Dissection Experience

Customizable images enliven presentations and quizzes for lecture or lab.

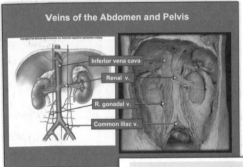

Veins of the Abdomen and Pelvis

Inferior vena cava

Renal v.

R. gonadal v.

Common iliac v.

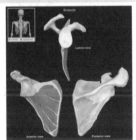

6. Which muscle's origin is highlighted in red?

a. Infraspinatus
b. *Subscapularis*
c. Teres minor
d. Teres major
e. Supraspinatus

Bonus: Which of the above muscles is not part of the rotator cuff?

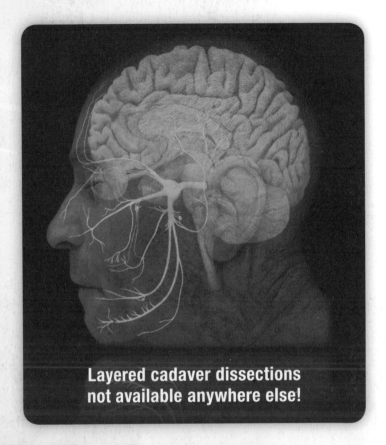

Layered cadaver dissections not available anywhere else!

www.aprevealed.com

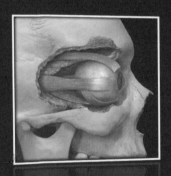  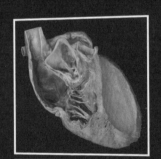 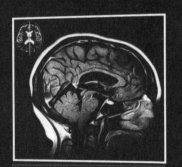

Portable cadavers can replace or enhance the laboratory experience.

Imaging
Correlate dissected anatomy with radiologic images like X-ray, MRI, and CT scans.

Self-Test
Gauge proficiency with quizzes and simulated practical exams.

Search
Type in a term to quickly locate any structure in the program.

McGraw Hill

second edition

HUMAN
Anatomy

Michael McKinley
Glendale Community College

Valerie Dean O'Loughlin
Indiana University

**McGraw-Hill
Higher Education**

Boston   Burr Ridge, IL   Dubuque, IA   New York   San Francisco   St. Louis
Bangkok   Bogotá   Caracas   Kuala Lumpur   Lisbon   London   Madrid   Mexico City
Milan   Montreal   New Delhi   Santiago   Seoul   Singapore   Sydney   Taipei   Toronto

 **McGraw-Hill Higher Education**

The McGraw·Hill Companies

HUMAN ANATOMY, SECOND EDITION

Published by McGraw-Hill, a business unit of The McGraw-Hill Companies, Inc., 1221 Avenue of the Americas, New York, NY 10020.
Copyright © 2008 by The McGraw-Hill Companies, Inc. All rights reserved. Previous edition © 2006. No part of this publication may be
reproduced or distributed in any form or by any means, or stored in a database or retrieval system, without the prior written consent of
The McGraw-Hill Companies, Inc., including, but not limited to, in any network or other electronic storage or transmission, or broadcast
for distance learning.

Some ancillaries, including electronic and print components, may not be available to customers outside the United States.

This book is printed on acid-free paper.

5 6 7 8 9 0 DOW/DOW 0 9

ISBN 978–0–07–296549–0
MHID 0–07–296549–5

Publisher: *Michelle Watnick*
Executive Editor: *Colin H. Wheatley*
Senior Developmental Editor: *Kristine A. Queck*
Marketing Manager: *Lynn M. Breithaupt*
Lead Project Manager: *Mary E. Powers*
Lead Production Supervisor: *Sandy Ludovissy*
Lead Media Project Manager: *Judi David*
Lead Media Producer: *John J. Theobald*
Senior Designer: *David W. Hash*
Cover Designer: *Elise Lansdon*
Cover Anatomy Art: *Kim E. Moss, Electronic Publishing Services Inc., NY*
Cover Photo: *Bruce Talbot/Getty Images, Inc.*
Lead Photo Research Coordinator: *Carrie K. Burger*
Photo Research: *Jerry Marshall*
Supplement Producer: *Mary Jane Lampe*
Compositor: *Electronic Publishing Services Inc., NY*
Typeface: *9.5/12 Slimbach*
Printer: *R. R. Donnelley Willard, OH*

The credits section for this book begins on page C-1 and is considered an extension of the copyright page.

To Jan, Renee, Ryan, and Shaun, and Janet Silver (the McKinley family).

To Bob and Erin (the O'Loughlin family).

And to Kris Queck and Laurel Shelton (our extended book family).

Thank you for all of your support, guidance, and patience with us throughout this project.

Library of Congress Cataloging-in-Publication Data

McKinley, Michael P.
  Human anatomy / Michael P. McKinley, Glendale, Valerie Dean O'Loughlin. — 2nd ed.
     p. cm.
  Includes index.
  ISBN 978–0–07–296549–0 — ISBN 0–07–296549–5 (hard copy : alk. paper)
     1. Human anatomy. I. O'Loughlin, Valerie Dean. II. Title.
QM23.2.M38 2008
611–dc22

                              2007020638

www.mhhe.com

MICHAEL McKINLEY received his undergraduate degree from the University of California, and both his M.S. and Ph.D. degrees from Arizona State University. In 1978, he accepted a postdoctoral fellowship at the University of California at San Francisco (UCSF) Medical School in the laboratory of Dr. Stanley Prusiner, where he worked for 12 years investigating prions and

prion-diseases. In 1980, he became a member of the anatomy faculty at the UCSF Medical School, where he taught medical histology for 10 years while continuing to do research on prions. During this time, he was an author or co-author of more than 80 scientific papers.

Since 1991, Mike has been a member of the biology faculty at Glendale Community College, where he teaches undergraduate anatomy and physiology,

general biology, and genetics. Between 1991 and 2000, in addition to teaching at Glendale Community College, he participated in Alzheimer disease research and served as director of the Brain Donation Program at the Sun Health Research Institute, while also teaching developmental biology and human genetics at Arizona State University, West. Mike's vast experience in histology, neuroanatomy, and cell biology greatly shaped the related content in *Human Anatomy*.

Mike is an active member of the Human Anatomy and Physiology Society (HAPS). He resides in Tempe, AZ, with his wife Jan.

VALERIE DEAN O'LOUGHLIN received her undergraduate degree from the College of William and Mary and her Ph.D. in biological anthropology from Indiana University. Since 1995, she has been a member of the Indiana University School of Medicine faculty, where she teaches human gross anatomy to first-year medical students and basic human anatomy to undergraduates. As part of her teaching, Valerie has performed numerous cadaver dissections and she drew heavily on this experience to ensure that both the narrative and the gross anatomy artwork in this book conform to standards typically seen in medical atlases and medical textbooks. However, she also made sure that the material is presented at a level that will not overwhelm the undergraduate reader.

Valerie's research interests span craniofacial growth and development, osteology, paleopathology, anatomy, educational research, and the scholarship of teaching and learning. In addition, she has prepared numerous web-based human embryology teaching modules. She has received numerous educational research grants as well as several

teaching awards, including a Teaching Excellence Recognition Award and a Trustee Teaching Award from Indiana University.

In 2007, Valerie received the American Association of Anatomists Basmajian Award for excellence in teaching and for her work in scholarship of education. Valerie is an active member of the American Association of Anatomists (AAA), the American Association of Physical Anthropologists (AAPA), and the Human Anatomy and Physiology Society (HAPS). She resides in Bloomington, IN, with her husband Bob and her daughter Erin.

Brief Contents

Contents

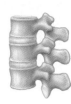

Chapter 9
Articulations 250

Chapter 10
Muscle Tissue and Organization 286

Chapter 11
Axial Muscles 320

Chapter 12
Appendicular Muscles 352

Chapter 17
Pathways and Integrative Functions   516

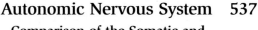

Chapter 18
Autonomic Nervous System   537

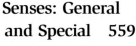

Chapter 19
Senses: General and Special   559

Chapter 20
Endocrine System   603

Chapter 21
Blood 635

Chapter 22
Heart 654

Chapter 23
Vessels and Circulation 681

Chapter 24
Lymphatic System 722

Chapter 25
Respiratory System 745

Chapter 26
Digestive System 776

Preface

Human anatomy is a fascinating field that has many layers of complexity. The subject is difficult to teach, and students can often be overwhelmed by its massive amount of material. In many respects, studying anatomy is similar to studying a foreign language because students must understand the vocabulary before they can apply the material. As many instructors know, textbook selection can either help or hinder student understanding. Throughout our teaching careers, we have examined and reviewed many textbooks. Some texts provide relatively accurate terminology and description but are too difficult for the average undergraduate to read. Other texts are easier to read but not as thorough or accurate in their discussions. We have strived to develop a text that is accurate and in-depth in its anatomic descriptions and yet easy to understand and full of pedagogical elements to help the student. This is the vision of *Human Anatomy*.

Audience

This textbook is designed for a one-semester human anatomy course, typically taken in the second or third year of college, for students in pre-allied health professions, nursing, exercise science, kinesiology, and/or other pre-professional health programs. It assumes the reader has no prior knowledge of biology or chemistry, and so the early chapters serve as a primer for the history of anatomy, biological terminology, and cell biology. This text provides all the background the introductory student needs to learn the basics of human anatomy.

What Makes This Book Special?

Although several human anatomy books exist in the market, a variety of features make this text different from the rest.

Superior Illustrations and a Quality Art Program

Anatomy is a visual subject, and one of the best ways a student can learn it is by studying beautiful, accurate drawings. We have been dismayed in the past to see texts in which sound anatomic discussions were accompanied by weak or inaccurate illustrations. One of our prime goals in producing this book was that the illustrations be just as accurate as the text. To meet this objective, we worked with an experienced team of certified medical illustrators to produce a collection of anatomic images unsurpassed by other anatomy texts. These images are not only beautiful but also as accurate as possible. We painstakingly scrutinized each rendering, relying on our experience in human gross anatomy, cadaver dissection, histology, and A&P—as well as trusted anatomic bibles such as Gray, Grant, Clemente, Netter, and a host of photographic atlases—to make sure the art matches life. Every illustration also went through an intensive peer review during which dozens of fellow instructors gave us pointed feedback on how to clarify concepts and make the drawings even more accurate—welcome assistance for our sometimes-weary eyes! Finally, we have carefully labeled the illustrations to coincide with coverage in the narrative to ensure that the pictures and words work together to tell a cohesive story. We challenge you to compare the artwork in this text with that in other human anatomy texts, and see which you and your students prefer.

Human Cadaver Photographs to Complement the Illustrations

Sometimes even the most beautiful art cannot prepare us for what anatomic structures look like in a real human being or for the normal variations that occur among individuals. Whenever possible, we have paired illustrations with human cadaver photographs to provide two valuable perspectives of key views: an artist's rendering that utilizes color and texture to make features stand out, and a photograph that demonstrates the appearance of real specimens. Furthermore, we have applied labels to complementary illustrations and photos so that they mirror each other whenever possible to make it easier for students to correlate structures between images. Christine Eckel of Salt Lake Community College tirelessly worked on the dissections and photographs of the cadavers. Her work is beautiful, and many of her dissections are presented in a way that is unparalleled in other texts. We suggest you turn to chapter 11 (Axial Muscles) and to chapter 15 (Brain and Cranial Nerves) and examine the photos. You will be impressed—and your students will appreciate their value as they are learning the laboratory material.

Writing Style: Blending Accuracy with Readability

Most, if not all, current undergraduate human anatomy textbooks are primarily "cut-down" versions of existing anatomy and physiology textbooks. Our text, *Human Anatomy*, was written exclusively for and with attention to the human anatomy course. Our text is not a "pared-down" version of an A&P text; we have designed it from the ground up to satisfy the needs of anatomy students and instructors.

Both authors have distinctive writing styles that, when combined in this text, provide the optimum balance between concise anatomic accuracy and user-friendly readability. We feel a text that is too condensed in its descriptions is more frustrating than helpful for students to use. Likewise, if a text is too verbose in its descriptions, students may feel they have read many pages that have said little. We have tried to strike a happy medium between these two extremes, so a student will feel that the text is easy to read and understand, while the instructor recognizes that the information is accurate, concise, and expertly written. We have been meticulous in our descriptions and level of accuracy.

In addition to making the text readable and accurate, we wanted to make it engaging and effective. To this end, we have incorporated many active learning techniques into the narrative. As we tell our students, you don't lose weight merely by watching an exercise program; you have to *do* the exercises in order to get results. Therefore, throughout our text, we have provided opportunities for the student to be an active learner, not just a passive reader. For example, students are encouraged to palpate structures on their bodies, perform basic experiments to test anatomic principles, and observe certain features on themselves. As the students perform these anatomy "exercises," their understanding will increase.

Themes and Distinctive Topic Approaches

Through our teaching experience, we have developed a few approaches that really seem to help students grasp certain topics or spark their interest. Thus, we have tried to incorporate these successful ideas from our own courses into our book.

Embryology

In many cases, a student can gain a complete understanding of adult anatomy only by first learning about the embryologic events that formed this anatomy. For this reason, we have placed an entire chapter on embryology (chapter 3) early in our text, as opposed to having a development chapter at the end of the book. In addition, "systems embryology" sections in each systems chapter (e.g., integumentary system, digestive system, etc.) provide a brief but thorough overview of the developmental processes for that particular system at a level that will not overwhelm the introductory student.

Forensic Anthropology

Many of our students are fascinated by crime shows on TV and love to learn how knowledge of anatomy can play a part in forensic analysis. With a Ph.D. in biological anthropology, Valerie shares this interest, and utilized her experience to craft the forensic applications in the skeletal system chapters. Chapters 6–8 feature discussions on such topics as epiphyseal plate fusion as a reliable indicator of age at death, sex differences in the skull, sex differences in the pelvis, and how morphologic changes in the pubic symphysis of the os coxae can be used to estimate age at death. These forensic applications are a great way to reinforce learning, and students will enjoy the "real-life" applications.

Surface Anatomy

Many of the students who take anatomy will become health-care professionals who use surface anatomy throughout their careers and need to know the importance of these landmarks. To best serve our student audience, we have given surface anatomy the coverage it deserves. Our chapter 13, Surface Anatomy, contains beautiful photographs and clear, concise text as well as numerous Clinical Views that illustrate the importance of the landmarks and how they are used daily in health care. Placing this chapter directly after the musculoskeletal chapters allows students to establish knowledge of the body's underlying framework before trying to understand surface landmarks.

Nervous System

In order to understand the workings of the nervous system, it is best to learn how the brain controls all aspects of the nervous system. Thus, in this text we examine the brain first, followed by a chapter comparing its similarities, differences, and relationships to the spinal cord. It seemed appropriate to use central nervous system terminology to describe the brain first and then the spinal cord. Additionally, because the nuclei of the cranial nerves are housed within the brain, we felt it made more sense to present the cranial nerves along with the brain.

Autonomic Nervous System

The autonomic nervous system is perhaps one of the most challenging topics in human anatomy. Why, then, do so many texts make a difficult topic even *more* difficult by presenting the sympathetic division first? We have seen in our own teaching experience that presenting the parasympathetic division (the relatively "easier" system) first increases the overall understanding of the autonomic nervous system. Thus, in chapter 18 (Autonomic Nervous System),

we discuss the parasympathetic division first, and follow up with a discussion of the more complex sympathetic division.

Arteries and Veins

We have been confused as to why other texts discuss all of the arteries in the body first, and then follow with a separate discussion of all of the veins. Presenting this material in such a fragmented fashion does not give students "the big picture." We feel that it makes much more sense to discuss blood flow in its entirety. For this reason, our text discusses arteries and veins in unison by region. For example, we present the arteries and veins of the upper limb together. This approach emphasizes to students that arteries often have corresponding veins and that both are responsible for the blood flow in a general region. We challenge you to compare our chapter 23 (Vessels and Circulation) with chapters from other texts. We predict that you and your students will appreciate our more unified presentation.

Reproductive System Homologues

Embryology has shown us that the female and male reproductive systems, and thus the homologues within those systems, originate from the same basic structures. An emphasis on homologues helps students grasp the similarities and differences between the female and male reproductive systems. Because the female reproductive system is the "basic" embryologic system (meaning that if no male hormonal influences occur in utero, the female pattern remains), we present the female reproductive system first, followed by the male reproductive system.

Accurate Terminology and Pronunciation Aids

The terms used in this text follow the standards set by the FCAT (Federative Committee on Anatomical Terminology) and published in *Terminologia Anatomica (TA)*. This reference is the international standard on which anatomic vocabulary should be based. In a few cases, *TA* terminology was not used because an alternative term was less confusing and more understandable for the student. In the case of an ambiguous term, *Stedman's Medical Dictionary* was also consulted. We have eliminated the use of eponyms as primary terms whenever possible. However, eponyms are given in italics so that the student and instructor can correlate an eponym with its proper anatomic term.

A large contributor to success in a human anatomy course is mastering the terminology. Students cannot properly learn anatomy if they cannot "talk the talk"—that is, pronounce the words and know what the words mean. Pronunciation guides and word origins are included throughout the book to teach students how to say the terms and give them helpful, memorable hints for decoding meaning. These vocabulary aids were derived from *Stedman's Medical Dictionary*.

Pedagogy

Learning human anatomy is often seen as an endeavor of rote memorization. In *Human Anatomy*, we have employed many pedagogical techniques that aim to take students beyond memorization and engage them in a thought-provoking discovery of facts that will lead to well-rounded understanding. Individuals learn in a variety of ways—some learn best by reading text, others by using visuals, and still others by studying information organized in tables. We have been careful to cover the concepts using all three of these media. These multifaceted concept presentations are then organized within a framework of pedagogical tools that help students build their knowledge base, challenge them to continue expanding their growing understanding of anatomy, and encourage them to actively apply the information they read. Question sets within each chapter and review activities at the end of each chapter provide a balanced combination of simple retention-based questions and more complex critical-thinking activities. Study Tip! boxes offer practical advice for understanding and remembering the material. Clinical View essays promote a deeper understanding of the material discussed in the text by demonstrating how basic concepts play out in disease processes. All of these pedagogical elements work together, sparking students to practice, remember, apply, and understand. The "Guided Tour" beginning on page xviii offers more specifics about the learning features in *Human Anatomy*.

What's New?

Although we have retained much of what made the first edition of *Human Anatomy* so successful, our second edition revisions were not superficial. New research findings, shifting terminology, technological advancements, and the evolving needs of students and instructors in the classroom require textbook authors to continually monitor and revise their content. To meet this demand, we have systematically revised the writing, artwork, and pedagogy using input from students and instructors to fine-tune our second edition.

Content and Writing Updates

Medical science is a dynamic and ever-changing field, with continual new discoveries and sometimes reversals in traditional ways of thinking about disease. We carefully scrutinized each chapter and updated material as new scientific advances became public. In addition, we read every word of this text from a student perspective and rewrote or reorganized passages as needed to facilitate better understanding of difficult topics. The many instructors who reviewed our first edition were helpful in this endeavor as well, pointing out spots where writing clarity or accuracy could be improved. We tried to incorporate all reviewer comments and suggestions when applicable and appropriate.

Illustration and Photo Updates

Every single figure in the book was carefully reviewed for labeling and anatomic accuracy and corrected when needed. In several cases, illustrations were reorganized to better present the material. Descriptive labels were added beneath most images to clarify the intent of the illustration. Terminology was updated in figures as needed to align with *Terminologia Anatomica,* and we carefully checked that items mentioned in the text were shown in the illustrations, and vice versa. Some figures were enlarged to better show important anatomic features. Some figures that were somewhat "static" in our first edition were redrawn as more dynamic process figures in the second edition to show particular events more clearly, and the associated descriptive steps were aligned more seamlessly with the text descriptions. Additional cadaver photographs have been added for clarity, and many of the micrographs have been replaced by images of superior quality.

Pedagogy Updates

Along with content and writing updates, we also enhanced the pedagogical aids in our second edition. Several students and faculty members provided us with new study tips that we incorporated into the text. We also streamlined the summaries for most chapters and revised many end-of-chapter questions to better test higher-order thinking skills. More "Developing Critical Reasoning" questions were also added to many chapters. We've tried to emphasize in this edition that anatomy is not just a subject for memorization—that students must also learn important concepts and applications in order to gain a complete understanding of the workings of the human body.

These overall enhancements, combined with the specific content updates throughout, make the second edition of *Human Anatomy* a revision we are proud to publish. The following list is by no means exhaustive, but it highlights some of the changes made in each chapter.

Chapter 1—A First Look at Anatomy   Expanded the discussion of the history of anatomy. Added posterior body cavities to table 1.4.

Chapter 2—The Cell: Basic Unit of Structure and Function   Reorganized table 2.2 (Components of the Cell). Clarified the terms "extracellular fluid" and "interstitial fluid." Converted figure 2.9 into a process figure showing the movement and packaging of materials in the Golgi apparatus. Rearranged the Life Cycle of the Cell section to define cell division before defining the cell cycle.

Chapter 3—Embryology   Included measurements of the zygote, blastocyst, and crown-rump lengths at major stages in embryologic development. Expanded the discussion of the effects of teratogens on organogenesis, and included a discussion of fetus susceptibility to a toxin and dose-dependent issues.

Chapter 4—Tissue Level of Organization   Per reviewer request, reorganized the discussion about glandular secretions into "anatomic" classifications (structure of the gland) and "physiologic" classifications (method of secretion). Added a new micrograph of osteons in compact bone (table 4.12). Relocated coverage of tissue repair and tissue death to chapter 5.

Chapter 5—Integumentary System   Clarified the description of calcitriol function. Rewrote the entire section on the stratum basale. Added new clinical information about sunless tanners. Incorporated sections on tissue repair and tissue death formerly included in chapter 4.

Chapter 6—Cartilage and Bone Connective Tissue   Included a discussion of epiphyseal arteries and veins. Expanded the discussion of the effects of certain hormones on bone growth and development. Included a new Clinical View on costochondritis. Revised figure 6.15 to better show the "dinner-fork deformity" of a Colles fracture and to more clearly differentiate spiral versus oblique fractures.

Chapters 7 and 8—Axial Skeleton and Appendicular Skeleton   Made some minor label and leader adjustments for better precision. Updated any terms that weren't already part of the accepted *TA* terminology.

Chapter 9—Articulations   Reorganized the presentation of movements at synovial joints to start with a discussion of simple movements (e.g., flexion and extension) and progress to lesser-known movements (e.g., abduction, adduction, and circumduction). Incorporated pronation and supination into the discussion of rotational movements. Renumbered and repositioned related figures, and reworded table 9.2 to coincide with the new organization.

Chapter 10—Muscle Tissue and Organization Moved the Exercise and Skeletal Muscle section to follow Skeletal Muscle Fiber Organization. Clarified the differences between G-actin and F-actin, the three functions of troponin, and the structure of I bands. Upgraded a series of sarcomere micrographs in figure 10.7.

Chapters 11 and 12—Axial Muscles and Appendicular Muscles Added the corrugator supercilii and levator palpebrae superioris to the discussion of facial muscles. Reorganized table 11.5 (pharyngeal and laryngeal muscles). Per reviewer requests, added new tables (table 11.11 and table 12.13) that summarize certain muscle actions.

Chapter 13—Surface Anatomy Revised the tracheotomy Clinical View and included a discussion of the complication of tracheal stenosis.

Chapter 14—Nervous Tissue Revised and clarified the description of myelination. Clarified artwork indicating impulse "input" and "output" throughout the chapter.

Chapter 15—Brain and Cranial Nerves Replaced "rostral" and "caudal" with "anterior" and "posterior" as primary terms. Reorganized the coverage of the cranial meninges from deep (pia mater) to superficial (dura mater), and then proceeded to the dural septa to emphasize that cranial dural septa are derived from dura mater. Expanded on specific hypothalamic nuclei and their individual functions.

Chapter 16—Spinal Cord and Spinal Nerves Revised the gross anatomy of the spinal cord section so that a lengthwise description of the cord comes first, followed by a cross-sectional description. Upgraded photos of the conus medullaris (figure 16.1c) and a transverse section of the spinal cord (figure 16.3b). Included a new Study Tip! (offered by a second edition reviewer) about somatic motor neurons. Added a brief mention of the sympathetic trunk and its ganglia (for those instructors who may not have time to assign the more detailed discussion of the autonomic nervous system in chapter 18).

Chapter 17—Pathways and Integrative Functions Added new tables (17.1 and 17.4) summarizing sensory and motor pathway neurons. Revised table 17.5 on principal motor spinal cord pathways.

Chapter 18—Autonomic Nervous System Reorganized the text within the Comparison of the Somatic vs. Autonomic Nervous System section to introduce similarities first, followed by differences.

Chapter 19—Senses: General and Special Clarified the difference between tonic and phasic receptors. Added information on primary and secondary odors, and added olfactory glands to figure 19.9. Added new Clinical Views addressing (1) the relationship between cochlear shape and function and (2) acoustic neuroma.

Chapter 20—Endocrine System Extensively revised the tables listing the hormones of the endocrine organs and added information on hormone targets, effects, and related disorders. Rewrote and clarified the section on anterior pituitary hormones and added the "FLAT PIG" mnemonic for remembering these hormones. Revamped and simplified illustrations of the pituitary gland throughout the chapter, and upgraded micrographs showing the anterior and posterior pituitary (figure 20.5). Added new information about oxytocin.

Chapter 21—Blood Added a description of acidosis and alkalosis. Per reviewer comments, revised the definition of hematocrit and compared the differences in how clinicians and basic scientists define this term.

Chapter 22—Heart Moved the overview of the cardiovascular system to precede the description of heart anatomy. Added a new section called "How the Heart Beats" to describe the characteristics of cardiac muscle tissue and the heart's conduction system. Added a section called "Tying It All Together: The Cardiac Cycle" to describe the steps in the cardiac cycle and summarize blood flow during the cycle. Revised the cardiac cycle art (figure 22.14) to show the cyclical pattern, and added the timing of various heart events.

Chapter 23—Vessels and Circulation Revised figures 23.19 and 23.20 to show the paired venous arches associated with the arterial arches. Per reviewer suggestions, included a new Study Tip! about arteries and veins.

Chapter 24—Lymphatic System Updated the AIDS clinical view and included recent information about the "pill-a-day" cocktail. Revised figure 24.5 (T-lymphocytes) to more clearly show the cells undergoing mitosis after being presented with an antigen. Added a cadaver photo of a lymph node to figure 24.10.

Chapter 25—Respiratory System Moved the discussion of pulmonary ventilation to precede the coverage of changes in thoracic wall dimensions during respiration. Revised and clarified the section describing the general organization and functions of the respiratory system. Added an explanation of the activity of the trachealis muscle during swallowing.

Chapter 26—Digestive System Revised the liver illustration (figure 26.18) to clarify the view orientation. Added coverage of celiac disease, bilirubin and jaundice, and Barrett esophagus to related Clinical Views.

Chapter 27—Urinary System Revised the explanation of the tissue layers surrounding the kidney. Rewrote and reorganized the description of the renal corpuscle. Moved the discussion of filtration membranes to precede the coverage of filtration. Added new clinical information and artwork about kidney transplants, as well as a discussion of incontinence related to aging.

Chapter 28—Reproductive System Added mention of the hormone inhibin to both the text and table 28.4. Revised figure 28.4 to clarify that not all ovarian follicles are present in the ovary at the same time, and reorganized the graphs of the ovarian and uterine cycles (figure 28.6). Relocated the discussion of the spermatic cord to fall before the discussion of testis anatomy and sperm development. Included information on interstitial cell androgen production. Added new clinical information on Implanon, Plan B, tubal ligation, and vasectomy as birth control methods; Gardasil and cervical cancer; and the association between circumcision and reduced HIV transmission.

Your Feedback Is Welcome!

We are dedicated to producing the best materials available to help students learn human anatomy and engender a love of this topic. Your suggestions for improving this textbook are always welcome!

Michael P. McKinley
Department of Biology
Glendale Community College
6000 W. Olive Avenue
Glendale, AZ 85302
michael.mckinley@gcmail.maricopa.edu

Valerie Dean O'Loughlin
Jordan Hall 010A
Medical Sciences
Indiana University
Bloomington, IN 47405
vdean@indiana.edu

Accurate and Engaging Illustrations

The brilliant illustrations in *Human Anatomy* bring the study of anatomy to life! Drawn by a team of medical illustrators, all figures have been carefully rendered to convey realistic, three-dimensional detail. Each drawing has been meticulously reviewed for accuracy and consistency, and precisely labeled to coordinate with the text discussions.

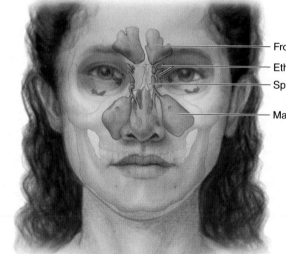

Frontal sinus
Ethmoidal sinuses
Sphenoidal sinus
Maxillary sinus

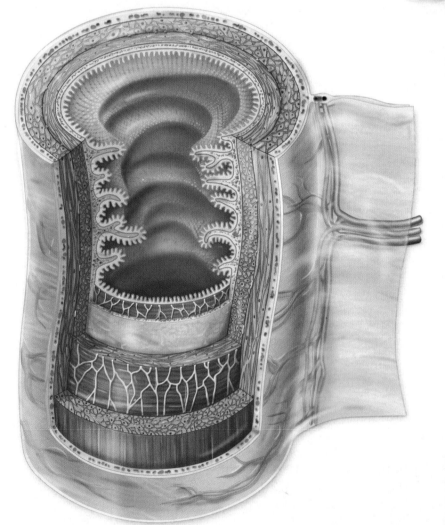

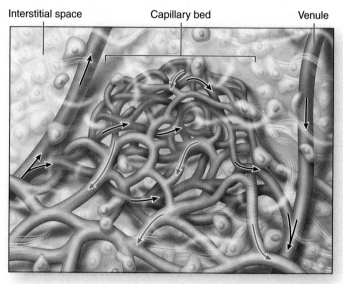

Interstitial space   Capillary bed   Venule

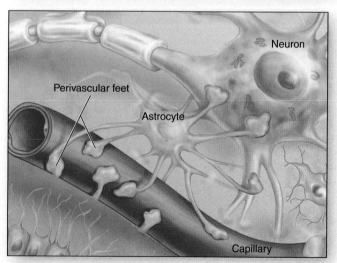

Neuron
Perivascular feet
Astrocyte
Capillary

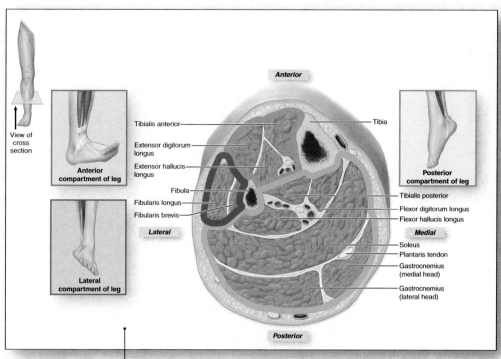

View Orientation

Reference diagrams clarify the view or plane an illustration represents.

Color-Coding

Many illustrations use color-coding to organize information and clarify concepts for visual learners.

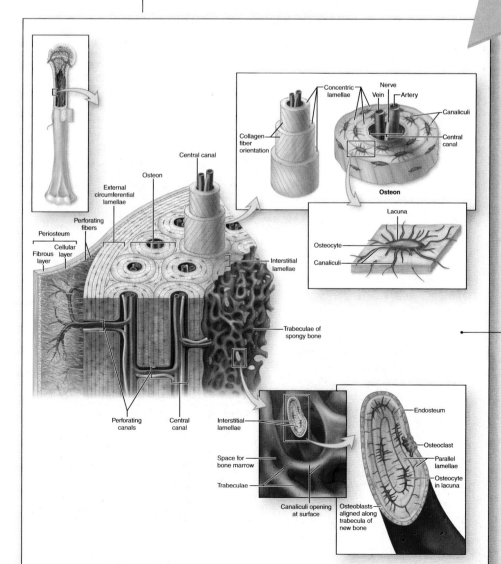

Multi-Level Perspective

Illustrations depicting complex structures connect macroscopic and microscopic views to show the relationships between increasingly detailed drawings.

Atlas-Quality Photographs

Human Anatomy features a beautiful collection of cadaver dissection images, bone photographs, surface anatomy shots, and histology micrographs. These detailed images capture the intangible characteristics of human anatomy that can only be conveyed in human specimens, and help familiarize students with the appearance of structures they will encounter in lab.

Complementary Views

Drawings are often paired with photographs to enhance visualization of structures. Labels on art and photos mirror each other whenever possible, making it easy to correlate structures between views.

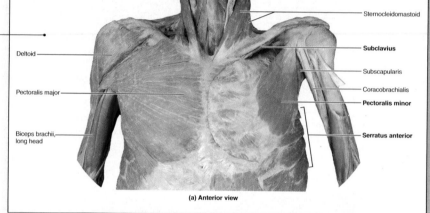

Superficial | Deep

Deltoid
Pectoralis major
Biceps brachii, long head

Sternocleidomastoid
Subclavius
Subscapularis
Coracobrachialis
Pectoralis minor
Serratus anterior

Deltoid
Pectoralis major
Biceps brachii, long head

Sternocleidomastoid
Subclavius
Subscapularis
Coracobrachialis
Pectoralis minor
Serratus anterior

(a) Anterior view

Cadaver Dissections

Expertly dissected specimens are preserved in richly colored photos that reveal incredible detail. Many unique views show relationships between anatomic structures from a new perspective.

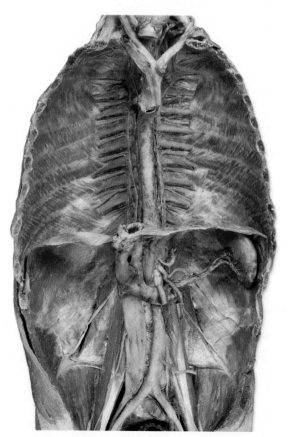

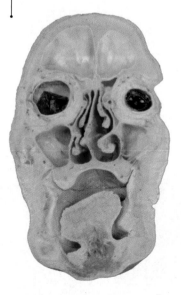

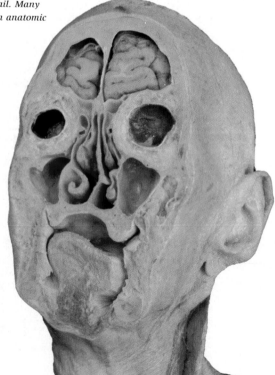

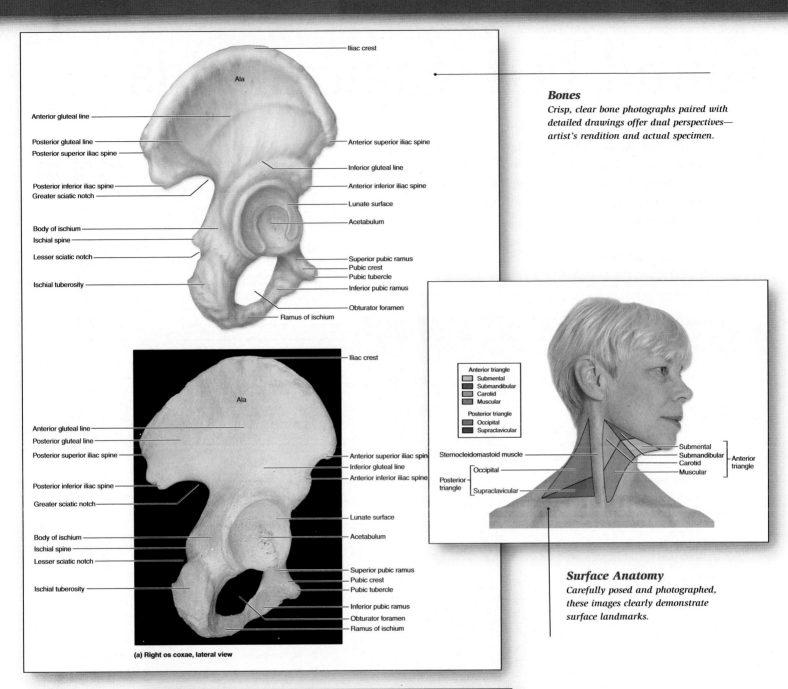

Iliac crest

Ala

Anterior gluteal line

Posterior gluteal line

Posterior superior iliac spine

Anterior superior iliac spine

Inferior gluteal line

Anterior inferior iliac spine

Posterior inferior iliac spine
Greater sciatic notch

Lunate surface

Acetabulum

Body of ischium
Ischial spine

Lesser sciatic notch

Superior pubic ramus
Pubic crest
Pubic tubercle
Inferior pubic ramus

Ischial tuberosity

Obturator foramen

Ramus of ischium

Iliac crest

Ala

Anterior gluteal line
Posterior gluteal line
Posterior superior iliac spine

Anterior superior iliac spin
Inferior gluteal line
Anterior inferior iliac spine

Posterior inferior iliac spine

Greater sciatic notch

Body of ischium
Ischial spine
Lesser sciatic notch

Lunate surface

Acetabulum

Superior pubic ramus
Pubic crest
Pubic tubercle

Ischial tuberosity

Inferior pubic ramus
Obturator foramen
Ramus of ischium

(a) Right os coxae, lateral view

Bones
Crisp, clear bone photographs paired with detailed drawings offer dual perspectives— artist's rendition and actual specimen.

Anterior triangle
Submental
Submandibular
Carotid
Muscular
Posterior triangle
Occipital
Supraclavicular

Sternocleidomastoid muscle

Posterior triangle

Occipital
Supraclavicular

Submental
Submandibular
Carotid
Muscular

Anterior triangle

Surface Anatomy
Carefully posed and photographed, these images clearly demonstrate surface landmarks.

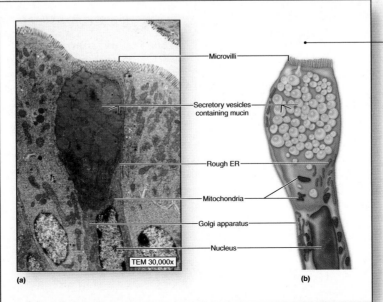

Microvilli

Secretory vesicles containing mucin

Rough ER

Mitochondria

Golgi apparatus

Nucleus

TEM 30,000x

(a)

(b)

Histology Micrographs
Light micrographs, as well as scanning and transmission electron micrographs, are used in conjunction with illustrations to present a true picture of microscopic anatomy. Magnifications provide a reference point for the sizes of the structures shown in the micrographs.

Guided Tour

Sound Pedagogical Aids

*H*uman Anatomy is built around a pedagogical framework designed to foster retention of facts and encourage the application of knowledge that leads to understanding. The learning aids in this book help organize studying, reinforce learning, and promote critical-thinking skills.

Chapter Outline
Each chapter begins with a page-referenced outline that provides a quick snapshot of the chapter contents and organization.

RESPIRATORY SYSTEM

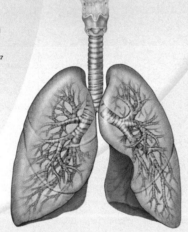

25

Respiratory System

Anatomy & Physiology | REVEALED
When applicable, icons indicate where content related to the chapter can be found on McGraw-Hill's Anatomy & Physiology | REVEALED software.

Key Topics
A brief list at the beginning of each section introduces the major concepts students should understand after completing the section. Reviewing these objectives before reading helps focus attention on critical information.

Epidermal Accessory Organs
Key topics in this section:
- Structure of nails
- Components of a hair and a hair follicle
- Growth, distribution, and replacement of hairs
- How hair changes throughout life
- Characteristics of sweat glands, sebaceous glands, and other glands found in the skin

Nails, hair, and sweat and sebaceous glands are derived from epidermis and are considered accessory organs, or appendages, of the integument. These structures originate from the invagination of the epidermis during embryological development; they are located in the dermis and may project through the epidermis to the surface. Both nails and hair are composed primarily of dead, keratinized cells.

Nails
Nails are scalelike modifications of the epidermis that form on the dorsal tips of the fingers and toes. They protect the exposed distal tips and prevent damage or distortion during jumping, kicking, catching, or grasping. Nails are hard derivatives from the stratum corneum layer of the epidermis. The cells that form the nails are densely packed and filled with parallel fibers of hard keratin.

Each nail has a pinkish **nail body** and a distal whitish **free edge** (figure 5.8a). Most of the nail body appears pink because of the blood flowing in the underlying capillaries. In contrast, the free edge of the nail appears white because there are no underlying capillaries. The **lunula** (loo'noo-lă; *luna* = moon) is the whitish semilunar area of the proximal end of the nail body. It appears whitish because a thickened underlying stratum basale obscures the underlying blood vessels. Along the lateral and proximal borders of the nail, portions of skin called **nail folds** overlap the nail so that the nail is recessed internal to the level of the surrounding epithelium and is bounded by a **nail groove**. The **eponychium** (ep-o-nik'ē-um; *epi* = upon, *onyx* = nail), also known as the **cuticle**, is a narrow

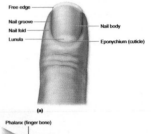

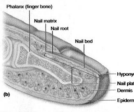

Figure 5.8
Structure of a Fingernail. Nails, the hard derivative of the stratum corneum, protect sensitive fingertips. *(a)* Surface view of a fingernail. *(b)* Sagittal section showing the internal details of a fingernail.

WHAT DID YOU LEARN?
4 What two main characteristics are used to classify epithelial tissues?
5 Why is one epithelium referred to as "pseudostratified"?
6 What are the two basic parts of a multicellular exocrine gland?
7 Why is epithelial cell regeneration important to the continued functioning of a holocrine gland?

What Did You Learn?
Review questions at the end of each section prompt students to test their comprehension of key concepts. These mini self-tests help students determine whether they have a sufficient grasp of the information before moving on to the next section of the chapter. Answers to the What Did You Learn? questions are provided on the McKinley/O'Loughlin Human Anatomy, 2e website at aris.mhhe.com.

Vocabulary Aids

Learning anatomy is, in many ways, like learning a new language. Human Anatomy stresses the use of proper anatomic terms, and includes vocabulary aids that help students master the terminology.

Support and Protection of the Brain

Key topics in this section:

- Characteristics of the cranial meninges and the cranial dural septa
- Origin, function, and pattern of cerebrospinal fluid circulation
- Structure of the blood-brain barrier and how it protects the brain

The brain is protected and isolated by multiple structures. The bony cranium provides rigid support, while protective connective tissue membranes called meninges surround, support, stabilize, and partition portions of the brain. Cerebrospinal fluid (CSF) acts as a cushioning fluid. Finally, the brain has a blood-brain barrier to prevent harmful materials from entering the bloodstream.

Cranial Meninges

The **cranial meninges** (mĕ-nin'jes, mĕ'nin-jēz; sing., *meninx*, men'ingks, mĕ'ninks; membrane) are three connective tissue layers that separate the soft tissue of the brain from the bones of the cranium, enclose and protect blood vessels that supply the brain, and contain and circulate cerebrospinal fluid. In addition, some parts of the cranial meninges form some of the veins that drain blood from the brain. From deep (closest to the brain) to superficial (farthest away from the brain), the cranial meninges are the pia mater, the arachnoid, and the dura mater (figure 15.4).

Pia Mater

Key terms are set in boldface where they are defined in the chapter, and many terms are included in the glossary at the end of the book. Pronunciation guides are included for difficult words.

The **pia mater** (pē'ă mah'ter, pī'ă mă'ter; *pia* = tender, delicate) is the innermost of the cranial meninges. It is a thin layer of delicate areolar connective tissue that is highly vascularized and tightly adheres to the brain, following every contour of the surface.

Arachnoid

The **arachnoid** (ă-rak'noyd), also called the *arachnoid mater* or the *arachnoid membrane*, lies external to the pia mater (figure 15.4). The term arachnoid means "resembling a spider web," and this meninx is so named because it is partially composed of a delicate web of collagen and elastic fibers, termed the **arachnoid trabeculae**. Immediately deep to the arachnoid is the **subarachnoid space**. The arachnoid trabeculae extend through this space from the arachnoid to the underlying pia mater. Between the arachnoid and the overlying dura mater is a potential space, the **subdural space**. The subdural space becomes an actual space if blood or fluid accumulates there, a condition called a subdural hematoma (see Clinical View).

Dura Mater

The **dura mater** (doo'ră mă'ter; *dura* = tough, *mater* = mother) is an external tough, dense irregular connective tissue layer composed of two fibrous layers. As its Latin name indicates, it is the strongest of the meninges. Within the cranium, the dura mater is composed of two layers. The **meningeal** (mĕ-nin'jē-ăl, men'in-jē'al) **layer** lies deep to the periosteal layer. The **periosteal** (per-ē-os'tē-ăl; *peri* = around, *osteon* = bone) **layer**, the more superficial layer, forms the periosteum on the internal surface of the cranial bones.

The meningeal layer is usually fused to the periosteal layer, except in specific areas where the two layers separate to form large, blood-filled spaces called **dural venous sinuses**. Dural venous sinuses are typically triangular in cross section, and unlike most other veins, they do not have valves to regulate venous blood flow. The dural venous sinuses are, in essence, large veins that drain blood from the brain and transport this blood to the internal jugular veins that help drain blood circulation to the head.

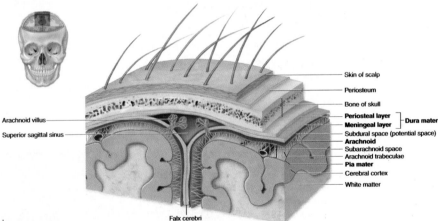

Arachnoid villus
Superior sagittal sinus

Skin of scalp
Periosteum
Bone of skull
Periosteal layer
Meningeal layer — Dura mater
Subdural space (potential space)
Arachnoid
Subarachnoid space
Arachnoid trabeculae
Pia mater
Cerebral cortex
White matter

Falx cerebri

Figure 15.4

Cranial Meninges. A coronal section of the head depicts the organization of the three meningeal layers: the dura mater, the arachnoid, and the pia mater. In the midline, folds of the inner meningeal layer of the dura mater form the falx cerebri, which partitions the two cerebral hemispheres. The inner meningeal layer and the outer periosteal layer sometimes separate to form the dural venous sinuses, such as the superior sagittal sinus (shown here), which drain blood away from the brain.

Because knowing the derivation of a term can enhance understanding and retention, word origins are given when relevant. Furthermore, a handy list of prefixes, suffixes, and combining forms is printed on the inside back cover as a quick reference for commonly used word roots.

WHAT DO YOU THINK?

❷ Try this experiment to determine the value of serous fluid: First, rub the palms of your hands quickly against one another. The sound you hear and the heat you feel are consequences of the friction being produced. Now put lotion (our version of serous fluid) on the palms of your hands and repeat the experiment. Do you still hear the noise and feel heat from your hands? What do you think would happen to your body organs if there were no serous fluid?

What Do You Think?

These critical-thinking questions actively engage students in application or analysis of the chapter material and encourage students to think more globally about the content. Answers to What Do You Think? questions are given at the end of each chapter, allowing students to evaluate the logic used to solve the problem.

Guided Tour

Study Tip!

The mnemonic "Never let monkeys eat bananas" is a simple way to recall the leukocytes in order of their relative abundance:

Never = **N**eutrophil (most abundant)

Let = **L**ymphocyte

Monkeys = **M**onocyte

Eat = **E**osinophil

Bananas = **B**asophil (least abundant)

Study Tips

Many anatomy instructors provide students with everyday analogies, mnemonics, and other useful tips to help them understand and remember the information. Study Tip! boxes throughout each chapter offer tried-and-tested practical learning strategies that students can apply as they read. These tips are not just useful—they can also be fun!

...d Circulation

...es harmful agents ...lood vessels. The ...he absorbed nutri- ...system of vessels ...r than distributing ...cular system. The ...rythrocyte destruction from the spleen, so that the liver can "recycle" some of these components.

The **hepatic portal vein** is the large vein that receives deoxygenated (oxygen-poor) but nutrient-rich blood from the gastrointestinal organs. Three main venous branches merge to form this vein: (1) The **inferior mesenteric vein,** a vertically positioned vein draining the distal part of the large intestine, receives blood from the superior rectal vein, sigmoid veins, and left colic vein. The inferior mesenteric vein typically (but not always) drains into the splenic vein. (2) The **splenic vein,** a horizontally positioned vein draining the spleen, receives blood from pancreatic veins, short gastric veins, and the right gastroepiploic vein. (3) The **superior mesenteric vein,** another vertically positioned vein on the right side of the body, drains the small intestine and part of the large intestine. It receives blood from the intestinal veins, pancreaticoduodenal veins, ileocolic vein, and right and middle colic veins. Some small veins, such as the left and right gastric veins, drain directly into the hepatic portal vein.

Figure 23.17 is a cadaver photo showing the arteries and veins of the posterior abdominal wall region. Note in this example that the inferior mesenteric vein drains into the superior mesenteric vein, not the splenic vein. This figure illustrates that the hepatic portal system can show great variation in some individuals.

Study Tip!

Although the pattern of the veins of the hepatic portal system can vary, together they typically resemble the side view of a chair. The front leg of the "chair" represents the inferior mesenteric vein, while the back leg represents the superior mesenteric vein. The seat of the chair is the splenic vein, while the back represents the hepatic portal vein.

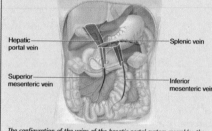

Hepatic portal vein

Splenic vein

Superior mesenteric vein

Inferior mesenteric vein

The configuration of the veins of the hepatic portal system resembles the side view of a chair.

The venous blood in the hepatic portal vein flows through the sinusoids of the liver for absorption, processing, and storage of nutrients. In these sinusoids, the venous blood mixes with some arterial blood entering the liver in the hepatic arteries. Thus, liver

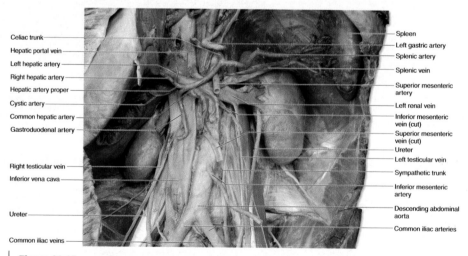

Celiac trunk
Hepatic portal vein
Left hepatic artery
Right hepatic artery
Hepatic artery proper
Cystic artery
Common hepatic artery
Gastroduodenal artery

Right testicular vein
Inferior vena cava

Ureter

Common iliac veins

Spleen
Left gastric artery
Splenic artery
Splenic vein
Superior mesenteric artery
Left renal vein
Inferior mesenteric vein (cut)
Superior mesenteric vein (cut)
Ureter
Left testicular vein
Sympathetic trunk
Inferior mesenteric artery
Descending abdominal aorta
Common iliac arteries

Figure 23.17

Major Vessels of the Posterior Abdominal Wall. In this cadaver photo, note that the inferior mesenteric vein varies from the "average" hepatic portal system pattern by draining into the superior mesenteric vein.

Clinical Coverage

Sometimes, an example of what can go wrong in the body helps crystallize understanding of the "norm." Clinical Views interspersed throughout each chapter provide insights into health or disease processes. Carefully checked by a clinician for accuracy with respect to patient care and the most recent treatments available, these clinical boxes expand upon topics covered in the text and provide relevant background information for students pursuing health-related careers.

CLINICAL VIEW

Gluteal Intramuscular Injections

The gluteal region is a preferred site for intramuscular (IM) injections because the gluteal muscles are quite thick and contain many blood vessels. However, health-care personnel must be careful not to accidentally inject the sciatic nerve or the superior and inferior gluteal vessels and nerves that supply the gluteal muscles. The sciatic nerve and the gluteal nerves and vessels are located primarily in the medial and inferior lateral part of the buttock. Therefore, the iliac crest is an important surface landmark for determining the safest place for a gluteal IM injection. Usually the injection is administered in the **superior lateral quadrant** of the buttock, about 5–7 centimeters inferior to the iliac crest. By placing the injection in the superior lateral quadrant, the healthcare worker can be reasonably certain of not accidentally piercing an important nerve or blood vessel.

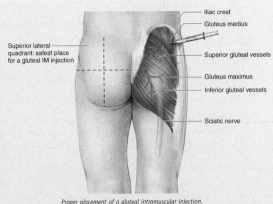

Proper placement of a gluteal intramuscular injection.

Clinical View

Interesting clinical sidebars reinforce or expand upon the facts and concepts discussed within the narrative.

CLINICAL VIEW: In Depth

Forensic Anthropology: Determining Age at Death

When the epiphyseal plates ossify, they fuse to and unite with the diaphysis. This process of epiphyseal plate ossification and fusion occurs in an orderly manner, and the timings of such fusions are well known. If an epiphyseal plate has not yet ossified, the diaphysis and epiphysis are still two separate pieces of bone. Thus, a skeleton that displays separate epiphyses and diaphyses (as opposed to whole fused bones) is that of a juvenile rather than an adult. Forensic anthropologists utilize this anatomic information to help determine the age of skeletal remains.

Fusion of an epiphyseal plate is progressive, and is usually scored as follows:

a. **Open** (no bony fusion or union between the epiphysis and the other bone end)
b. **Partial union** (some fusion between the epiphysis and the rest of the bone, but a distinct line of separation may be seen)
c. **Complete union** (all visible aspects of the epiphysis are united to the rest of the bone)

When determining the age at death from skeletal remains, the skeleton will be older than the oldest complete union and younger than the youngest open center. For example, if one epiphyseal plate that typically fuses at age 17 is completely united, but another plate that typically fuses at age 19 is open, the skeleton is that of a person between the ages of 17 and 19.

Current standards for estimating age based on epiphyseal plate fusion have primarily used male skeletal remains. Female epiphyseal plates tend to fuse approximately 1–2 years earlier than those of males, so this fact needs to be considered when estimating the age of a female skeleton. Further, population differences may exist with some epiphyseal plate unions. With these caveats in mind, the accompanying table lists standards for selected epiphyseal plate unions.

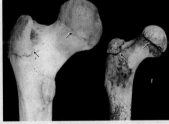

These two femurs came from individuals of different ages. (Left) In partial union (arrows), the epiphyses are partially fused. This individual likely was between the ages of 15 and 23 at death. (Right) No fusion has occurred between the epiphyses and the diaphysis (see arrows), a category called open. This individual likely was younger than 15 years of age.

Bone	Male Age at Epiphyseal Union (years)
Humerus, lateral epicondyle	11–16 (female: 9–13)
Humerus, medial epicondyle	11–16 (female: 10–15)
Humerus, head	14.5–23.5
Proximal radius	14–19
Distal radius	17–22
Distal fibula and tibia	14.5–19.5
Proximal tibia	15–22
Femur, head	14.5–23.5
Distal femur	14.5–21.5
Clavicle	19–30

Clinical View: In Depth

These boxed essays explore topics of clinical interest in detail. Subjects covered include pathologies, current research, treatments, forensics, and pharmacology.

CLINICAL TERMS

bradycardia (brad-ē-kar′dē-ă; *bradys* = slow) Slowing of the heartbeat, usually described as less than 50 beats per minute.
cardiomyopathy (kar′dē-ō-mī-op′ă-thē) Another term for disease of the myocardium; causes vary and include thickening of the ventricular septum (hypertrophy), secondary disease of the myocardium, or sometimes a disease of unknown cause.
endocarditis (en′dō-kar-dī′tis) Inflammation of the endocardium. Types include bacterial (caused by the direct invasion of bacteria), chorditis (affecting the chordae tendineae), infectious (caused by

microorganisms), mycotic (due to infection by fungi), and rheumatic (due to endocardial involvement as part of rheumatic heart disease).
ischemia (is-kē′mē-ă; *ischo* = to keep back) Inadequate blood flow to a structure caused by obstruction of the blood supply, usually due to arterial narrowing or disruption of blood flow.
myocarditis (mī′ō-kar-dī′tis) Inflammation of the muscular walls of the heart. This uncommon disorder is caused by viral, bacterial, or parasitic infections, exposure to chemicals, or allergic reactions to certain medications.

Clinical Terms

Selected clinical terms are defined at the end of each chapter.

End-of-Chapter Tools

A carefully devised set of learning aids at the end of each chapter helps students review the chapter content, evaluate their grasp of key concepts, and utilize what they have learned. Reading the chapter summary and completing the Challenge Yourself exercises is a great way to assess learning.

Chapter Summary Tables

Chapter summaries are presented in a concise, bulleted table format that provides a basic overview of each chapter. Page references make it easy to look up topics for review.

Challenge Yourself

This battery of matching, multiple choice, short answer, and critical-thinking questions is designed to test students on all levels of learning, from basic comprehension to synthesis of concepts. Answers to the Matching, Multiple Choice, and Developing Critical Reasoning questions are provided in the appendix. Answers to the Content Review questions are found on the McKinley/O'Loughlin Human Anatomy, 2e website at aris.mhhe.com.

Answers to What Do You Think?

The What Do You Think? questions are answered at the end of each chapter.

(Chapter Eight — Appendicular Skeleton 247)

CHAPTER SUMMARY

	■ The appendicular skeleton includes the bony supports (girdles) that attach the upper and lower limbs to the axial skeleton, as well as the bones of those limbs.
Pectoral Girdle 219	■ The pectoral girdle is composed of the clavicle and scapula.
	Clavicle 219
	■ The clavicle forms the collarbone.
	Scapula 219
	■ The scapula forms the "shoulder blade."
Upper Limb 223	■ Each upper limb contains a humerus, radius, ulna, 8 carpals, 5 metacarpals, and 14 phalanges.
	Humerus 223
	■ The head of the humerus articulates with the glenoid cavity of the scapula.
	■ The greater tubercle and lesser tubercle are important sites for muscle attachment. The trochlea and capitulum articulate with the radius and ulna at the elbow.
	Radius and Ulna 223
	■ The radius and ulna are the bones of the forearm.
	Carpals, Metacarpals, and Phalanges 228
	■ The carpal bones are the scaphoid, lunate, triquetrum, and pisiform (proximal row) and the trapezium, trapezoid, capitate, and hamate (distal row).

(Chapter One — A First Look at Anatomy 21)

CHALLENGE YOURSELF

Matching

Match each numbered item with the most closely related lettered item.

_____ 1. cranial   a. study of tissues

_____ 2. cytology   b. toward the tail

_____ 3. responsiveness   c. contains spinal cord

_____ 4. inguinal region   d. structural change in the body

_____ 5. caudal   e. study of organs of one system

_____ 6. development   f. thoracic cavity

_____ 7. vertebral cavity   g. detect and react to stimuli

_____ 8. histology   h. groin

_____ 9. mediastinum   i. toward the head

_____ 10. systemic anatomy   j. study of cells

Multiple Choice

Select the best answer from the four choices provided.

_____ 6. Which body cavity is located inferior to the diaphragm and superior to a horizontal line drawn between the superior edges of the hip bones?
a. abdominal cavity
b. thoracic cavity
c. pleural cavity
d. pelvic cavity

_____ 7. The term used when referring to a body structure that is below, or at a lower level than, another structure is
a. ventral.
b. medial.
c. inferior.
d. distal.

_____ 8. Which medical imaging technique uses modified x-rays to prepare three-dimensional cross-sectional "slices" of the body?
a. radiography
b. sonography
c. PET (positron emission tomography) scan
d. computed tomography (CT)

_____ 9. The _____ region is the "front" of the knee.

---

436 Chapter Fourteen Nervous Tissue

_____ 9. At an electrical synapse, presynaptic and postsynaptic membranes interface through
a. neurofibril nodes.
b. gap junctions.
c. telodendria.
d. neurotransmitters.

_____ 10. The epineurium is
a. a thick, dense irregular connective tissue layer enclosing the nerve.
b. a group of axons.
c. a delicate layer of areolar connective tissue.
d. a cellular layer of dense regular connective tissue.

Content Review

1. What are the three structural types of neurons? How do they compare to the three functional types of neurons?
2. What is the function of sensory neurons?
3. Identify the principal types of glial cells, and briefly discuss the function of each type.
4. How does the myelin sheath differ between the CNS and the PNS?
5. Describe the procedure by which a PNS axon may repair itself.
6. Describe the arrangement and structure of the three coverings that surround axons in ANS nerves.

7. Clearly distinguish among the following: a neuron, an axon, and a nerve.
8. What are the differences between electrical and chemical synapses? Which is the more common type of synapse in humans?
9. Discuss the similarities and differences between converging and parallel-after-discharge circuits.
10. What are the basic developmental events that occur during neurulation?

Developing Critical Reasoning

1. Over a period of 6 to 9 months, Marianne began to experience vision problems as well as weakness and loss of fine control of the skeletal muscles in her leg. Blood tests revealed the presence of antibodies (immune system proteins) that attack myelin. Beyond the presence of the antibodies, what was the cause of Marianne's vision and muscular difficulties?

2. Surgeons were able to reattach an amputated limb, sewing both the nerves and the blood vessels back together. After the surgery, which proceeded very well, the limb regained its blood supply almost immediately, but the limb remained motionless and the patient had no feeling in it for several months. Why did it take longer to reestablish innervation than circulation?

ANSWERS TO "WHAT DO YOU THINK?"

1. The term *visceral* refers to organs, especially thoracic and abdominal organs such as the heart, lungs, and gastrointestinal tract. Therefore, the parts of the sensory and motor nervous systems that innervate these viscera are called the visceral sensory and visceral motor (autonomic) nervous systems.

2. Tumors occur due to uncontrolled mitotic growth of cells. Since glial cells are mitotic and neurons typically are

nonmitotic, a "brain tumor" almost always develops from glial cells.

3. A myelinated axon takes up more space than an unmyelinated axon. There simply isn't enough space in the body to hold myelin sheaths for every axon. Thus, the body conserves this space by myelinating only the axons that must transmit nerve impulses very rapidly.

Visit the McKinley/O'Loughlin *Human Anatomy*, 2e website at aris.mhhe.com

Teaching and Learning Supplements

FOR INSTRUCTORS

Instructors can obtain teaching aids to accompany this textbook by visiting www.mhhe.com, calling 800-338-3987, or contacting a McGraw-Hill sales representative.

Textbook Website

McGraw-Hill's ARIS (Assessment, Review, and Instruction System) website, found at aris.mhhe.com, is a complete electronic homework and course management system. Available upon adoption of *Human Anatomy*, ARIS allows instructors to create and share course materials and assignments with colleagues—or disseminate them to students—with a few clicks of the mouse. Instructors can edit questions, import their own content, and create announcements and due dates for assignments. ARIS features automatic grading and reporting of easy-to-assign homework, quizzing, and testing. Once a student is registered in the course, all student activity within McGraw-Hill's ARIS is automatically recorded and available to the instructor through a fully integrated grade book that can be downloaded to Excel.

The instructors' ARIS site for McKinley/O'Loughlin also houses book-specific instructor materials, such as image files, question banks, and instructor's manual files. To access ARIS, request registration information from your McGraw-Hill sales representative.

Resource Library

Build instructional materials where-ever, when-ever, and how-ever you want! McGraw-Hill's Presentation Center is an online digital library containing assets such as illustrations, photographs, animations, PowerPoints, and other media files that can be used to create customized lectures, visually enhanced tests and quizzes, compelling course websites, or attractive printed support materials.

Access to your book, access to all books! The Presentation Center library includes thousands of assets from many McGraw-Hill titles. This ever-growing resource gives instructors the power to utilize assets specific to an adopted textbook as well as content from all other books in the library.

Nothing could be easier! Accessed from the instructor side of your textbook's ARIS website, Presentation Center's dynamic search engine allows you to explore by discipline, course, textbook chapter, asset type, or keyword. Simply browse, select, and download the files you need to build engaging course materials.

All assets are copyright McGraw-Hill Higher Education but can be used by instructors for classroom purposes.

Test Bank

A computerized test bank that uses testing software to quickly create customized exams is available for this text. The user-friendly program allows instructors to search for questions by topic or format, edit existing questions or add new ones, and scramble questions for multiple versions of the same test. Word files of the test bank questions are provided for those instructors who prefer to work outside the test-generator software.

Laboratory Manual

The *Human Anatomy Laboratory Manual*, by Christine Eckel of Salt Lake Community College and the University of Utah Medical School, is expressly written to supplement and expand upon content covered in the lecture course—not to repeat it. This hands-on learning tool guides students through human anatomy lab exercises using observation, touch, dissection, and practical activities such as sketching, labeling, and coloring. The manual focuses on human specimens, and also includes common animal dissections such as cow bone, cow eye, sheep brain, and sheep heart.

eInstruction

McGraw-Hill has partnered with eInstruction to bring the revolutionary Classroom Performance System (CPS) to the classroom. An instructor using this interactive system can administer questions electronically during class while students respond via hand-held remote-control keypads. Individual responses are logged into a grade book, and aggregated responses can be displayed in graphic form to provide immediate feedback on whether students understand a lecture topic or more clarification is needed. CPS promotes student participation, class productivity, and individual student accountability.

Course Delivery Systems

With help from our partners—WebCT, Blackboard, Top-Class, eCollege, and other course management systems—professors can take complete control of their course content. Course cartridges containing content from the ARIS textbook website, online testing, and powerful student tracking features are readily available for use within these platforms.

FOR STUDENTS

Students can order supplemental study materials by calling 800-262-4729 or by contacting their campus bookstore.

Anatomy & Physiology|REVEALED

This amazing multimedia tool is designed to help students learn and review human anatomy using cadaver specimens. Detailed cadaver photographs blended with a state-of-the-art layering technique provide a uniquely interactive dissection experience for the eleven body systems. *Anatomy & Physiology Revealed 2.0* features the following sections:

- **Dissection** Peel away layers of the human body to reveal structures beneath the surface. Structures can be pinned and labeled, just as in a real dissection lab. Each labeled structure is accompanied by detailed information and an audio pronunciation. Dissection images can be captured and saved.
- **Animation** Compelling animations demonstrate muscle actions, clarify anatomic relationships, or explain difficult concepts.
- **Histology** Labeled light micrographs presented with each body system allow students to examine tissues at their own pace.
- **Imaging** Labeled X-ray, MRI, and CT images familiarize students with the appearance of key anatomic structures as seen through different medical imaging techniques.
- **Self-Test** Challenging exercises let students test their ability to identify anatomic structures in a timed practical exam format or via traditional multiple choice. A results page provides analysis of test scores plus links back to all incorrectly identified structures for review.

- **Anatomy Terms** This visual glossary of general terms includes directional and regional terms, as well as planes and terms of movement.

Textbook Website

McGraw-Hill's ARIS (Assessment, Review, and Instruction System) for *Human Anatomy* at aris.mhhe.com offers students access to a vast array of online content to fortify the learning experience.

- **Text-Specific Study Tools** The McKinley/O'Loughlin ARIS site features quizzes, interactive learning games, and study tools tailored to coincide with each chapter of the text.
- **Online Tutoring** A 24-hour tutorial service moderated by qualified instructors means help is only an email away.
- **Course Assignments and Announcements** Students of instructors choosing to utilize McGraw-Hill's ARIS tools for course administration will receive a course code to log into their specific course for assignments.
- **Essential Study Partner** This collection of interactive study modules contains animations, learning activities, and quizzes designed to help students grasp complex concepts.

Virtual Anatomy Dissection Review

This multimedia program, created by John Waters of Pennsylvania State University and Melissa Janssen and Donna White of Collin County Community College, contains vivid, high-quality, labeled cat dissection photographs. Available online or on CD, the program helps students easily identify and compare cat structures with the corresponding human structures.

ACKNOWLEDGMENTS

Many people worked with us to produce this text. First and foremost, we are grateful for the tireless efforts of our Developmental Editor at McGraw-Hill, Kris Queck. Kris was a member of our "team" for the many years it took to bring the first edition to fruition, and has continued in this role with the second edition. Kris's attention to detail, her graphic artist's eye for making text and art visually pleasing as well as accurate, and her never-ending dedication to this project deserve many accolades. Kris is more than an excellent Developmental Editor; she is a good friend who worked "in the trenches" with us to produce the best book possible. Thank you, Kris!

This project was undertaken because of the vision and direction of Marty Lange, Vice President and Editor-in-Chief of McGraw-Hill Higher Education–Science, Engineering, and Math. We are grateful for his confidence in us as authors, and for his direction and support in the early development of this book.

Many others at McGraw-Hill have helped us bring the second edition of this text to market, and we wish to thank them all. Publisher Michelle Watnick and Sponsoring Editor Colin Wheatley attended to the details and behind-the-scenes planning. Marketing Manager Lynn Breithaupt helped promote our text and educate sales reps on its message. Project Manager Mary E. Powers helped us keep things moving through the various stages of production. Designer David Hash ensured that the design and art looked their best, and Photo Research Coordinator Carrie Burger managed the details of the photo program. Media Producer Jake Theobald, Media Project Manager Judi David, and Editorial Coordinator Ashley Zellmer saw to the details of creating the many ancillary materials that accompany our book. All of the professionals we have encountered throughout McGraw-Hill have been wonderful to work with, and we sincerely appreciate their efforts.

We were also fortunate to work with a number of individuals outside the McGraw-Hill organization who contributed their specific talents to various tasks. A very special thank you goes to Laurel Shelton, one of Mike's former students, who served as our administrative assistant for the first edition. Laurel provided us with a "student's eye" on the text, notifying us whenever she felt a student wouldn't understand our "technical-ese," and even offered some study tips and wording changes. We never could have done this without you, Laurel.

We are forever indebted to Kennie Harris, our copy editor for the first and second editions, who spent much time perfecting the descriptions to make them clear, concise, and consistent throughout all 28 chapters. Thank you, Kennie!

Christine Eckel of Salt Lake Community College worked tirelessly to produce the bone and cadaver dissection photos for this text. Christine has a wonderful artistic eye as well as a thorough knowledge of anatomy. Many thanks also to Jw Ramsey, Physiology Course Coordinator at Indiana University. His superb photography skills and meticulous photo preparation produced some of the best surface anatomy photos we have seen. In addition, we want to thank the numerous volunteers from Medical Sciences and the Bloomington, IN, community who served as models for our surface anatomy photos—your contributions are greatly appreciated. Al Telser of Northwestern University contributed numerous beautiful photomicrographs for this book and shot many images made to order. We are grateful to Dr. Mark Braun of Indiana University for both his thorough review of the text and his wonderfully written Clinical Views. Kudos to you, Mark! Frank Baker of Golden West College was our indispensable pronunciation and word root researcher. He spent many hours making sure each entry conformed to *Stedman's Medical Dictionary*.

One of the things we are most proud of is the beautiful, accurate, eye-catching artwork that graces the pages of this book. Electronic Publishing Services Inc. was the medical illustration firm that worked with us to create these wonderful drawings, as well as the carefully crafted page layouts. Eileen Mitchell, Kim Moss, Lisa Kinne, and the rest of the EPS team did their utmost to create exceptional artwork. We believe the EPS team has produced the most anatomically accurate and visually pleasing artwork of all the undergraduate anatomy texts currently on the market. We cannot thank EPS enough for their incredible efforts.

Numerous external reviewers and advisors evaluated the first edition and provided invaluable comments and suggestions that we utilized for the second edition revisions. We took everyone's comments to heart and incorporated as many of them as were reasonably possible. These reviewers ensured that this text was as accurate as possible, and for that we are grateful. They are recognized in the list that follows.

Finally, we could not have performed this effort were it not for the love and support of our families. The McKinley and O'Loughlin families provided us with the encouragement we needed, were forgiving when our book schedules made it seem as if we were working all the time, and made sacrifices along with us in order to see this project to fruition. Jan, Renee, Shaun, Ryan and Bob and Erin—thank you and we love you! We are blessed to have you.

Reviewers

Jonathan H. Anning
Slippery Rock University

Tamatha R. Barbeau
Francis Marion University

Debra J. Barnes
Contra Costa College

Fredric Bassett
Rose State College

Steven Bassett
Southeast Community College

Edward T. Bersu
University of Wisconsin

Leann Blem
Virginia Commonwealth University

Mary Bracken
Trinity Valley Community College

Alphonse R. Burdi
University of Michigan

Christine A. Byrd
Western Michigan University

David Cherney
Azusa Pacific University

Alexander G. Cheroske
Moorpark College

Jett S. Chinn
Cañada College

Roger D. Choate
Oklahoma City Community College

Harold Cleveland
University of Central Oklahoma

David F. Cox
Lincoln Land Community College

Paul V. Cupp, Jr.
Eastern Kentucky University

John M. Fitzsimmons
Michigan State University

Allan Forsman
East Tennessee State University

Carl D. Frailey
Johnson County Community College

Michael E. Fultz
Morehead State University

David R. Garris
University of Missouri–Kansas City

1

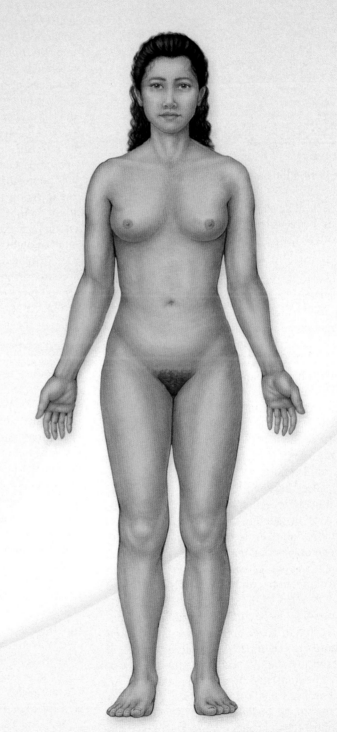

A First Look at Anatomy

You are about to embark on an exciting adventure in the world of human anatomy, investigating the structure and organization of an incredible machine, the human body. Human anatomy is an applied science that provides the basis for understanding health and physical performance. In this book, you will find that structure and function are inseparable, and you will discover what happens when the body works normally, as well as how it is affected by injury or disease.

Study Tip!

Throughout these chapters, boxed elements like this provide helpful analogies, mnemonics, and other "study tips" to help you better understand and learn the material. Look for these boxes throughout each chapter.

History of Human Anatomy

Key topics in this section:

- Early scientists' contributions to the field of human anatomy
- Significant technological developments that helped expand the study of human body structures and pass on that knowledge

For several centuries B.C., the main centers of the scientific world were in ancient Greece and Egypt. Around 400 B.C., the Greek physician Hippocrates developed a medical practice based on observations and studies of the human body. Hippocrates worked to accurately describe disease symptoms and thought that a physician should treat the body as a whole rather than as a collection of individual parts. Hippocrates is called the "Father of Medicine."

The ancient Egyptians had developed specialized knowledge in some areas of human anatomy, which they applied to efforts to mummify their deceased leaders. In Alexandria, Egypt, one of the great anatomy teachers in 300 B.C. was Herophilus, a Greek scientist who was the first to publicly dissect and compare human and animal bodies. Many of the early descriptions of anatomic structures were a result of his efforts. He is known as the "Father of Anatomy" because he based his conclusions (such as that blood vessels carry blood) on human dissection. The work of Herophilus greatly influenced Galen of Pergamum, who lived between 130 and 200 A.D. and was dubbed the "Prince of Physicians" because he stressed the importance of experimentation in medicine. Galen wrote many treatises, including *On the movement of the chest and of the lung*, *On anatomical procedure*, and *On the uses of the parts of the body of man*.

Advancements in anatomy were curtailed for almost a thousand years from 200 to 1200 A.D. Western Europeans had lost the anatomic treatises attributed to Galen. However, these works had been translated into Arabic by Islamic scholars. After 1200 A.D. Galen's treatises began to be translated from Arabic into Latin. In the mid-1200s the first European medical school was established in Italy at Salerno. There, human bodies were dissected in public. Importantly, in the mid-1400s, movable type and copperplate engraving were invented, thus providing a means for disseminating anatomic information on a larger scale. Just before 1500, in Padua, Italy, an anatomic theater opened and became the centerpiece for the study of human anatomy.

Illustrations became a way of recording anatomic findings and passing on that knowledge (**figure 1.1a**). Leonardo da Vinci began his study of the human body around 1500. He is considered one of the greatest anatomists and biological investigators of all time. Da Vinci became fascinated with the human body when he performed dissections to improve his drawing and painting techniques. In the mid-1500s, Andreas Vesalius, a Belgian physician and anatomist, began a movement in medicine and anatomy that was characterized by "refined observations." He organized the medical school classroom in a way that brought students close to the operating table. His dissections of the human body and descriptions of his findings helped correct misconceptions that had existed for 2000 years. Vesalius was called the "Reformer of Anatomy" because he promoted the idea of "living anatomy." His text, *De Humani Corporis Fabrica*, was the first anatomically accurate medical textbook, and the fine engravings in the book were produced from his personal sketches.

William Harvey was an Englishman who studied medicine at the University of Padua in Italy in the early 1600s, a time when this was the center for western European medical instruction. In 1628 he published a book, entitled *An Anatomical Study of the Motion of the Heart and of the Blood in Animals*, that described how blood was pumped from the heart to the body and then back to the heart. His ideas on recirculation formed the basis for modern efforts to study the heart and blood vessels. In a second publication, *Essays on the Generation of Animals*, Harvey established the basis for modern embryology.

A new art form for anatomy, called the preserved specimen, appeared in the late 1600s when anatomists began to collect bodies and body parts. Since these were real specimens, viewers of the exhibits containing these specimens were astonished.

In the 1700s the quality of anatomic illustrations improved dramatically with the simultaneous development of etching and engraving techniques along with mezzotint that provided beauty and texture. By the late 1700s to early 1800s, anatomists began to ensure that scientific illustrations were as accurate and realistic as possible by removing imaginative visual elements from artistic efforts.

Anatomists discovered in the early 1800s that cross sections obtained from frozen cadavers and parts of cadavers provided incredible insight into the complexity of the human body. The nature of the frozen specimens improved in the 1900s with advancements in this field, which came to be called cryotechnology. In the late 1980s the Visible Human Project began. Two donated bodies were deep-frozen in blue gelatin, and then cut into extremely thin cross sections from head to toe. Each newly exposed layer was photographed digitally for computer analysis.

Currently, a new technology to explore the wonders of human anatomy is sweeping the world in the form of Gunther von Hagens's "Body Worlds: The Anatomical Exhibition of Real Human Bodies." Von Hagens is a German anatomist who invented plastination, a unique technology that preserves specimens using reactive polymers. He has remarked that he observed specimens embedded in plastic and wondered, "Why not develop a way to force the plastic into the cells?" His technique has produced fantastic examples of preserved bodies for observation and study (figure 1.1b).

WHAT DID YOU LEARN?

1. What research method that is still used today formed the basis of our earliest knowledge about human body structure?

2. How did the invention of movable type and engraving techniques contribute to the science of human anatomy?

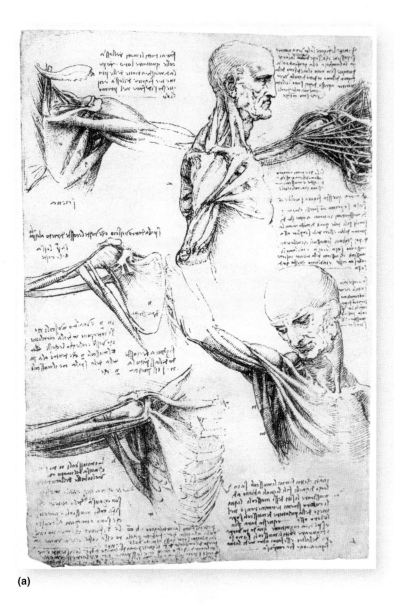

(a)

(b)

Figure 1.1

Aids for Anatomical Study.   (*a*) Early anatomists recorded the findings from their dissections of the human body by making detailed drawings. (*b*) Plastination is a recent technique that preserves body parts for further observation and study. Image taken from Body Worlds.

Study Tip!

The basic vocabulary used in anatomy is derived from Greek and Latin. Actively using this vocabulary will enhance your understanding and appreciation of normal body structure and function. Breaking a word into smaller parts can help you understand and remember its meaning. In this book, we frequently provide word derivations for new terms following their pronunciations. For example, in the case of histology, the study of tissues, we give (*histos* = web, tissue, *logos* = study). Many biological terms share some of the same prefixes, suffixes, and word roots, so learning the meanings of these can help you figure out the meanings of unfamiliar terms the first time you encounter them. A review of prefixes, suffixes, and word roots appears on the inside of the back cover of this book.

Definition of Anatomy

Key topics in this section:

- How anatomy differs from physiology
- Microscopic anatomy and its subdivisions
- Gross anatomy and some of its subdisciplines

Anatomy is the study of structure. The word anatomy is derived from Greek and means "to cut apart." Anatomists, scientists who study anatomy, examine the relationships among parts of the body as well as the structure of individual organs. Often the anatomy of specific body parts suggests their functions. The scientific discipline that studies the function of body structures is called **physiology**. A special relationship exists between anatomy and physiology because structure and function cannot be completely separated. The examples in **table 1.1** illustrate the differences and the interrelationships between anatomy (structure) and physiology (function).

The discipline of anatomy is an extremely broad field that can be divided into two general categories: microscopic anatomy and gross anatomy.

Microscopic Anatomy

Microscopic anatomy examines structures that cannot be observed by the unaided eye. For most such studies, scientists prepare individual cells or thin slices of some part of the body and examine them by microscope. Even so, there are limits to the magnification possible based on the sophistication of the equipment used.

Figure 1.2 illustrates how the microscope has evolved from the primitive form first developed in the seventeenth century to a modern microscope commonly found in anatomy labs today. Specialized subdivisions of microscopic anatomy are defined by the dimensional range of the material being examined. For example, **cytology** (sī-tol′ō-jē; *cyto* = cell, *logos* = study), or cellular anatomy, is the study of single

Table 1.1	Comparison of Anatomy and Physiology	
Anatomy		**Physiology**
The muscles of the thigh are composed of skeletal muscle tissue and receive innervation from somatic motor neurons. These muscles include the quadriceps and the hamstrings, which are designed to extend and flex the knee, respectively.		The muscles of the thigh are able to voluntarily contract and provide enough power to move the parts of the lower limbs during a foot race.
The wall of the small intestine contains two layers of smooth muscle: an inner circular layer and an outer longitudinal layer. The smooth muscle cells are spindle shaped and lack the striations seen in skeletal muscle.		The muscles of the intestinal wall contract slowly and involuntarily to gently squeeze and compress the internal chamber of the small intestine during digestion, processing, and absorption of ingested food.
The esophageal wall is composed of an innermost nonkeratinized stratified squamous epithelium, a middle layer of dense irregular connective tissue external to the epithelium, and an outer layer of muscle tissue (sometimes smooth muscle only, sometimes a combination of smooth and skeletal muscle tissue, and sometimes skeletal muscle only). The lumen (the inside opening of the esophagus) is thrown into folds.		The esophageal wall is designed to withstand the abrasive activities associated with swallowing food, and the muscle layers contract to propel food toward the stomach.
The walls of blood capillaries are composed of a thin epithelium called simple squamous epithelium. Some types of capillary walls also have fenestrations (openings) between the epithelial cells.		The structure of the capillary walls promotes nutrient and waste exchange between the blood and surrounding body fluid.

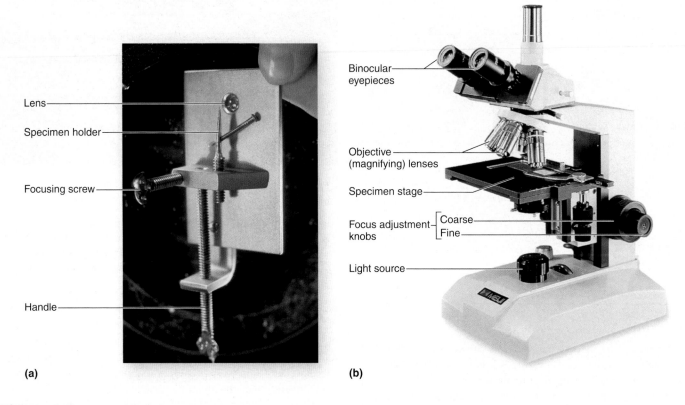

(a)   (b)

Figure 1.2

Microscopy. Scientists use the microscope to magnify objects and structures that cannot be seen by the unaided eye. (*a*) Brass replica of the first microscope, invented by Antoni van Leeuwenhoek. (*b*) A typical microscope used by students today.

body cells and their internal structures, while **histology** (his-tol′ō-jē; *histos* = web or tissue, *logos* = study) is the study of tissues. Histology takes a wider approach to microscopic anatomy by examining how groups of specialized cells and their products function for a common purpose.

Gross Anatomy

Gross anatomy, also called *macroscopic anatomy*, investigates the structure and relationships of large body parts that are visible to the unaided eye, such as the intestines, stomach, brain, heart, and kidneys. In these macroscopic investigations, preserved specimens or their parts are often cut open (dissected) for examination. There are several approaches to gross anatomy:

- **Comparative anatomy** examines the similarities and differences in the anatomy of species.
- **Developmental anatomy** investigates the changes in structure within an individual from conception through maturity.
- **Embryology** (em-brē-ol′ō-jē; *embryon* = young one) is concerned specifically with developmental changes occurring prior to birth.

- **Regional anatomy** examines all the structures in a particular region of the body as one complete unit—for example, the skin, connective tissue and fat, bones, muscles, nerves, and blood vessels of the neck.
- **Surface anatomy** examines both superficial anatomic markings and internal body structures as they relate to the skin covering them. Health-care providers use surface features to identify and locate specific bony processes at joints as well as to obtain a pulse or a blood sample from a patient.
- **Systemic anatomy** studies the gross anatomy of each system in the body. For example, studying the urinary system would involve examining the kidneys, where urine is formed, along with the organs of urine transport (ureters and urethra) and storage (urinary bladder).

Several specialized branches of anatomy focus on the diagnosis of medical conditions or the advancement of basic scientific research:

- **Pathologic** (path-ō-loj′-ik; *pathos* = disease) **anatomy** examines all anatomic changes resulting from disease.
- **Radiographic anatomy** studies the relationships among internal structures that may be visualized by specific scanning procedures, such as ultrasound, magnetic resonance imaging (MRI), or x-ray.
- **Surgical anatomy** investigates the anatomic landmarks used before and after surgery. For example, prior to back surgery, the location of the L_4 vertebra is precisely identified

by drawing an imaginary line between the hip bones. The intersection of this line with the vertebral column shows the location of L_4.

Although you might at first assume that the field of anatomy has already been completely described, it is not fixed. Anatomic studies are ongoing, and the success of the discipline depends upon precise observation, thorough description, and correct use of terminology. These tools are essential to your eventual mastery of the discipline.

WHAT DID YOU LEARN?

3 What is the relationship between anatomy and physiology?

4 What are some of the subdisciplines of gross anatomy?

Structural Organization of the Body

Key topics in this section:

- Levels of organization in the human body
- Characteristics of life
- The 11 organ systems of the body

Anatomists recognize several levels of increasingly complex organization in humans, as illustrated in **figure 1.3**. The simplest level of organization within the body is the **chemical level,** which is composed of atoms and molecules. **Atoms** are the smallest units

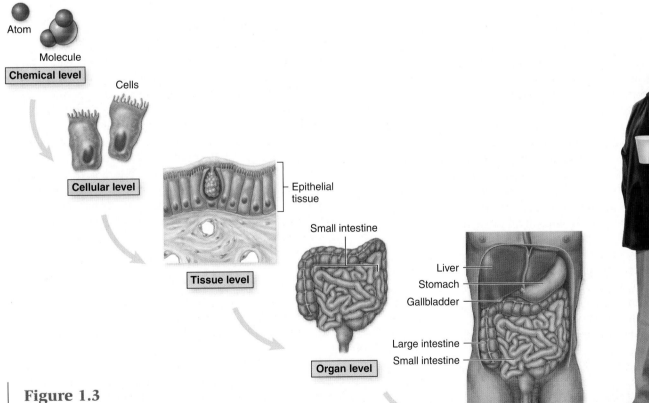

Atom

Molecule

Chemical level

Cells

Cellular level

Epithelial tissue

Tissue level

Small intestine

Organ level

Liver

Stomach

Gallbladder

Large intestine

Small intestine

Organ system level

Organismal level

Figure 1.3

Levels of Organization in the Human Body.   At each succeeding level, the structure becomes more complex.

of matter; two or more atoms combine to form a **molecule**, such as a protein, a water molecule, or a vitamin. Large molecules join in specific ways to form **cells**, the basic units of structure and function in organisms. At the **cellular level**, specialized structural and functional units called organelles permit all living cells to share certain common functions. The structures of cells vary widely, reflecting the specializations needed for their different functions. For example, a muscle cell may be very long and contain numerous organized proteins that aid in muscle contraction, whereas a blood cell is small, round, and flat, and designed to exchange respiratory gases quickly and effectively as it travels through the blood vessels.

Groups of similar cells with a common function form the next stage in the hierarchy, the **tissue level. Tissues** are precise organizations of similar cells that perform specialized functions. The four types of tissues and their general roles in the human body are: epithelial tissue (covers exposed surfaces and lines body cavities); connective tissue (protects, supports, and interconnects body parts and organs); muscle tissue (produces movement); and nervous tissue (conducts impulses for internal communication).

At the **organ level**, different tissue types combine to form an organ, such as the small intestine, brain, lungs, stomach, or heart. **Organs** contain two or more tissue types that work together to perform specific, complex functions. The small intestine, for example, has different structural and organizational relationships within its tissues that work together to process and absorb digested nutrients. Thus, the small intestine shown in figure 1.3 exhibits all four tissue types: an internal lining composed of simple columnar epithelium; a connective tissue layer that attaches the epithelium to an external layer of smooth muscle; and nervous tissue that innervates the organ.

The **organ system level** consists of related organs that work together to coordinate activities and achieve a common function. For example, several organs of the respiratory system (nose, pharynx, and trachea) collaborate to clean, warm, humidify, and conduct air from the atmosphere to the gas exchange surfaces in the lungs. Then special air sacs in the lungs allow exchange to occur between the respiratory gases from the atmosphere and the gases in the blood.

The highest level of structural organization in the body is the **organismal level**. All body systems function interdependently in a single living human being, the **organism**.

The importance of the interrelationships among structural levels of organization in the body becomes apparent when considering the devastating effects a gene mutation (the chemical level) may have on the body (the organismal level). For example, a common consequence of a specific genetic mutation in an individual's DNA is cystic fibrosis (discussed in a Clinical View in chapter 25). This disorder results when a defective or abnormal region in a molecule of DNA affects the normal function of cells in certain body organs. These cells are unable to transport salt across their membranes, thus disrupting the normal salt and water balance in the fluid covering these cells. Abnormal cellular function causes a corresponding failure in the functioning of the tissues composed of these abnormal cells, ultimately resulting in aberrant activity in the organ housing these tissues as well. Organ failure has devastating effects on organ system activities. It is apparent that as the structural level increases in complexity, the effects of a deviance or disruption magnify.

WHAT DO YOU THINK?

1 At which level of organization is the stomach? At which level is the digestive system?

Characteristics of Living Things

Life is neither defined by a single property nor exemplified by one characteristic only. The cell is the smallest structural unit that exhibits the characteristics of living things (organisms), and it is the smallest living portion of the human body. Several properties are common to all organisms, including humans:

- **Organization**. All organisms exhibit a complex structure and order. As mentioned earlier in this section, the human body has several increasingly complex levels of organization.
- **Metabolism**. All organisms carry out various chemical reactions, collectively termed **metabolism**. These chemical reactions include breaking down ingested nutrients into digestible particles, using the cells' own energy to perform certain functions, and contracting and relaxing muscles to move the body. Metabolic activities such as ingesting nutrients and expelling wastes enable the body to continue acquiring the energy needed for life's activities.
- **Growth and Development**. During their lifetime, organisms assimilate materials from their environment and exhibit increased size (growth) and increased specialization as related to form and function (development). As the human body grows in size, structures such as the brain become more complex and sophisticated.
- **Responsiveness**. All organisms sense and respond to changes in their internal or external environment. For example, a stimulus to the skin of the hand, such as extremely hot or cold temperature, causes a human to withdraw the hand from the stimulus, so as to prevent injury or damage.
- **Adaptation**. Over a period of time, an organism may alter an anatomic structure, physiologic process, or behavioral trait to increase its expected long-term reproductive success.
- **Regulation**. Control and regulatory mechanisms within an organism maintain a consistent internal environment, a state called **homeostasis** (hō′mē-ō-stā′sis; *homoios* = similar, *stasis* = standing). In a constantly changing environment, every organism must be able to maintain this "steady state." For example, when the body temperature rises, more blood is circulated near the surfaces of our limbs and digits (fingers and toes) to facilitate heat loss and a return to homeostasis.
- **Reproduction**. All organisms produce new cells for growth, maintenance, and repair. In addition, an organism produces sex cells (called gametes) that, under the right conditions, have the ability to develop into a new living organism (see chapter 3).

Introduction to Organ Systems

All organisms must exchange nutrients, gases, and wastes with their environment in order to carry on metabolism. Simple organisms exchange these substances directly across their surface membranes. Humans, by contrast, are complex, multicellular organisms that require sophisticated, specialized structures and mechanisms to perform the exchanges required for metabolic activities and the routine events of life. In humans, we commonly denote 11 organ systems, each composed of interrelated organs that work together to perform specific functions (**figure 1.4**). Thus, a human body maintains homeostasis, or internal equilibrium, through the intricate interworkings of all its organ systems. Subsequent chapters examine each of these organ systems in detail.

Figure 1.4

Organ Systems. Locations and major components of the 11 organ systems of the human body.

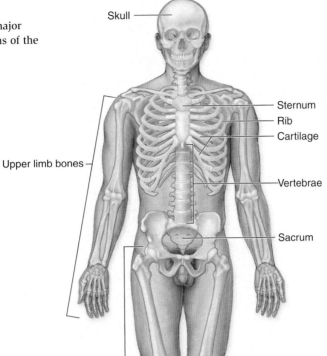

Skull

Sternum

Rib

Cartilage

Upper limb bones

Vertebrae

Sacrum

Lower limb bones

Knee joint

Skeletal System (Chapters 6–9)

Provides support and protection, site of hemopoiesis (blood cell production), stores calcium and phosphorus, provides sites for muscle attachments.

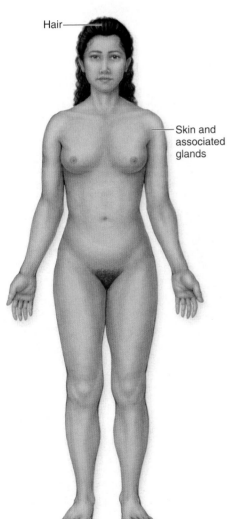

Hair

Skin and associated glands

Integumentary System (Chapter 5)

Provides protection, regulates body temperature, site of cutaneous receptors, synthesizes vitamin D, prevents water loss.

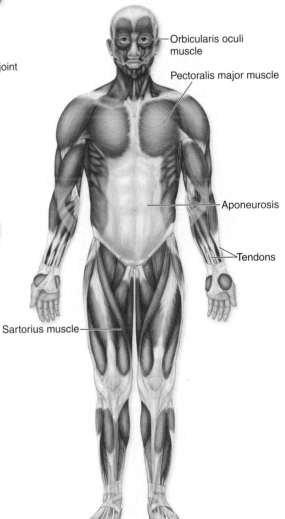

Orbicularis oculi muscle

Pectoralis major muscle

Aponeurosis

Tendons

Sartorius muscle

Muscular System (Chapters 10–12)

Produces body movement, generates heat when muscles contract.

Figure 1.4
Organ Systems. *(continued)*

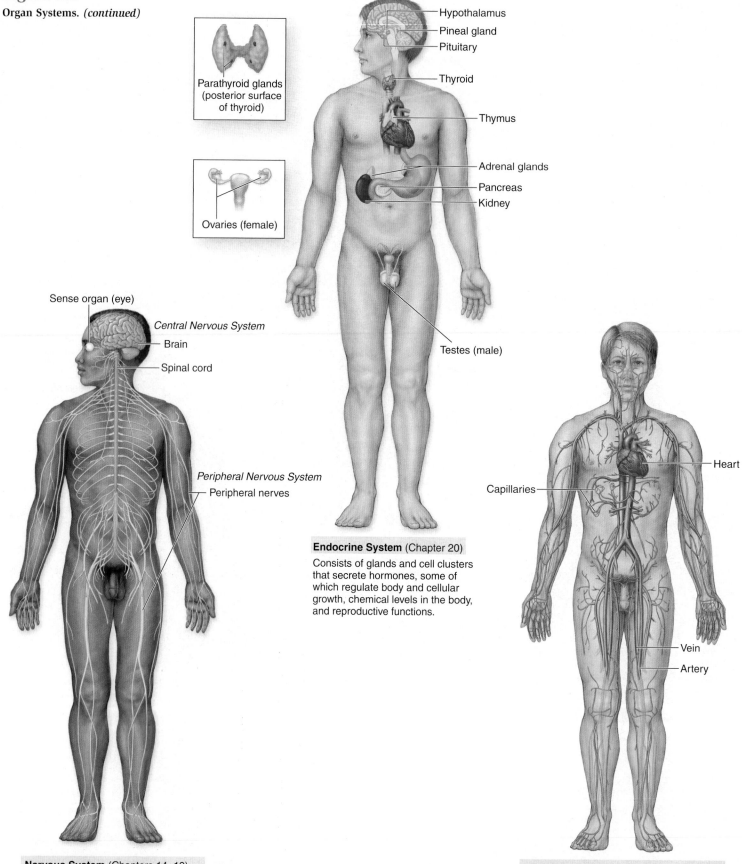

Parathyroid glands
(posterior surface
of thyroid)

Ovaries (female)

Hypothalamus
Pineal gland
Pituitary
Thyroid

Thymus

Adrenal glands
Pancreas
Kidney

Testes (male)

Endocrine System (Chapter 20)

Consists of glands and cell clusters
that secrete hormones, some of
which regulate body and cellular
growth, chemical levels in the body,
and reproductive functions.

Sense organ (eye)

Central Nervous System
Brain
Spinal cord

Peripheral Nervous System
Peripheral nerves

Nervous System (Chapters 14–19)

A regulatory system that controls body
movement, responds to sensory stimuli,
and helps control all other systems of
the body. Also responsible for
consciousness, intelligence, memory.

Heart

Capillaries

Vein
Artery

Cardiovascular System (Chapters 21–23)

Consists of the heart (a pump), blood, and
blood vessels; the heart moves blood
through blood vessels in order to distribute
hormones, nutrients, and gases, and pick up
waste products.

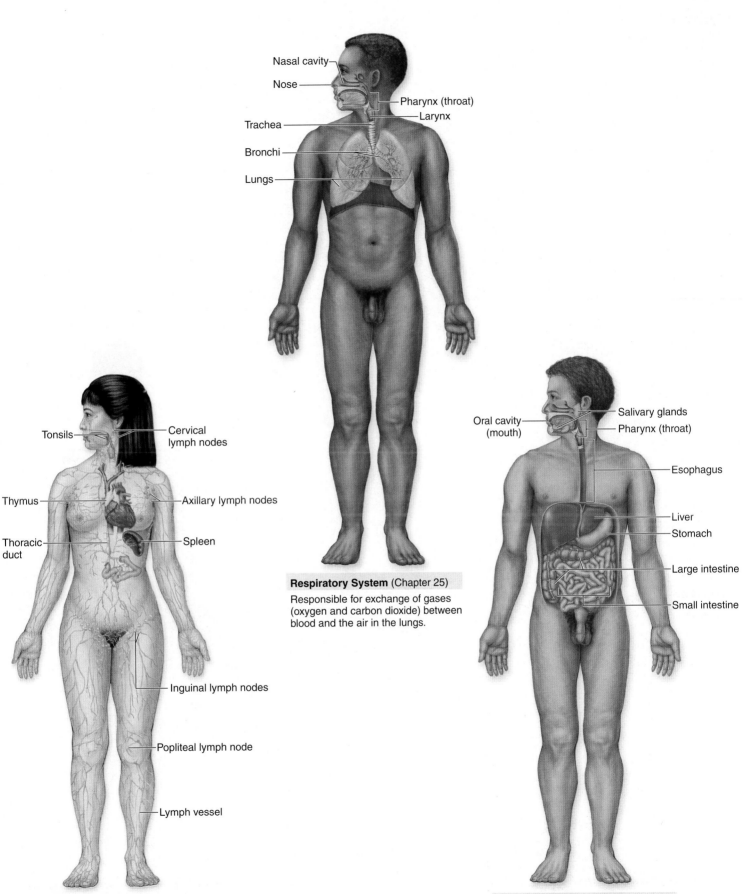

Nasal cavity
Nose
Pharynx (throat)
Larynx
Trachea
Bronchi
Lungs

Respiratory System (Chapter 25)

Responsible for exchange of gases
(oxygen and carbon dioxide) between
blood and the air in the lungs.

Tonsils
Cervical lymph nodes
Thymus
Axillary lymph nodes
Thoracic duct
Spleen
Inguinal lymph nodes
Popliteal lymph node
Lymph vessel

Lymphatic System (Chapter 24)

Transports and filters lymph (interstitial
fluid transported through lymph vessels)
and initiates an immune response when
necessary.

Oral cavity (mouth)
Salivary glands
Pharynx (throat)
Esophagus
Liver
Stomach
Large intestine
Small intestine

Digestive System (Chapter 26)

Mechanically and chemically digests
food materials, absorbs nutrients, and
expels waste products.

Figure 1.4

Organ Systems. *(continued)*

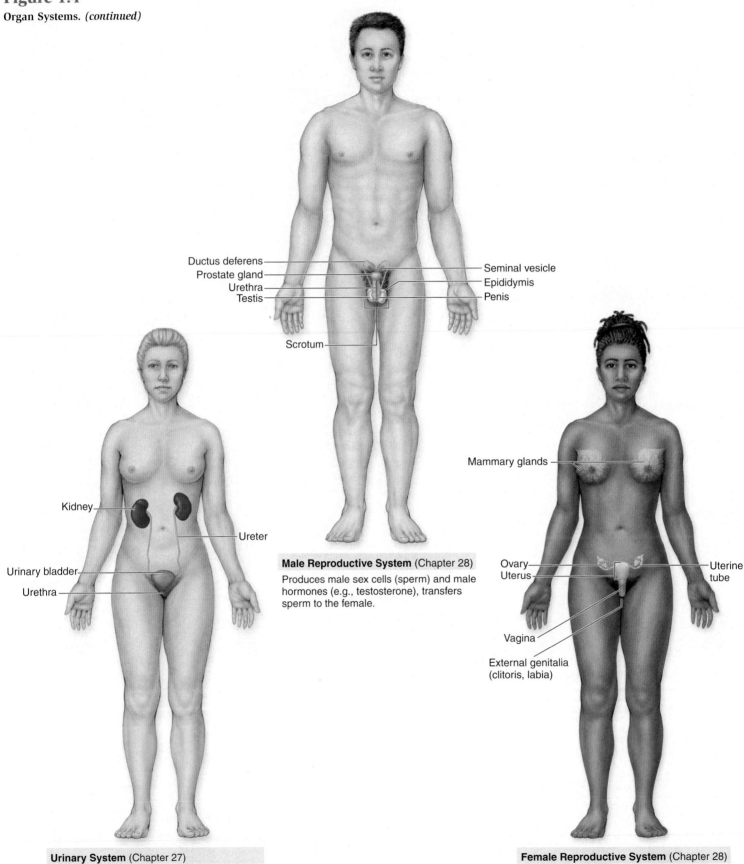

Ductus deferens

Prostate gland

Urethra

Testis

Seminal vesicle

Epididymis

Penis

Scrotum

Male Reproductive System (Chapter 28)

Produces male sex cells (sperm) and male hormones (e.g., testosterone), transfers sperm to the female.

Kidney

Ureter

Urinary bladder

Urethra

Urinary System (Chapter 27)

Filters the blood and removes waste products from the blood, concentrates waste products in the form of urine, and expels urine from the body.

Mammary glands

Ovary

Uterus

Uterine tube

Vagina

External genitalia (clitoris, labia)

Female Reproductive System (Chapter 28)

Produces female sex cells (oocytes) and female hormones (e.g., estrogen and progesterone), receives sperm from male, site of fertilization of oocyte, site of growth and development of embryo and fetus.

WHAT DID YOU LEARN?

5 Which level of organization consists of similar cells that work together to perform a common function?

6 List four characteristics common to all organisms.

Precise Language of Anatomy

Key topics in this section:

- Description and significance of the anatomic position
- Importance of the three common anatomic planes
- Terms that describe directions in the body
- Terms that describe major regions of the body
- Terms that identify the body cavities and their subdivisions
- The nine regions and four quadrants of the abdominopelvic cavity

All of us are interested in our own bodies, but we are often stymied by the seeming mountain of terminology that must be scaled before we can speak the language of anatomy correctly. For the sake of accuracy, anatomists must adhere to a set of proper terms, rather than the descriptive words of everyday conversation. For example, to properly describe human anatomic landmarks, we cannot use such common phrases as "in front of," "behind," "above," or "below." That is, it would be inaccurate to say, "The heart is above the stomach," because the heart appears to be "above" the stomach when a person is standing erect—but not when the person is lying on his or her back. Therefore, anatomists and health-care providers identify and locate body structures using descriptive terms based on the premise that the body is in the anatomic position, defined next.

Study Tip!

— You should always rely on two resource books while using this human anatomy text: *Stedman's Medical Dictionary,* which defines all medical terms, and *Terminologia Anatomica,* which uses the proper anatomic terms and organizes them for all the body systems. Cultivating a familiarity with these resources—and with the origins of terminology—will help you acquire the vocabulary necessary for succeeding in this discipline.

Anatomic Position

Descriptions of any region or body part require an initial point of reference and the use of directional indicators. In the **anatomic position**, an individual stands upright with the feet parallel and flat on the floor. The head is level, and the eyes look forward toward the observer. The arms are at either side of the body with the palms facing forward and the thumbs pointing away from the body **(figure 1.5)**. By visualizing the body in anatomic position, all observers have a common point of reference when describing and discussing its regions. All of the functional and directional terms used in this book refer to the body in anatomic position.

Sections and Planes

Anatomists refer to real or imaginary "slices" of the body, called sections or planes, in order to examine its internal anatomy and describe the position of one body part relative to another. The term *section* implies an actual cut or slice to expose the internal anatomy, while the word *plane* implies an imaginary flat surface passing through the body. The three major anatomic planes through the body or individual organs are the coronal, transverse, and midsagittal planes (figure 1.5).

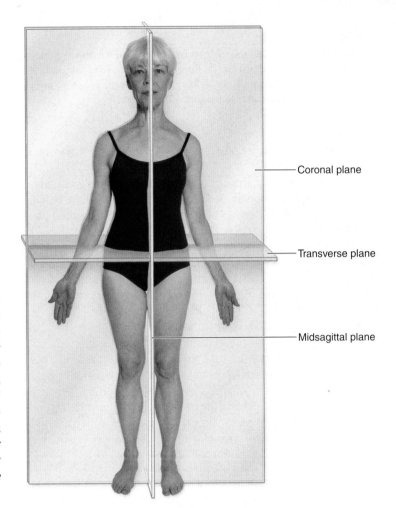

Figure 1.5

Anatomic Position and Body Planes. In the anatomic position, the body is upright, and the forearms are positioned with the palms facing forward. A plane is an imaginary surface that slices the body into specific sections. The three major anatomic planes of reference are the coronal, transverse, and midsagittal planes.

A **coronal** (kōr′ŏ-năl; *korone* = crown) **plane**, also called a *frontal plane*, is a vertical plane that divides the body into *anterior* (front) and *posterior* (back) parts. When a coronal plane is taken through the trunk, the anterior portion contains the chest, and the posterior portion contains the back.

A **transverse plane**, also called a *cross-sectional plane* or *horizontal plane*, cuts perpendicularly along the long axis of the body or organ. The body or organ is separated into both *superior* (upper) and *inferior* (lower) parts, and the relationship of neighboring organs at a particular level is revealed. Computed tomography (CT) scans provide transverse sectional images of the body for study (see Clinical View: In Depth at the end of this chapter).

A **midsagittal plane** (mid′saj′i-tăl; *sagittow* = arrow), or *median plane*, extends through the body or organ vertically and divides the structure into right and left halves. A plane that is parallel to the midsagittal plane, but either to the left or right of it, is termed a **sagittal plane**. Thus, a sagittal plane divides a structure into right and left portions that may or may not be equal. Although there is only one midsagittal plane, an infinite number of sagittal planes are possible. A midsagittal or sagittal plane is often used to show internal body parts, especially in the head and thoracic organs.

In addition to the coronal, transverse, and midsagittal planes, a minor plane, called the **oblique** (ob-lēk′) **plane**, passes through

Figure 1.6

Three-dimensional Reconstruction from Planes of Section.　Serial sections through an object are used to reconstruct its three-dimensional structure, as in these sections of the small intestine. Often a single section, such as the plane at the lower right of this figure, misrepresents the complete structure of the object.

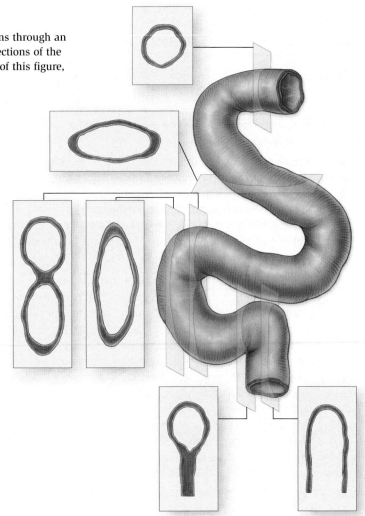

the specimen at an angle. (For an example, see **figure 1.6**, second section from the top.)

Interpreting body sections has become increasingly important for health-care professionals. Technical advances in medical imaging (described in Clinical View: In Depth at the end of this chapter) have produced spectacular sectional images. To determine the shape of any object within a section, we must be able to reconstruct its three-dimensional shape by observing many continuous sections. Just as sections of a curved, twisting tube may have significantly different appearances depending on where the section is taken, sectioning the body or an organ along different planes often results in very different views. For example, all sections (coronal, transverse, and midsagittal) through most regions of the abdominal cavity will exhibit multiple profiles of the long, twisted tube that is the small intestine. These sections will appear as circles, ovals, long tubes with parallel sides, a figure "8," and maybe a solid region because the section is through the wall only. Figure 1.6 shows the results of several possible sections through the small intestine. If you practice mentally converting two-dimensional images into three-dimensional shapes, your ability to assimilate anatomic information will advance quickly.

Anatomic Directions

Once the body is in the anatomic position, we can precisely describe the relative positions of various structures by using specific directional terms. Directional terms are precise and brief, and most of them have a correlative term that means just the opposite. **Table 1.2** and **figure 1.7** describe some important and commonly used

Table 1.2	**Anatomic Directional Terms**		
Direction	**Term**	**Meaning**	**Example**
Relative to front (belly side) or back (back side) of the body	Anterior	In front of; toward the front surface	The stomach is *anterior* to the spinal cord.
	Posterior	In back of; toward the back surface	The heart is *posterior* to the sternum.
	Dorsal	At the back side of the human body	The spinal cord is on the *dorsal* side of the body.
	Ventral	At the belly side of the human body	The umbilicus (navel, belly button) is on the *ventral* side of the body.
Relative to the head or tail of the body	Superior	Closer to the head	The chest is *superior* to the pelvis.
	Inferior	Closer to the feet	The stomach is *inferior* to the heart.
	Caudal	At the rear or tail end	The abdomen is *caudal* to the head.
	Cranial	At the head end	The head is *cranial* to the trunk.
Relative to the midline or center of the body	Medial	Toward the midline of the body	The lungs are *medial* to the shoulders.
	Lateral	Away from the midline of the body	The arms are *lateral* to the heart.
	Deep	On the inside, underneath another structure	Muscles are *deep* to the skin.
	Superficial	On the outside	The external edge of the kidney is *superficial* to its internal structure.
Relative to point of attachment of the appendage	Proximal	Closest to point of attachment to trunk	The elbow is *proximal* to the hand.
	Distal	Furthest from point of attachment to trunk	The wrist is *distal* to the elbow.

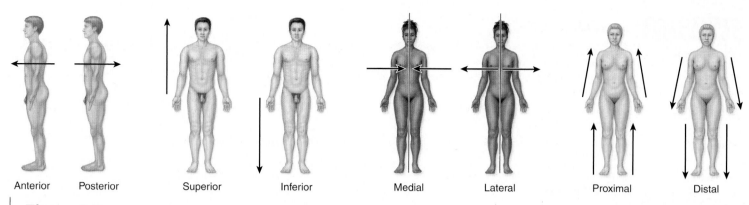

Figure 1.7

Directional Terms in Anatomy.   Directional terms precisely describe the location and relative relationships of body parts. See also table 1.2.

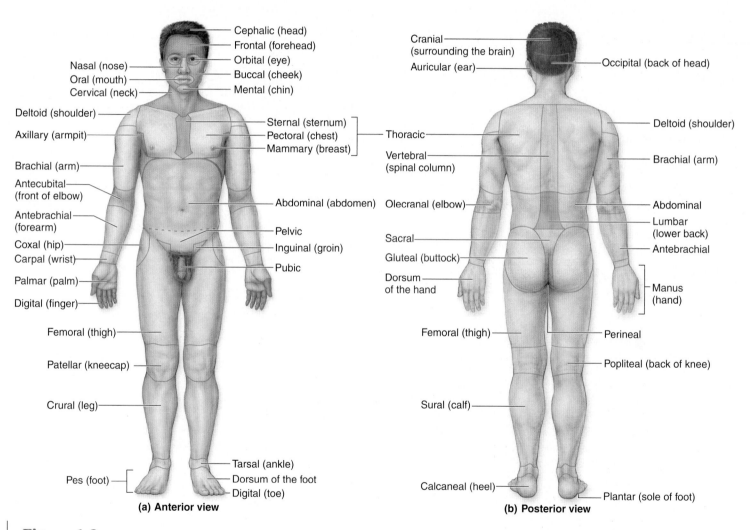

(a) Anterior view

(b) Posterior view

Figure 1.8

Regional Terms.   (*a*) Anterior and (*b*) posterior views identify the key regions of the body. Their common names appear in parentheses.

directional terms. Studying the table and the figure together will maximize your understanding of anatomic directions and aid your study of anatomy throughout the rest of this book.

Regional Anatomy

The human body is partitioned into two main regions, called the axial and appendicular regions. The **axial** (ak′sē-ăl) **region** includes the head, neck, and trunk; it forms the main vertical axis of the body. Our limbs, or appendages, attach to the body's axis and make up the

appendicular (ap′en-dik′ū-lăr) **region**. Several specific terms are used to identify the anatomic areas within these two regions. **Figure 1.8** and **table 1.3** identify the major regional terms and some additional minor ones. (Not all regions are shown in figure 1.8.)

Body Cavities and Membranes

Internal organs and organ systems are housed within separate enclosed spaces, or cavities. These cavities are named according to the bones that surround them or the organs they contain. For

Table 1.3	Human Body Regions			
Region Name	**Description**		**Region Name**	**Description**
Abdominal	Region inferior to the thorax (chest) and superior to the hip bones		Mental	Chin
Antebrachial	Forearm (the portion of the upper limb between the elbow and the wrist)		Nasal	Nose
Antecubital	Region anterior to the elbow; also known as the cubital region		Occipital	Posterior aspect of the head
Auricular	Ear (visible surface structures of the ear and the ear's internal organs)		Olecranal	Posterior of the elbow
Axillary	Armpit		Oral	Mouth
Brachial	Arm (the portion of the upper limb between the shoulder and the elbow)		Orbital	Eye
Buccal	Cheek		Palmar	Palm of the hand
Calcaneal	Heel of the foot		Patellar	Kneecap
Carpal	Wrist		Pelvic	Pelvis
Cephalic	Head		Perineal	Diamond-shaped region between the thighs that contains the anus and selected external reproductive organs
Cervical	Neck		Pes	Foot
Coxal	Hip		Plantar	Sole of the foot
Cranial	Skull		Pollex	Thumb
Crural	Leg (the portion of the lower limb between the knee and the ankle)		Popliteal	Area posterior to the knee
Deltoid	Shoulder		Pubic	Anterior region of the pelvis
Digital	Fingers or toes (also called phalangeal)		Radial	Lateral aspect of the forearm
Dorsal	Back		Sacral	Posterior region between the hip bones
Femoral	Thigh		Scapular	Shoulder blade
Fibular	Lateral aspect of the leg		Sternal	Anterior middle region of the thorax
Frontal	Forehead		Sural	Calf (posterior part of the leg)
Gluteal	Buttock		Tarsal	Root of the foot
Hallux	Great toe		Thoracic	Chest or thorax
Inguinal	Groin (sometimes used to indicate just the crease in the junction of the thigh with the trunk)		Tibial	Medial aspect of the leg
Lumbar	Relating to the loins, or the part of the back and sides between the ribs and pelvis		Ulnar	Medial aspect of the forearm
Mammary	Breast		Umbilical	Navel
Manus	Hand		Vertebral	Spinal column

purposes of discussion, the axial region is subdivided into two areas: the posterior aspect and the ventral cavity.

Posterior Aspect

The **posterior aspect** has two enclosed cavities **(figure 1.9a)**. A **cranial cavity** is formed by the cranium (specifically, the neurocranium) and houses the brain. A **vertebral** (ver′te-brăl) **canal** is formed by the individual bones of the vertebral column and contains the spinal cord. These two cavities are encased in bone and thus

are physically and developmentally different from the ventral cavity. Therefore, the parallel term *dorsal body cavity* is not used here.

Ventral Cavity

The **ventral cavity** arises from a space called the *coelom* that forms during embryonic development. The ventral cavity eventually becomes partitioned into a superior **thoracic** (thō-ras′ik) **cavity** and an inferior **abdominopelvic cavity** with the formation of the thoracic **diaphragm**, a muscular partition that develops between these cavities (figure 1.9).

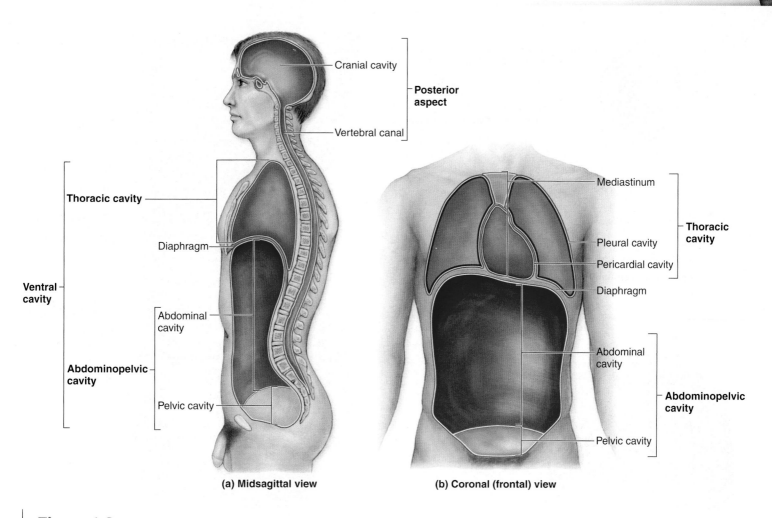

Cranial cavity

Posterior aspect

Vertebral canal

Thoracic cavity

Diaphragm

Ventral cavity

Abdominal cavity

Abdominopelvic cavity

Pelvic cavity

(a) Midsagittal view

Mediastinum

Thoracic cavity

Pleural cavity

Pericardial cavity

Diaphragm

Abdominal cavity

Abdominopelvic cavity

Pelvic cavity

(b) Coronal (frontal) view

Figure 1.9

Body Cavities.   The body is composed of two principal spaces: the posterior aspect and the ventral cavity. Many vital organs are housed within these spaces. (*a*) A midsagittal view shows both the posterior aspect and the ventral cavity. (*b*) A coronal view shows the relationship between the thoracic and abdominopelvic cavities within the ventral cavity.

Both the thoracic and abdominopelvic cavities are lined with thin **serous membranes**, which are composed of two layers. A **parietal layer** lines the internal surface of the body wall, while a **visceral layer** covers the external surface of organs (**viscera**) within the cavity. Between the parietal and visceral layers of the serous membrane is a thin **serous cavity** that is actually a *potential space*. A potential space is capable of becoming a larger cavity. A serous cavity contains a film of serous fluid that is secreted by the cells of the serous membranes. Serous fluid has the consistency of oil, and serves as a lubricant. In a living human, the organs (e.g., heart, lungs, stomach, and intestines) are moving and rubbing against each other and the body wall. This constant movement causes friction. The serous fluid's lubricant properties reduce this friction and help the organs move smoothly against both one another and the body wall.

WHAT DO YOU THINK?

❷ Try this experiment to determine the value of serous fluid: First, rub the palms of your hands quickly against one another. The sound you hear and the heat you feel are consequences of the friction being produced. Now put lotion (our version of serous fluid) on the palms of your hands and repeat the experiment. Do you still hear the noise and feel heat from your hands? What do you think would happen to your body organs if there were no serous fluid?

Figure 1.10a provides a helpful analogy for visualizing the serous membrane layers. The closed fist is comparable to an organ,

and the balloon is comparable to a serous membrane. When a fist is pushed into the wall of the balloon, the inner balloon wall that surrounds the fist is comparable to the visceral layer of the serous membrane. The outer balloon wall is comparable to the parietal layer of the serous membrane. The thin, air-filled space within the balloon, between the two "walls," is comparable to the serous cavity. Note that the organ is not *inside* the serous cavity; it is actually *outside* this cavity and merely covered by the visceral layer of the serous membrane!

Thoracic Cavity   The median space in the thoracic cavity is called the **mediastinum** (me′dē-as-tī′nŭm) (see figure 1.9*b*). It contains the heart, thymus, esophagus, trachea, and major blood vessels.

Within the mediastinum, the heart is enclosed by a two-layered serous membrane called the **pericardium** (see figure 1.10*b*). The **parietal pericardium** (per-i-kar′dē-ŭm; *peri* = around, *kardia* = heart) is the outermost layer and forms the sac around the heart; the **visceral pericardium** (also called *epicardium*; *epi* = upon) forms the heart's external surface. The **pericardial cavity** is the potential space between the parietal and visceral pericardia; it contains serous fluid.

The right and left sides of the thoracic cavity contain the lungs, which are lined by a two-layered serous membrane called the **pleura** (ploor′ă) (see figure 1.10*c*). The outer layer of this serous membrane is the **parietal** (pă-rī′ĕ-tăl) **pleura**; it lines the internal surface of the thoracic wall. The inner layer of this serous membrane is the **visceral**

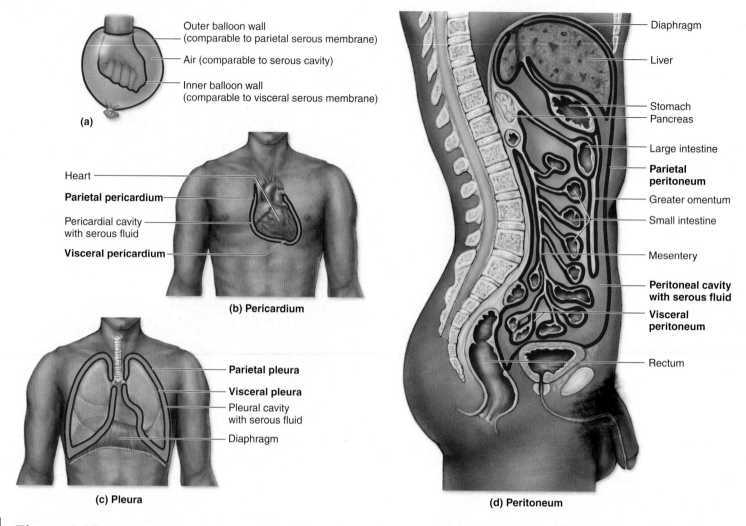

(a)

(b) Pericardium

(c) Pleura

(d) Peritoneum

Figure 1.10

Serous Membranes in the Thoracic and Abdominopelvic Body Cavities.   Serous membranes have two parts: the lining of the inside of the cavity (parietal layer) and the lining of the outside of an organ within the cavity (visceral layer). (*a*) The parietal and visceral serous membranes are similar to the inner and outer balloon walls that wrap around a fist, where the fist represents the body organ. (*b*) Parietal and visceral layers of the pericardium line the pericardial cavity around the heart. (*c*) Parietal and visceral layers of the pleura line the pleural cavity between the lungs and the chest wall. (*d*) Parietal and visceral layers of the peritoneum line the peritoneal cavity that lies between the abdominopelvic organs and the body wall.

pleura; it covers the external surface of the lung. The narrow, moist, potential space between the parietal and visceral layers is called the **pleural cavity**, and is the location of the lubricating serous fluid.

Abdominopelvic Cavity   The abdominopelvic cavity consists of an **abdominal cavity**, which is superior to an imaginary line drawn between the superior aspects of the hip bones, and a **pelvic cavity** that is inferior to this imaginary line. You can locate the division between these two cavities by palpating (feeling for) the superior ridges of your hip bones. The imaginary horizontal plane that rests on the superior ridge of each hip bone partitions these two cavities. The abdominal cavity contains most of the organs of the digestive system, as well as the kidneys and ureters of the urinary system. The organs of the pelvic cavity consist of the distal part of the large intestine, the urinary bladder and urethra, and the internal reproductive organs.

The **peritoneum** (per′i-tō-nē′ŭm; *periteino* = to stretch over) is a moist, two-layered serous membrane that lines the abdomi-

nopelvic cavity (see figure 1.10*d*). The **parietal peritoneum**, the outer layer of this serous membrane, lines the internal walls of the abdominopelvic cavity, whereas the **visceral peritoneum**, the inner layer of this serous membrane, ensheathes the external surfaces of most of the digestive organs. The potential space between these serous membrane layers in the abdominopelvic cavity is the **peritoneal cavity**, where the lubricating serous fluid is located.

Table 1.4 summarizes the characteristics of the body cavities.

Abdominopelvic Regions and Quadrants

In order to accurately describe organ location in the larger abdominopelvic cavity, anatomists and health-care professionals commonly partition the cavity into smaller, imaginary compartments. Nine compartments, called **abdominopelvic regions**, are delineated by using two transverse planes and two sagittal planes. The nine regions are arranged into three rows (superior, middle, and inferior) and three columns (left, middle, and right) (**figure 1.11***a*). Each region has a specific name:

Table 1.4	**Body Cavities**	
Posterior Aspect Cavities	**Description**	**Serous Membrane**
Cranial cavity	Formed by cranium; houses brain	None
Vertebral canal	Formed by vertebral column; contains spinal cord	None
Ventral Cavities	**Description**	**Serous Membrane**
THORACIC CAVITY	Chest cavity; bordered anteriorly and laterally by chest wall and inferiorly by diaphragm	
Mediastinum	Contains the pericardial cavity, thymus, trachea, esophagus, and major blood vessels	None
Pericardial	Contains the heart	Pericardium
Pleural	Contains the lungs	Pleura
ABDOMINOPELVIC CAVITY	Composed of two parts: abdominal and pelvic cavities	
Abdominal	Bordered superiorly by the diaphragm and inferiorly by a horizontal plane between the superior ridges of the hip bones. Associated with the abdominal viscera, including stomach, spleen, liver, pancreas, small intestine, most of large intestine, kidneys, ureters	Peritoneum
Pelvic	Region located between the hip bones and interior to a horizontal plane between the superior ridges of the hip bones. Associated with the pelvic viscera, including urinary bladder and urethra, internal reproductive organs, some of the large intestine	Peritoneum

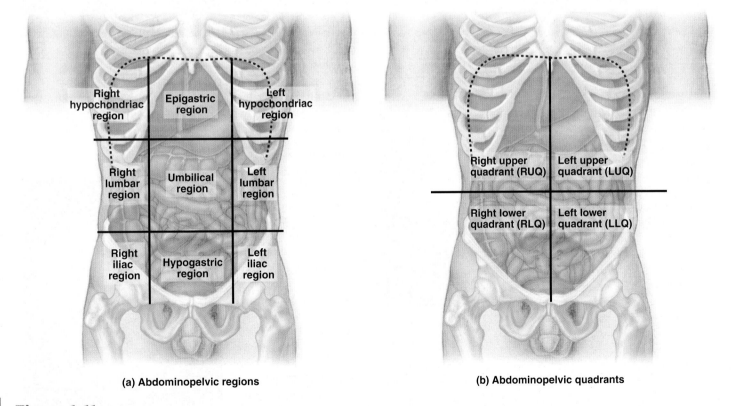

(a) Abdominopelvic regions

(b) Abdominopelvic quadrants

Figure 1.11

Abdominopelvic Regions and Quadrants. The abdominopelvic cavity can be subdivided into (*a*) nine regions or (*b*) four quadrants for purposes of description or identification.

- The **epigastric** (ep-i-gas′trik; *epi* = above, *gaster* = belly) **region**, the superior region in the middle column, typically contains part of the liver, part of the stomach, the duodenum, part of the pancreas, and both adrenal glands.
- The **umbilical** (ŭm-bil′i-kăl; *umbilicus* = navel) **region**, the middle region in the middle column, typically contains the transverse colon (middle part), part of the small intestine, and the branches of the blood vessels to the lower limbs.

- The **hypogastric** (hī-pō-gas′trik; *hypo* = under, *gaster* = belly) **region**, the inferior region in the middle column, typically contains part of the small intestine, the urinary bladder, and the sigmoid colon of the large intestine.
- The **right** and **left hypochondriac** (hī-pō-kon′drē-ak; *hypo* = under, *chondr* = cartilage) **regions** are the superior regions lateral to the epigastric region. The right hypochondriac region typically contains part of the liver, the gallbladder, and part

of the right kidney; the left hypochondriac region typically contains part of the stomach, the spleen, the left colic flexure of the large intestine, and part of the left kidney.

■ The **right** and **left lumbar regions** are the middle regions lateral to the umbilical region. The right lumbar region typically contains the ascending colon and the right colic flexure of the large intestine, the superior part of the cecum, part of the right kidney, and part of the small intestine; the left lumbar region contains the descending colon, part of the left kidney, and part of the small intestine.

■ The **right** and **left iliac** (il′ē-ak; *eileo* = to twist) **regions** are the inferior regions lateral to the hypogastric region. The right iliac region typically contains the inferior end of the cecum, the appendix, and part of the small intestine; the left iliac region contains the junction of parts of the colon as well as part of the small intestine.

Some health-care professionals prefer to partition the abdomen more simply into four quadrants (figure 1.11*b*). They use these areas to locate aches, pains, injuries, or other abnormalities.

CLINICAL VIEW: In Depth

Medical Imaging Procedures

To extend their ability to visualize internal body structures noninvasively (without inserting an instrument into the body), health-care professionals have taken advantage of various medical imaging techniques. Many of these techniques have quickly advanced health care and modern medicine. Some of the most common techniques are radiography, sonography, computed tomography, digital subtraction angiography, dynamic spatial reconstruction, magnetic resonance imaging, and positron emission tomography.

RADIOGRAPHY

Radiography (rā′dē-og′ră-fē; *radius* = ray, *grapho* = to write) is the primary method of obtaining a clinical image of a body part for diagnostic purposes. A beam of *x-rays*, which are a form of high-energy radiation, penetrates solid structures within the body. X-rays can pass through soft tissues, but they are absorbed by dense tissues, including bone, teeth, and tumors. Film images produced by x-rays passing through tissues leave the film lighter in the areas where the x-rays are absorbed. Hollow organs can be visualized by radiography if they are filled with a *radiopaque* (rā-dē-ō-pāk′; *radius* = ray, *opacus* = shady) substance that absorbs x-rays.

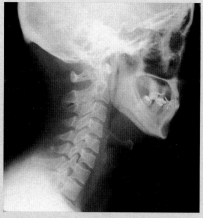
Radiograph (x-ray) of the head and neck.

Originally, x-rays got their name because they were an unknown type of radiation, but they are also called *roentgen rays* in honor of Wilhelm Roentgen, the German physicist who accidentally discovered them. The term x-ray also applies to the photograph (radiograph) made by this technique. Radiography is commonly used in dentistry, mammography, diagnosis of fractures, and chest examination. In terms of their disadvantages, x-rays sometimes produce images of overlapping organs, which can be confusing, and they are unable to reveal slight differences in tissue density.

SONOGRAPHY

The second most widely used imaging method is sonography (ultrasound). Generally, a technician slowly moves a small, handheld device across the body surface. This device produces high-frequency ultrasound waves and then receives signals that are reflected from internal organs. The image produced is called a *sonogram*. **Sonography** (sŏ-nog′ră-fi; *sonus* = sound, *grapho* =

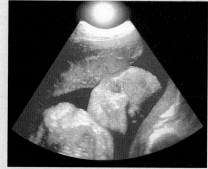
Sonogram of a fetus.

to write) is the method of choice in obstetrics, where a sonogram can show the location of the placenta and help evaluate fetal age, position, and development. Sonography avoids the harmful effects of x-rays, and the equipment is inexpensive and portable. Until recently, its primary disadvantage was that it did not produce a very sharp image, but technological advances are now markedly improving the images produced.

When radiography or sonography cannot produce the desired images, other more detailed (but much more expensive) imaging techniques are available.

COMPUTED TOMOGRAPHY (CT)

A *computed tomography (CT)* scan, previously termed a computerized axial tomographic (tō-mō-graf′ik; *tomos* = a section) (CAT) scan, is a more sophisticated application of x-rays. A patient is slowly moved through a doughnut-shaped machine while low-intensity x-rays are emitted on one side of the cylinder, passed through the body, collected by detectors, and then processed and analyzed by a computer. These signals produce an image of the body that is about the thickness of

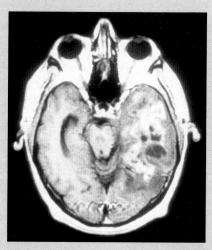
Computed tomographic (CT) scan of the head at the level of the eyes.

a dime. Continuous thin "slices" are used to reconstruct a three-dimensional image of a particular tissue or organ. By providing images of thin sections of the body, there is little overlap of organs and the image is much sharper than one obtained by a conventional x-ray. CT scanning is useful for identifying tumors, aneurysms, kidney stones, cerebral hemorrhages, and other abnormalities.

DIGITAL SUBTRACTION ANGIOGRAPHY (DSA)

Digital subtraction angiography (DSA) is a modified three-dimensional x-ray technique used primarily to observe blood vessels. It involves

Imaginary transverse and midsagittal planes pass through the umbilicus to divide the abdominopelvic cavity into these four quadrants: right upper quadrant (RUQ), left upper quadrant (LUQ), right lower quadrant (RLQ), and left lower quadrant (LLQ).

WHAT DID YOU LEARN?

7 What type of plane would separate the nose and mouth into superior and inferior structures?

8 If a physician makes an incision into a body cavity just superior to the diaphragm and inferior to the neck, what body cavity will be exposed?

9 Describe the location of the hypogastric region.

10 Use a directional term to describe the following:

The elbow is _____ to the wrist.

The neck is _____ to the shoulders.

taking radiographs both prior to and after injecting an opaque medium into the blood vessel. The computer compares the before and after images and removes the data from the before image from the data generated by the after image, thus leaving an image that indicates evidence of vessel blockages. DSA is useful in the procedure called **angioplasty** (an'jē-ō-plas-tē; *angos* = a vessel, *plastos* = formed), in which a physician directs a catheter through a blood vessel and puts a stent in the area where the vessel is blocked. The image produced by the DSA allows the physician to accurately guide the catheter to the blockage.

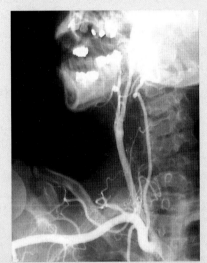

Digital subtraction angiography (DSA) shows three-dimensional images of blood vessels and normal changes in these vessels.

DYNAMIC SPATIAL RECONSTRUCTION (DSR)

Using modified CT scanners, a special technique called **dynamic spatial reconstruction (DSR)** provides two important pieces of medical information: (1) three-dimensional images of body organs, and (2) information about an organ's normal movement as well as changes in its internal volume. Unlike traditional static CT scans, DSR allows the physician to observe the movement of an organ. This type of observation, at slow speed or halted in time completely, has been invaluable for inspecting the heart and the flow of blood through vessels.

MAGNETIC RESONANCE IMAGING (MRI)

Magnetic resonance imaging (MRI), previously called nuclear magnetic resonance (NMR) imaging, was developed as a noninvasive technique to visualize soft tissues. The patient is placed in a prone position within a cylindrical chamber that is surrounded by a large electromagnet. The magnet generates a strong magnetic field that causes protons (hydrogen atoms) in the tissues to align. Thereafter, upon exposure to radio waves, the protons absorb additional energy and align in a different direction. The hydrogen atoms abruptly realign themselves to the magnetic field immediately after the radio waves are turned off. This results in the release of the atoms' excess energy at different rates, depending on the type of tissue. A computer analyzes the emitted energy to produce an image of the body. MRI is better than CT for distinguishing between soft tissues, such as the white and gray matter of the nervous system. However, dense structures (bone) do not show up well in MRI. Formerly,

another disadvantage of MRI was that patients felt claustrophobic while isolated within the closed cylinder. However, newer MRI technology has improved the hardware and lessened this effect. Recent advances in MRI, called functional MRI (fMRI), provide the means to map brain function based on local oxygen concentration differences in blood flow. Increased blood flow relates to brain activity and is detected by a decrease in deoxyhemoglobin (the form of hemoglobin lacking oxygen) in the blood.

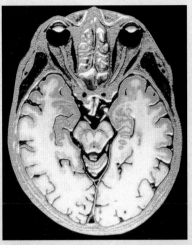

Magnetic resonance imaging (MRI) scan of the head at the level of the eyes.

POSITRON EMISSION TOMOGRAPHY (PET)

The **positron emission tomography (PET)** scan is used both to analyze the metabolic state of a tissue at a given moment in time and to determine which tissues are most active. The procedure begins with an injection of radioactively labeled glucose (sugar), which emits particles called positrons (like electrons, but with a positive charge). Collisions between a positron and an electron cause the release of gamma rays that can be detected by sensors and analyzed by computer. The result is a brilliant color image that shows which tissues were using the most glucose at that moment. In cardiology, the image can reveal the extent of damaged heart tissue. Because damaged heart tissue consumes little or no glucose, the tissue appears dark. Alternatively, PET scans have been used to illustrate activity levels in the brain. The PET scan is an example of nuclear medicine, which uses radioactive isotopes to form anatomic images of the body. Recently, PET scans have been used to detect whether certain cancers have metastasized throughout the body, because cancerous cells will take up more glucose and show up as a "hot spot" on the scan.

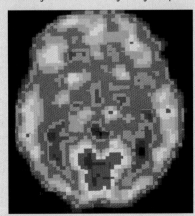

Positron emission tomography (PET) scan of the brain of an unmedicated schizophrenic patient. Red areas indicate high glucose use (metabolic activity). The visual center at the posterior region of the brain was especially active when the scan was made.

CLINICAL TERMS

abdominopelvic quadrants The four areas of the abdominopelvic cavity formed by passing one vertical and one horizontal plane through the umbilicus (navel).

abdominopelvic regions The nine areas in the abdominopelvic cavity formed by two transverse planes and two sagittal planes.

auscultation A diagnostic method that involves listening to the sounds produced by various body structures.

homeostasis State of equilibrium, or constant internal environment, in the body.

palpation Using the hands to detect organs, masses, or infiltration of a body part during a physical examination.

CHAPTER SUMMARY

History of Human Anatomy 2	■ The earliest studies of human anatomy date back to 400–300 B.C. and were based on evidence gleaned from dissection. ■ Anatomic studies revived in Europe during the Middle Ages, and advances in printing and engraving techniques led to Andreas Vesalius's illustrated and anatomically accurate textbook in the 1500s as well as to important books by William Harvey in the next century. ■ Later technological advances, including preserved specimens, cryotechnology, and plastination, have continued to improve and help disseminate knowledge about human body structure.
Definition of Anatomy 3	■ Anatomy is the study of the structure of individual body organs and their relationships to one another. ■ Physiology is the study of the functions of body structures. **Microscopic Anatomy 3** ■ Microscopic anatomy includes cytology, the study of cells, and histology, the study of tissues. **Gross Anatomy 4** ■ Gross anatomy includes numerous subdisciplines, such as regional anatomy, systemic anatomy, and surface anatomy. ■ Developmental anatomy investigates the changes in form that occur continuously from conception through physical maturity. Embryology is the study of the processes and developmental changes that occur prior to birth. ■ Some anatomic specialties important to health-care providers are pathologic anatomy, radiographic anatomy, and surgical anatomy. ■ The human body is organized in an increasingly complex series of interacting levels: the chemical level, the cellular level, the tissue level, the organ level, the organ system level, and the organismal level.
Structural Organization of the Body 5	**Characteristics of Living Things 6** ■ All living organisms exhibit several common properties: organization, metabolism, growth and development, responsiveness, adaptation, regulation, and reproduction. **Introduction to Organ Systems 6** ■ The organ systems of the body function together to maintain a constant internal environment, a state called homeostasis. ■ Clear, exact terminology accurately describes body structures and helps us identify and locate them.
Precise Language of Anatomy 11	**Anatomic Position 11** ■ The anatomic position is used as a standard reference point. **Sections and Planes 11** ■ Three planes describe relationships among the parts of the three-dimensional human body: the coronal (or frontal) plane, the transverse (cross-sectional or horizontal) plane, and the midsagittal plane. **Anatomic Directions 12** ■ Specific directional terms indicate relative body locations (see table 1.2). **Regional Anatomy 13** ■ Specific anatomic terms identify body regions (see table 1.3 and figure 1.8). **Body Cavities and Membranes 13** ■ Body cavities are spaces that enclose organs and organ systems. The posterior aspect of the body contains two cavities: the cranial cavity and the vertebral canal. The ventral cavity is separated into a superior thoracic cavity and an inferior abdominopelvic cavity. ■ The ventral cavity is lined by thin serous membranes. A parietal layer lines the internal body wall surface, and a visceral layer lines the external surface of the organs. ■ The thoracic cavity is composed of three separate spaces: a central space called the mediastinum, and two lateral spaces, the pleural cavities. ■ Within the mediastinum is a space called the pericardial cavity. ■ The abdominopelvic cavity is composed of two subdivisions: the abdominal cavity and the pelvic cavity. **Abdominopelvic Regions and Quadrants 16** ■ Regions and quadrants are two aids for describing locations of the viscera in the abdominopelvic area of the body. There are nine abdominopelvic regions and four abdominopelvic quadrants.

CHALLENGE YOURSELF

Matching

Match each numbered item with the most closely related lettered item.

_____ 1. cranial   a. study of tissues

_____ 2. cytology   b. toward the tail

_____ 3. responsiveness   c. contains spinal cord

_____ 4. inguinal region   d. structural change in the body

_____ 5. caudal   e. study of organs of one system

_____ 6. development   f. thoracic cavity

_____ 7. vertebral cavity   g. detect and react to stimuli

_____ 8. histology   h. groin

_____ 9. mediastinum   i. toward the head

_____ 10. systemic anatomy   j. study of cells

Multiple Choice

Select the best answer from the four choices provided.

_____ 1. Cutting a midsagittal section through the body separates the
    a. anterior and posterior portions of the body.
    b. superior and inferior portions of the body.
    c. dorsal and ventral portions of the body.
    d. right and left halves of the body.

_____ 2. Examination of superficial anatomic markings and internal body structures as they relate to the covering skin is called
    a. regional anatomy.
    b. surface anatomy.
    c. pathologic anatomy.
    d. systemic anatomy.

_____ 3. Which of the following regions corresponds to the forearm?
    a. cervical
    b. antebrachial
    c. femoral
    d. pes

_____ 4. The state of maintaining a constant internal environment is called
    a. reproduction.
    b. homeostasis.
    c. responsiveness.
    d. growth.

_____ 5. The _____ level of organization is composed of two or more tissue types that work together to perform a common function.
    a. cellular
    b. tissue
    c. organ
    d. organismal

_____ 6. Which body cavity is located inferior to the diaphragm and superior to a horizontal line drawn between the superior edges of the hip bones?
    a. abdominal cavity
    b. thoracic cavity
    c. pleural cavity
    d. pelvic cavity

_____ 7. The term used when referring to a body structure that is below, or at a lower level than, another structure is
    a. ventral.
    b. medial.
    c. inferior.
    d. distal.

_____ 8. Which medical imaging technique uses modified x-rays to prepare three-dimensional cross-sectional "slices" of the body?
    a. radiography
    b. sonography
    c. PET (positron emission tomography) scan
    d. computed tomography (CT)

_____ 9. The _____ region is the "front" of the knee.
    a. patellar
    b. popliteal
    c. pes
    d. inguinal

_____ 10. The subdiscipline of anatomy that examines structures not readily seen by the unaided eye is
    a. regional anatomy.
    b. microscopic anatomy.
    c. gross anatomy.
    d. pathologic anatomy.

Content Review

1. Distinguish between cytology and histology.

2. What properties are common to all living things?

3. List the levels of organization in a human, starting at the simplest level and proceeding to the most complex. Use arrows to connect the levels.

4. What are the organ systems in the human body?

5. Describe the body in the anatomic position. Why is the anatomic position used?

6. Describe the difference between the directional terms *superior* and *inferior.*

7. List the anatomic term that describes each of the following body regions: forearm, wrist, chest, armpit, thigh, and foot.

8. What are the two body cavities within the posterior aspect, and what does each cavity contain?

9. Describe the structure and the function of serous membranes in the body.

10. Describe which medical imaging techniques are best suited for examining soft tissues, and which are better suited for examining harder body tissues, such as bone.

Developing Critical Reasoning

1. If a person becomes ill and the symptoms indicate infection by a parasitic organism, treatment will depend upon correct diagnosis of the problem. What category of anatomic study would be most appropriate for identifying an infectious agent in the blood or muscle tissue? What kinds of effects would an infection in the blood or muscle tissue have?

2. Lynn was knocked off her bicycle during a race. She damaged a nerve in her antebrachial region, suffered an abrasion in her patellar region, and broke bones in her sacral and brachial regions. Explain where each of these injuries is located.

3. Your grandmother is being seen by a radiologist to diagnose a possible tumor in her small intestine. Explain to your grandmother what imaging techniques would best determine whether a tumor exists, and which imaging techniques would be inadequate for determining the placement of the tumor.

ANSWERS TO "WHAT DO YOU THINK?"

1. The stomach is at the organ level, while the digestive system is at the organ system level.

2. When you put lotion on your hands, the heat and the noise lessen considerably because friction is reduced. If the thoracic and abdominopelvic cavities didn't have the lubricating serous fluid, friction would build up, and you would feel pain whenever your organs moved. For example, the illness *pleurisy* (inflammation of the pleura) makes it very painful to breathe, because the pleura is inflamed and the serous fluid cannot lubricate the membranes.

Visit the McKinley/O'Loughlin *Human Anatomy,* 2e website at aris.mhhe.com

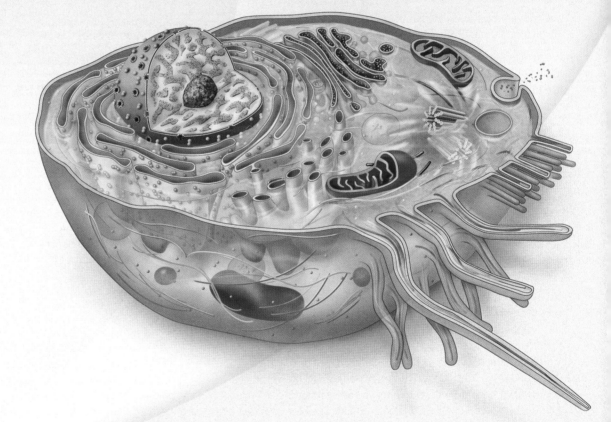

The Cell: Basic Unit of Structure and Function

Cells are the structural and functional units of all organisms, including humans. An adult human body contains about 75 trillion cells. Most cells are composed of characteristic parts that work together to allow them to perform specific body functions. There are approximately 200 different types of cells in the human body, but all of them share certain common characteristics:

■ All cells perform the general housekeeping functions necessary to sustain life. Each cell must obtain nutrients and other materials essential for survival from its surrounding fluids. Recall from chapter 1 that the total of all the chemical reactions that occur in cells is called metabolism.

■ Cells must dispose of the wastes they produce. If a cell didn't remove its waste products, this waste would build up in the cell and lead to its death.

■ The shape and integrity of a cell is maintained by both its internal contents and its surrounding membrane.

■ Most cells are capable of undergoing cell division to make more cells of the same type.

The Study of Cells

Key topics in this section:

■ Advantages and disadvantages of LM, TEM, and SEM
■ The relationship between structure and function in cells

The study of cells is called **cytology**. Throughout this chapter, we will examine the generalized structures and functions shared by all body cells. Subsequent chapters examine specialized cells and their unique functions.

Using the Microscope to Study Cells

The small size of cells is the greatest obstacle to determining their nature. Cells were discovered after microscopes were invented, and high-magnification microscopes are required to see the smallest human body cells. The dimensional unit often used to measure cell size is the micrometer (μm). One micrometer is equal to 1/10,000 of a centimeter (about 1/125,000 of an inch). For example, a red blood cell has a diameter of about 7–8 μm, whereas one of the largest human cells, an oocyte, has a diameter of about 120 μm. **Figure 2.1** compares the size of the smallest unit of structure in the human body (an atom) to various cell types as well as to other macroscopic structures, such as an ostrich egg and a human.

Microscopy is the use of the microscope. It has become a valuable asset in anatomic investigations. Most commonly used are the **light microscope (LM),** the **transmission electron microscope (TEM),** and the **scanning electron microscope (SEM).** Because specimen samples have no inherent contrast, they cannot be seen clearly under the microscope unless contrast is added. Therefore, colored-dye stains are used in light microscopy, and heavy-metal stain preparations are used in electron microscopy, which includes both TEM and SEM. **Figure 2.2** compares the images produced when each of these types of microscopes is used to examine the same specimen—in this case, the epithelium lining the respiratory tract.

The LM produces a two-dimensional image for study by passing visible light through the specimen. Glass lenses focus and

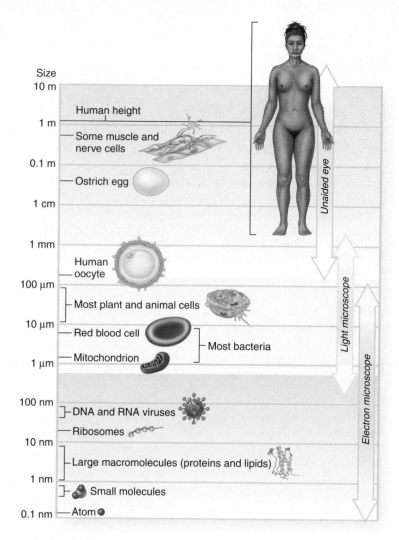

Figure 2.1

The Range of Cell Sizes. Most cells in the human body are between 1 micrometer (μm) and 100 μm in diameter. Individual cells are usually observed by light microscopy; subcellular structures are studied by electron microscopy.

magnify the image as it is projected toward the eye. Figure 2.2*a* shows the cellular structure of the epithelium as well as the hairlike structures (cilia) that project from its surface.

Electron microscopes use a beam of electrons to "illuminate" the specimen. Electron microscopes easily exceed the magnification obtained by light microscopy, but more importantly, they improve the resolution by more than a thousandfold. A TEM directs an electron beam through a thin-cut section of the specimen. The resultant two-dimensional image is focused onto a screen for viewing or onto photographic film to record the image. The TEM can show a close-up section of the cilia on the surface of the epithelial cells (figure 2.2*b*).

For a detailed three-dimensional study of the surface of the specimen, SEM analysis is the method of choice (figure 2.2*c*). Here,

Cilia

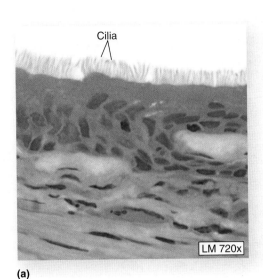

LM 720x

(a)

Cilia

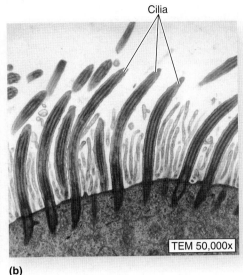

TEM 50,000x

(b)

Cilia

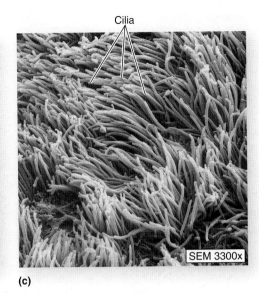

SEM 3300x

(c)

Figure 2.2

Microscopic Techniques for Cellular Studies.   Different techniques are used to investigate cellular anatomy. *(a)* A light microscope (LM) shows hairlike structures, termed cilia, that project from the free membrane surfaces of the cells lining the respiratory tract. *(b)* A transmission electron microscope (TEM) reveals the ultrastructure of the cilia on the same type of cells. *(c)* A scanning electron microscope (SEM) shows the three-dimensional image of the cilia-covered surface of the same type of cells.

the electron beam is moved across the surface of the specimen, and reflected electrons generate a surface-topography image captured on a television screen.

General Functions of Human Body Cells

Besides differing in size, cells also vary in shape, which may be flat, cylindrical, oval, or quite irregular. Often, cells' functions are reflected in either their size or their shape. Among the general functions of cells are the following:

- **Covering.** Epithelial cells form a sheet to cover surfaces. For example, skin cells cover the external body surface.
- **Lining.** Epithelial cells line the internal surfaces of our organs, such as the small intestine.
- **Storage.** Some body cells, such as hepatocytes (liver cells) and adipocytes (fat cells), store nutrients or energy reserves for the body.
- **Movement.** Muscle cells are composed of contractile proteins that cause the muscle to shorten (contract), thereby allowing movement to occur. Skeletal muscle cells attach to the skeleton so that when these cells contract, they move the skeleton. In contrast, when the muscle cells in the heart wall contract, they are able to pump blood throughout the body.
- **Connection.** Multiple cell types are found in connective tissues, which help connect and support other tissues. Fibroblast cells produce protein fibers that are found in ligaments, the connective tissue that binds bone to bone.
- **Defense.** Many cell types protect the body against pathogens or antigens (anything perceived as foreign in the body).

White blood cells (called leukocytes) are designed to recognize foreign material (antigens) and attack them. The process of attacking the foreign materials is called an immune response.

- **Communication.** Nerve cells (called neurons) transmit nerve impulses from one part of the body to another. The nerve impulse carries information between neurons within the nervous system, sensory information to the brain for processing, or motor information to make a muscle contract or a gland secrete.
- **Reproduction.** Some cells are designed solely to produce new individuals. For example, within the gonads, the sex cells (sperm and oocytes) are produced. They are specialized cells designed to join together and initiate the formation of a new individual. Additionally, within the bone marrow are stem cells that continuously produce new blood cells for the body.

Table 2.1 summarizes the types of cellular functions as they relate to cell structure. Now that we have mentioned that cells come in a variety of shapes and sizes and have different functions, let us examine the structures common to almost all cells.

WHAT DID YOU LEARN?

1 Describe an advantage of using TEM rather than LM to study intracellular structure.

2 What are some basic functions of human body cells?

Table 2.1	Selected Common Types of Cells and Their Functions				
Functional Category	**Example**	**Specific Functions**	**Functional Category**	**Example**	**Specific Functions**
Covering	Epidermal cells in skin	Protect outer surface of body	*Connection (attachment)*	Collagen (protein) fibers from fibroblasts	Form ligaments that attach bone to bone
Lining	Epithelial cells in small intestine	Regulate nutrient movement into body tissues	*Defense*	Lymphocytes	Produce antibodies to target antigens or invading cells
Storage	Fat cells Liver cells	Store lipid reserves Store carbohydrate nutrients as glycogen	*Communication*	Nerve cells	Send information between regions of the brain
Movement	Muscle cells of heart Skeletal muscle cells	Pump blood Move skeleton	*Reproduction*	Bone marrow stem cells Sperm and oocyte cells	Produce new blood cells Produce new individual

A Prototypical Cell

Key topics in this section:

- Characteristics of the plasma membrane, cytoplasm, and nucleus
- Contents of a prototypical cell

The generalized cell in **figure 2.3** isn't an actual body cell, but rather a representation of a cell that combines features of several different types of body cells. Almost all mature human cells share the same three basic constituents, which can be described in terms of the prototypical cell:

- **Plasma membrane.** The plasma membrane, sometimes called the cell membrane, forms the outer, limiting barrier separating the internal contents of the cell from the external environment.
- **Cytoplasm.** Cytoplasm (sī′tō-plazm; *kytos* = a hollow, *plasma* = a thing formed) is a general term for all cellular contents located between the plasma membrane and the nucleus. The three components of the cytoplasm are cytosol (a viscous fluid), inclusions (nonfunctional, temporary structures that store cellular products), and organelles (tiny structures that perform specific cellular functions).
- **Nucleus.** The nucleus (noo′klē-ŭs; *nux* = the kernel or inside of a thing) is the cell's control center. It controls protein synthesis (production of new proteins), and in so doing, it directs the functional and structural characteristics of the cell.

The next three sections of this chapter describe the contents and specific functions of the plasma membrane, the cytoplasm, and the nucleus. As you read these descriptions, it may help to refer to **table 2.2**, which summarizes this information.

WHAT DID YOU LEARN?

3 Briefly describe the three main constituents of a cell.

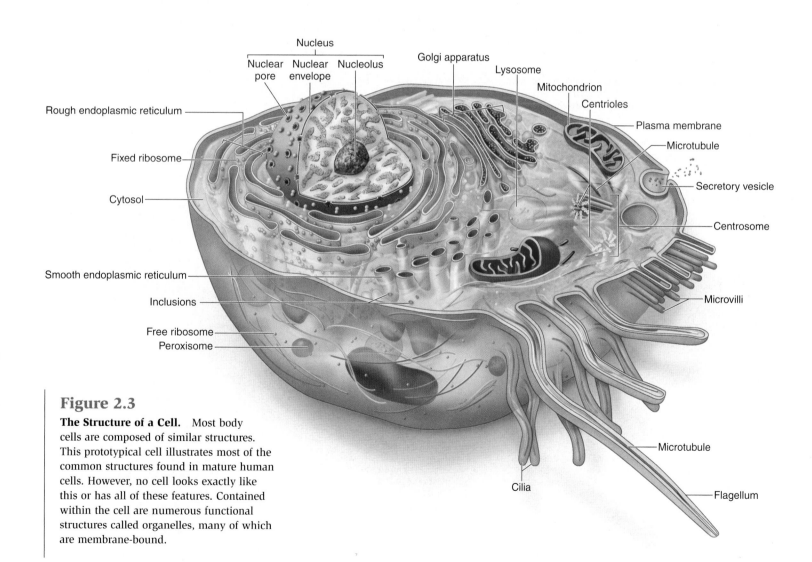

Figure 2.3

The Structure of a Cell. Most body cells are composed of similar structures. This prototypical cell illustrates most of the common structures found in mature human cells. However, no cell looks exactly like this or has all of these features. Contained within the cell are numerous functional structures called organelles, many of which are membrane-bound.

Table 2.2	Components of the Cell		
Component	**Structure**	**Function**	**Appearance**
MAJOR CELL COMPONENTS			
Plasma (cell) membrane	Phospholipid bilayer containing cholesterol and proteins (integral and peripheral) and some carbohydrates (externally)	Contains receptors for communication; forms intercellular connections; acts as physical barrier to enclose cell contents; regulates material movement into and out of the cell	Plasma membrane
Cytoplasm	Contains cytosol, a viscous fluid, and inclusions and organelles	Site of metabolic processes of the cell; stores nutrients and dissolved solutes	Cytoplasm, Cytosol, Organelles, Inclusions
Cytosol	Viscous fluid medium with dissolved solutes (ions, nutrients, proteins, carbohydrates, lipids, and other small molecules)	Provides support for organelles; serves as viscous medium through which diffusion occurs	
Organelles	Membrane-bound and non-membrane-bound structures that have unique functions and activities	Carry out specific metabolic activities of the cell	
Inclusions	Droplets of melanin, protein, glycogen granules, or lipid; usually non-membrane-bound	Store materials	
Nucleus	Surrounded by double membrane nuclear envelope (each membrane is a phospholipid bilayer); contains nucleolus and chromatin	Acts as cell control center; controls all genetic information (DNA); site of ribosome subunit assembly	Nucleus, Nuclear pores, Nuclear envelope, Nucleolus, Chromatin
Nuclear envelope	Double membrane boundary between cytoplasm and nuclear contents	Pores in envelope regulate exchange of materials with the cytoplasm	
Nuclear pores	Openings through the nuclear envelope	Allow for passage of materials between nucleus and cytoplasm	
Nucleolus (or nucleoli)	Spherical, dark-staining, dense granular region in the nucleus	Synthesizes rRNA and assembles ribosomes in the nucleus	
Chromatin and chromosomes	Filamentous association of DNA and histone proteins	Site of genes in the DNA	
MEMBRANE-BOUND ORGANELLES			
Smooth endoplasmic reticulum (smooth ER)	Interconnected network of membrane tubules and vesicles; no ribosomes	Synthesizes lipids; metabolizes carbohydrates; detoxifies drugs, alcohol	
Rough endoplasmic reticulum (rough ER)	Flattened intracellular network of membrane sacs called cisternae; ribosomes attached on cytoplasmic surface	Synthesizes proteins for secretion, new proteins for the plasma membrane, and lysosomal enzymes; transports and stores molecules	
Golgi apparatus	Stacked series of flattened, smooth membrane sacs with associated shuttle vesicles	Modifies, packages, and sorts newly synthesized proteins for secretion, inclusion in new plasma membrane, or lysosomal enzyme synthesis	

Table 2.2	Components of the Cell *(continued)*		
Component	**Structure**	**Function**	**Appearance**
MEMBRANE-BOUND ORGANELLES *(CONTINUED)*			
Lysosomes	Membrane sacs with digestive enzymes	Digest materials or microbes ingested by the cell; remove old/damaged organelles; self-destruct (autolyze)	
Peroxisomes	Membrane-enclosed sacs; usually contain large amounts of specific enzymes to break down harmful substances	Convert hydrogen peroxide formed during metabolism to water	
Mitochondria	Double membrane structures with cristae; fluid matrix contents at center	Synthesize most ATP during cellular respiration; "powerhouses of cell"	
NON-MEMBRANE-BOUND ORGANELLES			
Ribosomes	Dense cytoplasmic granules with two subunits (large and small); may be free in cytoplasm (free ribosomes) or bound to rough ER (fixed ribosomes)	Synthesize proteins for: 1. use in the cell (free ribosomes) 2. secretion, incorporation into plasma membrane, or lysosomes (fixed ribosomes)	Free ribosomes / Fixed ribosomes
Cytoskeleton	Organized network of protein filaments or hollow tubules throughout the cell	Provides structural support; facilitates cytoplasmic streaming, organelle and cellular motility, transport of materials, and chromosomal movement and cell division	Cytoskeleton / Intermediate filament / Microfilament / Microtubule
Microfilaments	Actin protein monomers formed into filaments	Maintain cell shape; aid in muscle contraction and intracellular movement; separate dividing cells	
Intermediate filaments	Various protein components	Provide structural support; stabilize cell junctions	
Microtubules	Hollow cylinders of tubulin protein; able to lengthen and shorten	Support cell; hold organelles in place; maintain cell shape and rigidity; direct organelle movement within cell and cell motility as cilia and flagella; move chromosomes at cell division	
Centrosome	Amorphous region adjacent to nucleus; contains a pair of centrioles	Organizes microtubules; participates in spindle formation during cell division	Centriole / Centrosome
Centrioles	Paired perpendicular cylindrical bodies; composed of microtubule triplets	Organize microtubules during cell division for movement of chromosomes	
Cilia	Short, membrane-attached projections containing microtubules; occur in large numbers on exposed membrane surfaces	Move fluid, mucus, and materials over the cell surface	Cilia
Flagellum	Long, singular membrane extension containing microtubules	Propels sperm cells in human male	Flagellum
Microvilli	Numerous thin membrane folds projecting from the free cell surface	Increase membrane surface area for increased absorption and/or secretion	Microvilli

Plasma Membrane

Key topics in this section:

- Structure of the plasma membrane
- Functions of selective permeability
- Specific types of passive and active transport

The **plasma membrane**, or *cell membrane*, forms the thin outer border of the cell. Also sometimes called the *plasmalemma* (plaz-mă-lem′ă; *plasma* = something formed, *lemma* = husk), the plasma membrane is a flexible and fluid molecular layer that separates the internal (intracellular) components of a cell from the external environment and extracellular materials. All materials that enter or leave the cell must pass across the plasma membrane. Therefore, the plasma membrane is a vital, selectively permeable barrier that functions as a "gatekeeper" to regulate the passage of gases, nutrients, and wastes between the internal and external environments. Selective permeability (sometimes called semipermeability) is essential to a cell's existence because it allows the entrance or exit of substances to be regulated or restricted.

Necessarily, the total surface area of the membrane must be extensive enough to permit all of these movements. As the cell grows larger, the surface area of the plasma membrane increases by square units, whereas the volume of cytoplasm within the cell increases by cubic units. It is possible that a cell may reach a point when it does not have the necessary area of membrane surface required to transport all of the materials it needs to maintain life processes. Thus, most cells necessarily remain small in order to acquire sufficient nutrients and dispose of their wastes.

Composition and Structure of Membranes

A plasma membrane is not a rigid layer of molecules. Rather, a typical plasma membrane is a fluid matrix composed of an approximately equal mixture, by weight, of lipids and proteins. While the lipids form the main structure of the plasma membrane, the proteins dispersed within it determine its primary function(s). In addition, the plasma membrane has an external carbohydrate (sugar) coat, called the **glycocalyx** (glī-kō-kā′liks; *glykys* = sweet, *kalyx* = husk). The following discussion explains how these components form the plasma membrane.

Lipids

Lipids are materials that are insoluble in water; examples are fats and oils, as well as steroids. The insolubility of the lipids within the plasma membrane ensures that the membrane will not simply "dissolve" when it comes in contact with water. The three types of lipids in the plasma membrane are phospholipids, cholesterol, and glycolipids.

Phospholipids   Most of the plasma membrane lipids are **phospholipids,** which contain both water-soluble and water-insoluble regions as well as the element phosphate. These molecules are called polar, meaning that a charge is distributed unevenly through the molecule so that one region has a positive charge and another region has a negative charge. Often these molecules are portrayed in the membrane as a balloon with two tails. The balloonlike, polar "head" is charged and hydrophilic ("water-loving," or attracted to water). The two "tails" are uncharged, nonpolar, and hydrophobic ("water-hating," or repelled by water). Because all phospholipid molecules have these two regions with different water association

properties, they readily associate to form two parallel sheets of phospholipid molecules lying tail-to-tail. The hydrophobic tails form the internal environment of the membrane, and their polar heads are directed outward. This basic structure of the plasma membrane is called the **phospholipid bilayer (figure 2.4)**. It ensures that **intracellular fluid (ICF)** (fluid within the cell) remains inside the cell, and **extracellular fluid (ECF)** (fluid outside the cell) remains outside. One type of ECF is **interstitial fluid**, the thin layer of fluid that bathes the external surface of a cell.

Cholesterol   **Cholesterol,** a type of lipid called a steroid, amounts to about 20% of the plasma membrane lipids. Cholesterol is scattered within the hydrophobic regions of the phospholipid bilayer, where it strengthens the membrane and stabilizes it at temperature extremes.

Glycolipids   **Glycolipids,** lipids with attached carbohydrate groups, form about 5% to 10% of the membrane lipids. They are located only on the outer layer of the membrane, where they are exposed to the extracellular fluid. The glycocalyx (the carbohydrate portion of the glycolipid molecule mentioned earlier) helps these molecules participate in cell–cell recognition, intracellular adhesion, and communication.

Proteins

The other common molecular structures within the plasma membrane are proteins. **Proteins** are complex, diverse molecules composed of chains of smaller molecules called amino acids. Proteins play various structural and functional roles within the cell and within the body. They make up about half of the plasma membrane by weight. Most of the membrane's specific functions are determined by its resident proteins. Plasma membrane proteins are of two types: integral and peripheral.

Integral proteins are embedded within, and extend across, the phospholipid bilayer. Some species of integral proteins act as membrane channels, providing a pore (hole) in the membrane through which specific substances pass. Other integral proteins, termed **receptors**, serve as binding sites for molecules outside of the cell. Hydrophobic regions within the integral proteins interact with the hydrophobic interior of the membrane. In contrast, the hydrophilic regions of the integral proteins are exposed to the aqueous environments on either side of the membrane.

Peripheral proteins are not embedded in the phospholipid bilayer. They are attached loosely to either the external or internal surface of the membrane, often to the exposed parts of the integral proteins. Peripheral proteins can "float" and move about the bilayer, much like a beach ball floating on the surface in a swimming pool.

Both integral and peripheral membrane proteins may serve as **enzymes,** which are also called catalysts. Enzymes are molecules

> ## Study Tip!
>
> Think of the glycoproteins and glycolipids as similar to your student ID card. This personal identification item supplies information about you and lets the school know you are supposed to be there. If a person doesn't have a student ID card, he or she is not allowed access to certain school facilities. Similarly, the glycoprotein and glycolipid molecules allow other cells in the body to recognize this cell and not confuse it with a foreign substance that must be destroyed.

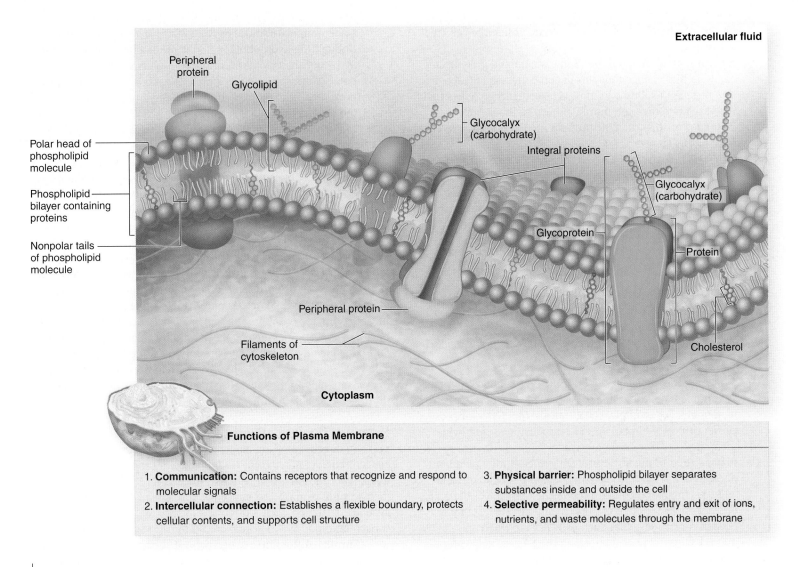

Functions of Plasma Membrane

1. **Communication:** Contains receptors that recognize and respond to molecular signals
2. **Intercellular connection:** Establishes a flexible boundary, protects cellular contents, and supports cell structure
3. **Physical barrier:** Phospholipid bilayer separates substances inside and outside the cell
4. **Selective permeability:** Regulates entry and exit of ions, nutrients, and waste molecules through the membrane

Figure 2.4
Structure of the Plasma Membrane.   The plasma membrane is a phospholipid bilayer with cholesterol and proteins scattered throughout and associated with its surfaces.

that are important for functional or metabolic activities in the cell because they change the rate of a reaction without being affected by the reaction itself. An enzyme is the equivalent of an electric starter for a barbecue grill; the starter can repeatedly ignite the fire in the grill because it is unchanged by the fire itself.

Many integral membrane proteins are **glycoproteins** (proteins with attached carbohydrate groups). They form about 90% of all the membrane molecules that have carbohydrates attached to their external surface. Together, the carbohydrate groups attached to both glycoproteins and the previously mentioned glycolipids form the fuzzy glycocalyx on the external surface of the plasma membrane.

WHAT DO YOU THINK?

1 What is the benefit to the cell of having a plasma membrane that is selectively permeable? What are some disadvantages of having a selectively permeable plasma membrane?

Protein-Specific Functions of the Plasma Membrane

The proteins of the plasma membrane perform a variety of important activities that promote its overall functions, including the following:

■ **Transport**. A transmembrane protein spans the plasma membrane completely. It has an internal hydrophobic region and hydrophilic regions at both the internal and external membrane surfaces. This protein assists the movement of a particular substance across the membrane. Sometimes the transport of material across the membrane requires cellular energy. A molecule called **ATP (adenosine triphosphate)** provides the energy for that transport. ATP releases energy when the bond that attaches its third phosphate to the rest of the molecule is broken.

■ **Intercellular connection**. Junctions form between some neighboring cells when proteins in the membranes of each cell attach. These junctions secure the cells to each other.

- **Anchorage for the cytoskeleton**. Cell shape is maintained by the attachment of structural proteins inside the cell (the cytoskeleton) to membrane proteins.
- **Enzyme (catalytic) activity**. Some membrane proteins are catalysts that change the rates of some metabolic reactions. The plasma membrane in most cells contains enzymes that increase the rate of ion movement across the membrane. Examples of such catalytic proteins are ion pumps, described later in this chapter.
- **Cell–cell recognition**. The carbohydrate components of both glycoproteins and glycolipids usually act as identification molecules that are specifically recognized by other cells.
- **Signal transduction**. Signal transduction is the transmission of a message from a molecule outside the cell to the inside of the cell. The cell then responds by changing its communication activities.

Transport Across the Plasma Membrane

As we've just discussed, the plasma membrane is selectively permeable, so it is able to regulate transport of materials into and out of the cell. The following factors influence membrane permeability:

- **Transport proteins**. Special integral membrane proteins attract specific molecules in both the internal and external environments of the cell and assist their transport across the membrane. For example, some transport proteins (also called carrier proteins) bind to specific carbohydrates and help them move across the membrane.
- **Plasma membrane structure**. Differences in the membrane phospholipids (both in the composition of the polar head and the length and composition of the tails) affect the ability of some molecules to cross that membrane. For example, because polar molecules such as water are small and able to interact with the phospholipids in the bilayer, they can pass through the phospholipid bilayer rapidly, while other polar molecules, such as simple sugars, cannot pass through the bilayer.
- **Concentration gradient**. Materials tend to move more rapidly when their concentrations are significantly different between two compartments. For example, if the intracellular fluid had a low concentration of a permeable substance, and the extracellular fluid had a high concentration of that substance, this substance would more easily pass through the membrane into the cell (where its original concentration was lower).
- **Ionic charge**. An **ion** (atom with a net negative or positive charge) may either be repulsed or attracted to the membrane structures. This ionic charge influences molecular movement across the membrane. For example, if the inside of the cell has a net negative charge, a negative ion outside the membrane might be repelled by the intracellular environment, whereas a positive ion might be attracted to the intracellular environment.
- **Lipid solubility**. Materials that are lipid-soluble easily dissolve through the phospholipid bilayer. Thus, lipid-soluble molecules can pass through the membrane more

easily than non-lipid-soluble molecules can. For example, small nonpolar molecules called fatty acids readily move through the hydrophobic interior of the phospholipid bilayer and enter the cytoplasm of the cell, whereas larger, charged polar molecules, such as simple sugars or amino acids, are prevented from moving through the hydrophobic region of the plasma membrane.

- **Molecular size**. Smaller molecules move across the plasma membrane readily, while larger molecules need special transport systems to move them across the membrane. For example, some small molecules and ions move continuously across the plasma membrane by passing between the molecules that form the fabric of the membrane.

Passive Transport

The processes of transporting substances across plasma membranes are classified as either passive or active. In **passive transport,** substances move across a plasma membrane without the expenditure of energy by the cell. Materials move along a concentration gradient, meaning that they flow from a region of higher concentration of the material to a region of lower concentration. Passive transport is similar to floating downstream with the current; no cellular energy (ATP) is needed for it to occur. Passive processes that move material across the plasma membrane include simple diffusion, osmosis, facilitated diffusion, and bulk filtration.

Simple Diffusion **Diffusion** (di-fū′zhun; *diffundo* = to pour in different directions) is the tendency of molecules to move down their concentration gradient; that is, the molecules move from a region of higher concentration to a region of lower concentration. This movement continues until the molecules are spread out evenly into the available space on each side of the membrane, at which point the concentration of this molecule is said to be at **equilibrium**.

Simple diffusion occurs when substances move across membranes unaided because they are either small or nonpolar, or because they are both. As a result of simple diffusion, a **net movement** of specific molecules or ions takes place from a region of their higher concentration to a region of their lower concentration. This net movement continues until all of those molecules are evenly distributed in the environment (the point of equilibrium). At this point, the concentration gradient no longer exists. However, molecular movement does not cease. Those molecules still move continuously in all directions at an equal rate. For example, there is no net movement of molecule "A" during its equilibrium, which means that one molecule "A" enters the cell for every molecule "A" that leaves the cell. An example of diffusion in the body is the movement of respiratory gases between the air sacs in the lungs and the blood vessels in the lungs. Oxygen continually moves from the lung air sacs into the blood, while carbon dioxide moves in the opposite direction. This movement guarantees that the blood will receive oxygen and eliminate carbon dioxide as part of normal respiration.

Osmosis **Osmosis** (os-mō′sis; *osmos* = a thrusting) is a special type of simple diffusion in which water diffuses from one side of the selectively permeable membrane to the other. The net movement of water across a semipermeable membrane continues from a region of high water concentration to a region of low water concentration until equilibrium is established. In the body, the movement of water between the blood and the extracellular fluid around cells occurs by osmosis.

Facilitated Diffusion   **Facilitated diffusion** requires the participation of specific transport proteins that help specific substances move across the plasma membrane. These substances are either large molecules or molecules that are insoluble in lipids. The molecule to be moved binds to the transport protein in the membrane. This binding helps alter the shape of both the transport protein and the molecule to be moved, thus permitting it to pass across the membrane. For example, glucose and some amino acids move across the membrane by this means. Facilitated diffusion differs from simple diffusion in that a specific transport protein is required. Thus, transport is aided by a protein.

Bulk Filtration   **Bulk filtration,** or *bulk movement,* involves the diffusion of solvents and solutes together across the selectively permeable membrane. **Solvents** are liquids that have substances called **solutes** dissolved in them. For example, water can be a solvent if it has a solute such as salt or sugar dissolved in it. An example of bulk filtration is when fluid and certain solutes are transported from the blood into the extracellular fluid. Bulk filtration works in this way: **Hydrostatic pressure** (hī-drō-stat′ik presh′ŭr) (fluid pressure exerted by blood pushing against the inside wall of a blood vessel) forces both water and small solutes from the blood across the plasma membranes of cells lining the blood vessel. Only smaller molecules (glucose) and ions (such as sodium [Na^+] and potassium [K^+]) can be forced across the membrane by hydrostatic pressure. The largest molecules (called macromolecules) and large solid particles in the solvent must be transported through the membrane by another process, which we examine next.

Active Transport

Active transport is the movement of a substance across a plasma membrane *against* a concentration gradient, so materials must be moved from an area of low concentration to an area of high concentration. Active transport is similar to swimming upstream against a current, where you must exert energy (swim) in order to move against the water flow. To move materials against their concentration gradient, active transport requires cellular energy in the form of ATP (adenosine triphosphate) and sometimes a transport protein as well. ATP is continually synthesized by mitochondria, cell structures described later in this chapter. Active transport methods include ion pumps and several processes collectively known as bulk transport.

Ion Pumps   Active transport processes that move ions across the membrane are called **ion pumps.** Ion pumps are a major factor in a cell's ability to maintain its internal concentrations of ions. One type of ion pump is the **sodium-potassium pump.** This transport mechanism is specifically called an exchange pump, because it moves one ion into the cell while simultaneously removing another type of ion from the cell (**figure 2.5**). For example, compared to their

surroundings, some human cells have much higher concentrations of potassium ions and much lower concentrations of sodium ions. The plasma membrane maintains these steep concentration gradient differences by continuously excluding sodium ions from the cell and moving potassium ions into the cell. Figure 2.5 shows the steps in this process. The cell must expend energy in the form of ATP to maintain these sodium and potassium levels.

Bulk Transport   Macromolecules, such as large proteins and polysaccharides, cannot move across the plasma membrane via ion pumps or even with the assistance of normal transport proteins. Instead, larger molecules or bulk structures move across the membrane via the active transport processes called exocytosis and endocytosis.

In **exocytosis** (ek′sō-sī-tō′sis; *exo* = outside, *kytos* = cell, *osis* = condition of), large molecules are secreted *from* the cell (**figure 2.6**). Typically, the material for secretion is packaged within intracellular transport **vesicles** (ves′i-kl; *vesica* = bladder), which move toward the plasma membrane. When the vesicle and plasma membrane come into contact, the lipid molecules of the vesicle and plasma membrane bilayers rearrange themselves so that the two membranes fuse. The fusion of these lipid bilayers requires the cell to expend energy in the form of ATP. Following fusion, the vesicle contents are released to the outside of the cell. An example of this process occurs in the pancreas, where cells release digestive enzymes into a pancreatic duct for transport to the small intestine.

CLINICAL VIEW

Cystic Fibrosis and Chloride Channels

The inherited disease cystic fibrosis (CF) involves defective plasma membrane proteins that affect chloride ion (Cl^-) channels in the membrane. These channels are transport proteins that use facilitated diffusion to move chloride ions across the plasma membrane. The genetic defect that causes CF results in the formation of abnormal chloride channel proteins in the membranes of cells lining the respiratory passageways and ducts in glands, such as the pancreas. The primary defect in these chloride channels results in an abnormal flow of chloride ions across the membrane, causing salt to be trapped within the cytoplasm of affected cells. Ultimately, the normal osmotic flow of water across the plasma membrane breaks down. The concentration of salt within the cytoplasm of these cells causes an increase in the osmotic flow of water into the cell, thereby resulting in thickening of the mucus in the respiratory passageways and the pancreatic ducts. The aggregation of thickened mucus plugs the airways of the lungs, leading to breathing problems and increasing the risk of infection. Therefore, a single genetic and biochemical defect in a transport protein produces significant health problems.

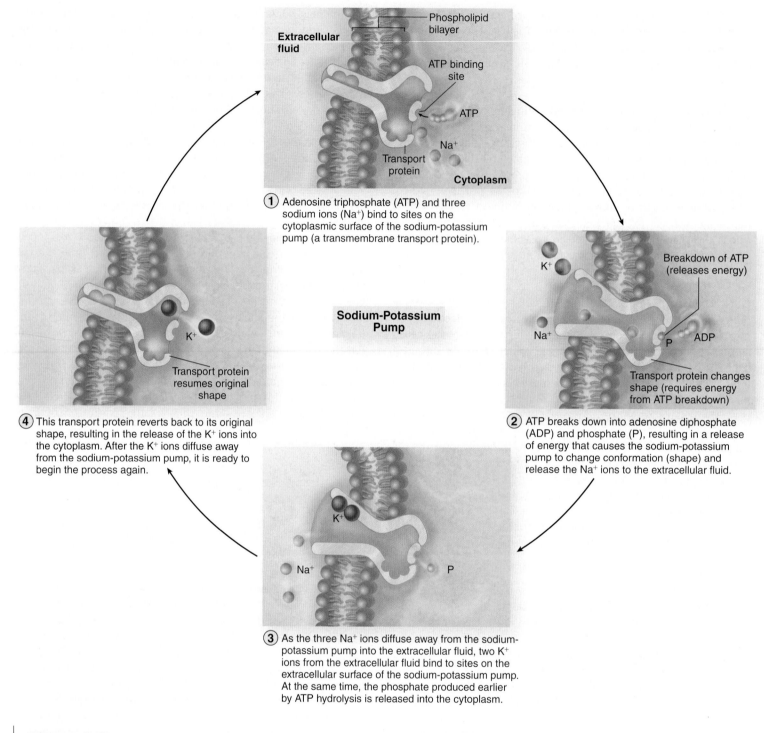

Sodium-Potassium Pump

1. Adenosine triphosphate (ATP) and three sodium ions (Na⁺) bind to sites on the cytoplasmic surface of the sodium-potassium pump (a transmembrane transport protein).

2. ATP breaks down into adenosine diphosphate (ADP) and phosphate (P), resulting in a release of energy that causes the sodium-potassium pump to change conformation (shape) and release the Na⁺ ions to the extracellular fluid.

3. As the three Na⁺ ions diffuse away from the sodium-potassium pump into the extracellular fluid, two K⁺ ions from the extracellular fluid bind to sites on the extracellular surface of the sodium-potassium pump. At the same time, the phosphate produced earlier by ATP hydrolysis is released into the cytoplasm.

4. This transport protein reverts back to its original shape, resulting in the release of the K⁺ ions into the cytoplasm. After the K⁺ ions diffuse away from the sodium-potassium pump, it is ready to begin the process again.

Figure 2.5

Sodium-Potassium Pump. A sodium-potassium pump has a transmembrane transport protein that uses energy to transport Na⁺ and K⁺ ions through the membrane from a region of low concentration to a region of high concentration. This continuous, active transport process can be broken down into four steps.

By contrast, large particulate substances and macromolecules are taken *into* the cell via **endocytosis** (en'dō-sī-tō'sis; *endon* = within). The steps of endocytosis are similar to those of exocytosis, only in reverse. In endocytosis, extracellular macromolecules and large particulate matter are packaged in a vesicle that forms at the cell surface for internalization into the cell. A small area of plasma membrane folds inward to form a pocket, or **invagination** (in-vaj'i-nā-shun; *in* = in, *vagina* = a sheath), which deepens and pinches off as the lipid bilayer fuses. This fusion of the lipid bilayer is the energy-expending step. A new intracellular vesicle is formed containing material that was formerly outside the cell. There are three types of endocytosis: phagocytosis, pinocytosis, and receptor-mediated endocytosis **(figure 2.7)**.

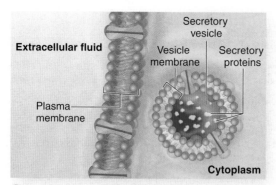

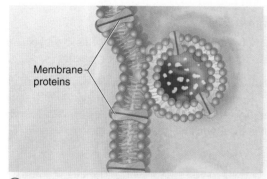

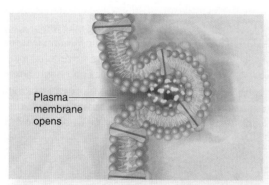

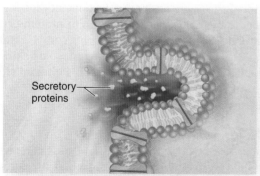

① Vesicle nears plasma membrane

② Fusion of vesicle membrane with plasma membrane

③ Exocytosis as plasma membrane opens externally

④ Release of vesicle components into the extracellular fluid and integration of vesicle membrane components into the plasma membrane

Figure 2.6

Exocytosis.   In exocytosis, the cell secretes bulk volumes of materials into the extracellular fluid.

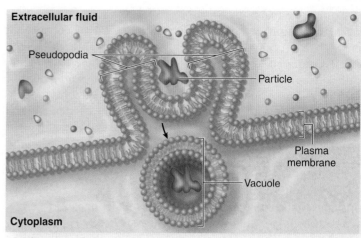

(a) Phagocytosis

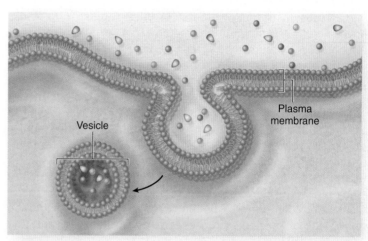

(b) Pinocytosis

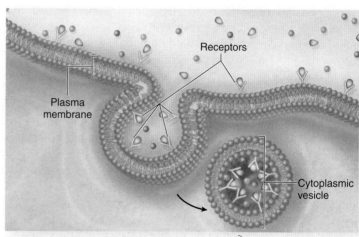

(c) Receptor-mediated endocytosis

Figure 2.7

Three Forms of Endocytosis.   Endocytosis is a process whereby the cell acquires materials from the extracellular fluid. (*a*) Phagocytosis occurs when membrane extensions, termed pseudopodia, engulf a particle and internalize it into a vacuole. (*b*) Pinocytosis is the incorporation of droplets of extracellular fluid into the cell via the formation of small vesicles. (*c*) In receptor-mediated endocytosis, receptors with specific molecules bound to them aggregate within the membrane, and then an invagination forms around them to create a cytoplasmic vesicle.

Phagocytosis (fag′ō-sī-tō′sis; *phago* = to eat, *kytos* = cell, *osis* = condition) means "cellular eating." It is a nonspecific process that occurs when a cell engulfs or captures a large particle external to the cell by forming membrane extensions, called **pseudopodia** (sing., pseudopodium; soo-dō-pō′dē-ŭm, -pō′dē-ă; *pous* = foot), or *false feet*, to surround the particle (figure 2.7*a*). Once the particle is engulfed by the pseudopodia, it is packaged within an enclosed membrane sac. If large enough, this sac is classified as a **vacuole** (vak′oo-ōl; *vacuum* = an empty space). The contents of the vacuole are broken down chemically (digested) after it fuses with a *lysosome* (lī′sō-sōm; *lysis* = a loosening, *soma* = body), which contains specific digestive enzymes that split large molecules into smaller ones. Only a few types of cells are able to perform phagocytosis. For example, phagocytosis occurs regularly when a white blood cell engulfs and digests a bacterium.

Pinocytosis (pin′ō-sī-tō′sis [or pī′nō-]; *pineo* = to drink, *kytos* = cell, *osis* = condition) is known as "cellular drinking." This process occurs when the cell internalizes a very small droplet of extracellular fluid into tiny internal vesicles (figure 2.7*b*). This process is nonspecific because all solutes dissolved in the droplet are taken into the cell. Most cells perform this type of transport across the membrane. Pinocytosis is similar to bulk filtration in that both types of transport move similar materials. However, it differs from bulk filtration because pinocytosis moves materials against a concentration gradient. An example of pinocytosis occurs within cells that form a capillary (tiny blood vessel) wall, where vesicles fill with a fluid droplet containing small solutes from the blood, carry this droplet to the other side of the cell, and then expel its contents outside the capillary wall.

Receptor-mediated endocytosis is the movement of specific molecules from the extracellular environment into a cell by way of a newly formed vesicle. This process begins when molecules in the extracellular fluid bind to their specific integral membrane protein receptors. (Recall that a membrane receptor is a protein that acts as a binding site for molecules outside the cell.) This process is different from the nonspecific transport mechanisms discussed earlier. It is considered a specific mechanism because the endocytosis is stimulated by the binding of the specific molecules to their specific membrane receptors. The receptor proteins then cluster in one region of the membrane to begin the process of endocytosis. The plasma membrane housing the bound specific molecules from the extracellular fluid folds inward to form a pocket, or invagination (figure 2.7*c*). This membrane pocket deepens and pinches off as the lipid bilayers fuse. The fusion of these lipid bilayers requires the cell to expend energy in the form of ATP. An example of receptor-mediated endocytosis occurs when human cells contain receptors that bind to and internalize cholesterol, which is required for new membrane synthesis. Cholesterol travels in our blood bound to proteins called low-density lipoproteins (LDL). LDL particles bind to LDL receptors in the membrane. Receptor-mediated endocytosis enables the cell to obtain bulk quantities of specific substances, even though those substances may not be very concentrated in the extracellular fluid.

Table 2.3 summarizes passive and active transport mechanisms.

WHAT DID YOU LEARN?

4 What types of lipids are found in the plasma membrane?

5 In general, what materials may cross a selectively permeable membrane?

6 What is diffusion?

7 Describe the process of osmosis.

8 Discuss the similarities between facilitated diffusion and receptor-mediated endocytosis.

Cytoplasm

Key topics in this section:

- Characteristics of the three parts of a cell's cytoplasm
- Structures and functions of cellular organelles

Cytoplasm is a nonspecific term for all of the materials contained within the plasma membrane and surrounding the nucleus. The cytoplasm includes three separate parts: cytosol, inclusions, and organelles (except the nucleus).

Cytosol

The **cytosol** (sī′tō-sol; *kytos* = cell, *sol* = abbrev. of soluble), also called the *cytoplasmic matrix* or *intracellular fluid*, is the viscous, syruplike fluid of the cytoplasm. It has a high water content and contains many dissolved solutes, including ions, nutrients, proteins, carbohydrates, lipids, and other small molecules. Many cytoplasmic proteins are the enzymes that act as catalysts in cellular reactions. The cytosol's carbohydrates and lipids serve as an energy source for the cell. Many of the small molecules in the cytosol are the building blocks of large macromolecules. For example, amino acids are small molecules dissolved in the cytosol that the cell uses to synthesize new proteins.

Inclusions

The cytosol of some cells contains **inclusions,** a large and diverse group of chemical substances that these cells store temporarily. Most inclusions are not bound by a membrane. Cellular inclusions include pigments, such as melanin; protein crystals; and nutrient stores, such as glycogen and triglycerides. **Melanin** (mel′ă-nin; *melas* = black), a stored pigment in some skin, hair, and eye cells, protects the body from the sun's ultraviolet light. **Glycogen** is a polysaccharide (a type of carbohydrate [sugar]) that is stored primarily in liver and skeletal muscle cells.

Organelles

Organelles (or′gă-nel; *organon* = organ, *elle* = the diminutive suffix), meaning "little organs," are complex, organized structures with unique, characteristic shapes. Each type of organelle performs a different function for the cell. Collectively, the specialized functions of all organelles are essential for normal cellular structure and activities. These unique structures permit the cell to carry on different activities simultaneously. This division of labor prevents interference between cellular activities and promotes maximal functional efficiency in the cell. Organelles assume specific roles in growth, repair, and cellular maintenance. The distribution and numbers of different types of organelles are determined by organelle function

Table 2.3	Transport Processes Across a Plasma Membrane		
Process	**Type of Movement**	**Energy Source**	**Example**
PASSIVE TRANSPORT	Movement of substance *along* a concentration gradient; ATP not required		
Simple diffusion	Unaided net movement of a substance due to molecular motion down its concentration gradient across selectively permeable membrane; continues until equilibrium is reached	Molecular movement	Exchange of oxygen and carbon dioxide between blood and body tissues
Osmosis	Diffusion of water across a selectively permeable membrane; direction is determined by relative solute concentrations; continues until equilibrium is reached	Molecular movement	Water in small kidney tubules moves across a cell barrier back into the blood from the tubular fluid that eventually forms urine
Facilitated diffusion	Movement of materials too large to pass through membrane channels; relies on transport proteins	Molecular movement requiring carrier assistance by a transport protein	Transport of glucose into cells
Bulk filtration	Bulk movement of solvents and solutes from an area of high concentration to an area of low concentration as a result of hydrostatic pressure differences across the membrane	Hydrostatic pressure	Transport of nutrients and fluids from the blood into body tissues
ACTIVE TRANSPORT	Movement of substances *against* a concentration gradient; requires ATP; requires assistance to move across the membrane, often by a transport protein and sometimes by bulk movement in vesicles.		
Ion pumps	Transport of ions across the membrane against a concentration gradient by transmembrane protein pumps	ATP	Sodium-potassium exchange pump
Bulk transport	Membrane vesicles form around materials for transport	ATP	
Exocytosis	Bulk movement of substances *out* of the cell by fusion of secretory vesicles with the plasma membrane	ATP	Release of digestive enzymes by pancreatic cells
Endocytosis	Bulk movement of substances *into* a cell by vesicles forming at the plasma membrane	ATP	
Phagocytosis	Type of endocytosis in which particulate materials move into a cell after being engulfed by pseudopodia at the cell surface; vesicles form at the inside of the plasma membrane; large sacs are called vacuoles	ATP	White blood cell engulfing a bacterium
Pinocytosis	Type of endocytosis in which plasma membrane folds inward to capture extracellular fluid droplet and its dissolved contents, forming a small new vesicle inside the cell	ATP	Formation of small vesicles in capillary wall to move fluid and small particulate materials out of the blood
Receptor-mediated endocytosis	Type of endocytosis in which specific molecule-receptor complexes in the plasma membrane stimulate the clustering of bound molecule-receptor complexes; vesicles containing specific molecules bound to receptors in the membrane stimulate internalization of the bound molecules	ATP	Uptake of cholesterol into cells

and vary among cells, depending upon the needs of the cells. Two categories of organelles are recognized: membrane-bound organelles and non-membrane-bound organelles.

Membrane-Bound Organelles

Some organelles are surrounded by a membrane and thus are called **membrane-bound organelles,** or *membranous organelles.* This membrane is similar to the plasma membrane surrounding the cell in that it is composed of a phospholipid bilayer with diverse associated proteins. Note that every membrane exhibits a unique protein-lipid composition, which confers a unique function(s) to that membrane. The membrane separates the organelle's contents from the cytosol so that the activities of the organelle can proceed without disrupting other cellular activities.

Membrane-bound organelles include the endoplasmic reticulum, the Golgi apparatus, lysosomes, peroxisomes, and mitochondria.

Endoplasmic Reticulum   The **endoplasmic reticulum** (re-tik'ū-lum; *rete* = net) **(ER)** is an extensive intracellular membrane network throughout the cytoplasm. ER is composed of two distinct regions that differ in structure and function: smooth endoplasmic reticulum (smooth ER, or SER) and rough endoplasmic reticulum (rough ER, or RER; **figure 2.8**). The amount of either kind of ER varies, depending on the specific functions of the cell.

Smooth ER is continuous with the rough ER. Because no ribosomes are attached to the smooth ER membranes, the walls have a smoother appearance than those of rough ER. Smooth ER resembles

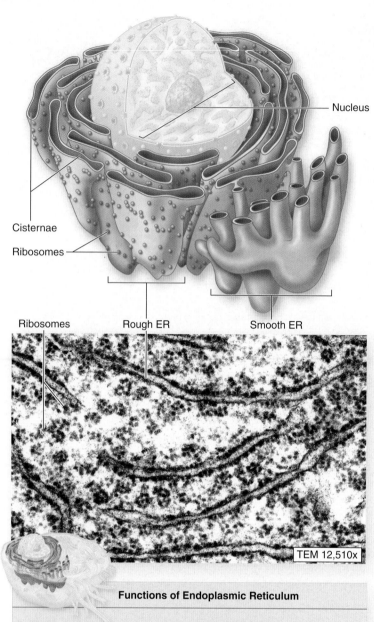

Nucleus

Cisternae

Ribosomes

| Ribosomes | Rough ER | Smooth ER |

TEM 12,510x

Functions of Endoplasmic Reticulum

1. **Synthesis:** Provides a place for chemical reactions
   a. Smooth ER is the site of lipid synthesis and carbohydrate metabolism
   b. Rough ER synthesizes proteins for secretion, incorporation into the plasma membrane, and as enzymes within lysosomes
2. **Transport:** Moves molecules through cisternal space from one part of the cell to another; sequestered away from the cytoplasm
3. **Storage:** Stores newly synthesized molecules
4. **Detoxification:** Smooth ER detoxifies both drugs and alcohol

Figure 2.8

Endoplasmic Reticulum (ER).   A drawing and TEM show that the ER is a membranous network of flattened membrane sacs (cisternae) and interconnected tubules that is continuous with the nuclear envelope. Smooth ER, which is not shown on this TEM, consists of even-surfaced, interconnected tubules, and it lacks associated ribosomes. Rough ER, by contrast, is composed of cisternal membranes with ribosomes attached to their cytoplasmic surfaces. However, the two types of ER are continuous.

multiple interconnected branches of tubules. The smooth ER of various cell types functions in diverse metabolic processes, including synthesis, transport, and storage of lipids; metabolism of carbohydrates; and detoxification of drugs, alcohol, and poisons. The amount of smooth ER is greater in cells that synthesize steroid hormones. In addition, the liver contains abundant amounts of smooth ER in order to process digested nutrients and detoxify drugs and alcohol.

Rough ER is responsible for producing, transporting, and storing proteins to be exported outside the cell, proteins to be incorporated into the plasma membrane, and enzymes that are housed within lysosomes. Rough ER consists of profiles of parallel membranes enclosing spaces called **cisternae** (sis'tern-ā; *cisterna* = cistern). Ribosomes are the small structures attached to the cytoplasmic sides (called faces) of these membranes. These ribosomes are called fixed ribosomes because they are attached to the membrane surface of the ER, thus forming the rough ER. These ribosomes synthesize the proteins targeted for cell export, insertion into the plasma membrane, or inclusion within a lysosome as a catalyst. As new proteins are synthesized by the fixed ribosomes, they pass through the membrane of the rough ER and enter its cisternae, where their original structure changes by either adding other molecules or removing part of what was originally synthesized. These modified proteins are packaged into small, enclosed membrane sacs that pinch off from the ER. These sacs, termed **transport vesicles** shuttle proteins from the rough ER to another organelle, the Golgi apparatus (discussed later) for further modification. For a seamless interaction and transition between organelles, transport vesicles, and the plasma membrane, the membranes of each structure have the same general lipid and protein composition and organization. However, as mentioned earlier, the molecules within these membranes also have some unique characteristics that are associated with the specific function(s) of each structure. The amount of rough ER is greater in cells producing large amounts of protein for secretion, such as a cell in the pancreas that secretes enzymes for digesting materials in the small intestine.

Golgi Apparatus   The **Golgi apparatus**, also called the *Golgi complex*, is a center for modifying, packaging, and sorting materials that arrive from the RER in transport vesicles. The Golgi apparatus is especially extensive and active in cells specialized for secretion.

The Golgi apparatus is composed primarily of a series of cisternae, which are arranged in a stack **(figure 2.9a)**. The edges of each sac bulge, and many small transport vesicles are clustered around the expanded edges of the individual sacs. The vesicles concentrated at the periphery of the Golgi apparatus are active in transporting and transferring material between the individual sacs of the Golgi apparatus as well as between the Golgi apparatus and other cellular structures.

The Golgi apparatus exhibits a distinct polarity: The membranes of the cisternae at opposite ends of a stack differ in thickness and molecular composition. These two poles of the Golgi apparatus are called the **receiving region** (or *cis-face*) and the **shipping region** (or *trans-face*), respectively. The diameter of the flattened sac is larger in the receiving region than in the shipping region. The products of the rough ER move through the Golgi apparatus via transport vesicles, going from the receiving region to the shipping region. Normally, materials move through the Golgi apparatus as shown in figure 2.9b and described here:

1. Newly synthesized proteins in the rough ER cisternae are sequestered into a transport vesicle.
2. The vesicle pinches off the ER and travels to the Golgi apparatus.
3. Newly arrived transport vesicles fuse with the receiving region of the Golgi apparatus.

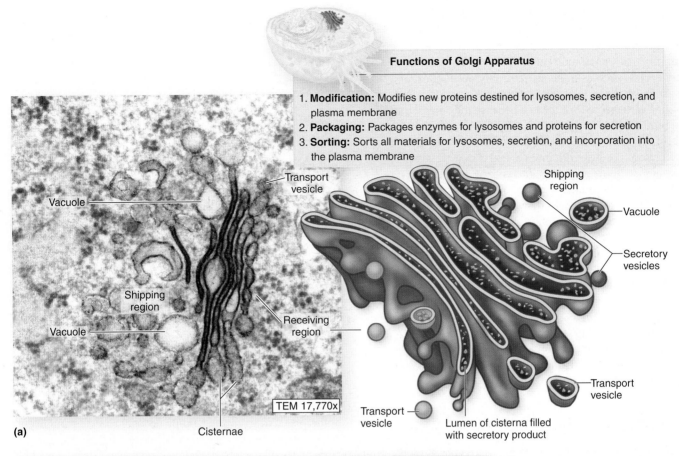

Functions of Golgi Apparatus

1. **Modification:** Modifies new proteins destined for lysosomes, secretion, and plasma membrane
2. **Packaging:** Packages enzymes for lysosomes and proteins for secretion
3. **Sorting:** Sorts all materials for lysosomes, secretion, and incorporation into the plasma membrane

(a)

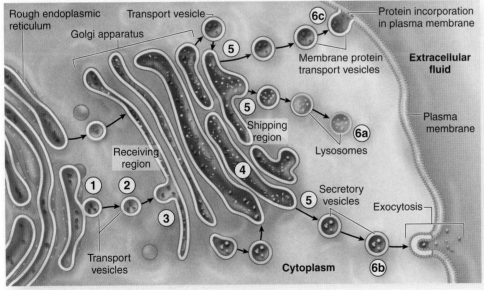

1. RER proteins in transport vesicle
2. Vesicle from RER moves to Golgi apparatus
3. Vesicle fuses with Golgi apparatus receiving region
4. Proteins are modified as they move through Golgi apparatus
5. Modified proteins are packaged in shipping region
6. Vesicles become either (a) lysosomes, (b) secretory vesicles that undergo exocytosis, or (c) plasma membrane

(b) Movement of materials through the Golgi apparatus

Figure 2.9

Golgi Apparatus.   Each Golgi apparatus is composed of several flattened membrane sacs (cisternae), with some associated transport vesicles at the periphery of these sacs. The arrangement of sacs exhibits both structural and functional polarity. *(a)* A TEM and a drawing provide different views of the Golgi apparatus along with a list of its functions. *(b)* The receiving region receives incoming transport vesicles from the rough ER; large vesicles carrying finished product exit the shipping region.

4. Protein modification occurs as the proteins are moved by transport vesicles sequentially through the Golgi apparatus cisternae from the receiving region to the shipping region.

5. Modified proteins are packaged in secretory vesicles.

6. Vesicles leaving the shipping region become (1) lysosomes, which contain proteins called digestive enzymes, (2) secretory vesicles that undergo exocytosis, or (3) new parts of the plasma membrane.

Lysosomes   **Lysosomes** (lī'sō-sōm; *lysis* = a loosening, *soma* = body) are membrane sacs formed by the Golgi apparatus (**figure 2.10**). Lysosomes contain enzymes used by the cell to digest waste products and ingested macromolecules. These enzymes break down large molecules, such as proteins, fats, polysaccharides, and nucleic acids, into smaller molecules. Lysosomes are sometimes referred to as the "garbagemen" of the cell because they digest and remove waste products.

Some substances digested by lysosomal enzymes enter the cell by endocytosis. Lysosomes fuse with internalized endocytic vesicles, and their enzymes combine with the internalized materials. The products resulting from these digestive activities are released from the lysosome into the cytosol, where they are recycled for various future uses. For example, a large protein is broken down into its component amino acids, which may be used to synthesize a new, different protein needed by that cell.

Lysosomes also remove the cell's damaged parts. An internal membrane encloses these damaged structures, and then it fuses with the lysosomes. Thus, old organelles are removed via a process called **autophagy** (aw-tōf'ă-jē; *autos* = self, *phago* = to eat). When a cell is damaged or dies, enzymes from all lysosomes are eventually released into the cell, resulting in the rapid digestion of the cell itself. This process is called **autolysis** (aw-tol'i-sis; *autos* = self, *lysis* = dissolution).

WHAT DO YOU THINK?

2   What would happen to a cell if it didn't contain any lysosomes (or if its lysosomes weren't functioning)? Would the cell be able to survive?

Tay-Sachs Disease

Tay-Sachs is a "lysosomal storage disease" that results in the buildup of fatty material in nerve cells. Healthy, properly functioning lysosomes are essential for the health of the cells and the whole body. Tay-Sachs disease occurs because one of the more than 40 different lysosomal enzymes is missing or nonfunctional. Lysosomes in affected individuals lack an enzyme that is needed to break down a complex membrane lipid. As a result, the complex lipid accumulates within cells. The cellular signs of Tay-Sachs disease are swollen lysosomes due to accumulation of the complex lipid that cannot be digested. Affected infants appear normal at birth, but begin to show signs of the disease by the age of 6 months. The nervous system bears the brunt of the damage. Paralysis, blindness, and deafness typically develop over a period of one or two years, followed by death by the age of 4. Unfortunately, there is no treatment or cure for this deadly disease.

Peroxisomes   **Peroxisomes** (per-ok'si-sōm) are membrane-enclosed sacs that are usually smaller in diameter than lysosomes (**figure 2.11**). They are formed by pinching off vesicles from the rough ER. Peroxisomes use oxygen to catalytically detoxify specific harmful substances either produced by the cell or taken into the cell. For example, the peroxisome is able to convert hydrogen peroxide (a toxic compound) that is sometimes produced by cells into water before it can damage

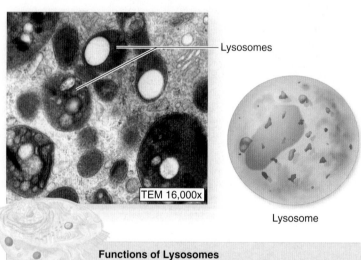

Functions of Lysosomes

1. **Digestion:** Digest all materials that enter cell by endocytosis
2. **Removal:** Remove worn-out or damaged organelles and cellular components; recycle small molecules for resynthesis (autophagy)
3. **Self-destruction:** Digest the remains (autolysis) after cellular death

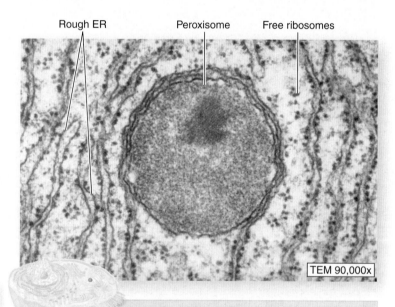

Function of Peroxisomes

Detoxification: Detoxify harmful substances; convert hydrogen peroxide to water; break down fatty acid molecules

Figure 2.10

Lysosomes.   A drawing and TEM show lysosomes, which are membrane-bound, spherical sacs in the cytoplasm of a cell. Lysosomes house enzymes for intracellular digestion, as well as performing the other functions listed here.

Figure 2.11

Peroxisomes.   A TEM shows a peroxisome in a cell. Peroxisomes are small, membrane-bound organelles that degrade harmful substances, including hydrogen peroxide, during cellular reactions. They also break down fatty acid molecules.

CLINICAL VIEW

Adrenoleukodystrophy (ALD)

Adrenoleukodystrophy is a rare inherited disorder that became widely known after the release of the movie *Lorenzo's Oil* in 1993. The movie chronicles the true story of Lorenzo Odone, a boy diagnosed with ALD, and his family's efforts to find a treatment and cure. ALD is caused when a membrane protein is missing in the peroxisome. In the healthy state, this protein transports into the peroxisome an enzyme that controls the breakdown of very-long-chain fatty acids, which are part of the neutral fats in our diets. When the enzyme cannot enter the peroxisome because the transport protein is missing, the peroxisomes cannot function normally, and so the very-long-chain fatty acids accumulate in cells of the central nervous system, eventually stripping these cells of their myelin covering. The absence of this myelin covering prevents the normal transmission of messages through the nerve cell, and the messages "short-circuit." The very-long-chain fatty acids also build up in the adrenal glands, causing them to malfunction.

ALD exists in several forms, but the most severe kind affects young boys between the ages of 4 and 10. Typically, the first signs of ALD are lethargy, weakness, and dizziness. Additionally, the patient's skin may darken, blood sugar levels decrease, heart rhythm is altered, and the levels of electrolytes in the body fluids become imbalanced. Control over the limbs deteriorates. In the severe form of ALD, the patient loses all motor function and becomes paralyzed. Eventually, the patient becomes blind, loses basic reflex actions, such as swallowing, and enters a vegetative state. Death often results.

There is no cure for ALD, but dietary modification (to reduce the amounts of very-long-chain fatty acids in the diet) and use of "Lorenzo's oil" (an oleic acid/rapeseed oil blend discovered by Lorenzo Odone's family) helps control the very-long-chain fatty acid buildup. Most recently, some research has indicated that statins (medicines that control cholesterol levels) may help prevent the buildup of the very-long-chain fatty acids. Researchers have also learned that the severity of the disease is reduced if new therapies are applied at an early age. In addition, new, noninvasive diagnostic techniques have been developed, and diagnosis has been further improved by recognizing different phenotypes (since ALD can be misdiagnosed as attention deficit/hyperactive disorder in some boys).

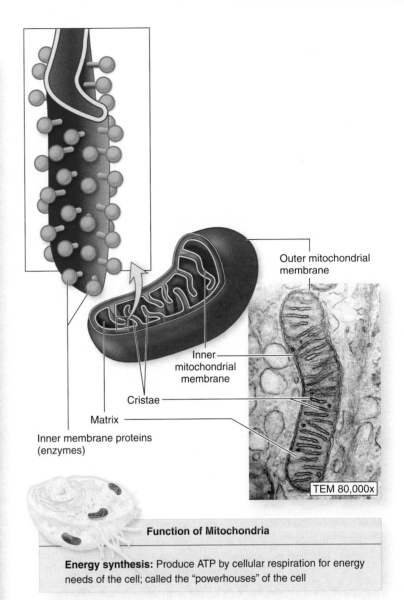

Outer mitochondrial membrane

Inner mitochondrial membrane

Cristae

Matrix

Inner membrane proteins (enzymes)

TEM 80,000x

Function of Mitochondria

Energy synthesis: Produce ATP by cellular respiration for energy needs of the cell; called the "powerhouses" of the cell

Figure 2.12

Mitochondria.   A drawing and TEM show the parts of a mitochondrion. Mitochondria are double-membrane-bound organelles that produce ATP for cellular work.

the cell. It does this using the enzyme catalase, which is a component of the peroxisome. Peroxisomes are most abundant in liver cells, where they break down fatty acids and detoxify some toxic materials, such as alcohol, that are absorbed in the digestive tract.

Mitochondria   Mitochondria (mī-tō-kon′drē-ă; sing., *mitochondrion*, mī-tō-kon′drē-on′; *mitos* = thread, *chondros* = granule) are organelles with a double membrane that are involved in producing large amounts of the cell's energy currency, ATP. For this reason, mitochondria are called the "powerhouses" of the cell. A mitochondrion is completely surrounded by an outer membrane, while a second, or inner membrane, is folded internally into the space at the center of the organelle. These folds, called **cristae** (kris′tă, -tē; *crista* = crest), increase the surface area that is exposed to the internal fluid contents, termed the **matrix (figure 2.12)**. Inner membrane proteins are on the cristae.

The number of mitochondria in a cell depends upon the cell's energy needs. Because mitochondria can self-replicate, the numbers of mitochondria are greater in cells that have a high energy demand. For example, muscle cells with a high rate of energy usage have a large number of mitochondria in their cytoplasm. Mitochondria numbers increase with increased demands for ATP. Additionally, mitochondria contain a small, unique fragment of DNA, the genetic material (described later in this chapter). In the mitochondria, this piece of DNA contains genes for producing mitochondrial proteins. Mitochondrial shape also varies among cells. Interestingly, the head of a sperm cell contains no mitochondria because that portion has no energy need. Instead, there are mitochondria in the midpiece of the sperm cell, the region responsible for propelling the sperm.

WHAT DO YOU THINK?

❸ While examining a cell by microscope, you observe that the cell has few mitochondria. What does this imply about the cell's energy requirements?

MELAS and Mitochondria

MELAS syndrome is a neurogenerative disorder named for its features: **M**itochondrial myopathy (mī-op′ă-thē), a weakness in muscle caused by reduced ATP production; **E**ncephalopathy, a brain disorder; **L**actic **A**cidosis, accumulation of lactic acid in tissues because of an inability to produce normal amounts of ATP; and **S**troke, impaired cerebral (brain) circulation. The abnormal mitochondrial function is the result of a single mutation in the mitochondrial DNA that makes affected individuals unable to synthesize some of the proteins needed for energy transactions. This mutation also leads to the elevated levels of lactic acid, brain pathology, and recurring strokes. The syndrome typically first presents with stroke (often between the ages of 4 and 15 years), a symptom that is followed by episodes of fatigue, developmental delays, and seizures. Low muscle tone and muscle weakness are common. Often patients have uncoordinated and numb hands or feet, as well as diabetes mellitus. MELAS is a progressive disorder that has a high rate of morbidity (illness) and mortality (death). There is no cure for MELAS, and drug therapies have been only minimally effective.

Non-Membrane-Bound Organelles

Organelles that are always in direct contact with the cytosol are called either **non-membrane-bound organelles** or *nonmembranous organelles.*

Ribosomes   **Ribosomes**   (rī′bō-sōm; *ribos* = reference to a s-carbon sugar, *soma* = body) are very small, dense granules that are responsible for protein production (synthesis). Each ribosome has a small subunit and a large subunit (**figure 2.13a**); the small subunit is about one-half the size of the large subunit. The parts of the subunits are formed in the nucleus, and the subunits are assembled within the cytosol at the time when a new protein is about to be synthesized. Once ribosomes are assembled, those that float freely within the cytosol of the cell are called **free ribosomes,** while those that are attached to the rough endoplasmic reticulum are called **fixed ribosomes** (figure 2.13b). Free ribosomes are responsible for the synthesis of proteins that remain within the cytosol of the cell. Fixed ribosomes produce proteins that are exported outside the cell, incorporated into the plasma membrane, or housed as enzymes within a new lysosome.

Cytoskeleton   The **cytoskeleton** is composed of protein subunits organized either as filaments or hollow tubes. The cytoskeleton has three separate components—microfilaments, intermediate filaments, and microtubules—which differ in their structures and functions (**figure 2.14**).

   Microfilaments (mī-krō-fil′ă-ment; *micros* = small) are the smallest components of the cytoskeleton. They are about 7 nanometers (nm) in diameter and are composed of thin protein filaments (actin proteins) organized into two intertwined strands. They form an interlacing network on the cytoplasmic side of the plasma membrane. Microfilaments help maintain cell shape, support changes in cell shape, participate in muscle contraction, separate the two cells formed during cell division, and facilitate cytoplasmic streaming, which is the movement of the cytoplasm associated with changing cell shape.

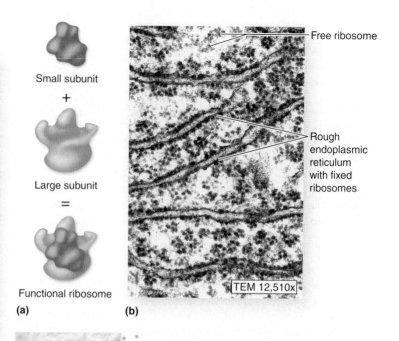

Small subunit

+

Large subunit

=

Functional ribosome

(a)

Free ribosome

Rough endoplasmic reticulum with fixed ribosomes

TEM 12,510x

(b)

Functions of Ribosomes

Protein synthesis:
1. Free ribosomes synthesize proteins for use within the cell
2. Fixed ribosomes synthesize proteins destined to be incorporated into the plasma membrane, exported from the cell, or housed within lysosomes

Figure 2.13

Ribosomes.   Ribosomes are small, dense, cytoplasmic granules where proteins are synthesized within the cell. (*a*) Ribosomes consist of both small and large subunits. (*b*) A TEM shows fixed and free ribosomes in the cell cytoplasm.

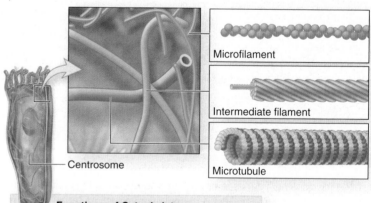

Microfilament

Intermediate filament

Centrosome

Microtubule

Functions of Cytoskeleton

1. **Structural:** Provides structural support to cell; stabilizes junctions between cells
2. **Movement:** Assists with cytosol streaming and cell motility; helps move organelles and materials throughout cell; helps move chromosomes during cell division

Figure 2.14

Cytoskeleton.   Filamentous proteins form the cytoskeleton, which helps give the cell its shape and coordinate cellular movements. The three cytoskeletal elements are microfilaments, intermediate filaments, and microtubules.

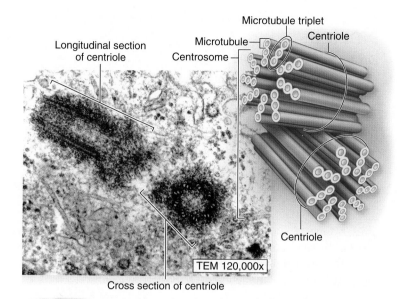

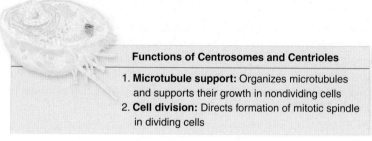

Functions of Centrosomes and Centrioles

1. **Microtubule support:** Organizes microtubules and supports their growth in nondividing cells
2. **Cell division:** Directs formation of mitotic spindle in dividing cells

Figure 2.15

Centrosome and Centrioles.   A drawing and TEM show that a region of the cytoplasm called the centrosome surrounds a centriole pair immediately adjacent to the nucleus.

Intermediate filaments are between 8 nm and 12 nm in diameter, and are more rigid than microfilaments. They support cells structurally and stabilize junctions between cells. Their protein component differs, depending upon the cells in which they are found in the body.

Microtubules (mī-krō-too′būl; *micros* = small, *tubus* = tube) are hollow tubules, about 25 nm in diameter, composed of long chains of a protein called tubulin. Microtubules radiate from the centrosome (discussed next) and help hold organelles in place, maintain cell shape and rigidity, direct organelle movement between different regions of the cell, provide a means of cell motility using structures called cilia or flagella, and move chromosomes during the process of cell division. Microtubules are not permanent structures, and they can be elongated or shortened as needed to complete their functions.

Centrosome and Centrioles   Closely adjacent to the nucleus in most cells is a nonmembranous, spherical structure called the **centrosome.** The matrix of this region is a microtubule organization center that supports the growth and elongation of microtubules. Within the region of the centrosome are a pair of cylindrical **centrioles** (sen′trē-ōl; *kentron* = a point, center) that lie perpendicular to one another. Each centriole is composed of nine sets of three closely aligned microtubules, called microtubule triplets, that are arranged in a circle **(figure 2.15)**. The centrioles replicate immediately prior to cell division (mitosis). During mitosis (described on page 47), they are responsible for organizing microtubules that are a part of the mitotic spindle. Some of these microtubules attach to chromosomes to facilitate their movement.

Cilia and Flagella   **Cilia** (sil′ē-ă; sing., *cilium*, sil′ē-ŭm; an eyelid) and **flagella** (flă-jel′ă; sing., *flagellum*, flă-jel′ŭm; a whip) are projections extending from the cell. They are composed of cytoplasm and supportive microtubules, and they are enclosed by the plasma membrane.

Cilia are usually found in large numbers on the exposed surfaces of certain cells **(figure 2.16a)**. For example, cells having cilia on their exposed surfaces line parts of the respiratory passageways. Here, these ciliated cells are always associated with mucin-secreting goblet cells. Mucus that is formed from the secreted mucin appears as a sticky film on the free surface of ciliated cells. The beating of the cilia moves the mucus and any adherent particulate material along the cell surface toward the throat, where it may then be expelled from the body.

Flagella are similar to cilia in basic structure; however, they are longer and usually appear alone (figure 2.16b). The function of

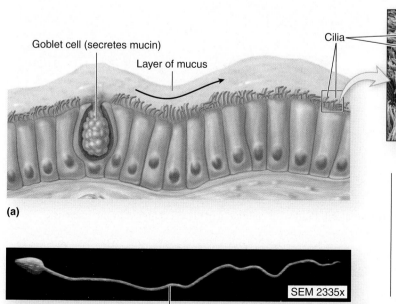

Figure 2.16

Cilia and Flagella.   Cilia and flagella are appendages extending from the surface of some cells. *(a)* Cilia usually occur in large numbers; they work together to move materials or fluids along the surface of a cell. *(b)* Flagella are longer than cilia, and usually occur as single appendages. In human sperm cells, the flagellum is the apparatus that enables the sperm to "swim."

a flagellum is to help propel or move an entire cell. In humans, the only example of a cell with a flagellum is the sperm cell.

Microvilli **Microvilli** are thin, microscopic projections extending from the surface of the plasma membrane. They are much smaller than cilia, much more densely packed together, and do not have powered movement (see figure 2.3). The main function of microvilli is to increase the surface area of the plasma membrane. In essence, these projections create a more extensive plasma membrane surface for molecules to travel across. Just as not all cells have cilia, not all cells have microvilli. Cells with microvilli occur throughout the small intestine, where increased surface area is needed to absorb digested nutrients.

WHAT DID YOU LEARN?

9 Describe the characteristics of the cytosol.

10 Describe the functions of lysosomes, mitochondria, and centrioles.

11 Contrast the fates of proteins synthesized on free ribosomes versus those synthesized on fixed ribosomes.

12 What is the function of cilia?

Nucleus

Key topics in this section:

- Contents and function of the nucleus
- Relationship between chromatin and chromosomes

The **nucleus** is the core, or the control center, of cellular activities. Usually, it is the largest structure within the cell, averaging about 5 μm to 7 μm in diameter (**figure 2.17**). Generally, its shape mirrors the shape of the cell. For example, a cuboidal cell has a spherical nucleus in the center of the cell, while a thin, elongated cell's nucleus is elongated in the same direction as the cell itself. Some cells contain uniquely shaped nuclei. For example, neutrophils, a type of white blood cell, have a multilobed nucleus—one that has three or more bulges.

The nucleus contains three basic structures: a nuclear envelope, nucleoli, and chromatin.

Nuclear Envelope

The nucleus is enclosed by a double membrane structure called the **nuclear envelope.** This boundary controls the entry and exit of materials between the nucleus and the cytoplasm. Each layer of the nuclear envelope is a phospholipid bilayer, similar in structure to the plasma membrane. The nuclear envelope has ribosomes attached to

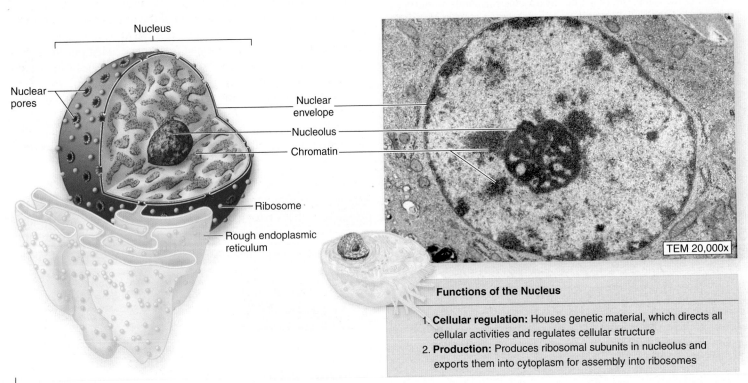

Functions of the Nucleus

1. **Cellular regulation:** Houses genetic material, which directs all cellular activities and regulates cellular structure
2. **Production:** Produces ribosomal subunits in nucleolus and exports them into cytoplasm for assembly into ribosomes

Figure 2.17
Nucleus. A drawing and TEM compare the structures of the nucleus within a cell. Control of cellular activities is centered in the nucleus.

its cytoplasmic surface, and it is continuous with the rough ER in the cytoplasm. **Nuclear pores** are open passageways that penetrate fused regions of the double membrane throughout the entire nuclear envelope. Nuclear pores allow the nuclear membrane to be selectively permeable and permit most ions and water-soluble molecules to shuttle between the nucleus and the cytoplasm.

Nucleoli

The cell nucleus may contain one or more dark-staining, usually spherical bodies called **nucleoli** (figure 2.17). (The singular term is *nucleolus* [noo-klē′ō-lŭs; pl., *nucleoli*, noo-klē′ō-lī].) Nucleoli are responsible for making the small and large subunits of ribosomes. These subunits are exported outside the nucleus into the cytoplasm, where they are then assembled to form ribosomes. You can think of the ribosomal subunits as puzzle pieces that are made in the nucleolus. Arrangement of the puzzle pieces into one complete puzzle (ribosome) occurs in the cytoplasm.

Not all cells contain a nucleolus. The presence and number of nucleoli indicate the protein synthetic activity of a cell. For example, nerve cells contain nucleoli because they produce many proteins. In contrast, sperm cells have no nucleoli because they produce no proteins.

DNA, Chromatin, and Chromosomes

The nucleus houses **deoxyribonucleic acid (DNA)**, an enormous macromolecule that contains the genetic material of the cell. The DNA within the nucleus, termed **nuclear DNA,** is much more complex than the DNA in mitochondria. DNA is organized into discrete units called **genes.** Genes provide the instructions for the production of specific proteins, and thereby direct all of the cell's activities.

DNA is in the shape of a double helix, or a ladder twisted into a spiral shape. The building blocks that form this double helix are called **nucleotides** (noo′klē-ō-tīd; *nucleus* = a little nut or kernel) **(figure 2.18a)**. A nucleotide contains a sugar (called a deoxyribose sugar), a phosphate molecule, and a nitrogen-containing base. There are four different types of nucleotides, each having one of four different bases: adenine (A), cytosine (C), guanine (G), and thymine (T). These nucleotides are arranged to form the unique double-helical shape of DNA. If you think of the DNA as a ladder, the sugar and phosphate components of the nucleotides

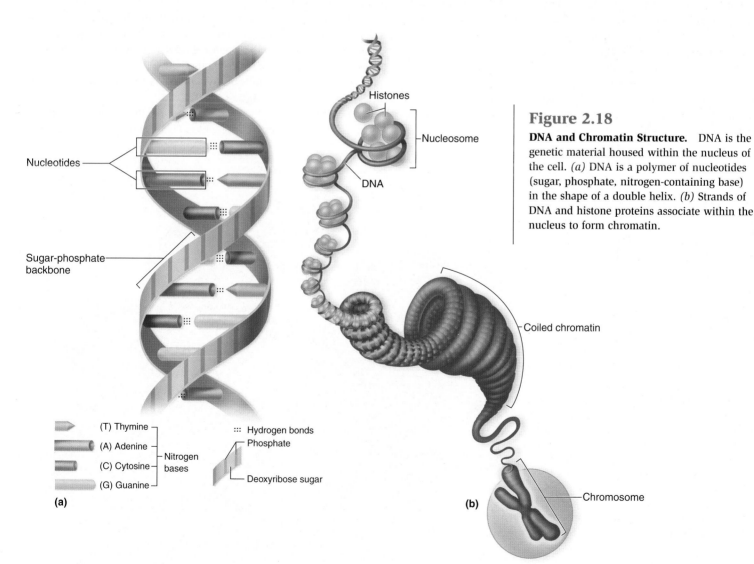

Figure 2.18

DNA and Chromatin Structure. DNA is the genetic material housed within the nucleus of the cell. *(a)* DNA is a polymer of nucleotides (sugar, phosphate, nitrogen-containing base) in the shape of a double helix. *(b)* Strands of DNA and histone proteins associate within the nucleus to form chromatin.

form the vertical "struts" of the ladder, while pairs of nucleotide bases interconnected by weak hydrogen bonds form the horizontal "rungs." Note that the base guanine only interconnects with the base cytosine, while the base adenine only pairs with the base thymine. The specific order of the bases in the nucleotides "codes for" specific proteins the body needs.

When a cell is not dividing, the DNA and its associated proteins are in the form of an unwound, finely filamented mass called **chromatin** (krō'ma-tin; *chroma* = color). Dark-staining chromatin in the nucleus of a nondividing cell is condensed chromatin. Other, light-staining regions of the nucleus contain chromatin that is uncoiled and spread out in fine strands of DNA and protein. When the cell is not dividing, the DNA remains unwound in fine, uncoiled chromatin, so that the genes within the DNA can direct the production of cellular proteins. This is not possible when the DNA is condensed and organized into a chromosome at the time of cell division.

Once the cell begins to divide, the chromatin rearranges itself in more precise and identifiable elongated structures called chromosomes. The **chromosome** (krō'mō-sōm; *chroma* = color, *soma* = body) is the most organized level of genetic material. Each chromosome contains a single, long molecule of DNA and associated proteins. Chromosomes become visible only when the cell is dividing. As a cell prepares for division, the DNA and protein in the chromatin coil, wrap, and twist to form the chromosomes, which resemble relatively short, thick rods. The long DNA double helix winds around a cluster of special nuclear proteins called **histones,** forming a complex known as a **nucleosome** (figure 2.18*b*). The degree of coiling of the DNA around the histone proteins ultimately determines the length and thickness of the chromosome.

Life Cycle of the Cell

Key topics in this section:

- Events that occur during interphase
- The phases of mitosis

Producing the trillions of cells that form a human body—and replacing the aging, damaged, or dead ones—requires continuous cell division. In cell division, one cell divides to produce two identical cells, called **daughter cells.** There are two types of cell division: mitosis and meiosis. **Mitosis** (mī-tō'sis; *mitos* = thread) is the cell division process that takes place in the **somatic cells,** which are all of the cells in the body except the sex cells. (Meiosis occurs in the sex cells, which give rise to sperm or oocytes ["eggs"], and is discussed in chapter 3.)

The events of cell division make up the **cell cycle.** The cell cycle has two phases: interphase and the mitotic (M) phase. **Interphase** is the time between cell divisions when the cell maintains and carries out normal metabolic activities and may also prepare for division. The **mitotic (M) phase** is the time when the cell divides into two cells **(figure 2.19).**

The lives of cells vary, depending on their specific type and their environment. For example, blood cells and epithelial skin cells

Figure 2.19

The Cell Cycle. A cell capable of division undergoes two general phases: interphase and the mitotic (M) phase. Interphase is a growth period that is subdivided into G_1, S, and G_2. Cell growth in preparation for division occurs during the G_1 and G_2 stages, and both cell growth and DNA replication occur during the S phase. The mitotic phase is composed of two processes: mitosis, during which the nucleus divides, and cytokinesis, when the cytoplasm divides.

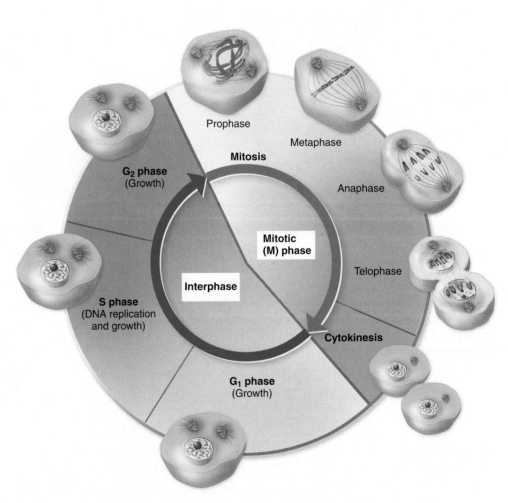

Prophase

Metaphase

Mitosis

G_2 **phase**
(Growth)

Anaphase

**Mitotic
(M) phase**

Telophase

Interphase

Cytokinesis

S phase
(DNA replication
and growth)

G_1 **phase**
(Growth)

are replaced frequently, so the cells that produce them undergo frequent cell division. Other cells, such as most nerve cells, undergo cell division infrequently or not at all. However, all somatic cells that divide go through the same stages, as described next.

Interphase

Most cells are in interphase during the majority of their lives. Interphase is a time when the cell appears to be resting because no overt activity is observed. However, while the cell carries on its normal activities, it may also be preparing for division. Interphase is a time for growth and making new cellular parts, replicating DNA and centrioles, and producing the proteins, RNA, and organelles needed for cell division. Interphase is divided into three distinct phases: G_1, S, and G_2 (figure 2.19).

G_1 Phase

During the G_1 **phase** (the first "growth" or gap stage), cells grow, produce new organelles, carry out specific metabolic activities, and produce proteins required for division. Near the end of G_1, the centrioles begin to replicate in preparation for cell division. Nondividing cells never finish G_1, and remain in a state of arrested development termed G_0. Most nerve cells appear to be in this state and do not enter cell division.

S Phase

The **S phase** ("synthesis" phase) is the next period of interphase for cells that will eventually be dividing. During this short phase, each DNA molecule replicates (makes an exact copy of itself) completely. Replication is in preparation for cell division and provides for the partitioning of all of the hereditary material of a parent cell into two identical daughter cells. A parent DNA molecule has two strands of DNA that are complementary, meaning that each base on one strand is paired with a specific partner: Adenine (A) pairs with thymine (T), and cytosine (C) pairs with guanine (G) (see figure 2.18a). The first steps in replication are the unwinding of the helix followed by the separation, or "unzipping," of the two strands of DNA in the parent molecule. Once separated, each parent strand serves as a template for the order of bases in the new complementary strand according to the base-pairing pattern just described. Each new DNA molecule now consists of one parent strand and one new strand.

G_2 Phase

The last part of interphase, called the G_2 **phase** (or the second "growth" or gap phase), is brief. During this phase, centriole replication is completed, organelle production continues, and enzymes needed for cell division are synthesized.

Mitotic (M) Phase

Cell division is necessary to provide the large number of cells essential for the growth and survival of a human. Cells divide at different rates through specific stages in their life cycle. Following interphase, cells enter the M (mitotic) phase **(figure 2.20)**. Two distinct events occur during this phase: **mitosis,** or division of the nucleus, followed by **cytokinesis** (sī'tō-ki-nē'sis; *kytos* = cell, *kinesis* = movement), division of the cytoplasm.

Mitotic cell division produces two daughter cells that are identical to the original (parent) cell. The nucleus divides such that the replicated DNA molecules of the original parent cell are apportioned into the two new daughter cells, with each receiving an identical copy of the DNA of the original cell.

Four consecutive phases take place during mitosis: prophase, metaphase, anaphase, and telophase. Each phase merges smoothly into the next in a nonstop process. The duration of mitosis varies according to cell type, but it typically lasts about 2 hours.

Prophase

Prophase is the first stage of mitosis (figure 2.20b). Chromatin becomes supercoiled into relatively short, dense chromosomes, which are more maneuverable during cell division than the long, delicate chromatin strands. Remember that the DNA replicated itself during interphase, so during prophase each chromosome (called a duplicated chromosome) contains two copies of its DNA. A duplicated chromosome consists of two genetically identical structures, called **sister chromatids.** Each sister chromatid is composed of an identical DNA double helix, and the two sister chromatids are joined together by proteins at a constricted region called the **centromere** (sen'trō-mēr; *kentron* = center, *meros* = part).

During prophase, the nucleolus breaks down and disappears. The chromosomes form a big puffy ball within the nucleus. Elongated microtubules called **spindle fibers** begin to grow from the centrioles, and this event pushes the two centriole pairs apart. Eventually, the centrioles come to lie at opposite poles of the cell. The end of prophase is marked by the dissolution of the nuclear envelope, which permits the chromosomes to move freely into and through the cytoplasm.

Metaphase

Metaphase occurs when the chromosomes line up along the equatorial plate of the cell (figure 2.20c). Spindle fibers grow from each centriole toward the chromosomes, and some attach to the centromere of each chromosome. The collection of spindle fibers extending from the centrioles to the chromosomes forms an oval-structured array termed the **mitotic spindle.** This arrangement remains in place until the next phase begins.

Anaphase

Anaphase begins as spindle fibers pull sister chromatids apart at the centromere (figure 2.20d). The spindle fibers shorten, and each "reels in" a chromatid, like a fishing line reeling in a fish. After the chromatids are pulled apart, each chromatid is called a **single-stranded chromosome,** as each forms its own unique centromere. Thus, a pair of single-stranded chromosomes is pulled apart from sister chromatids, and each migrates to the opposite end of the cell (cell pole). As each single-stranded chromosome migrates toward the cell pole, its centromere leads the way, and the arms of the chromosome trail behind.

Telophase

Telophase begins with the arrival of a group of single-stranded chromosomes at each cell pole (figure 2.20e). A new nuclear envelope forms around each set of chromosomes, and the chromosomes begin to uncoil and return to the form of dispersed threads of chromatin. The mitotic spindle breaks up and disappears. Each new nucleus forms nucleoli.

Telophase signals the end of nuclear division and it may overlap with cytokinesis, the division of the cytoplasm. A contractile ring of protein filaments at the periphery of the cell equator pinches the mother cell into two separate cells. The resulting **cleavage furrow** indicates where the cytoplasm is dividing. The two new daughter cells then enter the interphase of their life cycle, and the process continues.

Table 2.4 summarizes the events of the somatic cell cycle.

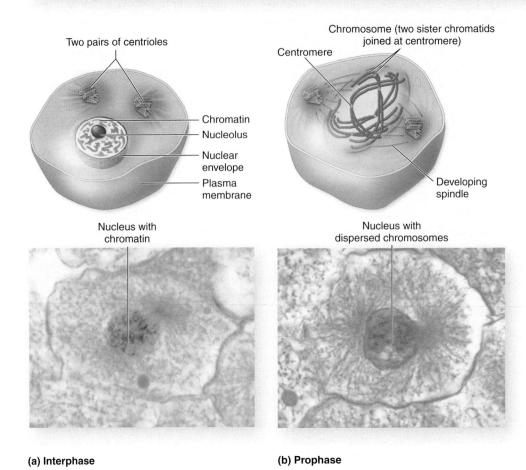

INTERPHASE AND MITOSIS

Figure 2.20

Interphase and Mitosis. Drawings and micrographs depict what happens inside a cell during the stages of *(a)* interphase and *(b–e)* mitotic cell division.

(a) Interphase

(b) Prophase

Table 2.4	Somatic Cell Cycle Events
Phase	**Cellular Events**
INTERPHASE	A time of normal metabolic activities with no noticeable change in either the cytoplasm or nucleus; cell is not dividing, and chromosomes are not visible by light microscopy
G_1 phase	First growth phase: Protein synthesis and metabolic activity occur; new organelles are produced; centriole replication begins at end of this phase
S phase	Nuclear DNA is replicated
G_2 phase	Second growth phase: Brief growth period for production of cell division enzymes; centriole replication finishes; organelle replication continues
MITOTIC (M) PHASE	Nuclear and cytoplasmic events produce two identical daughter cells from one parent cell
Mitosis	Division of the nucleus: Continuous series of nuclear events that distribute two sets of chromosomes into two daughter nuclei
Prophase	Chromatin threads appear due to coiling and condensation; elongated duplicated chromosomes consisting of identical sister chromatids become visible Nuclear envelope disappears at the end of this stage Nucleolus disappears Microtubules begin to form mitotic spindle Centrioles move toward opposing cell poles
Metaphase	Chromosomes line up at the equatorial plate of the cell Microtubules from the mitotic spindle attach to the centromeres of the chromosomes from the centrioles
Anaphase	Centromeres that held sister chromatid pairs together separate; they are now single-stranded chromosomes Identical pairs of single-stranded chromosomes are pulled toward opposite ends of the cell

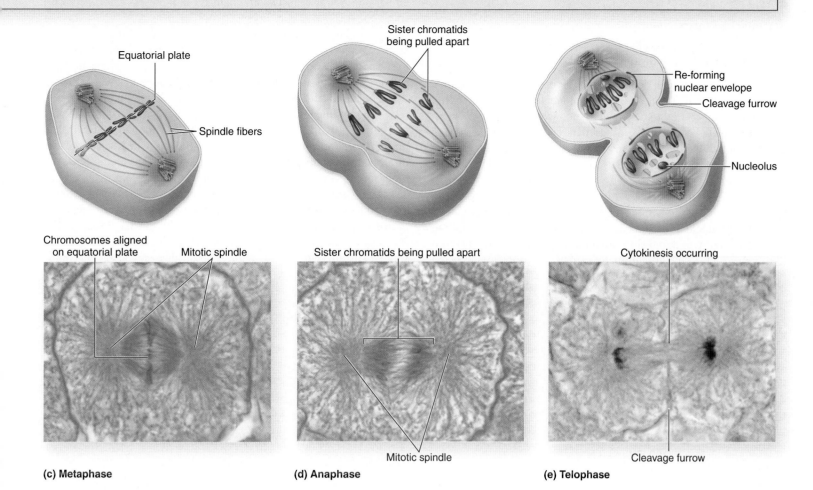

(c) Metaphase   (d) Anaphase   (e) Telophase

Table 2.4	Somatic Cell Cycle Events *(continued)*	
Phase	**Cellular Events**	
MITOTIC (M) PHASE	Nuclear and cytoplasmic events produce two identical daughter cells from one parent cell *(continued)*	
Telophase	Chromosomes arrive at cell poles and stop moving Nuclear envelope reappears, mitotic spindle disintegrates, chromosomes disappear and become thin chromatin threads within boundary of the new nuclear envelope Nucleoli reappear	
Cytokinesis	Usually begins before telophase ends; cleavage furrow is formed from a contractile ring of microfilaments; cytoplasm divides, completing the formation of two daughter cells	

Study Tip!

Use these study tips to help you remember some of the hallmark events that occur during each phase of mitosis:

- The **p** in **p**rophase stands for the **puffy** ball of chromosomes that forms in the center of the cell.
- The **m** in **m**etaphase stands for **middle:** During this phase, the chromosomes align along the **middle** of the cell.
- The **a** in **a**naphase stands for **apart:** During this phase, the sister chromatids are pulled **apart**.
- The **t** in **t**elophase stands for **two:** During this phase, **two** new cells begin to form as a cleavage furrow divides the cytoplasm.

WHAT DID YOU LEARN?

15 Observation shows that most cells are suspended in interphase for most of their lives. Identify the parts of interphase, and describe an event that occurs during each part.

16 List the stages of mitosis in order of occurrence. Describe a unique activity associated with each stage.

Aging and the Cell

Key topics in this section:

- Effects of aging on cells
- Two causes of cell death

Aging is a normal, continuous process that often exhibits obvious body signs. However, at the cellular level, changes within cells due to aging are neither obvious nor well understood. Often, reduced metabolic functions of normal cells have wide-ranging effects throughout the body, including cells' decreased ability to maintain homeostasis. These signs of aging reflect a reduced number of normal functional body cells, and may even suggest abnormal function in the remaining cells. Affected cells may exhibit alteration in either the structure or the number of specific organelles. For example, if mitochondrial function begins to fail, the cell's ability to synthesize ATP will diminish. Additionally, changes in the distribution and structure of the chromatin and chromosomes within the nucleus may occur. Often, both chromatin and chromosomes clump, shrink, or fragment as a result of repeated divisions.

Some cancers (e.g., prostate cancer) appear with greater frequency in elderly individuals. Cancer is essentially caused by cells that undergo uncontrolled cell division and fail to "turn off" the cell division process. Thus, as we age, the whole mechanism of cell division becomes more faulty, making cancers more prevalent. Further, pregnant women over the age of 35 are at greater risk for giving birth to a child with a birth defect than are younger pregnant women. One reason for this greater risk is that older women's sex cells (oocytes) are older, and their mechanisms for completing sex cell division and maturation may not operate properly.

Essentially, cells die by one of two mechanisms: (1) They are killed by harmful agents or mechanical damage, in a process called **necrosis** (ně-kro′sis; *nekrosis* = death), wherein the damage is irreversible and there is an inflammatory response, or (2) they are induced to commit suicide, a process of programmed cell death called **apoptosis** (ap′op-tō′sis; *apo* = off, *ptosis* = a falling). Cells in apoptosis exhibit nuclear changes (chromatin degradation), shrinkage in volume, and abnormal development in both organelle and plasma membrane structure.

Programmed cell death both promotes proper development and removes harmful cells. For example, in a human embryo, the proper development of fingers and toes begins with the formation of a paddlelike structure at the distal end of the developing limb. In order for our digits to form correctly, programmed death removes the cells and tissues between the true fingers and toes developing within this paddle structure. Additionally, programmed cell death sometimes destroys harmful cells, reducing potential health threats. For example, the cells of our immune system promote programmed cell death in some virus-infected cells to reduce the further spread of infection. Often, cells with damaged DNA appear to promote events leading to apoptosis, presumably to prevent these cells from causing developmental defects or becoming cancerous. Additionally, some cancer therapy treatments lead to apoptosis in certain types of cancer cells.

Precise control of cell division is required to maintain healthy, normal-functioning cells. The quality-control mechanisms inherent within normal cellular processes are meant to ensure continuous removal of unnecessary cells, old cells, or abnormal cells as normal aging progresses.

WHAT DID YOU LEARN?

17 What name is given to programmed cell death?

18 In general, what is the main characteristic of cancer?

CLINICAL VIEW: In Depth

Characteristics of Cancer Cells

Normal tissue development exhibits a balance between cell division and cell death. If this balance is upset and cells multiply faster than they die, abnormal growth results in a new cell mass called a neoplasm, or tumor. *Neoplasms* (nē′ō-plazm; *neos* = new, *plasma* = thing formed) are classified as benign or malignant, based on their cytologic and histologic features.

Benign (bē-nīn′; *benignus* = kind) neoplasms usually grow slowly and are confined within a connective tissue capsule. Cells within these tumors **dedifferentiate**—that is, they revert to a less specialized state and cause an increase in their own vascular supply to support their growth. These tumors are usually not lethal, but they have the potential to become life-threatening if they compress brain tissue, nerves, blood vessels, or airways.

Malignant (mă-lig′nănt; *maligno* = to do maliciously) neoplasms are unencapsulated, contain cells that dedifferentiate, increase their vascular supply, grow rapidly, and are able to spread easily to other organs by way of the blood or lymph, a phenomenon called *metastasis* (mě-tas′tă-sis; *meta* = in the midst of, *stasis* = a placing).

Cancer is the general term used to describe a group of diseases characterized by various types of malignant neoplasms. A *carcinogen* is any infectious agent or substance shown to cause changes within a normal cell that results in the formation of a cancer cell. Cancer cells resemble undifferentiated or primordial cell types. Generally, they do not mature before

they divide and are not capable of maintaining normal function. They use energy very inefficiently, and their growth comes at the expense of normal cells and tissues. The characteristics of cancer include the following:

- Cancer cells lose control of their cell cycle. Cell divide too frequently and grow out of control. A *mutagen* is any agent or factor that causes a change in genes; it may be responsible for stimulating the development of a cancerous cell.

- Cancer cells lose contact inhibition, meaning that they overgrow one another and lack the ability to stop growing and dividing when they crowd other cells.

- Cancer cells often exhibit dedifferentiation and revert to an earlier, less specialized developmental state.

- Cancer cells often produce chemicals that cause local blood vessel formation (a process called angiogenesis), resulting in increased blood vessels in the developing tumor.

- Cancer cells have the ability to squeeze into any space, a property called *invasiveness*. This permits cancer cells to leave their place of origin and travel elsewhere in the body.

- Cancer cells acquire the ability to metastasize—that is, spread to other organs in the body.

CLINICAL TERMS

anaplasia Obvious loss of cellular or structural differentiation and change in cells' orientation to each other and to blood vessels; seen in most malignant neoplasms.

dysplasia (dis-plă′zē-ă; *dys* = bad, *plasis* = a molding) Abnormal development of a tissue; a pathologic condition resulting in a change in the shape, size, and organization of adult cells; development of cellular and tissue elements that are not normal.

hyperplasia Increase in the normal number of cells within a tissue or organ; an excessive proliferation of normal cells; does not include tumor formation.

hypertrophy Generalized increase in the bulk or size of a part of an organ, not as a consequence of tumor formation.

malignant tumor An abnormal growth of cells that invades surrounding tissues.

metaplasia Abnormal transformation of a fully differentiated adult tissue into a differentiated tissue of another kind.

CHAPTER SUMMARY

The Study of Cells 24	▪ Cytology is the study of anatomy at the cellular level.
	Using the Microscope to Study Cells 24
	▪ Variations in magnification and resolution exist when comparing light microscopy (LM) and electron microscopy (TEM and SEM).
	General Functions of Human Body Cells 25
	▪ Cells vary in shape and size, often related to various cellular functions.
A Prototypical Cell 27	▪ A cell is surrounded by a thin layer of extracellular fluid. Interstitial fluid is a type of extracellular fluid forming a thin layer on the outside of the cell. Most mature human cells have an outer boundary called the plasma (cell) membrane, general cell contents termed cytoplasm, and a nucleus that serves as the cell's control center.
Plasma Membrane 30	▪ The plasma membrane acts as a gatekeeper to regulate movement of material into and out of the cell.
	Composition and Structure of Membranes 30
	▪ Plasma membranes are composed of an approximately equal mixture of lipids and proteins.
	▪ The primary membrane lipids are phospholipids, arranged as a bilayer.
	▪ Membrane proteins are of two types: integral proteins and peripheral proteins. Some integral membrane proteins have carbohydrate molecules attached to their external surfaces.
	▪ The glyocalyx is the carbohydrate component of the plasma membrane attached to either lipid (glycolipid) or protein (glycoprotein) components. It functions in cell–cell recognition and communication.
	Protein-Specific Functions of the Plasma Membrane 31
	▪ Plasma membrane proteins function in transport, intercellular attachment, cytoskeleton anchorage, catalytic (enzyme) activity, cell–cell recognition, and signal transduction.
	Transport Across the Plasma Membrane 32
	▪ Plasma membrane permeability is influenced by transport proteins, membrane structure, concentration gradient across the membrane, ionic charge, lipid solubility of materials, and molecular size.
	▪ Passive transport is the movement of a substance across a membrane at no energy cost to the cell; it includes diffusion (simple diffusion, osmosis, and facilitated diffusion) and bulk filtration.
	▪ All active transport processes require energy in the form of ATP. Two active processes are ion pumps and bulk transport in vesicles (exocytosis and endocytosis).
	▪ Bulk transport includes exocytosis, a mechanism to export packaged materials from the cell, and endocytosis, a mechanism by which materials are imported into the cell.
Cytoplasm 36	▪ The cytoplasm is all the material between the plasma membrane and the nucleus. It contains cytosol, inclusions, and organelles.
	Cytosol 36
	▪ Cytosol is a viscous intracellular fluid containing ions, nutrients, and other molecules necessary for cell metabolism.
	Inclusions 36
	▪ Inclusions are storage bodies in the cytoplasm.
	Organelles 36
	▪ Membrane-bound organelles include endoplasmic reticulum (both rough and smooth), the Golgi apparatus, lysosomes, peroxisomes, and mitochondria.
	▪ Non-membrane-bound organelles include ribosomes (both free and fixed), the cytoskeleton, the centrosome, and centrioles, cilia, flagella, and microvilli.

(continued on next page)

C H A P T E R   S U M M A R Y (*c o n t i n u e d*)

Nucleus 44	■ The nucleus is the cell's control center.
	Nuclear Envelope 44
	■ The nuclear envelope is a double membrane boundary surrounding the nucleus. Nuclear pores are openings that penetrate the nuclear envelope and permit direct communication with the cytosol.
	Nucleoli 45
	■ A nucleolus is a dark-staining, usually spherical body in the nucleus that produces the subunits that will form ribosomes.
	DNA, Chromatin, and Chromosomes 45
	■ Chromatin is the name of the fine, uncoiled strands of DNA in the nucleus. As the cell prepares to divide, the DNA strands begin to coil and wind to form large, microscopically identifiable structures termed chromosomes.
Life Cycle of the Cell 46	■ Cell division in somatic cells is called mitosis, and cell division in sex cells (sperm and oocytes) is called meiosis.
	Interphase 47
	■ Somatic cells spend the majority of their time in interphase, a time of maintenance and growth that occurs between cell divisions.
	Mitotic (M) Phase 47
	■ The division of the somatic cell nucleus is called mitosis, whereas the division of the cytoplasm following mitosis is called cytokinesis. Both mitosis and cytokinesis represent the mitotic phase.
	■ Four consecutive phases comprise mitosis: prophase, metaphase, anaphase, and telophase (see figure 2.20).
Aging and the Cell 50	■ Aging is a normal process that is often marked by changes in normal cells.
	■ Cells may be killed by harmful agents or mechanical damage, or they may undergo programmed cell death, called apoptosis.

C H A L L E N G E   Y O U R S E L F

Matching

Match each numbered item with the most closely related lettered item.

_____ 1. ribosomes   _____ 6. cytoskeleton

_____ 2. lysosomes   _____ 7. osmosis

_____ 3. peripheral proteins   _____ 8. S phase

_____ 4. Golgi apparatus   _____ 9. pinocytosis

_____ 5. exocytosis   _____ 10. nucleus

a. endocytosis of small amounts of fluid
b. organelle that sorts and packages molecules
c. diffusion of water across a semipermeable membrane
d. process of bulk export from the cell
e. responsible for synthesizing proteins
f. control center; stores genetic information
g. organelles housing digestive enzymes
h. not embedded in phospholipid bilayer
i. the time when DNA replication occurs
j. internal protein framework in cytoplasm

Multiple Choice

Select the best answer from the four choices provided.

_____ 1. When a cell begins to divide, its chromatin forms
    a. nucleoli.
    b. chromosomes.
    c. histones.
    d. None of these are correct.

_____ 2. Which of the following describes integral membrane proteins?
    a. They only permit water movement into or out of the cell.
    b. They only transport large proteins into the cell.
    c. They extend across the phospholipid bilayer.
    d. They are attached to the external plasma membrane surface.

_____ 3. Facilitated diffusion differs from active transport in that facilitated diffusion
    a. expends no ATP.
    b. moves molecules from an area of higher concentration to one of lower concentration.
    c. does not require a carrier protein for transport.
    d. moves molecules in vesicles across a semipermeable membrane.

_____ 4. Which plasma membrane structures serve in cell recognition and act as a "personal ID card" for the cell?
    a. integral proteins and peripheral proteins
    b. glycolipids and glycoproteins
    c. phospholipids and cholesterol
    d. cholesterol and integral proteins

_____ 5. _____ increase the outer surface area of the plasma membrane to increase absorption.
   a. Centrioles
   b. Cilia
   c. Microvilli
   d. Flagella

_____ 6. The major functions of the Golgi apparatus are
   a. diffusion and osmosis.
   b. detoxification of substances and removal of waste products.
   c. synthesis of new proteins for the cytoplasm.
   d. packaging, sorting, and modification of new molecules.

_____ 7. Interphase of the cell cycle consists of the following parts:
   a. prophase, metaphase, anaphase, and telophase.
   b. G_1, S, and G_2.
   c. mitosis and cytokinesis.
   d. All of these are correct.

_____ 8. The organelle that provides most of the ATP needed by the cell is the
   a. endoplasmic reticulum.
   b. mitochondrion.
   c. lysosome.
   d. Golgi apparatus.

_____ 9. During which phase of mitosis do the sister chromatids begin to move apart from each other at the middle of the cell?
   a. prophase
   b. metaphase
   c. anaphase
   d. telophase

_____ 10. A peroxisome uses oxygen to
   a. detoxify harmful substances.
   b. make ATP.
   c. help make proteins.
   d. package secretory materials.

Content Review

1. Describe the three main regions common to all cells, and briefly discuss the composition of each region.

2. Describe the structure and the function of the plasma membrane.

3. What is meant by passive transport of materials into a cell? Describe the passive processes by which substances enter and leave cells.

4. How does active transport differ from passive transport? What are the three specific forms of the active transport mechanism termed endocytosis?

5. Discuss the two categories of organelles and the main differences between these groups.

6. Compare and contrast the structure and functions of the SER and the RER.

7. Identify the three parts of the cytoskeleton, and describe the structure and function of each component.

8. What are the basic components of the nucleus, and what are their functions?

9. What is interphase? What role does it serve in the cell cycle?

10. Identify the phases of mitosis, and briefly discuss the events that occur during each phase.

Developing Critical Reasoning

1. You place some cells into a solution of unknown content and then observe them on a microscope slide. After a short period, all of the cells appear shrunken, and their plasma membranes look wrinkled. What took place, and why?

2. Why is it efficient for some organelles to be enclosed by a membrane similar to a plasma membrane?

ANSWERS TO "WHAT DO YOU THINK?"

1. A selectively permeable plasma membrane allows some materials to enter the cell and blocks the entry of other materials that may be detrimental to the cell. However, a selectively permeable plasma membrane may inadvertently block some beneficial material. In these cases, active transport methods (e.g., endocytosis) are needed to bring the material into the cell.

2. Most cells would not be able to function without lysosomes. Lysosomes are necessary for breaking down and removing waste products. If lysosomes do not function properly, the waste products build up in the cell and cause cell death.

3. The number of mitochondria is positively related to the metabolic activity of the cell. A cell with few mitochondria is probably not as active metabolically as a cell with numerous mitochondria.

Visit the McKinley/O'Loughlin _Human Anatomy,_ 2e website at aris.mhhe.com

3

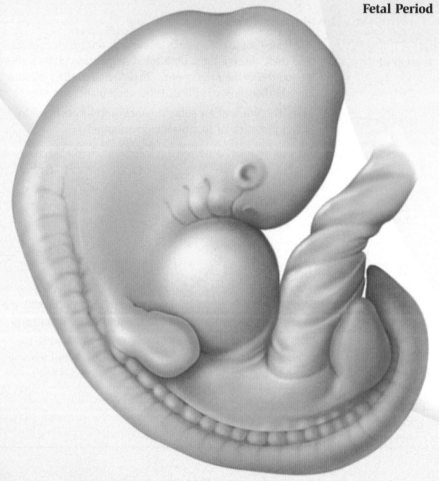

Embryology

ike all organisms, humans undergo development, a series of progressive changes that accomplishes two major functions: differentiation and reproduction. **Differentiation** leads to the formation and organization of all the diverse cell types in the body. **Reproduction** ensures that new individuals are produced from generation to generation. Development continues throughout the life of a human, but in this chapter we focus on the developmental events that occur prior to birth, a discipline known as **embryology** (em-brē-ol′ō-jē; *embryon* = a young one, *logos* = study).

Overview of Embryology

Key topics in this section:

- Major events of the three prenatal periods
- Processes that comprise embryogenesis

Embryology deals with the developmental events that occur during the **prenatal period**, the first 38 weeks of human development that begin with the fertilization of the secondary oocyte and end with birth.[1] The prenatal period is broken down into the following shorter periods:

- The **pre-embryonic period** is the first 2 weeks of development (the first 2 weeks after fertilization), when the single cell produced by fertilization (the zygote) becomes a spherical, multicellular structure (a blastocyst). This period ends when the blastocyst implants in the lining of the uterus.

- The **embryonic period** includes the third through eighth weeks of development. It is a remarkably active time during which rudimentary versions of the major organ systems appear in the body, which is now called an embryo.

- The **fetal** (fē′tal; *fetus* = offspring) **period** includes the remaining 30 weeks of development prior to birth, when the organism is called a fetus. During the fetal period, the fetus continues to grow, and its organs increase in complexity.

The developmental processes that occur in the pre-embryonic and embryonic periods are known collectively as **embryogenesis**. **Figure 3.1** shows the three stages of embryogenesis:

---

[1] Some physicians refer to a 40-week gestation period, or pregnancy. This time frame is measured from a woman's last period to the birth of the newborn. In this time frame, fertilization does not occur until week 2 (when a woman ovulates)! So why refer to a 40-week gestation? Physicians use this reference because a woman knows when her last period was, but she may not know the day she ovulated and had a secondary oocyte fertilized.

Figure 3.1

Developmental History of a Human.   The stages of development after fertilization through week 8 are known collectively as embryogenesis and its stages are separated into cleavage, gastrulation, and organogenesis. The fetal period occurs after week 8 until birth. Gametogenesis occurs in the sexually mature adult.

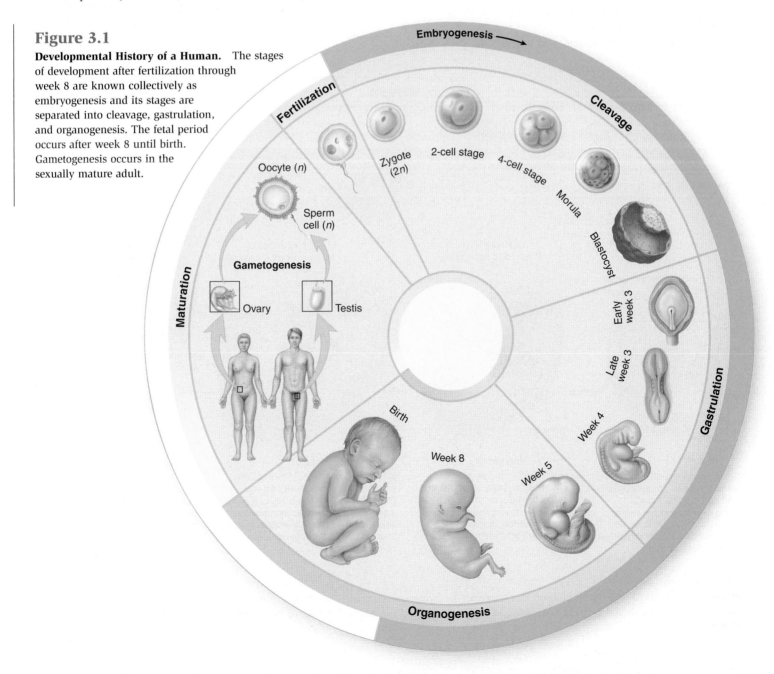

- **Cleavage.** The zygote divides by mitosis to form a multicellular structure called a blastocyst.
- **Gastrulation.** The blastocyst cells form three primary germ layers, which are the basic cellular structures from which all body tissues develop.
- **Organogenesis.** The three primary germ layers arrange themselves in ways that give rise to all organs in the body.

Following birth, an individual spends a great portion of his or her life undergoing maturation. During this stage, the body grows and develops, and the sex organs become mature. The sex organs (ovaries in the female, testes in the male) then begin to produce sex cells, or **gametes** (gam′ēt; *gameo* = to marry) through a process called **gametogenesis**.

Gametogenesis

Key topics in this section:

- The process of gametogenesis
- Events that occur during meiosis

Gametogenesis is necessary for the reproductive phase of development. When humans reproduce, they pass on their traits to a new individual. As mentioned in chapter 2, hereditary information is carried on chromosomes. Human somatic cells contain 23 pairs of chromosomes: 22 pairs of autosomes and one pair of sex chromosomes for a total of 46 chromosomes. **Autosomes** contain genetic information for most human characteristics, such as eye color, hair color, height, and skin pigmentation. A pair of similar autosomes are called **homologous chromosomes** (hō-mol′ō-gŭs; *homos* = same, *logos* = relation). The pair of **sex chromosomes** primarily determines whether an individual is female (she will have two X chromosomes) or male (he will have one X chromosome and one Y chromosome). One member of each pair of chromosomes (be they autosomes or sex chromosomes) is inherited from each parent. In other words, if you examined one of your body cells, you would discover that 23 of the chromosomes came from your mother, and the other 23 chromosomes in this same cell came from your father. A cell is said to be **diploid** (dip′loyd; *diploos* = double) if it contains 23 *pairs* of chromosomes. (A cell with pairs of chromosomes is designated as *2n*, as shown in figure 3.1.) In contrast, sex cells (either a secondary oocyte or a sperm cell) are **haploid** (hap′loyd; *haplos* = simple, *eidos* = appearance) because they contain 23 chromosomes only (and not 23 pairs of chromosomes). (A haploid number of chromosomes is designated as *n*.)

The process of gametogenesis begins with cell division, called meiosis. The sex cells produced in the female are secondary oocytes, while the sex cells produced in the male are sperm.

MEIOSIS I

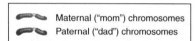

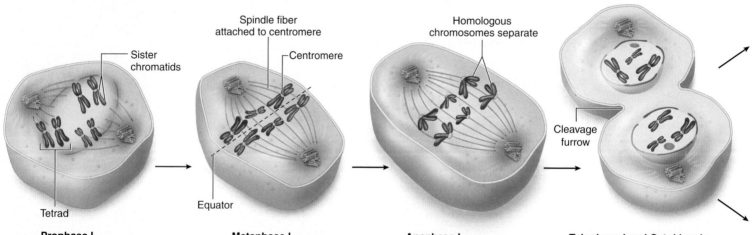

Prophase I
Homologous double-stranded chromosomes pair up (synapsis), and the pair forms a tetrad. Crossing over occurs between maternal ("mom") chromosomes and paternal ("dad") chromosomes, ensuring genetic diversity.

Metaphase I
Homologous double-stranded chromosomes line up above and below the equator of the cell, forming a double line of chromosomes. Spindle fibers attach to the chromosomes.

Anaphase I
Maternal and paternal pairs of chromosomes are separated and pulled to the opposite ends of the cell, a process called reduction division. Note that the sister chromatids remain attached in each double-stranded chromosome.

Telophase I and Cytokinesis
Nuclear division finishes and the nuclear envelopes re-form. The cytoplasm divides and two new cells are produced, each containing 23 chromosomes only. The chromosomes are still double-stranded.

Figure 3.2

Meiosis. Meiosis is a type of cell division that results in the formation of gametes (sex cells). In meiosis I, homologous chromosomes are separated after synapsis and crossing over occur. In meiosis II, sister chromatids are separated in a sequence of phases that resembles the steps of mitosis.

Meiosis

Meiosis (mī-ō'sis; *meiosis* = lessening) is a type of sex cell division that starts off with a diploid parent cell and produces haploid daughter cells. Mitosis (somatic cell division, described in chapter 2) and meiosis (sex cell division) differ in the following ways:

- Mitosis produces *two* daughter cells that are genetically *identical* to the parent cell. In contrast, meiosis produces *four* daughter cells that are genetically *different* from the parent cell.
- Mitosis produces daughter cells that are *diploid*, whereas meiosis produces daughter cells that are *haploid*.
- In meiosis, a process called **crossing over** occurs, whereby genetic material is exchanged between homologous chromosomes. Crossing over helps "shuffle the genetic deck of cards," so to speak. Thus, crossing over is a means of combining different genes from both parents on one of the homologous chromosomes. Crossing over does not occur in mitosis.

WHAT DO YOU THINK?

1 How does crossing over relate to genetic diversity among individual sex cells?

Meiosis begins with a diploid parent cell located in the gonad (testis or ovary). In this cell, 23 chromosomes came from the organism's mother (23 maternal or "mom" chromosomes), and 23 chromosomes came from the father (23 paternal or "dad" chromosomes). So this parent cell that is responsible for the production of gametes contains 23 *pairs* of chromosomes. In order for the organism to produce its own sex cells, this parent cell must divide by the process of meiosis.

Prior to meiosis is a cell phase known as interphase (discussed in chapter 2). During interphase, the DNA in each chromosome is replicated (duplicated) in the parent cell, resulting in identical or replicated chromosomes. These replicated chromosomes are **double-stranded chromosomes**, composed of two identical structures called **sister chromatids** (krō'mă-tid; *chromo* = color, *id* = two). Each sister chromatid in a double-stranded chromosome contains an identical copy of DNA. The sister chromatids are attached at a specialized region termed the **centromere**. Note that "double-stranded" does *not* mean the same as a "pair" of chromosomes. A double-stranded chromosome resembles a written letter **X** and is composed of two identical sister chromatids, whereas a homologous *pair* of chromosomes is composed of a maternal chromosome and a paternal chromosome of the same number. Therefore, after interphase, there are 23 pairs of double-stranded chromosomes.

Once the DNA is replicated in interphase, the phases of meiosis begin **(figure 3.2)**.

MEIOSIS II

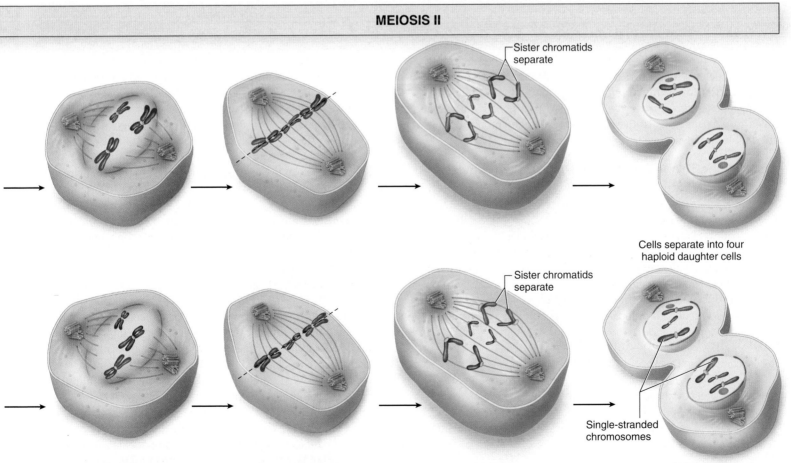

Sister chromatids separate

Sister chromatids separate

Cells separate into four haploid daughter cells

Single-stranded chromosomes

Prophase II	**Metaphase II**	**Anaphase II**	**Telophase II and Cytokinesis**
Nuclear envelope breaks down, and the chromosomes gather together. (There is no crossing over in Prophase II.)	Double-stranded chromosomes line up along the equator of the cell. Spindle fibers extend from the centrioles to the chromosomes.	Sister chromatids of each double-stranded chromosome are pulled apart at the centromere. Sister chromatids (now called single-stranded chromosomes) migrate to opposite ends of the cell.	Nuclear division finishes, and the nuclear envelopes re-form. The four new daughter cells that are produced each contain 23 single-stranded chromosomes only.

First Meiotic Prophase (Prophase I)

Homologous, double-stranded chromosomes in the parent cell form pairs. The process by which homologous chromosomes pair up is called **synapsis** (si-nap′sis; *syn* = together), and the actual pair of homologous chromosomes is called a **tetrad**. As the maternal and paternal chromosomes come close together, crossing over occurs. At this time, the homologous chromosomes exchange genetic material. A tiny portion of the genetic material in a sister chromatid of a maternal chromosome is exchanged with the same portion of genetic material transferred in a sister chromatid of a paternal chromosome. This shuffling of the genetic material ensures continued genetic diversity in new organisms.

First Meiotic Metaphase (Metaphase I)

The homologous pairs of double-stranded chromosomes line up above and below the equator, or middle, of the cell, forming a double line of chromosomes. This alignment of paired, double-stranded chromosomes is random with respect to whether the original maternal or paternal chromosome of a pair is on one side of the equator or the other. For example, some maternal chromosomes may be to the left of the equator, and other maternal chromosomes may be to the right. Spindle fibers formed by microtubules extend from centrioles at opposite ends of the cell and attach to the paired chromosomes.

First Meiotic Anaphase (Anaphase I)

Pairs of homologous chromosomes separate and are pulled to the opposite ends of the cell. For example, a maternal double-stranded chromosome may be pulled to one side of the cell, while the homologous paternal double-stranded chromosome is pulled to the opposite side. The process whereby maternal and paternal chromosome pairs are separated and move to opposite ends of the cell is referred to as **reduction division**. Note that the pairs of chromosomes are no longer together, because the members of each pair are being pulled to opposite ends of the cell. However, each chromosome is still double-stranded.

First Meiotic Telophase (Telophase I) and Cytokinesis

The chromosomes arrive at opposite ends of the cell, and a nuclear membrane re-forms around the chromosomes at each end of the cell. Then cleavage furrow forms in the cell, and the cell cytoplasm divides (cytokinesis) to produce two new cells. Each daughter cell contains 23 chromosomes only, but each of these chromosomes is double-stranded, meaning it is composed of two sister chromatids. These two cells must undergo further cell division so that the new cells will be composed of single-stranded chromosomes only. (Recall that a single-stranded chromosome contains only one chromatid.)

Second Meiotic Prophase (Prophase II)

The second prophase event resembles the prophase stage of mitosis. In each of the two new cells, the nuclear membrane breaks down, and the chromosomes collect together. However, crossing over does not occur in this phase because homologous chromosomes separated in anaphase I. (Crossing over occurs in the first meiotic prophase only.)

Second Meiotic Metaphase (Metaphase II)

The double-stranded chromosomes form a single line along the equator in the middle of the cell. Spindle fibers extend from the centrioles at the poles to the centromere of each double-stranded chromosome.

Second Meiotic Anaphase (Anaphase II)

The sister chromatids of each double-stranded chromosome are pulled apart at the centromere. Each chromatid, now called a single-stranded chromosome, is pulled to the opposite pole of the cell.

Second Meiotic Telophase (Telophase II) and Cytokinesis

The single-stranded chromosomes arrive at opposite ends of the cell. Nuclear membranes re-form, a cleavage furrow forms, and the cytoplasm in both cells divides, producing a total of four daughter cells. These daughter cells are haploid, because they contain 23 chromosomes only (not 23 pairs). These daughter cells mature into sperm (in males) or secondary oocytes (in females).

> ## Study Tip!
>
> Meiosis I (the first meiotic division) separates maternal and paternal *pairs* of chromosomes, while meiosis II (the second meiotic division) separates the remaining *double-stranded* chromosomes into *single-stranded* chromosomes. Also, meiosis II is very similar to mitosis. Thus, if you remember the steps of mitosis, you can figure out the steps of meiosis II.

Oocyte Development (Oogenesis)

In females, the sex cell produced is called the **secondary oocyte**, and the process of oocyte development is called **oogenesis** (ō-ō-jen′ĕ-sis; *oon* = egg, *genesis* = origin). This cell will have 22 autosomes and one X chromosome. Oogenesis is discussed in greater detail in chapter 28, but we provide a brief summary here.

The parent cells, or *stem cells*, that produce oocytes are called **oogonia** (ō-ō-gō′nē-ă), and they reside in the ovaries. Oogonia are diploid cells that undergo meiosis. In a female fetus, all the oogonia start the process of meiosis and form **primary oocytes** prior to birth. Primary oocytes are arrested in prophase I and remain this way until the female reaches puberty (i.e., begins monthly menstruation cycles). Then, each month, a number of primary oocytes begin to mature; usually only one becomes a secondary oocyte.

When the primary oocyte completes the first meiotic division (prophase I, metaphase I, anaphase I, and telophase I), two cells are produced. However, the division of the cytoplasm is grossly unequal. The cell we call the secondary oocyte receives the bulk of the cytoplasm and is the cell that is arrested in metaphase II. The diameter of the secondary oocyte varies, but is typically 100–120 micrometers (μm). The second cell, which receives only a tiny bit of the cytoplasm, is called a **polar body**. The polar body is a nonfunctional cell that eventually degenerates.

Thus, only the secondary oocyte has the potential to be fertilized. The secondary oocyte is **ovulated** (expelled from the ovary into the uterine tube) along with two other components surrounding the oocyte—cuboidal cells that form the **corona radiata** (kă-ro′nă rā-dē-ă′tă; radiate crown) and a thin ring of materials called the **zona pellucida** (pe-loo′sid-ă; *pellucid* = allowing the passage of light). The corona radiata and the zona pellucida form protective layers around the secondary oocyte.

The further development of the secondary oocyte varies, depending upon whether or not it is fertilized by a sperm. If the secondary oocyte is not fertilized, it degenerates about 24 hours after ovulation, still arrested in metaphase II. If the secondary oocyte is fertilized, it first finishes the process of meiosis. Two new cells are produced, and as before, the division of the cytoplasm is unequal. The cell that receives very little cytoplasm becomes another polar body and eventually degenerates. The cell that receives the majority of the cytoplasm becomes an **ovum** (ō′vŭm; egg). It is the ovum nucleus that combines with the sperm nucleus to produce the diploid fertilized cell, or zygote.

Typically, only one secondary oocyte is expelled (ovulated) from one of the two ovaries each month. Thus, during one month the left ovary matures and expels a secondary oocyte, and the next month the right ovary matures and expels its own secondary oocyte. In essence, the left and right ovaries "take turns." This is in stark contrast to sex cell production in males, whose bodies produce and release millions of gametes (sperm) throughout the entire month.

Sperm Development (Spermatogenesis)

In males, the sex cell produced is called a **sperm cell** (**sperm** or **spermatozoon**; pl., spermatozoa), and the process of sperm development is called **spermatogenesis**. Spermatogenesis is discussed in greater detail in chapter 28, but we provide a brief summary here.

The parent or stem cells that produce sperm are called **spermatogonia** (sper′mă-tō-gō′nē-ă; *sperma* = seed, *gone* = generation). Spermatogonia are diploid cells that reside in the male gonads, the testes. Each spermatogonium first divides by mitosis to make

CLINICAL VIEW

Nondisjunction

Abnormalities in chromosome number may originate during meiotic divisions. Normally, the two members of a homologous chromosome pair separate during meiosis I, and paired sister chromatids separate during the second meiotic division (meiosis II). Sometimes, however, separation fails (called **nondisjunction**), and both members of a homologous pair move into one cell, or both sister chromatids move into one cell. As a result of nondisjunction, one potential gamete receives two copies of a single chromosome and has 24 chromosomes, while the other potential gamete receives no copies of this same chromosome and has only 22 chromosomes. If either of these cells unites with a normal gamete with 23 chromosomes, the resulting individual will have either 47 chromosomes (**trisomy**) or 45 chromosomes (**monosomy**). Trisomy means the individual has three copies of a chromosome, while monosomy means an individual has only one copy of a chromosome.

A trisomy disorder is named according to the specific chromosome that has three copies. For example, in trisomy 18 an individual has three copies of chromosome 18. Although any chromosome may be affected by nondisjunction, the most well-known result is Down syndrome, also called trisomy 21. The cells of individuals with Down syndrome contain three copies of chromosome 21 instead of two. A person with trisomy 21 typically has the following characteristics: slight or moderate mental retardation, protruding tongue, epicanthic folds around the eyes, heart defects, and short stature. Many (but not all) cases of Down syndrome occur due to nondisjunction in the maternal line (in other words, the mother's sex cell did not undergo normal separation of chromosome 21). The incidence of Down syndrome increases with the mother's age, suggesting that nondisjunction problems may occur as the mother (and the mother's sex cells) age. However, there are many types of nondisjunction problems, and they may occur in either maternal or paternal sex cell lines.

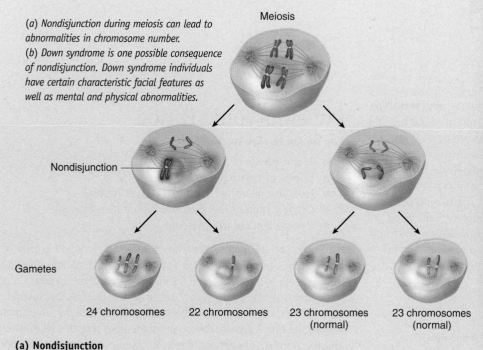

(*a*) Nondisjunction during meiosis can lead to abnormalities in chromosome number.
(*b*) Down syndrome is one possible consequence of nondisjunction. Down syndrome individuals have certain characteristic facial features as well as mental and physical abnormalities.

Meiosis

Nondisjunction

Gametes

24 chromosomes   22 chromosomes   23 chromosomes (normal)   23 chromosomes (normal)

(a) Nondisjunction

(b) Down syndrome (trisomy 21)

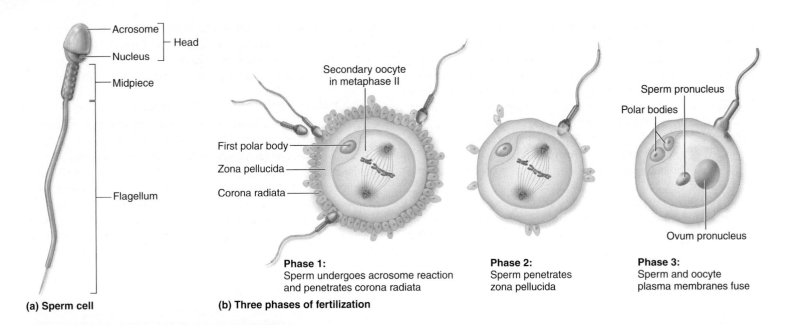

(a) Sperm cell

(b) Three phases of fertilization

Phase 1:
Sperm undergoes acrosome reaction and penetrates corona radiata

Phase 2:
Sperm penetrates zona pellucida

Phase 3:
Sperm and oocyte plasma membranes fuse

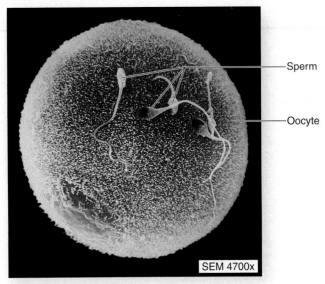

(c) Phase 1 of fertilization

Figure 3.3

Fertilization of a Secondary Oocyte in Humans.   A secondary oocyte is ovulated in a "developmentally arrested" state at metaphase II in meiosis. *(a)* Diagrammatic representation of a normal sperm cell. *(b)* Schematic representation of the three phases of fertilization. *(c)* Scanning electron micrograph of sperm cells in contact with the corona radiata surrounding a secondary oocyte.

an exact copy of itself, a new cell called a **primary spermatocyte**. Primary spermatocytes then undergo meiosis and produce haploid cells called **spermatids** (sper′mă-tid). Although spermatids contain 23 chromosomes only, they still must undergo further changes to form a sperm cell. In a process called **spermiogenesis** (sper′mē-ō-jen′ĕ-sis), the spermatids lose much of their cytoplasm and grow a long tail called a flagellum. The newly formed sperm cells are haploid cells that exhibit a distinctive head, a midpiece, and a tail, as shown in **figure 3.3a**. Thus, from a single spermatocyte, four new sperm are formed. Two of these sperm have 22 autosomes and one X chromosome, and two have 22 autosomes and one Y chromosome.

WHAT DID YOU LEARN?

1. What are two ways in which meiosis differs from mitosis?
2. What is crossing over, and during what phase of meiosis does it occur?
3. A secondary oocyte is arrested in what phase of meiosis?
4. What is the name of the stem cells that form mature sperm?

Pre-embryonic Period

Key topics in this section:

■ Major events of fertilization
■ Effects of cleavage
■ How the bilaminar germinal disc is formed
■ Organization of the extraembryonic membranes
■ Components of the placenta

The pre-embryonic period in human development begins with fertilization, when the male's sperm and the female's secondary oocyte unite to form a single diploid cell called the **zygote** (zī′gōt; *zygotes* = yoked). The zygote is the same size as the secondary oocyte, which typically is between 100 μm and 120 μm in diameter. Within the first 2 weeks, the zygote undergoes mitotic cell divisions, and the number of cells increases, forming a **pre-embryo**. The pre-embryonic stage of development spans the time from fertilization in the uterine tube through completion of implantation (burrowing and embedding) into the wall of the mother's uterus. **Table 3.1** traces the sequence of these events.

Table 3.1	Chronology of Events in Pre-embryonic Development		
Developmental Stage	**Time of Occurrence**	**Location**	**Events**
Fertilization Ovum pronucleus Sperm pronucleus ⊢— 120 µm —⊣	Within 12–24 hours after ovulation	Ampulla of uterine tube	Penetration of sperm into secondary oocyte; secondary oocyte completes meiosis and becomes an ovum; ovum and sperm pronuclei fuse
Zygote Nucleus ⊢— 120 µm —⊣	At the end of fertilization	Ampulla of uterine tube	Diploid cell produced when ovum and sperm pronuclei fuse
Cleavage ⊢— 120 µm —⊣ → ⊢— 120 µm —⊣ 4-cell stage 8-cell stage	30 hours to day 3 post fertilization	Uterine tube	Starting with zygote, cell division by mitosis occurs to increase cell number, but overall size of the structure remains constant
Morula ⊢— 120 µm —⊣	Days 3–4 post fertilization	Uterine tube	Structure formed resembles a solid ball of cells; 16 or more cells are present, but there is no change in diameter from original zygote
Blastocyst Embryoblast Trophoblast ⊢— 120 µm —⊣	Days 5–6	Uterus	Hollow ball of cells; outer ring of the ball formed by trophoblast cells; inner cell mass (embryoblast) is cell cluster inside blastocyst
Implantation Cytotrophoblast Embryoblast Syncytiotrophoblast	Begins late first week and is complete by end of second week	Functional layer of endometrium (inner lining) of uterus	Blastocyst adheres to uterine lining; trophoblast cells penetrate within functional layer of uterus, and together they start to form the placenta

Fertilization

Fertilization is the process whereby two sex cells fuse to form a new cell containing genetic material derived from both parents. Besides combining the male and female genetic material, fertilization restores the diploid number of chromosomes, determines the sex of the organism, and initiates cleavage (discussed later in this section). Fertilization occurs in the widest part of the uterine tube, called the ampulla. Following ovulation, the secondary oocyte remains viable in the female reproductive tract for no more than 24 hours, while sperm remain viable for an average of 3–4 days after ejaculation from the male.

Upon arrival in the female reproductive tract, sperm are not yet capable of fertilizing the secondary oocyte. Before they can successfully do so, sperm must undergo **capacitation** (kă-pas′i-tā′shun; *capacitas* = capable of), a period of conditioning. Capacitation takes place in the female reproductive tract, and typically lasts several hours. During this time, a glycoprotein coat and some proteins are removed from the sperm plasma membrane that overlies the acrosomal region of the sperm. The **acrosome** (ak′rō-sōm; *akros* = tip) is a membranous cap at the head of the sperm cell containing digestive enzymes that can break down the protective layers around the secondary oocyte. These enzymes are released when the sperm cell comes into contact with the secondary oocyte.

Normally, millions of sperm cells are deposited in the vagina of the female reproductive tract during intercourse. However, only a few hundred reach the secondary oocyte in the uterine tube. Many sperm leak out of the vagina, and some are not completely motile (able to swim). Other sperm do not survive the acidic environment of the vagina, and still more lose direction as they move through the uterus and get "churned" by its muscular contractions. Recall that each month, only one of the two uterine tubes contains the secondary oocyte. Sperm that travel into the uterine tube that does not contain the secondary oocyte die. Thus, while the male releases millions of sperm during sexual intercourse, only a few hundred have a chance at fertilization.

Once sperm reach the secondary oocyte, the race is on to see which sperm can fertilize the oocyte first. Only the first sperm to enter the secondary oocyte is able to fertilize it; the remaining sperm are prevented from penetrating the oocyte.

WHAT DO YOU THINK?

2 Rarely, two sperm may penetrate a secondary oocyte. Do you think this fertilized cell will survive for long? Why or why not?

Some causes of infertility (inability to achieve or maintain pregnancy) are due to immune system reactions related to the sperm or oocyte. Some men (and more rarely, some women) develop anti-sperm **antibodies**, which are substances that mark and target the sperm for destruction by the immune system. It is believed that men develop these antibodies against their sperm when the blood-testis barrier (a protective barrier between the testis and the blood vessels traveling through it) is breached, such as due to severe trauma to the testis or reversal of a vasectomy. In other immune-related fertility problems, the woman's body perceives the sperm and/or the fertilized oocyte as something foreign that must be destroyed. Researchers are examining ways to prevent these immune-related infertility problems from occurring.

The phases of fertilization are corona radiata penetration, zona pellucida penetration, and fusion of the sperm and oocyte plasma membranes (figure 3.3*b,c*).

Corona Radiata Penetration

The sperm cells that successfully reach the secondary oocyte release digestive enzymes from their acrosomes. These enzymes eat away (digest) the intercellular connections between the corona radiata cells, ultimately forming a passageway between the cells of the corona radiata. This release of enzymes from the acrosome is known as the **acrosome reaction**.

Zona Pellucida Penetration

Once the digestive enzymes from the acrosome of some sperm make a pathway through the corona radiata, other sperm that have now passed through this pathway also release these same enzymes to facilitate the penetration of the zona pellucida by sperm. After the first sperm cell successfully penetrates the zona pellucida and its nucleus enters the secondary oocyte, immediate changes occur to both the zona pellucida and the oocyte so that no other sperm can enter the oocyte. In essence, the zona pellucida hardens, preventing other sperm from binding to and ultimately digesting their way through this layer. This process is necessary to ensure that only one sperm cell fertilizes the oocyte.

On very rare occasions, two or more sperm cell nuclei simultaneously enter the secondary oocyte, a phenomenon called **polyspermy** (pol′ē-sper-mē; *polys* = many). Polyspermy is immediately fatal because it causes the fertilized oocyte to have 23 triplets (if two sperm enter) or 23 quadruplets (if three sperm enter) of chromosomes, instead of the normal 23 pairs of chromosomes.

Fusion of Sperm and Oocyte Plasma Membranes

When the sperm and oocyte plasma membranes come into contact, they immediately fuse. Only the nucleus of the sperm enters the cytoplasm of the secondary oocyte. Once the nucleus of the sperm enters the secondary oocyte, the secondary oocyte completes the second meiotic division and forms an ovum.

Following the completion of meiosis, the nucleus of the sperm cell and the nucleus of the ovum are called **pronuclei** (*pro* = before, precursor of) because they have a haploid number of chromosomes. These pronuclei come together and fuse, forming a single nucleus that contains a diploid number (23 pairs) of chromosomes. The single diploid cell formed is the zygote.

CLINICAL VIEW

Chromosomal Abnormalities and Their Effect on the Blastocyst

At first glance, the process of human development appears seamless, with few errors. However, abnormalities in chromosome number, shape, or form occur regularly. These abnormalities can occur during gametogenesis, fertilization, or cleavage. If the chromosomal abnormalities are severe enough, they result in the spontaneous abortion (miscarriage) of the blastocyst or embryo. Many of these spontaneous abortions occur early in pregnancy (within 2 to 3 weeks after fertilization), so a woman often spontaneously aborts without realizing she was ever pregnant.

Some estimates propose that approximately 50% of all pregnancies terminate as a result of spontaneous abortion; perhaps half of these are caused by chromosomal abnormalities in the developing organism. As a consequence, fewer organisms are stillborn or born with severe congenital malformations (birth defects). Thus, while 2–3% of all infants are born with some type of birth defect, this percentage would be much higher if not for the high frequency of spontaneous abortions very early in pregnancy.

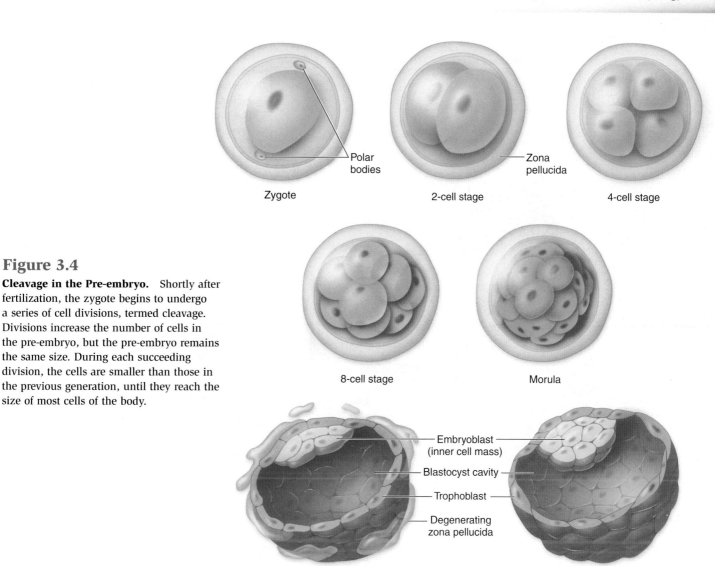

Figure 3.4

Cleavage in the Pre-embryo.   Shortly after fertilization, the zygote begins to undergo a series of cell divisions, termed cleavage. Divisions increase the number of cells in the pre-embryo, but the pre-embryo remains the same size. During each succeeding division, the cells are smaller than those in the previous generation, until they reach the size of most cells of the body.

Cleavage

Following fertilization, the zygote begins the process of becoming a multicellular organism. After the zygote divides once and reaches the 2-cell stage, a series of mitotic divisions, called **cleavage** (klēv′ij), results in an increase in cell number but not an increase in the overall size of the structure. The diameter of the structure remains about 120 μm, so the mitotic divisions produce greater numbers of smaller cells to fit in this structure. The structure will not increase in *size* until it implants in the uterine wall and derives a source of nourishment from the mother. **(figure 3.4)**. Before the 8-cell stage, cells are not tightly bound together, but after the third cleavage division, the cells become tightly compacted into a ball. The process by which contact between cells is increased to the maximum is called **compaction**. These cells now divide again, forming a 16-cell stage, the **morula** (mōr′oo-lă, mōr′ū; *morus* = mulberry). The cells of the morula continue to divide further.

Shortly after the morula enters the space (called the lumen) of the uterus, fluid begins to leak through the degenerating zona pellucida surrounding the morula. As a result, a fluid-filled cavity, called the **blastocyst cavity**, develops within the morula. The pre-embryo at this stage of development is known as a **blastocyst** (blas′tō-sist; *blastos* = germ), and it has two distinct components:

■ The **trophoblast** (trof′ō-blast; *trophe* = nourishment) is an outer ring of cells surrounding the fluid-filled cavity. These

cells will form the chorion, one of the extraembryonic membranes discussed later in this section.

■ The **embryoblast**, or **inner cell mass**, is a tightly packed group of cells located only within one side of the blastocyst. The embryoblast will form the embryo proper. These early cells are **pluripotent** (ploo-rip′ō-tent; *pluris* = multi, *potentia* = power), which means they have the power to differentiate into any cell or tissue type in the body.

An overview of fertilization and cleavage, including the movement of the pre-embryo from the uterine tube into the uterus, is given in **figure 3.5**.

Implantation

By the end of the first week after fertilization, the blastocyst enters the lumen of the uterus. The zona pellucida around the blastocyst begins to break down as the blastocyst prepares to invade the inner lining of the uterine wall, called the **endometrium**. The endometrium consists of a deeper **basal layer**, the *stratum basalis*, and a more superficial **functional layer**, the *stratum functionalis*. The blastocyst invades this functional layer. **Implantation** is the process by which the blastocyst burrows into and embeds within the endometrium.

The blastocyst begins the implantation process by about day 7 (the end of the first week of development), when trophoblast cells

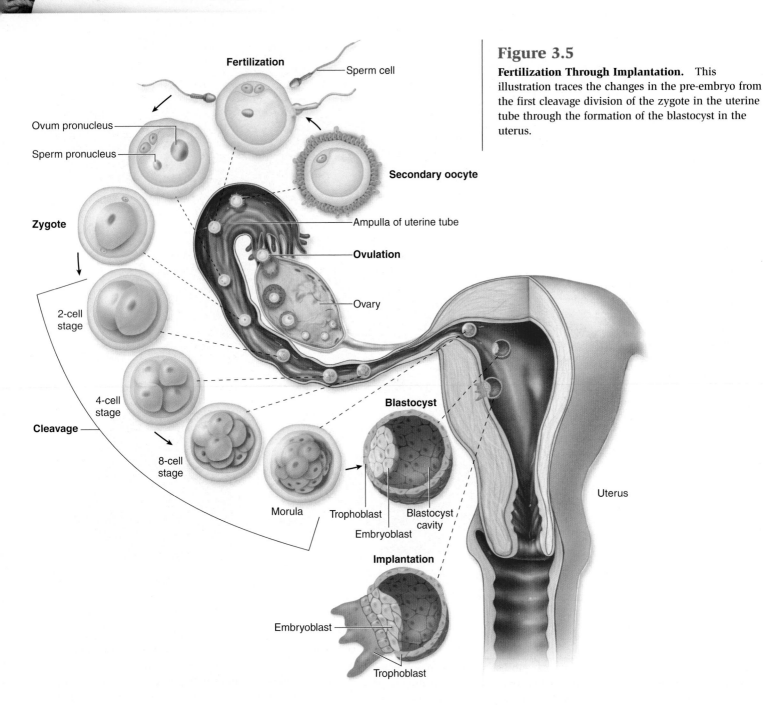

Fertilization

Sperm cell

Ovum pronucleus

Sperm pronucleus

Secondary oocyte

Zygote

Ampulla of uterine tube

Ovulation

Ovary

2-cell stage

4-cell stage

Blastocyst

Cleavage

8-cell stage

Uterus

Morula

Trophoblast   Blastocyst cavity

Embryoblast

Implantation

Embryoblast

Trophoblast

Figure 3.5

Fertilization Through Implantation. This illustration traces the changes in the pre-embryo from the first cleavage division of the zygote in the uterine tube through the formation of the blastocyst in the uterus.

begin to invade the functional layer of the endometrium (**figure 3.6**). Simultaneously, the trophoblast subdivides into two layers: a **cytotrophoblast** (sī-tō-trō′fō-blast; *kytos* = cell), which is the inner cellular layer of the trophoblast, and a **syncytiotrophoblast** (sin-sish′ē-ō-trō′fo-blast), which is the outer, thick layer of the trophoblast where no plasma membranes are visible. Over the next few days, the syncytiotrophoblast cells burrow into the functional layer of the endometrium and bring with them the rest of the blastocyst. By day 9, the blastocyst has completely burrowed into the uterine wall. Here, the blastocyst makes contact with the pools of nutrients in the uterine glands that supply the developing organism. Thus, implantation begins during the first week of development and is not complete until the second week.

Formation of the Bilaminar Germinal Disc

During the second week of development, as the blastocyst is undergoing implantation, changes also occur to the embryoblast portion of the blastocyst. By day 8 (the beginning of the second week of

CLINICAL VIEW

Human Chorionic Gonadotropin

The syncytiotrophoblast is responsible for producing a hormone called **human chorionic gonadotropin (hCG)**. This hormone signals other parts of the female reproductive system that fertilization and implantation have occurred, so the uterine lining should continue to grow and develop (rather than being shed as menstruation). By the end of the second week of development, sufficient quantities of hCG are produced to be detected in a woman's urine. The presence of hCG in urine indicates a woman is pregnant, and thus hCG is the basis for modern-day pregnancy tests. For the first 3 months of pregnancy, hCG levels remain high, but after that they decline. By this time, hCG is no longer needed because the placenta is producing its own hormones to maintain the pregnancy.

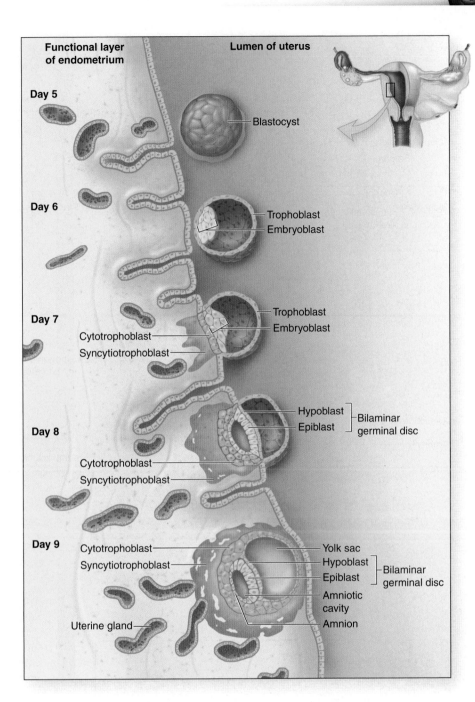

Figure 3.6

Implantation of the Blastocyst. The pre-embryo becomes a blastocyst in the uterine lumen prior to implantation. Contact between the blastocyst and the uterine wall begins the process of implantation about day 7. (The trophoblast of the implanting blastocyst differentiates into a cytotrophoblast and a syncytiotrophoblast shortly thereafter.) The implanting blastocyst makes contact with the maternal blood supply about day 9.

development), the cells of the embryoblast begin to differentiate into two layers. A layer of small, cuboidal cells adjacent to the blastocyst cavity is termed the **hypoblast** layer, and a layer of columnar cells adjacent to the amniotic cavity is called the **epiblast** layer (figure 3.6). Together, these layers form a flat disc termed a **bilaminar germinal disc**, or *blastodisc*.

Formation of Extraembryonic Membranes

The bilaminar germinal disc and trophoblast also produce extraembryonic membranes to mediate between them and the environment. These **extraembryonic membranes** are the yolk sac, amnion, and chorion **(figure 3.7)**. They first appear during the second week of development and continue to develop during the embryonic and fetal periods. They assist the embryo in vital functions such as nutrition, gas exchange, and removal and storage of waste materials. In addition, they protect the embryo by surrounding it with an aqueous environment.

The **yolk sac**, the first extraembryonic membrane to form, is formed from and continuous with the hypoblast layer. In humans, it does not store yolk, but it is an important site for early blood cell and blood vessel formation. The future gut tube (digestive system) maintains a connection with the yolk sac in the first trimester (first 3 months) of the pregnancy.

The **amnion** (am′nē-on; *amnios* = lamb) is a thin membrane that is formed from and continuous with the epiblast layer. The amnion eventually encloses the entire embryo in a fluid-filled sac called the **amniotic cavity** to prevent the embryo's desiccation. The amniotic membrane is specialized to secrete the amniotic fluid that bathes the embryo.

The **chorion** (kō′rē-on; membrane covering the fetus), the outermost extraembryonic membrane, is formed from the rapidly growing cytotrophoblast cells and syncytiotrophoblast. These cells blend with the functional layer of the endometrium and eventually form the placenta, the site of exchange between the embryo and the mother.

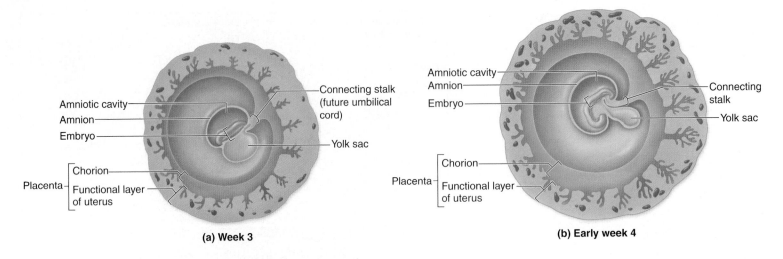

(a) Week 3

(b) Early week 4

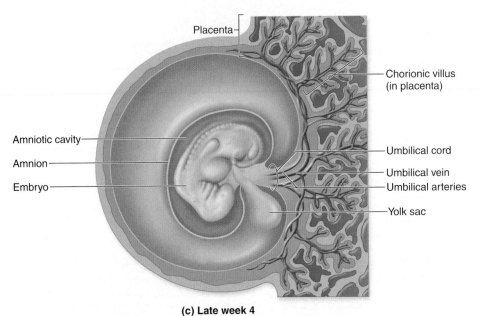

(c) Late week 4

Figure 3.7

Formation of Extraembryonic Membranes.
The extraembryonic membranes (amnion, chorion, and yolk sac) first appear during the second week of development. Their changes in growth and form are shown at *(a)* week 3, *(b)* early week 4, and *(c)* late week 4 of development.

Study Tip!

The *second week* of development may be thought of as the "period of twos," because many paired structures develop:

- ■ A *two-layered* (epiblast and hypoblast) germinal disc forms.
- ■ *Two membranes* (the yolk sac and the amnion) develop on either side of the bilaminar germinal disc.
- ■ The placenta develops from *two* components that merge (the chorion and the functional layer of the endometrium of the uterus).

Development of the Placenta

Recall that the blastocyst is approximately the same size as the initial zygote, but the blastocyst contains many more cells than the zygote. In order to develop into an embryo and fetus, the blastocyst must receive nutrients and respiratory gases from the maternal blood supply. The connection between the embryo or fetus and the mother is the richly vascular **placenta** (plă-sen'tă; a cake). The main functions of the placenta are:

- ■ Exchange of nutrients, waste products, and respiratory gases between the maternal and fetal bloodstreams.

- ■ Transmission of maternal antibodies (immune system substances that target viruses or bacteria) to the developing embryo or fetus.
- ■ Production of hormones (primarily estrogen and progesterone) to maintain and build the uterine lining.

The placenta begins to form during the second week of development. The fetal portion of the placenta develops from the chorion, while the maternal portion of the placenta forms from the functional layer of the uterus. The early organism is connected to the placenta via a structure called the **connecting stalk**. This connecting stalk eventually contains the umbilical arteries and veins that distribute blood through the embryo or fetus. The connecting stalk is the precursor to the future **umbilical cord**.

Figure 3.7 illustrates how the components of the placenta become better defined during the embryonic period. Stalklike structures called **chorionic villi** form from the chorion. The chorionic villi contain branches of the umbilical vessels. Adjacent to the chorionic villi is the functional layer of the endometrium, which contains maternal blood. Note that fetal blood and maternal blood do not mix; however, the bloodstreams are so close to one another that exchange of gases and nutrients can occur. Thus, the blood cells in the maternal tissue can pass along oxygen and nutrients to the fetal blood cells in the chorionic villi via the umbilical vein. Likewise, carbon dioxide and

CLINICAL VIEW

Regulation of Materials Along the Placental Barrier

The placenta may be thought of as a selectively permeable structure. Certain materials enter freely through the placenta into the fetal bloodstream, while other substances are effectively blocked. For example, respiratory gases and nutrients may freely cross the placental barrier, but certain microorganisms and large levels of maternal hormones are prevented from crossing this barrier into the developing fetus.

Unfortunately, a number of undesirable items *can* cross the placental barrier. Many viruses (such as HIV) and bacteria (such as *Treponema*, the bacterium that causes syphilis) can cross the placental barrier, infecting the fetus. Likewise, viruses such as rubella can cross the placental barrier and cause massive birth defects or death. Most drugs and alcohol can pass through the placental barrier as well, including anything from aspirin to barbiturates to heroin and cocaine. If a mother takes heroin or cocaine during her pregnancy, she can give birth to a baby who is addicted to these drugs and has mental and physical problems. Alcohol consumption can affect the developing fetus and cause a variety of physical and mental conditions, collectively known as *fetal alcohol syndrome*. The toxins from smoking (nicotine and carbon monoxide) can cross the placental barrier and cause low birth weight, among other problems.

Some fetuses may be more susceptible to materials that cross the placental barrier than other fetuses. In addition, the *dose* of the material crossing the placental barrier affects fetus susceptibility. These facts help explain why some newborns are strongly affected by materials that cross the placental barrier, while other newborns are relatively unaffected.

Prior to implantation, the blastocyst is not harmed by undesirable substances because it does not yet have a connection with the mother's uterine lining. However, once implantation begins and the placenta starts to form, the developing organism is exposed to most of the items to which the mother is exposed. For these reasons, pregnant females are strongly urged to quit smoking and to refrain from taking drugs and drinking alcohol during their pregnancies.

certain cellular waste products in the fetal blood may be passed from the fetal blood cells to the blood cells in the maternal tissue.

Although the placenta first forms during the pre-embryonic period, most of its growth and development occur during the fetal period. It takes about 3 months for the placenta to become fully formed and able to produce sufficient amounts of estrogen and progesterone to maintain and build the uterine lining. Within these first 3 months, a structure called the **corpus luteum** (in the ovary of the mother) produces the estrogen and progesterone. When the placenta matures, it resembles a disc in shape and adheres firmly to the wall of the uterus. Immediately after the baby is born, the placenta is also expelled from the uterus. The expelled placenta is often called the "afterbirth."

🔅 WHAT DID YOU LEARN?

5 How is a secondary oocyte different from an ovum?

6 What are some factors or events that can prevent sperm from reaching the secondary oocyte?

7 What is the name of the core of cells at one end of the blastocyst that will form the embryo proper?

8 What are the main functions of the placenta?

Embryonic Period

Key topics in this section:

- Process of gastrulation
- Nature of the three primary germ layers
- Major structures formed from each of the primary germ layers
- Steps involved in neurulation

The embryonic period begins with the establishment of the three primary germ layers through the process of gastrulation. Subsequent interactions and rearrangements among the cells of the three layers prepare for the formation of specific tissues and organs, a process called organogenesis. By the end of the embryonic period (week 8), the main organ systems have been established, and the major features of the external body form are recognizable. **Table 3.2** summarizes the events that occur during the embryonic period.

Table 3.2	Events in Embryonic Development	
Developmental Week		**Events**
Week 3 1.5 mm — Neural groove, Neural fold, Primitive streak		Primitive streak appears Three primary germ layers form Notochord develops Neurulation begins Length: 1.5 mm
Week 4 4.0 mm — Heart, Umbilical cord, Limb buds		Cephalocaudal and lateral folding produce a cylindrical embryo Basic human body plan is established Derivatives of the three germ layers begin to form Limb buds appear Crown-rump length: 4.0 mm
Weeks 5–8 30 mm		Head enlarges Eyes, ears, and nose appear Major organ systems are formed by the end of week 8 (although some may not be fully functional yet) Crown-rump length by the end of week 8: 30 mm

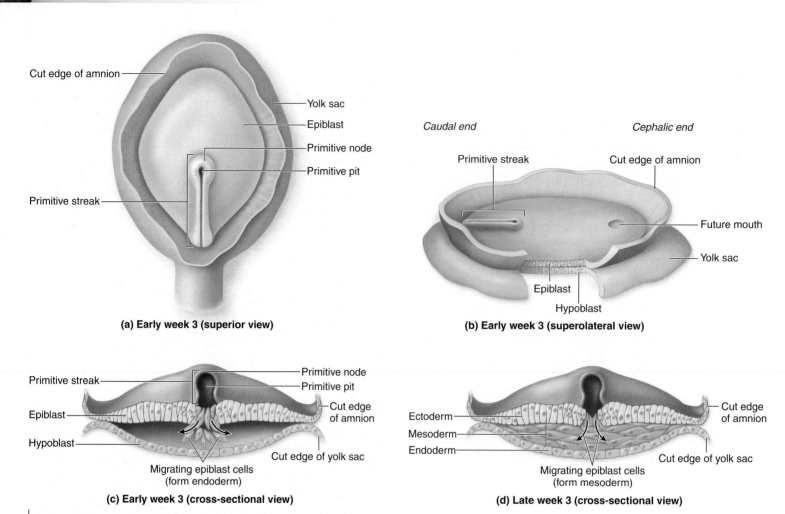

(a) Early week 3 (superior view)

Cut edge of amnion
Yolk sac
Epiblast
Primitive node
Primitive pit
Primitive streak

Caudal end   *Cephalic end*

(b) Early week 3 (superolateral view)

Primitive streak
Cut edge of amnion
Future mouth
Yolk sac
Epiblast
Hypoblast

(c) Early week 3 (cross-sectional view)

Primitive streak
Primitive node
Primitive pit
Epiblast
Cut edge of amnion
Hypoblast
Cut edge of yolk sac
Migrating epiblast cells
(form endoderm)

(d) Late week 3 (cross-sectional view)

Primitive node
Primitive pit
Cut edge of amnion
Ectoderm
Mesoderm
Endoderm
Cut edge of yolk sac
Migrating epiblast cells
(form mesoderm)

Figure 3.8

The Role of the Primitive Streak in Gastrulation.   (*a,b*) The primitive streak is a raised groove on the epiblast surface of the bilaminar germinal disc that appears early in the third week. (*c,d*) During gastrulation, epiblast cells migrate toward the primitive streak, where some become embryonic endoderm, and others form mesoderm between the epiblast and the new endoderm.

Gastrulation

Gastrulation (gas-troo-lā'shŭn; *gaster* = belly) occurs during the third week of development immediately after implantation, and is one of the most critical periods in the development of the embryo. Gastrulation is a process by which the cells of the epiblast migrate and form the three **primary germ layers**, which are the cells from which all body tissues develop. The three primary germ layers are called ectoderm, mesoderm, and endoderm. Once these three layers have formed, the developing trilaminar (three-layered) structure may be called an **embryo** (em'brē-o).

Gastrulation begins with formation of the **primitive streak**, a thin depression on the surface of the epiblast **(figure 3.8*a,b*)**. The cephalic (head) end of the streak, known as the **primitive node**, consists of a slightly elevated area surrounding a small **primitive pit**.

Cells detach from the epiblast layer and migrate through the primitive streak between the epiblast and hypoblast layers. This inward movement of cells is known as **invagination**. The layer of

cells that forms between these two layers becomes the primary germ layer known as **mesoderm** (mez'ō-derm; *meso* = middle, *derma* = skin). Other migrating cells eventually displace the hypoblast and form the **endoderm** (en'dō-derm; *endo* = inner). Cells remaining in the epiblast then form the **ectoderm** (ek'tō-derm; *ektos* = outside). Thus, the epiblast, through the process of gastrulation, is the source of the three primary germ layers, from which all body tissues and organs eventually derive (figure 3.8*c,d*).

> ## Study Tip!
> The *third week* of development produces an embryo with *three* primary germ layers: ectoderm, mesoderm, and endoderm.

Folding of the Embryonic Disc

The 3-week embryo is a flattened, disc-shaped structure. For this reason, the structure is also referred to as an **embryonic disc**

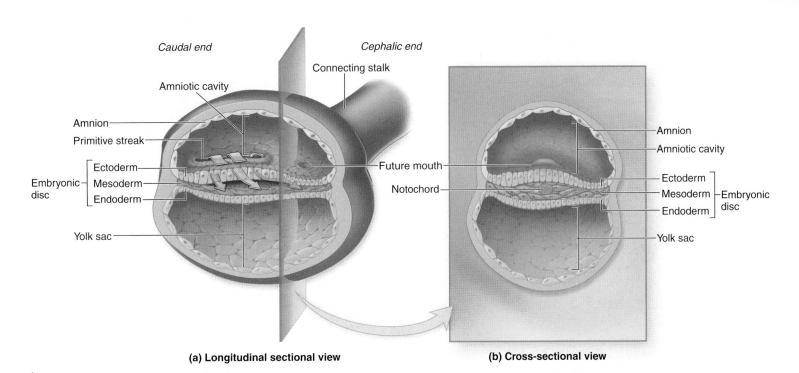

Caudal end   *Cephalic end*

Connecting stalk

Amniotic cavity

Amnion

Primitive streak

Embryonic disc — Ectoderm / Mesoderm / Endoderm

Yolk sac

Future mouth

(a) Longitudinal sectional view

Amnion

Amniotic cavity

Ectoderm / Mesoderm / Endoderm — Embryonic disc

Notochord

Yolk sac

(b) Cross-sectional view

Figure 3.9

Formation of the Embryonic Disc. Gastrulation produces a trilaminar embryonic disc that contains three primary germ layers: endoderm, mesoderm, and ectoderm.

(figure 3.9). So how does this flattened structure turn into a three-dimensional human?

The shape transformation begins during the late third and fourth weeks of development, when certain regions of the embryo grow faster than others. As a result of this differential growth, the embryonic disc starts to fold on itself and become more cylindrical. **Figure 3.10** illustrates the two types of folding that occur: cephalocaudal folding and transverse folding.

Cephalocaudal (sef′ă-lō-kaw′dăl) **folding** occurs in the cephalic (head) and caudal (tail) regions of the embryo. Essentially, the embryonic disc and amnion grow very rapidly, but the yolk sac does not grow at all. This differential growth causes the head and tail regions to fold on themselves.

Transverse folding (or *lateral folding*) occurs when the left and right sides of the embryo curve and migrate toward the midline. As these sides come together, they restrict and start to pinch off the yolk sac. Eventually, the sides of the embryonic disc fuse in the midline and create a cylindrical embryo. Thus, the ectoderm is now solely along the entire exterior of the embryo, while the endoderm is confined to the internal region of the embryo. As this midline fusion occurs, the yolk sac pinches off from most of the endoderm (with the exception of one small region of communication called the vitelline duct).

Thus, cephalocaudal folding helps create the future head and buttocks region of the embryo, while transverse folding creates a cylindrical trunk or torso region of the embryo. Let us now examine the specific derivatives of these primary germ layers.

Differentiation of Ectoderm

After the embryo undergoes cephalocaudal and transverse folding, the ectoderm is located on the external surface of the now-cylindrical embryo. The ectoderm is responsible for forming nervous system tissue as well as many externally placed structures, including the epidermis of the skin and epidermal derivatives such as hair and nails. The process of nervous system formation from the ectoderm is called neurulation.

Neurulation

A cylindrical structure of mesoderm, called the **notochord** (nō′tō-kōrd; *notos* = back, *chorde* = cord, string), forms immediately internal and parallel to the primitive streak. The notochord influences some of the overlying ectoderm to begin to form nervous tissue via a process called **induction**, in which one structure influences or induces another structure to change form. The inductive action that transforms a flat layer of ectodermal cells into a hollow nervous system tube is termed **neurulation** (noor-oo-lā′shun; *neuron* = nerve, *-ulus* = small one) **(figure 3.11)**.

In the third week of development, much of the ectoderm forms a thickened layer of cells called the **neural plate**. By the end of the third week, the lateral edges of this plate elevate to form **neural folds**, and the depression between the folds forms the **neural groove**. The neural folds approach each other gradually in the midline and fuse. Fusion of these folds produces a cylindrical **neural tube** (figure 3.11c). This fusion begins in the middle of the neural folds and proceeds in both cephalic and caudal directions.

Cephalocaudal folding

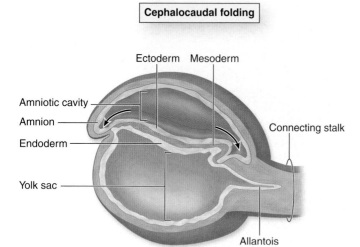

Week 3

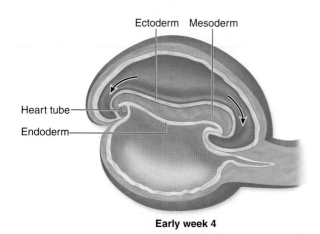

Early week 4

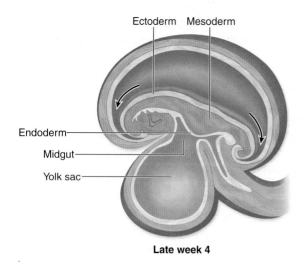

Late week 4

Transverse folding

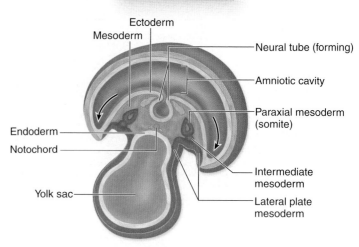

Late week 3

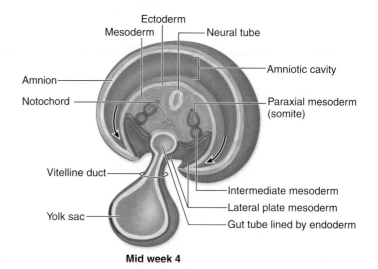

Mid week 4

Late week 4

Figure 3.10

Folding of the Embryonic Disc. During the third and fourth weeks of development, the flat embryo folds in both cephalocaudal and transverse (lateral) directions. The sequence of cephalocaudal folding is shown in the left column, while transverse folding is shown in the right column.

Superior views

Cross-sectional views

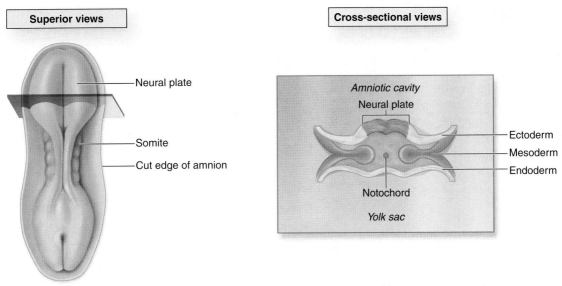

Neural plate

Somite

Cut edge of amnion

Amniotic cavity

Neural plate

Ectoderm

Mesoderm

Endoderm

Notochord

Yolk sac

(a) Mid week 3: neural plate forms

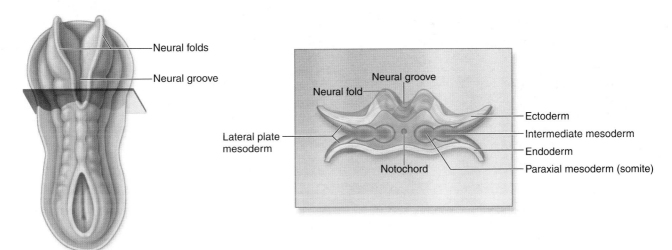

Neural folds

Neural groove

Lateral plate mesoderm

Neural groove

Neural fold

Ectoderm

Intermediate mesoderm

Endoderm

Notochord

Paraxial mesoderm (somite)

(b) Late week 3: neural folds and neural groove form

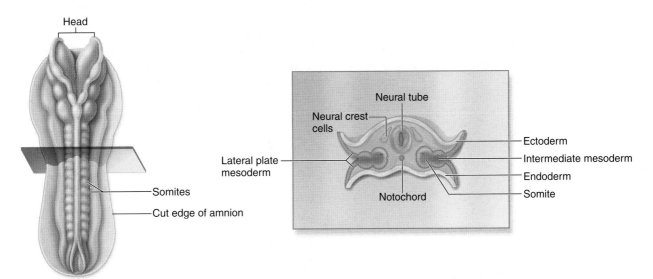

Head

Lateral plate mesoderm

Somites

Cut edge of amnion

Neural tube

Neural crest cells

Ectoderm

Intermediate mesoderm

Endoderm

Notochord

Somite

(c) Early week 4: neural folds fuse to form neural tube

Figure 3.11

Neurulation. Superior and cross-sectional views show the developing embryo during neural tube formation. *(a)* Ectoderm thickens in the midline, forming the neural plate. *(b)* The neural plate develops a depression called the neural groove and two elevations called the neural folds. *(c)* Neural folds fuse to form the neural tube, while neural crest cells pinch off the neural folds and migrate to various areas of the body.

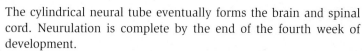

The cylindrical neural tube eventually forms the brain and spinal cord. Neurulation is complete by the end of the fourth week of development.

As the neural folds migrate toward each other and fuse, some cells along the lateral border of folds begin to dissociate from adjacent cells. Collectively, these cells are known as the **neural crest cells**. Neural crest cells migrate throughout the body and give rise to a vast, heterogeneous array of structures. Among the structures neural crest cells give rise to are melanocytes (pigment cells of the skin), the adrenal medulla (inner portion of the adrenal gland), some skeletal and muscular components of the head, spinal ganglia (specific nervous system structures), and a portion of the developing heart.

Not all ectodermal cells form the neural plate. The ectodermal cells covering the embryo after neurulation form the epidermis, the external layer of skin. Ectoderm also forms most exocrine glands, hair, nails, tooth enamel, and sensory organs. In general, ectoderm gives rise to those organs and structures that maintain contact with the outside world **(figure 3.12)**.

Differentiation of Mesoderm

Mesoderm subdivides into the following five categories:

- The tightly packed midline group of mesodermal cells, also called *chordamesoderm*, forms the notochord. The notochord serves as the basis for the central body axis and the axial skeleton, and induces the formation of the neural tube, as previously described.
- **Paraxial mesoderm** is found on both sides of the neural tube. The paraxial mesoderm then forms **somites** (sō'mīt; *soma* = body), which are blocklike masses responsible for the formation of the axial skeleton, most muscle (including limb musculature), and most of the cartilage, dermis, and connective tissues of the body (see figure 3.11).
- Lateral to the paraxial mesoderm are cords of **intermediate mesoderm**, which forms most of the urinary system and the reproductive system.
- The most lateral layers of mesoderm on both sides of the neural tube remain thin and are called the **lateral plate mesoderm**. These give rise to most of the components of the cardiovascular system, the lining of the body cavities, and all the connective tissue components of the limbs.
- The last region of mesoderm, called the **head mesenchyme** (mez'en-kīm), forms connective tissues and musculature of the face.

The derivatives of the mesoderm are listed and illustrated in figure 3.12.

Differentiation of Endoderm

Endoderm becomes the innermost tissue when the embryo undergoes transverse folding. Among the structures formed by embryonic endoderm are the linings of the digestive, respira-tory, and urinary tracts (figure 3.12). Endoderm also forms the thyroid gland, parathyroid glands, thymus, and portions of the palatine tonsils, as well as most of the liver, gallbladder, and pancreas.

WHAT DO YOU THINK?

3 If gastrulation did not occur properly and one of the primary germ layers wasn't formed, would the embryo be able to survive? Why or why not?

Organogenesis

Once the three primary germ layers have formed and the embryo has undergone cephalocaudal and transverse folding, the process of **organogenesis** (organ development) can begin. The upper and lower limbs attain their adult shapes, and the rudimentary forms of most organ systems have developed by week 8 of development.

By the end of the embryonic period, the embryo is slightly longer than 2.5 centimeters (1 inch), and yet it already has the outward appearance of a human. During the embryonic period, the embryo is particularly sensitive to **teratogens** (ter'ă-tō-jen; *teras* = monster, *gen* = producing), substances that can cause birth defects or the death of the embryo. Teratogens include alcohol, tobacco smoke, drugs, some viruses, and even some seemingly benign medications, such as aspirin. Because the embryonic period includes organogenesis, exposure to teratogens at this time can result in the malformation of some or all organ systems.

Although rudimentary versions of most organ systems have formed during the embryonic period, different organ systems undergo "peak development" periods at different times. For example, the peak development for limb maturation is weeks 4–8, while peak development of the external genitalia begins in the late embryonic period and continues through the early fetal period. Teratogens cause the most harm to an organ system during its peak development period. So, a drug such as thalidomide (which causes limb defects) causes the most limb development damage if taken by the mother during weeks 4–8.

The development of each organ system is discussed in detail at the end of later chapters. Thus, limb development is discussed at the end of chapter 8 (Appendicular Skeleton), and heart development is discussed at the end of chapter 22 (Heart).

WHAT DID YOU LEARN?

9 What is gastrulation?

10 What structure induces the process of neurulation?

11 Identify three structures that originate from ectoderm.

12 What structures in the embryo are derived from the somites?

13 What structures are formed from endoderm?

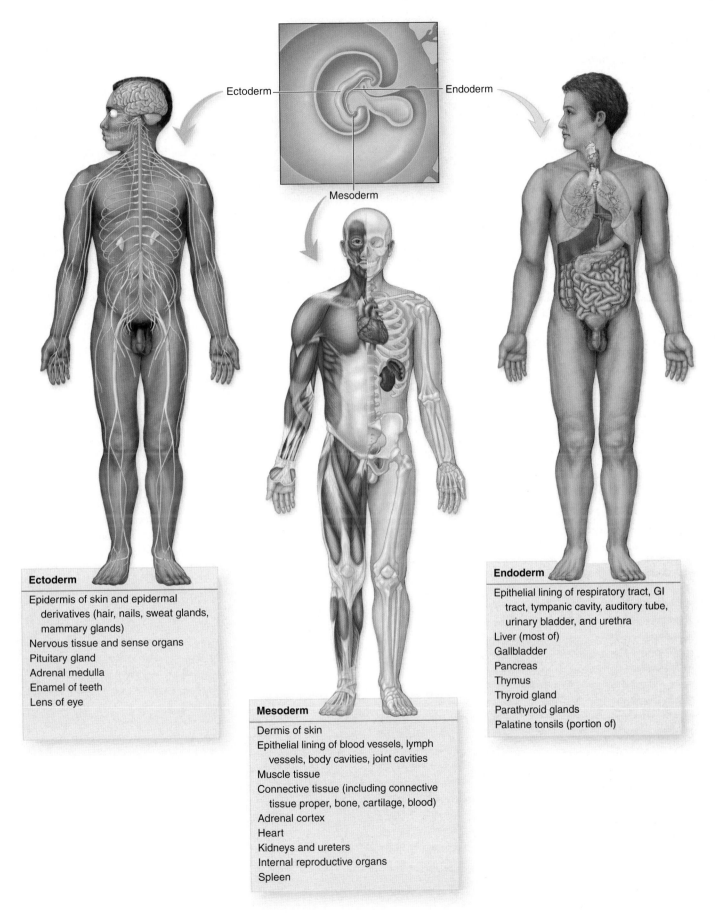

Ectoderm

Endoderm

Mesoderm

Ectoderm

Epidermis of skin and epidermal
  derivatives (hair, nails, sweat glands,
  mammary glands)
Nervous tissue and sense organs
Pituitary gland
Adrenal medulla
Enamel of teeth
Lens of eye

Mesoderm

Dermis of skin
Epithelial lining of blood vessels, lymph
  vessels, body cavities, joint cavities
Muscle tissue
Connective tissue (including connective
  tissue proper, bone, cartilage, blood)
Adrenal cortex
Heart
Kidneys and ureters
Internal reproductive organs
Spleen

Endoderm

Epithelial lining of respiratory tract, GI
  tract, tympanic cavity, auditory tube,
  urinary bladder, and urethra
Liver (most of)
Gallbladder
Pancreas
Thymus
Thyroid gland
Parathyroid glands
Palatine tonsils (portion of)

Figure 3.12
The Three Primary Germ Layers and Their Derivatives. Ectoderm, mesoderm, and endoderm give rise to all of the tissues in the body.

Fetal Period

Key topic in this section:

- Major events during the fetal stage of development

The fetal period extends from the beginning of the third month of development (week 9) to birth. It is characterized by maturation of tissues and organs and rapid growth of the body.

The length of the fetus is usually measured in centimeters, either as the crown-rump length (CRL) or the crown-heel length (CHL). Fetal length increases dramatically in months 3 to 5. The 2.5-centimeter embryo will grow in the fetal period to an average length of 53 centimeters (21 inches). Fetal weight increases steadily as well, although the weight increase is most striking in the last 2 months of pregnancy. The average weight of a full-term fetus ranges from 2.5 to 4.5 kilograms. The major events that occur during the fetal period are listed in **table 3.3**.

WHAT DID YOU LEARN?

14 The fetal period is characterized by which key events?

CLINICAL VIEW

Amniocentesis

Physicians can obtain information about the condition of the fetus, particularly the presence of certain abnormalities, by means of **amniocentesis** (am′nē-ō-sen-tē′sis; *kentesis* = puncture). This procedure is usually performed during the fourth month of pregnancy. About 5 to 10 milliliters (mL) of amniotic fluid (the fluid surrounding the developing fetus) are collected from within the mother's uterus using a hypodermic needle. The needle must be inserted in the abdominal wall, through the musculature, and then into the expanding uterus. The fluid sample contains cells shed by the developing embryo. Analysis of the embryo's chromosomes in these cells may reveal the presence of certain genetic diseases. For example, Down syndrome (trisomy 21) can be identified by detecting three copies of chromosome 21 instead of the normal two chromosomes.

CLINICAL VIEW: In Depth

Congenital Malformations

Congenital malformations, congenital anomalies, and birth defects are synonymous terms that describe structural, behavioral, functional, and metabolic disorders present at birth. The study of the causes of these disorders is **teratology** (ter-ă-tol′ŏ-jē). Major structural anomalies occur in 2–3% of live births. An additional 2–3% of anomalies are not immediately recognizable at birth, but become more apparent in children by age 5 years. Thus, the total percentage of anomalies that occur in live births and are detected by age 5 is 4–6%. Birth defects are the leading cause of infant mortality in the developed world, accounting for approximately 21% of all infant deaths.

In 40–60% of all birth defects, the cause is unknown. Of the remainder, genetic factors, such as chromosomal abnormalities and mutant genes, account for approximately 15%; environmental factors produce approximately 10%; and a combination of genetic and environmental influences produces 20–25%.

Minor anomalies occur in approximately 15% of newborns. These structural abnormalities, such as microtia (small ears), pigmented spots, and small eyelid openings, are not detrimental to the health of the individual but in some cases are associated with major defects. For example, infants with one minor anomaly have a 3% chance of having a major malformation; those with two minor anomalies have a 10% chance; and those with three or more minor anomalies have a 20% chance. Therefore, minor anomalies serve as diagnostic clues for more serious underlying defects. There are several types of anomalies.

- A **malformation** can occur during the formation of structures—that is, during the embryonic period (weeks 3 to 8 of development). These effects may include complete or partial absence of a structure or alterations in its normal configuration. Malformations may be caused by environmental or genetic factors acting independently or in concert. One example is atrial septal defect, a persistent hole in the wall between two chambers of the heart.

- A **disruption** results in morphological alterations of structures after their formation and is due to destructive processes. For example, some defects are produced by amniotic bands, tears in the amnion that may encircle part of the fetus, especially the limbs or digits. Amniotic bands may cause constrictions of the limbs or digits and amputations.

- A **deformation** is due to mechanical forces that mold a part of the fetus over a prolonged period of time. An example is clubfeet, caused by compression of the fetus in the uterus. Another example is plagiocephaly, where the skull is misshapen due to uterine constraint or premature fusion of some of the skull bones. Deformations often involve the muscular or skeletal systems and may be repairable or reversible postnatally.

- A **syndrome** refers to a group of anomalies occurring together that have a specific, common cause. This term indicates that a diagnosis has been made and that the risk of occurrence is known. An example is fetal alcohol syndrome, the result of alcoholic intake by the mother. Children with fetal alcohol syndrome tend to have mental deficiencies and characteristic facial features.

- An **association** refers to the nonrandom appearance of two or more anomalies that occur together more frequently than by chance alone, but for which the etiology (origin and cause) has not been determined. Examples include heart and blood vessel defects, retarded growth and development, and genital and urinary system anomalies. Associations are important because, although they do not constitute a diagnosis, recognition of one or more of the components promotes the search for others in the group.

Table 3.3	Fetal Stage of Development

Time Period	Major Events
Weeks 9–12 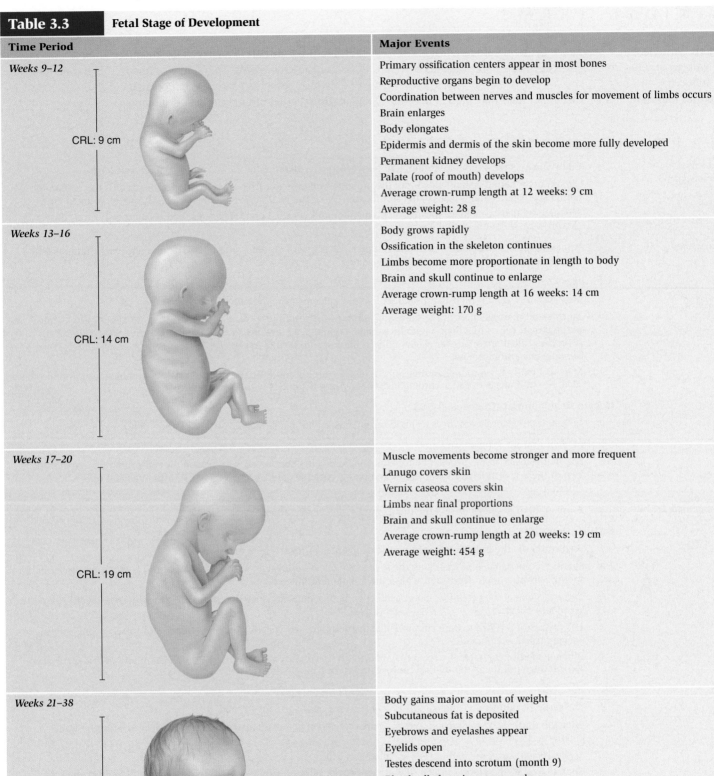CRL: 9 cm	Primary ossification centers appear in most bones Reproductive organs begin to develop Coordination between nerves and muscles for movement of limbs occurs Brain enlarges Body elongates Epidermis and dermis of the skin become more fully developed Permanent kidney develops Palate (roof of mouth) develops Average crown-rump length at 12 weeks: 9 cm Average weight: 28 g
Weeks 13–16 CRL: 14 cm	Body grows rapidly Ossification in the skeleton continues Limbs become more proportionate in length to body Brain and skull continue to enlarge Average crown-rump length at 16 weeks: 14 cm Average weight: 170 g
Weeks 17–20 CRL: 19 cm	Muscle movements become stronger and more frequent Lanugo covers skin Vernix caseosa covers skin Limbs near final proportions Brain and skull continue to enlarge Average crown-rump length at 20 weeks: 19 cm Average weight: 454 g
Weeks 21–38 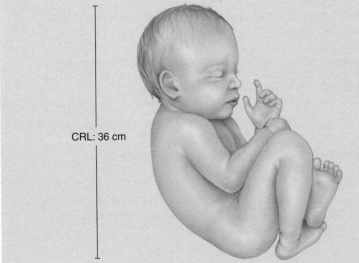CRL: 36 cm	Body gains major amount of weight Subcutaneous fat is deposited Eyebrows and eyelashes appear Eyelids open Testes descend into scrotum (month 9) Blood cells form in marrow only Average crown-rump length at 38 weeks: 36 cm Average *total* length at 38 weeks: 53 cm Average weight: 2.5–4.5 kg

CLINICAL TERMS

abortion Termination of pregnancy by premature removal of the embryo or fetus from the uterus; may be spontaneous or induced.

congenital anomaly A malformation or deformity present at birth.

ectopic pregnancy A pregnancy in which the embryo implants outside the uterus; commonly occurs in the uterine tube, in which case it is called a tubal pregnancy.

gestation Period of intrauterine development.

CHAPTER SUMMARY

Overview of Embryology 55	■ Embryology is the study of development between the fertilization of the secondary oocyte and birth. ■ The prenatal period is subdivided into the pre-embryonic period (the 2 weeks after fertilization), the embryonic period (between 3 and 8 weeks after fertilization), and the fetal period (the remaining 30 weeks prior to birth). ■ The process that takes place during the prenatal period is called embryogenesis, and it consists of three stages: cleavage, gastrulation, and organogenesis.
Gametogenesis 56	■ Human somatic cells contain a diploid number (23 pairs) of chromosomes: 22 pairs of autosomes (homologous chromosomes) and one pair of sex chromosomes. **Meiosis 57** ■ Meiosis is sex cell division that produces haploid gametes from diploid parent cells. ■ Mitosis and meiosis differ in the following ways: Mitosis produces two diploid cells that are genetically identical to the parent cell, while meiosis produces four haploid cells that are genetically different from the parent cell. A process called crossing over occurs only in meiosis and results in the exchange of genetic material between homologous chromosomes. ■ Meiosis involves two rounds of division. In meiosis I, crossing over occurs, and pairs of homologous chromosomes separate. In meiosis II, sister chromatids separate in the same way they do in mitosis. **Oocyte Development (Oogenesis) 58** ■ In females, oogenesis produces a single secondary oocyte, which has 22 autosomes and one X chromosome. **Sperm Development (Spermatogenesis) 59** ■ Spermatogenesis in the testes produces four haploid cells (sperm cells) from each diploid primary spermatocyte. Two of these sperm have 22 autosomes and one X chromosome, and two have 22 autosomes and one Y chromosome.
Pre-embryonic Period 60	■ A pre-embryo develops during the first 2 weeks after fertilization. **Fertilization 62** ■ Fertilization is the process whereby two sex cells fuse to form a new organism. ■ Sperm are not capable of fertilizing the secondary oocyte until they undergo capacitation, a conditioning period for sperm cells after their deposition within the female reproductive tract. ■ The acrosome at the tip of the sperm cell contains digestive enzymes that penetrate the protective layers around the secondary oocyte. ■ The phases of fertilization are corona radiata penetration, zona pellucida penetration, and fusion of the sperm and oocyte plasma membranes. ■ When a sperm penetrates the secondary oocyte, the secondary oocyte completes meiosis II and becomes an ovum. The pronuclei of the ovum and sperm fuse and form a single diploid cell called a zygote. **Cleavage 63** ■ The series of mitotic divisions of the zygote is called cleavage. ■ After the third cleavage division, the pre-embryo cells form a compressed ball of cells held together by tight junctions, a process called compaction. ■ At the 16-cell stage, the ball of cells is called a morula. ■ Upon arrival in the uterine lumen, the pre-embryo develops a single central fluid-filled cavity and is called a blastocyst. Cells within the blastocyst form the embryoblast (inner cell mass), which gives rise to the embryo proper. The outer ring of cells forms the trophoblast, which contributes to the placenta. **Implantation 63** ■ Implantation consists of attachment, cell changes in the trophoblast and uterine epithelium, and invasion of the endometrial wall. ■ The trophoblast subdivides into a cytotrophoblast (the inner cellular layer of the trophoblast) and a syncytiotrophoblast (the outer, thick layer). **Formation of the Bilaminar Germinal Disc 64** ■ During the implantation of the blastocyst into the endometrium of the uterus, cells of the embryoblast differentiate into two layers: the hypoblast and the epiblast. Together, these two layers form a flat disc called a bilaminar germinal disc.

Formation of Extraembryonic Membranes   65

■ The yolk sac is the site for the formation of the first blood cells and blood vessels.

■ The amnion is a thin ectodermal membrane that eventually encloses the entire embryo in a fluid-filled sac.

■ The chorion is the embryonic contribution to the placenta.

Development of the Placenta   66

■ The main functions of the placenta are the exchange of metabolic products and respiratory gases between the fetal and maternal bloodstreams, transmission of maternal antibodies, and hormone production.

Embryonic Period   67

■ The embryonic period extends from weeks 3 to 8.

Gastrulation   68

■ Gastrulation produces three primary germ layers: ectoderm, mesoderm, and endoderm.

Folding of the Embryonic Disc   68

■ The embryonic disc undergoes cephalocaudal and transverse folding, beginning late in the third week.

Differentiation of Ectoderm   69

■ The development of ectoderm into the entire nervous system is called neurulation.

■ Neurulation is an example of induction, a complex process whereby a structure stimulates a response from another tissue or group of cells.

■ Besides the nervous system, derivatives of the ectoderm include most exocrine glands, tooth enamel, epidermis, and the sense organs.

Differentiation of Mesoderm   72

■ The notochord is formed by chordamesoderm.

■ Paraxial mesoderm forms somites, which eventually give rise to the axial skeleton and most muscle, cartilage, dermis, and connective tissues.

■ Intermediate mesoderm forms most of the urinary system and the reproductive system.

■ Lateral plate mesoderm gives rise to most of the cardiovascular system, body cavity linings, and most limb structures.

■ Head mesenchyme forms connective tissues and musculature of the face.

Differentiation of Endoderm   72

■ Endoderm gives rise to the inner lining of the digestive, respiratory, and urinary tracts as well as the thyroid, parathyroid, and thymus glands, portions of the palatine tonsils, and most of the liver, gallbladder, and pancreas.

Organogenesis   72

■ Almost all of organogenesis (organ development) occurs during the embryonic period.

■ During the embryonic period, the embryo is very sensitive to teratogens, substances that can cause birth defects or the death of the embryo.

Fetal Period   74

■ The time from the beginning of the third month to birth is known as the fetal period. It is characterized by maturation of tissues and organs and rapid growth.

CHALLENGE YOURSELF

Matching

Match each numbered item with the most closely related lettered item.

_____ 1. amnion

_____ 2. paraxial mesoderm

_____ 3. gastrulation

_____ 4. neural tube

_____ 5. chorion

_____ 6. morula

_____ 7. intermediate mesoderm

_____ 8. hypoblast

_____ 9. blastocyst

_____ 10. zygote

a. fetal portion of placenta

b. formation of three primary germ layers

c. gives rise to most of urinary system

d. single cell produced by fertilization

e. fluid-filled membranous sac around fetus

f. forms brain and spinal cord

g. structure that implants in the uterus

h. forms muscle and the axial skeleton

i. solid ball of cells during cleavage

j. cell layer facing yolk sac in bilaminar germinal disc

Multiple Choice

Select the best answer from the four choices provided.

_____ 1. Fertilization of the secondary oocyte normally occurs in the
   a. ovary.
   b. uterine tube.
   c. uterus.
   d. vagina.

_____ 2. The outer layer of the blastocyst that attaches to the wall of the uterus at implantation is called the
   a. amnion.
   b. yolk sac.
   c. embryoblast.
   d. trophoblast.

_____ 3. At about day 3 after fertilization, the cells of the pre-embryo adhere tightly to each other and increase their surface contact in a process called
   a. implantation.
   b. compaction.
   c. gastrulation.
   d. neurulation.

_____ 4. Somites develop from
   a. paraxial mesoderm.
   b. intermediate mesoderm.
   c. lateral plate mesoderm.
   d. head mesenchyme.

_____ 5. During gastrulation, cells from the _____ layer of the bilaminar germinal disc migrate and form the three primary germ layers.
   a. notochord
   b. hypoblast
   c. epiblast
   d. mesoblast

_____ 6. An abnormal number of chromosomes in a cell following meiosis occurs as a result of
   a. synapsis.
   b. nondisjunction.
   c. crossing over.
   d. reduction division.

_____ 7. The cells of the embryoblast differentiate into the _____ and the _____.
   a. epiblast, hypoblast
   b. cytotrophoblast, syncytiotrophoblast
   c. amnioblast, epiblast
   d. epiblast, cytotrophoblast

_____ 8. Which of the following is not an extraembryonic membrane?
   a. amnion
   b. mesoderm
   c. chorion
   d. yolk sac

_____ 9. Capacitation occurs when sperm cells
   a. move through the uterine tubes.
   b. are mixed with secretions from the testes.
   c. are deposited within the female reproductive tract.
   d. are traveling through the penis.

_____ 10. The beginning of brain and spinal cord formation is termed
   a. nodal invagination.
   b. organogenesis.
   c. neurulation.
   d. gastrulation.

Content Review

1. Briefly describe the process of meiosis, mentioning a hallmark event that occurs during each phase.

2. What are the three phases of fertilization?

3. Describe the implantation of the blastocyst into the uterine wall.

4. What is the function of the chorion?

5. What important event occurs with the formation of the primitive streak?

6. Describe the formation of the primary germ layers.

7. Explain how the cylindrical body shape of the human embryo is derived from the flat embryonic disc.

8. List the five regions of the mesoderm, and identify some major body parts derived from each region.

9. Explain why teratogens are especially harmful to the developing organism during the embryonic period. What events occur during this period?

10. Describe the differences between the embryonic period and the fetal period.

Developing Critical Reasoning

1. Jennifer is a 37-year-old woman who is just over 3 months pregnant. She is the mother of three healthy children. Her obstetrician recommends that she be checked to see if her developing fetus is trisomic. What procedure is used to check for trisomy in the fetus? If a trisomic condition is detected, what is the most common cause, and how did it occur?

2. In the late 1960s, a number of pregnant women in Europe and Canada were prescribed a drug called thalidomide. Many of these women gave birth to children with amelia (no limbs) or meromelia (malformed upper and/or lower limbs). It was later discovered that thalidomide is a teratogen that can cause limb defects in an unborn baby. Based on this information, during what period of their pregnancy (pre-embryonic, embryonic, or fetal) do you think these women took thalidomide? During which of these periods would thalidomide cause the most harm to limb development?

3. A 22-year-old woman consumes large quantities of alcohol at a party and loses consciousness. Three weeks later, she misses her second consecutive period, and a pregnancy test is positive. Should she be concerned about the effects of her binge drinking episode on her baby?

ANSWERS TO "WHAT DO YOU THINK?"

1. Crossing over helps "shuffle" the genetic material between maternal and paternal chromosomes in the cell. Moving genetic material between these chromosomes forms genetically diverse sex cells that have the potential to produce a human being with its own unique array of traits.

2. If two sperm penetrate the secondary oocyte, the condition called polyspermy occurs. In this case, the fertilized cell contains 23 *triplets* of chromosomes, instead of the normal 23 pairs of chromosomes. In humans, no cell with 23 triplets of chromosomes can survive, so this fertilized cell will not survive.

3. If gastrulation did not occur properly and one of the primary germ layers wasn't formed, the embryo would not be able to survive. Each of the primary germ layers forms tissues that are vital for life, and without all of these tissues, there is no way the embryo could survive.

Visit the McKinley/O'Loughlin *Human Anatomy,* 2e website at aris.mhhe.com

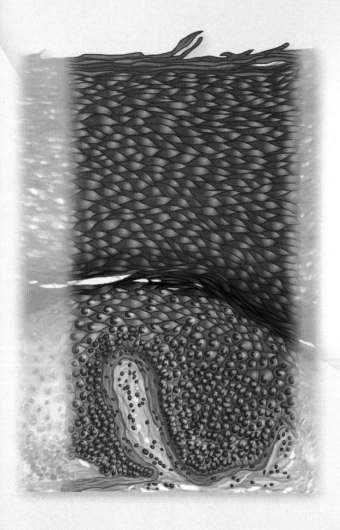

4

Tissue Level of Organization

The human body is composed of trillions of cells, which are organized into more complex units called tissues. **Tissues** are groups of similar cells and extracellular products that carry out a common function, such as providing protection or facilitating body movement. The study of tissues and their relationships within organs is called **histology**.

There are four principal types of tissues in the body: epithelial tissue, connective tissue, muscle tissue, and nervous tissue. Immediately following the connective tissue discussion, a section on Body Membranes has been inserted because these structures are composed of an epithelial sheet and an underlying connective tissue layer. Tissues are formed from the three primary germ layers (ectoderm, mesoderm, and endoderm). The four tissue types vary in terms of the structure of their specialized cells, the functions of these cells, and the presence of an **extracellular matrix** (mā′triks, mat′riks; *matrix* = womb) that is produced by the cells and surrounds them. The extracellular matrix is composed of varying amounts of water, protein fibers, and dissolved macromolecules. Its consistency ranges from fluid to quite solid. Epithelial, muscle, and nervous tissues have relatively little matrix between their cells. In contrast, connective tissue types contain varying amounts of extracellular matrix that exhibit differences in the volume of space occupied, the relative amounts of the extracellular matrix components, and the consistency (fluid to solid) of the extracellular matrix.

As we examine each of the four classes of tissues in this chapter, it may help you to refer to **table 4.1**, which summarizes their characteristics and functions. This chapter is a transition between chapter 2, which investigated the nature of cells, and later chapters, which examine tissue interactions in organs and organ systems.

Epithelial Tissue

Key topics in this section:

- Structure and function of each type of epithelial tissue
- Body locations where each type of epithelial tissue is found
- Specialized features of an epithelium
- Classification of exocrine glands

Epithelial (ep-i-thē′lē-ăl; *epi* = upon, *thēlē* = nipple) **tissue** covers or lines every body surface and all body cavities; thus it forms both the external and internal lining of many organs, and it constitutes the majority of glands. An **epithelium** (pl., epithelia) is composed of one or more layers of closely packed cells between two compartments having different components. There is little to no extracellular matrix between epithelial cells; additionally, no blood vessels penetrate an epithelium.

Characteristics of Epithelial Tissue

All epithelia exhibit several common characteristics:

- **Cellularity**. Epithelial tissue is composed almost entirely of cells. The cells of an epithelium are bound closely together by different types of intercellular junctions (discussed later). A minimal amount of extracellular matrix separates the cells in an epithelium.
- **Polarity**. Every epithelium has an **apical** (āp′i-kăl) **surface** (free, or top, surface), which is exposed either to the external environment or to some internal body space, and **lateral surfaces** having intercellular junctions. Additionally, each epithelium has a **basal** (bā′săl) **surface** (fixed, or bottom, surface) where the epithelium is attached to the underlying connective tissue.
- **Attachment**. At the basal surface of an epithelium, the epithelial layer is bound to a thin basement membrane, a complex molecular structure produced by both the epithelium and the underlying connective tissue.
- **Avascularity**. All epithelial tissues lack blood vessels. Epithelial cells obtain nutrients either directly across the apical surface or by diffusion across the basal surface from the underlying connective tissue.
- **Innervation**. Epithelia are richly innervated to detect changes in the environment at a particular body or organ surface region. Most nervous tissue is in the underlying connective tissue.

Table 4.1	Tissue Types			
Type	**General Characteristics**	**General Functions**	**Primary Germ Layer Derivative**	**Example Subtypes and Their Locations**
Epithelial tissue	Cellular, polar, attached, avascular, innervated, high regeneration capacity	Covers surfaces; lines insides of organs and body cavities	Ectoderm, mesoderm, endoderm	*Simple columnar epithelium:* Inner lining of digestive tract *Stratified squamous epithelium:* Epidermis of skin *Transitional epithelium:* Inner lining of urinary bladder
Connective tissue	Diverse types; all contain cells, protein fibers, and ground substance	Protects, binds together, and supports organs	Mesoderm	*Adipose connective tissue:* Fat *Dense regular connective tissue:* Ligaments and tendons *Dense irregular connective tissue:* Dermis of skin *Hyaline cartilage:* Articular cartilage in some joints *Fluid connective tissue:* Blood
Muscle tissue	Contractile; receives stimulation from nervous system and/or endocrine system	Facilitates movement of skeleton or organ walls	Mesoderm	*Skeletal muscle:* Muscles attached to bones *Cardiac muscle:* Muscle layer in heart *Smooth muscle:* Muscle layer in digestive tract
Nervous tissue	*Neurons:* Excitable, high metabolic rate, extreme longevity, nonmitotic *Glial cells:* Nonexcitable, mitotic	*Neurons:* Control activities, process information *Glial cells:* Support and protect neurons	Ectoderm	*Neurons:* Brain and spinal cord *Glial cells:* Brain and spinal cord

- **High regeneration capacity**. Because epithelial cells have an apical surface that is exposed to the environment, they are frequently damaged or lost by abrasion. However, damaged or lost epithelial cells generally are replaced as fast as they are lost because epithelia have a high regeneration capacity. The continual replacement occurs through the mitotic divisions of the deepest epithelial cells (called stem cells), which are found within the epithelium near its base.

Functions of Epithelial Tissue

Epithelia may have several functions, although no single epithelium performs all of them. These functions include:

- **Physical protection**. Epithelial tissues protect both exposed and internal surfaces from dehydration, abrasion, and destruction by physical, chemical, or biological agents.
- **Selective permeability**. All epithelial cells act as "gatekeepers," in that they regulate the movement of materials into and out of certain regions of the body. All substances that enter or leave the body must pass through the epithelium. Sometimes an epithelium exhibits a range of permeability; that is, it may be relatively impermeable to some substances, while at the same time promoting and assisting the passage of other molecules by absorption or secretion. The structure and characteristics of an epithelium may change as a result of applied pressure or stress; for example, walking around without shoes may increase the thickness of calluses on the bottom of the feet, which could alter or reduce the movement of materials across the epithelium.
- **Secretions**. Some epithelial cells, called exocrine glands, are specialized to produce secretions. Individual gland cells may be scattered among other cell types in an epithelium, or a large group of epithelial secretory cells may form a gland to produce specific secretions.
- **Sensations**. Epithelial tissues contain some nerve endings to detect changes in the external environment at their surface. These sensory nerve endings and those in the underlying connective tissue continuously supply information to the nervous system concerning touch, pressure, temperature, and pain. For example, receptors in the epithelium of the skin respond to pressure by stimulating adjacent sensory nerves. Additionally, several organs contain a specialized epithelium, called a neuroepithelium, that houses specific cells responsible for the senses of sight, taste, smell, hearing, and equilibrium.

WHAT DO YOU THINK?

1 Why do you think epithelial tissue does not contain blood vessels? Can you think of an epithelial function that could be compromised if blood vessels were running through the tissue?

Specialized Structure of Epithelial Tissue

Because epithelial tissues are located at all free surfaces in the body, they exhibit distinct structural specializations. An epithelium rests on a layer of connective tissue and adheres firmly to it. This secures the epithelium in place and prevents it from tearing. Between the epithelium and the underlying connective tissue is a thin extracellular layer called the **basement membrane**. The basement membrane consists of two specific layers: the basal lamina and the reticular lamina (**figure 4.1a**). The **basal lamina** contains collagen fibers as well as specific proteins and carbohydrates that are secreted by the cells of the epithelium. Cells in the connective tissue underlying the

epithelium secrete the **reticular lamina,** which contains protein fibers and both specific proteins and carbohydrates secreted by connective tissue cells. Together, these components of the basement membrane strengthen the attachment and form a selective molecular barrier between the epithelium and the underlying connective tissue.

The basement membrane has the following functions:

- Providing physical support for the epithelium.
- Anchoring the epithelium to the connective tissue.
- Acting as a barrier to regulate the movement of large molecules between the epithelium and the underlying connective tissue.

Intercellular Junctions

Epithelial cells are strongly bound together by specialized connections in the plasma membranes of their lateral surfaces called **intercellular junctions**. There are four types of junctions: tight junctions, adhering junctions, desmosomes, and gap junctions (figure 4.1b). Each of these types of junctions has a specialized structure.

Tight Junctions A **tight junction,** also called a *zonula* (zō′nū-lă) *occludens* ("occluding belt"), encircles epithelial cells near their apical surface and completely attaches each cell to its neighbors. Plasma membrane proteins among neighboring cells fuse, so the apical surfaces of the cells are tightly connected everywhere around the cell. This seals off the intercellular space and prevents substances from passing between the epithelial cells. The tight junction forces all materials to move *through*, rather than *between*, the epithelial cells in order to cross the epithelium. Thus, epithelial cells control whatever enters and leaves the body by moving across the epithelium. For example, in the small intestine, tight junctions prevent digestive enzymes that degrade molecules from moving between epithelial cells into underlying connective tissue.

Adhering Junctions An **adhering junction,** also called a *zonula adherens* ("adhesion belt"), is formed completely around the cell. This type of junction occurs when extensive zones of microfilaments extend from the cytoplasm into the plasma membrane, forming a supporting and strengthening belt within the plasma membrane that completely encircles the cell immediately adjacent to all of its neighbors. Typically, adhering junctions are located deep to the tight junctions; the anchoring of the microfilament proteins within this belt provides the only means of junctional support for the apical surface of the cell. The ultra-strong tight junctions are only needed near the apical surface and not along the entire length of the cell. Once neighboring cells are fused together by the tight junctions near the apical surface, the adhering junctions support the apical surface and provide for a small space between neighboring cells in the direction of the basal surface. Thus, the junction affords a passageway between cells for materials that have already passed through the apical surface of the epithelial cell and can then exit through the membranes on the lateral surface and continue their journey toward the basement membrane.

Desmosomes A **desmosome** (dez′mō-sōm; *desmos* = a band, *soma* = body), also called a *macula adherens* ("adhering spot"), is like a button or snap between adjacent epithelial cells. Each cell contributes half of the complete desmosome. It is a small region that holds cells together and provides resistance to mechanical stress at a single point, but it does not totally encircle the cell. In contrast to tight junctions, which encircle the cell to secure it to its neighbors everywhere around its periphery, the desmosome only attaches a cell to its neighbors at potential stress points. The neighboring cells are separated by a small space

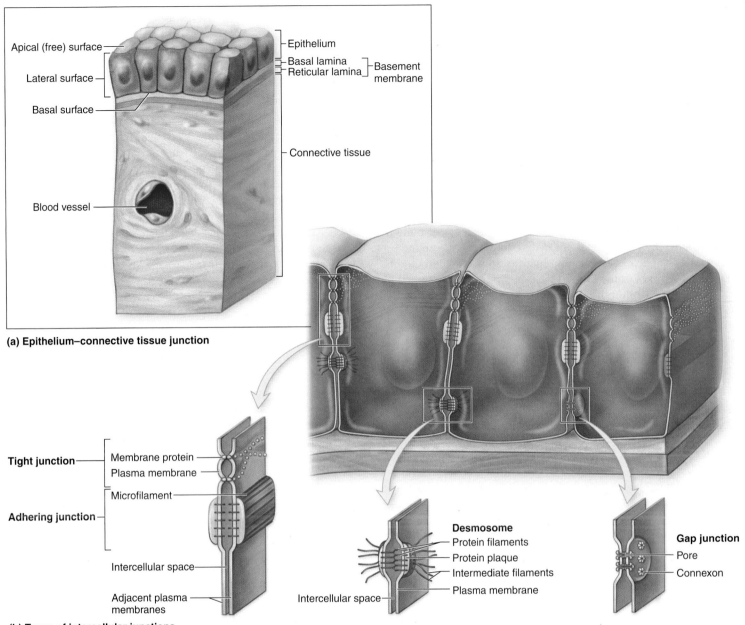

(a) Epithelium–connective tissue junction

Apical (free) surface
Lateral surface
Basal surface
Blood vessel

Epithelium
Basal lamina ⎤
Reticular lamina ⎦ Basement membrane
Connective tissue

Tight junction
  Membrane protein
  Plasma membrane

Adhering junction
  Microfilament
  Intercellular space
  Adjacent plasma membranes

Desmosome
  Protein filaments
  Protein plaque
  Intermediate filaments
  Plasma membrane
Intercellular space

Gap junction
  Pore
  Connexon

(b) Types of intercellular junctions

Figure 4.1

Polarity and Intercellular Junctions in an Epithelium.   An epithelium exhibits polarity and has intercellular junctions only on the lateral surfaces of its individual cells. *(a)* The apical surface is the free surface of the cell exposed to a body cavity, an organ lumen, or the exterior of the body. The basal surface of the cell adheres to the underlying connective tissue by a basement membrane. *(b)* The lateral surfaces of the cell contain intercellular junctions. Types of intercellular junctions are tight junctions, adhering junctions, desmosomes, and gap junctions.

that is spanned by a fine web of protein filaments. These filaments anchor into a thickened protein plaque located at the internal surface of the plasma membrane. On the cytoplasmic side of each plaque, intermediate filaments of the cytoskeleton penetrate the plaque to extend throughout the cell the support and strength supplied between the cells by the desmosome. The basal cells of epithelial tissue exhibit structures called **hemidesmosomes,** half-desmosomes that anchor them to the underlying basement membrane.

Gap Junctions   A **gap junction** is formed across the intercellular gap between neighboring cells. This gap (about 2 nanometers in length) is bridged by structures called **connexons** (kon-neks′on). Each connexon consists of six transmembrane proteins, arranged in a circular fashion to form a tiny, fluid-filled tunnel or **pore**. Gap junctions provide a direct passageway for small molecules traveling

between neighboring cells. Ions, glucose, amino acids, and other small solutes can pass directly from the cytoplasm of one cell into the neighboring cell through these channels. The flow of ions between cells coordinates such cellular activities as the beating of cilia. Gap junctions are also seen in certain types of muscle tissue, where they help coordinate contraction activities.

WHAT DID YOU LEARN?

❶ Describe the two layers of the basement membrane and the origin of each.

❷ Which intercellular junction ensures that epithelial cells act as "gatekeepers"?

❸ What type of intercellular junction provides resistance to mechanical stress at a single point?

Classification of Epithelial Tissue

The body contains many different kinds of epithelia, and the classification of each type is indicated by a two-part name. The first part of the name refers to the *number* of epithelial cell layers, and the second part describes the *shape* of the cells at the apical surface of the epithelium.

Classification by Number of Cell Layers

Epithelia may be classified based on number of cell layers as either simple or stratified (**figure 4.2a**). A **simple epithelium** is one cell layer thick, and all of these epithelial cells are in direct contact with the basement membrane. Often, the apical surface is covered by a thin layer of fluid or mucus to prevent desiccation and help protect the cells from abrasion or friction. A simple epithelium is found in areas where stress is minimal and where filtration, absorption, or secretion is the primary function. Such locations include the linings of the air sacs in the lungs, intestines, and blood vessels.

A **stratified epithelium** contains two or more layers of epithelial cells. Only the cells in the deepest (basal) layer are in contact with the basement membrane. A stratified epithelium resembles a brick wall, where the bricks in contact with the ground represent the basal layer and the bricks at the top of the wall represent the apical layer. The multiple cell layers of a stratified epithelium make it strong and capable of resisting stress and protecting underlying tissue. In contrast to a simple epithelium, a stratified epithelium is found in areas likely to be subjected to abrasive activities or mechanical stresses, where two or more layers of cells are better able to resist this wear and tear (e.g., the internal lining of the esophagus, pharynx, or vagina). Cells in the basal layer continuously regenerate as the cells in the more superficial layer are lost due to abrasion or stress.

Finally, a **pseudostratified** (soo′dō-strat′i-fĭd; *pseudes* = false, *stratum* = layer) **epithelium** looks layered (stratified) because the cells' nuclei are distributed at different levels between the apical and basal surfaces. But although all of these epithelial cells are attached to the basement membrane, some of them do not reach its apical surface. Those cells that do reach the apical surface often bear cilia to move mucus along the surface. This so-called ciliated pseudostratified epithelium lines the nasal cavity and the respiratory passageways.

Classification by Cell Shape

Epithelia are also classified by the shape of the cell at the apical surface. In a simple epithelium, all of the cells display the same shape. However, in a stratified epithelium there is usually a difference in cell shape between the basal layer and the apical layer. Figure 4.2*b* shows the three common cell shapes observed in epithelia: squamous, cuboidal, and columnar. (Note that the cells in this figure all appear hexagonal when looking at their apical surface, or "en face"; thus these terms describe the cells' shapes when viewed laterally, or from the side.)

Squamous (skwā′mŭs; *squamosus* = scaly) **cells** are flat, wide, and somewhat irregular in shape. The nucleus looks like a flattened disc. The cells are arranged like irregular, flattened floor tiles. **Cuboidal** (kū-boy′dăl; *kybos* = cube, *eidos* = resemblance) **cells** are about as tall as they are wide. The cells do not resemble perfect "cubes," because they do not have squared edges. The cell nucleus is spherical and located within the center of the cell. **Columnar** (kol′ŭm′năr; *columna* = column) **cells** are slender and taller than they are wide. The cells look like a group of hexagonal columns aligned next to each other. Each cell nucleus is oval and usually oriented lengthwise and located in the basal region of the cell.

Another shape that occurs in epithelial cells is called **transitional** (tran-zish′ŭn-ăl; *transitio* = to go across). These cells can readily change their shape or appearance depending upon how stretched the epithelium becomes. They are found where the epithelium cycles between distended and relaxed states, such as in the lining of the bladder, which fills with urine and is later emptied. When the transitional epithelium is in a relaxed state, the cells are described as *polyhedral*, which means "many-sided" and reflects the ranges in shape that are possible in this type of epithelium. When transitional epithelium is stretched, the surface cells resemble squamous cells.

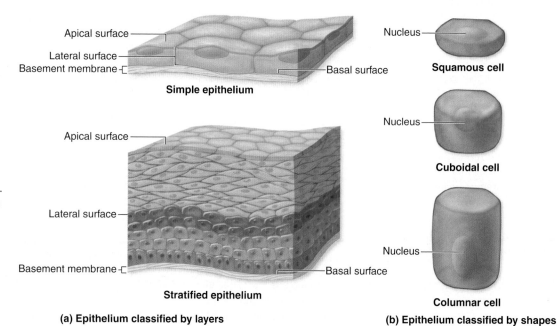

Figure 4.2

Classification of Epithelia. Two criteria are used to classify epithelia: the number of cell layers and the shape of the cell at the apical surface. (*a*) An epithelium is simple if it is one cell layer thick, and stratified if it has two or more layers of cells. (*b*) Epithelial cell shapes include squamous (thin, flattened cells), cuboidal (cells about as tall as they are wide), and columnar (cells taller than they are wide).

Apical surface — Lateral surface — Basement membrane — Basal surface

Simple epithelium

Apical surface — Lateral surface — Basement membrane — Basal surface

Stratified epithelium

(a) Epithelium classified by layers

Nucleus — **Squamous cell**

Nucleus — **Cuboidal cell**

Nucleus — **Columnar cell**

(b) Epithelium classified by shapes

Types of Epithelium

Using the classification system just described, epithelium can be broken down into the primary types shown in **table 4.2**. In this section, we describe the characteristics of these types of epithelium and show how each appears under the microscope.

Simple Squamous Epithelium

A **simple squamous epithelium** consists of a single layer of flattened cells (**table 4.3***a*). When viewed "en face," the irregularly shaped cells display a spherical to oval nucleus, and they appear tightly bound together in a mosaiclike pattern. Each squamous cell resembles a fried egg, with the nucleus representing the yolk.

This epithelium is extremely delicate and highly specialized to allow rapid movement of molecules across its surface by diffusion, osmosis, or filtration. Simple squamous epithelium is found only in protected regions where moist surfaces reduce friction and abrasion. For example, in the lining of the lung air sacs (alveoli), the thin epithelium is well suited for the exchange of oxygen and carbon dioxide between the blood and inhaled air. This type of epithelium is also found lining the lumen (inside space) of blood vessel walls, where it allows for rapid exchange of nutrients and waste between the blood and the interstitial fluid surrounding the blood vessels.

Simple squamous epithelia that line closed internal body cavities and all circulatory structures have special names. The simple squamous epithelium that lines the lumen of the blood and lymphatic vessels and the heart and its chambers is termed **endothelium** (en-dō-thē′-lē-ŭm; *endon* = within, *thele* = nipple). **Mesothelium** (mez-ō-thē′lē-ŭm; *mesos* = middle) is the simple squamous epithelium of the **serous membrane** (discussed in chapter 1) that lines the internal walls of the pericardial, pleural, and peritoneal cavities as well as the external surfaces of the organs within those cavities. Mesothelium gets its name from the primary germ layer mesoderm, from which it is derived.

Simple Cuboidal Epithelium

A **simple cuboidal epithelium** consists of a single layer of cells that are as tall as they are wide (table 4.3*b*). A spherical nucleus is located in the center of the cell. A simple cuboidal epithelium functions primarily to absorb fluids and other substances across its apical membrane and to secrete specific molecules. It forms the walls of kidney tubules, where it participates in the reabsorption of nutrients, ions, and water that are filtered out of the blood plasma. It also forms the ducts of exocrine glands, which secrete materials. Simple cuboidal epithelium covers the surface of the ovary and also lines the follicles of the thyroid gland.

Table 4.2	Types of Epithelium
Type	**Structure**
SIMPLE EPITHELIUM	One cell layer thick; all cells are tightly bound; all cells attach directly to the basement membrane
Simple squamous	One layer of flattened cells
Simple cuboidal	One layer of cells about as tall as they are wide
Simple columnar, nonciliated	One layer of nonciliated cells that are taller than they are wide; cells may contain microvilli
Simple columnar, ciliated	One layer of ciliated cells that are taller than they are wide
STRATIFIED EPITHELIUM	Two or more cell layers thick; only the deepest layer directly attaches to the basement membrane
Stratified squamous, keratinized	Many layers thick; cells in surface layers are dead, flat, and filled with the protein keratin
Stratified squamous, nonkeratinized	Many layers thick; no keratin in cells; surface layers are alive, flat, and moist
Stratified cuboidal	Two or more layers of cells; apical layer of cells is cuboidal-shaped
Stratified columnar	Two or more layers of cells; cells in apical layer are columnar-shaped
OTHER TYPES OF EPITHELIUM	Cell layers vary, from single to many
Pseudostratified columnar	One layer of cells of varying heights; all cells attach to basement membrane; ciliated form contains cilia and goblet cells; nonciliated form lacks cilia and goblet cells
Transitional	Multiple layers of polyhedral cells (when tissue is relaxed) or flattened cells (when tissue is distended); some cells may be binucleated

Table 4.3	Simple Epithelia

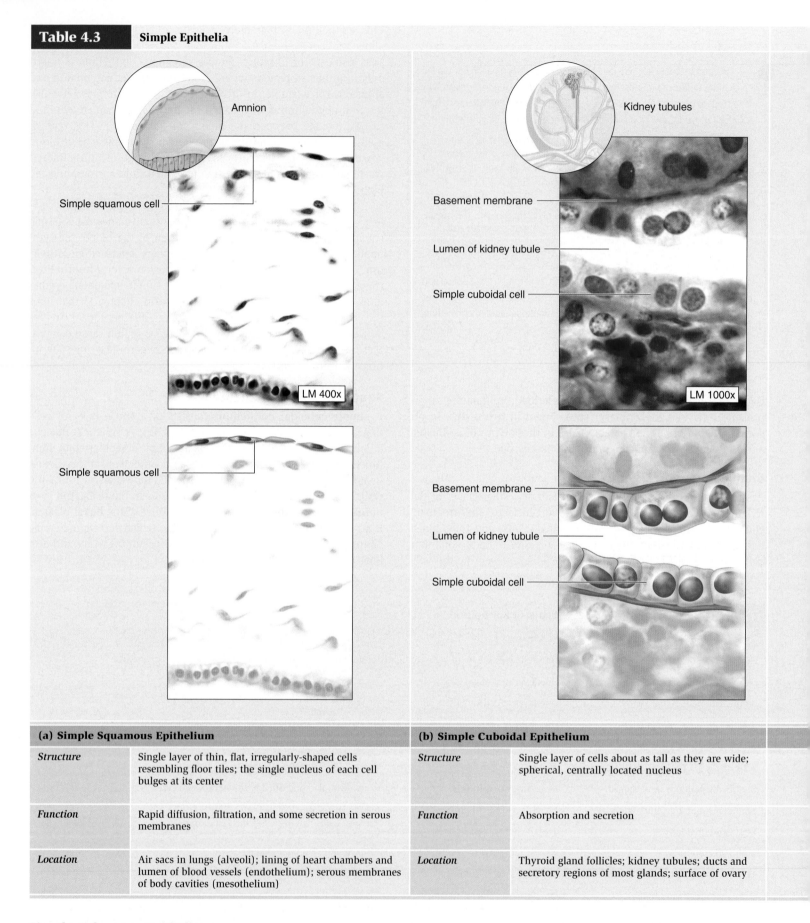

Amnion

Simple squamous cell

LM 400x

Simple squamous cell

Kidney tubules

Basement membrane

Lumen of kidney tubule

Simple cuboidal cell

LM 1000x

Basement membrane

Lumen of kidney tubule

Simple cuboidal cell

(a) Simple Squamous Epithelium

Structure	Single layer of thin, flat, irregularly-shaped cells resembling floor tiles; the single nucleus of each cell bulges at its center
Function	Rapid diffusion, filtration, and some secretion in serous membranes
Location	Air sacs in lungs (alveoli); lining of heart chambers and lumen of blood vessels (endothelium); serous membranes of body cavities (mesothelium)

(b) Simple Cuboidal Epithelium

Structure	Single layer of cells about as tall as they are wide; spherical, centrally located nucleus
Function	Absorption and secretion
Location	Thyroid gland follicles; kidney tubules; ducts and secretory regions of most glands; surface of ovary

Simple Columnar Epithelium

A **simple columnar epithelium** is composed of a single layer of tall, narrow cells. The nucleus is oval and located within the basal region of the cell. Active movement of molecules occurs across this type of epithelium by either absorption or secretion. Simple columnar epithelium has two forms; one type has no cilia, while the apical surface of the other type is lined with cilia.

Nonciliated simple columnar epithelium often contains **microvilli** and a scattering of unicellular glands called **goblet cells** (table 4.3c). Recall that microvilli are tiny, cytoplasmic projections on the

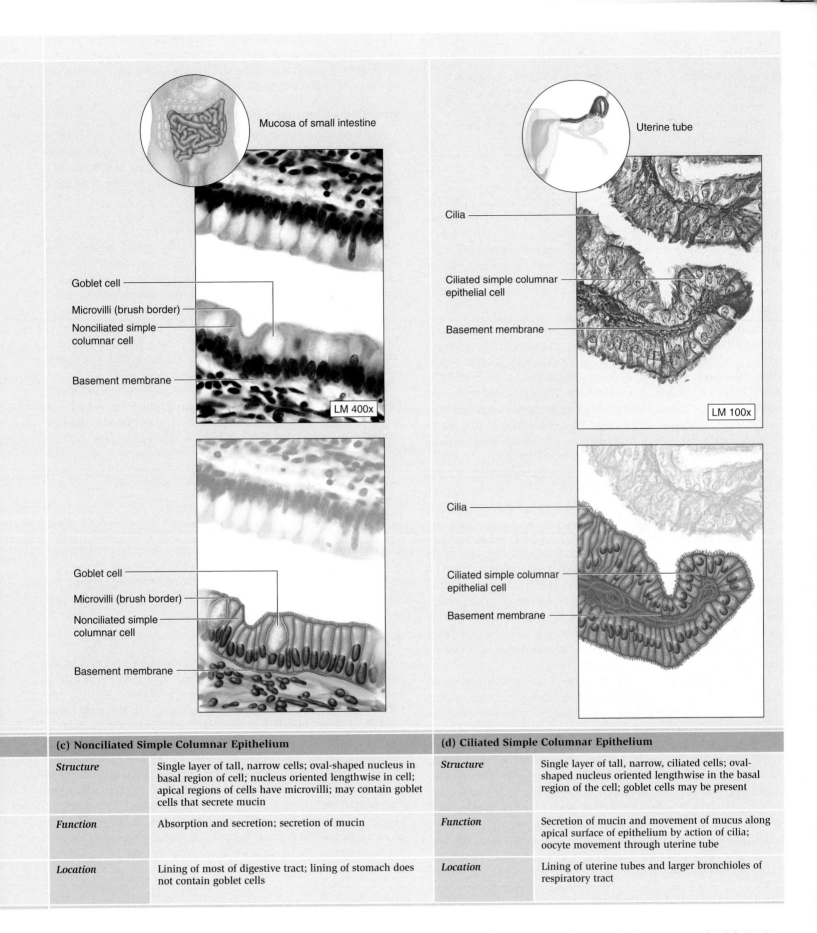

Mucosa of small intestine

Goblet cell

Microvilli (brush border)

Nonciliated simple columnar cell

Basement membrane

LM 400x

Goblet cell

Microvilli (brush border)

Nonciliated simple columnar cell

Basement membrane

Uterine tube

Cilia

Ciliated simple columnar epithelial cell

Basement membrane

LM 100x

Cilia

Ciliated simple columnar epithelial cell

Basement membrane

(c) Nonciliated Simple Columnar Epithelium	
Structure	Single layer of tall, narrow cells; oval-shaped nucleus in basal region of cell; nucleus oriented lengthwise in cell; apical regions of cells have microvilli; may contain goblet cells that secrete mucin
Function	Absorption and secretion; secretion of mucin
Location	Lining of most of digestive tract; lining of stomach does not contain goblet cells

(d) Ciliated Simple Columnar Epithelium	
Structure	Single layer of tall, narrow, ciliated cells; oval-shaped nucleus oriented lengthwise in the basal region of the cell; goblet cells may be present
Function	Secretion of mucin and movement of mucus along apical surface of epithelium by action of cilia; oocyte movement through uterine tube
Location	Lining of uterine tubes and larger bronchioles of respiratory tract

apical surface of the cell that increase the surface area for secretion and absorption. You cannot distinguish individual microvilli under the microscope; rather, the microvilli collectively appear as a darkened, fuzzy structure known as a **brush border**. Goblet cells secrete **mucin** (mū′sin; *mucus* = mucus), a glycoprotein that upon hydration (being mixed with water) forms mucus for lubrication. Nonciliated simple columnar epithelium lines most of the digestive tract, from the stomach to the anal canal.

In ciliated simple columnar epithelium, cilia project from the apical surfaces of the cells (table 4.3*d*). Mucus covers these apical

Study Tip!

— If you are having trouble distinguishing cilia from microvilli, recall that cilia appear under the light microscope like fine hairs extending from the apical surface of the cell, while microvilli are extensive folds of the plasma membrane that appear as a fuzzy, darkened brush border at the apical surface.

surfaces and is moved along by the beating of the cilia. Goblet cells typically are interspersed throughout this epithelium. This type of epithelium lines the luminal (internal) surface of the uterine tubes, where it helps move an oocyte from the ovary to the uterus. A ciliated simple columnar epithelium is also present in the bronchioles (smaller air tubes) of the lung.

Stratified Squamous Epithelium

A **stratified squamous epithelium** has multiple cell layers, and only the deepest layer of cells is in direct contact with the basement membrane. While the cells in the basal layers have a varied shape often described as polyhedral, the superficial cells at the apical surface display a flattened, squamous shape. Thus, stratified squamous epithelium is so named because of its multiple cell layers and the shape of its most superficial cells. This epithelium is adapted to protect underlying tissues from damage due to activities that are abrasive and cause friction. Stem cells in the basal layer continuously divide to produce a new stem cell and a committed cell that gradually moves toward the surface to replace the cells lost during protective activities. This type of epithelium exists in two forms: nonkeratinized and keratinized.

The cells in **nonkeratinized stratified squamous epithelium** remain alive all the way to its apical surface, and they are kept moist with secretions such as saliva or mucus. **Keratin,** a fibrous intracellular protein, is not present within the cells. Thus, since all of the cells are still alive, the flattened nuclei characteristic of squamous cells are visible even in the most superficial cells **(table 4.4a)**. Nonkeratinized stratified squamous epithelium lines the oral cavity (mouth), part of the pharynx (throat), the esophagus, the vagina, and the anus.

In **keratinized** (ker′ă-ti-nīzd; *keras* = horn) **stratified squamous epithelium**, the apical surface is composed of layers of cells that are dead; these cells lack nuclei and all organelles and are filled with tough, protective keratin. It is obvious that the superficial cells lack nuclei when they are viewed under the microscope (table 4.4b). New committed cells produced in the basal region of the epithelium migrate toward the apical surface. During their migration, they fill with keratin, lose their organelles and nuclei, and die. However, the keratin in these dead cells makes them very strong. Thus, there is a tradeoff with the appearance of keratin, in that the tissue becomes very strong, but the cells must die as a result. The epidermis (outer layer) of the skin consists of keratinized stratified squamous epithelium.

Stratified Cuboidal Epithelium

A **stratified cuboidal epithelium** contains two or more layers of cells, and the apical cells tend to be cuboidal in shape (table 4.4c).

This type of epithelium forms the walls of the larger ducts of most exocrine glands, such as the sweat glands in the skin. Although the function of stratified cuboidal epithelium is mainly protective, it also serves to strengthen the wall of gland ducts.

Stratified Columnar Epithelium

A **stratified columnar epithelium** is relatively rare in the body. It consists of two or more layers of cells, but only the apical surface cells are columnar in shape (table 4.4d). This type of epithelium is found in the large ducts of salivary glands and in the membranous segment of the male urethra.

Pseudostratified Columnar Epithelium

Pseudostratified columnar epithelium is so named because upon first glance, it appears to consist of multiple layers of cells. However, this epithelium is not really stratified, because all of its cells are in direct contact with the basement membrane. It may look stratified, but it is actually pseudostratified due to the fact that the nuclei are scattered at different distances from the basal surface but not all of the cells reach the apical surface **(table 4.5a)**. The columnar cells within this epithelium always reach the apical surface; the shorter cells are stem cells that give rise to the columnar cells.

There are two forms of pseudostratified columnar epithelium: **Pseudostratified ciliated columnar epithelium** has cilia on its apical surface, while **pseudostratified nonciliated columnar epithelium** lacks cilia. Both types of this epithelium perform protective functions. The ciliated form houses goblet cells, which secrete mucin that forms mucus. This mucus traps foreign particles and is moved along the apical surface by the beating of the cilia. Pseudostratified ciliated columnar epithelium lines much of the larger portions of the respiratory tract, including the nasal cavity, part of the pharynx (throat), the larynx (voice box), the trachea, and the bronchi. The cilia in this epithelium help propel dust particles and foreign materials away from the lungs and to the nose and mouth. In contrast, the nonciliated form of this epithelium has no goblet cells. It is a rare epithelium that occurs primarily in part of the male urethra and the epididymis.

Transitional Epithelium

A **transitional epithelium** varies in appearance, depending on whether it is in a relaxed or a stretched state (table 4.5b). In a relaxed state, the basal cells appear almost cuboidal, and the apical cells are large and rounded. During stretching, the transitional epithelium thins, and the apical cells continue to flatten, becoming almost squamous. In this distended state, it may be difficult to distinguish a transitional epithelium from a squamous epithelium. However, one distinguishing feature of transitional epithelium is the presence of a handful of **binucleated** (double-nucleus-containing) **cells.** This epithelium lines the urinary bladder, an organ that changes shape as it fills with urine. It also lines the ureters and the proximal part of the urethra. Transitional epithelium permits stretching and ensures that toxic urine does not seep into the underlying tissues and structures of these organs.

WHAT DO YOU THINK?

2 What types of epithelium are well suited for protection?

Table 4.4	**Stratified Epithelia**

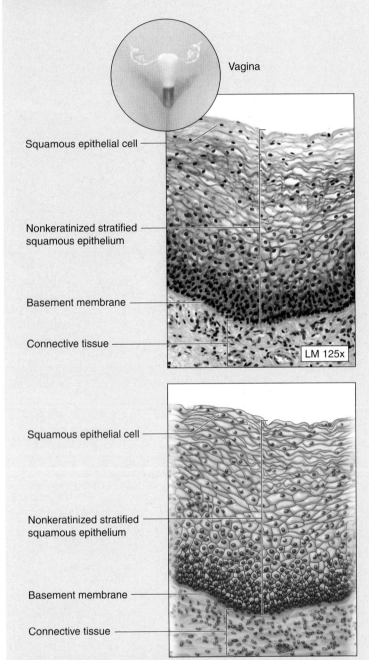

Vagina

Squamous epithelial cell

Nonkeratinized stratified squamous epithelium

Basement membrane

Connective tissue

LM 125x

Squamous epithelial cell

Nonkeratinized stratified squamous epithelium

Basement membrane

Connective tissue

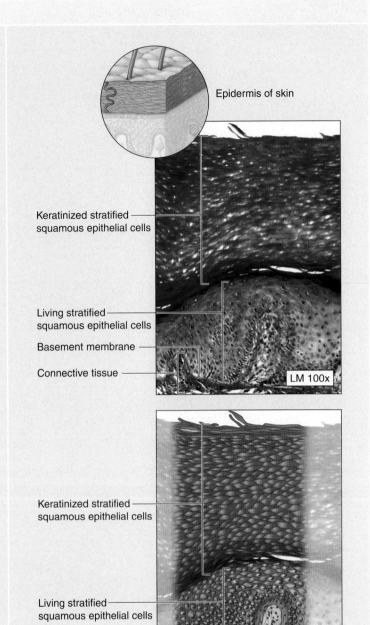

Epidermis of skin

Keratinized stratified squamous epithelial cells

Living stratified squamous epithelial cells

Basement membrane

Connective tissue

LM 100x

Keratinized stratified squamous epithelial cells

Living stratified squamous epithelial cells

Basement membrane

Connective tissue

(a) Nonkeratinized Stratified Squamous Epithelium		**(b) Keratinized Stratified Squamous Epithelium**	
Structure	Multiple layers of cells; basal cells typically are cuboidal or polyhedral, while apical (superficial) cells are squamous; surface cells are alive and kept moist	*Structure*	Multiple layers of cells; basal cells typically are cuboidal or polyhedral, while apical (superficial) cells are squamous; more superficial cells are dead and filled with the protein keratin
Function	Protection of underlying tissue	*Function*	Protection of underlying tissue
Location	Lining of oral cavity, part of pharynx, esophagus, vagina, and anus	*Location*	Epidermis of skin

(continued on next page)

Table 4.4	Stratified Epithelia *(continued)*

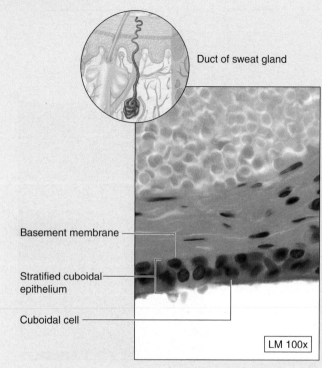

Duct of sweat gland

Basement membrane

Stratified cuboidal epithelium

Cuboidal cell

LM 100x

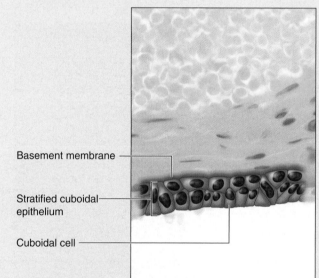

Basement membrane

Stratified cuboidal epithelium

Cuboidal cell

Columnar cell

Stratified columnar epithelium

Basement membrane

Connective tissue

Male urethra

LM 500x

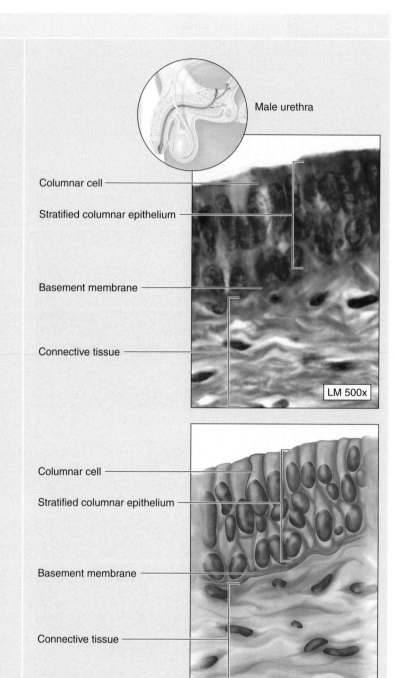

Columnar cell

Stratified columnar epithelium

Basement membrane

Connective tissue

(c) Stratified Cuboidal Epithelium		**(d) Stratified Columnar Epithelium**	
Structure	Two or more layers of cells; cells at the apical surface are cuboidal	*Structure*	Two or more layers of cells; cells at the apical surface are columnar
Function	Protection and secretion	*Function*	Protection and secretion
Location	Found in large ducts in most exocrine glands and in some parts of the male urethra	*Location*	Rare; found in large ducts of some exocrine glands and in some regions of the male urethra

Table 4.5	**Other Epithelia**

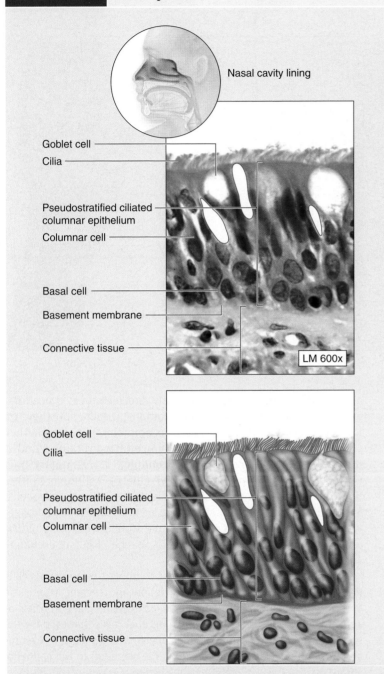

Nasal cavity lining

Goblet cell
Cilia
Pseudostratified ciliated columnar epithelium
Columnar cell
Basal cell
Basement membrane
Connective tissue
LM 600x

Goblet cell
Cilia
Pseudostratified ciliated columnar epithelium
Columnar cell
Basal cell
Basement membrane
Connective tissue

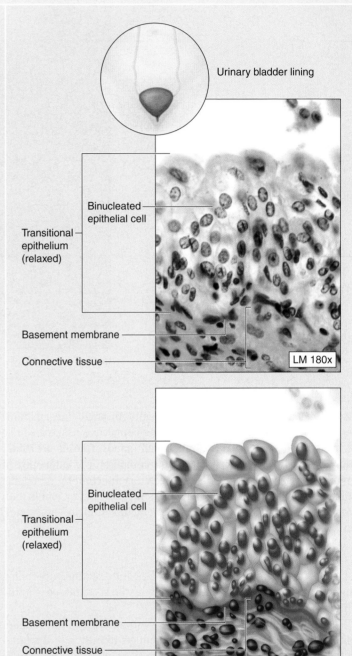

Urinary bladder lining

Binucleated epithelial cell
Transitional epithelium (relaxed)
Basement membrane
Connective tissue
LM 180x

Binucleated epithelial cell
Transitional epithelium (relaxed)
Basement membrane
Connective tissue

(a) Pseudostratified Columnar Epithelium		**(b) Transitional Epithelium**	
Structure	Single layer of cells with varying heights that appears multilayered; all cells connect to the basement membrane, but not all cells reach the apical surface. Ciliated form has goblet cells and cilia (*shown*); nonciliated form lacks goblet cells and cilia	*Structure*	Epithelial appearance varies, depending on whether the tissue is stretched or relaxed; shape of cells at apical surface changes; some cells may be binucleated
Function	Protection; ciliated form also involved in secretion of mucin and movement of mucus across surface by ciliary action	*Function*	Distention and relaxation to accommodate urine volume changes in bladder, ureters, and urethra
Location	Ciliated form lines most of respiratory tract, including nasal cavity, part of pharynx, larynx, trachea, bronchi. Nonciliated form is rare; lines epididymis and part of male urethra	*Location*	Lining of urinary bladder, ureters, and part of urethra

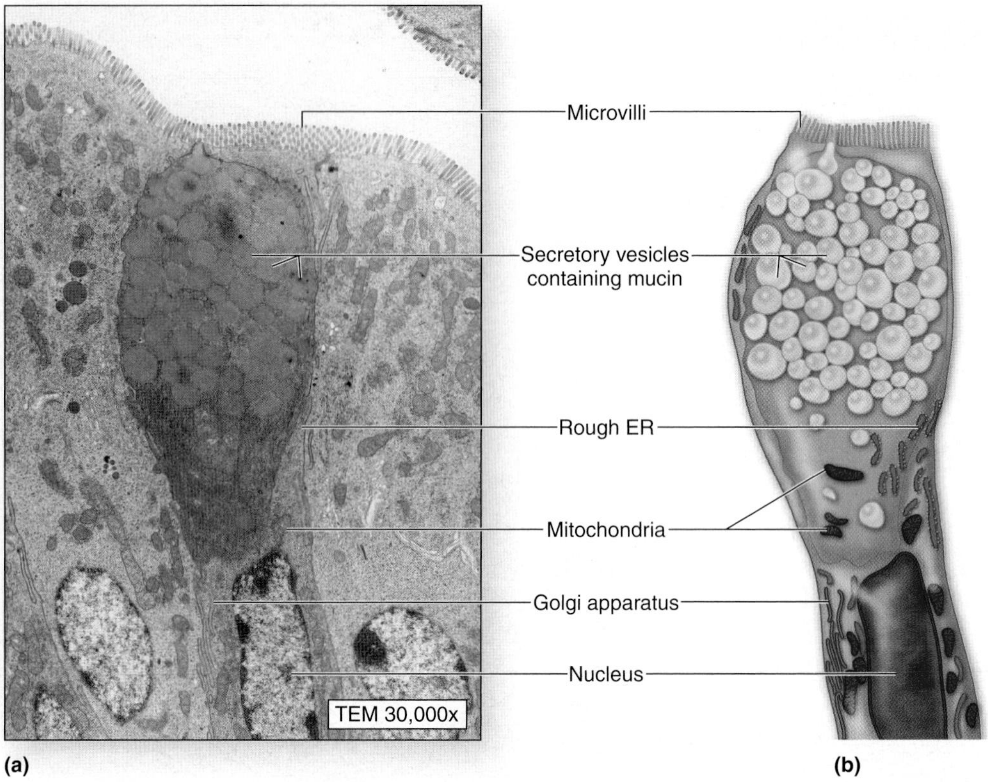

Figure 4.3

Goblet Cell: A Unicellular Exocrine Gland. *(a)* Photomicrograph and *(b)* diagram of a goblet cell in the small intestine.

Glands

As epithelial tissue develops in the embryo, small invaginations from this epithelium into the underlying connective tissue give rise to specialized secretory structures called glands. **Glands** are either individual cells or multicellular organs composed predominantly of epithelial tissue. Glands perform a secretory function by producing substances either for use elsewhere in the body or for elimination from the body. Glandular secretions include mucin, hormones, enzymes, and waste products.

Endocrine and Exocrine Glands

Glands are classified as either endocrine or exocrine, depending upon whether they have a duct connecting the secretory cells to the surface of an epithelium.

Endocrine (en′dō-krin; *endon* = within, *krino* = to separate) **glands** lack ducts and secrete their products directly into the interstitial fluid and bloodstream. The secretions of endocrine glands, called hormones, act as chemical messengers to influence cell activities elsewhere in the body. Endocrine glands are discussed in depth in chapter 20.

Exocrine (ek′sō-krin; *exo* = outside) **glands** typically originate from an invagination of epithelium that burrows into the deeper connective tissues. These glands usually maintain their contact with the epithelial surface by means of a **duct,** an epithelium-lined tube through which secretions of the gland are discharged onto the epithelial surface. This duct may secrete materials onto the surface of the skin (e.g., sweat from sweat glands or milk from mammary glands) or onto an epithelial surface lining an internal passageway (e.g., enzymes from the pancreas into the small intestine or saliva from the salivary glands into the oral cavity).

Exocrine Gland Structure

An exocrine gland may be unicellular or multicellular. A unicellular exocrine gland is an individual exocrine cell located within an epithelium that is predominantly nonsecretory. Unicellular exocrine glands typically do not contain a duct, and they are located close to the surface of the epithelium in which they reside. The most common type of unicellular exocrine gland is the goblet cell **(figure 4.3)**. For example, the respiratory tract is lined mainly by pseudostratified ciliated columnar epithelium, which also contains some mucin-secreting goblet cells. Mucus then coats the inner surface of the respiratory passageway to cover and protect its lining and to help warm, humidify, and cleanse the inhaled air before it reaches the gas exchange surfaces in the lungs.

Multicellular exocrine glands are composed of numerous cells that work together to produce a secretion and secrete it onto the surface of an epithelium. A multicellular exocrine gland consists of **acini** (as′i-nī; sing., as′i-nŭs; *acinus* = grape), sacs that produce the secretion, and one or more smaller ducts, which merge to eventually form a larger duct that transports the secretion to the epithelial surface **(figure 4.4)**. Acini are the secretory portions, while ducts are the conducting portions of these glands.

Most multicellular exocrine glands are enclosed within a fibrous **capsule**. Extensions of this capsule, called **septa** or *trabeculae*, partition the gland internally into compartments called **lobes**. Further subdivisions of the septa within each lobe form microscopic **lobules** (lob′ūl). The septa contain ducts, blood vessels, and nerves supplying the gland. The connective tissue framework of the gland is called the **stroma**. The stroma supports and organizes the **parenchyma** (pă-reng′ki-mă), the functional cells of the gland that produce and secrete the gland products. These cells are usually simple cuboidal or columnar epithelial cells. Multicellular exocrine glands are found in the mammary glands, pancreas, and salivary glands.

Classification of Exocrine Glands

Multicellular exocrine glands may be classified according to three criteria: (1) form and structure (*morphology*), which is

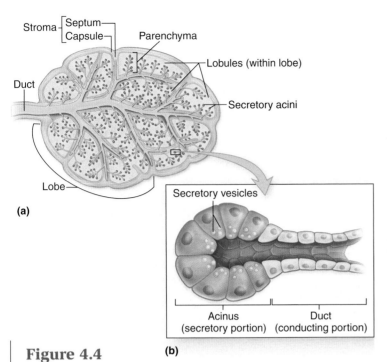

(a)

(b)

Figure 4.4

General Structure of Exocrine Glands.   *(a)* Exocrine glands have a connection called a duct that leads to an organ or body surface. Inside the gland, the duct branches repeatedly, following the connective tissue septa, until its finest divisions end on secretory acini. *(b)* The acinus is the secretory portion of the gland, and the duct is the conducting portion.

considered an anatomic classification, (2) type of secretion, and (3) method of secretion. The latter two are considered physiologic classifications.

Form and Structure   Based on the structure and complexity of their ducts, exocrine glands are considered either simple or compound. **Simple glands** have a single, unbranched duct; **compound glands** exhibit branched ducts.

Exocrine glands are also classified according to the shape or organization of their secretory portions. If the secretory portion and the duct are of uniform diameter, the gland is called **tubular**. If the secretory cells form an expanded sac, the gland is called **acinar** (as'i-nar). Finally, a gland with both secretory tubules and secretory acini is called a **tubuloacinar gland**. **Figure 4.5** shows the several types of exocrine glands as classified by morphology.

Secretion Types   Exocrine glands are classified by the nature of their secretions as serous glands, mucous glands, or mixed glands **Serous** (sēr'ŭs; *serum* = whey) **glands** produce and secrete a nonviscous, watery fluid, such as sweat, milk, tears, or digestive juices. This fluid carries wastes (sweat) to the surface of the skin, nutrients (milk) to a nursing infant, or digestive enzymes from the pancreas to the lumen of the small intestine. **Mucous** (mū'kŭs) **glands** secrete mucin, which forms mucus when mixed with water. Mucous glands are found in such places as the roof of the oral cavity and the surface of the tongue. **Mixed glands**, such as

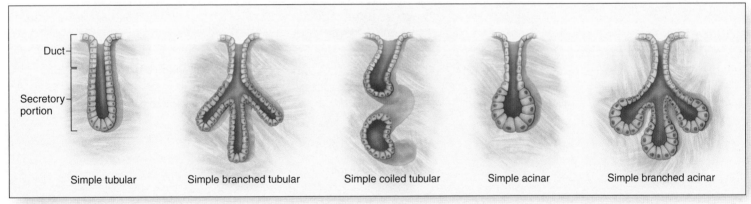

(a) Simple glands

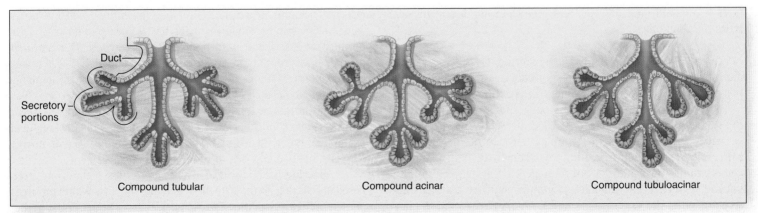

(b) Compound glands

Figure 4.5

Structural Classification of Multicellular Exocrine Glands.   *(a)* Simple glands have unbranched ducts, whereas *(b)* compound glands have ducts that branch. These glands also exhibit different forms: Tubular glands have secretory cells in a space with a uniform diameter, acinar glands have secretory cells arranged in saclike acini, and tubuloacinar glands have secretory cells in both types of regions.

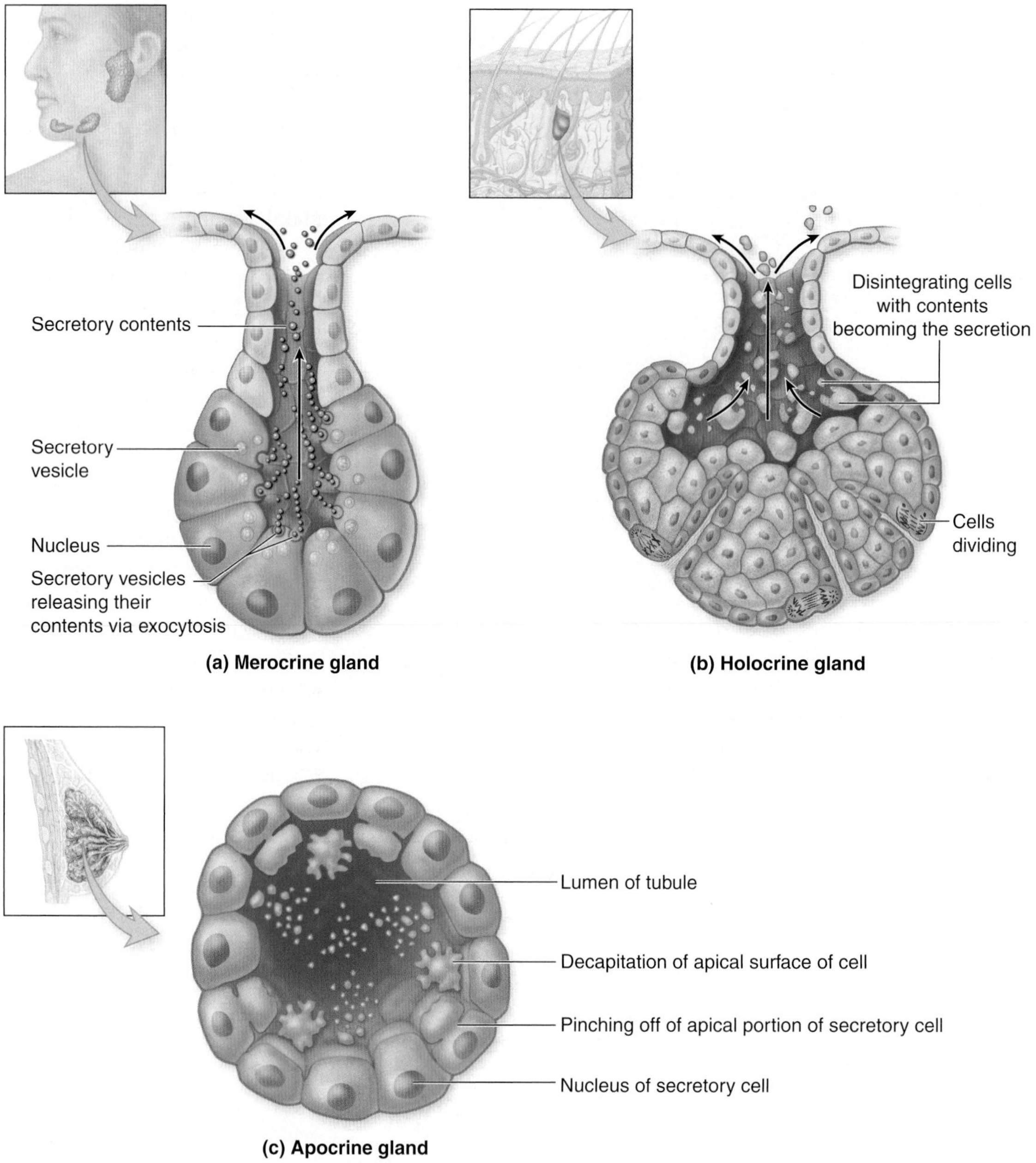

Figure 4.6

Modes of Exocrine Secretion. Exocrine glands use different processes to release their secretory products. *(a)* Merocrine glands secrete products by means of exocytosis at the apical surface of the secretory cells. *(b)* Holocrine gland secretion is produced through the destruction of the secretory cell. Lost cells are replaced by cell division at the base of the gland. *(c)* Apocrine gland secretion occurs with the "decapitation" of the apical surface of the cell and the subsequent release of secretory product and some cellular fragments.

Secretory contents

Secretory vesicle

Nucleus

Secretory vesicles releasing their contents via exocytosis

(a) Merocrine gland

Disintegrating cells with contents becoming the secretion

Cells dividing

(b) Holocrine gland

Lumen of tubule

Decapitation of apical surface of cell

Pinching off of apical portion of secretory cell

Nucleus of secretory cell

(c) Apocrine gland

the two pairs of salivary glands inferior to the oral cavity, contain both serous and mucous cells, and produce a mixture of the two types of secretions.

Secretion Methods Glands also can be classified by their mechanism of discharging secretory product as merocrine glands, holocrine glands, or apocrine glands **(figure 4.6)**.

Merocrine (mer′-ō-krin; *meros* = share) **glands** package their secretions in structures called secretory vesicles. The secretory vesicles travel to the apical surface of the glandular cell and release their secretion by exocytosis. The glandular cells remain intact and are not damaged in any way by producing the secretion. Lacrimal (tear) glands, salivary glands, some sweat glands, the exocrine glands of the pancreas, and the gastric glands of the stomach are examples of merocrine glands. Some merocrine glands are also called *eccrine glands*, to denote a type of sweat gland in the skin that is not connected to a hair follicle (see chapter 5).

Holocrine (hōl′ō-krin; *holos* = whole) **glands** are formed from cells that accumulate a product and then the entire cell disintegrates. Thus, a holocrine secretion is a mixture of cell fragments and the product the cell synthesized prior to its destruction. The ruptured, dead cells are continuously replaced by other epithelial cells undergoing mitosis. Without this regenerative capacity, holocrine glands would quickly lose all of their cells during their secretory activities. Holocrine secretions tend to be more viscous than merocrine secretions. The oil-producing glands (sebaceous glands) in the skin are an example of holocrine glands. (So the oily secretion you feel on your skin is actually composed of ruptured, dead cells!)

Apocrine (ap′ō-krin; *apo* = away from or off) **glands** are composed of cells that accumulate their secretory products within the apical portion of their cytoplasm. The secretion follows as this apical portion decapitates. The apical portion of the cytoplasm begins to pinch off into the lumen of the gland in order for the secretory product to be transported to the skin surface. Apocrine glands include the mammary glands and some sweat glands in the axillary and pubic regions.

WHAT DID YOU LEARN?

4 What two main characteristics are used to classify epithelial tissues?

5 Why is one epithelium referred to as "pseudostratified"?

6 What are the two basic parts of a multicellular exocrine gland?

7 Why is epithelial cell regeneration important to the continued functioning of a holocrine gland?

Connective Tissue

Key topics in this section:

■ Structure and function of connective tissue
■ Characteristics of embryonic connective tissue
■ Comparison of connective tissue proper, supporting connective tissue, and fluid connective tissue
■ Body locations where each type of connective tissue is found

Connective tissue is the most diverse, abundant, widely distributed, and microscopically variable of the tissues. Connective tissue is designed to support, protect, and bind organs. As its name implies, it is the "glue" that binds body structures together. The diversity of connective tissue is obvious when examining some of its types. Connective tissue includes the fibrous tendons and ligaments, body fat, the cartilage that connects the ends of ribs to the sternum, the bones of the skeleton, and the blood.

Characteristics of Connective Tissue

Although the types of connective tissue are diverse, all of them share three basic components: cells, protein fibers, and ground substance (**figure 4.7**). Their diversity is due to varying proportions of these components as well as to differences in the types and amounts of protein fibers.

Cells

Each type of connective tissue contains specific types of cells. For example, connective tissue proper contains fibroblasts, fat contains adipocytes, cartilage contains chondrocytes, and bone contains osteocytes. Most connective tissue cells are not in direct contact with each other, but are scattered throughout the tissue. This differs markedly from epithelial tissue, whose cells crowd closely together with little to no extracellular matrix surrounding them.

Protein Fibers

Most connective tissue contains protein fibers throughout. These fibers strengthen and support connective tissue. The type and abundance of these fibers indicate to what extent the particular connective tissue is responsible for strength and support. Three types of protein fibers are found in connective tissue: collagen fibers, which are strong and stretch-resistant; elastic fibers, which are flexible and resilient; and reticular fibers, which form an interwoven framework.

Ground Substance

Both the cells and the protein fibers reside within a material called ground substance. This nonliving material is produced by the connective tissue cells. It primarily consists of protein and carbohydrate molecules and variable amounts of water. The ground substance may be viscous (as in blood), semisolid (as in cartilage), or solid (as in bone). Together, the ground substance and the protein fibers form an **extracellular matrix**. Most connective tissues are composed primarily of an extracellular matrix, with relatively small proportions of cells.

WHAT DO YOU THINK?

3 Why does connective tissue contain fewer cells than epithelium? Can you think of some reasons related to the functions of connective tissue?

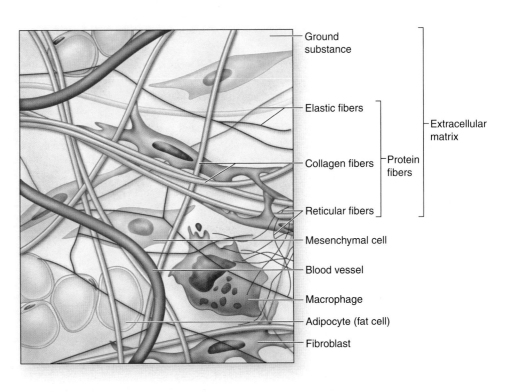

Figure 4.7

Connective Tissue Components and Organization.   Connective tissue is composed of cells and an extracellular matrix of protein fibers and ground substance.

Ground substance

Elastic fibers

Collagen fibers

Reticular fibers

Mesenchymal cell

Blood vessel

Macrophage

Adipocyte (fat cell)

Fibroblast

Protein fibers

Extracellular matrix

Functions of Connective Tissue

As a group, the many types of connective tissue perform a wide variety of functions, including the following:

- **Physical protection**. The bones of the cranium, sternum, and thoracic cage protect delicate organs, such as the brain, heart, and lungs; fat packed around the kidneys and at the posterior side of the eyes within the skull protects these organs.
- **Support and structural framework**. Bones provide the framework for the adult body and support the soft tissues; cartilage supports such body structures as the trachea, bronchi, ears, and nose; connective tissue sheets form capsules to support body organs such as the spleen and kidneys.
- **Binding of structures**. Ligaments bind bone to bone; tendons bind muscle to bone; dense irregular connective tissue binds skin to underlying muscle and bone.
- **Storage**. Fat is the major energy reserve in the body; bone is a large reservoir for calcium and phosphorus.
- **Transport**. Blood carries nutrients, gases, hormones, wastes, and blood cells between different regions of the body.
- **Immune protection**. Many connective tissue types contain white blood cells (leukocytes), which protect the body against disease and mount an immune response when the body is exposed to something foreign. A derivative of one type of leukocyte, called a macrophage, phagocytizes ("eats up") foreign materials. Additionally, the extracellular matrix is a viscous material that interferes with the movement and spread of disease-causing organisms.

Development of Connective Tissue

The primary germ layer mesoderm forms all connective tissues. There are two types of embryonic connective tissue: mesenchyme and mucous connective tissue. In the developing embryo, **mesenchyme** (mez'en-kīm; *mesos* = middle, *enkyma* = infusion) is the first type of connective tissue to emerge. It has star-shaped (stellate) or spindle-shaped mesenchymal cells dispersed within a gel-like ground substance that contains fine, immature protein fibers (**table 4.6a**). In fact, there is proportionately more ground substance than mesenchymal cells in this type of embryonic connective tissue. Mesenchyme is the source of all other connective tissues. Adult connective tissues often house numerous mesenchymal (stem) cells that support the repair of the tissue following damage or injury.

A second type of embryonic connective tissue is **mucous connective tissue** (*Wharton's jelly*). The immature protein fibers in mucous connective tissue are more numerous than those within mesenchyme (table 4.6b). Mucous connective tissue is located within the umbilical cord only.

Classification of Connective Tissue

The connective tissue types present after birth are classified into three broad categories: connective tissue proper, supporting

---

CLINICAL VIEW

What Are You Planning to Do with Your Baby's Umbilical Cord?

Years of medical research have led to an amazing discovery: A baby's blood contains stem cells that are the same as those found in a child's bone marrow, and these cells can be used to treat a variety of life-threatening diseases. What's more, these important stem cells are easy to collect. The leftover blood in the placenta and umbilical cord is a ready, and often discarded, source.

Cord blood can easily be harvested immediately following the birth of a baby. The specimen can be shipped to a cord blood bank for testing, processing, and storage. The cells are carefully banked in a cryogenic vault for optimal preservation should there be a future need. To date, conditions successfully treated with cord blood stem cells include lymphoma, leukemia, anemias resulting from severe bone marrow damage (especially complications of cancer chemotherapy), and sickle-cell disease.

Although this technology is hopeful, it is not without drawbacks. First, each cord blood sample contains relatively few stem cells. Although a method to increase the growth and number of stem cells is currently being investigated, at present the limited amount of cells available remains a problem. Second, harvesting and storing cord blood can be costly.

Donor registries are not yet available in all parts of the United States, and private banks are expensive. Presently, the typical charge for the initial processing at a private bank is as much as $1000, a cost not covered by insurance plans. And lifelong storage will certainly add to the overall expense.

The odds of a child needing a cord blood stem cell transplant are approximately 4 in 10,000, or about .04%. Certainly the odds of a person needing a transplant sometime during his or her life are much greater than just the .04% of childhood, but calculating the need over a typical adult life span is very difficult. However, the increasing risk with age, plus other possible applications of cord blood stem cells not yet discovered, seem to favor the banking of this valuable resource.

Table 4.6	Types of Embryonic Connective Tissue

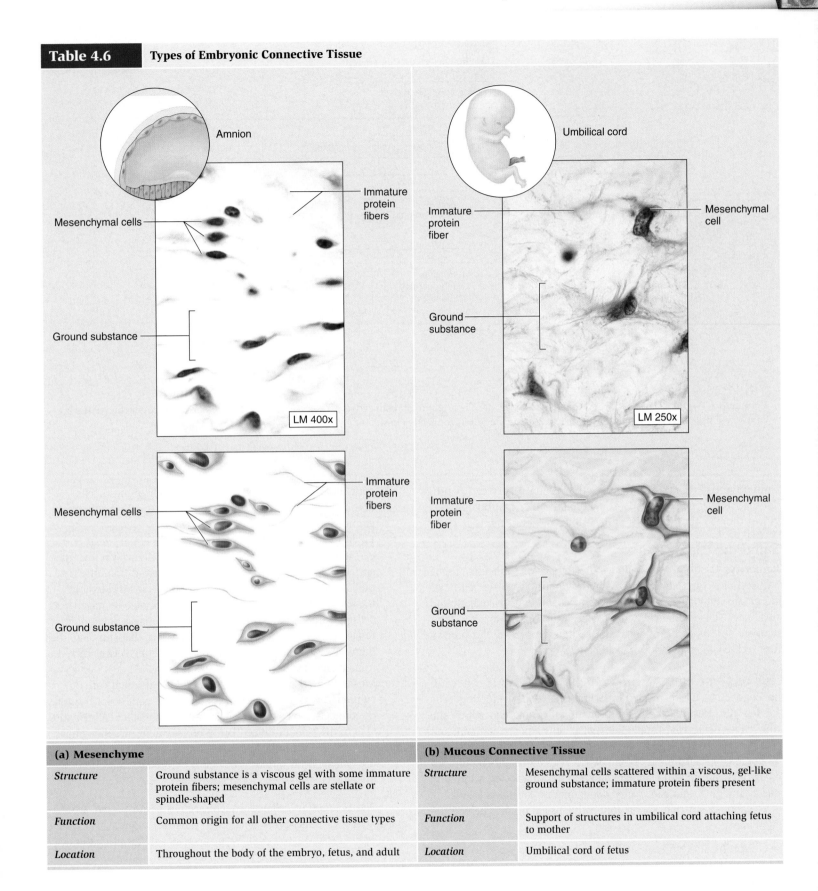

(a) Mesenchyme		(b) Mucous Connective Tissue	
Structure	Ground substance is a viscous gel with some immature protein fibers; mesenchymal cells are stellate or spindle-shaped	*Structure*	Mesenchymal cells scattered within a viscous, gel-like ground substance; immature protein fibers present
Function	Common origin for all other connective tissue types	*Function*	Support of structures in umbilical cord attaching fetus to mother
Location	Throughout the body of the embryo, fetus, and adult	*Location*	Umbilical cord of fetus

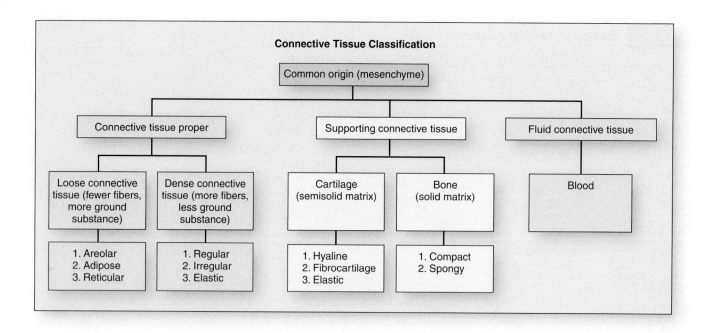

Figure 4.8

Connective Tissue Classification. Mesenchymal cells are the origin of all connective tissue cell types. The three classes of connective tissue are connective tissue proper, supporting connective tissue, and fluid connective tissue.

connective tissue, and fluid connective tissue. **Figure 4.8** provides an overview of these tissue types and the subcategories within them, each of which is described in detail next.

Connective Tissue Proper

Connective tissue proper includes those types of connective tissue that exhibit a variable mixture of both connective tissue cell types and extracellular protein fibers suspended within a viscous ground substance. These connective tissue types differ with respect to their numbers and types of cells and the relative properties and proportions of their fibers and ground substance.

Cells of Connective Tissue Proper Two classes of cells form the connective tissue proper: resident cells and wandering cells **(table 4.7)**. **Resident cells** are permanently contained within the connective tissue. These stationary cells help support, maintain, and repair the extracellular matrix. **Wandering cells** move throughout the connective tissue and are involved in immune protection and repair of damaged extracellular matrix. The number of wandering cells at any given moment varies depending on local conditions.

The resident cells of the connective tissue proper include the following types:

- **Fibroblasts** (fī′brō-blast; *fibra* = fiber, *blastos* = germ) are large, relatively flat cells with tapered ends. They are the most abundant resident cells in connective tissue proper. They produce the fibers and ground substance components of the extracellular matrix.
- **Adipocytes** (ad′i-pō-sīt; *adip* = fat) are also called *fat cells*. They often appear in small clusters within some types of

connective tissue proper. If a larger cluster of these cells dominates an area, the connective tissue is called adipose connective tissue.

- **Fixed macrophages** (mak′rō-fāj; *makros* = large, *phago* = to eat) are relatively large, irregular-shaped cells with numerous surface folds and projections. They are derived from one type of leukocyte (called a monocyte) and dispersed throughout the extracellular matrix, where they phagocytize damaged cells or pathogens. When the fixed macrophages encounter foreign materials, the cells release chemicals that stimulate the immune system and lure numerous wandering cells involved in body defense to the foreign materials.
- **Mesenchymal cells** are a type of embryonic stem cell contained within connective tissue. As a result of local injury or connective tissue damage, mesenchymal stem cells divide. One of the cells produced is the replacement mesenchymal cell, and the other becomes a committed cell that moves into the damaged or injured area and differentiates into the type of connective tissue cell that is needed.

The wandering cells of the connective tissue proper are primarily types of **leukocytes** (loo′kō-sīt; *leukos* = white), also called *white blood cells*. As you will learn in greater detail in chapter 21, there are several different types of leukocytes, and each type performs certain functions that help us overcome illness or fight foreign invaders in our bodies. Connective tissue proper contains specific types of wandering cells, including the following:

- **Mast cells**. These small, mobile cells contain a granule-filled cytoplasm. They are usually found close to blood vessels;

Table 4.7	Cells of Connective Tissue Proper	
Type of Cell	**Appearance**	**Function**
Resident Cells	Stationary	Maintain and repair extracellular matrix; store materials
Fibroblasts	Abundant, large, relatively flat cells, often with tapered ends	Produce fibers and ground substance of the extracellular matrix
Adipocytes	Fat cells with a single large lipid droplet; cellular components pushed to one side	Store lipid reserves
Fixed macrophages	Large cells derived from monocytes in blood; reside in extracellular matrix after leaving the blood	Phagocytize foreign materials
Mesenchymal cells	Stellate or spindle-shaped embryonic stem cells	Divide in response to injury to produce new connective tissue cells
Wandering Cells	Move through connective tissue spaces	Repair damaged extracellular matrix; active in immune response
Mast cells	Small cells with a granule-filled cytoplasm	Release histamine and heparin to stimulate local inflammation
Plasma cells	Small cells with a distinct nucleus derived from activated B-lymphocytes	Form antibodies that immobilize foreign substances, bacteria, viruses
Free macrophages	Mobile phagocytic cells formed from monocytes of the blood	Phagocytize foreign materials
Other leukocytes	White blood cells that enter connective tissue	Attack foreign materials (lymphocytes) or directly combat bacteria (neutrophils)

they secrete heparin to inhibit blood clotting, and histamine to dilate blood vessels and increase blood flow.

■ **Plasma cells**. When B-lymphocytes (a type of white blood cell) are activated by exposure to foreign materials, the cells mature into plasma cells. These cells are small "factories" that synthesize disease-fighting proteins called **antibodies** (an'tĭ-bod-ē; *anti* = against, *bodig* = corpus). Antibodies immobilize a foreign material and prevent it from causing further damage. Usually, plasma cells are found in the intestinal walls and in the spleen and lymph nodes.

■ **Free macrophages**. These mobile, phagocytic cells are formed from monocytes (a type of white blood cell) that migrate out of the bloodstream. They wander through connective tissue and engulf and destroy any bacteria, foreign particles, or damaged cells and debris they encounter.

■ **Other leukocytes**. In addition to the leukocytes just mentioned, other leukocytes migrate through the blood vessel walls into the connective tissue where they spend most of their time. The majority of these leukocytes are neutrophils, a type of white blood cell that seeks out and phagocytizes bacteria. The rest are lymphocytes, which attack and destroy foreign materials.

Fibers of Connective Tissue Proper As mentioned previously, the three types of protein fibers in connective tissue proper are collagen fibers, elastic fibers, and reticular fibers. Fibroblasts synthesize the components of all three fiber types, and then secrete these protein subunits into the interstitial fluid. The subunits combine or aggregate within the matrix and form the completed fiber.

Collagen (kol'lă-jen; *koila* = glue, *gen* = producing) **fibers** are long, unbranched extracellular fibers composed of the protein collagen. They are strong, flexible, and resistant to stretching. Collagen forms about 25% of the body's protein, making it the most abundant protein in the body. In fresh tissue, collagen fibers appear white, and thus they are often called white fibers. In tissue sections stained with hematoxylin and eosin to give contrast, they appear pink. In tissue sections, collagen forms coarse, sometimes wavy bundles. The parallel structure and arrangement of collagen bundles in tendons and ligaments allows them to withstand enormous forces in one direction.

Elastic (ĕ-las'tik; *elastreo* = drive) **fibers** contain the protein elastin and are thinner than collagen fibers. They stretch easily, branch, rejoin, and appear wavy. The coiled structure of elastin allows it to stretch and recoil like a rubber band when the deforming force is withdrawn. Elastic fibers permit the skin, lungs, and arteries to return to their normal shape after being stretched. Fresh elastic fibers have a yellowish color and are called yellow fibers. In tissue sections, elastic fibers are only visible when stained with special stains, such as Verhoff's stain, which makes elastic fibers appear black.

Reticular (re-tik'ū-lăr; *reticulum* = small net) **fibers** are thinner than collagen fibers. They contain the same protein subunits that collagen has, but their subunits are combined in a different way and they are coated with a glycoprotein (a protein with some carbohydrate attached to it). These fibers form a branching, interwoven framework that is tough but flexible. Reticular fibers are especially abundant in the **stroma**, a structural connective tissue framework in organs such as the lymph nodes, spleen, and liver. The meshlike arrangement of the reticular fibers permits them to physically support

Pathogenesis of Collagen

Collagen is an important protein that strengthens and supports almost all body tissues, especially the connective tissues. The pathogenesis (development of disease conditions) of certain connective tissue diseases may be traced to errors in collagen production. If the collagen does not form properly, the connective tissues are weak and subject to problems. Often these conditions are caused by a lack of dietary vitamin C (ascorbic acid), which is essential to collagen production. For example, **scurvy**, a disease caused by a deficiency in vitamin C, is marked by weakness, ulceration of gums with loss of teeth, and hemorrhages in mucous membranes and internal organs. Bone growth is abnormal, capillaries rupture easily, and wounds and fractures do not heal. All of these signs of scurvy are directly related to abnormal production of collagen.

Scurvy was especially prevalent many years ago among sailors who took long sea voyages and whose diets while at sea lacked vitamin C. These sailors eventually learned that bringing citrus fruits (such as limes and lemons) along on their voyages prevented scurvy. (This also explains how sailors received the nickname "limeys.") Nowadays, physicians try to treat collagen production disorders by promoting vitamin C supplementation in their patients' diets. Besides citrus fruits, foods high in vitamin C include broccoli, cauliflower, peppers, mustard greens, spinach, and tomatoes.

Marfan Syndrome

Marfan syndrome is a rare genetic disease of connective tissue that is characterized by skeletal, cardiovascular, and visual abnormalities. It is caused by an abnormal gene on chromosome 15.

Patients with Marfan syndrome are tall and thin. Their skeletal system deformities include abnormally long arms, legs, fingers, and toes; malformation of the thoracic cage and/or vertebral column as a result of excessive growth of ribs; and easily dislocated joints resulting from weak ligaments, tendons, and joint capsules. Cardiovascular system problems involve a weakness in the aorta and abnormal heart valves. Abnormalities in fibrillin, a protein that helps support blood vessels and other body structures, and in both collagen and elastin, are responsible for these clinical effects. Vision abnormalities develop because the thin fibers that hold the optic lens are weak, allowing the lens to slip out of place.

Patients usually exhibit symptoms of Marfan syndrome by age 10; those affected often die of cardiovascular problems before they reach 50 years of age. Several individuals have speculated that Abraham Lincoln suffered from Marfan syndrome because he exhibited many characteristics of the disease. Recently, however, researchers have stated there isn't enough conclusive evidence to prove the Lincoln-Marfan link.

Individual with Marfan syndrome.

organs and resist external forces that may damage the organ's cells and blood vessels.

Ground Substance of Connective Tissue Proper The cellular and fibrous components of the connective tissue proper are suspended within the ground substance, a colorless, featureless, viscous solution. Ground substance usually has a gelatinous, almost rubbery consistency due to the mixture of its component molecules, which vary both in their size and in their proportions of proteins and carbohydrates. The different molecules in the ground substance are called glycosaminoglycans, proteoglycans, and structural glycoproteins.

Categories of Connective Tissue Proper

Connective tissue proper is divided into two broad categories: loose connective tissue and dense connective tissue **(table 4.8)**. This classification is based on the relative proportions of cells, protein fibers, and ground substance.

Loose Connective Tissue **Loose connective tissue** contains relatively fewer cells and protein fibers than dense connective tissue. The protein fibers in loose connective tissue are loosely arranged rather than tightly packed together. Usually, this tissue occupies the spaces between and around organs. Loose connective tissues support the overlying epithelia and provide cushioning around organs, support and surround blood vessels and nerves, store lipids, and provide a medium for the diffusion of materials. Thus, loose connective tissues act as the body's "packing material." There are three types of loose connective tissue: areolar connective tissue, adipose connective tissue, and reticular connective tissue.

Table 4.8	Connective Tissue Proper		
Type	**Structure**	**Function**	**Location**
Loose Connective Tissue	Relatively fewer cells and fibers than in dense connective tissue; fibers are loosely arranged		
Areolar connective tissue	Fibroblasts; lesser amounts of collagen and elastic fibers; viscous ground substance	Binds and packs around organs	Surrounding nerves, vessels; subcutaneous layer
Adipose connective tissue	Adipocytes	Protects; stores fat; insulates	Subcutaneous layer; surrounding kidney and selected other organs
Reticular connective tissue	Meshwork of reticular fibers	Forms stroma of lymphatic organs	Stroma of spleen, liver, lymph nodes, bone marrow
Dense Connective Tissue	Higher proportion of fibers to ground substance; protein fibers densely packed together		
Dense regular connective tissue	Densely packed collagen fibers are parallel to direction of stress	Provides great strength and flexibility primarily in a single direction	Tendons and ligaments
Dense irregular connective tissue	Densely packed collagen fibers are interwoven; fibers are irregularly clumped together and project in all directions	Provides tensile strength in all directions	Dermis; capsules of organs
Elastic connective tissue	Elastic and collagen fibers are arranged irregularly	Provides framework and supports organs	Walls of large arteries

Areolar (ă-rē′ō-lăr) **connective tissue** is highly variable in appearance and the least specialized connective tissue in the body **(table 4.9a)**. It has a loosely organized array of collagen and elastic fibers and an abundant distribution of blood vessels. Areolar connective tissue contains all of the cell types of connective tissue proper, although the predominant cell is the fibroblast. A viscous ground substance occupies the spaces between fibers and accounts for most of the volume of areolar connective tissue. The ground substance cushions shocks, and the loosely organized fibers ensure that this type of connective tissue can be distorted without damage. Additionally, the elastic properties of this tissue promote independent movements. For instance, the dermis of the skin contains a superficial layer of areolar connective tissue, and thus tugging on the skin of the leg, for example, does not affect the underlying muscle.

Areolar connective tissue is found nearly everywhere in the body. It surrounds nerves, blood vessels, and individual muscle cells. It is also a major component of the subcutaneous layer deep to the skin.

Adipose connective tissue (commonly known as "fat") is a loose connective tissue composed primarily of cells called **adipocytes** (table 4.9b). Adipocytes usually range from 70 μm to 120 μm in diameter. In life, adipocytes are filled with lipid droplets. On a histology slide, the lipid has been extracted during preparation, so all that is left is the plasma membrane of the adipocyte, with the nucleus pushed to the side of a round, clear space looking much like a ring.

Adipose connective tissue serves as packing around structures and provides padding, cushions shocks, and acts as an insulator to slow heat loss through the skin. Adipose connective tissue is commonly found throughout the body in such diverse locations as a fat capsule surrounding the kidney, the pericardial and abdominopelvic cavities, and the subcutaneous layer.

Fat is a primary energy store for the body. The amount of stored fat fluctuates as the adipose cells either increase (called lipogenesis) or decrease (called lipolysis) their amount of stored fat. But although there is a constant turnover of the stored fat, an equilibrium is usually reached, and the amount of stored fat and the number of adipocytes are normally quite stable in an individual. Although adipocytes cannot divide, mesenchymal cells can provide additional fat cells if the body has excess nutrients. Thus, even after a surgical procedure to reduce the amount of body fat, such as liposuction, the mesenchymal stem cells may replace adipocytes to store excess fat in the body.

Reticular connective tissue contains a meshwork of reticular fibers, fibroblasts, and leukocytes (table 4.9c). This connective tissue forms the stroma of many lymphatic organs, such as the spleen, thymus, lymph nodes, and bone marrow.

Dense Connective Tissue

Dense connective tissue is composed primarily of protein fibers and has proportionately less ground substance than does loose connective tissue. Dense connective tissue is sometimes called *collagenous tissue* because collagen fibers are the dominant fiber type. There are three categories of dense connective tissue: (1) dense regular connective tissue, (2) dense irregular connective tissue, and (3) elastic connective tissue.

In **dense regular connective tissue,** collagen fibers are packed tightly and aligned parallel to an applied force. The parallel, wavy

Table 4.9 Connective Tissue Proper: Loose Connective Tissue

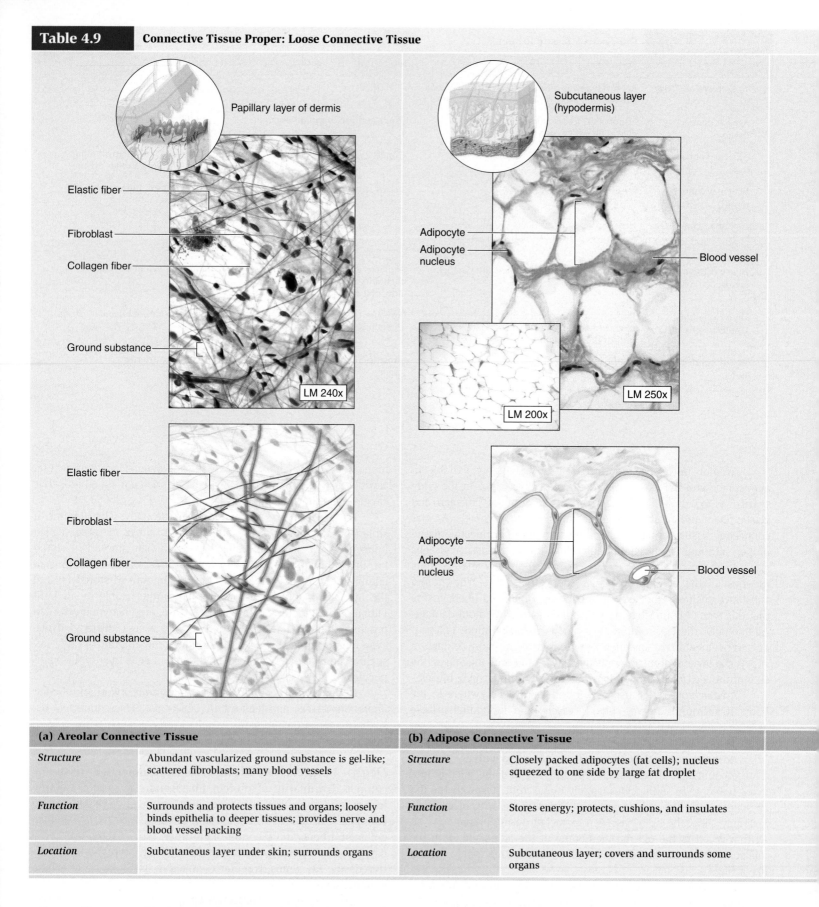

Papillary layer of dermis

Elastic fiber

Fibroblast

Collagen fiber

Ground substance

LM 240x

Elastic fiber

Fibroblast

Collagen fiber

Ground substance

Subcutaneous layer (hypodermis)

Adipocyte

Adipocyte nucleus

Blood vessel

LM 250x

LM 200x

Adipocyte

Adipocyte nucleus

Blood vessel

(a) Areolar Connective Tissue		(b) Adipose Connective Tissue	
Structure	Abundant vascularized ground substance is gel-like; scattered fibroblasts; many blood vessels	*Structure*	Closely packed adipocytes (fat cells); nucleus squeezed to one side by large fat droplet
Function	Surrounds and protects tissues and organs; loosely binds epithelia to deeper tissues; provides nerve and blood vessel packing	*Function*	Stores energy; protects, cushions, and insulates
Location	Subcutaneous layer under skin; surrounds organs	*Location*	Subcutaneous layer; covers and surrounds some organs

collagen fibers resemble lasagna noodles stacked one on top of another (**table 4.10a**). This tissue type is found in tendons and ligaments, where stress is applied in a single direction. Dense regular connective tissue has few blood vessels, and thus it takes a long time to heal following injury, since a rich blood supply is necessary for good healing.

In **dense irregular connective tissue,** individual bundles of collagen fibers extend in all directions in a scattered meshwork. These bundles of collagen fibers appear in clumps throughout the tissue, rather than arranged in parallel as seen in dense regular connective tissue (table 4.10b). Dense irregular connective tissue provides support and resistance to stress in multiple directions.

that supports and houses internal organs, such as the liver, kidneys, and spleen.

Elastic connective tissue has branching elastic fibers and more fibroblasts than loose connective tissue in addition to packed collagen fibers (table 4.10*c*). The elastic fibers provide resilience and the ability to deform and then return to normal shape. Examples of structures composed of elastic connective tissue are the vocal cords, the suspensory ligament of the penis, and some ligaments of the spinal column. Elastic connective tissue also is present as wavy sheets in the walls of large and medium arteries.

WHAT DO YOU THINK?

4 What type of connective tissue have you damaged when you sprain your ankle?

Study Tip!

Ask the following questions to help distinguish the types of connective tissue proper under the microscope:

1. Is the connective tissue loose or dense? Loose connective tissue has fewer protein fibers and relatively more ground substance. Dense connective tissue has more protein fibers and relatively little ground substance.

2. If the tissue is dense, are the protein fibers in clumps or in parallel? Protein fibers in clumps indicate dense irregular connective tissue. Protein fibers that run in parallel, like lasagna noodles stacked on top of one another, indicate dense regular connective tissue. Elastic connective tissue may resemble dense regular connective tissue, but its fibers are not as neatly arranged.

3. If the tissue is loose, what types of cells are present? Areolar connective tissue primarily contains fibroblasts, whereas adipose connective tissue contains adipocytes. The presence of numerous leukocytes may indicate reticular connective tissue.

Supporting Connective Tissue

Cartilage and bone are types of **supporting connective tissue** because they form a strong, durable framework that protects and supports the soft body tissues. The extracellular matrix in supporting connective tissue contains many protein fibers and a ground substance that ranges from semisolid to solid. In general, cartilage has a semisolid extracellular matrix while bone has a solid extracellular matrix.

Cartilage   **Cartilage** has a firm, gel-like extracellular matrix composed of both protein fibers and ground substance. Mature cartilage cells are called **chondrocytes** (kon'drō-sīt; *chondros* = gristle or cartilage, *cytos* = a hollow [cell]). They occupy small spaces called **lacunae** (lă-koo'ne; *lacus* = a hollow, a lake), within the extracellular matrix. The physical properties of cartilage vary with the extracellular matrix contents. Cartilage is stronger and more resilient than any previously discussed connective tissue type, and it provides more flexibility than bone. Collagen fibers within the matrix give cartilage its tensile strength; its resilience is attributed to elastic fibers and variations in the kinds and amounts of ground substance components, including water.

Cartilage is found in areas of the body that need support and must withstand deformation, such as the tip of the nose or the external part of the ear (auricle).

Chondrocytes produce a chemical that prevents blood vessel formation and growth within the extracellular matrix. Thus, mature cartilage is avascular, meaning without blood vessels. Therefore, the chondrocytes must exchange nutrients and waste

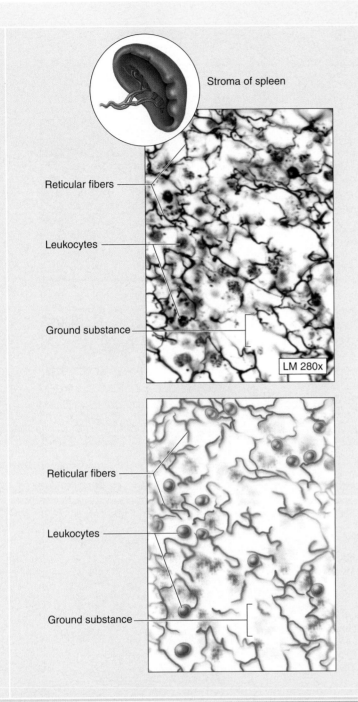

Stroma of spleen

Reticular fibers

Leukocytes

Ground substance

LM 280x

Reticular fibers

Leukocytes

Ground substance

(c) Reticular Connective Tissue

Structure	Ground substance is gel-like liquid; scattered arrangement of reticular fibers, fibroblasts, and leukocytes
Function	Provides supportive framework for spleen, lymph nodes, thymus, bone marrow
Location	Forms stroma of lymph nodes, spleen, thymus, bone marrow

An example of dense irregular connective tissue is the deep portion of the dermis, which lends strength to the skin and permits it to withstand applied forces from any direction. Dense irregular connective tissue also forms a supporting layer around cartilage (called the perichondrium) and around bone (called the periosteum), except at joints. In addition, it forms a thick, fibrous capsule

Table 4.10   Connective Tissue Proper: Dense Connective Tissue

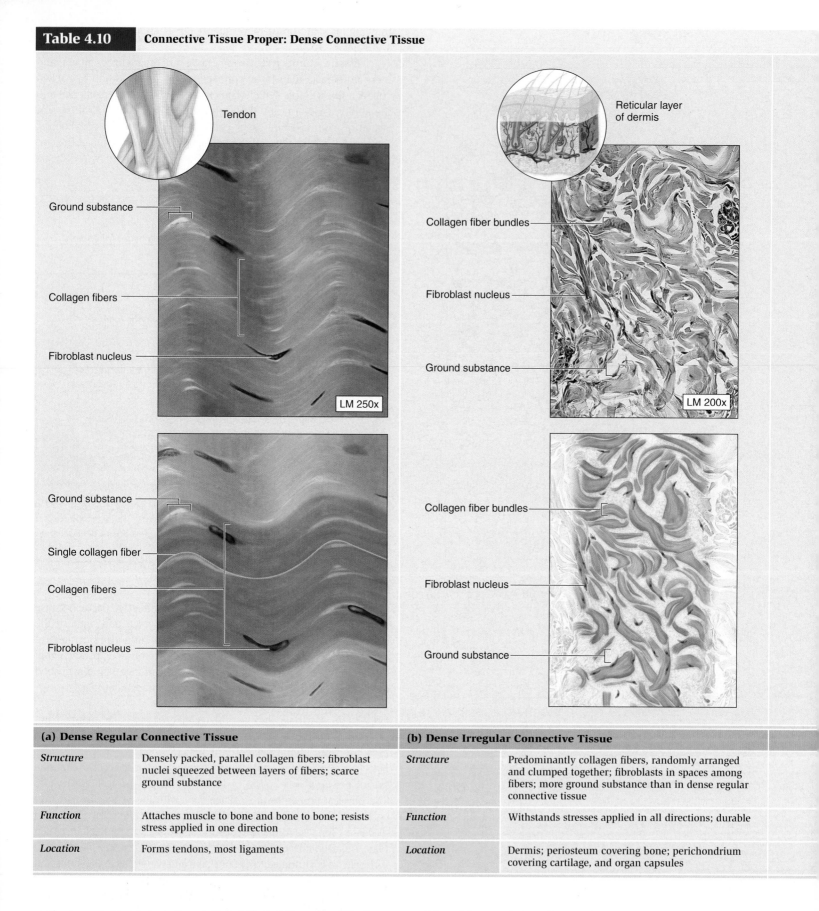

(a) Dense Regular Connective Tissue		(b) Dense Irregular Connective Tissue	
Structure	Densely packed, parallel collagen fibers; fibroblast nuclei squeezed between layers of fibers; scarce ground substance	*Structure*	Predominantly collagen fibers, randomly arranged and clumped together; fibroblasts in spaces among fibers; more ground substance than in dense regular connective tissue
Function	Attaches muscle to bone and bone to bone; resists stress applied in one direction	*Function*	Withstands stresses applied in all directions; durable
Location	Forms tendons, most ligaments	*Location*	Dermis; periosteum covering bone; perichondrium covering cartilage, and organ capsules

products with blood vessels outside of the cartilage by diffusion. Cartilage usually has a covering called the **perichondrium** (per-i-kon′drē-ŭm; *peri* = around, *chondros* = cartilage). Two distinct layers form the perichondrium: an outer, fibrous region of dense irregular connective tissue and an inner, cellular layer. The fibrous layer provides protection and mechanical support, and secures the

perichondrium to the cartilage and to other structures. The cellular layer contains stem cells (chondroblasts) necessary for the growth and maintenance of the cartilage.

Types of Cartilage   Three major types of cartilage are found in the body: hyaline cartilage, fibrocartilage, and elastic cartilage. The

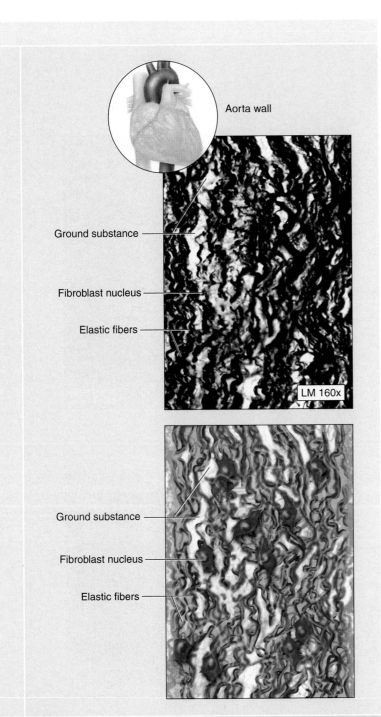

Aorta wall

Ground substance

Fibroblast nucleus

Elastic fibers

LM 160x

Ground substance

Fibroblast nucleus

Elastic fibers

(c) Elastic Connective Tissue

Structure	Predominantly freely branching elastic fibers; fibroblasts occupy some spaces between fibers
Function	Allows stretching of some organs
Location	Walls of elastic arteries; trachea; bronchial tubes; true vocal cords; suspensory ligaments of penis

cartilage types exhibit differences in density and dispersal of chondrocytes within the extracellular matrix.

Hyaline (hī′ă-lin, -lēn *hyalos* = glass) **cartilage** is the most common type of cartilage and also the weakest. It is named for its clear, glassy appearance under the microscope. The chondrocytes within their lacunae are irregularly scattered throughout the extracellular

matrix (**table 4.11***a*). However, the collagen within the matrix is not readily observed by light microscopy because it is primarily in the form of submicroscopic fibrils. Hyaline cartilage is surrounded by a perichondrium. If the hyaline cartilage tissue is stained by hematoxylin and eosin and then observed under the microscrope, the tissue resembles carbonated grape soda, where the lacunae represent the bubbles in the soda.

Hyaline cartilage has many functions in addition to its primary one of supporting soft tissue. It forms most of the fetal skeleton and is a model for most future bone growth. The cartilage at the articular ends of long bones allows the bones in a joint to move freely and easily. Hyaline cartilage is found in many other areas of the body, including the nose, trachea, most of the larynx, costal cartilage (the cartilage attached to the ribs), and the articular ends of long bones.

Fibrocartilage (fī-brō-kar′ti-lij; *fibro* = fiber) has numerous coarse, readily visible fibers in its extracellular matrix (table 4.11*b*). The fibers are arranged as irregular bundles between large chondrocytes. There is only a sparse amount of ground substance, and often the chondrocytes are arranged in parallel rows. The densely interwoven collagen fibers contribute to the extreme durability of this type of cartilage. It has no perichondrium.

Fibrocartilage is found in the intervertebral discs (circular structures between adjacent vertebrae), the pubic symphysis (a pad of cartilage between the anterior parts of the pelvic bones), and the menisci (C-shaped cartilage pads) of the knee joint. In these locations, fibrocartilage acts as a shock absorber and resists compression.

Elastic cartilage is so named because it contains numerous elastic fibers in its matrix (table 4.11*c*). The higher concentration of elastic fibers in this cartilage causes it to appear yellow in fresh sections. The chondrocytes of elastic cartilage are almost indistinguishable from those of hyaline cartilage. They are typically closely packed and surrounded by only a small amount of extracellular matrix. The elastic fibers are both denser and more highly branched in the central region of the extracellular matrix, where they form a weblike mesh around the chondrocytes within the lacunae. These fibers ensure that elastic cartilage is extremely resilient and flexible. Elastic cartilage is surrounded by a perichondrium.

Elastic cartilage is found in the epiglottis (a structure in the larynx that prevents swallowed food and fluids from entering the trachea) and in the external ear. You can see for yourself how flexible elastic cartilage is by performing this experiment: Fold your outer ear over your finger, hold for 10 seconds, and release. You will notice that your ear springs back to its original shape because the elastic cartilage resists the deformational pressure you applied. (This also explains why our ears aren't permanently misshapen if we sleep on them in an unusual way!)

Bone Bone connective tissue (or *osseous connective tissue*) makes up the mass of most of the body structures referred to as "bones." Bone is more solid than cartilage and provides greater support. Chapter 6 provides a detailed description of the histology of bone connective tissue.

About one-third of the dry weight of bone is composed of organic components (collagen fibers and different protein-carbohydrate molecules), and two-thirds consists of inorganic components (a mixture of calcium salts, primarily calcium phosphate). Bone derives its remarkable properties from its combination of components: Its organic portions provide some flexibility and

Table 4.11 — Supporting Connective Tissue: Cartilage

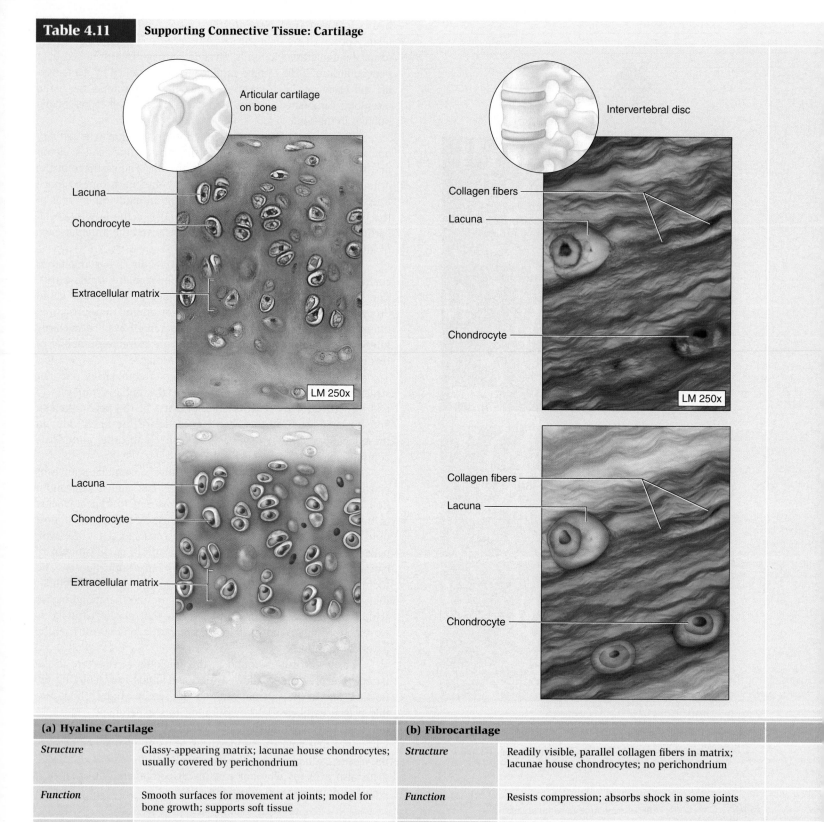

Articular cartilage on bone

Lacuna

Chondrocyte

Extracellular matrix

LM 250x

Lacuna

Chondrocyte

Extracellular matrix

Intervertebral disc

Collagen fibers

Lacuna

Chondrocyte

LM 250x

Collagen fibers

Lacuna

Chondrocyte

(a) Hyaline Cartilage		(b) Fibrocartilage	
Structure	Glassy-appearing matrix; lacunae house chondrocytes; usually covered by perichondrium	*Structure*	Readily visible, parallel collagen fibers in matrix; lacunae house chondrocytes; no perichondrium
Function	Smooth surfaces for movement at joints; model for bone growth; supports soft tissue	*Function*	Resists compression; absorbs shock in some joints
Location	Most of fetal skeleton; covers articular ends of long bones; costal cartilage; most of the larynx, trachea, nose	*Location*	Intervertebral discs; pubic symphysis; menisci of knee joints

tensile strength, and its inorganic portions provide compressional strength. The minerals are deposited onto the collagen fibers, resulting in a structure that is strong and durable but not brittle. Almost all bone surfaces (except for the surfaces of the joints of long bones) are covered by a dense irregular connective tissue

called the **periosteum** (per-ē-os′tē-ŭm; *osteon* = bone), which is similar to the perichondrium of cartilage.

There are two forms of bone connective tissue: compact bone and spongy bone. Both types are typically found in all bones of the body. **Compact bone** appears solid but is in fact perforated

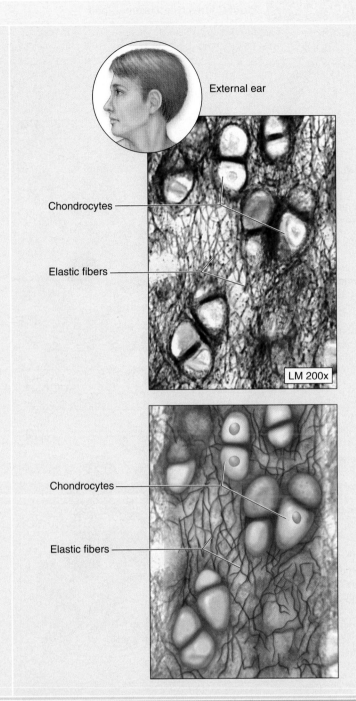

Chondrocytes

Elastic fibers

LM 200x

Chondrocytes

Elastic fibers

(c) Elastic Cartilage	
Structure	Contains abundant elastic fibers; elastic fibers form weblike mesh around lacunae; perichondrium present
Function	Maintains structure and shape while permitting extensive flexibility
Location	External ear; epiglottis of the larynx

Table 4.12	**Supporting Connective Tissue: Bone**

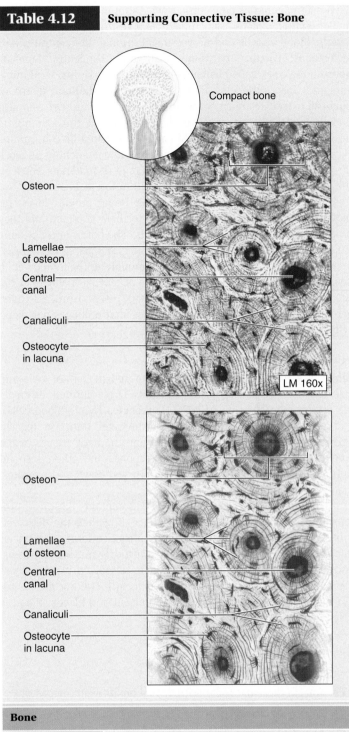

Compact bone

Osteon

Lamellae of osteon

Central canal

Canaliculi

Osteocyte in lacuna

LM 160x

Osteon

Lamellae of osteon

Central canal

Canaliculi

Osteocyte in lacuna

Bone	
Structure	*Compact bone:* Calcified matrix arranged in osteons (concentric lamellae arranged around a central canal containing blood vessels)
	Spongy bone: Lacks the organization of compact bone; contains macroscopic spaces; bone arranged in a meshwork pattern
Function	Supports soft structures; protects vital organs; provides levers for movement; stores calcium and phosphorus. Spongy bone is the site of hemopoiesis.
Location	Bones of the body

by a number of vascular canals. It usually forms the hard outer shell of the bone. **Spongy bone** (or *cancellous bone*) is located within the interior of a bone. Instead of being completely solid, spongy bone contains spaces, and the bone connective tissue forms a latticework structure that is very strong, yet lightweight.

This design allows our bones to be both strong and lightweight at the same time.

Compact bone has an ordered histologic pattern (**table 4.12**). It is formed from cylindrical structures called **osteons**, or *Haversian systems*. Osteons run parallel to the shafts of long bones.

Each osteon contains concentric rings of bone called **lamellae** (lă-mel'ē; *lamina* = plate), which encircle a **central canal** (*Haversian canal*). Blood vessels and nerves travel through the central canals of osteons. Lacunae between neighboring concentric lamellae house bone cells, called **osteocytes** (os'tē-ō-sīt). Diffusion of nutrients and waste products cannot occur through the hard matrix of bone, so osteocytes must communicate with one another, and ultimately with the blood vessels in the central canal, through minute passageways in the matrix called **canaliculi** (kan-ă-lik'ū-lī; *canalis* = canal). Together, all of the canaliculi form a branching network throughout compact bone for the exchange of materials between the blood vessels and the osteocytes within the lacunae.

Bone serves a variety of functions. Bones provide levers for movement when the muscles attached to them contract, and they protect soft tissues and vital body organs. The hard matrix of bone stores important minerals, such as calcium and phosphorus. Finally, many areas of spongy bone contain **hemopoietic** (hē'mō-poy-et'ik; *hemat* = blood) **cells**, which form a type of reticular connective tissue that is responsible for producing blood cells (a process called hemopoiesis). Thus, the connective tissue that produces our blood cells is stored within our spongy bone.

Fluid Connective Tissue

Blood is a fluid connective tissue composed in part of cells and cell fragments called formed elements. These formed elements are **erythrocytes** (red blood cells), **leukocytes** (white blood cells), and **platelets (table 4.13)**. The erythrocytes transport oxygen and carbon dioxide between the lungs and the body tissues, while some leukocytes mount an immune response and others respond to foreign pathogens such as bacteria, viruses, fungi, and parasites. Platelets are involved in blood clotting.

Besides the formed elements, blood contains dissolved protein fibers in a watery ground substance. Together, the dissolved protein fibers and the watery ground substance form an extracellular matrix called **plasma**. Plasma transports nutrients, wastes, and hormones throughout the body. The dissolved protein fibers are modified to become insoluble and form a clotting meshwork if a blood vessel or tissue becomes damaged and bleeds. Blood is discussed in greater detail in chapter 21.

WHAT DID YOU LEARN?

8 What is the extracellular matrix? What are its main components?

9 What three categories are used to classify connective tissue types?

10 Identify the three types of protein fibers in connective tissue proper.

11 Compare loose connective tissue to dense connective tissue with respect to fiber density and distribution, and amount of ground substance.

Body Membranes

Key topics in this section:

- Structures and functions of mucous, serous, cutaneous, and synovial membranes
- Body locations where the different types of membranes are found

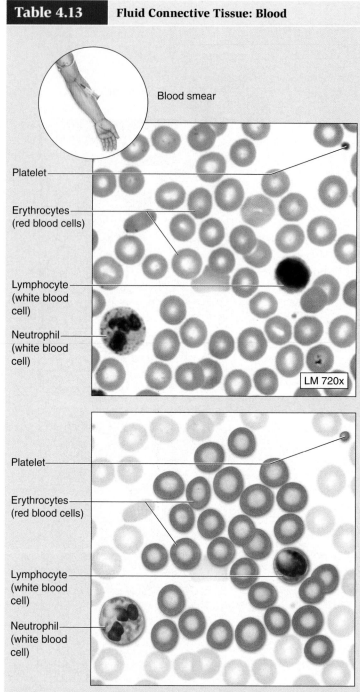

Table 4.13	Fluid Connective Tissue: Blood

Blood smear

Platelet

Erythrocytes (red blood cells)

Lymphocyte (white blood cell)

Neutrophil (white blood cell)

LM 720x

Platelet

Erythrocytes (red blood cells)

Lymphocyte (white blood cell)

Neutrophil (white blood cell)

Blood	
Structure	Contains erythrocytes, leukocytes, and platelets; soluble (dissolved) protein fibers and a watery ground substance form a fluid extracellular matrix called plasma
Function	Erythrocytes transport oxygen and some carbon dioxide. Leukocytes initiate and control immune response. Plasma contains clotting elements to stop blood loss. Platelets help with blood clotting. Plasma transports nutrients, wastes, and hormones throughout the body.
Location	Primarily within blood vessels (arteries, veins, capillaries) and the heart; leukocytes are also located in lymphatic organs and can migrate to infected or inflamed tissues in the body

The major cavities of the body have membranes that line both the internal surfaces of the cavities and the external surfaces of some of the viscera housed within those cavities. We discuss these membranes here because they consist of an epithelial sheet and an underlying connective tissue layer. The four types of body membranes are mucous, serous, cutaneous, and synovial membranes.

The two principal kinds of internal membranes are mucous and serous membranes. A **mucous membrane,** also called a *mucosa* (mū-kō'să), lines body passageways and compartments that eventually open to the external environment; these include the digestive, respiratory, reproductive, and urinary tracts. Mucous membranes perform absorptive, protective, and/or secretory functions. A mucous membrane is composed of an epithelium and underlying connective tissue called the lamina propria. Often, it is covered with a thin layer of mucus derived from goblet cells, multicellular mucous glands, or both. The mucus prevents the underlying layer of cells from drying out (a process called desiccation), provides lubrication, and traps bacteria and foreign particles to prevent them from invading the body.

A **serous membrane,** also termed a *serosa* (se-rō'să) is composed of a simple squamous epithelium called **mesothelium** and a thin underlying layer of loose connective tissue. The mesothelium is so named because it is derived from mesoderm. Serous membranes produce a thin, watery **serous fluid,** or *transudate* (tran'soo-dāt; *trans* = across, *sudo* = to sweat), which is derived from blood plasma. Serous membranes are composed of two parts: a **parietal layer** that lines the body cavity and a **visceral layer** that covers organs. The parietal and visceral layers are in close contact; a thin layer of serous fluid between them reduces the friction between their opposing surfaces. Examples of serous membranes include the pericardium, the peritoneum, and the pleura.

The largest body membrane is the **cutaneous** (kū-tā'nē-ŭs; *cutis* = skin) **membrane,** more commonly called the *skin.* The cutaneous membrane is composed of a keratinized stratified squamous epithelium (called the epidermis) and a layer of connective tissue (termed the dermis) upon which the epithelium rests (see chapter 5). It differs from the other membranes discussed so far in that it is relatively dry. Its many functions include protecting internal organs and preventing water loss.

Some joints of the skeletal system are lined by a fibrous **synovial** (si-nō'vē-ăl; *syn* = together, *ovum* = egg) **membrane,** which is composed of extensive areas of areolar connective tissue bounded by a superficial layer of squamous or cuboidal epithelial cells that lack a basement membrane. Some of the lining cells secrete a **synovial fluid** that reduces friction in the joint cavity and distributes nutrients to the cartilage on the joint surfaces of the bone.

WHAT DO YOU THINK?

5 What type of body membrane is found on the external surface of your forearm?

WHAT DID YOU LEARN?

12 What is the function of mucous membranes?

13 Distinguish between the parietal and visceral layers of the serous membrane.

Muscle Tissue

Key topics in this section:

- Structure and function of skeletal, cardiac, and smooth muscle
- Body locations where each type of muscle tissue is found

Muscle tissue is composed of specialized cells (fibers) that respond to stimulation from the nervous system by undergoing internal changes that cause them to shorten. As muscle tissue shortens, it exerts physical forces on other tissues and organs to produce movement; these movements include voluntary motion of body parts, blood circulation, respiratory activities, propulsion of materials through the digestive tract, and waste elimination. In order to perform these functions, muscle cells are very different from typical cells with respect to their cellular organization, cellular organelles, and other properties.

Classification of Muscle Tissue

The three histologic types of muscle in the body are skeletal muscle, cardiac muscle, and smooth muscle. The contraction mechanism is somewhat similar in all three, but they vary in their appearance, location, physiology, internal organization, and means of control by the nervous system. Specific details about the muscular system are discussed in chapter 10.

Skeletal Muscle Tissue

Skeletal muscle tissue is composed of cylindrical muscle cells called **muscle fibers (table 4.14a)**. Individual skeletal muscle cells are slender and often long (sometimes the length of the entire muscle). Such long cells need more than one nucleus to control and carry out all cellular functions, so each skeletal muscle fiber is multinucleated; some contain hundreds of nuclei. These multiple nuclei form when smaller embryonic muscle cells fuse early in the development of the skeletal muscle fiber. The nuclei in skeletal muscle fibers are located at the edge of the cell (called the periphery), immediately internal to the plasma membrane.

Skeletal muscle is described as striated and voluntary. Under the light microscope, the cells of skeletal muscle exhibit alternating light and dark bands, termed **striations,** (strī-ā'shŭn; *striatus* = furrow), that reflect the overlapping pattern of parallel thick and thin contractile protein filaments inside the cell. Additionally, skeletal muscle is considered **voluntary** because it usually does not contract unless stimulated by the somatic (voluntary) nervous system. Skeletal muscle attaches to the bones of the skeleton and also forms muscles associated with the skin, such as the muscles of facial expression and those forming body sphincters that help control waste removal. When skeletal muscles contract and relax, they produce heat for the body, which is why we become warmer when we "work out" at the gym or fitness center.

Cardiac Muscle Tissue

Cardiac muscle tissue is confined to the thick middle layer of the heart wall (called the myocardium). Macroscopically, cardiac muscle tissue resembles skeletal muscle in that both contain visible striations (table 4.14b). However, several obvious cellular differences distinguish the two types. First, the typical cardiac muscle cell is much shorter than a typical skeletal muscle fiber. Second, a cardiac muscle cell contains only one or two centrally

Table 4.14   Muscle Tissue

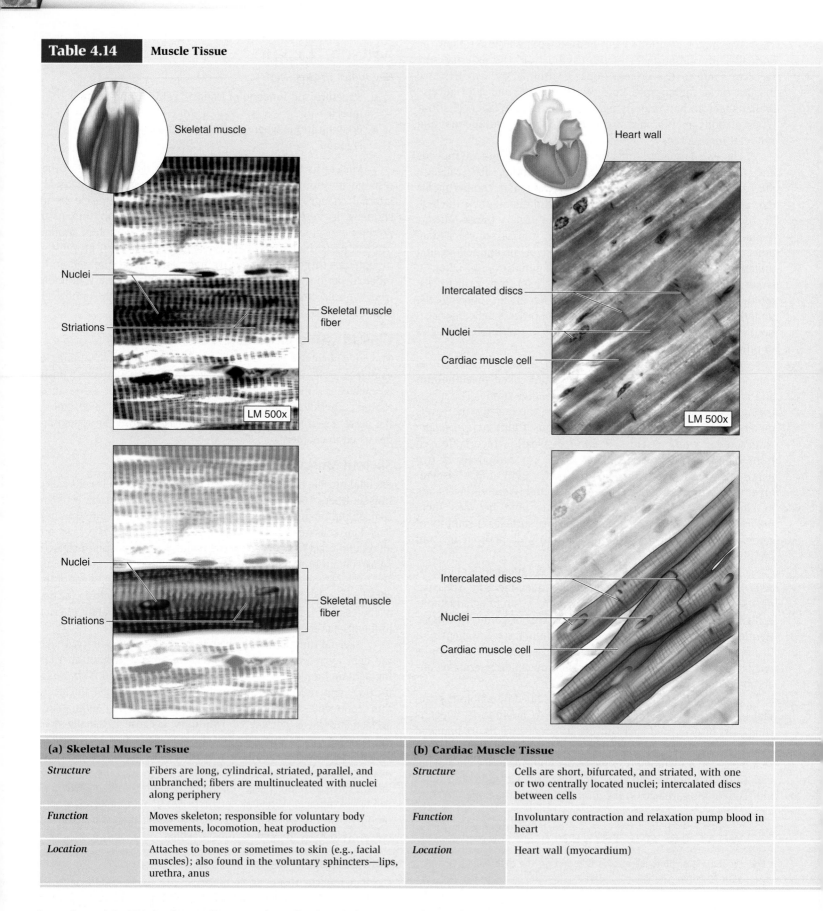

Skeletal muscle

Nuclei

Striations

Skeletal muscle fiber

LM 500x

Nuclei

Striations

Skeletal muscle fiber

Heart wall

Intercalated discs

Nuclei

Cardiac muscle cell

LM 500x

Intercalated discs

Nuclei

Cardiac muscle cell

(a) Skeletal Muscle Tissue		**(b) Cardiac Muscle Tissue**	
Structure	Fibers are long, cylindrical, striated, parallel, and unbranched; fibers are multinucleated with nuclei along periphery	*Structure*	Cells are short, bifurcated, and striated, with one or two centrally located nuclei; intercalated discs between cells
Function	Moves skeleton; responsible for voluntary body movements, locomotion, heat production	*Function*	Involuntary contraction and relaxation pump blood in heart
Location	Attaches to bones or sometimes to skin (e.g., facial muscles); also found in the voluntary sphincters—lips, urethra, anus	*Location*	Heart wall (myocardium)

located nuclei. Third, the cardiac muscle cell often bifurcates (branches), thus resembling a Y in shape. Finally, cardiac muscle cells are connected by **intercalated discs** (in-ter′kă-lā-ted; *intercalates* = inserted between), which are strong gap junctions between the cells. The intercalated discs promote the rapid transport of an electrical stimulus (nerve impulse) through many cardiac muscle cells at once, allowing the entire muscle wall to contract as a unit. When you view cardiac muscle tissue under the microscope, the intercalated discs appear as thick, dark lines between the cells.

the contraction. Thus, cardiac muscle tissue is both striated and involuntary.

Smooth Muscle Tissue

Smooth muscle tissue is so named because it lacks the striations seen in the other two types of muscle tissue, so the cells appear smooth (table 4.14c). Smooth muscle tissue is also called *visceral muscle tissue* because it is found in the walls of most viscera, such as the stomach, urinary bladder, and blood vessels. The contraction of smooth muscle helps propel and control the movement of material through these organs. Smooth muscle cells are fusiform (spindle-shaped), which means they are thick in the middle and tapered at their ends. The cells are also relatively short. Each cell has one centrally placed nucleus. Smooth muscle is considered involuntary because we do not have voluntary control over it. For example, you cannot voluntarily stop your stomach from digesting your food or your blood vessels from transporting your blood.

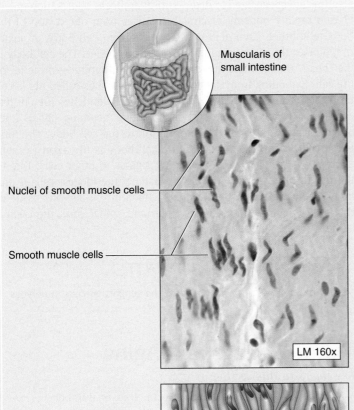

Muscularis of small intestine

Nuclei of smooth muscle cells

Smooth muscle cells

LM 160x

Nuclei of smooth muscle cells

Smooth muscle cells

(c) Smooth Muscle Tissue	
Structure	Cells are fusiform (spindle-shaped), short, nonstriated, and contain one centrally located nucleus
Function	Involuntary movements and motion; moves materials through internal organs
Location	Walls of hollow internal organs, such as vessels, airways, stomach, bladder, uterus

> ## Study Tip!
>
> Ask the following questions to help distinguish the three types of muscle tissue under the microscope:
>
> 1. What is the shape of the cell? Skeletal muscle fibers are long and cylindrical; smooth muscle cells are fusiform; and cardiac muscle cells are short and bifurcated.
>
> 2. Are the nuclei centrally located or at the periphery of the cell? Skeletal muscle nuclei are located at the periphery, while cardiac and smooth muscle cells have centrally located nuclei.
>
> 3. How many nuclei are present? A skeletal muscle fiber contains many nuclei at the periphery of the cylindrical fiber. A smooth muscle cell has one nucleus at the center of the cell. A cardiac muscle cell has one or two nuclei at its center.
>
> 4. Do the cells have striations? Smooth muscle cells have no striations, while cardiac and skeletal muscle cells do have striations.
>
> 5. Do you see intercalated discs? Only cardiac muscle has intercalated discs.

WHAT DID YOU LEARN?

14 What type of muscle tissue has long, cylindrical, multinucleated cells with obvious striations?

15 Why is smooth muscle referred to as involuntary?

Nervous Tissue

Key topics in this section:

- Structure and function of nervous tissue
- Body locations where nervous tissue is found

Nervous tissue is sometimes termed *neural tissue*. It consists of cells called **neurons** (noor'on), or *nerve cells*, and a larger number of different types of **glial cells** (or *supporting cells*) that support, protect, and provide a framework for neurons **(table 4.15)**. This tissue will be discussed in detail in chapter 14, but we provide a brief description here.

Cardiac muscle is responsible for the rhythmic heart contractions that pump blood throughout the blood vessels of the body. Cardiac muscle cells are considered **involuntary** because they do not require nervous system activity to initiate a contraction; instead, specialized cardiac muscle cells in the heart wall initiate

Table 4.15	**Nervous Tissue**

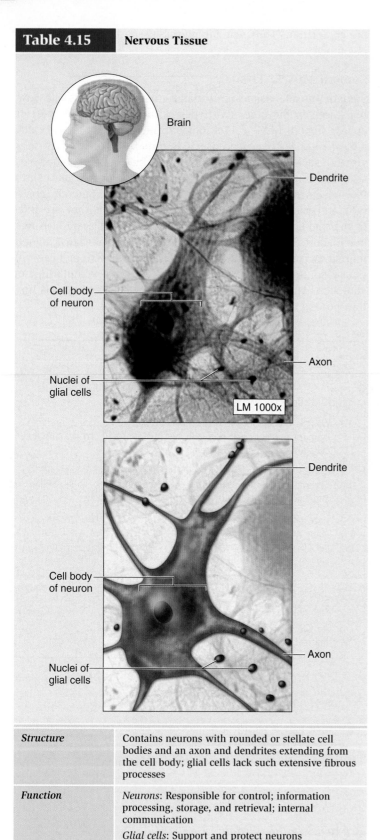

Structure	Contains neurons with rounded or stellate cell bodies and an axon and dendrites extending from the cell body; glial cells lack such extensive fibrous processes
Function	*Neurons*: Responsible for control; information processing, storage, and retrieval; internal communication *Glial cells*: Support and protect neurons
Location	Brain, spinal cord, and nerves

Characteristics of Neurons

Neurons are specialized to detect stimuli, process information quickly, and rapidly transmit electrical impulses from one region of the body to another. Each neuron has a prominent **cell body,** or soma, that houses the nucleus and most other organelles. The cell body is the "head" that controls the rest of the cell and produces proteins for the cell. Extending from the cell body are branches called nerve cell processes. The short, branched processes are **dendrites** (den′drītes; *dendritēs* = relating to a tree), which receive incoming signals from other cells and transmit the information to the cell body. The long nerve cell process extending from a cell body is the **axon** (ak′son; *axon* = axis), which carries outgoing signals to other cells. Due to the length of an axon, neurons are usually the longest cells in the body; some are longer than a meter. Much of the nervous tissue in the body is concentrated in the brain and spinal cord, the control centers for the nervous system.

WHAT DID YOU LEARN?

16 What general name is applied to the supporting cells in nervous tissue?

Tissue Change and Aging

Key topics in this section:

- How tissues may change in form, size, or number
- Changes that occur in tissues with age

Some tissue subtypes, although well established in an adult, may undergo changes.

Tissue Change

Sometimes a mature epithelium changes to a different form of mature epithelium, a phenomenon called **metaplasia** (met-ă-plā′zē-ă; *metaplasis* = transformation). Metaplasia may occur as an epithelium adapts to environmental conditions. For example, smokers typically experience metaplastic changes in the tracheal epithelium. The smoke and its by-products are the environmental stressors that change the normal pseudostratified ciliated columnar epithelium lining the trachea to a stratified squamous epithelium. Thus, the act of smoking causes metaplastic changes in the epithelium of the airway.

Tissues can grow or shrink in two ways: by a change in cell size or by a change in cell number. Increase in the size of existing cells is called **hypertrophy**; increase in the number of cells in the tissue due to mitosis is called **hyperplasia**. When growth proceeds out of control, a tumor that is composed of abnormal tissue develops, and the condition is termed **neoplasia** (nē-ō-plā′zē-ă; *neo* = new, *plasis* = molding).

Shrinkage of tissue by a decrease in either cell number or cell size is called **atrophy** (at′rō-fē). Atrophy may result from normal aging (senile atrophy) or from failure to use an organ (disuse atrophy). When people do not perform normal activities, their muscles exhibit disuse atrophy as the cells become smaller.

CLINICAL VIEW

Gangrene

Gangrene is the necrosis (death) of the soft tissues of a body part due to a diminished or obstructed arterial blood supply to that region. The body parts most commonly affected are the limbs, fingers, or toes. Gangrene may also occur as a consequence of either a bacterial infection or direct mechanical injury. Gangrene is a major complication for diabetics, who often suffer from diminished blood flow to their upper and lower limbs as a consequence of their disease. There are several different types of gangrene:

Intestinal gangrene usually occurs following an obstruction of the blood supply to the intestines. If the intestines are without sufficient blood, the tissue undergoes necrosis and gangrene. Untreated intestinal gangrene leads to death.

Dry gangrene is a form of gangrene in which the involved body part is desiccated, sharply demarcated, and shriveled, usually due to constricted blood vessels as a result of exposure to extreme cold. Dry gangrene can be a complication of frostbite or result from a variety of cardiovascular diseases that restrict blood flow,

primarily to the hands and feet, the areas most commonly affected by dry gangrene.

Wet gangrene is caused by a bacterial infection of tissues that have lost their blood and oxygen supply. The cells in the dying tissue rupture and release fluid (hence the name "wet" gangrene). The wet environment allows bacteria to flourish, and they often produce a foul-smelling pus. The most common bacteria associated with wet gangrene are *Streptococcus*, *Staphylococcus*, *Enterobacter*, and *Klebsiella*. Wet gangrene must be treated quickly with antibiotics and removal of the necrotic tissue.

Gas gangrene is often mistaken for wet gangrene. However, the bacteria typically associated with gas gangrene are *Clostridrium*, a type of bacterium that is called anaerobic because it can live and grow in the absence of oxygen. This type of gangrene usually affects muscle tissue. As the bacteria invade the necrotic tissue, a release of gases from the tissue produces gas bubbles. These bubbles make a crackling sound in the tissue, especially if the patient is moved. Symptoms of fever, pain, and edema (localized swelling) occur within 72 hours of the initial trauma to the region. The treatment for gas gangrene is similar to that for wet gangrene.

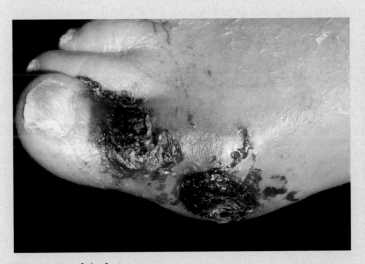

Dry gangrene of the foot.

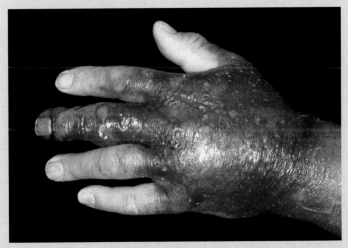

Gas gangrene in a recently amputated limb.

Tissue Aging

All tissues change as a result of aging. Proper nutrition, good health, normal circulation, and relatively infrequent wounds promote continued normal tissue functioning past middle age. Thereafter, the support, maintenance, and replacement of cells and extracellular matrix become less efficient. Physical damage, chemical changes, and physiologic changes can alter the structure and chemical composition of many tissues. For example, adequate intake of protein is required to enable the cells to continue synthesizing new proteins,

the body's structural and functional building blocks. As individuals age, epithelia become thinner, and connective tissues lose their pliability and resilience. Because the amount of collagen in the body declines, tissue repair and wound healing lose efficiency. Bones become brittle; joint pains and broken bones are common. Muscle and nervous tissue begin to atrophy. Diet and circulation problems contribute to these tissue declines. Eventually, cumulative losses from relatively minor damage or injury contribute to major health problems.

CLINICAL VIEW: In Depth

Tissue Transplantation

Grafting is the process of surgically transplanting healthy tissue to replace diseased, damaged, or defective tissue. The healthy tissue may be a person's own tissue or tissue from another individual or animal. The problem with using tissue from a different donor is that the patient's body may reject the tissue as "foreign." In essence, the body's immune system attacks the tissue because it recognizes that the tissue came from another body. There are four types of tissue grafts.

An **autograft** (au'tō-graft; *autos* = self, *graef* = implant) is a tissue transplant from one site on a person to a different site on the same person. Autografts are often performed with skin, as healthy skin from one part of the body is grafted to another part of the body where the skin has been damaged by burns or chemicals. Since an autograft uses a person's own tissue, the body does not reject the tissue as "foreign." However, autografts may not be feasible in certain situations, such as when the amount of skin damaged is so great that a transplant would not be possible. Most burn victims have damaged too much of their own skin to be able to provide autografts for all of their burned areas.

A **syngenetic** (sin-jĕ-net'ik; *syn* = together) **graft**, also called an *isograft*, is a tissue transplant from one person to a genetically identical person (i.e., an identical twin). It is unlikely that the body will reject the syngenetic graft because it came from a genetically identical individual. However, very few of us have an identical twin, so this type of graft is not possible for most people.

An **allograft** (al'ō-graft; *allos* = other) is the transplantation of tissue from a person who is not genetically identical. Many tissue types have been used as allografts, including skin, muscle, bone, and cartilage. In fact, the term allograft also refers to the transplantation of organs or parts of organs, such as heart valves, kidneys, and the liver. Orthopedists, physicians who treat musculoskeletal injuries, have frequently used musculoskeletal grafts from cadavers for such purposes as knee replacements or ligament reconstruction. In these cases, the bone, cartilage, and joint capsule from a cadaver are transplanted into another individual. These types of allografts are typically very successful. However, in 2002 the Food and Drug Administration (FDA) reported that some for-profit tissue banks had sent contaminated tis-sue samples to physicians. These tissues were contaminated with the bacterium *Clostridium*, which later infected the transplant recipients and caused deaths in some cases. The tissue banks in question came under investigation, and these tragic circumstances illustrated the potential complications of tissue allografts. In 2005, a New Jersey facility that was called a "tissue bank" stole tissue from cadavers awaiting burial or cremation. This tissue was implanted into patients. Forged death certificates and organ donor consent forms were used to try to legitimize this activity. In the spring of 2006, health professional organizations in Minnesota created a national model for the ethical procurement and use of human anatomic donations. This sets the stage for establishing best practices in the use of donated human organs.

Although most tissue allografts are successful, transplantation of entire organs is much more problematic. The patient and the organ donor must be as genetically similar as possible; traits such as blood type and other blood factors must "match." The closer the match, the less likely it is that the allograft will be rejected. The recipient of the transplanted organ(s) must take powerful immunosuppressant drugs, which help prevent the body from rejecting the organ. Unfortunately, these same drugs work by suppressing the immune system, making the transplant patient more susceptible to illness. Even with immunosuppressant drugs, rejection of allografts is common. Typically, graft rejection occurs after 15 to 25 days, when the transplant site has become infiltrated with graft-rejection cells that recognize and destroy the foreign cells.

A **heterograft** (hĕ'ter-ō-graft; *heteros* = other), also called a *xenograft* (zē'nō-graft; *xeno* = foreign), is a tissue transplant from an animal into a human being. For example, porcine (pig) and bovine (cow) tissue have been successfully used to replace heart valves, blood vessels, and bone. A chimpanzee kidney and a baboon heart have been transplanted into human patients. Porcine nervous tissue cells were transplanted into the brain of an individual with Parkinson disease in the hope that the healthy cells would stop or reverse the progress of the disease. Rejection of these animal tissues usually occurs frequently and quickly, which is not surprising, since tissue from a completely different species is being transplanted into a human. However, a few of these transplants have worked for a short time, and recent research is investigating the reasons for the lack of rejection of this tissue.

CLINICAL TERMS

adhesions Inflammatory bands that connect opposing serous surfaces.

biopsy Microscopic examination of tissue removed from the body for the purpose of diagnosing a disease. In a skin or muscle biopsy, a small piece of skin or muscle is removed, and the wound is sutured; in a needle biopsy, a tissue sample is removed from skin or an organ through an inserted needle; in an aspiration biopsy, cells are sucked into a syringe through a needle; in an endoscopic biopsy, a tissue section is taken by forceps in an endoscope within a hollow organ; in an open biopsy, a body cavity is opened for sample removal; and in an excisional biopsy, a lump is removed from a tissue or organ.

lesion (lē'zhŭn) Any localized wound, injury, or infection that affects tissue over a specific area rather than spread throughout the body.

liposuction A method of removing unwanted subcutaneous fat using a suction tube.

CHAPTER SUMMARY

<table>
<tr>
<td></td>
<td>■ There are four tissue types: epithelial tissue, connective tissue, muscle tissue, and nervous tissue.</td>
</tr>
<tr>
<td>Epithelial Tissue 81</td>
<td>■ Epithelial tissue covers the surface of the body, lines body cavities, and forms secretory structures called glands.</td>
</tr>
</table>

Characteristics of Epithelial Tissue   81

■ The characteristics of an epithelium include cellularity, polarity, attachment to a basement membrane, avascularity, innervation, and high regeneration ability.

■ Epithelial cells have an apical (free) surface, junctions on lateral membranes that bind neighboring cells, and a basal surface, which is closest to the basement membrane.

Functions of Epithelial Tissue   82

■ Epithelial tissues provide physical protection, control permeability, produce secretory products, and contain nerve cells that detect sensations. Gland cells are derived from epithelial cells and produce secretions.

Specialized Structure of Epithelial Tissue   82

■ The four types of epithelial cell junctions are tight junctions, adhering junctions, desmosomes, and gap junctions.

Classification of Epithelial Tissue   84

■ Epithelia are classified by two criteria: (1) number of cell layers, and (2) shape of apical surface cells.

■ A simple epithelium has only one cell layer overlaying the basement membrane. A stratified epithelium is two or more layers of cells thick, and only the deepest (basal) layer is in direct contact with the basement membrane. Pseudostratified columnar epithelium appears stratified but is not; all cells are in contact with the basement membrane.

Types of Epithelium   85

■ In a simple epithelium, the surface cells are thin and flat (squamous epithelium), about as tall as they are wide (cuboidal epithelium), or taller than they are wide (columnar epithelium).

■ The shape of transitional epithelium cells changes between relaxed and distended states.

Glands   92

■ Endocrine glands secrete hormones into the bloodstream. Exocrine glands secrete their products through ducts onto the epithelial surface.

■ Multicellular exocrine glands are classified by the structure of their ducts and the organization of the secretory portion of the gland.

■ Serous glands produce nonviscous, watery fluids; mucous glands secrete mucin that forms mucus; and mixed glands produce both types of secretions.

<table>
<tr>
<td>Connective Tissue 95</td>
<td>■ Connective tissue binds, protects, and supports the body organs.</td>
</tr>
</table>

Characteristics of Connective Tissue   95

■ Connective tissue contains cells, protein fibers, and a ground substance. The protein fibers and ground substance together form the extracellular matrix.

Functions of Connective Tissue   96

■ Connective tissue provides physical protection, support and structural framework, binding of structures, storage, transport, and immune protection.

Development of Connective Tissue   96

■ All connective tissues are derived from two types of embryonic connective tissue, mesenchyme and mucous connective tissue.

Classification of Connective Tissue   96

■ Loose connective tissue has a high volume of ground substance; it is easily distorted and serves to cushion shocks.

■ Dense connective tissue consists primarily of large amounts of extracellular protein fibers.

■ Supporting connective tissue (cartilage and bone) provides support and protection to the soft tissues and organs of the body.

■ Blood is a fluid connective tissue. Its cells are called formed elements, and the dissolved protein fibers and watery ground substance form an extracellular matrix called plasma.

<table>
<tr>
<td>Body Membranes 108</td>
<td>■ Mucous membranes line cavities that communicate with the exterior.</td>
</tr>
</table>

■ Serous membranes line internal cavities and are delicate, moist, and very permeable.

■ The external body surface is covered by the cutaneous membrane, which is dry, keratinized, and relatively thick.

■ Synovial membranes line the inner surface of synovial joint cavities.

(continued on next page)

CHAPTER SUMMARY (continued)

Muscle Tissue 109	■ Muscle tissue is composed of muscle cells, sometimes termed muscle fibers, which are capable of contractions resulting in cellular shortening along their longitudinal axes and producing movement, either of the skeleton or specific body parts.
	Classification of Muscle Tissue 109
	■ Skeletal muscle tissue is composed of long, multinucleated, cylindrical fibers that are striated and voluntary.
	■ Cardiac muscle tissue is located within the wall of the heart. It is composed of branched, short cells with one or two centrally located nuclei. It is striated and involuntary.
	■ Smooth muscle tissue is found in the walls of organs; it has short, tapered cells that are nonstriated and involuntary.
Nervous Tissue 111	■ Nervous tissue is composed of two specific cell types: neurons and glial cells. Neurons receive stimuli and transmit impulses in response. Glial cells interact with each other to form an extensive supporting framework for neurons and nervous tissue. Additionally, glial cells help provide nutrient support to the neuron.
	Characteristics of Neurons 112
	■ Neurons have a prominent cell body, dendrites, and a long process called the axon.
Tissue Change and Aging 112	**Tissue Change 112**
	■ Metaplasia is a change from one mature epithelial type to another in response to injury or stress.
	■ Hypertrophy is an increase in cell size, whereas hyperplasia is an increase in cell number.
	Tissue Aging 113
	■ When tissues age, repair and maintenance become less efficient, and the structure and chemical composition of many tissues are altered.

CHALLENGE YOURSELF

Matching

Match each numbered item with the most closely related lettered item.

_____ 1. smooth muscle

_____ 2. merocrine secretion

_____ 3. ground substance

_____ 4. simple columnar epithelium

_____ 5. goblet cell

_____ 6. dense regular connective tissue

_____ 7. endothelium

_____ 8. cardiac muscle

_____ 9. dense irregular connective tissue

_____ 10. avascular

a. a characteristic of all epithelia

b. contains intercalated discs

c. lines the small intestine lumen

d. scattered arrangement of protein fibers

e. part of extracellular matrix

f. unicellular exocrine gland

g. parallel arrangement of protein fibers

h. salivary glands, for example

i. lines blood vessel lumen

j. has no striations

Multiple Choice

Select the best answer from the four choices provided.

_____ 1. Which type of tissue contains a calcified ground substance and is specialized for structural support?
a. muscle tissue
b. nervous tissue
c. areolar connective tissue
d. bone connective tissue

_____ 2. What is the predominant cell type in areolar connective tissue?
a. mesenchymal cell
b. fibroblast
c. adipocyte
d. satellite cell

_____ 3. Preventing desiccation and providing surface lubrication within a body cavity are the functions of _____ membranes.
a. cutaneous
b. mucous
c. serous
d. synovial

_____ 4. Which of the following is a correct statement about a simple epithelium?
a. It protects against mechanical abrasion.
b. It may contain the protein keratin.
c. It is adapted for diffusion and filtration.
d. It is formed from multiple layers of epithelial cells.

_____ 5. Which of the following is not a function of an epithelium?
a. It is selectively permeable.
b. It serves as a packing and binding material.
c. The cells can produce secretory products.
d. It is designed for physical protection.

6. Which connective tissue type is composed of cells called chondrocytes and may be surrounded by a covering called perichondrium?
   a. cartilage
   b. dense irregular connective tissue
   c. bone
   d. areolar connective tissue

7. Aging effects on tissue include which of the following?
   a. Tissue is less able to maintain itself.
   b. Tissue has a decreased ability to repair itself.
   c. Epithelium becomes thinned.
   d. All of these are correct.

8. Which epithelial tissue type lines the trachea (air tube)?
   a. simple columnar epithelium
   b. pseudostratified ciliated columnar epithelium
   c. simple squamous epithelium
   d. stratified squamous epithelium

9. Which muscle type consists of long, cylindrical, striated cells with multiple nuclei located at the periphery of the cell?
   a. smooth muscle
   b. cardiac muscle
   c. skeletal muscle
   d. All of these are correct.

10. A gland that releases its secretion by exocytosis into secretory vesicles is called a _____ gland.
    a. apocrine
    b. merocrine
    c. holocrine
    d. All of these are correct.

Content Review

1. What are some common characteristics of all types of epithelium?

2. Describe the types of intercellular junctions between epithelial cells and where each is located.

3. List the epithelial type that is found: (a) lining the lumen of the stomach, (b) lining the oral cavity, (c) lining the urinary bladder, and (d) lining the tiny air sacs of the lungs.

4. What are the three secretion methods of exocrine glands, and how does each method work?

5. What characteristics are common to all connective tissues?

6. What are the main structural differences between dense regular and dense irregular connective tissue?

7. In what regions of the body would you expect to find hyaline cartilage, fibrocartilage, and elastic cartilage, and why would these supporting connective tissues be located in these regions?

8. Name the four types of body membranes, and cite a location of each type.

9. A significant structural feature in the microscopic study of cardiac muscle cells is the presence of gap junctions between neighboring cells. Why are these junctions so important?

10. What are the similarities and differences between skeletal muscle, cardiac muscle, and smooth muscle?

Developing Critical Reasoning

1. During a microscopy exercise in the anatomy laboratory, a student makes the following observations about a tissue section: (1) The section contains some different types of scattered protein fibers—that is, they exhibit different widths, some are branched, some are long and unbranched, and their staining characteristics differ (some are observed only with specific stains). (2) Several cell types with different morphologies are scattered throughout the section, but these cells are not grouped tightly together. (3) The observed section has some "open spaces"—that is, places between cells and the observed fibers in the section that appear clear with no recognizable features. What type of tissue is the student observing? Where might this tissue be found in the body?

2. Your father is suffering from painful knee joints. He has been told that he either has the early stages of arthritis or some inherent joint problems. His friend recommends that he take a chemical supplement with his meals (chondroitin sulfate), which has been shown to help some people with joint aches and pains. This supplement stimulates growth and recovery of degenerated cartilage on the surfaces of bones in joints. Based on your knowledge of connective tissues, do you think the chondroitin sulfate supplements could help your father's knee problems?

ANSWERS TO "WHAT DO YOU THINK?"

1. If epithelium contained blood vessels, the "gatekeeper" function of selective permeability would be compromised. Materials would be able to enter the body by entering the bloodstream without passing through the epithelium.

2. All types of stratified epithelium (stratified squamous, stratified columnar, stratified cuboidal) and transitional epithelium are suited for protection, because they have multiple layers of cells.

3. Connective tissue has fewer cells because it contains other materials, such as protein fibers and ground substance. The number of protein fibers in connective tissue is related to the strength and support the connective tissue gives. The ground substance can serve as a packing and binding material and can suspend the cells and protein fibers.

4. You have damaged dense regular connective tissue when you sprain your ankle.

5. A cutaneous membrane is found on the external surface of your forearm.

Visit the McKinley/O'Loughlin *Human Anatomy,* 2e website at aris.mhhe.com

5

INTEGUMENTARY SYSTEM

Integumentary System

The **integument** (in-teg'ū-ment; *integumentum* = a covering) is the skin that covers your body. Skin is also known as the **cutaneous** (kū-tā'nē-ŭs) **membrane**, or *cutaneous layer*. The **integumentary** (in-teg-ū-men'tă-rē) **system** consists of the skin and its derivatives—nails, hair, sweat glands, and sebaceous glands. We are most conscious of this highly visible and over-examined body system, because it characterizes our self-image and reflects our emotions. Our skin is a vulnerable barrier to the outside world; it is subjected to trauma, harmful chemicals, pollutants, microbes, and damaging sunlight. Still, it usually remains strong and pliable, is easily cleaned, is self-renewing, and serves as a visual indicator of our physiology and health. Changes in the color of the skin may reflect body disorders or anomalies; skin changes or lesions sometimes indicate systemic infections or diseases. The scientific study and treatment of the integumentary system is called **dermatology** (der-mă-tol'ō-jē; *derma* = skin, *logos* = study).

Structure and Function of the Integument

Key topics in this section:

- General structure of the integument
- Varied functions of the integument

The integument, or skin, is the body's largest organ. Although the skin is not as complex as most other organs, it does consist of different tissue types that collectively perform specific activities. Its surface is covered by an epithelium that protects underlying body layers. The connective tissues that underlie the epithelium contain blood vessels, which provide nutrients to the epithelial cells and give strength and resilience to the skin. Smooth muscle controls blood vessel diameter and hair position for these integumentary structures. Finally, nervous tissue supports and monitors sensory receptors in the skin, which provide information about touch, pressure, temperature, and pain.

Integument Structure

The integument covers the entire body surface, an area that ranges between about 1.5 and 2.0 square meters (m^2) and accounts for 7% to 8% of the body weight. Its thickness ranges between 1.5 and 4 millimeters (mm) or more, depending on body location. The integument consists of two distinct layers: a layer of stratified squamous epithelium called the epidermis, and a deeper layer of dense irregular connective tissue called the dermis **(figure 5.1)**. Deep to the dermis is a layer composed of areolar and adipose connective tissue called the subcutaneous layer, or hypodermis. The subcutaneous layer is not part of the integumentary system; however, it is described in this

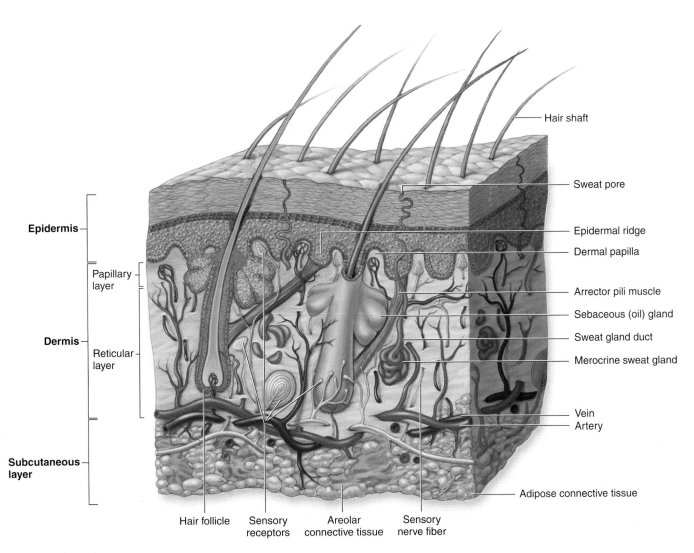

Figure 5.1

Layers of the Integument. A diagrammatic sectional view through the integument shows the relationship of the cutaneous membrane (skin) to the underlying subcutaneous layer.

chapter because it is closely involved with both the structure and function of the skin.

The integument meets the mucous membranes within the nostrils, lips, anus, urethral opening, and vaginal opening. At these sites, the transition is seamless, and the epithelial defenses remain intact and functional.

Integument Functions

The integument is more than just a wrapping around the body. It serves many varied functions, including protection, prevention of water loss, temperature regulation, metabolic regulation, immune defense, sensory reception, and excretion.

Protection

The skin acts as a physical barrier that protects the entire body from physical injury, trauma, bumps, and scrapes. It also offers protection against harmful chemicals, toxins, microbes, and excessive heat or cold. Paradoxically, it can absorb certain chemicals and drugs (such as estrogen from a birth control patch or nicotine from a nicotine patch). Thus, the skin is said to be **selectively permeable** because some materials are able to pass through it while others are effectively blocked. The epidermis is designed to withstand stresses and regenerate itself continuously throughout a person's lifetime. The skin also protects deeper tissues from solar radiation, especially ultraviolet rays. When exposed to the sun, the melanocytes become more active and produce more melanin, thus giving the skin a darker, tanned look. Even when you get a sunburn, the deeper tissues (muscles and internal organs) remain unaffected.

Prevention of Water Loss

The epidermis is water resistant and helps prevent unnecessary water loss. (If the skin were not water resistant, each time you took a bath you would swell up like a sponge as your skin absorbed water!) Water cannot easily enter or exit the skin, unless it is specifically secreted by the sweat glands. The skin also prevents the water within the body cells and in the extracellular (the fluid outside of cells) from "leaking out." When the skin is severely burned, a primary danger is dehydration, because the individual has lost the protective skin barrier, and water can escape from body tissues.

Although the integument is water resistant, it is not entirely waterproof. Some interstitial fluids slowly escape through the epidermis to the surface, where they evaporate into the surrounding air, a process called **transepidermal water loss (TEWL).** Approximately 500 milliliters (ml) (approximately 1 pint) of water is lost daily by evaporation of moisture from the skin or from respiratory passageways during breathing. **Insensible perspiration** is the release of water vapor from sweat glands under "normal" circumstances when we are not sweating. In contrast, **sensible perspiration** is sweating.

Temperature Regulation

Body temperature is influenced by vast capillary networks and sweat glands in the dermis. When the body is too warm and needs to dissipate heat, the diameter of the blood vessels in the dermis enlarges to permit more blood flow through the dermis and sweat glands release fluid onto the skin surface. As relatively more blood flows through these dermal vessels, the warmth from the blood dissipates through the skin, and the body cools off by evaporation of the sweat. Conversely, when the body is cold and needs to conserve heat, the blood vessels in the dermis constrict to reduce blood flow. In an effort to conserve heat, more blood is shunted to deeper body tissues, and relatively less blood flows in the dermal blood vessels.

Metabolic Regulation

Vitamin D₃ is a cholesterol derivative synthesized from **cholecalciferol** (kŏ′lē-kal-sif′er-ol), which is produced by some epidermal cells when they are exposed to ultraviolet radiation. **Calcitriol** (kal-si-trī′ol) is synthesized from the cholecalciferol by some endocrine cells in the kidney. Calcitriol, the active form of vitamin D₃, is a hormone that promotes calcium and phosphorus absorption from ingested materials across the wall of the small intestine. Thus, the synthesis of vitamin D₃ is important in regulating the levels of calcium and phosphate in the blood. As little as 15 minutes of sunlight a day will provide your body with its daily vitamin D requirement!

Immune Defense

The epidermis contains a small population of immune cells. These immune cells, called **epidermal dendritic** (den-drit′ik) **cells,** or *Langerhans cells*, play an important role in initiating an immune response by phagocytizing pathogens that have penetrated the epidermis and also against epidermal cancer cells.

Sensory Reception

The skin contains numerous sensory receptors. These receptors are associated with nerve endings that detect heat, cold, touch, pressure, texture, and vibration. For example, **tactile cells** (or *Merkel cells*) are large, specialized epithelial cells that stimulate specific sensory nerve endings when they are distorted by fine touch or pressure. Because your skin is responsible for perceiving many stimuli, it needs different sensory receptor types to detect, distinguish, and interpret these stimuli.

Excretion by Means of Secretion

Skin exhibits an excretory function when it secretes substances from the body during sweating. Sweating, or sensible perspiration, occurs when the body needs to cool itself off. Notice that sweat sometimes feels "gritty" because of the waste products being secreted onto the skin surface. These substances include water, salts, and urea, a nitrogen-containing waste product of body cells. In addition, the skin contains sebaceous glands that secrete an oily material called sebum, which lubricates the skin surface and hair.

WHAT DID YOU LEARN?

1 What are the two major layers of the integument and the components of each?

2 What is the relationship between exposure to sunlight and the body's need for vitamin D?

WHAT DO YOU THINK?

1 During the Industrial Revolution, as children spent little time outdoors and most of their time working in factories, increasing numbers of them developed a bone disorder called rickets. Rickets is caused by inadequate vitamin D. Based on your knowledge of skin function, why do you think these children developed rickets?

Epidermis

Key topics in this section:

- Arrangement and functions of the epidermal strata
- Epidermal variations in thickness, color, and markings

The epithelium of the integument is called the **epidermis** (ep-i-derm′is; *epi* = on, *derma* = skin). The epidermis is a keratinized, stratified squamous epithelium. Like other epithelia, the epidermis is avascular, and it acquires its nutrients through diffusion from the underlying dermis.

Epidermal Strata

Careful examination of the epidermis, from the basement membrane to its surface, reveals several layers, or strata. From deep to superficial, these layers are the stratum basale, the stratum spinosum, the stratum granulosum, the stratum lucidum (found in thick skin only), and the stratum corneum (**figure 5.2**). The first three strata listed are composed of living keratinocytes, and last two strata contain dead keratinocytes.

Stratum Basale

The deepest epidermal layer is the **stratum basale** (strat′ŭm bah-sā′lē) (also known as the *stratum germinativum* or *basal layer*). This single layer of cells ranges from cuboidal to low columnar in appearance. It is tightly attached to an underlying basement membrane that separates the epidermis from the connective tissue of the adjacent dermis. Three types of cells occupy the stratum basale (figure 5.2b):

1. **Keratinocytes** (ke-rat′i-nō-sīt; *keras* = horn) are the most abundant cell type in the epidermis and are found throughout all epidermal strata. The stratum basale is dominated by large keratinocyte stem cells, which divide to provide both replacement stem cells and new keratinocytes that replace the dead keratinocytes shed from the surface. Their name is derived from their role in the synthesis of the protein **keratin** (ker′ă-tin) in the epidermal cells of the skin. Keratin is a family of fibrous structural proteins that are both tough and insoluble. Fibrous keratin molecules can twist and intertwine around each other to form helical intermediate filaments of the cytoskeleton (see chapter 2). The keratins found in epidermal cells of the skin are called *cytokeratins*. Their structure in these cells gives skin its strength and makes the epidermis almost waterproof.

2. **Melanocytes** (mel′ă-nō-sīt; *melano* = black) have long, branching cytoplasmic processes and are scattered among the keratinocytes of the stratum basale. These processes transfer pigment granules, called **melanosomes** (mel′ă-nō-sōmes), into the keratinocytes within the basal layer and sometimes within more superficial layers. This pigment (black, brown, or yellow-brown) accumulates around the nucleus of the keratinocyte and shields the DNA within the nucleus from ultraviolet radiation. The darker tones of the skin result from melanin being produced by the melanocytes and from the darkening of melanin already present upon exposure to ultraviolet light.

3. **Tactile cells** are few in number and found scattered among the cells within the stratum basale. Tactile cells are sensitive to touch, and when compressed, they release chemicals that stimulate sensory nerve endings, providing information about objects touching the skin.

Stratum Spinosum

Several layers of polygonal keratinocytes form the **stratum spinosum** (spī-nō′sŭm), or *spiny layer*. Each time a keratinocyte stem cell in the stratum basale divides, the daughter cell that will differentiate into the new epidermal cell is pushed toward the external surface from the stratum basale. Once this new cell enters the stratum spinosum, the cell begins to differentiate into a nondividing, highly specialized keratinocyte. Sometimes the deepest cells in this layer still undergo mitosis to help replace epidermal cells that exfoliate from the epidermal surface. The nondividing keratinocytes in the stratum spinosum attach to their neighbors by many intercellular junctions called desmosomes (described in chapter 4). The process of preparing epidermal tissue for observation on a microscope slide shrinks the cytoplasm of the cells in the stratum spinosum. Because the cytoskeletal elements and desmosomes remain intact, the shrunken stratum spinosum cells resemble miniature porcupines attached to their neighbors. These bridges between neighboring cells provide a spiny appearance, explaining the name of the layer.

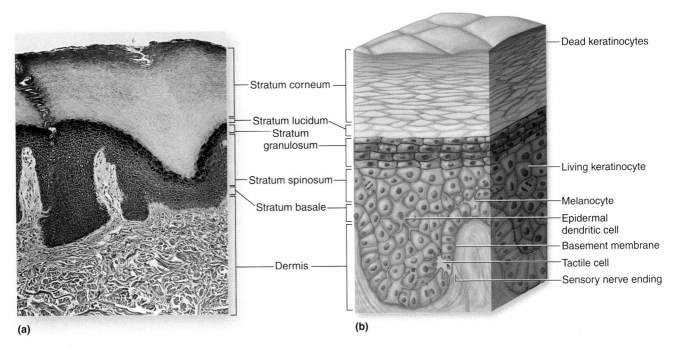

(a)   **(b)**

Figure 5.2

Epidermal Strata. *(a)* Photomicrograph and *(b)* diagram compare the order and relationships of the epidermal strata in thick skin.

In addition to the keratinocytes, the stratum spinosum also contains the fourth epidermal cell type, the epidermal dendritic cells (figure 5.2b). Epidermal dendritic cells are immune cells that help fight infection in the epidermis. These cells are often present but not easily identifiable in both the stratum spinosum and the more superficial stratum granulosum. Their phagocytic activity initiates an immune response to protect the body against pathogens that have penetrated the superficial layers of the epidermis as well as against epidermal cancer cells.

Stratum Granulosum

The **stratum granulosum** (gran-ū-lō′sum), or *granular layer*, consists of three to five layers of keratinocytes superficial to the stratum spinosum. Within this stratum begins a process called **keratinization** (ker′ă-tin-i-zā′shŭn), by which the keratinocytes fill up with the protein keratin. Several significant events occur during keratinization. As the cells pass through the stratum granulosum and true keratin filaments (intermediate filaments of the cytoskeleton) begin to develop, the cells become thinner and flatter. Their membranes thicken and become less permeable. The nucleus and all organelles disintegrate, and the cells start to die. Subsequently, the dehydrated material left within the cells forms a tightly interlocked layer of keratin fibers sandwiched between thickened phospholipid membranes. Keratinization is not complete until the cells reach the more superficial epidermal layers. A fully keratinized cell is dead (because it has neither a nucleus nor organelles), but it is strong because it contains keratin.

Stratum Lucidum

The **stratum lucidum** (lū′sĭ-dum), or *clear layer*, is a thin, translucent region about two to three cell layers thick that is superficial to the stratum granulosum. This stratum is found only in thick skin, such as the palms of the hands and the soles of the feet. Cells occupying this layer appear pale and featureless, and have indistinct boundaries. The keratinocytes within this layer are flattened and filled with the protein **eleidin** (ē-lē′ĭ-din), an intermediate product in the process of keratin maturation.

Stratum Corneum

The **stratum corneum** (kōr′nē-ŭm; *corneus* = horny, or *hornlike layer*), is the most superficial layer of the epidermis. It is the stratum you see when you look at your skin. The stratum corneum consists of about 20–30 layers of dead, scaly, interlocking keratinized cells called corneocytes (kōr′nē-ō-sīt). The dead cells are **anucleate** (lacking a nucleus) and tightly packed together.

A keratinized (or cornified) epithelium contains large amounts of keratin. After keratinocytes are formed from stem cells within the stratum basale, they change in structure and in their relationship to their neighbors as they move through the different strata until they eventually reach the stratum corneum and are sloughed off from its external surface. Migration of the keratinocyte to the stratum corneum from the stratum basale occurs during the first 2 weeks of the keratinocyte's life. The dead, keratinized cells usually remain for an additional 2 weeks in the exposed stratum corneum layer, providing a barrier for cells deeper in the epidermis before they are shed, washed away, or removed by abrasion. Overall, keratinocytes are present for about 1 month following their formation.

The normally dry stratum corneum presents a thickened surface unsuitable for the growth of many microorganisms. Additionally, some secretions onto the surface of the epidermis from exocrine glands help prevent the growth of microorganisms on the epidermis, thus supporting its barrier function.

Study Tip!

In your anatomy lab, you may be asked to identify a specific epidermal stratum. Answer the following questions to help identify these strata.

1. Is the epidermal stratum near the free surface of the epithelium or closer to the basal surface? Remember, the stratum corneum forms the free surface, while the stratum basale forms the deepest epidermal layer.

2. What is the shape of the cells? The stratum basale contains cells that are cuboidal to low columnar in shape, the stratum spinosum contains polygonal cells, and the stratum lucidum and stratum corneum contain squamous cells.

3. Do the keratinocytes have a nucleus, or are they anucleate (lacking a nucleus)? When the keratinocytes are still alive (as in the strata basale, spinosum, and granulosum), you will be able to see nuclei in the keratinocytes. The stratum lucidum and stratum corneum layers contain anucleate keratinocytes.

4. How many layers of cells are in the stratum? The stratum basale has only one layer of cells, and the stratum corneum contains 20–30 layers of cells. The other layers contain about 2–5 layers of cells.

5. Does the cytoplasm of the cells contain visible dark granules? If the answer is yes, you likely are looking at the stratum granulosum.

Variations in the Epidermis

The epidermis exhibits variations among different body regions within a single individual, as well as differences between individuals. The epidermis varies in thickness, coloration, and skin markings.

Thick Skin Versus Thin Skin

Over most of the body, the skin ranges from 1 mm to 2 mm in thickness. Skin is classified as either thick or thin based on the number of strata in the epidermis and the relative thickness of the epidermis, rather than the thickness of the entire integument (**figure 5.3**).

CLINICAL VIEW

Transdermal Administration of Drugs

Some drugs may be administered through the skin, a process called **transdermal administration**. Drugs that are soluble either in oils or lipid-soluble carriers may be administered transdermally by affixing a patch containing the drug to the skin surface. These drugs slowly penetrate the epidermis and are absorbed into the blood vessels of the dermis. Transdermal patches are especially useful because they release a continual, slow amount of the drug over a relatively long period of time. The epidermal barrier requires that the concentration of the drug in the patch be relatively high. There are transdermal patches that contain nicotine (to help people quit smoking), estrogen (for hormone replacement therapy [HRT] or birth control), or nitroglycerin (to prevent heart attack). These patches are advantageous because the patient is not required to ingest daily medication.

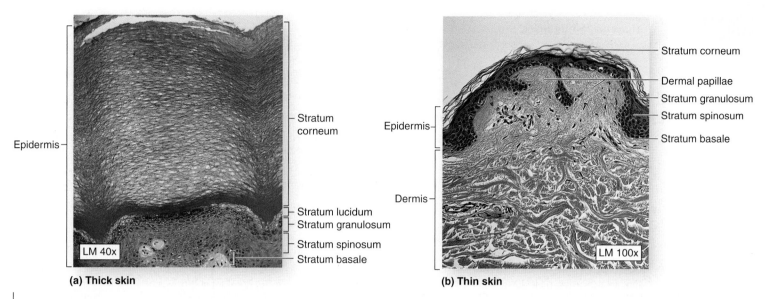

(a) Thick skin

Epidermis

Stratum corneum

Stratum lucidum
Stratum granulosum
Stratum spinosum
Stratum basale

LM 40x

(b) Thin skin

Stratum corneum
Dermal papillae
Stratum granulosum
Stratum spinosum
Stratum basale

Epidermis

Dermis

LM 100x

Figure 5.3

Thick Skin and Thin Skin.   The stratified squamous epithelium of the epidermis varies in thickness, depending upon the region of the body in which it is located. *(a)* Thick skin contains all five epidermal strata and covers the soles of the feet and the palms of the hands. *(b)* Thin skin covers most body surfaces; it lacks a stratum lucidum.

Thick skin is found on the palms of the hands, the soles of the feet, and corresponding surfaces of the fingers and toes. All five epidermal strata occur in thick skin. Thick skin ranges between 400 and 600 micrometers (μm) thick. Thick skin contains sweat glands, but no hair follicles or sebaceous glands.

Thin skin covers most of the body. The epidermis lacks the stratum lucidum, so it has only four layers. Thin skin contains the following accessories: hair follicles, sebaceous glands, and sweat glands. The epidermis of thin skin is only 75 μm to 150 μm thick.

WHAT DO YOU THINK?

2 Why does thick skin lack hair follicles and sebaceous glands? Think about the body locations of thick skin and how the presence of hair follicles and sebaceous glands might interfere with the job of thick skin in those areas.

Skin Color

Normal skin color results from a combination of hemoglobin, melanin, and carotene. **Hemoglobin** (hē-mō-glō′bin; *haima* = blood) is an oxygen-binding protein present within red blood cells. Upon binding oxygen, hemoglobin exhibits a bright red color, giving blood vessels in the dermis a bright reddish tint that is most easily observed in the skin of lightly pigmented individuals.

Melanin (mel′ă-nin) is a pigment produced and stored in cells called melanocytes (**figure 5.4**; see figure 5.2*b*). This pigment is synthesized from the amino acid tyrosine, and its production requires the enzyme tyrosinase. There are two types of melanin, *eumelanin* and *pheomelanin*, and they occur in various ratios of yellow, reddish, tan, brown, and black shades. Melanin is transferred in membrane-bound vesicles from melanocytes to keratinocytes in the stratum basale. The keratinocytes that receive the melanin are

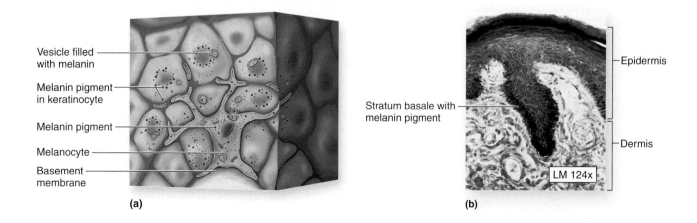

Vesicle filled with melanin

Melanin pigment in keratinocyte

Melanin pigment

Melanocyte

Basement membrane

(a)

Stratum basale with melanin pigment

Epidermis

Dermis

LM 124x

(b)

Figure 5.4

Production of Melanin by Melanocytes.   Melanin gives a yellow to tan to brown color to the skin. *(a)* Vesicles in melanocytes transport the melanin pigment to the keratinocytes, where the pigment surrounds the nucleus. *(b)* Melanin is incorporated into the cells of the stratum basale.

Table 5.1	Abnormal Skin Colors	
Name	**Description**	**Cause**
Albinism	Hair is white, skin is pale, irises of eyes are pink	Lack of melanin production; inherited recessive condition in which enzyme needed to synthesize melanin is nonfunctional, so melanocytes cannot produce melanin
Bronzing	Skin appears golden brown, copper, or bronze in color	Glucocorticoid hormone deficiency in the adrenal cortex; Addison disease
Cyanosis	Skin appears blue as a result of oxygen deficiency in circulating blood	Airway obstruction, emphysema, or respiratory arrest; also results from exposure to cold weather or from cardiac arrest with slow blood flow
Erythema	Skin appears abnormally red	Exercise, sunburn, excess heat, emotions (anger or embarrassment) resulting in increased blood flow in dilated blood vessels in the dermis
Hematoma	A bruise (visible pool of clotted blood) is observable through the skin	Usually due to trauma (a blow to the skin); also may be indicative of hemophilia or a nutritional or metabolic disorder
Jaundice	Skin and sclera (white of the eyes) appear yellow	Elevated levels of bilirubin in the blood; often occurs when normal liver function is disrupted, and in premature infants whose liver function is not yet sufficient
Pallor	Skin appears ashen, pale due to white collagen fibers housed within the dermis	Decreased blood flow to the skin; occurs as a result of low blood pressure, cold temperature, emotional stress, severe anemia, or circulatory shock

displaced toward the stratum corneum, and thus melanocyte activity affects the color of the entire epidermis.

All people have about the same number of melanocytes. However, melanocyte activity and the color of the melanin produced by these cells varies among individuals and races, resulting in different skin tones. Darker-skinned individuals have melanocytes that produce relatively more melanin than do those of lighter-skinned individuals. Further, these more active melanocytes tend to package and send melanin to cells in the more superficial epidermal layers, such as the stratum granulosum. The amount of melanin in the skin is determined by both heredity and light exposure. Melanin pigment surrounds the keratinocyte nucleus, where it absorbs **ultraviolet (UV) radiation** in sunlight,

thus preventing damage to nuclear DNA. Exposure to UV light both darkens melanin already present and stimulates melanocytes to make more melanin.

Carotene (kar'ō-tēn) is a yellow-orange pigment that is acquired in the body by eating various yellow-orange vegetables, such as carrots, corn, and squash. Normally, carotene accumulates inside keratinocytes of the stratum corneum and within the subcutaneous fat. In the body, carotene is converted into vitamin A, which has an important function in normal vision. Additionally, carotene has been implicated in reducing the number of potentially dangerous molecules formed during normal metabolic activity and in improving immune cell number and activity.

Table 5.1 describes some abnormalities in skin color.

CLINICAL VIEW

UV Radiation, Sunscreens, and Sunless Tanners

The sun generates three forms of ultraviolet radiation: UVA (ultraviolet A), UVB (ultraviolet B), and UVC (ultraviolet C). The wavelength of UVA ranges between 320 and 400 nanometers (nm), that of UVB ranges between 290 and 320 nm, and the peak output of UVC occurs at 253 nm. In contrast, visible light ranges begin at about 400 nm (the deepest violet). UVC rays are absorbed by the upper atmosphere and do not reach the earth's surface, while UVA and UVB rays can affect individuals' skin color.

UVA light is commonly termed "tanning rays," and UVB is often called "burning sun rays." Many tanning salons claim to provide a "safe" tan because they use only UVA rays. However, UVA rays can cause burning as well as tanning, and they also inhibit the immune system. Both UVA and UVB rays are believed to initiate skin cancer. Thus, there is no such thing as a "healthy" suntan.

Sunscreens are lotions that contain materials to help protect the skin from UVA and UVB rays. Sunscreens can help protect against skin cancer, but *only* if they are used correctly. Many people do not follow the directions, so they have a false sense of security when they apply sunscreen. First, sunscreen must be applied liberally over all exposed body surfaces, and reapplied after entering the water or perspiring. Second, it is important to use a sunscreen that has a high enough **SPF (sun protection factor).**

SPF is a number determined experimentally by exposing subjects to a light spectrum that mimics noontime sun. Some of the subjects wear sunscreen while others do not. The amount of light that induces redness in sunscreen-protected skin, divided by the amount of light that induces redness in unprotected skin, equals the SPF. For example, a sunscreen with an SPF of 15 will delay the onset of a sunburn in a person who would otherwise burn in 10 to 150 minutes. Thus, a sunscreen with an SPF of 15 will keep the skin from burning 15 times longer than if the skin is unprotected. However, it is never safe to assume that a sunscreen will protect you completely from the sun's harmful rays.

Sunless tanners create a tanned, bronzed skin without UV light exposure. There are many types of sunless tanners, but the most effective ones contain **dihydroxyacetone (DHA)** as their active ingredient. DHA is a colorless sugar derived from glyercin. Its effects on the skin were first discovered by the Germans in the 1920s, when they saw that accidentally spilling DHA on the skin produced darkening. When applied to the epidermis, DHA interacts with the amino acids in the cells to produce a darkened, brown color. Since only the most superfical epidermal cells are affected, the color change is temporary, lasting about 5 to 7 days. There are other sunless tanners on the market that contain other chemicals, but they do not appear to be as effective. It is important to note that sunless tanners contain no sunscreen and offer no protection against UV rays. Thus, individuals who use sunless tanners should also apply sunscreen to protect their skin.

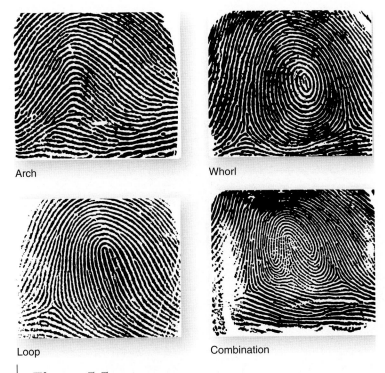

Arch

Whorl

Loop

Combination

Figure 5.5

Friction Ridges of Thick Skin. Friction ridges form fingerprints, palm prints, and toe prints. Shown here are four basic fingerprint patterns.

Skin Markings

A **nevus** (nē′vŭs; pl., nē′vī; *naevus* = mole, birthmark), commonly called a **mole** is a harmless, localized overgrowth of melanin-forming cells. Almost everyone is born with a few nevi, and some people have as many as 20 or more. On very rare occasions, a nevus may become malignant, typically as a consequence of excessive UV light exposure. Thus, nevi should be monitored for changes that may suggest malignancy. **Freckles** are yellowish or brown spots that represent localized areas of excessive melanocyte activity, not an increase in melanocyte numbers. A freckle's degree of pigmentation varies and depends on both sun exposure and heredity.

A **hemangioma** (he-man′jē-ō′mă; *angio* = vessel, *oma* = tumor) is a congenital anomaly that results in skin discoloration due to blood vessels that proliferate and form a benign tumor. Capillary hemangiomas or "strawberry-colored birthmarks," appear in the skin as bright red to deep purple nodules that usually disappear in childhood. Cavernous hemangiomas, sometimes called "port-wine stains," involve larger dermal blood vessels and may last a lifetime.

The contours of the skin surface follow ridge patterns, varying from small, conical pegs (in thin skin) to complex arches and whorls (in thick skin) called **friction ridges**. Friction ridges are found on the fingers, palms, soles, and toes **(figure 5.5)**. These ridges are formed from large folds and valleys of both the dermis and epidermis. Friction ridges increase friction so that objects do not slip easily from our hands and our feet do not slip on the floor when we walk. Friction ridges can leave noticeable prints on touched surfaces, commonly called "fingerprints." Because each individual has a unique pattern of friction ridges, fingerprints have become a valuable identification tool for law enforcement. Medical applications are possible as well (see Clinical View: Dermatoglyphics).

WHAT DO YOU THINK?

❸ Why are people's attempts to change their recognizable fingerprints usually not successful?

WHAT DID YOU LEARN?

❸ Why is the stratum spinosum important in maintaining the integrity of the skin?

❹ Briefly describe the process of keratinization. Where does it begin? Why is it important?

❺ What normal skin accessories are not present in thick skin?

❻ How do melanocytes help protect the skin?

Dermis

Key topics in this section:

- Organization and function of the layers of the dermis
- Nerve and blood supply to the dermis

The **dermis** (der′mis) lies deep to the epidermis and ranges in thickness from 0.5 mm to 3.0 mm. This connective tissue layer of the integument is composed of cells of the connective tissue proper and primarily of collagen fibers, although both elastic and reticular fibers are also present. Other components of the dermis are blood vessels, sweat glands, sebaceous glands, hair follicles, nail roots, sensory nerve endings, and smooth muscle tissue. There are two major regions of the dermis: a superficial papillary layer and a deeper reticular layer **(figure 5.6)**.

Figure 5.6

Layers of the Dermis. The dermis is composed of a papillary layer and a reticular layer.

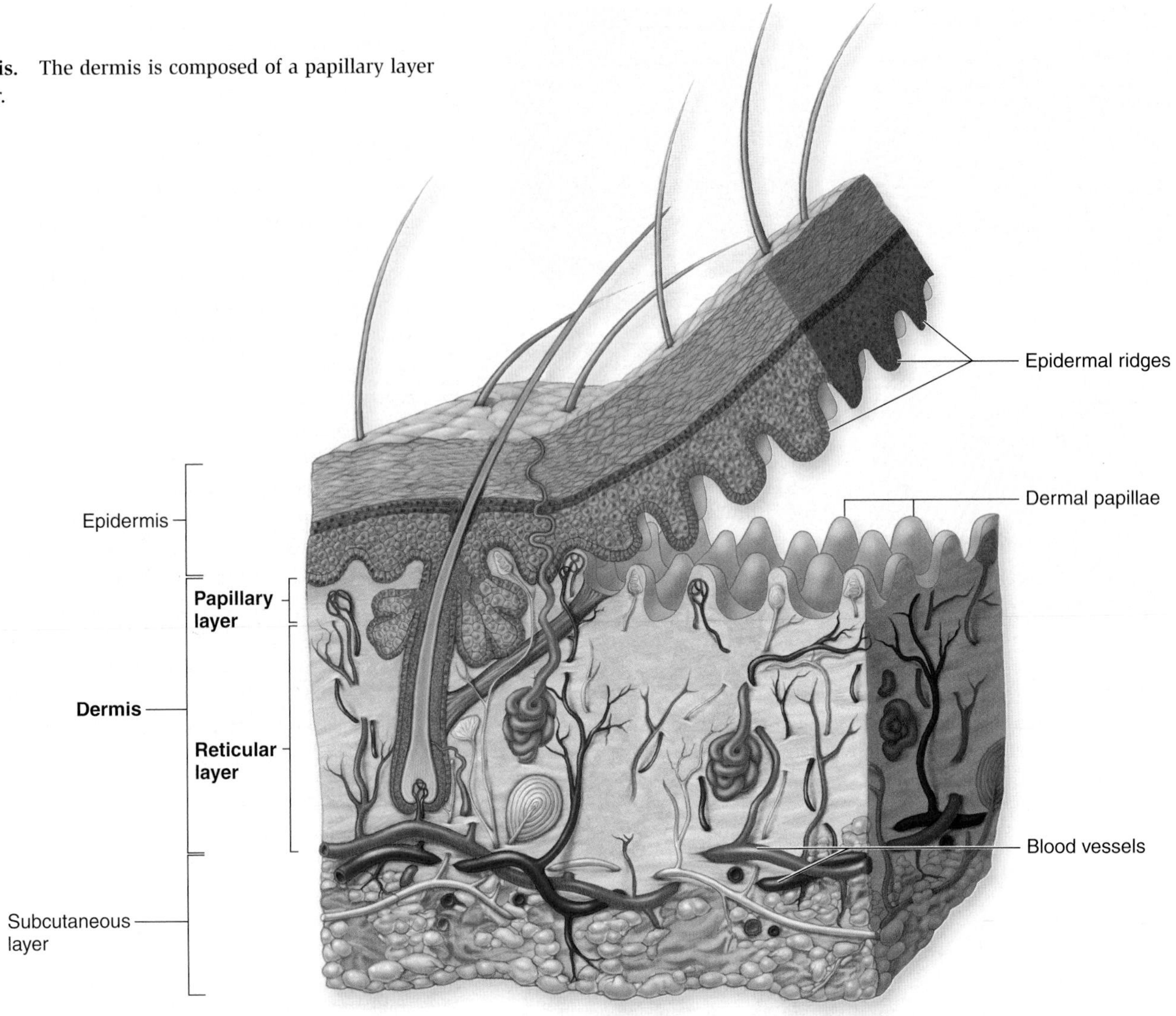

Papillary Layer of the Dermis

The **papillary** (pap′i-lār-ē) **layer** is the superficial region of the dermis directly adjacent to the epidermis. It is composed of areolar connective tissue, and it derives its name from the projections of the dermis toward the epidermis called **dermal papillae** (der′măl pă-pil′ē; sing., papilla; a nipple). The dermal papillae interlock with deep projections of epidermis called **epidermal ridges**. Together, the epidermal ridges and dermal papillae increase the area of contact between the epidermis and dermis and connect these layers. Each dermal papilla contains the capillaries that supply nutrients to the cells of the epidermis. It also houses sensory receptors, such as some of the receptors shown in figure 5.1, that continuously monitor touch on the surface of the epidermis. Chapter 19 discusses tactile receptors in detail.

Reticular Layer of the Dermis

The **reticular layer** forms the deeper, major portion of the dermis and extends from the thin, overlying papillary layer to the underlying subcutaneous layer. The reticular layer consists primarily of dense irregular connective tissue through which large bundles of collagen fibers project in all directions. These fibers are interwoven into a meshwork that surrounds the structures in the dermis, such as hair follicles, sebaceous glands, sweat glands, nerves, and blood

vessels. The word reticular in the name of this layer means "network" and refers to the meshwork of collagen fibers. These interwoven collagen fiber bundles obscure any distinct boundary between the papillary and reticular layers. Additionally, collagen fibers extend internally from the reticular layer of the dermis into the underlying subcutaneous layer.

Stretch Marks, Wrinkles, and Lines of Cleavege

Together, collagen fibers and elastic fibers in the dermis contribute to the observed physical characteristics of the skin. Whereas collagen fibers impart tensile strength, elastic fibers allow some stretch and contraction in the dermis during normal movement. Stretching of the skin, which may occur as a result of excessive weight gain or pregnancy, often exceeds the skin's elastic capabilities. When the skin is stretched beyond its capacity, some collagen fibers are torn and result in stretch marks, called **striae** (strī′ē; *stria* = furrow). Both the flexibility and thickness of the dermis are diminished by the effects of exposure to UV light and by aging. These causative agents may result in either sagging skin or increased wrinkles.

At specific body locations, the majority of the collagen and elastic fibers in the skin are oriented in parallel bundles. The specific orientation of dermal fiber bundles is a result of the direction of applied stress during routine movement; therefore, the alignment of

Figure 5.7

Lines of Cleavage. Lines of cleavage (tension lines) partition the skin and indicate the predominant direction of underlying collagen fibers in the reticular layer of the dermis.

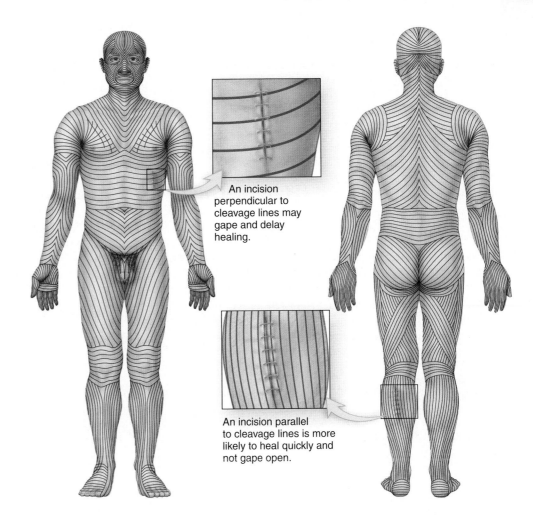

An incision perpendicular to cleavage lines may gape and delay healing.

An incision parallel to cleavage lines is more likely to heal quickly and not gape open.

the bundles functions to resist stress. **Lines of cleavage** (or tension lines) in the skin identify the predominant orientation of collagen fiber bundles **(figure 5.7)**. These are clinically and surgically significant because any procedure resulting in a cut at right angles to a cleavage line is usually pulled open due to the recoil from cut elastic fibers. This often results in slow healing and increased scarring. In contrast, a cut parallel to a cleavage line usually remains closed, resulting in faster healing. Therefore, surgical procedures should be planned to allow for these lines of cleavage, thus ensuring rapid healing and preventing scarring.

Innervation and Blood Supply

Nerve fibers are extensively dispersed throughout the dermis, a property called innervation. Nerve fibers in the skin monitor sensory receptors in the dermis and epidermis, and they also control both blood flow and gland secretion rates. Tactile corpuscles and tactile cells perceive touch sensations and work with a variety of other sensory nerve endings in the skin. This rich innervation allows us to be very aware of our surroundings and to differentiate among the different kinds of sensory signals from receptors in the skin.

Recall that all epithelia, including the epidermis, are avascular. Therefore, blood vessels within the dermis must supply nutrients to the living cells in the epidermis as well as to all structures in the dermis. The largest of these blood vessels lie along the border between the reticular layer of the dermis and the subcutaneous layer. Smaller vessels branch into the dermis to supply the hair follicles, sweat glands, sensory receptors, and other structures housed

there. The smallest arterial vessels connect to capillary loops within the dermal papillae. Eventually, these capillaries drain into small vessels, forming a vessel network that merges into small veins draining the dermis.

Dermal blood vessels have an important role in regulating body temperature and blood pressure. **Vasoconstriction** (vā′sō; *vas* = a vessel) means that the diameters of the vessels narrow, so relatively less blood can travel through them. Therefore, relatively more blood must travel in blood vessels that are deeper internal to the skin. The net effect of vasoconstriction of the dermal blood vessels is a shunting of blood *away* from the periphery of the body. If the body is cold, the dermal blood vessels vasoconstrict to conserve heat in the blood. This is why we are paler when we are exposed to cold temperatures.

Conversely, **vasodilation** of the dermal blood vessels means that the diameter of the vessels increases, so relatively more blood can travel through them. As more blood is shunted to these superficial blood vessels, the heat from the blood may be more easily dissipated through the skin. If the body is too warm, the dermal blood vessels vasodilate so more blood can travel close to the surface and excess heat can be lost. This additional blood flow in the dermis gives a more reddish or pinkish hue to the skin. Thus, your face may become flushed when you exercise because your dermal blood vessels are dilated in an attempt to release the excess heat you generated while working out. Because blood volume typically remains constant, any increase in circulation to the skin results in a decrease in circulation to other organs.

WHAT DID YOU LEARN?

7 Briefly describe the structure of epidermal ridges and dermal papillae. What is the importance of each, and how do they interact?

8 What is indicated by the lines of cleavage in the skin, and why are they medically important?

9 Why must the circulation to the skin be closely regulated?

Subcutaneous Layer (Hypodermis)

Key topic in this section:

■ Structure and function of the subcutaneous layer

Deep to the integument is the **subcutaneous** (sŭb-kū-tā′nē-ŭs; *sub* = beneath; *cutis* = skin) **layer,** also called the *hypodermis* or *superficial fascia.* It is not considered a part of the integument. This layer consists of both areolar connective tissue and adipose connective tissue (see figure 5.1). In some locations of the body, adipose connective tissue predominates, and the subcutaneous layer is called **subcutaneous fat**. The connective tissue fibers of the reticular layer are extensively interwoven with those of the subcutaneous layer to stabilize the position of the skin and bind it to the underlying tissues. The subcutaneous layer pads and protects the body and its parts, acts as an energy reservoir, and provides thermal insulation.

Drugs are often injected into the subcutaneous layer because its excessive vascular network promotes rapid absorption. Normally, the subcutaneous layer is thicker in women than in men, and its regional distribution also differs between the sexes. Adult males accumulate subcutaneous fat primarily at the neck, upper arms, abdomen, along the lower back, and over the buttocks, whereas adult females accumulate subcutaneous fat primarily in the breasts, buttocks, hips, and thighs.

Table 5.2 reviews the layers of the integument and the subcutaneous layer.

WHAT DID YOU LEARN?

10 What are some functions of the subcutaneous layer?

Table 5.2	Layers of the Integument and the Subcutaneous Layer	
Layer	**Specific Sublayers**	**Structure**
INTEGUMENT: EPIDERMIS		
	Stratum corneum	Most superficial layer of epidermis; 20–30 layers of dead, flattened, anucleate, keratin-filled keratinocytes called corneocytes
	Stratum lucidum	2–3 layers of anucleate, dead cells; only seen in thick skin (e.g., palms, soles)
	Stratum granulosum	3–5 layers of keratinocytes with distinct granules in the cytoplasm: keratinization begins in this layer
	Stratum spinosum	Several layers of keratinocytes attached to neighbors by desmosomes; epidermal dendritic cells present
	Stratum basale	Deepest, single layer of cuboidal to low columnar cells in contact with basement membrane; mitosis occurs here; contains keratinocytes, melanocytes, and tactile cells
INTEGUMENT: DERMIS		
	Papillary layer	More superficial layer of dermis; composed of areolar connective tissue; contains dermal papillae
	Reticular layer	Deeper layer of dermis; dense irregular connective tissue surrounding blood vessels, hair follicles, nerves, sweat glands, and sebaceous glands
SUBCUTANEOUS LAYER		
		Not considered part of the integument; deep to dermis; composed of areolar connective tissue and adipose connective tissue

Epidermal Accessory Organs

Key topics in this section:

- Structure of nails
- Components of a hair and a hair follicle
- Growth, distribution, and replacement of hairs
- How hair changes throughout life
- Characteristics of sweat glands, sebaceous glands, and other glands found in the skin

Nails, hair, and sweat and sebaceous glands are derived from epidermis and are considered accessory organs, or appendages, of the integument. These structures originate from the invagination of the epidermis during embryological development; they are located in the dermis and may project through the epidermis to the surface. Both nails and hair are composed primarily of dead, keratinized cells.

Nails

Nails are scalelike modifications of the epidermis that form on the dorsal tips of the fingers and toes. They protect the exposed distal tips and prevent damage or distortion during jumping, kicking, catching, or grasping. Nails are hard derivatives from the stratum corneum layer of the epidermis. The cells that form the nails are densely packed and filled with parallel fibers of hard keratin.

Each nail has a pinkish **nail body** and a distal whitish **free edge (figure 5.8a)**. Most of the nail body appears pink because of the blood flowing in the underlying capillaries. In contrast, the free edge of the nail appears white because there are no underlying capillaries. The **lunula** (loo'noo-lă; *luna* = moon) is the whitish semilunar area of the proximal end of the nail body. It appears whitish because a thickened underlying stratum basale obscures the underlying blood vessels. Along the lateral and proximal borders of the nail, portions of skin called **nail folds** overlap the nail so that the nail is recessed internal to the level of the surrounding epithelium and is bounded by a **nail groove**. The **eponychium** (ep-o-nik'ē-um; *epi* = upon, *onyx* = nail), also known as the **cuticle**, is a narrow band of epidermis that extends from the margin of the nail wall onto the nail body.

The nail body covers a layer of epidermis called the **nail bed**, which contains only the deeper, living cell layers of the epidermis (figure 5.8b). The **nail root** is the proximal part of the nail embedded in the skin. At the nail root, the nail bed thickens to form the **nail matrix**, which is the actively growing part of the nail. The **hyponychium** (hī-po-nik'ē-um; *hypo* = below) is a region of thickened stratum corneum over which the free nail edge projects. Together, the nail root, the nail body, and the free edge make up the **nail plate**.

Hair

Hair is found almost everywhere on the body except the palms of the hands, the sides and soles of the feet, the lips, the sides of the fingers and toes, and portions of the external genitalia. Most of the hairs on the human body are on the general body surface rather than the head. The general structure of hair and its relationship to the integument are shown in **figure 5.9**.

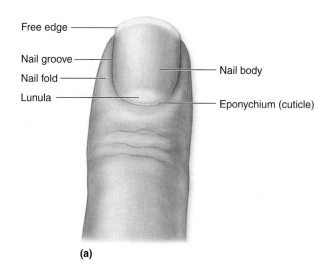

(a)

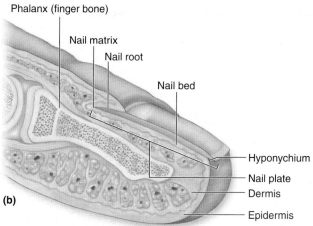

(b)

Figure 5.8

Structure of a Fingernail. Nails, the hard derivative of the stratum corneum, protect sensitive fingertips. *(a)* Surface view of a fingernail. *(b)* Sagittal section showing the internal details of a fingernail.

Hair Type and Distribution

A single hair is also called a *pilus*. It has the shape of a slender filament, and is composed of keratinized cells growing from hair follicles that extend deep into the dermis, often projecting into the underlying subcutaneous layer. Differences in hair density are due primarily to differences in its texture and pigmentation.

During our lives, we produce three kinds of hair: lanugo, vellus, and terminal hair. **Lanugo** (lă-noo'gō) is a fine, unpigmented, downy hair that first appears on the fetus in the last trimester of development. At birth, most of the lanugo has been replaced by similarly fine, unpigmented or lightly pigmented hair called **vellus** (vel'ŭs; *vellus* = fleece). Vellus is the primary human hair and is found on the upper and lower limbs. **Terminal hair** is usually coarser, pigmented, and longer than vellus. It grows on the scalp, and is also the hair of eyebrows and eyelashes. At puberty, terminal hair replaces vellus in the axillary (ak'sil-ār-e; underarm) and pubic regions. Additionally, it forms the beard on the faces of males, as well as on their arms, legs, and trunk.

CLINICAL VIEW

Nail Disorders

Changes in the shape, structure, or appearance of the nails may indicate the existence of a disease that affects metabolism throughout the body. In fact, the state of a person's fingernails and toenails can be indicative of his or her overall health. Nails are subject to various disorders.

Brittle nails are prone to vertical splitting and separation of the nail plate layers at the free edge. Overexposure to water or to certain household chemicals can cause brittle nails, because these substances dry out the nails. Keeping the nails moisturized and limiting exposure to water and chemicals can alleviate brittle nails.

An **ingrown nail** occurs when the edge of a nail digs into the skin around it. This condition is first characterized by pain and inflammation. Any nail may be affected, but the great toenail is the most common site. Some ingrown toenails, if left untreated, can cause infection. The most common causes of ingrown nails are too-tight shoes and improperly trimmed nails (e.g., cutting the nails too short or cutting them in a rounded fashion, instead of straight across).

Onychomycosis (on'i-kō-mī-kō'sis; *onych* = nail, *mykes* = fungus, *osis* = condition) is also known as a *fungal infection*. Fungal infections account for about half of all nail disorders. These infections occur in nails constantly exposed to warmth and moisture, such as toenails in overly warm shoes or fingernails on hands that are constantly in warm water (e.g., washing dishes). The fungus starts to grow under the nail and eventually causes a yellowish discoloration, thickened nail, and brittle, cracked edges (figure *a*). Fungus infections can result in permanent damage to the nail or spread of the infection. Treatment involves taking oral fungal medications for long periods of time (a minimum of 6 to 12 weeks, and in some cases up to a year) in order to eradicate the fungal infection.

Bacterial and viral infections can also affect the nails. To treat a bacterial infection, oral antibiotics are administered.

Yellow nail syndrome occurs when growth and thickening of the nail slows or stops completely. As nail growth slows, the nails become yellowish or sometimes greenish (figure *b*). Yellow nail syndrome is often, but not always, an outward sign of respiratory disease, such as chronic bronchitis.

In **spoon nails**, or *koilonychia* (koy-lō-nik'ē-ă; *koilos* = hollow), nails are malformed so that the outer surfaces are concave instead of convex (figure *c*). Spoon nails frequently are a sign of iron deficiency. Treating the iron deficiency should alleviate the condition.

Beau's lines run horizontally across the nail and indicate a temporary interference with nail growth at the time this portion of the nail was formed (figure *d*). Severe illness or injury can cause Beau's lines. Beau's lines may also be seen in individuals suffering from chronic malnutrition.

Vertical ridging of the nails is common and usually does not indicate any serious medical problem. The condition occurs more frequently as we get older.

In the condition called **half-and-half**, a transverse line forms on the nail to partition it into a distal brown or pink region and a proximal dull white region. Half-and-half is the result of *uremia*, excess nitrogen waste in the blood.

Hapalonychia (hap'ă-lō-nik'ē-ă; *hapalos* = soft) is a condition in which the free edge of the nail bends and breaks as a result of nail thinning.

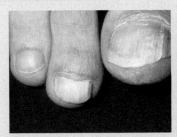

(a) Onychomycosis

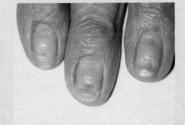

(b) Yellow nail syndrome

(c) Spoon nails

(d) Beau's lines

Hair Structure and Follicles

There are three recognizable zones along the length of a hair:

1. The **hair bulb** consists of epithelial cells and is a swelling at the base where the hair originates in the dermis. The epithelium at the base of the bulb surrounds a small **hair papilla**, which is composed of a small amount of connective tissue containing tiny blood vessels and nerves.
2. The **root** is the hair within the follicle internal to the skin surface.
3. The **shaft** is that portion of the hair that extends beyond the skin surface.

The root and shaft consist of dead epithelial cells, while the hair bulb contains living epithelial cells. Thus, it doesn't hurt to get a haircut because the hairstylist is cutting dead cells. In constrast, it hurts to pull a hair out by its root, because you are disturbing the live portion of the hair.

Hair production involves a specialized type of keratinization that occurs in the hair **matrix**. Basal epithelial cells near the center of the hair matrix divide, producing daughter cells that are gradually pushed toward the surface. The **medulla**, not found in all hair types, is a remnant of the soft core of the matrix. It is composed of loosely arranged cells and air spaces, and contains flexible, soft keratin. Several layers of flattened cells closer to the outer surface of the developing hair form the relatively hard **cortex**. Hair stiffness is derived from the hard keratin contained within the cortex. Multiple cell layers around the cortex form the **cuticle** (kū'ti-kl), which coats the hair. The free edges of cuticle cells are directed externally.

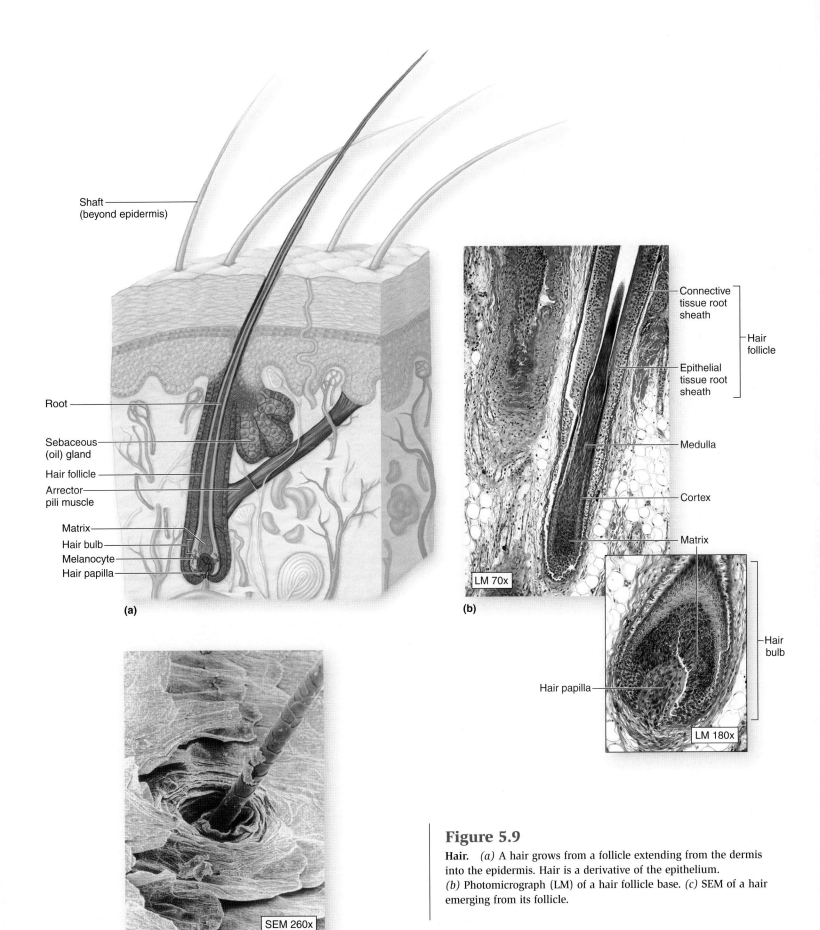

Shaft
(beyond epidermis)

Root

Sebaceous
(oil) gland

Hair follicle

Arrector
pili muscle

Matrix

Hair bulb

Melanocyte

Hair papilla

(a)

Connective
tissue root
sheath

Hair
follicle

Epithelial
tissue root
sheath

Medulla

Cortex

Matrix

LM 70x

(b)

Hair
bulb

Hair papilla

LM 180x

SEM 260x

(c)

Figure 5.9

Hair.   *(a)* A hair grows from a follicle extending from the dermis
into the epidermis. Hair is a derivative of the epithelium.
(b) Photomicrograph (LM) of a hair follicle base. *(c)* SEM of a hair
emerging from its follicle.

The hair root extends from the hair bulb to the region where the hair shaft is completely mature. The entire hair root lies internal to the skin, and the hair shaft extends from the hair root to the exposed tip the hair. The hair shaft's size, shape, and color can be highly variable.

The **hair follicle** (fol'i-kl; *folliculus* = a small sac) is an oblique tube that surrounds the root hair. The follicle always extends into the dermis and sometimes into the subcutaneous layer. The cells of the follicle walls are organized into two principal concentric layers: an outer **connective tissue root sheath**, which originates from the dermis, and an inner **epithelial tissue root sheath**, which originates from the epidermis (figure 5.9b).

The epithelial tissue root sheath is composed of two parts: an internal root sheath and an external root sheath. The **internal root sheath** is produced by peripheral cells of the matrix. It surrounds both the hair and the deep part of the shaft. This layer does not extend the entire length of the follicle because its cells are quickly destroyed. The **external root sheath** extends between the skin surface and the hair matrix. In general, it contains the same epidermal cell layers as the skin surface. However, all of the cells resemble those of the stratum basale where this sheath joins the hair matrix.

Extending from the dermal papillae are thin ribbons of smooth muscle that are collectively called the **arrector pili** (ă-rek'tōr pī'lī; *rectus* = to raise up, *pilus* = hair) **muscle**. The arrector pili is usually stimulated in response to an emotional state, such as fear or rage, or exposure to cold temperatures. Upon stimulation, the arrector pili contracts, pulling on the follicles and elevating the hairs, to produce "goose bumps."

Functions of Hair

The millions of hairs on the surface of the human body have important functions, including:

- **Protection.** The hair on the head protects the scalp from sunburn and injury. Hairs within the nostrils protect the respiratory system by preventing inhalation of large foreign particles, while those within the external ear canal protect the ear from insects and foreign particles. Eyebrows and eyelashes protect the eyes.
- **Heat retention.** Hair on the head prevents the loss of conducted heat from the scalp to the surrounding air. Individuals who have lost their scalp hair lose much more heat than those who have a full head of hair. The scalp is the only place on the body where the hair is thick enough to retain heat.
- **Facial expression.** The hairs of the eyebrows function primarily to enhance facial expression.
- **Sensory reception.** Hairs have associated touch receptors (hair root plexuses) that detect light touch.
- **Visual identification.** Hair characteristics are important in determining species, age, and sex, and in identifying individuals.
- **Chemical signal dispersal.** Hairs help disperse **pheromones,** which are chemical signals involved in attracting members of the opposite sex and in sex recognition. After pheromones are secreted by selected sweat glands, such as those in the axillary and pubic regions, they are released onto the hairs in these areas.

Hair Color

Hair color is a result of the synthesis of melanin in the matrix adjacent to the papillae. Variations in hair color reflect genetically determined differences in the structure of the melanin. Additionally, environmental and hormonal factors may influence the color of the hair. As people age, pigment production decreases, and thus hair becomes lighter in color. Gray hair results from the gradual reduction of melanin production within the hair follicle, while white hair signifies the lack of pigment entirely. Usually, hair color changes gradually.

Hair Growth and Replacement

A hair in the scalp normally grows about one-third of a millimeter per day for 2–5 years, and may attain a length of about a meter. Normally, it enters a dormant phase of 3 to 4 months after this growth phase. A new hair begins to grow inside the follicle internal to the older hair. The older hair is pushed out and eventually falls from the follicle.

The hair growth rate and the duration of the hair growth cycle vary; however, the scalp normally loses between 10 and 100 hairs per day. Continuous losses that exceed 100 hairs per day often indicate a health problem. Sometimes hair loss may be temporary as a result of one or more of the following factors: exposure to drugs, dietary factors, radiation, high fever, or stress. Thinning of the hair, called **alopecia** (al-ō-pē'shē-ă; *alopekia* = a disease like fox mange), can occur in both sexes, usually as a result of aging. In **diffuse hair loss**, a condition that is both dramatic and distressing, hair is shed from all parts of the scalp. Women primarily suffer from this condition, which may be due to hormones, drugs, or iron deficiency.

In males, the condition called **male pattern baldness** causes loss of hair first from only the crown region of the scalp rather than uniformly. It is caused by a combination of genetic and hormonal influences. At puberty, the testes begin secreting large quantities of male sex hormones (primarily testosterone). As one effect of sex hormone production, males develop a typical pattern of underarm hair, facial hair, and chest hair. The relevant gene for male pattern baldness has two alleles, one for uniform hair growth and one for baldness. The baldness allele is dominant in males and is expressed only in the presence of a high level of testosterone. In men who are either heterozygous or homozygous for the baldness allele, testosterone causes the terminal hair of the scalp to be replaced by thinner vellus, beginning on the top of the head and later at the sides. In females, the baldness allele is recessive. This is a *sex-influenced trait*, in which an allele is dominant in one sex (males) and recessive in the other (females). Changes in the level of the sex hormones circulating in the blood can affect hair development on the scalp, causing a shift from terminal hair to vellus production.

Exocrine Glands of the Skin

The skin houses two types of exocrine glands: **sweat (sudoriferous) glands** and **sebaceous glands (figure 5.10a)**. Sweat glands produce a watery solution that performs several specific functions. Sebaceous glands produce an oily material that coats hair shafts and the epidermal surface (see chapter 4). **Table 5.3** compares the types of glands found in the skin.

Sweat Glands

The two types of sweat glands in the skin are **merocrine sweat glands** and **apocrine sweat glands**. These sweat glands have a coiled, tubular secretory portion located either in the reticular layer of the dermis,

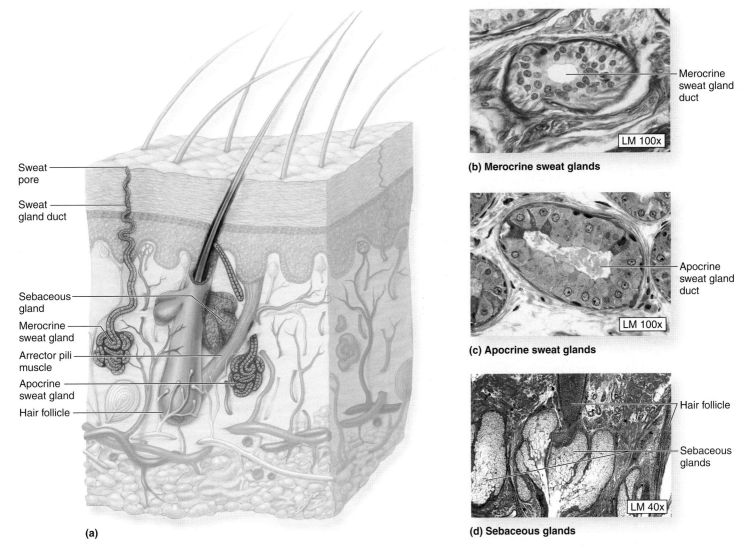

Sweat pore

Sweat gland duct

Sebaceous gland

Merocrine sweat gland

Arrector pili muscle

Apocrine sweat gland

Hair follicle

(a)

Merocrine sweat gland duct

LM 100x

(b) Merocrine sweat glands

Apocrine sweat gland duct

LM 100x

(c) Apocrine sweat glands

Hair follicle

Sebaceous glands

LM 40x

(d) Sebaceous glands

Figure 5.10

Exocrine Glands of the Skin. *(a)* The integument contains sweat glands and sebaceous glands. *(b)* Merocrine sweat glands have a duct with a narrow lumen that opens onto the skin surface through a pore. *(c)* Apocrine sweat glands exhibit a duct with a large lumen to convey secretion products into a hair follicle. *(d)* The cells of sebaceous glands are destroyed during the release of their oily secretion into the follicle.

Table 5.3	Glands of the Skin	
Type	**Location**	**Products Secreted/Description**
SWEAT (SUDORIFEROUS) GLANDS		
Apocrine glands	Distributed in axillary, anal, areolar, and pubic regions	Produces viscous, complex secretion; secretion influenced by hormones; may act in signaling/communication
Merocrine glands	Distributed throughout body, except external genitalia, nipples, and lips; especially prevalent on palms, soles, and forehead	Produce nonviscous, watery secretion; controlled by nervous system; provide some antibacterial protection; function in thermoregulation and excretion; flush surface of epidermis
SEBACEOUS GLANDS		
Sebaceous glands	Associated with hair follicles	Produce lipid material called sebum, which coats epidermis and shaft of hair; provide lubrication and antibacterial activity
OTHER GLANDS		
Ceruminous glands	External acoustic meatus	Cerumen
Mammary glands	Breasts	Milk to nourish offspring

or in the subcutaneous layer. A **sweat gland duct** carries the secretion to the surface of the epidermis (in a merocrine gland) or into a hair follicle (in an apocrine gland). The opening of the sweat gland duct on the epidermal surface is an indented region called a **sweat pore**.

Both types of sweat glands contain **myoepithelial cells**. These specialized cells are sandwiched between the secretory gland cells and the underlying basement membrane. In response to sympathetic nervous system stimulation, myoepithelial cells contract to squeeze the gland, causing it to discharge its accumulated secretions into the duct.

WHAT DO YOU THINK?

4 The sympathetic nervous system is the part of the nervous system that can be activated when we are frightened or nervous. What would you expect to happen to sweat gland production and secretion when we are experiencing these emotions?

Merocrine Sweat Glands Merocrine sweat glands are simple, coiled, tubular glands that release their secretion onto the surface of the skin. They are the most numerous and widely distributed sweat glands in the body. The adult integument contains between 3 and 4 million merocrine sweat glands. The palms of the hands, the soles of the feet, and the forehead have the highest numbers of these glands; some estimates suggest that the palm of each hand houses about 500 merocrine glands per square centimeter (or about 3000 glands per square inch). Merocrine sweat glands are controlled by the nervous system. The secretory portion of the gland is housed within the dermis or the subcutaneous layer; the conducting portion of the gland is an undulating or coiled duct leading to a sweat pore on the skin surface (figure 5.10*b*).

The clear secretion produced by merocrine glands is termed sweat, or sensible perspiration. It begins as a protein-free filtrate of blood plasma. Sweat is approximately 99% water and 1% other chemicals, including some electrolytes (primarily sodium and chloride), metabolites (lactic acid), and waste products (urea and ammonia). It is the sodium chloride that gives sweat a salty taste.

Some of the functions of merocrine sweat glands include:

- **Thermoregulation.** The major function of merocrine sweat glands is to help regulate body temperature through evaporation of fluid from the skin. The secretory activity of these glands is regulated by neural controls. In very hot weather or while exercising, a person may lose as much as a liter of perspiration each hour, and thus dangerous fluid and electrolyte losses are possible.
- **Secretion.** Merocrine sweat gland secretions help rid the body of excess water and electrolytes. In addition, the secretions may help eliminate some types of ingested drugs.
- **Protection.** Merocrine sweat gland secretions provide some protection against environmental hazards both by diluting harmful chemicals and by preventing the growth of microorganisms.

Apocrine Sweat Glands Apocrine sweat glands are simple, coiled, tubular glands that release their secretions into hair follicles at the armpits (axillae), around the nipples (areola), in the groin (pubic region), and around the anus (anal region). Originally, these glands were named apocrine because their cells were thought to secrete their product by an apocrine mechanism (meaning that the apical portion of the cell's cytoplasm pinches off and, along with cellular components of the apical region, becomes the secretory product) (see chaper 4). Now, researchers have shown that both apocrine and merocrine sweat glands produce their secretions by exocytosis. However, the secretory portion of an apocrine gland has a much larger lumen than that of a merocrine gland (figure 5.10*c*), so these glands continue to be called apocrine glands. The secretion they produce is viscous, cloudy, and composed of proteins and lipids that are acted upon by bacteria, producing a distinct, noticeable odor. (Underarm deodorant is designed to mask the odor from the secretory product of these glands.) Secretion is influenced by hormones and may function in both signaling and communication. These sweat glands become active and produce secretory product after puberty.

Sebaceous Glands

Sebaceous glands are holocrine glands that discharge an oily, waxy secretion called **sebum** (sē′bŭm), usually into a hair follicle (figure 5.10*a,d*). Sebum acts as a lubricant to keep the skin and hair from becoming dry, brittle, and cracked. Several sebaceous glands may open onto a single follicle by means of one or more short ducts. Sebaceous glands are relatively inactive during childhood; however, they are activated during puberty in both sexes, when the production of sex hormones begins to increase.

Sebum has bactericidal (bacteria-killing) properties. However, under some conditions, bacteria can cause an infection within the sebaceous gland and produce a local inflammation called **folliculitis** (fo-lik-ū-lī′tis). A blocked duct in a sebaceous gland often develops into a distinctive abscess called a **furuncle** (fū′rŭng-kl; *furunculus* = a petty thief), or *boil*. A furuncle is usually treated by lancing (cutting it open) to facilitate normal drainage and healing.

Other Integumentary Glands

Some specialized glands of the integument are restricted to specific locations. Two important examples are the ceruminous glands and the mammary glands.

Ceruminous (sĕ-roo′mi-nŭs; *cera* = wax) **glands** are modified sweat glands located only in the external acoustic meatus, where their secretion mixes with both sebum and exfoliated keratinocytes to form waterproof earwax called **cerumen** (sĕ-roo′men). They are simple, coiled, tubular glands with ducts leading to the surface of the skin. Ceruminous glands differ from sweat glands in that their coils have a very large lumen and their gland cells contain many pigment granules and liquid droplets. Cerumen, together with tiny hairs along the ear canal, helps trap foreign particles or small insects and keeps them from reaching the eardrum. Cerumen also lubricates the external acoustic meatus and eardrum.

The **mammary glands** of the breasts are modified apocrine sweat glands. Both males and females have mammary glands, but these glands only become functional in pregnant females, when they produce a secretion (milk) that nourishes offspring. The development of the gland and the production of its secretory products are controlled by a complex interaction between gonadal and pituitary hormones. The structure and function of mammary glands are discussed in chapter 28.

WHAT DID YOU LEARN?

11 Why does the lunula of the nail have a whitish appearance?

12 What stimulates the arrector pili muscle to contract?

13 Compare and contrast merocrine and apocrine sweat gland secretions.

14 What do sebaceous glands secrete?

CLINICAL VIEW

Acne and Acne Treatments

The term **acne** (ak'-nē) describes plugged sebaceous ducts. Acne may become abundant beginning at puberty, because increases in sex hormone levels stimulate sebaceous gland secretion, making the pores more prone to blockage. Acne is prevalent during the teenage years, although any age group (including people in their 30s and 40s) can have acne.

The types of acne lesions include:

- **Comedo** (kom'ē-dō; pl., comedones). A sebaceous gland's ducts plugged with sebum. An open comedo is called a **blackhead**, because the plugged material has a dark appearance. A closed comedo is called a **whitehead**, because the top surface is whitish in color.
- **Papule** (pap'ūl) and **pustule** (pūs'chool). Dome-shaped lesions filled with a mixture of white blood cells, dead skin cells, and bacteria.
- **Nodule** (nod'ūl). Similar to a pustule, but extending into the deeper skin layers. Nodules can be prone to scarring.
- **Cyst**. A fluid-filled nodule that can become severely inflamed and painful, and can lead to scarring.

Many medicinal treatments are available for acne, depending on its type and severity. The effectiveness of the following medications varies from individual to individual:

- **Benzoyl peroxide**. Used as a treatment for mild acne for decades, benzoyl peroxide has antibacterial properties and appears to decrease the secretion of some sebum components. Excessive use can cause the skin to become overly dry.

- **Salicylic acid**. Salicylic (sal-i-sil'ik) acid helps unclog pores and appears to affect the rate of skin cell shedding. Like benzoyl peroxide, excessive use can cause overdrying of the skin.
- **Topical and oral antibiotics**. Because many forms of acne are filled with bacteria (called *P. acne*), the use of an antibiotic helps prevent acne outbreaks. Common prescription antibiotic treatments for acne include doxycycline, tetracycline, or erythromycin.
- **Topical retinoids**. Retinoids (e.g., Retin-A) are essentially vitamin A-like compounds that are effective against acne. These prescription medications help normalize the abnormal growth and death of cells in the sebaceous glands. Topical retinoids can cause redness, dryness, and peeling in the treated areas.
- **Systemic retinoids**. Systemic retinoids (e.g., Accutane) are similar to topical retinoids, but are taken orally. These prescription medications are very effective for treating even the most severe forms of acne. Unfortunately, systemic retinoids are associated with major birth defects, so women who are pregnant or thinking of becoming pregnant should not take these drugs. In fact, females who take this medication are required to use multiple forms of birth control so as to prevent an unintended pregnancy.

Other acne treatments include light chemical skin peels and comedo extraction (surgical removal of the comedones by a dermatologist). If severe acne is not treated, it can lead to permanent scarring. In addition, constant "picking" at acne can leads to scars. Thus, dermatologists strongly recommend that individuals refrain from picking blemishes.

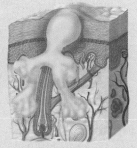

| Normal hair follicle | Blackhead (open comedo) | Whitehead (closed comedo) | Papule | Nodule |

Integument Repair and Regeneration

Key topics in this section:

- How burns affect the integument
- Treatment of burns

The components of the integumentary system exhibit a tremendous ability to respond to stresses, trauma, and damage. Repetitive mechanical stresses applied to the integument stimulate mitotic activity in the stem cells of the stratum basale, resulting in thickening of the epidermis and improved ability to withstand stress. For example, walking about without shoes causes the soles of the feet to thicken, thus providing more protection for the underlying tissues.

Damaged tissues are normally repaired in one of two ways. **Regeneration** replaces damaged or dead cells with the same cell type and restores organ function. When regeneration is not possible because part of the organ is too severely damaged or its cells lack the capacity to divide, the body fills in the gap with scar tissue. This process, known as **fibrosis**, effectively binds the "broken" parts back together. The replacement scar tissue is produced by fibroblasts and composed primarily of collagen fibers. Although fibrosis

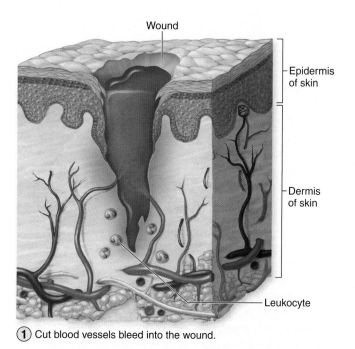

① Cut blood vessels bleed into the wound.

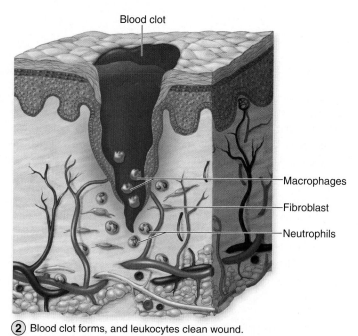

② Blood clot forms, and leukocytes clean wound.

③ Blood vessels regrow, and granulation tissue forms.

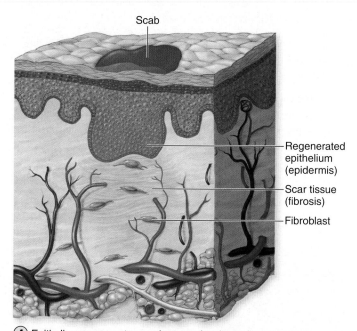

④ Epithelium regenerates, and connective tissue fibrosis occurs.

Figure 5.11

Stages in Wound Healing. Cut blood vessels in tissue initiate the multistep process of wound healing.

restores some structure it does not restore function. Fibrosis is the repair response in tissues subjected to severe injuries or burns.

Both regeneration and fibrosis may occur in the healing of damaged skin. **Figure 5.11** illustrates the following stages in wound healing:

1. Cut blood vessels initiate bleeding into the wound. The blood brings clotting proteins, platelets, numerous white blood cells, and antibodies to the site. The clotting proteins and platelets stop the bleeding, while the white blood cells and antibodies clean the wound and fight any infection that may have been introduced.

2. A blood clot forms, temporarily patching the edges of the wound together and acting as a barrier to prevent the entry

of pathogens into the body. Internal to the clot, macrophages and neutrophils (two types of leukocytes) clean the wound of cellular debris.

3. The cut blood vessels regenerate and grow in the wound. A soft mass deep in the wound becomes **granulation** (gran′ū-lā′shŭn) **tissue,** a vascular connective tissue that initially forms in a healing wound. Macrophages within the wound begin to remove the clotted blood. Fibroblasts produce new collagen in the region.

4. Regeneration of the epidermis occurs due to division of epithelial cells at the edge of the wound. These new epithelial cells migrate over the wound, creeping internally to the now superficial remains of the clot (the scab). The connective tissue is replaced by fibrosis.

CLINICAL VIEW: In Depth

Burns

Burns are a major cause of accidental death, primarily as a result of their effects on the skin. They are usually caused by heat, radiation, harmful chemicals, sunlight, or electrical shock. The immediate threat to life results primarily from fluid loss, infection, and the effects of burned, dead tissue. Burns are classified based on the depth of tissue involvement. First- and second-degree burns are called *partial-thickness burns;* third-degree burns are called *full-thickness burns.*

First-degree burns involve only the epidermis and are characterized by redness, pain, and slight edema (swelling) (figure *a*). An example is a mild sunburn. Treatment involves immersing the burned area in cool water or applying cool, wet compresses; this is sometimes followed by covering the burn with sterile, nonadhesive bandages. The healing time averages about 3 to 5 days, and typically no scarring results.

Second-degree burns involve the epidermis and part of the dermis. The skin appears red, tan, or white and is blistered and painful (figure *b*). Examples include very severe sunburns (characterized by blisters) or scalding from hot liquids or chemicals. Treatment is similar to that for first-degree burns. Care must be taken not to break the blisters because breaking would increase the risk of infection. In addition, ointments should not be applied to the blisters because ointments could cause heat to be retained in the burned area. Elevating burned limbs is recommended to prevent swelling. Healing takes approximately 2 to 4 weeks, and slight scarring may occur.

Third-degree burns involve the epidermis, dermis, and subcutaneous layer, which often are destroyed (figure *c*). Third-degree burns are usually caused by contact with corrosive chemicals or fire, or by prolonged contact with extremely hot water. Dehydration is a major concern with a third-degree burn, because the entire portion of skin has been lost, and water cannot be retained in the area. Third-degree burn victims must be aggressively treated for dehydration, or they may die. In addition, patients are typically given antibiotics because the risk of infection is very great. Treatment may vary slightly, depending upon what caused the burn. Most third-degree burns require hospitalization.

With third-degree burns, regeneration of the integument may occur from the edge only, due to the absence of dermis. Skin grafting is usually needed for patients with third-degree burns, since the entire dermis and its vasculature are destroyed and regeneration is limited. A **skin graft** is a piece of skin transplanted from one part of the body to another to cover a destroyed area. Skin grafts help prevent infection and dehydration in the affected area, and they also help minimize abnormal connective tissue fibrosis and disfigurement.

The first step in determining the severity of burns involves a careful assessment of the body surface area (TBSA). Physicians calculate the body surface area that has been burned by using the **rule of nines**. For example, head and neck = 9%, each arm = 9%, anterior thorax = 18%, posterior thorax = 18%, each leg = 18%, and the perineum = 1%. The amount of surface area involved affects the treatment plan.

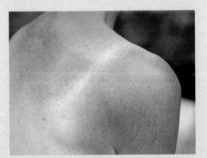

(a) **First-degree burn**

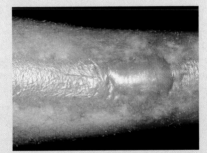

(b) **Second-degree burn**

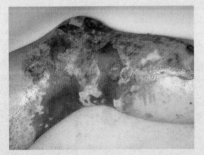

(c) **Third-degree burn**

The skin repair and regeneration process is not rapid. The wider and deeper the surface affected, the longer it takes for skin to be repaired. Additionally, the area under repair is usually more susceptible to complications due to fluid loss and infection. As the severity of damage increases, the repair and regeneration ability of the skin is strained, and a return to its original condition becomes much less likely. Some integumentary system components are not repaired following demage; these include hair follicles, exocrine glands, nerve cells, and muscle fibers.

WHAT DID YOU LEARN?

15 What is the source of new epidermal cells and new dermal cells in the repair of the integument?

Aging of the Integument

Key topics in this section:

- Changes that occur in the skin during aging
- Warning signs and characteristics of skin cancer

Although some people develop acne when they enter puberty, most skin problems do not become obvious until an individual reaches middle age. Eventually, all components of the integumentary system are affected by age in the following ways:

- As an individual ages, the skin repair processes take longer to complete because of the reduced number and activity of stem cells. Skin repair and regeneration that take 3 weeks in a healthy young person often take twice that time for a person in his or her 70s. Additionally, the reduced stem cell activity in the epidermis results in thinner skin that is less likely to protect against mechanical trauma.
- Collagen fibers in the dermis decrease in number and organization, and elastic fibers lose their elasticity. Also years of making particular facial expressions (e.g., squinting, smiling) produce crease lines in the integument. As a result, the skin forms wrinkles and becomes less resilient.
- The skin's immune responsiveness is diminished by a decrease in the number and efficiency of epidermal dendritic cells. This decreased immune response may be related to the

appearance of longer dendritic processes and the decrease in molecules on their surface that recognize pathogens.

- Skin becomes drier and sometimes scaly because decreased sebaceous gland activity diminishes the amounts of natural skin lubricants.
- A decrease in melanocytes causes altered skin pigmentation. As a result, hair becomes gray or white, and sensitivity to sun exposure increases. Often, exposure to the sun or other forms of UV light leads to an increase in skin pigmentation in certain body areas. The resulting flat, brown or black spots are called liver spots, although they are unrelated to the liver.
- Sweat production diminishes as a result of decreased sweat gland activity.
- The dermal blood vessels lose some permeability as a result of decreased elasticity. Blood supply to the dermis is reduced as the extent of blood vessel distribution decreases. These vascular changes, along with glandular changes associated with aging, lead to impaired thermoregulation.
- Hair follicles either produce thinner hairs or stop production entirely.
- Integumentary production of vitamin D_3 decreases. If vitamin D_3 levels diminish significantly, the body is unable to absorb calcium and phosphorus from the digestive tract. The resulting declines in calcium and phosphorus concentrations affect muscle activity and bone density.
- Chronic overexposure to UV rays can damage the DNA in epidermal cells and accelerate aging.

Skin Cancer

Chronic overexposure to UV rays can damage the DNA in epidermal cells and accelerate aging. This overexposure is the predominant factor in the development of nearly all skin cancers. Skin cancer is the most common type of cancer. It occurs most frequently on the head and neck regions, followed by other regions commonly exposed to the sun. Fair-skinned individuals, especially those who experienced severe sunburns as children, are most at risk for skin cancer. However, skin cancer can arise in anyone of any age. Individuals should use sunscreen regularly and avoid prolonged exposure to the sun (see Clinical View earlier in this chapter). An individual should regularly and thoroughly inspect his or her skin for any changes, such as an increase in the number or size of moles or the appearance of new skin lesions. In addition, an examination by a dermatologist should be part of everyone's routine health check-up. **Table 5.4** describes the three main types of skin cancer.

Table 5.4	Skin Cancer
BASAL CELL CARCINOMA	
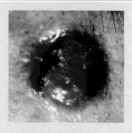	■ Most common type of skin cancer ■ Least dangerous type because it seldom metastasizes ■ Originates in stratum basale ■ First appears as small, shiny elevation that enlarges and develops central depression with pearly edge ■ Usually occurs on face ■ Treated by surgical removal of lesion
SQUAMOUS CELL CARCINOMA	
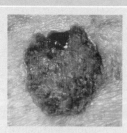	■ Arises from keratinocytes of stratum spinosum ■ Lesions usually appear on scalp, ears, lower lip, or back of hand ■ Early lesions are raised, reddened, scaly; later lesions form concave ulcers with elevated edges ■ Treated by early detection and surgical removal of lesion ■ May metastasize to other parts of the body
MALIGNANT MELANOMA	
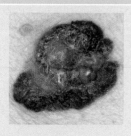	■ Most deadly type of skin cancer due to aggressive growth and metastasis ■ Arises from melanocytes, usually in a preexisting mole ■ Individuals at increased risk include those who have had severe sunburns, especially as children. ■ Characterized by change in mole diameter, color, shape of border, and symmetry ■ Survival rates improved by early detection and surgical removal of lesion. ■ Advanced cases (metastasis of disease) are difficult to cure and are treated with chemotherapy, interferon therapy, and radiation therapy.

The usual signs of melanoma may be easily remembered using the ABCD rule. Report any of the following changes in a birthmark or mole to your physician:

A = **Asymmetry:** One-half of a mole or birthmark does not match the other.
B = **Border:** Edges are notched, irregular, blurred, or ragged.
C = **Color:** Color is not uniform; differing shades (usually brown or black and sometimes patches of white, blue, or red) may be observed.
D = **Diameter:** Affected area is larger than 6 mm (about 1/4 inch) or is growing larger.

Development of the Integumentary System

Key topics in this section:

- Development of the integument from surface ectoderm and mesoderm
- Development of epidermal derivatives

The structures of the integumentary system are derived from the ectodermal and mesodermal germ layers. The ectoderm is the origin of the epidermis, while the mesoderm gives rise to the dermis.

Integument Development

By the end of the seventh week of development, the surface ectoderm is composed of a simple cuboidal epithelium. These epithelial cells divide, grow, and form a layer of squamous epithelium that flattens and becomes a covering layer called the **periderm** and an underlying **basal layer** (**figure 5.12**). The basal layer will form the stratum basale and all other epidermal layers. By the eleventh week of development, the cells of the basal layer from an intermediate layer of skin, and by the twenty-first week, the stratum corneum forms. Also by this time, the friction ridges have formed. During the fetal period, the periderm is eventually sloughed off, and these sloughed off cells mix with sebum secreted by the sebaceous glands, producing a waterproof coating called the **vernix caseosa**. The vernix caseosa protects the skin of the fetus.

Although keratinocytes are formed from epidermal cells, melanocytes originate from specialized neural crest cells called **melanoblasts**, which arise from the ectoderm that also forms nervous tissue. Melanoblasts migrate to the future epidermis. Melanoblasts differentiate into melanocytes about 40–50 days after fertilization and thereafter begin to produce the pigment melanin.

The dermis is derived from mesoderm. During the embryonic period, this mesoderm becomes mesenchyme, and it occupies a zone internal to the ectoderm. At about 11 weeks, the mesenchymal cells begin to form the components of the dermis. The formation of collagen and elastic fibers causes folding at the boundary of the overlying epidermis and dermis, resulting in the formation of dermal papillae. Blood vessels start to form in the dermis, and by the end of the first trimester, the primary vascular pattern in the dermis is present.

Nail Development

Fingernails and toenails start to form in the tenth week of development. These nails form from thickened ridges of epithelium called **nail fields** at the tip of each digit. The nail fields are surrounded by folds of epidermis called nail folds. The proximal nail fold grows over the nail field and becomes keratinized, forming the nail plate. The fingernails reach the tips of the fingers by 32 weeks, while the toenails become fully formed by about 36 weeks. Infants born prematurely may not have fully formed fingernails and toenails.

Hair Development

Hair development is illustrated in **figure 5.13a**. Hair follicles must be present before hair can form. Hair follicles begin to appear between 9 and 12 weeks of development as pockets of cells called **hair buds** that invade the dermis from the overlying stratum basale of the epidermis. These buds differentiate into a hair bulb, hair papillae, sebaceous glands, and other structures associated with hair follicles. The **hair papilla** is formed from differentiating mesenchymal cells located around the epithelial cells of the hair bulb. Eventually, hair grows due to continuous mitotic activity in the epithelial cells of the hair bulb. These hairs do not become easily recognizable in the fetus until about the twentieth week, when they appear as lanugo. These very fine, soft hairs help hold the vernix caseosa on the skin. They are replaced by vellus after birth.

Sebaceous and Sweat Gland Development

Both sweat and sebaceous glands develop from the stratum basale of the epidermis (figure 5.13b). These glands originate from epidermis but start to grow and burrow into the underlying dermis. Sweat glands appear at about 20 weeks on the palms and soles and later in other regions. The secretory portion of these glands coils as the gland develops within the dermis. Sebaceous glands typically develop as epidermal outgrowths from the sides of a developing hair follicle. Sebaceous glands start to form sebum during the fetal period. As previously described, this sebum mixes with the cells of the sloughed off periderm to form the vernix caseosa.

Mammary Gland Development

A **primary mammary bud** first appears during the sixth week of development as an epidermal growth into the underlying dermal

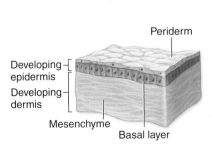

Developing epidermis
Developing dermis
Periderm
Mesenchyme
Basal layer

7 – 8 weeks

Periderm
Melanoblast

11 – 12 weeks

Vernix caseosa
Epidermis
Dermis
Melanocyte

Birth

Figure 5.12

Integument Development. Skin structure becomes increasingly complex in the period from 7 weeks to birth.

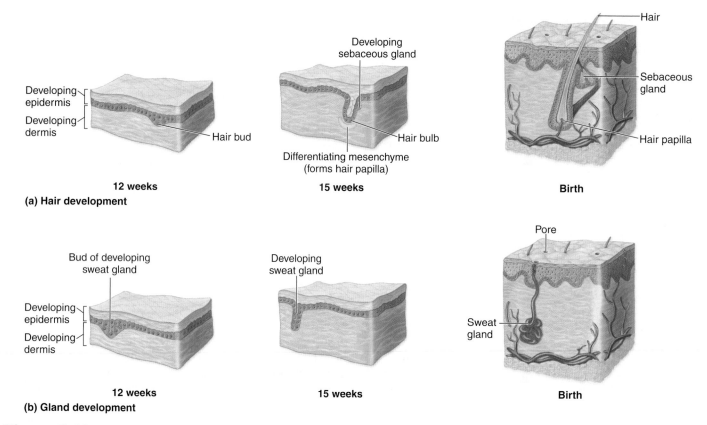

(a) Hair development

Developing epidermis
Developing dermis
Hair bud
12 weeks

Developing sebaceous gland
Hair bulb
Differentiating mesenchyme (forms hair papilla)
15 weeks

Hair
Sebaceous gland
Hair papilla
Birth

Bud of developing sweat gland
Developing epidermis
Developing dermis
12 weeks

Developing sweat gland
15 weeks

Pore
Sweat gland
Birth

(b) Gland development

Figure 5.13

Hair and Gland Development.　Comparison of the development of the hair and the glands of the skin at 12 weeks, at 15 weeks, and at birth.

Figure 5.14

Mammary Gland Development. Development of the mammary glands at 6 weeks, 16 weeks, and 28 weeks.

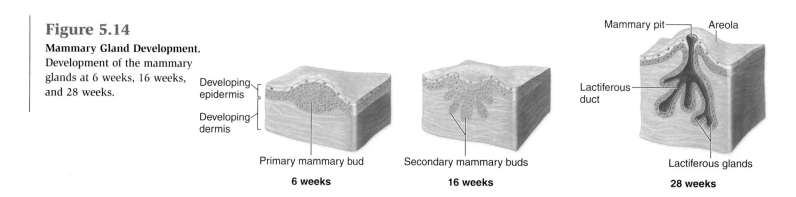

Developing epidermis
Developing dermis
Primary mammary bud
6 weeks

Secondary mammary buds
16 weeks

Mammary pit
Areola
Lactiferous duct
Lactiferous glands
28 weeks

layer (**figure 5.14**). At about 16 weeks (the fourth month) of development, each primary mammary bud branches to form **secondary mammary buds**, which branch and elongate. Later in the fetal period, these mammary buds develop lumina (internal openings) that eventually form milk ducts. The fat and the connective tissue of the mammary gland are formed from the nearby mesenchyme within the dermis.

Late in the fetal period, the mammary gland develops an external epidermal depression called the **mammary pit**. The mammary pit forms the center around which the nipple tissue will grow. The developing milk ducts open up into this pit. At birth, the mammary glands remain underdeveloped. At puberty, female mammary glands grow and differentiate due to increased levels of female sex hormones.

CLINICAL TERMS

athlete's foot　A fungal infection of the skin, especially between the toes; causes itching, redness, and peeling. Also called tinea pedis.

blister　A thin-walled, fluid-filled sac either internal to or within the epidermis; caused by a burn or by excessive friction.

cold sore　Small, fluid-filled blister that is sensitive and painful to the touch; associated with the lips and the mucosa of the oral cavity; caused by herpes simplex type I virus, which infects nerve cells that supply the skin. Also called a *fever blister*.

dandruff　Flaking of the epidermis of the scalp, resulting in white or gray scales in the hair.

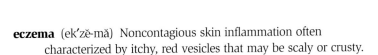

eczema (ek′zĕ-mă) Noncontagious skin inflammation often characterized by itchy, red vesicles that may be scaly or crusty.

hives Eruption of reddish, raised areas on the skin, usually accompanied by extreme itching; causes include certain foods, specific drugs, or stress. Also called urticaria.

impetigo (im-pe-tī′gō) A contagious, pus-forming bacterial infection of the skin; fluid-filled vesicles form and then rupture, forming a yellow crust.

keloid (kē′loyd) Excess scar tissue caused by collagen formation during healing; often painful and tender.

pruritis (proo-rī′tŭs) Irritating, itching condition of the skin that may be caused by infection or exposure to various irritants, such as chemicals, cleaning solutions, or mites.

psoriasis (sō-rī′ă-sis) Chronic inflammatory condition characterized by lesions with dry, silvery scales, usually on the scalp, elbows, and knees.

wart A growth of epidermal cells that forms a roughened projection from the surface of the skin; caused by human papillomavirus.

C H A P T E R   S U M M A R Y

	■ The integumentary system consists of the skin (integument) and its derivatives (nails, hair, sweat glands, and sebaceous glands).
Structure and Function of the Integument 119	■ The integument is the body's largest organ.
	Integument Structure 119
	■ The integument contains a superficial, stratified squamous epithelium called the epidermis, and a deeper, dense irregular connective tissue layer called the dermis.
	■ Deep to the dermis is the subcutaneous layer, or hypodermis, which is not part of the integumentary system.
	Integument Functions 120
	■ The integument's functions include providing mechanical protection and a physical barrier, protecting against water loss, regulating temperature, aiding metabolism, contributing to immune defense, perceiving sensations, and excreting wastes through secretion.
Epidermis 120	■ Four distinct cell types are found within the epidermis: keratinocytes melanocytes, tactile cells, and epidermal dendritic cells.
	Epidermal Strata 121
	■ The stratum basale is a single cell layer of stem cells adjacent to the basement membrane separating the epidermis from the dermis.
	■ The stratum spinosum contains multiple layers of keratinocytes attached together by desmosomes.
	■ The stratum granulosum is composed of three to five layers of keratinocytes. The process of keratinization begins here.
	■ The stratum lucidum is a thin, translucent layer of anucleate cells superficial to the stratum granulosum; it occurs only in the thick skin of the palms and soles.
	■ The stratum corneum has numerous layers of dead, scaly, interlocking keratinized cells.
	Variations in the Epidermis 122
	■ Normal skin color is a result of a combination of hemoglobin in the blood of the dermis and variable quantities of the pigments melanin and carotene.
Dermis 125	■ Components of the dermis include blood vessels, sweat glands, sebaceous glands, hair follicles, nail roots, sensory nerve endings, and muscular tissue.
	Papillary Layer of the Dermis 126
	■ The papillary layer is composed of areolar connective tissue. Epidermal ridges interdigitate with dermal papillae at the boundary between the epidermis and dermis to interlock these layers and increase the area of contact between them.
	Reticular Layer of the Dermis 126
	■ The reticular layer lies deep to the papillary layer; it consists of dense irregular connective tissue.
	Stretch Marks, Wrinkles, and Lines of Cleavage 126
	■ Skin stretching due to weight gain or pregnancy causes stretch marks, called striae.
	■ Lines of cleavage in the skin indicate the predominant direction of the underlying bundles of collagen fibers.
	Innervation and Blood Supply 127
	■ Vasoconstriction of dermal blood vessels causes decreased circulation to the skin and a corresponding conservation of heat in the blood. Vasodilation of dermal blood vessels causes increased circulation to the skin and loss of excess heat.

(continued on next page)

CHAPTER SUMMARY (*continued*)

Subcutaneous Layer (Hypodermis) 128	■ The subcutaneous layer consists of areolar connective tissue and adipose connective tissue.
Epidermal Accessory Organs 129	■ The nails, hair, and sweat and sebaceous glands are epidermal derivatives that are considered accessory organs of the integument.

Nails 129

■ Nails are formed from stratum corneum; they protect the exposed distal tips of the fingers and toes.

Hair 129

■ Hairs project beyond the skin surface almost everywhere except over the palms of the hands, the sides and soles of the feet, the lips, the sides of the fingers and toes, and portions of the external genitalia.

■ Hair functions include protection, thermoregulation, facial expression, sensory reception, visual identification, and dispersal of pheromones.

Exocrine Glands of the Skin 132

■ Merocrine sweat glands produce a thin, watery secretion called sweat (sensible perspiration).

■ Apocrine sweat glands produce a thick secretion that becomes odorous after exposure to bacteria on the skin surface.

■ Sebaceous glands discharge an oily sebum into hair follicles by holocrine secretion.

■ Ceruminous glands housed within the external ear canal are modified sweat glands; they produce a waxy product called cerumen.

■ Mammary glands are modified apocrine sweat glands that produce milk to nourish a newborn infant.

Integument Repair and Regeneration 135	■ The skin can regenerate even after considerable damage, including trauma due to burns.
	■ Severe damage to the dermis and accessory structures of the skin cannot be repaired. Often, fibrous scar tissue forms, and a graft is required.
Aging of the Integument 137	■ Changes to the skin due to aging include slower regeneration and repair, decreased numbers of collagen fibers and melanocytes, diminished immune responsiveness and sweat production, and increased dryness.

Skin Cancer 138

■ UV rays from the sun pose the greatest risk for this most common type of cancer.

Development of the Integumentary System 139	■ The epidermis is derived from the ectoderm, and the dermis is derived from the mesoderm.

Integument Development 139

■ Surface ectoderm forms a covering called periderm and an underlying basal layer.

Nail Development 139

■ Nails form from thickened epithelial ridges called nail fields.

Hair Development 139

■ Hair follicles form from hair buds that differentiate into hair bulbs, hair papillae, and sebaceous glands.

Sebaceous and Sweat Gland Development 139

■ Both types of glands originate from the stratum basale of the epidermis.

Mammary Gland Development 139

■ A primary mammary bud develops as an epidermal outgrowth in the underlying dermis: each bud branches to form secondary mammary buds.

CHALLENGE YOURSELF

Matching

Match each numbered item with the most closely related lettered item.

_____ 1. integument

_____ 2. fingernails

_____ 3. keratin

_____ 4. tactile cells

_____ 5. melanocytes

_____ 6. keratinocytes

_____ 7. epidermal dendritic cell

_____ 8. subcutaneous layer

_____ 9. reticular layer

_____ 10. arrector pili

a. smooth muscle attached to hair follicle

b. most numerous epidermal cell

c. a phagocytic cell (active in immune response)

d. layer deep to dermis

e. formed from stratum corneum

f. receptors for touch

g. composed of epidermis and dermis

h. dense irregular connective tissue

i. fibrous protein in epidermis

j. pigment-forming cells

Multiple Choice

Select the best answer from the four choices provided.

_____ 1. " Strawberry-colored birthmarks" are also called
a. cavernous hemangiomas.
b. freckles.
c. capillary hemangiomas.
d. erythema.

_____ 2. The layer of the epidermis in which cells begin the process of keratinization is the
a. stratum corneum.
b. stratum basale.
c. stratum lucidum.
d. stratum granulosum.

_____ 3. The sweat glands that communicate with skin surfaces only in the axillary, areolar, pubic, and anal regions are
a. apocrine glands.
b. merocrine glands.
c. sebaceous glands.
d. All of these are correct.

_____ 4. Which of the following is not a function of the integument?
a. acts as a physical barrier
b. stores calcium in the dermis
c. regulates temperature through vasoconstriction and vasodilation of dermal blood vessels
d. participates in immune defense

_____ 5. Which of the following layers contains areolar connective tissue and dermal papillae?
a. reticular layer
b. subcutaneous layer
c. papillary layer
d. epidermis

_____ 6. Melanin is
a. an orange-yellow pigment that strengthens the epidermis.
b. a pigment that accumulates inside keratinocytes.
c. a protein fiber found in the dermis.
d. a pigment that gives the characteristic color to hemoglobin.

_____ 7. The layer of squamous epithelium that forms by the seventh week of development to give rise to the integument is the
a. mesenchyme.
b. periderm.
c. basal layer.
d. sebaceous layer.

_____ 8. The cells in a hair follicle that are responsible for forming hair are the
a. papillary cells.
b. matrix cells.
c. medullary cells.
d. cortex cells.

_____ 9. Which epidermal cell type is responsible for detecting touch sensations?
a. keratinocyte
b. melanocyte
c. tactile cell
d. epidermal dendritic cell

_____ 10. Water loss due to evaporation of interstitial fluid through the surface of the skin is termed
a. latent perspiration.
b. sensible perspiration.
c. active perspiration.
d. insensible perspiration.

Content Review

1. What effect does the protein keratin have on both the appearance and the function of the integument?

2. Describe two ways in which the skin helps regulate body temperature.

3. List the layers of the epidermis from deep to superficial and compare their structure.

4. Identify and distinguish among the three types of hair produced during a person's lifetime.

5. List and discuss the three zones along the length of a hair.

6. How do apocrine and merocrine sweat glands differ in structure and function?

7. Describe how the skin is involved in vitamin D production.

8. Briefly discuss the origin and function of sebum.

9. Describe the four steps in wound repair of the integument.

10. What are some effects of aging on the integument?

Developing Critical Reasoning

1. When you are outside on a cold day, your skin is much paler than normal. Later, you enter a warm room, and your face becomes flushed. What are the reasons for these changes in color?

2. Teri is a 14-year-old with a bad case of acne. Explain the probable cause of Teri's skin condition.

3. As a young man, John spent every summer afternoon at the pool for many years. As he approached the age of 50, his skin was quite wrinkled, and he discovered some suspicious growths on his face. He visited a dermatologist, who removed these growths. What were the growths, and what probably caused them?

A N S W E R S   T O   " W H A T   D O   Y O U   T H I N K ? "

1. The children were not getting enough vitamin D in their diet, and they were spending all of their daylight hours indoors. Since the children weren't exposed to much sunlight, their skin could not synthesize vitamin D from the UV rays of the sun. Without adequate amounts of vitamin D, the children succumbed to rickets.

2. Thick skin is found on the palms of the hands and the soles of the feet. Secretions from sebaceous glands would make these areas slippery, which would interfere with grasping objects and walking. The presence of hair in these areas would interfere with these same functions.

3. Fingerprints are formed from folds of both epidermal and dermal tissue, so in order to physically "change" or remove fingerprints, you would have to destroy or damage both layers. This would be very painful, and permanent scarring or malformation would likely result.

4. When we are frightened or nervous, the sympathetic nervous system is stimulated, which in turn stimulates the sweat gland to produce and release sweat. This is why our palms and other body regions become sweaty in nervous or frightening situations.

5. The person who has never used sunscreen is more likely to get wrinkles. Constant, unprotected exposure to UV rays can cause wrinkling and increased aging of the skin, and it also increases the risk of developing skin cancer.

Visit the McKinley/O'Loughlin *Human Anatomy,* 2e website at aris.mhhe.com

OUTLINE

SKELETAL SYSTEM

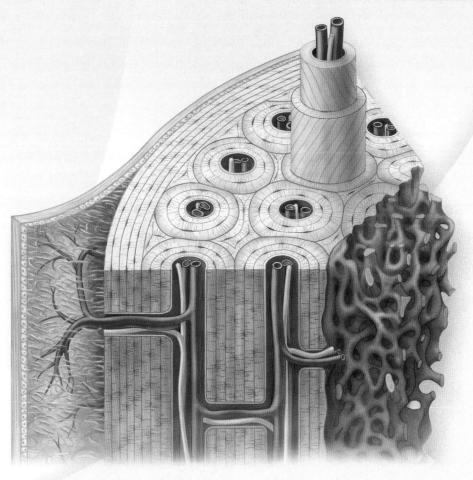

Cartilage and Bone Connective Tissue

Mention of the skeletal system conjures up images of dry, lifeless bones in various sizes and shapes. But the **skeleton** (skel′ĕ-ton; *skeletos* = dried) is much more than a supporting framework for the soft tissues of the body. The skeletal system is composed of dynamic living tissues; it interacts with all of the other organ systems and continually rebuilds and remodels itself. Our skeletal system includes the bones of the skeleton as well as cartilage, ligaments, and other connective tissues that stabilize or connect the bones. Bones support our weight and interact with muscles to produce precisely controlled movements. This interaction permits us to sit, stand, walk, and run. Further, our bones serve as vital reservoirs for calcium and phosphorus. Before concentrating on bone connective tissue, we first examine the cartilage components of the skeleton.

Cartilage Connective Tissue

Key topics in this section:

- Characteristics and functions of cartilage
- Structure, function, and distribution of hyaline cartilage, fibrocartilage, and elastic cartilage
- Interstitial and appositional growth of cartilage

Cartilage connective tissue is found throughout the human body **(figure 6.1)**. Cartilage is a semirigid connective tissue that is weaker than bone, but more flexible and resilient (see chapter 4). As with all connective tissue types, cartilage contains a population of cells scattered throughout a matrix of protein fibers embedded within a gel-like ground substance. **Chondroblasts** (kon′drō-blast; *chondros* = grit or gristle, *blastos* = germ) are the cells that produce the matrix of cartilage. Once they become encased within the matrix they have produced and secreted, the cells are called **chondrocytes** (kon′drō-sīt; *cyte* = cell) and occupy small spaces called **lacunae**. These mature cartilage cells maintain the matrix and ensure that it remains healthy and viable. Mature cartilage is avascular (not penetrated by blood vessels).

Functions of Cartilage

Cartilage has three major functions in the body:

- Supporting soft tissues. For example, C-shaped hyaline cartilage rings in the trachea support the connective tissue and musculature of the tracheal wall, and flexible elastic cartilage supports the fleshy, external part of the ear called the auricle (aw′ri-kl; *auris* = ear).

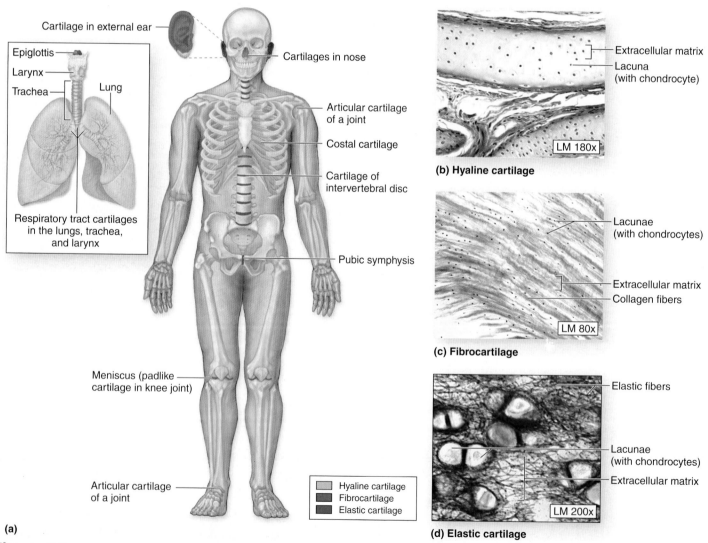

(a)

(b) Hyaline cartilage

(c) Fibrocartilage

(d) Elastic cartilage

Figure 6.1

Distribution of Cartilage in an Adult.   *(a)* Three types of cartilage are found within an adult. Photomicrographs show *(b)* hyaline cartilage, *(c)* fibrocartilage, and *(d)* elastic cartilage.

- Providing a gliding surface at **articulations** (joints), where two bones meet.
- Providing a model for the formation of most of the bones in the body. Beginning in the embryonic period, this cartilage serves as a "rough draft" for bone that is later replaced by bone tissue.

Types of Cartilage

The human body has three types of cartilage: hyaline cartilage, fibrocartilage, and elastic cartilage (see chapter 4).

Hyaline (hī′ă-lin; *hyalos* = glass) **cartilage** is the most abundant type of cartilage. It is found in the trachea, portions of the larynx, the articular (joint) cartilage on bones, epiphyseal plates (discussed later in this chapter), and the fetal skeleton. It provides support through flexibility and resilience, and its extracellular matrix has a translucent appearance, with no clearly visible collagen fibers, when viewed in microscopic section (figure 6.1*b*). Most hyaline cartilage is surrounded by a dense connective tissue covering called **perichondrium**.

Fibrocartilage (fī-brō-kar′ti-lij) has an extracellular matrix with numerous thick collagen fibers that help resist both tensile (stretching) and compressional (compaction) forces (figure 6.1*c*). Fibrocartilage can act as a shock absorber, and is located in regions of the body where these strengths are required, including the intervertebral discs (pads of fibrocartilage between the vertebrae), the menisci of the knee (pads of fibrocartilage between the tibia and femur), and the pubic symphysis (a pad of fibrocartilage between the two pubic bones). Fibrocartilage lacks a perichondrium because stress applied at the surface of the fibrocartilage would quickly destroy this layer.

Elastic cartilage contains highly branched elastic fibers (elastin) within its extracellular matrix (figure 6.1*d*). Elastic cartilage is typically found in regions requiring a highly flexible form of support, such as the auricle of the ear, the external auditory canal (canal in the ear where sound waves travel), and the epiglottis (part of the larynx). Elastic cartilage is surrounded by a perichondrium.

Growth Patterns of Cartilage

Cartilage grows in two ways. Growth from within the cartilage itself is termed **interstitial** (in-ter-stish′ăl) **growth**. Growth along the cartilage's outside edge, or periphery, is called **appositional** (ap-ō-zish′ŭn-ăl) **growth (figure 6.2)**.

Interstitial Growth

Interstitial growth occurs through a series of steps:

1. Chondrocytes housed in lacunae undergo mitotic cell division.
2. Following cell division, the two new cells occupy a single lacuna.
3. As the cells begin to synthesize and secrete new cartilage matrix, they are pushed apart and now reside in their own lacunae.
4. The new individual cells within their own lacunae are called chondrocytes. New matrix has been produced internally, and thus interstitial growth has occurred.

Appositional Growth

Appositional growth also occurs through a series of defined steps:

1. Stem cells at the internal edge of the perichondrium begin to divide, forming new stem cells and committed cells.
2. The committed cells differentiate into chondroblasts.
3. These chondroblasts, located at the periphery of the old cartilage, begin to produce and secrete new cartilage matrix.

As a result, they push apart and become chondrocytes, each occupying its own lacuna.
4. The new matrix has been produced peripherally, and thus appositional growth has occurred.

During early embryonic development, both interstitial and appositional cartilage growth occur simultaneously. However, interstitial growth declines rapidly as the cartilage matures because the cartilage becomes semirigid as it matures, and the matrix is no longer able to expand. Further growth can occur only at the periphery of the tissue, so later growth is primarily appositional. Once the cartilage is fully mature, new cartilage growth typically stops entirely. From this point on, cartilage growth usually occurs only after injury to the cartilage.

WHAT DID YOU LEARN?

❶ Identify the three types of cartilage. How do they differ with respect to their locations in the body?

❷ Compare and contrast interstitial and appositional growth of cartilage. In older cartilage, which type of growth predominates?

Bone

Key topic in this section:

- Functions of bone

The bones of the skeleton are complex, dynamic organs containing all tissue types. Their primary component is **bone connective tissue**, also called *osseous* (os′ē-ŭs) *connective tissue* (see chapter 4). In addition, they contain connective tissue proper (periosteum), cartilage connective tissue (articular cartilage), smooth muscle tissue (forming the walls of blood vessels that supply bone), fluid connective tissue (blood), epithelial tissue (lining the inside opening of blood vessels), and nervous tissue (nerves that supply bone). The matrix of bone connective tissue is sturdy and rigid due to deposition of minerals in the matrix, a process called **calcification** (kal′si-fi-kā′shŭn), or *mineralization*.

> ## Study Tip!
>
> You can do a quick overnight experiment to demonstrate what would happen to our body shape if the composition of our bones changed. Obtain the "wishbone" (fused clavicles) from a chicken or game hen and observe its physical characteristics. Next, place the bone in a glass container of vinegar. Let it stand overnight, and then examine the bone. You should see the following changes: (1) The bone is darker because the acid in the vinegar has dissolved the calcium phosphate in the bone, and (2) the bone is somewhat limp like a wet noodle because it has lost its strength due to the removal of the calcium phosphate from the bone.

Functions of Bone

Bone connective tissue and the bones that compose the skeletal system perform several basic functions: support and protection, movement, hemopoiesis, and storage of mineral and energy reserves.

Support and Protection

Bones provide structural support and serve as a framework for the entire body. Bones also protect many delicate tissues and organs from injury and trauma. The rib cage protects the heart and lungs,

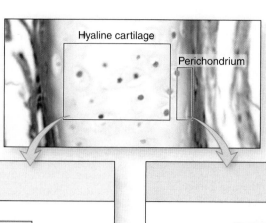

Interstitial Growth

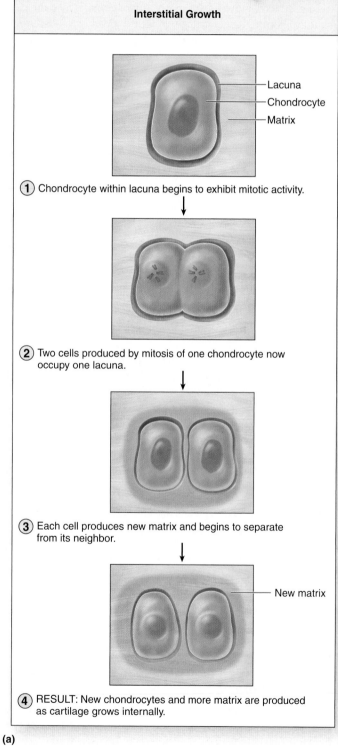

- Lacuna
- Chondrocyte
- Matrix

(1) Chondrocyte within lacuna begins to exhibit mitotic activity.

(2) Two cells produced by mitosis of one chondrocyte now occupy one lacuna.

(3) Each cell produces new matrix and begins to separate from its neighbor.

- New matrix

(4) RESULT: New chondrocytes and more matrix are produced as cartilage grows internally.

(a)

Appositional Growth

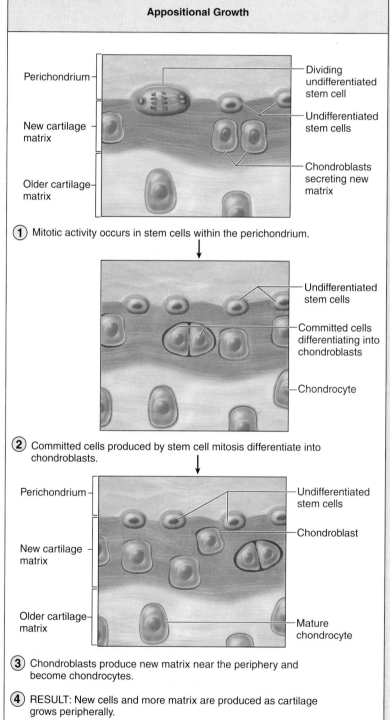

Perichondrium —

New cartilage matrix —

Older cartilage matrix —

- Dividing undifferentiated stem cell
- Undifferentiated stem cells
- Chondroblasts secreting new matrix

(1) Mitotic activity occurs in stem cells within the perichondrium.

- Undifferentiated stem cells
- Committed cells differentiating into chondroblasts
- Chondrocyte

(2) Committed cells produced by stem cell mitosis differentiate into chondroblasts.

Perichondrium —

New cartilage matrix —

Older cartilage matrix —

- Undifferentiated stem cells
- Chondroblast
- Mature chondrocyte

(3) Chondroblasts produce new matrix near the periphery and become chondrocytes.

(4) RESULT: New cells and more matrix are produced as cartilage grows peripherally.

(b)

Figure 6.2

Formation and Growth of Cartilage. Cartilage grows either from within (interstitial growth) or at its edge (appositional growth). *(a)* In interstitial growth, chondrocytes within lacunae divide to form two chondroblasts; these cells grow, begin to produce new matrix, and push apart from each other, forming two new chondrocytes. *(b)* In appositional growth, cartilage grows when stem cells at the internal edge of the perichondrium divide. Differentiation of committed cells into chondroblasts results in the formation of new cartilage matrix and the differentiation of these cells into chondrocytes within the inner layer of the perichondrium.

the cranial bones enclose and protect the brain, the vertebrae enclose the spinal cord, and the pelvis cradles some digestive, urinary, and reproductive organs.

Movement

Individual groups of bones serve as attachment sites for skeletal muscles, other soft tissues, and some organs. Muscles attached to the bones of the skeleton contract and exert a pull on the skeleton, which then functions as a series of levers. The bones of the skeleton can alter the direction and magnitude of the forces generated by the skeletal muscles. Potential movements range from powerful contractions needed for running and jumping to delicate, precise movements required to remove a splinter from the finger.

Hemopoiesis

The process of blood cell production is called **hemopoiesis** (hē′mō-poy-ē′sis; *haima* = blood, *poiesi* = making). Blood cells are produced in a connective tissue called **red bone marrow**, which is located in some spongy bone. Red bone marrow contains stem cells that form all of the formed elements in the blood.

The locations of red bone marrow differ between children and adults. In children, red bone marrow is located in the spongy bone of most of the bones of the body. As children mature into adults, much of the red bone marrow degenerates and turns into a fatty tissue called **yellow bone marrow**. As a result, adults have red bone marrow only in selected portions of the axial skeleton, such as the flat bones of the skull, the vertebrae, the ribs, the sternum (breastbone), and the ossa coxae (hip bones). Adults also have red bone marrow in the proximal epiphyses of each humerus and femur.

Storage of Mineral and Energy Reserves

More than 90% of the body's reserves of the minerals calcium and phosphate are stored and released by bone. Calcium is an essential mineral for such body functions as muscle contraction, blood clotting, and nerve impulse transmission. Phosphate is needed for ATP utilization, among other things. When calcium or phosphate is needed by the body, some bone connective tissue is broken down, and the minerals are released into the bloodstream. In addition, potential energy in the form of lipids is stored in yellow bone marrow, which is located in the shafts of long bones.

WHAT DID YOU LEARN?

❸ Briefly describe at least four functions of bone.

Classification and Anatomy of Bones

Key topics in this section:

- Characteristics of long, short, flat, and irregular bones
- Gross anatomy of a long bone
- Microscopic anatomy of compact bone and spongy bone

Bones of the human skeleton occur in various shapes and sizes, depending on their function. The four classes of bone as determined by shape are long bones, short bones, flat bones, and irregular bones (**figure 6.3**).

Long bones have a greater length than width. These bones have an elongated, cylindrical shaft (diaphysis). This is the most common bone shape. Long bones are found in the upper limb

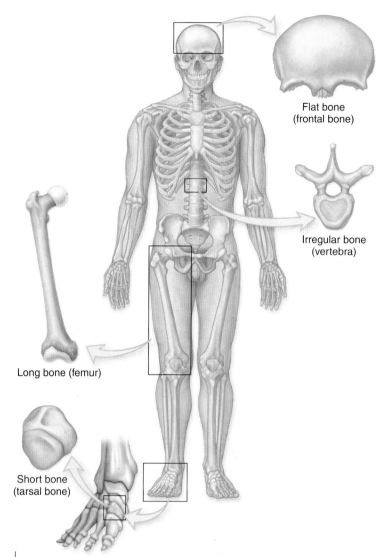

Figure 6.3

Classification of Bone by Shape. Four different classes of bone are recognized according to shape: long, short, flat, and irregular.

(namely, the arm, forearm, palm, and fingers) and lower limb (thigh, leg, sole of the foot, and toes). Long bones vary in size; the small bones in the fingers and toes are long bones, as are the larger tibia and fibula of the lower limb.

Short bones have a length nearly equal to their width. The external surfaces of short bones are covered by compact bone, and their interior is composed of spongy bone. Examples of short bones include the carpals (wrist bones) and tarsals (bones in the foot). *Sesamoid bones*, which are tiny, seed-shaped bones along the tendons of some muscles, are also classified as short bones. The patella (kneecap) is the largest sesamoid bone.

Flat bones are so named because they have flat, thin surfaces. These bones are composed of roughly parallel surfaces of compact bone with a layer of internally placed spongy bone. They provide extensive surfaces for muscle attachment and protect underlying soft tissues. Flat bones form the roof of the skull, the scapulae (shoulder blades), the sternum (breastbone), and the ribs.

Irregular bones have elaborate, complex shapes and do not fit into any of the preceding categories. The vertebrae, ossa coxae (hip bones), and several bones in the skull, such as the ethmoid and sphenoid bones, are examples of irregular bones.

WHAT DO YOU THINK?

1 Why is the rib classified as a flat bone instead of a long bone? Describe the features and functions of flat bones and compare these to the features and functions of a rib.

General Structure and Gross Anatomy of Long Bones

Long bones, the most common bone shape in the body, serve as a useful model of bone structure. Two examples of long bones are the femur (thigh bone) and the humerus (arm bone) **(figure 6.4)**. A typical long bone contains the following parts:

- One of the principal gross features of a long bone is its shaft, or **diaphysis** (dī-af′i-sis; pl., *diaphyses*, dī-af′i-sēz; growing

between). The elongated, usually cylindrical diaphysis provides for the leverage and major weight support of a long bone.

- At each end of a long bone is an expanded, knobby region called the **epiphysis** (e-pif′i-sis; pl., *epiphyses*, e-pif′i-sēz; *epi* = upon, *physis* = growth). The epiphysis is enlarged to strengthen the joint and provide added surface area for bone-to-bone articulation as well as tendon and ligament attachment. It is composed of an outer layer of compact bone and an inner layer of spongy bone. A **proximal epiphysis** is the end of the bone closest to the body trunk, and a **distal epiphysis** is the end farthest from the trunk.

- The **metaphysis** (mĕ-taf′i-sis) is the region in a mature bone sandwiched between the diaphysis and the epiphysis. In a

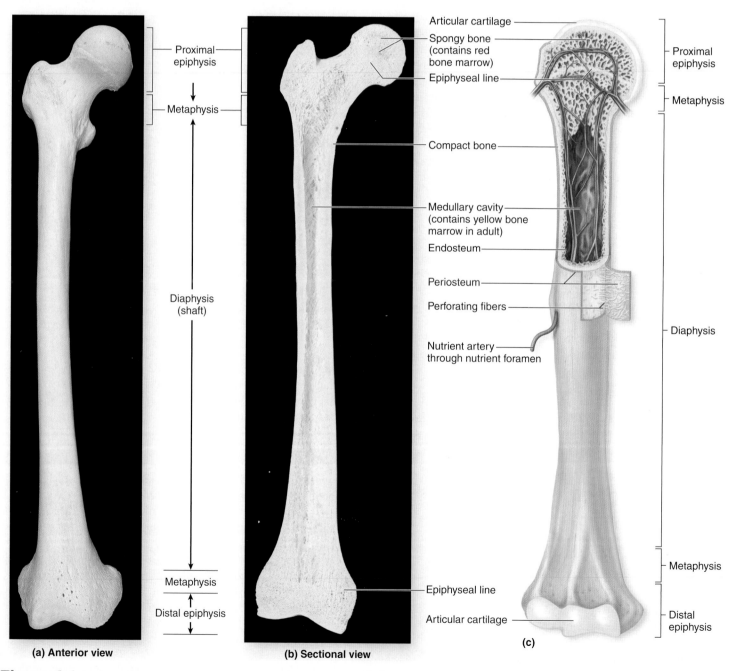

(a) Anterior view

(b) Sectional view

(c)

Figure 6.4

Gross Anatomy of a Long Bone.   Long bones support soft tissues in the limbs. The femur, the bone of the thigh, is shown in both *(a)* anterior and *(b)* sectional views. *(c)* A typical long bone, such as the humerus, contains both compact and spongy bone.

growing bone, this region contains the epiphyseal (growth) plate, thin layers of hyaline cartilage that provide for the continued lengthwise growth of the diaphysis. In adults, the remnant of the epiphyseal plate is a thin layer of compact bone called the **epiphyseal line**.

- The thin layer of hyaline cartilage covering the epiphysis at a joint surface is called **articular cartilage**. This cartilage helps reduce friction and absorb shock in movable joints.
- The hollow, cylindrical space within the diaphysis is called the **medullary cavity** (*marrow cavity*). In adults, it contains yellow bone marrow.
- The **endosteum** (en-dos′tē-ŭm; *endo* = within, *osteon* = bone) is an incomplete layer of cells that covers all internal surfaces of the bone, such as the medullary cavity. The endosteum contains osteoprogenitor cells, osteoblasts, and osteoclasts **(figure 6.5)**, and is active during bone growth, repair, and remodeling.
- A tough sheath called **periosteum** (per-ē-os′tē-ŭm; *peri* = around) covers the outer surface of the bone, except for the areas covered by articular cartilage. Periosteum is made of dense irregular connective tissue and consists of an outer fibrous layer and an inner cellular layer (figure 6.5). The periosteum is anchored to the bone by numerous strong collagen fibers called **perforating fibers,** which run perpendicular to the diaphysis. The periosteum protects the bone from surrounding structures, anchors blood vessels and nerves to the surface of the bone, and provides stem cells (osteoprogenitor cells and osteoblasts) for bone width growth and fracture repair.

WHAT DID YOU LEARN?

④ What are the four classes of bone in terms of shape? Into which group would the os coxae (hip bone) be placed?

⑤ What is the difference between the diaphysis and the epiphysis?

Cells of Bone

Four types of cells are found in bone connective tissue: osteoprogenitor cells, osteoblasts, osteocytes, and osteoclasts **(figure 6.6)**.

Osteoprogenitor (os′tē-ō-prō-jen′i-ter; *osteo* = bone) **cells** are stem cells derived from mesenchyme. When they divide, they produce another stem cell and a "committed cell" that matures to become an osteoblast. These stem cells are located in both the periosteum and the endosteum.

Osteoblasts (*blast* = germ) are formed from osteoprogenitor stem cells. Often, osteoblasts exhibit a somewhat cuboidal structure. They secrete the initial semisolid, organic form of bone matrix called **osteoid** (os′tē-oyd; *eidos* = resemblance). Osteoid later calcifies and hardens as a result of calcium salt deposition. Osteoblasts produce new bone, and once osteoblasts become entrapped in the matrix they produce and secrete, they differentiate into osteocytes.

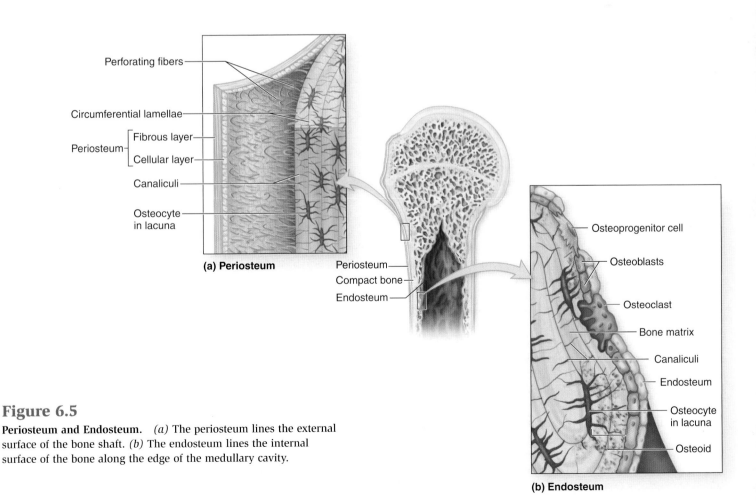

Figure 6.5

Periosteum and Endosteum. *(a)* The periosteum lines the external surface of the bone shaft. *(b)* The endosteum lines the internal surface of the bone along the edge of the medullary cavity.

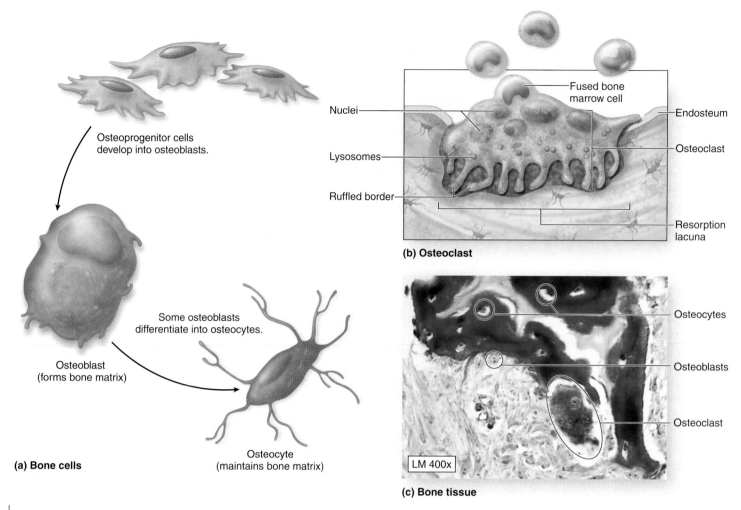

Figure 6.6

Types of Cells in Bone Connective Tissue.   Four different types of cells are found in bone connective tissue. *(a)* Osteoprogenitor cells develop into osteoblasts, many of which differentiate to become osteocytes. *(b)* Bone marrow cells fuse to form osteoclasts. *(c)* A photomicrograph shows osteoblasts, osteocytes, and an osteoclast.

Osteocytes (*cyt* = cell) are mature bone cells derived from osteoblasts that have become entrapped in the matrix they secreted. They reside in small spaces within the matrix called **lacunae**. Osteocytes maintain the bone matrix and detect mechanical stress on a bone. This information is communicated to osteoblasts, and may result in the deposition of new bone matrix at the surface.

Osteoclasts (os′-tē-ō-klast; *klastos* = broken) are large, multi-nuclear, phagocytic cells. They appear to be derived from fused bone marrow cells similar to those that produce monocytes (described in chapter 21). These cells exhibit a ruffled border where they contact the bone, which increases their surface area exposure to the bone. An osteoclast is often located within or adjacent to a depression or pit on the bone surface called a **resorption lacuna** (*Howship's lacuna*). Osteoclasts are involved in an important process called **bone resorption** that takes place as follows: Osteoclasts secrete hydrochloric acid, which dissolves the mineral parts (calcium and phosphate) of the bone matrix, while lysosomes within the osteoclasts secrete enzymes that dissolve the organic part of the matrix (described in the next section). The release of the stored calcium and phosphate from the bone matrix is called **osteolysis** (os-tē-ol′i-sis; *lysis* = dissolution, loosening). The liberated calcium and phosphate ions enter the tissue fluid and then the blood.

Osteoclasts remove matrix and osteoblasts add to it, maintaining a delicate balance. Osteoblast and osteoclast activity may be affected by hormonal levels (discussed at the end of the chapter), the body's need for calcium and/or phosphorus, and gravitational or mechanical stressors to bone. For example, when a person wears orthodontic braces, osteoblasts and osteoclasts work together to modify the tooth-jaw junction, in response to the mechanical stress applied by the braces to the teeth and jaw. If osteoclasts resorb the bone to remove calcium salts at a faster rate than osteoblasts produce matrix to stimulate deposition, bones lose mass and become weaker; in contrast, when osteoblast activity outpaces osteoclast activity, bones have a greater mass.

Composition of the Bone Matrix

The matrix of bone connective tissue has both organic and inorganic components. About one-third of bone mass is composed of organic components, including cells, collagen fibers, and ground substance. The collagen fibers give a bone tensile strength by resisting stretching and twisting, and contribute to its overall flexibility. The ground substance is the semisolid material that suspends and supports the collagen fibers. The inorganic components of the bone provide its compressional strength. Calcium phosphate, $Ca_3(PO_4)_2$, accounts for most of the inorganic components of bone. Calcium phosphate and calcium hydroxide interact to form crystals of **hydroxyapatite** (hī-drok′sē-ap-ă-tīt), which is $Ca_{10}(PO_4)_6(OH)_2$. These crystals deposit around the collagen fibers in the extracellular matrix, leading to hardening of the matrix. The crystals also incorporate other salts,

Osteitis Deformans

Osteitis deformans (Paget disease of bone) was first described by Sir James Paget in 1877. The disease results from a disruption in the balance between osteoclast and osteoblast function. It is characterized by excessive bone resorption (excessive osteoclast activity) followed by excessive bone deposition (excessive osteoblast activity). The resulting bone is structurally unstable and immature.

In osteitis deformans, the osteoclasts are anatomically and physiologically abnormal; they are five times larger than normal and may contain 20 or more nuclei (compared to about 3–5 nuclei in normal osteoclasts). These larger osteoclasts resorb bone at a higher rate than normal. In response to this excessive bone resorption, the osteoblasts (which are normal-sized) deposit additional bone, but this new bone is poorly formed, making it more susceptible to deformation and fractures.

Osteitis deformans most commonly occurs in the bones of the pelvis, skull, vertebrae, femur (thigh bone), and tibia (leg bone). Initial symptoms include bone deformity and pain. Eventually, the lower limb bones may be bowed, and the skull often becomes thicker and enlarged. Biochemical tests can measure the level of osteoclast activity. There is no cure for osteitis deformans, but medications can reduce bone pain and bone resorption by osteoclasts.

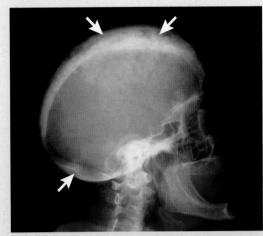

Lateral x-ray of a skull with Paget disease. White arrows indicate areas of excessive bone deposition.

such as calcium carbonate, and ions, such as sodium, magnesium, sulfate, and fluoride, in the process of calcification.

Comparison of Compact and Spongy Bone

Two types of bone connective tissue are present in most of the bones of the body: **compact bone** (also called *dense* or *cortical bone*) and **spongy bone** (also called *cancellous* or *trabecular bone*). As their names imply, compact bone is solid and relatively dense, whereas spongy bone appears more porous, like a sponge. The arrangement of compact bone and spongy bone components differs at the microscopic level. Spongy bone forms an open lattice of narrow plates of bone, called **trabeculae** (tră-bek′ū-lē; sing., *trabecula*, tră-bek′ū-lă; *trabs* = a beam). In a long bone, compact bone forms the solid external walls of the bone, and spongy bone is located internally, primarily within the epiphyses. In a flat bone of the skull, the spongy bone, also called **diploë** (dip′lō-ē; *diplous* = double), is sandwiched between two layers of compact bone (**figure 6.7**).

Compact Bone Microscopic Anatomy   Compact bone has an organized structure when viewed under the microscope. A cylindrical **osteon** (os′tē-on; bone), or *Haversian system*, is the basic functional and structural unit of mature compact bone. Osteons run parallel to the diaphysis of the long bone. An osteon is a three-dimensional structure that has several components (**figures 6.8, 6.9a,b**).

- The **central canal** (*Haversian canal*) is a cylindrical channel that lies in the center of the osteon and runs parallel to it. Traveling within the central canal are the blood vessels and nerves that supply the bone.
- **Concentric lamellae** (lă-mel′-ē; sing., *lamella*, lă-mel′ă; *lamina* = plate, leaf) are rings of bone connective tissue that surround the central canal and form the bulk of the osteon. The numbers of concentric lamellae vary among osteons. Each lamella contains collagen fibers oriented in one direction; adjacent lamellae contain collagen fibers oriented in alternating directions. In other words, if one lamella has collagen fibers directed superiorly and to the right, the next

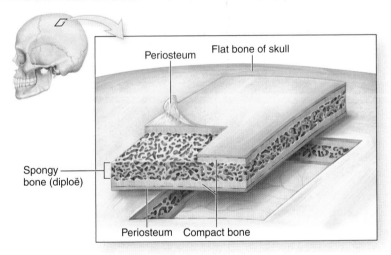

Figure 6.7

Flat Bones Within the Skull.   These bones are composed of two layers of compact bone, with a region of spongy bone (diploë) sandwiched between. Both layers of compact bone are covered by periosteum.

lamella will have collagen fibers directed superiorly and to the left. This alternating collagen fiber direction gives bone part of its strength and resilience.

- **Osteocytes** are housed in lacunae and are found between adjacent concentric lamellae.
- **Canaliculi** (kan-ă-lik′ū-lī; sing., *canaliculus*, kan-ă-lik′ū-lŭs; *canalis* = canal) are tiny, interconnecting channels within the bone connective tissue that extend from each lacuna, travel through the lamellae, and connect to other lacunae and the central canal. Canaliculi house osteocyte cytoplasmic projections that permit intercellular contact and communication. Thus, nutrients, minerals, gases, and wastes can travel through these passageways between the central canal and the osteocytes.

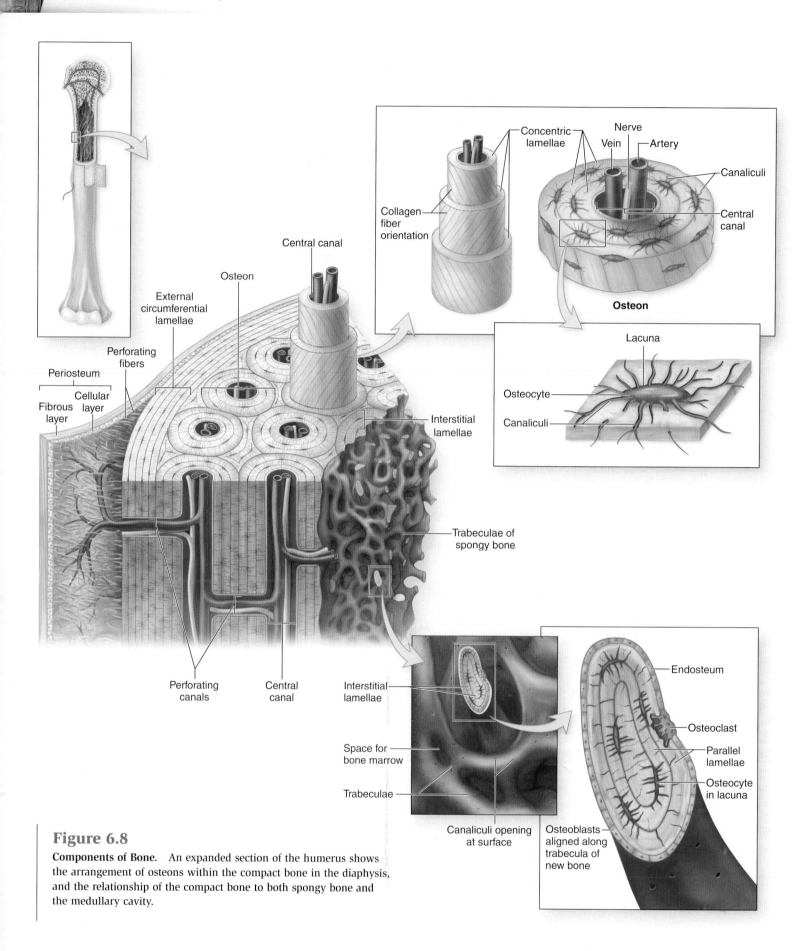

Figure 6.8

Components of Bone. An expanded section of the humerus shows the arrangement of osteons within the compact bone in the diaphysis, and the relationship of the compact bone to both spongy bone and the medullary cavity.

Osteon

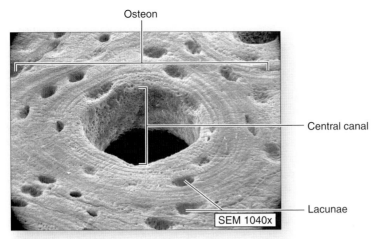

Central canal

SEM 1040x

Lacunae

(a) Compact bone

Lacuna
(with osteocyte)   Osteon

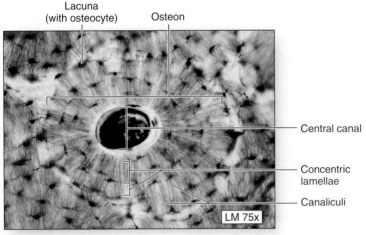

Central canal

Concentric
lamellae

Canaliculi

LM 75x

(b) Compact bone

Osteoblasts   Spongy   Red bone
bone   marrow

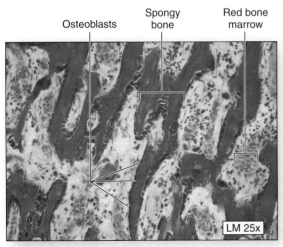

LM 25x

(c) Spongy bone

Figure 6.9

Microscopic Anatomy of Bone. *(a)* SEM and *(b)* light micrograph of osteons in a cross section of bone. *(c)* Light micrograph of spongy bone.

Several other structures are found in compact bone, but are not part of the osteon proper, including the following (see figure 6.8):

- **Perforating canals** (*Volkmann canals*) resemble central canals in that they also contain blood vessels and nerves. However, perforating canals run perpendicular to the central canals and help connect multiple central canals, thus creating a vascular and innervation connection among the multiple osteons.
- **Circumferential lamellae** are rings of bone immediately internal to the periosteum of the bone (**external circumferential lamellae**) or internal to the endosteum (**internal circumferential lamellae**). These two distinct regions appear during the original formation of the bone. Both external and internal circumferential lamellae run the entire circumference of the bone itself (hence, their name).
- **Interstitial lamellae** are the leftover parts of osteons that have been partially resorbed. They often look like a "bite" has been taken out of them. The interstitial lamellae are incomplete and typically have no central canal.

Spongy Bone Microscopic Anatomy   Spongy bone contains no osteons (figure 6.9c). Instead, the trabeculae of spongy bone are composed of **parallel lamellae.** Between adjacent lamellae are osteocytes resting in lacunae, with numerous canaliculi radiating from the lacunae. Nutrients reach the osteocytes by diffusion through canaliculi that open onto the surfaces of the trabeculae. Note that the trabeculae often form a meshwork of crisscrossing bars and plates of bone pieces. This structure provides great resistance to stresses applied in many directions by distributing the stress throughout the entire framework. As an analogy, visualize the jungle gym climbing apparatus on a children's playground. It is capable of supporting the weight of numerous children whether they are distributed throughout its structure or all localized in one area. This is accomplished because stresses and forces are distributed throughout the structure.

WHAT DO YOU THINK?

2 Long bones typically contain both compact bone and spongy bone. What benefit does spongy bone provide? Why wouldn't you want compact bone throughout the entire bone?

WHAT DID YOU LEARN?

6 What are some of the organic and inorganic components of bone?

7 If the activity of osteoblasts exceeds the activity of osteoclasts, how is the mass of the bone affected?

8 Compare the following spaces in bone: central canal, canaliculi, and lacunae. How are they similar and different? Where is each type located?

Ossification

Key topics in this section:

- Intramembranous ossification and endochondral ossification
- Components of bone that enable it to grow and be remodeled

Ossification (os′i-fi-kā′shŭn; *os* = bone, *facio* = to make), or *osteogenesis* (os′tē-ō-jen′ĕ-sis; *osteo* = bone, *genesis* = beginning), refers to the formation and development of bone connective tissue. Ossification begins in the embryo and continues as the skeleton grows during childhood and adolescence. Even after the adult bones have formed, ossification continues, as will be described later in this section. By the eighth through twelfth weeks of development, the skeleton begins forming from either thickened condensations of mesenchyme or a hyaline cartilage model of bone. Thereafter, these models are replaced by hard bone.

Intramembranous Ossification

Intramembranous (in′tră-mem′brā-nŭs) **ossification** literally means "bone growth within a membrane," and is so named because the thin layer of mesenchyme in these areas is sometimes referred to as a membrane. Intramembranous ossification is also sometimes called *dermal ossification*, because the mesenchyme that is the source of these bones is in the area of the future dermis. Recall from chapter 4 that mesenchyme is an embryonic connective tissue that has mesenchymal cells and abundant ground substance. Intramembranous ossification produces the flat bones of the skull, some of the facial bones (zygomatic bone, maxilla), the mandible (lower jaw), and the central part of the clavicle (collarbone). It begins when mesenchyme becomes thickened and condensed with a dense supply of blood capillaries, and continues in several steps (**figure 6.10**):

1. **Ossification centers form within thickened regions of mesenchyme**. Beginning at the eighth week of development, some cells in the thickened, condensed mesenchyme divide, and the committed cells that result then differentiate into osteoprogenitor cells. Some osteoprogenitor cells become osteoblasts, which secrete the semisolid organic components of the bone matrix called osteoid. Multiple **ossification centers** develop within the thickened mesenchyme as the number of osteoblasts increases.

2. **Osteoid undergoes calcification**. Osteoid formation is quickly followed by initiation of the process of calcification, as calcium salts are deposited onto the osteoid and then crystallize (solidify). Both organic matrix formation and calcification occur simultaneously at several sites within the condensed mesenchyme. When calcification entraps osteoblasts within lacunae in the matrix, the entrapped cells become osteocytes.

3. **Woven bone and its surrounding periosteum form**. Initially, the newly formed bone connective tissue is

immature and not well organized, a type called **woven bone**, or *primary bone*. This woven bone is eventually replaced by **lamellar bone**, or *secondary bone*. The mesenchyme that still surrounds the woven bone begins to thicken and eventually organizes to form the periosteum. The bone continues to grow, and new osteoblasts are trapped in the expanding bone. Additional osteoblasts are continually produced as mesenchymal cells grow and develop. Newly formed blood vessels also branch throughout this region. The calcified trabeculae and intertrabecular spaces are composed of spongy bone.

4. **Lamellar bone replaces woven bone, as compact bone and spongy bone form.** Lamellar bone replaces the trabeculae of woven bone. On the internal and external surfaces, spaces between the trabeculae are filled, and the bone becomes compact bone. Internally, the trabeculae are modified slightly and produce spongy bone. The typical structure of a flat cranial bone results: two external layers of compact bone with a layer of spongy bone in between.

Endochondral Ossification

Endochondral (en-dō-kon′drăl; *endo* = within, *chondral* = cartilage) **ossification** begins with a hyaline cartilage model and produces most of the other bones of the skeleton, including those of the upper and lower limbs, the pelvis, the vertebrae, and the ends of the clavicle. Long bone development in the limb is a good example of this process, which takes place in the following steps (**figure 6.11**):

1. **The fetal hyaline cartilage model develops.** During the eighth to twelfth week of development, chondroblasts secrete cartilage matrix, and a hyaline cartilage model forms. Within this cartilage model, the chondroblasts have become chondrocytes trapped within lacunae. A perichondrium surrounds the cartilage.

2. **Cartilage calcifies, and a periosteal bone collar forms.** Within the center of the cartilage model (future diaphysis), chondrocytes start to *hypertrophy* (enlarge) and *resorb* (eat away) some of the surrounding cartilage matrix, producing larger holes in the matrix. As these chondrocytes enlarge, the cartilage matrix begins to calcify. Chondrocytes in this region die and disintegrate because nutrients cannot diffuse to them through this calcified matrix. The result is a calcified cartilage shaft with large holes in the place where chondrocytes once were.

   As the cartilage in the shaft is calcifying, blood vessels grow toward the cartilage and start to penetrate the perichondrium around the shaft. Stem cells within the perichondrium divide to form osteoblasts. As the osteoblasts develop and this supporting connective tissue becomes highly vascularized, the perichondrium becomes a periosteum. The osteoblasts within the internal layer of the periosteum start secreting a layer of osteoid around the calcified cartilage shaft. The osteoid hardens and forms a **periosteal bone collar** around this shaft.

3. **The primary ossification center forms in the diaphysis.** A growth of capillaries and osteoblasts, called a *periosteal bud*, extends from the periosteum into the core of the cartilage shaft, invading the spaces left by the chondrocytes. The remains of the calcified cartilage serve as a template on which osteoblasts begin to produce osteoid. This region, where bone replaces cartilage in the center of the diaphysis of the

Intramembranous Ossification

(1) Ossification centers form within thickened regions of mesenchyme.

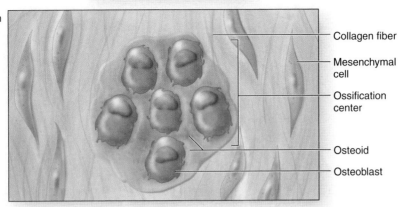

Collagen fiber

Mesenchymal cell

Ossification center

Osteoid

Osteoblast

(2) Osteoid undergoes calcification.

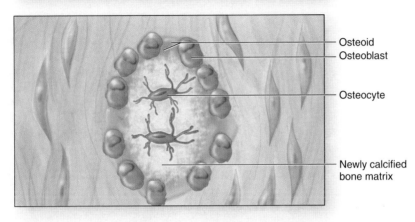

Osteoid
Osteoblast

Osteocyte

Newly calcified bone matrix

Figure 6.10

Intramembranous Ossification. A flat bone in the skull forms from mesenchymal cells in a series of continuous steps.

(3) Woven bone and surrounding periosteum form.

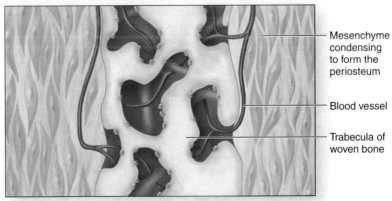

Mesenchyme condensing to form the periosteum

Blood vessel

Trabecula of woven bone

(4) Lamellar bone replaces woven bone, as compact and spongy bone form.

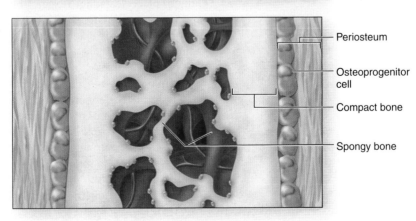

Periosteum

Osteoprogenitor cell

Compact bone

Spongy bone

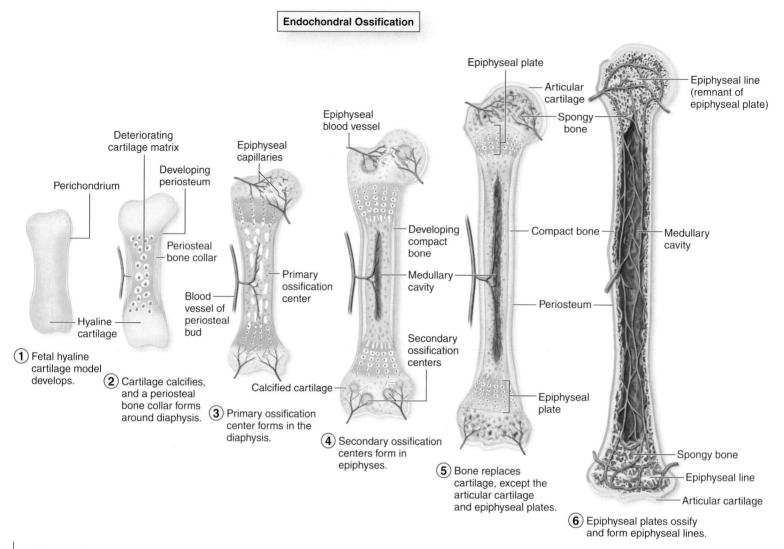

Endochondral Ossification

① Fetal hyaline cartilage model develops.

② Cartilage calcifies, and a periosteal bone collar forms around diaphysis.

③ Primary ossification center forms in the diaphysis.

④ Secondary ossification centers form in epiphyses.

⑤ Bone replaces cartilage, except the articular cartilage and epiphyseal plates.

⑥ Epiphyseal plates ossify and form epiphyseal lines.

Figure 6.11

Endochondral Ossification. Endochondral ossification of a long bone occurs in progressive stages. Bone growth is complete when each epiphyseal plate has ossified and the epiphyseal line has formed. Depending on the bone, epiphyseal plate ossification occurs between the ages of 10 and 25 years.

hyaline cartilage model, is called the **primary ossification center** because it is the first major center of bone formation. Bone development extends in both directions toward the epiphyses from the primary ossification center. Healthy bone tissue quickly replaces the calcified, degenerating cartilage in the shaft. Most, but not all, primary ossification centers have formed by the twelfth week of development.

4. **Secondary ossification centers form in the epiphyses.** The same basic process that formed the primary ossification center occurs later in the epiphyses. Beginning around the time of birth, the hyaline cartilage in the center of each epiphysis calcifies and begins to degenerate. Epiphyseal blood vessels and osteoprogenitor cells enter each epiphysis. **Secondary ossification centers** form as bone replaces calcified cartilage. Note that not all secondary ossification centers form at birth; some form later in childhood. As the secondary ossification centers form, osteoclasts resorb some bone matrix within the diaphysis, creating a hollow medullary cavity.

5. **Bone replaces cartilage, except the articular cartilage and epiphyseal plates.** By late stages of bone development, almost all of the hyaline cartilage is replaced by bone. At this point, hyaline cartilage is found only as articular cartilage on the articular surface of each epiphysis, and as a region called the **epiphyseal** (ep-i-fiz′ē-ăl) **plate,** sandwiched between the diaphysis and the epiphysis.

6. **Epiphyseal plates ossify and form epiphyseal lines.** As the bone reaches its adult size, each epiphyseal plate ossifies. Eventually, the only remnant of each epiphyseal plate is an internal thin line of compact bone called an **epiphyseal line.** Depending upon the bone, most epiphyseal plates fuse between the ages of 10 and 25. (The last epiphyseal plates to ossify are those of the clavicle in the late 20s.)

Although adult bone size has been reached, the bone continues to reshape itself throughout a person's lifetime in a constant process of bone resorption and deposition called bone remodeling (discussed later in this section).

Study Tip!

Endochondral bone growth is a tough process to learn and understand. Before trying to remember every single detail, first learn these basics:

1. A hyaline cartilage model of bone forms.

2. Bone first replaces hyaline cartilage in the diaphysis.

3. Later, bone replaces hyaline cartilage in the epiphyses.

4. Eventually, bone replaces hyaline cartilage everywhere, except the epiphyseal plates and articular cartilage.

5. By a person's late 20s, the epiphyseal plates have ossified, and lengthwise bone growth is complete.

WHAT DO YOU THINK?

3 Why does endochondral bone formation involve so many complex steps? Instead of having the hyaline cartilage model followed by the separate formation of the diaphysis and epiphyses, why can't bone simply be completely formed in the fetus?

Epiphyseal Plate Morphology

Recall that the epiphyseal plate is a layer of hyaline cartilage at the boundary between an epiphysis and the diaphysis. The epiphyseal plate exhibits five distinct microscopic zones, which are continuous from the first zone (zone 1) nearest the epiphysis to the last zone (zone 5) nearest the diaphysis **(figure 6.12)**.

1. **Zone of resting cartilage**. This zone is farthest from the medullary cavity of the diaphysis and nearest the epiphysis. It

CLINICAL VIEW: In Depth

Forensic Anthropology: Determining Age at Death

When the epiphyseal plates ossify, they fuse to and unite with the diaphysis. This process of epiphyseal plate ossification and fusion occurs in an orderly manner, and the timings of such fusions are well known. If an epiphyseal plate has not yet ossified, the diaphysis and epiphysis are still two separate pieces of bone. Thus, a skeleton that displays separate epiphyses and diaphyses (as opposed to whole fused bones) is that of a juvenile rather than an adult. Forensic anthropologists utilize this anatomic information to help determine the age of skeletal remains.

Fusion of an epiphyseal plate is progressive, and is usually scored as follows:

a. **Open** (no bony fusion or union between the epiphysis and the other bone end)

b. **Partial union** (some fusion between the epiphysis and the rest of the bone, but a distinct line of separation may be seen)

c. **Complete union** (all visible aspects of the epiphysis are united to the rest of the bone)

When determining the age at death from skeletal remains, the skeleton will be older than the oldest complete union and younger than the youngest open center. For example, if one epiphyseal plate that typically fuses at age 17 is completely united, but another plate that typically fuses at age 19 is open, the skeleton is that of a person between the ages of 17 and 19.

Current standards for estimating age based on epiphyseal plate fusion have primarily used male skeletal remains. Female epiphyseal plates tend to fuse approximately 1–2 years earlier than those of males, so this fact needs to be considered when estimating the age of a female skeleton. Further, population differences may exist with some epiphyseal plate unions. With these caveats in mind, the accompanying table lists standards for selected epiphyseal plate unions.

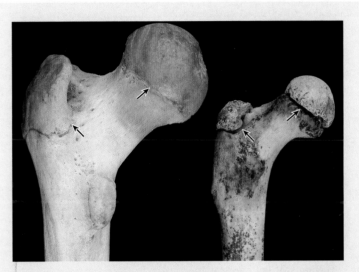

These two femurs came from individuals of different ages. (Left) *In partial union* (arrows), *the epiphyses are partially fused. This individual likely was between the ages of 15 and 23 at death.* (Right) *No fusion has occurred between the epiphyses and the diaphysis* (see arrows), *a category called open. This individual likely was younger than 15 years of age.*

Bone	Male Age at Epiphyseal Union (years)
Humerus, lateral epicondyle	11–16 (female: 9–13)
Humerus, medial epicondyle	11–16 (female: 10–15)
Humerus, head	14.5–23.5
Proximal radius	14–19
Distal radius	17–22
Distal fibula and tibia	14.5–19.5
Proximal tibia	15–22
Femur, head	14.5–23.5
Distal femur	14.5–21.5
Clavicle	19–30

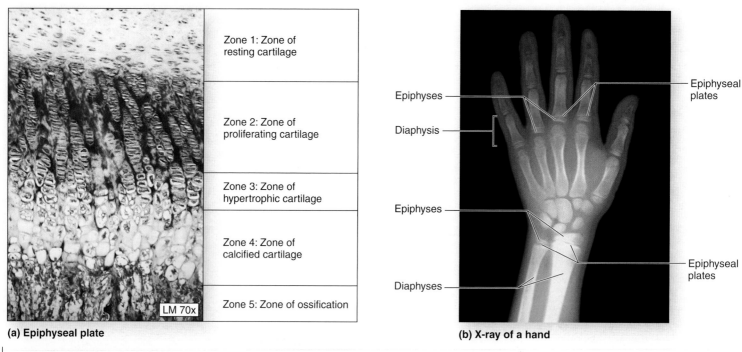

Figure 6.12

Epiphyseal Plate. *(a)* In a growing long bone, the epiphyseal plate, located at the boundary between the diaphysis and the epiphysis, exhibits five distinct but continuous zones. Zones 1–4 are cartilage, while zone 5 is bone. *(b)* An x-ray of a child's hand shows the cartilaginous epiphyseal plates as dark lines between the epiphysis and the diaphysis of long bones.

is composed of small chondrocytes distributed throughout the cartilage matrix, and resembles mature, healthy hyaline cartilage. This region secures the epiphysis to the epiphyseal plate.

2. **Zone of proliferating cartilage.** Chondrocytes in this zone undergo rapid mitotic cell division, enlarge slightly, and become aligned like a stack of coins into longitudinal columns of flattened lacunae.

3. **Zone of hypertrophic cartilage**. Within this zone, chondrocytes cease dividing and begin to hypertrophy (enlarge) greatly. The walls of the lacunae become thin as the chondrocytes resorb matrix during their hypertrophy.

4. **Zone of calcified cartilage**. This narrow zone of cartilage is only a few cells thick. Minerals are deposited in the matrix between the columns of lacunae; this calcification kills the chondrocytes and makes the matrix appear opaque.

5. **Zone of ossification**. The walls break down between lacunae in the columns, forming longitudinal channels. These spaces are invaded by capillaries and osteoprogenitor cells from the medullary cavity. New matrix of bone is deposited on the remaining calcified cartilage matrix.

Growth of Bone

As with cartilage growth, a long bone's growth in length is called interstitial growth, and its growth in diameter or thickness is called appositional growth. Interstitial growth occurs within the epiphyseal plate as chondrocytes undergo mitotic cell division in zone 2 and chondrocytes hypertrophy in zone 3. These activities combine to push the zone of resting cartilage toward the epiphysis, while new bone is being produced at the same rate in zone 5, resulting in increased bone length. The epiphyseal plate maintains its thickness as it is pushed away from the center of the shaft. At maturity, the rate of epiphyseal cartilage production slows, and the rate of osteo-

blast activity accelerates. As a result, the epiphyseal plate becomes narrower, until it ultimately disappears, and interstitial growth completely stops. The appearance of the remnant epiphyseal line signals the termination of lengthwise growth of the bone.

Appositional growth occurs within the periosteum (**figure 6.13**). In this process, osteoblasts in the inner cellular layer of the periosteum lay down bone matrix in layers parallel to the surface, called external circumferential lamellae. These lamellae are analogous to tree rings: As they increase in number, the structure widens. Thus, the bone becomes wider as new bone is laid down at the periphery. As this new bone is being laid down, osteoclasts along the medullary cavity resorb bone matrix, creating an expanding medullary cavity. The combined effects of bone growth at the periphery and bone resorption within the medullary cavity transform an infant bone into a larger version called an adult bone.

Bone Remodeling

Bone continues to grow and renew itself throughout life. The continual deposition of new bone tissue and the removal (resorption) of old bone tissue is called **bone remodeling**. Bone remodeling helps maintain calcium and phosphate levels in body fluids, and can be stimulated by stress on a bone (e.g., bone fracture, or exercise that builds up muscles that attach to bone). This ongoing process occurs at both the periosteal and endosteal surfaces of a bone. It either modifies the architecture of the bone or changes the total amount of minerals deposited in the skeleton. Prior to and throughout puberty, the formation of bone typically exceeds its resorption. In young adults, the processes of formation and resorption tend to occur at about the same rate. However, they become disproportionate in older adults when resorption of bone exceeds its formation.

It is estimated that about 20% of the adult human skeleton is replaced yearly. However, bone remodeling does not occur at the

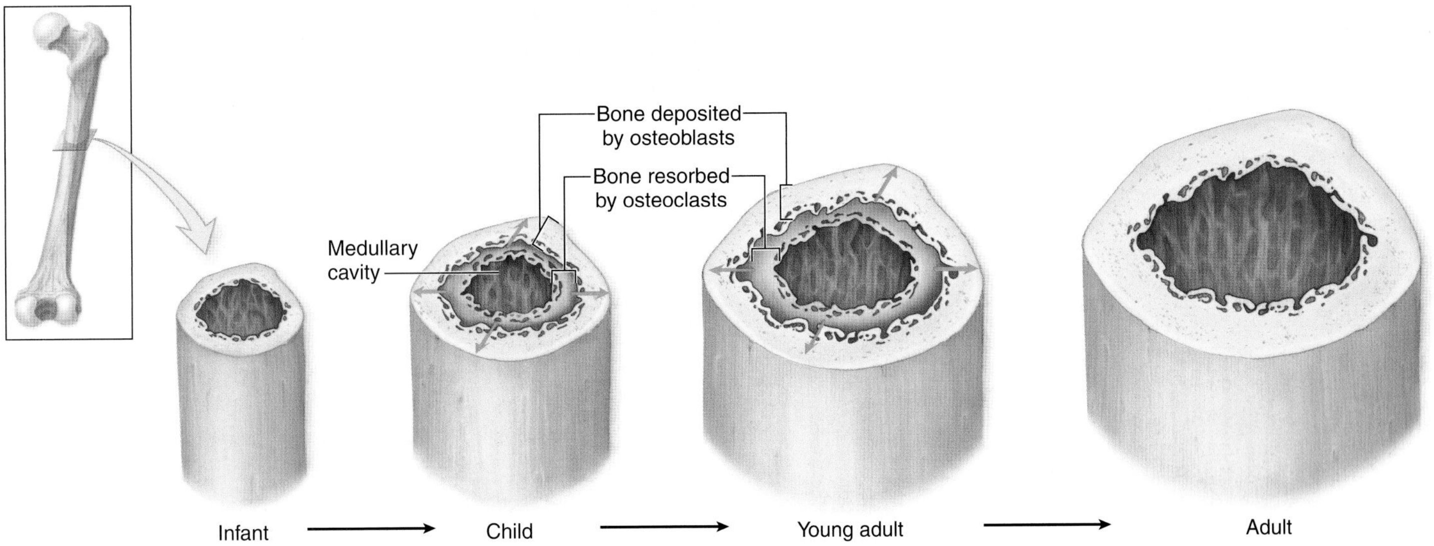

Bone deposited
by osteoblasts

Bone resorbed
by osteoclasts

Medullary
cavity

Infant ⟶ Child ⟶ Young adult ⟶ Adult

Figure 6.13

Appositional Bone Growth.   A bone increases in diameter as new bone is added to the surface. At the same time, some bone may be removed from the inner surface to enlarge the marrow cavity.

same rate everywhere in the skeleton. For example, the compact bone in our skeleton is replaced at a slower rate than the spongy bone. The distal part of the femur (thigh bone) is replaced every 4 to 6 months, while the diaphysis of this bone may not be completely replaced during an individual's lifetime.

Blood Supply and Innervation

Bone is highly vascularized (meaning it is supplied by many blood vessels), especially in regions containing red bone marrow. Blood vessels enter bones from the periosteum. A typical long bone such as the humerus has four major sets of blood vessels **(figure 6.14)**.

Nutrient blood vessels, called the **nutrient artery** and the **nutrient vein**, supply the diaphysis of a long bone. Typically, only one nutrient artery enters and one nutrient vein leaves the bone via a nutrient foramen in the bone. These vessels branch and extend along the length of the shaft toward the epiphyses and into the central canal of osteons within compact bone and the marrow cavity.

Metaphyseal blood vessels (**metaphyseal arteries** and **metaphyseal veins**) provide the blood supply to the diaphyseal side of the epiphyseal plate, which is the region where new bone ossification forms bone connective tissue to replace epiphyseal plate cartilage.

Epiphyseal arteries and **epiphyseal veins** provide the blood supply to the epiphyses of the bone. In early childhood, the cartilaginous epiphyseal plate separates the epiphyseal and metaphyseal vessels. However, once an epiphyseal plate ossifies and becomes an epiphyseal line, the epiphyseal vessels and metaphyseal vessels anastomose (interconnect) through channels formed in the epiphyseal line (see figure 6.14 for examples).

Periosteal blood vessels (**periosteal arteries** and **periosteal veins**) provide blood to the external circumferential lamellae and the superficial osteons within the compact bone at the external edge of the bone. These vessels and the accompanying periosteal nerves penetrate the diaphysis and enter the perforating canals at many locations.

Nerves that supply bones accompany blood vessels through the nutrient foramen and innervate the bone as well as its periosteum, endosteum, and marrow cavity. These are mainly sensory nerves that signal injuries to the skeleton.

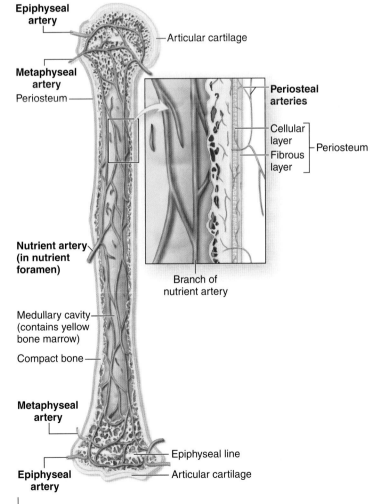

Figure 6.14

Arterial Supply to a Mature Bone.   Four major sets of blood vessels supply the humerus, a long bone: nutrient arteries and veins, metaphyseal arteries and veins, epiphyseal arteries and veins, and periosteal arteries and veins.

Achondroplastic Dwarfism

Achondroplasia (ā-kon-drō-plā'zē-ă) is characterized by abnormal conversion of hyaline cartilage to bone. The most common form is **achondroplastic dwarfism**, in which the long bones of the limbs stop growing in childhood, while the other bones usually continue to grow normally. Thus, an individual with achondroplastic dwarfism is short in stature but generally has a large head. Often the forehead is prominent, and the nose is flat at the bridge. Those affected may have bowlegs and *lordosis* (exaggerated curvature of the lumbar spine). Most individuals are about 4 feet tall. Their intelligence and life span are within normal range.

Achondroplastic dwarfism results from a failure of chondrocytes in the second and third zones of the epiphyseal plate (see figure 6.12a) to multiply and enlarge, leading to inadequate endochondral ossification. Most cases result from a spontaneous mutation during DNA replication. Thus, even parents who are of normal height and have no family history of dwarfism may have a child with achondroplastic dwarfism. Children of an achondroplastic dwarfism parent also may inherit the disorder. This is because it is an autosomal dominant condition, meaning that a child may inherit only one defective gene from a parent (as opposed to having both genes defective) in order to express the condition. This condition differs from pituitary dwarfism, which results when the pituitary gland produces insufficient growth hormone or none at all. In pituitary dwarfism, the growth of all the bones is stunted, so the individual is short in stature but has normal proportions throughout the skeletal system.

Table 6.1	Effects of Hormones and Vitamins on Bone Maintenance and Growth
HORMONES	
Growth hormone	Stimulates liver to produce the hormone somatomedin, which causes cartilage proliferation at epiphyseal plate and resulting bone elongation; too little growth hormone results in short stature in the child
Thyroid hormone	Stimulates bone growth by stimulating metabolic rate of osteoblasts; too little thyroid hormone results in short stature
Calcitonin	Promotes calcium deposition in bone and inhibits osteoclast activity
Parathyroid hormone	Increases blood calcium levels by encouraging bone resorption by osteoclasts
Sex hormones (estrogen and testosterone)	Stimulate osteoblasts; promote epiphyseal plate growth and closure
Glucocorticoids	If levels are chronically too high, bone resorption occurs and significant bone mass is lost
VITAMINS	
Vitamin A	Activates osteoblasts
Vitamin C (ascorbic acid)	Promotes collagen production
Vitamin D	Promotes absorption of calcium and phosphate into blood; helps with calcification of bone

WHAT DID YOU LEARN?

9 What is intramembranous ossification? What bones form by this process?

10 Identify the locations of the primary and secondary ossification centers in a long bone.

11 How could a physician determine whether a patient had reached full height by examining x-rays of his or her bones?

12 Name the five zones in an epiphyseal plate and the characteristics of each.

Maintaining Homeostasis and Promoting Bone Growth

Key topics in this section:

- Effects of hormones, vitamins, and exercise on bone maintenance
- Steps in the healing of bone fractures

Bone growth and maintenance normally depend upon both hormones and vitamins **(table 6.1)**.

Effects of Hormones

Hormones control and regulate growth patterns in bone by altering the rates of osteoblast and osteoclast activity. **Growth hormone**, also called *somatotropin* (sō'mă-tō-trō'pin), is produced by the anterior pituitary gland. It affects bone growth by stimulating the formation of another hormone, *somatomedin* (sō'mă-tō-mē'din), which is produced

by the liver. Somatomedin directly stimulates growth of cartilage in the epiphyseal plate. **Thyroid hormone**, secreted by the thyroid gland, stimulates bone growth by influencing the basal metabolic rate of bone cells. Together, growth hormone and thyroid hormone, if maintained in proper balance, regulate and maintain normal activity at the epiphyseal plates until puberty. If a child's growth hormone and/or thyroid hormone levels are chronically too low, then bone growth is adversely affected, and the child will be short in stature.

Another thyroid gland hormone is **calcitonin** (kal-si-tō'nin; *calx* = lime, *tonos* = stretching), which is secreted in response to high levels of calcium in the blood. Calcitonin encourages calcium deposition from blood into bone and inhibits osteoclast activity.

Parathyroid hormone is secreted and released by the parathyroid glands in response to reduced calcium levels in the blood. Ultimately, parathyroid hormone increases the blood calcium levels, so other body tissues can utilize this calcium. Parathyroid hormone stimulates osteoclasts to resorb bone and thereby increase calcium levels in the blood.

Sex hormones (**estrogen** and **testosterone**), which begin to be secreted in great amounts at puberty, dramatically accelerate bone growth. Sex hormones increase the rate of bone formation by osteoblasts in ossification centers within the epiphyseal plate, resulting in increased length of long bones and increased height. The appearance of high levels of sex hormones at puberty also signals the beginning of the end for growth at the epiphyseal plate. Eventually, more bone is produced at the epiphyseal plate than the cartilage within the plate can support. As a result, the thickness of the epiphyseal plate cartilage begins to diminish, and eventually it disappears altogether, leaving behind the epiphyseal line. Older individuals (who have a normal reduction in sex hormones) also may experience a decrease in bone mass as they age.

Finally, *abnormal* amounts of certain hormones can affect bone maintenance and growth. As mentioned earlier, chronically low levels of growth hormone and/or thyroid hormone in a child inhibit bone growth and result in short stature. Another example are the **glucocorticoids**, a group of hormones produced by the adrenal cortex. Normal glucocorticoid levels tend not to have any major effects on bone growth or mass. However, if glucocorticoid levels are chronically too high, they stimulate bone resorption and can lead to significant loss of bone mass.

Effects of Vitamins

A continual dietary source of vitamins is required for normal bone growth. For example, **vitamin A** activates osteoblasts, while **vitamin C** is required for normal synthesis of collagen, the primary organic

---

CLINICAL VIEW

Rickets

Rickets is a disease caused by a vitamin D deficiency in childhood and characterized by overproduction and deficient calcification of osteoid. Due to the lack of vitamin D, the digestive tract is unable to absorb calcium and phosphorus, minerals needed for the hardening of the osteoid during the formation of bone.

Rickets usually develops in children, and results in bones that are poorly calcified and exhibit too much flexibility. Patients with rickets acquire a bowlegged appearance as their weight increases and the bones in their legs bend. In addition to skeletal deformities, rickets is characterized by disturbances in growth, hypocalcemia (an abnormally low level of calcium in the blood), and sometimes tetany (cramps and muscle twitches), usually caused by low blood calcium. The condition is often accompanied by irritability, listlessness, and generalized muscular weakness. Fractures frequently occur in patients with rickets.

During the Industrial Revolution, the incidence of rickets increased as children were forced to work indoors in factories. These children had little exposure to sunlight and were usually malnourished as well. (Recall from chapter 5 that the body can manufacture its own vitamin D when the integument is exposed to sunlight.) Rickets continues to occur in some developing nations, and recently the incidence has increased in urban areas of the United States. Researchers have discovered that these children spend much of their time indoors and typically do not drink enough milk, opting for soft drinks instead. So, unfortunately, a disease that is easily preventable is making a comeback in the United States due to poor dietary and lifestyle habits among the nation's youth.

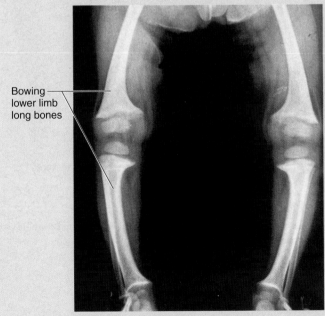

Bowing lower limb long bones

Radiograph of a 10-month-old with rickets.

CLINICAL VIEW

Bone Scans

Bone scans are tests that can detect bone pathologies sooner than standard x-rays, while exposing the patient to only a fraction of the radiation of a normal x-ray. The patient is injected intravenously with a small amount of a radioactive tracer compound that is absorbed by bone. A scanning camera then detects and measures the radiation emitted from the bone. This information is converted into a diagram or photograph that can be read like an x-ray. In these films, normal bone tissue is a consistent gray color, while darker areas are "hot spots" indicating increased metabolism, and lighter areas are "cold spots" indicating decreased metabolism. Abnormalities that can be detected by a bone scan include fractures, decalcification of bone, osteomyelitis, degenerative bone disease, and Paget disease. Bone scans are also used to determine whether cancer has metastasized to bone, to identify bone infections, to monitor the progress of bone grafts and degenerative bone disorders, to evaluate unexplained bone pain or possible fracture, and to monitor response to therapy of a cancer that has spread to bone.

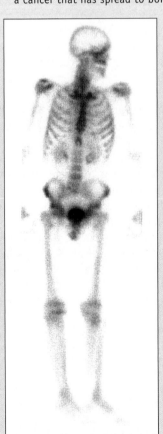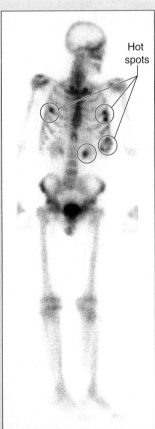

Hot spots

(a) Normal bone scan   **(b) Abnormal scan with numerous hot spots**

component in the bone matrix. **Vitamin D** stimulates the absorption and transport of calcium and phosphate ions into the blood. It also is necessary for the calcification of bone. As calcium and phosphate levels rise in the blood, calcitonin is secreted, which encourages the deposition of these minerals into bone.

Effects of Exercise

Mechanical stress, in the form of exercise, is required for normal bone remodeling. In response to mechanical stress, bone has the ability to increase its strength over a period of time by increasing the amounts of mineral salts deposited and collagen fibers synthesized. Stress also increases the production of the hormone calcitonin, which helps inhibit bone resorption by osteoclasts and encourage bone deposition by osteoblasts.

Mechanical stresses that significantly affect bone result from repeated skeletal muscle contraction and gravitational forces. Typically, the bones of athletes become noticeably thicker as a result of repetitive and stressful exercise. Weight-bearing activities, such as weight lifting or walking, help build and retain bone mass. In contrast, lack of mechanical stress weakens bone through both demineralization of the bone matrix and reduction of collagen formation. For example, if a person has a fractured bone in a cast or is bedridden, the mass of the unstressed bone decreases in the immobilized limbs. While in space, astronauts must exercise so that the lesser gravity won't weaken their bones.

Research has shown that regular weight-bearing exercise can increase total bone mass in adolescents and young adults prior to its inevitable reduction later in life. In fact, recent studies have shown that even 70- and 80-year-olds who perform moderate weight training can increase their bone mass.

Fracture Repair

Bone has great strength, and yet it may break as a result of unusual stress or a sudden impact. Breaks in bones, called **fractures,** are classified in several ways. A **stress fracture** is a thin break caused by recent increased physical activity in which the bone experiences repetitive loads (e.g., as seen in some runners). Stress fractures tend to occur in the weight-bearing bones (e.g., pelvis and lower limb). A **pathologic fracture** usually occurs in bone that has been weakened by disease, such as when the vertebrae fracture in someone with osteoporosis (a bone condition discussed in the next section). In a **simple fracture**, the broken bone does not penetrate the skin, while in a **compound fracture**, one or both ends of the broken bone pierce the overlying skin and body tissues. **Table 6.2** shows the different classifications of fractures, and **figure 6.15** illustrates some of the most common types.

The healing of a simple fracture takes about 2 to 3 months, whereas a compound fracture takes longer to heal. Fractures heal much more quickly in young children (average healing time, 3 weeks) and become slower to heal as we age. In the elderly, the normal thinning and weakening of bone increases the incidence of fractures, and some severe fractures never heal without surgical intervention. Bone fracture repair can be described as a series of steps (**figure 6.16**):

1. **A fracture hematoma forms**. A bone fracture tears blood vessels inside the bone and within the periosteum, causing bleeding, and then a **fracture hematoma** forms from the clotted blood.

2. **A fibrocartilaginous (soft) callus forms**. Regenerated blood capillaries infiltrate the fracture hematoma due to an increase in osteoblasts in both the periosteum and

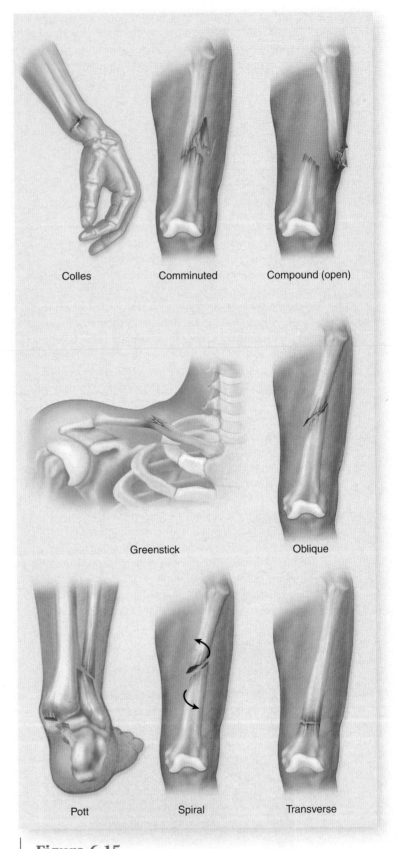

Colles   Comminuted   Compound (open)

Greenstick   Oblique

Pott   Spiral   Transverse

Figure 6.15

Types of Bone Fractures. Selected bone fractures listed in table 6.2 are illustrated here.

Table 6.2	Classification of Bone Fractures		
Fracture	**Description**	**Fracture**	**Description**
Avulsion	Complete severing of a body part (typically a toe or finger)	Impacted	One fragment of bone is firmly driven into the other
Colles	Fracture of the distal end of the lateral forearm bone (radius); produces a "dinner fork" deformity	Incomplete	Partial fracture that extends only partway across the bone
Comminuted	Bone is splintered into several small pieces between the main parts	Linear	Fracture is parallel to the long axis of the bone
Complete	Bone is broken into two or more pieces	Oblique	Diagonal fracture at an angle between linear and transverse
Compound (open)	Broken ends of the bone protrude through the skin	Pathologic	Weakening of a bone caused by disease processes (e.g., cancer)
Compression	Bone is squashed (may occur in a vertebra during a fall)	Pott	Fracture at the distal end of the tibia, fibula, or both
Depressed	Broken part of the bone forms a concavity (as in skull fracture)	Simple (closed)	Bone does not break through the skin
Displaced	Fractured bone parts are out of anatomic alignment	Spiral	Fracture spirals around axis of long bone; results from twisting stress
Epiphyseal	Epiphysis is separated from the diaphysis at the epiphyseal plate	Stress	Thin fractures due to repeated, stressful impact such as running. (These fractures are sometimes difficult to see on x-rays, and a bone scan may be necessary to accurately identify their presence.)
Greenstick	Partial fracture; one side of bone breaks—the other side is bent	Transverse	Fracture at right angles to the long axis of the bone
Hairline	Fine crack in which sections of bone remain aligned (common in skull)		

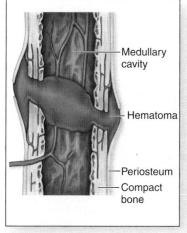

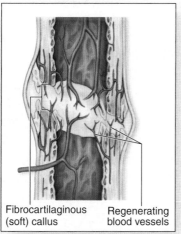

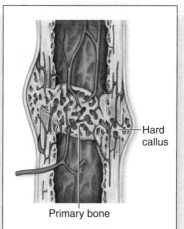

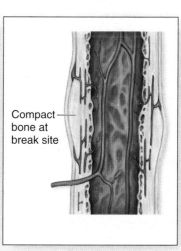

① A fracture hematoma forms.   ② A fibrocartilaginous (soft) callus forms.   ③ A hard (bony) callus forms.   ④ The bone is remodeled.

Figure 6.16

Fracture Repair.   The repair of a bone fracture occurs in a series of steps.

the endosteum near the fracture site. First, the fracture hematoma is reorganized into an actively growing connective tissue called a **procallus**. Fibroblasts within the procallus produce collagen fibers that help connect the broken ends of the bones. Chondroblasts in the newly growing connective tissue form a dense regular connective tissue associated with the cartilage. Eventually, the procallus becomes a **fibrocartilaginous (soft) callus** (kal′ŭs; hard skin). The fibrocartilaginous callus stage lasts at least 3 weeks.

3. **A hard (bony) callus forms**. Within a week, osteoprogenitor cells in areas adjacent to the fibrocartilaginous callus become osteoblasts and produce trabeculae of primary bone. The fibrocartilaginous callus is then replaced by this bone, which forms a **hard (bony) callus**. The trabeculae of the hard callus continue to grow and thicken for several months.

4. **The bone is remodeled**. Remodeling is the final phase of fracture repair. The hard callus persists for at least 3 to 4 months as osteoclasts remove excess bony material

from both exterior and interior surfaces. Compact bone replaces primary bone. The fracture usually leaves a slight thickening of the bone (as detected by x-ray); however, in many instances healing occurs with no obvious thickening.

WHAT DID YOU LEARN?

13 What are the effects of growth hormone and parathyroid hormone on bone growth and/or bone mass?

14 Which vitamins help regulate bone growth?

15 A _____ fracture is diagnosed when the bone has broken through the skin.

Bone Markings

Key topic in this section:

■ Anatomic terms that describe the surface features of bone

Distinctive **bone markings**, or *surface features*, characterize each bone in the body. Projections from the bone surface mark the point where tendons and ligaments attach. Sites of articulation between adjacent bones are smooth, flat areas. Depressions, grooves, and tunnels through bones indicate sites where blood vessels and nerves either lie alongside or penetrate the bone. Anatomists use specific terms to describe these elevations and depressions (**figure 6.17**).

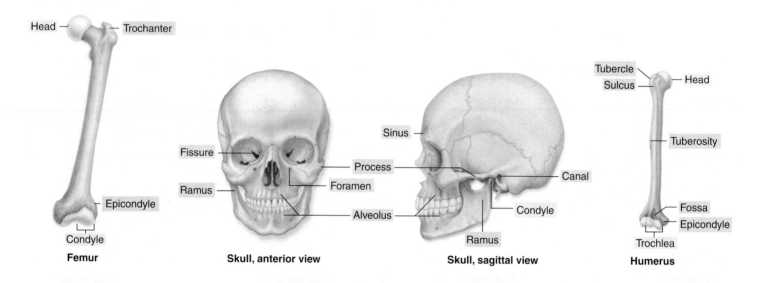

Femur   Skull, anterior view   Skull, sagittal view   Humerus

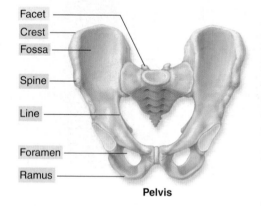

Pelvis

Figure 6.17

Bone Markings. Specific anatomic terms describe the characteristic identification marks for bones.

General Structure	Anatomical Term	Description
Articulating surfaces	Condyle	Large, smooth, rounded articulating oval structure
	Facet	Small, flat, shallow articulating surface
	Head	Prominent, rounded epiphysis
	Trochlea	Smooth, grooved, pulley-like articular process
Depressions	Alveolus (pl., *alveoli*)	Deep pit or socket in the maxillae or mandible
	Fossa (pl., *fossae*)	Flattened or shallow depression
	Sulcus	Narrow groove
Projections for tendon and ligament attachment	Crest	Narrow, prominent, ridgelike projection
	Epicondyle	Projection adjacent to a condyle
	Line	Low ridge
	Process	Any marked bony prominence
	Ramus (pl., *rami*)	Angular extension of a bone relative to the rest of the structure
	Spine	Pointed, slender process
	Trochanter	Massive, rough projection found only on the femur
	Tubercle	Small, round projection
	Tuberosity	Large, rough projection
Openings and spaces	Canal	Passageway through a bone
	Fissure	Narrow, slitlike opening through a bone
	Foramen (pl., *foramina*)	Rounded passageway through a bone
	Sinus	Cavity or hollow space in a bone

Knowing the names of bone markings will help you learn about specific bones in chapters 7 and 8. For example, knowing that foramen means "hole" or "passageway" will tell you to look for a hole when trying to find the foramen magnum on the skull. Likewise, you can usually correctly assume that any smooth, oval prominence on a bone is called a condyle. Refer back to figure 6.17 frequently for assistance in learning the bones and their features.

For professional criminologists, pathologists, and anthropologists, bones can tell an intricate anatomic story. Bone markings on skeletal remains indicate where the soft tissue components once were, often allowing an individual's height, age, sex, and general appearance to be determined.

WHAT DID YOU LEARN?

16 What is the anatomic name for a narrow, slitlike opening through a bone?

Aging of the Skeletal System

Key topic in this section:

- Changes in bone architecture and bone mass related to age

Aging affects bone connective tissue in two ways. First, the tensile strength of bone decreases due to a reduced rate of protein synthesis, which in turn results in decreased ability to produce the organic portion of bone matrix. Consequently, the percentage of inorganic minerals in the bone matrix increases, and the bones of the skeleton become brittle and susceptible to fracture.

Second, bone loses calcium and other minerals (demineralization). The bones of the skeleton become thinner and weaker, resulting in insufficient ossification, a condition called **osteopenia** (os'tē-ō-pē'nē-ă; *penia* = poverty). Aging causes all people to become slightly osteopenic. This reduction in bone mass may begin as early as 35–40 years of age, when osteoblast

---

CLINICAL VIEW

Osteoporosis

Osteoporosis, meaning "porous bones," is a disease that results in decreased bone mass and microarchitectural changes that lead to weakened bones that are prone to fracture. Both bone matrix and calcium are lost, particularly in metabolically active spongy bone. The occurrence of osteoporosis is greatest among the elderly, especially Caucasian women, and the severity is closely linked to age and the onset of menopause. Postmenopausal women are at risk because (1) women have less bone mass than men, (2) women begin losing bone mass earlier and faster in life (sometimes as early as 35 years of age), and (3) postmenopausal women no longer produce significant amounts of estrogen, which appears to help protect against osteoporosis by stimulating bone growth. Although the condition does affect men, osteoporosis in men is typically less severe than in women for the reasons just mentioned.

As a result of osteoporosis, the incidence of fractures increases, most frequently in the wrist, hip, and vertebral column, and usually as the result of a normal amount of stress. Wrist fractures occur at the distal end of the radius (Colles fracture), and hip fractures usually occur at the neck of the femur. The weight-bearing regions of the vertebrae lose spongy bone and are more easily compressed, leading to a loss of height and sometimes to compression fractures of the vertebral bodies.

Although diagnosis and monitoring of osteoporosis have been simplified, a cure remains elusive. The best treatment seems to be prevention. Young adults should maintain good nutrition and physical activity to ensure adequate bone density, thus allowing for the normal, age-related loss later in life. Calcium supplements may help maintain bone health, but by themselves will not stimulate new bone growth.

Medical treatments involve two strategies: (1) slowing the rate of bone loss and (2) attempting to stimulate new bone growth. Formerly, because of the link between estrogen and bone growth, hormone replacement therapy (HRT) was widely used to slow bone degeneration in postmenopausal women. Unfortunately, new studies have linked estrogen supplementation to increased risk of cardiovascular (heart and blood vessel) problems, such as stroke and heart attacks, as well as increases in blood clots in the lung (pulmonary emboli). These significant complications of HRT have substantially limited its usefulness in therapy and prevention.

A new class of medications, the bisphosphonates, have shown great promise in slowing the progression of osteoporosis. Examples of bisphosphonates include alendronate (brand name: Fosamax), pamidronate (Aredia), and risedronate (Actonel). These drugs work by interfering with osteoclast function and thus retarding the removal of bone during remodeling. Since bone remodeling goes on all the time, even in people with osteoporosis, slowing osteoclast-driven bone destruction even a little can help preserve, and even add to, bone mass.

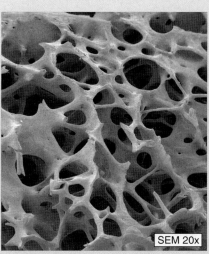

SEM 20x

(a) Normal bone

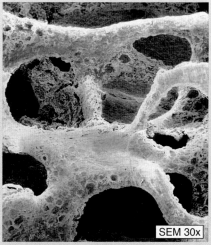

SEM 30x

(b) Osteoporotic bone

activity declines while osteoclast activity continues at previous levels. During each successive decade of their lives, women lose roughly more of their skeletal mass than do men. Different parts of the skeleton are affected unequally. Vertebrae, jaw bones, and epiphyses lose large amounts of mass, resulting in reduced height, loss of teeth, and fragile limbs. A significant percentage of older women and a smaller proportion of older men suffer from **osteoporosis** (os'tē-ō-pō-rō'sis; *poros* = pore, *osis* = condition), in which bone mass becomes reduced enough to compromise normal function (see Clinical View).

 WHAT DO YOU THINK?

4 What major differences might we expect when comparing bone composition in a 65-year-old man with that of his 13-year-old granddaughter?

CLINICAL TERMS

chondroma Benign (noncancerous) tumor derived from cartilage cells.

chondrosarcoma Malignant (cancerous) tumor derived from cartilage cells.

hyperostosis Excessive formation of bone tissue.

osteogenesis imperfecta Also known as "brittle bone disease"; Inherited condition that affects collagen fiber distribution and organization. It occurs due to impaired osteoblast function, and results in abnormal bone growth, brittle bones, continuing deformation of the skeleton, and increased susceptibility to fracture.

osteosarcoma The most common and malignant bone sarcoma; arises from bone-forming cells (osteoblasts) and chiefly affects the ends of long bones.

osteoma Benign tumor in lamellar bone, often in the jaw or the skull.

osteomalacia Vitamin D deficiency disease in adults characterized by gradual softening and bending of the bones as a result of decreased mineral content; although bone mass is still present, it is demineralized.

osteomyelitis Infection and inflammation within both the bone marrow and neighboring regions of the bone.

CHAPTER SUMMARY

	■ The skeletal system is composed of bones, cartilage that supports the bones, and ligaments that bind together, support, and stabilize bones.
Cartilage Connective Tissue 146	■ Cartilage contains cells embedded within a matrix of protein fibers and a gel-like ground substance.
	Functions of Cartilage 146
	■ Cartilage provides support for soft tissues, a sliding surface for bone, and a model for formation of most of the bones of the body.
	Types of Cartilage 147
	■ Hyaline cartilage, the most common type, has a distinct, glassy appearance and is widely distributed.
	■ Fibrocartilage contains thick collagen fibers to help resist compression and tension.
	■ Elastic cartilage contains numerous highly branched elastic fibers to provide flexibility to structures.
	Growth Patterns of Cartilage 147
	■ Cartilage growth includes both interstitial growth (growth from within preexisting cartilage) and appositional growth (growth around the periphery of cartilage).
Bone 147	■ Bones are organs that contain multiple tissue types, the most abundant being bone (osseous) connective tissue.
	Functions of Bone 147
	■ Bone performs the following functions: support and protection, movement, hemopoiesis, and storage of minerals and energy.
Classification and Anatomy of Bones 149	■ Bones are categorized by shape as long, short, flat, or irregular.
	General Structure and Gross Anatomy of Long Bones 150
	■ A long bone contains the following parts: diaphysis, epiphyses, metaphysis, articular cartilage, medullary cavity, periosteum, and endosteum.
	■ Osteoblasts synthesize and secrete osteoid, the matrix of bone prior to its calcification.
	■ Osteocytes are mature bone cells that reside in lacunae.
	■ Osteoclasts are large, multinucleated cells involved in bone resorption.
	■ An osteon is the basic unit of structure and function of mature compact bone.
	■ An osteon contains a central canal that houses blood vessels and nerves, concentric bone layers called lamellae, osteocytes in lacunae, and thin channels called canaliculi.
Ossification 156	■ Ossification is the process of bone connective tissue formation.
	Intramembranous Ossification 156
	■ In intramembranous ossification, bone forms from mesenchyme.
	Endochondral Ossification 156
	■ Endochondral ossification uses a hyaline cartilage model that is gradually replaced by newly formed osseous tissue.

Ossification 156	**Epiphyseal Plate Morphology 159** ▪ The epiphyseal plate contains five zones where cartilage grows and is replaced by bone. **Growth of Bone 160** ▪ Lengthwise bone growth is called interstitial growth, while a bone increases in diameter through appositional growth at the periosteum. ▪ The continual deposition of new bone tissue and resorption of old bone tissue is called bone remodeling. **Blood Supply and Innervation 161** ▪ Four categories of blood vessels develop to supply a typical bone: nutrient vessels, metaphyseal vessels, epiphyseal vessels, and periosteal vessels.
Maintaining Homeostasis and Promoting Bone Growth 162	**Effects of Hormones 162** ▪ Growth hormone, thyroid hormone, calcitonin, and sex hormones stimulate bone growth. ▪ Parathyroid hormone stimulates osteoclast activity. **Effects of Vitamins 163** ▪ Vitamins A and C are essential for bone growth and remodeling. Vitamin D is needed for calcium and phosphorus absorption and calcification of bone. **Effects of Exercise 164** ▪ Stress in the form of exercise strengthens bone tissue by increasing the amounts of mineral salts deposited and collagen fibers synthesized. **Fracture Repair 164** ▪ A fracture is a break in a bone that can usually be healed if portions of the blood supply, endosteum, and periosteum remain intact.
Bone Markings 166	▪ Specific names denote bone markings such as projections, elevations, depressions, and passageways.
Aging of the Skeletal System 167	▪ Due to aging, the tensile strength of bone decreases, and bone loses calcium and other minerals (demineralization).

CHALLENGE YOURSELF

Matching

Match each numbered item with the most closely related lettered item.

_____ 1. flat bone of skull

_____ 2. osteon

_____ 3. spongy bone

_____ 4. epiphysis

_____ 5. osteoid

_____ 6. parathyroid hormone

_____ 7. endosteum

_____ 8. osteoclasts

_____ 9. vitamin D

_____ 10. hydroxyapatite

a. end of a long bone

b. formed by intramembranous ossification

c. organic components of bone matrix

d. stimulates osteoclasts to become active

e. lines medullary cavity

f. calcium phosphate/hydroxide crystals

g. responsible for bone resorption

h. increases calcium absorption in intestine

i. formed from trabeculae

j. contains concentric lamellae

Multiple Choice

Select the best answer from the four choices provided.

_____ 1. The immature cells that produce osteoid are called
   a. osteocytes.
   b. osteoblasts.
   c. osteoclasts.
   d. osteons.

_____ 2. Hyaline cartilage is found in all of the following structures *except* the
   a. trachea.
   b. larynx.
   c. pubic symphysis.
   d. fetal skeleton.

_____ 3. A small space within compact bone housing an osteocyte is termed a
   a. lamella.
   b. lacuna.
   c. canaliculus.
   d. medullary cavity.

_____ 4. Endochondral ossification begins with a _____ model of bone.
   a. dense regular connective tissue
   b. hyaline cartilage
   c. fibrocartilage
   d. elastic cartilage

_____ 5. Production of new bone _____ as a result of increased sex hormone production at puberty.
   a. is not affected
   b. slows down
   c. increases slowly
   d. increases rapidly

_____ 6. An epiphyseal line appears when
   a. epiphyseal plate growth has ended.
   b. epiphyseal plate growth is just beginning.
   c. growth in bone diameter is just beginning.
   d. a primary ossification center first develops.

_____ 7. The condition of inadequate ossification that may accompany aging and is a result of reduced calcification is called
   a. osteopenia.
   b. osteomyelitis.
   c. osteitis.
   d. osteosarcoma.

_____ 8. A fracture of the distal end of the radius that produces a characteristic "dinner fork" deformity is a _____ fracture.
   a. displaced
   b. Colles
   c. Pott
   d. stress

_____ 9. The femur is an example of a
   a. flat bone.
   b. long bone.
   c. irregular bone.
   d. short bone.

_____ 10. A large, rough projection of a bone is termed a
   a. fossa.
   b. tuberosity.
   c. ramus.
   d. tubercle.

Content Review

1. Identify the three types of cartilage, describing the extracellular matrix of each type.

2. Describe the structure of the periosteum, and list its functions.

3. Describe the characteristics of articular cartilage, the medullary cavity, and endosteum in a long bone.

4. Describe the microscopic anatomy of compact bone.

5. Why is spongy bone able to withstand stress in an area such as the expanded end of a long bone?

6. What is ossification? What is the difference between intramembranous and endochondral ossification?

7. List the steps involved in endochondral ossification.

8. List the four types of arteries that are found in a long bone, and what portions of the bone each artery supplies.

9. Discuss the effect of exercise on bone mass.

10. What are the steps in fracture repair?

Developing Critical Reasoning

1. Marty fell off his skateboard and suffered a broken leg. A cast was put on the leg for 6 weeks. After the bone healed and the cast was removed, an enlarged, bony bump remained at the region of the fracture. Eventually, this enlargement disappeared, and the leg regained its normal appearance. What happened from the time the cast was removed until the enlargement disappeared?

2. Elise is 14 and lives in an apartment in the city. She does not like outdoor activities, so she spends most of her spare time watching TV, playing video games, drinking soft drinks, and talking to friends on the phone. One afternoon, Elise tries to run down the stairs while talking on the phone, and falls, breaking her leg. Although she appears healthy, her leg takes longer to heal than expected. What might cause the longer healing time?

3. Connor is a healthy, active 7-year-old who fell while climbing on a bar apparatus in the playground, breaking his forearm near the wrist. The doctor told Connor's father that the fracture would require insertion of screws to align ends of the broken bones and ensure proper growth in the future. Why was the physician taking special care with the healing of this fracture?

A N S W E R S   T O   " W H A T   D O   Y O U   T H I N K ? "

1. Ribs are best classified as flat bones because they have parallel surfaces of compact bone with internally placed spongy bone. Flat bones tend to be relatively thin (like the rib) and are both light and strong. In contrast, long bones typically have a cylindrical diaphysis, while ribs are flattened.

2. Spongy bone is lighter and able to withstand stresses applied from many directions. In addition, hemopoietic tissue resides in the spaces within some spongy bone. Compact bone is very strong but weighs more than spongy bone. A bone made entirely of compact bone would be too heavy to move and too metabolically expensive to maintain, partly due to the increased musculature necessary to move it.

3. The numerous complex steps in endochondral bone formation ensure that a working bone may be formed for a newborn and later develop into a working adult bone. Having a bone collar, epiphyseal plates, and constant bone remodeling ensures that the bone can reshape itself, grow in both width and length, and develop a medullary cavity so that it will not weigh too much.

4. The 13-year-old will likely have several active epiphyseal plates (indicating that the bones are still growing in length), while the 65-year-old's bones will have stopped growing in length. Typically, the 65-year-old will have less bone mass and be at greater risk for osteopenia than the 13-year-old. (However, staying active will help the 65-year-old maintain bone mass and help ward off osteoporosis.)

Visit the McKinley/O'Loughlin _Human Anatomy_, 2e website at aris.mhhe.com

SKELETAL SYSTEM

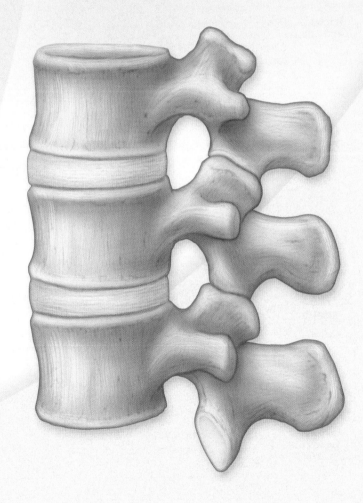

Axial Skeleton

The bones of the skeleton form an internal framework to support soft tissues, protect vital organs, bear the body's weight, and help us move. Without a bony skeleton, we would collapse into a formless mass. Typically, there are 206 bones in an adult skeleton, although this number varies in some individuals. A larger number of bones appear to be present at birth, but the total number decreases with growth and maturity as some separate bones fuse. Bones differ in size, shape, weight, and even composition, and this diversity is directly related to the skeleton's many functions.

The skeletal system is divided into two parts: the axial skeleton and the appendicular skeleton. The **axial skeleton** is so named because it is composed of the bones along the central axis of the body, which we commonly divide into three regions—the skull, the vertebral column, and the thoracic cage **(figure 7.1)**. The **appendicular skeleton** consists of the bones of the appendages (upper and lower limbs), as well as the bones that hold the limbs to the trunk of the body (the pectoral and pelvic girdles). The axial skeleton is the topic of this chapter; in chapter 8, we discuss the appendicular skeleton.

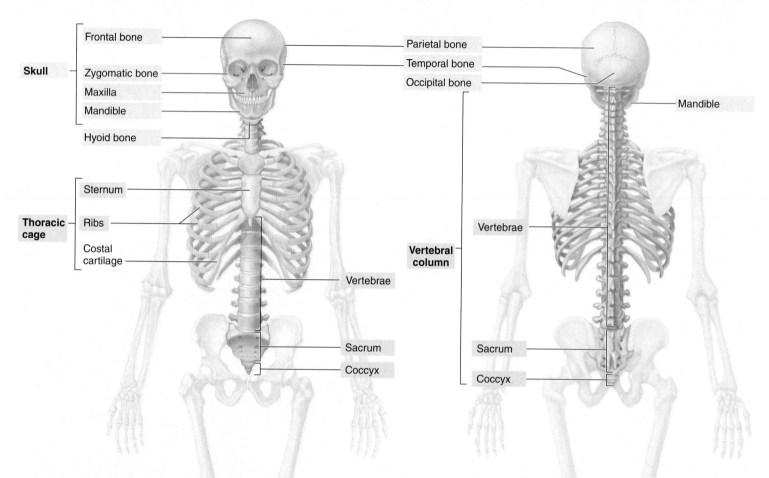

Bones of the Axial Skeleton (80)				
Skull (22)	Cranial bones (8) Frontal bone (1), parietal bones (2), temporal bones (2), occipital bone (1), sphenoid bone (1), ethmoid bone (1)	**Vertebral column (26)**	Vertebrae (24) Cervical vertebrae (7), thoracic vertebrae (12), lumbar vertebrae (5)	
	Facial bones (14) Zygomatic bones (2), lacrimal bones (2), nasal bones (2), vomer (1), inferior nasal conchae (2), palatine bones (2), maxillae (2), mandible (1)		Sacrum (1)	
			Coccyx (1)	
Associated bones of the skull (7)	Auditory ossicles (6) Malleus (2), incus (2), stapes (2) Hyoid bone (1)	**Thoracic cage (25)**	Sternum (1)	
			Ribs (24)	

(a) Anterior view   (b) Posterior view

Figure 7.1

Axial Skeleton. *(a)* Anterior and *(b)* posterior views show the axial skeleton, which is composed of the skull, vertebral column, and thoracic cage. A table summarizes the bones of the axial regions.

The main function of the axial skeleton is to form a framework that supports and protects the organs. The axial skeleton also houses special sense organs (the organs for hearing, balance, taste, smell, and vision) and provides areas for the attachment of skeletal muscles. Additionally, the spongy bone of most of the axial skeleton contains hemopoietic tissue, which is responsible for blood cell formation.

We begin our examination of the axial skeleton by discussing its most complex structure, the skull.

Study Tip!

Many bones have the same names as the body regions where they are found. Before you begin learning about the bones of the axial skeleton, it may help you to review table 1.2 (Anatomic Directional Terms) and table 1.3 (Human Body Regions). We will be using these terms as we discuss various features of bones in the next few chapters.

Skull

Key topics in this section:

- Description of the cranial and facial bones of the skull
- Locations of the sutures between cranial bones
- Structure of the nasal complex and the paranasal sinuses
- Identification of the three auditory ossicles
- Structure of the hyoid bone

The **skull** is composed of both cranial and facial bones **(figure 7.2)**. **Cranial bones** form the rounded **cranium** (krā′nē-um; *kranion* = skull), which completely surrounds and encloses the brain.[1] Eight bones make up the cranium: the unpaired ethmoid, frontal, occipital, and sphenoid bones, and the paired parietal and temporal bones. These bones also provide attachment sites for several jaw, head, and neck muscles. Touch the top, sides, and back of your head; these parts of your skull are cranial bones. **Facial bones** form the face. They also protect the entrances to the digestive and respiratory systems as well as providing attachment sites for facial muscles. Touch your cheeks, your jaws, and the bridge of your nose; these bones are facial bones.

The skull contains several prominent cavities **(figure 7.3)**. The largest cavity is the **cranial cavity**, which encloses, protects, and supports the brain and has an adult volume of approximately 1300 to 1500 cubic centimeters. The skull also has several smaller cavities, including the orbits (eye sockets), the oral cavity (mouth), the nasal (nā′zal; *nasus* = nose) cavity, and the paranasal sinuses.

WHAT DID YOU LEARN?

1 What are the two groups of skull bones?

---

[1] *Osteologists (scientists who study bones) define the cranium as the entire skull minus the mandible. In this text, we use the term* cranium *to denote the bones of the braincase only.*

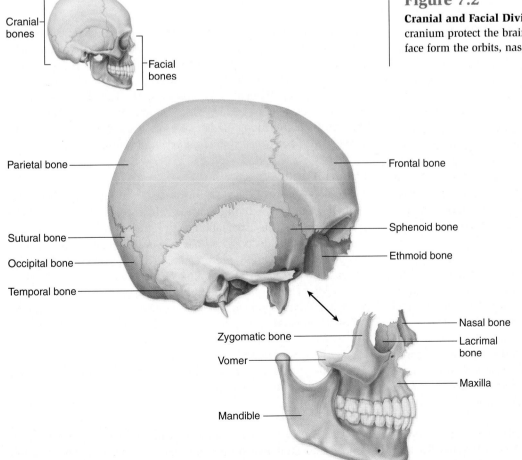

Cranial bones

Facial bones

Parietal bone

Sutural bone

Occipital bone

Temporal bone

Zygomatic bone

Vomer

Mandible

Frontal bone

Sphenoid bone

Ethmoid bone

Nasal bone

Lacrimal bone

Maxilla

Figure 7.2

Cranial and Facial Divisions of the Skull. The bones of the cranium protect the brain. The bones of both the cranium and the face form the orbits, nasal cavity, and mouth.

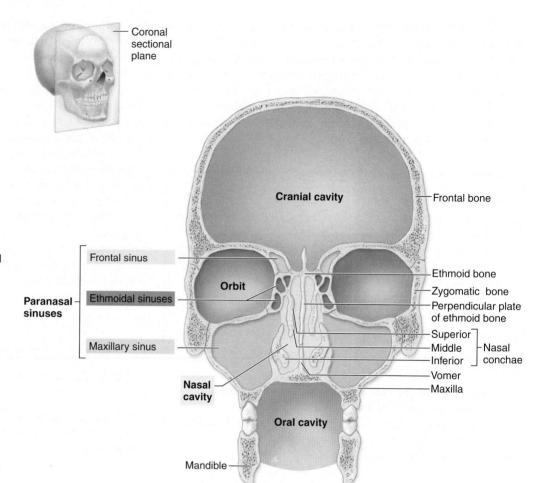

Figure 7.3

Major Cavities of the Skull. A coronal section diagram identifies the cavities within the skull.

Views of the Skull and Landmark Features

The skull is composed of multiple bones that exhibit complex shapes, and each bone articulates with at least one other. In order to best understand the complex nature of the skull, we first examine the skull as a whole and learn which bones are best seen from a particular view. Note that only some major features will be mentioned in this section. Later in the chapter, we examine the individual skull bones in detail.

A cursory glance at the skull reveals numerous bone markings, such as canals, fissures, and foramina, which are passageways for blood vessels and nerves. The major foramina and canals of the cranial and facial bones are listed in **table 7.1**. Refer to this table as we examine the skull from various directions. (This table also will be important when we study individual blood vessels and nerves in later chapters.)

Anterior View

An anterior view best shows several major bones of the skull (**figure 7.4**). The *frontal bone* forms the forehead. Put your hand on your forehead; you are feeling your frontal bone. The left and right *orbits* (eye sockets) are formed from a complex articulation of multiple skull bones. Within the orbits are two large openings, called the *superior orbital fissure* and the *inferior orbital fissure*. Superior to the orbits are the *superciliary arches*, otherwise known as the brow ridges. The left and right *nasal bones* form the bony "bridge" of the nose. Superior to the nasal bones and between the orbits is a landmark area called the *glabella*.

The left and right *maxillae* fuse to form most of the upper jaw and the lateral boundaries of the nasal cavity. The maxillae also help form the floor of each orbit. Inferior to each orbit, within each

maxilla, is an *infraorbital foramen*, which conducts blood vessels and nerves to the face. The lower jaw is formed by the *mandible*. The prominent "chin" of the mandible is called the *mental protuberance*.

An anterior view also shows the nasal cavity. Its inferior border is marked by a prominent *anterior nasal spine*. The thin ridge of bone that subdivides the nasal cavity into left and right halves helps form the *nasal septum*. Along the lateral walls of the nasal cavity are two scroll-shaped bones called the *inferior nasal conchae*.

Superior View

The superior view of the skull in **figure 7.5a** primarily shows four of the cranial bones: the frontal bone, both *parietal bones*, and the *occipital bone*. The articulation between the frontal and parietal bones is the *coronal suture*, so named because it runs along a coronal plane. The *sagittal suture* connects the left and right parietal bones almost exactly in the midline of the skull.

Along the posterior one-third of the sagittal suture are either a single *parietal foramen* or paired *parietal foramina*. This foramen conducts tiny emissary veins from the veins of the brain to the veins of the scalp. The number of parietal foramina can vary in individuals and between left and right sides of the same skull. The superior part of the *lambdoid suture* represents the articulation of the occipital bone with both parietal bones.

Posterior View

The posterior view of the skull in **figure 7.5b** shows the occipital bone and its lambdoid suture, as well as portions of the parietal and temporal bones. The *external occipital protuberance* is a bump on the

Table 7.1	Passageways Within the Skull	
Passageway	**Location**	**Structures That Pass Through**
CRANIAL BONES		
Carotid canal	Petrous part of temporal bone	Internal carotid artery
Cribriform foramina	Cribriform plate of ethmoid bone	Olfactory nerves (CN I)
Foramen lacerum	Between petrous part of temporal bone, sphenoid bone, and occipital bone	None
Foramen magnum	Occipital bone	Vertebral arteries; spinal cord, accessory nerves (CN XI)
Foramen ovale	Greater wing of sphenoid bone	Mandibular branch of trigeminal nerve (CN V_3)
Foramen rotundum	Greater wing of sphenoid bone	Maxillary branch of trigeminal nerve (CN V_2)
Foramen spinosum	Greater wing of sphenoid bone	Middle meningeal vessels
Hypoglossal canal	Anteromedial to occipital condyle of occipital bone	Hypoglossal nerve (CN XII)
Inferior orbital fissure	Junction of maxilla, sphenoid, and zygomatic bones	Infraorbital nerve (branch of CN V_2)
Jugular foramen	Between temporal bone and occipital bone (posterior to carotid canal)	Internal jugular vein; glossopharyngeal nerve (CN IX), vagus nerve (CN X), and accessory nerve (CN XI)
Mastoid foramen	Posterior to mastoid process of temporal bone	Mastoid emissary veins
Optic canal	Posteromedial part of orbit in lesser wing of sphenoid bone	Optic nerve (CN II)
Stylomastoid foramen	Between mastoid and styloid processes of temporal bone	Facial nerve (CN VII)
Superior orbital fissure	Posterior part of orbit between greater and lesser wings of sphenoid bone	Ophthalmic veins; oculomotor nerve (CN III), trochlear nerve (CN IV), ophthalmic branch of trigeminal nerve (CN V_1), and abducens nerve (CN VI)
Supraorbital foramen	Supraorbital margin of orbit in frontal bone	Supraorbital artery; supraorbital nerve (branch of CN V_1)
FACIAL BONES		
Greater and lesser palatine foramina	Palatine bone	Palatine vessels; greater and lesser palatine nerves (branches of CN V_2)
Incisive foramen	Posterior to incisor teeth in hard palate of maxilla	Branches of nasopalatine nerve (branch of CN V_2)
Infraorbital foramen	Inferior to orbit in maxilla	Infraorbital artery; infraorbital nerve (branch of CN V_2)
Lacrimal groove	Lacrimal bone	Nasolacrimal duct
Mandibular foramen	Medial surface of ramus of mandible	Inferior alveolar blood vessels; inferior alveolar nerve (branch of CN V_3)
Mental foramen	Inferior to second premolar on anterolateral surface of mandible	Mental blood vessels; mental nerve (branch of CN V_3)

back of the head. Palpate the back of your head; males tend to have a prominent, pointed external occipital protuberance, while females have a more subtle, rounded protuberance. Within the lambdoid suture there may be one or more *sutural (Wormian) bones*.

Lateral View

The lateral view of the skull in **figure 7.6** shows the following skull bones: one *parietal bone*, one *temporal bone*, one *zygomatic bone*, one *maxilla*, the *frontal bone*, the *mandible*, and portions of the *occipital bone*. The tiny *lacrimal bone* articulates with the nasal bone anteriorly and with the ethmoid bone posteriorly. A portion of the *sphenoid bone* articulates with the frontal, parietal, and temporal bones. The region called the **pterion** (tĕ′rē-on; *ptéron* = wing), circled on figure 7.6, represents the H-shaped set of sutures of these four articulating bones.

The temporal process of the zygomatic bone and the zygomatic process of the temporal bone fuse to form the *zygomatic arch*. Put your fingers along the bony prominences ("apples") of your cheeks and move your fingers posteriorly to your ears; you are feeling the zygomatic arch. The zygomatic arch terminates superior to the point where the mandible articulates with the *mandibular fossa* of the temporal bone. This articulation is called the *temporomandibular joint*. By putting your finger anterior to your external ear opening and opening your jaw, you can feel that joint moving. The external ear opening overlies the *external acoustic meatus* of the skull. Posterior to this canal is the *mastoid process*, the bump you feel posterior and inferior to your external ear opening.

Sagittal Sectional View

Cutting the skull along a sagittal sectional plane reveals bones that form the nasal cavity and the endocranium (**figure 7.7**). The cranial cavity is formed from a complex articulation of the frontal, parietal, temporal, occipital, ethmoid, and sphenoid bones. Vessel impressions on the internal surface of the skull show up clearly. The *frontal sinus* (a space in the frontal bone) and the *sphenoidal sinus* (a space in the sphenoid bone) are visible.

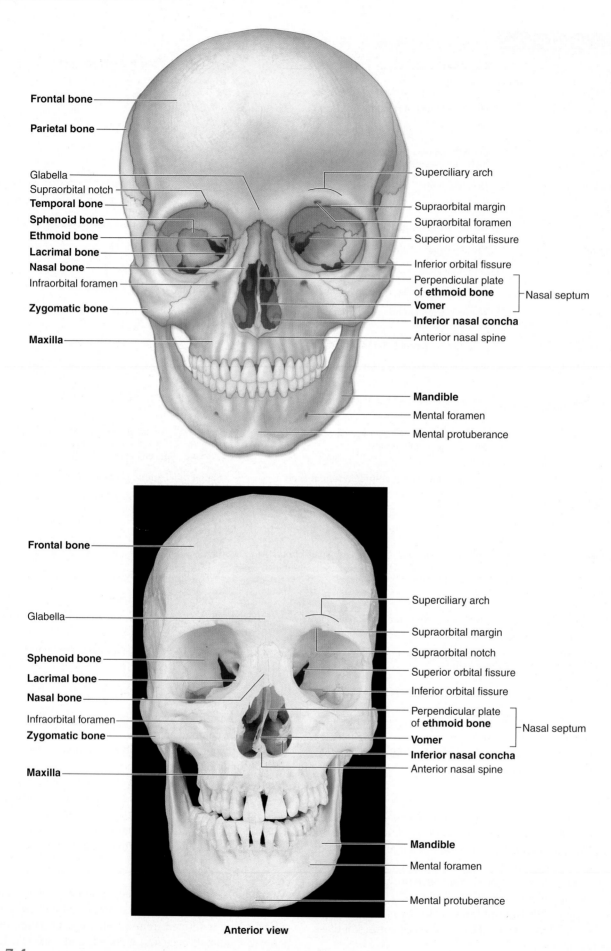

Frontal bone

Parietal bone

Glabella
Supraorbital notch
Temporal bone
Sphenoid bone
Ethmoid bone
Lacrimal bone
Nasal bone
Infraorbital foramen

Zygomatic bone

Maxilla

Superciliary arch

Supraorbital margin
Supraorbital foramen
Superior orbital fissure

Inferior orbital fissure

Perpendicular plate
of **ethmoid bone**
Vomer
Inferior nasal concha
Anterior nasal spine

Mandible
Mental foramen
Mental protuberance

Nasal septum

Frontal bone

Glabella

Sphenoid bone

Lacrimal bone

Nasal bone

Infraorbital foramen
Zygomatic bone

Maxilla

Superciliary arch

Supraorbital margin
Supraorbital notch
Superior orbital fissure
Inferior orbital fissure

Perpendicular plate
of **ethmoid bone**
Vomer
Inferior nasal concha
Anterior nasal spine

Mandible
Mental foramen
Mental protuberance

Nasal septum

Anterior view

Figure 7.4

Anterior View of the Skull. The frontal bone, nasal bones, maxillae, and mandible are prominent in this view.

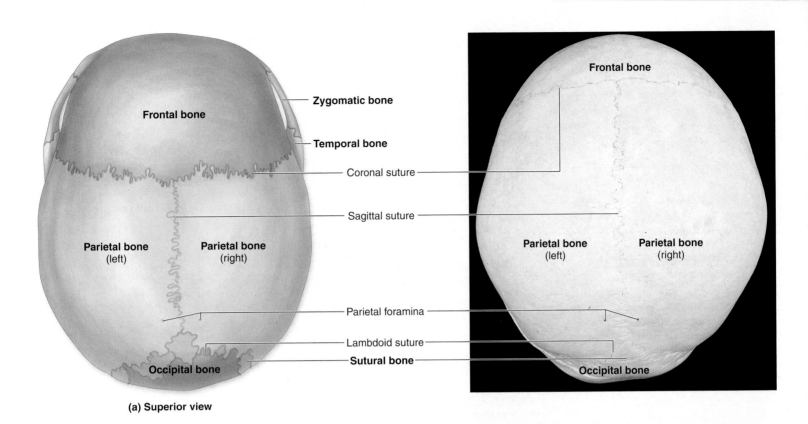

(a) Superior view

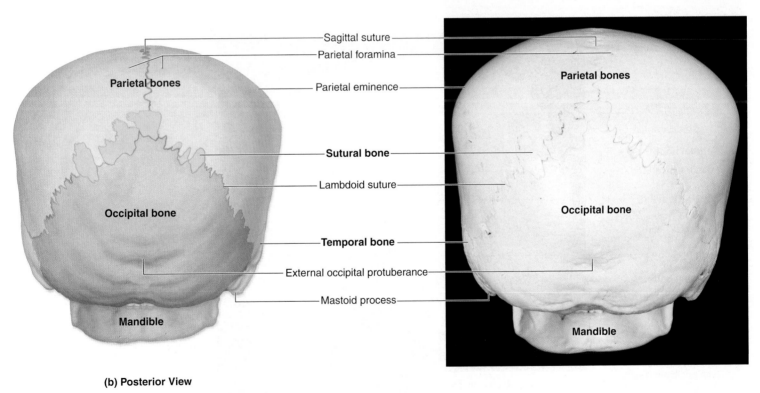

(b) Posterior View

Figure 7.5

Superior and Posterior Views of the Skull. *(a)* The superior aspect of the skull shows the major sutures and some of the bones of the skull. *(b)* The posterior view is dominated by the occipital and parietal bones.

A sagittal sectional view also shows the bones that form the nasal septum more clearly. The *perpendicular plate* forms the superior portion of the nasal septum, while the *vomer* forms the inferior portion. The ethmoid bone serves as a "wall" between the anterior floor of the cranial cavity and the roof of the nasal cavity. The *maxillae* and *palatine bones* form the hard palate, which acts as both the floor of the nasal cavity and part of the roof of the mouth.

Move your tongue along the roof of your mouth; you are palpating the maxillae anteriorly and the palatine bones posteriorly.

Inferior (Basal) View

In an inferior (basal) view, the skull looks a bit complex, with all of its foramina and weird-shaped bone features **(figure 7.8)**. However, you will soon be able to recognize and distinguish some landmark features.

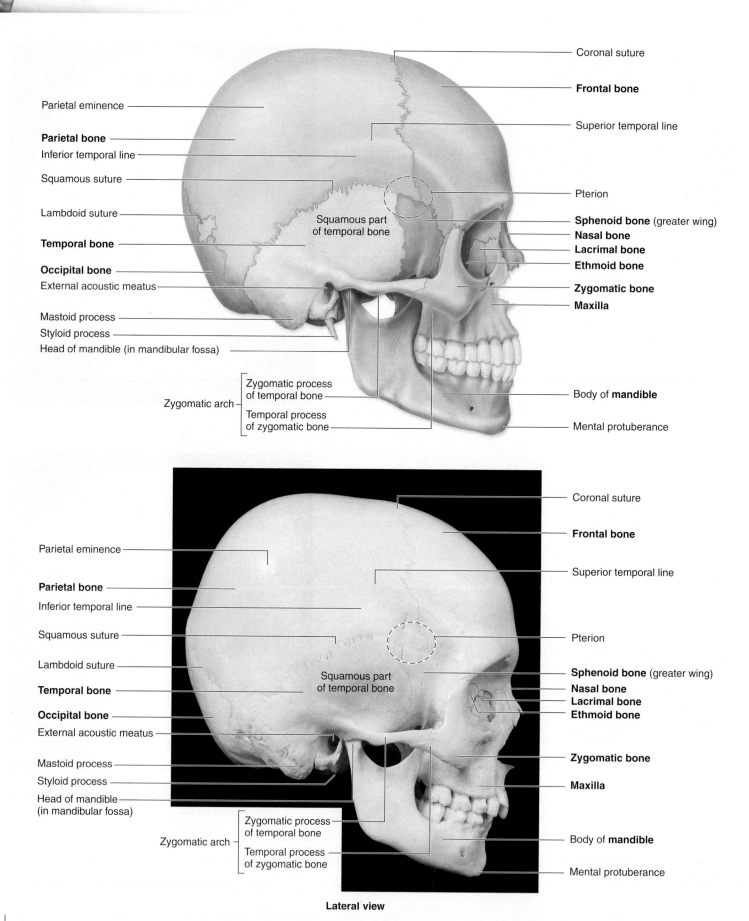

Coronal suture

Frontal bone

Superior temporal line

Parietal eminence

Parietal bone

Inferior temporal line

Squamous suture

Lambdoid suture

Pterion

Squamous part
of temporal bone

Sphenoid bone (greater wing)
Nasal bone
Lacrimal bone
Ethmoid bone

Temporal bone

Occipital bone

Zygomatic bone
Maxilla

External acoustic meatus

Mastoid process

Styloid process

Head of mandible (in mandibular fossa)

Zygomatic arch

Zygomatic process
of temporal bone

Temporal process
of zygomatic bone

Body of **mandible**

Mental protuberance

Coronal suture

Frontal bone

Superior temporal line

Parietal eminence

Parietal bone

Inferior temporal line

Squamous suture

Lambdoid suture

Pterion

Squamous part
of temporal bone

Sphenoid bone (greater wing)
Nasal bone
Lacrimal bone
Ethmoid bone

Temporal bone

Occipital bone

Zygomatic bone

External acoustic meatus

Maxilla

Mastoid process

Styloid process

Head of mandible
(in mandibular fossa)

Zygomatic arch

Zygomatic process
of temporal bone

Temporal process
of zygomatic bone

Body of **mandible**

Mental protuberance

Lateral view

Figure 7.6

Lateral View of the Skull.   The parietal, temporal, zygomatic, frontal, and occipital bones, as well as the maxilla and mandible, are prominent in this view.

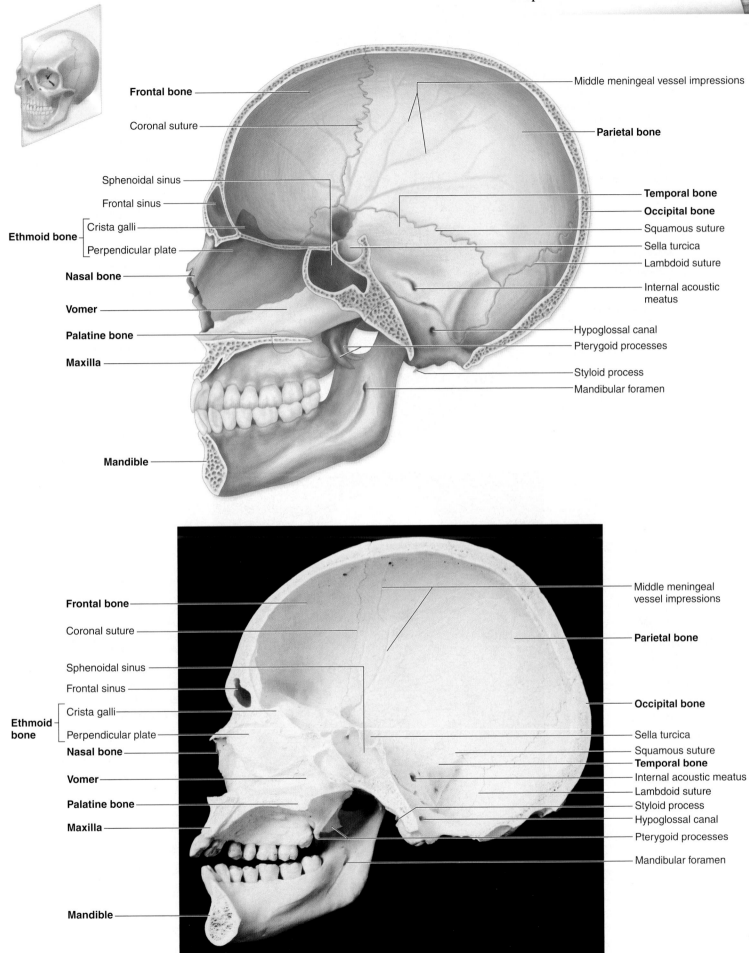

Frontal bone

Coronal suture

Middle meningeal vessel impressions

Parietal bone

Sphenoidal sinus

Frontal sinus

Temporal bone

Occipital bone

Ethmoid bone
- Crista galli
- Perpendicular plate

Squamous suture

Sella turcica

Lambdoid suture

Nasal bone

Internal acoustic meatus

Vomer

Palatine bone

Hypoglossal canal

Pterygoid processes

Maxilla

Styloid process

Mandibular foramen

Mandible

Frontal bone

Coronal suture

Middle meningeal vessel impressions

Parietal bone

Sphenoidal sinus

Frontal sinus

Occipital bone

Ethmoid bone
- Crista galli
- Perpendicular plate

Sella turcica

Squamous suture

Nasal bone

Temporal bone

Internal acoustic meatus

Vomer

Lambdoid suture

Styloid process

Palatine bone

Hypoglossal canal

Maxilla

Pterygoid processes

Mandibular foramen

Mandible

Sagittal section

Figure 7.7

Sagittal Section of the Skull.   Features such as the perpendicular plate of the ethmoid bone, the vomer, and the frontal and sphenoidal sinuses, as well as the internal relationships of the skull bones, are best seen in this view.

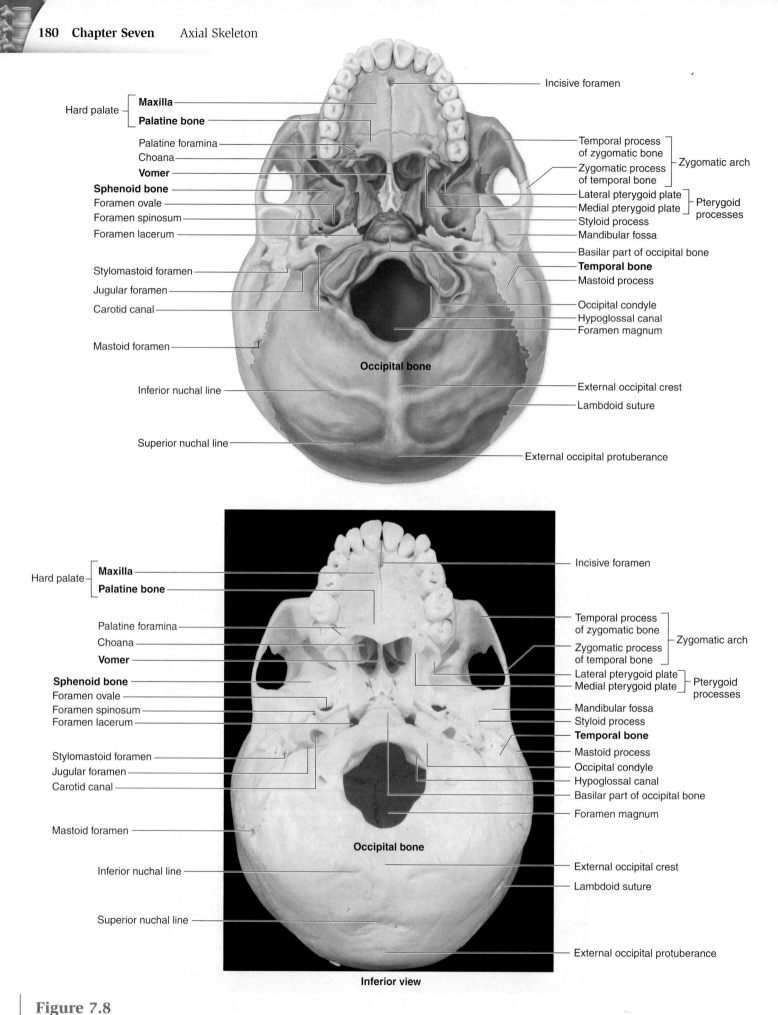

Incisive foramen

Hard palate ⎡ **Maxilla**
⎣ **Palatine bone**

Palatine foramina
Choana
Vomer
Sphenoid bone
Foramen ovale
Foramen spinosum
Foramen lacerum

Stylomastoid foramen
Jugular foramen
Carotid canal

Mastoid foramen

Temporal process
of zygomatic bone ⎤
⎦ Zygomatic arch
Zygomatic process
of temporal bone
Lateral pterygoid plate ⎤
Medial pterygoid plate ⎦ Pterygoid processes
Styloid process
Mandibular fossa
Basilar part of occipital bone
Temporal bone
Mastoid process
Occipital condyle
Hypoglossal canal
Foramen magnum

Occipital bone

Inferior nuchal line

Superior nuchal line

External occipital crest
Lambdoid suture

External occipital protuberance

Incisive foramen

Hard palate ⎡ **Maxilla**
⎣ **Palatine bone**

Palatine foramina
Choana
Vomer
Sphenoid bone
Foramen ovale
Foramen spinosum
Foramen lacerum

Stylomastoid foramen
Jugular foramen
Carotid canal

Mastoid foramen

Temporal process
of zygomatic bone ⎤
⎦ Zygomatic arch
Zygomatic process
of temporal bone
Lateral pterygoid plate ⎤
Medial pterygoid plate ⎦ Pterygoid processes
Mandibular fossa
Styloid process
Temporal bone
Mastoid process
Occipital condyle
Hypoglossal canal
Basilar part of occipital bone
Foramen magnum

Occipital bone

Inferior nuchal line

Superior nuchal line

External occipital crest
Lambdoid suture

External occipital protuberance

Inferior view

Figure 7.8

Inferior View of the Skull.   The hard palate, sphenoid bone, parts of the temporal bone, and the occipital bone with its foramen magnum are readily visible when the mandible is removed in this view.

Craniosynostosis

Sutures in the skull allow the cranium to grow and expand during childhood. In adulthood, when cranial growth has stopped, the sutures fuse and are obliterated. **Craniosynostosis** (krā′nē-ō-sin′os-tō′sis) refers to the premature fusion or closing of one or more of these cranial sutures. If this premature fusion occurs early in life or in utero, skull shape is dramatically affected. If not surgically treated, a craniosynostotic individual often grows up with an unusual craniofacial shape.

The morphological effects of craniosynostosis (i.e., the changes in head shape) are referred to as **craniostenosis** (krā′nē-ō-sten-ō′sis). For example, if the sagittal suture fuses prematurely, a condition called **sagittal synostosis**, the skull cannot grow and expand laterally as the brain grows, and compensatory skull growth occurs in an anterior-posterior fashion. A child with sagittal synostosis develops a very elongated, narrow skull shape, a condition called **scaphocephaly** or *dolicocephaly*. **Coronal synostosis** refers to premature fusion of the coronal suture, which causes the skull to be abnormally short and wide.

Craniosynostosis appears to have multiple causes, including genetics, teratogens (a drug or other agent that can cause birth defects), and environmental factors. Many people with craniosynostosis have no complications other than the unusual skull shape. Those who do experience complications may have increased intracranial pressure (leading to headache and seizures if severe), optic nerve compression, and mental retardation (due to restricted brain growth).

To limit and correct unusual skull shape, craniosynostosis must e surgically corrected as soon as feasibly possible, preferably within the first year of life. In a procedure called a **craniectomy** (krā′nē-ek′tō-mē), the fused suture is incised and opened. In more severe craniosynostosis cases, larger pieces of cranial bones may be cut and fit differently to reshape the skull. After surgery, the child may also be fitted with a "molding helmet" to help the skull bones grow and develop along desired trajectories. The earlier in life a child receives treatment, the greater are his or her chances of having a more normal skull shape.

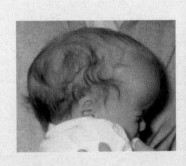
Sagittal synostosis.

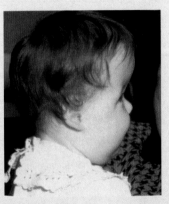
Coronal synostosis.

The most anterior structure is the *hard palate*. On the posterior aspect of either side of the palate are the *pterygoid processes* of the sphenoid bone. Adjacent to these structures are the internal openings of the nasal cavity, called the *choanae*.

Between the mandibular fossa and the pterygoid processes are several **foramina** (fō-ram′i-nă; sing., *foramen*, fo′rā′men; *forare* = to bore) and canals. Closest to the pterygoid process is the *foramen ovale*, and lateral to this is the *foramen spinosum*. Posterior and lateral to these foramina is the **jugular** (jŭg′ū-lar; *jugulum* = throat) **foramen**, which is a space between the temporal and occipital bones. The entrance to the *carotid canal* is anteromedial to the jugular foramen. The *foramen lacerum* (anteromedial to the carotid canal) extends between the occipital and temporal bones. This opening is closed off by connective tissue in a living individual.

The largest foramen of all is the foramen magnum, literally meaning "big hole." Through this opening, the spinal cord enters the cranial cavity and becomes continuous with the brainstem. On either side of the foramen magnum are the rounded occipital condyles, which articulate with the vertebral column.

Internal View of Cranial Base

When the superior part of the skull is cut and removed, the internal view of the cranial base **(figure 7.9)** reveals the *frontal bone*, the most anteriorly located bone. It surrounds the delicate *cribriform plate* of the ethmoid bone. Posterior to the frontal bone are the *lesser wings* and the *greater wings* of the sphenoid bone. The *temporal bones* from the lateral regions of the cranial base, while the occipital bone forms its posterior aspect.

Many of the foramina labeled on the inferior view of the skull can also be seen from this internal view, but some new openings are visible as well. For example, left and right *optic canals* are located

Study Tip!

In your lab (and with your instructor's permission), put colored pipe cleaners through the cranial foramina to observe how they travel through the skull. For example, if you put a pipe cleaner through the carotid canal, you will see that it bends as it travels through the temporal bone to open at the base of the skull.

in the lesser wings of the sphenoid. The *internal acoustic meatus* is located more posteriorly in the temporal bone.

Sutures

Sutures (soo′choor; *sutura* = a seam) are immovable joints that form the boundaries between the cranial bones (see figures 7.5–7.8). Dense regular connective tissue seals cranial bones firmly together at a suture. Different types of sutures are distinguished by the margins between the bones, which often have intricate interlocking forms, like puzzle pieces, and form a strong union, or articulation.

There are numerous sutures in the skull, each with a specific name. Many of the smaller sutures are named for the bones or features they interconnect. For example, the *occipitomastoid suture* connects the occipital bone with the portion of the temporal bone that houses the mastoid process. Here we discuss only the four largest sutures—the coronal, lambdoid, sagittal, and squamous sutures:

- The **coronal** (kō-rō′nal; *coron* = crown) **suture** extends across the superior surface of the skull along a coronal (or frontal) plane. It represents the articulation between the anterior frontal bone and the more posterior parietal bones.

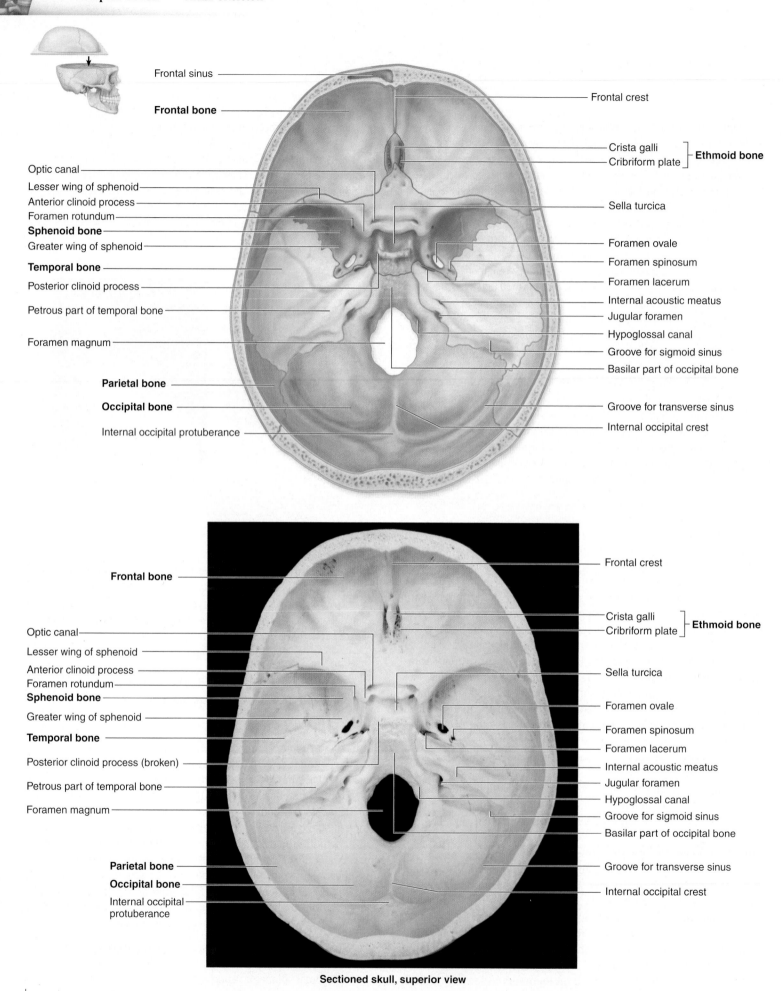

Frontal sinus
Frontal bone
Frontal crest
Crista galli
Cribriform plate
} **Ethmoid bone**

Optic canal
Lesser wing of sphenoid
Anterior clinoid process
Foramen rotundum
Sphenoid bone
Greater wing of sphenoid
Temporal bone
Posterior clinoid process
Petrous part of temporal bone
Foramen magnum

Sella turcica
Foramen ovale
Foramen spinosum
Foramen lacerum
Internal acoustic meatus
Jugular foramen
Hypoglossal canal
Groove for sigmoid sinus
Basilar part of occipital bone

Parietal bone
Occipital bone
Internal occipital protuberance

Groove for transverse sinus
Internal occipital crest

Frontal bone
Frontal crest
Crista galli
Cribriform plate
} **Ethmoid bone**

Optic canal
Lesser wing of sphenoid
Anterior clinoid process
Foramen rotundum
Sphenoid bone
Greater wing of sphenoid
Temporal bone
Posterior clinoid process (broken)
Petrous part of temporal bone
Foramen magnum

Sella turcica
Foramen ovale
Foramen spinosum
Foramen lacerum
Internal acoustic meatus
Jugular foramen
Hypoglossal canal
Groove for sigmoid sinus
Basilar part of occipital bone

Parietal bone
Occipital bone
Internal occipital protuberance

Groove for transverse sinus
Internal occipital crest

Sectioned skull, superior view

Figure 7.9

Superior View of the Skull. In this horizontal section, the frontal, ethmoid, sphenoid, temporal, and occipital bones are prominent.

- The **lambdoid** (lam'doyd) **suture** extends like an arc across the posterior surface of the skull, articulating with the parietal bones and the occipital bone. It is named for the Greek letter "lambda," which its shape resembles.
- The **sagittal** (saj'i-tăl; *sagitta* = arrow) **suture** extends between the superior midlines of the coronal and lambdoid sutures. It is in the midline of the cranium (along the midsagittal plane) and is the articulation between the right and left parietal bones.
- A **squamous** (skwā'mus; *squama* = scale) **suture** on each side of the skull articulates the temporal bone and the parietal bone of that side. The squamous (flat) part of the temporal bone typically "overlaps" the parietal bone.

One common variation in sutures is the presence of **sutural bones** (*Wormian bones*) (see figures 7.5 and 7.6). Sutural bones typically are small, ranging in size from a tiny pebble to a quarter, but they can be much larger. Any suture may have sutural bones, but they are most common and numerous in the lambdoid suture. Sutural bones represent independent bone ossification centers. Researchers do not know why sutural bone incidence varies among individuals, but most suspect a combination of genetic and environmental influences.

In our adult years, the sutures typically disappear as the adjoining bones fuse. This fusion starts internally (endocranially) and is followed by fusion on the skull's external (ectocranial) surface. Although the timing of suture closure can be highly variable, the coronal suture typically is the first to fuse, usually in the late 20s to early 30s, followed by the sagittal suture and then the lambdoid suture (usually in the 40s). The squamous suture usually does not fuse until late adulthood (60+ years), or it may not fuse at all. Osteologists can estimate the approximate age at death of an individual by examining the extent of suture closure in the skull.

WHAT DID YOU LEARN?

2 Which three skull sutures can be seen from a superior view of the skull? Which bones articulate at these sutures?

Bones of the Cranium

The eight bones of the cranium collectively form a rigid protective structure for the brain. The cranium consists of a roof and a base. The roof, called the **calvaria** (kal-vā'rē-ă), or *skullcap*, is composed of the squamous part of the frontal bone, the parietal bones, and the squamous part of the occipital bone. The **base** of the cranium is composed of portions of the ethmoid, sphenoid, occipital, and temporal bones. Some skulls in the anatomy lab have had their calvariae cut away, making the distinction between the calvaria and base easier to distinguish. Each bone of the cranium has specific surface features (**table 7.2**).

> ## Study Tip!
>
> As you learn about the individual skull bones, be sure to review figures 7.4 through 7.9 to see how each bone fits within the various views of the whole skull. Comparing the individual bone images with the whole skull images will help you better understand the complex articulations among all of the cranial and facial bones.

Frontal Bone

The **frontal bone** forms part of the calvaria, the forehead, and the roof of the orbits (**figure 7.10**). During development, the cranial bones (including the frontal bone) form as a result of the fusion of separate ossification centers, an event that may not occur until after birth. The frontal bone is formed from two major, separate ossification centers. Soon after birth, the left and right sides of the developing frontal bone

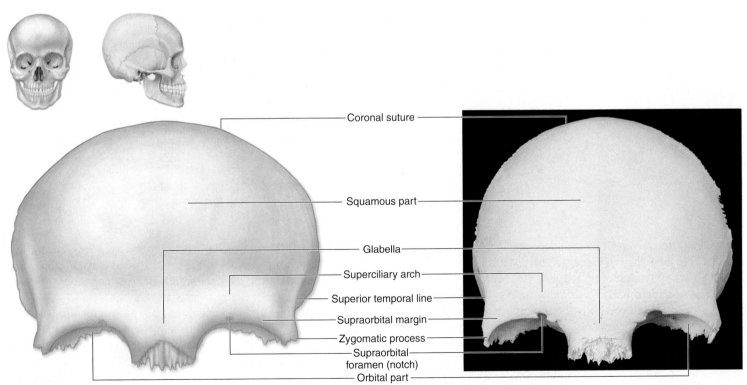

Coronal suture
Squamous part
Glabella
Superciliary arch
Superior temporal line
Supraorbital margin
Zygomatic process
Supraorbital foramen (notch)
Orbital part

Frontal bone, anterior view

Figure 7.10

Anterior View of the Frontal Bone. The frontal bone forms the forehead and part of the orbits.

Table 7.2	**Cranial Bones and Selected Features**	
Bone(s) and Associated Passageways	**Bone Boundaries Within the Skull**	**Selected Features and Their Functions**
Frontal bone Supraorbital foramen	Forms superior and anterior parts of skull, part of anterior cranial fossa and orbit	**Frontal crest:** Attachment site for meninges to help stabilize brain within skull **Frontal sinuses:** Lighten bone, moisten inhaled air, and give resonance to voice **Orbital part:** Forms roof of orbit **Squamous part:** Attachment of scalp muscles **Supraorbital margin:** Forms protective superior border of orbit
Parietal bones	Each forms most of lateral and superior walls of skull	**Inferior and superior temporal lines:** Attachment site for temporalis muscle **Parietal eminence:** Forms rounded prominence on each side of skull

Temporal bones
  Carotid canal
  External acoustic meatus
  Internal acoustic meatus
  Mastoid foramen
  Stylomastoid foramen
	Each forms inferolateral wall of the skull; forms part of middle cranial fossa; has three parts—petrous, squamous, and tympanic	**Mandibular fossa:** Articulates with mandible
		Mastoid air cells: Lighten mastoid process
		Mastoid process: Attachment site of some neck muscles to extend or rotate head
		Petrous part: Protects sensory structures in inner ear
		Styloid process: Attachment site for hyoid bone ligaments and muscles
		Squamous part: Attachment site of some jaw muscles
		Zygomatic process: Articulates with zygomatic bone to form zygomatic arch

Occipital bone
  Foramen magnum
  Hypoglossal canal
  Jugular foramen (with temporal bone)
	Forms posteroinferior part of skull, including most of the posterior cranial fossa; forms part of base of skull	**External occipital crest:** Attachment site for ligaments
		External occipital protuberance: Attachment of muscles that move head
		Inferior and superior nuchal lines: Attachment of neck ligaments and muscles
		Occipital condyles: Articulate with first cervical vertebra (atlas)

Sphenoid bone
  Foramen lacerum (with temporal and occipital bones)
  Foramen ovale
  Foramen rotundum
  Foramen spinosum
  Optic canal
  Superior orbital fissure
	Forms part of base of skull, posterior part of eye orbit, part of anterior and middle cranial fossae	**Body:** Houses sphenoidal sinuses
		Sella turcica: Houses pituitary gland
		Optic canals: House optic nerves (CN II)
		Medial and lateral pterygoid plates: Attachment of muscles of the jaw
		Lesser wings: Form part of anterior cranial fossa; contain optic canal
		Greater wings: Form part of middle cranial fossa and orbit
		Sphenoidal sinuses: Moisten inhaled air and give resonance to voice

Ethmoid bone Cribriform foramina	Forms part of the anterior cranial fossa; part of nasal septum; roof and lateral walls of nasal cavity; part of medial wall of eye orbit	**Crista galli:** Attachment site for cranial dural septa to help stabilize brain within skull **Ethmoidal labyrinths:** Contain the ethmoidal sinuses and nasal conchae **Ethmoidal sinuses:** Lighten bone, moisten inhaled air, and give resonance to voice **Nasal conchae (superior and middle):** Increase airflow turbulence in nasal cavity so air can be adequately moistened and cleaned by nasal mucosa **Orbital plate:** Forms part of medial wall of the orbit **Perpendicular plate:** Forms superior part of nasal septum

are united by the **metopic** (me-tō′pik, me-top′ik; *metopon* = forehead) **suture**. This suture usually fuses and disappears by age 2, although a trace of it persists in some adult skulls.

The **squamous part** of the frontal bone is the vertical flattened region. The squamous part ends at the **supraorbital margins**, each of which forms the superior ridge of the orbit. The midpoint of each supraorbital margin contains a single **supraorbital foramen**, or **notch**. Superior to the supraorbital margins are the **superciliary** (soo-per-sil′ē-ār-ē; *super* = above, *cilium* = eyelid) **arches**, which are the brow ridges. Male skulls tend to have more pronounced superciliary arches than do female skulls. The part of the frontal bone sandwiched between the superciliary arches is the **glabella** (glă-bel′ă; *glabellus* = smooth).

The **orbital part** of the frontal bone is the smooth, inferior portion that forms the roof of the orbit. Lateral to each orbital part is the **zygomatic process** of the frontal bone, which articulates with the frontal process of the zygomatic bone.

Within the frontal bone is a pair of **frontal sinuses** (see figure 7.9). The frontal sinuses usually start to appear after age 6 and become more fully developed after age 10. Some people never develop these sinuses at all.

On the internal surface of the frontal bone is a midline elevation of bone called the **frontal crest**. The frontal crest serves as a point of attachment for the falx cerebri, a protective connective tissue sheet that helps support the brain.

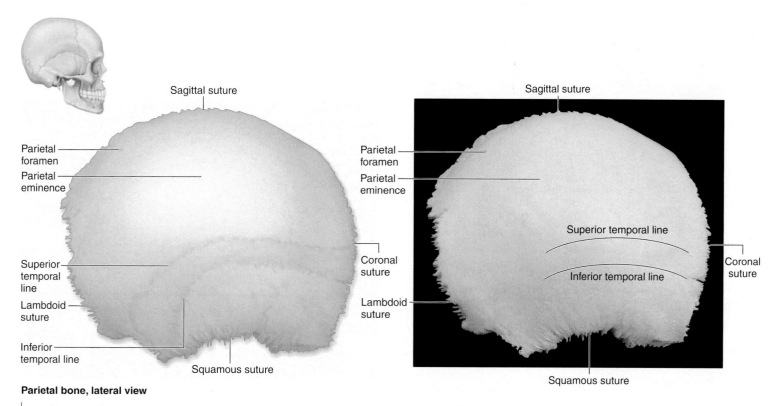

Sagittal suture

Parietal foramen

Parietal eminence

Superior temporal line

Lambdoid suture

Inferior temporal line

Squamous suture

Coronal suture

Parietal bone, lateral view

Sagittal suture

Parietal foramen

Parietal eminence

Superior temporal line

Inferior temporal line

Lambdoid suture

Coronal suture

Squamous suture

Figure 7.11

Lateral View of the Parietal Bone.   The parietal bones form the lateral aspects of the skull.

Parietal Bones

The right and left **parietal** (pă-rī′ĕ-tăl; *paries* = wall) **bones** form the lateral walls and roof of the cranium **(figure 7.11)**. Each parietal bone is bordered by four sutures that unite it to the neighboring bones. A **parietal foramen** sometimes occurs in the posterior one-third of the parietal bone, adjacent to the sagittal suture. A tiny emissary vein travels through this opening, connecting the venous sinuses with the veins of the scalp. On the lateral surface, each parietal bone exhibits a pair of faint ridges called the **superior** and **inferior temporal lines**. These lines arc across the surface of the parietal and frontal bones. They mark the attachment site of the large, fan-shaped temporalis muscle that closes the mouth. Superior to these lines, the rounded, smooth parietal surface is called the **parietal eminence**. The internal surfaces of the parietal bones exhibit many grooves that accommodate some of the blood vessels within the cranium.

Temporal Bones

The paired **temporal bones** form the inferior lateral walls and part of the floor of the cranium **(figure 7.12)**. Each temporal bone has a complex structure composed of three parts: the petrous, squamous, and tympanic parts.

The thick **petrous** (pet′rŭs; *patra* = a rock) **part** of the temporal bone houses sensory structures of the inner ear that provide information about hearing and balance. In figure 7.12*b*, observe the **internal acoustic meatus** (mē-ā′tŭs; a passage) (also called either the *internal auditory meatus* or *internal auditory canal*). It provides a passageway for nerves and blood vessels to and from the inner ear. A **groove for the sigmoid** (sig′moyd; *sigma* = the letter S, *eidos* = resemblance) **sinus** runs along the inferior surface of the petrous region. The sigmoid sinus is a venous sinus (vein) that drains blood from the brain.

Externally, the prominent bulge on the inferior surface of the temporal bone is the **mastoid** (mas′toyd; *masto* = breast) **process**, an anchoring site for muscles that move the neck. Rather than being solid bone, it is filled with many small, interconnected air cells (called mastoid air cells) that communicate with the middle ear.

On the posteroinferior surface of the temporal bone, a variable **mastoid foramen** opens near the mastoid process. Tiny emissary veins travel through this foramen to connect the venous sinuses inside the cranium with the veins on the scalp. A thin, pointed projection of bone, called the **styloid** (stī′loyd; *stylos* = pillar post) **process**, serves as an attachment site for several hyoid and tongue muscles. The **stylomastoid foramen** lies between the mastoid process and the styloid process (see figure 7.8). The facial nerve (CN VII) extends through the stylomastoid foramen to innervate the facial muscles. The **carotid canal** (ka-rot′id; *karoo* = to put to sleep) is medial to the styloid process and transmits the internal carotid artery.

The **squamous part**, or *squama*, is the lateral flat surface of the temporal bone immediately inferior to the squamous suture (see figure 7.12). Immediately inferior to the squamous part, a prominent **zygomatic** (zī′gō-mat′ik; *zygoma* = a joining, a yoke) **process** curves laterally and anteriorly to unite with the **temporal process** of the **zygomatic bone**. The union of these processes forms the **zygomatic arch** (see figures 7.6 and 7.8). Each temporal bone articulates with the mandible inferior to the base of both zygomatic processes in a depression called the **mandibular** (man-dib′ū-lăr) **fossa**. Anterior to the mandibular fossa is a bump called the **articular tubercle**.

Immediately posterolateral to the mandibular fossa is the **tympanic** (tim-pan′ik; *tympanon* = drum) **part**, a small, bony ring surrounding the entrance to the **external acoustic meatus**, or *external auditory canal* (see figure 7.12).

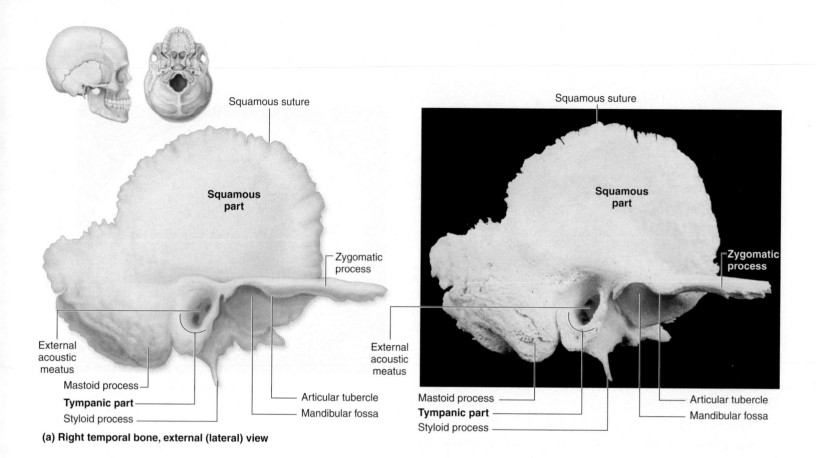

(a) Right temporal bone, external (lateral) view

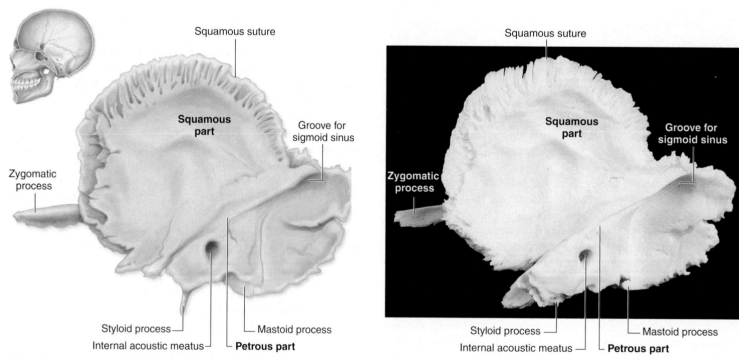

(b) Right temporal bone, internal (medial) view

Figure 7.12

Temporal Bone. The temporal bone is located adjacent to the ear. *(a)* External (lateral) view and *(b)* internal (medial) views of the right temporal bone are shown.

Occipital Bone

The occipital bone is subdivided into a flattened **squamous part**, which forms the posterior region of the skull, and a median **basilar part**, which forms a portion of the base of the cranium **(figure 7.13)**. Within the basilar part of the occipital bone is a large, circular open-

ing called the **foramen magnum**, and lateral to this foramen are smooth knobs called **occipital** (ok-sip′i-tăl; *occiput* = back of head) **condyles**. The skull articulates with the first cervical vertebra at the occipital condyles. When you nod "yes," you are moving the occipital condyles against the vertebra. At the anteromedial edge of

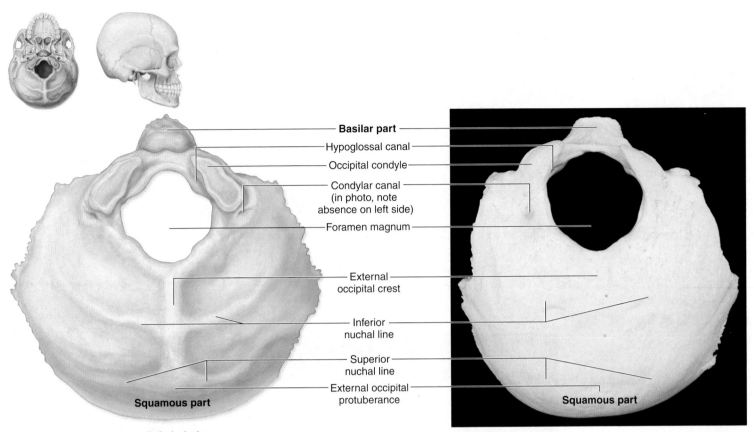

(a) Occipital bone, external (inferior) view

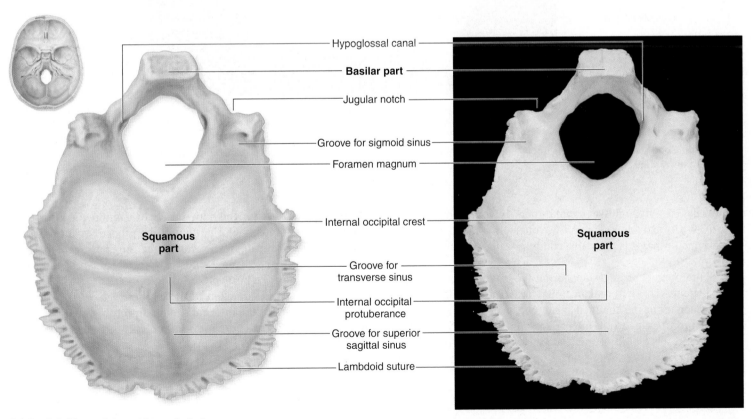

(b) Occipital bone, internal (superior) view

Figure 7.13

Occipital Bone.   The occipital bone forms the posterior portion of the skull. (*a*) External (inferior) view shows the nuchal lines and the external occipital protuberance. (*b*) Internal (superior) view shows the internal occipital protuberance and grooves for the venous sinuses.

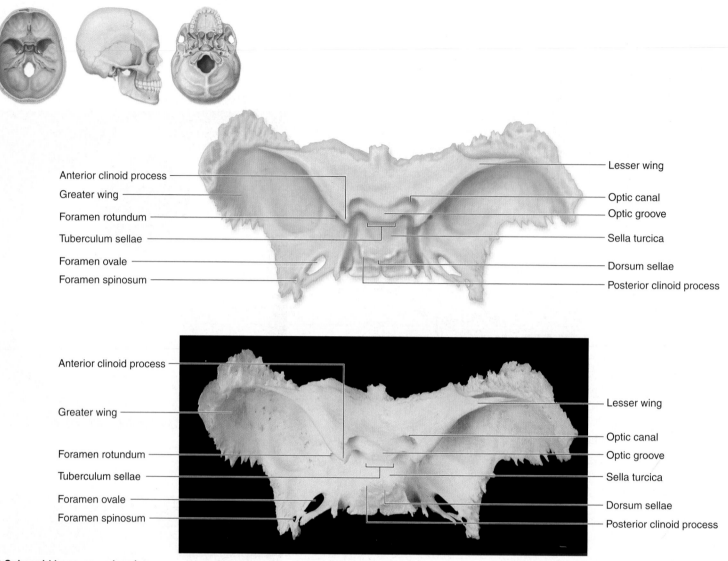

(a) Sphenoid bone, superior view

Figure 7.14

Sphenoid Bone. *(a)* Superior and *(b)* posterior views show that the sphenoid bone is a butterfly-shaped bone that forms the centerpiece of the base of the cranium.

each condyle is a **hypoglossal canal** through which the hypoglossal nerve (CN XII) extends to supply the tongue muscles. Posterior to each occipital condyle is a variable **condylar canal**, which transmits a vein.

Some prominent ridges appear on the external surface of the occipital bone. The **external occipital crest** projects in a posterior direction from the foramen magnum, ending in the **external occipital protuberance** (prō-too′ber-ans). Intersecting the external occipital crest are two horizontal ridges, the **superior** and **inferior nuchal** (noo′kăl) **lines**. These ridges are attachment sites for ligaments and neck muscles. Males have larger and more robust nuchal lines, because males tend to have larger muscles and ligaments.

The portion of the occipital bone that helps form the jugular foramen is called the **jugular notch** (figure 7.13*b*). The concave internal surface of the occipital bone closely follows the contours of the brain. Additionally, there are grooves formed from the impressions of, and named for, the venous sinuses within, the cranium. For

example, the **groove for the superior sagittal sinus**, the **groove for the transverse sinus**, and a portion of the **groove for the sigmoid sinus** represent the impressions that the superior sagittal sinus, transverse sinus, and sigmoid sinus make on the internal surface of the occipital bone, respectively. Also on the internal surface of the occipital bone, at the junction of the left and right grooves for the transverse sinuses, is the **internal occipital protuberance**. An **internal occipital crest** extends from the protuberance to the posterior border of the foramen magnum. This crest is a site of attachment for the falx cerebelli, a connective tissue sheet that helps support the cerebellum of the brain.

Sphenoid Bone

The **sphenoid** (sfē′noyd; wedge-shaped) **bone** has a complex shape, resembling that of a butterfly **(figure 7.14)**. It is often referred to as a "bridging bone," or the "keystone of the skull," because it unites the cranial and facial bones and articulates with almost every

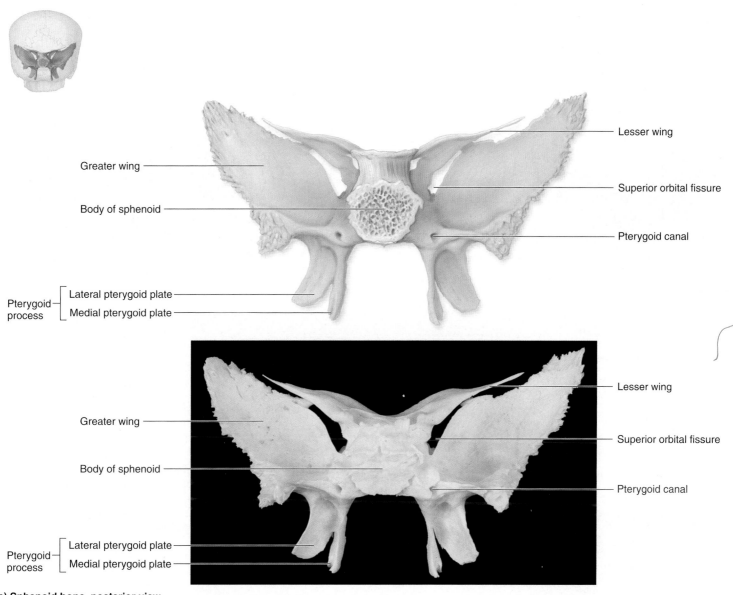

Greater wing

Body of sphenoid

Pterygoid process
- Lateral pterygoid plate
- Medial pterygoid plate

Lesser wing

Superior orbital fissure

Pterygoid canal

Greater wing

Body of sphenoid

Pterygoid process
- Lateral pterygoid plate
- Medial pterygoid plate

Lesser wing

Superior orbital fissure

Pterygoid canal

(b) Sphenoid bone, posterior view

other bone of the skull **(figure 7.15)**. Medially, it has a thick **body**, which contains the **sphenoidal sinuses**. Laterally, it extends to form the **greater** and **lesser wings**. Although the sphenoid is relatively large, much of it is hidden by more superficial bones.

The pituitary gland is suspended inferiorly from the brain into a prominent midline depression between the greater and lesser wings. This depression is termed the **hypophyseal fossa**, and the bony enclosure is called the **sella turcica** (sel′ă, saddle; tur′si-kă, Turkish) (see figure 7.14). The sella turcica houses the pituitary gland. On either side of the sella turcica are projections called the **anterior clinoid** (klĭ′noyd; *kline* = bed) **processes** and the **posterior clinoid processes**. The anterior border of the sella turcica is formed by the **tuberculum sellae**; the posterior border is formed by the **dorsum sellae**.

Anterior to the sella turcica, a shallow, transverse depression called the **optic** (op′tik; *ops* = eye) **groove** crosses the superior surface of the sphenoid bone. An **optic canal** (or *foramen*) is located at either end of this groove. The optic nerves (CN II) that carry

visual information from the eyes to the brain travel through these canals. On either side of the sella turcica, the **foramen rotundum** (rō-tŭn′dum; round), the **foramen ovale** (ō-văl′ē; oval), and the **foramen spinosum** (spĭ-nō′sŭm) penetrate the greater wings of the sphenoid bone. These openings carry blood vessels to the meninges around the brain and nerves to structures of the orbit, face, and jaws. A **pterygoid canal** is located on either side of the body and transmits nerves. The **pterygoid** (ter′i-goyd; *pteryx* = winglike) **processes** are vertical projections that begin at the boundary between the greater and lesser wings. Each pterygoid process forms a pair of **medial** and **lateral pterygoid plates**, which provide the attachment surfaces for some muscles that move the lower jaw and soft palate.

Ethmoid Bone

The irregularly shaped **ethmoid** (eth′moyd; *ethmos* = sieve) **bone** is positioned between the orbits (figure 7.15). It forms the anteromedial floor of the cranium, the roof of the nasal cavity, part of the medial wall of each orbit, and part of the nasal septum.

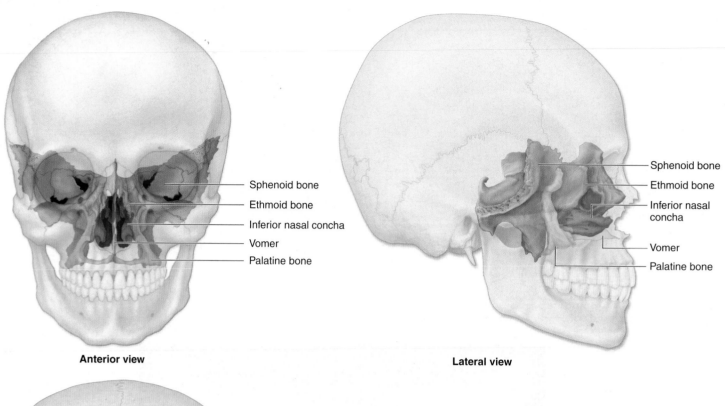

Sphenoid bone
Ethmoid bone
Inferior nasal concha
Vomer
Palatine bone

Anterior view

Sphenoid bone
Ethmoid bone
Inferior nasal concha
Vomer
Palatine bone

Lateral view

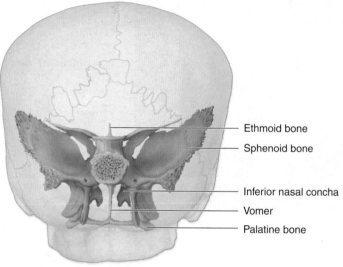

Ethmoid bone
Sphenoid bone

Inferior nasal concha
Vomer
Palatine bone

Posterior view

Figure 7.15

Articulations of the Sphenoid and Ethmoid Bones. Several bones of the skull, such as the ethmoid and sphenoid, are primarily located deep to other bones and can be seen only when the skull bones are disarticulated. This figure illustrates the positioning of these internal bones, relative to the externally placed skull bones.

The superior part of the ethmoid bone exhibits a thin midsagittal elevation called the **crista** (kris′tă; crest) **galli** (gal′lē; cock) **(figure 7.16)**. This bony crest is the point of attachment for the falx cerebri, a membranous sheet that helps support the brain. Immediately lateral to each side of the crista galli, the horizontal **cribriform** (krib′ri-fōrm; *cribrum* = sieve) **plate** has numerous perforations called the **cribriform foramina**. These foramina provide passageways for the olfactory nerves (CN I).

The paired **ethmoidal labyrinths** (*lateral masses*) contain tiny spaces called the **ethmoidal sinuses**, which open into both sides of the nasal cavity. The smooth part of each ethmoidal labyrinth is called the **orbital plate**, and forms part of the medial wall of the orbit. The ethmoidal labyrinths are partially composed of thin, scroll-like bones called the **superior** and the **middle nasal conchae** (kon′kē; sing., *concha*, kon′kă; shell). The inferior, midline projection of the ethmoid bone is called the **perpendicular plate**, and forms the superior part of the nasal septum.

Cranial Fossae

The contoured floor of the cranial cavity exhibits three curved depressions called the **cranial fossae (figure 7.17)**. Their surfaces contain depressions for parts of the brain, grooves for blood vessels, and numerous foramina.

The **anterior cranial fossa** is the shallowest of the three depressions. It is formed by the frontal bone, the ethmoid bone, and the lesser wings of the sphenoid bone. The anterior cranial fossa houses the frontal lobes of the cerebral hemispheres. It ranges from the internal surface of the inferior part of the frontal bone (anteriorly) to the posterior edge of the lesser wings of the sphenoid bone (posteriorly).

The **middle cranial fossa** is inferior and posterior to the anterior cranial fossa. It ranges from the posterior edge of the lesser wings of the sphenoid bone (anteriorly) to the anterior part of the petrous part of the temporal bone (posteriorly). This fossa is formed by the parietal, sphenoid, and temporal bones. It houses the temporal lobes of the cerebral hemispheres and part of the brainstem.

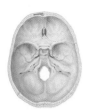

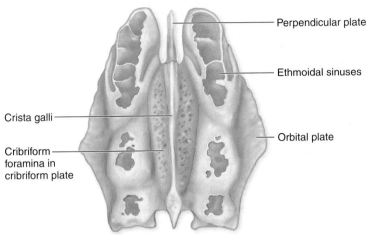

Perpendicular plate

Ethmoidal sinuses

Crista galli

Orbital plate

Cribriform foramina in cribriform plate

(a) Ethmoid bone, superior view

Figure 7.16

Ethmoid Bone.   This irregularly shaped bone forms part of the orbital wall, the anteromedial floor of the cranium, and part of the nasal cavity and nasal septum. *(a)* Superior and *(b)* anterior views show the ethmoid bone.

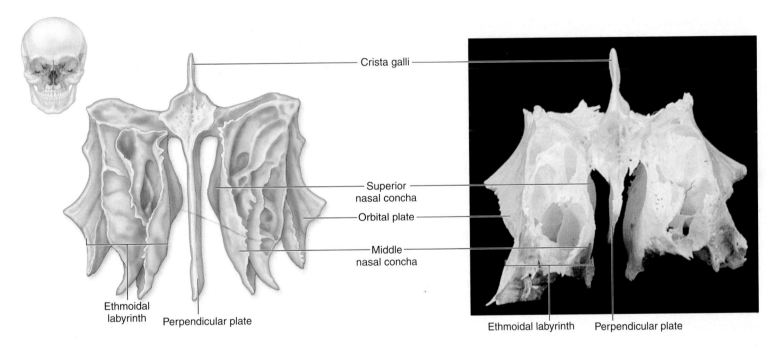

Crista galli

Superior nasal concha

Orbital plate

Middle nasal concha

Ethmoidal labyrinth   Perpendicular plate

Ethmoidal labyrinth   Perpendicular plate

(b) Ethmoid bone, anterior view

The **posterior cranial fossa** is the most inferior and posterior cranial fossa. It extends from the posterior part of the petrous part of the temporal bones to the internal posterior surface of the skull. The posterior fossa is formed by the occipital, temporal, and parietal bones. This fossa houses the cerebellum and part of the brainstem.

WHAT DID YOU LEARN?

❸ What are the three parts of the temporal bone? In which part is the mastoid process located?

❹ What are the functions of the superior and inferior nuchal lines? How do these lines differ in male versus female skulls?

❺ What is the sella turcica, its function, and which bone has this feature?

Bones of the Face

The **facial bones** (see figure 7.2) give shape and individuality to the face, form part of the orbit and nasal cavities, support the teeth, and provide for the attachment of muscles involved in facial expression and mastication. There are 14 facial bones, including the

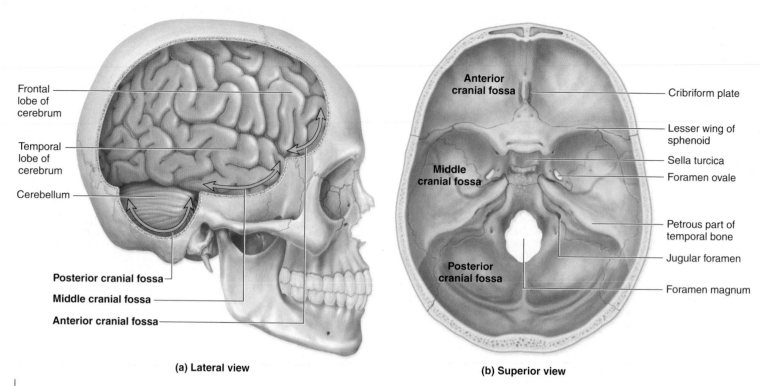

(a) Lateral view

(b) Superior view

Figure 7.17

Cranial Fossae. *(a)* Lateral and *(b)* superior views show the three levels of depression in the cranium (anterior, middle, and posterior) that parallel the contours of the ventral surface of the brain.

paired zygomatic bones, lacrimal bones, nasal bones, inferior nasal conchae, palatine bones, and maxillae, and the unpaired vomer and mandible (**table 7.3**).

Zygomatic Bones

The **zygomatic bones**, commonly referred to as the "cheekbones," form part of the lateral wall of each orbit and the cheeks (**figure 7.18**). A prominent **zygomatic arch** is formed by the articulation of the **temporal process** of each zygomatic bone with the zygomatic process of each temporal bone (see figure 7.6). The bone also has a **maxillary** (mak′si-lār-ē) **process**, which articulates with the zygomatic process of the maxilla, and a **frontal process**, which articulates with the frontal bone. The **orbital surface** of the zygomatic bone forms the lateral wall of the orbit.

Lacrimal Bones

The small, paired **lacrimal** (lak′ri-măl; *lacrima* = a tear) **bones**, form part of the medial wall of each orbit (see figure 7.4). A small, depressed inferior opening called the **lacrimal groove** provides a passageway for the nasolacrimal duct, which drains tears into the nasal cavity.

Nasal Bones

The paired **nasal bones** form the bridge of the nose (see figure 7.4). The medial edge of each maxilla articulates with the lateral edge of a nasal bone. The nasal bones are often fractured by blows to the nose.

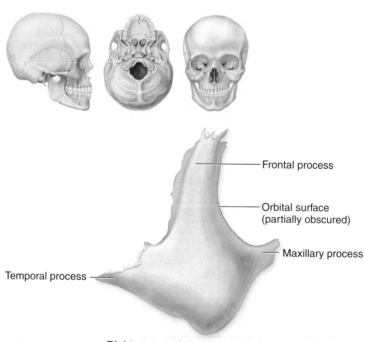

Right zygomatic bone, lateral view

Figure 7.18

Zygomatic Bone. The zygomatic bone forms the cheek and part of the lateral wall of the orbit.

Table 7.3	Facial Bones and Selected Features	
Bone and Associated Passageways	**Description and Boundaries of Bone**	**Selected Features and Their Functions**
Zygomatic bones	Each forms the cheek and lateral part of the orbit	**Frontal process:** Articulates with frontal bone **Maxillary process:** Articulates with maxilla **Temporal process:** Articulates with temporal bone to form zygomatic arch
Lacrimal bones	Each forms part of the medial wall of the orbit	**Lacrimal groove:** Contains nasolacrimal duct
Nasal bones	Each forms the anterosuperior bridge of the nose	
Vomer	Forms inferior and posterior parts of nasal septum	**Ala:** Articulates with the sphenoid bone **Vertical plate:** Articulates with perpendicular plate of ethmoid
Inferior nasal conchae	Curved bones that project from lateral walls of the nasal cavity	Increase airflow turbulence in nasal cavity
Palatine bones Greater and lesser palatine foramina	Each forms posterior part of hard palate; forms small part of nasal cavity and orbit wall	**Horizontal plate:** Forms posterior part of palate **Perpendicular plate:** Forms part of nasal cavity and orbit

Maxillae
   Incisive foramen
   Infraorbital foramen
	Each forms anterior portion of face; forms upper jaw and parts of the hard palate, inferior parts of orbits, and part of the walls of nasal cavity	**Alveolar process:** Houses the teeth
		Frontal process: Forms part of lateral aspect of nasal bridge
		Infraorbital margin: Forms inferolateral border of orbit
		Maxillary sinus: Lightens bone
		Palatine process: Forms most of bony palate
		Zygomatic process: Articulates with zygomatic bone

Mandible
   Mandibular foramen
   Mental foramen
	Forms the lower jaw	**Alveolar process:** Houses the teeth
		Coronoid process: Attachment of temporalis muscle
		Head of mandible: Articulates with temporal bone
		Mental protuberance: Forms the chin
		Mylohyoid line: Attachment site for mylohyoid muscle

Vomer

The **vomer** (vō′mer; plowshare) has a triangular shape, and when viewed laterally, resembles a farming plow **(figure 7.19)**. It articulates along its midline with both the maxillae and the palatine bones. Its curved, thin, horizontal projection, called the **ala**, meaning "wing," articulates superiorly with the sphenoid bone. The **vertical plate** of the vomer articulates with the perpendicular plate of the ethmoid bone. Anteriorly, both the vomer and the perpendicular plate of the ethmoid bone form the bony nasal septum.

Inferior Nasal Conchae

The **inferior nasal conchae** are located in the inferolateral wall of the nasal cavity (see figure 7.15). They are similar to the superior and middle nasal conchae of the ethmoid bone in that they help create turbulence in inhaled air. However, inferior nasal conchae are separate bones, while the middle and superior nasal conchae are parts of the ethmoid bone.

Palatine Bones

The **palatine** (pal′a-tin) **bones** are small bones with a distinct L shape. They form part of the hard palate, nasal cavity, and eye orbit **(figure 7.20)**. The posterior portion of the hard palate is formed by the **horizontal plate** of the palatine bone, which articulates anteriorly with the palatine process of the maxilla. Greater and lesser **palatine foramina** perforate this horizontal plate (see figure 7.8). Nerves to the palate and upper teeth travel through these palatine formina. The

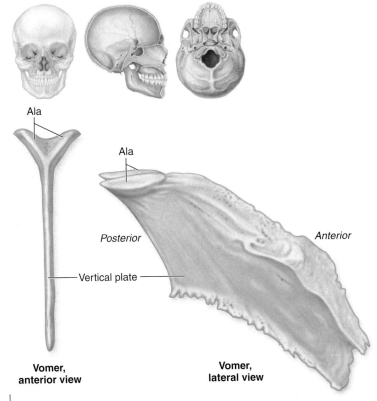

Ala

Ala

Posterior

Anterior

Vertical plate

Vomer, anterior view

Vomer, lateral view

Figure 7.19

Vomer. The vomer forms the inferior part of the nasal septum.

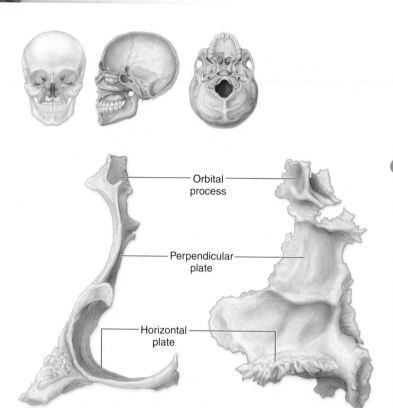

Figure 7.20 labels:
- Orbital process
- Perpendicular plate
- Horizontal plate

Right palatine bone, anterior view **Right palatine bone, medial view**

Figure 7.20

Palatine Bone. The L-shaped palatine bone forms part of the nasal complex and the hard palate.

perpendicular plate forms part of the lateral wall of the nasal cavity. The most superior part of the perpendicular plate has an **orbital process** that forms a small part of the medial floor of the orbit.

Maxillae

The paired **maxillae** (mak-sil'ē; sing., *maxilla,* mak-sil'ă; jawbone), also called *maxillary bones,* form the central part of the facial skeleton **(figure 7.21)**. Left and right maxillae unite to form the upper jaw. Together, the united maxillae form a prominent **anterior nasal spine** along the inferior surface of the nasal cavity. Each maxilla contains an **infraorbital margin** and an inferior **orbital surface**. A large **infraorbital foramen** provides passage for a blood vessel and nerve (infraorbital artery and nerve). Within the orbit, this foramen extends along the **infraorbital groove**. The inferior portions of the maxillae contain the **alveolar** (al-vē'ō-lăr) **processes** that house the upper teeth.

Most of the hard palate is formed anteriorly by horizontal medial extensions of both maxillae, called **palatine processes** (see figure 7.21*b*). Near the anterior margin of the fused palatine processes, immediately posterior to the teeth called incisors, is an **incisive foramen**. This foramen is a passageway for branches of the nasopalatine nerve. Lateral to the nasal cavity, each maxilla contains

a large, spacious cavity called the **maxillary sinus**. Laterally, each maxilla articulates with a zygomatic bone via a **zygomatic process**. Superiorly, the maxillae articulate with the frontal bones via **frontal processes** (see figure 7.21*a*).

CLINICAL VIEW

Cleft Lip and Cleft Palate

Several embryonic structures must grow toward each other and fuse to form a normal upper lip and palate. If some of the embryonic structures fail to fuse properly, an opening called a *cleft* can result. Incomplete fusion of the medial and lateral nasal prominences and the maxillary process results in a split upper lip, called **cleft lip**, extending from the mouth to the side of one nostril. Cleft lip may be unilateral (occurring on one side only) or bilateral (occurring on both sides). Cleft lip appears in 1 per 1000 births and tends to be more common in males. The etiology of cleft lip is multifactorial, in that both genetic and environmental factors appear to contribute to the condition.

Another anomaly that can develop is **cleft palate**, a congenital fissure in the midline of the palate. Normally, the palatine processes of the maxillae and palatine bones join between the tenth and twelfth weeks of embryonic development. A cleft palate results when the left and right bones fuse incompletely or do not fuse at all. The opening in the palate varies from tiny to large; in very severe cases, the palate doesn't form at all. In the more severe cases, children experience swallowing and feeding problems because food can easily travel from the oral cavity into the nasal cavity. Cleft palate occurs in about 1 per 2500 births and tends to be more common in females. Like cleft lip, the etiology of cleft palate is multifactorial. Cleft palate sometimes occurs in conjunction with cleft lip.

Both the position and extent of the cleft lip or cleft palate determine how speech and swallowing are affected. Early treatment by oral and facial surgeons often yields excellent results.

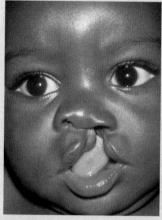

Cleft lip.

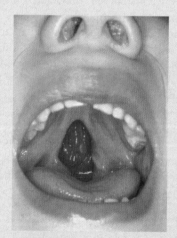

Cleft palate.

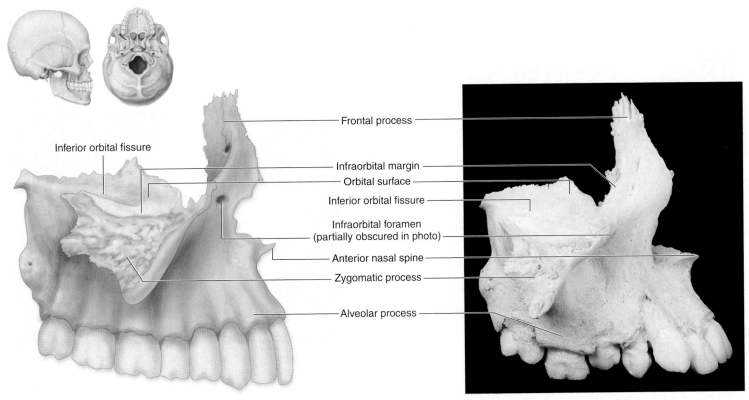

Frontal process

Inferior orbital fissure

Infraorbital margin
Orbital surface
Inferior orbital fissure
Infraorbital foramen
(partially obscured in photo)
Anterior nasal spine
Zygomatic process

Alveolar process

(a) Right maxilla, lateral view

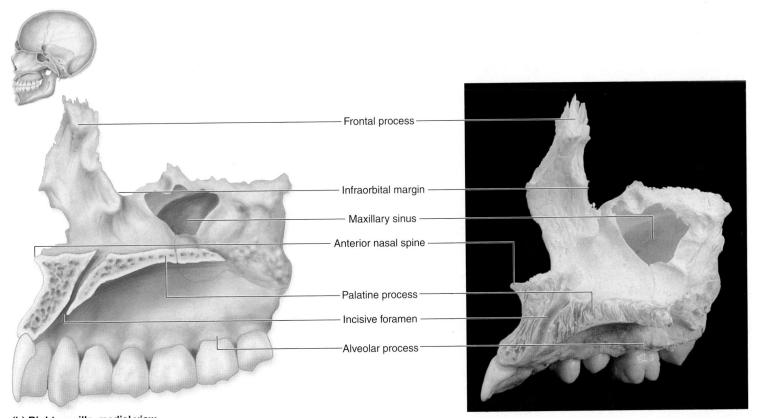

Frontal process

Infraorbital margin

Maxillary sinus

Anterior nasal spine

Palatine process

Incisive foramen

Alveolar process

(b) Right maxilla, medial view

Figure 7.21

Maxilla.   Both maxillae form the upper jaw. *(a)* Lateral and *(b)* medial views show the right maxilla.

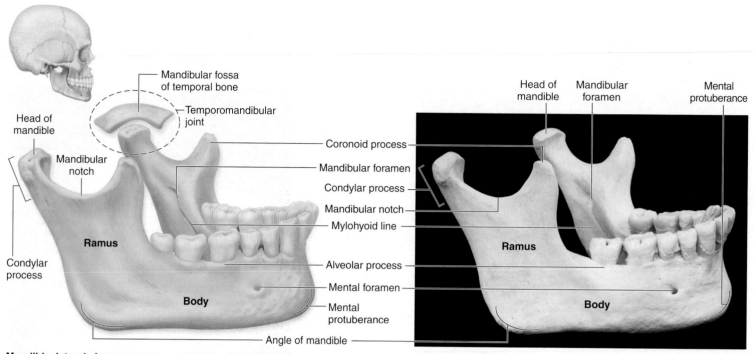

Mandible, lateral view

Figure 7.22

Mandible. The mandible forms the lower jaw, and its articulation with the temporal bone forms the temporomandibular joint.

Mandible

The **mandible** (man′di-bl) forms the entire lower jaw (**figure 7.22**). It supports the inferior teeth and provides attachment for the muscles of mastication. The mandible has a horizontal **body** and two vertical-to-oblique ascending posterior regions called the **rami** (rā′mī; sing., *ramus*, rā′mŭs). The teeth are supported by the **alveolar process** of the mandibular body. Each ramus intersects the body at a "corner" called the **angle of the mandible**. The point of the chin is the **mental** (*mentum* = chin) **protuberance**. On the anterolateral surface of the body, a **mental foramen** penetrates the body on each side of the chin to provide a passageway for nerves and blood vessels.

An **alveolar process** covers both the alveoli and the roots of the teeth medially in the lower jaw. On the medial wall of each ramus, at the **mylohyoid** (mī′lō-hī-oyd; *myle* = molar teeth) **line**, the mylohyoid muscle inserts to support the tongue and the floor of the mouth. At the posterosuperior end of each mylohyoid line, a prominent **mandibular foramen** provides a passageway for blood vessels and nerves that innervate the inferior teeth.

The posterior projection of each mandibular ramus, called the **condylar** (kon′di-lăr) **process**, terminates at the **head of the mandible**, also called the *mandibular condyle* (kon′dīl). Each articulation of the head of the mandible with the mandibular fossa of the temporal bone is called the **temporomandibular joint (TMJ)**, a mobile joint that allows us to move the lower jaw when we talk or chew. The anterior projection of the ramus, termed the **coronoid** (kōr′ŏ-noyd; *korone* = a crown) **process**, is the insertion point for the temporalis muscle, a powerful muscle involved in closing the mouth. The U-shaped depression between the two processes is called the **mandibular notch**.

Nasal Complex

The **nasal complex** is composed of bones and cartilage that enclose the nasal cavity and the paranasal sinuses. These bones are shown in both sagittal section and coronal section, as in **figure 7.23**.

- The roof, or superior border, of the nasal complex is formed by the cribriform plate of the ethmoid bone and parts of the frontal and sphenoid bones.
- The floor, or inferior border, is formed by the palatine processes of the maxillae and the horizontal plates of the palatine bones.
- The lateral walls are formed by the ethmoid bone, maxillae, inferior nasal conchae, the perpendicular plates of the palatine bones, and the lacrimal bones.

Most of the anterior walls of the nasal cavity are formed by cartilage and the soft tissues of the nose, but the bridge of the nose is supported by the maxillae and the nasal bones.

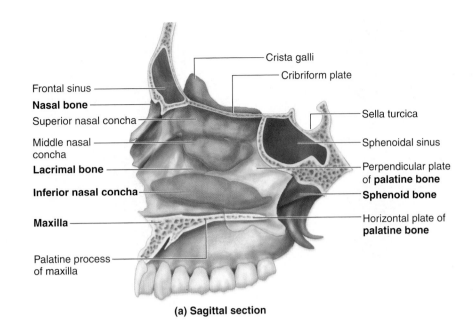

Crista galli
Cribriform plate
Frontal sinus
Nasal bone
Superior nasal concha
Sella turcica
Middle nasal concha
Sphenoidal sinus
Lacrimal bone
Perpendicular plate of **palatine bone**
Inferior nasal concha
Sphenoid bone
Maxilla
Horizontal plate of **palatine bone**
Palatine process of maxilla

(a) Sagittal section

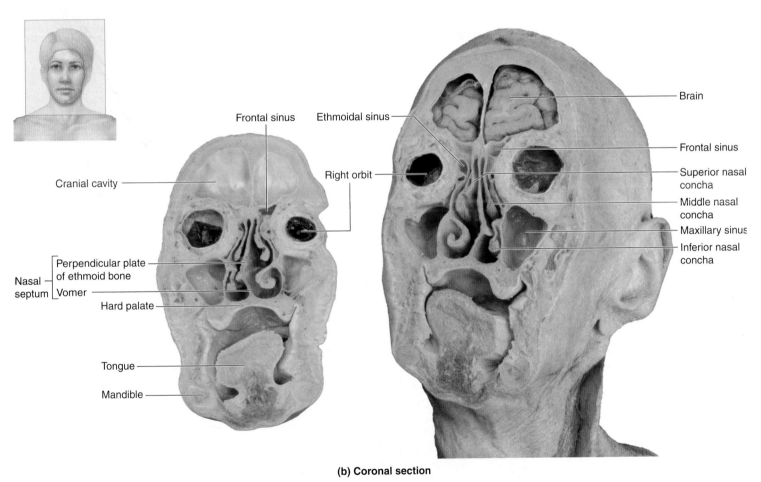

Brain
Frontal sinus
Ethmoidal sinus
Frontal sinus
Cranial cavity
Superior nasal concha
Right orbit
Middle nasal concha
Maxillary sinus
Inferior nasal concha
Perpendicular plate of ethmoid bone
Nasal septum
Vomer
Hard palate
Tongue
Mandible

(b) Coronal section

Figure 7.23

Nasal Complex. Multiple skull bones form the intricate nasal complex. *(a)* A sagittal section shows the right side of the nasal complex. *(b)* Cadaver photo of coronal sections through the head shows the nasal complex.

Paranasal Sinuses

We have already described the ethmoidal, frontal, maxillary, and sphenoidal sinuses in connection with the bones where they are found. As a group, these air-filled chambers that open into the nasal cavities are called the **paranasal sinuses** (sī′nŭs; cavity, hollow) **(figure 7.24)**. The sinuses have a mucous lining that helps to humidify and warm inhaled air. Additionally, the sinus spaces in some skull bones cause these skull bones to be lighter, and also provide resonance to the voice.

Orbital Complex

The bony cavities called orbits enclose and protect the eyes and the muscles that move them. The **orbital complex** consists of multiple

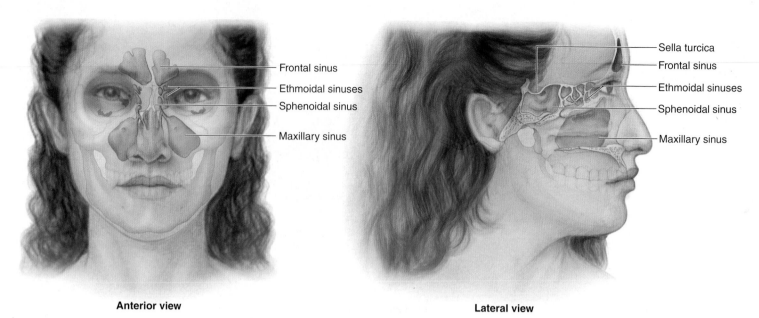

Anterior view

Lateral view

Sella turcica
Frontal sinus
Ethmoidal sinuses
Sphenoidal sinus
Maxillary sinus

Frontal sinus
Ethmoidal sinuses
Sphenoidal sinus
Maxillary sinus

Figure 7.24

Paranasal Sinuses. The paranasal sinuses are air-filled chambers within the frontal, ethmoid, and sphenoid bones and the maxillae. They act as extensions of the nasal cavity.

bones that form each orbit **(figure 7.25)**. The borders of the orbital complex are as follows:

- The roof of the orbit is formed from the orbital part of the frontal bone and the lesser wing of the sphenoid bone.
- The floor of the orbit is formed primarily by the orbital surface of the maxilla, although the zygomatic bone and orbital process of the palatine bone also contribute a portion.

- The medial wall is formed from the frontal process of the maxilla, the lacrimal bone, and the orbital plate of the ethmoid bone.
- The lateral wall of the orbit is formed from the orbital surface of the zygomatic bone, the greater wing of the sphenoid bone, and the zygomatic process of the frontal bone.
- The posterior wall of the orbit is formed primarily from the sphenoid bone.

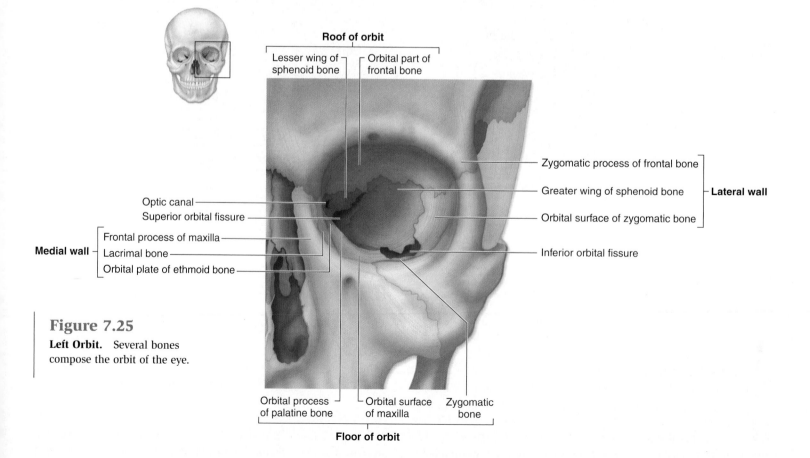

Roof of orbit

Lesser wing of sphenoid bone

Orbital part of frontal bone

Zygomatic process of frontal bone

Greater wing of sphenoid bone

Lateral wall

Orbital surface of zygomatic bone

Optic canal
Superior orbital fissure

Inferior orbital fissure

Medial wall — Frontal process of maxilla
Lacrimal bone
Orbital plate of ethmoid bone

Orbital process of palatine bone

Orbital surface of maxilla

Zygomatic bone

Floor of orbit

Figure 7.25

Left Orbit. Several bones compose the orbit of the eye.

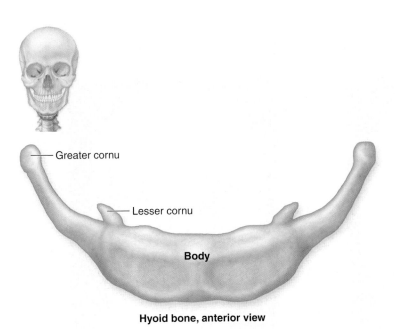

— Greater cornu

— Lesser cornu

Body

Hyoid bone, anterior view

Figure 7.26

Hyoid Bone.   The hyoid bone is not in direct contact with any other bone of the skeleton.

Bones Associated with the Skull

The auditory ossicles and the hyoid bone are bones of the axial skeleton associated with the skull. Three tiny ear bones called **auditory ossicles** (os′i-kl) are housed within the petrous part of each temporal bone. These bones—the **malleus** (mal′ē-us), the **incus** (ink′kus), and the **stapes** (stā′pēz)—are discussed in depth in chapter 19.

The **hyoid bone** is a slender, curved bone located inferior to the skull between the mandible and the larynx (voice box) (**figure 7.26**). It does not articulate with any other bone in the skeleton. The hyoid has a midline **body** and two hornlike processes, the **greater cornu** (kōr′noo; pl., *cornua*, kor′noo-ă; horn) and the **lesser cornu**. The cornua and body serve as attachment sites for tongue and larynx muscles and ligaments.

WHAT DID YOU LEARN?

6 The hard palate is composed of what bones and their parts?

7 Identify the bones in which the paranasal sinuses are located.

8 What bones form the lateral wall of the orbit?

9 Identify the auditory ossicles.

Sex Differences in the Skull

Key topic in this section:

■ Comparison of male and female skulls.

As mentioned briefly throughout the chapter, human male and female skulls display differences in general shape and size, a phenomenon known as *sexual dimorphism*. Typical "female" features are gracile (delicate, small), while "male" features tend to be more robust (larger, sturdier, bulkier). **Table 7.4** summarizes the sex differences in the skull.

It is possible to determine the sex of an individual from skeletal remains by examining the skull, but there are some caveats to keep in mind. First, the features of the skull (and those of other skeletal remains as well) vary from population to population. For example, some male Asian skeletal remains may be less robust than those of, say, female Native Americans. Further, it is difficult (and in many cases, impossible) to determine the sex of infant and juvenile remains, since skull characteristics appear "female" until well after puberty.

The most accurate method of determining sex is to look at multiple features on the skeleton and make a judgment based on the majority of features present. For example, if a skull displays two female-like characteristics and four male-like characteristics, the skull will likely be classified as male. If your anatomy lab uses real skulls, use table 7.4 to determine the sex of the skull you are studying.

WHAT DID YOU LEARN?

10 What are some features that differ between female and male skulls?

WHAT DO YOU THINK?

1 It is difficult to determine the sex of a young child's skull because both male and female young adult skulls appear female-like. What factors cause those features to change in males by adulthood?

Aging of the Skull

Key topic in this section:

■ Comparison of fetal, child, and adult skulls

Although many centers of ossification are involved in the formation of the skull, fusion of the centers produces a smaller number of composite bones as development proceeds. For example, the ethmoid bone forms from three separate ossification centers, and the occipital bone forms from four separate ossification centers. At birth, some of these ossification centers have not yet fused, so an infant initially has two frontal bone elements, four occipital bone elements, and a number of sphenoid and temporal elements.

The shape and structure of cranial elements differ in the skulls of infants and adults, causing variations in proportions and size. The most significant growth in the skull occurs before age 5, when the brain is still growing and exerting pressure against the internal surface of the developing skull bones. Brain growth is 90–95% complete by age 5, at which time cranial bone growth is also nearly complete, and the cranial sutures are almost fully developed. Note that the skull grows at a much faster rate than the rest of the body. Thus, the cranium of a young child is relatively larger than that of an adult. **Figure 7.27** shows lateral and superior views of a neonatal (infant) cranium.

The infant's cranial bones are connected by flexible areas of dense regular connective tissue, and in some regions the brain is covered only by this connective tissue sheet, since the bones aren't yet big enough to fully surround the brain. The regions between the cranial bones are thickened, fibrous membrane remnants that are not yet ossified, called **fontanelles** (fon′tă-nel′; little spring; sometimes spelled *fontanels*). Fontanelles are sometimes referred to as

Table 7.4 — Sex Differences in the Skull

View	Female Skull	Male Skull
Anterior View		
Lateral View		

Skull Feature	Female Skull Characteristic	Male Skull Characteristic
General Size and Appearance	More gracile and delicate	More robust (big and bulky), more prominent muscle markings
Nuchal Lines and External Occipital Protuberance	External surface of occipital bone is relatively smooth, with no major bony projections	Well-demarcated nuchal lines and prominent bump or "hook" for external occipital protuberance
Mastoid Process	Relatively small	Large, may project inferior to external acoustic meatus
Squamous Part of Frontal Bone	Usually more vertically oriented and rounded than in males	Exhibits a sloping angle
Supraorbital Margin	Thin, sharp border	Thick, rounded, blunt border
Superciliary Arches	Little or no prominence	More prominent and bulky
Mandible (general features)	Smaller and lighter	Larger, heavier, more robust
Mental Protuberance (chin)	More pointed and triangular-shaped, less forward projection	Squarish, more forward projection
Mandibular Angle	Typically greater than 125 degrees	Flared, less obtuse, less than 125 degrees (typically about 90 degrees)
Sinuses	Smaller in total volume	Larger in total volume
Teeth	Relatively smaller	Relatively larger

the "soft spots" on a baby's head. When a baby passes through the birth canal, the cranial bones overlap at these fontanelles in order to ease the baby's passage. Newborns frequently have a "cone-shaped" head due to this temporary deformation, but by a few days after birth, the cranial bones have returned to their normal position.

The fontanelles are present until many months after birth, when skull bone growth finally starts to keep pace with brain growth. The small **mastoid** and **sphenoidal fontanelles** close relatively quickly, compared to the larger posterior and anterior fontanelles. The **posterior fontanelle** normally closes around 9 months of age, while the larger **anterior fontanelle** doesn't close until about 15 months of age. It is not uncommon to see rhythmic pulsations of the blood vessels internal to these fontanelles.

Although the skull may come close to its adult size by age 5, it still undergoes many more changes in subsequent years. The maxillary sinus becomes a bit more prominent beginning at age 5, and by age 10 the frontal sinus is becoming well formed. Later, the cranial sutures start to fuse and ossify. As a person ages, the teeth start to wear down from use, a process called *dental attrition*. Finally, if an individual loses some or all of his teeth, the alveolar processes of the maxillae and mandible regress, become less prominent, and eventually disappear.

WHAT DID YOU LEARN?

11 What are the two largest fontanelles, and when do they disappear?

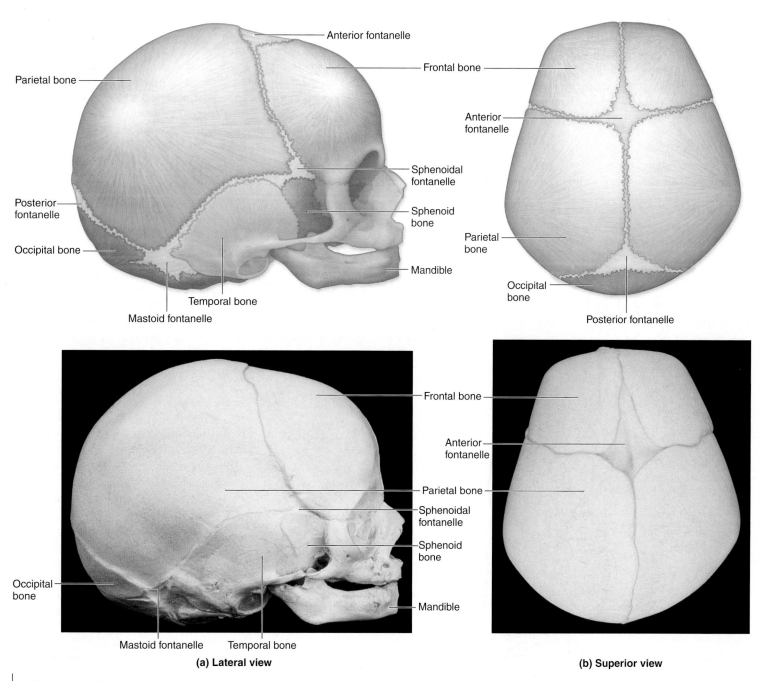

(a) Lateral view   **(b) Superior view**

Figure 7.27

Fetal Skull. *(a)* Lateral and *(b)* superior views show the flat bones in an infant skull, which are separated by fontanelles. These allow for the flexibility of the skull during birth and the growth of the brain after birth.

Vertebral Column

Key topics in this section:

■ Functions of the vertebral column
■ General structure of the vertebral column
■ The parts of a typical vertebra
■ Comparison of the vertebrae in each region

The adult **vertebral column** is composed of 26 bones, including 24 individual **vertebrae** (ver′tĕ-brē; sing., *vertebra*, ver′tĕ-bră) and the fused vertebrae that form both the sacrum and the coccyx. Each vertebra (except the first and the last) articulates with one superior vertebra and one inferior vertebra. The vertebral column has several functions:

■ Providing vertical support for the body.
■ Supporting the weight of the head.
■ Helping to maintain upright body position.
■ Helping to transfer axial skeletal weight to the appendicular skeleton of the lower limbs.
■ Housing and protecting the delicate spinal cord and providing a passageway for spinal nerves that connect to the spinal cord.

Divisions of the Vertebral Column

The vertebral column is partitioned into five regions **(figure 7.28)**. Vertebrae are identified by a capital letter that denotes their region, followed by a numerical subscript that indicates their sequence, going from superior to inferior.

■ Seven **cervical** (ser′vĭ-kal; *cervix* = neck) **vertebrae** (designated C_1–C_7) form the bones of the neck. The first cervical vertebra (C_1) articulates superiorly with the occipital condyles of the occipital bone of the skull. The seventh cervical vertebra (C_7) articulates inferiorly with the first thoracic vertebra.

■ Twelve **thoracic vertebrae** (designated T_1–T_{12}) form the superior regions of the back, and each articulates laterally with one or two pairs of ribs. The twelfth thoracic vertebra articulates inferiorly with the first lumbar vertebra.

■ Five **lumbar vertebrae** (L_1–L_5) form the inferior concave region ("small") of the back; L_5 articulates inferiorly with the sacrum.

■ The **sacrum** (sā′krŭm) is formed from five sacral vertebrae (S_1–S_5), which fuse into a single bony structure by the mid to late 20s. The sacrum articulates with L_5 superiorly and with the first coccygeal vertebra inferiorly. In addition, the sacrum articulates laterally with the two ossa coxae (hip bones).

■ The **coccyx** (kok′siks), commonly called the "tailbone," is formed from four **coccygeal vertebrae** (Co_1–Co_4) that start to unite during puberty. The first coccygeal vertebra (Co_1) articulates with the inferior end of the sacrum. When a person is much older, the coccyx may also fuse to the sacrum.

CLINICAL VIEW

Spinal Curvature Abnormalities

Distortion of the normal spinal curvature may be caused by poor posture, disease, congenital defects in the structure of the vertebrae, or weakness or paralysis of muscles of the trunk. There are three main spinal curvature deformities: kyphosis, lordosis, and scoliosis.

Kyphosis (kĭ-fō′sis) is an exaggerated thoracic curvature that is directed posteriorly, producing a "hunchback" look. Kyphosis often results from osteoporosis, but it also occurs in individuals who experience any of the following: a vertebral compression fracture that affects the anterior region of the vertebral column, osteomalacia (a disease in which adult bones become demineralized), heavy weight lifting during adolescence, abnormal vertebral growth, or chronic contractions in muscles that insert on the vertebrae.

Lordosis (lōr-dō′sis) is an exaggerated lumbar curvature, often called "swayback," that is observed as a protrusion of the abdomen and buttocks. Lordosis may have the same causes as kyphosis, or it may result from the added abdominal weight associated with pregnancy or obesity.

Scoliosis (skō-lē-ō′sis) is the most common spinal curvature deformity. It may affect one or more of the movable vertebrae, but it occurs most often in the thoracic region, especially among adolescent females. Scoliosis is an abnormal lateral curvature that sometimes results during development when both the vertebral arch and the body either fail to form or form incompletely on one side of a vertebra. It also can be caused by unilateral muscular paralysis, or spasm, in the back.

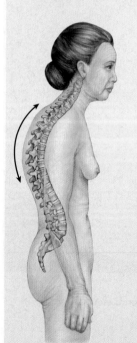
Kyphosis ("hunchback")

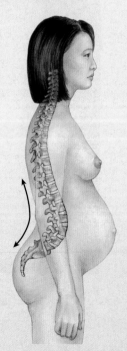
Lordosis ("swayback")

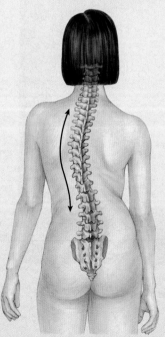
Scoliosis

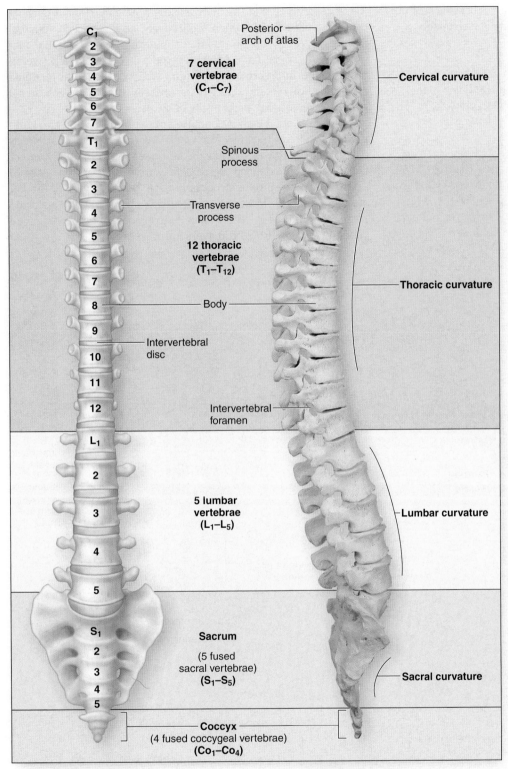

Posterior arch of atlas

7 cervical vertebrae (C₁–C₇)

Cervical curvature

Spinous process

Transverse process

12 thoracic vertebrae (T₁–T₁₂)

Thoracic curvature

Body

Intervertebral disc

Intervertebral foramen

L₁

5 lumbar vertebrae (L₁–L₅)

Lumbar curvature

Sacrum

(5 fused sacral vertebrae) (S₁–S₅)

Sacral curvature

Coccyx (4 fused coccygeal vertebrae) (Co₁–Co₄)

(a) Anterior view

(b) Right lateral view

Figure 7.28

Vertebral Column. (a) Anterior and (b) right lateral views show the regions and curvatures in the vertebral column.

Spinal Curvatures

The vertebral column has some flexibility because it is not straight and rigid. When viewed from a lateral perspective, the adult vertebral column has four **spinal curvatures**: the **cervical curvature**, **thoracic curvature**, **lumbar curvature**, and **sacral curvature**. These spinal curvatures better support the weight of the body when standing than a straight spine could.

The spinal curvatures appear sequentially during fetal, newborn, and child developmental stages. The **primary curves** are the thoracic and sacral curvatures, which appear late in fetal develop-

ment. These curves are also called *accommodation curves* because they accommodate the thoracic and abdominopelvic viscera. In the newborn, only these primary curves are present, and the vertebral column is C-shaped.

The **secondary curves** are the cervical and lumbar curvatures and appear after birth. These curves arch anteriorly and are also known as *compensation curves* because they help shift the trunk weight over the legs. The cervical curvature appears around 3–4 months of age, when the child is first able to hold up its head without support. The lumbar curvature appears by the first year of life, when the child is learning to stand and walk. These curves become accentuated as the child becomes more adept at walking and running.

WHAT DO YOU THINK?

2 Which do you think is better able to support the weight of the body—a completely straight vertebral column or a vertebral column with spinal curvatures? Why?

Vertebral Anatomy

Most vertebrae share common structural features **(figure 7.29)**. The anterior region of each vertebra is a rounded or cylindrical **body**, also called a *centrum* (pl., *centra*), which is the weight-bearing structure of almost all vertebra. Posterior to the vertebral body is the **vertebral arch**, also called the *neural arch*. Together,

the vertebral arch and the body enclose a roughly circular opening called the **vertebral foramen**. Collectively, all the stacked vertebral foramina form a superior-to-inferior directed **vertebral canal** that contains the spinal cord. Lateral openings between adjacent vertebrae are the **intervertebral foramina**. The intervertebral foramina provide a horizontally directed passageway through which spinal nerves travel to other parts of the body.

The vertebral arch is composed of two pedicles and two laminae. The **pedicles** (ped′ĭ-kl; *pes* = foot) originate from the posterolateral margins of the body, while the **laminae** (lam′ĭ-nē; sing., *lamina*, lam′ĭ-nă; layer) extend posteromedially from the posterior edge of each pedicle. A **spinous process** projects posteriorly from the left and right laminae. Most of these spinous processes can be palpated through the skin of the back. Lateral projections on both sides of the vertebral arch are called **transverse processes**.

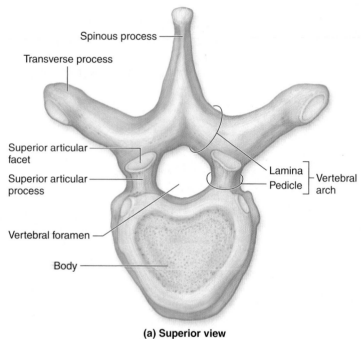

(a) Superior view

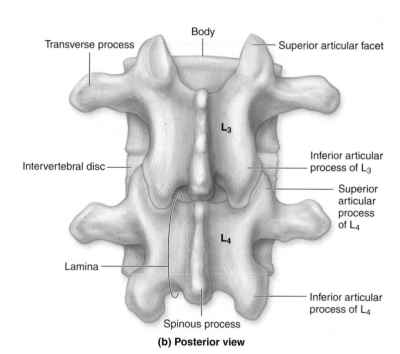

(b) Posterior view

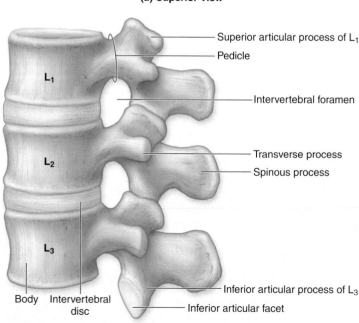

(c) Lateral view

Figure 7.29

Vertebral Anatomy. *(a)* Superior view of a thoracic vertebra. *(b)* Articulation between lumbar vertebrae, posterior view. *(c)* Articulation between lumbar vertebrae, lateral view.

Each vertebra has **articular processes** on both its superior and inferior surfaces that project from the junction between the pedicles and laminae. The **inferior articular processes** of each vertebra articulate with the **superior articular processes** of the vertebra immediately inferior to it. Each articular process has a smooth surface, called an **articular facet** (fas′et, fă-set′).

The vertebral bodies are interconnected by ligaments. Adjacent vertebral bodies are separated by pads of fibrocartilage called the **intervertebral** (in-ter-ver′te-bral) **discs**. Intervertebral discs are composed of an outer ring of fibrocartilage, called the **anulus fibrosus** (an′ū-lŭs fī-brō′sŭs), and an inner circular region, called the **nucleus pulposus**. The nucleus pulposus has a high water content, giving it a gelatinous consistency. Intervertebral discs make up approximately one-quarter of the entire vertebral column. They act as shock absorbers between the vertebral bodies, and also allow the vertebral column to bend. For example, when you bend your torso anteriorly, the intervertebral discs are compressed at the bending (anterior) surface and pushed out toward the opposite (posterior) surface.

Intervertebral discs are able to withstand a certain amount of compression. Over the course of a day, as body weight and gravity act on the vertebral column, the intervertebral discs become compressed and flattened. But while a person sleeps, lying horizontally, the intervertebral discs are able to spring back to their original shape.

In general, the vertebrae are smallest near the skull, and become gradually larger moving inferiorly through the body trunk as weight-bearing increases. Thus, the cervical vertebrae are the smallest, followed by the thoracic, lumbar, and sacral vertebrae.

Although vertebrae are divided into regions, there are no anatomically discrete "cutoffs" between the regions. For example, the most inferior cervical vertebra has some structural similarities to the most superior thoracic vertebra, since the two vertebrae are adjacent to one another. Likewise, the most inferior thoracic vertebra may look similar to the first lumbar vertebra. Despite this, there are basic characteristics that distinguish different types of vertebrae. We discuss these characteristics in the next sections. **Table 7.5** compares the characteristics of the cervical, thoracic, and lumbar vertebrae.

Cervical Vertebrae

The cervical vertebrae, C_1–C_7, are the most superiorly placed vertebrae (table 7.5a). They extend inferiorly from the occipital bone of the skull through the neck to the thorax. Since cervical vertebrae support only the weight of the head, their vertebral bodies are relatively small and light.

The body of a typical cervical vertebra (C_3–C_6) is relatively small compared to its foramen. The superior surface of a cervical vertebral body is concave from side to side, and it exhibits a superior slope from the posterior edge to the anterior edge. The spinous process is relatively short, usually less than the diameter of the vertebral foramen. The tip of each process, other than C_7, is usually bifurcated (bifid), meaning that the posterior end of the spinous process appears to be split in two. The transverse processes of the first six (and sometimes the seventh) cervical vertebrae are unique in that they contain prominent, round **transverse foramina**, which provide a protective bony passageway for the vertebral arteries and veins supplying the brain.

CLINICAL VIEW

Herniated Discs

Certain twisting and flexing motions of the vertebral column can injure the intervertebral discs. The cervical and lumbar intervertebral discs are the most common discs to be injured, because the vertebral column has a great deal of mobility in these regions, and the lumbar region bears increased weight. Intervertebral discs in the thoracic part of the vertebral column tend not to be injured because this part of the vertebral column is less mobile and more stable due to its articulation with the ribs.

A **herniated** (her′nē-ā-ted) **disc** occurs when the gelatinous nucleus pulposus protrudes into or through the anulus fibrosus. This herniation produces a "bulging" of the disc posterolaterally into the vertebral canal and pinches the spinal cord and/or nerves of the spinal cord. The symptoms of a herniated disc vary, depending on the location of the herniation. Cervical herniated discs can cause neck pain and pain down the upper limb, since the nerves that supply the upper limb originate in this region of the spinal cord. Muscle weakness in the upper limb may also occur. The most common cervical disc ruptures are between vertebrae C_5 and C_6 or C_6 and C_7. Lumbar herniated discs frequently cause low back pain. If the disc starts to pinch nerve fibers, the patient may feel pain down the entire lower limb, a condition known as **sciatica**. The most common lumbar disc rupture is between vertebrae L_4 and L_5.

Treatment options for a herniated disc vary. Conservative approaches include "wait-and-see" if the disc heals on its own, nonsteroidal anti-inflammatory drugs (NSAIDS) such as ibuprofen, steroid drugs, and physical therapy. If conservative treatments fail and the patient is still in severe pain, surgical treatments include **microdiscectomy**, a microsurgical technique whereby the herniated portion of the disc is removed, or **discectomy**, a more invasive technique in which the laminae of the nearby

vertebrae and the back muscles are incised before removing the herniated portions of the disc. Most recently, artificial discs made of synthetic material have been developed to replace herniated discs. Currently, only a few medical centers offer disc replacement, and individuals who want to explore this option typically must be part of a clinical trial.

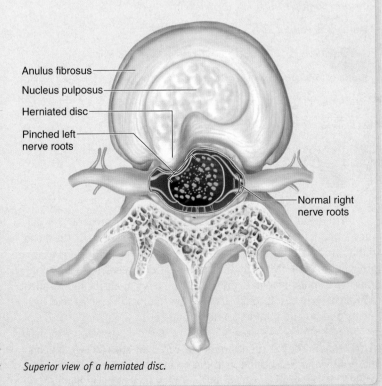

Anulus fibrosus
Nucleus pulposus
Herniated disc
Pinched left nerve roots
Normal right nerve roots

Superior view of a herniated disc.

Table 7.5 Characteristic Features of Cervical, Thoracic, and Lumbar Vertebrae

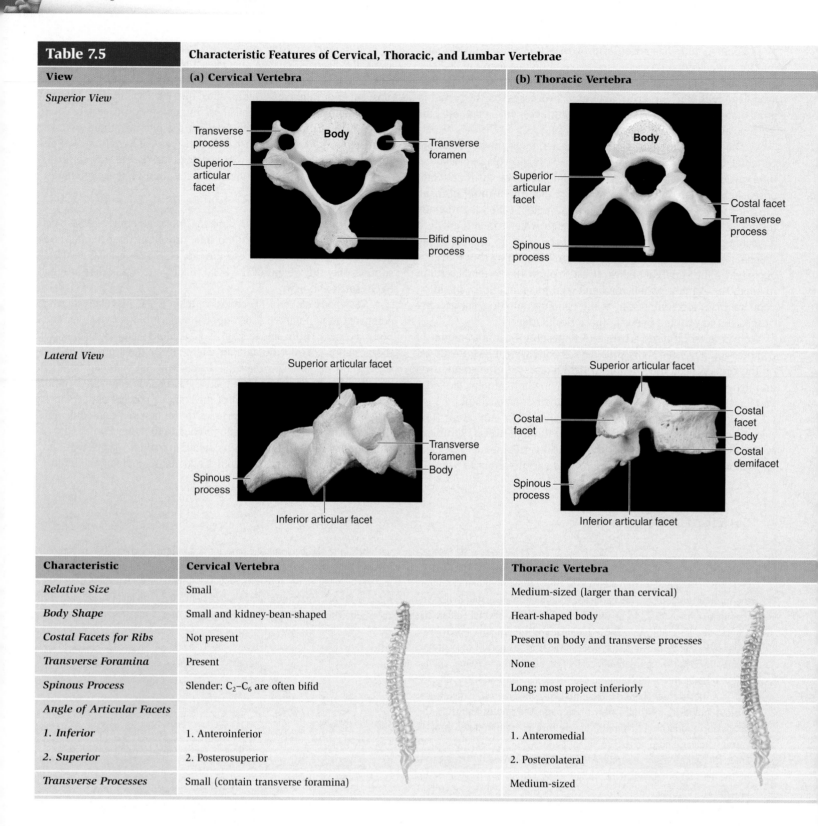

View	(a) Cervical Vertebra	(b) Thoracic Vertebra
Superior View	Transverse process · Body · Superior articular facet · Transverse foramen · Bifid spinous process	Body · Superior articular facet · Costal facet · Transverse process · Spinous process
Lateral View	Superior articular facet · Spinous process · Transverse foramen · Body · Inferior articular facet	Superior articular facet · Costal facet · Costal facet · Body · Costal demifacet · Spinous process · Inferior articular facet

Characteristic	Cervical Vertebra	Thoracic Vertebra
Relative Size	Small	Medium-sized (larger than cervical)
Body Shape	Small and kidney-bean-shaped	Heart-shaped body
Costal Facets for Ribs	Not present	Present on body and transverse processes
Transverse Foramina	Present	None
Spinous Process	Slender: C_2–C_6 are often bifid	Long; most project inferiorly
Angle of Articular Facets		
1. Inferior	1. Anteroinferior	1. Anteromedial
2. Superior	2. Posterosuperior	2. Posterolateral
Transverse Processes	Small (contain transverse foramina)	Medium-sized

The head is a large, heavy structure that is precariously balanced upon the cervical vertebrae. Small muscles keep the head stable. However, if the body changes position suddenly—for example, due to a fall or the impact from a car crash—these "balancing" muscles cannot stabilize the head. A cervical dislocation called **whiplash** may result, characterized by injury to muscles and ligaments and potential injury to the spinal cord.

Atlas (C_1)

The first cervical vertebra, called the **atlas** (at′las), supports the head via its articulation with the occipital condyles of the occipital bone. This vertebrae is named for the Greek mythological figure Atlas, who carried the world on his shoulders. The articulation between the occipital condyles and the atlas, called the atlanto-occipital joint, permits us to nod our heads "yes." The atlas is readily distinguished from the other vertebrae because it lacks a body and a spinous process. Instead, the atlas has lateral masses that are connected by semicircular **anterior** and **posterior arches**, each containing slight protuberances, the **anterior** and **posterior tubercles** (too′ber-kl) (**figure 7.30a**). The atlas has depressed, oval **superior articular facets** that articulate with the occipital condyles of the skull. The atlas also has **inferior articular facets** that articulate with the superior articular facets of the axis. Finally, the atlas has

(c) Lumbar Vertebra

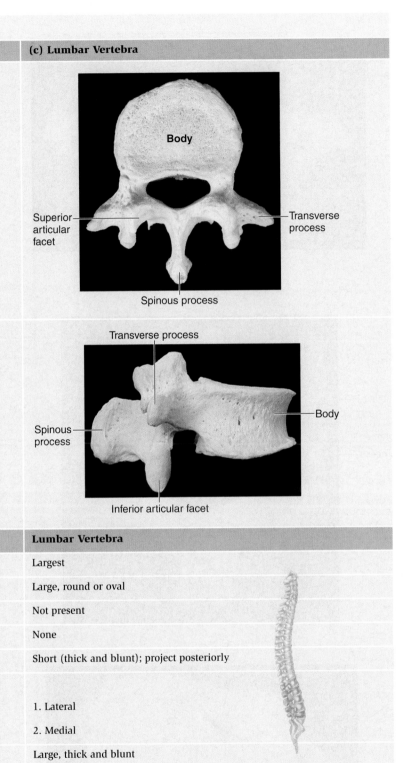

Body	
Superior articular facet	Transverse process
Spinous process	

Transverse process

Spinous process — — Body

Inferior articular facet

Lumbar Vertebra

Largest

Large, round or oval

Not present

None

Short (thick and blunt); project posteriorly

1. Lateral

2. Medial

Large, thick and blunt

an articular **facet for dens** on its anterior arch where it articulates with the dens of the axis.

Axis (C₂) During development, the body of the atlas fuses to the body of the second cervical vertebra, called the **axis** (ak′sis) (figure 7.30*b*). This fusion produces the most distinctive feature of the axis, the prominent **dens**, or *odontoid* (ō-don′toyd; *odont* = tooth) *process*. The dens rests in the **articular facet for dens** of the atlas, where it is held in place by a transverse ligament. The dens acts as a pivot for the rotation of both the atlas and the skull. This articulation between the atlas and axis, called the atlantoaxial joint, permits us to shake our heads "no" (figure 7.30*c*).

Since both the dens and the spinal cord occupy the vertebral foramen at the level of the axis, any trauma that dislocates the dens often results in severe injury. For example, an impact to the head or even severe shaking of a child can dislocate the dens and cause severe damage to the spinal cord. In an adult, a severe blow at or near the base of the skull is often equally dangerous because dislocation of the atlantoaxial joint can force the dens into the base of the brain, with fatal results.

Vertebra Prominens (C₇) The seventh cervical vertebra represents a transition to the thoracic vertebral region and has some features of thoracic vertebrae. The spinous process of C₇ is typically not bifurcated, and it is much larger and longer than the spinous processes of the other cervical vertebrae. This large spinous process is easily seen and palpated through the skin, sometimes appearing as a slight protrusion between the shoulder blades and inferior to the neck. Thus, C₇ is called the **vertebra prominens** (prom′i-nens; prominent).

Thoracic Vertebrae

There are 12 thoracic vertebrae, designated T₁–T₁₂, and each vertebra articulates with the ribs (table 7.5*b*). The thoracic vertebrae lack the mobility of the other vertebrae due to their stabilizing articulation with the ribs. The thoracic vertebrae also lack the transverse foramina and bifid spinous processes of the cervical vertebrae, but they have their own distinctive characteristics. A thoracic vertebra has a heart-shaped body that is larger and more massive than the body of a cervical vertebra. Its spinous process is relatively pointed and long; in some thoracic vertebrae, it angles sharply in an inferior direction.

Thoracic vertebrae are distinguished from all other types of vertebrae by the presence of **costal facets** or **costal demifacets** (dem′ē; half) on the lateral side of the body and on the sides of the transverse processes. A costal facet is a circular depression that articulates with the entire head or tubercle of the rib, while a costal demifacet is a semicircular depression that articulates with either the superior or inferior edge of the head of the rib. The head of the rib articulates with the costal facet on the body of the thoracic vertebra. The tubercle of the rib articulates with the costal facets on the transverse processes of the vertebra.

The thoracic vertebrae vary slightly in terms of their transverse costal facets. Vertebrae T₁–T₁₀ have transverse costal facets on their transverse processes; T₁₁ and T₁₂ lack these transverse costal facets because the eleventh and twelfth ribs do not have tubercles (and thus do not articulate with the transverse processes). The costal facets on the *bodies* of the thoracic vertebrae also display variations:

- The body of vertebra T₁ bears a full costal facet for the first rib and a demifacet for the second rib.
- The bodies of vertebrae T₂–T₈ have two demifacets each: one on the superior edge of the body, and the other on the inferior edge of the body.
- The body of vertebra T₉ has only a superior demifacet for the articulation with the ninth rib.
- The bodies of vertebrae T₁₀–T₁₂ have a single whole facet to articulate with their respective ribs.

Lumbar Vertebrae

The largest vertebrae are the lumbar vertebrae. A typical lumbar vertebra body is thicker than those of all the other vertebrae, and its superior and inferior surfaces are oval rather than heart-shaped (table 7.5*c*). The lumbar vertebrae are distinguished by the features they lack: A lumbar vertebra has neither transverse foramina nor costal facets. The transverse processes are thin and project dorsolaterally. The spinous

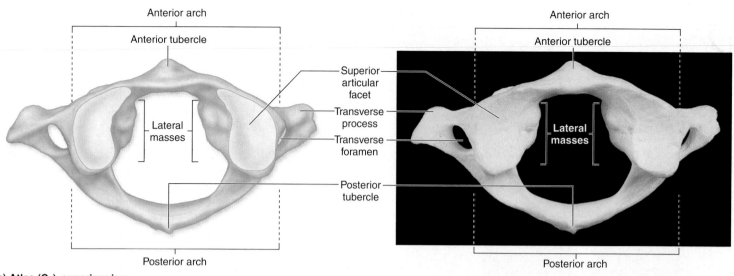

(a) Atlas (C₁), superior view

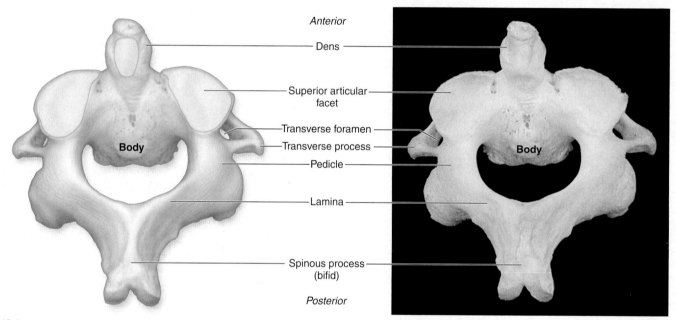

(b) Axis (C₂), posterosuperior view

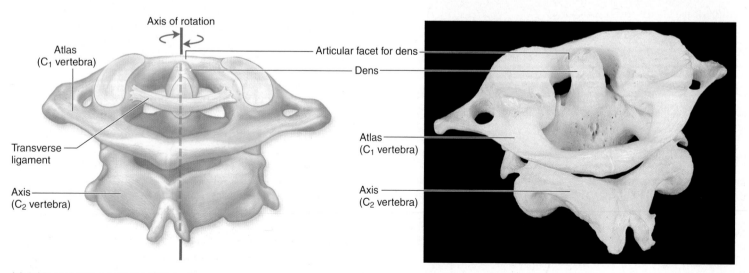

(c) Atlas and axis, posterosuperior view

Figure 7.30

Cervical Vertebrae C₁ and C₂. The atlas (C₁) and the axis (C₂) differ in structure from a typical cervical vertebra. (*a*) Superior view of the atlas. (*b*) Posterosuperior view of the axis. (*c*) The articulation of the atlas and axis, called the atlantoaxial joint, allows partial rotation of the atlas.

processes are thick and project dorsally, unlike the thoracic vertebrae spinous processes, which are long, slender, and point inferiorly.

The lumbar vertebrae bear most of the weight of the body. The thick spinous processes provide extensive surface area for the attachment of inferior back muscles that reinforce or adjust the lumbar curvature.

WHAT DO YOU THINK?

3 You are given a vertebra to identify. It has transverse foramina and a bifid spinous process. Is this a cervical, thoracic, or lumbar vertebra?

Sacrum

The sacrum is an anteriorly curved, somewhat triangular bone that forms the posterior wall of the pelvic cavity. The **apex** of the sacrum is a narrow, pointed portion of the bone that projects inferiorly, whereas the bone's broad superior surface forms its **base**. The lateral sacral curvature is more pronounced in males than in females.

The sacrum is composed of the five fused sacral vertebrae **(figure 7.31)**. These vertebrae start to fuse shortly after puberty and are usually completely fused between ages 20 and 30. The horizontal lines of fusion that remain are called **transverse ridges**. Superiorly, the sacrum articulates with L_5 via a pair of **superior articular processes**, and inferiorly it articulates with the coccyx.

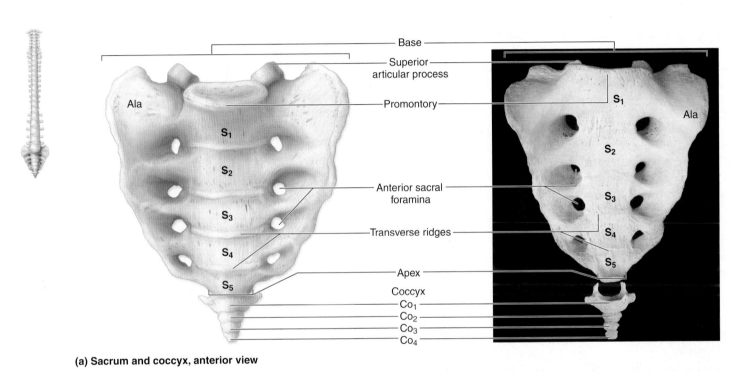

(a) Sacrum and coccyx, anterior view

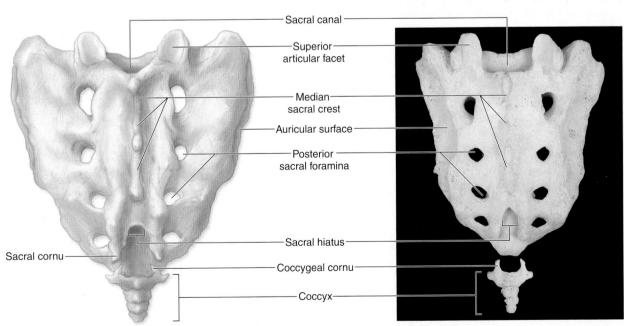

(b) Sacrum and coccyx, posterior view

Figure 7.31

Sacrum and Coccyx. The sacrum is formed by the fusion of five sacral vertebrae, and the coccyx is formed by the fusion of four coccygeal vertebrae. *(a)* Anterior and *(b)* posterior views show the sacrum and the coccyx.

The vertebral canal becomes much narrower and continues through the sacrum as the **sacral canal**. The sacral canal terminates in an inferior opening called the **sacral hiatus** (hī-ā′tŭs; *hio* = to yawn). The sacral hiatus represents an area where the laminae of the last sacral vertebra failed to fuse. On either side of the sacral hiatus are bony projections called the **sacral cornua**.

The anterosuperior edge of the first sacral vertebra bulges anteriorly into the pelvic cavity and is called the **promontory**. Four anterior lines cross the anterior surface of the sacrum, marking the lines of fusion of the sacral vertebrae. The paired **anterior sacral foramina** permit the passage of nerves to the pelvic organs. A dorsal ridge, termed the **median sacral crest**, is formed by the fusion of the spinous processes of individual sacral vertebrae. Also on the dorsal surface of the sacrum are four pairs of openings for spinal nerves, called the **posterior sacral foramina**. On each lateral surface of the sacrum is the **ala** (meaning "wing"). On the lateral surface of the ala is the **auricular surface**, which marks the site of articulation with the os coxae of the pelvic girdle, forming the strong, nearly immovable **sacroiliac** (sā-krō-il′ē-ak) **joint**.

Coccyx

Four small coccygeal vertebrae fuse to form the **coccyx** (figure 7.31*a*). The individual vertebrae begin to fuse by about age 25. The coccyx is an attachment site for several ligaments and some muscles. The first and second coccygeal vertebrae have unfused vertebral arches and transverse processes. The prominent laminae of the first coccygeal vertebrae are known as the **coccygeal cornua**, which curve to meet the sacral cornua. Fusion of the coccygeal vertebrae is not complete until adulthood. In males, the coccyx tends to project anteriorly, but in females it tends to project more inferiorly, so as not to obstruct the birth canal. In very old individuals, the coccyx may fuse with the sacrum.

WHAT DID YOU LEARN?

12 Identify the five vertebral regions in order, from superior to inferior.

13 If an athlete suffers a hairline fracture at the base of the dens, what bone is fractured, and where is it located?

14 Compare the locations and functions of transverse foramina, intervertebral foramina, and the vertebral foramen.

Thoracic Cage

Key topic in this section:

■ General structure of the sternum and the ribs

The bony framework of the chest is called the **thoracic cage**; it consists of the thoracic vertebrae posteriorly, the ribs laterally, and the sternum anteriorly **(figure 7.32)**. The thoracic cage acts as a protective framework around vital organs, including the heart, lungs, trachea, and esophagus. It also provides attachment points for many muscles supporting the pectoral girdles (the bones that hold the upper limb in place), the chest, the neck, the shoulders, the back, and the muscles involved in respiration.

Sternum

The adult **sternum** (ster′nŭm; *sternon* = the chest), also called the "breastbone," is a flat bone that forms in the anterior midline of the thoracic wall. Its shape has been likened to that of a sword. The sternum is composed of three parts: the manubrium, the body, and the xiphoid process.

The **manubrium** (mă-noo′brē-ŭm) is the widest and most superior portion of the sternum (the "handle" of the bony sword). Two **clavicular notches** articulate the sternum with the left and right clavicles. The shallow superior indentation between the clavicular notches is called the **suprasternal notch** (or *jugular notch*). A single pair of **costal notches** represent articulations for the first ribs' costal cartilages.

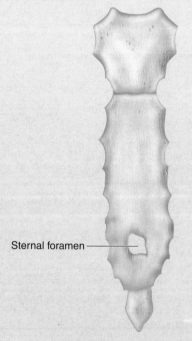

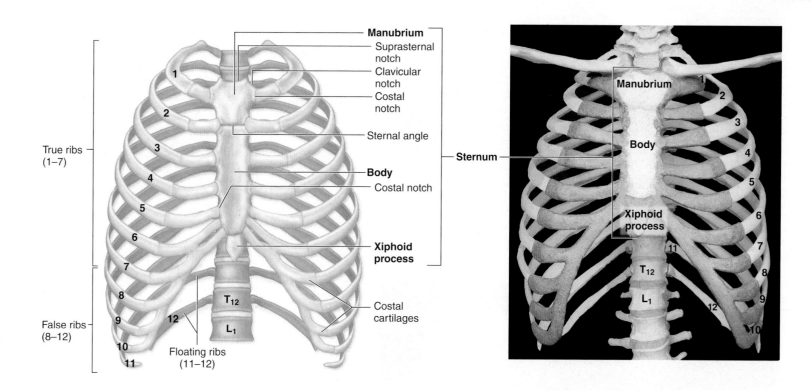

Figure 7.32

Thoracic Cage. Drawing and photograph show anterior views of the bones of the thoracic cage, which protect and enclose the organs in the thoracic cavity.

The **body** (or *gladiolus*) is the longest part of the sternum and forms its bulk (the "blade" of the bony sword). Individual costal cartilages from ribs 2–7 are attached to the body at indented articular costal notches. The body and the manubrium articulate at the **sternal angle,** a horizontal ridge that may be palpated under the skin. The sternal angle is an important landmark in that the costal cartilages of the second ribs attach there; thus, it may be used to count the ribs.

The **xiphoid** (zi'foyd; *xiphos* = sword) **process** represents the very tip of the "sword blade." This small, inferiorly pointed projection is cartilaginous and often doesn't ossify until after age 40. The connection of the xiphoid process to the body of the sternum may be broken by an impact or strong pressure. The resulting internal projection of bone can severely damage the heart or liver.

Ribs

The ribs are elongated, curved, flattened bones that originate on or between the thoracic vertebrae and end in the anterior wall of the thorax (figure 7.32). Both males and females have the same number of ribs—12 pairs. Ribs 1–7 are called **true ribs**. At the anterior body wall, the true ribs connect individually to the sternum by separate cartilaginous extensions called **costal** (kos'tăl; *costa* = rib) **cartilages.** The smallest true rib is the first.

Ribs 8–12 are called **false ribs** because their costal cartilages do not attach directly to the sternum. The costal cartilages of ribs 8–10 fuse to the costal cartilage of rib 7 and thus indirectly articulate with the sternum. The last two pairs of false ribs (ribs 11 and 12) are called **floating ribs** because they have no connection with the sternum.

The vertebral end of a typical rib articulates with the vertebral column at the **head** (or *capitulum*). The articular surface of the head is divided into **superior** and **inferior articular facets** by an interarticular **crest (figure 7.33a)**. The surfaces of these facets articulate with the costal facets on the bodies of the thoracic vertebrae. The **neck** of the rib lies between the head and the tubercle. The **tubercle** (or *tuberculum*) of the rib has an articular facet for the costal facet on the transverse process of the thoracic vertebra. Figure 7.33b,c illustrates how most of the ribs articulate with the thoracic vertebrae. Rib 1 articulates with vertebra T_1. The head of the rib articulates at a costal facet on the body, and the tubercle of the rib articulates at a transverse costal facet on the transverse process of T_1.

Ribs 2–9 articulate with vertebrae T_2–T_9. Each of these vertebrae has two demifacets on the lateral side of its body. The superior articular facet on the head of the rib articulates with the more superior vertebra, and the inferior articular facet articulates with the more inferior

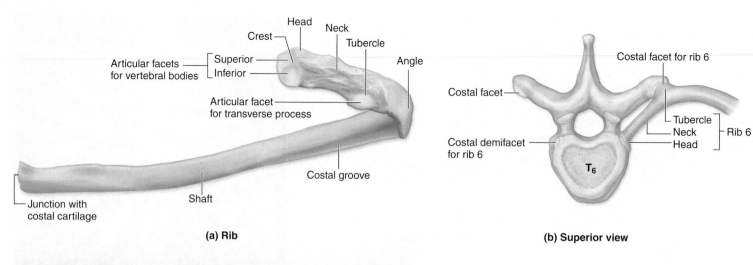

(a) Rib

(b) Superior view

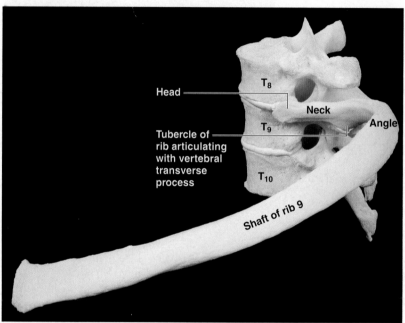

(c) Lateral view

Figure 7.33

Rib Anatomy and Articulation with Thoracic Vertebrae.
Paired ribs attach to thoracic vertebrae posteriorly and extend
anteroinferiorly to the anterior chest wall. (*a*) Features of ribs
2–10. (*b*) Superior and (*c*) lateral views show the articulation
of a rib with a vertebra.

vertebra. For example, the superior articular facet on the head of rib
2 articulates with the inferior costal demifacet on the body of T_1, and
the inferior articular facet on the head of rib 2 articulates with the
superior costal demifacet on the body of T_2. The tubercle of each rib
articulates with the transverse costal facet on the transverse process of
each vertebra. For example, the tubercle of rib 3 articulates with the
transverse costal facet on the transverse process of T_3.

Ribs 10–12 articulate with vertebrae T_{10}–T_{12}. Each of these vertebrae has a whole costal facet on the lateral body to articulate with
the head of its respective rib. Vertebra T_{10} also has transverse costal
facets on its transverse processes to articulate with the tubercle of
each rib 10. Ribs 11 and 12 do not have tubercles, so there are no
costal facets on the transverse processes of T_{11} and T_{12}.

The **angle** (*border*) of the rib indicates the site where the tubular **shaft** (or *body*) begins curving anteriorly toward the sternum. A
prominent **costal groove** along its inferior internal border marks the
path of nerves and blood vessels to the thoracic wall.

WHAT DID YOU LEARN?

15 What are the three components of the sternum, and
what ribs articulate directly with the sternum?

16 The tubercle of a rib articulates with what specific vertebral
feature?

Aging of the Axial Skeleton

Key topic in this section:

■ How the axial skeleton changes as we grow and mature

The general changes in the axial skeleton that take place with age have been described throughout this chapter. As mentioned before, many bones fuse when we age. Also, skeletal mass and density become reduced, and bones often become more porous and brittle, a condition known as osteoporosis (see chapter 6). Osteoporotic bones are susceptible to fracture, which is why elderly individuals are at greater risk for bone fractures.

In addition, articulating surfaces deteriorate, contributing to arthritic conditions. Whereas a younger individual has smooth, well-formed articular surfaces, those of an older individual may be rough, worn away, or covered with bony, spurlike growths, making movement at these surfaces painful and difficult. These changes begin in early childhood and continue throughout life.

Development of the Axial Skeleton

Key topic in this section:

■ Major events in skeletal development prior to birth

As mentioned in chapter 6, bone forms by either intramembranous ossification within a mesenchyme layer or endochondral ossification from hyaline cartilage models. **Figure 7.34** shows which bones are formed by which type of ossification.

The following bones of the skull are formed by intramembranous ossification: the flat bones of the skull (e.g., parietal, frontal, and part of the occipital bones), the zygomatic bones, the maxillae, and the mandible. Most of the bones at the base of the cranium (e.g., the sphenoid, part of the temporal bone, and part of the occipital bone) are formed by endochondral ossification. These bones become rather well formed by 12–20 weeks of development.

Almost all of the remaining bones of the skeleton form through endochondral ossification. (The exception is the clavicle, which is formed from a central membranous ossification center while its ends are formed from endochondral ossification centers.)

The sternum develops from left and right cartilaginous **sternal bars (figure 7.35)**. The sternal bars meet along the midline and start to fuse, beginning in the seventh week. Fusion commences at the superior end and finishes at the inferior end by the ninth week. Within these sternal bars, multiple bony ossification centers will develop.

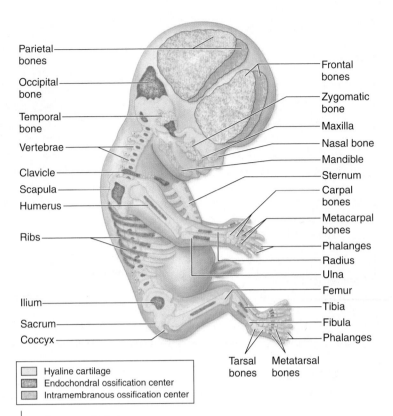

Hyaline cartilage
Endochondral ossification center
Intramembranous ossification center

Figure 7.34

Development of the Axial Skeleton.   Many centers of ossification for the axial skeleton are readily observed by the tenth week of development.

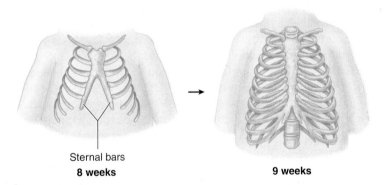

Sternal bars
8 weeks

9 weeks

Figure 7.35

Sternum Development.   The sternum forms from the fusion of sternal bars.

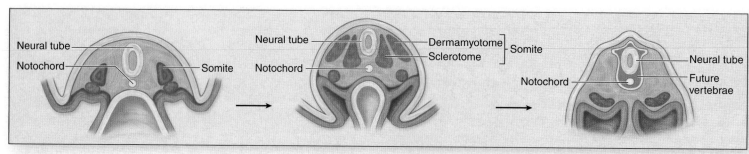

(a) Week 4: Sclerotome portions of somites surround the neural tube and form the future vertebrae and ribs.

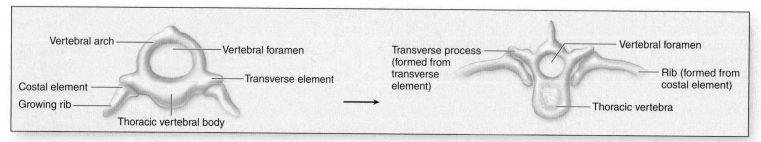

(b) Week 5: Transverse elements of thoracic vertebrae form transverse processes, while costal elements of thoracic vertebrae form ribs.

Figure 7.36

Rib and Vertebrae Development. Ribs and vertebrae form from portions of somites called sclerotomes.

Rib and vertebrae development are intertwined, as shown in **figure 7.36**. Recall from chapter 3 that blocks of mesoderm called **somites** are located on either side of the developing neural tube. A portion of each somite, called a **sclerotome** (sklēr′ō-tōm), separates from the dermamyotome portion of the somite and is the origin of the vertebrae and ribs. The sclerotomes start to surround the neural tube during the fourth week of development. The sclerotomes later become cartilaginous and form portions of the vertebrae, including the **costal element** (or *costal process*) and the **transverse element**. For most vertebrae, the costal element and transverse element fuse to form the transverse process of the vertebra. However, in the thoracic region of the spinal cord, the costal element and transverse element remain separate. The transverse element in the developing thoracic vertebra forms the transverse process. The costal elements of the thoracic vertebrae elongate and form the ribs. This elongation of the costal elements starts during the fifth week of development, and ossification of the ribs occurs during the fetal period.

Within the cartilage of the developing vertebrae, multiple bony ossification centers form. At birth, only the body and the vertebral arch have ossified; it isn't until puberty that other secondary ossification centers, such as those in the tips of spinous and transverse processes, form. All secondary ossification centers fuse with the primary vertebral ossification centers by about age 25.

CLINICAL VIEW

Variations in Rib Development

Variations in rib development are not uncommon. For example, in one out of every 200 people, the costal element of the seventh cervical vertebra elongates and forms a rudimentary **cervical rib**. Cervical ribs may compress the artery and nerves extending toward the upper limb, producing tingling or pain. If a cervical rib continuously produces these symptoms, it is usually removed surgically. Less commonly, an extra pair of ribs may form from the costal elements of the first lumbar vertebra. These ribs tend to be asymptomatic.

Some individuals lack a pair of twelfth ribs, because their costal elements from the twelfth thoracic vertebra failed to elongate. Finally, **bifid ribs** occur in 1.2% of the world's population (and up to 8.4% of Samoans). A bifid rib splits into two separate portions when it reaches the sternum. Like most other variations in rib development, bifid ribs are typically asymptomatic and may be discovered only incidentally on a chest x-ray.

CLINICAL TERMS

plagiocephaly Asymmetric or unusual head shape, caused either by premature suture closure or long-term positional pressures placed on the skull.

sinusitis Inflammation of the mucous membrane of one or more paranasal sinuses.

spinal fusion Medical procedure used to stabilize part of the vertebral column; bone chips are inserted surgically as a tissue graft after a vertebra has been fractured or a disc prolapsed.

CHAPTER SUMMARY

Skull 173

- The axial skeleton includes the bones of the skull, the vertebral column, and the thoracic cage.
- The appendicular skeleton includes the bones of the pectoral and pelvic girdles and the upper and lower limbs.
- Cranial bones enclose the cranial cavity. Facial bones protect and support the entrances to the digestive and respiratory systems.

Views of the Skull and Landmark Features   174

- Anterior, superior, posterior, lateral, sagittal sectional, inferior (basal), and internal views show specific bones, foramina, processes, and bone landmarks.

Sutures   181

- Sutures are immovable joints that form boundaries between skull bones.

Bones of the Cranium   183

- The frontal bone forms the forehead and the superior region of the orbit.
- Paired parietal bones form part of the superolateral surfaces of the cranium.
- Paired temporal bones are on the lateral sides of the cranium.
- The occipital bone forms part of the base of the skull.
- The sphenoid bone contributes to the cranial base.
- The ethmoid bone forms part of the orbit and the roof of the nasal cavity.

Bones of the Face   191

- The paired zygomatic bones form the cheeks.
- The paired lacrimal bones are located in the anteromedial portion of each orbit.
- The paired nasal bones form the bridge of the nose.
- The vomer forms the inferior portion of the nasal septum.
- The paired inferior nasal conchae attach to the lateral walls of the nasal cavity.
- The paired palatine bones form the posterior portion of the hard palate.
- The paired maxillae form the upper jaw and most of the hard palate.
- The mandible is the lower jaw.

Nasal Complex   196

- The nasal complex is composed of bones and cartilage that enclose the nasal cavities and the paranasal sinuses.

Paranasal Sinuses   197

- Paranasal sinuses are hollow cavities in the maxillae, ethmoid, frontal, and sphenoid bones that connect with the nasal cavity.

Orbital Complex   197

- Seven bones form the orbit: the maxilla, frontal, lacrimal, ethmoid, sphenoid, palatine, and zygomatic bones.

Bones Associated with the Skull   199

- Auditory ossicles (malleus, incus, and stapes) are three tiny ear bones housed in each temporal bone.
- The hyoid bone serves as a base for the attachment of several tongue and larynx muscles.

Sex Differences in the Skull   199

- Female skulls tend to be more gracile, have more pointed (versus squared-off) chins, and have sharper orbital rims.
- Male skulls tend to be more robust and have larger features as well as squared-off chins.

Aging of the Skull   199

- Fontanelles permit the skulls of infants and young children to expand as the brain grows.

Vertebral Column   202

- The vertebral column is composed of 26 vertebrae.

Divisions of the Vertebral Column   202

- There are 7 cervical vertebrae, 12 thoracic vertebrae, 5 lumbar vertebrae, the sacrum, and the coccyx.

(continued on next page)

CHAPTER SUMMARY *(continued)*

Vertebral Column (continued) 202	**Spinal Curvatures 203** ■ The adult spinal column exhibits four curvatures. The thoracic and sacral curvatures are called primary (accommodation) curves, and the cervical and lumbar curvatures are termed secondary (compensation) curves. **Vertebral Anatomy 204** ■ A typical vertebra has a body and a posterior vertebral arch. The vertebral arch is formed by pedicles and laminae. The vertebral foramen houses the spinal cord. ■ Between adjacent vertebrae are fibrocartilaginous intervertebral discs. ■ Cervical vertebrae have transverse foramina and bifid spinous processes. ■ Thoracic vertebrae have heart-shaped bodies, long spinous processes, costal facets on the body, and transverse processes that articulate with the ribs. ■ The lumbar vertebrae are the most massive.
Thoracic Cage 210	■ The skeleton of the thoracic cage is composed of the thoracic vertebrae, the ribs, and the sternum. **Sternum 210** ■ The sternum consists of a superiorly placed manubrium, a middle body, and an inferiorly placed xiphoid process. **Ribs 211** ■ Ribs 1–7 are called true ribs, and ribs 8–12 are called false ribs (while ribs 11–12 are also known as floating ribs).
Aging of the Axial Skeleton 213	■ Skeletal mass and density are often reduced with age, and articulating surfaces deteriorate, leading to arthritis.
Development of the Axial Skeleton 213	■ The flat bones of the skull are formed from intramembranous ossification, whereas almost all other bones of the skull are formed from endochondral ossification. ■ The sternum forms from two cartilaginous sternal bars that start to fuse during the eighth week of development. ■ Vertebrae and ribs are formed from the sclerotomes of developing somites.

CHALLENGE YOURSELF

Matching

Match each numbered item with the most closely related lettered item.

_____ 1. supraorbital foramen   a. mandible

_____ 2. foramen magnum   b. frontal bone

_____ 3. petrous part   c. maxillae

_____ 4. sella turcica   d. cervical vertebrae

_____ 5. cribriform plate   e. occipital bone

_____ 6. mental protuberance   f. sternum

_____ 7. transverse foramina   g. thoracic vertebrae

_____ 8. costal demifacets   h. temporal bone

_____ 9. xiphoid process   i. ethmoid bone

_____ 10. upper jaw   j. sphenoid bone

Multiple Choice

Select the best answer from the four choices provided.

_____ 1. Which bones form the hard palate?
    a. mandible and maxillae
    b. palatine bones and mandible
    c. palatine bones and maxillae
    d. maxillae only

_____ 2. The bony portion of the nasal septum is formed by the
    a. perpendicular plate of the ethmoid bone and vomer.
    b. perpendicular plate of the ethmoid bone only.
    c. nasal bones and perpendicular plate of the ethmoid bone.
    d. vomer and sphenoid bones.

_____ 3. The mandible articulates with the _____ bone.
    a. occipital
    b. frontal
    c. temporal
    d. parietal

_____ 4. Some muscles that control the tongue and larynx are attached to the
    a. maxillae.
    b. cervical vertebrae.
    c. hyoid bone.
    d. malleus bone.

_____ 5. The frontal and parietal bones articulate at the _____ suture.
    a. coronal
    b. sagittal
    c. lambdoid
    d. squamous

_____ 6. The compression of an infant's skull bones at birth is facilitated by spaces between unfused cranial bones called
   a. ossification centers.
   b. fortanelles.
   c. foramina.
   d. fossae.

_____ 7. All of the following are openings in the sphenoid *except* the
   a. foramen rotundum.
   b. hypoglossal canal.
   c. foramen spinosum.
   d. optic canal.

_____ 8. Each temporal bone articulates with the
   a. frontal, temporal, occipital, and parietal bones only.
   b. frontal, zygomatic, occipital, parietal and sphenoid bones.
   c. occipital, zygomatic, sphenoid, and parietal bones, and the mandible.
   d. frontal, occipital, temporal, sphenoid, and parietal bones.

_____ 9. Most _____ vertebrae have a long spinous process that is angled inferiorly.
   a. cervical
   b. thoracic
   c. lumbar
   d. sacral

_____ 10. The clavicles articulate with the _____ of the sternum.
   a. manubrium
   b. xiphoid process
   c. body
   d. angle

Content Review

1. Explain the primary difference between a facial bone and a cranial bone.

2. What are sutures, and how do they affect skull shape and growth?

3. With which bones does the occipital bone articulate?

4. What are the boundaries of the middle cranial fossa?

5. Compare the superior, middle, and inferior nasal conchae. Are they part of another bone? Where in the nasal complex are they found?

6. Identify the seven bones that form the orbit, and discuss their arrangement.

7. What are the functions of the paranasal sinuses?

8. Identify the first two cervical vertebrae, describe their unique structures, and discuss the functions these vertebrae perform in spinal mobility.

9. Identify the region of the vertebral column that is most likely to experience a herniated disc, and discuss the causes of this problem.

10. Describe similarities and differences among true, false, and floating ribs.

Developing Critical Reasoning

1. Two patients see a doctor with complaints about lower back pain. The first is a construction worker who lifts bulky objects every day, and the second is an overweight teenager. Is there a common cause for these complaints? What might the doctor recommend?

2. Paul viewed his newborn daughter through the nursery window at the hospital and was distressed because the infant's skull was badly misshapen. A nurse told him not to worry—the shape of the infant's head would return to normal in a few days. What caused the misshapen skull, and what anatomic feature of the neonatal skull allows it to return to a more rounded shape?

3. A forensic anthropologist was asked to determine the sex of a skull found at a crime scene. How would she be able to discern this information?

ANSWERS TO "WHAT DO YOU THINK?"

1. Male sex hormones and increased growth beginning at puberty cause the skull to become more robust, with more prominent features and a more squared-off jaw.

2. A completely straight vertebral column would not be as well adapted for weight-bearing as a vertebral column with spinal curvatures. The spinal curvatures support the weight of the body better by bringing that weight in line with the body axis and thus helping us walk upright. (Compare the spinal curvatures of a human to those of an animal that does not normally walk upright, such as a chimpanzee.)

3. A cervical vertebra typically has transverse foramina and a bifid spinous process.

Visit the McKinley/O'Loughlin *Human Anatomy,* 2e website at aris.mhhe.com

8

Anatomy &
Physiology | REVEALED®
aprevealed.com

SKELETAL SYSTEM

Appendicular Skeleton

One benefit of space travel is that it has helped advance our knowledge of human anatomy. We are excited and amazed by video showing astronauts in space running on treadmills, riding stationary bicycles, jumping rope, and doing various other exercises. For astronauts—and for all of us—exercise is essential for maintaining bone mass and strength as well as muscle tone and strength. When we exercise, our contracting muscles apply stress to the bones to which they are attached, thereby strengthening the bone and ultimately preventing it from becoming thin and brittle. Regular exercise prevents degenerative changes in the skeleton and helps us avoid serious health problems later in life.

When we exercise, we move the bones of the **appendicular skeleton,** which includes the bones of the upper and lower limbs, and the girdles of bones that hold and attach the upper and lower limbs to the axial skeleton **(figure 8.1)**. The pectoral girdle consists of bones that hold the upper limbs in place, while the pelvic girdle consists of bones that hold the lower limbs in place. In this chapter, we examine the specific components of the appendicular skeleton and explore their interactions with other systems, such as the muscular and cardiovascular systems. Be sure to review figure 6.17 regarding bone feature names before you proceed.

Study Tip!

Many bones and bony features may be palpated (felt) underneath the skin. As you hold a bone in lab, try to palpate the same bone on your own body. In this way, you will understand how the bone is positioned, how it associates with other bones, and how it moves in a living body. In effect, you can use your body as a "bone study guide."

Pectoral Girdle

Key topics in this section:

- Bones of the pectoral girdle and their functions
- Bone surface features in the pectoral girdle

The left and right **pectoral** (pek'tŏ-răl; *pectus* = breastbone) **girdles** (ger'dl) articulate with the trunk, and each supports one upper limb. A pectoral girdle consists of two bones: the clavicle (collarbone) and the scapula (shoulder blade). The pectoral girdle also provides attachment sites for many muscles that move the limb, and it promotes upper limb mobility in two ways: (1) Because the scapula is not directly attached to the axial skeleton, it moves freely across the posterior surface of the thorax, permitting the arm to move with it, and (2) the shallow cavity of the shoulder joint permits a wide range of movement of the upper limb.

WHAT DO YOU THINK?

1 Why is the clavicle commonly known as the "collarbone"? Based on this layman's term, can you figure out where the clavicle is located in your body?

Clavicle

The **clavicle** (klav'i-kl; *clavis* = key) is an S-shaped bone that extends between the manubrium of the sternum and the acromion of the scapula **(figure 8.2)**. It is the only direct connection between the pectoral girdle and the axial skeleton. Its **sternal end** (medial end) is roughly pyramidal in shape and articulates with the manubrium of the sternum, forming the sternoclavicular joint. The **acromial end** (lateral end) of the clavicle is broad and flattened. The acromial end articulates with the acromion of the scapula, forming the acromioclavicular joint.

You can palpate your own clavicle by first locating the superior aspect of your sternum and then moving your hand laterally. The curved bone you feel under your skin, and close to the collar of your shirt, is your clavicle.

The superior surface of the clavicle is relatively smooth, but the inferior surface is marked by grooves and ridges for muscle and ligament attachment. On the inferior surface, near the acromial end, is a rough tuberosity called the **conoid** (kō'noyd; *konoeides* = cone-shaped) **tubercle.** The inferiorly located prominence at the sternal end of the clavicle is called the **costal tuberosity**.

Scapula

The **scapula** (skap'ū-lă) is a broad, flat, triangular bone that forms the "shoulder blade" **(figure 8.3)**. You can palpate your scapula by putting your hand on your superolateral back region and moving your upper limb; the bone you feel moving is the scapula. Several large projections extend from the scapula and provide surface area for muscle and ligament attachments. The **spine** of the scapula is a ridge of bone on the posterior aspect of the scapula. It is easily palpated under the skin. The spine is continuous with a larger, posterior process called the **acromion** (ă-krō'mē-on; *akron* = tip, *omos* = shoulder), which forms the bony tip of the shoulder. Palpate the superior region of your shoulder; the prominent bump you feel is the acromion. The acromion articulates with the acromial end of the clavicle. The **coracoid** (kōr'ă-koyd; *korakodes* = like a crow's beak) **process** is the smaller, more anterior projection.

CLINICAL VIEW

Fracture of the Clavicle

The clavicle can fracture relatively easily because it is not strong and cannot resist stress. In addition, the sternoclavicular joint is incredibly strong, so if stress is placed on both the clavicle and the joint, the clavicle will fracture before the joint is damaged. A direct blow to the middle part of the clavicle, a fall onto the lateral border of the shoulder, or use of the arms to brace against a forward fall is often stress enough to fracture the clavicle. Because the clavicle has an anterior and posterior curve along its length between the medial and lateral edges, severe stress to the mid-region of the bone usually results in an anterior fracture. A posterior fracture may be more serious because bone splinters can penetrate the subclavian artery and vein, which lie immediately posterior and inferior to the clavicle and are the primary blood vessels supplying the upper limb.

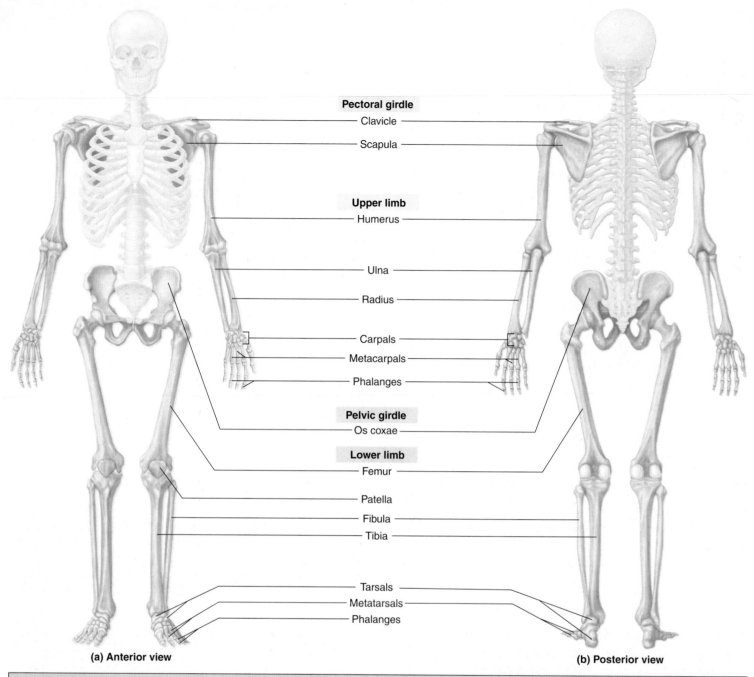

(a) Anterior view

(b) Posterior view

Bones of the Appendicular Skeleton (63 bones per each side of the body, 126 bones total)			
Pectoral girdles (2 bones per each girdle, 4 bones total)	Clavicle (2)	**Pelvic girdles (1 bone per each girdle, 2 bones total)**	Os coxae (2) Ilium, ischium, and pubis bones fuse in early adolescence
	Scapula (2)		
Upper limbs (30 bones per each upper limb, 60 bones total)	Humerus (2)	**Lower limbs (30 bones per each lower limb, 60 bones total)**	Femur (2)
	Radius (2)		Patella (2)
	Ulna (2)		Tibia (2)
	Carpals (16) Scaphoid (2), lunate (2), triquetrum (2), pisiform (2), trapezium (2), trapezoid (2), capitate (2), hamate (2)		Fibula (2)
			Tarsals (14) Calcaneus (2), talus (2), navicular (2), cuboid (2), medial cuneiform (2), intermediate cuneiform (2), lateral cuneiform (2)
	Metacarpals (10)		
	Phalanges (28) Proximal phalanx (10), middle phalanx (8), distal phalanx (10)		Metatarsals (10)
			Phalanges (28) Proximal phalanx (10), middle phalanx (8), distal phalanx (10)

Figure 8.1

Appendicular Skeleton.   (*a*) Anterior and (*b*) posterior views show the pectoral and pelvic girdles and the bones of the upper and lower limbs, all of which make up the appendicular skeleton. A table summarizes the bones of each region.

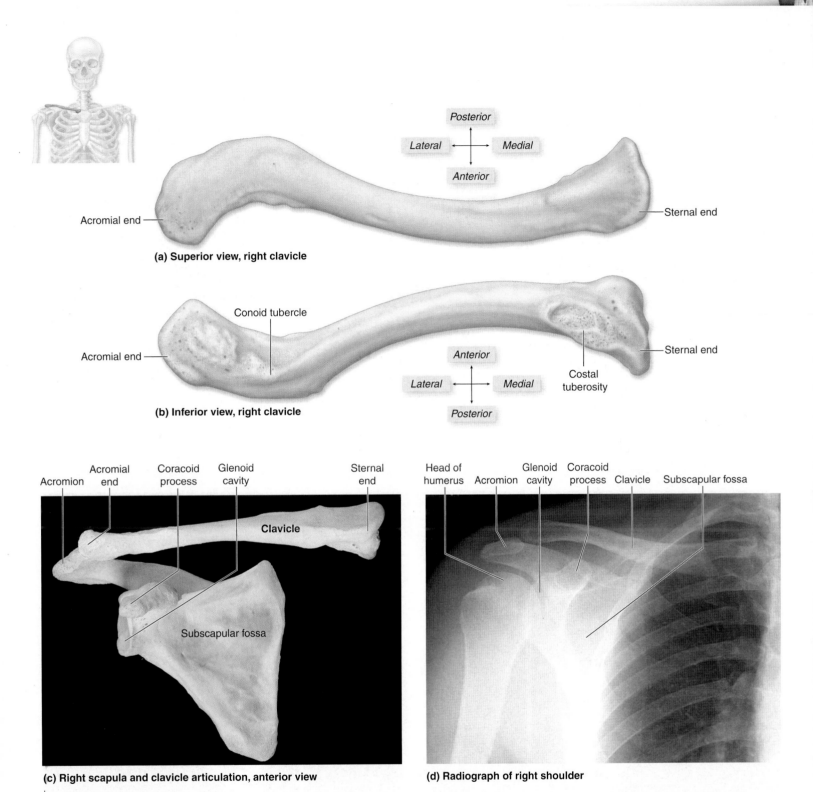

(a) Superior view, right clavicle

(b) Inferior view, right clavicle

(c) Right scapula and clavicle articulation, anterior view

(d) Radiograph of right shoulder

Figure 8.2

Clavicle.   The S-shaped clavicle is the only direct connection between the pectoral girdle and the axial skeleton. *(a)* Superior and *(b)* inferior views of the right clavicle. *(c)* Anterior view of an articulated right clavicle and scapula. *(d)* A radiograph of an articulated right clavicle and scapula.

The triangular shape of the scapula forms three sides, or borders. The **superior border** is the horizontal edge of the scapula superior to the spine of the scapula; the **medial border** (also called the *vertebral border*) is the edge of the scapula closest to the vertebrae; and the **lateral border** (also called the *axillary border*) is closest to the axilla (armpit). A conspicuous **suprascapular notch**

(which in some individuals is a **suprascapular foramen**) in the superior border provides passage for the suprascapular nerve.

Between these borders are the superior, inferior, and lateral angles. The **superior angle** is the pointed part of the scapula between the superior and medial borders, while the **inferior angle** is located between the medial and lateral borders. The

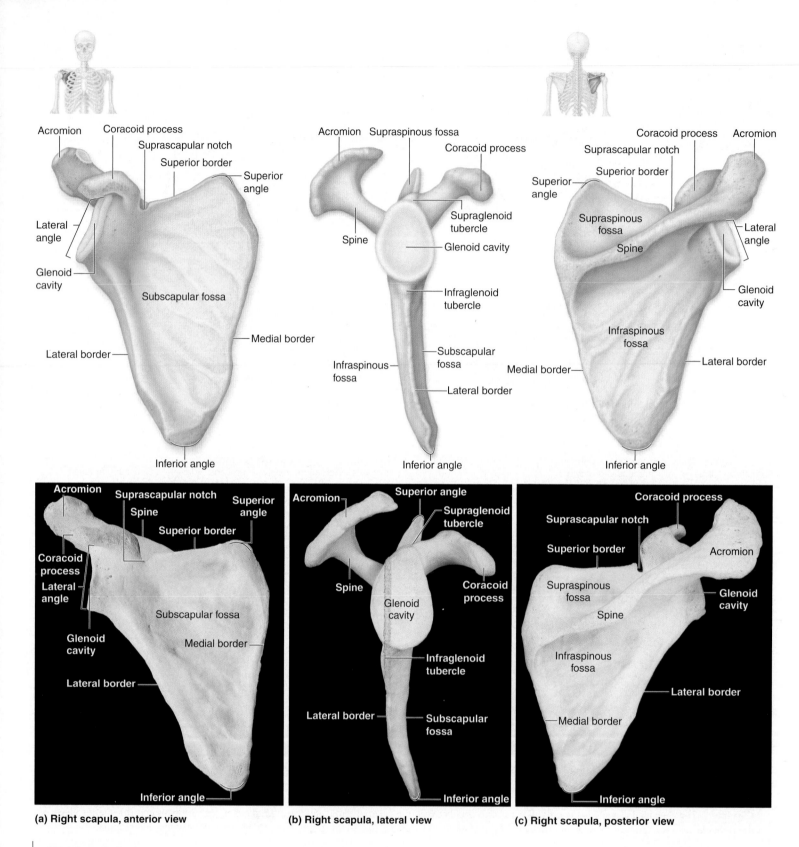

(a) Right scapula, anterior view

(b) Right scapula, lateral view

(c) Right scapula, posterior view

Figure 8.3

Scapula.   The upper limb articulates with the pectoral girdle at the scapula, as shown in *(a)* anterior, *(b)* lateral, and *(c)* posterior views.

lateral angle is composed primarily of the cup-shaped, shallow **glenoid** (glēn'oyd; resembling a socket) **cavity**, or *glenoid fossa*, which articulates with the humerus, the bone of the arm. Tubercles (too'ber-kl) on the superior and inferior edges of the glenoid cavity serve as attachment sites for the muscles that position the shoulder and arm. Near the superior edge of the glenoid cavity is the

supraglenoid tubercle, and near the inferior edge is the **infraglenoid tubercle**.

The scapula has several flattened regions of bone that provide surfaces for the attachment of some of the rotator cuff muscles, which help stabilize and move the shoulder joint. The broad, relatively smooth, anterior surface of the scapula is called the subscapular (sŭb-skap′ū-lăr; *sub* = under) **fossa** (fos′ă; pl., *fossae*, fos′ē). It is slightly concave and relatively featureless. A large muscle called the subscapularis overlies this fossa. The spine subdivides the posterior surface of the scapula into two shallow depressions, or fossae. The depression superior to the spine is the **supraspinous** (soo-pră-spī′nŭs; *supra* = above) **fossa**; inferior to the spine is a broad, extensive surface called the **infraspinous fossa.** The supraspinatus and infraspinatus muscles, respectively, occupy these fossae.

WHAT DID YOU LEARN?

1 Which scapular angle contains the glenoid cavity?

Upper Limb

Key topic in this section:
- Bones of the upper limb and their prominent markings.

The upper limb is composed of many long and some short bones, which articulate to provide great movement. Each upper limb contains a total of 30 bones:

- 1 humerus, located in the brachium region
- 1 radius and 1 ulna, located in the antebrachium region
- 8 carpal bones, which form the wrist
- 5 metacarpal bones, which form the palm of the hand
- 14 phalanges, which form the fingers

Humerus

The **humerus** (hū′mer-ŭs) is the longest and largest upper limb bone **(figure 8.4)**. Its proximal end has a hemispherical **head** that articulates with the glenoid cavity of the scapula. Adjacent to the head are two tubercles. The prominent **greater tubercle** is positioned more laterally and helps form the rounded contour of the shoulder. The **lesser tubercle** is smaller and located more anteromedially. Between the two tubercles is the **intertubercular sulcus** (or *bicipital sulcus*), a depression that contains the tendon of the long head of the biceps brachii muscle.

Between the tubercles and the head of the humerus is the **anatomical neck,** an almost indistinct groove that marks the location of the former epiphyseal plate. The **surgical neck** is a narrowing of the bone immediately distal to the tubercles, at the transition from the head to the shaft. This feature is called the "surgical" neck because it is a common fracture site.

The **shaft** of the humerus has a roughened area, termed the **deltoid** (del′toyd; *deltoeides* = like the Greek letter Δ) **tuberosity** (too′ber-os′i-tē), which extends along its lateral surface for about half the length of the humerus. The deltoid muscle of the shoulder attaches to this roughened surface. The **radial groove** (or *spiral groove*) is located adjacent to the deltoid tuberosity and is where the radial nerve and some blood vessels travel.

Together, the bones of the humerus, radius, and ulna form the elbow joint (figure 8.4*b,c*). The **medial** and **lateral epicondyles** (ep-i-kon′dīl; *epi* = upon, *kondylos* = a knuckle) are bony side projections on the distal humerus that provide surfaces for muscle attachment. Palpate both sides of your elbow; the bumps you feel are the medial and lateral epicondyles. Traveling posterior to the medial epicondyle is the ulnar nerve, which supplies many intrinsic hand muscles.

The distal end of the humerus also has two smooth, curved surfaces for articulation with the bones of the forearm. The rounded **capitulum** (kă-pit′ū-lŭm; *caput* = head) is located laterally and articulates with the head of the radius. The pulley-shaped **trochlea** (trok′lē-ă; *trochileia* = a pulley) is located medially and articulates with the trochlear notch of the ulna. Additionally, the distal end of the humerus exhibits three depressions, two on its anterior surface and one on its posterior surface. The anterolaterally placed **radial fossa** accommodates the head of the radius, while the anteromedially placed **coronoid** (kōr′ŏ-noyd; *korone* = a crow, *eidos* = resembling) **fossa** accommodates the coronoid process of the ulna. The posterior depression called the **olecranon** (ō-lek′ră-non; *olene* = ulna, *kranion* = head) **fossa** accommodates the olecranon of the ulna.

Radius and Ulna

The radius and ulna are the bones of the forearm **(figure 8.5)**. In anatomic position, these bones are parallel, and the **radius** (rā′dē-ŭs; spoke of a wheel, ray) is located more laterally. The proximal end of the radius has a distinctive disc-shaped **head** that articulates with the capitulum of the humerus. A narrow **neck** separates the radial head from the **radial tuberosity** (or *bicipital tuberosity*). The radial tuberosity is an attachment site for the biceps brachii muscle.

The **shaft** of the radius curves slightly and leads to a wide distal end where there is a laterally placed **styloid** (stī′loyd; *stylos* = pillar, post) **process**. This bony projection can be palpated on the lateral side of the wrist, just proximal to the thumb. On the distal medial surface of the radius is an **ulnar notch,** where the medial surface of the radius articulates with the distal end of the ulna (figure 8.5*c*).

The **ulna** (ŭl′nă; *olene* = elbow) is the longer, medially placed bone of the forearm. At the proximal end of the ulna, a C-shaped **trochlear notch** interlocks with the trochlea of the humerus. The posterosuperior aspect of the trochlear notch has a prominent projection called the **olecranon**. The olecranon articulates with the olecranon fossa of the humerus and forms the posterior "bump" of the elbow. (Palpate your posterior elbow; the bump you feel is the olecranon.) The inferior lip of the trochlear notch, called the **coronoid process,** articulates with the humerus at the coronoid fossa. Lateral to the coronoid process, a smooth, curved **radial notch** accommodates the head of the radius and helps form the proximal radioulnar joint (figure 8.5*b*). Also at the proximal end of this bone is the **tuberosity of ulna**. At the distal end of the ulna, the shaft narrows and terminates in a knoblike **head** that has a posteromedial **styloid process.** The styloid process of the ulna may be palpated on the medial ("little finger" side) of the wrist.

Figure 8.4

Humerus and Elbow Joint.   The right humerus
is shown in (a) anterior and (d) posterior views.
(b) Anterior and (c) posterior views of the
elbow joint, which is formed by the humerus
articulating with the radius and ulna.

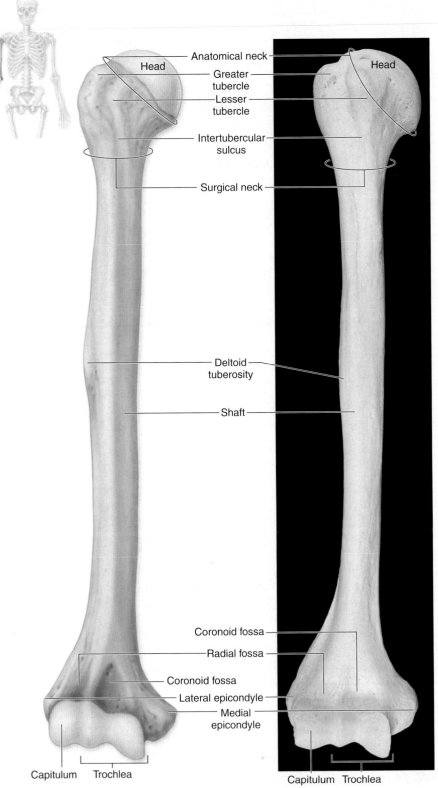

Head

Anatomical neck

Greater
tubercle

Lesser
tubercle

Intertubercular
sulcus

Surgical neck

Head

Deltoid
tuberosity

Shaft

Coronoid fossa

Radial fossa

Coronoid fossa

Lateral epicondyle

Medial
epicondyle

Capitulum   Trochlea

Capitulum Trochlea

(a) Right humerus, anterior view

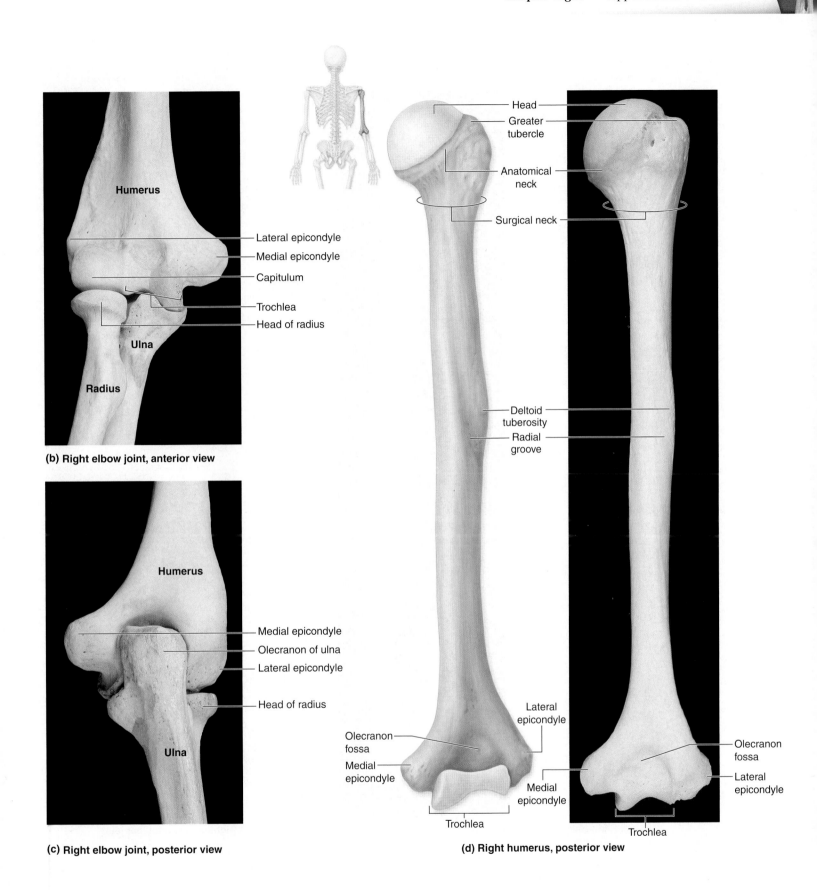

(b) Right elbow joint, anterior view

Humerus

Lateral epicondyle
Medial epicondyle
Capitulum
Trochlea
Head of radius

Ulna

Radius

(c) Right elbow joint, posterior view

Humerus

Medial epicondyle
Olecranon of ulna
Lateral epicondyle

Head of radius

Ulna

Head
Greater tubercle
Anatomical neck
Surgical neck

Deltoid tuberosity
Radial groove

Lateral epicondyle

Olecranon fossa
Medial epicondyle

Medial epicondyle

Trochlea

Olecranon fossa
Lateral epicondyle

Trochlea

(d) Right humerus, posterior view

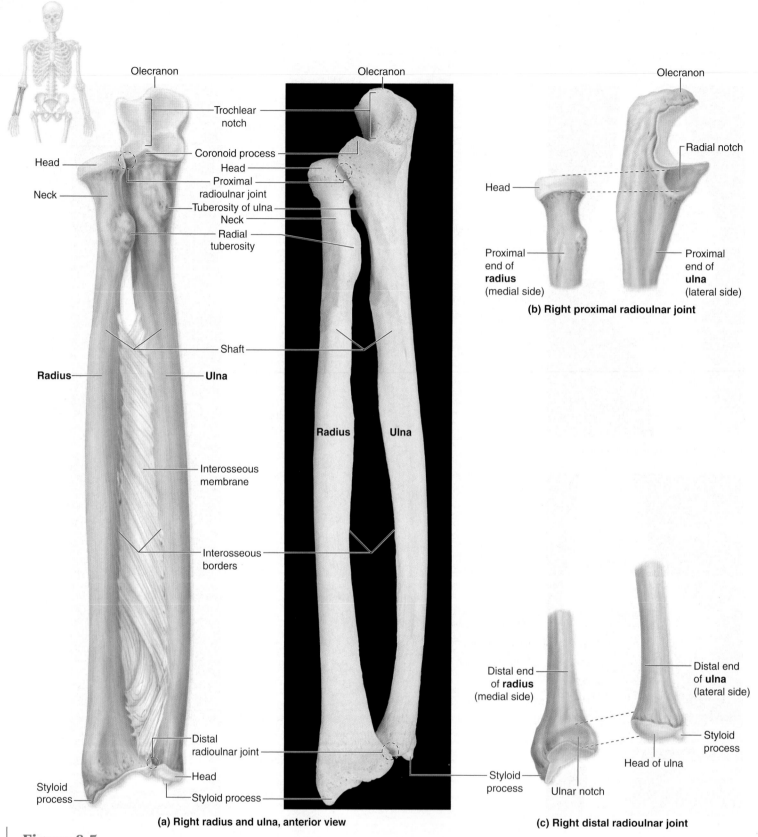

Olecranon

Trochlear notch

Coronoid process

Head

Proximal radioulnar joint

Tuberosity of ulna

Neck

Radial tuberosity

Head

Neck

Shaft

Radius

Ulna

Interosseous membrane

Interosseous borders

Distal radioulnar joint

Head

Styloid process

Styloid process

Styloid process

(a) Right radius and ulna, anterior view

Radius

Ulna

Olecranon

Head

Radial notch

Proximal end of **radius** (medial side)

Proximal end of **ulna** (lateral side)

(b) Right proximal radioulnar joint

Olecranon

Distal end of **radius** (medial side)

Distal end of **ulna** (lateral side)

Styloid process

Head of ulna

Styloid process

Ulnar notch

(c) Right distal radioulnar joint

Figure 8.5

Radius and Ulna. *(a)* Anterior view of the bones of the right forearm along with *(b, c)* the proximal and distal radioulnar joints. *(d)* Supination and *(e)* pronation of the right forearm. *(f)* Posterior view of the right forearm.

Both the radius and the ulna exhibit **interosseous borders,** which face each other; the ulna's interosseous border faces laterally, while the radius's interosseous border faces medially. These interosseous borders are connected by an **interosseous membrane** (*interosseous ligament*), composed of dense regular connective tissue, that helps keep the radius and ulna a fixed distance apart

from one another and provides a pivot of rotation for the forearm. The bony joints that move during this rotation are the proximal and distal radioulnar joints (figure 8.5*b,c*).

In anatomic position, the palm of the hand is facing anteriorly, and the bones of the forearm are said to be in **supination** (soo'pi-nă'shŭn) (figure 8.5*d*). Note that the radius and the ulna are parallel with one

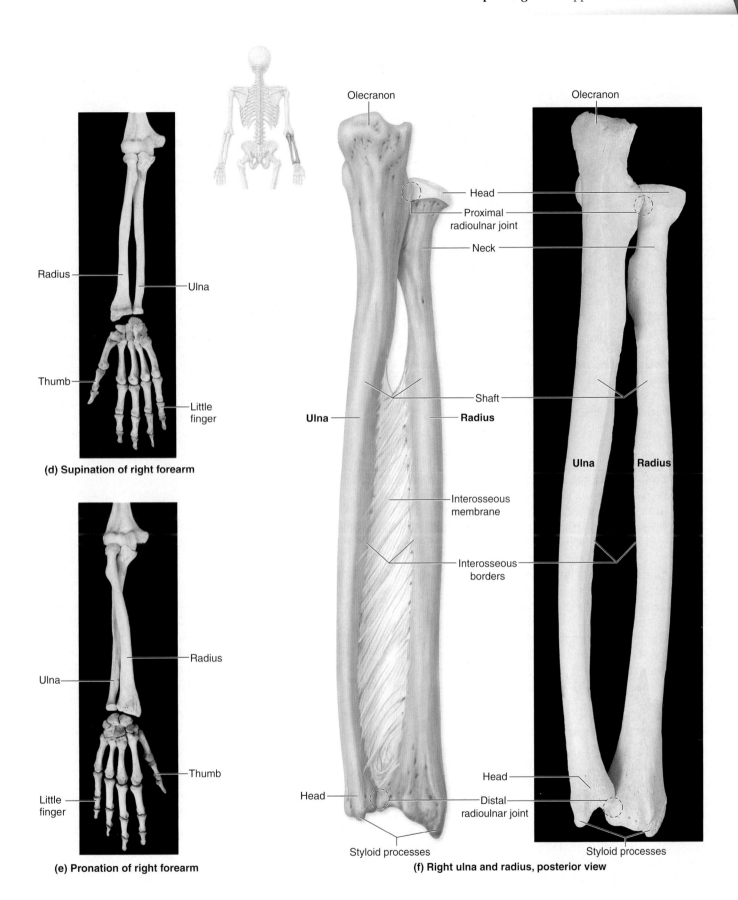

(d) Supination of right forearm

(e) Pronation of right forearm

(f) Right ulna and radius, posterior view

another. If you stand in anatomic position, so that you can view your own forearm, the radius is on the lateral (thumb) side of the forearm, while the ulna is on the medial (little finger) side of the forearm.

Pronation (prō-nā'shŭn) of the forearm requires that the radius cross over the ulna and that both bones pivot along the

Study Tip!

No matter what the position of the forearm (whether pronated or supinated), the distal end of the radius is always near the thumb, and the distal end of the ulna is always on the side of the little finger.

Colles Fracture

A **Colles** (cōl′) **fracture** is a fracture of the distal radius (see figure 6.15). This type of fracture typically occurs when a person extends a hand (and thus the forearm is pronated) while trying to break a fall. The force of the fall on the outstretched hand fractures the distal radius, just proximal to the wrist. The force can be transmitted via the interosseous membrane to the ulna and may also result in a distal ulna fracture. Colles fractures are very common in adults, especially in elderly individuals who suffer from osteoporosis.

The common symptoms of a Colles fracture are pain and swelling just proximal to the wrist and weakness in the affected hand. In addition, when viewed from the side, the wrist is not straight, but has a bend and angle similar to the shape of a dinner fork, because the broken distal part of the radius overrides the proximal part. A Colles fracture can be diagnosed with an x-ray.

Treatment typically requires immobilizing the affected bones with a splint or cast. A fracture involving multiple pieces of broken bone, called a comminuted fracture, may require surgical intervention with internally placed pins, screws, or plates.

interosseous membrane (figure 8.5e). When the forearm is pronated, the palm of the hand is facing posteriorly. Now pronate your own forearm; you can sometimes palpate the radius and feel it criss-crossing over the ulna. In this position, the head of the radius is still along the lateral side of the elbow, but the distal end of the radius has crossed over and is the more medial structure.

When an individual has the upper limbs extended and forearms supinated, note that the bones of the forearm may angle laterally from the elbow joint. This positioning is referred to as the "carrying angle" of the elbow, and this angle measures approximately 5 to 15 degrees. The carrying angle positions the bones of the forearms such that the forearms will clear the hips during walking (and as the forearm bones swing during the process). Females have wider carrying angles than males, presumably because they have wider hips than males.

Carpals, Metacarpals, and Phalanges

The bones that form the wrist and hand are the carpals, metacarpals, and phalanges **(figure 8.6)**. The **carpals** (kar′păl) are small, short bones that form the wrist. They are arranged in two rows (a proximal row and a distal row) of four bones each. These small bones allow for the multiple movements possible at the wrist. The proximal row of carpal bones, listed from lateral to medial, are the **scaphoid** (skaf′oyd; *skaphe* = boat), **lunate** (loo′nāt; *luna* = moon), **triquetrum** (trī-kwē′trŭm; *triquetrus* = three-cornered), and **pisiform** (pis′i-fōrm; *pisum* = pea, *forma* = appearance). The bones of the distal row of

carpal bones are the most laterally placed **trapezium** (tra-pē′zē-ŭm; *trapeza* = table), **trapezoid** (trap′ĕ-zoyd), **capitate** (kap′i-tāt; *caput* = head), and **hamate** (ha′māt; *hamus* = hook).

WHAT DO YOU THINK?

2 Why does each wrist have so many carpal bones (eight)? How does the number of carpal bones relate to the amount of movement in the wrist? Would your wrist be as freely movable if you had just one or two large carpal bones?

Scaphoid Fractures

The scaphoid bone is one of the more commonly fractured carpal bones. A fall on the outstretched hand may cause the scaphoid to fracture into two separate pieces. When this happens, only one of the two pieces maintains its blood supply. Usually, blood vessels are torn on the proximal part of the scaphoid, resulting in **avascular necrosis**, death of the bone tissue in that area due to inadequate blood supply. Scaphoid fractures take quite a while to heal properly due to this complication. Additionally, avascular necrosis may cause the patient to develop degenerative joint disease of the wrist.

The bones in the palm of the hand are called **metacarpals** (met′ă-kar′păl; *meta* = after, *karpus* = wrist). Five metacarpal bones articulate with the distal carpal bones and support the palm. Roman numerals I–V denote the metacarpal bones, with metacarpal I located at the base of the thumb, and metacarpal V at the base of the little finger.

The bones of the digits are the **phalanges** (fă-lan′jēz; sing., *phalanx*, fā′langks; line of soldiers). There are three phalanges in each of the second through fifth fingers and two phalanges only in the thumb, also known as the **pollex** (pol′eks; thumb), for a total of 14 phalanges per hand. The **proximal phalanx** articulates with the head of a metacarpal, while the **distal phalanx** is the bone in the very tip of the finger. The **middle phalanx** of each finger lies between the proximal and distal phalanges; however, a middle phalanx is not present in the thumb.

WHAT DID YOU LEARN?

2 What is the location and the purpose of the intertubercular sulcus of the humerus?

3 What bone of the forearm articulates with the trochlea of the humerus?

4 Describe the structure of the head of the radius, and discuss its functions.

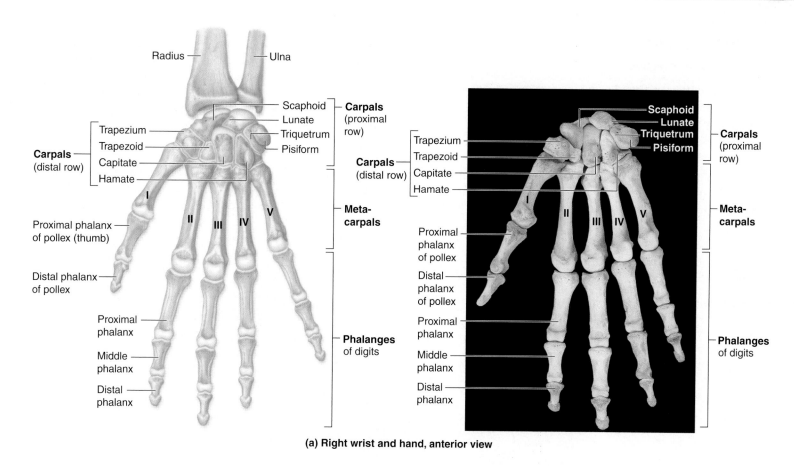

(a) Right wrist and hand, anterior view

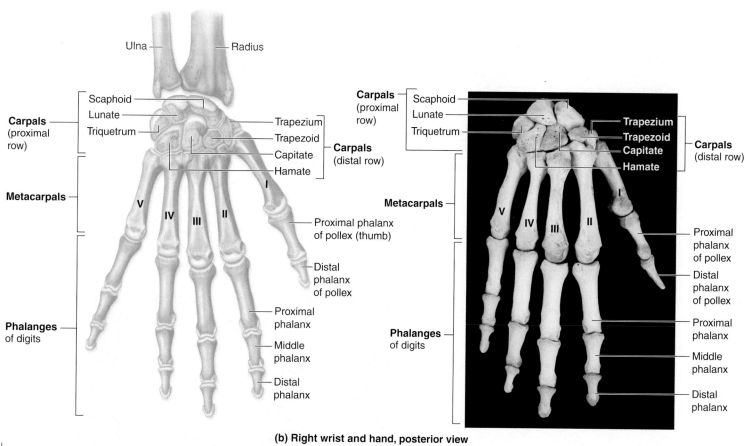

(b) Right wrist and hand, posterior view

Figure 8.6

Bones of the Carpals, Metacarpals, and Phalanges.   Diagrams and photos compare the carpal bones, which form the wrist, and the metacarpals and phalanges, which form the hand. *(a)* Anterior (palmar) and *(b)* posterior views of the right wrist and hand.

Pelvic Girdle

Key topics in this section:

- Bones of the pelvic girdle and their prominent surface features
- How each bone contributes to the pelvic girdle's strength and function
- Comparison of male and female pelves

The adult **pelvis** (pel'vis; pl., *pelves*, pel'vēz; basin) is composed of four bones: the sacrum, the coccyx, and the right and left **ossa coxae** (os'ă cox'ē; sing., *os coxae*; hip bone) **(figure 8.7)**. The pelvis protects and supports the viscera in the inferior part of the ventral body cavity. In contrast, the term **pelvic girdle** refers to the left and right ossa coxae only. The radiograph in **figure 8.8** illustrates how the pelvis articulates with each bone of the thigh (femur). Note that the head of each femur fits snugly into the acetabulum of each os coxae. When a person is standing upright, the pelvis is tipped slightly anteriorly.

Os Coxae

The os coxae is commonly referred to as the "hip bone" (and sometimes as the *coxal bone* or the *innominate bone*). Each os coxae is formed from three separate bones: the ilium, the ischium,

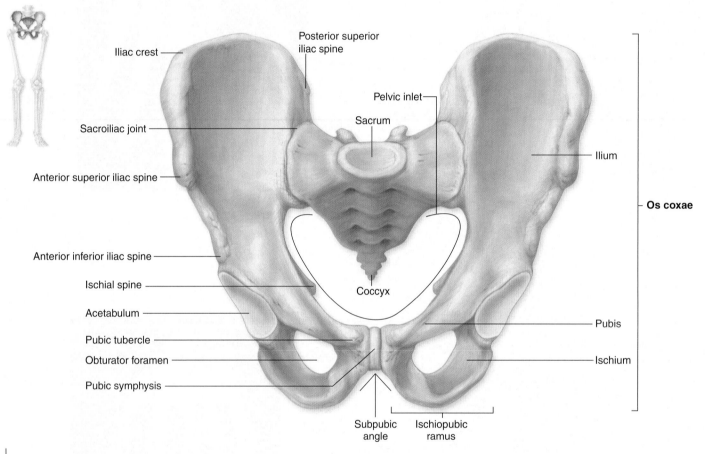

Figure 8.7

Pelvis. The complete pelvis consists of the two ossa coxae, the sacrum, and the coccyx.

Figure 8.8

Pelvis and Femur Articulation. A radiograph shows an anterior view of the articulation between the pelvis and the femur.

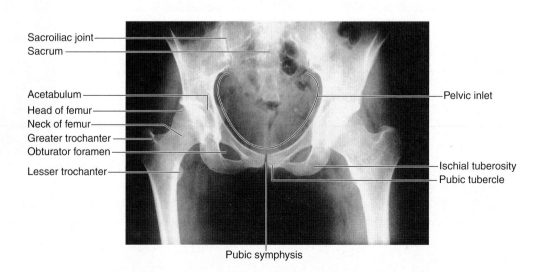

and the pubis (**figure 8.9**). These three bones fuse between the ages of 13 and 15 years to form the single os coxae. Each os coxae articulates posteriorly with an auricular surface of the sacrum at the sacroiliac joint. The femur articulates with a deep, curved depression on the lateral surface of the os coxae called the **acetabulum** (as-ĕ-tab'ū-lŭm; shallow cup). The acetabulum contains a smooth curved surface, called the **lunate surface**, which is C-shaped and articulates with the femoral head. The three bones that form the os coxae (ilium, ischium, and pubis) all contribute a portion to the acetabulum. Thus, the acetabulum represents a region where these bones have fused.

The largest of the three coxal bones is the **ilium** (il'ē-ŭm; groin, flank), which forms the superior region of the os coxae and the largest portion of the acetabular surface. The wide, fan-shaped portion of the ilium is called the **ala** (ā'lă; wing). The ala terminates inferiorly at a ridge called the **arcuate line** (ar'kū-āt; *arcuatus* = bowed) on the medial surface of the ilium. On the medial side of the ala is a depression termed the **iliac fossa**. From a lateral view, an observer sees **anterior, posterior,** and **inferior gluteal lines** that are attachment sites for the gluteal muscles of the buttock. The posteromedial side of the ilium exhibits a large, roughened area called the **auricular** (aw-rik'ū-lăr; *auris* = ear) **surface,** where the ilium articulates with the sacrum at the sacroiliac joint.

The superior-most ridge of the ilium is the **iliac crest**. Palpate the posterosuperior edges of your hips; the ridge of bone you feel on each side is the iliac crest. The iliac crest arises anteriorly from a projection called the **anterior superior iliac spine** and extends posteriorly to the **posterior superior iliac spine**. Located inferiorly to the ala of the ilium are the **anterior inferior iliac spine** and the **posterior inferior iliac spine**. The posterior inferior iliac spine is adjacent to a prominent **greater sciatic** (sī-at'ik; *sciaticus* = hip joint) **notch,** through which the sciatic nerve travels to the lower limb.

The ilium fuses with the **ischium** (is'kē-ŭm; *ischion* = hip) near the superior and posterior margins of the acetabulum. The ischium accounts for the posterior two-fifths of the acetabular surface. Posterior to the acetabulum, the prominent triangular **ischial** (is'kē-ăl) **spine** projects medially. The bulky bone superior to the ischial spine is called the ischial **body**. The **lesser sciatic notch** is a semicircular depression inferior to the ischial spine. The posterolateral border of the ischium is a roughened projection called the **ischial tuberosity**. The ischial tuberosities are also called the "sits bones" by some health professionals and fitness instructors, because they support the weight of the body when seated. If you palpate your buttocks while in a sitting position, you can feel the large ischial tuberosities. An elongated **ramus** (rā'mŭs; pl., *rami,* rā'mē) of the ischium extends from the ischial tuberosity toward its anterior fusion with the pubis.

The **pubis** (pew'bŭs) fuses with the ilium and ischium at the acetabulum. The ischial ramus fuses anteriorly with the **inferior pubic ramus** to form the **ischiopubic ramus** (see figure 8.7). The **superior pubic ramus** originates at the anterior margin of the acetabulum. The **obturator** (ob'too-rā-tŏr; *obturo* = to occlude) **foramen** is a space in the os coxae that is encircled by both pubic and ischial rami. A roughened ridge called the **pubic crest** is located on the anterosuperior surface of the superior ramus, and it ends at the **pubic tubercle**. The pubic tubercle is an attachment site for the inguinal ligament. A roughened area on the anteromedial surface of the pubis, called the **symphysial surface** (sim'fĭ-sis; growing

together), denotes the site of articulation between the pubic bones. On the medial surface of the pubis, the **pectineal** (pek-tin'ē-ăl; ridged or relating to the pubis) **line** originates and extends diagonally across the pubis to merge with the arcuate line.

WHAT DO YOU THINK?

3 Compare and contrast the glenoid cavity of the scapula with the acetabulum of the os coxae. Which girdle maintains stronger, more tightly fitting bony connections with its respective limbs—the pectoral girdle or the pelvic girdle?

True and False Pelves

The **pelvic brim** is a continuous oval ridge that extends from the pubic crest, pectineal line, and arcuate line to the rounded inferior edges of the sacral ala and promontory. This pelvic brim helps subdivide the entire pelvis into a true pelvis and a false pelvis (**figure 8.10**). The **true pelvis** lies *inferior* to the pelvic brim. It encloses the pelvic cavity and forms a deep bowl that contains the pelvic organs. The **false pelvis** lies *superior* to the pelvic brim. It is enclosed by the ala of the iliac bones. It forms the inferior region of the abdominal cavity and houses the inferior abdominal organs.

The pelvis also has a superior and an inferior opening, and each has clinical significance. The **pelvic inlet** is the superiorly positioned space enclosed by the pelvic brim. In other words, the pelvic brim is the bony oval *ridge* of bone, whereas the pelvic inlet is the *space* surrounded by the pelvic brim. The pelvic inlet is the opening at the boundary between the true pelvis and the false pelvis.

The **pelvic outlet** is the inferiorly placed opening bounded by the coccyx, the ischial tuberosities, and the inferior border of the pubic symphysis. In males, the ischial spines sometimes project into the pelvic outlet, thereby narrowing the diameter of this outlet. In contrast, female ischial spines rarely project into the pelvic outlet. The pelvic outlet is covered with muscles and skin and forms the body region called the perineum (per'i-nē'ŭm). The width and size of the pelvic outlet is especially important in females, because the opening must be wide enough to accommodate the fetal head during childbirth.

Sex Differences Between the Female and Male Pelves

Although it is possible to determine the sex of a skeleton by examining the skull (see chapter 7), the most reliable indicator of sex is the pelvis, primarily the ossa coxae. The ossa coxae are the most sexually dimorphic bones of the body due to the requirements of pregnancy and childbirth in females. For example, the female pelvis is shallower and wider than the male pelvis in order to accommodate an infant's head as it passes through the birth canal. Some of these differences are obvious, such as that males have narrower hips than females do. But we can find many other differences by examining the shapes and orientations of the pelvic bones. For example, the female ilium flares more laterally, while the male ilium projects more superiorly, which is why males typically have narrower hips. Since the female pelvis is wider, the acetabulum projects more laterally, and the greater sciatic notch is much wider as well. In contrast, the male acetabulum projects more anteriorly, and the male greater sciatic notch is much narrower and U-shaped. Females tend to have a **preauricular sulcus,** which is a depression/groove between the greater sciatic notch and the sacroiliac articulation. Males tend not

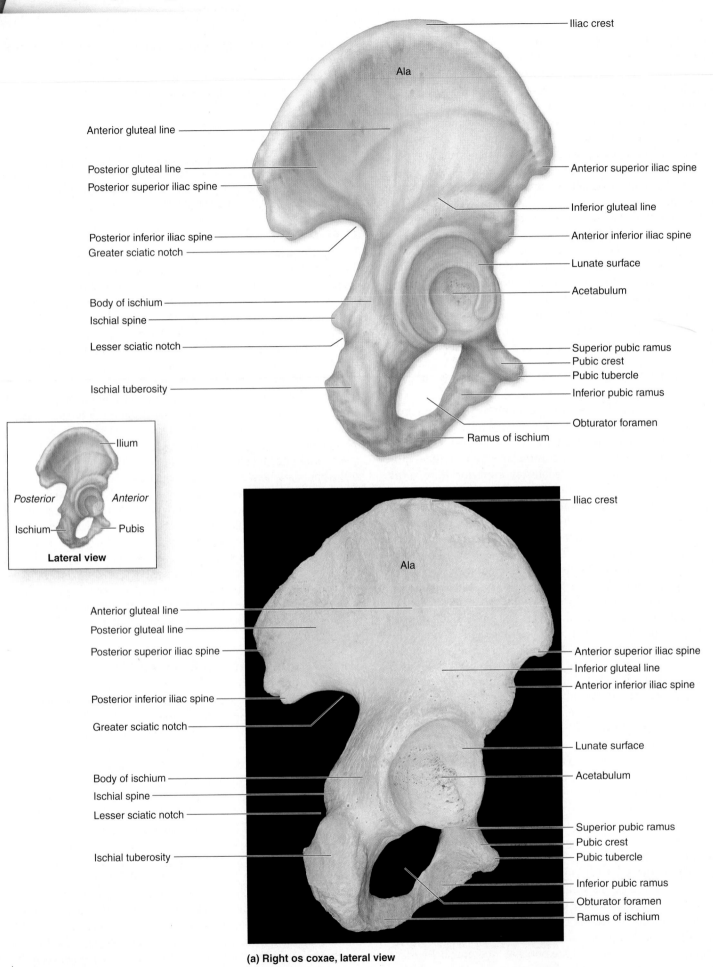

Iliac crest

Ala

Anterior gluteal line

Posterior gluteal line
Posterior superior iliac spine

Anterior superior iliac spine

Inferior gluteal line

Posterior inferior iliac spine
Greater sciatic notch

Anterior inferior iliac spine

Lunate surface

Acetabulum

Body of ischium
Ischial spine

Lesser sciatic notch

Superior pubic ramus
Pubic crest
Pubic tubercle

Ischial tuberosity

Inferior pubic ramus

Obturator foramen

Ramus of ischium

Ilium

Posterior *Anterior*

Ischium Pubis

Lateral view

Iliac crest

Ala

Anterior gluteal line

Posterior gluteal line

Posterior superior iliac spine

Anterior superior iliac spine
Inferior gluteal line
Anterior inferior iliac spine

Posterior inferior iliac spine

Greater sciatic notch

Lunate surface

Acetabulum

Body of ischium
Ischial spine

Lesser sciatic notch

Superior pubic ramus
Pubic crest
Pubic tubercle

Ischial tuberosity

Inferior pubic ramus

Obturator foramen
Ramus of ischium

(a) Right os coxae, lateral view

Figure 8.9

Os Coxae. Each os coxae of the pelvic girdle is formed by the fusion of three bones: an ilium, an ischium, and a pubis. Diagrams and photos show the features of these bones and their relationships in *(a)* lateral and *(b)* medial views.

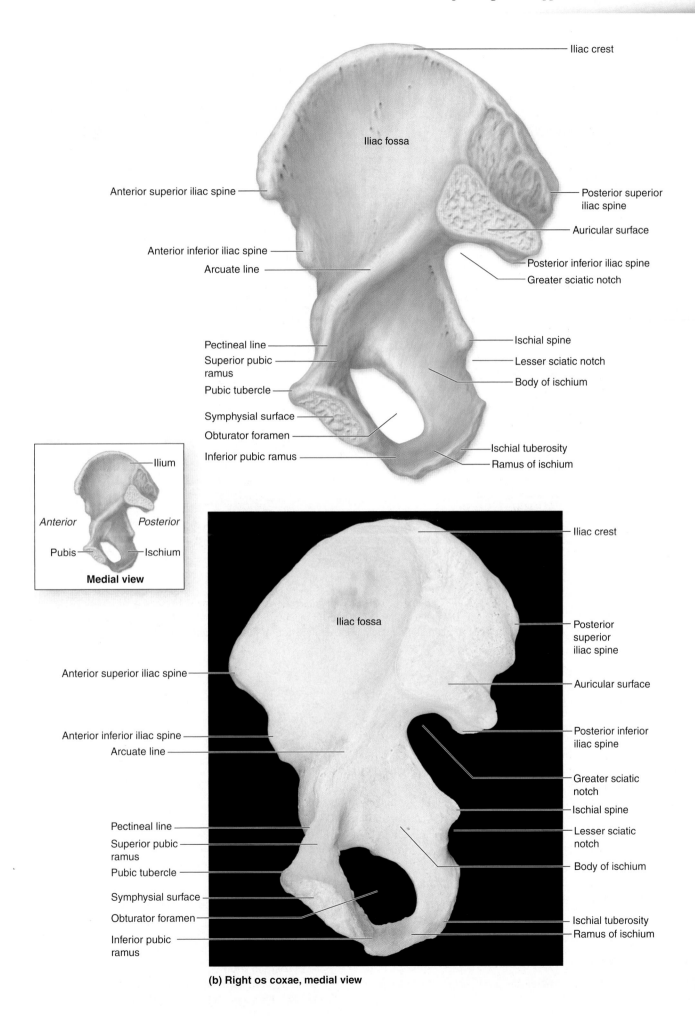

Iliac crest

Iliac fossa

Anterior superior iliac spine

Posterior superior iliac spine

Auricular surface

Anterior inferior iliac spine

Posterior inferior iliac spine

Arcuate line

Greater sciatic notch

Pectineal line

Ischial spine

Superior pubic ramus

Lesser sciatic notch

Pubic tubercle

Body of ischium

Symphysial surface

Obturator foramen

Ischial tuberosity

Inferior pubic ramus

Ramus of ischium

Ilium

Anterior *Posterior*

Pubis

Ischium

Medial view

Iliac crest

Iliac fossa

Anterior superior iliac spine

Posterior superior iliac spine

Auricular surface

Anterior inferior iliac spine

Posterior inferior iliac spine

Arcuate line

Greater sciatic notch

Ischial spine

Pectineal line

Lesser sciatic notch

Superior pubic ramus

Body of ischium

Pubic tubercle

Symphysial surface

Obturator foramen

Ischial tuberosity

Inferior pubic ramus

Ramus of ischium

(b) Right os coxae, medial view

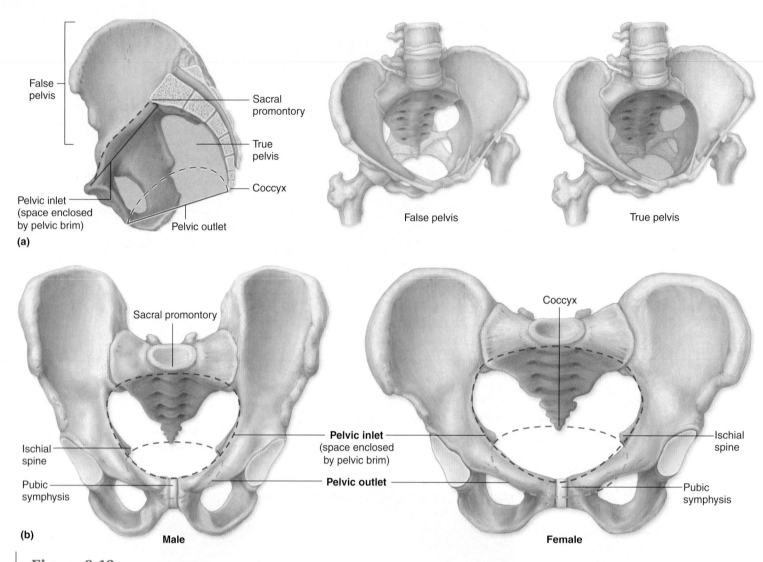

Figure 8.10

Features of the Pelvis.   The pelvic brim is the oval bony ridge that subdivides the pelvis into a true pelvis and a false pelvis. The pelvic inlet is the space enclosed by the pelvic brim, whereas the pelvic outlet is the inferior opening in the true pelvis. *(a)* Medial and anterolateral views of the true and false pelves. *(b)* Anterosuperior view of male and female pelves, demonstrating the sex differences between the pelvic inlet and outlet.

to have this sulcus. The sacrum is usually shorter and wider in females. The coccyx projects more vertically in males, whereas the female coccyx has a posterior tilt.

The body of the pubis in females is much longer and almost rectangular in shape, compared to the shorter, triangular male pubic body. The **subpubic angle** (or *pubic arch*) is the angle formed when the left and right pubic bones are aligned at their pubic symphyses. Since females have much longer pubic bones, the corresponding subpubic angle is much wider and more convex, usually much greater than 100 degrees. The male pubic arch is much narrower and typically does not extend past 90 degrees.

Several significant differences between the female and male pelves are listed and illustrated in **table 8.1**.

WHAT DID YOU LEARN?

5 What three bones fuse to form each os coxae?

6 What is the difference between the pelvic inlet and the pelvic outlet?

Lower Limb

Key topic in this section:
- Bones of the lower limb and their prominent markings

The arrangement and numbers of bones in the lower limb are similar to those of the upper limb. However, since the bones of the lower limb are adapted for weight-bearing and locomotion, they may be shaped somewhat differently and articulate differently than the comparable bones of the upper limb. Each lower limb contains a total of 30 bones:

- 1 femur, located in the femoral region
- 1 patella (kneecap), located in the patellar region
- 1 tibia and 1 fibula, located in the crural region
- 7 tarsal bones, which form the bones of the ankle and proximal foot
- 5 metatarsal bones, which form the arched part of the foot
- 14 phalanges, which form the toes

Table 8.1	Sex Differences Between the Female and Male Pelves	
View	**Female Pelvis**	**Male Pelvis**
Medial View	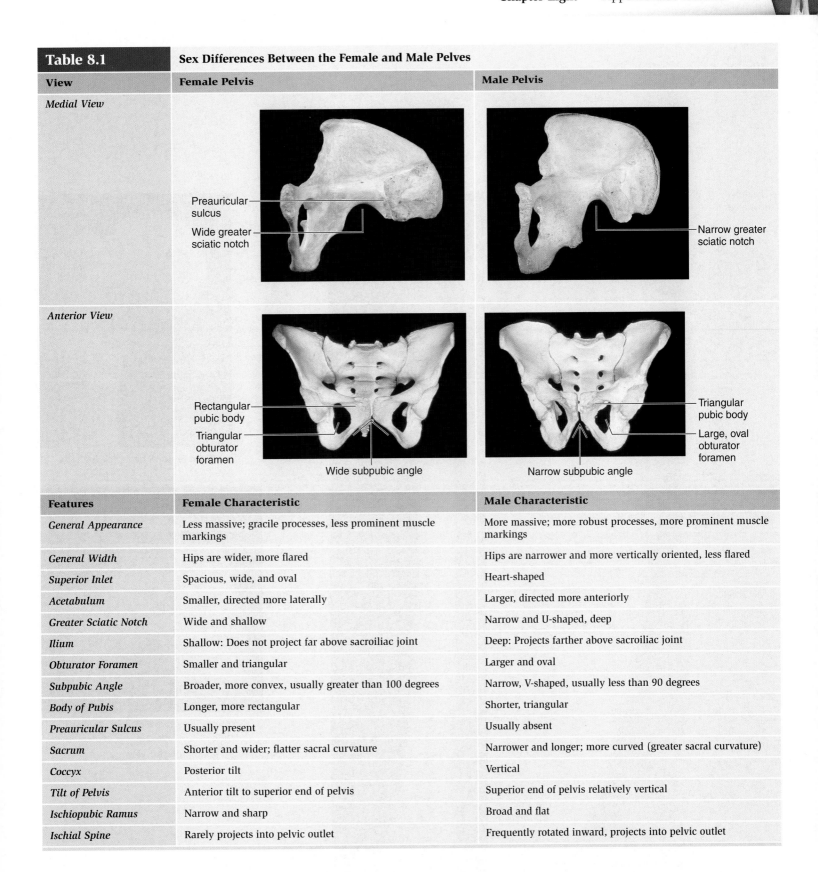	
Anterior View		

Features	**Female Characteristic**	**Male Characteristic**
General Appearance	Less massive; gracile processes, less prominent muscle markings	More massive; more robust processes, more prominent muscle markings
General Width	Hips are wider, more flared	Hips are narrower and more vertically oriented, less flared
Superior Inlet	Spacious, wide, and oval	Heart-shaped
Acetabulum	Smaller, directed more laterally	Larger, directed more anteriorly
Greater Sciatic Notch	Wide and shallow	Narrow and U-shaped, deep
Ilium	Shallow: Does not project far above sacroiliac joint	Deep: Projects farther above sacroiliac joint
Obturator Foramen	Smaller and triangular	Larger and oval
Subpubic Angle	Broader, more convex, usually greater than 100 degrees	Narrow, V-shaped, usually less than 90 degrees
Body of Pubis	Longer, more rectangular	Shorter, triangular
Preauricular Sulcus	Usually present	Usually absent
Sacrum	Shorter and wider; flatter sacral curvature	Narrower and longer; more curved (greater sacral curvature)
Coccyx	Posterior tilt	Vertical
Tilt of Pelvis	Anterior tilt to superior end of pelvis	Superior end of pelvis relatively vertical
Ischiopubic Ramus	Narrow and sharp	Broad and flat
Ischial Spine	Rarely projects into pelvic outlet	Frequently rotated inward, projects into pelvic outlet

Study Tip!

As you learn the bones of the lower limb, compare and contrast them with their corresponding upper limb bones. For example, compare the femur with the humerus. Review how the two are similar, and then determine what features differ between them. This method will help you better remember and understand the bones and their features.

Femur

The **femur** (fē′mŭr; thigh) is the longest bone in the body as well as the strongest and heaviest **(figure 8.11)**. The nearly spherical **head** of the femur articulates with the pelvis at the acetabulum. A tiny ligament connects the acetabulum to a depression in the head of the femur, called the **fovea** (fō′vē-ă; a pit), or *fovea capitis* (kăp′i-tĭs; head). Distal to the head, an elongated, constricted **neck** joins the **shaft** of the femur at an angle. This results in a medial angling of the femur, which brings the knees closer to the midline.

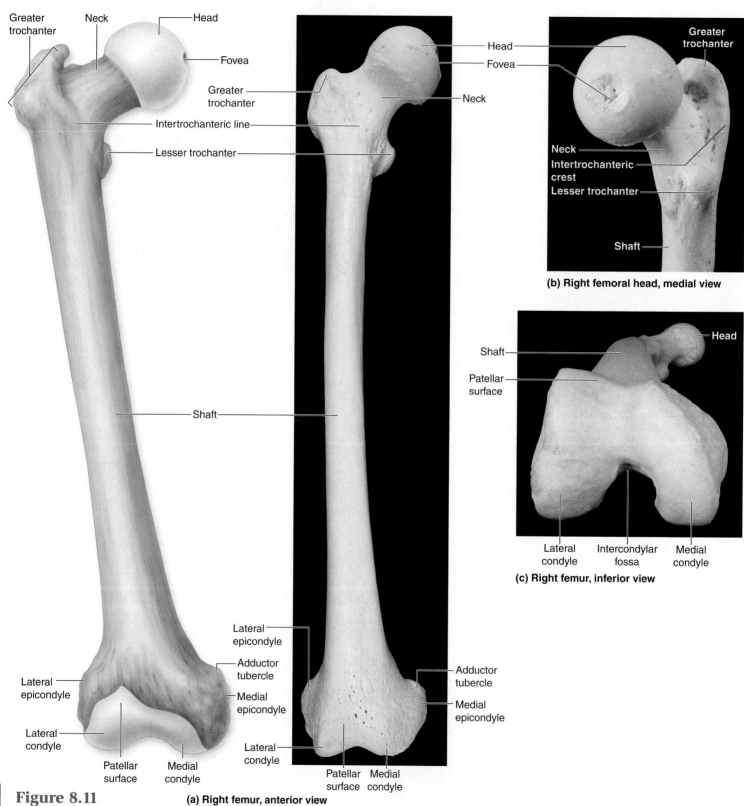

Greater trochanter — Neck — Head

Fovea

Greater trochanter

Intertrochanteric line

Lesser trochanter

Shaft

Lateral epicondyle

Adductor tubercle

Medial epicondyle

Lateral epicondyle

Lateral condyle

Patellar surface Medial condyle

Lateral condyle

Patellar surface Medial condyle

Head

Fovea

Neck

Adductor tubercle

Medial epicondyle

Lateral condyle

Patellar surface Medial condyle

(a) Right femur, anterior view

Head

Fovea

Greater trochanter

Neck

Intertrochanteric crest

Lesser trochanter

Shaft

(b) Right femoral head, medial view

Shaft

Patellar surface

Head

Lateral condyle Intercondylar fossa Medial condyle

(c) Right femur, inferior view

Figure 8.11

Femur. The femur is the bone of the femoral region. *(a)* Diagram and photo show an anterior view of the right femur. *(b)* Superomedial, *(c)* inferior, and *(d)* posterior views of the right femur.

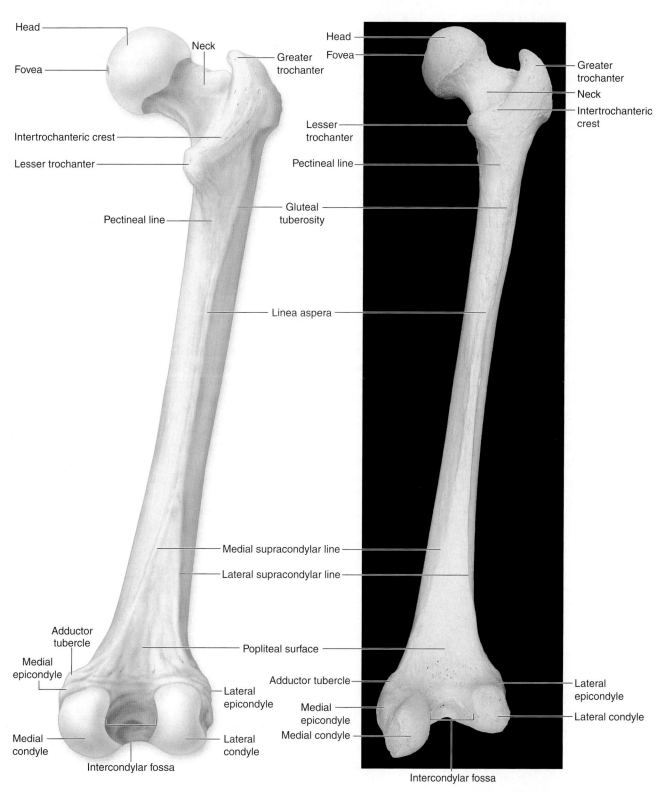

(d) Right femur, posterior view

Two massive, rough processes originate near the proximal end of the femur and serve as insertion sites for the powerful hip muscles. The **greater trochanter** (trō-kan′ter; a runner) projects laterally from the junction of the neck and shaft. Stand up and palpate your lateral thigh, near the hip joint; the bony projection you feel is the greater trochanter. A **lesser trochanter** is located on the femur's posteromedial surface. The greater and lesser trochanters are connected on the posterior surface of the femur by a thick oblique ridge of bone called the **intertrochanteric** (in′ter-trō-kan-tār′ik) **crest**. Anteriorly, a raised **intertrochanteric line** extends between the two trochanters and marks the distal edge of the hip joint capsule. Inferior to the intertrochanteric crest, the **pectineal line** marks the attachment of the pectineus muscle, while the **gluteal** (gloo′tē-ăl; *gloutos* = buttock) **tuberosity** marks the attachment of the gluteus maximus muscle.

The prominent feature on the posterior surface of the shaft is an elevated, midline ridge called the **linea aspera** (lin′ē-ă as′pĕ-ră; rough line). This ridge denotes the attachment site for many thigh muscles. The gluteal tuberosity and pectineal line merge proximally to the linea aspera. Distally, the linea aspera branches into **medial** and **lateral supracondylar lines**. A flattened triangular area, called the **popliteal** (pop-lit′ē-ăl; *poples* = ham of knee) **surface,** is circumscribed by these ridges and an imaginary line between the distal epicondyles. The medial supracondylar ridge terminates in the **adductor tubercle,** a rough, raised projection that is the site of attachment for the adductor magnus muscle.

On the distal, inferior surface of the femur are two smooth, oval articulating surfaces called the **medial** and **lateral condyles** (kon′dīl). Superior to each condyle are projections called the **medial** and **lateral epicondyle,** respectively. When you flex your knee, you can palpate these epicondyles in the thigh on the sides of your knee joint. The medial and lateral supracondylar lines terminate at these epicondyles. On the distal posterior surface of the femur, a deep **intercondylar fossa** separates the two condyles. Both condyles continue from the posterior surface to the anterior surface, where their articular faces merge, producing an articular surface with elevated lateral borders. This smooth anteromedial depression, called the **patellar surface,** is the place where the patella articulates with the femur.

Patella

The **patella** (pa-tel′ă; *patina* = shallow disk), or kneecap, is a large, roughly triangular sesamoid bone located within the tendon of the quadriceps femoris muscle **(figure 8.12)**. The patella allows the tendon of the quadriceps femoris to glide more smoothly, and it protects the knee joint. The superior **base** of the patella is broad, and its inferior **apex** is pointed. The patella may be easily palpated along the anterior surface of the knee. The posterior aspect of the patella has an **articular surface** that articulates (connects) with the patellar surface of the femur.

Tibia and Fibula

Anatomists identify the part of the lower limb between the knee and the ankle as the crural region, or leg. The skeleton of the leg has two parallel bones, the thick, strong tibia and a slender fibula **(figure 8.13)**. These two bones are connected by an **interosseous membrane** composed of dense regular connective tissue, which extends between their **interosseous borders**. The interosseous membrane stabilizes the relative positions of the tibia and fibula, and additionally provides a pivot of minimal rotation for these two bones.

The **tibia** (tib′ē-ă; large shinbone) is the medially placed bone and the only weight-bearing bone of the crural region. Its broad, superior head has two relatively flat surfaces, the **medial** and **lateral condyles,** which articulate with the medial and lateral condyles of the femur, respectively. Separating the medial and lateral condyles of the tibia is a prominent ridge called the **intercondylar eminence** (em′i-nens). On the proximal posterolateral side of the tibia is a **fibular articular facet** where the head of the fibula articulates to form the **superior** (or *proximal*) **tibiofibular joint**.

The rough anterior surface of the tibia near the medial and lateral condyles is the **tibial tuberosity,** which can be palpated just inferior to the patella and marks the attachment site for the patellar ligament. The **anterior border** (or *margin*) is a ridge that extends distally along the anterior tibial surface from the tibial tuberosity. This crest can be readily felt through the skin and is commonly referred to as the "shin."

The tibia narrows distally, but at its medial border, it forms a large, prominent process called the **medial malleolus** (ma-lē′ō-lŭs; *malleus* = hammer). Palpate the medial side of your ankle; the bump you feel is your medial malleolus. On the distal posterolateral side of the tibia is a **fibular notch,** where the fibula articulates and

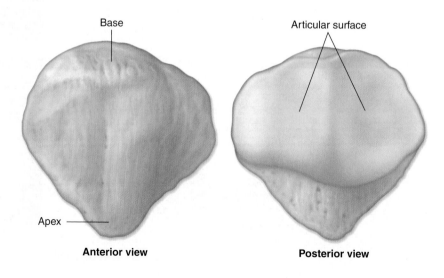

Figure 8.12

Patella. The patella is a sesamoid bone located within the tendon of the quadriceps femoris muscle. These views show the right patella.

Base

Articular surface

Apex

Anterior view

Posterior view

forms the **inferior** (or *distal*) **tibiofibular joint**. On the inferior distal surface of the tibia is the smooth **inferior articular surface** for the talus, one of the tarsal bones.

The **fibula** (fib′ū-lă; buckle, clasp) is the long, thin, laterally placed bone of the leg. It has expanded proximal and distal ends. The fibula does not bear any weight, but several muscles originate from it. Along the lateral edge of the tibia, the fibula articulates with the surface of the tibia. The rounded, knoblike **head** of the fibula is slightly inferior and posterior to the lateral condyle of the tibia. On the head, the smooth **articular facet** articulates with the tibia. Distal to the fibular head is the **neck** of the fibula, followed by its **shaft**. Although the fibula does not bear or transfer weight, its distal tip, called the **lateral malleolus**, extends laterally to the ankle joint, where it provides lateral stability. Palpate the lateral side of your ankle; the bump you feel is your lateral malleolus.

WHAT DO YOU THINK?

4 The medial and lateral malleoli of the leg are similar to what bony features of the forearm?

Tarsals, Metatarsals, and Phalanges

The bones that form the ankle and foot are the tarsals, metatarsals, and phalanges **(figure 8.14)**. The seven **tarsals** (tar′săl; *tarsus* = flat surface) of the ankle and proximal foot are similar to the eight carpal bones of the wrist in some respects, although their shapes and arrangement are different from those of their carpal bone counterparts. The tarsal bones are thoroughly integrated into the structure of the foot because they help the ankle bear the body's weight.

The largest tarsal bone is the **calcaneus** (kal-kā′nē-ŭs; heel), which forms the heel. Its posterior end is a rough, knob-shaped projection that is the point of attachment for the calcaneal (Achilles) tendon extending from the strong muscles on the posterior side of the leg. The superior-most and second-largest tarsal bone is the **talus** (tā′lŭs; ankle bone). The superior aspect of the talus articulates with the articular surface of the tibia. The **navicular** (nă-vik′ū-lăr; *navis* = ship) **bone** is on the medial side of the ankle. The talus, calcaneus, and navicular are considered the proximal row of tarsal bones.

The distal row is formed by a group of four tarsal bones. The three **cuneiform** (kū′nē-i-fŏrm; *cuneus* = wedge) **bones** are wedge-shaped bones with articulations between them, positioned anterior to the navicular bone. They are named according to their position: **medial cuneiform, intermediate cuneiform,** and **lateral cuneiform bones.** The cuneiform bones articulate proximally with the anterior surface of the navicular bone. The laterally placed **cuboid** (kū′boyd; *kybos* = cube) **bone** articulates at its medial surface with the lateral cuneiform and the calcaneus. The distal surfaces of the cuboid bone and the cuneiform bones articulate with the metatarsal bones of the foot.

WHAT DO YOU THINK?

5 What are some similarities and differences between the carpal bones and the tarsal bones?

The **metatarsal** (met′ă-tar′săl) **bones** of the foot are five long bones similar in arrangement and name to the metacarpal bones of the hand. They form the sole of the foot and are identified with Roman numerals I–V, proceeding medially to laterally across the sole (figure 8.14). Metatarsals I through III articulate with the three cuneiform bones, while metatarsals IV and V articulate with the cuboid bone. Distally, each metatarsal bone articulates with a proximal phalanx. At the head of the first metatarsal are two tiny sesamoid bones, which insert on the tendons of the flexor hallucis brevis muscle and help these tendons move more freely.

The bones of the toes (like the bones of the fingers) are called **phalanges.** The toes contain a total of 14 phalanges. The great toe is the **hallux** (hal′ŭks; *hallex* = great toe), and it has only two phalanges (proximal and distal); each of the other four toes has three phalanges (proximal, middle, and distal).

Arches of the Foot

Normally, the sole of the foot does not rest flat on the ground. Rather, the foot is arched, which helps it support the weight of the body and ensures that the blood vessels and nerves on the sole of the foot are not pinched when we are standing. The three arches of the foot are the medial longitudinal, lateral longitudinal, and transverse arches **(figure 8.15)**.

The **medial longitudinal arch** (*arcus* = bow) extends from the heel to the great toe. It is formed from the calcaneus, talus, navicular, and cuneiform bones and from metatarsals I–III. The medial longitudinal arch is the highest of the three arches. The medial longitudinal arch prevents the medial side of the foot from touching the ground and gives our footprint its characteristic shape; note that when you make a footprint, the medial side of the foot does not contribute to the print (figure 8.15*d*).

The **lateral longitudinal arch** is not as high as the medial longitudinal arch, so the lateral part of the foot *does* contribute to a footprint. This arch extends between the little toe and the heel, and it is formed from the calcaneus and cuboid bones and from metatarsals IV and V. The lateral longitudinal arch elevates the lateral edge of the foot slightly to help redistribute some of the body weight among the cuboid and calcaneal bones and metatarsals IV and V.

The **transverse arch** runs perpendicular to the longitudinal arches. It is formed from the distal row of tarsals (cuboid and cuneiforms) and the bases of all five metatarsals. Note in figure 8.15*c* that the medial part of the transverse arch is higher than the lateral part. This is because the medial longitudinal arch (found along the medial side of the transverse arch) is higher than the lateral longitudinal arch (found along the lateral side of the transverse arch).

The shape of the foot arches is maintained primarily by the foot bones themselves. These bones are shaped so that they can interlock and support their weight in an arch, much like the wedge-shaped blocks of an arched bridge can support the bridge without other mechanical supports. Secondarily, strong ligaments attach to the bones and contracting muscles pull on the tendons, thereby helping to maintain the shape of the foot arches.

WHAT DID YOU LEARN?

7 Where is the interosseous membrane of the leg located? What are its functions?

8 What are the names of the tarsal bones? Which tarsal bone articulates with the leg, and which tarsal bone articulates with the metatarsals of the foot?

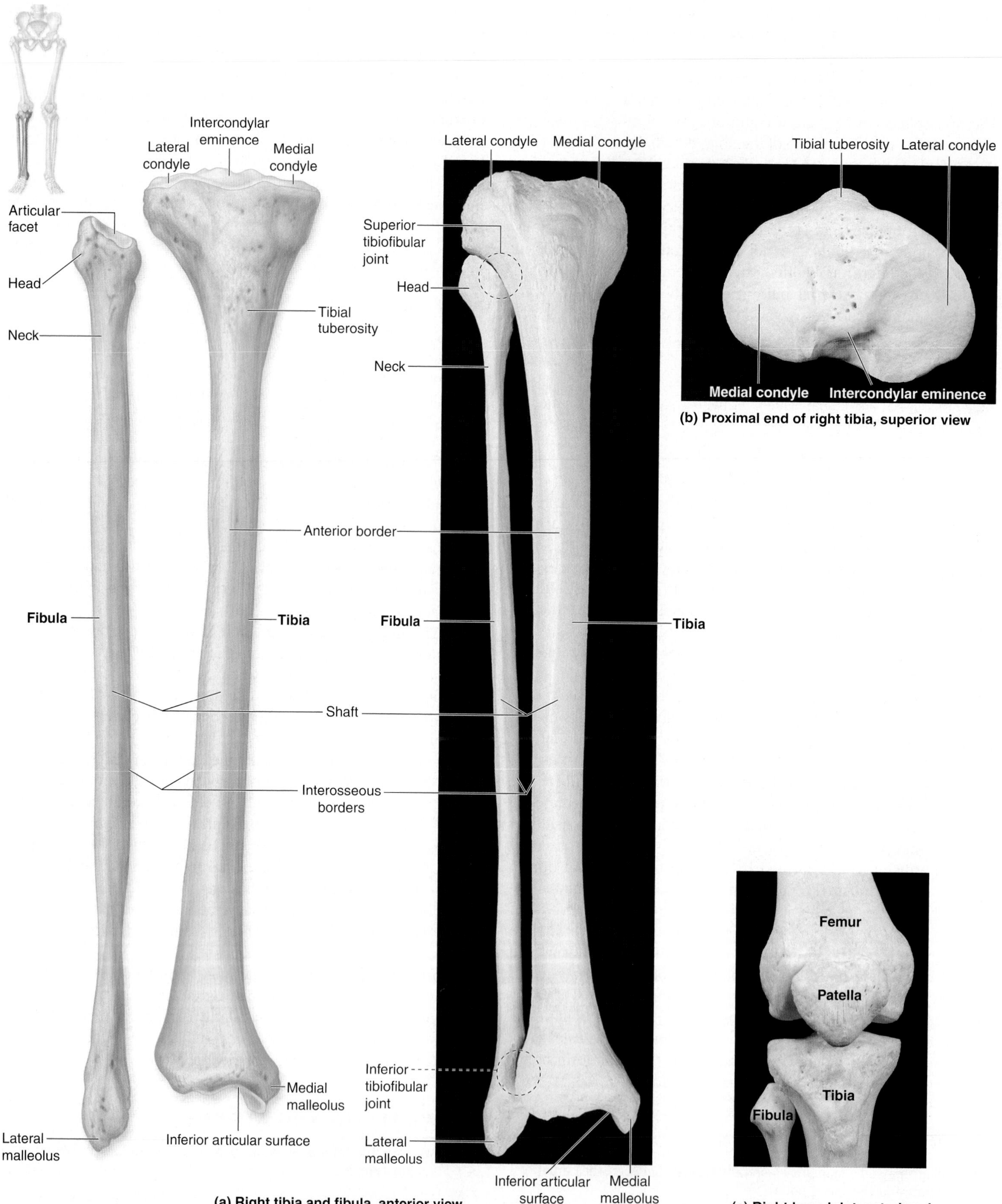

Figure 8.13

Intercondylar eminence

Lateral condyle
Medial condyle

Articular facet

Head

Neck

Fibula

Tibia

Shaft

Interosseous borders

Lateral malleolus

Medial malleolus

Inferior articular surface

Tibial tuberosity

(a) Right tibia and fibula, anterior view

Lateral condyle
Medial condyle

Superior tibiofibular joint

Head

Neck

Anterior border

Fibula

Tibia

Inferior tibiofibular joint

Lateral malleolus

Inferior articular surface
Medial malleolus

Tibial tuberosity
Lateral condyle

Medial condyle
Intercondylar eminence

(b) Proximal end of right tibia, superior view

Femur

Patella

Tibia

Fibula

(c) Right knee joint, anterior view

Tibia, Fibula, and Knee Joint. The tibia and fibula are the bones of the crural (leg) region. *(a)* Diagram and photo show an anterior view of the right tibia and fibula. *(b)* Proximal end of the right tibia. *(c)* Anterior view of the right knee joint. *(d)* Posterior view of the right tibia and fibula. *(e, f)* Posterior and *(g)* lateral views of the right knee joint. Note that the fibula does not directly participate in the knee joint proper.

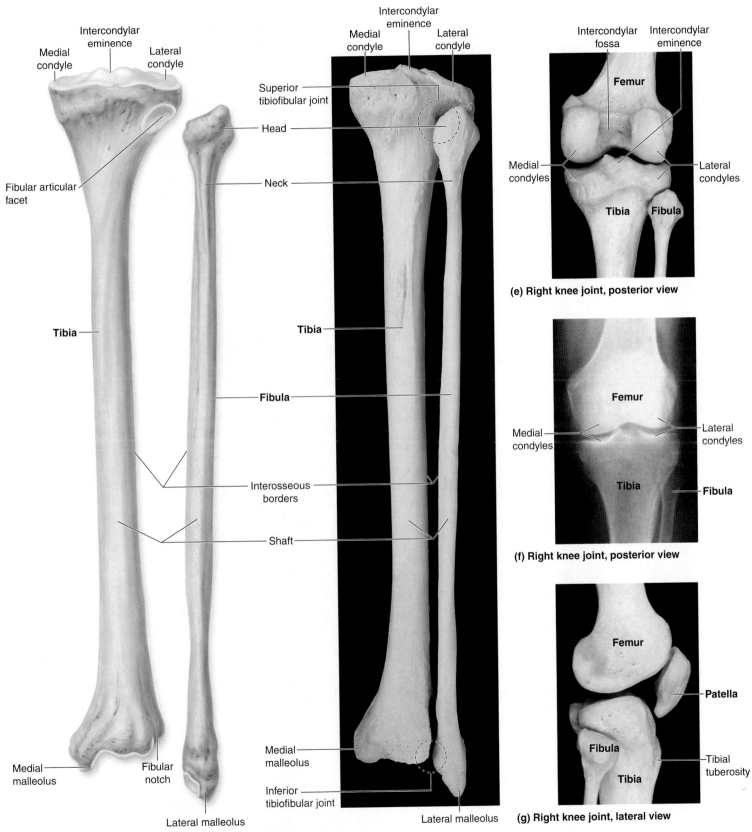

Intercondylar
eminence

Medial
condyle

Lateral
condyle

Fibular articular
facet

Tibia

Medial
malleolus

Fibular
notch

Lateral malleolus

Head

Neck

Interosseous
borders

Shaft

(d) Right tibia and fibula, posterior view

Intercondylar
eminence

Medial
condyle

Lateral
condyle

Superior
tibiofibular joint

Tibia

Fibula

Medial
malleolus

Inferior
tibiofibular joint

Lateral malleolus

Intercondylar
fossa

Intercondylar
eminence

Femur

Medial
condyles

Lateral
condyles

Tibia

Fibula

(e) Right knee joint, posterior view

Femur

Medial
condyles

Lateral
condyles

Tibia

Fibula

(f) Right knee joint, posterior view

Femur

Patella

Fibula

Tibia

Tibial
tuberosity

(g) Right knee joint, lateral view

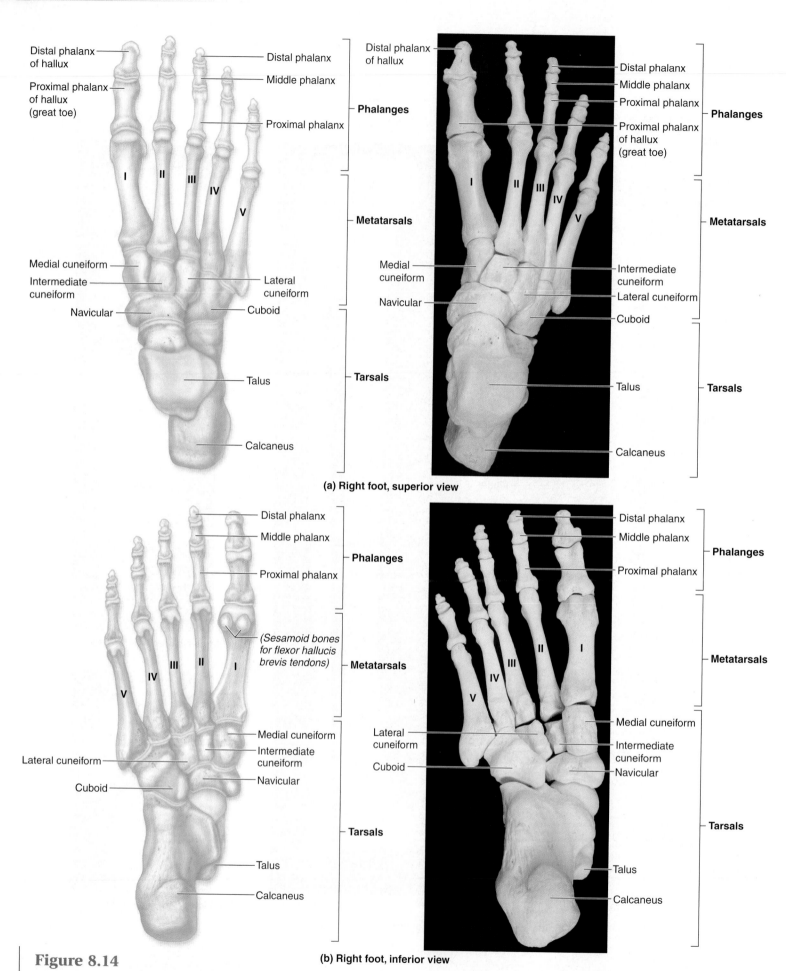

Figure 8.14

Bones of the Tarsals, Metatarsals, and Phalanges.   Tarsal bones form the ankle and proximal foot, metatarsals form the arched sole of the foot, and phalanges form the toes. Diagrams and photos show (*a*) superior and (*b*) inferior views of the right foot.

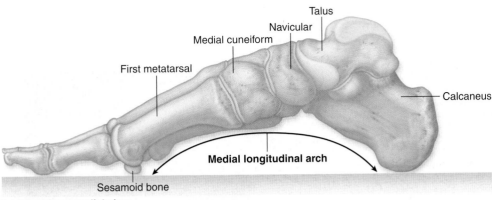

(a) Right foot, medial view

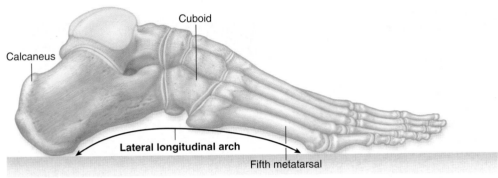

(b) Right foot, lateral view

Figure 8.15

Arches of the Foot. The foot's two longitudinal arches and one transverse arch allow for better weight support. *(a)* Medial longitudinal arch. *(b)* Lateral longitudinal arch. *(c)* Transverse arch as seen in cross-sectional view. *(d)* A footprint illustrates the placement of the longitudinal arches.

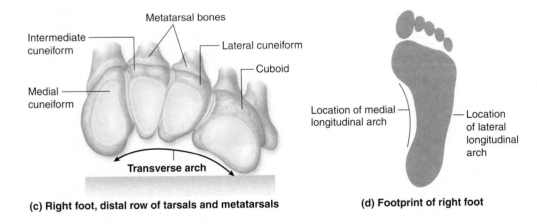

(c) Right foot, distal row of tarsals and metatarsals

(d) Footprint of right foot

Aging of the Appendicular Skeleton

Key topic in this section:

- How the appendicular skeleton changes as we grow older

As we age, skeletal mass and density decline, while erosion and porosity increase, potentially resulting in osteoporosis. Bones become more brittle and susceptible to fracture. Articulating surfaces deteriorate, contributing to osteoarthritis. Changes in the skeleton begin in childhood and continue throughout life. For example, most of the epiphyseal plates fuse between the ages of 10 and 25 years. Degenerative changes in the normal skeleton, such as a reduction in mineral content, don't begin until middle age. Measurable loss of calcium in men begins by age 45, but may start in some women as early as age 35.

The os coxae is not only a reliable indicator of sex, but it also can provide a good estimate of a skeleton's age at death. In particular, the pubic symphysis undergoes age-related changes. The pubic symphysis appears roughened or billowed in the teens and early 20s. Thereafter, the symphysis flattens and loses its billowing. In the 30s and 40s, the pubic symphysis develops a well-defined rim. Finally, as a person gets older, it begins to develop concavities and arthritic changes.

Development of the Appendicular Skeleton

Key topic in this section:

- Events that occur during development of the appendicular skeleton

The appendicular skeleton begins to develop during the fourth week, when **limb buds** appear as small ridges along the lateral sides of the embryo. The upper limb buds appear early in the fourth

CLINICAL VIEW: In Depth

Pathologies of the Foot

Some medical problems associated with the foot include bunions, pes cavus, talipes equinovarus, pes planus, and metatarsal stress fractures. These pathologies have various causes.

A **bunion** (bŭn'yŭn; *buigne* = bump) is a localized swelling at either the dorsal or medial region of the first metatarsophalangeal joint (the joint between the first metatarsal and the proximal phalanx of the great toe). It looks like a bump on the foot near the great toe, and causes that toe to point toward the second toe instead of in a purely anterior direction. Other effects of bunions include bone spurs, which are knobby, abnormal projections from the bone surface that increase tension forces in the nearby tendon; inflammation of a tiny fluid-filled sac (bursa) that acts as a gliding surface to reduce friction around tendons to the great toe; and calluses, thickening of the surface layer of the skin (called hyperkeratosis), usually in response to pressure. Bunions are usually caused by wearing shoes that fit too tightly, and are among the most common foot problems.

Pes cavus (pes că'vus), or *clawfoot*, is characterized by excessively high longitudinal arches. In addition, the joints between the metatarsals and proximal phalanges are overly extended, and the joints between the different phalanges are bent so that they appear clawed. Pes cavus is often seen in patients with neurological disorders (such as poliomyelitis) or muscular disorders (such as atrophy of leg muscles).

Talipes (tal'i-pēz; *talus* = ankle, *pes* = foot) **equinovarus** (ē-kwī-nō-vā'rŭs; *equus* = horse, *varus* = bent inward) is commonly referred to as *congenital clubfoot*. This foot deformity occurs in 0.1% of births, sometimes when there isn't enough room in the womb. In this condition, the feet are permanently inverted (the soles of the feet are twisted medially), and the ankles are plantar flexed (the soles of the feet are twisted more inferiorly), as if the patient were trying to stand on tiptoe. Treatment for mild cases may consist of applying casts or adhesive tape immediately after birth. More severe cases require surgery and corrective shoes.

Pes planus (pla'nus), commonly known as *flat feet*, is a foot deformity in which the medial longitudinal arch is flattened (or "fallen") so that the entire sole touches the ground. Pes planus is often caused by excessive weight, postural abnormalities, or weakened supporting tissue. Individuals who spend most of their day standing may have slightly fallen arches by the end of the day, but with proper rest their arches can return to normal shape. A custom-designed arch support is usually prescribed to treat pes planus. A congenital variation develops *in utero* (during gestation) when the navicular bone articulates with the dorsal side of the talus, thus fixing the talus in a plantar flexed position.

A common foot injury is a **metatarsal stress fracture.** This injury usually results when repetitive pressure or stress on the foot causes a small crack to develop in the outer surface of the bone. The second and third metatarsals are most often involved, although any metatarsal can be affected. Often the individual has no recollection of any injury, and the fracture may not become apparent on x-rays until a few weeks later. Runners are especially prone to this injury because they put repetitive stress on their feet. Extended rest and wearing either stiff or well-cushioned shoes are required for healing.

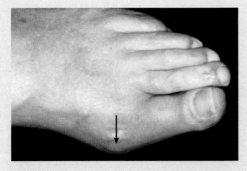

A bunion (arrow).

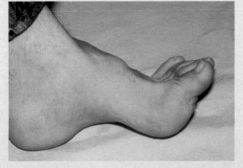

Pes cavus.

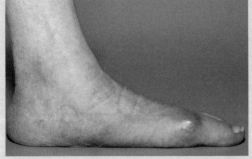

Talipes equinovarus (congenital clubfoot).

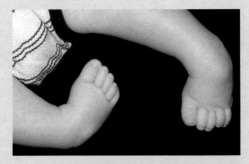

Pes planus.

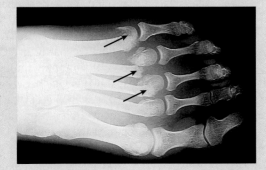

Metatarsal stress fractures (arrows).

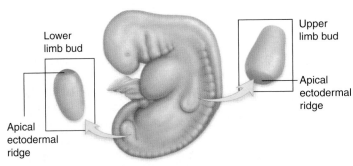

(a) Week 4: Upper and lower limb buds form.

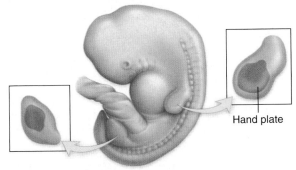

(b) Week 5: Hand plate forms.

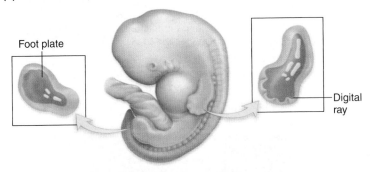

(c) Week 6: Digital rays appear in hand plate. Foot plate forms.

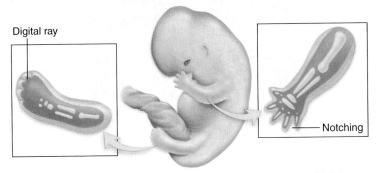

(d) Week 7: Notching develops between digital rays of hand plate. Digital rays appear in foot plate.

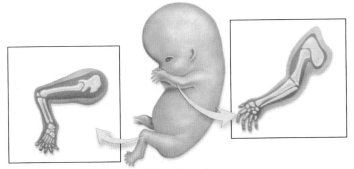

(e) Week 8: Separate fingers and toes formed.

Figure 8.16

Development of the Appendicular Skeleton. The upper and lower limbs develop between weeks 4 and 8. Upper limb development precedes corresponding lower limb development by 2 to 4 days.

week (approximately day 26), while the lower limb buds appear a few days later (day 28) **(figure 8.16)**. In general, the development of upper and lower limbs is similar. However, upper limb development precedes corresponding lower limb development by about 2 to 4 days. The upper and lower limbs form *proximodistally*, meaning that the more proximal parts of the limbs form first (in weeks 4–5), while the more distal parts differentiate later.

Early limb buds are composed of lateral plate mesoderm and covered by a layer of ectoderm. The lateral plate mesoderm later forms the bones, tendons, cartilage, and connective tissue of the limb, while the ectoderm forms the epidermis and the epidermal derivatives. The musculature of the limbs forms from somitic mesoderm that migrates to the developing limbs during the fifth week of development.

At the apex of each limb bud, part of the ectoderm forms an elevated thickening called the **apical ectodermal** (ek-tō-der′măl) **ridge,** which plays a role in the differentiation and elongation of the limb. By mechanisms not completely understood, this ridge "signals" the underlying tissue to form the various components of the limb. Experiments on animals indicate that the limb fails to develop if the apical ectodermal ridge is removed. Thus, this ridge is vital for limb development and differentiation.

Initially, the limb buds are cylindrical. By the early fifth week, the distal portion of the upper limb bud forms a rounded, paddle-shaped **hand plate,** which later becomes the palm and fingers. In the lower limb bud, a corresponding **foot plate** forms during the sixth week. These plates develop longitudinal thickenings called **digital rays,** which eventually form the digits. The digital rays in the hand plate appear in the late sixth week, and the foot digital rays appear during the early seventh week. The digital rays are initially connected by intermediately placed tissue, which later undergoes programmed cell death (apoptosis). Thus, as this intermediate tissue dies, notching occurs between the digital rays, and separate digits are formed. This process occurs in the seventh week and is complete by the eighth week for both the fingers and the toes.

As the limb buds enlarge, bends appear where the future elbow and shoulder joints will develop. During the late seventh through early eighth weeks, the upper limb rotates laterally, so that the elbows are directed posteriorly, while the lower limb rotates medially, so that the knees are directed anteriorly. By week 8 of development, primary ossification centers begin to form in each bone. Hyaline cartilage is gradually replaced by bony tissue via the process of endochondral ossification (see chapter 6). By week 12, the shafts of the limb bones are rapidly ossifying, but other developing bones remain as cartilage. All of these bones continue to develop throughout the fetal period and well into childhood.

WHAT DID YOU LEARN?

9 What is the function of the apical ectodermal ridge?

10 During what week are separate fingers and toes formed?

CLINICAL VIEW

Limb Malformations

Limb and finger malformations may occur due to genetic or environmental influences. Some limb and finger malformations include the following:

- **Polydactyly** (pol-ē-dak'ti-lē; *poly* = many, *daktylos* = finger) is the condition of having extra digits. This trait may be either unilateral (occurring on one hand or foot only) or bilateral (occurring on both). Polydactyly tends to run in families and appears to have a genetic component.

- **Ectrodactyly** (ek-trō-dak'ti-lē; *ectro* = congenital absence of a part) is the absence of a digit. Like polydactyly, ectrodactyly runs in families.

- **Syndactyly** (sin-dak'ti-lē; *syn* = together) refers to "webbing" or abnormal fusion of the digits. It occurs when the intermediate tissue between the digital rays fails to undergo normal programmed cell death. In some cases, there is merely extra tissue between the digits, while in more severe cases, two or more digits are completely fused. Several genes have been implicated in syndactyly, although certain drugs taken by the pregnant mother can cause this condition as well.

- **Amelia** (ă-mē'lē-ă; *a* = without, *melos* = a limb) refers to the complete absence of a limb, whereas **meromelia** (mer-ō-mē'lē-ă; *mero* = part) refers to the partial absence of a limb.

- **Phocomelia** (fō-kō-mē'lē-ă; *phoke* = a seal) refers to a short, poorly formed limb that resembles the flipper of a seal.

A notable instance of limb malformation involved the drug **thalidomide**, which was first marketed in Europe in 1954 as a non-barbiturate sleep aid. Physicians later discovered that the drug also helped quell nausea, and so some prescribed it for their pregnant patients who were experiencing morning sickness. Most believed the drug to be free of side effects. The popularity of thalidomide increased, and soon it was approved for use in over 20 countries (although not in the United States).

In the late 1950s and early 1960s, the incidence of limb malformations in Europe and Canada skyrocketed. Many children were born with limbs shaped like seal flippers (phocomelia) or with no limbs at all (amelia). Other less severe malformations, such as syndactyly, were occurring more frequently as well. Medical detective work soon linked the increase in limb malformations to thalidomide. Researchers found that thalidomide binds to particular regions of chromosomal DNA, effectively "locking up" specific genes and preventing their expression. Among the genes most affected in the fetus were those responsible for blood vessel growth. In the absence of an adequate vascular network, limb bud formation is disrupted. It was discovered that if a pregnant female took thalidomide during weeks 4–8 of embryonic development (the time when the limbs are at their critical stage of development), there was a much greater chance of limb formation being severely disrupted.

Thalidomide was taken off the market in the 1960s, but recently it has made a comeback because it has been shown to be an excellent anti-inflammatory agent and especially effective in reducing the more devastating effects of leprosy. Although thalidomide doesn't kill the causative *Mycobacterium leprae* organism, it provides symptomatic relief while specific antibiotics destroy the bacteria. Thalidomide is also being used as an anti-cancer drug. Because it stops the production of new blood vessels, researchers hope that thalidomide can stop the growth and spread of cancer as well. Already, thalidomide has been approved for the treatment of multiple myeloma (a type of cancer of bone marrow cells). Thalidomide has also been approved for treating the symptoms of AIDS and lupus (an autoimmune disease). However, although thalidomide can be a valuable drug, under no conditions should a pregnant woman ever take it. Thalidomide is the classic example of how teratogens can affect the delicate cycle of embryonic development, and why females of childbearing age should be sure they aren't pregnant before taking any medication.

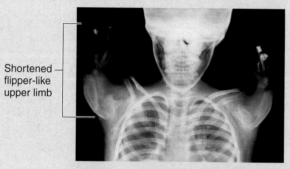

Shortened flipper-like upper limb

Radiograph of a child with phocomelia.

CLINICAL TERMS

hip pointer   Bruising of the soft tissues and bone associated with the anterior superior iliac spine.

lateral humeral epicondylitis   Inflammation of the tissues surrounding the lateral epicondyle of the humerus; also called *tennis elbow*.

patellar dislocation   Displacement of the patella as the result of a blow to the knee or a forceful, unnatural movement of the leg; the patella usually slips to one side.

patellofemoral syndrome   Condition in which the patella doesn't track or align properly on the femur. Females are more prone to this condition because their hips are wider, and thus their femurs flare at a wider angle, affecting the knee joint as well. Due to a weakness of the vastus medialis portion of the quadriceps muscle, the patella is pulled laterally. Patients can alleviate the resulting knee pain by performing specific directed exercises to strengthen the vastus medialis muscle.

CHAPTER   SUMMARY

	■ The appendicular skeleton includes the bony supports (girdles) that attach the upper and lower limbs to the axial skeleton, as well as the bones of those limbs.
Pectoral Girdle 219	■ The pectoral girdle is composed of the clavicle and scapula. **Clavicle 219** ■ The clavicle forms the collarbone. **Scapula 219** ■ The scapula forms the "shoulder blade."
Upper Limb 223	■ Each upper limb contains a humerus, radius, ulna, 8 carpals, 5 metacarpals, and 14 phalanges. **Humerus 223** ■ The head of the humerus articulates with the glenoid cavity of the scapula. ■ The greater tubercle and lesser tubercle are important sites for muscle attachment. The trochlea and capitulum articulate with the radius and ulna at the elbow. **Radius and Ulna 223** ■ The radius and ulna are the bones of the forearm. **Carpals, Metacarpals, and Phalanges 228** ■ The carpal bones are the scaphoid, lunate, triquetrum, and pisiform (proximal row) and the trapezium, trapezoid, capitate, and hamate (distal row). ■ Five metacarpal bones form the bones in the palm of the hand. ■ The phalanges are the finger bones. Four of the fingers contain three phalanges; the pollex (thumb) has only two.
Pelvic Girdle 230	■ The pelvic girdle consists of two ossa coxae. The pelvis is composed of the two ossa coxae, the sacrum, and the coccyx. **Os Coxae 230** ■ Each os coxae forms through the fusion of an ilium, an ischium, and a pubis. The acetabulum is the socket that articulates with the head of the femur. **True and False Pelves 231** ■ The pelvic brim is an oval ridge of bone that divides the entire pelvis into a true (inferior) pelvis and a false (superior) pelvis. **Sex Differences Between the Female and Male Pelves 231** ■ The shapes and orientations of the pelvic bones are very different in females and males.
Lower Limb 234	■ The lower limb is composed of the femur, patella, tibia, fibula, 7 tarsals, 5 metatarsals, and 14 phalanges. **Femur 235** ■ The femur has a rounded head and an elongated neck. The greater and lesser trochanters are projections near the head. ■ The medial and lateral condyles articulate with the condyles of the tibia. **Patella 238** ■ The patella is the kneecap. **Tibia and Fibula 238** ■ In the leg, the tibia is medially located. Its medial malleolus forms the medial bump of the ankle. ■ The fibula is the lateral leg bone. Its lateral malleolus forms the lateral bump of the ankle. **Tarsals, Metatarsals, and Phalanges 239** ■ The seven tarsal bones are the calcaneus, talus, navicular, three cuneiforms, and the cuboid. ■ When we stand, our weight is transferred along the longitudinal and transverse arches of the foot.
Aging of the Appendicular Skeleton 243	■ Some age-related changes in the skeleton are due to maturation and further development, while others reflect deterioration of bone tissue.
Development of the Appendicular Skeleton 243	■ The limbs first appear as limb buds during the fourth week. In general, lower limb development lags behind upper limb development by 2 to 4 days.

CHALLENGE YOURSELF

Matching

Match each numbered item (bone feature or bone description) with the most closely related lettered item (the bone having that feature).

_____ 1. lateral malleolus   a. tibia

_____ 2. supraspinous fossa   b. fibula

_____ 3. tarsal bone   c. ulna

_____ 4. capitulum   d. lunate

_____ 5. radial notch   e. clavicle

_____ 6. acetabulum   f. femur

_____ 7. lesser trochanter   g. scapula

_____ 8. medial malleolus   h. talus

_____ 9. sternal end   i. os coxae

_____ 10. carpal bone   j. humerus

Multiple Choice

Select the best answer from the four choices provided.

_____ 1. The female pelvis typically has which of the following characteristics?

    a. narrow, U-shaped greater sciatic notch.

    b. wide subpubic angle, greater than 100 degrees

    c. short, triangular pubic body

    d. smaller, heart-shaped pelvic inlet

_____ 2. The posterior surface depression at the distal end of the humerus is the

    a. intercondylar fossa.

    b. coronoid fossa.

    c. olecranon fossa.

    d. intertubercular groove.

_____ 3. The spine of the scapula separates which two fossae?

    a. supraspinous, subscapular

    b. subscapular, infraspinous

    c. infraspinous, supraspinous

    d. supraspinous, glenoid

_____ 4. The femur articulates with the tibia at the femur's

    a. linea aspera.

    b. medial and lateral condyles.

    c. head of the femur.

    d. greater trochanter of the femur.

_____ 5. The bony feature palpated on the dorsolateral side of the wrist is the

    a. styloid process of radius.

    b. head of ulna.

    c. pisiform bone.

    d. radial tuberosity.

_____ 6. Identify the bone that articulates with the os coxae at the acetabulum.

    a. sacrum

    b. humerus

    c. femur

    d. tibia

_____ 7. Which of the following is a carpal bone?

    a. cuneiform

    b. cuboid

    c. trapezium

    d. talus

_____ 8. When sitting upright, you are resting on your

    a. pubic bones.

    b. ischial tuberosities.

    c. sacroiliac joints.

    d. iliac crest.

_____ 9. The two prominent bumps you can palpate on the sides of your ankle are the

    a. head of the fibula and the tibial tuberosity.

    b. calcaneus and cuboid.

    c. medial malleolus and lateral malleolus.

    d. styloid processes.

_____ 10. The glenoid cavity articulates with which bone or bone feature?

    a. clavicle

    b. head of the humerus

    c. acromion process

    d. trochlea of the humerus

Content Review

1. Compare the anatomic and functional features of the pectoral and pelvic girdles.

2. Identify and describe the borders of the scapula.

3. What is the difference between the anatomical neck and the surgical neck of the humerus?

4. Name and describe the placement of the eight carpal bones of the wrist.

5. How do the glenoid cavity and the acetabulum differ?

6. When do the ilium, ischium, and pubis fuse to form the os coxae? What features do each of these bones contribute to the os coxae?

7. Distinguish between the true and false pelves. What bony landmark separates the two?

8. What is the function of the slender leg bone called the fibula?

9. Discuss the functions of the arches of the foot.

10. Discuss the development of the limbs. What primary germ layers form the limb bud? List the major events during each week of limb development.

Developing Critical Reasoning

1. A female in her first trimester of pregnancy sees her physician. She suffers from lupus and has read that the drug thalidomide has shown remarkable promise in treating the symptoms. Should the physician prescribe the drug for her at this time? Why or why not?

2. Forensic anthropologists are investigating portions of a human pelvis found in a cave. How can they tell the sex, relative age, and some physical characteristics of the individual based on the pelvis alone?

3. A young male wishes to enlist in the Army. During his physical, the physician tells him he has pes planus and will not be able to enlist because of it. What is pes planus, and why does the Army not accept individuals who have it?

ANSWERS TO "WHAT DO YOU THINK?"

1. The clavicle is called the "collarbone" because the collar of a shirt rests over this bone.

2. Eight cube-shaped carpal bones allow a great deal more movement than just one or two large carpal bones could because movement can occur between each joint among two or more bones. Having eight carpal bones results in many intercarpal joints where movement may occur, and hence many more possible movements.

3. The glenoid cavity of the scapula is flatter and shallower than the deep, curved acetabulum of the os coxae. The pelvic girdle (both ossa coxae) maintains stronger, more tightly fitting bony connections with the lower limbs than the pectoral girdle (scapula and clavicle) does with the upper limbs.

4. The medial and lateral malleoli of the leg bones are analogous to the styloid processes of the radius and ulna. Both sets of bony features produce the bumps felt along the ankle and the wrist.

5. Both the carpal and the tarsal bones are short bones. The multiple bones (8 carpal bones, 7 tarsal bones) allow for a range of movement at the intercarpal or intertarsal joints. However, the wrist is more freely movable than the ankle and proximal foot, because the foot is adapted for weight bearing. The tarsal bones are larger and bulkier than the carpal bones.

Visit the McKinley/O'Loughlin *Human Anatomy,* 2e website at aris.mhhe.com

9

Anatomy & Physiology | **REVEALED**
aprevealed.com

SKELETAL SYSTEM

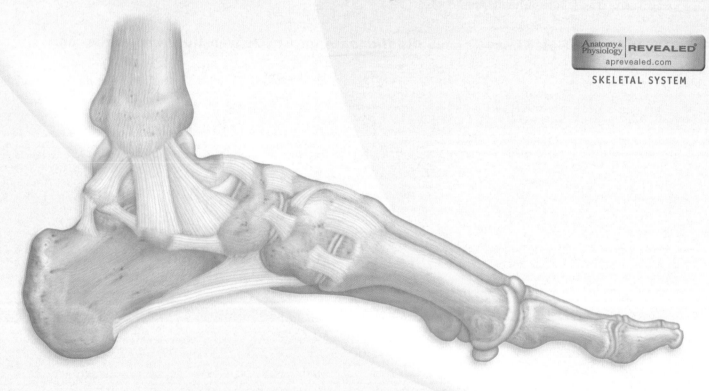

Articulations

Our skeleton protects vital organs and supports soft tissues. Its marrow cavity is the source of new blood cells. When it interacts with the muscular system, the skeleton helps the body move. Bones are too rigid to bend; however, they meet at joints, which anatomists call articulations. In this chapter, we examine how bones articulate and sometimes the bones still allow some freedom of movement, depending on the shapes and supporting structures of the various joints.

Articulations (Joints)

Key topics in this section:

- General structure of articulations
- How degree of movement is determined at a joint
- Structural and functional classifications of joints

A **joint**, or **articulation** (ar-tik-ū-lā′shŭn), is the place of contact between bones, between bone and cartilage, or between bones and teeth. Bones are said to **articulate** with each other at a joint. The scientific study of joints is called **arthrology** (ar-throl′ō-jē; *arthron* = joint, *logos* = study).

> ## Study Tip!
>
> You can figure out the names of most joints by piecing together the names of the bones that form them. For example, the *glenohumeral* joint is where the glenoid cavity of the scapula meets the head of the humerus, and the *sternoclavicular* joint is where the manubrium of the sternum articulates with the sternal end of the clavicle.

The motion permitted at a joint ranges from no movement (e.g., where some skull bones interlock at a suture) to extensive movement (e.g., at the shoulder, where the arm connects to the scapula). The structure of each joint determines its mobility and its stability. There is an inverse relationship between mobility and stability in articulations. The more mobile a joint is, the less stable it is. In contrast, if a joint is immobile, it is correspondingly more stable. **Figure 9.1** illustrates the "tradeoff" between mobility and stability for various joints.

Classification of Joints

Joints are categorized structurally on the basis of the type of connective tissue that binds the articulating surfaces of the bones, and whether a space occurs between the articulating bones:

- A **fibrous** (fī′brŭs) **joint** occurs where bones are held together by dense regular (fibrous) connective tissue.
- A **cartilaginous** (kar-ti-laj′i-nŭs; *cartilago* = gristle) **joint** occurs where bones are joined by cartilage.
- A **synovial** (si-nō′vē-ăl) **joint** has a fluid-filled, joint cavity that separates the articulating surfaces of the bones. The articulating surfaces are enclosed within a capsule, and the bones are also joined by various ligaments.

Joints may also be classified functionally based on the extent of movement they permit:

- A **synarthrosis** (sin′ar-thrō′sis; pl., sin′ar-thrō′sēz; *syn* = joined together) is an immobile joint.
- An **amphiarthrosis** (am′fi-ar-thrō′sis; pl., -sēz; *amphi* = around) is a slightly mobile joint.
- A **diarthrosis** (dī-ar-thrō′sis; pl., -sēz; *di* = two) is a freely mobile joint.

The following discussion of articulations is based on their structural classification, with functional categories included as appropriate. As you read about the various types of joints, it may help you to refer to the summary of joint classifications in **table 9.1**.

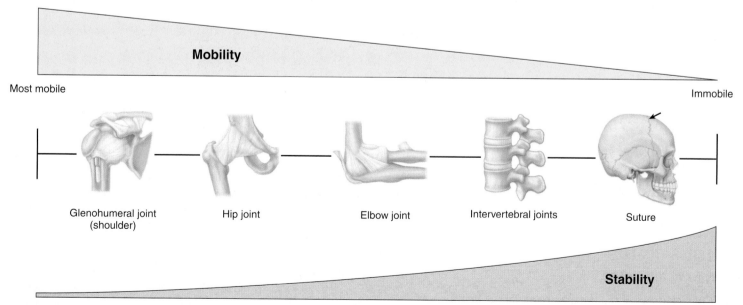

Mobility

Most mobile   Immobile

Glenohumeral joint (shoulder)	Hip joint	Elbow joint	Intervertebral joints	Suture

Stability

Very unstable   Most stable

Figure 9.1

Relationship Between Mobility and Stability in Joints. In every joint, there is a "tradeoff" between the relative amounts of mobility and stability. The more mobile the joint, the less stable it is. Conversely, the more stable the joint, the less mobile it is. Note how the glenohumeral (shoulder) joint is very mobile but not very stable, while a suture is immobile and yet very stable.

Table 9.1		Joint Classifications		
Structural Classification	**Structural Characteristics**	**Structural Category**	**Example**	**Functional Classification**
Fibrous	Dense regular connective tissue holds together the ends of bones and bone parts; no joint cavity	Gomphosis: Periodontal membranes hold tooth to bony jaw	Tooth to jaw	Synarthrosis (immobile)
		Suture: Dense regular connective tissue connects skull bones	Lambdoid suture (connects occipital and parietal bones)	Synarthrosis (immobile)
		Syndesmosis: Dense regular connective tissue fibers (interosseous membrane) between bones	Articulation between radius and ulna, and between tibia and fibula	Amphiarthrosis (slightly mobile)
Cartilaginous	Pad of cartilage is wedged between the ends of bones; no joint cavity	Synchondrosis: Hyaline cartilage plate between bones	Epiphyseal plates in growing bones; costochondral joints	Synarthrosis (immobile)
		Symphysis: Fibrocartilage pad between bones	Pubic symphysis; intervertebral disc articulations	Amphiarthrosis (slightly mobile)
Synovial	Ends of bones covered with articular cartilage; joint cavity separates the articulating bones; enclosed by an articular capsule, lined by a synovial membrane; contains synovial fluid	**Uniaxial** Plane joint: Flattened or slightly curved faces slide across one another Hinge joint: Permits angular movements in a single plane Pivot joint: Permits rotation only	Plane joint: Intercarpal joints, intertarsal joints Hinge joint: Elbow joint Pivot joint: Atlantoaxial joint	Diarthrosis (freely mobile)
		Biaxial Condylar joint: Oval articular surface on one bone closely interfaces with a depressed oval surface on another bone Saddle joint: Saddle-shaped articular surface on one bone closely interfaces with saddle-shaped surface on another bone	Condylar joint: MP (metacarpophalangeal) joints Saddle joint: Articulation between carpal and first metacarpal bone	Diarthrosis (freely mobile)
		Multiaxial (triaxial) Ball-and-socket joint: Round head of one bone rests within cup-shaped depression in another bone	Ball-and-socket joint: Glenohumeral joint, hip joint	Diarthrosis (freely mobile)

Fibrous Joints

Key topics in this section:

- Characteristics of the three types of fibrous joints
- Some locations of gomphoses, sutures, and syndesmoses in the body

Articulating bones are joined by dense regular connective tissue in fibrous joints. Most fibrous joints are immobile or only slightly mobile. Fibrous joints have no joint cavity (space between the articulating bones). The three types of fibrous joints are gomphoses, sutures, and syndesmoses **(figure 9.2)**.

Gomphoses

A **gomphosis** (gom-fō′sis; pl., -sēz; *gomphos* = bolt, *osis* = condition) resembles a "peg in a socket." The only gomphoses in the human body are the articulations of the roots of individual teeth with the sockets of the mandible and the maxillae. A tooth is held firmly in place by a fibrous **periodontal** (per′ē-ō-don′tăl; *peri* = around, *odous* = tooth) **membrane**. This joint is functionally classified as a synarthrosis.

The reasons orthodontic braces can be painful and take a long time to correctly position the teeth are related to the gomphosis architecture. The orthodontist's job is to reposition these normally immobile joints through the use of bands, rings, and braces. In response to these mechanical stressors, osteoblasts and osteoclasts work together to modify the alveolus, resulting in the remodeling of the joint and the slow repositioning of the teeth.

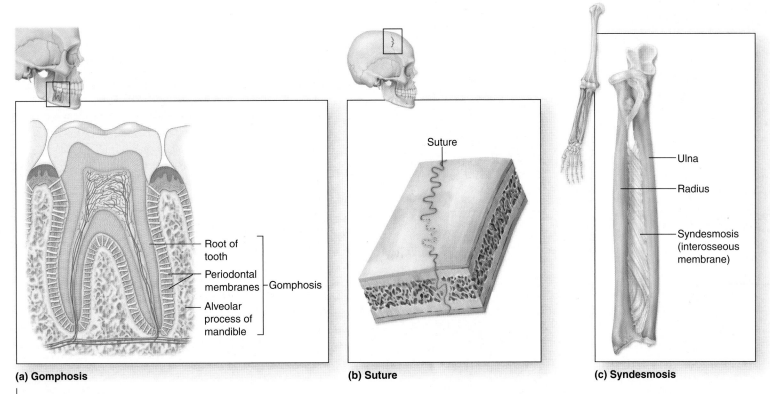

(a) Gomphosis

Root of tooth

Periodontal membranes — Gomphosis

Alveolar process of mandible

(b) Suture

Suture

(c) Syndesmosis

Ulna

Radius

Syndesmosis (interosseous membrane)

Figure 9.2

Fibrous Joints.   Dense regular connective tissue binds the articulating bones in fibrous joints to prevent or severely restrict movement. *(a)* A gomphosis is the immobile joint between a tooth and the jaw. *(b)* A suture is an immobile joint between bones of the skull. *(c)* A syndesmosis permits slight mobility between the radius and the ulna.

Sutures

Sutures (soo′choor; *sutura* = a seam) are immobile fibrous joints (synarthoses) that are found only between certain bones of the skull. Sutures have distinct, interlocking, usually irregular edges that both increase their strength and decrease the number of fractures at these articulations. In addition to joining bones, sutures permit the skull to grow as the brain increases in size during childhood. In an older adult, the dense regular connective tissue in the suture becomes ossified, fusing the skull bones together. When the bones have completely fused across the suture line, these obliterated sutures become **synostoses** (sin-os-tō′sēz; sing., -sis).

Syndesmoses

Syndesmoses (sin′dez-mō′sēz; sing., -sis; *syndesmos* = a fastening) are fibrous joints in which articulating bones are joined by long strands of dense regular connective tissue only. Because syndesmoses allow for slight mobility, they are classified as amphiarthroses. Syndesmoses are found between the radius and ulna, and between the tibia and fibula. The shafts of the two articulating bones are bound side by side by a broad ligamentous sheet called an **interosseous membrane** (or *interosseous ligament*). The interosseous membrane provides a pivot point where the radius and ulna (or the tibia and fibula) can move against one another.

WHAT DID YOU LEARN?

❸ Describe the three types of fibrous joints, and name a place in the body where each type is found.

Cartilaginous Joints

Key topics in this section:

- Characteristics of the two types of cartilaginous joints
- Some locations of synchondroses and symphyses in the body

The articulating bones in cartilaginous joints are attached to each other by cartilage. These joints lack a joint cavity. The two types of cartilaginous joints are synchondroses and symphyses **(figure 9.3)**.

Synchondroses

An articulation in which bones are joined by hyaline cartilage is called a **synchondrosis** (sin′kon-drō′sis; pl., -sēz; *chondros* = cartilage). Functionally, all synchondroses are immobile and thus are classified as synarthroses. The hyaline cartilage of epiphyseal plates in children forms synchondroses that bind the epiphyses and the diaphysis of long bones. When the hyaline cartilage stops growing, bone replaces the cartilage, and a synchondrosis no longer exists.

The spheno-occipital synchondrosis is found between the body of the sphenoid and the basilar part of the occipital bone. This synchondrosis fuses between 18 and 25 years of age, making it a useful tool for assessing the age of a skull.

Another synchondrosis is the attachment of the first rib to the sternum by costal cartilage (called the first sternocostal joint). Here, the first rib and its costal cartilage (formed from hyaline

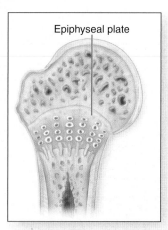

Epiphyseal plate

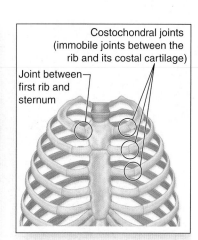

Costochondral joints
(immobile joints between the
rib and its costal cartilage)

Joint between
first rib and
sternum

(a) Synchondroses (contain hyaline cartilage)

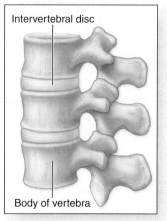

Intervertebral disc

Body of vertebra

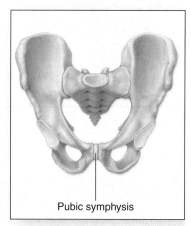

Pubic symphysis

(b) Symphyses (contain fibrocartilage)

Figure 9.3

Cartilaginous Joints. Articulating bones are joined by cartilage. *(a)* Synchondroses are immobile joints that occur in an epiphyseal plate in a long bone and in the joint between a rib and the sternum. *(b)* Symphyses are amphiarthroses and occur in the intervertebral discs and the pubic symphysis.

cartilage) are united firmly to the manubrium of the sternum to provide stability to the rib cage. A final example of synchondroses are the **costochondral** (kos-tō-kon'drăl; *costa* = rib) **joints,** the joints between each bony rib and its respective costal cartilage. (Note that the costochondral joints are different from the articulation between the sternum and the costal cartilage of ribs 2–7, which is a synovial joint, not a synchondrosis.)

WHAT DO YOU THINK?

① Why is a synchondrosis a synarthrosis? Why would you want a synchondrosis to be immobile?

Symphyses

A **symphysis** (sim'fi-sis; pl., -sēz; growing together) has a pad of fibrocartilage between the articulating bones. The fibrocartilage resists compression and tension stresses and acts as a resilient shock absorber. All symphyses are amphiarthroses, meaning that they allow slight mobility.

One example of a symphysis is the pubic symphysis, which is located between the right and left pubic bones. In pregnant

Costochondritis

Costochondritis (kos-tō-kon-drī' tis; *itis* = inflammation) refers to inflammation and irritation of the costochondral joints, resulting in localized chest pain. Any costochondral joint may be affected, although the joints for ribs 2–6 are those most commonly injured. The cause of costochondritis is usually unknown, but some documented causes include repeated minor trauma to the chest wall (e.g., from forceful repeated coughing during a respiratory infection or overexertion during exercise) and bacterial or viral infection of the joints themselves. Some backpackers who do not use the chest brace have experienced bouts of costochondritis.

The most common symptom of costochondritis is localized chest pain, typically following exertion or a respiratory infection. The pain may be mistaken for that caused by a myocardial infarction (heart attack), and thus may cause needless anxiety for the patient. Sitting, lying on the affected side, and increased mental stress can exacerbate symptoms. Costochondritis is not a medical emergency and may be treated with NSAIDs (nonsteroidal anti-inflammatory drugs, such as aspirin). With proper rest and treatment, symptoms typically disappear after several weeks.

females, the pubic symphysis becomes more mobile to allow the pelvis to change shape slightly as the fetus passes through the birth canal.

Other examples of symphyses are the intervertebral joints, where the bodies of adjacent vertebrae are both separated and united by intervertebral discs. These intervertebral discs allow only slight movements between the adjacent vertebrae; however, the collective movements of all the intervertebral discs afford the spine considerable flexibility.

WHAT DID YOU LEARN?

④ Describe a symphysis. In what functional category is this type of joint placed, and why?

Synovial Joints

Key topics in this section:

- General anatomy of synovial joints and their accessory structures
- Classes of synovial joints based on the shapes of the joint surfaces and the types of movement permitted
- Dynamic movements at synovial joints

Synovial joints are freely mobile articulations. Unlike the joints previously discussed, the bones in a synovial joint are separated by a space called a joint cavity. Most of the commonly known joints in the body are synovial joints, including the glenohumeral (shoulder) joint, the temporomandibular joint, the elbow joint, and the knee joint. Functionally, all synovial joints are classified as diarthroses, since all are freely mobile. Often, the terms *diarthrosis* and *synovial joint* are equated.

General Anatomy of Synovial Joints

All types of synovial joints have several basic features: an articular capsule, a joint cavity, synovial fluid, articular cartilage, ligaments, and nerves and blood vessels (**figure 9.4**).

Each synovial joint is composed of a double-layered capsule called the **articular** (ar-tik′ū-lăr) **capsule** (or *joint capsule*). The outer layer is called the **fibrous layer**, while the inner layer is a **synovial membrane** (or *synovium*). The fibrous layer is formed from dense regular connective tissue, and it strengthens the joint to prevent the bones from being pulled apart. The synovial membrane is composed primarily of areolar connective tissue, covers all the internal joint surfaces not covered by cartilage, and lines the articular capsule.

Only synovial joints house a **joint cavity** (or *articular cavity*), a space that contains a small amount of synovial fluid. The cavity permits separation of the articulating bones. The articular cartilage and synovial fluid within the joint cavity reduce friction as bones move at a synovial joint.

Lining the joint cavity is the synovial membrane, which secretes a viscous, oily **synovial fluid**. Synovial fluid is composed of secretions from synovial membrane cells and a filtrate from blood plasma. Synovial fluid has three functions:

1. Synovial fluid lubricates the articular cartilage on the articulating bones (in the same way that oil in a car engine lubricates the moving engine parts).
2. Synovial fluid nourishes the articular cartilage's chondrocytes. The relatively small volume of synovial fluid must be circulated continually to provide nutrients and remove wastes to these cells. Whenever movement occurs at a synovial joint, the combined compression and re-expansion of the articular cartilage circulate the synovial fluid into and out of the cartilage matrix.
3. Synovial fluid acts as a shock absorber, distributing stresses and force evenly across the articular surfaces when the pressure in the joint suddenly increases.

All articulating bone surfaces in a synovial joint are covered by a thin layer of hyaline cartilage called **articular cartilage**. This cartilage reduces friction in the joint during movement, acts as a spongy cushion to absorb compression placed on the joint, and prevents damage to the articulating ends of the bones. This special hyaline cartilage lacks a perichondrium. Mature cartilage is avascular, so it does not have blood vessels to bring nutrients to and remove waste products from the tissue. The repetitious compression/relaxation that occurs during exercise is vital to the articular cartilage's well-being because the accompanying pumping action enhances its nutrition and waste removal.

Ligaments (lig′ă-ment; *ligamentum* = a band) are composed of dense regular connective tissue. Ligaments connect one bone to another bone and strengthen and reinforce most synovial joints. **Extrinsic ligaments** are outside of and physically separate from the articular capsule, whereas **intrinsic ligaments** represent thickenings of the articular capsule itself. Intrinsic ligaments include *extracapsular ligaments* outside the articular capsule and *intracapsular ligaments* within the articular capsule.

Tendons (ten′dŏn; *tendo* = extend) are not part of the synovial joint itself. Like a ligament, a tendon is composed of dense regular connective tissue. However, whereas a ligament binds bone to bone, a tendon attaches a muscle to a bone. When a muscle contracts, the tendon from that muscle moves the bone to which it is attached, thus creating movement at the joint. Tendons help stabilize joints because they pass across or around a joint to provide mechanical support, and sometimes they limit the range or amount of movement permitted at a joint.

All synovial joints have numerous sensory **nerves** and **blood vessels** that innervate and supply the articular capsule and associated ligaments. The sensory nerves detect painful stimuli in the joint and report on the amount of movement and stretch in the joint. By monitoring stretching at a joint, the nervous system can detect changes in our posture and adjust body movements.

In addition to the main structures just described, synovial joints usually have the following accessory structures: bursae, fat pads, and tendons.

A **bursa** (ber′să; pl., *bursae*, ber′sē; a purse) is a fibrous, saclike structure that contains synovial fluid and is lined by a synovial membrane (**figure 9.5a**). Bursae are found around most synovial joints and also where bones, ligaments, muscles, skin, or tendons overlie each other and rub together. Bursae may be either connected to the joint cavity or completely separate from it. They are designed to alleviate the friction resulting from the various body movements, such as a tendon or ligament rubbing against bone. An elongated bursa called a **tendon sheath** wraps around tendons where there may be excessive friction. Tendon sheaths are especially common in the confined spaces of the wrist and ankle (figure 9.5b).

Fat pads are often distributed along the periphery of a synovial joint. They act as packing material and provide some protection for the joint. Often fat pads fill the spaces that form when bones move and the joint cavity changes shape.

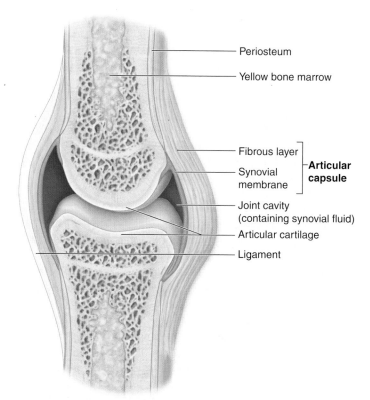

Periosteum

Yellow bone marrow

Fibrous layer ⎫
Synovial membrane ⎬ **Articular capsule**

Joint cavity (containing synovial fluid)

Articular cartilage

Ligament

Typical synovial joint

Figure 9.4

Synovial Joints.   All synovial joints are diarthroses, and they permit a wide range of motion.

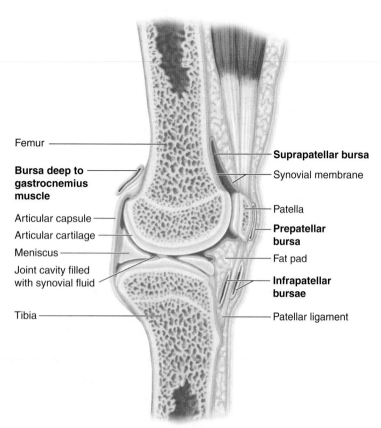

(a) Bursae of the knee joint, sagittal section

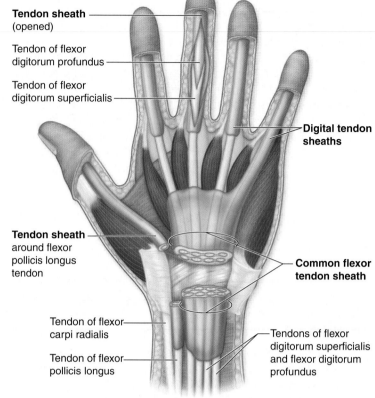

(b) Tendon sheaths of wrist and hand, anterior view

Figure 9.5

Bursae and Tendon Sheaths.   Synovial-fluid-filled structures called bursae and tendon sheaths reduce friction where ligaments, muscles, tendons, and bones rub together. *(a)* The knee joint contains a number of bursae. *(b)* The wrist and hand contain numerous tendon sheaths (blue).

CLINICAL VIEW

"Cracking Knuckles"

Cracking or popping sounds often result when people pull forcefully on their fingers. Stretching or pulling on a synovial joint causes the joint volume to immediately expand and the pressure on the fluid within the joint to decrease, so that a partial vacuum exists within the joint. As a result, the gases dissolved in the fluid become less soluble, and they form bubbles, a process called *cavitation*. When the joint is stretched to a certain point, the pressure in the joint drops even lower, so the bubbles in the fluid burst, producing a popping or cracking sound. (Similarly, displaced water in a sealed vacuum tube makes this sound as it hits against the glass wall.) It typically takes about 25 to 30 minutes for the gases to dissolve back into the synovial fluid. You cannot crack your knuckles again until these gases dissolve. Contrary to popular belief, cracking your knuckles does *not* cause arthritis.

Types of Synovial Joints

Synovial joints are classified by the shapes of their articulating surfaces and the types of movement they allow. Movement of a bone at a synovial joint is best described with respect to three intersecting perpendicular planes or axes:

■ A joint is said to be **uniaxial** (yū-nē-ak′sē-ăl; *unus* = one) if the bone moves in just one plane or axis.

■ A joint is **biaxial** (bī-ak′sē-ăl; *bi* = double) if the bone moves in two planes or axes.

■ A joint is **multiaxial** (or **triaxial** [trī-ak′sē-ăl; *tri* = three]) if the bone moves in multiple planes or axes.

Note that all synovial joints are diarthroses, although some are more mobile than others. From least mobile to most freely mobile, the six specific types of synovial joints are plane joints, hinge joints, pivot joints, condylar joints, saddle joints, and ball-and-socket joints **(figure 9.6)**.

A **plane** (*planus* = flat) **joint**, also called a *planar* or *gliding joint*, is the simplest synovial articulation and the least mobile type of diarthrosis. This type of synovial joint is also known as a *uniaxial* joint because only side-to-side movements are possible. The articular surfaces of the bones are flat, or planar. Examples of plane joints include the intercarpal and intertarsal joints (the joints between the cube-shaped carpal and tarsal bones).

A **hinge joint** is a uniaxial joint in which the convex surface of one articulating bone fits into a concave depression on the other bone. Movement is confined to a single axis, like the hinge of a door. An example is the elbow joint. The trochlear notch of the ulna fits directly into the trochlea of the humerus, so the forearm can be moved only anteriorly toward the arm or posteriorly away from the arm. Other hinge joints occur in the knee and the finger (interphalangeal (IP)) joints.

A **pivot** (piv′ŏt) **joint** is a uniaxial joint in which one articulating bone with a rounded surface fits into a ring formed by a ligament and another bone. The first bone rotates on its longitudinal axis relative to the second bone. An example is the proximal radioulnar

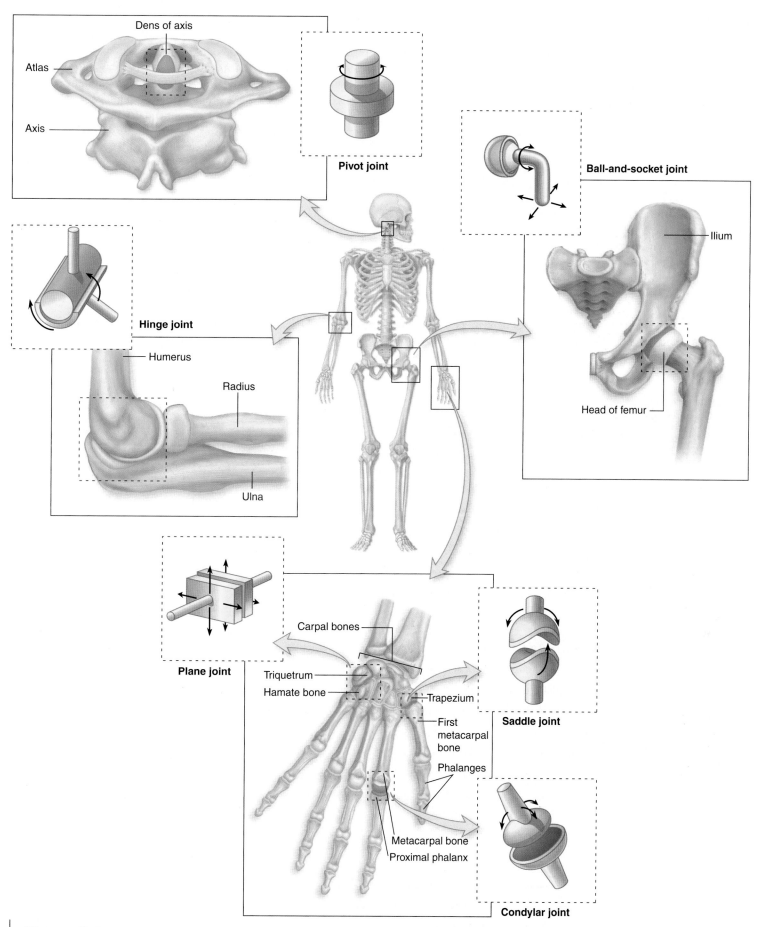

Figure 9.6

Types of Synovial Joints. These six types of synovial joints permit specific types of movement.

joint, where the rounded head of the radius pivots along the ulna and permits the radius to rotate. Another example is the atlantoaxial joint between the first two cervical vertebrae. The rounded dens of the axis fits snugly against an articular facet on the anterior arch of the atlas. This joint pivots when you shake your head "no."

Condylar (kon′di-lar) **joints**, also called *condyloid* or *ellipsoid joints*, are biaxial joints with an oval, convex surface on one bone that articulates with a concave articular surface on the second bone. Biaxial joints can move in two axes, such as back-and-forth and side-to-side. Examples of condylar joints are the metacarpophalangeal (MP) (met′ă-kar′pō-fă-lan′jē-ăl) joints of fingers 2 through 5. The MP joints are commonly referred to as "knuckles." Examine your hand and look at the movements along the MP joints; you can flex and extend the fingers at this joint (that is one axis of movement). You also can move your fingers apart from one another and move them closer together, which is the second axis of movement.

A **saddle joint** is so named because the articular surfaces of the bones have convex and concave regions that resemble the shape of a saddle. It allows a greater range of movement than either a condylar or hinge joint. The carpometacarpal joint of the thumb (between the trapezium and the first metacarpal) is an example of a saddle joint. This joint permits the thumb to move toward the other fingers so that we can grasp objects.

Ball-and-socket joints are multiaxial joints in which the spherical articulating head of one bone fits into the rounded, cuplike socket of a second bone. Examples of these joints are the hip joint and the glenohumeral joint. The multiaxial nature of these joints permits movement in three axes. Move your arm at your shoulder, and observe the wide range of movements that can be produced. This is why the ball-and-socket joint is considered the most freely mobile type of synovial joint.

WHAT DO YOU THINK?

2 If a ball-and-socket joint is more mobile than a plane joint, which of these two joints is more *stable*?

Movements at Synovial Joints

Four types of motion occur at synovial joints: gliding, angular, rotational, and special movements (motions that occur only at specific joints) **(table 9.2)**.

Table 9.2	Movements at Synovial Joints	
Movement	**Description**	**Opposing Movement[1]**
Gliding Motion	Two opposing articular surfaces slide past each other in almost any direction; the amount of movement is slight	None
Angular Motion	The angle between articulating bones increases or decreases	
Flexion	The angle between articulating bones decreases; usually occurs in the sagittal plane	Extension
Extension	The angle between articulating bones increases; usually occurs in the sagittal plane	Flexion
Hyperextension	Extension movement continues past the anatomic position	Flexion
Lateral flexion	The vertebral column moves in either lateral direction along a coronal plane	None
Abduction	Movement of a bone away from the midline; usually in the coronal plane	Adduction
Adduction	Movement of a bone toward the midline; usually in the coronal plane	Abduction
Circumduction	A continuous movement that combines flexion, abduction, extension, and adduction in succession; the distal end of the limb or digit moves in a circle	None
Rotational Motion	A bone pivots around its own longitudinal axis	None
Pronation	Rotation of the forearm whereby the palm is turned posteriorly	Supination
Supination	Rotation of the forearm whereby the palm is turned anteriorly	Pronation
Special Movements	Types of movement that don't fit in the previous categories	
Depression	Movement of a body part inferiorly	Elevation
Elevation	Movement of a body part superiorly	Depression
Dorsiflexion	Ankle joint movement whereby the dorsum of the foot is brought closer to the anterior surface of the leg	Plantar flexion
Plantar flexion	Ankle joint movement whereby the sole of the foot is brought closer to the posterior surface of the leg	Dorsiflexion
Inversion	Twisting motion of the foot that turns the sole medially or inward	Eversion
Eversion	Twisting motion of the foot that turns the sole laterally or outward	Inversion
Protraction	Anterior movement of a body part from anatomic position	Retraction
Retraction	Posterior movement of a body part from anatomic position	Protraction
Opposition	Special movement of the thumb across the palm toward the fingers to permit grasping and holding of an object	Reposition

[1] Some movements (e.g., circumduction) do not have an opposing movement.

Gliding Motion

Gliding is a simple movement in which two opposing surfaces slide slightly back-and-forth or side-to-side with respect to one another. In a gliding motion, the angle between the bones does not change, and only limited movement is possible in any direction. Gliding motion typically occurs along plane joints.

Angular Motion

Angular motion either increases or decreases the angle between two bones. These movements may occur at many of the synovial joints; they include the following specific types: flexion and extension, hyperextension, lateral flexion, abduction and adduction, and circumduction.

Flexion (flek′shŭn; *flecto* = to bend) is movement in an anterior-posterior (AP) plane of the body that *decreases* the angle between the articulating bones. Bones are brought closer together as the angle between them decreases. Examples include bending your fingers toward your palm to make a fist, bending your forearm toward your arm at the elbow, flexion at the shoulder when you raise an arm anteriorly, and flexion of the neck when you bend your head anteriorly to look down at your feet. The opposite of flexion is **extension** (eks-ten′shŭn; *extensio* = a stretching out), which is movement in an anterior-posterior plane that *increases* the angle

between the articulating bones. Extension is a straightening action that usually occurs in the sagittal plane of the body. Straightening your arm and forearm until the upper limb projects directly away from the anterior side of your body or straightening your fingers after making a clenched fist are examples of extension. Flexion and extension of various body parts are illustrated in **figure 9.7a–d.**

Hyperextension (hī′per-eks-ten′shŭn; *hyper* = above normal) is the extension of a joint beyond 180 degrees. For example, if you extend your arm and hand with the palm facing inferiorly, and then raise the back of your hand as if admiring a new ring on your finger, the wrist is hyperextended. If you glance up at the ceiling while standing, your neck is hyperextended.

Lateral flexion occurs when the trunk of the body moves in a coronal plane laterally away from the body. This type of movement occurs primarily between the vertebrae in the cervical and lumbar regions of the vertebral column (figure 9.7e).

Abduction (ab-dŭk′shŭn), which means to "move away," is a lateral movement of a body part *away from* the body midline. Abduction occurs when either the arm or the thigh is moved laterally away from the body midline. Abduction of either the fingers or the toes means that you spread them apart, away from the longest digit, which is acting as the midline. Abducting the wrist (also known

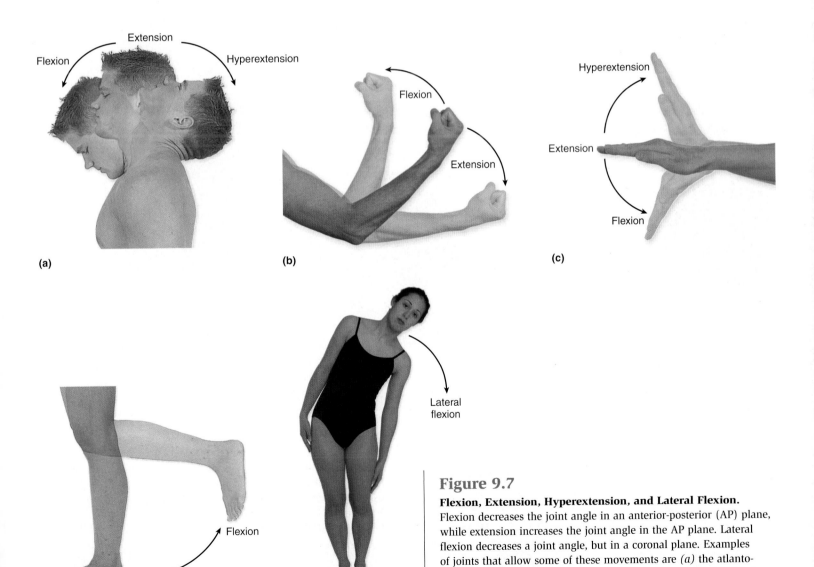

(a)   **(b)**   **(c)**   **(d)**   **(e)**

Figure 9.7

Flexion, Extension, Hyperextension, and Lateral Flexion.
Flexion decreases the joint angle in an anterior-posterior (AP) plane, while extension increases the joint angle in the AP plane. Lateral flexion decreases a joint angle, but in a coronal plane. Examples of joints that allow some of these movements are *(a)* the atlanto-occipital joint, *(b)* the elbow joint, *(c)* the radiocarpal joint, *(d)* the knee joint, and *(e)* the intervertebral joints.

as *radial deviation*) involves pointing the hand and fingers laterally, away from the body. The opposite of abduction is **adduction** (ad-dŭk'shŭn), which means to "move toward," and is the medial movement of a body part *toward* the body midline. Adduction occurs when you bring your raised arm or thigh back toward the body midline, or in the case of the digits, toward the midline of the hand. Adducting the wrist (also known as *ulnar deviation*) involves pointing the hand and fingers medially, toward the body. Abduction and adduction of various body parts are shown in **figure 9.8**.

Circumduction (ser-kŭm-dŭk'shŭn; *circum* = around, *duco* = to draw) is a sequence of movements in which the proximal end of an appendage remains relatively stationary while the distal end makes a circular motion **(figure 9.9)**. The resulting movement makes an imaginary cone shape. For example, when you draw a circle on the blackboard, your shoulder remains stationary while your hand moves. The tip of the imaginary cone is the stationary shoulder, while the rounded "base" of the cone is the circle the hand makes. Circumduction is a complex movement that occurs as a result of a continuous sequence of flexion, abduction, extension, and adduction.

WHAT DO YOU THINK?

❸ When sitting upright in a chair, are your hip and knee joints flexed or extended?

Rotational Motion

Rotation is a pivoting motion in which a bone turns on its own longitudinal axis **(figure 9.10)**. Rotational movement occurs at the atlantoaxial joint, which pivots when you rotate your head to gesture "no." Some limb rotations are described as either away from the median plane or toward it. For example, **lateral rotation** (or *external rotation*) turns the anterior surface of the femur or humerus laterally, while **medial rotation** (or *internal rotation*) turns the anterior surface of the femur or humerus medially.

Pronation (prō-nā'shŭn) is the medial rotation of the forearm so that the palm of the hand is directed posteriorly or inferiorly. The radius and ulna are crossed to form an X. **Supination** (soo'pi-nā'shŭn; *supinus* = supine) occurs when the forearm rotates laterally so that the palm faces anteriorly or superiorly, and the radius is parallel with the ulna. In the anatomic position, the forearm is supinated. Figure 9.10*d* illustrates pronation and supination.

Special Movements

Some motions occur only at specific joints and do not readily fit into any of the functional categories previously discussed. These **special movements** include: depression and elevation, dorsiflexion and plantar flexion, inversion and eversion, protraction and retraction, and opposition.

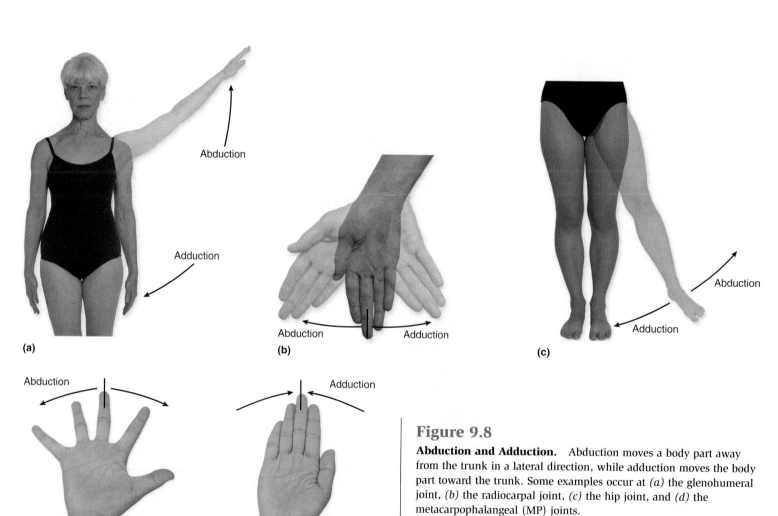

(a) Abduction / Adduction

(b) Abduction / Abduction / Adduction

(c) Abduction / Adduction

(d) Abduction / Adduction

Figure 9.8

Abduction and Adduction. Abduction moves a body part away from the trunk in a lateral direction, while adduction moves the body part toward the trunk. Some examples occur at *(a)* the glenohumeral joint, *(b)* the radiocarpal joint, *(c)* the hip joint, and *(d)* the metacarpophalangeal (MP) joints.

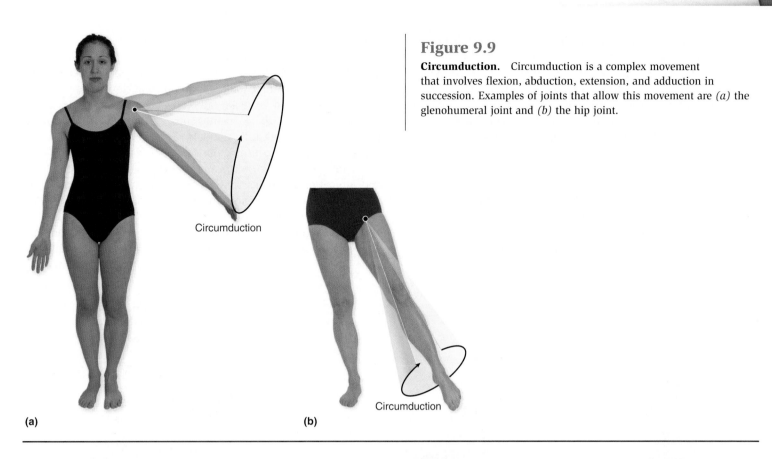

Figure 9.9

Circumduction. Circumduction is a complex movement that involves flexion, abduction, extension, and adduction in succession. Examples of joints that allow this movement are *(a)* the glenohumeral joint and *(b)* the hip joint.

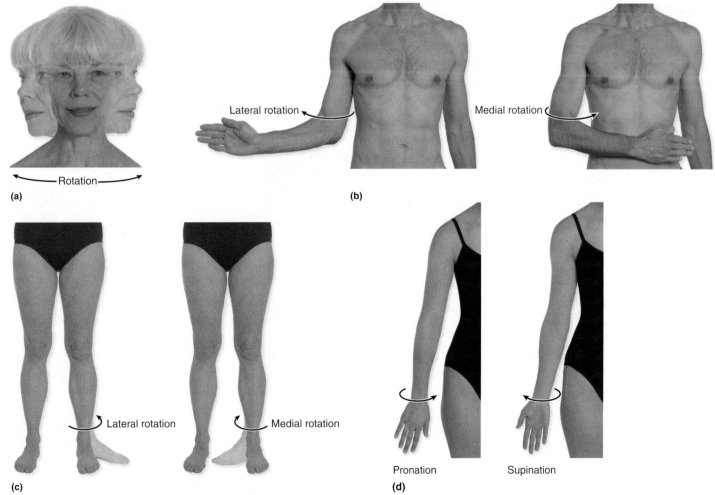

Figure 9.10

Rotational Movements. Rotation allows a bone to pivot on its longitudinal axis. Examples of joints that allow this movement are *(a)* the atlantoaxial joint, *(b)* the glenohumeral joint, and *(c)* the hip joint. *(d)* Pronation and supination occur at the forearm.

Depression (*de* = away or down, *presso* = to press) is the inferior movement of a part of the body. Examples of depression include the movement of the mandible while opening your mouth to chew food and the movement of your shoulders in an inferior direction. **Elevation** is the superior movement of a body part. Examples of elevation include the superior movement of the mandible while closing your mouth at the temporomandibular joint and the movement of the shoulders in a superior direction (shrugging your shoulders). **Figure 9.11a** illustrates depression and elevation at the glenohumeral joint.

Dorsiflexion and plantar flexion are limited to the ankle joint (figure 9.11*b*). **Dorsiflexion** (dōr-si-flek′shŭn) occurs when the talocrural (ankle) joint is bent such that the superior surface of the foot and toes moves toward the leg. This movement occurs when you dig in your heels, and it prevents your toes from scraping the

ground when you take a step. In **plantar** (plan′tăr; *planta* = sole of foot) **flexion**, movement at the talocrural joint permits extension of the foot so that the toes point inferiorly. When a ballerina is standing on her tiptoes, her ankle joint is in full plantar flexion.

Inversion and eversion are movements that occur at the intertarsal joints of the foot only (figure 9.11*c*). In **inversion** (in-ver′zhŭn; turning inward), the sole of the foot turns medially. In **eversion** (ē-ver′zhŭn; turning outward), the sole turns to face laterally. (Note: Some orthopedists and runners use the terms pronation and supination when describing foot movements as well, instead of using inversion and eversion. Inversion is foot supination, whereas eversion is foot pronation).

Protraction (prō-trak′shŭn; to draw forth) is the anterior movement of a body part from anatomic position, as when moving your jaw anteriorly at the temporomandibular joint or hunching

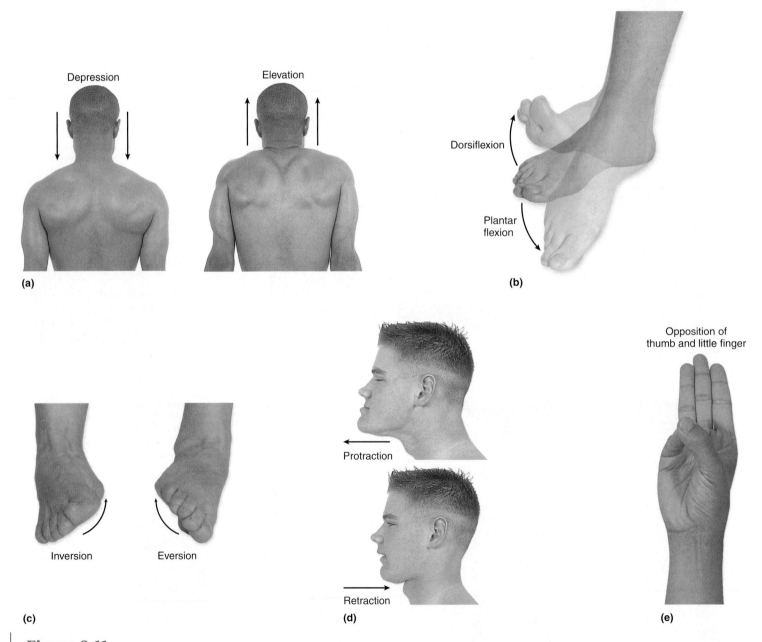

Figure 9.11

Special Movements Allowed at Synovial Joints. *(a)* Depression and elevation at the glenohumeral joint. *(b)* Dorsiflexion and plantar flexion at the talocrural joint. *(c)* Inversion and eversion at the intertarsal joints. *(d)* Protraction and retraction at the temporomandibular joint. *(e)* Opposition at the carpometacarpal joints.

your shoulders anteriorly by crossing your arms. In the latter case, the clavicles move anteriorly due to movement at both the acromioclavicular and sternoclavicular joints. **Retraction** (rē-trak′shŭn; to draw back) is the posteriorly directed movement of a body part from anatomic position. Figure 9.11*d* illustrates protraction and retraction at the temporomandibular joint.

At the carpometacarpal joint, the thumb moves toward the palmar tips of the fingers as it crosses the palm of the hand. This movement is called **opposition** (op′pō-si′shŭn) (figure 9.11*e*). It enables the hand to grasp objects and is the most distinctive digital movement in humans. The opposite movement is called reposition.

> ## Study Tip!
>
> When your mother tells you to "pull your shoulders back and stand up straight," you are *retracting* your shoulders. Conversely, when you are slumped forward in a chair, your shoulders are protracted.

💡 WHAT DID YOU LEARN?

5 What are the basic characteristics of all types of synovial joints?

6 Compare the structure and motion permitted in saddle and condylar joints.

7 Describe the following types of movements, and give an example of each: (a) flexion, (b) circumduction, (c) opposition.

Selected Articulations in Depth

Key topic in this section:

■ Characteristics of the major articulations of the axial and appendicular skeletons

In this section, we examine the structure and function of the more commonly known articulations of the axial and appendicular skeletons. For the axial skeleton, we present in-depth descriptions of the temporomandibular joint and the intervertebral articulations. **Table 9.3** summarizes the main features of these two areas and also provides comparable information about the other major joints of the axial skeleton.

Joints of the Axial Skeleton
Temporomandibular Joint

The **temporomandibular** (tem′pŏ-rō-man-dib′ū-lăr) **joint (TMJ)** is the articulation formed at the point where the head of the mandible articulates with the articular tubercle of the temporal bone anteriorly and the mandibular fossa posteriorly. This small, complex articulation is the only mobile joint between skull bones. The temporomandibular joint has several unique anatomic features **(figure 9.12)**. A loose **articular capsule** surrounds the joint and promotes an extensive range of motion. The TMJ is poorly stabilized, and thus a forceful anterior or lateral movement of the mandible can result in partial or complete dislocation of the mandible. The joint contains an **articular disc**, which is a thick pad of fibrocartilage separating the articulating bones and extending horizontally to divide the joint cavity into two separate chambers. As a result, the temporomandibular joint is really two synovial joints—one between the temporal bone and the articular disc, and a second between the articular disc and the mandible.

Several ligaments support this joint. The **sphenomandibular ligament** (an extracapsular ligament) is a thin band that extends anteriorly and inferiorly from the sphenoid to the medial surface of the mandibular ramus. The **stylomandibular ligament** (an extracapsular ligament) is a thick band that extends from the styloid process of the temporal bone to the mandibular angle. The **temporomandibular ligament** (or *lateral ligament*) is composed of two short bands on the lateral portion of the articular capsule. These

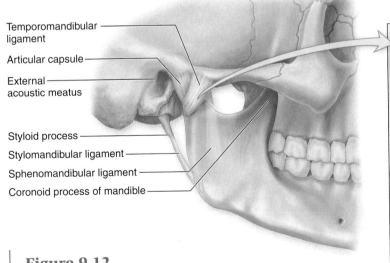

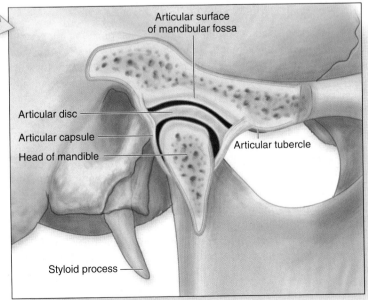

Figure 9.12

Temporomandibular Joint.   The articulation between the head of the mandible and the mandibular fossa of the temporal bone exhibits a wide range of movements.

Table 9.3		Axial Skeleton Joints			
	Joint	**Articulation Components**	**Structural Classification**		
	Suture	Adjacent skull bones	Fibrous joint		
	Temporomandibular	Head of mandible and mandibular fossa of temporal bone	Synovial (hinge, plane) joints		
		Head of mandible and articular tubercle of temporal bone			
	Atlanto-occipital	Superior articular facets of atlas and occipital condyles of occipital bone	Synovial (condylar) joint		
	Atlantoaxial	Anterior arch of atlas and dens of axis	Synovial (pivot) joint		
	Intervertebral	Vertebral bodies of adjacent vertebrae	Cartilaginous joint (symphysis) between vertebral bodies; synovial (plane) joint between articular processes		
		Superior and inferior articular processes of adjacent vertebrae			
	Vertebrocostal	Facets of heads of ribs and bodies of adjacent thoracic vertebrae and intervertebral discs between adjacent vertebrae	Synovial (plane) joint		
		Articular part of tubercles of ribs and facets of transverse processes of thoracic vertebrae			
	Lumbosacral	Body of the fifth lumbar vertebra and base of the sacrum	Cartilaginous joint (symphysis) between lumbar body and base of sacrum; synovial (plane) joint between articular facets		
		Inferior articular facets of fifth lumbar vertebra and superior articular facets of first sacral vertebra			
	Sternocostal	Sternum and first seven pairs of ribs	Cartilaginous joint (synchondrosis) between sternum and first ribs; synovial (plane) joint between sternum and ribs 2–7		

Diagram labels (left): Suture, Temporomandibular, Atlanto-occipital, Atlantoaxial, Intervertebral, Vertebrocostal, Lumbosacral, Sternocostal

bands extend inferiorly and posteriorly from the articular tubercle to the mandible.

The temporomandibular joint exhibits hinge, gliding, and some pivot joint movements. It functions like a hinge during jaw depression and elevation while chewing. It glides slightly forward during protraction of the jaw for biting, and glides slightly from side to side to grind food between the teeth during chewing.

Intervertebral Articulations

Intervertebral articulations occur between the bodies of the vertebrae, as well as between the superior and inferior articular processes of adjacent vertebrae. All of the vertebral bodies, between the axis (C_2) and the sacrum, are separated and cushioned by pads of fibrocartilage called **intervertebral discs (figure 9.13).** Each intervertebral disc consists of two components: (1) an anulus fibrosus and (2) a nucleus pulposus. The **anulus fibrosus** is the tough outer layer of fibrocartilage that covers each intervertebral disc. The anulus

fibrosus contains collagen fibers that attach the disc to the bodies of adjacent vertebrae. The anulus fibrosus also shares connections with many of the ligaments that run along the bodies of the vertebrae. The **nucleus pulposus** is the inner gelatinous core of the disc and is primarily composed of water, with some scattered reticular and elastic fibers.

Two factors compress the substance of the nucleus pulposus and displace it in every direction—movement of the vertebral column and the weight of the body. However, as humans age, water is gradually lost from the nucleus pulposus within each disc. Thus, over time the discs become less effective as a cushion, and the chances for vertebral injury increase. Loss of water by the discs also contributes to the shortening of the vertebral column, which accounts for the characteristic decrease in height that occurs with advanced age.

Several ligaments stabilize the vertebral column by supporting vertebrae through attachments to either their bodies or their

Functional Classification	Description of Movement
Synarthrosis	None allowed
Diarthrosis	Depression, elevation, lateral displacement, protraction, retraction, slight rotation
Diarthrosis	Extension and flexion of the head; slight lateral flexion of head to sides
Diarthrosis	Head rotation
Amphiarthrosis between vertebral bodies; diarthrosis between articular processes	Extension, flexion, lateral flexion of vertebral column
Diarthrosis	Some slight gliding
Amphiarthrosis between body and base; diarthrosis between articular facets	Extension, flexion, lateral flexion of vertebral column
Synarthrosis between sternum and first ribs; diarthrosis between sternum and ribs 2–7	No movement between sternum and first ribs; some gliding movement permitted between sternum and ribs 2–7

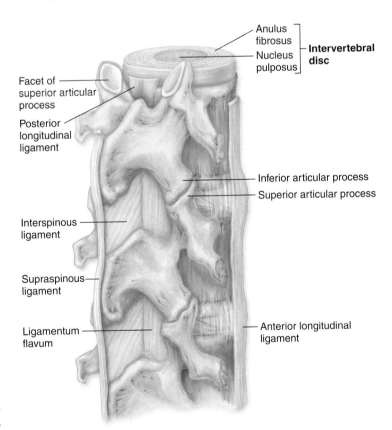

Figure 9.13

Intervertebral Articulations. Vertebrae articulate with adjacent vertebrae at both their superior and inferior articular processes. Intervertebral discs separate the bodies of adjacent vertebrae.

processes. The **anterior longitudinal ligament** is a thick, sturdy ligament that attaches vertebral bodies and intervertebral discs at their anterior surfaces. The **posterior longitudinal ligament** attaches the posterior aspects of the vertebral bodies and discs. It is much thinner than the anterior longitudinal ligament, and it runs within the vertebral canal. Multiple **interspinous ligaments** interconnect the spinous processes of adjacent vertebrae. Their angled fibers merge with the supraspinous ligament. The **supraspinous ligament** interconnects the tips of the spinous processes from C_7 to the sacrum. The **ligamentum** (lig′ă-men′tŭm) **nuchae** (noo′kē; back of neck) is the part of the supraspinous ligament that extends between C_7 and the base of the skull. The ligamentum nuchae is very thick and sturdy, and helps stabilize the skull on the cervical vertebrae. The **ligamentum flavum** (flā′vŭm) connects the laminae of adjacent vertebrae.

A second type of intervertebral articulation occurs at the synovial joints between adjacent superior and inferior articular processes.

The articular facets of the superior and inferior articular processes form plane joints that permit restricted gliding movements. An articular capsule surrounds these articular processes.

The movement possible between a single set of vertebrae is limited. However, when you add the movements of all the intervertebral joints of all the vertebrae together, an entire range of movements becomes possible, including flexion, extension, lateral flexion, and some rotation.

WHAT DO YOU THINK?

④ In which position does the anterior longitudinal ligament become taut—flexion or extension?

Joints of the Pectoral Girdle and Upper Limbs

Table 9.4 lists the features of the major joints of the pectoral girdle and upper limbs. Here, we provide an in-depth examination of several of these joints.

Sternoclavicular Joint

The **sternoclavicular** (ster′nō-kla-vik′ū-lăr) **joint** is a saddle joint formed by the articulation between the manubrium of the sternum and the sternal end of the clavicle **(figure 9.14)**. An **articular disc** partitions the sternoclavicular joint into two parts and creates two separate joint cavities. As a result, a wide range of movements is possible, including elevation, depression, and circumduction.

Support and stability are provided to this articulation by the fibers of the articular capsule. The **anterior sternoclavicular ligament** and the **posterior sternoclavicular ligament** reinforce the capsule. In addition, two extracapsular ligaments also help strengthen the joint: (1) The clavicle is attached to the first rib by the strong, wide **costoclavicular ligament**. This ligament stabilizes the joint and prevents dislocation of the shoulder when the shoulder is elevated. (2) The **interclavicular ligament** runs along the sternal notch and attaches to each clavicle. It reinforces the superior regions of the adjacent capsules. This design makes the sternoclavicular

Table 9.4		Pectoral Girdle and Upper Limb Joints	
	Joint	Articulation Components	
	Sternoclavicular	Sternal end of clavicle, manubrium of sternum, and first costal cartilage	
	Acromioclavicular	Acromial end of clavicle and acromion of scapula	
	Glenohumeral	Glenoid cavity of scapula and head of humerus	
	Humeroulnar	Trochlea of humerus and trochlear notch of ulna	
	Humeroradial	Capitulum of humerus and head of radius	
	Radioulnar	**Proximal joint:** Head of radius and radial notch of ulna	
		Distal joint: Distal end of ulna and ulnar notch of radius	
	Radiocarpal	Distal end of radius; lunate, scaphoid, and triquetrum	
	Intercarpal	Adjacent bones in proximal row of carpal bones	
		Adjacent bones in distal row of carpal bones	
		Adjacent bones between proximal and distal rows (midcarpal joints)	
	Carpometacarpal	Thumb: Trapezium (carpal bone) and first metacarpal	
		Other digits: Carpals and metacarpals II–V	
	Metacarpophalangeal (MP joints, "knuckles")	Head of metacarpals and bases of proximal phalanges	
	Interphalangeal (IP joints)	Heads of proximal and middle phalanges with bases of middle and distal phalanges, respectively	

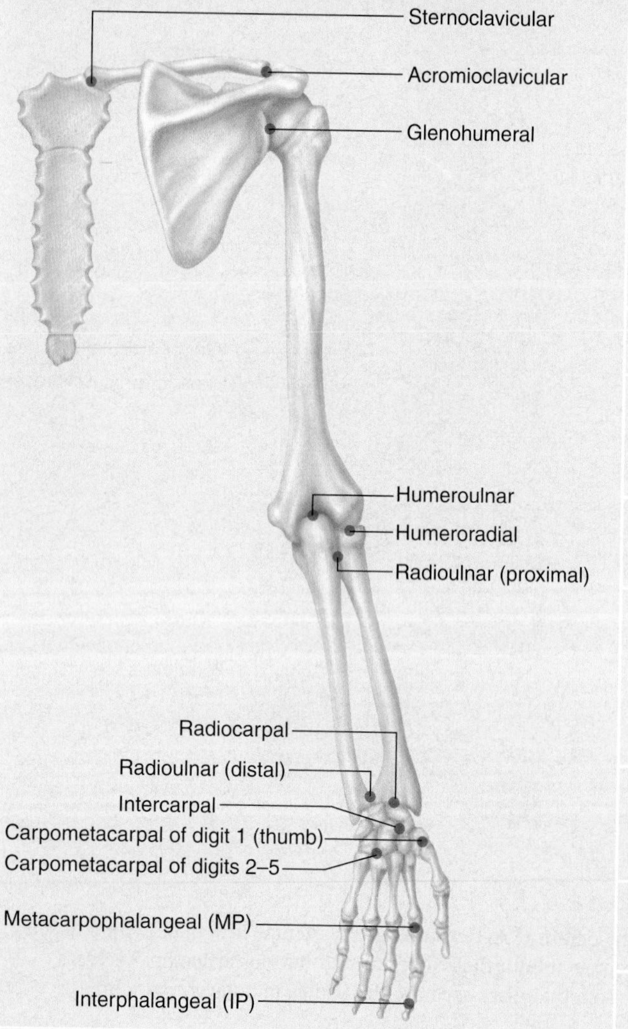

Labels on figure:
Sternoclavicular
Acromioclavicular
Glenohumeral
Humeroulnar
Humeroradial
Radioulnar (proximal)
Radiocarpal
Radioulnar (distal)
Intercarpal
Carpometacarpal of digit 1 (thumb)
Carpometacarpal of digits 2–5
Metacarpophalangeal (MP)
Interphalangeal (IP)

joint very stable. If a person falls on an outstretched hand so that force is applied to the joint, the clavicle will fracture before this joint ever dislocates.

Acromioclavicular Joint

The **acromioclavicular** (ă-krō′mē-ō-kla-vik′ū-lăr) **joint** is a plane joint between the acromion and the acromial end of the clavicle **(figure 9.15)**. A fibrocartilaginous **articular disc** lies within the joint cavity between these two bones. This joint works with both the sternoclavicular joint and the glenohumeral joint to give the upper limb a full range of movement.

Several ligaments provide great stability to this joint. The articular capsule is strengthened superiorly by an **acromioclavicular ligament**. In addition, a very strong **coracoclavicular** (kōr′ă-kō-kla-vik′ū-lăr) **ligament** binds the clavicle to the coracoid process of the scapula. The coracoclavicular ligament is responsible for most of the stability of the joint, because it indirectly prevents the clavicle from losing contact with the acromion. If this ligament is torn (as occurs

in severe shoulder separations; see Clinical View), the acromion and clavicle no longer align properly.

Glenohumeral (Shoulder) Joint

The **glenohumeral** (glē′nō-hū′mer-ăl) **joint** is commonly referred to as the shoulder joint. It is a ball-and-socket joint formed by the articulation of the head of the humerus and the glenoid cavity of the scapula (figure 9.15). It permits the greatest range of motion of any joint in the body, and so it is also the most unstable joint in the body and the one most frequently dislocated.

The fibrocartilaginous **glenoid labrum** encircles and covers the surface of the glenoid cavity. A relatively loose articular capsule attaches to the surgical neck of the humerus. The glenohumeral joints has several major ligaments. The **coracoacromial** (kōr′ă-kō-ă-krō′mē-ăl) **ligament** extends across the space between the coracoid process and the acromion. The large **coracohumeral** (kōr′ă-kō-hū′mer-ăl) **ligament** is a thickening of the superior part of the joint capsule. It runs from the coracoid process to the humeral head. The **glenohumeral ligaments** are three thickenings of the anterior portion of the articular capsule. These ligaments are often indistinct or absent and provide only minimal support. The **transverse humeral ligament** is a narrow sheet that extends between the greater and lesser tubercles of the humerus. In addition, the **tendon of the long head of biceps brachii** travels within the articular capsule and helps stabilize the humeral head in the joint.

Structural Classification	Functional Classification	Description of Movement
Synovial (saddle)	Diarthrosis	Elevation, depression, circumduction
Synovial (plane)	Diarthrosis	Gliding of scapula on clavicle
Synovial (ball-and-socket)	Diarthrosis	Abduction, adduction, circumduction, extension, flexion, lateral rotation, and medial rotation of arm
Synovial (hinge)	Diarthrosis	Extension and flexion of forearm
Synovial (hinge)	Diarthrosis	Extension and flexion of forearm
Synovial (pivot)	Diarthrosis	Rotation of radius with respect to the ulna
Synovial (condylar)	Diarthrosis	Abduction, adduction, circumduction, extension, and flexion of wrist
Synovial (plane)	Diarthrosis	Gliding
Synovial (saddle) at thumb; synovial (plane) at other digits	Diarthrosis	Abduction, adduction, circumduction, extension, flexion, and opposition at thumb; gliding at other digits
Synovial (condylar)	Diarthrosis	Abduction, adduction, circumduction, extension, and flexion of phalanges
Synovial (hinge)	Diarthrosis	Extension and flexion of phalanges

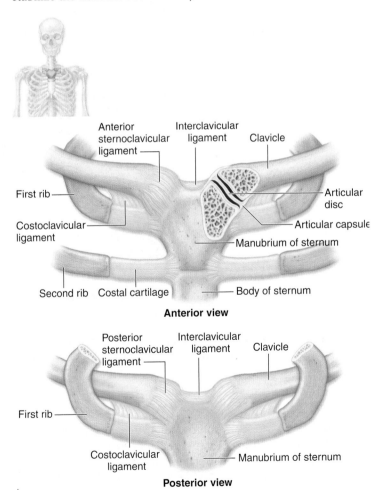

Anterior view

Posterior view

Figure 9.14

Sternoclavicular Joint. The sternoclavicular joint helps stabilize movements of the entire shoulder.

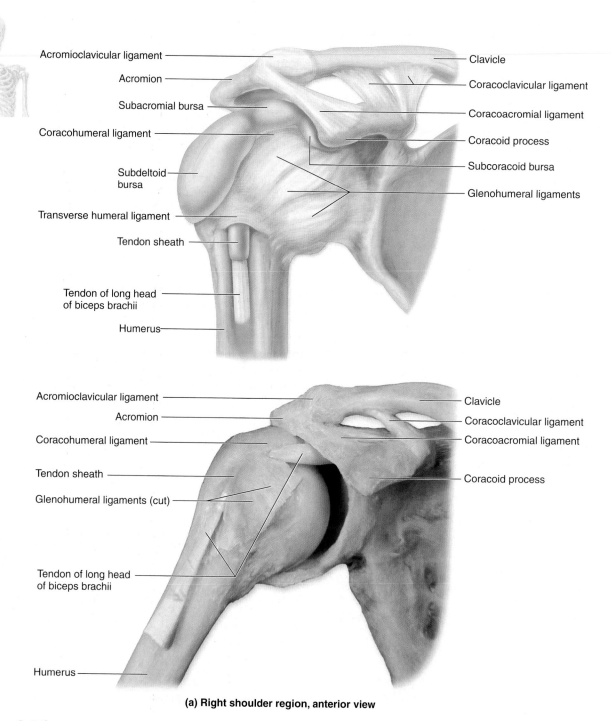

Acromioclavicular ligament

Acromion

Subacromial bursa

Coracohumeral ligament

Subdeltoid bursa

Transverse humeral ligament

Tendon sheath

Tendon of long head of biceps brachii

Humerus

Clavicle

Coracoclavicular ligament

Coracoacromial ligament

Coracoid process

Subcoracoid bursa

Glenohumeral ligaments

Acromioclavicular ligament

Acromion

Coracohumeral ligament

Tendon sheath

Glenohumeral ligaments (cut)

Tendon of long head of biceps brachii

Humerus

Clavicle

Coracoclavicular ligament

Coracoacromial ligament

Coracoid process

(a) Right shoulder region, anterior view

Figure 9.15

Acromioclavicular and Glenohumeral Joints. *(a)* Anterior diagrammatic view and cadaver photo of both joints on the right side of the body. *(b)* Right lateral view and *(c)* right coronal section show the articulating bones and supporting structures at the shoulder.

Ligaments of the glenohumeral joint strengthen the joint only minimally. Most of the joint's strength is due to the **rotator cuff** muscles surrounding it. The rotator cuff muscles (infraspinatus, subscapularis, supraspinatus, and teres minor) work as a group to hold the head of the humerus in the glenoid cavity. The tendons of these muscles encircle the joint (except for the inferior portion) and fuse with the articular capsule. Because the inferior portion of the joint lacks rotator cuff muscles, this area is weak and is the most likely site of injury.

Bursae help decrease friction at the specific places on the shoulder where both tendons and large muscles extend across the articular capsule. The shoulder has a relatively large number of bursae. The **subacromial** (sŭb-ă-krō'mē-ăl) **bursa** prevents rubbing between the acromion and the articular capsule. The **subcoracoid**

Tendon of long head of biceps brachii
Acromioclavicular tendon
Supraspinatus tendon
Acromion
Infraspinatus tendon
Subacromial bursa
Teres minor muscle
Glenoid cavity
Glenoid labrum
Articular capsule

Coracoacromial ligament
Clavicle
Coracoclavicular ligaments
Coracoid process
Subcoracoid bursa
Subscapularis muscle
Subscapular bursa
Glenohumeral ligaments

(b) Right lateral view

Acromion
Acromioclavicular joint
Clavicle
Articular disc
Supraspinatus tendon
Synovial membrane
Glenoid cavity of scapula
Glenoid labrum
Articular capsule

Tendon of long head of biceps brachii
Subdeltoid bursa
Deltoid muscle
Humerus

(c) Right coronal section

CLINICAL VIEW

Shoulder Separation

The term **shoulder separation** refers to a dislocation of the acromioclavicular joint. **Dislocation** (dis-lō-kā′shŭn; *dis* = apart, *locatio* = placing) is a joint injury in which the articulating bones have separated. This injury often results from a hard blow to the joint, as when a hockey player is "slammed into the boards." Shoulder separation is also common in wrestlers. The symptoms of a shoulder separation include:

- Tenderness and edema (swelling) in the area of the joint
- Surface deformity at the acromioclavicular joint; since the bones are displaced, the acromion is very prominent and appears more pointed.

- Pain when the arm is abducted more than 90 degrees, the position at which significant movement occurs between the separated clavicular and acromial surfaces.

Acromioclavicular dislocations are graded according to severity. In the most severe injury, the joint is completely dislocated, and the coracoclavicular ligament is torn. Since the coracoclavicular ligament provides most of the stability to this joint, damage to it means the bones will not stay in alignment. The coracoclavicular ligament must be surgically repaired in order for the bones of the joint to remain fixed in place.

bursa prevents contact between the coracoid process and the articular capsule. The **subdeltoid bursa** and the **subscapular bursa** allow for easier movements of the deltoid and supraspinatus muscles, respectively.

Elbow Joint

The **elbow joint** is a hinge joint composed primarily of two articulations: (1) the humeroulnar joint, where the trochlear notch of the ulna articulates with the trochlea of the humerus, and (2) the humeroradial joint, where the capitulum of the humerus articulates with the head of the radius. These joints are enclosed within a single articular capsule **(figure 9.16)**.

The elbow is an extremely stable joint for several reasons. First, the articular capsule is fairly thick, and thus effectively protects the articulations. Second, the bony surfaces of the humerus and ulna interlock very well, and thus provide a solid bony support. Finally, multiple strong supporting ligaments help reinforce

the articular capsule. Remember that there is a tradeoff between stability and mobility in a joint. Thus, while the elbow joint is very stable, it is not as mobile as some other joints, such as the glenohumeral joint.

The elbow joint has two main supporting ligaments. The **radial collateral ligament** (or *lateral collateral ligament*) is responsible for stabilizing the joint at its lateral surface; it extends around the head of the radius between the anular ligament and the lateral epicondyle of the humerus. The **ulnar collateral ligament** (or *medial collateral ligament*) stabilizes the medial side of the joint and extends from the medial epicondyle of the humerus to the coronoid process of the ulna, and posteriorly to the olecranon. In addition, an **anular** (an′ū-lăr; *anulus* = ring) **ligament** surrounds the neck of the radius and binds the proximal head of the radius to the ulna. The anular ligament helps hold the head of the radius in place.

Despite the support from the capsule and ligaments, the elbow joint is subject to damage from a severe impact or unusual

Dislocation of the Glenohumeral Joint

Because the glenohumeral joint is very mobile and yet unstable, dislocations are very common. Glenohumeral dislocations usually occur when a fully abducted humerus is struck hard—for example, when a quarterback is hit as he is about to release a football, or when a person falls on an outstretched hand.

The following sequence of events occurs in a glenohumeral dislocation:

1. Immediately after the initial blow, the head of the humerus pushes into the inferior part of the articular capsule. (Recall the inferior part of the capsule is relatively weak and not protected by muscle tendons as the other surfaces of the capsule are.)
2. The head of the humerus tears the inferior part of the capsule and dislocates the humerus, so that the humerus lies inferior to the glenoid cavity.

3. Once the humeral head has become dislocated from the glenoid cavity, the anterior thorax (chest) muscles pull on the head superiorly and medially, causing the humeral head to lie just inferior to the coracoid process.

The result is that the shoulder appears flattened and "squared-off," because the humeral head is dislocated anteriorly and inferiorly to the glenohumeral joint capsule.

Some glenohumeral dislocations can be repaired by "popping" the humerus back into the glenoid cavity. More severe dislocations may need surgical repair.

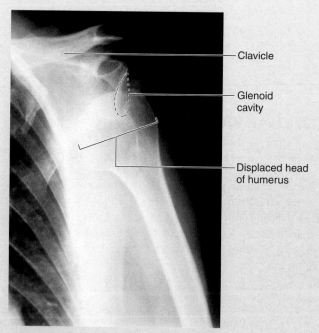

(b) Radiograph of a glenohumeral dislocation

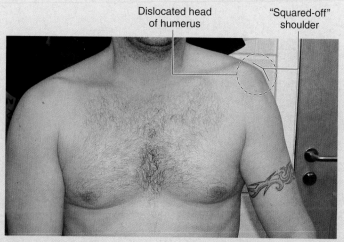

(a) Dislocated glenohumeral joint

stress. For example, if you fall on an outstretched hand with your elbow joint partially flexed, the posterior stress on the ulna combined with contractions of muscles that extend the elbow may break the ulna at the center of the trochlear notch. Sometimes dislocations result from stresses to the elbow. This is particularly true when growth is still occurring at the epiphyseal plate, so children and teenagers may be prone to humeral epicondyle dislocations or fractures.

Radiocarpal (Wrist) Joint

The **radiocarpal** (rā′dē-ō-kar′păl) **joint**, also known as the wrist joint, is an articulation among the three proximal carpal bones (scaphoid, lunate, and triquetrum), the distal articular surface

of the radius, and a fibrocartilaginous **articular disc (figure 9.17)**. This articular disc separates the ulna from the radiocarpal joint (which is why the ulna is not considered part of this joint). The entire wrist complex is ensheathed by an articular capsule that has reinforcing broad ligaments to support and stabilize the carpal bone positions. The radiocarpal joint is a condylar articulation that permits flexion, extension, adduction, abduction, and circumduction, but no rotation. Rotational movements (in the form of supination and pronation) occur at the distal and proximal radioulnar joints.

Additional movements in the carpus region are made possible by **intercarpal articulations**, which are plane joints that permit gliding movements between the individual carpal bones.

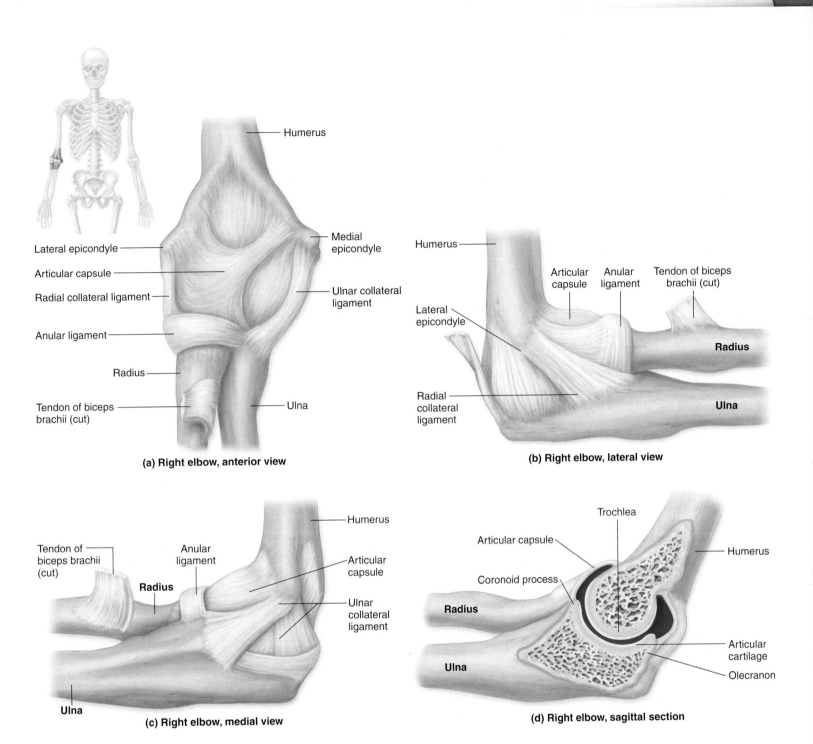

(a) Right elbow, anterior view

Humerus

Lateral epicondyle

Articular capsule

Radial collateral ligament

Anular ligament

Radius

Tendon of biceps brachii (cut)

Medial epicondyle

Ulnar collateral ligament

Ulna

(b) Right elbow, lateral view

Humerus

Articular capsule

Anular ligament

Tendon of biceps brachii (cut)

Lateral epicondyle

Radial collateral ligament

Radius

Ulna

(c) Right elbow, medial view

Tendon of biceps brachii (cut)

Anular ligament

Radius

Humerus

Articular capsule

Ulnar collateral ligament

Ulna

(d) Right elbow, sagittal section

Trochlea

Articular capsule

Coronoid process

Radius

Ulna

Humerus

Articular cartilage

Olecranon

Figure 9.16

Elbow Joint.   The elbow joint is a hinge joint. The right elbow is shown here in (*a*) anterior view, (*b*) lateral view, (*c*) medial view, and (*d*) sagittal section.

WHAT DID YOU LEARN?

8 Describe the structural arrangement and function of the anulus fibrosus and the nucleus pulposus in an intervertebral disc.

9 Which ligaments limit the relative mobility of the clavicle?

10 Which structures are the primary stabilizers of the glenohumeral joint?

11 What is the function of the anular ligament in the elbow joint?

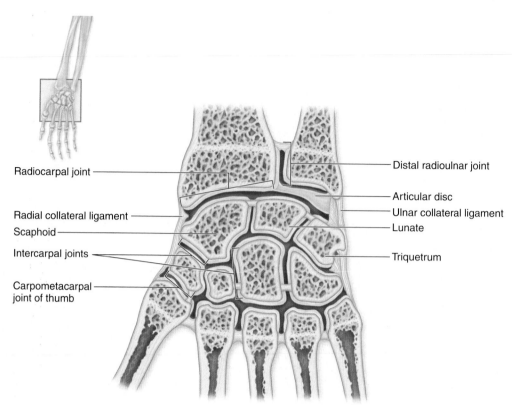

Figure 9.17

Radiocarpal Joint. A right coronal section depicts the condylar articulation between the radius and three proximal carpal bones.

Radiocarpal joint

Radial collateral ligament

Scaphoid

Intercarpal joints

Carpometacarpal joint of thumb

Distal radioulnar joint

Articular disc

Ulnar collateral ligament

Lunate

Triquetrum

Right radiocarpal joint, coronal section

CLINICAL VIEW

Subluxation of the Head of the Radius

The term **subluxation** refers to an incomplete dislocation, in which the contact between the bony joint surfaces is altered, but they are still in partial contact. In subluxation of the head of the radius, the head is pulled out of the anular ligament. Laymen's terms for this injury include "pulled elbow," "nursemaid's elbow," or "slipped elbow." This injury occurs commonly and almost exclusively in children (typically those younger than age 5), because a child's anular ligament is thin and the head of the radius is not fully formed. Thus, it is much easier for the head of the radius to be pulled out of the anular ligament. After age 5, both the ligament and the radial head are more fully formed, and the risk of this type of injury lessens dramatically.

A classic example of subluxation of the head of the radius occurs when a parent or caregiver suddenly pulls on a child's pronated forearm, and the child, resisting, puts his or her upper limb in a flexed and partially pronated position. As the child resists moving and the adult pulls on the upper limb, the head of the radius pulls out of the anular ligament. The child later complains of pain on the lateral side of the elbow, where a prominent "bump" (caused by the subluxated radial head) also appears. Luckily, treatment is simple: The pediatrician applies posteriorly placed pressure to the head of the radius while slowly supinating and extending the child's forearm. This movement literally "screws" the radial head back into the anular ligament. In most cases, this manual treatment brings immediate relief. A child who has had this injury may be more likely to reinjure this articulation prior to age 5.

Joints of the Pelvic Girdle and Lower Limbs

Table 9.5 lists the features of the major joints of the pelvic girdle and lower limbs. Here, we provide an in-depth examination of several of these joints.

Hip (Coxal) Joint

The **hip joint**, also called the *coxal joint*, is the articulation between the head of the femur and the relatively deep, concave acetabulum of the os coxae (**figure 9.18**). A fibrocartilaginous **acetabular labrum** further deepens this socket. The hip joint's more extensive bony architecture is therefore much stronger and more stable than

that of the glenohumeral joint. Conversely, the hip joint's increased stability means that it is less mobile than the glenohumeral joint. The hip joint must be more stable (and thus less mobile) because it supports the body weight.

The hip joint is secured by a strong articular capsule, several ligaments, and a number of powerful muscles. The articular capsule extends from the acetabulum to the trochanters of the femur, enclosing both the femoral head and neck. This arrangement prevents the head from moving away from the acetabulum. The ligamentous fibers of the articular capsule reflect around the neck of the femur. These reflected fibers, called **retinacular** (ret-i-nak'ū-lăr; = a band) **fibers,**

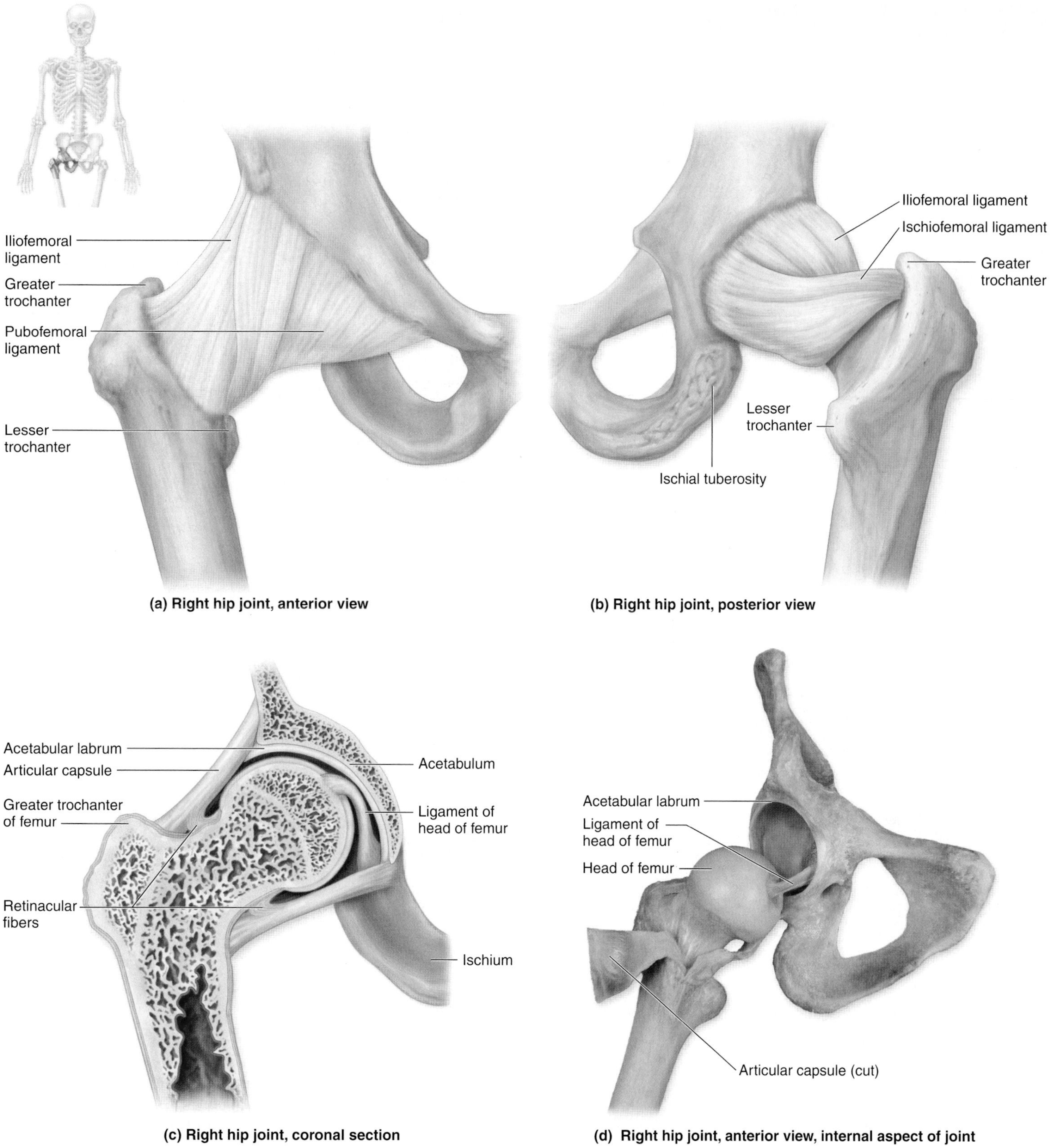

(a) Right hip joint, anterior view

Iliofemoral ligament
Greater trochanter
Pubofemoral ligament
Lesser trochanter

(b) Right hip joint, posterior view

Iliofemoral ligament
Ischiofemoral ligament
Greater trochanter
Lesser trochanter
Ischial tuberosity

(c) Right hip joint, coronal section

Acetabular labrum
Articular capsule
Greater trochanter of femur
Retinacular fibers
Acetabulum
Ligament of head of femur
Ischium

(d) Right hip joint, anterior view, internal aspect of joint

Acetabular labrum
Ligament of head of femur
Head of femur
Articular capsule (cut)

Figure 9.18

Hip Joint. The hip joint is formed by the head of the femur and the acetabulum of the os coxae. The right hip joint is shown in *(a)* anterior view, *(b)* posterior view, and *(c)* coronal section. *(d)* Cadaver photo of the hip joint, with the articular capsule cut to show internal structures.

Table 9.5 — **Pelvic Girdle and Lower Limb Joints**

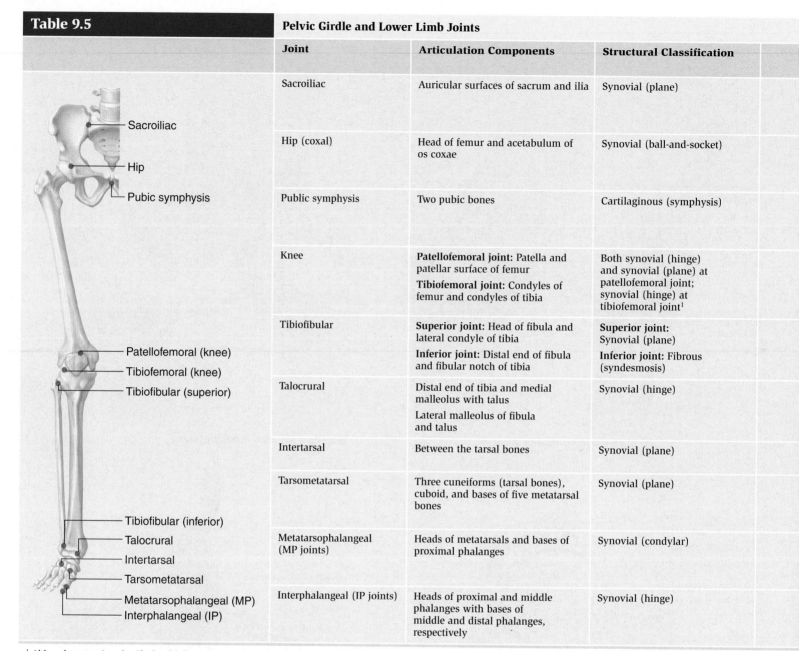

Joint	Articulation Components	Structural Classification	
Sacroiliac	Auricular surfaces of sacrum and ilia	Synovial (plane)	
Hip (coxal)	Head of femur and acetabulum of os coxae	Synovial (ball-and-socket)	
Public symphysis	Two pubic bones	Cartilaginous (symphysis)	
Knee	**Patellofemoral joint:** Patella and patellar surface of femur **Tibiofemoral joint:** Condyles of femur and condyles of tibia	Both synovial (hinge) and synovial (plane) at patellofemoral joint; synovial (hinge) at tibiofemoral joint[1]	
Tibiofibular	**Superior joint:** Head of fibula and lateral condyle of tibia **Inferior joint:** Distal end of fibula and fibular notch of tibia	**Superior joint:** Synovial (plane) **Inferior joint:** Fibrous (syndesmosis)	
Talocrural	Distal end of tibia and medial malleolus with talus Lateral malleolus of fibula and talus	Synovial (hinge)	
Intertarsal	Between the tarsal bones	Synovial (plane)	
Tarsometatarsal	Three cuneiforms (tarsal bones), cuboid, and bases of five metatarsal bones	Synovial (plane)	
Metatarsophalangeal (MP joints)	Heads of metatarsals and bases of proximal phalanges	Synovial (condylar)	
Interphalangeal (IP joints)	Heads of proximal and middle phalanges with bases of middle and distal phalanges, respectively	Synovial (hinge)	

[1] Although anatomists classify the tibiofemoral joint as a hinge joint, some kinesiologists and exercise scientists prefer to classify the tibiofemoral joint as a modified condylar joint.

provide additional stability to the capsule. Traveling through the retinacular fibers are retinacular arteries (branches of the deep femoral artery), which supply almost all of the blood to the head and neck of the femur.

The articular capsule is reinforced by three spiraling intracapsular ligaments. The **iliofemoral** (il′ē-ō-fem′ŏ-răl) **ligament** is a Y-shaped ligament that provides strong reinforcement for the anterior region of the articular capsule. The **ischiofemoral** (is-kē-ō-fem′ŏ-răl) **ligament** is a spiral-shaped, posteriorly located ligament. The **pubofemoral** (pū′bō-fem′ŏ-răl) **ligament** is a triangular thickening of the capsule's inferior region. All of these spiraling ligaments become taut when the hip joint is extended, so the hip joint is most stable

in the extended position. Try this experiment: Flex your hip joint, and try to move the femur; you may notice a great deal of mobility. Now extend your hip joint (stand up), and try to move the femur. Because those ligaments are taut, you don't have as much mobility in the joint as you did when the hip joint was flexed.

Another tiny ligament, the **ligament of head of femur**, also called the *ligamentum teres*, originates along the acetabulum. Its attachment point is the center of the head of the femur. This ligament does not provide much strength to the joint; rather, it typically contains a small artery that supplies the head of the femur.

The combination of a deep bony socket, a strong articular capsule, supporting ligaments, and muscular padding gives

Functional Classification	Description of Movement
Diarthrosis	Slight gliding; more movement during pregnancy and childbirth
Diarthrosis	Abduction, adduction, circumduction, extension, flexion, medial and lateral rotation of thigh
Amphiarthrosis	Very slight movements; more movement during childbirth
Diarthrosis	Extension, flexion, lateral rotation of leg in flexed position, slight medial rotation
Amphiarthrosis	Slight rotation of fibula during dorsiflexion of foot
Diarthrosis	Dorsiflexion and plantar flexion
Diarthrosis	Eversion and inversion of foot
Diarthrosis	Slight gliding
Diarthrosis	Abduction, adduction, circumduction, extension, and flexion of proximal phalanges
Diarthrosis	Extension and flexion of phalanges

CLINICAL VIEW

Fracture of the Femoral Neck

Fracture of the femoral neck is a common and complex injury. Although this injury is often referred to as a "fractured hip," the os coxae isn't broken, just the femoral neck. When the femoral neck breaks, the pull of the lower limb muscles causes the leg to rotate laterally and shorten by several inches. Fractures of the femoral neck are of two types: intertrochanteric and subcapital.

Intertrochanteric fractures of the femoral neck occur distally to or outside the hip articular capsule—in other words, these fractures are *extracapsular*. The fracture line runs between the greater and lesser trochanters. This type of injury typically occurs in younger and middle-aged individuals, and usually in response to trauma.

Subcapital fractures (or *intracapsular fractures*) of the femoral neck occur within the hip articular capsule, very close to the head of the femur itself. This type of fracture usually occurs in elderly people whose bones have been weakened by osteoporosis. The weakened femur is more susceptible to fracture, and when the femoral neck fractures, the elderly individual usually falls.

Subcapital fractures result in tearing of the retinacular fibers and the retinacular arteries that supply the head and neck of the femur. The ligament to the head of the femur may be torn as well. As a result, the head and neck of the femur lose their blood supply. If a bone doesn't have an adequate blood supply, it develops **avascular necrosis**, which is death of the bone tissue due to lack of blood. Avascular necrosis of the femoral head and neck is a common complication in subcapital fractures. Frequently, hip replacement surgery is needed, whereby a metal femoral head and neck replace the dying bone. This surgery is not without risk, and many elderly patients do not survive.

The knee joint has an articular capsule that encloses only the medial, lateral, and posterior regions of the knee joint. The articular capsule does not cover the anterior surface of the knee joint; rather, the quadriceps femoris muscle tendon passes over the anterior surface. The patella is embedded within this tendon, and the **patellar ligament** extends beyond the patella and continues to its attachment on the tibial tuberosity of the tibia. Thus, there is no single unified capsule in the knee, nor is there a common joint cavity.

On either side of the joint are two collateral ligaments that become taut on extension and provide additional stability to the joint. The **fibular collateral ligament** (*lateral collateral ligament*) reinforces the lateral surface of the joint. This ligament runs from the femur to the fibula and prevents hyperadduction of the leg at the knee. (In other words, it prevents the leg from moving too far medially relative to the thigh.) The **tibial collateral ligament** (*medial collateral ligament*) reinforces the medial surface of the knee joint. This ligament runs from the femur to the tibia and prevents hyperabduction of the leg at the knee. (In other words, it prevents the leg from moving too far laterally relative to the thigh.) This ligament is attached to the medial meniscus of the knee joint as well, so

the hip joint its stability. Movements possible at the hip joint include flexion, extension, abduction, adduction, rotation, and circumduction.

Knee Joint

The **knee joint** is the largest and most complex diarthrosis of the body (**figure 9.19**). It is primarily a hinge joint, but when the knee is flexed, it is also capable of slight rotation and lateral gliding. Structurally, the knee is composed of two separate articulations: (1) The **tibiofemoral** (tib-ē-ō-fem′ŏ-răl) **joint** is between the condyles of the femur and the condyles of the tibia. (2) The **patellofemoral joint** is between the patella and the patellar surface of the femur.

Figure 9.19

Knee Joint. This joint is the largest and most complex diarthrosis of the body. *(a)* Anterior superficial, *(b)* sagittal section, *(c)* anterior deep, and *(d)* posterior deep views reveal the complex interrelationships among the parts of the right knee.

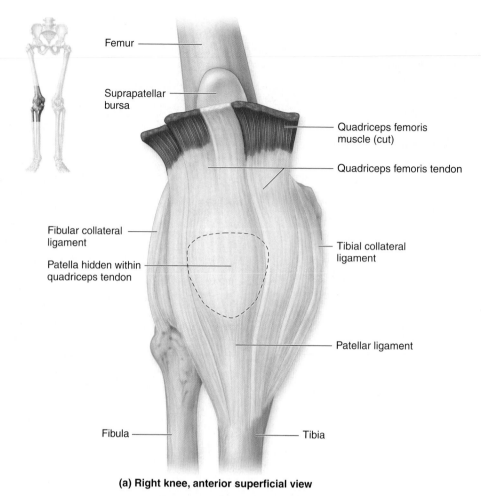

Femur

Suprapatellar bursa

Quadriceps femoris muscle (cut)

Quadriceps femoris tendon

Fibular collateral ligament

Patella hidden within quadriceps tendon

Tibial collateral ligament

Patellar ligament

Fibula

Tibia

(a) Right knee, anterior superficial view

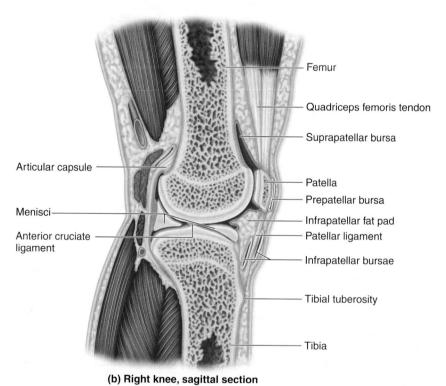

Femur

Quadriceps femoris tendon

Suprapatellar bursa

Articular capsule

Patella

Prepatellar bursa

Menisci

Infrapatellar fat pad

Anterior cruciate ligament

Patellar ligament

Infrapatellar bursae

Tibial tuberosity

Tibia

(b) Right knee, sagittal section

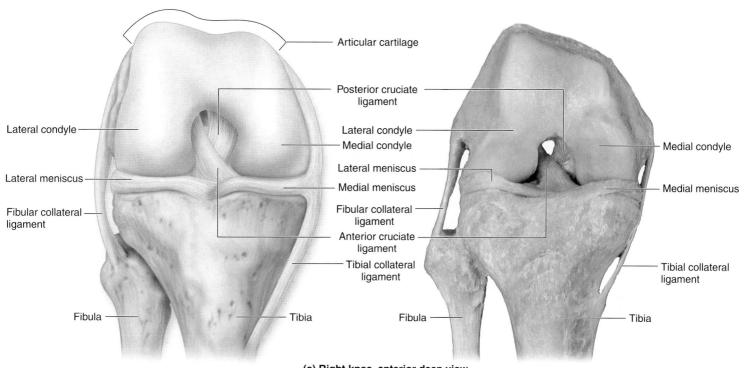

Articular cartilage

Posterior cruciate ligament

Lateral condyle

Lateral condyle
Medial condyle

Lateral meniscus

Lateral meniscus

Medial meniscus

Medial meniscus

Fibular collateral ligament

Fibular collateral ligament

Anterior cruciate ligament

Tibial collateral ligament

Tibial collateral ligament

Fibula

Tibia

Fibula

Tibia

(c) Right knee, anterior deep view

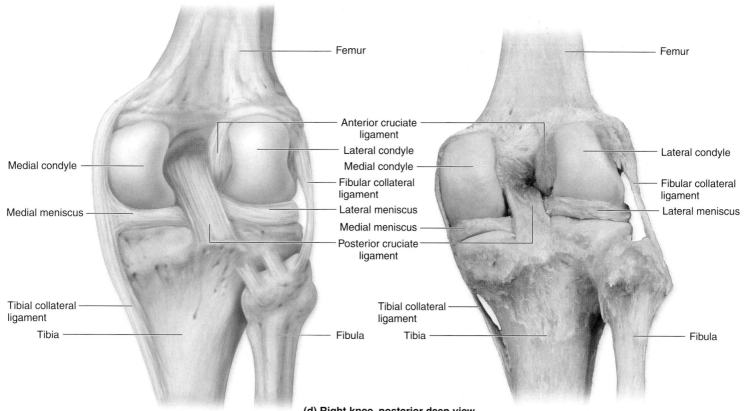

Femur

Femur

Anterior cruciate ligament

Medial condyle

Lateral condyle
Medial condyle

Lateral condyle

Fibular collateral ligament

Fibular collateral ligament

Medial meniscus

Lateral meniscus

Lateral meniscus

Medial meniscus

Tibial collateral ligament

Posterior cruciate ligament

Tibial collateral ligament

Tibia

Fibula

Tibia

Fibula

(d) Right knee, posterior deep view

an injury to the tibial collateral ligament usually affects the medial meniscus.

Deep to the articular capsule and within the knee joint itself are a pair of C-shaped fibrocartilage pads located on the condyles of the tibia. These pads, called the **medial meniscus** and the **lateral meniscus**, partially stabilize the joint medially and laterally, act as cushions between articular surfaces, and continuously change shape to conform to the articulating surfaces as the femur moves.

Also deep to the articular capsule of the knee joint are two **cruciate** (kroo′shē-āt) **ligaments**, which limit the anterior and posterior movement of the femur on the tibia. These ligaments cross each other in the form of an X, hence the name "cruciate" (which means "cross"). The **anterior cruciate ligament (ACL)** runs from the posterior femur to the anterior side of the tibia. When the knee is extended, the ACL is pulled tight and prevents hyperextension. The ACL prevents the tibia from moving too far anteriorly on the femur. The **posterior cruciate ligament (PCL)** runs from the anteroinferior femur to the posterior side of the tibia.

The PCL becomes taut on flexion, and so it prevents hyperflexion of the knee joint. The PCL also prevents posterior displacement of the tibia on the femur.

"Locking" the Knee Humans are bipedal animals, meaning that they walk on two feet. An important aspect of bipedal locomotion is the ability to "lock" the knees in the extended position and stand erect for long periods without tiring the leg muscles. At full extension, the tibia rotates laterally so as to tighten the anterior cruciate ligament and squeeze the meniscus between the tibia and femur. Muscular contraction by the popliteus muscle unlocks the knee joint. Contraction of this muscle causes a slight rotational movement between the tibia and the femur.

Talocrural (Ankle) Joint

The **talocrural joint,** or ankle joint, is a highly modified hinge joint that permits dorsiflexion and plantar flexion, and includes two articulations within one articular capsule. One of these articulations

---

CLINICAL VIEW: *In Depth*

Knee Ligament Injuries

Although the knee is capable of bearing much weight and has numerous strong supporting ligaments, it is highly vulnerable to injury, especially among athletes. The knee is susceptible to both horizontal and rotational stress, most commonly when struck either from the lateral or posterior aspect while slightly flexed. Because the knee is reinforced by tendons and ligaments only, ligamentous injuries to the knee are very common.

The tibial collateral ligament is frequently injured when the leg is forcibly abducted at the knee. For example, if a person's knee is hit on the lateral side, the leg is hyperabducted, and the tibial collateral ligament is strained and frequently torn. Because the tibial collateral ligament is attached to the medial meniscus, the medial meniscus may be injured as well.

Injury to the fibular collateral ligament can occur if the medial side of the knee is struck, resulting in hyperadduction of the leg at the knee. This type of injury is fairly rare, in part because the fibular collateral ligament is very strong and also because medial blows to the knee are not common.

The anterior cruciate ligament (ACL) can be injured when the leg is hyperextended—for example, if a runner's foot hits a hole. Because the ACL is rather weak compared to the other knee ligaments, it is especially prone to injury. ACL injury often occurs in association with another ligament injury. To test for ACL injury, a physician gently tugs anteriorly on the tibia. In this so-called "anterior drawer test," too much forward movement indicates an ACL tear.

Posterior cruciate ligament (PCL) injury may occur if the leg is hyperflexed or if the tibia is driven posteriorly on the femur. PCL injury occurs rarely, because this ligament is rather strong. To test for PCL injury, a physician gently pushes on the tibia. In this "posterior drawer test," too much posterior movement indicates a PCL tear.

The **unhappy triad** of injuries refers to a triple ligamentous injury of the tibial collateral ligament, medial meniscus, and anterior cruciate ligament. This is the most common type of football injury. It occurs when

a player is illegally "clipped" by a lateral blow to the knee, and the leg is forcibly abducted and laterally rotated. If the blow is severe enough, the tibial collateral ligament tears, followed by tearing of the medial meniscus, since these two structures are connected. The force that tears the tibial collateral ligament and the medial meniscus is thus transferred to the ACL. Because the ACL is relatively weak, it tears as well.

The treatment of ligamentous knee injuries depends upon the severity and type of injury. Conservative treatment involves immobilizing the knee for a period of time to rest the joint. Surgical treatment can include repairing the torn ligaments or replacing the ligaments with a graft from another tendon or ligament (such as the quadriceps tendon). Rehabilitation of the knee also requires strengthening the muscles and tendons that surround the knee, so they can provide additional support to the joint.

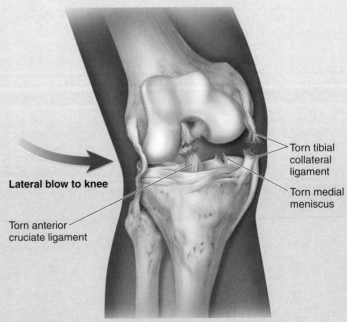

Lateral blow to knee

Torn tibial collateral ligament

Torn medial meniscus

Torn anterior cruciate ligament

"Unhappy triad" of injuries to the right knee.

is between the distal end of the tibia and the talus, and the other is between the distal end of the fibula and the lateral aspect of the talus **(figure 9.20)**. The medial and lateral malleoli of the tibia and fibula, respectively, form extensive medial and lateral margins and prevent the talus from sliding side-to-side.

The talocrural joint includes several distinctive anatomic features. An articular capsule covers the distal surfaces of the tibia, the medial malleolus, the lateral malleolus, and the talus. A multipart **deltoid ligament** (or *medial ligament*) binds the tibia to the foot on the medial side. This ligament prevents overeversion of the foot. The deltoid ligament is incredibly strong and rarely tears; in fact, it will pull the medial malleolus off the tibia before it ever ruptures! A much thinner, multipart **lateral ligament** binds the fibula to the foot on the lateral side. This ligament prevents overinversion of the foot. It is not as strong as the deltoid

ligament, and is prone to sprains and tears. Two **tibiofibular** (tib-ē-ō-fib′ū-lăr) **ligaments** (**anterior** and **posterior**) bind the tibia to the fibula.

Joints of the Foot

Four types of synovial joints are found in the foot: intertarsal joints, tarsometatarsal joints, metatarsophalangeal joints, and interphalangeal joints **(figure 9.21)**. **Intertarsal joints** are the articulations between the tarsal bones. Some of these joints go by specific names (e.g., talonavicular joint, calcaneocuboid joint). It is at the intertarsal joints that inversion and eversion of the foot occur.

The articulations between the tarsal and metatarsal bones form the **tarsometatarsal** (tar-sō-met′ă-tar′săl) **joints**. These are plane articulations that permit some twisting and limited side-to-side movements. The medial, intermediate, and lateral cuneiform

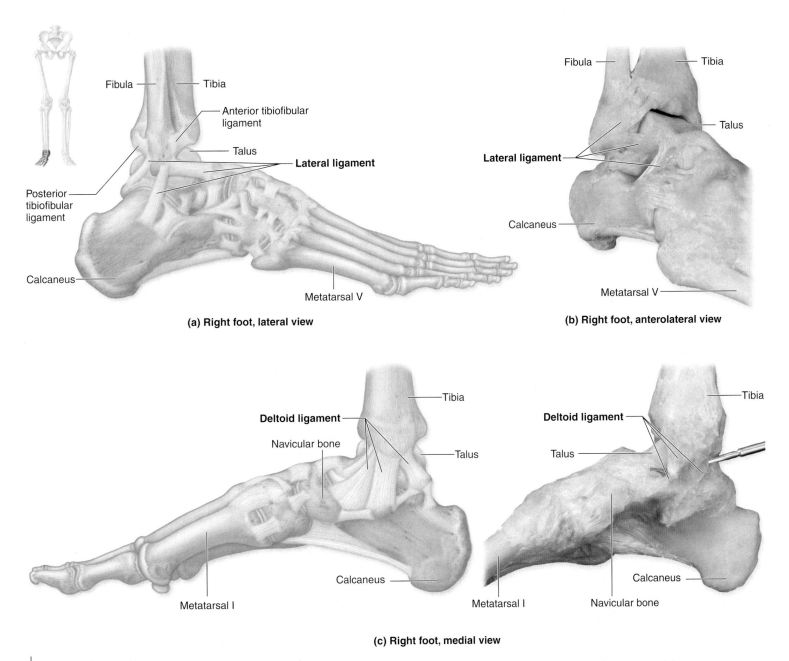

(a) Right foot, lateral view

(b) Right foot, anterolateral view

(c) Right foot, medial view

Figure 9.20

Talocrural Joint. *(a)* Lateral, *(b)* anterolateral, and *(c)* medial views of the right foot show that the talocrural joint contains articulations among the tibia, fibula, and talus. This joint permits dorsiflexion and plantar flexion only.

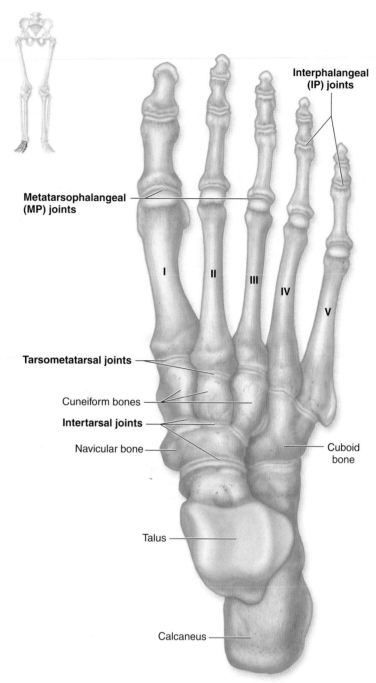

Interphalangeal (IP) joints

Metatarsophalangeal (MP) joints

I   II   III   IV   V

Tarsometatarsal joints

Cuneiform bones

Intertarsal joints

Navicular bone

Cuboid bone

Talus

Calcaneus

Right foot, superior view

Figure 9.21

Joints of the Foot.   The intertarsal, tarsometatarsal, metatarsophalangeal (MP), and interphalangeal (IP) joints help move the toes and foot.

bones articulate with the first three metatarsals. The fourth and fifth metatarsals articulate with the cuboid.

The **metatarsophalangeal** (met′ă-tar′sō-fă-lan′jē-ăl) **joints**, also called the **MP joints**, are between the metatarsals and the phalanges of the toes. These are condylar joints, and they permit limited abduction and adduction of the toes, as well as flexion and extension.

Finally, the **interphalangeal (IP) joints** occur between individual phalanges. Each interphalangeal joint is a hinge joint that permits flexion and extension only.

Ankle Sprains and Pott Fractures

A **sprain** is a stretching or tearing of ligaments, without fracture or dislocation of the joint. An ankle sprain results when the foot is twisted, almost always due to *overinversion*. Fibers of the lateral ligaments are either stretched (in mild sprains) or torn (in more severe sprains), producing localized swelling and tenderness anteroinferior to the lateral malleolus. *Overeversion* sprains rarely occur due to the strength of the deltoid ligament. Remember from chapter 4 that ligaments are composed of dense regular connective tissue, which is poorly vascularized. Tissue that is poorly vascularized takes a long time to heal, and that is the case with ankle sprains. They are also prone to reinjury.

If overeversion *does* occur, the injury that usually results is called a **Pott fracture** (see chapter 6). If the foot is overeverted, it pulls on the deltoid ligament, which is very strong and doesn't tear. Instead, the pull on the deltoid ligament can avulse (pull off) the medial malleolus of the tibia. The force from the injury then continues to move the talus laterally, since the medial malleolus can no longer restrict side-to-side movements of the ankle. As the talus moves laterally and puts force on the fibula, the fibula fractures as well (usually at its distal end or by the lateral malleolus). Thus, both the tibia and the fibula fracture in this injury, and yet the deltoid ligament remains intact.

WHAT DID YOU LEARN?

12   Which ligaments support the hip joint?

13   List the intracapsular ligaments of the knee joint, and discuss their function.

14   Compare the deltoid and lateral ligaments of the ankle joint. Which of these ligaments is stronger? What types of injuries are associated with these ligaments?

Disease and Aging of the Joints

Key topic in this section:

- Effects of aging on the joints

During a person's lifetime, the joints are subjected to extensive wear and tear. A joint's size, flexibility, and shape are affected and modified by use. Active joints develop larger and thicker capsules, and the supporting ligaments and bones increase in size.

Prior to the closure of the epiphyseal plates in early adulthood, some injuries to a young person may result in subluxation or fracture of an epiphysis, with potential adverse effects on the future development and health of the joint. After the epiphyseal plates close, injuries at the epiphyses typically result in sprains.

Arthritis is a disease that involves damage to articular cartilage (see Clinical View: In Depth). The primary problem that develops in an aging joint is osteoarthritis, also known as degenerative arthritis. The cause of the damage may vary, but it usually results from cumulative wear and tear at the joint surface.

Just as the strength of a bone is maintained by continual application of stress, the health of joints is directly related to moderate exercise. Exercise compresses the articular cartilages, causing synovial fluid to be squeezed out of the cartilage and then pulled

CLINICAL VIEW: In Depth

Arthritis

Arthritis (ar-thrī′tis) is a group of inflammatory or degenerative diseases of joints that occur in various forms. Each form presents the same symptoms: swelling of the joint, pain, and stiffness. It is the most prevalent crippling disease in the United States. Some common forms of arthritis are gouty arthritis, osteoarthritis, and rheumatoid arthritis.

Gouty arthritis is typically seen in middle-aged and older individuals, and is more common in males. Often called "gout," this disease occurs as a result of an increased level of uric acid (a normal cellular waste product) in the blood. This abnormal level causes urate crystals to accumulate in the blood, synovial fluid, and synovial membranes. The body's inflammatory response to the urate crystals results in joint pain. Gout usually begins with an attack on a single joint (often in the great toe), and later progresses to other joints. Eventually, gouty arthritis may immobilize joints by causing fusion between the articular surfaces of the bones. Often, nonsteroidal anti-inflammatory drugs (NSAIDs) are used to alleviate symptoms and reduce the inflammation.

Osteoarthritis is the most common type of arthritis. This chronic degenerative joint condition is termed "wear-and-tear arthritis" because repeated use of a joint gradually wears down the articular cartilage, much like the repeated use of a pencil eraser wears down the eraser. If the cartilage is worn down enough, osteoarthritis results. Eventually, bone rubs against bone, causing abrasions on the bony surfaces. Without the protective articular cartilage, movements at the joints become stiff and painful. The joints most affected by osteoarthritis are those of the fingers, knuckles, hips, knees, and shoulders. The nonsynovial joints between the vertebral bodies are also susceptible to osteoarthritis, especially in the cervical and lumbar regions. Osteoarthritis is typically seen in older individuals, although more and more athletes are experiencing arthritis at an earlier age due to the repetitive stresses placed on their joints. NSAIDs are used to alleviate the symptoms of osteoarthritis.

Rheumatoid (roo′mă-toyd) **arthritis** is typically seen in younger and middle-aged adults, and is much more prevalent in women. It presents with pain and swelling of the joints, muscle weakness, osteoporosis, and assorted problems with both the heart and the blood vessels.

Rheumatoid arthritis is an *autoimmune disorder* in which the body's immune system targets its own tissues for attack. Although the cause of this reaction is unknown, it often follows infection by certain bacteria and viruses that have surface molecules similar to molecules normally present in the joints. When the body's immune system is stimulated to attack the foreign molecules, it also destroys its own joint tissue, thus initiating the autoimmune disorder.

Rheumatoid arthritis starts with synovial membrane inflammation. Fluid and white blood cells leak from small blood vessels into the joint cavity, causing an increase in synovial fluid volume. As a consequence, the joint swells, and the inflamed synovial membrane thickens; eventually, the articular cartilage and, often, the underlying bone become eroded. Scar tissue later forms and ossifies, and bone ends fuse together, immobilizing the joint. Medications that help suppress the immune system (e.g., prednisone) are frequently used to alleviate the symptoms of rheumatoid arthritis.

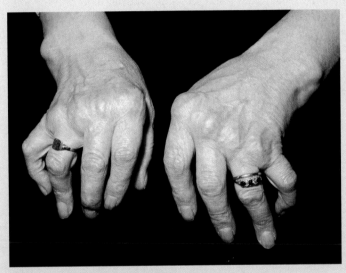

(a) Hands with rheumatoid arthritis

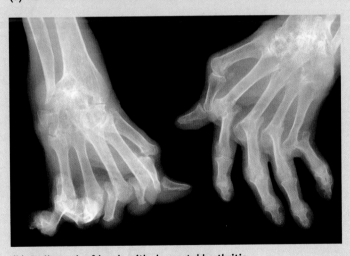

(b) Radiograph of hands with rheumatoid arthritis

back inside the cartilage matrix. This flow of fluid gives the chondrocytes within the cartilage the nourishment required to maintain their health. Joints become stronger, and the rate of degeneration of the articular cartilage is reduced. Exercise also strengthens the muscles that support and stabilize the joint. However, extreme exercise should be avoided, because it aggravates potential joint problems and may worsen osteoarthritis. Athletes such as baseball player Nolan Ryan and Olympic figure skater Dorothy Hamill have experienced osteoarthritis at an early age due to extreme exercise in their youth.

WHAT DID YOU LEARN?

15 What is osteoarthritis, and what causes it?

Development of the Joints

Key topic in this section:

- How joints develop in the embryo

Joints start to form by the sixth week of development and become better differentiated during the fetal period. Some of the mesenchyme around the developing bones develops into the connective tissues of the articulations. For example, in the area of future fibrous joints, the mesenchyme around the developing bones differentiates into dense regular connective tissue, and later joins the developing bones together. In cartilaginous joints, the mesenchyme differentiates into either fibrocartilage or hyaline cartilage.

The development of the synovial joints is more complex than that of fibrous and cartilaginous joints (**figure 9.22**). The mesenchyme around the articulating bones differentiates into the components of a synovial joint. The most *laterally* placed mesenchyme forms the articular capsule and supporting ligaments of the joint. Just medial to this region, the mesenchyme forms the synovial membrane, which then starts secreting synovial fluid into the joint cavity. The *centrally* located mesenchyme differentiates in one of three ways, depending upon the type of synovial joint:

1. The centrally located mesenchyme is resorbed, and a **free joint cavity** forms. Examples of free joint cavities are the interphalangeal joints of the fingers and toes.

2. The centrally located mesenchyme forms incomplete cartilaginous rings or blocks called **menisci**, which serve as shock absorbers in joints. The knee joint is a synovial joint that contains menisci.

3. The centrally located mesenchyme condenses and forms a cartilaginous **articular disc** within the joint cavity. The articular disc assists the movement of the articulating bones. Examples of synovial joints with articular discs are the sternoclavicular joint, the acromioclavicular joint, and the radiocarpal joint.

Differentiation of the centrally located mesenchyme occurs by about the twelfth week of development, and the entire joint continues to differentiate throughout the fetal period.

WHAT DID YOU LEARN?

16 Joints start to form during which week of development?

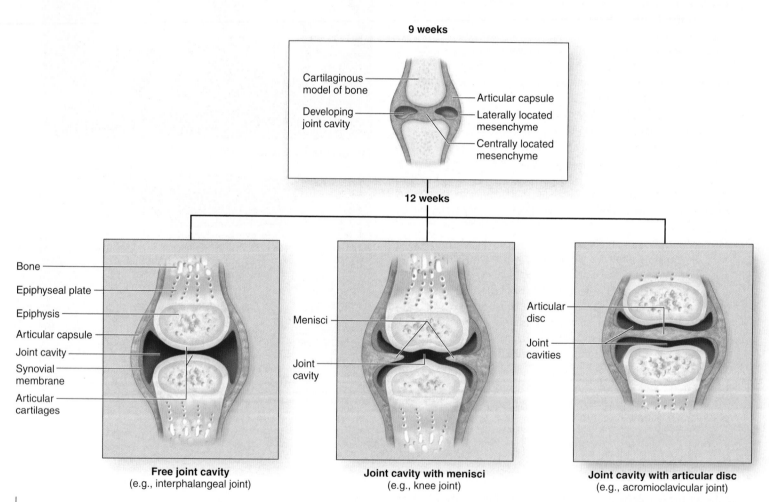

Figure 9.22

Development of the Synovial Joints. By 9 weeks of development, as the future bones are developing, a primitive model of the synovial joint cavities forms. Over the next several weeks, the synovial joints continue to differentiate. By 12 weeks, some have formed free joint cavities (e.g., the interphalangeal joints) while others have formed menisci (e.g., the knee joint). Still other joints have an articular disc (e.g., the acromioclavicular joint) that separates the articulating bones.

CLINICAL TERMS

ankylosis (ang'ki-lō'sis) Stiffening of a joint due to the union of fibers or bones across the joint as the result of a disease.

arthralgia (ar-thral'jē-ă) Joint-associated pain that is not usually inflammatory.

arthroplasty (ar'thrō-plas-tē) Construction of an artificial joint to provide relief from or to correct advanced degenerative arthritis.

arthrosis Condition pertaining to an articulation; a noninflammatory disease of a joint.

bursitis Inflammation of a bursa.

chondromalacia (kon'drō-mă-lā'shē-ă) **patellae** Softening of the articular cartilage of the patella; sometimes considered a subtype of patellofemoral syndrome.

rheumatism (roo'mă-tizm) Any one of various conditions exhibiting joint pain or other symptoms of articular origin; often associated with muscular or skeletal system problems.

synovitis (sin-ō-vī'tis) Inflammation of the synovial membrane of a joint.

tenosynovitis (ten'ō-sin-ō-vī'tis) Inflammation of a tendon or its sheath.

CHAPTER SUMMARY

Articulations (Joints) 251	■ Articulations occur where bones interact. Joints differ in structure, function, and the amount of movement they allow, which may be extensive, slight, or none at all.
	Classification of Joints 251
	■ There are three structural categories of joints: fibrous, cartilaginous, and synovial.
	■ The three functional categories of joints are synarthroses, which are immobile; amphiarthroses, which are slightly mobile; and diarthroses, which are freely mobile.
Fibrous Joints 252	■ In fibrous joints, articulating bones are interconnected by dense regular connective tissue.
	Gomphoses 252
	■ A gomphosis is a synarthrosis between the tooth and either the mandible or the maxillae.
	Sutures 253
	■ A suture is a synarthrosis that tightly binds bones of the skull. Closed sutures are called synostoses.
	Syndesmoses 253
	■ A syndesmosis is an amphiarthrosis, and the bones are connected by interosseous membranes.
Cartilaginous Joints 253	■ In cartilaginous joints, articulating bones are attached to each other by cartilage.
	Synchondroses 253
	■ A synchondrosis is a synarthrosis where hyaline cartilage is wedged between articulating bones.
	Symphyses 254
	■ A symphysis is an amphiarthrosis, and has a disc of fibrocartilage wedged between the articulating bones.
Synovial Joints 254	■ Synovial joints are diarthroses.
	General Anatomy of Synovial Joints 255
	■ Synovial joints contain an articular capsule, a joint cavity, synovial fluid, articular cartilage, ligaments, and nerves and blood vessels.
	Types of Synovial Joints 256
	■ The six types of synovial joints are plane, hinge, pivot, condylar, saddle, and ball-and-socket.
	Movements at Synovial Joints 258
	■ Motions that occur at synovial joints include gliding, angular, rotational, and special.
Selected Articulations in Depth 263	■ In each articulation, unique features of the articulating bones support only the intended movement.
	Joints of the Axial Skeleton 263
	■ The temporomandibular joint is an articulation between the head of the mandible and the mandibular fossa of the temporal bone.
	■ Vertebrae articulate between their bodies as well as between the inferior and superior articular processes.
	Joints of the Pectoral Girdle and Upper Limbs 266
	■ The sternoclavicular joint is a saddle joint between the manubrium of the sternum and the sternal end of the clavicle.
	■ The acromioclavicular joint is a plane synovial joint between the acromion and the acromial end of the clavicle.
	■ The glenohumeral joint is a ball-and-socket joint between the glenoid cavity of the scapula and the head of the humerus.
	■ The elbow is a hinge joint.
	■ The radiocarpal joint involves the distal radius and three proximal carpal bones.

(continued on next page)

CHAPTER SUMMARY (continued)

Selected Articulations in Depth (continued) 263	**Joints of the Pelvic Girdle and Lower Limbs 272**
	■ The hip joint is a ball-and-socket joint between the head of the femur and the acetabulum of the os coxae.
	■ The knee joint is primarily a hinge joint, but is capable of slight rotation and gliding.
	■ The talocrural joint is a hinge joint that permits dorsiflexion and plantar flexion.
	■ Intertarsal joints occur between the tarsal bones.
	■ Tarsometatarsal joints are the articulations between the tarsal and metatarsal bones.
	■ Metatarsophalangeal joints are the articulations between the metatarsals and the proximal phalanges.
	■ Interphalangeal joints occur between individual phalanges.
Disease and Aging of the Joints 280	■ Osteoarthritis is a common joint problem that occurs with aging.
Development of the Joints 281	■ Joints begin to form during week 6 of development.

CHALLENGE YOURSELF

Matching

Match each numbered item with the most closely related lettered item.

_____ 1. joint between sternum and clavicle

_____ 2. joint between tooth and jaw

_____ 3. joint angle is increased in an AP plane

_____ 4. bursa

_____ 5. palm faces posteriorly

_____ 6. standing on tiptoe

_____ 7. intervertebral disc

_____ 8. articulation among tibia, fibula, and talus

_____ 9. menisci

_____ 10. ligament of head of femur

a. talocrural joint

b. plantar flexion

c. gomphosis

d. hip joint

e. located in knee joint

f. has anulus fibrosus and nucleus pulposus

g. pronation

h. extension

i. sternoclavicular joint

j. sac filled with synovial fluid

Multiple Choice

Select the best answer from the four choices provided.

_____ 1. The greatest range of mobility of any joint in the body is found in the
   a. knee joint.
   b. hip joint.
   c. glenohumeral joint.
   d. elbow joint.

_____ 2. The movement of the foot that turns the sole laterally is called
   a. dorsiflexion.
   b. inversion.
   c. eversion.
   d. plantar flexion.

_____ 3. A _____ is formed when two bones previously connected in a suture fuse.
   a. gomphosis
   b. synostosis
   c. symphysis
   d. syndesmosis

_____ 4. The ligament that helps to maintain the alignment of the condyles between the femur and tibia and to limit the anterior movement of the tibia on the femur is the
   a. tibial collateral ligament.
   b. posterior cruciate ligament.
   c. anterior cruciate ligament.
   d. fibular collateral ligament.

_____ 5. The glenohumeral joint is primarily stabilized by the
   a. coracohumeral ligament.
   b. glenohumeral ligaments.
   c. rotator cuff muscles that move the humerus.
   d. scapula.

_____ 6. In a biaxial articulation,
   a. movement can occur in all three planes.
   b. only circumduction occurs.
   c. movement can occur in two planes.
   d. movement can occur in only one plane.

_____ 7. A metacarpophalangeal (MP) joint, which has oval articulating surfaces and permits movement in two planes, is what type of synovial joint?
   a. condylar
   b. plane
   c. hinge
   d. saddle

_____ 8. The ligament that is not associated with the intervertebral joints is the
   a. anterior longitudinal ligament.
   b. pubofemoral ligament.
   c. ligamentum flavum.
   d. supraspinous ligament.

_____ 9. Which of the following is a function of synovial fluid?
   a. lubricates the joint
   b. provides nutrients for articular cartilage
   c. absorbs shock within the joint
   d. All of these are correct.

_____ 10. All of the following movements are possible at the radiocarpal joint except
   a. circumduction.
   b. abduction.
   c. flexion.
   d. rotation.

Content Review

1. Discuss the factors that influence both the stability and the mobility of a joint. What is the relationship between a joint's mobility and its stability?

2. Describe the structural differences between fibrous joints and cartilaginous joints.

3. Describe all joints that are functionally classified as synarthroses.

4. Discuss the origin and function of synovial fluid within a synovial joint.

5. Compare a hinge joint and a pivot joint with respect to structure, function, and location within the body.

6. Describe and compare the movements of abduction, adduction, pronation, and supination.

7. Describe the basic anatomy of the glenohumeral joint.

8. What are the main supporting ligaments of the elbow joint?

9. How do the tibia and talus maintain their correct positioning in the talocrural joint?

10. What is the primary age-related change that can occur in a joint?

Developing Critical Reasoning

1. During soccer practice, Erin tripped over the outstretched leg of a teammate and fell directly onto her shoulder. She was taken to the hospital in excruciating pain. Examination revealed that the head of the humerus had moved inferiorly and anteriorly into the axilla. What happened to Erin in this injury?

2. While Lucas and Omar were watching a football game, a player was penalized for "clipping," meaning that he had hit an opposing player on the lateral knee, causing hyperabduction at the knee joint. Lucas asked Omar what the big deal was about "clipping." What joint is most at risk, and what kind of injuries can occur if a player gets "clipped"?

ANSWERS TO "WHAT DO YOU THINK?"

1. A synchondrosis is designed to be a synarthrosis because as the bone ends are growing, they must not be allowed to move in relation to one another. For example, if the epiphysis and diaphysis move along the epiphyseal plate, bone growth is compromised and the bone becomes misshapen.

2. The plane joint, the least mobile of the two joints, must be more stable than the ball-and-socket joint.

3. When sitting upright in a chair, both the hip and knee joints are flexed.

4. The anterior longitudinal ligament becomes taut during extension, so it is taut when we are standing and more relaxed when we are sitting.

Visit the McKinley/O'Loughlin *Human Anatomy*, 2e website at aris.mhhe.com

MUSCULAR SYSTEM

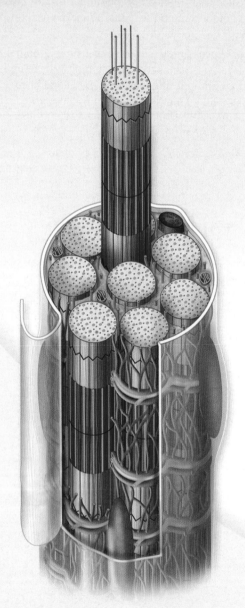

10

Anatomy &
Physiology **REVEALED**
aprevealed.com

MUSCULAR SYSTEM

Muscle Tissue and Organization

When most of us hear the word "muscle," we think of the muscles that move the skeleton. These muscles are considered organs because they are composed not only of muscle tissue, but also of epithelial, connective, and nervous tissue. Over 700 skeletal muscles have been named and together they form the **muscular system.**

However, skeletal muscles are not the only places where muscle tissue is found. In fact, muscle tissue is distributed almost everywhere in the body and is responsible for the movement of materials within and throughout the body. Without this vital tissue, the food we eat would not be propelled along the gastrointestinal tract, the waste products we produce would not be rhythmically expelled, and our blood would not be pumped to body tissues.

In order to understand how muscle tissue performs these functions, this chapter examines its gross and microscopic anatomy, with emphasis on skeletal muscle tissue. The chapter also briefly reviews the characteristics of the other two types of muscle tissue (cardiac and smooth), which were introduced in chapter 4.

Properties of Muscle Tissue

Key topic in this section:

- Four unique properties of muscle tissue

Because of their potentially extraordinary length, skeletal *muscle cells* are often referred to as **muscle fibers**. All muscle tissue is composed of muscle cells and exhibits certain common properties:

- **Excitability.** Excitability is equated with responsiveness. In other words, muscle cells are very responsive to input from stimuli. When a muscle cell is stimulated by the nervous system, by stretching, or by other stimuli in the environment, it responds by initiating electrical changes that sweep across its entire plasma membrane and spark internal events leading to muscle contraction.
- **Contractility.** Stimulation of muscle cells generates tension within the cell (contraction), which may cause the cell to shorten. This shortening results in either a pull on bones of the skeleton or the movement of specific body parts.
- **Elasticity.** A contracted muscle cell recoils to its resting length when the applied tension is removed. Thus, elasticity is not the muscle's ability to stretch, but its ability to return to its original length when tension is released.
- **Extensibility.** A muscle cell must be capable of extending in length in response to the contraction of opposing muscle cells. For example, when you flex your elbow joint, you are contracting the biceps brachii on the anterior side of your arm and extending the triceps brachii on the posterior side of your arm. The reverse is true when you extend your elbow joint. During these movements, muscle cells undergo alternating cycles of contraction and extension.

WHAT DID YOU LEARN?

1 What are the properties of muscle tissue?

Characteristics of Skeletal Muscle Tissue

Key topics in this section:

- The many roles of skeletal muscle in the body
- Levels of organization in a skeletal muscle
- How muscles are attached to other body structures
- Components of muscle fibers

Each skeletal muscle is an organ composed of the four types of tissue: epithelial, connective, muscle, and nervous tissue. A skeletal muscle is striated (marked by stripes or bands) and usually attached to one or more bones. Individual skeletal muscles vary widely in shape, ranging from elongated and flat to thick and triangular, or even circular. A single muscle may be composed of thousands of fibers, and each fiber is as long as the muscle itself. A small skeletal muscle fiber in a muscle in the toe may have a length of about 100 micrometers (μm) and a diameter of about 10 μm. In contrast, fibers in the muscle on the anterior side of the arm may extend up to 35 centimeters (cm) and have a diameter of about 100 μm (i.e., the same thickness as a fine strand of hair).

Functions of Skeletal Muscle Tissue

Skeletal muscles perform the following functions:

- **Body movement.** Bones of the skeleton move when muscles contract and pull on the tendons by which the muscles are attached to the bones. The integrated functioning of muscles, bones, and joints can produce both highly coordinated movements such as running and relatively simple, localized movements such as underlining a word in a book.
- **Maintenance of posture.** Contraction of specific skeletal muscles stabilizes joints and helps maintain the body's posture or position. Postural muscles contract continuously when a person is awake to prevent collapse. (Note that when a person falls asleep in, say, a lecture hall, the head droops forward and the mouth rolls open.)
- **Temperature regulation.** Energy is required for muscle tissue contraction, and heat is always produced as a waste product of this energy usage. Most of this heat maintains our normal body temperature. Note that when you exercise, you feel warmer; the heat you feel is produced by your working muscles. Likewise, you shiver when you are cold because your muscles are contracting and relaxing in order to produce heat.
- **Storage and movement of materials.** Circular muscle bands, called **sphincters** (sfingk'ter; *sphinkter* = a band), contract at the openings, or **orifices** (or'i-fis; *orificium* = opening), of the gastrointestinal and urinary tracts. These sphincters may be voluntarily closed (to store the material within an organ) or opened (to facilitate movement of materials).
- **Support.** Skeletal muscle is sometimes arranged in flat sheets or layers, such as along the walls of the abdominal cavity and the floor of the pelvic cavity. These sheets of muscle protect the organs and support their weight within the abdominopelvic cavity. In fact, the primary function of your abdominal muscles is to hold your abdominal organs in place!

Gross Anatomy of Skeletal Muscle

In general, a skeletal muscle is composed of numerous skeletal muscle fibers, blood vessels and nerves, and connective tissue sheets that surround the muscle fibers and connect the muscle to bone. While an individual muscle fiber runs the *length* of a muscle, it takes groupings of many muscle fibers to form the *width* of a muscle. If you look at a cross section **(figure 10.1)**, each skeletal muscle is composed of **fascicles** (fas'i-kl; *fascis* = bundle), which are bundles of muscle fibers. Muscle fibers, in turn, contain cylindrical structures called myofibrils, which are composed of myofilaments. **Table 10.1** illustrates the levels of organization in a skeletal muscle.

Connective Tissue Components

Three concentric layers of connective tissue, composed of collagen and elastic fibers, encircle each individual muscle fiber, groups of muscle fibers, and the entire muscle itself. These layers provide protection, sites for distribution of blood vessels and nerves, and a means of attachment to the skeleton. The three connective tissue layers are the endomysium, the perimysium, and the epimysium.

The **endomysium** (en'dō-mis'ē-ŭm, -miz'ē-ŭm; *endon* = within, *mys* = muscle) is the innermost connective tissue layer. The endomysium is a delicate, areolar connective tissue layer that surrounds and

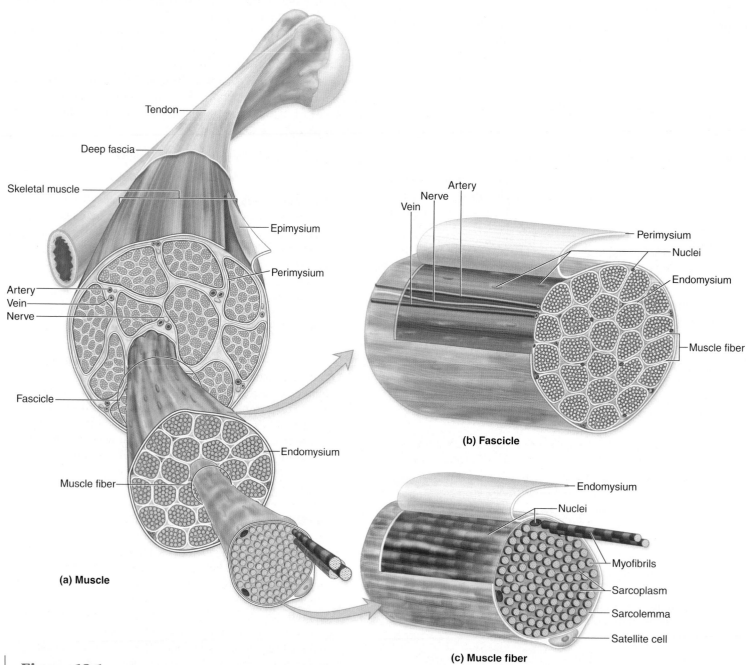

Figure 10.1

Structural Organization of Skeletal Muscle. (*a*) A skeletal muscle is ensheathed within a connective tissue layer called the epimysium. (*b*) Each fascicle (bundle of muscle fibers) is wrapped within a connective tissue layer called the perimysium. (*c*) Each muscle fiber is surrounded by a delicate connective tissue layer termed the endomysium.

Table 10.1 Organizational Levels of Skeletal Muscle

<table>
<tr><th>Level</th><th>Appearance</th><th>Description</th><th>Connective Tissue Covering</th></tr>
<tr><td>Muscle</td><td>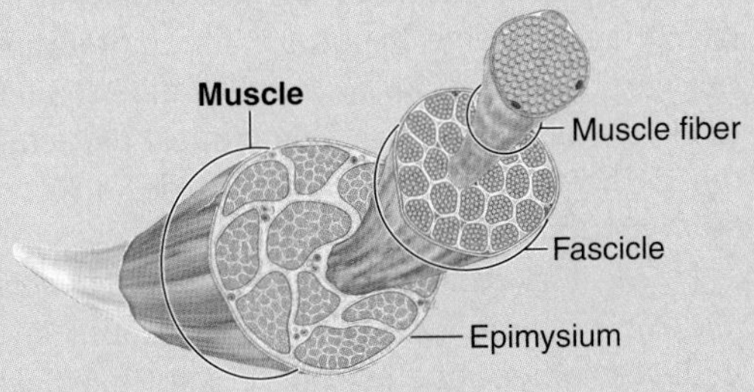
Muscle
Muscle fiber
Fascicle
Epimysium</td><td>Multiple fascicles housing many muscle fibers, connective tissue coverings, blood vessels, nerve fibers</td><td>Epimysium</td></tr>
<tr><td>Fascicle</td><td>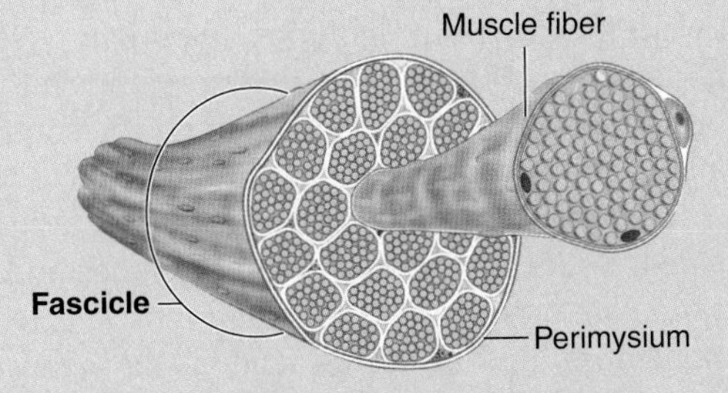
Muscle fiber
Fascicle
Perimysium</td><td>A bundle of muscle fibers separated from other bundles of fibers by a dense irregular connective tissue covering</td><td>Perimysium</td></tr>
<tr><td>Muscle Fiber (muscle cell)</td><td>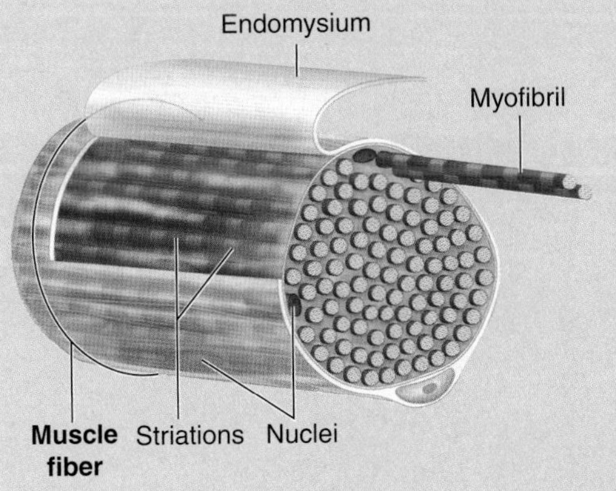
Endomysium
Myofibril
Muscle fiber
Striations
Nuclei</td><td>Elongated, multinucleated, cylindrical fiber (cell); contains myofibrils, separated from other fibers by delicate layer of areolar connective tissue; exhibits striations</td><td>Endomysium</td></tr>
<tr><td>Myofibril</td><td>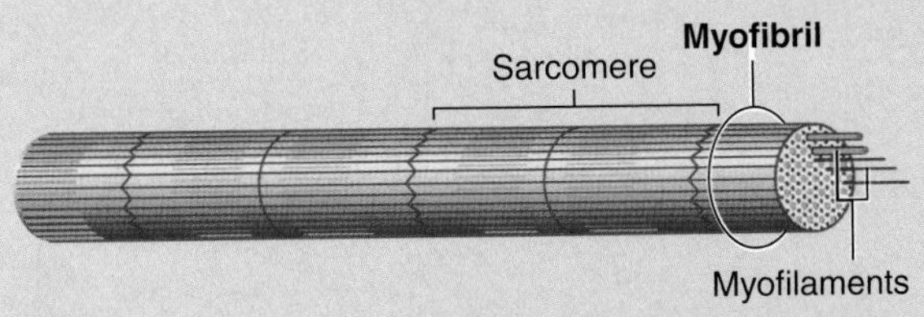
Sarcomere
Myofibril
Myofilaments</td><td>Long, cylindrical contractile element within muscle fiber; as long as the muscle fiber itself; composed of myofilaments; exhibits striations</td><td>None</td></tr>
<tr><td>Myofilaments</td><td>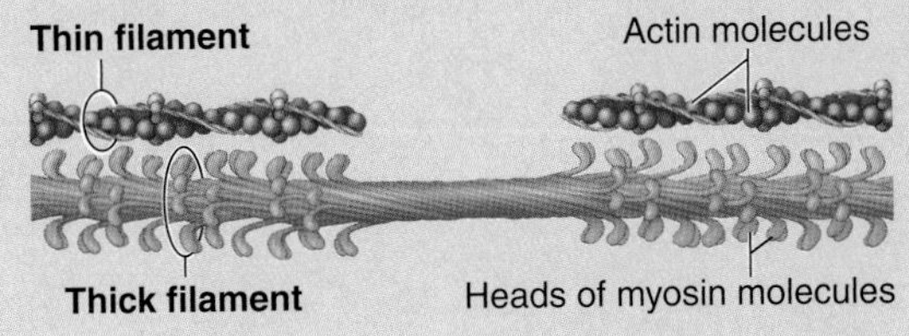
Thin filament
Actin molecules
Thick filament
Heads of myosin molecules</td><td>Short contractile proteins of two types: thick (composed of myosin) and thin (composed of actin, tropomyosin, and troponin)</td><td>None</td></tr>
</table>

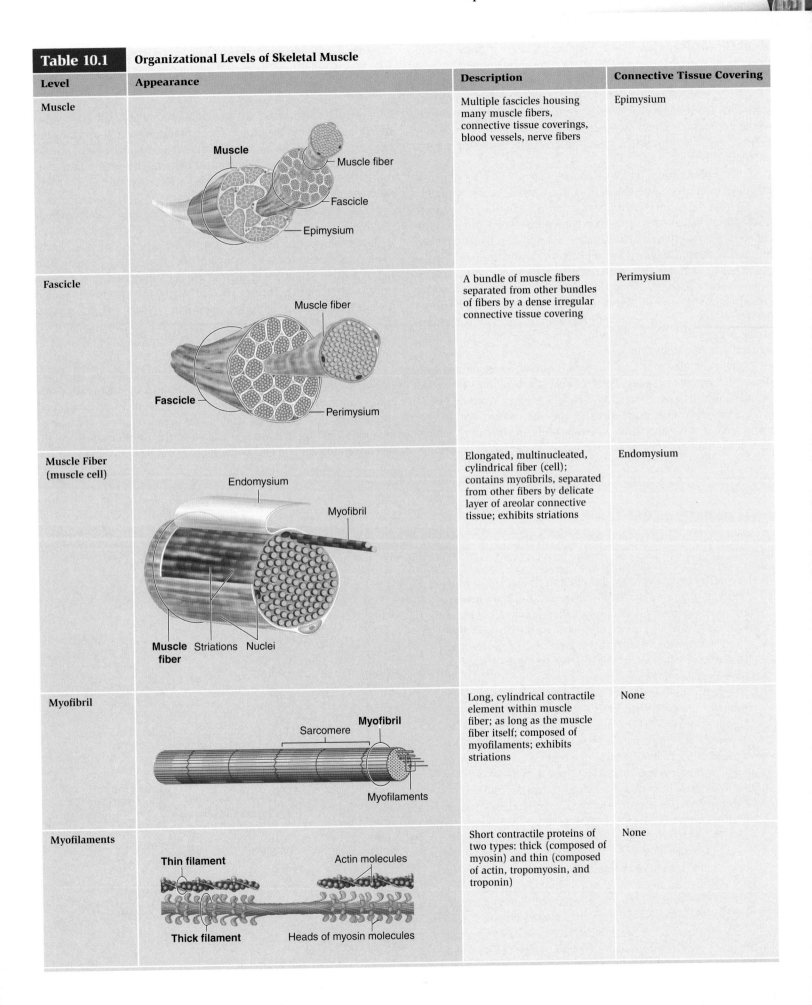

electrically insulates each muscle fiber. It has reticular fibers to help bind together neighboring muscle fibers and support capillaries near these fibers.

The **perimysium** (per-i-mis′ē-ŭm, -miz′ē-ŭm; *peri* = around) surrounds the fascicles. The dense irregular connective tissue sheath of the perimysium contains extensive arrays of blood vessels and nerves (called neurovascular bundles) that branch to supply each individual fascicle.

The **epimysium** (ep-i-mis′ē-ŭm; *epi* = upon) is a layer of dense irregular connective tissue that surrounds the whole skeletal muscle.

These three connective tissue layers are ensheathed by **deep fascia** (fash′ē-ă; band or filler), an expansive sheet of dense irregular connective tissue that separates individual muscles, binds together muscles with similar functions, and forms sheaths to help distribute nerves, blood vessels, and lymphatic vessels, and to fill spaces between muscles. Deep fascia is also called *visceral* or *muscular fascia*. The deep fascia is deep or internal to a layer called the **superficial fascia** (also called the *subcutaneous layer*). The superficial fascia is composed of areolar and adipose connective tissue that separates muscle from skin.

The next time you go to the supermarket, examine a cut of steak, and compare it to skeletal muscle organization. An entire steak is equivalent to at least one skeletal muscle: The fat and connective tissue encircling the steak is the epimysium. Within the steak itself, a whitish, marbled texture is the perimysium, which encircles bundles of muscle fibers (fascicles). You won't be able to distinguish individual muscle fibers or the endomysium in the steak. (We hope you haven't been eating dinner as you have read this section—you may never eat steak again!)

Muscle Attachments   At the ends of a muscle, the connective tissue layers merge to form a fibrous **tendon,** which attaches the muscle to bone, skin, or another muscle. Tendons usually have a thick, cordlike structure. Sometimes, the tendon forms a thin, flattened sheet, termed an **aponeurosis** (ap′ō-noo-rō′sis; *apo* = from, *neuron* = sinew).

Most skeletal muscles extend between bones and cross at least one mobile joint. Upon contraction, one of the bones moves while the other bone usually remains fixed. The less mobile attachment of a muscle is called its **origin** (ōr′i-jin; *origo* = source). The more mobile attachment of the muscle is its **insertion** (in-ser′shŭn; *insero* = to plant in) **(figure 10.2)**. Usually, the insertion is pulled toward the origin. In the limbs, the origin typically lies proximal to the insertion. For example, the biceps brachii muscle originates on the scapula and inserts on the radius. The contraction of this muscle pulls the forearm toward the shoulder.

Sometimes, neither the origin nor the insertion can be determined easily by either movement or position. In this case, other criteria are used. For example, if a muscle extends between a broad aponeurosis and a narrow tendon, the aponeurosis is considered the origin, and the tendon is attached to the insertion. If there are several tendons at one end of the muscle and just one tendon at the

---

CLINICAL VIEW

Tendonitis

Tendonitis refers to inflammation of either a tendon or a synovial sheath surrounding the tendon. (The suffix *-itis* indicates an inflammatory process.) Tendonitis typically results from overuse of the tendon, as when a computer operator spends too much time typing on the keyboard, so that the finger tendons are affected. The synovial sheath around the tendon becomes inflamed, resulting in swelling and pain.

Tendonitis can also be due to age-related changes or to an autoimmune condition such as rheumatoid arthritis. Although the cause of age-related tendonitis is not well understood, aging is known to cause a person's tendons to lose their elasticity and ability to glide smoothly. The resulting limitation of movement is thought to be instrumental in developing age-related tendonitis. In rheumatoid arthritis, a person's immune system reacts against his or her own body. The same inflammatory process that damages the articular joints can injure the synovial sheaths of the tendons.

The treatment of tendonitis begins with giving the affected limb or joint an extended rest. Icing the area helps relieve the acute symptoms, and use of nonsteroidal anti-inflammatory agents (aspirin, ibuprofen, etc.) reduces the inflammation. In the long run, stretching and strengthening the muscles in the affected area is beneficial. A physical therapist can often devise a measured exercise program to avoid future occurrences. In rare cases, tendonitis may have an anatomic cause. For example, if a tendon does not have a smooth and direct "glide-path," it is more likely to become inflamed during use. In this situation, surgical realignment of the tendon may be necessary.

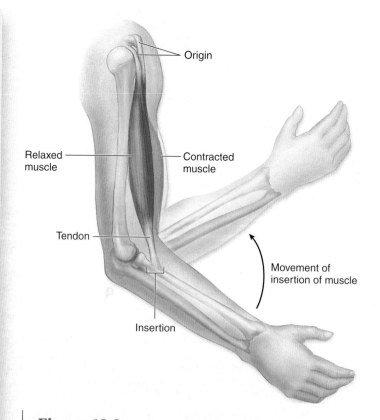

Figure 10.2

Muscle Origin and Insertion.   The origin is the less mobile point of attachment of the muscle, whereas the insertion is more mobile, as shown in this view of the biceps brachii muscle.

other end, each of the multiple tendons is considered an origin, and the single tendon is considered the insertion.

Blood Vessels and Nerves

An extensive network of blood vessels and nerve fibers extends through both the epimysium and the perimysium. The blood vessels deliver to the muscle fibers nutrients and oxygen needed for the production of ATP (adenosine triphosphate). They also remove waste products produced by the muscle fibers.

Skeletal muscles are classified as **voluntary muscles** because they are controlled by the somatic (voluntary) nervous system and we can voluntarily move our skeletal muscles. The neurons (nerve cells) that stimulate muscle contraction are called motor neurons and are said to innervate muscle fibers within the muscle. Each motor neuron has a long extension called an **axon** (ak'son; axis), or nerve fiber. An axon of a motor neuron transmits a nerve impulse to a muscle fiber. The axon travels through the epimysium and perimysium, and enters the endomysium, where it delivers a nerve impulse to an individual muscle fiber. The junction between the axon and the muscle fiber is called a neuromuscular junction, and will be discussed later in this chapter.

Microscopic Anatomy of Skeletal Muscle

Skeletal muscle fibers have many of the same components as the typical cells described in chapter 2, but some of them are named differently **(table 10.2)**.

For instance, the plasma membrane of a skeletal muscle fiber is called the **sarcolemma** (sar'kō-lem'ă; *sarx* = flesh, *lemma* = husk), and the cytoplasm of the muscle fiber is called the **sarcoplasm** (sar'kō-plazm) **(figure 10.3)**. A typical skeletal muscle fiber contains abundant mitochondria (approximately 300 mitochondria per muscle fiber), since the fibers have a great demand for energy. They also contain lysosomes and other cellular organelles. Two main structures are unique to muscle fibers: T-tubules and sarcoplasmic reticulum.

Transverse (trans-vers'; *trans* = across, *versus* = to turn) **tubules,** or **T-tubules,** are deep invaginations of the sarcolemma that extend into the sarcoplasm of skeletal muscle fibers as a network of narrow membranous tubules. Muscle impulses generated when a nerve impulse reaches the sarcolemma travel rapidly along the sarcolemma and then spread internally along the membranes of the T-tubule network. The passage of this impulse helps stimulate and coordinate muscle contractions, as we will see later.

Skeletal muscle fibers contain another internal membrane complex called the **sarcoplasmic** (sar-kō-plaz'mik) **reticulum** (re-tik'ū-lŭm; *rete* = a net) **(SR)** that is similar to the smooth endoplasmic reticulum of other cells. The sarcoplasmic reticulum stores calcium ions needed to initiate muscle contraction. Some parts of the sarcoplasmic reticulum run parallel to the muscle fiber, whereas other parts appear as blind sacs perpendicular to the fiber's length. These blind sacs of the sarcoplasmic reticulum are called **terminal cisternae** (sis-ter'nē; sing., sis-ter'nă). Pairs of terminal cisternae are immediately adjacent to each T-tubule; they are the reservoirs and specific sites for calcium ion release to initiate muscle contraction, and they interact with the T-tubules during muscle contraction. Together,

Table 10.2	Muscle Fiber Components	
Structure	**Description**	**Function**
Muscle Fiber	Single muscle cell	Metabolic activities; contraction
Sarcolemma	Plasma membrane of a muscle fiber	Surrounds muscle fiber and regulates entry and exit of materials
Sarcoplasm	Cytoplasm of a muscle fiber	Site of metabolic processes for normal muscle fiber activities
Sarcoplasmic reticulum	Smooth endoplasmic reticulum in a muscle fiber	Stores calcium ions needed for muscle contraction
Terminal cisternae	Expanded ends of the sarcoplasmic reticulum that are in contact with the transverse tubules	Site of calcium ion release to promote muscle contraction
Transverse tubule (T-tubule)	Narrow, tubular extensions of the sarcolemma into the sarcoplasm, contacting the terminal cisternae; wrapped around myofibrils	Quickly transports a muscle impulse from the sarcolemma throughout the entire muscle fiber
Myofibrils	Organized bundles of myofilaments; cylindrical structures as long as the muscle fiber itself.	Contain myofilaments that are responsible for muscle contraction
Thick filament	Fine protein myofilament composed of bundles of myosin (about 11 nm in diameter)	Bind to thin filament and cause contraction
Thin filament	Fine protein myofilament composed of actin, troponin, and tropomyosin (about 5–6 nm in diameter)	Thick filaments bind to it and cause contraction
Actin	Double-stranded contractile protein	Binding site for myosin to shorten a sarcomere
Tropomyosin	Double-stranded regulatory protein	Covers the active sites on actin, preventing myosin from binding to actin when muscle fiber is at rest
Troponin	Regulatory protein that holds tropomyosin in place and anchors to actin	When calcium ions bind to one of its subunits, troponin changes shape, causing the tropomyosin to move off the actin active site, and this permits myosin binding to actin
Titin	Filaments of an elastic protein	Help return myofilaments to resting position after contraction; maintain positions of myofilaments in sarcomere

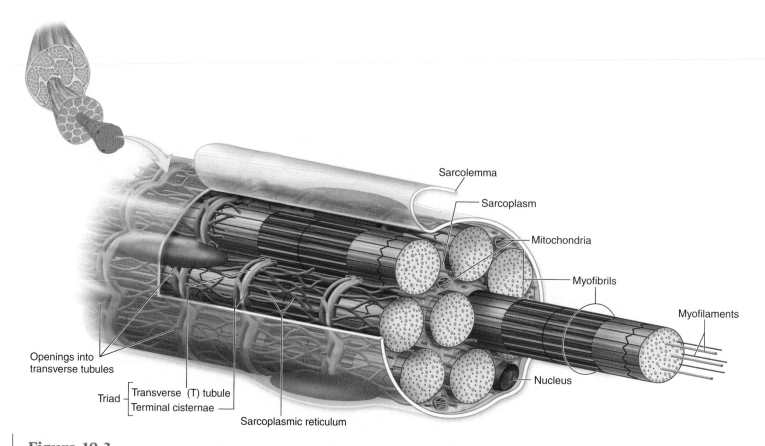

Figure 10.3

Formation, Structure, and Organization of a Skeletal Muscle Fiber.   The internal organization of a muscle fiber includes both thick and thin myofilaments, myofibrils, mitochondria, sarcoplasmic reticulum, and triads.

the two terminal cisternae and the centrally placed T-tubule form a structure called a **triad.**

Recall from chapter 4 that skeletal muscle fibers are multinucle-ated. This increase in the number of nuclei occurs during development when groups of embryonic cells, termed **myoblasts** (mī′ō-blast; *blastos* = germ), fuse to form single skeletal muscle fibers (**figure 10.4**). During this fusion process, each myoblast nucleus contributes to the eventual total number of nuclei. Some myoblasts do not fuse with muscle fibers during development. These embryonic-like cells remain in adult skeletal muscle tissue as **satellite** (sat′ĕ-līt) **cells.** If a skeletal muscle is injured, some satellite cells may be stimulated to differentiate and assist in its repair and regeneration.

Myofibrils and Myofilaments

The sarcoplasm of a skeletal muscle fiber contains hundreds to thousands of long, cylindrical structures termed **myofibrils** (mī-ō-fī′bril). Each myofibril is about 1–2 micrometers in diameter and extends the length of the entire muscle fiber (see figure 10.3). During contraction, the myofibrils shorten as their component proteins change position. Because myofibrils are attached to the ends of the muscle fiber, the shortening of the myofibrils during a contraction causes the fiber to shorten. Myofibrils consist of bundles of short **myofilaments** (mī-ō-fil′ă-ment; *filum* = thread). Whereas a single myofibril runs the length of the muscle fiber, it takes many successive groupings of myofilaments to run the entire length of a myofibril.

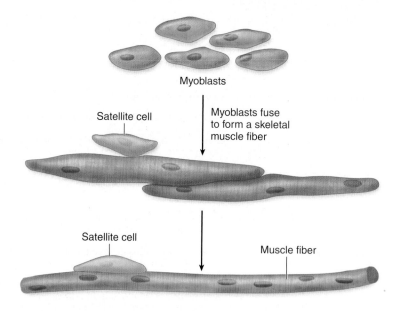

Figure 10.4

Development of Skeletal Muscle.   Embryonic muscle cells called myoblasts fuse to form a single skeletal muscle fiber. Satellite cells are myoblasts that do not go on to form the skeletal muscle fiber. Instead, satellite cells remain with postnatal skeletal muscle tissue and assist in repair of muscles.

Mitochondrial Myopathies

Mitochondria produce adenosine triphosphate (ATP), the source of energy for cell activities. Muscle fibers house numerous mitochondria because each fiber must produce abundant amounts of ATP for muscle contraction.

Mutations in either nuclear or mitochondrial DNA are known to cause mitochondrial myopathies in muscles. The term *myopathy* implies an abnormality or disorder of the muscle fiber that is not inflammatory in nature. Mitochondrial myopathies may result in either an inconsistent supply of ATP or an inability to produce ATP at the rate needed by the muscle fiber. Often, this leads to muscle weakness. It is as if the body has "engine problems." If the motor of your car suddenly worked on only half its cylinders, it would not matter how much gas you put in the car, it could only run at low speeds. Similarly, with a mitochondrial

myopathy, cells simply cannot produce the amount of ATP necessary to carry on normal activities.

Mitochondrial myopathies typically occur in childhood or young adulthood, and are manifest by varying degrees of skeletal muscle dysfunction. The most significantly affected muscles are the extraocular muscles, which are responsible for eye movement, and the muscles of the pectoral and pelvic girdles. Double vision resulting from lack of coordinated eye movements, failing upper body strength, and problems with gait and posture are common complaints.

Mitochondrial myopathies may also develop in older individuals. For example, the well-known athlete Greg LeMond won the greatest cycling race in the world, the Tour de France, in 1986, 1989, and 1990. But after developing a type of muscular dystrophy as a result of mitochondrial myopathy, he was left no choice but to leave the sport in 1994.

Thin and Thick Filaments   The bundles of myofilaments are classified as thin myofilaments (usually simply called thin filaments) and thick myofilaments (usually called thick filaments) **(figure 10.5)**.

Thin filaments are only about 5–6 nanometers in diameter. They are primarily composed of two strands of the protein **actin** twisted around each other to form a helical shape. In each helical strand of actin, many small, spherical molecules are connected to form a long filament resembling a string of beads. Each spherical molecule is called G (globular) actin, and each filament composed of a strand of G-actin molecules is called F (filamentous) actin. Two regulatory proteins, **tropomyosin** (trō-pō-mī'ō-sin) and **troponin** (trō'pō-nin), are part of the thin filaments. The tropomyosin molecule is a short, thin, twisted filament that covers small sections of the actin strands. The troponin has three functions: Structurally, it (1) attaches to actin to anchor itself in place, and (2) attaches to tropomyosin to hold it in place over the surface of the actin; (3) functionally, troponin provides a binding site for calcium ions.

In contrast, **thick filaments** are about twice as large as thin filaments, with a diameter of about 11 nanometers. Thick filaments are assembled from bundles of the protein **myosin**. Each myosin molecule in a thick filament consists of two strands; each strand has a free, globular head and an attached, elongated tail. The myosin molecules are oriented on either end of the thick filament so that the long tails point toward the center of the filament and the heads point toward the edges of the filament and project outward toward the surrounding thin filaments. During a contraction, myosin heads form **crossbridges** by binding thick filaments to actin in the thin filaments.

As mentioned previously, skeletal muscle has striations. This striated appearance is due to size and density differences between thick filaments and thin filaments. Under the light microscope, two differently shaded bands are observed. The dark bands, called **A bands,** contain the entire thick filament. At either end of a thick filament is a light band region occupied by thin filaments that extend into the A band between the stacked thick filaments. The light bands, called **I bands,** contain thin filaments but no thick filaments. In addition to the thin filaments in I bands, there are protein filaments called **titin** (tī'tin), which play a role in muscle elasticity, control of thick filament assembly, and passive stiffness generated in the muscle.

Within both the A bands and the I bands, other important structures can be clearly observed using an electron microscope:

- The **H zone** (also called the *H band*) is a light, central region in the A band. It is lighter shaded because only thick filaments are present—that is, there are no thin filaments overlapping the thick filaments in the H zone in a relaxed muscle fiber. At maximal contraction, the thin filaments are pulled into this zone, and the H zone disappears.

- The **M line** is a thin transverse protein meshwork structure in the center of the H zone of a relaxed fiber. It serves as an attachment site for the thick filaments and keeps the thick filaments aligned during contraction and relaxation.

- The **Z disc** (also called the *Z line*) is a thin transverse protein structure in the center of the I band that serves as an attachment site for thin filament ends. Although the Z disc is circular, when viewed "head-on," only the edge of the circle is visible, so it sometimes looks like a line. **Connectins** (kon-nek'tins) are Z disc proteins that anchor and interconnect the thin filament ends at either end of a sarcomere (discussed next).

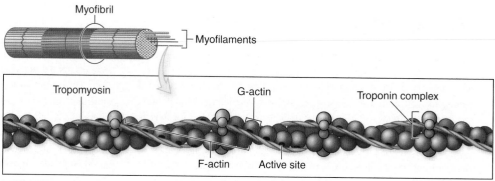

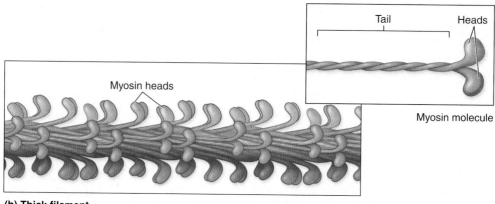

(a) Thin filament

Figure 10.5

Molecular Structure of Thin and Thick Filaments. Contractile proteins form myofilaments within myofibrils. *(a)* A thin filament is composed of actin, tropomyosin, and troponin proteins. *(b)* A thick filament consists of myosin protein molecules. *(c)* Comparison of thick and thin filaments.

(b) Thick filament

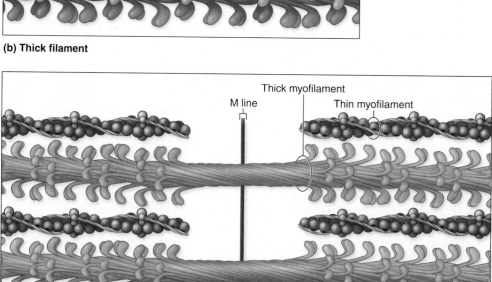

(c) Comparison of thick and thin filaments

Organization of a Sarcomere

A **sarcomere** (sar′kō-mēr; *meros* = part) is the *functional contractile unit* of a skeletal muscle fiber. A sarcomere is defined as the distance from one Z disc to the next adjacent Z disc. Myofibrils contain multiple Z discs; thus, there are numerous sarcomeres in each myofibril. Imagine that the sarcomeres are arranged within a myofibril like a stack of coins, where the width of each coin represents a single sarcomere and the opposing faces of adjacent coins form the Z disc between adjacent sarcomeres. Each sarcomere shortens as the muscle fiber contracts.

An individual sarcomere is shown in **figure 10.6**, and its components are listed in **table 10.3**. Notice that thick filaments are positioned at the center of the sarcomere. A cross section through the lateral parts of the A band reveals the relative sizes, arrangements, and organization of thick and thin filaments. In

cross section, each thin filament is sandwiched by three thick filaments that form a triangle at its periphery, and similarly each thick filament when viewed in cross section is surrounded by six thin filaments.

WHAT DID YOU LEARN?

2 List and describe the connective tissue components of skeletal muscle.

3 Describe the difference between the origin and the insertion of a skeletal muscle. Where is each typically located?

4 Compare and contrast myofibrils and myofilaments.

5 What proteins are in thick filaments and thin filaments?

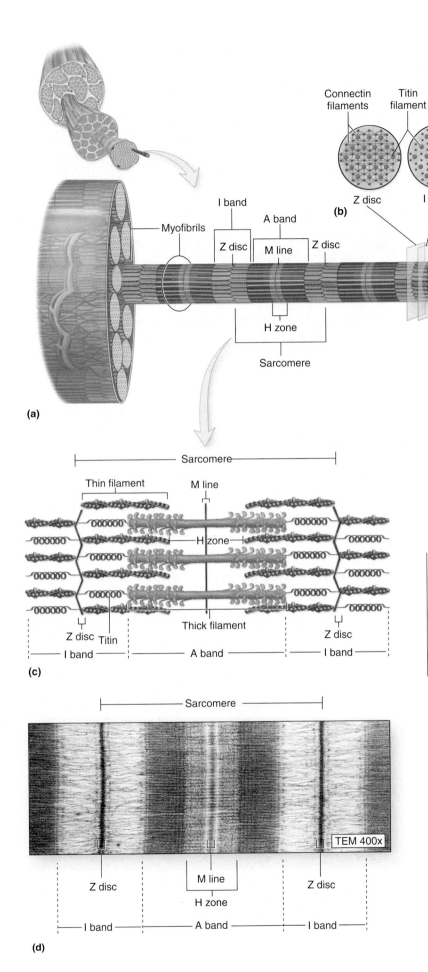

Connectin filaments   Titin filament   Thin filament   Thick filament   Thick filament   Thin filament

Z disc   I band   M line   H zone   A band

(b)

Myofibrils

I band
A band
Z disc   M line   Z disc
H zone
Sarcomere

(a)

Sarcomere

Thin filament   M line

H zone

Z disc   Titin
Thick filament
I band   A band   I band

Z disc

(c)

Sarcomere

TEM 400x

Z disc   M line   Z disc
H zone
I band   A band   I band

(d)

Figure 10.6

Structure of a Sarcomere.   *(a)* The microscopic arrangement of bands and zones within myofibrils in a muscle fiber defines a sarcomere. *(b)* Cross sections of a myofibril reveal the different arrangements of myofilaments in each region of the sarcomere. *(c)* Diagram and *(d)* electron micrograph compare the organization of a sarcomere.

Table 10.3	Sarcomere and Its Components	
Structure	**Description**	**Does it shorten/ narrow when the muscle contracts?**
Sarcomere	Functional contractile unit of skeletal muscle; the distance between two adjacent Z discs	Yes
A band	Dark band in the middle of the sarcomere; composed of entire thick filaments and on its lateral end regions of overlapping thin filaments	No
H zone	Lighter region in the middle of the A band; contains thick filaments only	Yes
M line	A thin transverse protein meshwork that appears as a dark protein disc in center of H zone where thick filaments attach	No
I band	Light band containing thin filaments only and titin proteins	Yes
Z disc	Dark proteins called connectins in the center of the I band where thin filaments attach	No

Contraction of Skeletal Muscle Fibers

Key topics in this section:

- Structure of a neuromuscular junction
- Process of skeletal muscle contraction
- Structure and function of a motor unit
- Comparison of isometric and isotonic contraction

A contracting skeletal muscle fiber typically shortens as all of its sarcomeres shorten in length. Ultimately, **tension** is exerted on the portion of the skeleton where the muscle is attached. The thick and thin protein filaments in sarcomeres interact to cause muscle contraction. The mechanism for contraction is explained by the sliding filament theory.

The Sliding Filament Theory

According to the **sliding filament theory,** when a muscle contracts, thick and thin filaments slide past each other, and the sarcomere shortens **(figure 10.7)**. The following changes occur within a sarcomere during a contraction:

- The width of the A band remains constant, but the H zone disappears.
- The Z discs in one sarcomere move closer together.

- The sarcomere narrows or shortens in length.
- The I bands narrow or shorten in length.

Thick and thin filaments maintain their same length, whether the muscle is relaxed or contracted. However, during muscle fiber contraction, the relative position between the thick and thin filaments within the sarcomeres changes markedly. Thick filaments in neighboring sarcomeres move closer together, as do the thin filaments on either end of one sarcomere.

> ## Study Tip!
>
> To help remember what shortens (and what stays the same length) during muscle contraction, try this experiment: Interweave your second through fifth fingers together—they represent the thick and thin filaments. Your thumbs (pointing upward) represent the Z discs. The distance between your thumbs is the sarcomere. Now slide your fingers toward each other. The distance between your thumbs (the sarcomere) shortens. Your thumbs (the Z discs) and your fingers (the thick and thin filaments) remain the same length. This illustrates that the filaments do not change *length*—only the *distance* between the Z discs is affected.

Neuromuscular Junctions

Earlier we noted that motor neuron activity stimulates skeletal muscle contraction. Muscle contraction begins when a nerve impulse stimulates an impulse in a muscle fiber. Each muscle fiber is controlled by one motor neuron. The motor neuron transmits the effect of a nerve impulse to the muscle fiber at a **neuromuscular** (noor-ō-mŭs′kū-lăr) **junction,** the point where a motor neuron meets a skeletal muscle fiber **(figure 10.8)**. A neuromuscular junction has the following components:

- The **synaptic** (si-nap′tik; *syn* = with or together, *hapto* = to clasp) **knob** of the neuron is an expanded tip of an axon. When it nears the sarcolemma of a muscle fiber, it expands further to cover a relatively large surface area of the sarcolemma. A nerve impulse travels through the axon to the synaptic knob.
- The synaptic knob cytoplasm houses numerous **synaptic vesicles** (small membrane sacs) filled with molecules of the neurotransmitter **acetylcholine** (as-e-til-kō′lēn) **(ACh).**
- The **motor end plate** is a specialized region of the sarcolemma. It has folds and indentations to increase the membrane surface area covered by the synaptic knob.
- The **synaptic cleft** is a narrow space separating the synaptic knob and the motor end plate.
- **ACh receptors** in the motor end plate act like doors that normally are closed. ACh is the only "key" to open these receptor doors.
- The enzyme **acetylcholinesterase** (as′e-til-kō-lin-es′ter-ās) **(AChE)** resides in the synaptic cleft and rapidly breaks down molecules of ACh that are released into the synaptic cleft. Thus, AChE is needed so that ACh will not continuously stimulate the muscle.

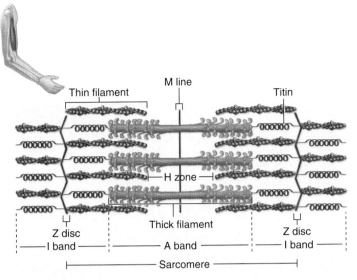

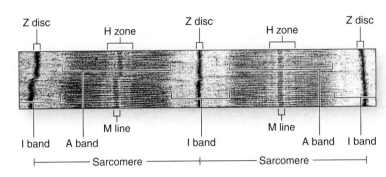

(a) Relaxed muscle
Sarcomere, I band, and H zone at a relaxed length.

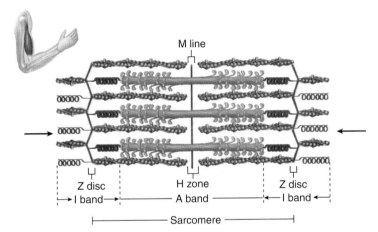

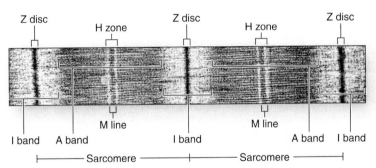

(b) Partially contracted muscle
Thick and thin filaments start to slide past one another. The sarcomere, I band, and H zone are narrower and shorter.

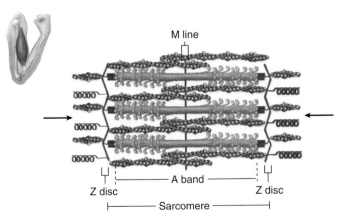

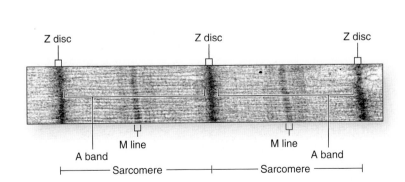

(c) Fully contracted muscle
The H zone and I band disappear, and the sarcomere is at its shortest length. Remember the lengths of the thick and thin filaments do not change.

Figure 10.7

Sliding Filament Theory of Contraction.   Diagrams and micrographs compare the changes in the striations in skeletal muscle fibers according to the sliding filament model. (*a*) In their relaxed state, the sarcomere, I band, and H zone are at their expanded length. (*b*) The Z discs at the lateral edges of the sarcomere are drawn closer together during contraction as they move toward the edges of the thick filaments in the A band. (*c*) The H zone and I bands narrow or shorten and may disappear altogether.

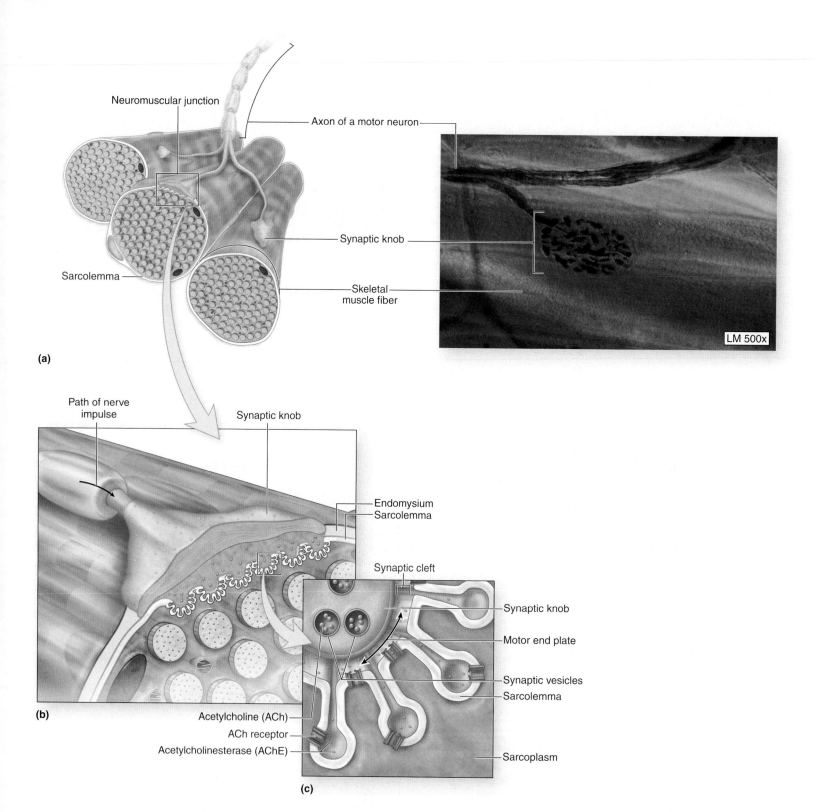

Neuromuscular junction

Axon of a motor neuron

Synaptic knob

Sarcolemma

Skeletal muscle fiber

LM 500x

(a)

Path of nerve impulse

Synaptic knob

Endomysium

Sarcolemma

Synaptic cleft

Synaptic knob

Motor end plate

Synaptic vesicles

Sarcolemma

Acetylcholine (ACh)

ACh receptor

Acetylcholinesterase (AChE)

Sarcoplasm

(b)

(c)

Figure 10.8

Neuromuscular Junction. *(a)* Diagram and light micrograph show the synaptic knob of an axon meeting a muscle fiber to form a neuromuscular junction. *(b, c)* Progressively enlarged views illustrate the components of a neuromuscular junction.

CLINICAL VIEW

Muscular Paralysis and Neurotoxins

Many toxins interfere with some of the processes that occur at the neuromuscular junction, resulting in muscular paralysis. Two of the conditions that can result are tetanus and botulism.

Tetanus (tet'ă-nŭs; *tetanos* = compulsive tension) is a form of spastic paralysis caused by a toxin produced by the bacterium *Clostridium tetani* only under anaerobic (lack of oxygen) conditions. The toxin blocks the release of an inhibitory neurotransmitter called glycine in the spinal cord, resulting in overstimulation of the muscles and excessive muscle contraction. *C. tetani* is a common soil organism that can transform into a spore, a form able to survive for months or years.

Unfortunately, *C. tetani* spores are very common in the environment and can be infective if introduced into a wound. Released by the bacteria, the toxin is absorbed into the bloodstream and circulates to the nervous system where it initiates tetanic (spasmodic) muscle contractions. This condition is potentially life-threatening, and so

we routinely immunize ourselves against it. Unfortunately, one injection ("tetanus shot") does not confer life-long immunity, so periodic "booster" shots are needed.

Botulism (bot'ū-lizm) is a potentially fatal muscular paralysis caused by a toxin produced by the bacterium *Clostridium botulinum*. The toxin prevents the release of acetylcholine (ACh) at neuromuscular junctions and leads to muscular paralysis. Like *C. tetani*, *C. botulinum* is common in the environment and also produces its toxin only under anaerobic conditions. Most cases of botulism poisoning result from ingesting the toxin in canned foods that were not processed at temperatures high enough to kill the botulism spores.

In 2002, the Food and Drug Administration (FDA) approved the injection of botulinum toxin type A (Botox) to remove and lessen the appearance of skin wrinkles. The toxin paralyzes or weakens the injected muscle by blocking the release of acetylcholine. Botox is reportedly effective for up to 120 days, and the FDA recommends that injections be administered no more than once every 3 months.

Physiology of Muscle Contraction

The arrival of a nerve impulse at the synaptic knob causes synaptic vesicles to release acetylcholine into the synaptic cleft (**figure 10.9**). ACh attaches to receptors in the motor end plate. This causes the receptor to open, allowing sodium (Na$^+$) ions to enter the muscle fiber. This ion movement changes the voltage (potential) across the sarcolemma, and a muscle impulse is initiated. The muscle impulse travels along the sarcolemma and into the muscle fiber via the T-tubules. The muscle impulse continues to spread throughout the muscle fiber as long as ACh keeps the receptors open. Usually ACh is quickly broken down and removed from the receptor by acetylcholinesterase.

Interaction Between the T-tubules and Terminal Cisternae

Recall that T-tubules distribute the muscle impulse throughout the inside of the muscle fiber, and they are sandwiched by terminal cisternae, which are reservoirs storing the calcium ions required for muscle contraction. Spread of a muscle impulse along the T-tubule membrane causes calcium ions to leak out of the terminal cisternae into the sarcoplasm of the muscle fiber. These calcium ions diffuse throughout the sarcoplasm and attach to the troponin in the thin filaments.

Interactions Between Thick and Thin Filaments

Contraction of a muscle fiber requires that the myosin heads in the thick filament bind to active sites on G-actin molecules within the thin filaments (see figure 10.5). When the muscle fiber is in a relaxed state, the tropomyosin molecules cover these active sites, preventing interaction between thick filaments and thin filaments. However, the nerve impulse arriving at the muscle fiber ultimately generates an influx of calcium ions into the sarcoplasm of the muscle fiber from the sarcoplasmic reticulum. Some calcium ions bind to troponin subunits with a binding site for calcium, causing the entire troponin molecule to change shape. As the troponin changes shape, it simultaneously moves the tropomyosin molecule to which it is attached, thus exposing the active sites on the G-actin molecules. Myosin is now able

to bind to actin. When the stimulation from the nerve impulse ceases, calcium ions are sequestered back into the sarcoplasmic reticulum, and the troponin-tropomyosin complex moves back to its original conformation where the tropomyosin once again blocks active sites on actin.

The Mechanism of Sliding After the myosin heads bind to thin filaments, myofilament sliding begins. As crossbridges form, the myosin heads pivot toward the center of the sarcomere. This action pulls the thin filaments toward the sarcomere center, causing the Z discs to move closer together as the sarcomere shortens. When the myosin head finishes pivoting, the crossbridge detaches and returns to its original cocked position, ready to repeat the cycle of "attach, pivot, detach, and return."

The mechanism of sliding is analogous to climbing a rope in the gym. Imagine that your arms are the thick filament, the rope is the thin filament, the big "knots" in the rope are the active sites, and your hands are the myosin heads. Your hands grab a knot in the rope (an "active site" on a G-actin), and you pull yourself up the rope (just as the head of a thick filament attaches onto the active site of a thin filament and pulls the thin filament toward the center of the sarcomere). Then, your hands move alternately to the next knot in the rope, and you continue to pull yourself higher. As you lift yourself up the rope, you are performing a series of "attach, pivot, detach, and return."

Energy to drive this myosin movement is provided in the form of ATP. Myosin head attachment and pivoting do not require energy, but ATP is needed in order for the myosin head crossbridge to detach from actin. Energy is released to power the detachment when ATP binds to the myosin head and is broken down into ADP (adenosine diphosphate) and P (phosphate). The return of the fiber to its resting length is completely passive and results from the pull of antagonistic muscles and from elastic forces in the contracting muscle fibers.

Muscle Contraction: A Summary

The events of muscle contraction have been called "excitation-contraction coupling," meaning that the stimulation of a muscle fiber by a nerve impulse results in a series of events that culminates

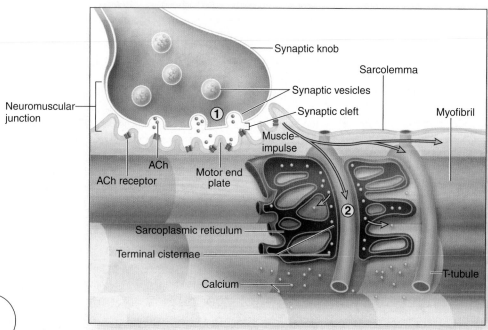

① A nerve impulse triggers release of ACh from the synaptic knob into the synaptic cleft. ACh binds to ACh receptors in the motor end plate of the neuromuscular junction, initiating a muscle impulse in the sarcolemma of the muscle fiber.

② As the muscle impulse spreads quickly from the sarcolemma along T-tubules, calcium ions are released from terminal cisternae into the sarcoplasm.

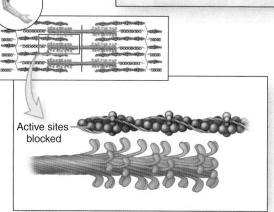

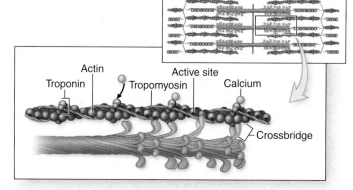

⑤ When the impulse stops, calcium ions are actively transported into the sarcoplasmic reticulum, tropomyosin re-covers active sites, and filaments passively slide back to their relaxed state.

③ Calcium ions bind to troponin. Troponin changes shape, moving tropomyosin on the actin to expose active sites on actin molecules of thin filaments. Myosin heads of thick filaments attach to exposed active sites to form crossbridges.

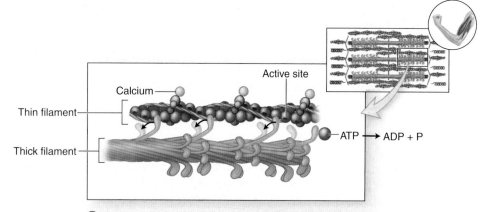

④ Myosin heads pivot, moving thin filaments toward the sarcomere center. ATP binds myosin heads and is broken down into ADP and P. Myosin heads detach from thin filaments and return to their pre-pivot position. The repeating cycle of *attach–pivot–detach–return* slides thick and thin filaments past one another. The sarcomere shortens and the muscle contracts. The cycle continues as long as calcium ions remain bound to troponin to keep active sites exposed.

Figure 10.9

Events in Muscle Contraction.

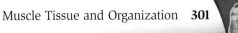

in muscle fiber contraction. These events are summarized in figure 10.9 and in the following list:

1. A nerve impulse causes ACh release at a neuromuscular junction. ACh binds receptors on the motor end plate, initiating a muscle impulse.
2. The muscle impulse spreads quickly along the sarcolemma and into the muscle fiber along T-tubule membranes, causing calcium ions to be released into the sarcoplasm.
3. Calcium ions bind to troponin, causing tropomyosin to move and expose active sites on actin. Myosin heads attach to the actin and form crossbridges.
4. Myosin heads go through cyclic "attach, pivot, detach, return" events as the thin filaments are pulled past the thick filaments. ATP is required to detach the myosin heads and complete the sequence of cyclic events. The sarcomere shortens, and the muscle contracts. The cyclic events continue as long as calcium ions remain bound to the troponin.
5. Calcium ions are moved back into the sarcoplasmic reticulum by ATP-driven ion pumps to reduce calcium concentration in the sarcoplasm, leading to relaxation. Termination of the muscle impulse results in the passive sliding of myofilaments back to their original state.

Motor Units

We have examined how a motor neuron initiates the process of muscle contraction in a single muscle fiber. However, a single motor neuron typically controls numerous muscle fibers in a muscle. This motor neuron has a neuromuscular junction with each muscle fiber

it controls. Thus, a **motor unit** is composed of a single motor neuron and all of the muscle fibers it controls **(figure 10.10)**.

A motor unit typically controls only some of the muscle fibers in an entire muscle. Most muscles have many motor units, which means that many motor neurons are needed to innervate (supply) an entire muscle. Thus, a large muscle such as the biceps brachii of the arm has numerous motor units, each controlling a portion of the total number of fibers in the muscle.

An inverse relationship exists between the size of the motor unit and the degree of control provided. Overall, the smaller the motor unit, the finer the control. For example, since precise control is essential in the muscles that move the eye, motor neurons innervating eye muscles may control only two or three muscle fibers. In contrast, power-generating muscles in our lower limbs require less precise control, so a single motor neuron controls several thousand individual muscle fibers.

Each muscle fiber obeys the **all-or-none principle,** which states that a muscle fiber either contracts completely or does not contract at all. When a motor unit is stimulated, all its fibers contract at the same time. The total force exerted by the muscle depends on the number of activated motor units. If more motor units are activated, more muscle fibers contract and greater force is exerted. Movements that require less force need fewer activated motor units. Thus, while the muscle fibers obey the all-or-none principle, the force and precision of muscle movement can be varied, depending on how many muscle fibers and motor units are activated.

WHAT DO YOU THINK?

1 What is the benefit of having multiple motor units in a skeletal muscle? Why wouldn't you simply want one motor unit to innervate all of the muscle fibers?

CLINICAL VIEW

Rigor Mortis

At death, circulation ceases, and all body tissues are immediately deprived of oxygen and nutrients. But even though there is no brain function and no cardiac or respiratory activity, some tissues continue to "live" for as long as an hour as they metabolize stored energy reserves. From a physiologic standpoint, death is not so much an event, as a process. Consider what happens to the skeletal muscles.

Within a few hours after the heart stops beating, ATP levels in skeletal muscle fibers have been completely exhausted, thus preventing myosin head detachment from actin. At the same time, as a consequence of ATP exhaustion in skeletal muscle fibers, the sarcoplasmic reticulum loses its ability to recall calcium ions from the sarcoplasm. As a result, the calcium ions already present in the sarcoplasm, as well as those that continue to leak out of the sarcoplasmic reticulum, trigger a sustained contraction in the fiber. Because no ATP is available, the crossbridges between thick and thin filaments cannot detach. All skeletal muscles lock into a contracted position, and the deceased individual becomes rigid. This physiologic state, termed **rigor mortis** (rig'er mōr'tis), continues for about

15 to 24 hours. Rigor mortis then disappears because lysosomal enzymes are released within the muscle fibers, causing autolysis (self-destruction and breakdown) of the myofibrils. As the muscle tissue breaks down, the muscles of the deceased person again become flaccid (relaxed). Forensic pathologists often use the development and resolution of rigor mortis to establish an approximate time of death. Because a number of factors affect the rate of development and resolution of rigor mortis, environmental conditions need to be taken into consideration. For example, a warmer body will develop and resolve rigor mortis much more quickly than a body of normal temperature, because the elevated temperature increases the rate of the processes involved in rigor mortis. This means that people who die while they have a fever, or those exposed to hot environmental conditions, will develop rigor mortis more quickly than average. Conversely, a person who dies and remains in the cold will develop rigor mortis more slowly than normal. Therefore, the pathologist must know the ambient (existing) temperature and general conditions where the body was found to accurately establish a time of death. The following chart provides rough guidelines for estimating the death interval, assuming average body temperature and average ambient temperature.

Death Interval	Body Temperature	Stiffness
Dead less than 3 hours	Warm	No stiffness
Dead 3–8 hours	Warm, but cooling	Developing stiffness
Dead 8–24 hours	Ambient temperature	Stiff, but resolving
Dead 24–36 hours	Ambient temperature	No stiffness

Figure 10.10

A Motor Unit. Each skeletal muscle fiber is innervated (supplied) by a single axon from a motor neuron. In a motor unit, a motor neuron axon branches to innervate a number of muscle fibers within a muscle.

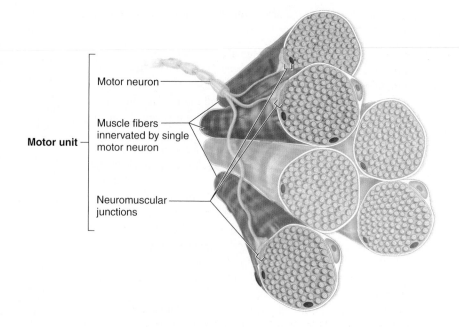

Motor neuron

Muscle fibers innervated by single motor neuron

Motor unit

Neuromuscular junctions

Muscle Tone

Some motor units are always active, even when a muscle is at rest. The motor units cause the muscle to become tense, but their contractions do not produce enough tension to cause movement. **Muscle tone** is the resting tension in a skeletal muscle. Usually, motor units are stimulated randomly so that the attached tendon maintains a constant tension, but individual muscle fibers are afforded some relaxation time. This resting muscle tone stabilizes the position of bones and joints.

As stimulation begins and then increases in an active muscle, the fibers within the muscle begin to contract, generating tension in the muscle. The two types of muscle contraction are isometric and isotonic (**figure 10.11**).

During an **isometric** (ī-sō-met′rik; *iso* = same, *metron* = measure) **contraction,** the length of the muscle does not change

because the tension produced by this contracting muscle never exceeds the resistance (load). For example, when a person tries to lift an extremely heavy object (e.g., pick up the back of a car off the ground), the arm muscles are isometrically contracted. Tension is generated, but it is not great enough to move the load.

In an **isotonic** (ī-sō-ton′ik; *tonos* = tension) **contraction,** the tension produced equals or is greater than the resistance, and then the muscle fibers shorten, resulting in movement. An example of isotonic contraction occurs when a person lifts a grocery bag out of the car. The tension generated in the appropriate muscles equals the load, and then the muscle fibers shorten. Most of the movements performed by individuals working out on weight machines produce isotonic contractions. Isotonic contractions are of two types. **Concentric contractions** actively shorten a muscle, as when

Figure 10.11

Isometric Versus Isotonic Contraction. *(a)* When a muscle is isometrically contracted, its length does not change, but the muscle is tensed. *(b)* Isotonic muscle contraction causes the muscle to shorten and the body part to move.

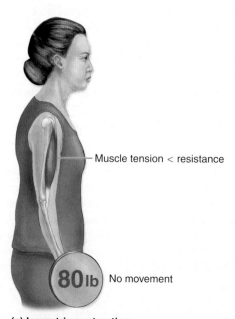

Muscle tension < resistance

80 lb No movement

(a) Isometric contraction
Muscle tension is less than the resistance. Although tension is generated, the muscle does not shorten, and no movement occurs.

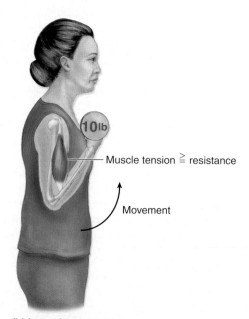

10 lb

Muscle tension ≥ resistance

Movement

(b) Isotonic contraction
Muscle tension equals or is greater than the resistance. The muscle shortens, and movement occurs.

CLINICAL VIEW

Hypertonia and Hypotonia

Increases or decreases in muscle tone result from disorders of the nervous system or electrolyte imbalances in our body fluids. If our nervous system malfunctions, appropriate stimulating signals may not reach the muscles that need to be contracted, leading to changes in muscle tone. On the other hand, if the concentration of electrolytes (such as Na^+) changes in the cellular or interstitial fluid, the generation of a muscle impulse may be adversely affected—that is, you could generate an impulse unintentionally or be unable to generate one when you want to.

Hypertonia (hī-per-tŏ'nē-ă; *hyper* = above, over) is an increase in muscle tone or stiffness that presents as either spasticity or rigidity.

Spasticity is associated with an increase in tendon reflexes and certain pathologic reflexes, such as the Babinski sign, in which the great toe extends when the sole of the foot is stroked. **Rigidity** refers to an increase in muscle tone but does not involve reflexes. Alteration in the balance between neuron stimulation and muscle fiber inhibition leads to stiffness and irregular, jerky movements.

Hypotonia (hī-pō-tŏ'nē-ă; *hypo* = under), on the other hand, involves decreased or lost muscle tone. Hypotonic muscles are flaccid and appear flattened. Flaccid paralysis is characterized by loss of muscle tone and tendon reflexes, along with atrophy and degeneration of muscles. There are many causes of hypotonia, some involving disorders of motor neurons and others involving the muscles themselves.

we lift a baby from a crib. The load (weight of the baby) is less than the maximum tension that can be generated by the muscles in the arm; thus, the muscle shortens and the baby is raised up. **Eccentric contractions** actively lengthen a muscle when we are placing the baby back into the crib. The muscles in our arm are active while lengthening as we lay the baby down gently rather than just letting her "fall" into the crib.

Simply put, an isotonic contraction of the arm muscles allows you to lift a book from the table, while an isometric contraction of the arm muscles occurs when you push against a wall.

WHAT DID YOU LEARN?

6 Describe the structural relationship between a nerve fiber and a muscle fiber at the neuromuscular junction.

7 What is the function of calcium ions in skeletal muscle contraction?

8 Describe how thick and thin filaments interact during a muscle contraction.

9 Describe the difference between isometric and isotonic contractions, and give an example of each.

Types of Skeletal Muscle Fibers

Key topics in this section:

- Characteristics of the three types of skeletal muscle fibers
- Types of movements facilitated by slow, intermediate, and fast muscle fibers

Skeletal muscles must perform a wide variety of actions. The range of actions is determined by the types of fibers that form the muscle. The three types of skeletal muscle fibers are slow (type I, slow oxidative), intermediate (type II-A, fast aerobic), and fast (type II-B, fast anaerobic). Their characteristics vary **(table 10.4)**. Each

Table 10.4	Structural and Functional Characteristics of Different Types of Skeletal Muscle Fibers		
Fiber Characteristic	**Slow Fibers (Type I, Slow Oxidative)**	**Intermediate Fibers (Type IIa, Fast Aerobic)**	**Fast Fibers (Type IIb, Fast Anaerobic)**
ATP Use	Slowly	Quickly	Quickly
Capacity to Make ATP	Aerobic	Aerobic	Anaerobic
Concentration of Capillaries	Extensive	Extensive	Sparse
Color of Fibers	Red	Red	White (pale)
Contraction Velocity	Slow	Fast	Fast
Resistance to Fatigue	Highest	High	Low
Fiber Distribution	Found in greatest abundance in muscles of the trunk, especially postural muscles	Found in greatest abundance in muscles of the lower limbs	Found in greatest abundance in muscles of the upper limbs
Fiber Diameter	Smallest	Intermediate	Largest
Number of Mitochondria	Many	Many	Few
Amount of Myoglobin	Large	Large	Small
Primary Fiber Function	Endurance (e.g., marathon running), maintaining posture	Medium duration, moderate movement (e.g., walking, biking)	Short duration, intense movement (e.g., sprinting, lifting weights)

CLINICAL VIEW

Anabolic Steroids and Performance-Enhancing Compounds

Anabolic (an-ă-bol′ik) **steroids** are synthetic substances that mimic the actions of natural testosterone. To date, over 100 compounds have been developed with anabolic properties, but they all require a prescription for legal use in the United States. Anabolic steroids have only a few accepted medical uses—among them, the treatment of delayed puberty, certain types of impotence, and the wasting condition associated with HIV infection and other diseases. Because anabolic steroids stimulate the manufacture of muscle proteins, these compounds have become popular with some athletes as performance enhancers.

The stimulation of excessive muscle development requires relatively large doses of anabolic steroids, and the extra strength and speed gained come at a price. Many devastating side effects are associated with extended anabolic steroid use, including increased risk of heart disease and stroke, kidney damage, aggressive behavior, liver tumors, testicular atrophy, acne, and personality aberrations. Because anabolic steroids mimic the effects of testosterone, female athletes who use them often experience menstrual irregularities, growth of facial hair, and in extreme circumstances atrophy of the uterus and mammary glands; sterility has even been reported. Adding to these problems is the route of administration. Because many of these steroid preparations must be injected, the improper use or sharing of needles raises the possibility of transferring a disease such as AIDS or hepatitis. Therefore, the use of anabolic steroids as performance enhancers has been widely banned.

In addition to anabolic steroids, athletes have used other substances and some hormones, such as GH and IGF-1, to increase muscle performance. Compounds such as creatine, branch-chain amino acids, and ephedrine have also been touted as increasing muscle size and power, but most research indicates this is not the case. In fact, ephedrine has been banned after being linked to death in some individuals taking it.

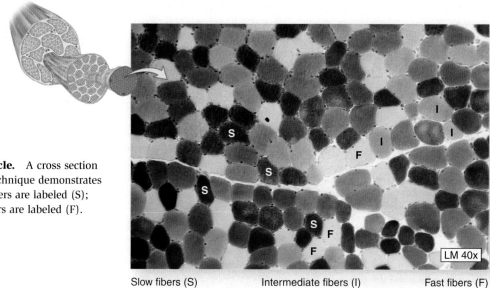

Figure 10.12

Comparison of Fiber Types in Skeletal Muscle. A cross section of a skeletal muscle using a specific staining technique demonstrates the three types of fibers in the muscle. Slow fibers are labeled (S); intermediate fibers are labeled (I); and fast fibers are labeled (F).

Slow fibers (S)   Intermediate fibers (I)   Fast fibers (F)

skeletal muscle typically contains a percentage of each of these muscle fiber types, as shown in **figure 10.12**.

Slow fibers are typically small in diameter—about half the diameter of fast fibers. They contract more slowly than fast fibers, often taking two or three times longer to contract after stimulation. These fibers are specialized to continue contracting for extended periods of time. A similar effort by a fast muscle would result in fatigue. The vascular supply to slow muscle fibers is more extensive than the network of capillaries around fast muscle fibers; thus, the supply of nutrients and oxygen to the slow fibers is markedly increased. Slow fibers are also called *red fibers* because they contain the red pigment **myoglobin** (mī-ō-glō′bin). Myoglobin is a globular, oxygen-binding, reddish-appearing protein that is structurally related to hemoglobin, the oxygen-binding protein in

erythrocytes. Therefore, resting slow muscle fibers contain substantial oxygen reserves that can be mobilized during a contraction to provide the needed ATP. An additional advantage of slow fibers is their relatively large number of mitochondria. This permits slow muscle fibers to produce a greater amount of ATP than fast muscle fibers while contractions are under way, making the fiber less dependent on anaerobic metabolism. Both slow and intermediate fibers require oxygen to produce ATP, so the metabolic reactions within these fibers are termed *aerobic*.

Intermediate fibers exhibit properties that are somewhere between those of slow fibers and fast fibers. The intermediate fibers contract faster than the slow fibers and slower than the fast fibers. Histologically, intermediate fibers resemble fast fibers; however, they have a greater resistance to fatigue.

Fast fibers are large in diameter, and they contain large glycogen reserves, densely packed myofibrils, and relatively few mitochondria. They are also called *white fibers* because they are pale in color due to their lack of myoglobin.

The majority of skeletal muscle fibers in the body are fast fibers. The name "fast" suggests the instantaneous nature of their contraction, which usually occurs 0.01 second or less after they have been stimulated. Fast-fiber muscles produce powerful contractions because they contain a large number of sarcomeres, and the tension produced by a muscle fiber is directly proportional to the number of sarcomeres it contains. These contractions use vast quantities of ATP. Their prolonged activity is supported primarily by **anaerobic** (an-ār-ō'bik; *an* = without, *aer* = air, *bios* = life) metabolic reactions—those that do not require oxygen. As a result, fast fibers fatigue rapidly. (Note: Oxygen is required for the production of the maximum amount of ATP.)

WHAT DO YOU THINK?

2 The next time you sit down to a turkey dinner, you may be asked if you prefer the "white meat" or the "dark meat." What can you say about the diameter of the muscle fibers of each type of meat? Which would be fast fibers and which would be slow fibers? In life, which type of muscle would have had the greatest resistance to fatigue?

Distribution of Slow, Intermediate, and Fast Fibers

The relative number of muscle fibers in each of the slow, intermediate, and fast types varies in different skeletal muscles. Usually, skeletal muscles have a diverse number and composition of fiber types. However, within a single motor unit, all fibers belong to the same type. Some muscles have no slow fibers, while others have no fast fibers. For example, slow fibers dominate many back and calf muscles, which contract almost continually to help us maintain an upright posture. In contrast, there are no slow fibers in the muscles of the eye and hand, where swift but brief contractions are required.

The relative number of slow fibers compared to fast fibers in each muscle is determined by a person's genes, and their proportions determine an individual's endurance. For example, distance runners who have higher proportions of slow muscle fibers in their leg muscles are able to outperform runners who have a greater number of fast muscle fibers in their leg muscles.

Additionally, an individual has better proficiency in performing repeated contractions under aerobic conditions if he or she has a greater percentage of slow fibers in specific muscles. However, for brief periods of intense activity, such as sprinting or weight-lifting, the individual with a higher percentage of fast muscle fibers has the advantage.

The proportion of intermediate fibers changes with physical conditioning; if used repeatedly for endurance events, fast fibers can develop the appearance and functional capabilities of intermediate fibers. Thus, physical conditioning may lead to muscle hypertrophy, which enables athletes to improve both strength and endurance.

WHAT DID YOU LEARN?

10 List the three types of skeletal muscle fibers, and describe the anatomic and functional characteristics of each.

Skeletal Muscle Fiber Organization

Key topic in this section:

■ Four organizational patterns in fascicles

As mentioned earlier in this chapter, bundles of muscle fibers termed fascicles lie parallel to each other within each muscle. However, the organization of fascicles in different muscles often varies. There are four different patterns of fascicle arrangement: circular, convergent, parallel, and pennate **(table 10.5)**.

Circular Muscles

The muscle fibers in a **circular muscle** are concentrically arranged around an opening or recess. A circular muscle is also called a sphincter, because contraction of the muscle closes off the opening. Circular muscles are located at entrances and exits of internal passageways. An example is the orbicularis oris muscle that encircles the opening of the mouth.

Parallel Muscles

The fascicles in a **parallel muscle** run parallel to its long axis. Each muscle fiber in this type of skeletal muscle exhibits the functional characteristics of the entire parallel muscle. Parallel muscles have a central body, called the belly, or *gaster*. This muscle shortens when it contracts, and its body increases in diameter. Parallel muscles have high endurance but are not as strong as other muscle types. Examples of parallel muscles include the rectus abdominis (an anterior abdominal muscle that forms the "six-pack" of a well-sculpted abdomen), the biceps brachii of the arm, and the masseter that moves the mandible.

Convergent Muscles

A **convergent** (kon-ver'jent) **muscle** has widespread muscle fibers that converge on a common attachment site. This attachment site may be a single tendon, a tendinous sheet, or a slender band of collagen fibers known as a **raphe** (rā'fē; *rhaphe* = seam). These muscle fibers are often triangular in shape, resembling a broad fan with a tendon at the tip. A convergent muscle is versatile—that is, the direction of its pull can be modified merely by activating a single group of muscle fibers at any one time. However, when the fibers in a convergent muscle all contract at once, they do not pull as hard on the tendon as a parallel muscle of the same size because the muscle fibers on opposite sides of the tendon are not working together; rather, they are pulling in different directions. An example of a convergent muscle is the pectoralis major of the chest.

Pennate Muscles

Pennate (pen'āt; *penna* = feather) **muscles** are so named because their tendons and muscle fibers resemble a large feather. Pennate muscles have one or more tendons extending through their body, and the fascicles are arranged at an oblique angle to the tendon. Because pennate muscle fibers pull at an angle to the tendon, this type of muscle does not move its tendons as far as parallel muscles move their tendons. However, a contracting pennate muscle generates more tension than does a parallel muscle of the same size, and thus it is stronger.

There are three types of pennate muscles:

■ In a **unipennate muscle,** all of the muscle fibers are on the same side of the tendon. The extensor digitorum, a long muscle that extends the fingers, is a unipennate muscle.

Table 10.5	Skeletal Muscle Architecture	
Pattern of Muscle Fibers	**Description**	**Example**
Circular	Fibers arranged concentrically around an opening Functions as a sphincter to close a passageway or opening (e.g., orbits, mouth, anus)	Orbicularis oris
Parallel	Fascicles are parallel to the long axis of the muscle Body of muscle increases in diameter with contraction High endurance, not very strong	Rectus abdominis
Convergent	Triangular muscle with common attachment site Direction of pull of muscle can be changed Does not pull as hard as equal-sized parallel muscle	Pectoralis major
Pennate	Muscle body has one or more tendons Fascicles at oblique angle to tendon Pulls harder than a parallel muscle of equal size *Unipennate:* all muscle fibers on the same side of the tendon *Bipennate:* muscle fibers on both sides of the tendon *Multipennate:* tendon branches within the muscle	Extensor digitorum (unipennate) Rectus femoris (bipennate) Deltoid (multipennate)

- A **bipennate muscle,** the most common type, has muscle fibers on both sides of the tendon. The palmar and dorsal interosseous muscles that attach to the metacarpals are composed of bipennate muscle that helps adduct and abduct the digits.
- A **multipennate muscle** has branches of the tendon within the muscle. The triangular deltoid that covers the superior surface of the shoulder joint is a multipennate muscle.

WHAT DID YOU LEARN?

11 What are the four main patterns of skeletal muscle fiber organization?

Exercise and Skeletal Muscle

Key topic in this section:

- Characteristics of muscle atrophy and muscle hypertrophy

Muscle building is a time-consuming endeavor that requires proper diet, specific workout regimens, and sufficient rest and recovery times between workouts. In contrast, the lack of sufficient exercise is detrimental to the health of skeletal muscle.

Muscle Atrophy

Atrophy (at′rō-fē; *a* = without, *trophe* = nourishment) is a wasting of tissue that results in a reduction in muscle size, tone, and power. If a skeletal muscle experiences markedly reduced stimulation, it loses both mass and tone. The muscle becomes flaccid, and its fibers decrease in size and become weaker. Even a temporary reduction in muscle use can lead to muscular atrophy. For example, after a cast has been worn, the limb muscles exhibit reduced tone and size. Individuals who suffer damage to the nervous system or are paralyzed by spinal injuries gradually lose muscle tone and size in the areas affected. Although the muscle atrophy is initially reversible, dead or dying muscle fibers are not replaced. When extreme atrophy occurs, the loss of gross muscle function is permanent. For these reasons, physical therapy is required for patients who suffer temporary loss of mobility.

Muscle Hypertrophy

An increase in muscle fiber size is called **hypertrophy.** Hypertrophy does not result in an increase in muscle fiber number, which would be termed *hyperplasia.* However, it does result in an increase in the number of myofibrils per fiber. Muscle size may be increased by exercising. The repetitive, exhaustive stimulation of muscle fibers results in more mitochondria, larger glycogen reserves, and an increased ability to produce ATP. Ultimately, each muscle fiber develops more myofibrils, and each myofibril contains a larger number of myofilaments. An athlete who competes as a bodybuilder or weight lifter exhibits hypertrophied muscular development.

WHAT DID YOU LEARN?

12 What anatomic changes occur in a muscle fiber when it undergoes hypertrophy?

Levers and Joint Biomechanics

Key topics in this section:

- Characteristics of the three types of levers in the body
- How interacting skeletal muscles can initiate or prevent movement

If a skeletal muscle crosses a joint, its contractions may change the angular relationships of the bones in the joint, resulting in body movement. The location and nature of the muscular connection to the skeleton influences the speed, range, and force of movement.

When analyzing muscle contraction, anatomists often compare it to the mechanics of a lever; this practice of applying mechanical principles to biology is known as **biomechanics.** A **lever** (lev′er, lĕ′ver; to lift) is an elongated, rigid object that rotates around a fixed point called the **fulcrum** (ful′krŭm). A seesaw is a familiar example of a lever. Levers have the ability to change the speed and distance of movement produced by a force, the direction of an applied force, and the force strength. Rotation occurs when an effort applied to one point on the lever exceeds a **resistance** located at some other point. The part of a lever from the fulcrum to the point of effort is called the **effort arm,** and the part from the fulcrum to the point of resistance is the **resistance arm.** In the body, a long bone acts as a lever, a joint serves as the fulcrum, and the effort is generated by a muscle attached to the bone.

Classes of Levers

Three classes of levers are found in the human body: first-class, second-class, and third-class **(figure 10.13)**.

First-Class Levers

A **first-class lever** has a fulcrum in the middle, between the effort and the resistance. An example of a first-class lever is a pair of scissors. The effort is applied to the handle of the scissors while the resistance is at the cutting end of the scissors. The fulcrum (pivot for movement) is along the middle of the scissors, between the handle and the cutting ends. In the body, an example of a first-class lever is the atlanto-occipital joint of the neck, where the muscles on the posterior side of the neck pull inferiorly on the nuchal lines of the skull and oppose the tendency of the head to tip anteriorly.

Second-Class Levers

The resistance in a **second-class lever** is between the fulcrum and the applied effort. A common example of this type of lever is lifting the handles of a wheelbarrow, allowing it to pivot on its wheel at the opposite end and lift a load in the middle. The load weight is the resistance, and the upward lift on the handle is the effort. A small force can balance a larger weight in this type of lever, because the force is always farther from the fulcrum than the resistance. In the body, second-class levers are rare, but one example occurs when the foot is depressed (plantar flexion) so that a person can stand on tiptoe. The contraction of the calf muscle causes a pull superiorly by the calcaneal tendon attached to the heel (calcaneus).

Third-Class Levers

In a **third-class lever,** an effort is applied between the resistance and the fulcrum, as when picking up a small object with a pair of forceps. Third-class levers are the most common levers in the body. One example is found at the elbow, where the fulcrum is the joint between the humerus and ulna, the effort is applied by the biceps brachii muscle, and the resistance is provided by any weight in the hand or by the weight of the forearm itself. In addition, the mandible acts as a third-class lever when you bite with your incisors on a piece of food. The temporomandibular joint is the fulcrum, and the temporalis muscle exerts the effort, while the resistance is the item of food being bitten.

WHAT DO YOU THINK?

3 What type of lever is the knee joint?

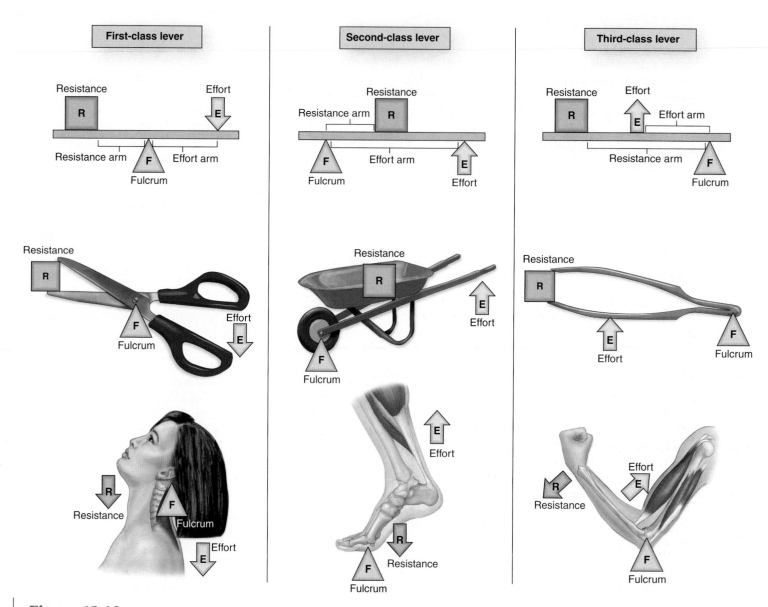

Figure 10.13

Classes of Levers.   This comparison of the three classes of levers shows: *(top row)* relative position of each lever; *(middle row)* a mechanical example of each, and *(bottom row)* an anatomic example of each.

Actions of Skeletal Muscles

Skeletal muscles generally do not function in isolation; rather, they work together to produce movements. Muscles are grouped according to their primary actions into three types: agonists, antagonists, and synergists.

An **agonist** (ag′on-ist; *agon* = a contest), also called a *prime mover*, is a muscle that contracts to produce a particular movement, such as extending the forearm. The triceps brachii of the posterior arm is an agonist that causes forearm extension.

An **antagonist** (an-tag′ŏ-nist; *anti* = against) is a muscle whose actions oppose those of the agonist. If the agonist produces extension, the antagonist produces flexion. The contraction of the agonist stretches the antagonist, and vice versa. As this movement occurs, the stretched muscle usually does not relax completely. Instead, the tension within the muscle being stretched is adjusted to control the speed of the movement and ensure that it is smooth. For example, when the triceps brachii acts as an agonist to extend the forearm, the biceps brachii on the anterior side of the humerus acts as an antagonist to stabilize the movement and produce the opposing action, which is flexion of the forearm.

A **synergist** (sin′er-jist; *ergon* = work) is a muscle that assists the agonist in performing its action. The contraction of a synergist usually either contributes to tension exerted close to the insertion of the muscle or stabilizes the point of origin. Usually, synergists are most useful at the start of a movement when the agonist is stretched and cannot exert much power. Examples of synergistic muscles are the biceps brachii and the brachialis muscles of the arm. Both muscles work synergistically (together) to flex the elbow joint. Synergists may also assist an agonist by *preventing* movement at a joint and thereby stabilizing the origin of the agonist. In this case, these synergistic muscles are called fixators.

WHAT DID YOU LEARN?

13 Contrast an agonist and a synergist.

The Naming of Skeletal Muscles

Key topic in this section:

- How muscle names incorporate appearance, location, function, orientation, and unusual features

Usually, the names of skeletal muscles provide clues to their identification. As **figure 10.14** shows, skeletal muscles are named according to the following criteria:

- **Muscle action.** Names that indicate the primary function or movement of the muscle include *flexor, extensor,* and *pronator.* These are such common actions that the names almost always contain other clues to the appearance or location of the muscle. For example, the pronator teres is a long, round muscle responsible for pronating the forearm.

- **Specific body regions.** The rectus femoris is on the thigh *(femur),* and the tibialis anterior is on the *anterior* surface of the *tibia.* Muscles that can be observed at the body surface are often termed *superficialis* (soo′per-fish-ē-ā′lis) or *externus* (eks-ter′nŭs). In contrast, deeper or more internally placed muscles may have names such as *profundus* (prō-fŭn′dŭs; deep) or *internus.*

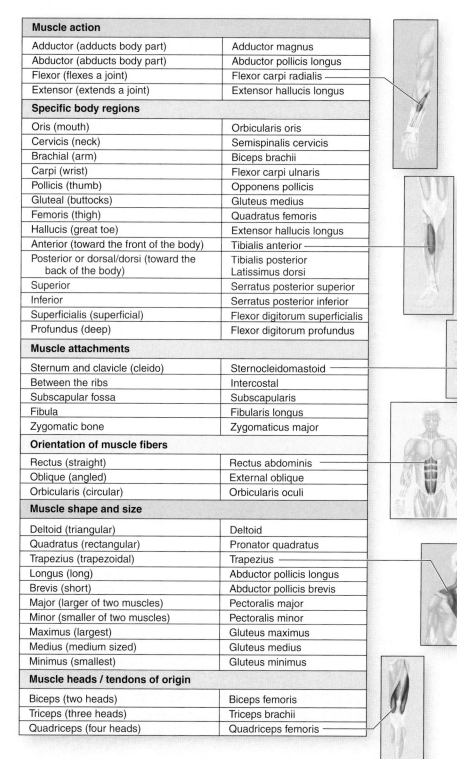

Figure 10.14

Muscle Naming. Muscles are named according to a variety of features.

Muscle action	
Adductor (adducts body part)	Adductor magnus
Abductor (abducts body part)	Abductor pollicis longus
Flexor (flexes a joint)	Flexor carpi radialis
Extensor (extends a joint)	Extensor hallucis longus
Specific body regions	
Oris (mouth)	Orbicularis oris
Cervicis (neck)	Semispinalis cervicis
Brachial (arm)	Biceps brachii
Carpi (wrist)	Flexor carpi ulnaris
Pollicis (thumb)	Opponens pollicis
Gluteal (buttocks)	Gluteus medius
Femoris (thigh)	Quadratus femoris
Hallucis (great toe)	Extensor hallucis longus
Anterior (toward the front of the body)	Tibialis anterior
Posterior or dorsal/dorsi (toward the back of the body)	Tibialis posterior / Latissimus dorsi
Superior	Serratus posterior superior
Inferior	Serratus posterior inferior
Superficialis (superficial)	Flexor digitorum superficialis
Profundus (deep)	Flexor digitorum profundus
Muscle attachments	
Sternum and clavicle (cleido)	Sternocleidomastoid
Between the ribs	Intercostal
Subscapular fossa	Subscapularis
Fibula	Fibularis longus
Zygomatic bone	Zygomaticus major
Orientation of muscle fibers	
Rectus (straight)	Rectus abdominis
Oblique (angled)	External oblique
Orbicularis (circular)	Orbicularis oculi
Muscle shape and size	
Deltoid (triangular)	Deltoid
Quadratus (rectangular)	Pronator quadratus
Trapezius (trapezoidal)	Trapezius
Longus (long)	Abductor pollicis longus
Brevis (short)	Abductor pollicis brevis
Major (larger of two muscles)	Pectoralis major
Minor (smaller of two muscles)	Pectoralis minor
Maximus (largest)	Gluteus maximus
Medius (medium sized)	Gluteus medius
Minimus (smallest)	Gluteus minimus
Muscle heads / tendons of origin	
Biceps (two heads)	Biceps femoris
Triceps (three heads)	Triceps brachii
Quadriceps (four heads)	Quadriceps femoris

- **Muscle attachments.** Many muscle names identify their origins, insertions, or other prominent attachments. In this case, the first part of the name indicates the origin and the second part the insertion. For example, the sternocleidomastoid has origins on the *sternum* and clavicle (*cleido*) and an insertion on the *mastoid* process of the temporal bone.
- **Orientation of muscle fibers.** The rectus abdominis muscle is named for its lengthwise-running muscle fibers; *rectus* means "straight." Similarly, names such as *oblique* or *obliquus* (ob-lĭ′kwŭs) indicate muscles with fibers extending at an oblique angle to the longitudinal axis of the body.
- **Muscle shape and size.** Examples of shape in muscle names include: deltoid, shaped like a triangle; orbicularis, a circle; rhomboid, a rhombus; and trapezius, a trapezoid. Short muscles are called *brevis* (brev′is); long muscles are called *longus* (lon′gŭs) or *longissimus* (lon-jĭs′ĭ-mŭs; longest). *Teres* (ter′ēz, te′rēz) muscles are both long and round. Large muscles are called *magnus* (mag′nŭs; big), *major* (mā′jĕr; bigger), or *maximus* (măk′sĭ-mŭs; biggest). Small muscles are called *minor* (mī′nor; smaller) or *minimus* (mī′nĭ-mus; smallest).
- **Muscle heads/tendons of origin.** Some muscles are named after specific features—for example, how many tendons of origin or how many muscle bellies or heads each contains. A *biceps* muscle has two tendons of origin, a *triceps* muscle has three heads/tendons, and a *quadriceps* muscle has four heads/tendons. So the quadriceps femoris muscle is a muscle of the thigh that has four muscle heads/tendons of origin.

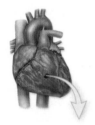

WHAT DID YOU LEARN?

14 Give at least four criteria used to determine how muscles are named.

15 The gluteus maximus muscle gets its name from which categories for naming muscles?

Characteristics of Cardiac and Smooth Muscle

Key topic in this section:

- Similarities and differences among the three types of muscle tissue

In addition to skeletal muscle, which we have just described in depth, two other types of muscle are found in the body: cardiac muscle and smooth muscle. Here we briefly review their characteristics and compare them to skeletal muscle.

Cardiac Muscle

Cardiac muscle cells are individual muscle cells arranged in thick bundles within the heart wall. These cells are striated like skeletal muscle fibers, but shorter and thicker, and they have only one or two nuclei. Cardiac muscle cells form Y-shaped branches and join to adjacent muscle cells at junctions termed **intercalated** (in-ter′kă-lā-ted; *inter* = between, *calarius* = inserted) **discs (figure 10.15).**

Cardiac muscle cells are **autorhythmic,** meaning that the individual cells can generate a muscle impulse without nervous stimulation. This feature of cardiac muscle cells is responsible for our repetitious, rhythmic heartbeat (see chapter 22 for an in-depth discussion). The autonomic nervous system controls the rate of rhythmic contraction of cardiac muscle.

Cardiac muscle cells are dependent for their contraction upon calcium ions. Most of the calcium ions that stimulate the contraction of the cardiac muscle cells originate within the interstitial fluid bathing these cells, because the terminal cisternae of their sarcoplasmic reticulum are less well developed and thus store fewer calcium ions than do skeletal muscle fibers.

Figure 10.15

Cardiac Muscle.   Cardiac muscle is found only in the heart walls. *(a)* Cardiac muscle cells branch and are connected by intercalated discs. *(b)* A section through cardiac muscle cells illustrates the distribution of some of their internal structures.

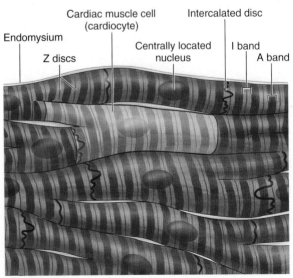

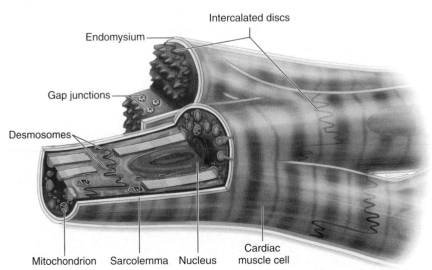

(a)   **(b)**

Cardiac muscle uses aerobic respiration almost exclusively. Cardiac muscle cells contain a large number of mitochondria to generate the ATP required for their unceasing work.

Smooth Muscle

Smooth muscle is composed of short muscle cells that have a fusiform shape (widest at the middle of the cell and tapered at each end) **(figure 10.16)**. They have a single, centrally located nucleus. Although smooth muscle cells have both thick and thin filaments, they are not precisely aligned, so no visible striations or sarcomeres are present. Z discs are absent from smooth muscle cells; instead, thin filaments are attached to **dense bodies** by elements of the cytoskeleton. Dense bodies are small concentrations of protein scattered throughout the sarcoplasm and on the inner face of the sarcolemma. The sarcoplasmic reticulum is sparse, and transverse tubules and troponin are absent. The calcium needed to activate smooth muscle contraction originates in the interstitial fluid around the cell.

Two major differences are obvious when comparing the mechanism of contraction between smooth muscle and skeletal muscle. Both involve a calcium-binding protein called calmodulin. When the smooth muscle cells are inactive, no calcium enters from the interstitial fluid to bind to calmodulin; thus, smooth cells remain quiescent. However, when the cells are stimulated, calcium ions enter from the interstitial fluid, and the following events result: (1) Smooth muscle cells have a unique protein called myosin light-chain that may exist in either a phosphorylated (with phosphate attached) or nonphosphorylated state. The phosphorylated state is a necessary first step for the onset of contraction in the smooth muscle cell. Calmodulin regulates the addition of phosphate to myosin light-chain. (2) Although troponin is lacking in smooth muscle cells, cytoplasmic calcium levels still regulate contractile activity when calmodulin binds to a protein called caldesmon, which then regulates the movement of tropomyosin from the myosin binding sites in thin filaments.

Smooth muscle contraction is slow, resistant to fatigue, and usually sustained for an extended period of time. These are impor-

tant characteristics because smooth muscle in the walls of viscera and blood vessels must sustain an extended contraction without undergoing fatigue. Smooth muscle cells are resistant to fatigue because their ATP requirements are markedly reduced compared to the amount needed by skeletal muscle.

Smooth muscle contraction is under involuntary control. For example, we cannot voluntarily contract the smooth muscle in the walls of our digestive organs. Thus, smooth muscle cells contract and our stomach seems to "growl" at an inappropriate time. As with cardiac muscle, the autonomic nervous system is one way that the contraction of smooth muscle is controlled.

Skeletal, cardiac, and smooth muscle are compared in **table 10.6**.

💡 WHAT DID YOU LEARN?

16 Describe smooth muscle cell structure. What are dense bodies?

Aging and the Muscular System

Key topic in this section:

■ Effects of aging on skeletal muscle

In a person's mid-30s, a slow, progressive loss of skeletal muscle mass begins as a direct result of increasing inactivity. The size and power of all muscle tissues also decrease. Often, the lost muscle mass is replaced by either adipose or fibrous connective tissue. Aging affects the muscular system in several specific ways.

First, the number of myofibrils and myofilaments in existing muscle fibers decreases. This muscular atrophy results in a decrease in muscle fiber volume. Second, skeletal muscle fibers decrease in diameter, usually due to a loss of myofibrils within the fiber. The consequences of decreased fiber diameter include a cutback in oxygen storage capacity because of less myoglobin, reduced glycogen reserves, and a decreased ability to produce ATP. Overall, muscle strength and endurance are impaired, and the individual has a tendency to fatigue quickly. Decreased cardiovascular performance often accompanies aging; thus, increased circulatory supply to active muscles occurs much more slowly in the elderly as they begin to exercise.

In addition, a person's tolerance for exercise lowers with age. Decreased muscle strength is compounded by a tendency toward rapid fatigue. The atrophy of muscle secondarily results in less tension generated by that muscle. Consequently, the muscle fatigues more readily while performing a specific amount of work.

As a person grows older, muscle tissue has a reduced capacity to recover from disease or injury. The number of satellite cells in skeletal muscle steadily decreases, and scar tissue often forms due to diminished repair capabilities. Finally, the elasticity of skeletal muscle also decreases as muscle mass is often replaced by dense regular (fibrous) connective tissue, a process called **fibrosis** (fī-brō′sis). The increasing amounts of dense regular connective tissue decrease the flexibility of muscle; an increase in collagen fibers can restrict movement and circulation.

All of us will eventually experience a decline in muscular performance, irrespective of our lifestyle or exercise patterns. But we can improve our chances of being in good shape later in life by striving for physical fitness early in life. Other benefits of regular exercise include strengthening our bones and increasing or maintaining bone mass, controlling body weight, and generally improving the quality of life. Exercise that is regular and moderate, rather than extremely demanding, is the most helpful type.

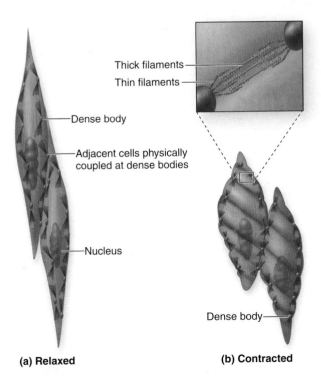

Thick filaments
Thin filaments

Dense body

Adjacent cells physically coupled at dense bodies

Nucleus

Dense body

(a) Relaxed   **(b) Contracted**

Figure 10.16

Smooth Muscle. Smooth muscle is located in the walls of hollow organs and blood vessels. *(a)* Relaxed smooth muscle cells appear more elongated than *(b)* contracted smooth muscle cells.

Table 10.6	**Muscle Tissue Types: General Comparisons**			
Muscle Tissue	**Appearance and Shape of Cell**	**Location of Muscle**	**Cell Length/ Cell Diameter**	**Autorhythmicity/Contraction Speed/Nervous System Control**
Skeletal Muscle	Long, cylindrical fiber; multiple peripheral nuclei internal to sarcolemma; striated Muscle fiber Nuclei Striations Striations Nuclei Muscle (dark vs. light) fiber	Attached to bones (usually via tendons) or to subcutaneous layer	*Length:* Long (100 µm–30 cm) *Diameter:* Large (10–100 µm)	*Autorhythmicity:* No *Contraction speed:* Fast *Nervous system control:* Voluntary
Cardiac Muscle	Short, branched cell with 1 or 2 centrally located nuclei; intercalated discs between fibers; striated Muscle fiber Striations Muscle Intercalated cell Nuclei Striations discs	Only in heart wall	*Length:* Short (50–100 µm) *Diameter:* Large (about 15 µm diameter)	*Autorhythmicity:* Yes *Contraction speed:* Moderate *Nervous system control:* Involuntary
Smooth Muscle	Spindle-shaped cell; a single centrally located nucleus; no striations Muscle cell Nucleus Muscle cell Nucleus	Found in walls of hollow organs (e.g., intestines, blood vessels); and in iris and ciliary body of eye	*Length:* Short (50–200 µm) *Diameter:* Small (5–10 µm)	*Autorhythmicity:* In some types only *Contraction speed:* Slow *Nervous system control:* Involuntary

[1] Recent research indicates that cardiac muscle may have very limited regenerative capacity in certain situations.

Connective Tissue Components	Calcium Source/ Junctions Between Cells/ Neuromuscular Junctions	Presence of Myofibrils	Regeneration Capacity	Sarcomere Protein Organization	Sarcoplasmic Reticulum/ Transverse Tubules
Three layers: epimysium, perimysium, endomysium 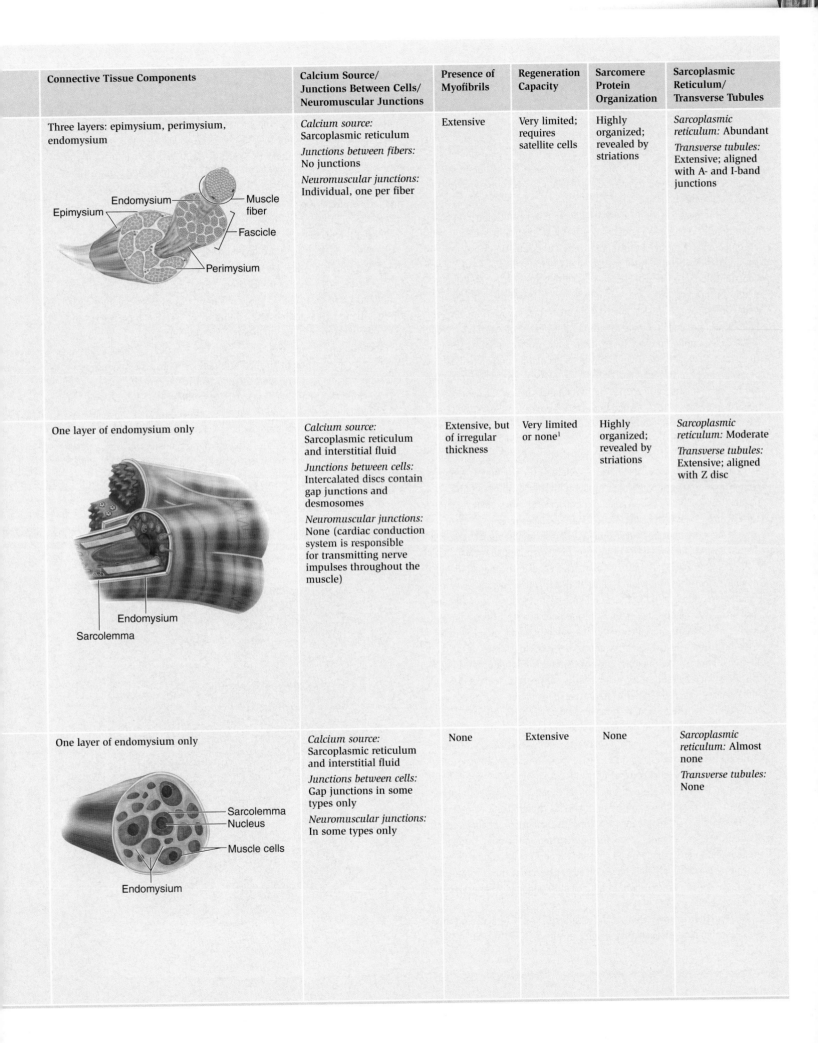	*Calcium source:* Sarcoplasmic reticulum *Junctions between fibers:* No junctions *Neuromuscular junctions:* Individual, one per fiber	Extensive	Very limited; requires satellite cells	Highly organized; revealed by striations	*Sarcoplasmic reticulum:* Abundant *Transverse tubules:* Extensive; aligned with A- and I-band junctions
One layer of endomysium only	*Calcium source:* Sarcoplasmic reticulum and interstitial fluid *Junctions between cells:* Intercalated discs contain gap junctions and desmosomes *Neuromuscular junctions:* None (cardiac conduction system is responsible for transmitting nerve impulses throughout the muscle)	Extensive, but of irregular thickness	Very limited or none[1]	Highly organized; revealed by striations	*Sarcoplasmic reticulum:* Moderate *Transverse tubules:* Extensive; aligned with Z disc
One layer of endomysium only	*Calcium source:* Sarcoplasmic reticulum and interstitial fluid *Junctions between cells:* Gap junctions in some types only *Neuromuscular junctions:* In some types only	None	Extensive	None	*Sarcoplasmic reticulum:* Almost none *Transverse tubules:* None

Labels in first image: Epimysium, Endomysium, Muscle fiber, Fascicle, Perimysium

Labels in second image: Endomysium, Sarcolemma

Labels in third image: Sarcolemma, Nucleus, Muscle cells, Endomysium

CLINICAL VIEW: In Depth

Neuromuscular Diseases

Disease or damage affecting somatic motor neurons, neuromuscular junctions, or muscle fibers often results in abnormal muscle functions, collectively termed *neuromuscular disease.* Some of these noninfectious disorders are fibromyalgia, muscular dystrophy, myasthenia gravis, and myofascial pain syndrome.

Fibromyalgia (fī-brō-mī-al′jă; *algos* = pain) presents as a syndrome of chronic severe pain involving both the muscles and the skeleton, and accompanied by fatigue, morning stiffness, and sometimes psychological depression. No single cause has been determined, but a number of contributing factors have been identified, including alterations in brain neurotransmitters, stress and anxiety, sleep disturbances, and possibly endocrine abnormalities. Although the pain associated with this condition is body-wide, it most commonly affects the neck or lower back. About 1–2% of the population has this syndrome, and women are affected more frequently than men. Treatment consists of a combination of medication and self-care. Nonsteroidal anti-inflammatory medications (aspirin, ibuprofen, etc.) are useful for relieving pain and stiffness. Antidepressants are also known to help some people with this condition. Proven self-care measures include stress reduction, regular exercise, reduction or elimination of caffeine intake, and smoking cessation.

Muscular dystrophy (dis′trō-fē) is a collective term for several hereditary diseases in which the skeletal muscles degenerate, lose strength, and are gradually replaced by adipose and fibrous connective tissue. In a vicious cycle, the new connective tissues impede blood circulation, which further accelerates muscle degeneration.

Duchenne muscular dystrophy (DMD) is the most common form of the illness. It results from the expression of a sex-linked recessive allele. DMD is almost exclusively a disease of males and occurs in about 1 in 3500 live births. For these individuals, muscular difficulties become apparent in early childhood. Walking is a problem; the child falls frequently and has difficulty standing up again. The hips are affected first, followed by the lower limbs, and eventually the abdominal and spinal muscles. Muscular atrophy causes shortening of the muscles, which results in postural abnormalities such as scoliosis (lateral curvature of the spine). DMD is an incurable disease, with patients confined to a wheelchair by adolescence. The DMD patient rarely lives beyond the age of 20, and death typically results from respiratory or heart complications.

Recent advances in gene therapy research may result in a treatment for DMD within the next few years. Using an altered virus that carries the gene for dystrophin, an essential protein for muscle function that is missing in DMD patients, investigators have successfully stimulated the manufacture of the protein in laboratory mice that lack this gene. Although these early results provide great hope for treatment of humans, they are also cause for concern. Scientists recognize that the ability to deliver genes to every muscle fiber in the body through a single injection may also provide a means for performance enhancement in athletes. Athletic officials have voiced concern that the current doping practice using steroids would appear old-fashioned if gene doping were to become possible.

Myasthenia gravis (mī-as-thē′nē-ă; *astheneia* = weakness) **(MG)** is an autoimmune disease that occurs in about 1 in 10,000 people,

primarily women between 20 and 40 years of age. Antibodies attack the neuromuscular junctions, binding ACh receptors together into clusters. The abnormally clustered ACh receptors are removed from the muscle fiber sacrolemma by endocytosis, thus significantly diminishing the number of receptors on the muscle fiber. The resulting decreased ability to stimulate the muscle causes rapid fatigue and muscle weakness. Eye and facial muscles are often attacked first, resulting in double vision and drooping eyelids. These symptoms are usually followed by swallowing problems, limb weakness, and overall low physical stamina.

Some patients with MG have a normal life span, while others die quickly from paralysis of the respiratory muscles. Treatment often involves the use of cholinesterase inhibitors, immunosuppressive drugs, and thymectomy (removal of the thymus gland, which is abnormal in these patients). The cholinesterase inhibitors prevent ACh destruction by acetylcholinesterase, allowing for prolonged stimulation of the remaining ACh receptors in the muscle fiber membrane. Immunosuppressive drugs decrease the production of antibodies directed against the ACh receptors and thereby slow the attack on the neuromuscular junction.

Myofascial (mī-ō-fash′ē-ăl) **pain syndrome (MPS)** is a common disorder usually associated with excessive use of the postural muscles. Pain results when bands of muscle fibers tighten and then twitch after the overlying skin is stimulated. This "trigger point" stimulation of the pain is a widely recognized feature of MPS. Occasionally, the trigger points can produce autonomic nervous system changes, resulting in localized flushing of the skin, sweating, and even "goose bumps." Unlike some other neuromuscular disorders, MPS does not cause fatigue and morning stiffness. As many as 50% of people, between the ages of 30 and 60, may be affected by this disease. Massages, muscle stretching, and nonsteroidal anti-inflammatory drugs are effective in treating this condition.

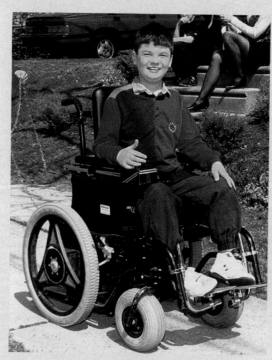

Child with Duchenne muscular dystrophy.

Development of the Muscular System

Key topic in this section:

■ Developmental events in muscle formation

This brief examination of the development of the muscular system focuses primarily on skeletal muscle development. Skeletal muscle tissue formation is initiated during the fourth week of development from blocks of paraxial mesoderm that form structures called **somites** (sō′mīt). The first somites appear in the cranial portion of the embryo at about day 20; new somites are formed caudally at regular intervals until the forty-fourth pair of somites appears on about day 30.

Cells within a somite differentiate into three distinct regions: (1) The **sclerotome** (sklĕr′ō-tōm; *skleros* = hard, *tome* = a cutting) separates from the rest of the somite and gives rise to the vertebral skeleton. (2) The **dermatome** (der′mă-tōm; *derma* = skin) forms the connective tissue of the skin. (3) The **myotome** (mī′ō-tōm) gives rise to the skeletal muscles. Myoblasts within the myotome begin rapid mitotic division (see figure 10.4). The number of cells continues to increase as myoblasts migrate and fuse into multinucleated **myotubules**. At about 9 weeks, primitive myofilaments extend through the myotubules. Growth in length continues via the addition of more myoblasts.

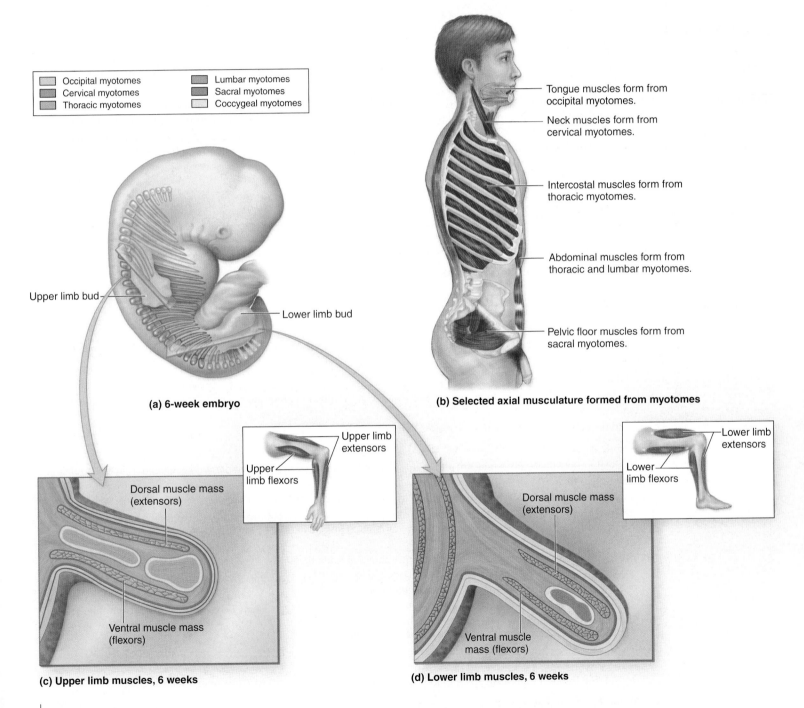

(a) 6-week embryo

(b) Selected axial musculature formed from myotomes

(c) Upper limb muscles, 6 weeks

(d) Lower limb muscles, 6 weeks

Figure 10.17

Development of Skeletal Muscles. *(a)* The distribution of embryonic myotomes at about 6 weeks. *(b)* Axial muscles form from adjacent segmental myotomes. *(c)* Upper limb musculature forms from cervical and thoracic myotomes that migrate into the upper limb beginning in the fifth week of development. The myotomes split into two separate groups that later form the flexors and the extensors of the upper limb. *(d)* Lower limb musculature forms from lumbar and sacral myotomes that migrate into the lower limb beginning in the fifth week of development. The myotomes split into two separate groups that will form the flexors and extensors of the lower limb.

The myotomes give rise to virtually all skeletal muscles, except some muscles of the head (**figure 10.17**). The axial musculature (muscles that attach to the axial skeleton) forms from adjacent myotomes. For example, myotomes in the thoracic and lumbar regions form the lower back muscles and the abdominal muscles. Some cervical myotomes form neck muscles, where the cervical vertebrae are found.

During the fifth week of development, portions of some of the somites migrate into the developing upper and lower limbs. The upper limb musculature forms from some cervical and thoracic myotomes. The lower limb musculature forms from some lumbar and sacral myotomes that migrate into the developing limb. As these myotomes migrate, the myotome group is subdivided into a **ventral muscle mass** (so named because this mass of muscles is close to the ventral, or belly, side of the embryo) and a **dorsal muscle mass** (located closer to the dorsal side of the embryo). The ventral muscle mass will form the future flexors of that limb. The biceps brachii muscle of the anterior arm is a ventral mass muscle, because this muscle flexes the elbow joint. The dorsal muscle mass will form the future extensors of that limb. The triceps brachii muscle of the posterior arm is a dorsal mass muscle, and this muscle extends the elbow joint. These muscle mass groups differentiate into individual skeletal muscles during the embryonic period.

CLINICAL TERMS

"charley horse" Localized pain or muscle stiffness following the contusion of a muscle.

myoma (mī-ō′mă) Benign tumor of muscular tissue.

myomalacia (mī-ō-mă-lā′shē-ă) Pathologic softening of muscular tissue.

spasm (spazm) Sudden, involuntary contraction of one or more muscles.

strain To injure or tear a muscle as a result of either overuse or improper use.

tic Repeated contraction of certain muscles, resulting in stereotyped, individualized actions that can be voluntarily suppressed for only brief periods (e.g., clearing the throat, pursing the lips, or excessive blinking).

CHAPTER SUMMARY

	■ Muscle tissue moves materials within and throughout the body.
	■ The types of muscle tissue are skeletal muscle, cardiac muscle, and smooth muscle.
Properties of Muscle Tissue 287	■ Muscles exhibit excitability, contractility, elasticity, and extensibility.
Characteristics of Skeletal Muscle Tissue 287	■ Individual skeletal muscles vary in shape and are striated.

Functions of Skeletal Muscle Tissue   287

■ Skeletal muscles produce body movement, maintain posture and body temperature, store and move materials, and provide support.

Gross Anatomy of Skeletal Muscle   288

■ Muscles are ensheathed by three connective tissue layers: an inner endomysium, a middle perimysium, and an outer epimysium.

■ Tendons or aponeuroses attach muscle ends to other structures.

Microscopic Anatomy of Skeletal Muscle   291

■ Skeletal muscle fibers are cylindrical and multinucleated, and have a plasma membrane (sarcolemma) that surrounds the cytoplasm (sarcoplasm).

■ T-tubules extend as sarcolemma invaginations into the sarcoplasm and conduct impulses for muscle contraction.

■ The sarcoplasmic reticulum (SR) is a specialized internal membrane system that stores calcium ions needed for muscle contraction. The calcium ions are housed in the terminal cisternae of the SR. A T-tubule together with the terminal cisternae at both its sides is called a triad.

■ Muscle fibers are filled with myofibrils that house thin and thick protein filaments sometimes called myofilaments.

■ Thin filaments are composed of actin (a contractile protein) and two regulatory proteins (tropomyosin and troponin). Thick filaments are made of myosin (a contractile protein).

■ Striations in a muscle fiber are alternating dark bands (A bands) and light bands (I bands) that result from thin and thick filament differences. The H zone is a light region in the middle of the A band. The Z disc is a protein attachment site for filaments in the middle of the I band; it appears as a thin, dark line.

■ The functional contractile unit of a skeletal muscle fiber is a sarcomere that extends between adjacent Z discs. Sarcomeres stack in each myofibril.

■ The M line is a thin transverse protein meshwork structure in the center of the H zone of a relaxed fiber.

Contraction of Skeletal Muscle Fibers 296

The Sliding Filament Theory   296

■ Thick and thin filaments slide past each other during a muscle contraction. A sarcomere shortens, but filament length does not change.

■ Nerve impulses initiate impulses in skeletal muscle fibers.

Neuromuscular Junctions   296

■ The neuromuscular junction is the structure where axons of neurons transmit nerve impulses to muscle fibers.

Contraction of Skeletal Muscle Fibers (continued)	**Physiology of Muscle Contraction 299** ■ Nerve impulses in axons cause muscle impulses in the sarcolemma; these spread into T-tubules, causing calcium ions to be released from the terminal cisternae. ■ Calcium ions bind troponin, resulting in tropomyosin movement away from the surface of actin. ■ Myosin heads attach to actin (form crossbridges), pivot toward the center of the sarcomere, detach from actin, and recock to their original position. **Muscle Contraction: A Summary 299** ■ (1) A muscle impulse is generated; (2) calcium ions are released into the sarcoplasm; (3) calcium ions bind to troponin, causing tropomyosin to move off actin; (4) myosin heads attach, pivot, detach, and return, causing protein filaments to slide by each other; (5) the cycle continues as long as calcium ions bind to troponin. Rest occurs when muscle impulses cease. **Motor Units 301** ■ A motor unit is composed of a motor neuron and the muscle fibers it controls. ■ Isometric contractions result in no muscle shortening; isotonic contractions result in muscle shortening.
Fixed Types of Skeletal Muscle Fibers 303	■ The three muscle fiber types are slow, intermediate, and fast. **Distribution of Slow, Intermediate, and Fast Fibers 305** ■ Muscle function determines the presence of fiber types. Postural muscles have more slow fibers, whereas the muscles that move the eyes have more fast fibers.
Skeletal Muscle Fiber Organization 305	■ Muscles are classified according to fiber arrangements. **Circular Muscles 305** ■ Circular muscles are composed of concentric fibers, and are found at entrances and exits from the body or in internal passageways. **Parallel Muscles 305** ■ In parallel muscles, fascicles are oriented parallel to the muscle fibers' long axes. **Convergent Muscles 305** ■ Convergent muscles are widespread fibers joined at a single attachment site. **Pennate Muscles 305** ■ Pennate muscles are composed of one or more tendons extending through the muscle. Fascicles are oblique to the tendons. Subtypes are unipennate, bipennate, and multipennate.
Exercise and Skeletal Muscle 307	■ Exercise and proper diet are required to increase skeletal muscle size. **Muscle Atrophy 307** ■ Muscles become flaccid and atrophied if stimulation is inadequate. **Muscle Hypertrophy 307** ■ Muscular hypertrophy is an increase in muscle fiber size. Repetitive stimulation causes more myofibrils to develop.
Levers and Joint Biomechanics 307	■ Levers rotate around a fixed point to affect speed of movement, distance moved, direction of force, and strength of force. **Classes of Levers 307** ■ Three classes of levers are defined by the arrangement of the fulcrum, resistance point, and effort arm. Third-class levers are the most common type of lever in the body. **Actions of Skeletal Muscles 308** ■ An agonist is a muscle that contracts to produce a particular movement. The antagonist opposes the agonist. The synergist assists the agonist.
The Naming of Skeletal Muscles 309	■ Muscle names indicate muscle action, body regions, attachment sites, orientation, shape, size, heads, and tendon origins.
Characteristics of Cardiac and Smooth Muscle 310	**Cardiac Muscle 310** ■ Cardiac muscle is involuntary. Cardiac muscle cells occur in the wall of the heart and are autorhythmic. Intercalated discs are junctions between cardiac muscle cells. **Smooth Muscle 311** ■ Smooth muscle is involuntary; it has short, fusiform, nonstriated cells.
Aging and the Muscular System 311	■ The aging process decreases the elasticity and diameter of skeletal muscle fibers, and reduces their ability to recover from disease or injury. Exercise tolerance decreases as the body ages.
Development of the Muscular System 315	■ The development of skeletal muscle tissue is initiated during the fourth week of development. Blocks of mesoderm called somites form on either side of the neural tube. The myotome portion of the somites forms most skeletal muscle in the body.

CHALLENGE YOURSELF

Matching

Match each numbered item with the most closely related lettered item.

_____ 1. perimysium

_____ 2. sarcolemma

_____ 3. I band

_____ 4. insertion

_____ 5. sarcomere

_____ 6. fast fiber

_____ 7. muscle tone

_____ 8. antagonist

_____ 9. circular muscle

_____ 10. myosin

a. type of muscle fiber that fatigues easily

b. muscle fiber plasma membrane

c. muscle whose function opposes agonist

d. connective tissue covering a fascicle

e. functional contractile unit of a skeletal muscle fiber

f. protein in thick filaments

g. sarcomere region with thin filaments only

h. muscle that surrounds an opening

i. the resting tension within a muscle

j. the more mobile attachment of a muscle

Multiple Choice

Select the best answer from the four choices provided.

_____ 1. The unit of muscle structure that is composed of bundles of myofibrils, enclosed within a sarcolemma, and surrounded by a connective tissue covering called endomysium is a
   a. myofibril.
   b. fascicle.
   c. myofilament.
   d. muscle fiber.

_____ 2. During the contraction of a muscle fiber, myofibrils
   a. lengthen.
   b. remain unchanged.
   c. increase in diameter.
   d. shorten.

_____ 3. In a convergent muscle, the fibers are
   a. oblique to the tendon of the muscle.
   b. concentrically arranged.
   c. parallel to the long axis of the muscle.
   d. widespread over a broad area and joined at a common attachment site.

_____ 4. The plasma membrane of a skeletal muscle fiber is called the
   a. sarcoplasmic reticulum.
   b. sarcolemma.
   c. sarcoplasm.
   d. sarcomysium.

_____ 5. In a skeletal muscle fiber, a triad is composed of
   a. myosin, actin, and myofibrils.
   b. one A band, one H zone, and one I band.
   c. one transverse tubule and two terminal cisternae.
   d. myofilaments, myofibrils, and sarcomeres.

_____ 6. During development, the _____ of a somite gives rise to skeletal muscle.
   a. dermatome
   b. scleroblast
   c. myotome
   d. satellite cell

_____ 7. What is a synaptic knob?
   a. receptor for neurotransmitter in the neuromuscular junction
   b. membrane sac holding the neurotransmitter
   c. expanded tip of an axon at the neuromuscular junction
   d. the space that separates the neuron from the muscle fiber

_____ 8. The bundle of dense regular connective tissue that attaches a skeletal muscle to bone is called a(n)
   a. tendon.
   b. ligament.
   c. endomysium.
   d. fascicle.

_____ 9. The muscle that assists the agonist is called the
   a. antagonist.
   b. synergist.
   c. prime mover.
   d. None of these are correct.

_____ 10. Which of the following changes in skeletal muscles is associated with aging?
   a. Muscle fibers increase in diameter.
   b. Muscles become more elastic.
   c. Muscle fibers increase their glycogen reserves.
   d. The number of satellite cells in muscle decreases.

Content Review

1. Compare the roles of the three concentric layers of connective tissue wrappings in the organization of skeletal muscle.

2. Describe the structure of tendons and aponeuroses, and discuss their purpose in the body.

3. Describe the structural relationships among the following in a resting skeletal muscle fiber: A band, H zone, M line.

4. Describe the changes that occur in the length of thick and thin filaments, I bands, and the sarcomere during muscle contraction.

5. Describe in your own words the events that occur during skeletal muscle contraction.

6. Explain why the ratio of motor neurons to muscle fibers is greater in muscles that control eye movement than in postural muscles of the leg.

7. Briefly describe the differences between muscle atrophy and muscle hypertrophy. What might cause each condition?

8. Explain why athletes who excel at short sprints probably have fewer slow fibers in their leg muscles.

9. Explain what the following names tell us about muscles: longus, extensor, triceps, rectus, superficialis.

10. Describe the structure of intercalated discs in cardiac muscle cells, and briefly discuss their functions.

Developing Critical Reasoning

1. Paul broke his left radius playing football, and the doctor placed his forearm in a cast for 6 weeks. When the cast was removed, his left forearm was noticeably thinner and less well developed than his right forearm. How can you explain the noticeable decrease in mass in the left forearm?

2. Tessa and her laboratory partners were given the following assignment: Explain how the structure of myofilaments is related to their function, and then describe how the sliding filament theory allows for the shortening of a muscle fiber. If they asked for your help, how would you answer these questions?

3. Savannah and her 59-year-old grandfather spent a long day gardening. Pulling weeds, raking, and planting left both of them a bit sore. However, Savannah's grandfather was more stiff and sore, and it took him much longer to recover than Savannah. Why did Savannah and her grandfather react to this exertion differently?

A N S W E R S   T O   " W H A T   D O   Y O U   T H I N K ? "

1. If a muscle has multiple motor units, you can control the force and precision of the muscle contraction. The greater the number of motor units activated, the more forceful the contraction, because more muscle fibers are contracting. Likewise, a muscle can perform more precise movements if the motor units are controlling only a few muscle fibers.

2. The "white meat" of turkey contains large-diameter, fast muscle fibers that, in life, would have low resistance to fatigue. In contrast, the "dark meat" contains small-diameter, slow muscle fibers that would have great resistance to fatigue.

3. The knee joint is a third-class lever, because the fulcrum (bony knee joint) and the resistance (movement of the leg) are at either end, and the effort is applied by the patellar ligament located between the two.

Visit the McKinley/O'Loughlin *Human Anatomy*, 2e website at aris.mhhe.com

11

MUSCULAR SYSTEM

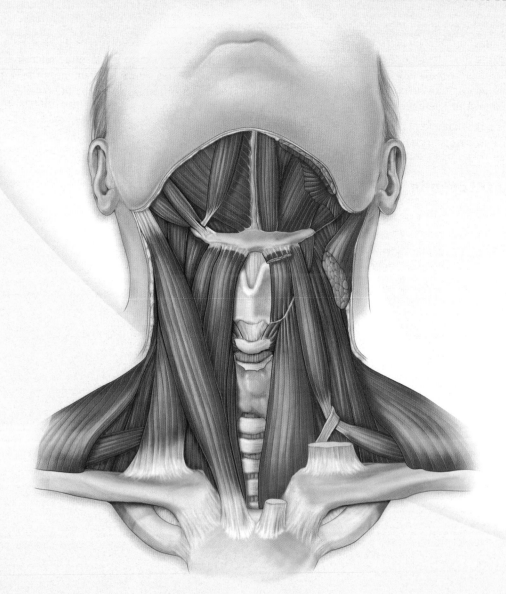

Axial Muscles

The artificial partitioning of the skeletal system into axial and appendicular divisions provides a useful guideline for subdividing the muscular system as well. **Axial muscles** have both their origins and insertions on parts of the axial skeleton. Axial muscles support and move the head and spinal column, function in nonverbal communication by affecting facial features, move the lower jaw during chewing, assist in food processing and swallowing, aid breathing, and support and protect the abdominal and pelvic organs. The axial muscles are not responsible for stabilizing or moving the pectoral or pelvic girdles or their attached limbs; those are functions of the appendicular muscles (see chapter 12). Some muscles of both divisions are shown in **figure 11.1**.

Study Tip!

— The following suggestions may help you learn the muscles, whether axial or appendicular:

- The axial muscles may be organized into groups based on their location or function. Learning the muscles in groups is easiest.

- When studying a particular muscle, try to palpate it on yourself. Contract the muscle to sense its action and identify its regional location.

- Repeat the name of a muscle aloud to become familiar with its name. It is easier to remember and spell terms that you can pronounce.

- Associate visual images from models, cadavers, a photographic atlas, or dissected animals with muscle names. Connecting a muscle name with direct observation will help you remember it.

- Locate the origins and insertions of muscles on an articulated skeleton. This helps you visualize the locations of muscles and understand how they produce particular actions.

- Learn the derivation of each muscle name because it usually describes the muscle's action, location, number of heads, orientation of muscle fibers, shape, or size.

Refer to figure 10.14 for examples of how some muscles are named.

The axial muscles are organized into five groups based on their location:

- Muscles of the head and neck
- Muscles of the vertebral column
- Muscles of respiration
- Muscles of the abdominal wall
- Muscles of the pelvic floor

The discussion in this chapter has been organized according to these specific groups. For each group, tables provide descriptions of the muscles as well as information about their action, origin, insertion, and innervation. (Note: The word *innervation* refers to the nerve(s) that supplies a muscle and stimulates it to contract. For further information about the nerves listed in the tables, see chapters 15 and 16.)

Muscles of the Head and Neck

Key topics in this section:

- Major muscles involved in facial expression
- Extrinsic muscles of the eye and how each affects eye movement
- Muscles of mastication and how each affects mandibular movement
- Movements of the tongue and comparison of its extrinsic and intrinsic muscles
- Muscles of the pharynx and their function in swallowing
- Organization and distribution of the muscles of the anterior neck
- Muscles involved in the major movements of the head and neck

The muscles of the head and neck are separated into several specific groups, based on their location or general functions. Almost all of these muscles (except for a few muscles of the anterior neck) originate on either the skull or the hyoid bone.

Muscles of Facial Expression

The muscles of facial expression have their origin in the superficial fascia or on the skull bones **(figure 11.2)**. These muscles insert into the superficial fascia of the skin, so when they contract, they contort the skin, causing it to move. Most of these muscles are innervated by the seventh cranial nerve (CN VII), the facial nerve.

The **epicranius** is composed of the **occipitofrontalis muscle** and a broad **epicranial aponeurosis,** also called the *galea aponeurotica.* The **frontal belly** of the occipitofrontalis is superficial to the frontal bone on the forehead. When this muscle contracts, it raises the eyebrows and wrinkles the skin of the forehead. The **occipital belly** of the occipitofrontalis covers the posterior side of the head. When this muscle contracts, it retracts the scalp slightly. Deep to the frontal belly is the **corrugator supercilii.** This muscle draws the eyebrows together and creates vertical wrinkle lines above the nose. The **orbicularis oculi** consists of circular muscle fibers that surround the orbit. When this muscle contracts, the eyelid closes, as when you wink, blink, or squint. The **levator palpebrae superioris** (discussed in detail in chapter 19, and shown in figure 19.10) elevates the upper eyelid when you open your eyes.

Several muscles of facial expression are associated with the nose. The **nasalis** elevates the corners of the nostrils. When you "flare your nostrils," you are using the nasalis muscles. If you wrinkle your nose in distaste after smelling a foul odor, you have used your **procerus** muscle. This muscle is continuous with the frontalis muscle, and it runs over the bridge of the nose, where it produces transverse wrinkles when it contracts.

The mouth is the most expressive part of the face, and not surprisingly the muscles in that area are very diverse. The **orbicularis oris** consists of muscle fibers that encircle the opening of the mouth. When this muscle contracts, you close your mouth. When you "pucker up for a kiss," you are using this muscle. The **depressor labii inferioris** does what its name suggests—it pulls the lower lip inferiorly. The **depressor anguli oris** is considered the "frown" muscle,

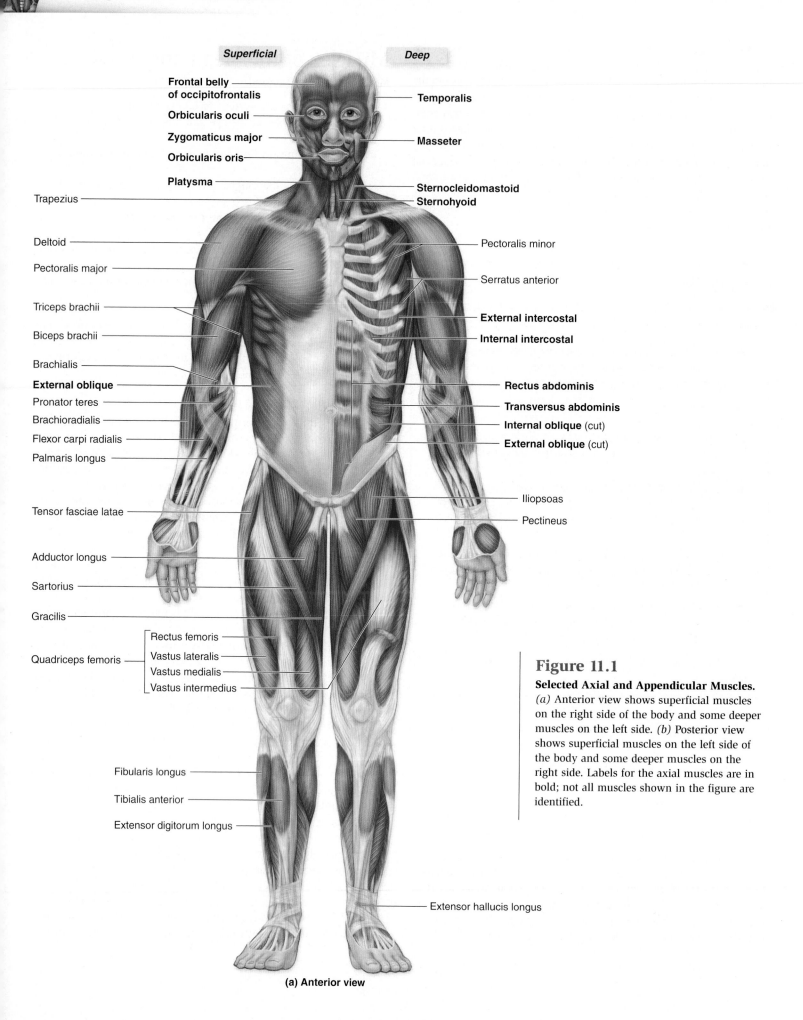

Superficial

Deep

Frontal belly
of occipitofrontalis

Orbicularis oculi

Zygomaticus major

Orbicularis oris

Platysma

Temporalis

Masseter

Sternocleidomastoid
Sternohyoid

Trapezius

Deltoid

Pectoralis major

Triceps brachii

Biceps brachii

Brachialis

External oblique

Pronator teres

Brachioradialis

Flexor carpi radialis

Palmaris longus

Tensor fasciae latae

Adductor longus

Sartorius

Gracilis

Quadriceps femoris ─ [Rectus femoris
Vastus lateralis
Vastus medialis
Vastus intermedius

Fibularis longus

Tibialis anterior

Extensor digitorum longus

Pectoralis minor

Serratus anterior

External intercostal

Internal intercostal

Rectus abdominis

Transversus abdominis

Internal oblique (cut)

External oblique (cut)

Iliopsoas

Pectineus

Extensor hallucis longus

Figure 11.1

Selected Axial and Appendicular Muscles.
(a) Anterior view shows superficial muscles
on the right side of the body and some deeper
muscles on the left side. *(b)* Posterior view
shows superficial muscles on the left side of
the body and some deeper muscles on the
right side. Labels for the axial muscles are in
bold; not all muscles shown in the figure are
identified.

(a) Anterior view

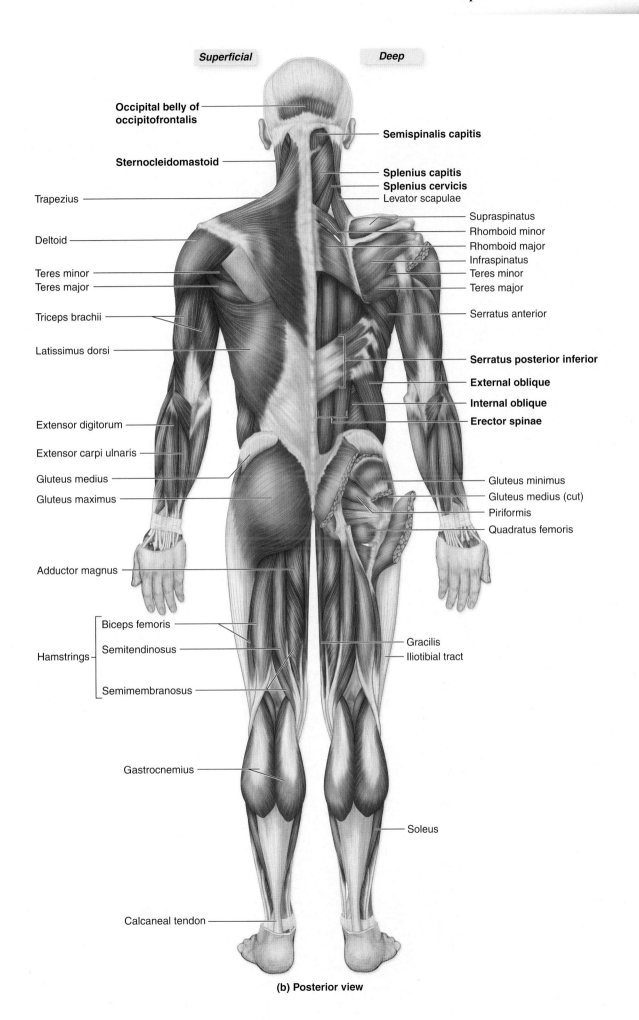

Superficial   *Deep*

Occipital belly of
occipitofrontalis

Semispinalis capitis

Sternocleidomastoid

Splenius capitis
Splenius cervicis
Levator scapulae

Trapezius

Supraspinatus
Rhomboid minor
Rhomboid major
Infraspinatus
Teres minor
Teres major

Deltoid

Teres minor
Teres major

Serratus anterior

Triceps brachii

Latissimus dorsi

Serratus posterior inferior

External oblique

Internal oblique

Erector spinae

Extensor digitorum

Extensor carpi ulnaris

Gluteus medius

Gluteus maximus

Gluteus minimus
Gluteus medius (cut)
Piriformis
Quadratus femoris

Adductor magnus

Biceps femoris

Gracilis
Iliotibial tract

Hamstrings   Semitendinosus

Semimembranosus

Gastrocnemius

Soleus

Calcaneal tendon

(b) Posterior view

Superficial

Deep

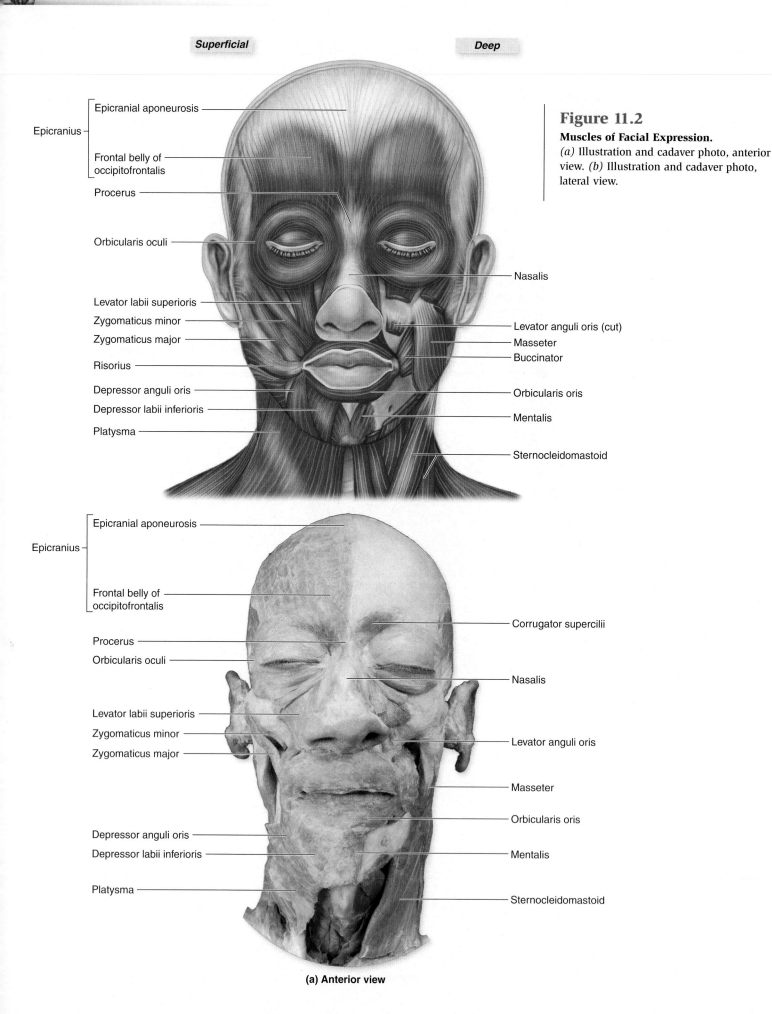

Epicranius
- Epicranial aponeurosis
- Frontal belly of occipitofrontalis

Procerus

Orbicularis oculi

Levator labii superioris

Zygomaticus minor

Zygomaticus major

Risorius

Depressor anguli oris

Depressor labii inferioris

Platysma

Nasalis

Levator anguli oris (cut)

Masseter

Buccinator

Orbicularis oris

Mentalis

Sternocleidomastoid

Figure 11.2

Muscles of Facial Expression.
(a) Illustration and cadaver photo, anterior view. *(b)* Illustration and cadaver photo, lateral view.

Epicranius
- Epicranial aponeurosis
- Frontal belly of occipitofrontalis

Procerus

Orbicularis oculi

Levator labii superioris

Zygomaticus minor

Zygomaticus major

Depressor anguli oris

Depressor labii inferioris

Platysma

Corrugator supercilii

Nasalis

Levator anguli oris

Masseter

Orbicularis oris

Mentalis

Sternocleidomastoid

(a) Anterior view

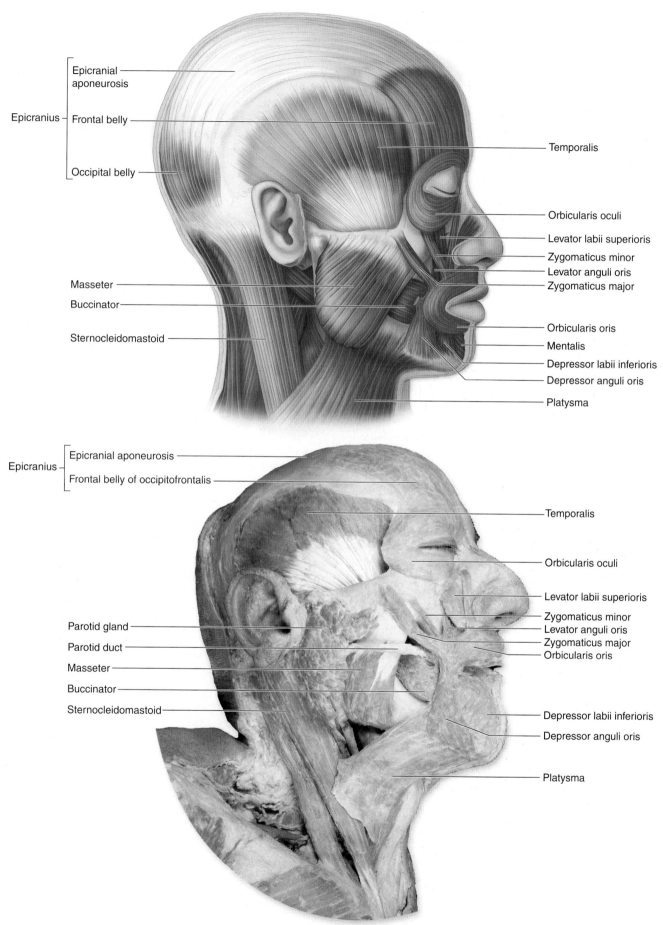

Epicranial aponeurosis

Epicranius

Frontal belly

Occipital belly

Temporalis

Orbicularis oculi

Levator labii superioris

Zygomaticus minor

Levator anguli oris

Zygomaticus major

Masseter

Buccinator

Sternocleidomastoid

Orbicularis oris

Mentalis

Depressor labii inferioris

Depressor anguli oris

Platysma

Epicranius

Epicranial aponeurosis

Frontal belly of occipitofrontalis

Temporalis

Orbicularis oculi

Levator labii superioris

Zygomaticus minor

Levator anguli oris

Zygomaticus major

Orbicularis oris

Parotid gland

Parotid duct

Masseter

Buccinator

Sternocleidomastoid

Depressor labii inferioris

Depressor anguli oris

Platysma

(b) Lateral view

because it pulls the corners of the mouth inferiorly. (Note, however, that it takes more muscles than this one to produce a frown!)

In contrast, some muscles of the mouth elevate part or all of the upper lips. The **levator labii superioris** pulls the upper lip superiorly, as if a person is sneering or snarling. The **levator anguli oris** pulls the corners of the mouth superiorly and laterally. The **zygomaticus major** and **zygomaticus minor** work with the levator anguli oris muscles. You use these when you smile. The **risorius** pulls the corner of the lips laterally; you use this muscle if you make a closed-mouth smile.

The **mentalis** attaches to the lower lip, and when it contracts, it protrudes the lower lip (as when a person "pouts"). The **platysma** tenses the skin of the neck and pulls the lower lip inferiorly. If you stand in front of a mirror and tense the skin of your neck, you can see these thin muscles bulging out.

The **buccinator** compresses the cheek against the teeth when we chew (and is the reason our cheeks don't bulge like a squirrel's cheeks when we eat). Infants use the buccinator when they suckle at the breast. Some trumpet players (such as Dizzy Gillespie) have stretched out their buccinator muscles, allowing their cheeks to be "puffy" with air when they play the trumpet.

Table 11.1 summarizes the attachments and movements of the muscles of facial expression. **Figure 11.3** illustrates how these muscles produce some of the more characteristic expressions.

WHAT DO YOU THINK?

1 What muscles of facial expression must contract in order for you to smile?

CLINICAL VIEW

Idiopathic Facial Nerve Paralysis (Bell Palsy)

Unilateral paralysis of the muscles of facial expression is termed **facial nerve paralysis**. This condition results from either disease or injury to the facial nerve (CN VII). If the cause of the condition is unknown, doctors refer to it as *idiopathic* (id′ē-ō-path′ik; *idios* = one's own, *pathos* = suffering) *facial nerve paralysis*, or **Bell Palsy**. Recent studies indicate a frequent link between CN VII paralysis and herpes simplex 1 viral infection. Facial nerve paralysis is also associated with exposure to cold temperatures, and is commonly seen in individuals who sleep with one side of their head facing an open window. Another possible cause of facial nerve paralysis is compression of the facial nerve by an adjacent blood vessel. Whatever the underlying cause, the nerve becomes inflamed and compressed within the narrow stylomastoid foramen.

The facial nerve innervates all but one of the muscles of facial expression, so if its function becomes impaired, the muscles on the same side of the face are paralyzed. The patient may be unable to wrinkle the forehead (paralyzed occipitofrontalis muscle), pucker the lips (paralyzed orbicularis oris), or close the eyelid on the affected side (paralyzed orbicularis oculi). These symptoms lead to other problems as well. For example, if the affected individual cannot close his or her eye, the eye becomes dry, possibly damaging the cornea. If unable to close the mouth, the person drools, and the mucous membranes of the mouth become parched.

Treatment of facial nerve paralysis usually means alleviating the symptoms. Doctors often use prednisone (a type of steroid) to reduce the inflammation and swelling of the nerve. If herpes simplex infection is suspected, an antiviral medication called Acyclovir is also given. Paralysis of the orbicularis oculi may require eyedrops to combat the symptoms of dry eye, and sometimes it's necessary to patch the affected eye to keep it closed while the patient is sleeping.

Like its underlying cause, recovery from idiopathic facial nerve paralysis is mysterious. Over 50% of all patients experience a complete, spontaneous recovery within 30 days of their first symptoms. For others, recovery may take longer, while still others may never recover. Current statistics indicate that the recovery rate for idiopathic facial nerve paralysis averages about 80%, and does not appear to be related to its treatment.

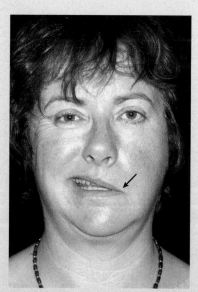

Facial nerve (CN VII) paralysis on the left side of the face. Note the drooping left side of the mouth (arrow) and the lack of contraction by the left orbicularis oculi when the woman tries to smile.

Table 11.1	Muscles of Facial Expression		
Region/Muscle	**Action**	**Origin/Insertion**	**Innervation**
SCALP			
Epicranius (ep′ĭ-krā′nē-us) *epi* = over *cran* = skull		Composed of an epicranial aponeurosis and the occipitofrontalis muscle	
Frontal belly of occipitofrontalis (ok-sip′i-tō-fron-tă′lis) *front* = forehead	Moves scalp, eyebrows; wrinkles skin of forehead	O: Frontal bone I: Epicranial aponeurosis	CN VII (facial nerve)
Occipital belly of occipitofrontalis *occipito* = base of skull	Retracts scalp	O: Superior nuchal line I: Epicranial aponeurosis	CN VII (facial nerve)
NOSE			
Nasalis (nā′ză-lis) *nasus* = nose	Compresses bridge and depresses tip of nose; elevates corners of nostrils	O: Maxillae and alar cartilage of nose I: Dorsum of nose	CN VII (facial nerve)
Procerus (prō-sē′rŭs) *procerus* = long	Moves and wrinkles nose	O: Nasal bone and lateral nasal cartilage I: Aponeurosis at bridge of nose and skin of forehead	CN VII (facial nerve)
MOUTH			
Buccinator (buk′sĭ-nā′tōr) *bucco* = cheek	Compresses cheek; holds food between teeth during chewing	O: Alveolar processes of mandible and maxillae I: Orbicularis oris	CN VII (facial nerve)
Depressor anguli oris (dē-pres′ōr ang′ŭ-lī ōr′ŭs) *depressor* = depresses *angul* = angle *or* = mouth	Draws corners of mouth inferiorly and laterally ("frown" muscle)	O: Body of mandible I: Skin at inferior corner (angle) of mouth	CN VII (facial nerve)
Depressor labii inferioris (dē-pres′ōr lā′bē-ī in-fēr′ē-ōr-is) *labi* = lip *infer* = below	Draws lower lip inferiorly	O: Body of mandible lateral to midline I: Skin at inferior lip	CN VII (facial nerve)
Levator anguli oris (lē-vā′tor, le-vā′ter ang′ŭ-lī ōr′ŭs) *leva* = raise	Draws corners of mouth superiorly and laterally ("smile" muscle)	O: Lateral maxilla I: Skin at superior corner of mouth	CN VII (facial nerve)
Levator labii superioris (lē-vā′tor, le-vā′ter lā′bē-ī sū-pěr′ē-ōr-is)	Opens lips; raises and furrows the upper lip ("Elvis" lip snarl)	O: Zygomatic bone; maxilla I: Skin and muscle of superior lip	CN VII (facial nerve)
Mentalis (men-tā′lis) *ment* = chin	Protrudes lower lip ("pout"); wrinkles chin	O: Central mandible I: Skin of chin	CN VII (facial nerve)
Orbicularis oris (ōr-bik′ŭ-lā′ris ōr′is) *orb* = circular *or* = mouth	Compresses and purses lips ("kiss" muscle)	O: Maxilla and mandible; blend with fibers from other facial muscles I: Encircling mouth; skin and muscles at angles to mouth	CN VII (facial nerve)
Risorius (ri-sōr′ē-ŭs) *risor* = laughter	Draws corner of lip laterally; tenses lips; synergist of zygomaticus	O: Deep fascia associated with masseter muscle I: Skin at angle of mouth	CN VII (facial nerve)
Zygomaticus major (zī′gō-mat′i-kŭs) *zygomatic* = cheekbone *major* = greater	Elevates corner of the mouth ("smile" muscle)	O: Zygomatic bone I: Skin at superolateral edge of mouth	CN VII (facial nerve)
Zygomaticus minor (zī′gō-mat′i-kŭs) *minor* = lesser	Elevates corner of the mouth ("smile" muscle)	O: Zygomatic bone I: Skin of superior lip	CN VII (facial nerve)

(continued on next page)

Table 11.1	Muscles of Facial Expression *(continued)*		
Region/Muscle	**Action**	**Origin/Insertion**	**Innervation**
EYE			
Corrugator supercilii (kōr′ŭ-gā-ter soo′per-sil′ĕ-ī) *corrugo* = to wrinkle *cilium* = eyelid	Pulls eyebrows inferiorly and medially; creates vertical wrinkles above nose	O: Medial end of superciliary arch I: Skin superior to supraorbital margin and superciliary arch	CN VII (facial nerve)
Levator palpebrae superioris (see fig. 19.10) (le-vă′ter pal-pē′bră soo-pēr′ē-ōr-ĭs) *levo* = to lift *palpebra* = eyelid	Elevates superior eyelid	O: Lesser wing of sphenoid bone I: Superior tarsal plate and skin of superior eyelid	CN III (oculomotor nerve)
Orbicularis oculi (ōr-bik′ŭ-lā′ris ok′ū-lī) *orb* = circular *ocul* = eye	Closes eye; produces winking, blinking, squinting ("blink" muscle)	O: Medial wall or margin of orbit I: Skin surrounding eyelids	CN VII (facial nerve)
NECK			
Platysma (plă-tiz′mă) *platy* = flat	Pulls lower lip inferiorly; tenses skin of neck	O: Fascia of deltoid and pectoralis major muscles and acromion of scapula I: Skin of cheek and mandible	CN VII (facial nerve)

Extrinsic Eye Muscles

The **extrinsic eye muscles,** often called *extraocular muscles*, move the eyes. They are termed extrinsic because they originate within the orbit and insert onto the white outer surface of the eye, called the sclera. There are six extrinsic eye muscles: the rectus muscles (medial, lateral, inferior, and superior) and the oblique muscles (inferior and superior) **(figure 11.4)**.

The rectus eye muscles have their origin from a **common tendinous ring** in the orbit. These muscles insert on the *anterior* part of the sclera of the eye, and are named according to which side of the eye they are located at (medial, lateral, inferior, or superior).

The **medial rectus** attaches to the anteromedial surface of the eye and pulls the eye medially (adducts the eye). It is innervated by CN III (oculomotor nerve). The **lateral rectus** attaches to the anterolateral surface of the eye and pulls the eye laterally (abducts the eye). This muscle is innervated by CN VI (abducens). (Note that this nerve's name tells you what muscle it innervates—the eye muscle that abducts the eye). The **inferior rectus** attaches to the anteroinferior part of the sclera. The inferior rectus pulls the eye inferiorly (as when you look down) and medially (as when you look at your nose). The **superior rectus** is located superiorly and attaches to the anterosuperior part of the sclera. The superior rectus pulls the eye superiorly (as when you look up) and medially (as when you look at your nose). The inferior and superior rectus muscles are innervated by CN III. Figure 11.4*d* illustrates that the superior and inferior rectus muscles do not pull directly parallel to the long axis of the eye; that is why both muscles also move the eye slightly in the medial direction.

The oblique eye muscles originate from within the orbit and insert on the *posterolateral* part of the sclera of the eye. The

Depressor anguli oris
 (frown)

Orbicularis oculi
 (blink/close eyes)

Zygomaticus major
 (smile)

Orbicularis oris
 (close mouth/kiss)

Frontal belly of occipitofrontalis
 (wrinkle forehead, raise eyebrows)

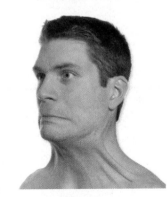

Platysma
 (tense skin of neck)

Figure 11.3

Surface Anatomy of Some Muscles of Facial Expression. These muscles permit complex expressions that are often used as a means of communication.

inferior oblique elevates the eye and turns the eye laterally. Since this muscle attaches to the inferior *posterior* part of the eye, contracting this muscle pulls the posterior part of the eye inferiorly (but *elevates* the *anterior* part of the eye). This muscle is innervated by CN III. The **superior oblique** depresses the eye and turns the eye laterally. This muscle passes through a pulley-like loop, called the **trochlea,** in the anteromedial orbit. This muscle attaches to the superior *posterior* part of the eye, so contracting this muscle pulls the posterior part of the eye superiorly (but *depresses* the *anterior* surface of the eye). This muscle is innervated by CN IV (trochlear). (Note that this nerve's name is derived from the trochlea that holds the superior oblique in place.)

Table 11.2 compares the extrinsic muscles of the eye.

> ## Study Tip!
> Remembering the innervation of the eye muscles can be difficult. Use the following "chemical formula" to help you learn the eye muscle innervation:
> $$[(SO_4)(LR_6)]_3$$
> In other words, the superior oblique **(SO)** is innervated by cranial nerve IV **(4)**, the lateral rectus **(LR)** is innervated by cranial nerve VI **(6)**, and the rest of the eye muscles are innervated by cranial nerve III **(3)**.

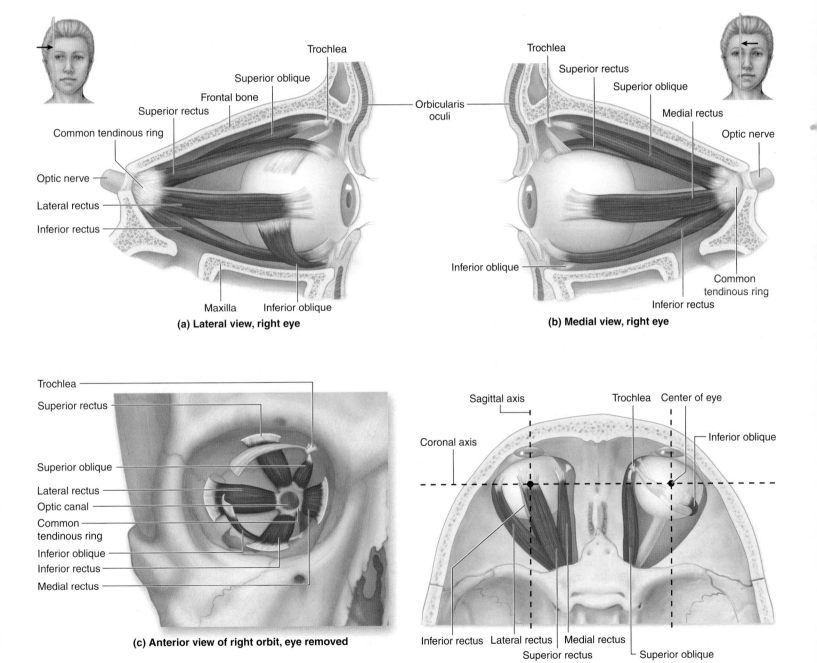

(a) Lateral view, right eye

(b) Medial view, right eye

(c) Anterior view of right orbit, eye removed

(d) Superior view

Figure 11.4

Extrinsic Muscles of the Eye.   The extrinsic eye muscles control movements of the eye. *(a)* The insertions for the lateral, superior, and inferior rectus, as well as both the inferior and superior oblique muscles, appear prominently in a lateral view of the right eye. *(b)* The medial rectus muscle appears prominently in a medial view of the right eye. *(c)* Most of the eye muscles originate from a common tendinous ring, shown here in an anterior view of the right orbit. *(d)* A superior view of the left and right orbits illustrates the insertion differences between the rectus and oblique muscles, and how these differences affect their movement of the eye.

Table 11.2	Extrinsic Eye Muscles		
Group/Muscle	**Action**	**Origin/Insertion**	**Innervation**
RECTUS MUSCLES			
Medial rectus (mē´dē-ăl rek´tus) *rectus* = straight	Moves eye medially (adducts eye)	O: Common tendinous ring I: Anteromedial surface of eye	CN III (oculomotor nerve)
Lateral rectus (lat´er-ăl rek´tus)	Moves eye laterally (abducts eye)	O: Common tendinous ring I: Anterolateral surface of eye	CN VI (abducens nerve)
Inferior rectus (in-fē´rē-ōr rek´tus)	Moves eye inferiorly (depresses eye) and medially (adducts eye)	O: Common tendinous ring I: Anteroinferior surface of eye	CN III (oculomotor nerve)
Superior rectus (soo-pěr´ē-ōr rek´tus)	Moves eye superiorly (elevates eye) and medially (adducts eye)	O: Common tendinous ring I: Anterosuperior surface of eye	CN III (oculomotor nerve)
OBLIQUE MUSCLES			
Inferior oblique (in-fē´rē-ōr ob-lěk´) *obliquus* = slanting	Moves eye superiorly (elevates eye) and laterally (abducts eye)	O: Anterior orbital surface of maxilla I: Posteroinferior, lateral surface of eye	CN III (oculomotor nerve)
Superior oblique (soo-pěr´ē-ōr ob-lěk´)	Moves eye inferiorly (depresses eye) and laterally (abducts eye)	O: Sphenoid bone I: Posterosuperior, lateral surface of eye	CN IV (trochlear nerve)

CLINICAL VIEW

Strabismus

When the eyes are improperly aligned, the condition is called **strabismus** (stra-biz´mŭs; *strabismos* = a squinting). The misalignment means the eyes are not working synchronously to transmit a stereoscopic view to the brain. With each eye sending a slightly different image, the brain becomes confused and ignores one of the images. The ignored eye becomes weaker and weaker over time, resulting in a condition termed "lazy eye." If uncorrected, the lazy eye loses visual acuity, a condition termed **strabismic amblyopia** (amblē-ō´pē-ă; *amblys* = dull, *ops* = eye).

Causes of strabismus include birth injuries, diseases localized to the eye or its bony orbit, improper attachment of the extrinsic eye muscles, and heredity. Two forms of strabismus are recognized. *External strabismus* occurs when the oculomotor nerve (CN III) is injured, so that the affected eye moves laterally while at rest but cannot move medially and inferiorly. Conversely, *internal strabismus* occurs when the abducens nerve (CN VI) is injured. The affected eye moves medially but cannot move laterally.

WHAT DID YOU LEARN?

1. What are the origins for all the muscles of facial expression?
2. List the extrinsic eye muscles, and describe the function of each muscle.
3. The corners of the mouth are pulled inferiorly into a frown position by the contraction of what muscle?

Muscles of Mastication

The term **mastication** (mas-ti-kā´shŭn; *masticatus* = to chew) refers to the process of chewing. These muscles move the mandible at the temporomandibular joint. There are four paired muscles of mastication: the temporalis, the masseter, and the lateral and medial pterygoids (**figure 11.5**). The muscles of mastication are innervated by the mandibular division of CN V (trigeminal nerve).

The **temporalis** (or *temporal* muscle) is a broad, fan-shaped muscle that extends from the temporal lines of the skull and inserts on the coronoid process of the mandible. It elevates and retracts (pulls posteriorly) the mandible. You can palpate the temporalis by placing your fingers along your temple (lateral skull at same level of orbits) as you open and close your mouth. The muscle you feel contracting is the temporalis.

The **masseter** elevates and protracts (pulls anteriorly) the mandible. It is the most powerful and important of the masticatory muscles. This short, thick muscle is superficial to the temporalis. You can feel the contraction of the masseter by palpating near the angle of the mandible as you open and close your mouth.

The **lateral** and **medial pterygoid** muscles arise from the lateral pterygoid plates of the sphenoid bone and insert on the mandible. Both pterygoids protract the mandible and move it from side to side during chewing. These movements maximize the efficiency of the teeth while chewing or grinding foods of various consistencies. The medial pterygoid also elevates the mandible.

Table 11.3 summarizes the characteristics of the muscles of mastication.

Muscles That Move the Tongue

The tongue is an agile, highly mobile organ. It consists of **intrinsic muscles** that curl, squeeze, and fold the tongue during chewing and speaking. Thus, the tongue itself acts like a big muscle.

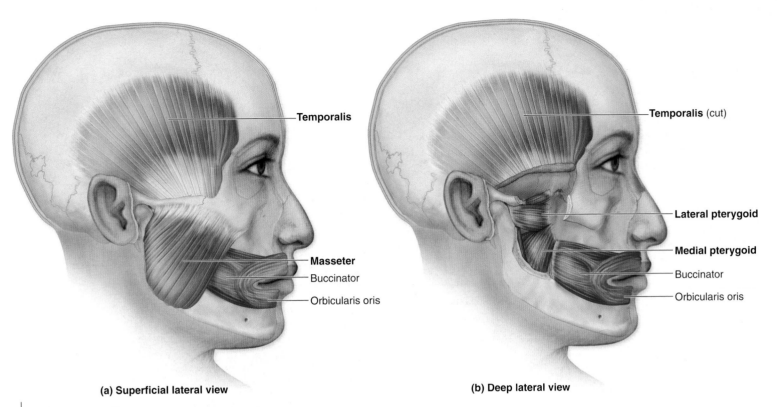

(a) Superficial lateral view   (b) Deep lateral view

Figure 11.5

Muscles of Mastication.   *(a)* Superficial and *(b)* deep lateral views of the muscles of mastication (shown in bold), which move the mandible.

The **extrinsic muscles** of the tongue have their origin on other head and neck structures and insert on the tongue. The extrinsic muscles end in the suffix *-glossus*, meaning "tongue" **(figure 11.6)**. These extrinsic tongue muscles are used in various combinations to accomplish the precise, complex, and delicate tongue movements required for proper speech. Additionally, they manipulate food within the mouth in preparation for swallowing. Most of these muscles are innervated by CN XII, the hypoglossal nerve.

The left and right **genioglossus** muscles have their origin on the mandible and protract the tongue. You use these muscles when you stick out your tongue. The left and right **styloglossus** muscles originate on the styloid processes of the temporal bone. These muscles elevate and retract the tongue (pull the tongue posteriorly, back into the mouth). The left and right **hyoglossus** muscles originate at the hyoid bone and insert on the sides of the tongue. These muscles depress and retract the tongue. The left and right **palatoglossus** muscles originate on the soft palate and elevate the posterior portion of the tongue.

Table 11.3	**Muscles of Mastication**		
Muscle	**Action**	**Origin/Insertion**	**Innervation**
Temporalis (tem-pō-rā′lis) *tempora* = pertaining to temporal bone	Elevates and retracts mandible	O: Superior and inferior temporal lines I: Coronoid process of mandible	CN V$_3$ (trigeminal nerve, mandibular division)
Masseter (ma′se-ter) *maseter* = chewer	Elevates and protracts mandible; prime mover of jaw closure	O: Zygomatic arch I: Coronoid process, lateral surface and angle of mandible	CN V$_3$ (trigeminal nerve, mandibular division)
Medial pterygoid (mē′dē-ăl ter′i-goyd)	Elevates and protracts mandible; produces side-to-side movement of mandible	O: Maxilla, palatine, and medial surface of lateral pterygoid plate I: Medial surface of mandibular ramus	CN V$_3$ (trigeminal nerve, mandibular division)
Lateral pterygoid (lat′er-ăl ter′i-goyd) *pterygoid* = winglike	Protracts mandible; produces side-to-side movement of mandible	O: Greater wing of sphenoid and lateral surface of lateral pterygoid plate I: Condylar process of mandible	CN V$_3$ (trigeminal nerve, mandibular division)

Figure 11.6

Muscles That Move the Tongue. Extrinsic tongue muscles (shown in bold) originate on structures other than the tongue and insert onto it to allow gross tongue movement.

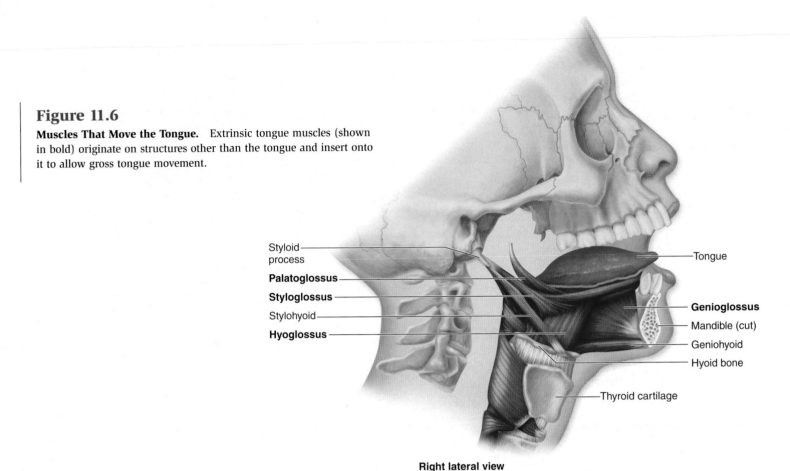

Right lateral view

Table 11.4 summarizes the characteristics of the muscles that move the tongue.

Muscles of the Pharynx

The **pharynx,** commonly called the "throat," is a funnel-shaped tube that lies posterior to and extends inferiorly from both the oral and nasal cavities. Several muscles help form this muscular tube or attach to it and aid in swallowing

(**figure 11.7**). Most pharyngeal muscles are innervated by CN X (vagus nerve).

The primary pharyngeal muscles are the **pharyngeal constrictors (superior, middle,** and **inferior).** When a bolus of food enters the pharynx, these muscles contract sequentially to initiate swallowing and force the bolus inferiorly into the esophagus. Other pharyngeal muscles help elevate or tense the palate when swallowing. These muscles are summarized in **table 11.5.**

Table 11.4	Muscles That Move the Tongue		
Muscle	**Action**	**Origin/Insertion**	**Innervation**
Genioglossus (jĕ′nĭ-ō-glos′ŭs) *geni* = chin *glossus* = tongue	Protracts tongue	O: Mental spines of mandible I: Inferior region of tongue; hyoid bone	CN XII (hypoglossal nerve)
Styloglossus (stī′lō-glos′ŭs) *stylo* = pertaining to styloid process of temporal bone	Elevates and retracts tongue	O: Styloid process of temporal bone I: Sides and inferior aspect of tongue	CN XII (hypoglossal nerve)
Hyoglossus (hī′ō-glos′ŭs) *hyo* = pertaining to hyoid bone	Depresses and retracts tongue	O: Hyoid bone I: Inferolateral side of tongue	CN XII (hypoglossal nerve)
Palatoglossus (pal-ă-tō-glos′ŭs) *palato* = palate	Elevates posterior part of tongue	O: Anterior surface of soft palate I: Side and posterior aspect of tongue	CN X (vagus nerve) via pharyngeal plexus of nerves

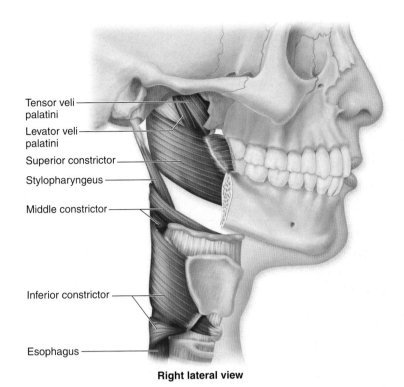

Tensor veli palatini

Levator veli palatini

Superior constrictor

Stylopharyngeus

Middle constrictor

Inferior constrictor

Esophagus

Right lateral view

Figure 11.7

Pharyngeal Constrictors, Palate Muscles, and Laryngeal Elevators.
A right lateral view reveals some of the muscles that constrict the pharynx when swallowing, move the palate, and elevate the larynx (palatopharyngeus and salpingopharyngeus not shown).

Muscles of the Anterior Neck

The muscles of the anterior neck are divided into the **suprahyoid muscles,** which are superior to the hyoid bone, and the **infrahyoid muscles,** which are inferior to the hyoid bone (**figure 11.8**). The **suprahyoid muscles** are associated with the floor of the mouth. In general, these muscles act as a group to elevate the hyoid bone during swallowing or speaking. Some of these muscles perform additional functions: The **digastric** has two bellies, anterior and posterior. One belly extends from the mental protuberance to the hyoid, and the other continues from the hyoid to the mastoid portion of the temporal bone. The two bellies are united by an intermediate tendon that is held in position by a fascia sling (fibrous loop). In addition to elevating the hyoid bone, this muscle can also depress the mandible.

Table 11.5	Muscles of the Pharynx[1]		
Region/Muscle	**Action**	**Origin/Insertion**	**Innervation**
PALATE MUSCLES			
Levator veli palatini (lĕ-vā′tor, le-vā′ter vel′ī pal′ă-tē′nī) *levator* = elevates *velum* = veil	Elevates soft palate when swallowing	O: Petrous part of temporal bone I: Soft palate	CN X (vagus nerve)
Tensor veli palatini (ten′sōr vel′ī pal′ă-tē′nī) *tensus* = to stretch	Tenses soft palate and opens auditory tube when swallowing or yawning	O: Sphenoid bone; region around auditory tube I: Soft palate	CN V$_3$ (trigeminal nerve, mandibular division)
PHARYNGEAL CONSTRICTORS			
Superior constrictor (kon-strik′ter, -tor) *constringo* = to draw together	Constricts pharynx in sequence to force bolus into esophagus; superior is innermost	O: Pterygoid process of sphenoid bone; medial surface of mandible I: Posterior median raphe (muscle fiber union from both sides)	CN X (vagus nerve) via branches of pharyngeal plexus
Middle constrictor	Constricts pharynx in sequence	O: Hyoid bone I: Posterior median raphe	CN X (vagus nerve) via branches of pharyngeal plexus
Inferior constrictor	Constricts pharynx in sequence; inferior is outermost	O: Thyroid and cricoid cartilage I: Posterior median raphe	CN X (vagus nerve) via branches of pharyngeal plexus
LARYNGEAL (VOICE BOX) ELEVATORS			
Palatopharyngeus (păl′ă-tō-far-in′-jē-ŭs) *pharynx* = pharynx	Elevates pharynx and larynx	O: Soft palate I: Side of pharynx and thyroid cartilage of larynx	CN X (vagus nerve) via branches of pharyngeal plexus
Salpingopharyngeus (sal-ping′gō-făr-in′jē-ŭs) *salpinx* = trumpet	Elevates pharynx and larynx	O: Auditory tube I: Blends with palatopharyngeus on lateral wall of pharynx	CN X (vagus nerve) via branches of pharyngeal plexus
Stylopharyngeus (stī′lō-far-in′jē-ŭs) *stylo* = styloid process	Elevates pharynx and larynx	O: Styloid process of temporal bone I: Side of pharynx and thyroid cartilage of larynx	CN IX (glossopharyngeal nerve) via branches of pharyngeal plexus

[1] Only the pharyngeal constrictors are discussed in the text.

Superficial

Deep

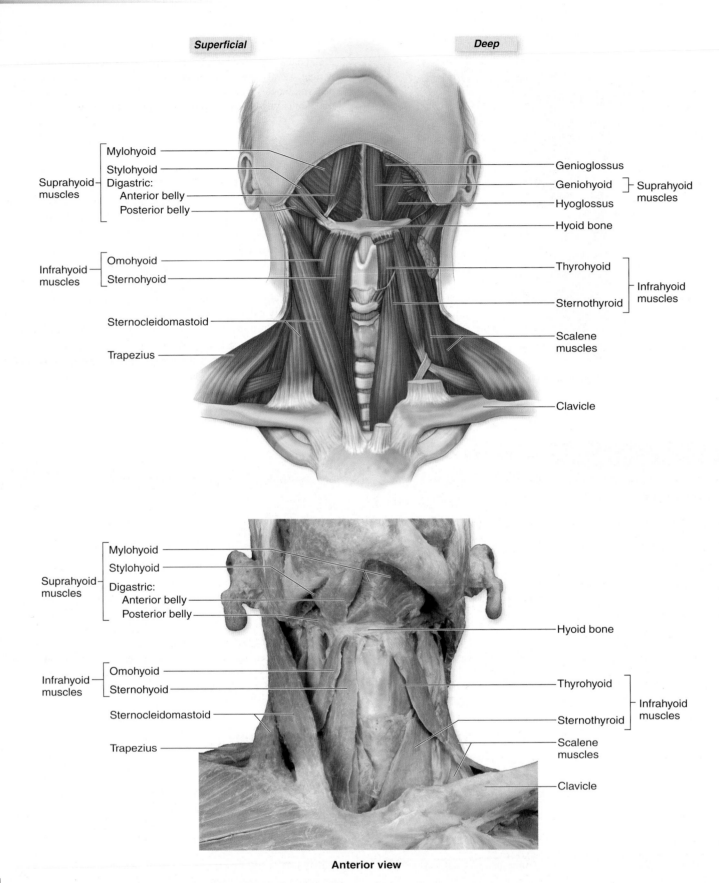

Suprahyoid muscles
- Mylohyoid
- Stylohyoid
- Digastric:
  - Anterior belly
  - Posterior belly

Infrahyoid muscles
- Omohyoid
- Sternohyoid

Sternocleidomastoid

Trapezius

Genioglossus
Geniohyoid } Suprahyoid muscles
Hyoglossus
Hyoid bone

Thyrohyoid
Sternothyroid } Infrahyoid muscles

Scalene muscles

Clavicle

Suprahyoid muscles
- Mylohyoid
- Stylohyoid
- Digastric:
  - Anterior belly
  - Posterior belly

Infrahyoid muscles
- Omohyoid
- Sternohyoid

Sternocleidomastoid

Trapezius

Hyoid bone

Thyrohyoid

Sternothyroid } Infrahyoid muscles

Scalene muscles

Clavicle

Anterior view

Figure 11.8

Muscles of the Anterior Neck.   An illustration and a cadaver photo show the anterior neck muscles, which move the hyoid bone and the thyroid cartilage. Superficial muscles are shown on the right side, while deeper muscles are shown on the left side.

The **geniohyoid** originates from the mental spines of the mandible and inserts on the hyoid bone. This muscle elevates the hyoid bone. The broad, flat **mylohyoid** provides a muscular floor to the mouth. When this muscle contracts, it both elevates the hyoid bone and raises the floor of the mouth. The muscle fibers of the left and right mylohyoid are aligned in a V shape. The **stylohyoid** originates from the styloid process of the skull and inserts on the hyoid. Upon contraction, it elevates the hyoid bone, causing the floor of the oral cavity to elongate during swallowing.

As swallowing ends, the **infrahyoid muscles** contract to influence the position of the hyoid bone and the larynx. In general, these muscles either depress the hyoid bone or depress the thyroid cartilage of the larynx. The **omohyoid** contains two thin muscle bellies anchored in place by a connective tissue "sling." This muscle is lateral to the sternohyoid and extends from the superior border of the scapula and inserts on the hyoid, where it depresses the hyoid bone. The **sternohyoid** extends from the sternum to the hyoid, where it depresses the hyoid bone. The **sternothyroid** is deep to the sternohyoid. It extends from the sternum to the thyroid cartilage of the larynx. It depresses the thyroid cartilage to return it to its original position after swallowing. The **thyrohyoid** extends from the thyroid cartilage of the larynx to the hyoid. It depresses the hyoid bone and elevates the thyroid cartilage to close off the larynx during swallowing. In addition, the omohyoid, sternohyoid, and thyrohyoid help anchor the hyoid so the digastric can depress the mandible.

Table 11.6 summarizes the characteristics of the muscles of the anterior neck.

WHAT DO YOU THINK?

2 Since muscles frequently are named for their attachment sites, what do you think the prefix *omo* in "omohyoid" means?

Muscles That Move the Head and Neck

Muscles that move the head and neck originate on the vertebral column, the thoracic cage, and the pectoral girdle, and insert on bones of the cranium (**figure 11.9**; see figure 11.8).

Anterolateral Neck Muscles

The anterolateral neck muscles flex the head and/or neck. The main muscles in this group are the sternocleidomastoid and the three scalenes.

The **sternocleidomastoid** is a thick, cordlike muscle that extends from the sternum and clavicle to the mastoid process posterior to the ear. Contraction of both sternocleidomastoid muscles (called **bilateral contraction**) flexes the neck. Contraction of just one sternocleidomastoid muscle (called **unilateral contraction**) results in lateral flexion of the neck and rotation of the head to the opposite side. Thus, if the left sternocleidomastoid muscle contracts, it rotates the head to the right side of the body. The three **scalene muscles**

Table 11.6	Muscles of the Anterior Neck		
Region/Muscle	**Action**	**Origin/Insertion**	**Innervation**
SUPRAHYOID MUSCLES			
Digastric (dī-gas′trik) *di* = two *gaster* = belly	Depresses mandible; elevates hyoid bone	O: Anterior belly, mandible near mental protuberance; posterior belly, mastoid process I: Hyoid bone via fascia sling	Anterior belly: CN V$_3$ (trigeminal nerve, mandibular division) Posterior belly: CN VII (facial nerve)
Geniohyoid (jě′nī-ō-hī′-oyd) *hyoid* = hyoid bone	Elevates hyoid bone	O: Mental spines of mandible I: Hyoid bone	First cervical spinal nerve (C1) via CN XII (hypoglossal nerve)
Mylohyoid (mī′lō-hī′oyd) *myle* = molar	Elevates hyoid bone; elevates floor of mouth	O: Mylohyoid line of mandible I: Hyoid bone	CN V$_3$ (trigeminal nerve, mandibular division)
Stylohyoid (stī′lō-hī′oyd)	Elevates hyoid bone	O: Styloid process of temporal bone I: Hyoid bone	CN VII (facial nerve)
INFRAHYOID MUSCLES			
Omohyoid (ō′mō-hī′oyd) *omo* = shoulder	Depresses hyoid bone; fixes hyoid during opening of mouth	O: Superior border of scapula I: Hyoid bone	Cervical spinal nerves C1–C3 through ansa cervicalis (from cervical plexus)
Sternohyoid (ster′nō-hī′oyd) *sterno* = sternum	Depresses hyoid bone	O: Manubrium of sternum and medial end of clavicle I: Hyoid bone	Cervical spinal nerves C1–C3 through ansa cervicalis (from cervical plexus)
Sternothyroid (ster′nō-thī′royd) *thyro* = thyroid cartilage	Depresses thyroid cartilage of larynx	O: Posterior surface of manubrium of sternum I: Thyroid cartilage of larynx	Cervical spinal nerves C1–C3 through ansa cervicalis (from cervical plexus)
Thyrohyoid (thī′rō-hī′oyd)	Depresses hyoid bone and elevates thyroid cartilage of larynx	O: Thyroid cartilage of larynx I: Hyoid bone	First cervical spinal nerve C1 via CN XII (hypoglossal nerve)

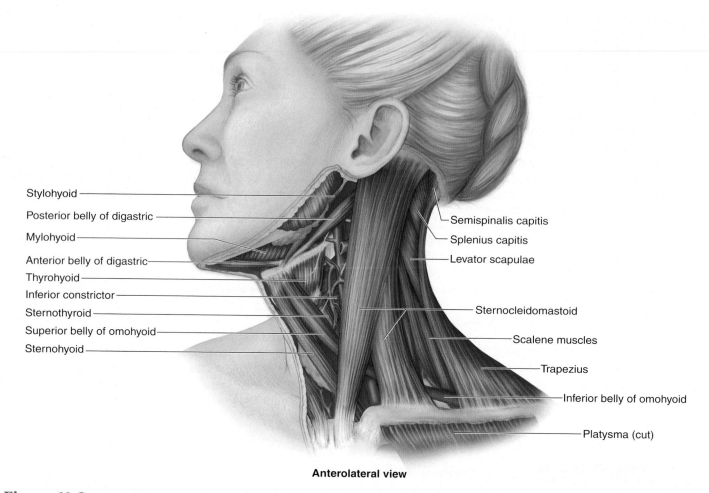

Stylohyoid
Posterior belly of digastric
Mylohyoid
Anterior belly of digastric
Thyrohyoid
Inferior constrictor
Sternothyroid
Superior belly of omohyoid
Sternohyoid

Semispinalis capitis
Splenius capitis
Levator scapulae

Sternocleidomastoid
Scalene muscles
Trapezius
Inferior belly of omohyoid
Platysma (cut)

Anterolateral view

Figure 11.9

Muscles That Move the Head and Neck.   Anterolateral muscles collectively flex the neck, while posterior neck muscles extend the head and/or neck.

(**anterior, middle,** and **posterior**) work with the sternocleidomastoid to flex the neck. In addition, the scalene muscles elevate the first and second ribs during forced inhalation.

Posterior Neck Muscles

Several muscles work together to extend the head and/or neck (**figure 11.10**). The trapezius attaches to the skull and helps extend the head and/or neck. The primary function of the trapezius is to help move the pectoral girdle, so it is discussed in greater detail in chapter 12. When the left and right **splenius capitis, splenius cervicis, semispinalis capitis,** and **longissimus capitis** muscles bilaterally contract, they extend the neck. Unilateral contraction turns the head and neck to the same side. A group of muscles called the suboccipital muscles includes the **obliquus capitus superior, obliquus capitus inferior, rectus capitis posterior major,** and **rectus capitis**

posterior minor. The obliquus muscles turn the head to the same side, while the rectus muscles extend the head and neck.

Table 11.7 summarizes the characteristics of the muscles of the head and neck.

WHAT DID YOU LEARN?

4 What movements do the medial and lateral pterygoids perform?

5 Which muscle protracts the tongue?

6 List the suprahyoid muscles. What is their common function?

7 The unilateral contraction of what muscle causes lateral flexion of the neck and rotation of the head to the opposite side?

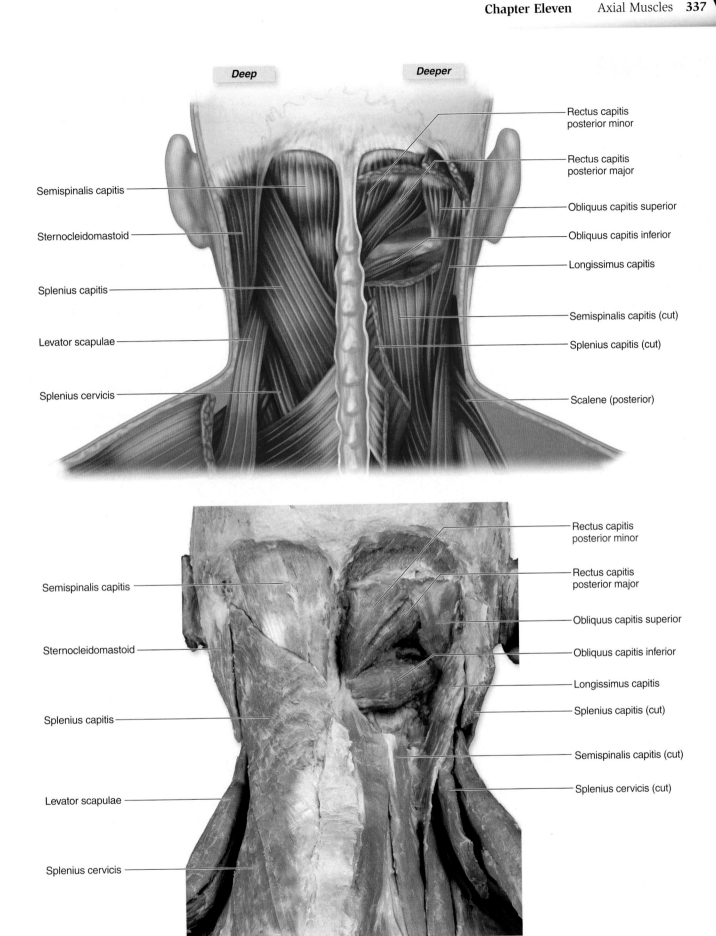

Deep   **Deeper**

Semispinalis capitis

Sternocleidomastoid

Splenius capitis

Levator scapulae

Splenius cervicis

Rectus capitis
posterior minor

Rectus capitis
posterior major

Obliquus capitis superior

Obliquus capitis inferior

Longissimus capitis

Semispinalis capitis (cut)

Splenius capitis (cut)

Scalene (posterior)

Semispinalis capitis

Sternocleidomastoid

Splenius capitis

Levator scapulae

Splenius cervicis

Rectus capitis
posterior minor

Rectus capitis
posterior major

Obliquus capitis superior

Obliquus capitis inferior

Longissimus capitis

Splenius capitis (cut)

Semispinalis capitis (cut)

Splenius cervicis (cut)

Posterior view

Figure 11.10

Posterior Neck Muscles.   An illustration and a cadaver photo show the deep and deeper muscles that extend and rotate the head and neck.

Table 11.7	Muscles That Move the Head and Neck		
Muscle	**Action**	**Origin/Insertion**	**Innervation**
Sternocleidomastoid (ster′nō-klī′dō-măs′toyd) *sterno* = sternum *cleido* = clavicle *masto* = mastoid process	Unilateral action[1]: Lateral flexion, rotation of head to opposite side Bilateral action[2]: Flexes neck	O: Manubrium and sternal end of clavicle I: Mastoid process	CN XI (accessory nerve)
Scalene muscles (anterior, middle, posterior) (see also table 11.9) (skā′lēnz) *scalene* = uneven	Flex neck (when 1st rib is fixed); elevate 1st and 2nd ribs during forced inhalation when neck is fixed	O: Transverse processes of cervical vertebrae I: Superior surface of 1st and 2nd ribs	Cervical spinal nerves
Splenius capitis and **cervicis** (splē′nē-ŭs ka′pĭ-tis) (ser′vi-sis) *splenion* = bandage	Unilateral action: Turns head to same side Bilateral action: Extends head/neck	O: Ligamentum nuchae I: Occipital bone and mastoid process of temporal bone	Cervical spinal nerves
Longissimus capitis (lon-jis′i-mŭs ka′pĭ-tis) *longissimus* = longest *caput* = head	Unilateral action: Turns (rotates) head toward same side Bilateral action: Extends head/neck	O: Transverse process of T_1–T_4 and articular processes of C_4–C_7 vertebrae I: Mastoid process	Cervical and thoracic spinal nerves
Obliquus capitis superior (ob-lī′kŭs ka′pĭ-tis soo-pēr′ē-ŏr)	Turns head to same side	O: Transverse process of atlas I: Inferior nuchal line	Suboccipital nerve (posterior ramus of 1st cervical spinal nerve)
Obliquus capitis inferior (ob-lī′kŭs ka′pĭ-tis in-fē′rē-ŏr) *obliquus* = slanting	Turns head to same side	O: Spinous process of axis I: Transverse process of atlas	Suboccipital nerve (posterior ramus of 1st cervical spinal nerve)
Rectus capitis posterior major (rek′tŭs ka′pĭ-tis pos-tēr′ē-ŏr)	Extends head/neck	O: Spinous process of axis I: Inferior nuchal line of occipital bone	Suboccipital nerve (posterior ramus of 1st cervical spinal nerve)
Rectus capitis posterior minor	Extends head/neck	O: Posterior tubercle of atlas I: Inferior nuchal line of occipital bone	Suboccipital nerve (posterior ramus of 1st cervical spinal nerve)

[1] *Unilateral action* means only one muscle (either the left or right muscle) is contracting.

[2] *Bilateral action* means both the left and right muscles are contracting together.

Muscles of the Vertebral Column

Key topic in this section:

- Muscles involved in the movements of the vertebral column

The muscles of the vertebral column are very complex; they have multiple origins and insertions, and they exhibit extensive overlap (**figure 11.11**). All of these muscles are covered by the most superficial back muscles, which actually move the upper limb, including the trapezius and the latissimus dorsi.

Note that the "neck" is actually the cervical portion of the vertebral column. Thus, the muscles discussed previously in connection with neck extension (splenius cervicis, splenius capitis, longissimus capitis, semispinalis capitis) extend the *cervical portion* of the vertebral column.

The **erector spinae** function to maintain posture and to help an individual stand erect. When the left and right erector spinae muscles contract together, they extend the vertebral column. If the erector spinae muscles on only one side contract, the vertebral column flexes laterally toward that same side.

The erector spinae muscles are organized into three groups; a series of multipart, overlapping muscles compose each of these groups. These muscles share a common tendinous insertion from the posterior part of the iliac crest, posterior sacrum, and spinous processes of the lumbar vertebrae. The muscles are named based on the body region with which they are associated.

- The **iliocostalis group** is the most laterally placed of the three erector spinae components. It is composed of three parts: cervical, thoracic, and lumbar.
- The **longissimus group** is medial to the iliocostalis group. The longissimus muscle group inserts on the transverse processes of the vertebrae. The longissimus group is composed of three parts: capitis, cervical, and thoracic.
- The **spinalis group** is the most medially placed of the erector spinae muscles. The spinalis muscle fibers insert on the spinous processes of the vertebrae (hence, the name of this muscle group). The spinalis group is composed of cervical and thoracic parts. The cervical part originates from the C_7 spinous process.

Deep to the erector spinae, a group of muscles collectively called the **transversospinalis** muscles connect and stabilize the vertebrae (**figure 11.12**). There are several specific muscles in this group (**table 11.8**). In addition, minor deep back muscles called **interspinales** and **intertransversarii** assist the transversospinalis muscles with moving the vertebral column.

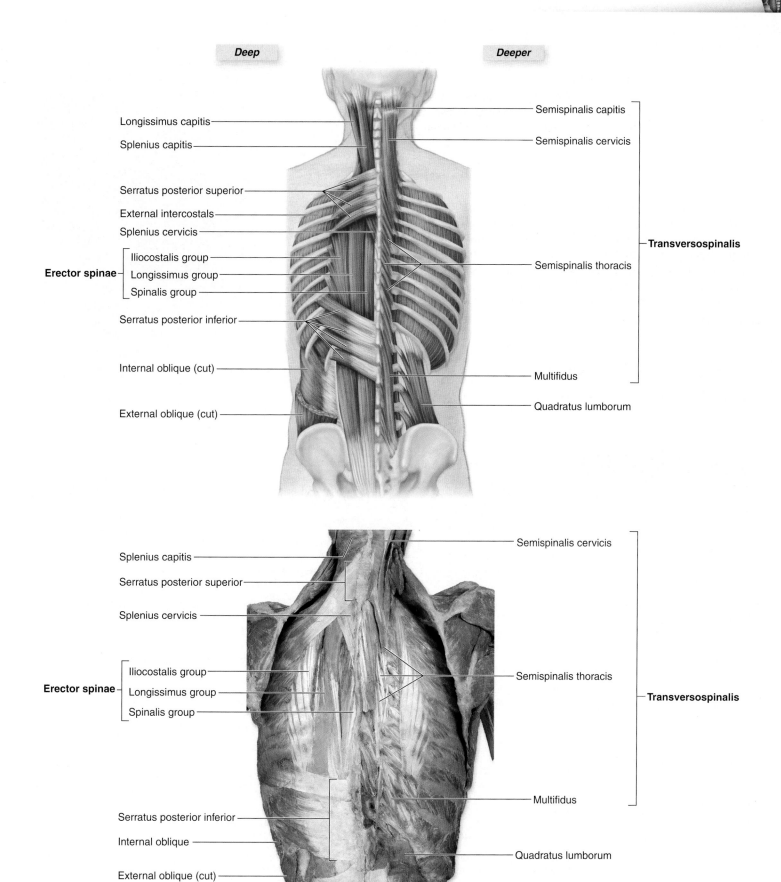

Deep

Deeper

Longissimus capitis

Splenius capitis

Serratus posterior superior

External intercostals

Splenius cervicis

Erector spinae
- Iliocostalis group
- Longissimus group
- Spinalis group

Serratus posterior inferior

Internal oblique (cut)

External oblique (cut)

Semispinalis capitis

Semispinalis cervicis

Transversospinalis

Semispinalis thoracis

Multifidus

Quadratus lumborum

Splenius capitis

Serratus posterior superior

Splenius cervicis

Erector spinae
- Iliocostalis group
- Longissimus group
- Spinalis group

Serratus posterior inferior

Internal oblique

External oblique (cut)

Semispinalis cervicis

Semispinalis thoracis

Transversospinalis

Multifidus

Quadratus lumborum

Posterior view

Figure 11.11

Deep Muscles of the Vertebral Column.   An illustration and a cadaver photo show the muscles that affect, modify, and stabilize the positions of the vertebral column, neck, and ribs.

Figure 11.12

Transversospinalis Muscles and Minor Deep Back Muscles. These muscles affect and modify the positions of the vertebral column.

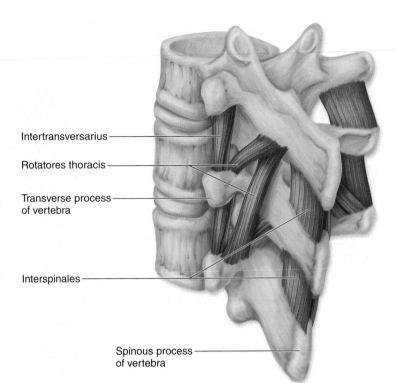

Intertransversarius

Rotatores thoracis

Transverse process of vertebra

Interspinales

Spinous process of vertebra

A final pair of muscles helps move the vertebral column. The **quadratus lumborum** muscles are located primarily in the lumbar region (see figure 11.11). When the left and right quadratus lumborum muscles bilaterally contract, they extend the vertebral column. When either the left or right quadratus lumborum muscle unilaterally contracts, it laterally flexes the vertebral column.

Table 11.8 summarizes the characteristics of the muscles of the vertebral column.

WHAT DID YOU LEARN?

8 Describe the erector spinae, and briefly discuss their function.

Table 11.8	Muscles of the Vertebral Column		
Group/Muscle	**Action**	**Origin/Insertion**	**Innervation**
ERECTOR SPINAE			
Iliocostalis group (il-ē-ō-kos-tā′lis) *ilio* = ilium *cost* = rib	Extends neck and vertebral column; maintains posture	O: Tendon from posterior part of iliac crest, posterior sacrum, and lumbar spinous processes I: Angles of ribs; transverse processes of cervical vertebrae	Cervical, thoracic, and lumbar spinal nerves
Longissimus group (lon-jis′i-mŭs) *longissimus* = longest	Extends neck and vertebral column and rotates head; maintains posture	O: Tendon from posterior part of iliac crest, posterior sacrum, and lumbar spinous processes I: Mastoid process of temporal bone and transverse processes of cervical and thoracic vertebrae	Cervical and thoracic spinal nerves
Spinalis group (spī-nā′lis) *spin* = spine	Extends neck and vertebral column; maintains posture	O: Lumbar spinous processes (thoracic part) and C_7 spinous process (cervical part) I: Spinous process of axis and thoracic vertebrae	Cervical and thoracic spinal nerves
TRANSVERSOSPINALIS GROUP			
Multifidus (mul-tif′i-dŭs) *multus* = much *findo* = to cleave	Extends vertebral column; rotates vertebral column toward opposite side	O: Sacrum and transverse processes of each vertebra I: Spinous processes of vertebrae located 2–4 segments superior to origin	Cervical, thoracic, and lumbar spinal nerves
Rotatores (rō-tā′tōrz) *rotatus* = to revolve	Extends vertebral column; rotates vertebral column toward opposite side	O: Transverse processes of each vertebra I: Spinous process of immediately superior vertebra	Cervical, thoracic, and lumbar spinal nerves
Semispinalis group (sem′ē-spī-nā′lis)	Bilateral action: Extends vertebral column/neck Unilateral action: Laterally flexes vertebral column/neck	O: Transverse processes of C_4–T_{12} vertebrae I: Occipital bone and spinous processes of cervical and thoracic vertebrae	Cervical and thoracic spinal nerves

Table 11.8	Muscles of the Vertebral Column *(continued)*		
Group/Muscle	**Action**	**Origin/Insertion**	**Innervation**
MINOR DEEP BACK MUSCLES			
Interspinales (in-ter-spī-nā′lēz) *inter* = between	Extends vertebral column; rotates vertebral column to opposite side	O: Spinous processes of each vertebra I: Spinous processes of more superior vertebrae	Cervical, thoracic, and lumbar spinal nerves
Intertransversarii (in-ter-trans′ver-sǎr′ē-ī)	Lateral flexion of vertebral column	O: Transverse processes of each vertebra I: Transverse process of more superior vertebrae	Cervical, thoracic, and lumbar spinal nerves
SPINAL EXTENSORS AND LATERAL FLEXORS			
Quadratus lumborum (kwah-drā′tūs lŭm-bōr′ŭm) *quad* = four-sided *lumb* = lumbar region	Bilateral action: Extends vertebral column Unilateral action: Laterally flexes vertebral column	O: Iliac crest and iliolumbar ligament I: Last rib; transverse processes of lumbar vertebrae	Thoracic and lumbar spinal nerves

Muscles of Respiration

Key topic in this section:

■ Muscles of respiration and their functions

The process of respiration involves **inhalation** and **exhalation.** When an individual inhales, several muscles contract to increase the dimensions of the thoracic cavity to allow the lungs to fill with air. When an individual exhales, some respiratory muscles contract and others relax, collectively decreasing the dimensions of the thoracic cavity and forcing air out of the lungs.

The muscles of respiration are on the anterior and posterior surfaces of the thorax. These muscles are covered by more superficial muscles (such as the pectoral muscles, trapezius, and latissimus dorsi) that move the upper limb. Two posterior thorax muscles assist with respiration. These muscles are located deep to the trapezius and latissimus dorsi, but superficial to the erector spinae muscles. The **serratus posterior superior** attaches to ribs 2–5 (see figure 11.11) and elevates these ribs during inspiration, thereby increasing the lateral dimensions of the thoracic cavity. The **serratus posterior inferior** attaches to ribs 8–12 and depresses those ribs during expiration.

Several groups of anterior thorax muscles change the dimensions of the thorax during respiration **(figure 11.13)**. The scalene muscles (discussed previously with other neck muscles) help elevate the first and second ribs during forced inspiration, thereby increasing the dimensions of the thoracic cavity.

The **external intercostals** extend inferomedially from the superior rib to the adjacent inferior rib. The external intercostals assist in expanding the thoracic cavity by elevating the ribs during inhalation. This movement is like lifting a bucket handle—that is, as the bucket handle (rib) is elevated, its distance from the center of the

CLINICAL VIEW

Paralysis of the Diaphragm

Injury to critical parts of the brain, spinal cord, or phrenic nerves can result in the loss of diaphragmatic innervation and cause paralysis. When the diaphragm becomes paralyzed, it cannot contract, and thus no air is exchanged in the lungs. The patient cannot breathe, and death is inevitable unless artificial breathing measures are implemented.

The most common cause of diaphragmatic paralysis today is spinal cord injury at or superior to the fourth cervical vertebra, where the motor neurons that innervate the diaphragm are located. In years past, infection with the polio virus was a common cause of diaphragmatic paralysis, as the viral infection destroyed brain and spinal cord motor neurons. Prior to the development of the modern ventilator (a device to help the patient breathe) and other respiratory assistance techniques, the polio patient with diaphragmatic paralysis was placed in a device known as an *iron lung*. This was a chamber in which the air pressure surrounding the patient was cyclically decreased to facilitate inhalation and then increased to facilitate exhalation. Lying prone,

with only the head extending from this large, tubular apparatus, a patient spent weeks or even months while the iron lung supplied the needed respiratory assistance.

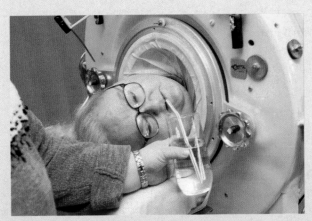

A patient in an iron lung due to diaphragm paralysis.

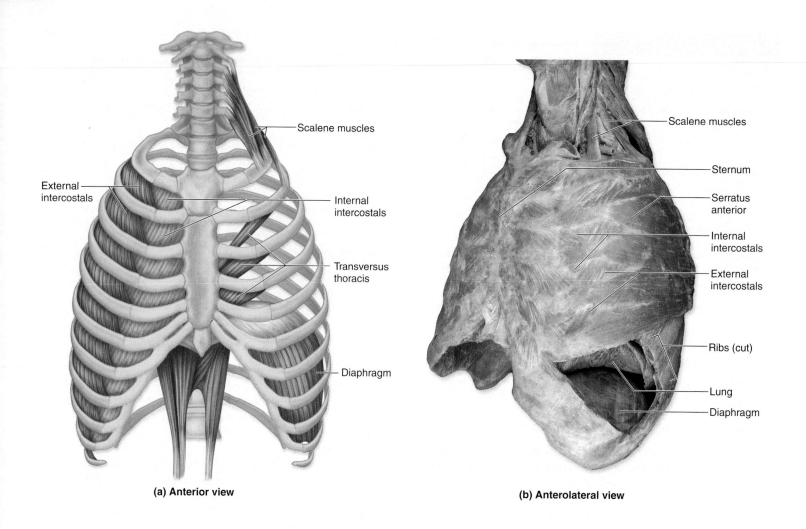

Scalene muscles

External intercostals

Internal intercostals

Transversus thoracis

Diaphragm

(a) Anterior view

Scalene muscles

Sternum

Serratus anterior

Internal intercostals

External intercostals

Ribs (cut)

Lung

Diaphragm

(b) Anterolateral view

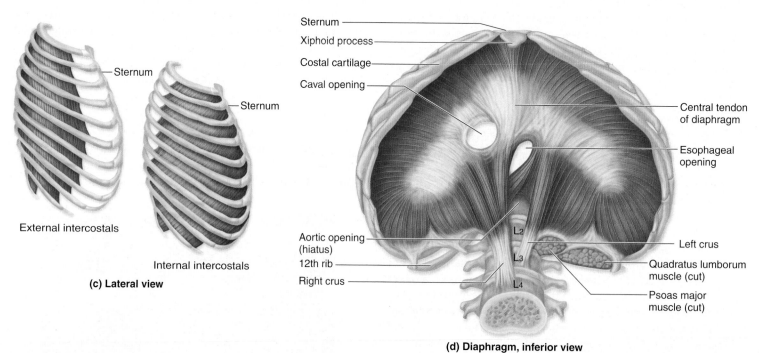

Sternum

Sternum

External intercostals

Internal intercostals

(c) Lateral view

Sternum

Xiphoid process

Costal cartilage

Caval opening

Aortic opening (hiatus)

12th rib

Right crus

L2

L3

L4

Central tendon of diaphragm

Esophageal opening

Left crus

Quadratus lumborum muscle (cut)

Psoas major muscle (cut)

(d) Diaphragm, inferior view

Figure 11.13

Muscles of Respiration.   These skeletal muscles contract rhythmically to alter the size of the thoracic cavity and facilitate respiration. *(a)* Anterior view. *(b)* A cadaver photo provides an anterolateral view, with the inferior ribs cut to expose the thoracic cavity and the superior surface of the diaphragm. *(c)* Lateral views demonstrate fiber directions of the external and internal intercostals. *(d)* Inferior view of the diaphragm.

Table 11.9	Muscles of Respiration			
Muscle	**Action**	**Description**	**Origin/Insertion**	**Innervation**
Serratus posterior superior (sĕr-ā′tŭs pos-tēr-ē-ōr soo-pēr′ē-ōr) *serratus* = a saw	Elevates ribs during inhalation	Thin, superior intermediate back muscle; splits into 4 separate muscle segments	O: Spinous processes of C_7–T_3 vertebrae I: Lateral borders of ribs 2–5	Thoracic spinal nerves
Serratus posterior inferior (sĕr-ā′tŭs pos-tēr′ē-ōr infē′rē-ōr)	Depresses ribs during exhalation	Thin, inferior intermediate back muscle; splits into 4 separate muscle segments	O: Spinous processes of T_{11}–L_2 vertebrae I: Inferior borders of ribs 8–12	Thoracic spinal nerves
Scalene muscles (anterior, middle, posterior) (see table 11.7 for description)				
External intercostals (eks-ter′năl in′ter-kos′talz) *inter* = between *cost* = rib	Elevates ribs during inhalation	11 pairs of oblique fibers between ribs; project anteroinferiorly	O: Inferior border of superior rib I: Superior border of inferior rib	Thoracic spinal nerves
Internal intercostals (in-ter′năl in′ter-kos′talz)	Depresses ribs during forced exhalation; antagonistic to external intercostals	11 pairs of oblique fibers between ribs; project posteroinferiorly	O: Superior border of inferior rib I: Inferior border of superior rib	Thoracic spinal nerves
Transversus thoracis (trans-ver′sŭs thō-ra′sis)	Depresses ribs during exhalation	Assist in decreasing diameter of thoracic cavity	O: Posterior surface of xiphoid process and inferior region of sternum I: Costal cartilages 2–6	Thoracic spinal nerves
Diaphragm (dī′ă-fram) *dia* = across *phragm* = partition	Contraction causes flattening of diaphragm (moves inferiorly), and thus expansion of thoracic cavity; increases pressure in abdominopelvic cavity	Dome-shaped, broad muscle; separates thoracic and abdominopelvic cavities	O: Inferior internal surface of ribs 7–12; xiphoid process of sternum and costal cartilages of inferior 6 ribs; lumbar vertebrae I: Central tendon	Phrenic nerves (C3–C5)

bucket (thorax) increases. Thus, contraction of the external intercostals increases the transverse dimensions of the thoracic cavity. The **internal intercostals** lie deep to the external intercostals, and their muscle fibers are at right angles to the external intercostals. The internal intercostals depress the ribs, but only during forced exhalation; normal exhalation takes no active muscular effort.

A small **transversus thoracis** extends across the inner surface of the thoracic cage and inserts on ribs 2–6. It helps depress the ribs.

Finally, the **diaphragm** is an internally placed, dome-shaped muscle that forms a partition between the thoracic and abdominopelvic cavities. It is the most important muscle associated with breathing. The muscle fibers of the diaphragm converge from its margins toward a fibrous **central tendon,** a strong aponeurosis that is the insertion tendon for all peripheral muscle fibers of the diaphragm. When the diaphragm contracts, the central tendon is pulled inferiorly toward the abdominopelvic cavity, thereby increasing the vertical dimensions of the thoracic cavity. As it compresses the abdominopelvic cavity, it also increases intra-abdominal pressure, an event that is necessary for urination, defecation, and childbirth. Beyond respiration, diaphragm movements are also important in helping return venous blood to the heart from the inferior half of the body.

Table 11.9 summarizes the characteristics of the muscles of respiration.

WHAT DO YOU THINK?

❸ After you've eaten a very large meal, it is sometimes difficult to take a big, deep breath. Why is it more difficult to breathe deeply with a full GI tract?

WHAT DID YOU LEARN?

❾ Compare the functions of the external intercostals and the internal intercostals.

❿ Identify the muscle of respiration that partitions the thoracic and abdominopelvic cavities. What is the name of the structure to which all fibers of this muscle converge?

Muscles of the Abdominal Wall

Key topic in this section:

- Organization and function of the muscles of the abdominal wall

The anterolateral wall of the abdomen is reinforced by four pairs of muscles that collectively compress and hold the abdominal organs in place: the external oblique, internal oblique, transversus abdominis, and rectus abdominis (**figure 11.14**). These muscles

Superficial   **Deep**

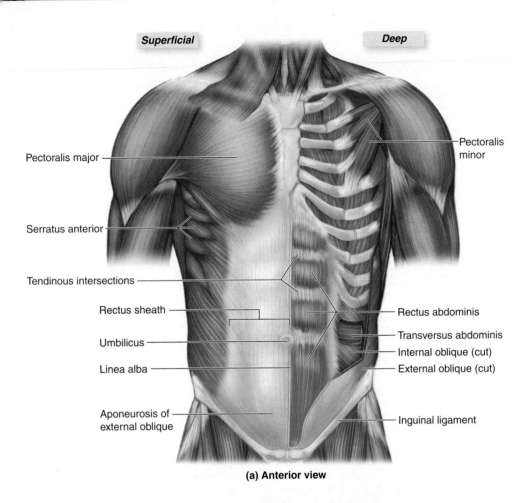

Pectoralis major

Serratus anterior

Tendinous intersections

Rectus sheath

Umbilicus

Linea alba

Aponeurosis of
external oblique

Pectoralis
minor

Rectus abdominis

Transversus abdominis

Internal oblique (cut)

External oblique (cut)

Inguinal ligament

(a) Anterior view

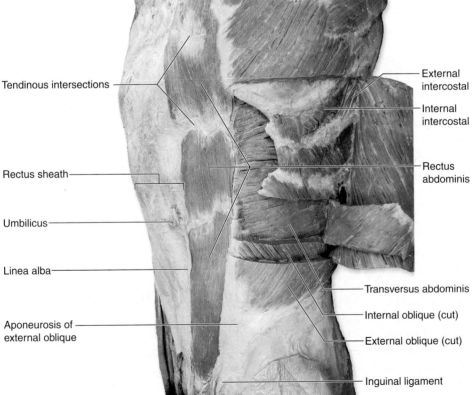

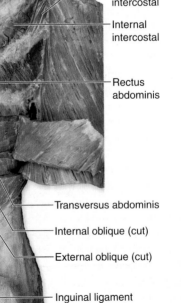

Tendinous intersections

Rectus sheath

Umbilicus

Linea alba

Aponeurosis of
external oblique

External
intercostal

Internal
intercostal

Rectus
abdominis

Transversus abdominis

Internal oblique (cut)

External oblique (cut)

Inguinal ligament

(b) Anterolateral view

External oblique

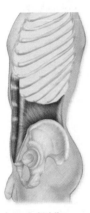

Internal oblique
and
rectus abdominis

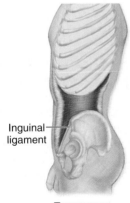

Inguinal
ligament

Transversus
abdominis

(c)

Figure 11.14

Muscles of the Abdominal Wall. The
abdominal muscles compress abdominal
contents and flex the vertebral column. (a) An
illustration provides an anterior view of some
superficial and deep muscles. (b) A cadaver
photo provides an anterolateral view of the
muscles of the abdominal wall. (c) Diagrams
show some individual abdominal muscles,
ranging from superficial to deep.

Table 11.10	Muscles of the Abdominal Wall			
Muscle	**Action**	**Description**	**Origin/Insertion**	**Innervation**
External oblique (eks-ter′năl ob-lēk′)	Unilateral action[1]: Lateral flexion of vertebral column; rotation of vertebral column to opposite side Bilateral action[2]: Flexes vertebral column and compresses abdominal wall	Large superficial muscle sheet; fibers project inferomedially; forms inguinal ligament	O: External and inferior borders of the inferior 8 ribs I: Linea alba by a broad aponeurosis; some to iliac crest	Spinal nerves T8–T12, L1
Internal oblique (in-ter′nal ob-lēk′)	Unilateral action: Lateral flexion of vertebral column; rotation of vertebral column to opposite side Bilateral action: Flexes vertebral column and compresses abdominal wall	Middle muscle sheet, deep to external oblique and superficial to transversus abdominis; fibers primarily project superomedially	O: Lumbar fascia, inguinal ligament, and iliac crest I: Linea alba, pubic crest, inferior rib surfaces (last 4 ribs); costal cartilages of ribs 8–10	Spinal nerves T8–T12, L1
Transversus abdominis (trans-ver′sŭs ab-dom′i-nis)	Unilateral action: Lateral flexion of vertebral column Bilateral action: Flexes vertebral column; compresses abdominal wall	Deepest of the 3 muscle sheets; fibers project horizontally	O: Iliac crest, cartilages of inferior 6 ribs; lumbar fascia; inguinal ligament I: Linea alba and pubic crest	Spinal nerves T8–T12, L1
Rectus abdominis (rek′tŭs ab-dom′i-nis)	Flexes vertebral column; compresses abdominal wall	Paired anterior and medial surface muscles; extend from thoracic cage to pubis; wrapped by aponeuroses of lateral muscles; segmented by three perpendicular tendinous intersections	O: Superior surface of pubis near symphysis I: Xiphoid process of sternum; inferior surfaces of ribs 5–7	Spinal nerves T7–T12

[1] *Unilateral action* means only one muscle (either the left or right muscle) is contracting.

[2] *Bilateral action* means both the left and right muscles are contracting together.

also work together to flex and stabilize the vertebral column. When these muscles unilaterally contract, they laterally flex the vertebral column.

The muscle fibers of the superficial **external oblique** are directed inferomedially. The external oblique is muscular along the lateral abdominal wall and forms an aponeurosis as it projects anteriorly. Inferiorly, the aponeurosis of the external oblique forms a strong, cordlike **inguinal ligament** that extends from the anterior superior iliac spine to the pubic tubercle. Immediately deep to the external oblique is the **internal oblique.** Its muscle fibers project superomedially, which is at right angles to the external oblique. Like the external oblique, this muscle forms an aponeurosis as it projects anteriorly. Unilaterally, the external and internal oblique also rotate the vertebral column to the opposite side of the contracting muscle.

The deepest muscle is the **transversus abdominis,** whose fibers project transversely across the abdomen. The **rectus abdominis** is a long, straplike muscle that extends vertically the entire length of the anteromedial abdominal wall between the sternum and the pubic symphysis. It is partitioned into four segments by three fibrous **tendinous intersections,** which form the traditional "six-pack" of a muscular, toned abdominal wall. The rectus abdominis is enclosed within a fibrous sleeve called the **rectus sheath,** which is formed from the aponeuroses of the external oblique, internal oblique, and transversus abdominis muscles. The left and right rectus sheaths are connected by a vertical fibrous strip termed the **linea alba.**

Table 11.10 summarizes the characteristics of the muscles of the abdominal wall.

Study Tip!

Knowing the direction of the oblique and intercostal muscle fibers can help you identify these specific muscles on models and cadavers:

■ The fibers of the external intercostals and external oblique muscles run in the same direction—inferomedially. This is the same direction that you put your hands in your pockets.

■ The fibers of the internal intercostals and internal oblique muscles run perpendicular (in the opposite direction) to the external muscles—superomedially.

We have seen that multiple muscles may work together to perform a common function. For example, several neck muscles and back muscles work together to extend the vertebral column. Learning muscles in groups according to common function helps most students assimilate the anatomy information. **Table 11.11** summarizes the actions of various axial muscles and groups them according to common function. Note that a muscle that has multiple functions is listed in more than one group.

WHAT DID YOU LEARN?

11 Identify the muscles of the abdominal wall.

Table 11.11	**Muscle Actions on the Axial Skeleton**					
Extend the Head, Neck, and/or Vertebral Column	**Flex the Head, Neck, and/or Vertebral Column**	**Laterally Flex the Vertebral Column**	**Rotate the Head and/or Neck to One Side**	**Elevate the Ribs**	**Depress the Ribs**
Splenius muscles[2]	Sternocleidomastoid[2]	Quadratus lumborum[1]	Sternocleidomastoid[1]	Serratus posterior superior	Serratus posterior inferior
Erector spinae[2] (iliocostalis, longissimus, spinalis)	Scalenes[2]	External oblique[1]	Splenius muscles[1]	External intercostals	Internal intercostals
Quadratus lumborum[2]	External oblique[2]	Internal oblique[1]	Longissimus capitis[1]	Scalene muscles (1st and 2nd ribs only)	Transversus thoracis
Transversospinalis group[2]	Internal oblique[2]	Transversus abdominis[1]	Obliquus capitis inferior[1]		
Minor deep back muscles[2]	Transversus abdominis[2]		Obliquus capitis superior[1]		
Rectus capitis posterior major and minor[2]	Rectus abdominis[2]				

[1] Unilateral action of muscles

[2] Bilateral action

Muscles of the Pelvic Floor

Key topic in this section:

- Muscles that form the pelvic floor and perineum

The floor of the pelvic cavity is formed by three layers of muscles and associated fasciae, collectively known as the **pelvic diaphragm.** (The term *diaphragm* refers to a muscle or group of muscles that covers or partitions an opening.) The pelvic diaphragm extends from the ischium and pubis of the ossa coxae across the pel-vic outlet to the sacrum and coccyx. These muscles collectively form the pelvic floor and support the pelvic viscera (**figure 11.15**).

The **coccygeus** pulls the coccyx anteriorly after its poste-rior deflection during defecation or childbirth. The **external anal sphincter** (located within a region called the anal triangle) assists in defecation. The largest and most important collection of muscles in the pelvic floor is the **levator ani.** It supports the pelvic viscera and functions as a sphincter at the anorectal junction, urethra, and vagina. The levator ani is formed by the **iliococcygeus,** the **pubo-coccygeus,** and the **puborectalis.** The puborectalis muscle forms a

---

CLINICAL VIEW: In Depth

Hernias

The condition in which a portion of the viscera protrudes through a weakened point of the muscular wall of the abdominopelvic cavity is called a **hernia** (her'nē-ă; rupture). A significant medical problem may develop if the herniated portion of the intestine swells, becoming trapped. Blood flow to the trapped segment may diminish, causing that portion of the intestine to die. This condition, called a *strangulated intestinal hernia,* is very painful and can be life-threatening if not treated promptly.

Two common types of hernias are inguinal hernias and femoral hernias. An **inguinal hernia** is the most common type of hernia to require treatment. The inguinal region is one of the weakest areas of the abdominal wall. Within this region is a canal (inguinal canal) that allows the passage of the spermatic cord in males, and a smaller structure in females called the *round ligament* of the uterus. The inguinal canal, or the **superficial inguinal ring** associated with it, is often the site of a rupture or separation of the abdominal wall. Males are more likely to develop inguinal hernias than females, because their inguinal canals and superficial inguinal rings are larger to allow room for the spermatic cord. Rising pressure in the abdominopelvic cavity, as might develop while straining to lift a heavy object, provides the force to push

a segment of the small intestine into the canal. There are two types of inguinal hernia:

- In a *direct inguinal hernia,* the loop of small intestine protrudes directly through the superficial inguinal ring, but not through the entire length of the inguinal canal, and creates a bulge in the lower anterior abdominal wall. This type of hernia is typically seen in middle-aged males with poorly developed abdominal muscles and protruding abdomens.

- In an *indirect inguinal hernia,* the herniation travels through the entire inguinal canal and may even extend all the way into the scrotum, since the path of the herniation follows the path of the spermatic cord. This type of hernia tends to occur in younger males or male children who have a congenital anomaly called *patent process vaginalis,* in which the embryonic path taken by the testis into the scrotum fails to regress.

A **femoral hernia** occurs in the upper thigh, just inferior to the inguinal ligament, originating in a region called the femoral triangle. The medial part of the femoral triangle is relatively weak and prone to stress injury, thus allowing a loop of small intestine to protrude. Women more com-monly develop femoral hernias because of the greater width of their femoral triangle, which equates to the wider hip span of the female anatomy.

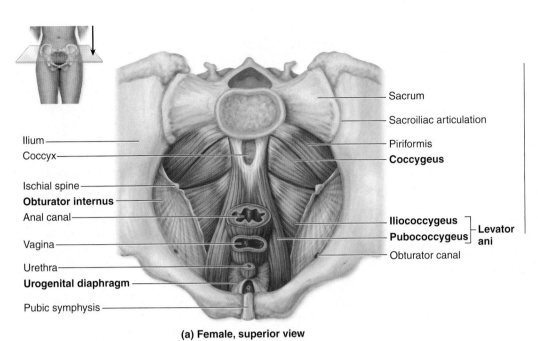

(a) **Female, superior view**

Figure 11.15

Muscles of the Pelvic Floor.   The pelvic cavity floor is composed of muscle layers that form the urogenital and anal triangles, extend across the pelvic outlet, and support the organs in the pelvic cavity (puborectalis not shown). (*a*) Superior view of the female pelvic cavity. (*b, c*) Inferior views show male and female perineal regions, respectively.

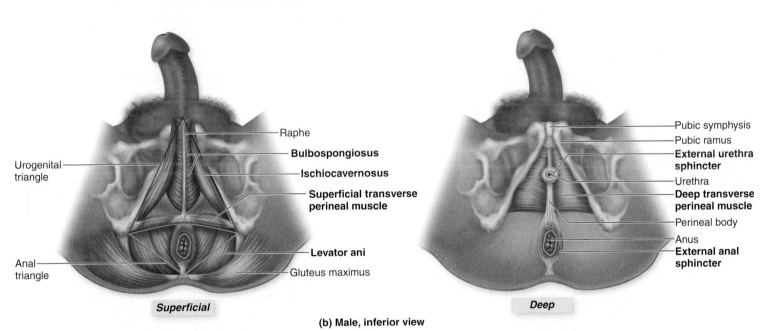

(b) **Male, inferior view**

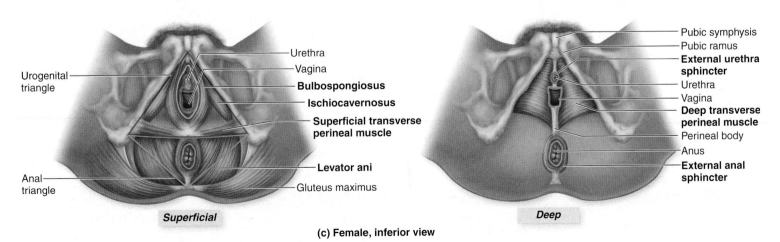

(c) **Female, inferior view**

circular "sling" around the anorectal junction. When this muscle is contracted, it increases the bend or angle of the anorectal junction. When an individual wishes to defecate, the puborectalis muscle must be consciously relaxed in order to decrease the bend of the anorectal junction and allow for the easy passage of feces through the rectum.

The diamond-shaped region between the lower appendages is called the **perineum.** The perineum has four significant bony landmarks: the pubic symphysis anteriorly, the coccyx posteriorly, and both ischial tuberosities laterally. A transverse line drawn between the ischial tuberosities partitions the perineum into an anterior **urogenital triangle** that contains the external genitalia and urethra, and a posterior **anal triangle** that contains the anus (figure 11.15b,c).

The urogenital triangle is subdivided into a superficial layer and a deep layer. The **superficial layer** of the urogenital triangle is composed of three muscles: **bulbospongiosus, ischiocavernosus,** and **superficial transverse perineal (table 11.12).** The ischiocavernosus inserts on the pubic symphysis, while the bulbospongiosus and superficial transverse perineal muscles insert on a tendinous central structure called the **perineal body** (or *central tendon of the perineum*). The **deep layer** of the urogenital triangle is composed of two muscles: the **deep transverse perineal** and the **external urethral sphincter.** These muscles are collectively referred to as the **urogenital diaphragm,** because they serve as a partition for the urogenital portion of the pelvic floor.

Table 11.12 summarizes the characteristics of the muscles of the pelvic floor.

CLINICAL VIEW

Episiotomy

An **episiotomy** (e-piz-ē-ot'ō-mē; *epicios* = vulvar, *otomy* = to cut) is a surgical incision made in the perineal skin and soft tissues between the vagina and the anus during childbirth to prevent tearing of the mother's tissues and to minimize fetal injury. It has long been believed that a clean surgical incision heals more rapidly and effectively than a laceration or tear.

The episiotomy has been a very common medical procedure in the United States, but lately its routine use has come into question. Large studies have indicated that serious perineal lacerations may be more common in women who have had an episiotomy. Furthermore, in some European countries where the procedure is rarely done, no significant increase in serious maternal or fetal complications has been seen. Because of general public concern, the medical community has begun to examine the unwarranted and routine use of the episiotomy. Medical studies now indicate that, at most, 25–30% of women would benefit from the procedure, and that the selective use of episiotomy may be a better health-care option. Women approaching childbirth are advised to talk in advance with their health-care provider about the necessity of having an episiotomy.

Table 11.12	Muscles of the Pelvic Floor		
Group/Muscle	**Action**	**Origin/Insertion**	**Innervation**
ANAL TRIANGLE			
Coccygeus (kok-si'jē-ŭs) *coccy* = coccyx	Forms pelvic floor and supports pelvic viscera	O: Ischial spine I: Lateral and inferior borders of sacrum	Spinal nerves S4–S5
External anal sphincter (eks-ter'năl ă'năl sfingk'ter) *anal* = referring to anus *sphin* = squeeze	Closes anal opening; must relax in order to defecate	O: Perineal body I: Encircles anal opening	Pudendal nerve (S2–S4)
Levator ani (lē-vă'tor, le-vă'-ter ă'nī) *levator* = raises *ani* = anus		Group of muscles that form the anterior and lateral parts of the pelvic diaphragm	
Iliococcygeus (il'ē-ō-kok-si'jē-ŭs) *ilio* = ilium	Forms pelvic floor and supports pelvic viscera	O: Pubis and ischial spine I: Coccyx and median raphe	Pudendal nerve (S2–S4)
Pubococcygeus (pū'bō-kok-si'jē-ŭs) *pubes* = pubis	Forms pelvic floor and supports pelvic viscera	O: Pubis and ischial spine I: Coccyx and median raphe	Pudendal nerve (S2–S4)
Puborectalis (pū-bō-rek'tăl-is) *rectal* = rectum	Supports anorectal junction; must relax in order to defecate	O: Pubis and ischial spine I: Coccyx and median raphe	Pudendal nerve (S2–S4)
UROGENITAL TRIANGLE			
SUPERFICIAL LAYER			
Bulbospongiosus (female) (bul'bō-spŭn'jē-ō'sŭs) *bulbon* = bulb *spongio* = sponge	Narrows vaginal opening; compresses and stiffens clitoris	O: Sheath of collagen fibers at base of clitoris I: Perineal body	Pudendal nerve (S2–S4)
Bulbospongiosus (male)	Ejects urine or semen; compresses base of penis; stiffens penis	O: Sheath of collagen fibers at base of penis I: Median raphe and perineal body	Pudendal nerve (S2–S4)

Table 11.12	Muscles of the Pelvic Floor *(continued)*		
Group/Muscle	**Action**	**Origin/Insertion**	**Innervation**
UROGENITAL TRIANGLE			
SUPERFICIAL LAYER			
Ischiocavernosus (ish′ē-ō-kav′er-nō′sŭs) *ischi* = hip *caverna* = hollow chamber	Assists erection of penis or clitoris	O: Ischial tuberosities and ischial ramus I: Pubic symphysis	Pudendal nerve (S2–S4)
Superficial transverse perineal muscle (soo-per-fish′ăl trans-vers′ per-i-nē′-ăl)	Supports pelvic organs	O: Ramus of ischium I: Perineal body	Pudendal nerve (S2–S4)
DEEP LAYER (UROGENITAL DIAPHRAGM)			
Deep transverse perineal muscle (dēp trans-vers′ per-i-nē′ăl)	Supports pelvic organs	O: Ischial ramus I: Median raphe of urogenital diaphragm	Pudendal nerve (S2–S4)
External urethral sphincter (eks-ter′năl ū-rē′thrăl sfingk′ter) *sphin* = squeeze	Constricts urethra to voluntarily inhibit urination	O: Rami of ischium and pubis I: Median raphe of urogenital diaphragm	Pudendal nerve (S2–S4)

CLINICAL TERMS

linea nigra Condition sometimes seen in pregnant females. The linea alba of the rectus sheath darkens (hence the term "nigra"), forming a line that extends along the midline of the abdomen.

rectus sheath separation Separation of the left and right rectus muscles/sheaths due to great expansion of the abdomen (as occurs in some pregnancies). If severe, the separation may have to be repaired surgically.

CHAPTER SUMMARY

	■ The axial muscles attach to components of the axial skeleton, whereas the appendicular muscles stabilize or move components of the appendicular skeleton.
	■ Axial musculature functions include supporting and positioning the head, vertebral column, and thoracic cage; controlling movements associated with respiration; and forming part of the floor of the pelvic cavity.
Muscles of the Head and Neck 321	■ Muscles of the head and neck are separated into groups based on their specific activities.
	Muscles of Facial Expression 321
	■ The muscles of facial expression arise from the skull and often attach to the skin.
	Extrinsic Eye Muscles 328
	■ The six extrinsic eye muscles attach to the eye and control the movements and position of the eyes.
	Muscles of Mastication 330
	■ The muscles of mastication move the mandible during chewing.
	Muscles That Move the Tongue 330
	■ The muscles of the tongue are divided into intrinsic muscles, which function during chewing and speaking, and extrinsic muscles, which function during food manipulation, swallowing, and some speech-related activities.
	Muscles of the Pharynx 332
	■ Muscles of the pharynx function in swallowing.
	Muscles of the Anterior Neck 333
	■ Anterior neck muscles are the suprahyoid muscles, superior to the hyoid bone, and the infrahyoid muscles, inferior to the hyoid bone. These move the hyoid bone or thyroid cartilage during swallowing or speaking.
	Muscles That Move the Head and Neck 335
	■ Muscles that move the head and neck originate on the vertebral column, the thoracic cage, and the pectoral girdle, and insert on bones of the cranium.
Muscles of the Vertebral Column 338	■ Deep back muscles extend the vertebral column.

(continued on next page)

CHAPTER SUMMARY (continued)

Muscles of Respiration 341	■ The muscles of respiration are located on the anterior and posterior surfaces of the thorax and are covered by more superficial muscles. Their contraction either increases or decreases the size of the thoracic cavity.
Muscles of the Abdominal Wall 343	■ The muscles of the abdominal wall compress the abdomen, help hold the abdominal organs in place, and assist the stabilization and lateral flexion of the vertebral column.
Muscles of the Pelvic Floor 346	■ The muscles of the pelvic floor extend from the pubis and ischium anteriorly to the sacrum and coccyx posteriorly. They support the pelvic cavity organs and control the evacuation of waste materials from the digestive and urinary systems.

CHALLENGE YOURSELF

Matching

Match each numbered item with the most closely related lettered item.

_____ 1. platysma

_____ 2. buccinator

_____ 3. lateral rectus

_____ 4. temporalis

_____ 5. levator ani

_____ 6. digastric

_____ 7. external intercostal

_____ 8. styloglossus

_____ 9. zygomaticus major

_____ 10. spinalis group

a. moves eye laterally

b. elevates and retracts tongue

c. elevates and retracts mandible

d. tenses skin of neck

e. extends vertebral column

f. elevates angles of mouth

g. compresses cheeks

h. supports pelvic floor and viscera

i. depresses mandible

j. elevates ribs

Multiple Choice

Select the best answer from the four choices provided.

_____ 1. The geniohyoid muscle
a. depresses the hyoid bone and larynx.
b. elevates the floor of the mouth.
c. elevates the hyoid bone.
d. depresses the larynx.

_____ 2. When the left and right _____ contract, they flex the neck.
a. sternocleidomastoid
b. longissimus group
c. splenius
d. rectus abdominis

_____ 3. When this large muscle contracts, the vertical dimensions of the thoracic cavity increase.
a. external intercostal
b. internal intercostal
c. diaphragm
d. transversus thoracis

_____ 4. Which of the following is *not* a muscle within the urogenital triangle?
a. bulbospongiosus
b. coccygeus
c. superficial transverse perineal
d. ischiocavernosus

_____ 5. The muscle that does *not* cause some lateral movement in the eye is the
a. inferior rectus.
b. inferior oblique.
c. lateral rectus.
d. superior oblique.

_____ 6. Which muscle allows you to stick out your tongue?
a. palatoglossus
b. genioglossus
c. lateral pterygoid
d. hyoglossus

_____ 7. Each of these muscles can *laterally* flex the vertebral column, except the
a. external oblique.
b. transversus abdominis.
c. spinalis.
d. internal oblique.

_____ 8. Which muscle is *not* involved in extending the head or neck?
a. rectus capitis posterior major
b. longissimus capitis
c. sternocleidomastoid
d. splenius cervicis

_____ 9. One function of the transversus abdominis muscle is to
a. elevate the ribs.
b. compress the abdominal wall.
c. extend the vertebral column.
d. increase the dimensions of the thoracic cavity.

_____ 10. Which muscle protrudes the lower lip (as when you "pout")?
a. risorius
b. levator labii superioris
c. mentalis
d. zygomaticus major

Content Review

1. Describe which muscles of facial expression you use to (a) smile and (b) frown.

2. Compare and contrast the functions of the extrinsic muscles of the tongue.

3. Discuss why the eye moves slightly medially during the contraction of either the superior or inferior rectus muscle.

4. Discuss the effect of contracting the three pharyngeal constrictors during swallowing.

5. Distinguish between suprahyoid and infrahyoid muscles, and describe the functions of each group.

6. Describe the differences in action between bilateral and unilateral contraction of the splenius muscles.

7. Describe the functions of the thoracic diaphragm.

8. What is the effect of contracting the abdominal oblique muscles?

9. What structures form the rectus sheaths, and how do the left and right rectus sheaths relate to the linea alba?

10. What are the general functions of the pelvic diaphragm muscles, and what specific muscles form the pelvic diaphragm?

Developing Critical Reasoning

1. Albon is a 45-year-old male who characterizes himself as a "couch potato." He exercises infrequently and has a rounded abdomen ("beer belly"). While helping a friend move some heavy furniture, he felt a sharp pain deep within his abdominopelvic cavity. An emergency room resident told Albon that he had suffered an inguinal hernia. What is this injury, how did it occur, and how might Albon's poorly developed abdominal musculature have contributed to it?

2. While training on the balance beam, Pat slipped during her landing from a back flip and fell, straddling the beam. Although only slightly sore from the fall, she became concerned when she suddenly lost the ability to completely control her urination. What might have happened to Pat's pelvic floor structures during the fall?

ANSWERS TO "WHAT DO YOU THINK?"

1. The zygomaticus major, zygomaticus minor, levator anguli oris, risorius, and levator labii superioris all contract when you smile.

2. Since the omohyoid attaches to the scapula, the prefix *omo-* means "shoulder."

3. When you breathe deeply, the diaphragm contracts and pushes down on the GI tract (abdominal viscera). If these viscera are bulging with food, the diaphragm has difficulty contracting fully, making it hard to take deep breaths.

Visit the McKinley/O'Loughlin *Human Anatomy*, 2e website at aris.mhhe.com

12

MUSCULAR SYSTEM

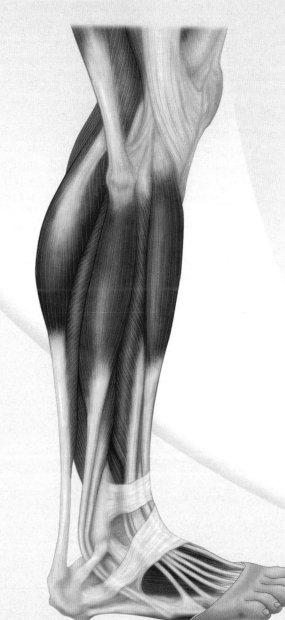

Appendicular Muscles

he appendicular muscles control the movements of the upper and lower limbs, and stabilize and control the movements of the pectoral and pelvic girdles. These muscles are organized into groups based on their location in the body or the part of the skeleton they move. Beyond their individual activities, these muscles also work in groups that are either synergistic or antagonistic. Refer to figure 10.14 to review how muscles are named, and recall the first Study Tip! from chapter 11 that gives suggestions for learning the muscles.

Muscles That Move the Pectoral Girdle and Upper Limb

Key topics in this section:

- Major movements of the pectoral girdle and upper limb, and the muscles involved
- Muscles that move the scapula and their actions
- Muscles of the glenohumeral joint and how each affects the movement of the humerus

- Muscles that move the elbow joint
- Muscles of the forearm, wrist joint, fingers, and thumb

Muscles that move the pectoral girdle and upper limbs are organized into specific groups: (1) muscles that move the pectoral girdle, (2) muscles that move the glenohumeral joint/arm, (3) arm and forearm muscles that move the elbow joint/forearm, (4) forearm muscles that move the wrist joint, hand, and fingers, and (5) intrinsic muscles of the hand. Some of these muscles are superficial, and others are deep.

Muscles That Move the Pectoral Girdle

The muscles of the pectoral girdle originate on the axial skeleton and insert on the scapula and clavicle **(figures 12.1** and **12.2)**. These muscles both stabilize the scapula and move it to increase the arm's angle of movements. Some of the superficial muscles of the thorax are grouped together according to the scapular movement they direct: elevation, depression, protraction, or retraction **(figure 12.3)**. The muscles that move the pectoral girdle are classified according to their location in the thorax as either anterior or posterior thoracic muscles.

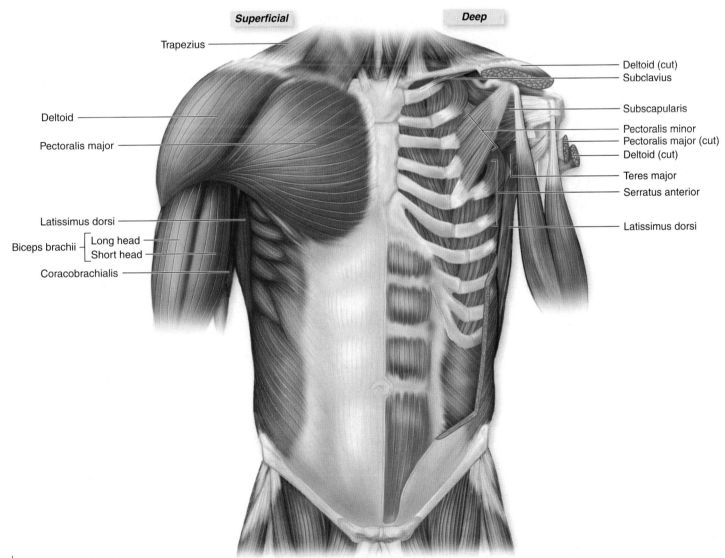

Anterior view

Figure 12.1

Anterior Muscles Associated with the Proximal Upper Limb. This anterior view compares some components of both the axial and appendicular musculature. Only those muscles that move the upper limb are labeled. Superficial muscles are shown on the right side of the body, and deep muscles are shown on the left side.

Study Tip!

When studying appendicular muscle function, remember these two basic rules:

1. If a muscle crosses over or spans a joint, it must move that joint. For example, since the biceps brachii crosses over the elbow joint, it must move the elbow joint.

2. Conversely, if a muscle doesn't cross over or span a joint, it cannot move that joint. For example, the deltoid is found in the shoulder, and it does not cross over the wrist joint. Therefore, there is no possible way the deltoid can move the wrist joint!

If you can visualize where a muscle is located in your body, you can usually figure out what type of movement the muscle performs.

The anterior thoracic muscles are the pectoralis minor, serratus anterior, and subclavius (shown in **figure 12.4a**). The **pectoralis minor** is a thin, flat, triangular muscle deep to the pectoralis major. The muscle helps depress and protract (pull anteriorly) the scapula. When your shoulders are hunched forward, the pectoralis minor muscle is contracting. The **serratus anterior** is a large, flat, fan-shaped muscle positioned between the ribs and the scapula. Its name is derived from the saw-toothed (serrated) appearance of its origins on the ribs. This muscle is the prime mover (agonist) in scapula protraction, and thus works with the pectoralis minor. It is also the primary muscle that helps stabilize the scapula against the posterior side of the rib cage and is a powerful superior rotator of the scapula by moving the glenoid cavity superiorly, as occurs when you abduct the upper limb. The **subclavius** is a small, cylindrical muscle named for its location inferior to the clavicle. It extends from the first rib to the clavicle, and its main action is to stabilize and depress the clavicle.

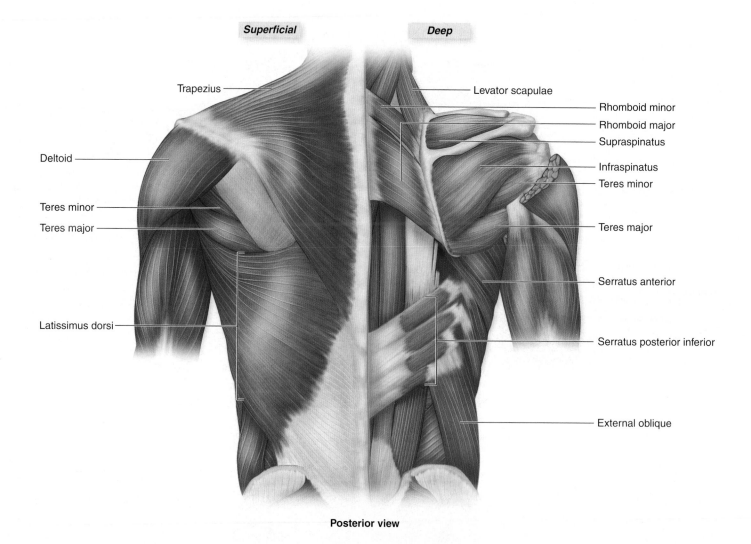

Posterior view

Figure 12.2

Posterior Muscles Associated with the Proximal Upper Limb.   This posterior view compares some components of both the axial and appendicular musculature. Only those muscles that move the upper limb are labeled. Superficial muscles are shown on the left, and deep muscles are shown on the right.

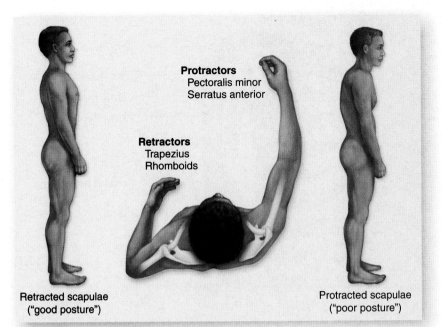

Protractors
Pectoralis minor
Serratus anterior

Retractors
Trapezius
Rhomboids

Retracted scapulae
("good posture")

Protracted scapulae
("poor posture")

(a) Retraction and protraction of scapula

Figure 12.3

Actions of Some Thoracic Muscles on the Scapula.
Individual muscles may contribute to different, multiple
actions. *(a)* The scapula can be retracted or protracted. When
you are standing upright and have good posture, your scapulae
are retracted. Conversely, poor posture demonstrates scapular
protraction. *(b)* Muscles that elevate and depress the scapula.
(c) Muscles that rotate the scapula.

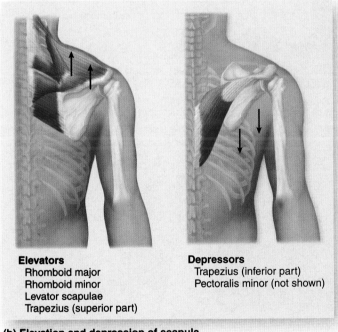

Elevators
Rhomboid major
Rhomboid minor
Levator scapulae
Trapezius (superior part)

Depressors
Trapezius (inferior part)
Pectoralis minor (not shown)

(b) Elevation and depression of scapula

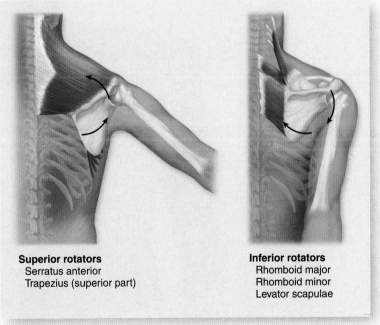

Superior rotators
Serratus anterior
Trapezius (superior part)

Inferior rotators
Rhomboid major
Rhomboid minor
Levator scapulae

(c) Superior and inferior rotation of scapula

The posterior thoracic muscles are the levator scapulae, rhomboid major, rhomboid minor, and trapezius (shown in figure 12.4*b*).
The **levator scapulae** is a narrow, elongated muscle that is deep to
both the trapezius and sternocleidomastoid muscles. It originates
from multiple heads on the transverse processes of the cervical vertebrae and inserts on the superior angle of the scapula. As its name
implies, its primary action is to elevate the scapula. It can also inferiorly rotate the scapula so that the glenoid cavity points inferiorly.

Both the **rhomboid major** and the **rhomboid minor** are located
deep to the trapezius. These rhomboid muscles are parallel bands
that run inferolaterally from the vertebrae to the scapula. Often they
are indistinct from each other. They help elevate and retract (adduct)

the shoulder, as when you are standing up straight. The rhomboid
muscles also inferiorly rotate the scapula.

The **trapezius** is a large, flat, diamond-shaped muscle that
extends from the skull and vertebral column to the pectoral girdle
laterally. In general, the trapezius can elevate, depress, retract, or
rotate the scapula, depending upon which fibers of the muscle are
actively contracting. The superior fibers of the trapezius elevate and
superiorly rotate the scapula. The middle fibers work with the rhomboid muscles to retract the scapula, while the inferior fibers depress
the scapula.

Table 12.1 summarizes the characteristics of the thoracic
muscles that move the pectoral girdle.

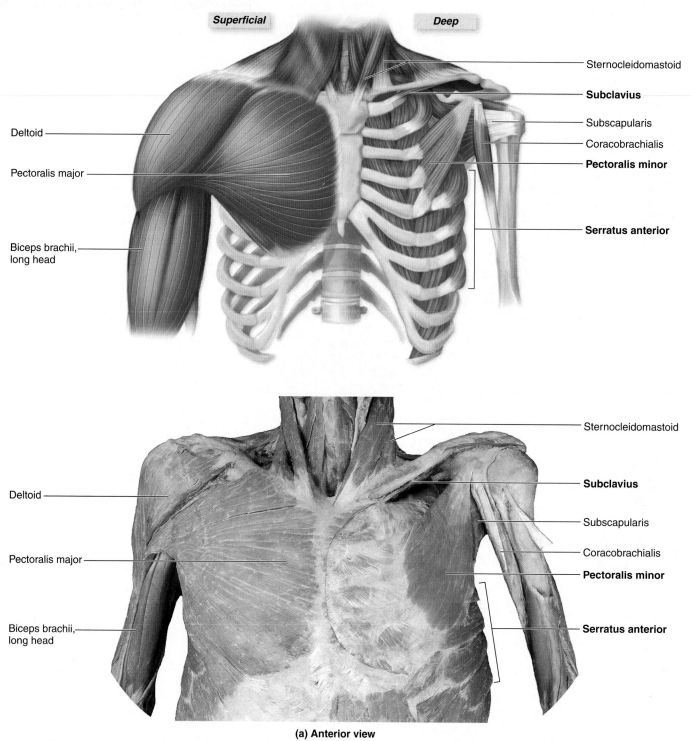

Superficial   Deep

Sternocleidomastoid

Subclavius

Subscapularis

Coracobrachialis

Pectoralis minor

Serratus anterior

Deltoid

Pectoralis major

Biceps brachii, long head

Sternocleidomastoid

Subclavius

Subscapularis

Coracobrachialis

Pectoralis minor

Serratus anterior

Deltoid

Pectoralis major

Biceps brachii, long head

(a) Anterior view

Figure 12.4

Muscles That Move the Pectoral Girdle and the Glenohumeral Joint/Arm. Illustrations and cadaver photos show (*a*) anterior and (*b*) posterior views of the muscles whose primary function is to move the pectoral girdle (scapula or clavicle), labeled in bold. Muscles that attach to the pectoral girdle but primarily move the arm are labeled but not in bold. Superficial muscles are shown on the right side of the body and deep muscles on the left.

CLINICAL VIEW

Paralysis of the Serratus Anterior Muscle ("Winged Scapula")

The serratus anterior muscle receives its innervation from the long thoracic nerve, which travels inferiorly along the anterolateral chest wall. Because of its location, the long thoracic nerve is occasionally damaged or cut during surgical removal of the breast (radical mastectomy). Damage to this nerve causes paralysis of the serratus anterior muscle. Recall that the serratus

anterior is the primary protractor of the scapula, and under normal circumstances, its pull on the scapula is counterbalanced by the posterior thoracic (superficial back) muscles. Paralysis of the serratus anterior muscle leaves the posterior thoracic muscles unopposed. If a patient with this condition puts both arms anteriorly on a wall and then pushes, the scapula on the injured side will poke posteriorly like a bird's wing, a classic sign known as the "winged scapula." The scapula on the unaffected side will remain in the expected anatomic position, close to the thoracic wall.

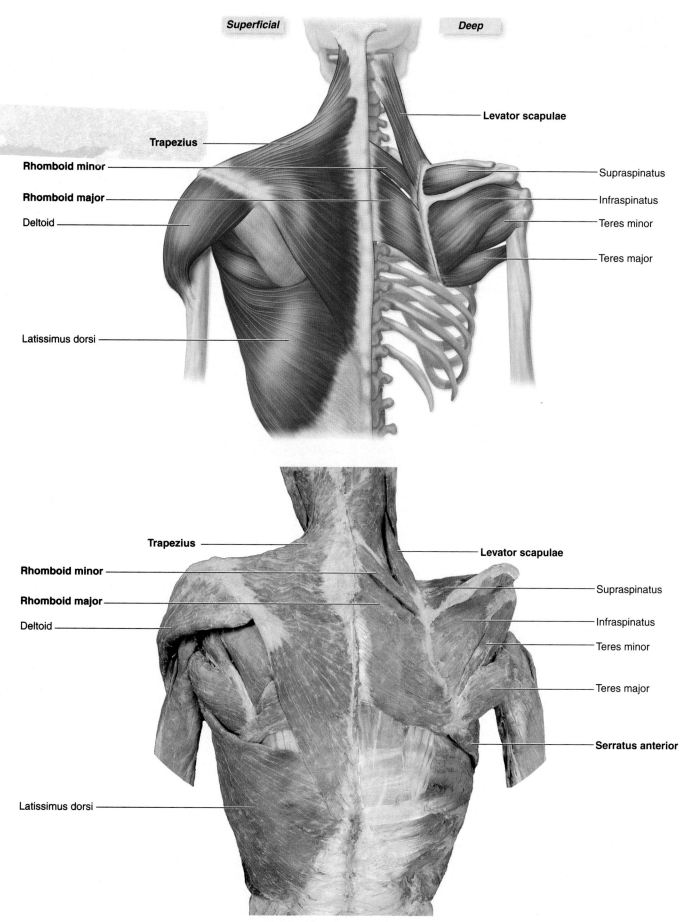

Superficial

Deep

Levator scapulae

Trapezius

Rhomboid minor

Rhomboid major

Deltoid

Latissimus dorsi

Supraspinatus

Infraspinatus

Teres minor

Teres major

Trapezius

Levator scapulae

Rhomboid minor

Supraspinatus

Rhomboid major

Infraspinatus

Deltoid

Teres minor

Teres major

Serratus anterior

Latissimus dorsi

(b) Posterior view

Table 12.1	Thoracic Muscles That Move the Pectoral Girdle		
Group/Muscle	**Actions**	**Origin/Insertion**	**Innervation**
ANTERIOR MUSCLES			
Pectoralis minor (pek′tō-ra′lis mī′ner) *pectus* = chest	Protracts and depresses scapula	O: Ribs 3–5 I: Coracoid process of scapula	Medial pectoral nerve (C8–T1)
Serratus anterior (ser-ā′tŭs an-tēr′ē-ōr) *serratus* = saw	Prime mover in scapula protraction; superiorly rotates scapula (so glenoid cavity moves superiorly); stabilizes scapula	O: Ribs 1–8, anterior and superior margins I: Medial border of scapula, anterior surface	Long thoracic nerve (C5–C7)
Subclavius (sŭb-klā′vē-ŭs) *sub* = under *clav* = clavicle	Stabilizes and depresses clavicle	O: Rib 1 I: Inferior surface of clavicle	Nerve to subclavius (C5–C6)
POSTERIOR MUSCLES			
Levator scapulae (lē-vā′tor, le-vā′ter skap′ū-lé) *levator* = raises	Elevates scapula; inferiorly rotates scapula (pulls glenoid cavity inferiorly)	O: Transverse processes of C_1–C_4 I: Superior part of medial border of scapula	Cervical nerves (C3–C4) and dorsal scapular nerve (C5)
Rhomboid major (rom′boyd mā′jŏr) *rhomboid* = diamond-shaped	Elevates and retracts (adducts) scapula; inferiorly rotates scapula	O: Spinous processes of T_2–T_5 I: Medial border of scapula from spine to inferior angle	Dorsal scapular nerve (C5)
Rhomboid minor	Elevates and retracts (adducts) scapula; inferiorly rotates scapula	O: Spinous processes of C_7–T_1 I: Medial border of scapula superior to spine	Dorsal scapular nerve (C5)
Trapezius (tra-pē′zē-ŭs) *trapezion* = irregular four-sided figure	Superior fibers: Elevate and superiorly rotate scapula Middle fibers: Retract scapula Inferior fibers: Depress scapula	O: Occipital bone (superior nuchal line); ligamentum nuchae; spinous processes of C_7–T_{12} I: Clavicle; acromion process and spine of scapula	Accessory nerve (CN XI)

Muscles That Move the Glenohumeral Joint/Arm

The phrases "moving the glenohumeral joint" and "moving the arm or humerus" mean the same thing. A movement such as flexion of the arm requires movement at the glenohumeral joint. Throughout this text, we refer to both the joint (where the movement is occurring) and the body region (that is being moved) to minimize any confusion you may have.

The glenohumeral joint is crossed by 10 muscles that insert on the arm (humerus) (see figure 12.4). Two of these muscles—the latissimus dorsi and the pectoralis major—are called axial muscles because they originate on the axial skeleton. The **latissimus dorsi** is a broad, triangular muscle located on the inferior part of the back. Often, it is referred to as the "swimmer's muscle," because many of its actions are required for certain swimming strokes. It is the prime arm extensor, and also adducts and medially rotates the arm. The **pectoralis major** is a large, thick, fan-shaped muscle that covers the superior part of the thorax. It is the principal flexor of the arm, and also adducts and medially rotates the arm.

The latissimus dorsi and pectoralis major muscles are the primary attachments of the arm to the trunk, and they are the prime movers of the glenohumeral joint. These muscles are antagonists with respect to arm flexion and arm extension. However, these same two muscles work together (synergistically) when performing other movements, such as adducting and medially rotating the humerus.

The triceps brachii, discussed in detail with the muscles that move the elbow joint, also participates in the glenohumeral joint. Specifically, it is the long head of the triceps brachii that spans the glenohumeral joint due to its origin on the infraglenoid tubercle of the scapula and insertion on the olecranon of the ulna. This part of the muscle helps extend and adduct the arm. The long head of the biceps brachii originates on the supraglenoid tubercle of the scapula and assists in flexing the arm.

Study Tip!

Generally speaking, muscles that originate anterior to the glenohumeral joint flex the arm (move it anteriorly), and those that originate posterior to the joint extend the arm (move it posteriorly).

The seven remaining muscles that move the humerus at the glenohumeral joint are termed the scapular muscles, because they originate entirely on the scapula. These muscles include the deltoid, coracobrachialis, teres major, and the four rotator cuff muscles. The **deltoid** is a thick, powerful muscle that functions as a prime abductor of the arm and forms the rounded contour of the shoulder. Note that the fibers of the deltoid originate from three different points, and these different fiber groups all perform different functions: (1) The anterior fibers flex and medially rotate the arm. (2) The middle fibers abduct the arm; in fact, the deltoid is the prime abductor of the arm. (3) The posterior fibers extend and laterally rotate the arm. The **coracobrachialis** works as a synergist to the pectoralis major in flexing and adducting the arm. The **teres major** works synergistically with the latissimus dorsi by extending, adducting, and medially rotating the arm.

Four **rotator cuff muscles** (subscapularis, supraspinatus, infraspinatus, and teres minor) provide strength and stability to the glenohumeral joint **(figure 12.5)**. These muscles attach the scapula

to the humerus. The specific movements of each muscle are best learned when equating them to pitching a ball:

- The **subscapularis** is used when you wind up for a pitch. It medially rotates the arm.
- The **supraspinatus** is used when you start to execute the pitch, by fully abducting the arm.
- The **infraspinatus** and **teres minor** help slow down the pitching arm upon completion of the pitch. These two muscles adduct and laterally rotate the arm.

Table 12.2 summarizes the characteristics of the muscles that move the glenohumeral joint and arm.

WHAT DO YOU THINK?

1 Which rotator cuff muscle tends to suffer the most injuries, and why?

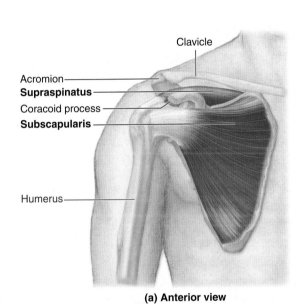

(a) Anterior view

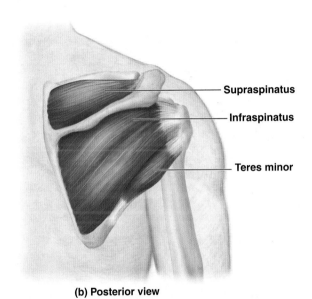

(b) Posterior view

Subscapularis

Supraspinatus

Infraspinatus and teres minor

(c) Movement of rotator cuff muscles

Figure 12.5

Rotator Cuff Muscles.   The rotator cuff muscles reinforce the glenohumeral joint and secure the head of the humerus in the glenoid cavity.
(a) The subscapularis is best seen in an anterior view of the right shoulder. *(b)* The supraspinatus, infraspinatus, and teres minor are best seen in a posterior view. *(c)* The subscapularis medially rotates the humerus (as when winding up for a pitch), the supraspinatus abducts the humerus (as when executing the pitch), and the infraspinatus and teres minor laterally rotate the humerus (as when completing the pitch and slowing down the pitching arm).

Table 12.2	**Muscles That Move the Glenohumeral Joint/Arm**		
Group/Muscle	**Action**	**Origin/Insertion**	**Innervation**
MUSCLES ORIGINATING ON AXIAL SKELETON			
Latissimus dorsi (lă-tis'ĭ-mŭs dōr'sī) *latissimus* = widest *dorsi* = back	Prime mover of arm extension; also adducts and medially rotates arm ("swimmer's muscle")	O: Spinous processes of T_7–T_{12}; ribs 8–12; iliac crest; thoracolumbar fascia I: Intertubercular groove of humerus	Thoracodorsal nerve (C6–C8)
Pectoralis major (pek'tō-ră'lis mā'jŏr)	Prime mover of arm flexion; also adducts and medially rotates arm	O: Medial clavicle; costal cartilages of ribs 2–6; body of sternum I: Lateral part of intertubercular groove of humerus	Lateral pectoral (C5–C7) and medial pectoral (C8–T1) nerves
MUSCLES ORIGINATING ON SCAPULA			
Deltoid (del'toyd) *delta* = triangular	Anterior fibers: Flex and medially rotate arm Middle fibers: Prime mover of arm abduction Posterior fibers: Extend and laterally rotate arm	O: Acromial end of clavicle; acromion and spine of scapula I: Deltoid tuberosity of humerus	Axillary nerve (C5–C6)
Coracobrachialis (kōr'ă-kō-brā-kē-a'lis) *coraco* = coracoid *brachi* = arm	Adducts and flexes arm	O: Coracoid process of scapula I: Middle medial shaft of humerus	Musculocutaneous nerve (C5–C6 fibers)
Teres major (ter'ēz, tĕr'ēz, mā'jŏr) *teres* = round	Extends, adducts, and medially rotates arm	O: Inferior lateral border and inferior angle of scapula I: Lesser tubercle and intertubercular groove of humerus	Lower subscapular nerve (C5–C6)
Triceps brachii (long head) (trī'seps brā'kē-ī) *triceps* = three heads	Extends and adducts arm	O: Infraglenoid tubercle of scapula I: Olecranon process of ulna	Radial nerve (C5–C7 axons)
Biceps brachii (long head) (bī'seps)	Flexes arm	O: Supraglenoid tubercle of scapula I: Radial tuberosity and bicipital aponeurosis	Musculocutaneous nerve (C5–C6 fibers)
Rotator Cuff Muscles (rō-tā'tŏr kŭf) *rotatio* = to revolve	Collectively, these 4 muscles stabilize the glenohumeral joint		
Subscapularis (sŭb-skap-ū-lār'ris) *sub* = under	Medially rotates arm	O: Subscapular fossa of scapula I: Lesser tubercle of humerus	Upper and lower subscapular nerves (C5–C6)
Supraspinatus (soo-pră-spī-nā'tŭs) *supra* = above, over *spin* = spine	Abducts arm	O: Supraspinous fossa of scapula I: Greater tubercle of humerus	Suprascapular nerve (C5–C6)
Infraspinatus (in-fră-spī-nā'tŭs) *infra* = below	Adducts and laterally rotates arm	O: Infraspinous fossa of scapula I: Greater tubercle of humerus	Suprascapular nerve (C5–C6)
Teres minor	Adducts and laterally rotates arm	O: Upper dorsal lateral border of scapula (superior to teres major origin) I: Greater tubercle of humerus	Axillary nerve (C5–C6)

Table 12.3	Summary of Muscle Actions at the Glenohumeral Joint/Arm				
Abduction	**Adduction**	**Extension**	**Flexion**	**Lateral Rotation**	**Medial Rotation**
Deltoid (middle fibers)	**Latissimus dorsi**	**Latissimus dorsi**	**Pectoralis major**	**Infraspinatus**	**Subscapularis**
Supraspinatus	**Pectoralis major**	**Deltoid (posterior fibers)**	**Deltoid (anterior fibers)**	**Teres minor**	Deltoid (anterior fibers)
	Coracobrachialis	Teres major	Coracobrachialis	Deltoid (posterior fibers)	Latissimus dorsi
	Teres major	Long head of triceps brachii	(Long head of biceps brachii)		Pectoralis major
	Teres minor				Teres major
	Infraspinatus				

Boldface indicates a prime mover; others are synergists. Parentheses around an entire muscle name indicate only a slight effect.

Study Tip!

The best way to remember the appendicular muscles is to group muscles that have similar functions. Note that a muscle that has multiple functions is in more than one group. The muscles that move the arm at the glenohumeral joint are grouped in **table 12.3** according to different types of actions. We recommend that you copy the columns multiple times and then test your knowledge by trying to write out all of the muscles in a group without looking at your notes. If you can list them all, you truly remember the information!

WHAT DID YOU LEARN?

1 What muscles are you using when you protract the scapula?

2 What is the primary action of the levator scapulae?

3 What muscles cause medial rotation of the arm?

4 How can the deltoid extend and flex the arm?

Arm and Forearm Muscles That Move the Elbow Joint/Forearm

When you move the elbow joint, you move the bones of the forearm. Thus, the term "flexing the elbow joint" is synonymous with "flexing the forearm." Keep this in mind as we discuss the muscles that move the elbow joint and forearm.

The muscles in limbs are organized into **compartments,** which are surrounded by deep fascia. Each compartment houses functionally related skeletal muscles, as well as their associated nerves and blood vessels. The muscles of the arm may be subdivided into an **anterior compartment** and a **posterior compartment (figure 12.6)**. The anterior compartment primarily contains elbow flexors, and the posterior compartment contains elbow extensors, so these compartments are also called the *flexor compartment* and the *extensor compartment,* respectively.

On the anterior side of the humerus are the principal flexors of the forearm: the biceps brachii and the brachialis (**figure 12.7**). The **biceps brachii** is a large, two-headed muscle on the anterior surface

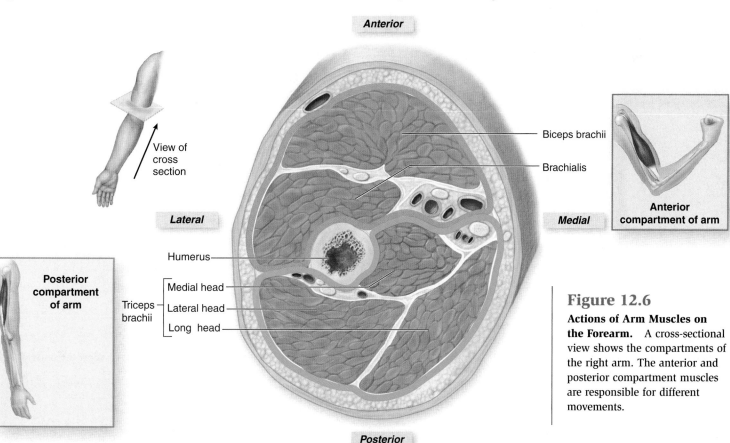

Figure 12.6

Actions of Arm Muscles on the Forearm.   A cross-sectional view shows the compartments of the right arm. The anterior and posterior compartment muscles are responsible for different movements.

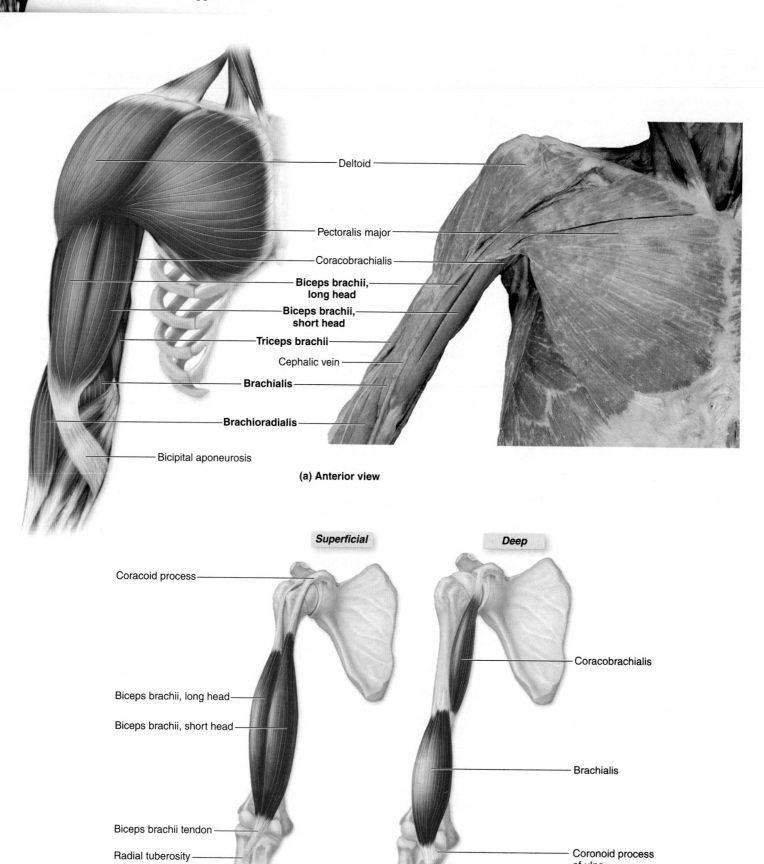

(a) Anterior view

Deltoid

Pectoralis major

Coracobrachialis

Biceps brachii, long head

Biceps brachii, short head

Triceps brachii

Cephalic vein

Brachialis

Brachioradialis

Bicipital aponeurosis

Superficial

Deep

Coracoid process

Biceps brachii, long head

Biceps brachii, short head

Coracobrachialis

Brachialis

Biceps brachii tendon

Radial tuberosity

Coronoid process of ulna

(b) Anterior muscles

Figure 12.7

Anterior Muscles That Move the Elbow Joint/Forearm. *(a)* Illustration and cadaver photo of the right arm and shoulder show the muscles that produce movements at the elbow joint, labeled in bold. *(b)* Superficial and deep anterior arm muscles.

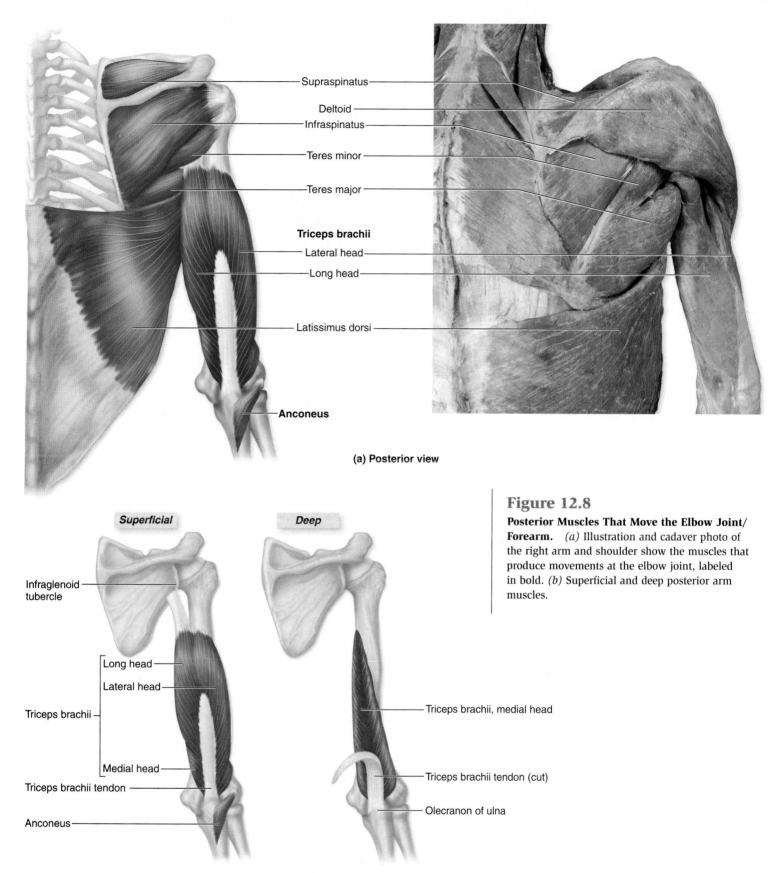

Figure 12.8

Posterior Muscles That Move the Elbow Joint/ Forearm. *(a)* Illustration and cadaver photo of the right arm and shoulder show the muscles that produce movements at the elbow joint, labeled in bold. *(b)* Superficial and deep posterior arm muscles.

of the humerus. The biceps brachii flexes the forearm and is a powerful supinator of the forearm when the elbow is flexed. (An example of this supination movement occurs when you tighten a screw with your right hand.) The tendon of the long head of the biceps brachii crosses the shoulder joint, and so this muscle helps flex the humerus as well (albeit weakly). The **brachialis** is deep to the biceps brachii and lies on the anterior surface of the humerus. It is the most power-

ful flexor of the forearm at the elbow. The **brachioradialis** is a prominent muscle on the lateral surface of the forearm. It is a synergist in forearm flexion, effective primarily when the prime movers of forearm flexion have already partially flexed the elbow.

The posterior compartment of the arm contains two muscles that extend the forearm at the elbow: the triceps brachii and the anconeus **(figure 12.8)**. The **triceps brachii** is the large, three-headed muscle on

Figure 12.9

Forearm Muscles That Supinate and Pronate. A view of the right upper limb shows the supinator muscle supinates the forearm, while the pronator teres and pronator quadratus pronate the forearm. (The biceps brachii, an arm muscle not shown here, also supinates the forearm.)

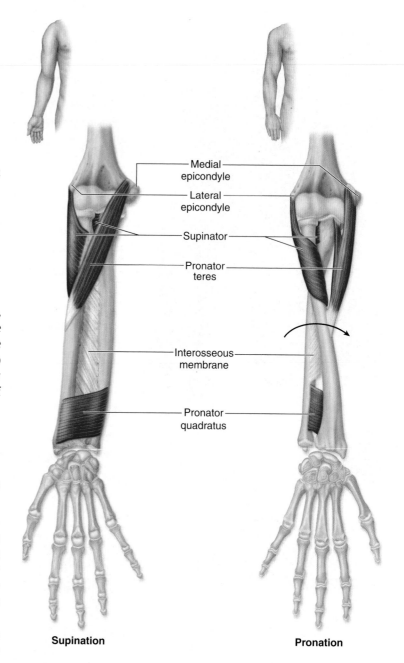

the posterior surface of the arm. It is the prime extensor of the forearm, and so its action is antagonistic to that of the biceps brachii. Only the long head of the triceps brachii crosses the glenohumeral joint, where it helps extend the humerus. All three parts of this muscle merge to form a common insertion on the olecranon of the ulna. A weak elbow extensor is the small **anconeus** that crosses the posterolateral region of the elbow.

WHAT DO YOU THINK?

2 The brachialis is on the anterior surface of the arm. Without looking at the muscle tables, determine whether this muscle flexes or extends the elbow joint. How did you reach your conclusion?

Some forearm muscles pronate or supinate the forearm (**figure 12.9**). As their names imply, both the **pronator teres** and the **pronator quadratus** rotate the radius across the surface of the ulna to pronate the forearm. These muscles are located in the anterior compartment of the forearm. They are antagonistic to the **supinator** in the posterior compartment of the forearm. The supinator works synergistically with the biceps brachii to supinate the forearm.

Table 12.4 summarizes the characteristics of the muscles that move the forearm, and **table 12.5** groups them according to common function. By learning these muscles as groups, you will have a better understanding of how they work together to perform specific functions.

Forearm Muscles That Move the Wrist Joint, Hand, and Fingers

Most muscles in the forearm move the hand at the wrist and/or the fingers. These muscles are called extrinsic muscles of the wrist and hand, because the muscles originate on the forearm, not the wrist or hand. Palpate your own forearm; it is bigger near the elbow because the bellies of these forearm muscles form the bulk of this region. Moving toward the wrist, the forearm thins because there are no longer big muscle bellies, but rather the long tendons that project from these muscles.

Deep fascia partitions the forearm muscles into an anterior (flexor) compartment and a posterior (extensor) compartment (**figure 12.10**). Most of the **anterior compartment** muscles originate

on the medial epicondyle of the humerus via a common flexor tendon. Muscles in the anterior compartment of the forearm tend to flex the wrist, the metacarpophalangeal (MP) joints, and/or the interphalangeal (IP) joints of the fingers. Most of the **posterior compartment** muscles originate on the lateral epicondyle of the humerus via a common extensor tendon. Muscles in the posterior compartment of the forearm tend to extend the wrist, the MP joints, and/or the IP joints.

Note that not all anterior forearm muscles cause flexion. Both the pronator teres and the pronator quadratus, discussed previously, are located in the anterior compartment of the forearm, although their primary function is pronation. Likewise, the supinator muscle is in the posterior compartment of the forearm, even though its primary function is supination.

Table 12.4 — Muscles That Move the Forearm

Muscle	Action	Origin/Insertion	Innervation
FLEXORS (ANTERIOR ARM)			

Biceps brachii (bī′seps brā′kē-ī)
biceps = two heads
  Long head
  Short head
	Flexes forearm, powerful supinator of forearm
	Long head flexes arm
		O: Long head: supraglenoid tubercle of scapula
		Short head: coracoid process of scapula
		I: Radial tuberosity and bicipital aponeurosis
			Musculocutaneous nerve (C5–C6 fibers)

Brachialis (brā′kē-al′is)	Primary flexor of forearm	O: Distal anterior surface of humerus I: Tuberosity and coronoid process of ulna	Musculocutaneous nerve (C5–C6 fibers)
Brachioradialis (brā′kē-ō-rā′dē-al′is)	Flexes forearm	O Lateral supracondylar ridge of humerus I: Styloid process of radius	Radial nerve (C6–C7 fibers)
EXTENSORS (POSTERIOR ARM)			

Triceps brachii (trī′seps brā′kē-ī)
  Long head
  Lateral head
  Medial head
	Primary extensor of forearm
	Long head of triceps also extends and adducts arm
		O: Long head: infraglenoid tubercle of scapula
		Lateral head: posterior humerus above radial groove
		Medial head: posterior humerus below radial groove
		I: Olecranon of ulna
			Radial nerve (C5–C7 fibers)

Anconeus (ang-kō′nē-ŭs) *ankon* = elbow	Extends forearm	O: Lateral epicondyle of humerus I: Olecranon of ulna	Radial nerve (C6–C8 fibers)
PRONATORS (ANTERIOR FOREARM MUSCLES)			
Pronator quadratus (prō-nā′tōr kwah-drā′tŭs)	Pronates forearm	O: Distal 1/4 of ulna I: Distal 1/4 of radius	Median nerve (C8–T1 fibers)
Pronator teres (prō-nā′tōr ter′ēz)	Pronates forearm	O: Medial epicondyle of humerus and coronoid process of ulna I: Lateral surface of radius	Median nerve (C6–C7 fibers)
SUPINATOR (POSTERIOR FOREARM MUSCLE)			
Supinator (soo′pi-nā-tōr)	Supinates forearm	O: Lateral epicondyle of humerus and ulna distal to radial notch I: Anterolateral surface of radius distal to radial tuberosity	Radial nerve (C6–C8 fibers)

Table 12.5 — Summary of Muscle Actions at the Elbow Joint/Forearm

Extension	Flexion	Pronation	Supination
Triceps brachii	**Brachialis**	**Pronator teres**	**Biceps brachii**
(Anconeus)	Biceps brachii	**Pronator quadratus**	Supinator
	Brachioradialis		

Boldface indicates a prime mover; others are synergists. Parentheses indicate only a slight effect.

The tendons of forearm muscles typically are surrounded by tendon (synovial) sheaths and held adjacent to the skeletal elements by strong fascial structures. At the wrist, the deep fascia of the forearm forms thickened, fibrous bands termed retinacula. The retinacula help hold the tendons close to the bone and prevent the tendons from "bowstringing" outward. The palmar (anterior) surface of the carpal bones is covered by the **flexor retinaculum (figure 12.11a)**. Flexor tendons of the digits and the median nerve

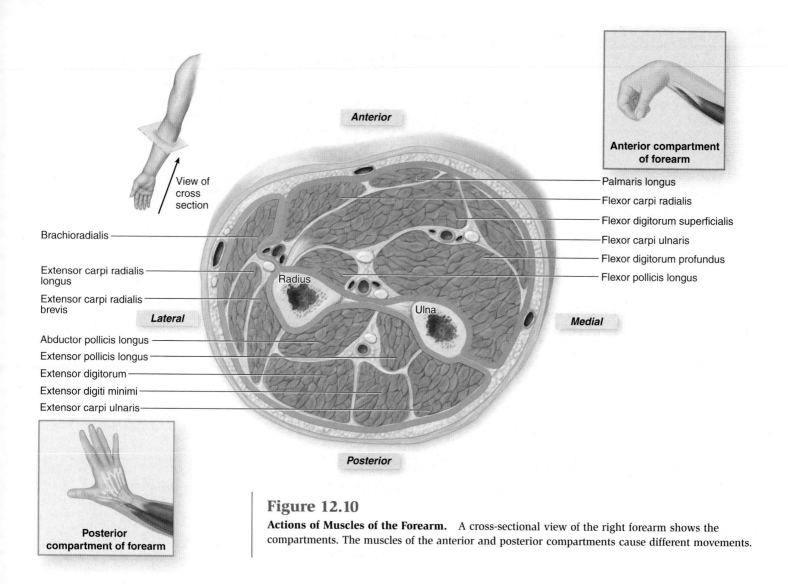

Figure 12.10

Actions of Muscles of the Forearm. A cross-sectional view of the right forearm shows the compartments. The muscles of the anterior and posterior compartments cause different movements.

pass through the tight space between the carpal bones and the flexor retinaculum, which is called the **carpal tunnel**. The **extensor retinaculum** is superficial to the dorsal surface of the carpal bones. Extensor tendons of the wrist and digits pass between the carpal bones and the extensor retinaculum.

The muscles of the **anterior compartment** of the forearm may be subdivided into a superficial layer, an intermediate layer,

Lateral Epicondylitis ("Tennis Elbow")

Lateral epicondylitis (ep'i-kon-di-li'tis), or **"tennis elbow,"** is a painful condition resulting from trauma or overuse of the common extensor tendon of the posterior forearm muscles. Although the pain is perceived as coming from the elbow joint, it actually arises from the lateral epicondyle of the humerus, the attachment site of the common extensor tendon. Lateral epicondylitis most often results from the repeated forceful contraction of the forearm extensors. These are the muscles used to extend the hand at the wrist, as when pulling a heavy object from an overhead shelf, shoveling snow, or hitting a backhand shot in tennis.

and a deep layer. The superficial and intermediate muscles originate from the common flexor tendon that attaches to the medial epicondyle of the humerus. The deep layer of muscles originates directly on the forearm.

The **superficial layer** of anterior forearm muscles is arranged from the lateral to the medial surface of the forearm in the following order: pronator teres (described previously with muscles that move the forearm), flexor carpi radialis, palmaris longus, and flexor carpi ulnaris (figure 12.11*a*).

The **flexor carpi radialis** extends diagonally across the anterior surface of the forearm. Its tendon is prominent on the lateral side of the forearm. This muscle flexes the wrist and abducts the hand at the wrist. The **palmaris longus** is absent in approximately 10% of all individuals. This narrow, superficial muscle on the anterior surface of the forearm weakly assists in wrist flexion. On the anteromedial side of the forearm, the **flexor carpi ulnaris** is positioned to both flex the wrist and adduct the hand at the wrist.

You can determine the positioning of the three superficial muscles of the anterior forearm and the pronator teres muscle on your own body by performing the exercise shown in **figure 12.12**. Wrap your thumb around the medial epicondyle of the other arm, so your thumb is positioned posterior to the elbow. Align your little finger along the medial border of your forearm. The natural placement of your four fingers, from your index finger to your little

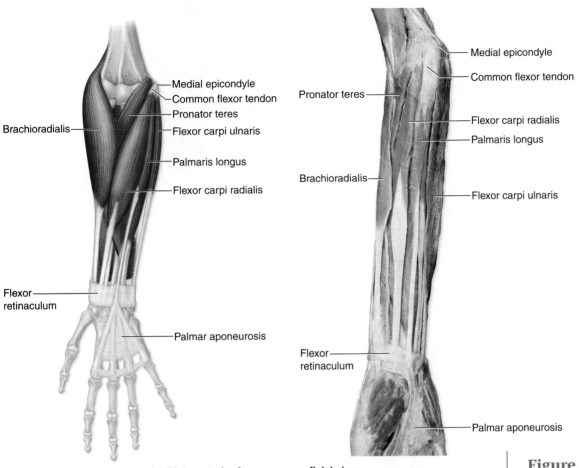

Brachioradialis

Medial epicondyle
Common flexor tendon
Pronator teres
Flexor carpi ulnaris
Palmaris longus
Flexor carpi radialis

Flexor retinaculum

Palmar aponeurosis

Pronator teres

Brachioradialis

Flexor retinaculum

Medial epicondyle
Common flexor tendon
Flexor carpi radialis
Palmaris longus
Flexor carpi ulnaris

Palmar aponeurosis

(a) Right anterior forearm, superficial view

Figure 12.11

Anterior Forearm Muscles.   The anterior forearm muscles pronate the forearm or flex the wrist and fingers. They may be subdivided into superficial, intermediate, and deep groups. *(a)* Illustration and cadaver photo show the superficial muscles of the right anterior forearm. *(b)* Intermediate and *(c)* deep muscles of the right anterior forearm.

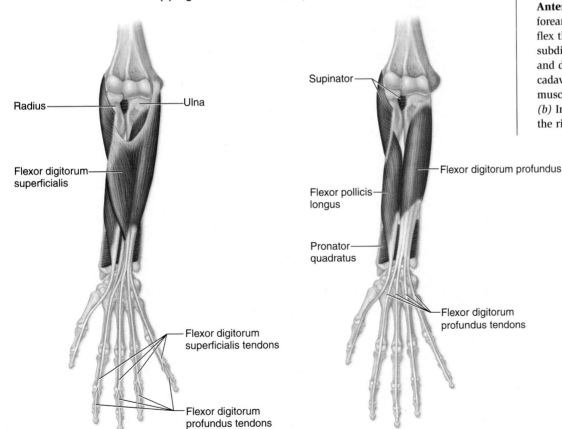

Radius

Ulna

Flexor digitorum superficialis

Flexor digitorum superficialis tendons

Flexor digitorum profundus tendons

(b) Right anterior forearm, intermediate view

Supinator

Flexor digitorum profundus

Flexor pollicis longus

Pronator quadratus

Flexor digitorum profundus tendons

(c) Right anterior forearm, deep view

CLINICAL VIEW

Carpal Tunnel Syndrome

The flexor retinaculum extends from the medial carpal bones to the lateral carpal bones, and the space between the carpal bones and the flexor retinaculum is the carpal tunnel. Through this narrow tunnel, numerous long muscle tendons extend to the fingers from the forearm. Accompanying these tendons is the median nerve, which innervates the skin on the lateral region of the hand and the muscles that move the thumb.

Any compression of either the median nerve or the tendons in the tunnel results in **carpal tunnel syndrome.** A common cause is inflammation of any component in the carpal tunnel—for example, swollen tendons as a result of overuse. Workers who repeatedly flex either their fingers or wrists, such as typists and computer programmers, experience this condition. In addition, females in their last trimester of pregnancy

may develop carpal tunnel syndrome as increased water retention results in compression within the carpal tunnel.

Carpal tunnel syndrome is characterized by pain and **paresthesia** (par-es-thē′zē-ă; *aisthesis* = sensation), which is the feeling of "pins and needles." Sometimes, there is more extensive sensory loss as well as motor loss in the muscles of the hand supplied by the median nerve. The median nerve supplies the muscles of the thumb, so in severe cases of carpal tunnel syndrome, these muscles may atrophy as their nerve supply is diminished. Treatment of the syndrome includes supporting the hand in a splint and administering anti-inflammatory drugs (e.g., ibuprofen or prescription medicine). In severe, chronic cases that do not respond to more conservative treatment, a surgeon can incise the flexor retinaculum and open the carpal tunnel, relieving the pressure.

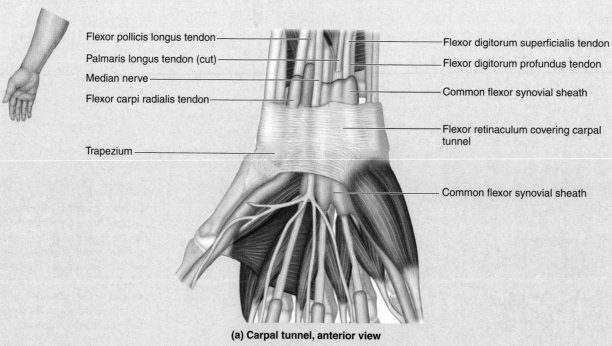

Flexor pollicis longus tendon
Palmaris longus tendon (cut)
Median nerve
Flexor carpi radialis tendon
Trapezium

Flexor digitorum superficialis tendon
Flexor digitorum profundus tendon
Common flexor synovial sheath
Flexor retinaculum covering carpal tunnel
Common flexor synovial sheath

(a) Carpal tunnel, anterior view

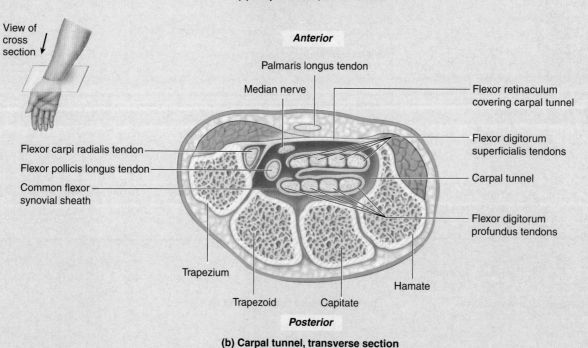

View of cross section

Anterior

Palmaris longus tendon
Median nerve

Flexor retinaculum covering carpal tunnel
Flexor digitorum superficialis tendons
Carpal tunnel
Flexor digitorum profundus tendons

Flexor carpi radialis tendon
Flexor pollicis longus tendon
Common flexor synovial sheath

Trapezium
Trapezoid   Capitate
Hamate

Posterior

(b) Carpal tunnel, transverse section

(Left hand covers medial epicondyle)

Figure 12.12

Positioning of the Superficial Anterior Forearm Muscles. By positioning the left hand at the medial epicondyle of the right humerus, fingers 2–5 lay in the approximate position of the superficial muscles of the anterior forearm.

finger, overlies the placement of the pronator teres, flexor carpi radialis, palmaris longus, and flexor carpi ulnaris, respectively.

The **intermediate layer** in the anterior compartment of the forearm contains a single muscle (see figure 12.11*b*), the **flexor digitorum superficialis.** This muscle splits into four tendons that each insert on the middle phalanges of fingers 2–5. This muscle crosses over the wrist, MP joints, and PIP (proximal interphalangeal) joints of fingers 2–5; thus, it flexes all of these joints. Since the flexor digitorum superficialis does not cross over the DIP (distal interphalangeal) joints of these fingers, it cannot move the DIP joints.

The **deep layer** of the forearm anterior compartment muscles includes the flexor pollicis longus (lateral side) and the flexor digitorum profundus (medial side). Deep to both of these muscles is the pronator quadratus muscle, discussed previously with muscles that pronate the forearm (see figure 12.11*c*).

The **flexor pollicis longus** attaches to the distal phalanx of the thumb and flexes the MP and IP joints of the thumb. In addition, because this muscle crosses the wrist joint, it can weakly flex the wrist. The **flexor digitorum profundus** lies deep to the flexor digitorum superficialis. This muscle splits into four tendons that insert on the distal phalanges of fingers 2–5. At the level of the middle phalanges, the tendons of the flexor digitorum superficialis split to allow the flexor digitorum profundus tendons to pass to the tips of the fingers. The flexor digitorum profundus flexes the wrist, MP joints, PIP joints, and DIP joints of fingers 2–5.

Muscles of the **posterior compartment** of the forearm are primarily wrist and finger extensors. An exception is the supinator, which helps supinate the forearm. The posterior compartment muscles may be subdivided into a superficial layer and a deep layer.

The **superficial layer** of posterior forearm muscles originates from a common extensor tendon on the lateral epicondyle of the humerus (**figure 12.13***a*). These muscles are positioned laterally to medially as follows:

- The **extensor carpi radialis longus** is a long, tapered muscle that is medial to the brachioradialis. It extends the wrist and abducts the hand at the wrist.
- The **extensor carpi radialis brevis** works synergistically with the extensor carpi radialis longus.
- The **extensor digitorum** splits into four tendons that insert on the distal phalanges of fingers 2–5. It extends the wrist, MP joints, PIP joints, and DIP joints of fingers 2–5.

Anatomic Snuffbox

The **anatomic snuffbox** is a triangular region on the posterolateral side of the hand, just proximal to the thumb. This region is bounded by the three tendons of the deep posterior compartment muscles of the forearm: (1) abductor pollicis longus, (2) extensor pollicis brevis, and (3) extensor pollicis longus. This area is termed the anatomic snuffbox because in historical times, such as during the American Revolution, people put finely ground tobacco called snuff in this little depression and then inhaled it. The floor of the snuffbox is formed by the scaphoid bone. This fact has diagnostic importance, because a person who fractures the scaphoid bone experiences extreme localized tenderness in the region of the anatomic snuffbox.

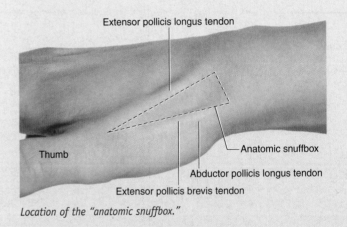

Location of the "anatomic snuffbox."

- The **extensor digiti minimi** attaches to the distal phalanx of finger 5. It works with the extensor digitorum to extend the little finger.
- On the medial surface of the posterior forearm, the **extensor carpi ulnaris** inserts on the fifth metacarpal bone, where it acts to extend the wrist and adduct the hand.

The **deep layer** originates directly on the posterior forearm and inserts on the wrist or hand (figure 12.13*b*). These muscles are arranged from lateral to medial in the following order: supinator (previously described), abductor pollicis longus, extensor pollicis brevis, extensor pollicis longus, and extensor indicis. These muscles weakly extend the wrist, and perform the following other functions.

- The **abductor pollicis longus** inserts on the first metacarpal. It abducts the thumb.
- The **extensor pollicis brevis** lies immediately medial to the abductor pollicis longus. The extensor pollicis brevis attaches to the proximal phalanx of the thumb and helps extend the MP joint of the thumb.
- The **extensor pollicis longus** extends the MP and IP joints of the thumb.
- The **extensor indicis** is the most medial muscle of the deep posterior compartment. It extends the MP, PIP, and DIP joints of the index finger (finger 2).

Table 12.6 summarizes the characteristics of the muscles that move the wrist joint, hand, and fingers.

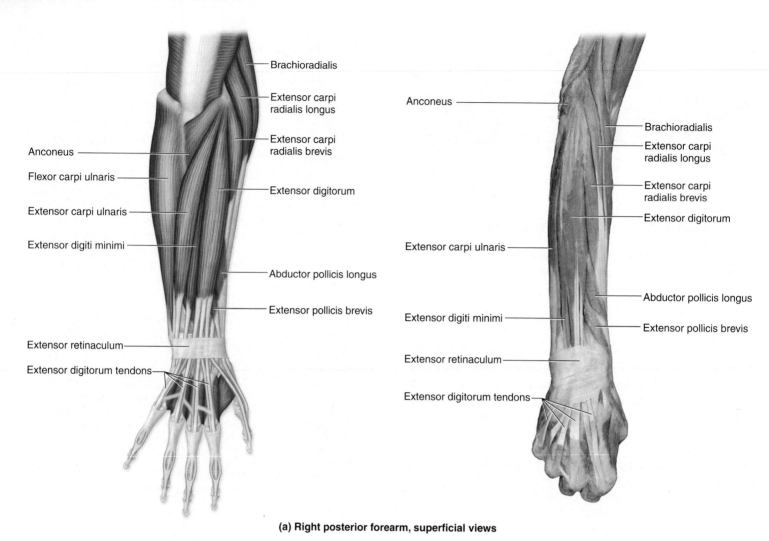

Brachioradialis

Extensor carpi radialis longus

Extensor carpi radialis brevis

Anconeus

Flexor carpi ulnaris

Extensor carpi ulnaris

Extensor digiti minimi

Extensor digitorum

Abductor pollicis longus

Extensor pollicis brevis

Extensor retinaculum

Extensor digitorum tendons

Anconeus

Brachioradialis

Extensor carpi radialis longus

Extensor carpi radialis brevis

Extensor digitorum

Extensor carpi ulnaris

Extensor digiti minimi

Extensor retinaculum

Extensor digitorum tendons

Abductor pollicis longus

Extensor pollicis brevis

(a) Right posterior forearm, superficial views

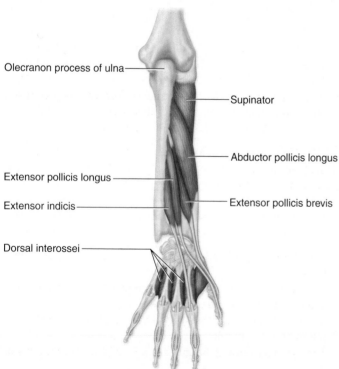

Olecranon process of ulna

Supinator

Abductor pollicis longus

Extensor pollicis longus

Extensor indicis

Extensor pollicis brevis

Dorsal interossei

(b) Right posterior forearm, deep views

Figure 12.13

Posterior Forearm Muscles. The posterior forearm muscles supinate the forearm or extend the wrist or fingers. They may be subdivided into *(a)* superficial and *(b)* deep groups, as shown in these views of the right forearm.

Table 12.6	Forearm Muscles That Move the Wrist Joint, Hand, and Fingers		
Group/Muscle	**Action**	**Origin/Insertion**	**Innervation**
ANTERIOR MUSCLES: SUPERFICIAL			
Pronator teres (described in table 12.4)			
Flexor carpi radialis (flek′ser kar′pī rā-dē-āl′is) *carpi* = wrist	Flexes wrist and abducts hand	O: Medial epicondyle of humerus I: Base of metacarpals II and III	Median nerve (C6–C7 fibers)
Palmaris longus (pawl-mar′is lon′gŭs)	Weak wrist flexor	O: Medial epicondyle of humerus I: Flexor retinaculum and palmar aponeurosis	Median nerve (C6–C7 fibers)
Flexor carpi ulnaris (ŭl-nar′is)	Flexes wrist and adducts hand	O: Medial epicondyle of humerus; olecranon and posterior surface of ulna I: Pisiform and hamate bones; base of metacarpal V	Ulnar nerve (C8–T1)
ANTERIOR MUSCLES: INTERMEDIATE			
Flexor digitorum superficialis (dij′i-tōr′ŭm soo′per-fish-ē-ā′lis) *superficial* = close to surface	Flexes wrist, 2nd–5th MP joints, and PIP joints	O: Medial epicondyle of humerus, coronoid process of ulna I: Middle phalanges of fingers 2–5	Median nerve (C6–C7 fibers)
ANTERIOR MUSCLES: DEEP			
Flexor pollicis longus (pol′i-sis lon′gŭs) *pollex* = thumb	Flexes MP joint of thumb, IP joint of thumb; weakly flexes wrist	O: Anterior shaft of radius; interosseous membrane I: Distal phalanx of thumb	Median nerve (C6–C7 fibers)
Flexor digitorum profundus (prō-fŭn′dŭs) *profound* = deep	Flexes wrist, 2nd–5th MP joints, PIP joints, and DIP joints	O: Anteromedial surface of ulna; interosseous membrane I: Distal phalanges of fingers 2–5	Lateral 1/2 of muscle innervated by median nerve (C6–C8 fibers), medial 1/2 of muscle innervated by ulnar nerve (C8 fibers)
Pronator quadratus (described in table 12.4)			
POSTERIOR MUSCLES: SUPERFICIAL			
Extensor carpi radialis longus (eks-ten′ser)	Extends wrist, abducts hand	O: Lateral supracondylar ridge of humerus I: Base of metacarpal II	Radial nerve (C6–C7 fibers)
Extensor carpi radialis brevis (brev′is) *brevis* = short	Extends wrist, abducts hand	O: Lateral epicondyle of humerus I: Base of metacarpal III	Radial nerve (C6–C7 fibers)
Extensor digitorum (dij′i-tōr′ŭm)	Extends wrist, extends 2nd–5th MP joints, PIP joints, and DIP joints	O: Lateral epicondyle of humerus I: Distal and middle phalanges of fingers 2–5	Radial nerve (C6–C8 fibers)
Extensor digiti minimi (dij′i-tī mi′nī-mī) *digitus minimus* = little finger	Extends wrist, MP, and PIP joints of finger 5	O: Lateral epicondyle of humerus I: Proximal phalanx of finger 5	Radial nerve (C6–C8 fibers)
Extensor carpi ulnaris	Extends wrist, adducts hand	O: Lateral epicondyle of humerus; posterior border of ulna I: Base of metacarpal V	Radial nerve (C6–C8 fibers)

(continued on next page)

Table 12.6	Forearm Muscles That Move the Wrist Joint, Hand, and Fingers *(continued)*		
Group/Muscle	**Action**	**Origin/Insertion**	**Innervation**
POSTERIOR MUSCLES: DEEP			
Abductor pollicis longus (ab-dŭk′ter, -tōr)	Abducts thumb, extends wrist (weakly)	O: Proximal dorsal surfaces of radius and ulna; interosseous membrane I: Lateral edge of metacarpal I	Radial nerve (C6–C8 fibers)
Extensor pollicis brevis	Extends MP joints of thumb, extends wrist (weakly)	O: Posterior surface of radius; interosseous membrane I: Proximal phalanx of thumb	Radial nerve (C6–C8 fibers)
Extensor pollicis longus	Extends MP and IP joints of thumb, extends wrist (weakly)	O: Posterior surface of ulna; interosseous membrane I: Distal phalanx of thumb	Radial nerve (C6–C8 fibers)
Extensor indicis (in′di-sis) *index* = forefinger	Extends MP, PIP, and DIP joints of finger 2, extends wrist (weakly)	O: Posterior surface of ulna; interosseous membrane I: Tendon of extensor digitorum	Radial nerve (C6–C8 fibers)
Supinator (described in table 12.4)			

MP = metacarpophalangeal, PIP = proximal interphalangeal, DIP = distal interphalangeal

Intrinsic Muscles of the Hand

The intrinsic muscles of the hand are small muscles that both originate and insert on the hand; they are housed entirely within the palm (figure 12.14). These muscles are divided into three groups: (1) The **thenar group** forms the thick, fleshy mass (thenar eminence) at the base of the thumb. (2) The **hypothenar group** forms a smaller fleshy mass (hypothenar eminence) at the base of the little finger. (3) The **midpalmar group** occupies the space between the first two groups.

The thenar and hypothenar groups contain smaller muscles:

- Small flexors (**flexor pollicis brevis** in the thenar group and **flexor digiti minimi brevis** in the hypothenar group) flex the thumb and the little finger, respectively.
- Abductors (**abductor pollicis brevis** in the thenar group and **abductor digiti minimi** in the hypothenar group) abduct the thumb and little finger, respectively.
- Opponens muscles (**opponens pollicis** in the thenar group and **opponens digiti minimi** in the hypothenar group) assist in the opposition of the thumb and little finger, respectively.

The midpalmar group contains twelve muscles that are partitioned into the following subgroups: lumbricals, dorsal interossei, palmar interossei, and adductor pollicis. The **lumbrical muscles** are four worm-shaped muscles. These muscles flex the MP joints and at the same time extend the PIP and DIP joints of fingers 2–5. The **dorsal interossei** are four deep bipennate muscles located between the metacarpals. They flex the MP joints and at the same time extend the PIP and DIP joints of fingers 2–5. In addition, the dorsal interossei abduct fingers 2–5. The **palmar interossei** are three small muscles that insert on fingers 2, 4, and 5 and adduct the fingers. In addition, these muscles work with the lumbricals and dorsal interossei to flex the MP joints and at the same time extend the PIP and DIP joints of fingers 2–5. The **adductor pollicis** is sometimes incorrectly classified as a palmar interosseous muscle. As its name suggests, this muscle adducts the thumb.

> ## Study Tip!
> To remember the functions of the palmar and dorsal interosseous muscles, use this mnemonic:
>
> **PAD-DAB**
>
> (**P**almar interossei **AD**duct the fingers, while **D**orsal interossei **AB**duct the fingers.)

The intrinsic muscles of the hand are summarized in **table 12.7** and a summary of muscle actions at the wrist and hand are listed in **table 12.8**.

WHAT DID YOU LEARN?

5 Identify the muscles that rotate (pronate or supinate) the forearm.

6 What muscles are flexors of the forearm?

7 What are the actions of the extensor carpi radialis muscles?

8 Identify the intrinsic muscles of the hand that cause abduction of the fingers.

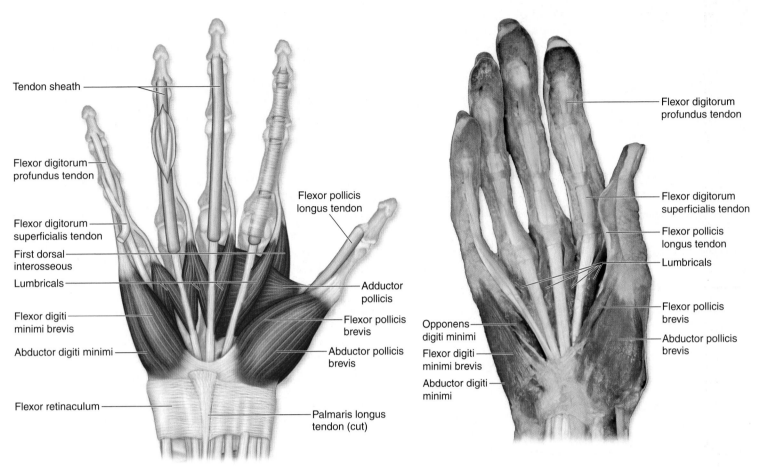

(a) Right hand, superficial palmar view

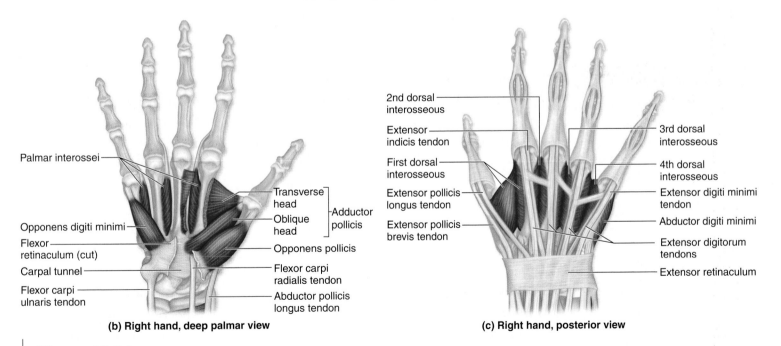

(b) Right hand, deep palmar view

(c) Right hand, posterior view

Figure 12.14

Intrinsic Muscles of the Hand.   These muscles allow the fine, controlled movements necessary for such activities as writing, typing, and playing a guitar. *(a)* Palmar (anterior) views of the superficial muscles of the right hand. *(b)* Palmar view of the deep muscles. *(c)* Posterior (dorsal) view of the superficial muscles.

Table 12.7 — Intrinsic Muscles of the Hand

Group/Muscle	Action	Origin/Insertion	Innervation
THENAR GROUP			
Flexor pollicis brevis	Flexes thumb	O: Flexor retinaculum, trapezium I: Proximal phalanx of thumb	Median nerve (C8–T1 fibers)
Abductor pollicis brevis	Abducts thumb	O: Flexor retinaculum, scaphoid, trapezium I: Lateral side of proximal phalanx of thumb	Median nerve (C8–T1 fibers)
Opponens pollicis (ō-pō′nens) *opponens* = to place against	Opposition of thumb	O: Flexor retinaculum, trapezium I: Lateral side of metacarpal I	Median nerve (C8–T1 fibers)
HYPOTHENAR GROUP			
Flexor digiti minimi brevis	Flexes finger 5	O: Hamate bone, flexor retinaculum I: Proximal phalanx of finger 5	Ulnar nerve (C8–T1)
Abductor digiti minimi	Abducts finger 5	O: Pisiform bone, tendon of flexor carpi ulnaris I: Proximal phalanx of finger 5	Ulnar nerve (C8–T1)
Opponens digiti minimi	Opposition of finger 5	O: Hamate bone, flexor retinaculum I: Metacarpal bone V	Ulnar nerve (C8–T1)
MIDPALMAR GROUP			
Lumbricals (lŭm′bri-kălz) *lumbricus* = earthworm	Flexes 2nd–5th MP joints and extends 2nd–5th PIP and DIP joints	O: Tendons of flexor digitorum profundus I: Dorsal tendons on fingers 2–5	Median nerve (lateral two lumbricals 1, 2) and ulnar nerve (medial two lumbricals 3, 4)
Dorsal interossei (dōr′săl in′ter-os′ē-ī) *interossei* = between bones	Abducts fingers 2–5; flexes MP joints 2–5, and extends PIP and DIP joints	O: Adjacent, opposing faces of metacarpals I: Dorsal tendons on fingers 2–5	Ulnar nerve (C8–T1)
Palmar interossei (pal′mer)	Adducts fingers 2–5; flexes MP joints 2–5, and extends PIP and DIP joints	O: Metacarpal bones II, IV, V I: Sides of proximal phalanx bases for fingers 2, 4, and 5	Ulnar nerve (C8–T1)
Adductor pollicis	Adducts thumb	O: Oblique head: capitate bone, bases of metacarpals II, III Transverse head: metacarpal III I: Medial side of proximal phalanx of thumb	Ulnar nerve (C8–T1)

Table 12.8 — Summary of Muscle Actions at the Wrist and Hand

Hand Abduction	Hand Adduction	Wrist Extension	Wrist Flexion
Flexor carpi radialis	Extensor carpi ulnaris	Extensor digitorum	Flexor carpi radialis
Extensor carpi radialis brevis	Flexor carpi ulnaris	Extensor carpi radialis brevis	Flexor carpi ulnaris
Extensor carpi radialis longus		Extensor carpi radialis longus	Flexor digitorum superficialis
		Extensor carpi ulnaris	Flexor digitorum profundus
		(Extensor indicis)	(Palmaris longus)
		(Extensor pollicis longus)	(Flexor pollicis longus)
		(Extensor pollicis brevis)	
		(Abductor pollicis longus)	

Finger Abduction	Finger Adduction	IP Joint Extension	IP Joint Flexion
Dorsal interossei	Palmar interossei	Extensor digitorum	Flexor digitorum profundus
Abductor pollicis longus	Adductor pollicis	Extensor indicis	Flexor digitorum superficialis
Abductor pollicis brevis		Extensor pollicis brevis	Flexor pollicis longus
Abductor digiti minimi		Extensor pollicis longus	Flexor pollicis brevis
		Extensor digiti minimi	Flexor digiti minimi
		Lumbricals	
		Dorsal interossei	
		Palmar interossei	

Parentheses indicate only a slight effect.

Muscles That Move the Pelvic Girdle and Lower Limb

Key topics in this section:

- Major movements at the pelvic girdle and lower limb, and the muscles involved
- Muscles that move the thigh, and their organization into movement groups
- Muscles that move the leg, ankle, foot, and toes

The most powerful and largest muscles in the body are those of the lower limb. Several of these muscles cross and act upon two joints—the hip joint and the knee joint.

Muscles That Move the Hip Joint/Thigh

Note that in the subsequent discussion the phrases "moving the thigh" and "moving the hip joint" mean the same thing. The **fascia lata,** the deep fascia of the thigh, encircles the thigh muscles like a supportive stocking and tightly binds them. The fascia lata partitions the thigh muscles into compartments, each with its own blood and nerve supply. The anterior compartment muscles either extend the knee or flex the thigh. The muscles of the medial compartment act as adductors of the thigh. The muscle in the lateral compartment abducts the thigh. Most muscles of the posterior compartment act as both flexors of the knee and extensors of the thigh. Some of these muscles also abduct the thigh. We discuss the muscles that move the thigh first.

Most muscles that act on the thigh originate on the os coxae. These muscles stabilize the highly mobile hip joint and support the body during standing and walking. A majority of the muscles that move the thigh at the hip joint originate on the os coxae and insert on the femur.

Multiple muscles insert on the anterior femur and flex the thigh (**figure 12.15a**): The **psoas major** and the **iliacus** have different origins, but they share the common insertion at the lesser

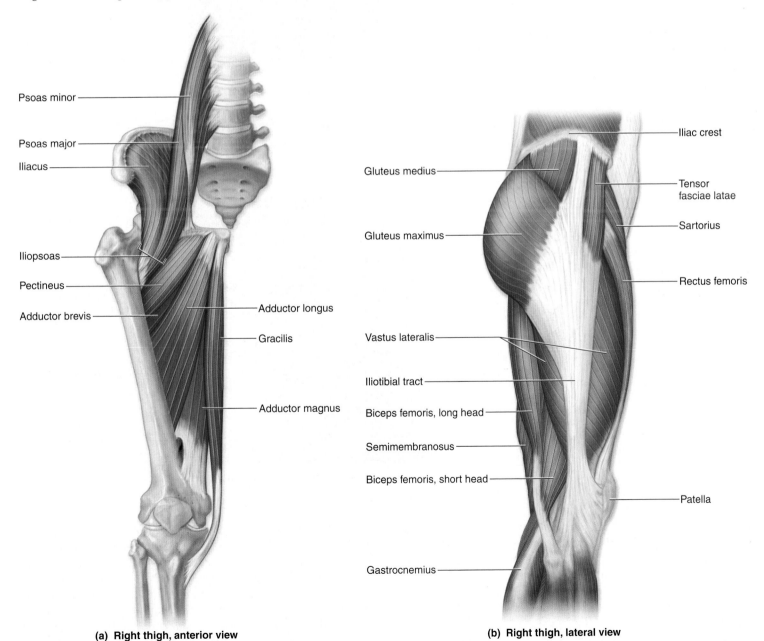

(a) Right thigh, anterior view

(b) Right thigh, lateral view

Figure 12.15

Muscles That Act on the Hip and Thigh. *(a)* Anterior, *(b)* lateral, and *(c)* deep posterior views of the right thigh. Most muscles that act on the thigh (femur) originate from the os coxae. *(continues on next page)*

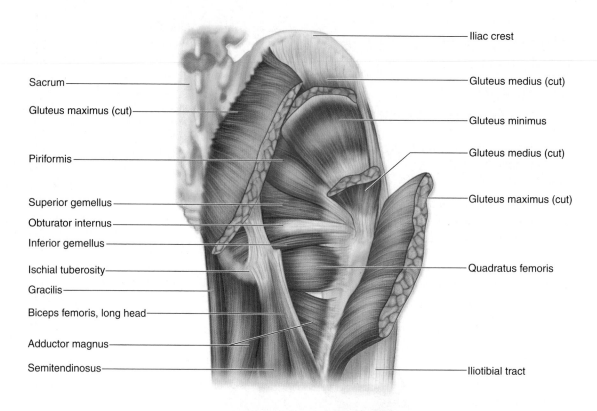

Iliac crest

Gluteus medius (cut)

Sacrum

Gluteus maximus (cut)

Gluteus minimus

Gluteus medius (cut)

Piriformis

Gluteus maximus (cut)

Superior gemellus

Obturator internus

Inferior gemellus

Ischial tuberosity

Gracilis

Quadratus femoris

Biceps femoris, long head

Adductor magnus

Semitendinosus

Iliotibial tract

Figure 12.15

Muscles That Act on the Hip and Thigh. *(continued)*

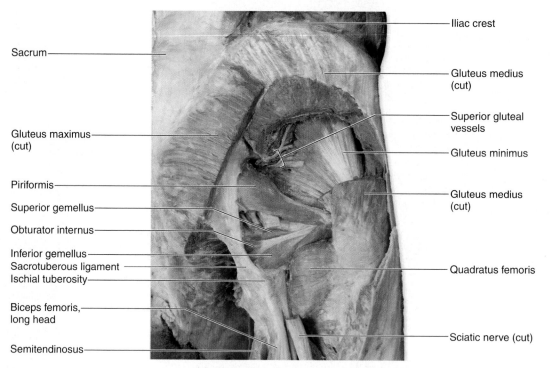

Iliac crest

Sacrum

Gluteus medius (cut)

Superior gluteal vessels

Gluteus maximus (cut)

Gluteus minimus

Piriformis

Superior gemellus

Gluteus medius (cut)

Obturator internus

Inferior gemellus

Sacrotuberous ligament

Ischial tuberosity

Quadratus femoris

Biceps femoris, long head

Semitendinosus

Sciatic nerve (cut)

(c) Right thigh, deep posterior view

trochanter of the femur. Collectively, the two muscles merge and insert on the femur as the **iliopsoas.** Together, these muscles work synergistically to flex the thigh. A long, thin muscle called the **sartorius** crosses over the anterior thigh and helps flex the thigh. The rectus femoris also flexes the thigh and is examined later in this chapter in connection with the thigh muscles that move the knee joint and leg.

Six muscles are located in the medial compartment of the thigh. Most of these muscles adduct the thigh, and some of them perform

additional functions. The **adductor longus, adductor brevis, gracilis,** and **pectineus** also flex the thigh. A fifth muscle, the **adductor magnus,** also extends and laterally rotates the thigh. The **obturator externus** does not adduct the thigh, but it laterally rotates the thigh.

On the lateral thigh is a single muscle called the **tensor fasciae latae** (figure 12.15b). It attaches to a lateral thickening of the fascia lata, called the **iliotibial tract** (or *iliotibial band*), which extends from the iliac crest to the lateral condyle of the tibia. The tensor fasciae latae abducts and medially rotates the thigh.

The posterior muscles that move the thigh include three gluteal muscles and the "hamstring" muscle group (figure 12.15*b,c*). The **gluteus maximus** is the largest and heaviest of the three gluteal muscles and one of the largest muscles in the body. It is the chief extensor of the thigh, and it laterally rotates the thigh. Deep to the gluteus maximus is the **gluteus medius,** a powerful abductor of the thigh. This muscle also medially rotates the thigh. Intramuscular injections are often given in this muscle. The smallest of the gluteal muscles is the **gluteus minimus.** It lies deep to the gluteus medius, with which it works to abduct and medially rotate the thigh.

Deep to the gluteal muscles are a group of muscles that laterally rotate the thigh and the hip joint, as when the legs are crossed with one ankle resting on the knee. These muscles are organized from superior to inferior within the posterior thigh as the **piriformis, superior gemellus, obturator internus, inferior gemellus,** and **quadratus femoris.**

Finally, the posterior thigh contains a group of muscles that are collectively referred to as the **hamstrings** because a ham is strung up by these muscles while being smoked. The hamstring muscles are the biceps femoris, semimembranosus, and semitendinosus. These muscles share a common origin on the ischial tuberosity of the os coxae, and insert on the leg. Thus, these muscles move both the thigh and the knee. Their primary thigh movement is extension. These muscles will be discussed again when we discuss movement at the knee joint and leg.

Table 12.9 summarizes the characteristics of the muscles that move the hip joint and thigh, and **table 12.10** groups these muscles according to their common actions on the hip joint and thigh.

WHAT DID YOU LEARN?

9 What two muscles insert on the iliotibial tract?

10 What muscles adduct the thigh?

11 Which muscles laterally rotate the thigh?

12 Identify the muscles that extend the thigh.

Table 12.9	Muscles That Move the Hip Joint/Thigh		
Group/Muscle	**Action**	**Origin/Insertion**	**Innervation**
ANTERIOR THIGH COMPARTMENT (THIGH FLEXORS)			
Psoas major (sō'as mā'jŏr) *psoa* = loin muscle	Flexes thigh	O: Transverse processes and bodies of T_{12}–L_5 vertebrae I: Lesser trochanter of femur with iliacus	Branches of lumbar plexus (L2–L3)
Iliacus (il-ī'ă-kŭs) *iliac* = ilium	Flexes thigh	O: Iliac fossa I: Lesser trochanter of femur with psoas major	Femoral nerve (L2–L3 fibers)
Sartorius (sar-tōr'ē-ŭs) *sartor* = tailor	Flexes thigh and rotates thigh laterally; flexes leg and rotates leg medially	O: Anterior superior iliac spine I: Tibial tuberosity, medial side	Femoral nerve (L2–L3 fibers)
Rectus femoris (rek'tŭs fem'ō-ris) *rectus* = straight *femoris* = femur	Flexes thigh, extends leg	O: Anterior inferior iliac spine I: Quadriceps tendon to patella and then patellar ligament to tibial tuberosity	Femoral nerve (L2–L4)
MEDIAL THIGH COMPARTMENT (THIGH ADDUCTORS)			
Adductor longus (a-dŭk'ter, -tōr) *adduct* = to move toward midline	Adducts thigh; flexes thigh	O: Pubis near pubic symphysis I: Linea aspera of femur	Obturator nerve (L2–L4)
Adductor brevis	Adducts thigh; flexes thigh	O: Inferior ramus and body of pubis I: Upper third of linea aspera of femur	Obturator nerve (L2–L3 fibers)
Gracilis (gră-cil'is) *gracilis* = slender	Adducts and flexes thigh; flexes leg	O: Inferior ramus and body of pubis I: Upper medial surface of tibia	Obturator nerve (L2–L4)
Pectineus (pek-ti'nē-us) *pectin* = comb	Adducts thigh; flexes thigh	O: Pectineal line of pubis I: Pectineal line of femur	Femoral nerve (L2–L4) or obturator nerve (L2–L4)
Adductor magnus (mag'nŭs) *magnus* = large	Adducts thigh; adductor part of muscle flexes thigh; hamstring part of muscle extends and laterally rotates thigh	O: Inferior ramus of pubis and ischial tuberosity I: Hamstring part: linea aspera of femur Adductor part: adductor tubercle of femur	Adductor part: Obturator nerve (L2–L4) Hamstring part: Tibial division of sciatic nerve (L2–L4 fibers)
Obturator externus (ob'too-rā-tŏr eks-ter'nŭs) *obturator* = any structure that occludes an opening *externus* = outside	Laterally rotates thigh	O: Margins of obturator foramen and obturator membrane I: Trochanteric fossa of posterior femur	Obturator nerve (L3–L4 fibers)

(continued on next page)

Table 12.9	Muscles That Move the Hip Joint/Thigh *(continued)*		
Group/Muscle	**Action**	**Origin/Insertion**	**Innervation**
LATERAL THIGH COMPARTMENT (THIGH ABDUCTOR)			
Tensor fasciae latae (ten'sŏr fash'ă lā'tē) *tensor* = to make tense *fascia* = band *lata* = wide	Abducts thigh; medially rotates thigh	O: Iliac crest and lateral surface of anterior superior iliac spine I: Iliotibial band	Superior gluteal nerve (L4–S1)
GLUTEAL GROUP			
Gluteus maximus (gloo-tē'ŭs mak'si-mŭs) *glutos* = buttock *maximus* = largest	Extends thigh; laterally rotates thigh	O: Iliac crest, sacrum, coccyx I: Iliotibial tract of fascia lata; linea aspera and gluteal tuberosity of femur	Inferior gluteal nerve (L5–S2)
Gluteus medius (mē'dē-ŭs) *medius* = middle	Abducts thigh; medially rotates thigh	O: Posterior iliac crest; lateral surface between posterior and anterior gluteal lines I: Greater trochanter of femur	Superior gluteal nerve (L4–S1)
Gluteus minimus (min'i-mŭs) *minimus* = smallest	Abducts thigh; medially rotates thigh	O: Lateral surface of ilium between inferior and anterior gluteal lines I: Greater trochanter of femur	Superior gluteal nerve (L4–S1)
DEEP MUSCLES OF THE GLUTEAL REGION (LATERAL THIGH ROTATORS)			
Piriformis (pir'i-fōr'mis) *pirum* = pear *forma* = form	Laterally rotates thigh	O: Anterolateral surface of sacrum I: Greater trochanter	Nerve to piriformis (S1–S2)
Superior gemellus (jē-mel'ŭs) *gemin* = twin, double	Laterally rotates thigh	O: Ischial spine and tuberosity I: Obturator internus tendon	Nerve to obturator internus (L5–S1)
Obturator internus (in-ter'nŭs) *internus* = inside	Laterally rotates thigh	O: Posterior surface of obturator membrane; margins of obturator foramen I: Greater trochanter	Nerve to obturator internus (L5–S1)
Inferior gemellus	Laterally rotates thigh	O: Ischial tuberosity I: Obturator internus tendon	Nerve to quadratus femoris (L5–S1)
Quadratus femoris	Laterally rotates thigh	O: Lateral border of ischial tuberosity I: Intertrochanteric crest of femur	Nerve to quadratus femoris (L5–S1)
POSTERIOR THIGH (HAMSTRING) COMPARTMENT (THIGH EXTENSORS AND LEG FLEXORS)			
Biceps femoris Long head Short head	Extends thigh (long head only); flexes leg (both long head and short head); laterally rotates leg	O: Long head: ischial tuberosity Short head: linea aspera of femur I: Head of fibula	Long head: tibial division of sciatic nerve (L4–S1 fibers) Short head: common fibular division of sciatic nerve (L5–S1 fibers)
Semimembranosus (sem'ē-mem-bră-nō'sŭs) *semi* = half *membranosus* = membrane	Extends thigh and flexes leg; medially rotates leg	O: Ischial tuberosity I: Posterior surface of medial condyle of tibia	Tibial division of sciatic nerve (L4–S1 fibers)
Semitendinosus (sem'ē-ten-di-nō'sŭs) *tendinosus* = tendon	Extends thigh and flexes leg; medially rotates leg	O: Ischial tuberosity I: Proximal medial surface of tibia	Tibial division of sciatic nerve (L4–S1 fibers)

Table 12.10	Summary of Muscle Actions at the Hip Joint/Thigh				
Abduction	**Adduction**	**Extension**	**Flexion**	**Lateral Rotation**	**Medial Rotation**
Gluteus medius	Adductor brevis, longus, magnus	**Gluteus maximus**	**Iliopsoas**	Adductor magnus (hamstring part)	Gluteus medius
Gluteus minimus	Gracilis	Adductor magnus (hamstring part)	Adductor brevis, longus, magnus (adductor part)	Gluteus maximus	Gluteus minimus
Tensor fasciae latae	Pectineus	Biceps femoris (long head)	Pectineus	Sartorius	Tensor fasciae latae
		Semimembranosus	Sartorius	Obturator externus	
		Semitendinosus	Rectus femoris	Obturator internus	
			Gracilis	Piriformis	
				Superior gemellus	
				Inferior gemellus	
				Quadratus femoris	

Boldface indicates a prime mover; others are synergists.

Thigh Muscles That Move the Knee Joint/Leg

The muscles that act on the knee form most of the mass of the thigh. Muscles in the thigh are separated by deep fascia into anterior, medial, and posterior compartments (**figure 12.16**).

The **anterior (extensor) compartment** of the thigh is composed of the large **quadriceps femoris,** the prime mover of knee extension and the most powerful muscle in the body. The quadriceps femoris is a composite muscle with four heads, as shown on **figure 12.17**:

■ The **rectus femoris** is on the anterior surface of the thigh; this muscle originates on the os coxae, and so it also flexes the thigh.

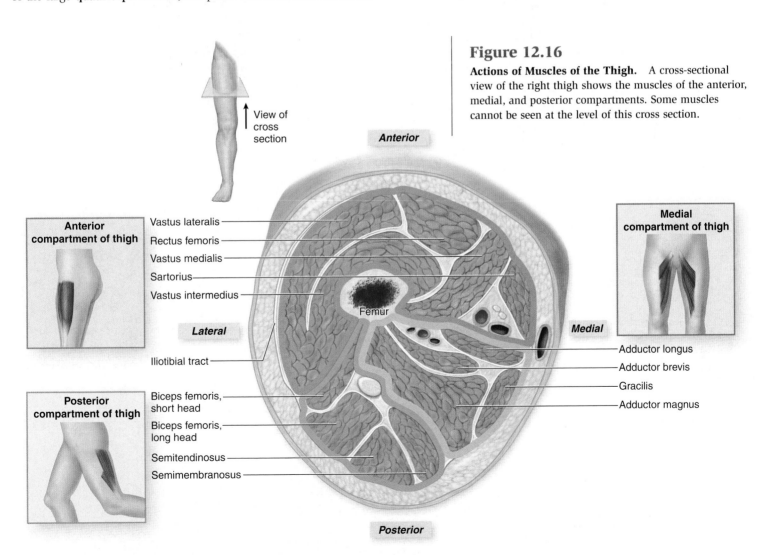

Figure 12.16

Actions of Muscles of the Thigh. A cross-sectional view of the right thigh shows the muscles of the anterior, medial, and posterior compartments. Some muscles cannot be seen at the level of this cross section.

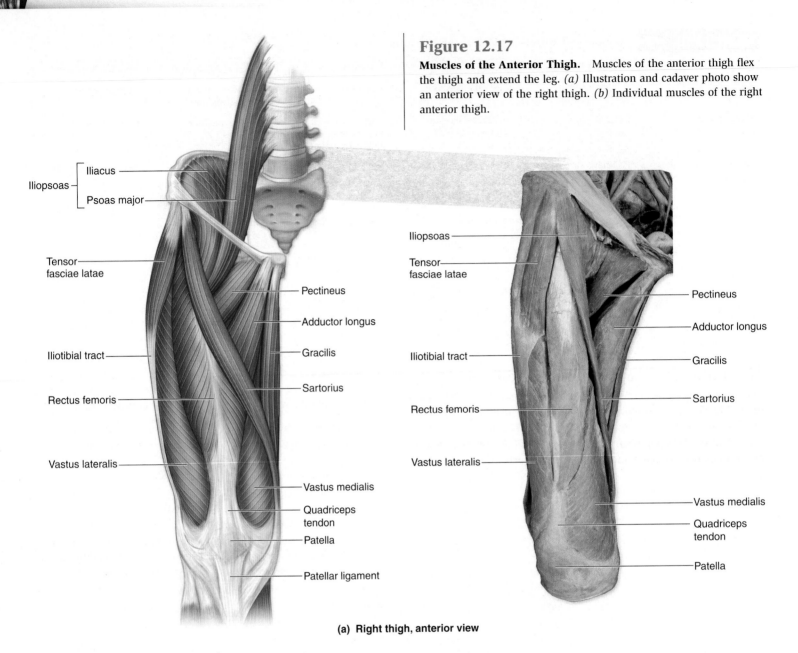

Figure 12.17

Muscles of the Anterior Thigh.   Muscles of the anterior thigh flex the thigh and extend the leg. *(a)* Illustration and cadaver photo show an anterior view of the right thigh. *(b)* Individual muscles of the right anterior thigh.

(a) Right thigh, anterior view

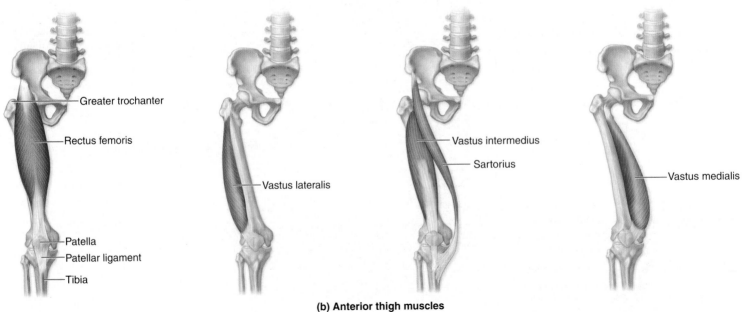

(b) Anterior thigh muscles

- The **vastus lateralis** forms the anterolateral surface of the thigh.
- The **vastus medialis** forms the anteromedial surface of the thigh.
- The **vastus intermedius** is positioned deep to the rectus femoris and sandwiched between the other two vastus muscles.

All four muscles converge on a single **quadriceps tendon,** which extends to the patella and then continues inferiorly as the **patellar ligament** and inserts on the tibial tuberosity. The patella becomes encased in this tendon and ligament. The quadriceps femoris is the great extensor muscle of the leg. It extends the knee when you stand up, take a step, or kick a ball, and it is very important in running, because it acts with the iliopsoas to flex the hip while the leg is off the ground.

The anterior compartment of the thigh contains another muscle worth noting. The long, straplike **sartorius** projects obliquely across the anterior surface of the thigh from the superolateral to the inferomedial side. It acts on both the hip and knee joints, flexing and laterally rotating the thigh while flexing and medially rotating the leg. This muscle is the longest in the body and is nicknamed the "tailor's muscle" because it helps us sit cross-legged, as tailors used to do.

The **medial (adductor) compartment** of the thigh is named for the muscles that adduct the thigh (see figures 12.15*a* and 12.17*a*). The **gracilis** is part of this compartment and also flexes the leg, since it spans the knee joint.

The **posterior (flexor) compartment** of the thigh contains the three hamstring muscles discussed previously **(figure 12.18)**. These muscles also flex the leg. The **biceps femoris** is a two-headed muscle that inserts on the lateral side of the leg. This muscle also can laterally rotate the leg when the leg is flexed. The long head of the biceps femoris originates on the ischial tuberosity with the semimembranosus and semitendinosus. The short head of the biceps femoris originates on the linea aspera of the femur. The short head cannot move the thigh, but it does help the other hamstring

muscles in flexing the leg. The **semimembranosus** is deep to the semitendinosus. It originates from the ischial tuberosity and attaches to the medial side of the leg. The **semitendinosus** is superficial to the semimembranosus and is attached to the medial leg. The semimembranosus and semitendinosus also medially rotate the leg when the leg is flexed.

Finally, several leg muscles span the knee joint and work to flex the leg. These muscles (gastrocnemius, plantaris, and popliteus) are discussed in the next section, as we examine muscles of the leg.

Table 12.11 summarizes the characteristics of the muscles that move the knee joint and leg.

WHAT DO YOU THINK?

3 Recall that if a muscle spans a joint, it must move the joint (and conversely, if a muscle doesn't span a joint, it cannot move that joint). Based on this rule, would you expect the iliopsoas to flex the knee joint/leg? Why or why not?

CLINICAL VIEW

Lower Limb Muscle Injuries

The muscle groups in the lower limbs are prone to injury, especially in people who are physically active. Two examples of such injuries are groin pull and strained (or pulled) hamstrings.

A **groin pull** is caused by tearing, stretching, or straining the proximal attachments of the medial muscles of the thigh—the adductor muscles of the leg and/or the iliopsoas muscle. This type of injury most frequently results from activities that involve rapid accelerations, as are called for in football, baseball, tennis, running, and soccer.

Strained or **pulled hamstrings** are common in athletes who perform quick starts and stops, run very fast, or sustain sudden lateral or medial stress to the knee joint. The violent muscular exertion required to perform these running feats sometimes causes the tendinous origins of the hamstrings to be avulsed (torn away) from their attachment on the ischial tuberosity. The biceps femoris is especially susceptible to this type of stress injury. Contusions (bruising), blood vessel rupture, pain, hematoma formation (accumulation of blood in soft tissue), and tearing of muscle fibers may accompany hamstring damage. To prevent this type of injury, experts recommend that athletes "warm up" and perform stretching exercises prior to running.

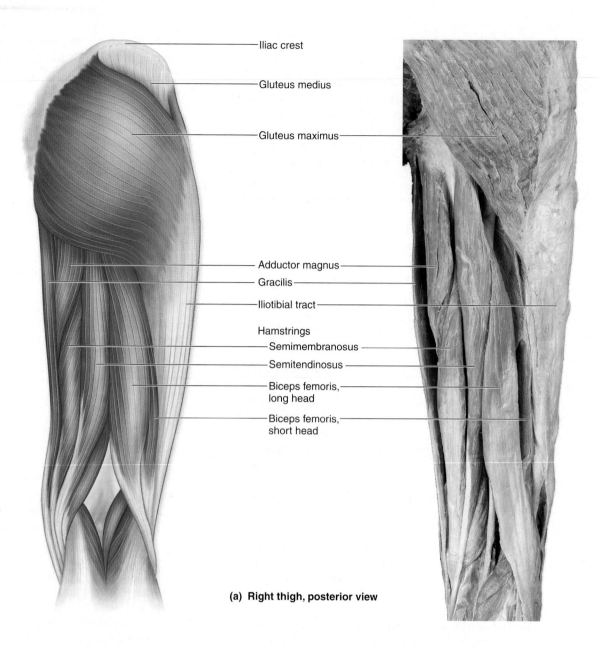

- Iliac crest
- Gluteus medius
- Gluteus maximus
- Adductor magnus
- Gracilis
- Iliotibial tract
- Hamstrings
  - Semimembranosus
  - Semitendinosus
  - Biceps femoris, long head
  - Biceps femoris, short head

(a) Right thigh, posterior view

Figure 12.18

Muscles of the Gluteal Region and Posterior Thigh. Muscles of the posterior thigh extend the thigh and flex the leg. *(a)* Illustration and cadaver photo show the gluteal and posterior muscles of the right thigh. *(b)* Individual muscles that extend the thigh are shown in bold. Note that the short head of biceps femoris does not participate in thigh extension.

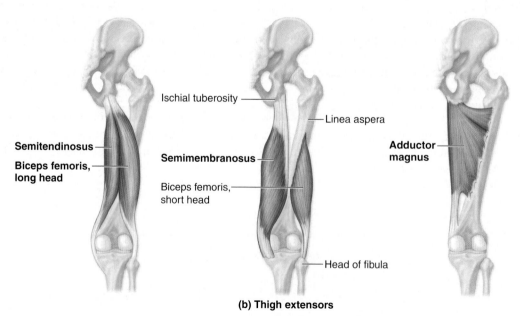

- **Semitendinosus**
- **Biceps femoris, long head**
- Ischial tuberosity
- Linea aspera
- **Semimembranosus**
- Biceps femoris, short head
- Head of fibula
- **Adductor magnus**

(b) Thigh extensors

Table 12.11	Thigh Muscles That Move the Knee Joint/Leg		
Group/Muscle	**Action**	**Origin/Insertion**	**Innervation**
LEG EXTENSORS (ANTERIOR THIGH MUSCLES)			
Quadriceps Femoris			
Rectus femoris	Extends leg; flexes thigh	O: Anterior inferior iliac spine I: Quadriceps tendon to patella and then patellar ligament to tibial tuberosity	Femoral nerve (L2–L4)
Vastus intermedius (vas′tŭs inter-mē′dē-ŭs) *vastus* = great *intermedius* = intermediate	Extends leg	O: Anterolateral surface of femur I: Quadriceps tendon to patella and then patellar ligament to tibial tuberosity	Femoral nerve (L2–L4)
Vastus lateralis (lat-er-ăl′is)	Extends leg	O: Greater trochanter and linea aspera I: Quadriceps tendon to patella and then patellar ligament to tibial tuberosity	Femoral nerve (L2–L4)
Vastus medialis (mē-dē-ăl′is)	Extends leg	O: Intertrochanteric line and linea aspera of femur I: Quadriceps tendon to patella and then patellar ligament to tibial tuberosity	Femoral nerve (L2–L4)
LEG FLEXORS			
Sartorius	Flexes thigh and rotates thigh laterally; flexes leg and rotates leg medially	See table 12.9	
Gracilis	Flexes and adducts thigh; flexes leg	See table 12.9	
Hamstrings			
(Biceps femoris, semimembranosus, semitendinosus)	Extend thigh and flex leg; rotate leg laterally	See table 12.9	

Leg Muscles

The muscles that move the ankle, foot, and toes are housed within the leg and are called the **crural muscles.** Some of these muscles also help flex the leg. The deep fascia partitions the leg musculature into three compartments (anterior, lateral, and posterior), each with its own nerve and blood supply **(figure 12.19).**

Anterior compartment leg muscles dorsiflex the foot and/or extend the toes **(figure 12.20).** The **extensor digitorum longus** sends four long tendons to attach to the dorsal surface of toes 2–5. This muscle dorsiflexes the foot and extends toes 2–5. The **extensor hallucis longus** sends a tendon to the dorsum of the great toe (hallux), and so it dorsiflexes the foot and extends the great toe. The **fibularis tertius** (or *peroneus tertius*) extends from the extensor digitorum longus muscle. It dorsiflexes and weakly everts the foot. The **tibialis anterior** is the primary dorsiflexor of the foot at the ankle.

This muscle attaches to the medial plantar side of the foot, so it also inverts the foot. Analogous to the wrist, tendons of the muscles within the anterior compartment are held tightly against the ankle by multiple deep fascia thickenings, collectively referred to as the **extensor retinaculum.**

The **lateral compartment** leg muscles contain two synergistic muscles that are very powerful evertors of the foot and weak plantar flexors **(figure 12.21).** The long, flat **fibularis longus** (or *peroneus longus*) is a superficial lateral muscle that covers the fibula. Its tendon attaches to the plantar side of the foot on the base of metatarsal I and the medial cuneiform. The **fibularis brevis** (or *peroneus brevis*) lies deep to the fibularis longus. Its tendon inserts onto the base of the fifth metatarsal.

The **posterior compartment** of the leg is composed of seven muscles that are separated into superficial and deep groups

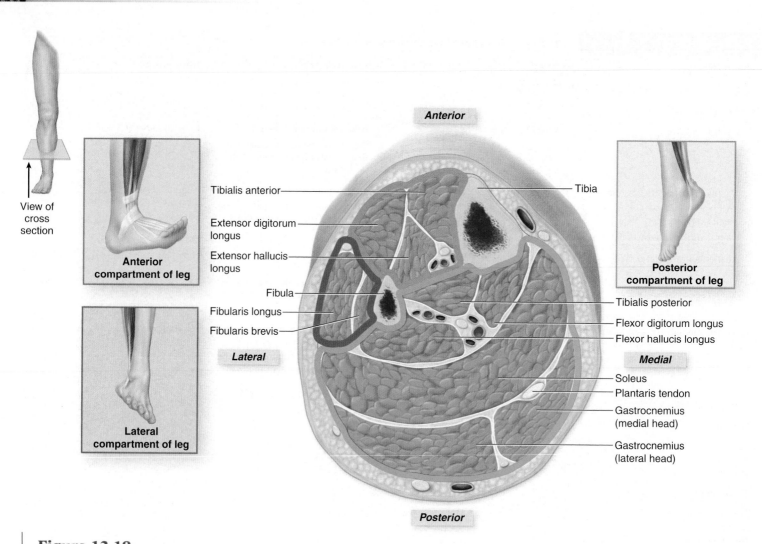

View of cross section

Anterior compartment of leg

Lateral compartment of leg

Anterior

Tibialis anterior

Extensor digitorum longus

Extensor hallucis longus

Fibula

Fibularis longus

Fibularis brevis

Lateral

Tibia

Posterior compartment of leg

Tibialis posterior

Flexor digitorum longus

Flexor hallucis longus

Medial

Soleus

Plantaris tendon

Gastrocnemius (medial head)

Gastrocnemius (lateral head)

Posterior

Figure 12.19

Actions of Muscles of the Leg. A cross-sectional view of the right leg shows the muscles of the anterior, lateral, and posterior compartments, which perform different movements.

(figure 12.22). The superficial muscles and most of the deep muscles plantar flex the foot at the ankle. The **superficial layer** of the posterior compartment contains three muscles: gastrocnemius, soleus, and plantaris. The **gastrocnemius** is the most superficial muscle. It has two thick muscle bellies, the lateral head and the medial head, that collectively form the prominence on the posterior part of the leg often referred to as the "calf." This muscle spans both the knee and the ankle joints; it flexes the leg and plantar flexes the foot. The **soleus** is a broad, flat muscle deep to the gastrocnemius that resembles a flat fish. This muscle plantar flexes the foot. The **plantaris** is a small muscle that is absent in some individuals. It projects obliquely between the gastrocnemius and soleus muscles. This muscle is a weak leg flexor and plantar flexor of the foot.

The gastrocnemius and soleus are collectively known as the **triceps surae,** and together they are the most powerful plantar flexors of all of the leg muscles. These two muscles share a common tendon of insertion, the **calcaneal tendon** (or *Achilles tendon*).

The **deep layer** of the posterior compartment contains four muscles. The **flexor digitorum longus** attaches to the distal phalanges of toes 2–5, plantar flexes the foot, and flexes the MP, PIP, and DIP joints of toes 2–5. The **flexor hallucis longus** originates on the fibula, and yet its tendon travels medially and runs along the plantar side of the foot to attach to the distal phalanx of the great toe. This muscle plantar flexes the foot and flexes the great toe. The **tibialis posterior** is the deepest of the posterior compartment muscles. It plantar flexes and inverts the foot. The **popliteus** forms the floor of the popliteal fossa, and acts to flex the leg. This muscle also medially rotates the tibia slightly to "unlock" the fully extended knee joint. This muscle originates and inserts in the popliteal region, so it only moves the knee, not the foot.

Table 12.12 lists the characteristics of the muscles that move the leg. **Table 12.13** groups muscles according to their action on the leg. Note that many thigh and leg muscles are involved with leg flexion.

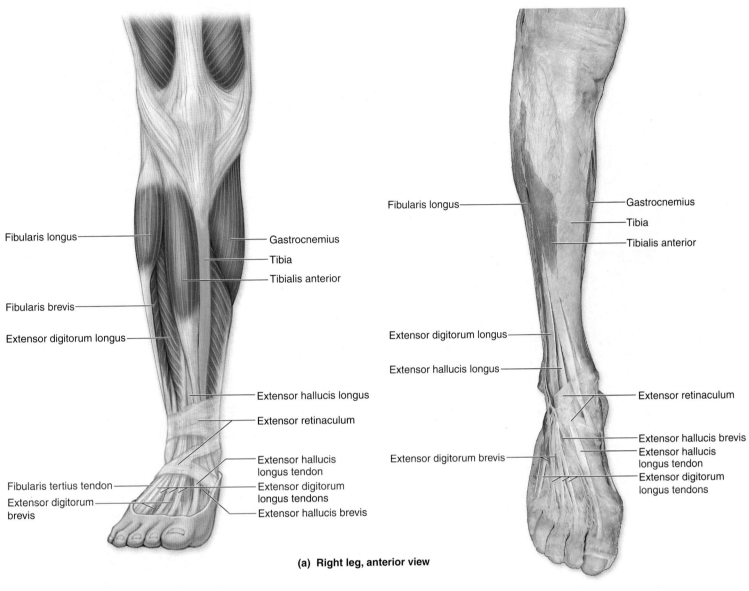

Fibularis longus

Gastrocnemius

Tibia

Tibialis anterior

Fibularis brevis

Extensor digitorum longus

Extensor hallucis longus

Extensor retinaculum

Fibularis tertius tendon

Extensor hallucis longus tendon

Extensor digitorum brevis

Extensor digitorum longus tendons

Extensor hallucis brevis

Fibularis longus

Gastrocnemius

Tibia

Tibialis anterior

Extensor digitorum longus

Extensor hallucis longus

Extensor retinaculum

Extensor hallucis brevis

Extensor hallucis longus tendon

Extensor digitorum brevis

Extensor digitorum longus tendons

(a) Right leg, anterior view

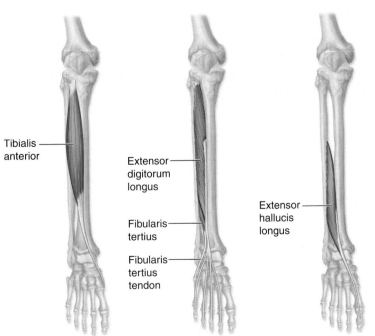

Tibialis anterior

Extensor digitorum longus

Fibularis tertius

Fibularis tertius tendon

Extensor hallucis longus

(b) Anterior leg muscles

Figure 12.20

Muscles of the Anterior Leg.   The anterior muscles of the leg dorsiflex the foot and extend the toes. (*a*) Illustration and cadaver photo show an anterior view of the right leg. (*b*) Individual muscles of the right anterior leg.

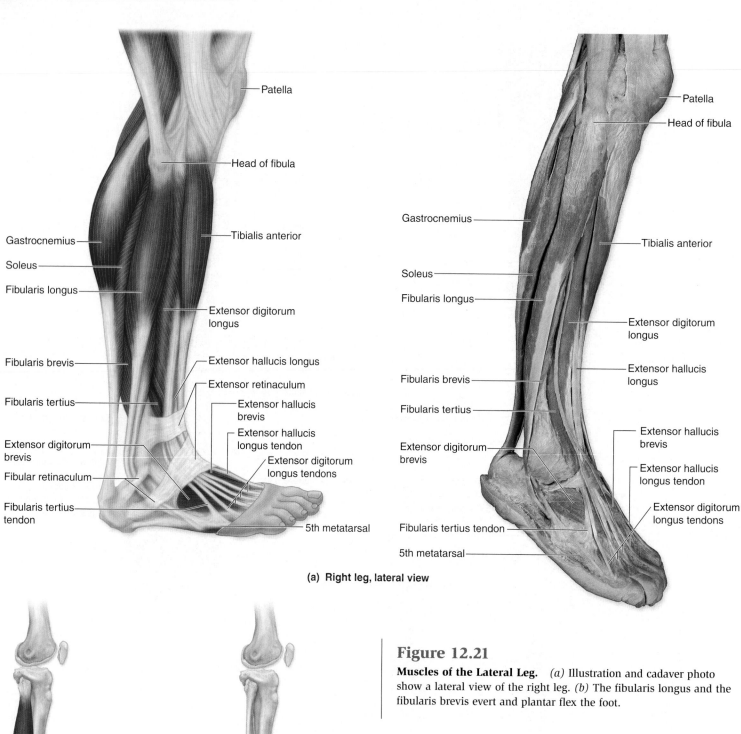

Patella

Head of fibula

Tibialis anterior

Gastrocnemius

Soleus

Fibularis longus

Extensor digitorum longus

Fibularis brevis

Extensor hallucis longus

Fibularis tertius

Extensor retinaculum

Extensor hallucis brevis

Extensor hallucis longus tendon

Extensor digitorum brevis

Extensor digitorum longus tendons

Fibular retinaculum

Fibularis tertius tendon

5th metatarsal

Patella

Head of fibula

Gastrocnemius

Tibialis anterior

Soleus

Fibularis longus

Extensor digitorum longus

Extensor hallucis longus

Fibularis brevis

Fibularis tertius

Extensor hallucis brevis

Extensor digitorum brevis

Extensor hallucis longus tendon

Extensor digitorum longus tendons

Fibularis tertius tendon

5th metatarsal

(a) Right leg, lateral view

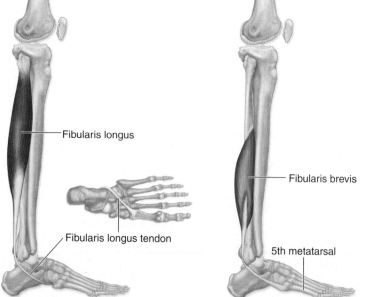

Fibularis longus

Fibularis brevis

Fibularis longus tendon

5th metatarsal

(b) Lateral leg muscles

Figure 12.21

Muscles of the Lateral Leg. *(a)* Illustration and cadaver photo show a lateral view of the right leg. *(b)* The fibularis longus and the fibularis brevis evert and plantar flex the foot.

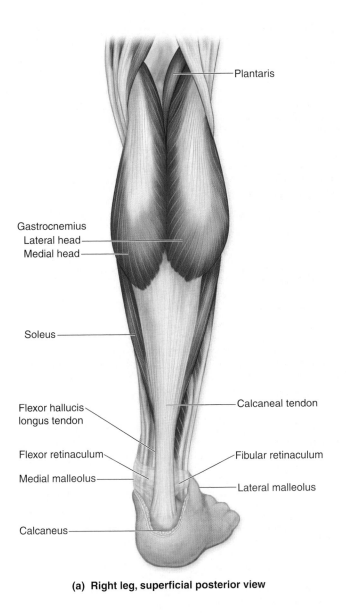

Plantaris

Gastrocnemius
Lateral head
Medial head

Soleus

Flexor hallucis
longus tendon

Flexor retinaculum

Medial malleolus

Calcaneus

Calcaneal tendon

Fibular retinaculum

Lateral malleolus

(a) Right leg, superficial posterior view

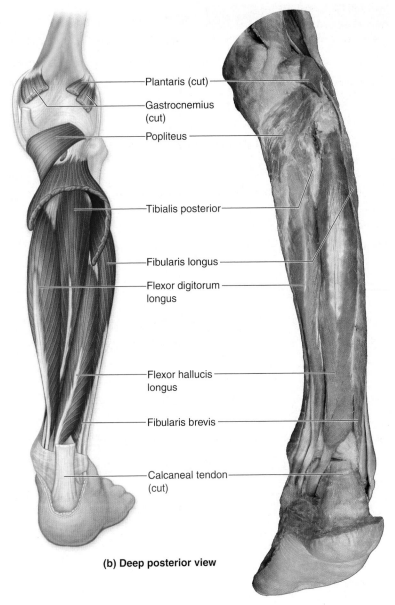

Plantaris (cut)

Gastrocnemius
(cut)

Popliteus

Tibialis posterior

Fibularis longus

Flexor digitorum
longus

Flexor hallucis
longus

Fibularis brevis

Calcaneal tendon
(cut)

(b) Deep posterior view

Tibia

Tibialis
posterior

Interosseous
membrane

Tarsal and
metatarsal
bones

Popliteus

Flexor digitorum
longus

Distal
phalanges of
toes 2–5

Fibula

Flexor
hallucis
longus

Distal
phalanx
of hallux

(c) Deep posterior leg muscles

Figure 12.22

Muscles of the Posterior Leg.   The posterior muscles of the leg
plantar flex the foot and flex the toes. *(a)* Superficial and *(b)* deep
views of the posterior right leg. *(c)* Selected individual muscles of the
deep posterior compartment.

Table 12.12	Leg Muscles		
Group/Muscle	**Action**	**Origin/Insertion**	**Innervation**
ANTERIOR COMPARTMENT (DORSIFLEXORS AND TOE EXTENSORS)			
Extensor digitorum longus	Extends toes 2–5; dorsiflexes foot	O: Lateral condyle of tibia; anterior surface of fibula; interosseous membrane I: Distal phalanges of toes 2–5	Deep fibular nerve (L4–S1)
Extensor hallucis longus (hal′ū-sis) *hallux* = great toe	Extends great toe (1); dorsiflexes foot	O: Anterior surface of fibula; interosseous membrane I: Distal phalanx of great toe (1)	Deep fibular nerve (L4–S1)
Fibularis tertius (fib-ū-lar′is ter′shē-ŭs) *fibularis* = fibula *tertius* = third	Dorsiflexes and weakly everts foot	O: Anterior distal surface of fibula; interosseous membrane I: Base of metatarsal V	Deep fibular nerve (L5–S1)
Tibialis anterior (tib-ē-a′lis)	Dorsiflexes foot; inverts foot	O: Lateral condyle and proximal shaft of tibia; interosseous membrane I: Metatarsal I and first (medial) cuneiform	Deep fibular nerve (L4–S1)
LATERAL COMPARTMENT (EVERTORS AND WEAK PLANTAR FLEXORS)			
Fibularis longus	Everts foot; weak plantar flexor	O: Head and superior 2/3 of shaft of fibula; lateral condyle of tibia I: Base of metatarsal I; medial cuneiform bone	Superficial fibular nerve (L5–S2)
Fibularis brevis	Everts foot; weak plantar flexor	O: Midlateral shaft of fibula I: Base of metatarsal V	Superficial fibular nerve (L5–S2)
POSTERIOR COMPARTMENT (PLANTAR FLEXORS, FLEXORS OF THE LEG AND TOES)			
Superficial Layer			
Triceps surae			

Gastrocnemius (gas-trok-nē′mē-ŭs)
  gaster = belly
  kneme = leg
	Flexes leg; plantar flexes foot	O: Superior posterior surfaces of lateral and medial condyles of femur
		I: Calcaneus (via calcaneal tendon)
			Tibial nerve (L4–S1 fibers)

Soleus (sō-lē′ŭs) *soleus* = flat fish	Plantar flexes foot	O: Head and proximal shaft of fibula; medial border of tibia I: Calcaneus (via calcaneal tendon)	Tibial nerve (L4–S1 fibers)
Plantaris (plan-tār′is) *planta* = sole of foot	Weak leg flexor and plantar flexor	O: Lateral supracondylar ridge of femur I: Posterior region of calcaneus	Tibial nerve (L4–S1 fibers)
Deep Layer			
Flexor digitorum longus	Plantar flexes foot; flexes MP, PIP, and DIP joints of toes 2–5	O: Posteromedial surface of tibia I: Distal phalanges of toes 2–5	Tibial nerve (L5–S1 fibers)
Flexor hallucis longus	Plantar flexes foot; flexes MP and IP joints of great toe (1)	O: Posterior inferior 2/3 of fibula I: Distal phalanx of great toe (1)	Tibial nerve (L5–S1 fibers)
Tibialis posterior	Plantar flexes foot; inverts foot	O: Fibula, tibia, and interosseous membrane I: Metatarsals II–IV; navicular bone; cuboid bone; all cuneiforms	Tibial nerve (L5–S1 fibers)
Popliteus (pop-li-tē′ŭs) *poplit* = back of knee	Flexes leg; medially rotates tibia to unlock the knee	O: Lateral condyle of femur I: Posterior, proximal surface of tibia	Tibial nerve (L4–L5 fibers)

Table 12.13	Summary of Muscle Actions at the Knee Joint/Leg	
Extension	**Flexion**	
Quadriceps femoris:	Sartorius	
Rectus femoris	Gracilis	
Vastus lateralis	Adductor longus, brevis, magnus	
Vastus intermedius	Biceps femoris	
Vastus medialis	Semimembranosus	
	Semitendinosus	
	Gastrocnemius	
	Popliteus	
	(Plantaris)	

Parentheses indicate only a slight effect.

Intrinsic Muscles of the Foot

The intrinsic muscles of the foot both originate and insert within the foot. They support the arches and move the toes to aid locomotion. Most of these muscles are comparable to the intrinsic muscles of the hand, meaning that they have similar names and locations. However, the intrinsic muscles of the foot rarely perform all the precise movements their names suggest.

The intrinsic foot muscles form a dorsal group and a plantar group. The dorsal group contains only two muscles: the extensor hallucis brevis and the extensor digitorum brevis (see figures 12.20 and 12.21). The **extensor hallucis brevis** extends the MP joint of the great toe. The **extensor digitorum brevis** is deep to the tendons

Plantar Fasciitis

Plantar fasciitis (fas-ē-ī′tis) is an inflammation caused by chronic irritation of the plantar aponeurosis (fascia). Most often, the inflammation is greatest at the origin of the plantar aponeurosis on the calcaneus bone. This condition is frequently caused by overexertion that stresses the plantar fascia, but it may also be related to age, since loss of elasticity in the aponeurosis, which is known to occur with age, seems to be a factor in developing this condition. Other factors associated with plantar fasciitis include weight-bearing activities (lifting heavy objects, jogging, or walking), excessive body weight, improperly fitting shoes, and poor biomechanics (wearing high-heeled shoes or having flat feet). Because plantar fasciitis can occur as a consequence of repetitively pounding the feet on the ground, this condition has become the most common cause of heel pain in runners.

of the extensor digitorum longus, and it extends the MP, PIP, and DIP joints of toes 2–4.

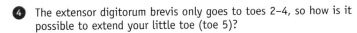

WHAT DO YOU THINK?

4 The extensor digitorum brevis only goes to toes 2–4, so how is it possible to extend your little toe (toe 5)?

The plantar surface of the foot is supported by the **plantar aponeurosis** formed from the deep fascia of the foot. This aponeurosis

Compartment Syndrome

Cross sections of the limbs reveal that the musculature is surrounded by deep fascia. This deep fascia is tough and unyielding; it wraps around the muscle like support hose, and it groups the muscles into compartments. Thus, the compartments of the leg (anterior, lateral, and posterior) are subdivided by this fascia. Sometimes problems develop within these compartments.

Shin splints, also called *shin splint syndrome*, refers specifically to soreness or pain somewhere along the length of the tibia, usually on the inferior portion. Causes of the syndrome include one or more of the following: (1) stress fractures of the tibia; (2) tendonitis involving muscles of the anterior compartment muscles of the leg, often the tibialis anterior muscle; and (3) inflammation of the periosteum, called periostitis. Shin splints often occurs in runners or joggers who are either new to the sport or poorly conditioned. Athletes who run on exceptionally hard surfaces (concrete, asphalt) or wear running shoes that do not properly support the foot are susceptible to shin splints. Some healthcare professionals consider shin splints a compartment syndrome of the anterior compartment of the leg.

Generally, **compartment syndrome** is a condition in which the blood vessels within a limb compartment become compressed as a result

of inflammation and swelling secondary to muscle strain, contusion, or overuse. For example, compartment syndrome can occur in an individual who suddenly embarks on an intensive exercise regimen. More severe compartment syndrome can occur due to trauma to the limb compartment, such as a bone fracture or rupture of a blood vessel.

The syndrome develops when the overworked muscles start to swell, compressing other structures within the compartment. Since the connective tissue sheets forming the compartment boundaries are tight and cannot stretch, any accumulating fluid or blood increases pressure on the muscles, nerves, and blood vessels. Both the circulatory supply and the nerves of the compartment become compressed and compromised. The reduced blood flow (known as **ischemia** [is-kē′mē-ă; *isch* = to keep back, *haima* = blood]) leads to **hypoxia** (hī-pok′sē-ă; *hypo* = under, *oxo* = oxygen), which is a lack of oxygen within the compartment. If the blood flow is not restored, this situation can lead to death of nerves within 2 hours and death of skeletal muscle within 6 hours. Although the damaged nerves may regenerate after the compartment syndrome has been resolved, the loss of muscle cells is irreversible. Mild cases of compartment syndrome may be treated by immobilizing and resting the affected limb. In more severe cases, the fascia may have to be incised (cut) to relieve the pressure and decompress the affected compartment.

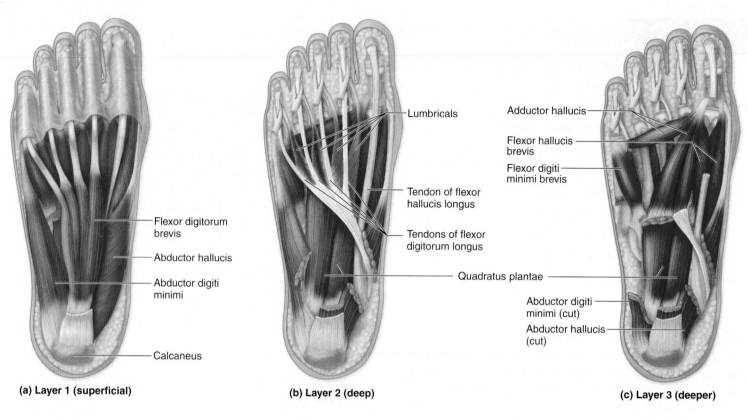

Flexor digitorum brevis

Abductor hallucis

Abductor digiti minimi

Calcaneus

(a) Layer 1 (superficial)

Lumbricals

Tendon of flexor hallucis longus

Tendons of flexor digitorum longus

Quadratus plantae

Abductor digiti minimi (cut)

Abductor hallucis (cut)

(b) Layer 2 (deep)

Adductor hallucis

Flexor hallucis brevis

Flexor digiti minimi brevis

(c) Layer 3 (deeper)

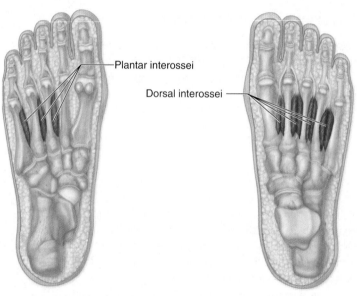

Plantar interossei

Dorsal interossei

(d) Layer 4 (deepest), plantar view

(e) Layer 4 (deepest), dorsal view

Figure 12.23

Plantar Intrinsic Muscles of the Foot. These muscles move the toes. *(a)* Superficial, *(b)* deep, and *(c)* deeper layers of the intrinsic muscles of the right foot. *(d)* Plantar and *(e)* dorsal views of the deepest layers.

extends between the phalanges of the toes and the calcaneus. It also encloses the plantar muscles of the foot.

The plantar muscles are grouped into four layers **(figure 12.23)**. The first layer of muscles is the most superficial. It includes the **flexor digitorum brevis** that attaches to the middle phalanges of the toes, so it can flex the MP and PIP joints (but *not* the DIP joints) of toes 2–5. Also included in this layer are the **abductor hallucis,** which abducts the great toe (1), and the **abductor digiti minimi,** which abducts the small toe (5).

The second layer (deep to the first) consists of the thick, medial **quadratus plantae,** which attaches to the tendons of the

flexor digitorum longus. Note how the flexor digitorum longus tendons attach to the toes at an angle. The quadratus plantae pulls on the slanted flexor digitorum longus tendons in a posterior fashion so that toes 2–5 may be flexed properly, not at an angle. In addition, this layer contains the four **lumbrical muscles,** small muscles that attach to the tendons of the flexor digitorum longus and serve to flex the MP joints and extend the DIP and PIP joints of toes 2–5.

The third layer (deeper) contains the **adductor hallucis,** which adducts the great toe (1). Also in this layer are the medial **flexor hallucis brevis** and the lateral **flexor digiti minimi brevis,**

Table 12.14	Intrinsic Muscles of the Foot		
Group/Muscle	**Action**	**Origin/Insertion**	**Innervation**
DORSAL SURFACE (TOE EXTENSORS)			
Extensor hallucis brevis	Extends MP joint of great toe (1)	O: Calcaneus and inferior extensor retinaculum I: Proximal phalanx of great toe (1)	Deep fibular nerve (S1–S2 fibers)
Extensor digitorum brevis	Extends MP, PIP, and DIP joints of toes 2–4	O: Calcaneus and inferior extensor retinaculum I: Middle phalanges of toes 2–4	Deep fibular nerve (S1–S2 fibers)
PLANTAR SURFACE (TOE FLEXORS, ADDUCTORS, ABDUCTORS)			
Layer 1 (Superficial)			
Flexor digitorum brevis	Flexes MP and PIP joints of toes 2–5	O: Calcaneus I: Middle phalanges of toes 2–5	Medial plantar nerve (S2–S3)
Abductor hallucis	Abducts great toe (1)	O: Calcaneus I: Medial side of proximal phalanx of great toe (1)	Medial plantar nerve (S2–S3)
Abductor digiti minimi	Abducts toe 5	O: Calcaneus (inferior surface tuberosity) I: Lateral side of proximal phalanx of toe 5	Lateral plantar nerve (S2–S3)
Layer 2 (Deep)			
Quadratus plantae (plan′tē) *planta* = sole of foot	Pulls on flexor digitorum longus tendons to flex toes 2–5	O: Calcaneus, long plantar ligament I: Tendons of flexor digitorum longus	Lateral plantar nerve (S2–S3)
Lumbricals	Flexes MP joints and extends PIP and DIP joints of toes 2–5	O: Tendons of flexor digitorum longus I: Tendons of extensor digitorum longus	Medial plantar nerve (1st lumbrical); lateral plantar nerve (2nd–4th lumbricals)
Layer 3 (Deeper)			
Adductor hallucis	Adducts great toe (1)	O: Transverse head: capsules of MP joints III–V Oblique head: bases of metatarsals II–IV I: Lateral side of proximal phalanx of great toe (1)	Lateral plantar nerve (S2–S3)
Flexor hallucis brevis	Flexes MP joint of great toe (1)	O: Cuboid and lateral (3rd) cuneiform bones I: Proximal phalanx of great toe (1)	Medial plantar nerve (S2–S3)
Flexor digiti minimi brevis	Flexes MP joint of toe 5	O: Metatarsal V I: Proximal phalanx of toe 5	Lateral plantar nerve (S2–S3)
Layer 4 (Deepest)			
Dorsal interossei	Abducts toes	O: Adjacent sides of metatarsals I: Sides of proximal phalanges of toes 2–4	Lateral plantar nerve (S2–S3)
Plantar interossei	Adducts toes	O: Sides of metatarsals III–V I: Medial side of proximal phalanges of toes 3–5	Lateral plantar nerve (S2–S3)

which flex the great toe (1) and the small toe (5), respectively. The fourth layer (deepest) consists of four **dorsal interossei** and three **plantar interossei.** The dorsal interossei abduct the toes, while the plantar interossei adduct the toes.

Table 12.14 summarizes the characteristics of the intrinsic muscles of the foot, and **table 12.15** groups the leg and intrinsic foot muscles according to their common actions on the foot.

WHAT DID YOU LEARN?

13 What are the basic functions of the hamstring muscles?

14 Identify the muscles that flex the knee joint/leg.

15 What is the action of the muscles in the lateral compartment of the leg?

16 Identify the intrinsic muscles of the foot that extend the toes.

Table 12.15		**Summary of Leg and Foot Muscle Actions at the Foot and Toes**	
FOOT			
Dorsiflexion	**Plantar Flexion**	**Eversion**	**Inversion**
Tibialis anterior	**Gastrocnemius**	**Fibularis longus**	**Tibialis posterior**
Extensor digitorum longus	**Soleus**	**Fibularis brevis**	**Tibialis anterior**
(Extensor hallucis longus)	Flexor digitorum longus	(Fibularis tertius)	
(Fibularis tertius)	Flexor hallucis longus		
	Tibialis posterior		
	(Fibularis longus)		
	(Fibularis brevis)		
	(Plantaris)		
TOES			
Extension	**Flexion**	**Abduction**	**Adduction**
Extensor digitorum longus	Flexor digitorum longus	Abductor hallucis	Adductor hallucis
Extensor hallucis longus	Flexor hallucis longus	Abductor digiti minimi	Plantar interossei
Extensor digitorum brevis	Flexor digitorum brevis	Dorsal interossei	
Extensor hallucis brevis	Flexor hallucis brevis		
	Flexor digiti minimi brevis		

Boldface indicates a prime mover; others are synergists. Parentheses indicate only a slight effect.

CLINICAL TERMS

adductor strain Extreme thigh abduction results in pulled adductor muscles, especially the adductor longus.
electromyography (ē-lek′trō-mī-og′ră-fē) Recording of electrical activity generated in a muscle for diagnostic purposes.

hallux valgus (hal′ŭks văl′gus; *valgus* = turned outward) Deviation of the tip of the great toe to the medial side of the foot away from the second toe.

CHAPTER SUMMARY

- The appendicular muscles stabilize and help move the pectoral and pelvic girdles, and move the upper and lower limbs.

Muscles That Move the Pectoral Girdle and Upper Limb 353

- Five groups of muscles are associated with pectoral girdle and upper limb movement: muscles that move (1) the pectoral girdle, (2) the glenohumeral joint/arm, (3) the elbow joint/forearm, (4) the wrist joint, hand, and fingers, and (5) the intrinsic muscles of the hand.

Muscles That Move the Pectoral Girdle 353

- Anterior thoracic muscles protract, rotate, and/or depress the scapula; one anterior thoracic muscle depresses the clavicle.
- The posterior thoracic muscles elevate, retract, and/or rotate the scapula.

Muscles That Move the Glenohumeral Joint/Arm 358

- The pectoralis major flexes the arm, and the latissimus dorsi extends it, while both adduct and medially rotate it.
- Seven scapular muscles move the humerus, individually or together causing abduction, adduction, extension, flexion, lateral rotation, or medial rotation of the arm.
- Collectively, the subscapularis, supraspinatus, infraspinatus, and teres minor are called the rotator cuff muscles. They provide strength and stability to the glenohumeral joint, and move the humerus.

Arm and Forearm Muscles That Move the Elbow Joint/Forearm 361

- The principal flexors are on the anterior side of the arm, and the principal extensors are on the posterior side of the arm.
- Muscles that move the forearm are the biceps brachii (flexes and supinates) and the triceps brachii (extends).
- The pronator teres and pronator quadratus cause pronation only of the forearm; the supinator opposes this movement.

Muscles That Move the Pectoral Girdle and Upper Limb (continued) 353	**Forearm Muscles That Move the Wrist Joint, Hand, and Fingers 364** ■ Strong fibrous bands, called the flexor retinaculum and the extensor retinaculum, hold tendon sheaths of the forearm muscles close to the skeletal elements of the wrist. ■ Superficial muscles of the forearm function coordinately to flex the wrist; some also cause abduction and adduction of the wrist. The intermediate and deep anterior forearm muscles flex the wrist and the MP and IP joints. ■ Extension of the wrist and fingers is provided by the extensor muscles of the forearm. Some also cause abduction or adduction of the hand. The deep posterior compartment muscles are primarily wrist and finger extensors. **Intrinsic Muscles of the Hand 372** ■ Intrinsic muscles of the hand are small muscles entirely within the hand and divided into three groups: (1) the thenar group, (2) the hypothenar group, and (3) the midpalmar group between the thenar and hypothenar groups.
Muscles That Move the Pelvic Girdle and Lower Limb 375	■ Four groups of muscles are associated with the pelvis and lower limb: muscles that move (1) the hip joint/thigh, (2) the knee joint/leg, and (3) leg muscles, as well as (4) the intrinsic muscles of the foot. **Muscles That Move the Hip Joint/Thigh 375** ■ Muscles that act on the thigh (femur) originate on the surface of the ossa coxae (pelvis) and insert on the femur. ■ Anterior compartment muscles flex the thigh. ■ There are three gluteal muscles: The gluteus maximus extends and laterally rotates the thigh; the gluteus medius and minimus are powerful abductors of the thigh. ■ The deep gluteal muscles contain the lateral rotators of the thigh. ■ The posterior compartment (hamstring) muscles extend the thigh. ■ Medial compartment muscles adduct the thigh, and most also flex the thigh. ■ The lateral compartment muscle (tensor fasciae latae) abducts the thigh. **Thigh Muscles That Move the Knee Joint/Leg 379** ■ The quadriceps femoris extends the leg. ■ The sartorius muscle flexes both the thigh and the leg. ■ Some medial thigh muscles also flex the leg. ■ The extensors of the thigh and flexors of the leg, termed the hamstrings, occupy the posterior compartment.
	Leg Muscles 383 ■ The anterior compartment muscles dorsiflex the foot and extend the toes. One muscle also inverts the foot. ■ Lateral compartment muscles evert and plantar flex the foot. ■ Posterior compartment muscles plantar flex the foot. ■ Deep posterior compartment muscles either flex the leg or plantar flex the foot, and either flex the toes or invert the foot. **Intrinsic Muscles of the Foot 389** ■ Dorsal muscles extend the toes. ■ The four layers of plantar muscles flex, extend, abduct, and/or adduct the toes.

C H A L L E N G E   Y O U R S E L F

Matching

Match each numbered item with the most closely related lettered item.

_____ 1. serratus anterior   a. elevates scapula

_____ 2. rhomboid major   b. protracts scapula

_____ 3. teres minor   c. adducts and flexes thigh

_____ 4. deltoid   d. connective tissue band

_____ 5. pronator teres   e. plantar flexes foot

_____ 6. extensor retinaculum   f. extends leg

_____ 7. quadriceps femoris   g. prime abductor of humerus

_____ 8. pectineus   h. dorsiflexes foot

_____ 9. soleus   i. laterally rotates humerus

_____ 10. tibialis anterior   j. pronates forearm

Multiple Choice

Select the best answer from the four choices provided.

_____ 1. The dorsal interossei muscles in the hand
a. adduct fingers 2–5.   c. flex the PIP and DIP joints.
b. abduct fingers 2–5.   d. extend the MP joints.

_____ 2. The contraction of the _____ causes medial rotation of the thigh.
a. pectineus   c. gluteus minimus
b. obturator externus   d. gracilis

_____ 3. Muscles in the anterior compartment of the leg
a. evert the foot.
b. dorsiflex the foot and extend the toes.
c. plantar flex the foot.
d. flex the toes.

_____ 4. All of the following muscles flex the forearm *except* the
   a. brachialis.
   b. biceps brachii.
   c. brachioradialis.
   d. anconeus.

_____ 5. The quadriceps femoris is composed of which of the following muscles?
   a. biceps femoris, rectus femoris, vastus lateralis, and gracilis
   b. vastus lateralis, vastus medialis, rectus femoris, and vastus intermedius
   c. semimembranosus, semitendinosus, and biceps femoris
   d. popliteus, gracilis, and sartorius

_____ 6. Thumb opposition is caused by contraction of the _____ muscle.
   a. flexor digiti minimi brevis
   b. opponens pollicis
   c. extensor pollicis longus
   d. adductor pollicis

_____ 7. The _____ flexes the knee and causes a slight medial rotation to "unlock" the knee joint.
   a. sartorius
   b. soleus
   c. tensor fasciae latae
   d. popliteus

_____ 8. Eversion of the foot is caused by the contraction of the _____ muscle.
   a. soleus
   b. plantaris
   c. fibularis brevis
   d. gastrocnemius

_____ 9. Which muscles originate on the ischial tuberosity and extend the thigh plus flex the leg?
   a. adductor muscles
   b. fibularis muscles
   c. hamstring muscles
   d. quadriceps muscles

_____ 10. The _____ causes plantar flexion of the foot.
   a. iliopsoas
   b. gastrocnemius
   c. fibularis tertius
   d. vastus intermedius

Content Review

1. The trapezius can perform what types of movements?

2. What movements are possible at the glenohumeral joint, and which muscles perform each of these movements?

3. Identify the compartments of the arm, the muscles in each compartment, and their function.

4. Compare and contrast the flexor digitorum superficialis and the flexor digitorum profundus; where does each insert, how are their tendons interrelated, and what muscle actions do they perform?

5. What is the primary function of the retinacula in a limb?

6. What muscles are responsible for thigh extension? Which of these is the prime mover of thigh extension?

7. Which muscles abduct the thigh, and where are they located?

8. What leg muscles are contracted when you are sitting at your desk?

9. What leg muscles allow a ballet dancer to rise up and balance on her toes?

10. Which muscles are responsible for foot inversion, and what muscles are antagonists to those that invert the foot?

Developing Critical Reasoning

1. Edmund suffers an injury to the anterior compartment of his thigh that results in paralysis of the muscles of this compartment. What muscle movements will be compromised?

2. After falling while skateboarding, Karen had surgery on her elbow. During her recovery, she must visit the physical therapist to improve muscle function around the elbow. Develop a series of exercises that may improve all of Karen's elbow movements, and determine which muscles are being helped by each exercise.

3. Why is it more difficult for Eric to lift a heavy weight when his forearm is pronated, than when it is in the supine position?

A N S W E R S   T O   " W H A T   D O   Y O U   T H I N K ? "

1. The supraspinatus muscle is the most commonly injured rotator cuff muscle, due in part to its location in the narrow space between the acromion and the humerus. Repetitive arm motions impinge (pinch) the muscle in this narrow space, and may eventually lead to tears of the muscle or tendon.

2. The brachialis is an anterior arm muscle. Since anterior arm muscles tend to flex the elbow joint, we can surmise that the brachialis flexes the elbow joint.

3. The iliopsoas cannot flex the leg, because this muscle does not cross over the knee joint.

4. Remember that there are *leg* muscles that also move the toes. In this case, the extensor digitorum longus (a leg muscle) attaches to toes 2–5 and helps move them all.

Visit the McKinley/O'Loughlin *Human Anatomy, 2e* website at aris.mhhe.com

MUSCULAR SYSTEM

Surface Anatomy

13

Imagine this scenario: An unconscious patient has been brought to the emergency room. Although the patient cannot tell the ER physician what is wrong or "where it hurts," the doctor can assess some of the injuries by observing surface anatomy, including:

- Locating pulse points to determine the patient's heart rate and pulse strength.
- Palpating the bones under the skin to determine if a fracture has occurred.
- Passively moving the limbs to observe potential damage to muscles and tendons.
- Examining skeletal and muscular landmarks to discover whether joints are dislocated.

Examination of surface anatomy must often substitute for interviewing the patient, and when the health-care professional is a keen observer, it may be very accurate in assessing illness or injury.

A Regional Approach to Surface Anatomy

Key topics in this section:

- Importance of surface anatomy in learning about internal structures
- How surface anatomy studies help us diagnose and treat disease

Surface anatomy is a branch of gross anatomy that examines shapes and markings on the surface of the body as they relate to deeper structures. An understanding of surface anatomy is essential for locating and identifying anatomic structures prior to studying internal gross anatomy. Health-care personnel use surface anatomy to help diagnose medical conditions and to treat patients, as when taking a pulse, inserting a needle or tube, or performing physical therapy. You have already begun your study of surface anatomy; each time we have asked you to palpate a part of your body and feel for a structure, you have examined your *own* surface anatomy.

Health-care professionals rely on four techniques when examining surface anatomy. Using **visual inspection,** they directly observe the structure and markings of surface features. Through **palpation** (pal-pā′shŭn) (feeling with firm pressure or perceiving by the sense of touch), they precisely locate and identify anatomic features under the skin. Using **percussion** (per-kŭsh′ŭn), they tap firmly on specific body sites to detect resonating vibrations. And via **auscultation** (aws-kŭl-tā′shŭn), they listen to sounds emitted from organs.

In our discussion of surface anatomy in this chapter, the illustrations include some structures that we have discussed previously and other features yet to be discussed. We strongly suggest that you return to this chapter often as you explore and examine other body systems in subsequent chapters. In addition, before you begin, refer back to chapter 1 and review the discussion of body region names and abdominopelvic regions and quadrants.

Study Tip!

When preparing for an anatomy exam, your best study aid is your own body. Palpating surface anatomy features on yourself will help you recall these features on an exam and, more importantly, give you a richer understanding of human anatomy.

Head Region

Key topic in this section:

- Surface features of the cranial and facial regions

The **head** is the most complex and highly integrated region of the body because it houses the brain, which communicates with and controls all of the body systems. The head is structurally and developmentally divided into the cranium and the face. **Figure 13.1** shows the regions and many of the surface anatomy structures of the head and neck.

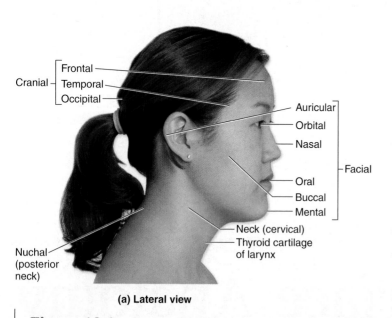

(a) Lateral view

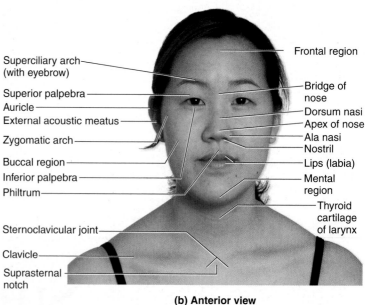

(b) Anterior view

Figure 13.1

Head and Neck. The major regions of the head and neck are shown in *(a)* lateral view, while specific features are shown in *(b)* anterior view.

Cranium

The **cranium** (also called the **cranial region** or *braincase*) is covered by the scalp, which is composed of skin and subcutaneous tissue. The cranium can be subdivided into three regions, each having prominent surface anatomy features.

The **frontal region** of the cranium is the forehead. Covering this region is the **frontal belly** of the **occipitofrontalis** muscle. The frontal region terminates at the **superciliary arches.** You can feel these bony elevations immediately inferior to your eyebrows.

Laterally, the scalp covers the sides of the skull in each **temporal region** and terminates just superior to the ear. The **temporalis** muscle is attached at the temporal region, and is easily palpable when the jaw is repeatedly clenched. Running over the temporalis muscle is the **superficial temporal artery.** You can feel the pulse of this artery just posterior to the orbits and anterior to the auricle of your ear.

The posterior part of the cranium is the **occipital** (ok-sip′i-tăl) **region.** In the center of that region is the **external occipital protuberance,** a rounded or pointed projection (see figures 7.5, 7.8). In chapter 7, you learned that males tend to have a more prominent, pointed external occipital protuberance than females. Palpate your own external occipital protuberance. Is it small and rounded, or somewhat larger and pointed?

WHAT DO YOU THINK?

1 Can you name some other facial muscles not already mentioned above that can be palpated easily under the skin?

Face

The **face** is divided into five regions: auricular, orbital, nasal, oral, and mental.

The **auricular** (aw-rik′ū-lăr; *auris* = ear) **region** is composed of the visible surface structures of the ear as well as the ear's internal organs, which function in hearing and maintaining equilibrium. The **auricle** (aw′ri-kl), or *pinna*, is the fleshy part of the external ear. Within the auricle is a tubular opening called the **external acoustic meatus.** The **mastoid process** is posterior and inferior to the auricle. Palpate your mastoid process; it should feel like a bony bump immediately posteroinferior to the ear.

The **orbital** (or *ocular*) **region** includes the **eyes** and associated structures. Most orbital region surface features protect the eye. **Eyebrows** protect against sunlight and potential mechanical damage to the eyes; **eyelids** (**palpebrae;** pal′pē-brē) close reflexively to protect against objects moving near the eye; and eyelashes prevent airborne particles from contacting the eyeball. The **superior palpebral fissure,** or upper eyelid crease, is palpated easily on most individuals, although Asians do not have a superior palpebral fissure.

The **nasal region** contains the nose. The firm, narrow part of the nose that projects anteriorly between the eyes is the **bridge;** it is formed by the union of the nasal bones. Anteroinferior to the bridge is the fleshy part of the nose, called the **dorsum nasi** (nā′zē; *nasus* = nose). Farther anteroinferiorly is the tip of the nose, called the **apex.** The **nostrils,** or **external nares** (nā′res; sing., *naris*, na′ris), are the paired openings into the nose. The **ala nasi** (wing of the nose) forms the flared posterolateral margin of each nostril.

The **oral region** is inferior to the nasal region; it includes the **buccal** (cheek) **region,** the fleshy upper and lower **lips** (**labia**),

CLINICAL VIEW

Lip Color as a Diagnostic Tool

Lip color is the collective result of a combination of pigments that contribute to a person's skin color, most notably melanin, hemoglobin, and carotene.

The pigment **melanin** has two subtypes: *Eumelanin* is black, and *pheomelanin* is typically slightly yellow in low concentrations but slightly reddish in high concentrations. The ratio of these types of melanin determines the skin and lip color. **Hemoglobin,** an oxygen-binding protein in red blood cells, contributes a red or pink hue. **Carotene,** a yellow-orange pigment found in carrots, sweet potatoes, and squash, contributes those hues to the skin and lips.

Variations in lip color result from the combinations and amounts of these pigments. However, lip color is also affected by our environment and state of health. For example, cold weather causes our lips to appear "blue" because blood (and its reddish-colored hemoglobin) is being shunted away from the superficial lips and toward deeper body structures in order to conserve heat. Low body temperature for other reasons, as well as several health conditions, also cause "blue lips." For instance, a patient who has anemia, pneumonia, emphysema, or certain disorders of the cardiovascular system could exhibit blueness or discoloration of the lips.

and the structures of the **oral cavity** (mouth) that can be observed when the mouth is open. Look in the mirror and observe the vertical depression between your nose and upper lip; this is called the **philtrum** (fil′trŭm; *philtron* = a love charm).

The **buccal** (bŭk′al) **region** refers to the cheek. Within this region is the buccinator muscle. Palpate the superolateral region of your cheek and locate your **zygomatic bone** and the **zygomatic arch.**

Finally, continuing in the inferior direction, look in the mirror and observe the **mental region,** which contains the **mentum,** or *chin*. Usually, the mentum tends to be pointed and almost triangular in females, while males tend to have a "squared-off" mentum.

WHAT DID YOU LEARN?

1 Identify at least two surface features of the orbital region that protect the eye.

2 Identify the narrow, bony, superior part of the nasal region between the eyes.

Neck Region

Key topics in this section:

- Palpable structures in the regions of the neck
- The triangles of the neck and the structures they contain

The **neck,** also called the *cervical region* or *cervix* (ser′viks), is a complex region that connects the head to the trunk. The spinal cord, nerves, trachea, esophagus, and major vessels traverse this highly flexible area. In addition, the neck contains the larynx (voice box) and several important glands. For purposes of discussion, the neck can be subdivided into anterior, posterior, and lateral regions.

The anterior region of the neck has several palpable landmarks, including the larynx, trachea, and sternal notch. The **larynx** (lar'ingks; voice box), found in the middle of the anterior neck, is composed of multiple cartilages. Its largest cartilage is the **thyroid cartilage,** which you can palpate as the big bulge on the anterior side of your neck. In males, the larynx has a noticeably pointed **laryngeal prominence** (commonly known as the "Adam's apple") that may be visualized and palpated easily. (Females do not have a noticeably pointed laryngeal prominence.) Inferior to the larynx are the cricoid cartilage and **trachea** (air tube). The neck terminates at the **suprasternal notch** of the manubrium and the left and right clavicles. Palpate your anterior neck region: Moving inferiorly along the neck, you first feel the prominent thyroid cartilage, then the trachea with its hard, cartilaginous rings, and eventually the suprasternal notch and the clavicles.

The posterior neck region is also referred to as the **nuchal region** (see figure 13.1a). This region houses the spinal cord, cervical vertebrae, and associated structures. You can easily palpate the spinous process of the **vertebra prominens** (C$_7$), especially during neck flexion. Palpate your nuchal region; the bump you feel at the inferior boundary of this region is the vertebra prominens. As you move your fingers superiorly along the midline of the neck, you can palpate the **ligamentum nuchae,** a thick ligament that extends from C$_7$ to the nuchal lines of the skull.

The left and right lateral portions of the neck contain the **sternocleidomastoid muscles,** which become prominent when a person turns his or her head to one side. Each sternocleidomastoid muscle partitions each side of the neck into two clinically important triangles, an anterior triangle and a posterior triangle (**figure 13.2**). Each of these triangles houses important structures that extend through the neck, and these triangles are further subdivided into smaller triangles.

The **anterior triangle** lies anterior to the sternocleidomastoid muscle and inferior to the mandible. It is subdivided into four smaller triangles: the submental, submandibular, carotid, and muscular triangles.

The **submental triangle,** the most superiorly placed of the four triangles, is posteroinferior to the chin in the midline of the neck and partially bounded by the anterior belly of the digastric muscle. It contains some **cervical lymph nodes** and tiny veins. When you are ill, these lymph nodes (as well as some glands) enlarge and become tender. A physician palpates these to determine if you have an infection that has been detected by immune cells within these nodes.

CLINICAL VIEW

Tracheotomy

The **tracheotomy** (trā-kē-ot'ō-mē; *tome* = incision) is one of the oldest surgical procedures in medicine. A tracheotomy is performed when a patient requires extended ventilatory (breathing) assistance based on one of three recognized indications:

- The presence of an upper airway obstruction due to a foreign body, trauma, swelling, etc.
- Difficulty breathing due to advanced pneumonia, emphysema, or severe chest wall injury
- Respiratory paralysis, as may result from head injury, polio, or tetanus infection

Understanding the surface anatomy of the neck is critical to performing a tracheotomy correctly. Typically, the physician makes a skin incision about 1 to 1.5 centimeters superior to the suprasternal notch. Care must be taken not to damage the anterior jugular veins, and sometimes the thyroid must be incised in the midline and divided in order to gain access to the trachea. Retractors are used to separate the subcutaneous tissue and expose the trachea. Then the surgeon makes an incision in the trachea between the third and fourth tracheal rings to allow the insertion of a tracheotomy tube; this opening is called a *tracheostomy*. Once the tube has been taped into place, the patient's breathing bypasses the nasal cavity and larynx.

The tracheotomy is an important and often life-saving procedure, but it is not without risks. A misplaced incision in the anterior neck can lead to serious damage of the larynx and possibly even a fatal hemorrhage. Other potential complications include infection at the site, aspiration of foreign matter directly into the lungs, or tracheal stenosis (a narrowing of the trachea at the incision site due to scar tissue formation).

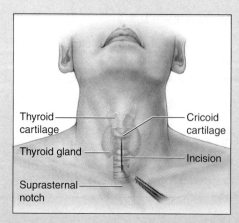

① Incision is made superior to suprasternal notch. Thyroid may have to be cut as well.

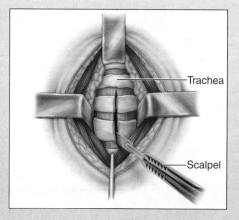

② Retractors separate the tissue, and an incision is made through the third and fourth tracheal rings.

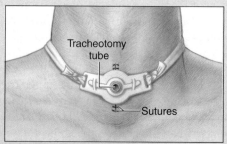

③ A tracheotomy tube is inserted, and the remaining incision is sutured closed.

Tracheotomy procedure.

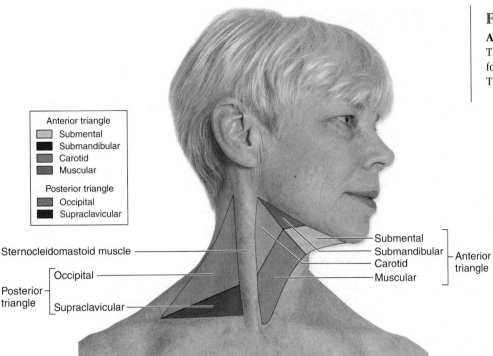

Figure 13.2
Anterior and Posterior Triangles of the Neck.
The sternocleidomastoid muscle is the landmark
for separating the anterior and posterior triangles.
These triangles are further subdivided as shown.

Anterior triangle
☐ Submental
■ Submandibular
■ Carotid
■ Muscular

Posterior triangle
■ Occipital
■ Supraclavicular

Sternocleidomastoid muscle ———

Posterior ⎱ ⎰Occipital ———
triangle ⎰ ⎱Supraclavicular ———

Submental
Submandibular ⎤
Carotid   ⎬ Anterior
Muscular   ⎦ triangle

The **submandibular triangle** is inferior to the mandible
and posterolateral to the submental triangle. It is bounded by the
mandible and the bellies of the digastric muscle. If you palpate this
triangle, you can feel the **submandibular gland,** which is the bulge
inferior to the mandible.

The **carotid** (kă-rŏt'id) **triangle** is bounded by the sternocleidomastoid, omohyoid, and posterior digastric muscles. Palpate
this triangle until you feel the strong pulsation of an artery; that is
the **common carotid artery.** This triangle also contains the internal
jugular vein and some cervical lymph nodes, which may be easily
palpated here as well.

The **muscular triangle** is the most inferior of the four triangles. It contains the **sternohyoid** and **sternothyroid muscles,** as
well as the lateral edges of the larynx and the **thyroid gland.** Try
to palpate the thyroid gland here. Also in this triangle are cervical
lymph nodes. (Cervical lymph nodes are present throughout the
neck, as we have indicated.)

The **posterior triangle** is in the lateral region of the neck,
posterior to the sternocleidomastoid muscle, superior to the clavicle
inferiorly, and anterior to the trapezius muscle. This triangle is subdivided into two smaller triangles: the occipital and supraclavicular
triangles.

The **occipital triangle** is the larger and more posteriorly
placed of the two triangles. It is bounded by the omohyoid, trapezius, and sternocleidomastoid muscles. This important triangle
contains the **external jugular vein** (which may be visible internal to
the skin), the accessory nerve, the brachial plexus (a mass of nerves
that innervates the upper limbs), and some lymph nodes.

The **supraclavicular** (soo-pră-kla-vik'ū-lăr) **triangle** also
goes by the names *omoclavicular* and *subclavian.* It is bounded
by the clavicle, omohyoid, and sternocleidomastoid muscles. It

contains part of the subclavian vein and artery as well as some
lymph nodes.

WHAT DID YOU LEARN?

3 Describe two structures contained within the anterior region of
the neck.

4 What muscle divides each lateral region of the neck into anterior
and posterior triangles?

Trunk Region

Key topics in this section:

- Surface features of the thorax, abdomen, and back
- Auscultation sites in the thorax and abdominopelvic region

The **trunk,** or *torso,* is partitioned into the **thorax** (chest),
the **abdominopelvic region,** and the **back.** The surface anatomy of
the trunk is particularly important in determining the location and
condition of the viscera. However, some of the surface features may
be obscured due to the age, sex, or body weight of an individual.

Thorax

The thorax is the superior portion of the trunk sandwiched between
the neck superiorly and the abdomen inferiorly. It consists of the
chest and the "upper back." On the anterior surface of the chest are
the two dominating surface features of the thorax—the clavicles and
the sternum—as well as several other important surface anatomy
landmarks **(figure 13.3).**

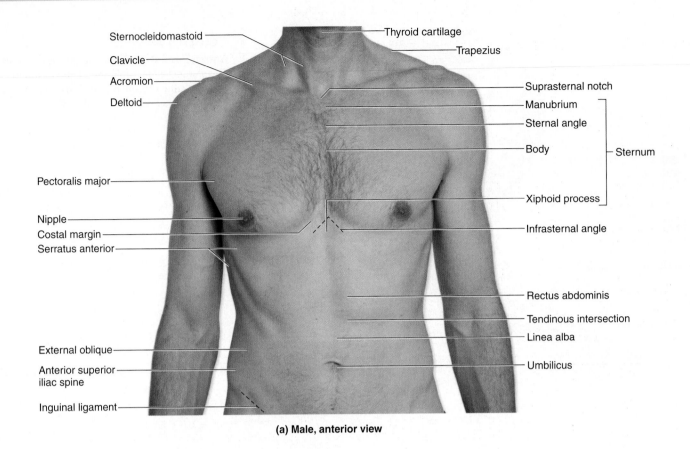

(a) Male, anterior view

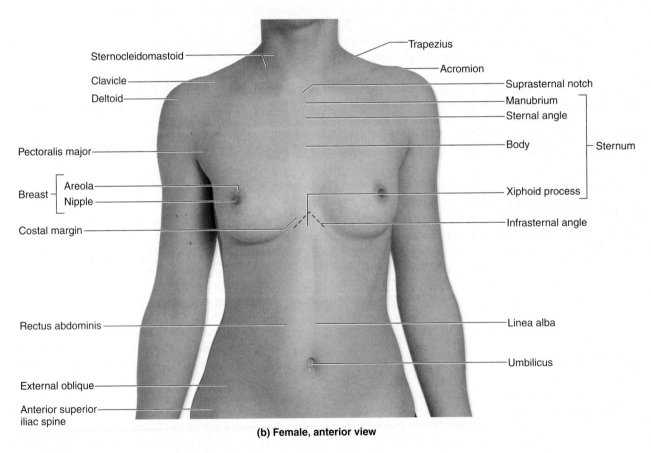

(b) Female, anterior view

Figure 13.3

Anterior Trunk Region. Anterior views show selected surface anatomy landmarks in *(a)* a male and *(b)* a female.

CLINICAL VIEW

Surface Anatomy and CPR

All adults should learn the life-saving technique known as **cardiopulmonary resuscitation (CPR).** CPR is a combination of rescue breathing and chest compressions given to an individual who is in *cardiac arrest,* meaning that the heart has stopped working. Classes are available for those who want to learn proper CPR techniques and obtain CPR certification. Below is a brief summary about CPR, but please note that this summary is not comprehensive and *not* a substitute for proper CPR certification:

- After dialing 9-1-1, first make sure the victim's airway is open. Look, listen, and feel for breathing. Remove any foreign material from the mouth, and if the person is not breathing, give two full rescue breaths.

- If no breathing, coughing, or movement (and thus no normal signs of circulation) result in response to the two rescue breaths, chest compressions should be performed by (1) palpating the xiphoid process and placing two fingers there; (2) placing the heel of the hand superior to the two fingers so that the hand rests on the body of the sternum; and (3) pressing down about 2 inches on the body of the sternum, the place where the heart will receive maximum benefit.

- Specific ratios of chest compression to rescue breathing must be performed, depending upon whether the patient is a child or an adult.

CPR classes provide instruction and further details about this procedure. CPR guidelines are updated and changed periodically, so individuals need to renew their certification on a yearly basis. Become CPR certified: Someone's life may depend on what you know!

 ① Palpate xiphoid process

 ② Place hand on body of sternum

 ③ Start chest compressions

Chest compression steps in CPR.

The paired **clavicles** and the **suprasternal notch** represent the border between the thorax and the neck. Feel the clavicles on the superior anterior surface where they extend between the base of the neck on the right and left sides laterally to the shoulders. The left and right **costal margins** of the rib cage form the inferior boundary of the thorax. The **infrasternal angle** *(subcostal angle)* is where the costal margins join to form an inverted V at the xiphoid process. On a thin person, many of the **ribs** can be seen. Most of the ribs (with the exception of the first ribs) can be palpated.

The **sternum** is palpated readily as the midline bony structure in the thorax. Its three components (the **manubrium,** the **body,** and the **xiphoid process**) may also be palpated. The **sternal angle** can be felt as an elevation between the manubrium and the body. The sternal angle is clinically important because it is at the level of the costal cartilage of the second rib, where it is often used as a landmark for counting the ribs.

Each **breast** is located lateral to the sternum. The breast has a projection, the **nipple,** surrounded by a circular, colored region called the **areola.**

Abdominopelvic Region

The abdominopelvic region is the portion of the trunk that lies inferior to the rib cage. Surface anatomy features in this region may be difficult to palpate in some obese people. Some of the features in this region are shown in figure 13.3, and the abdominopelvic regions and quadrants are shown in figure 1.10.

On the anterior surface of the abdomen, the **umbilicus** *(navel)* is the prominent depression (if you have an "innie") or projection (if you have an "outie") in the midline of the abdominal wall. Also in the midline of the abdominal anterior surface is the **linea alba,** a tendinous structure that extends inferiorly from the xiphoid process to the pubic symphysis. If you observe the anteroinferior surface of the abdomen, you can readily palpate the **pubic bones** in the pubic region, underneath the pubic hair.

Additionally, the abdominopelvic region houses several other readily observed structures. The left and right **rectus abdominis** muscles and their **tendinous intersections** are observable in individuals with well-developed abdominal musculature. These muscles and intersections are referred to as "six-pack abs." The superior aspect of the ilium (iliac crest) terminates anteriorly at the **anterior superior iliac spine.** Attached to the anterior superior iliac spine is the **inguinal ligament,** which forms the inferior boundary of the abdominal wall. The inguinal ligament terminates on a small anterior, rounded projection on the pubis called the **pubic tubercle.** Superior to the medial portion of the inguinal ligament is the **superficial inguinal ring.** This "ring" is actually a superficial opening in the inferior anterior abdominal wall via the **inguinal canal,** and represents a weak spot in the wall. Although this ring may not be readily seen, a physician can palpate the ring and the inguinal canal to detect an inguinal hernia.

Testing for Inguinal Hernias

An **inguinal hernia** is a protrusion of intestine through a weak spot in the anterior abdominal wall. This weak spot typically is the superficial inguinal ring, located superolateral to the pubic tubercle. (Inguinal hernias are discussed in detail in the Clinical View: In Depth in chapter 11.)

A physician must have a thorough knowledge of surface anatomy in order to test for an inguinal hernia. First, the physician must find the superficial inguinal ring by locating both the inguinal ligament (found inferior to the ring) and the pubic tubercle (just inferior and medial to the ring). Once the ring is located, the physician inserts a finger in the depression formed by the superficial inguinal ring and into the inguinal canal, and asks the patient to turn his head and "cough." (Having the subject turn his head simply ensures that the patient doesn't cough in the physician's face.) The act of coughing increases intra-abdominal pressure, and would encourage a portion of intestine to poke through the ring if there was a problem. While the patient coughs, the physician palpates the superficial inguinal ring to make sure no intestine is protruding through it.

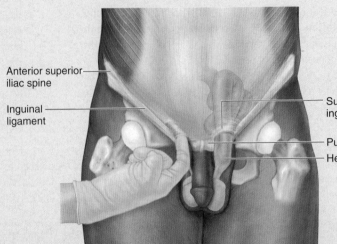

Anterior superior iliac spine

Inguinal ligament

Superficial inguinal ring

Pubic tubercle

Herniated intestine

Testing for a potential inguinal hernia on the right side. Note the herniated intestine protruding through the left superficial inguinal ring.

Back

The surface anatomy features of the back appear in **figure 13.4**. At the superior midline surface of the back near the interface between the neck and back, the vertebra prominens is readily palpated, as discussed previously. Moving inferiorly from the vertebra prominens in the midline, some vertebral spinous processes can be observed. The entire series of vertebral spines is visible when the vertebral column is flexed. When the back is extended, some of the inferior vertebral spines are obscured; instead, a vertically oriented indentation called the **median furrow** is all that can be seen along the inferior midline of the back.

Several prominent features of the posterior scapula are observed and easily palpated in some individuals, including the **lateral** and **medial borders**. The **spine of the scapula** is covered by the trapezius muscle, but it still may be palpated, especially when the back is flexed. The **triangle of auscultation** is a region bordered by three muscles: the rhomboid major, trapezius, and latissimus dorsi. When an individual flexes his or her back, this triangle becomes larger, and the sixth intercostal space becomes subcutaneous (lies directly internal to the skin). Thus, at this site a physician can hear respiratory sounds more readily through a stethoscope, without their being muffled by the muscles.

Finally, the **iliac crests** mark the superior surface of the ossa coxae. They originate along the abdominal wall and continue along the inferior border of the back. When the superiormost points of the iliac crests are palpated, drawing a horizontal line through them bisects the spinous process of the L_4 vertebra. (The iliac crests may be difficult to palpate in obese people.)

Surface Anatomy and Lumbar Puncture

A **lumbar puncture**, sometimes called a *spinal tap*, is a procedure in which a needle is inserted into the vertebral canal for the purpose of giving an anesthetic or testing the fluid around the spinal cord for evidence of infection or hemorrhage. This procedure requires careful and precise use of surface anatomy features.

Because the adult spinal cord typically ends at the level of the L_1 vertebra (see chapter 16), a lumbar puncture must be performed inferior to this level to ensure that the spinal cord is not pierced by the needle. The typical location is between the L_3 and L_4 vertebrae. To locate this region, the physician must first palpate the iliac crests, the superiormost points of which are level with the spinous process of the L_4 vertebra. Once the physician has properly palpated the iliac crests, she can draw an imaginary horizontal line to the level of the L_4 vertebra to identify the correct spinous process. She can then insert the lumbar puncture needle either directly above or directly below the spinous process when the vertebral column is flexed.

WHAT DID YOU LEARN?

5 What is the sternal angle, and what is its clinical significance?

6 The most superior levels of the iliac crests are at the same level as what bony structure? Why is this clinically important?

7 List the muscles that border the triangle of auscultation. What is the clinical significance of this triangle?

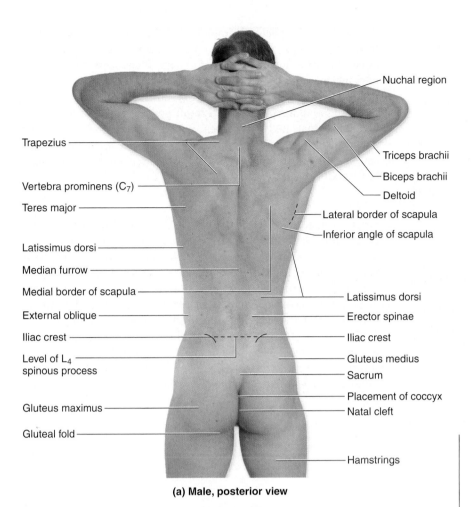

Nuchal region

Trapezius

Vertebra prominens (C₇)

Teres major

Latissimus dorsi

Median furrow

Medial border of scapula

External oblique

Iliac crest

Level of L₄
spinous process

Gluteus maximus

Gluteal fold

Triceps brachii

Biceps brachii

Deltoid

Lateral border of scapula

Inferior angle of scapula

Latissimus dorsi

Erector spinae

Iliac crest

Gluteus medius

Sacrum

Placement of coccyx

Natal cleft

Hamstrings

(a) Male, posterior view

Figure 13.4

Posterior Trunk Region.   Posterior views show selected surface anatomy landmarks in *(a)* a male and *(b)* a female.

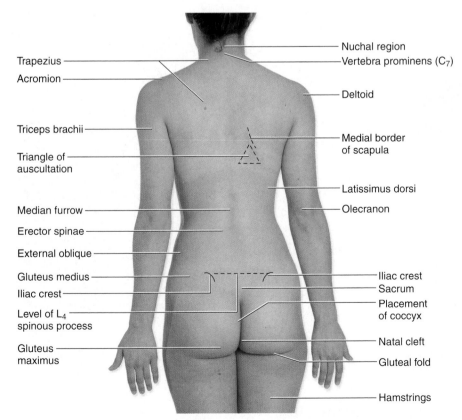

Trapezius

Acromion

Triceps brachii

Triangle of
auscultation

Median furrow

Erector spinae

External oblique

Gluteus medius

Iliac crest

Level of L₄
spinous process

Gluteus
maximus

Nuchal region

Vertebra prominens (C₇)

Deltoid

Medial border
of scapula

Latissimus dorsi

Olecranon

Iliac crest

Sacrum

Placement
of coccyx

Natal cleft

Gluteal fold

Hamstrings

(b) Female, posterior view

Shoulder and Upper Limb Region

Key topics in this section:

- Surface features of the shoulder and upper limb
- Clinically relevant features of the axilla, the cubital fossa, and the wrist

The anatomy of the shoulder and upper limb region is clinically important because of frequent trauma to these body regions. Additionally, vessels of the upper limb are often used as pressure sites and as sites for drawing blood, providing nutrients and fluids, and administering medicine.

Shoulder

The scapula, clavicle, and proximal part of the humerus collectively form the shoulder, and portions of each of these bones are important surface landmarks in this region (see figures 13.3 and 13.4). Anteriorly, the **clavicle** and **acromion** of the scapula may be observed and palpated. The acromion helps form the rounded, superior projection on your shoulder. The rounded curve of the shoulder is formed by the thick **deltoid muscle,** which is a frequent site for intramuscular injections.

Axilla

The **axilla** (ak'sil'ă), commonly called the armpit, is clinically important because of the nerves, axillary blood vessels, and lymph nodes located there. The **pectoralis major** forms the fleshy **anterior axillary fold,** which acts as the anterior border of the axilla. The **latissimus dorsi** and **teres major** muscles form the fleshy **posterior axillary fold,** which is the posterior border of the axilla (**figure 13.5**). Palpate your own axilla to locate the posterior and anterior axillary folds.

The **axillary** (ak'sil-ār-ē) **lymph nodes** are also in this region and may be palpated easily. These lymph nodes may become tender and swollen due to an infection, or they may become hard if breast cancer cells spread to them. Also palpable in this region is the pulse of the **axillary artery.**

Arm

Several structures are clearly visible in the arm, also called the **brachium** (brā'kē-ŭm), which extends from the shoulder to the elbow on the upper limb. On the anterior side of the arm, the **cephalic vein** is evident in muscular individuals as it traverses along the anterolateral border of the entire upper limb. This vein terminates in a small surface depression, bordered by the deltoid and pectoralis major muscles, called the **clavipectoral triangle** (or *deltopectoral triangle*) (**figure 13.6**). The **basilic vein** is sometimes evident along the medial side of the upper limb. The **brachial artery** becomes subcutaneous along the medial side of the brachium, and its pulse may be detected there. This region is clinically important in measuring blood pressure.

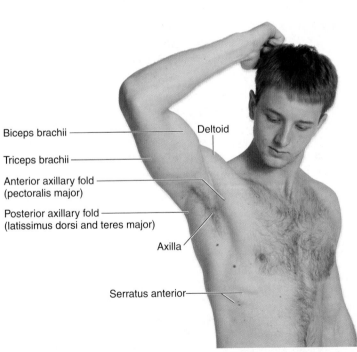

Biceps brachii

Triceps brachii

Anterior axillary fold
(pectoralis major)

Posterior axillary fold
(latissimus dorsi and teres major)

Deltoid

Axilla

Serratus anterior

Anterolateral view

Figure 13.5

Axilla and Trunk.   An anterolateral view of the axilla and trunk in an adult male.

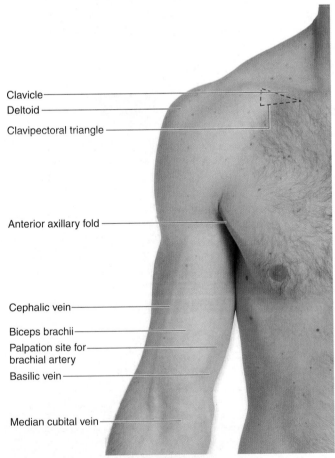

Clavicle

Deltoid

Clavipectoral triangle

Anterior axillary fold

Cephalic vein

Biceps brachii

Palpation site for brachial artery

Basilic vein

Median cubital vein

Right brachium, anterior view

Figure 13.6

Arm.   Anterior view shows the right shoulder and arm region in a male.

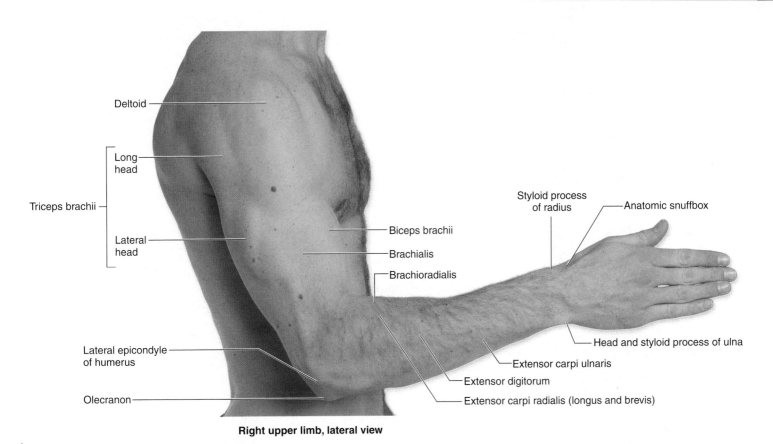

Deltoid

Long head

Triceps brachii

Lateral head

Styloid process of radius

Anatomic snuffbox

Biceps brachii

Brachialis

Brachioradialis

Lateral epicondyle of humerus

Olecranon

Head and styloid process of ulna

Extensor carpi ulnaris

Extensor digitorum

Extensor carpi radialis (longus and brevis)

Right upper limb, lateral view

Figure 13.7

Upper Limb.   Lateral view shows the right upper limb in a male.

The **biceps brachii** muscle becomes prominent when the elbow is flexed (**figure 13.7**). Located on the anterior surface of the elbow region, the **cubital fossa** is a depression within which the **median cubital vein** connects the basilic and cephalic veins (figure 13.6). The cubital fossa is a common site for **venipuncture** (removal of blood from a vein).

The bulk of the posterior surface of the brachium is formed by the **triceps brachii** muscle. Three bony prominences are readily identified in the distal region of the brachium near the elbow (**figure 13.8**; see also figure 13.7). The **lateral epicondyle** of the **humerus** is a rounded lateral projection at the distal end of the humerus. The **olecranon** (ō-lek′ră-non) of the ulna is palpated easily along the posterior aspect of the elbow. The **medial epicondyle** of the humerus is more prominent and may be easily palpated. The **ulnar nerve** traverses posterior and inferior to the medial epicondyle. When we hit our "funny bone," we actually are hitting or pinching the ulnar nerve as it travels posterior to the medial epicondyle.

Forearm

The radius, the ulna, and the muscles that control hand movements form the forearm, or **antebrachium** (an-te-brā′kē-ŭm) (figure 13.8). Palpate your own forearm. Note that the proximal part of the forearm is bulkier, due to the fleshy bellies of the forearm muscles. The **head** of the radius may be palpated just distal to the lateral epicondyle of the humerus, especially when you pronate and supinate your forearm. As you palpate distally, the forearm becomes thinner because you are palpating the tendons of these muscles. The **styloid**

process of the radius is readily palpable as the lateral bump along the wrist, while the **head** and **styloid process** of the ulna collectively form the medial prominence of the wrist.

The **pulse of the radial artery** may be detected between the distal tendons of the flexor carpi radialis and the brachioradialis (**figure 13.9**). The **pulse of the ulnar artery** is a bit more difficult to find. You can locate it by feeling for the medial bump in your hand that is the **pisiform bone** and then placing your fingers immediately lateral to the pisiform bone.

The tendons of the **extensor pollicis brevis, abductor pollicis longus,** and **extensor pollicis longus** muscles mark the boundary of the triangular **anatomic snuffbox.** You can palpate the pulse of the radial artery here as well. In addition, you can palpate the **scaphoid bone** in this region. (See also the chapter 12 Clinical View, "Anatomic Snuffbox," on page 369.)

Hand

The most obvious surface anatomy features of the hand involve flexion creases, fingerprints, and fingernails (figure 13.9). Anteriorly, flexion creases are seen for the **metacarpophalangeal (MP), proximal interphalangeal (PIP),** and **distal interphalangeal (DIP) joints.** Palpate the **thenar** (thē′nar) **eminence** (em′i-nens), the thickened, muscular region the hand that forms the base of the thumb. The **hypothenar eminence** is the medial region of the palm immediately proximal to the little finger.

The tendons that extend to each of the fingers from the **extensor digitorum** muscle are readily observed on the posterior side of

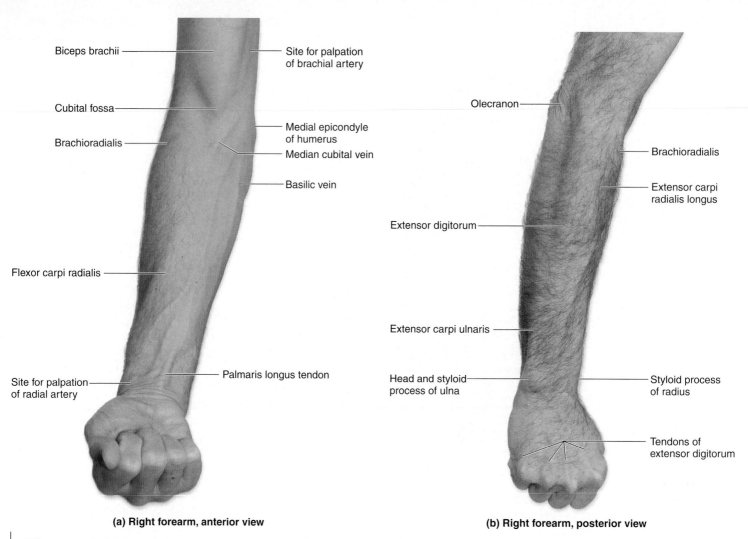

(a) Right forearm, anterior view

(b) Right forearm, posterior view

Figure 13.8

Forearm and Hand. *(a)* Anterior and *(b)* posterior views show the right forearm in a male.

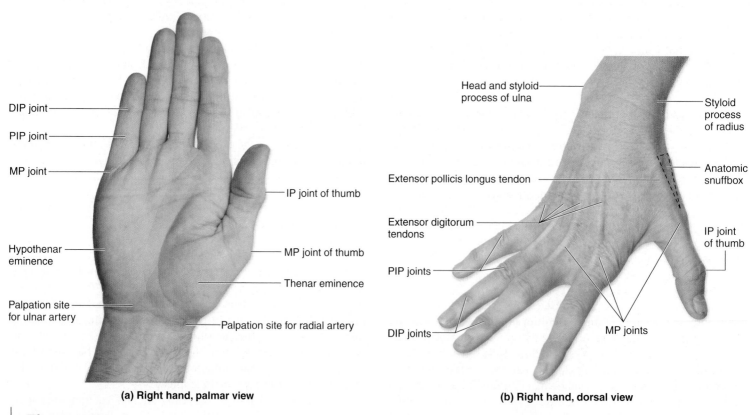

(a) Right hand, palmar view

(b) Right hand, dorsal view

Figure 13.9

Wrist and Hand. *(a)* Palmar and *(b)* dorsal views of the right hand illustrate many surface anatomy features.

the hand when the digital joints are extended. Also on the posterior side of the hand, the MP joints ("knuckles") are formed by the distal ends of metacarpal bones II–V. Palpate each **phalanx** (fā'langks) and all the interphalangeal joints.

WHAT DID YOU LEARN?

8 What muscles form the anterior and posterior axillary folds?

9 Discuss the location of the cubital fossa, and describe what is found in this depression.

10 Where is the pulse of the brachial artery taken?

Lower Limb Region

Key topics in this section:

- Surface features of the lower limb
- Clinical importance of the femoral triangle and the arches of the foot

The massive bones and strong muscles of the lower limbs are weight bearing important with respect to locomotion, our ability to move from place to place.

Gluteal Region

The buttock could have been included in our discussion of the posterior region of the pelvis, but instead we discuss it here as the **gluteal** (gloo'tē-ăl; *gloutos* = buttock) **region.** Several surface anatomy features are in this region (see figure 13.4).

The inferior border of the **gluteus maximus** muscle forms the **gluteal fold.** The **natal cleft** extends vertically to separate the buttocks into two prominences. In the inferior portion of each buttock, an **ischial** (is'kē-ăl) **tuberosity** can be palpated; these tuberosities support body weight while seated. The gluteus maximus muscle and some fat form most of the inferolateral "fleshy" part of the buttock. The **gluteus medius** muscle may be palpated only in the superolateral portion of each buttock. The **sciatic nerve** isn't easily palpable, but knowledge of its location in the buttock region is essential for health-care providers who must give gluteal intramuscular injections. The sciatic nerve originates in the inferior medial quadrant of the buttock and travels inferiorly to the lower limb.

WHAT DO YOU THINK?

2 What muscle functions would be impaired if you accidentally pierced or injured the sciatic nerve?

Thigh

Many muscular and bony features are readily identified in the thigh, which extends between the hip and the knee on each lower limb **(figure 13.10)**. An extremely important element of thigh surface anatomy is a region called the **femoral triangle.** The femoral triangle is a depression inferior to the inguinal ligament and on the anteromedial surface in the superior portion of the thigh. It is bounded superiorly by the inguinal ligament, laterally by the sartorius muscle, and medially by the adductor longus muscle. The femoral artery, vein, and nerve travel through this region, making it

---

CLINICAL VIEW

Gluteal Intramuscular Injections

The gluteal region is a preferred site for intramuscular (IM) injections because the gluteal muscles are quite thick and contain many blood vessels. However, health-care personnel must be careful not to accidentally inject the sciatic nerve or the superior and inferior gluteal vessels and nerves that supply the gluteal muscles. The sciatic nerve and the gluteal nerves and vessels are located primarily in the medial and inferior lateral part of the buttock. Therefore, the iliac crest is an important surface landmark for determining the safest place for a gluteal IM injection. Usually the injection is administered in the **superior lateral quadrant** of the buttock, about 5–7 centimeters inferior to the iliac crest. By placing the injection in the superior lateral quadrant, the healthcare worker can be reasonably certain of not accidentally piercing an important nerve or blood vessel.

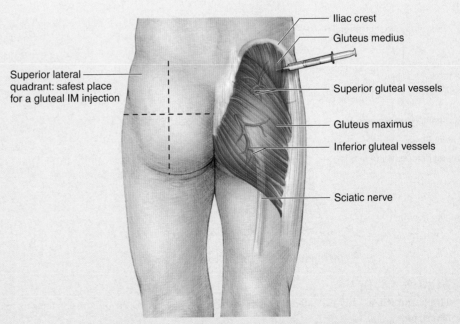

Proper placement of a gluteal intramuscular injection.

an important arterial pressure point for controlling lower limb hemorrhage. Now focus your attention on the distal part of your anterior thigh, and try to palpate three parts of the **quadriceps femoris** as they approach the knee.

Still on the anterior side of the thigh, four obvious skeletal features can be observed and palpated: (1) The **greater trochanter** is palpated on the superior lateral surface of the thigh; (2) the **patella** is located easily within the patellar tendon; and (3) the **lateral** and (4) **medial epicondyles** of both the femur and tibia are indentified and palpated at each knee. On the lateral side of the thigh, the tendinous **iliotibial tract** may be palpated.

The posterior side of the thigh has the tendinous attachments of the **hamstring** muscles. By flexing your knee, you can readily palpate these tendons along the posterior aspect of the knee joint. Also in this part of the thigh near the knee, observe the depression on the posterior part of the knee joint, called the **popliteal** (pop-lit′ē-ăl, pop-li-tē′ăl) **fossa** (pl., *fossae*, fos′ē). This is often the site of vascular problems in the elderly due to reduced muscle tone and inactivity that lead to incompetent valves in the veins, poor circulation, and blood clots.

Leg

Several skeletal features are observable as part of the surface anatomy of the leg, which extends from the knee to the foot **(figure 13.11)**. From an anterior perspective, palpate the **tibial tuberosity** immediately inferior to the knee joint. Then, moving inferiorly, feel the **anterior border** of the tibia (called the "shin"), which is subcutaneous and palpable along its length. At the distal end of the leg, observe and palpate the **medial malleolus** and **lateral malleolus** along the sides of the ankle.

On the posterior side of the leg, palpate the strong **calcaneal tendon** (*Achilles tendon*) along the posteroinferior leg. Then try to palpate the **pulse of the posterior tibial artery,** posteroinferior to the medial malleolus of the tibia.

Next, observe the leg first from a lateral perspective and then from a medial perspective. The superior part of the lateral leg is the **head of the fibula.** Palpate this bony projection. Then look distally and observe the lateral malleolus that you palpated when observing the leg in anterior view. The **small saphenous vein** is easily seen along the lateral malleolus, traveling superiorly along the posterior part of the leg. Switch over to a medial view and observe the medial malleolus

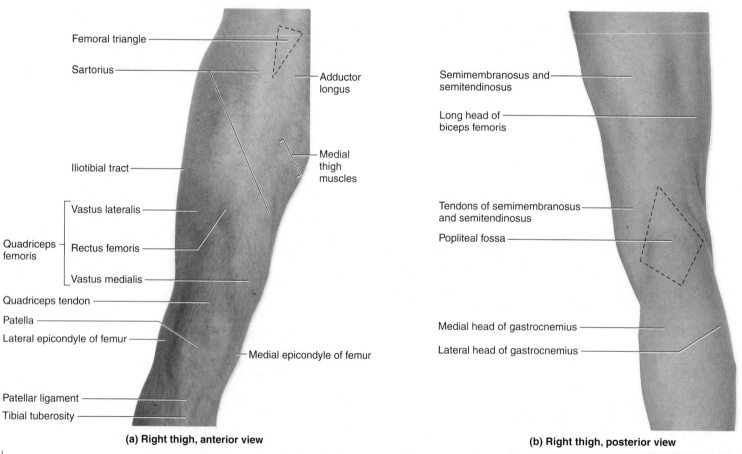

(a) Right thigh, anterior view

Femoral triangle
Sartorius
Adductor longus
Medial thigh muscles
Iliotibial tract
Vastus lateralis
Quadriceps femoris
Rectus femoris
Vastus medialis
Quadriceps tendon
Patella
Lateral epicondyle of femur
Medial epicondyle of femur
Patellar ligament
Tibial tuberosity

(b) Right thigh, posterior view

Semimembranosus and semitendinosus
Long head of biceps femoris
Tendons of semimembranosus and semitendinosus
Popliteal fossa
Medial head of gastrocnemius
Lateral head of gastrocnemius

Figure 13.10

Thigh and Knee. *(a)* An anterior view of the right thigh reveals the quadriceps femoris and patella, while *(b)* a posterior view illustrates the hamstrings.

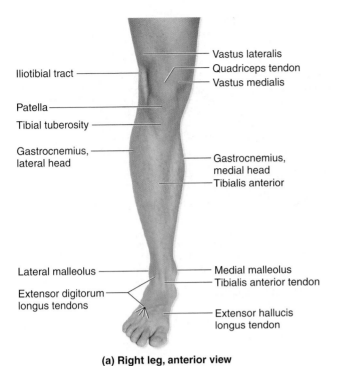

Vastus lateralis
Quadriceps tendon
Vastus medialis
Iliotibial tract
Patella
Tibial tuberosity
Gastrocnemius, lateral head
Gastrocnemius, medial head
Tibialis anterior
Lateral malleolus
Medial malleolus
Tibialis anterior tendon
Extensor digitorum longus tendons
Extensor hallucis longus tendon

(a) Right leg, anterior view

Figure 13.11

Leg. *(a)* Anterior, *(b)* posterior, and *(c)* lateral views of the right leg show the prominent surface landmarks.

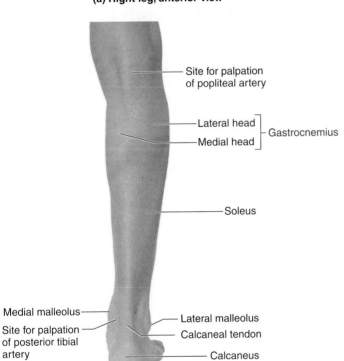

Site for palpation of popliteal artery
Lateral head
Medial head
Gastrocnemius
Soleus
Medial malleolus
Site for palpation of posterior tibial artery
Lateral malleolus
Calcaneal tendon
Calcaneus

(b) Right leg, posterior view

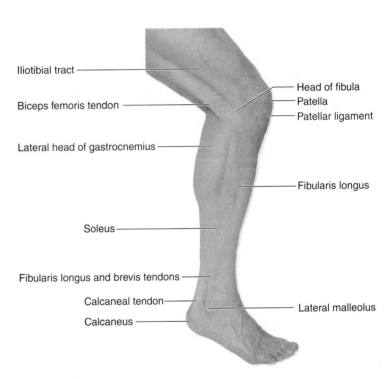

Iliotibial tract
Head of fibula
Patella
Patellar ligament
Biceps femoris tendon
Lateral head of gastrocnemius
Fibularis longus
Soleus
Fibularis longus and brevis tendons
Calcaneal tendon
Lateral malleolus
Calcaneus

(c) Right leg, lateral view

you palpated earlier. Another superficial vein of the leg is the **great saphenous vein,** seen subcutaneously on the medial side of the leg.

Foot

In most individuals, numerous surface features of the foot can be clearly observed (**figure 13.12**; see figure 13.11*a*). On the superior side of the foot, also called the *dorsum,* palpate the tendons for the **tibialis anterior, extensor digitorum longus,** and **extensor hallucis**

longus. Each of these tendons is clearly observed, especially when the foot is dorsiflexed. Move back toward the ankle, and palpate the **navicular bone** along the dorsum. You may feel the **pulse of the dorsalis pedis artery** either over the navicular (the medial dorsal side of the foot) or along the dorsal interspace between the first and second toes. Physicians check the pulse of this artery for possible compartment syndrome in the anterior leg (see Clinical View: In Depth, chapter 12, page 389).

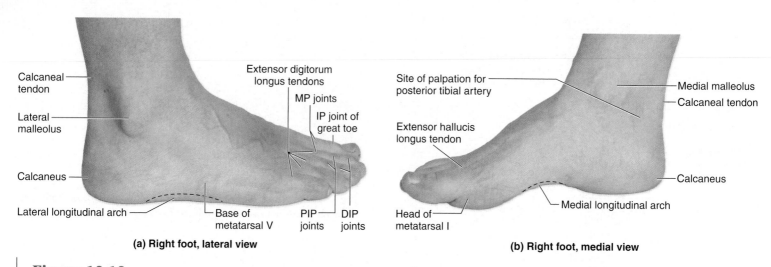

(a) Right foot, lateral view

(b) Right foot, medial view

Figure 13.12

Ankle and Foot. *(a)* Lateral and *(b)* medial views of the right ankle and foot show the prominent surface landmarks.

Both sides of the foot have distinct surface anatomy structures. Along the lateral surface, palpate the **base of metatarsal V.** The tendon for the fibularis brevis attaches here. Then observe the **lateral longitudinal arch,** the curvature along the lateral side of the foot. The **phalanges, metatarsophalangeal (MP) joints, PIP** and **DIP joints,** and **toenails** are obvious surface landmarks readily observed when viewing either the lateral side or the dorsum of the foot. The medial surface of the foot clearly illustrates the high, arched **medial longitudinal arch.** At the distal end of the medial longitudinal arch, the **head of metatarsal I** appears as a rounded prominence.

WHAT DID YOU LEARN?

11 Why is the femoral triangle an important element of the surface anatomy of the thigh?

12 What two superficial veins are observed on the surface of the leg?

13 What tendons may be easily seen along the dorsum of the foot?

CLINICAL TERM

femoral hernia Herniation of the intestines through the medial part of the femoral triangle.

CHAPTER SUMMARY

A Regional Approach to Surface Anatomy 396	■ Visual inspection, palpation, percussion, and auscultation are the primary tools used when examining surface anatomy.
Head Region 396	■ The head is composed of the cranial and facial regions.
	Cranium 397
	■ The cranium is composed of frontal, temporal, and occipital regions.
	Face 397
	■ The facial region is composed of auricular, orbital, nasal, oral, and mental regions.
Neck Region 397	■ The sternocleidomastoid muscle subdivides each lateral region of the neck into anterior and posterior triangles.
Trunk Region 399	■ The trunk is divided into the thorax, abdominopelvic region, and back.
	Thorax 399
	■ Palpable structures of the thorax include the sternum, sternal angle, and ribs.

Trunk Region (continued)	**Abdominopelvic Region 401** ■ The important surface anatomy features of the abdominopelvic region include the linea alba, the umbilicus, and the inguinal ligaments. **Back 402** ■ The triangle of auscultation is an area where breathing sounds may be heard clearly with a stethoscope.
Shoulder and Upper Limb Region 404	■ The surface anatomy of the shoulder and upper limb is used for drawing blood, providing nutrients and fluids, and administering medicine. **Shoulder 404** ■ The scapula, the clavicle, and the humerus are palpable contributors to the structure of the shoulder. **Axilla 404** ■ The axilla, or armpit, is important clinically because of the blood vessels, nerves, and lymph nodes located there. **Arm 404** ■ The pulse of the brachial artery may be felt along the medial surface of the arm. **Forearm 405** ■ The styloid process of the radius and the head and styloid process of the ulna are readily palpable landmarks of the forearm. **Hand 405** ■ The thenar and hypothenar eminences are easily palpated on the anterior surface of the hand.
Lower Limb Region 407	■ The massive bones and strong muscles of the lower limbs are weight-bearers and serve in locomotion. **Gluteal Region 407** ■ The gluteal muscles form the prominences of the buttocks. **Thigh 407** ■ The femoral triangle and popliteal fossa are clinically important surface landmarks. **Leg 408** ■ The pulse of the posterior tibial artery is palpated posteroinferiorly from the medial malleolus. **Foot 409** ■ The pulse of the dorsalis pedis artery may be palpated on the medial dorsal side of the foot above the navicular bone, or along the dorsal interspace between the first and second toes.

CHALLENGE YOURSELF

Matching

Match each numbered item with the most closely related lettered item.

_____ 1. latissimus dorsi

_____ 2. cubital fossa

_____ 3. neck

_____ 4. auricle

_____ 5. umbilicus

_____ 6. scapula

_____ 7. deltoid

_____ 8. L$_4$ spinous process

_____ 9. sternal angle

_____ 10. philtrum

a. common place for an intramuscular injection

b. costal cartilage of second ribs attach here

c. venipuncture performed here

d. its spine may be palpated on the back

e. lumbar puncture performed near here

f. contains the carotid triangle

g. forms part of posterior axillary fold

h. vertical depression inferior to nose and superior to lips

i. fleshy part of external ear

j. structure on the abdomen

Multiple Choice

Select the best answer from the four choices provided.

_____ 1. Which of these can be palpated in the nuchal region of the neck?
   a. hyoid bone
   b. larynx
   c. cervical vertebrae
   d. trachea

_____ 2. Which muscle forms the anterior axillary fold?
   a. latissimus dorsi
   b. pectoralis major
   c. biceps brachii
   d. teres major

_____ 3. Which nerve is pinched when you "hit your funny bone"?
   a. ulnar nerve
   b. brachial nerve
   c. cephalic nerve
   d. radial nerve

_____ 4. Which of the following surface features may not be observed easily on obese people?
   a. philtrum
   b. auricle
   c. iliac crests
   d. natal cleft

_____ 5. An arterial pulse in the neck is best detected at the
   a. carotid triangle.
   b. submandibular triangle.
   c. submental triangle.
   d. supraclavicular triangle.

_____ 6. The costal margin is the inferior edge of the
   a. sternum.
   b. clavicle.
   c. rib cage.
   d. linea alba.

_____ 7. Eyebrows are located on the
   a. ala nasi.
   b. external occipital protuberance.
   c. superciliary arches.
   d. superior palpebral fissure.

_____ 8. Which artery can be palpated between the tendons of the flexor carpi radialis and the brachioradialis?
   a. popliteal
   b. radial
   c. brachial
   d. femoral

_____ 9. The great and small saphenous veins are located in the
   a. forearm.
   b. foot.
   c. leg.
   d. neck.

_____ 10. The triangle of auscultation is formed by all of the following muscles *except* the
   a. trapezius.
   b. rhomboid minor.
   c. latissimus dorsi.
   d. rhomboid major.

Content Review

1. Identify the five regions of the face, and specify a surface feature of each region.

2. What is the vertebra prominens, and where is it located?

3. Identify the two major triangles of the neck, and describe the structural subdivisions of each triangle. Also, identify an important structure in each triangle.

4. Where is the superficial inguinal ring, and where can it be palpated? What is the ring's clinical significance?

5. Why must a physician know surface anatomy of the back in order to perform a lumbar puncture? What key surface anatomy features are used to do this procedure correctly?

6. What prominent features may be palpated at or near the elbow?

7. Identify and describe the tendons that can be observed or palpated along the anterior surface of the wrist.

8. Discuss and describe the surface features that form the boundaries of a buttock.

9. Describe the location of the popliteal fossa, and discuss the clinical importance of this region.

10. Describe the anatomic locations where the following could be observed: (a) the posterior tibial arteries; (b) the greater trochanter; (c) the medial malleolus and lateral malleolus; and (d) the tendinous attachments of the hamstring muscles.

Developing Critical Reasoning

1. Marcie went to the doctor and received an intramuscular injection in her right gluteal region. Afterwards, Marcie had partial paralysis and lack of sensation in her right leg. What may have happened to Marcie during this injection?

2. Javier was hit hard in the lateral thoracic region. His doctor told him he had fractured his right sixth rib. How was the doctor able to determine which rib was fractured? What surface anatomy feature did he use to count the ribs?

3. When Louisa was sick with the flu, the doctor palpated her neck. What specific neck structures was the doctor palpating, and how do these structures relate to Louisa's infection?

ANSWERS TO "WHAT DO YOU THINK?"

1. Most of the muscles of facial expression may be palpated under the skin, including the orbicularis oculi, orbicularis oris, zygomaticus major, and platysma. In addition, you can palpate the masseter muscle, a muscle of mastication.

2. The sciatic nerve supplies the hamstrings, all leg muscles, and all muscles of the foot, so none of these muscles would work properly if the sciatic nerve were injured.

Visit the McKinley/O'Loughlin *Human Anatomy*, 2e website at aris.mhhe.com

OUTLINE

NERVOUS SYSTEM

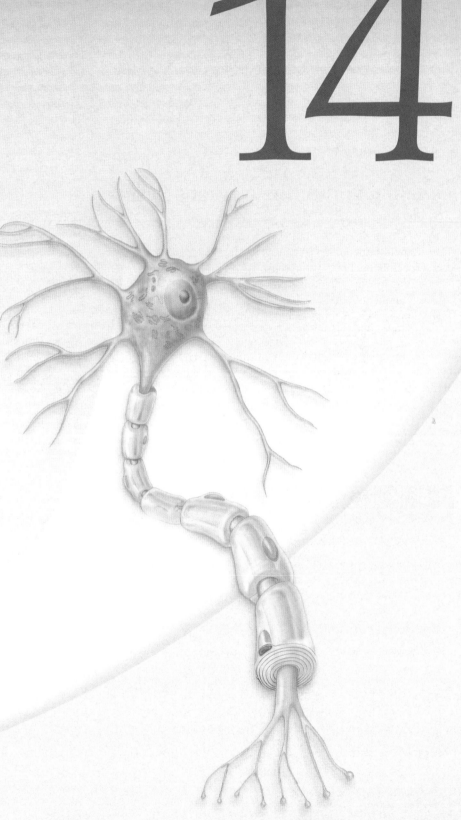

14

Nervous Tissue

Throughout the day, your body perceives and responds to multiple sensations. You smell spring flowers, feel the touch of a hand on your shoulder, and perceive your limbs moving. You control multiple muscle movements in order to walk, talk to the person sitting next to you, and hold this textbook. Other muscle movements occur without your voluntary input: Your heart beats, your stomach churns to digest your breakfast, and you jump at the sound of a honking horn. All of these sensations and muscle movements are interpreted and controlled by your **nervous system.** The nervous system is composed of all tissue types, but primarily of **nervous tissue**—neurons and glial cells (see chapter 4). This chapter introduces the study of the nervous system by first describing its overall organization and then investigating the components of nervous tissue.

Organization of the Nervous System

Key topics in this section:

- Organs of the CNS and PNS
- General functions of the nervous system
- Specific functions of the sensory and motor nervous systems
- Comparison of somatic sensory and visceral sensory components
- Comparison of somatic motor and autonomic (visceral) motor components

As the body's primary communication and control system, the nervous system is extremely complex. In order to describe its interacting structures and functions, anatomists and physiologists have devised specialized terms and organizational systems. For example, the nervous system may be divided into either structural or functional categories, as shown in **table 14.1**. However, always keep in mind that such artificial divisions are merely intended to simplify discussion—there is only one nervous system.

Structural Organization: Central and Peripheral Nervous Systems

Based on its anatomic components, the nervous system consists of two subdivisions: the central nervous system and the peripheral nervous system (**figure 14.1**). The **central nervous system (CNS)** is composed of the **brain** and **spinal cord.** The brain is protected and enclosed within the skull, while the spinal cord is housed and protected within the vertebral canal. The **peripheral** (pĕ-rif′-ĕ-răl) **nervous system (PNS)** includes the **cranial nerves** (nerves that extend from the brain), **spinal nerves** (nerves that extend from the spinal cord), and **ganglia** (gang′glē-ă; sing., *ganglion* = swelling), which are clusters of neuron cell bodies located outside the CNS.

Functional Organization: Sensory and Motor Nervous Systems

Together, the CNS and PNS perform three general functions:

- **Collecting information.** Specialized PNS structures called **receptors** (dendrite endings of sensory neurons or cells) detect changes in the internal or external environment and pass them on to the CNS as sensory input (discussed in chapter 19).
- **Processing and evaluating information.** After processing sensory input, the CNS determines what, if any, response is required.
- **Responding to information.** After selecting an appropriate response, the CNS initiates specific nerve impulses (rapid movements of an electrical charge along the neuron's plasma membrane), called motor output. **Motor output** travels

Table 14.1	Structural and Functional Divisions of the Nervous System	
Nervous System Organization	**Anatomic Components**	**Description**
STRUCTURAL DIVISIONS		
Central nervous system (CNS)	Brain and spinal cord	Command center of nervous system that integrates and processes nervous information
Peripheral nervous system (PNS)	Nerves and ganglia	Projects information to and receives information from CNS; mediates some reflexes
FUNCTIONAL DIVISIONS		
Sensory nervous system	Some CNS and PNS components (including sensory neurons)	Consists of all axons that transmit a nerve impulse *from* a peripheral structure *to* the CNS; includes pain, touch, temperature, and pressure ("input" information)
Somatic sensory		Transmits input from skin, fascia, joints, and skeletal muscle
Visceral sensory		Transmits input from viscera
Motor nervous system	Some CNS and PNS components (including motor neurons)	Consists of all axons that transmit a nerve impulse *from* the CNS *to* a muscle or gland ("output" information)
Somatic motor (somatic nervous system; SNS)		Voluntary control of skeletal muscle
Autonomic motor (autonomic nervous system; ANS)		Involuntary control of smooth muscle, cardiac muscle, and glands

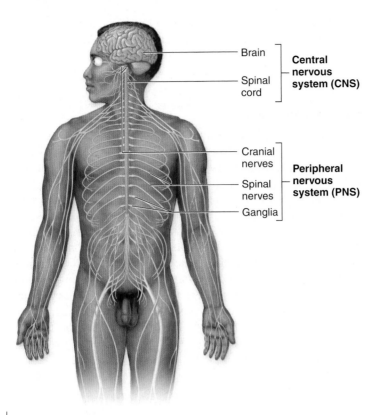

Figure 14.1

Organization of the Nervous System.   The central nervous system (CNS) is composed of the brain and spinal cord. The peripheral nervous system (PNS) is composed of cranial nerves, spinal nerves, and ganglia.

through structures of the PNS to **effectors** (the cells that receive impulses from motor neurons: muscles or glands).

There are two functional divisions of the nervous system: the sensory nervous system and the motor nervous system **(figure 14.2)**.

⚡ WHAT DO YOU THINK?

1 Why is the term *visceral* sometimes used to describe certain parts of the sensory and motor nervous systems?

Sensory Nervous System

The **sensory** (or *afferent;* af′er-ent) **nervous system** is responsible for receiving sensory information *from* receptors and transmitting this information *to* the CNS. (The term *afferent* means "inflowing," which indicates that nerve impulses are transmitted to the CNS.) Thus, the sensory nervous system is responsible for **input.** The sensory nervous system contains both PNS and CNS components: Nerves of the PNS transmit the sensory information, and certain parts of the brain and spinal cord in the CNS interpret this information.

The sensory nervous system has two components: somatic sensory and visceral sensory. The **somatic sensory** components are the general somatic senses—touch, pain, pressure, vibration, temperature, and proprioception (sensing the position or movement of joints and limbs)—and the special senses (taste, vision, hearing, balance, and smell). These functions are considered *voluntary* because we have some control over them and we tend to be conscious of them. The **visceral sensory** components transmit nerve impulses from blood vessels and viscera to the CNS. The visceral senses primarily include temperature and stretch (of muscles of the organ

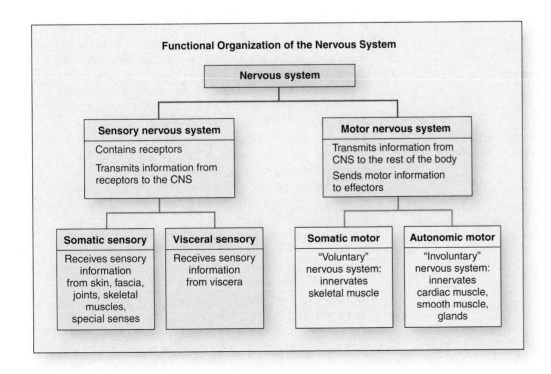

Figure 14.2

Functional Organization of the Nervous System.   The nervous system is functionally divided into a sensory nervous system and a motor nervous system. Both of these parts of the nervous system contain somatic and visceral components.

wall). These functions are said to be *involuntary* because most of the time you do not have voluntary control over them and are not conscious of them. However, you may become aware of visceral sensations when they are extreme—for example, if you have eaten too much and your stomach is bloated.

Motor Nervous System

The **motor** (or *efferent*; ef'er-ent) **nervous system** is responsible for transmitting motor impulses *from* the CNS *to* muscles or glands. (The term *efferent* means "conducting outward," which indicates that nerve impulses are transmitted from the CNS.) Thus, the motor nervous system is responsible for **output**. The motor nervous system contains both CNS and PNS components: Parts of the brain and spinal cord (CNS) initiate nerve impulses, which travel through motor nerves that in turn transmit these impulses to effector organs.

The motor division is subdivided into somatic motor and autonomic motor components. The **somatic motor** component (**somatic nervous system; SNS**) conducts nerve impulses from the CNS to the skeletal muscles, causing them to contract. The somatic motor division is often called the *voluntary nervous system* because the contractions of the skeletal muscles are under conscious control; for example, you exert voluntary control over your leg muscles as you press on the accelerator of your car.

The **autonomic motor** component is often called the **autonomic nervous system (ANS).** Because it innervates internal organs and regulates smooth muscle, cardiac muscle, and glands without our control, it is also known as the *visceral motor system* or the *involuntary nervous system.* For example, we cannot voluntarily make our hearts stop beating, nor can we prevent our stomachs from growling. The autonomic nervous system has two further subdivisions—parasympathetic and sympathetic—which we examine in chapter 18.

💡 WHAT DID YOU LEARN?

1 Together, what three functions do the CNS and PNS perform?

2 Compare and contrast the meanings of afferent and efferent.

Cytology of Nervous Tissue

Key topics in this section:

- Basic features common to all neurons
- Structural and functional classifications of neurons
- Types of glial cells and comparison of their morphology and function

Two distinct cell types form nervous tissue: neurons, which are excitable cells that are able to generate, transmit, and receive nerve impulses, and glial cells, which are nonexcitable cells that support and protect the neurons.

Neurons

The basic structural unit of the nervous system is the **neuron** (noor'on). Neurons conduct nerve impulses from one part of the body to another. They have several special characteristics:

- Neurons have a high metabolic rate. Their survival depends on continuous and abundant supplies of glucose and oxygen.
- Neurons have extreme longevity. Most neurons formed during fetal development are still functional in very elderly individuals.

- Neurons typically are nonmitotic (unable to divide and produce new neurons). During the fetal development of neurons, mitotic activity is lost, except possibly in certain areas of the brain (see Clinical View: "New Neurons in Adults").

Neuron Structure

Neurons come in all shapes and sizes, but all neurons share certain basic structural features (**figure 14.3**). A typical neuron has a cell body. Projecting from the cell body are processes called dendrites and an axon.

The **cell body**, also called a *soma*, serves as the neuron's control center and is responsible for receiving, integrating, and sending nerve impulses. The cell body is enclosed by a plasma membrane and contains cytoplasm surrounding a **nucleus**. The nucleus contains a prominent **nucleolus**, reflecting the high metabolic activity of neurons, which require the production of many proteins. Numerous mitochondria are present within this cytoplasm to produce the large amounts of ATP needed by the neuron. Large numbers of free ribosomes and rough ER produce proteins for the active neuron. Together, both free and bound ribosomes go by two names: **chromatophilic** (krō-mă-tō-fil'ik; *chromo* = color; *phileo* = to love) **substance**, because they stain darkly with basic dyes, or *Nissl bod-*

CLINICAL VIEW

New Neurons in Adults?

For years, prevailing medical wisdom has maintained that the number of neurons you have shortly after birth is your supply for a lifetime. Recent studies, however, have shown that this is not always the case. Researchers investigating the hippocampus of the brain, the region involved in memory processing (described in chapter 15), have found that mature neurons are indeed "terminally" differentiated, meaning they lack the ability to divide and produce daughter cells. However, there now appears to be a population of immature progenitor cells in the hippocampus that are called neural stem cells. These stem cells were once thought to give rise only to new glial cells in adults, but it is now clear that under special circumstances they can mature into neurons. What's more, the new neurons appear able to incorporate themselves, at least to some degree, into the brain circuitry. Researchers have learned that the surrounding glial cells provide the chemical signals that direct a stem cell down the path of nerve cell maturation.

Although excited by these new data, researchers remain divided as to whether the new cells truly function in the same way as the ones that were present all along. It has yet to be conclusively shown that new hippocampal neurons can make all the necessary connections needed to function as fully integrated parts of the brain. One problem hindering medical application of this new information is that stem cells apparently do not mature to neurons uniformly throughout the adult brain. Only a few regions seem to be so lucky, and the hippocampus is one. Even so, understanding the mechanisms that drive this process in the hippocampus greatly expands our knowledge of brain function and may help explain how the brain can continue to function for so many years. Clinicians hope that research in this field will lead to therapies for conditions that cause the loss of neurons.

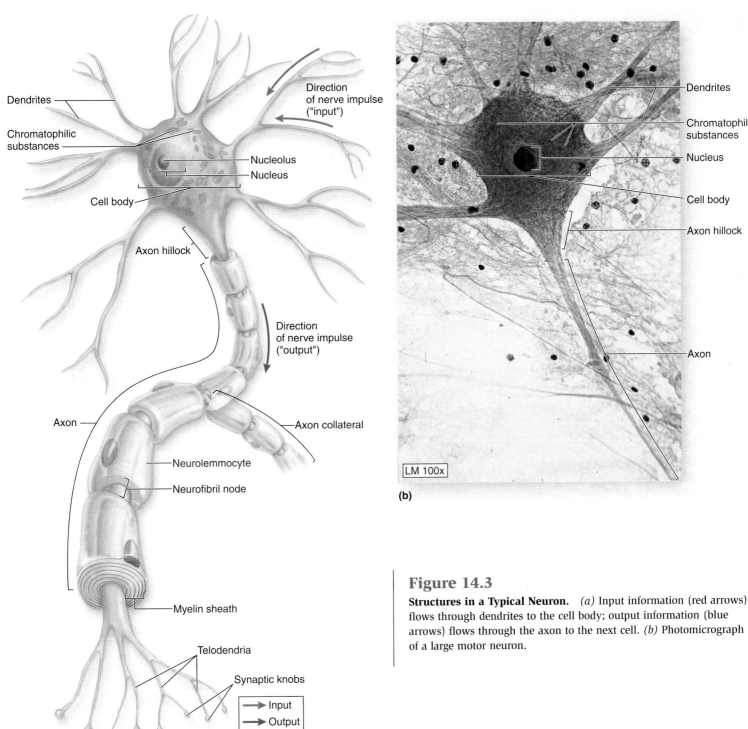

Dendrites

Chromatophilic substances

Direction of nerve impulse ("input")

Nucleolus

Nucleus

Cell body

Axon hillock

Axon

Axon collateral

Direction of nerve impulse ("output")

Neurolemmocyte

Neurofibril node

Myelin sheath

Telodendria

Synaptic knobs

→ Input
→ Output

(a)

Dendrites

Chromatophilic substances

Nucleus

Cell body

Axon hillock

Axon

LM 100x

(b)

Figure 14.3

Structures in a Typical Neuron. *(a)* Input information (red arrows) flows through dendrites to the cell body; output information (blue arrows) flows through the axon to the next cell. *(b)* Photomicrograph of a large motor neuron.

ies, because they were first described by the German microscopist Franz Nissl. Cytologists believe that the chromatophilic substance together with dendrites and cell bodies account for the gray color of the gray matter, as seen in brain and spinal cord areas containing collections of neuron cell bodies.

Dendrites (den′drīt; *dendrites* = relating to a tree) tend to be shorter, smaller processes that branch off the cell body. Some neurons have only one dendrite, while others have many. Dendrites conduct nerve impulses toward the cell body; in essence, they receive input and then transfer it to the cell body for processing. The more dendrites a neuron has, the more nerve impulses that neuron can receive from other cells.

The typically longer nerve cell process emanating from the cell body is the **axon** (ak′son; *axon* = axis), sometimes called a *nerve fiber*.

Neurons have either one axon or no axon at all (neurons with only dendrites and no axons are called **anaxonic** [an-aks′on; *an* = without]). Most neurons, however, have a single axon. The axon transmits a nerve impulse away from the cell body toward another cell; in essence, the axon transmits output information to other cells. The axon connects to the cell body at a triangular region called the **axon hillock** (hil′lok). Unlike the rest of the cell body, the axon hillock is devoid of chromatophilic substance, and so it lacks those dark-staining regions when viewed under the

Table 14.2 — Parts of a Neuron

Category/Structure	Description
Neuron	Structural and functional cell of the nervous system; sometimes called a nerve cell
Cell body	Nucleus and surrounding cytoplasm of a neuron (excluding its dendrites and axon)
Dendrites	Neuron processes that conduct information *to* the cell body ("input")
Axon	Neuron process that conducts nerve impulses *away* from the cell body ("output")
Axon hillock	Triangular region connecting axon to cell body
Axon collaterals	Side branches of an axon
Telodendria	Fine terminal branches of an axon or axon collateral
Synaptic knobs	Slightly expanded regions at the tips of telodendria

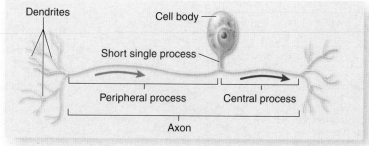

(a) Unipolar neuron

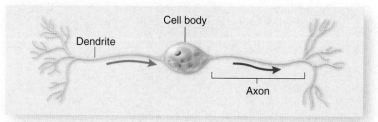

(b) Bipolar neuron

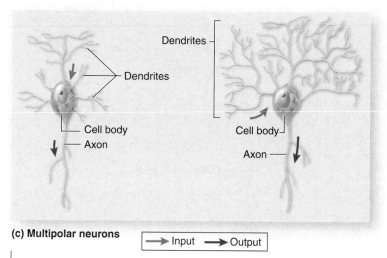

(c) Multipolar neurons

→ Input → Output

Figure 14.4

Structural Classification of Neurons. Neurons can be classified according to the number of processes extending directly from the cell body. *(a)* A unipolar neuron has a single process that divides into a peripheral process and a central process. *(b)* A bipolar neuron has two processes. *(c)* A multipolar neuron has three or more processes.

microscope. Although an axon remains relatively unbranched for most of its length, it may give rise to a few side branches called **axon collaterals.** Most axons and their collaterals branch extensively at their distal end into an array of fine terminal extensions called **telodendria** (tel-ō-den'drĭa; sing., telodendrion; *telos* = end), or *axon terminals*. The extreme tips of these fine extensions are slightly expanded regions called **synaptic** (si-nap'tik) **knobs** (also called *end bulbs* or *terminal boutons*).

Special terms denote other internal structures of a neuron. The cytoplasm within the cell body is called the **perikaryon** (per-i-kar'ē-on; *peri* = around, *karyon* = kernel), although some anatomists use that term to describe the whole cell body. The microtubules that form the cytoskeleton are called neurotubules. **Neurofilaments** (noor-ō-fil'ă-ment; *filamentum* = thread) are intermediate filaments that aggregate to form bundles called **neurofibrils** (noor-ō-fī'bril; *fibrilla* = fiber). Neurofibrils extend as a complex network into both the dendrites and axons, where their tensile strength provides support for these processes. **Table 14.2** reviews some of the terms used to describe neuron structures.

Neuron Classification

Neurons vary widely in morphology and location. They can be classified according to either their structure or their function (**table 14.3**).

Structural Classification   Structurally, neurons are classified into three types, based on the number of neuron processes emanating directly from the cell body: unipolar, bipolar, or multipolar (**figure 14.4**).

Unipolar neurons have a single, short neuron process that emerges from the cell body and branches like a T. These neurons are also called *pseudounipolar* (soo'dō-oo-nē-pō-lăr; *pseudo* = false, *uni* = one) because they start out as bipolar neurons during development, but their two processes fuse into a single process. The naming of the branched processes in unipolar neurons has been a source of confusion as it relates to the common definitions of dendrites and axons. It seems most appropriate to call the short, multiple-branched receptive endings dendrites. The combined **peripheral process** (from dendrites to the cell body) and **central process** (from the cell body into the CNS) together denote the axon, because these processes generate and conduct impulses and are often myelinated.

Bipolar neurons have two neuron processes that extend from the cell body—one axon and one dendrite. These neurons are relatively uncommon in humans and primarily limited to some of the special senses. For example, bipolar neurons are located in the olfactory epithelium of the nose and in the retina of the eye.

Multipolar neurons are the most common type of neuron. Multiple neuron processes—many dendrites and a single axon—extend from the cell body. Examples of multipolar neurons include motor neurons that innervate muscle and glands.

Functional Classification   Functionally, neurons are classified as one of three types according to the direction the nerve impulse travels relative to the CNS: sensory neurons, motor neurons, or interneurons (**figure 14.5**).

Sensory neurons, or *afferent neurons*, transmit nerve impulses *from* sensory **receptors** *to* the CNS. These neurons are specialized to detect changes in their environment called **stimuli** (sing., *stimulus*). Stimuli can be in the form of touch, pressure, heat, light, or chemicals. Most sensory neurons are unipolar,

Table 14.3	Structural and Functional Classifications of Neurons	
Structural Classification	**Description**	**Functional Example**
Unipolar neuron	Common type of sensory neuron; single short cell process extends directly from the cell body and looks like a T as a result of the fusion of two processes into one long axon	Most sensory neurons (detect stimuli in the form of touch, pressure, temperature, or chemicals)
Bipolar neuron	Relatively uncommon; two nerve cell processes extend directly from the cell body	Some special sense neurons (e.g., in olfactory epithelium of nose, retina of eye)
Multipolar neuron	Most common type of neuron; multiple nerve cell processes extend from cell body; typically one axon and many dendrites	Interneurons, motor neurons
Functional Classification	**Description**	**Structural Example**
Sensory	Conducts nerve impulses *from* body *to* CNS	Most sensory neurons are unipolar; a few (e.g., some in olfactory epithelium and retina) are bipolar
Motor	Conducts nerve impulses *from* CNS *to* muscles or glands	Multipolar
Interneuron	Found only in CNS; facilitates communication between motor and sensory neurons	Multipolar

although a few are bipolar (e.g., those in the olfactory epithelium of the nose and the retina of the eye, as previously mentioned). The cell bodies of unipolar sensory neurons are located outside the CNS and housed within structures called posterior (dorsal) root ganglia.

Motor neurons, or *efferent neurons,* transmit nerve impulses *from* the CNS *to* muscles or glands. They are called motor neurons because most of them extend to muscle cells, and the nerve impulses they transmit cause these cells to contract. The muscle and gland cells that receive nerve impulses from motor neurons are called effectors, because their stimulation produces a response or effect. The cell bodies of most motor neurons lie in the spinal cord, whereas the axons primarily travel in cranial or spinal nerves to muscles and glands. All motor neurons are multipolar.

Interneurons, or *association neurons,* lie entirely within the CNS and are multipolar structures. They receive nerve impulses from many other neurons and carry out the integrative function of the nervous system—that is, they retrieve, process, and store information and "decide" how the body responds to stimuli. Thus, interneurons facilitate communication between sensory and motor neurons. Figure 14.5 shows a sensory neuron transmitting stimuli (sensory information) to an interneuron, which then processes that information and signals the appropriate motor neuron(s) to transmit a nerve impulse to the muscle. Interneurons outnumber all other neurons in both their total number and different types; it is estimated that 99% of our neurons are interneurons. The number of interneurons activated during processing or storing increases dramatically with the complexity of the response.

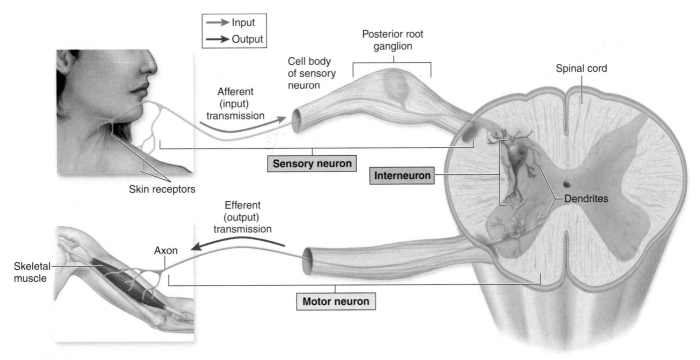

Figure 14.5

Functional Classification of Neurons.   Sensory neurons carry afferent (input) signals to the central nervous system. Interneurons process information in the CNS. Motor neurons transmit efferent (output) impulses from the CNS to effectors.

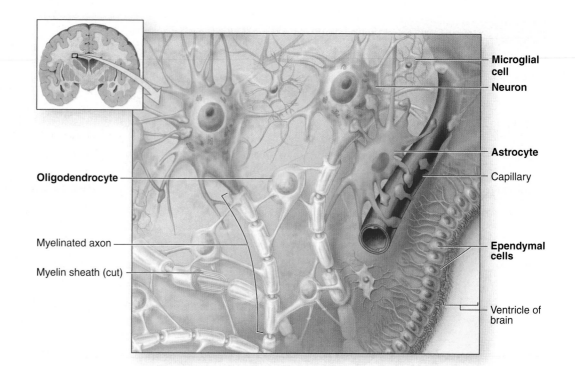

Figure 14.6

Cellular Organization of Nervous Tissue Within the CNS. The four types of glial cells in the CNS are shown in relationship to a neuron.

Glial Cells

Glial (glī′ăl) **cells**, sometimes referred to as *neuroglia* (noo-rog′lē-a; *glia* = glue) occur within both the CNS and the PNS. Glial cells differ from neurons in that they are smaller and capable of mitosis. Glial cells do not transmit nerve impulses, but they do assist neurons with their functions. Collectively, the glial cells physically protect and help nourish neurons, and provide an organized, supporting framework for all the nervous tissue. During development, glial cells form the framework that guides young migrating neurons to their destinations.

Glial cells far outnumber neurons. The nervous tissue of a young adult may contain 35 to 100 billion neurons and 100 billion to 1 trillion glial cells. Collectively, glial cells account for roughly half the volume of the nervous system.

WHAT DO YOU THINK?

2 If a person has a "brain tumor," is it more likely to have developed from neurons or from glial cells? Why?

Glial Cells of the CNS

Four types of glial cells are found in the central nervous system: astrocytes, ependymal cells, microglial cells, and oligodendrocytes **(figure 14.6)**. They can be distinguished on the basis of size, intracellular organization, and the presence of specific cytoplasmic processes **(table 14.4)**.

Astrocytes **Astrocytes** (as′trō-sīt; *astron* = star) exhibit a star-like shape due to many projections from their surface **(figure 14.7a)**.

Table 14.4	Glial Cells	
Cell Type	**Appearance**	**Functions**
CENTRAL NERVOUS SYSTEM		
Astrocyte	Large cell with numerous cell processes; in contact with neurons and capillaries; most common type of glial cell	Helps form the blood-brain barrier Regulates tissue fluid composition Provides structural support and organization to CNS Replaces damaged neurons Assists with neuronal development
Ependymal cell	Simple cuboidal epithelial cell lining cavities in brain and spinal cord; cilia on apical surface	Lines ventricles of brain and central canal of spinal cord Assists in production and circulation of CSF
Microglial cell	Small cells with slender branches from cell body; least common type of glial cell	Defends against pathogens Removes debris Phagocytizes wastes
Oligodendrocyte	Rounded, bulbous cell with slender cytoplasmic extensions; extensions wrap around CNS axons	Myelinates and insulates CNS axons Allows faster nerve impulse conduction through the axon
PERIPHERAL NERVOUS SYSTEM		
Satellite cell	Flattened cell clustered around neuronal cell bodies in a ganglion	Protects and regulates nutrients for cell bodies in ganglia
Neurolemmocyte	Flattened cell wrapped around a portion of an axon in the PNS	Myelinates and insulates PNS axons Allows for faster nerve impulse conduction through the axon

CNS Glial Cells

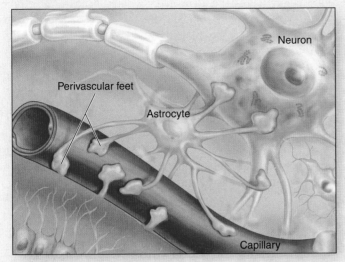

(a) Astrocyte

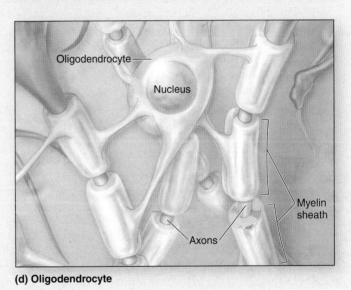

(b) Ependymal cells

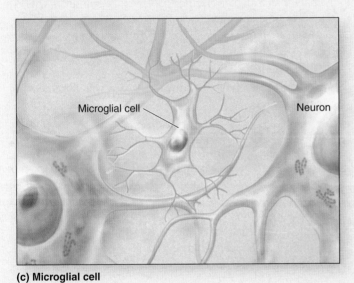

(c) Microglial cell

(d) Oligodendrocyte

PNS Glial Cells

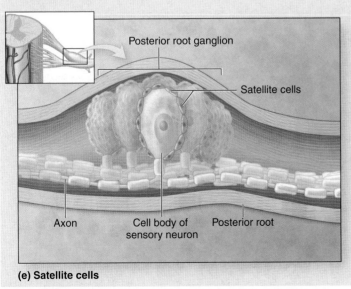

(e) Satellite cells

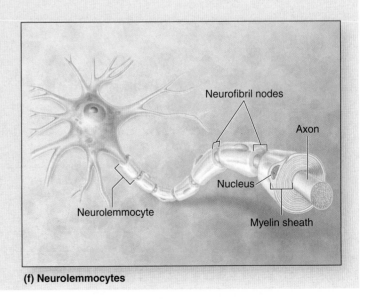

(f) Neurolemmocytes

Figure 14.7

Glial Cells. *(a)* Astrocytes have perivascular feet (only a few are shown here to appreciate their morphology) that wrap completely around capillaries in the CNS. *(b)* Ependymal cells line the fluid-filled spaces in the brain and spinal cord. *(c)* Microglial cells phagocytize damaged neurons and cellular debris. *(d)* Oligodendrocytes myelinate axons in the CNS. *(e)* In the PNS, satellite cells surround neuron cell bodies in ganglia, such as the posterior root ganglion. *(f)* Neurolemmocytes myelinate axons in the PNS.

These numerous cell processes touch both capillary walls and different parts of neurons. Astrocytes are the most abundant glial cell in the CNS, and they constitute over 90% of the nervous tissue in some areas of the brain. Their functions include:

- **Helping form the blood-brain barrier.** Ends of astrocyte processes called perivascular feet wrap completely around and cover the outer surface of capillaries in the brain. Together, the perivascular feet and the brain capillaries, which are less "leaky" than other capillaries in the body, contribute to a **blood-brain barrier (BBB)** that strictly controls substances entering the nervous tissue in the brain from the bloodstream. This blood-brain barrier protects the delicate brain from toxins (such as certain waste products and drugs in the blood), but allows needed nutrients to pass through. Sometimes this barrier is detrimental; for example, some medications are not allowed to exit the capillaries and enter the nervous tissue in the brain.

- **Regulating tissue fluid composition.** Astrocytes help regulate the chemical composition of the interstitial fluid within the brain by controlling movement of molecules from the blood to the interstitial fluid.

- **Forming a structural network.** The cytoskeleton in astrocytes strengthens and organizes nervous tissue in the CNS.

- **Replacing damaged neurons.** When neurons are damaged and die, the space they formerly occupied is often filled by cells produced by astrocyte division, a process termed *astrocytosis*.

- **Assisting neuronal development.** Astrocytes help direct the development of neurons in the fetal brain by secreting chemicals that regulate the connections between neurons.

Ependymal Cells   **Ependymal** (ep-en′di-măl) **cells** are cuboidal epithelial cells that line the internal cavities (ventricles) of the brain and the central canal of the spinal cord (figure 14.7*b*). These cells have slender processes that branch extensively in order to make contact with other glial cells in the surrounding nervous tissue. Ependymal cells and nearby blood capillaries together form a network called the choroid (ko′royd) plexus. The choroid plexus produces cerebrospinal fluid (CSF), a clear liquid that bathes the CNS and fills its internal cavities. The ependymal cells have cilia on their apical surfaces that help circulate the CSF. (Chapter 15 describes ependymal cells, the choroid plexus, and CSF in more detail.)

Microglial Cells   **Microglial** (mī-krog′le-ăl; *micros* = small) **cells** represent the smallest percentage of CNS glial cells; some estimates of their prevalence are as low as 5%. Microglial cells are typically small cells that have slender branches extending from the main cell body (figure 14.7*c*). They wander through the CNS and replicate in response to an infection. They perform phagocytic activity and remove debris from dead or damaged nervous tissue. Thus, the activities of microglial cells resemble those of the macrophages of the immune system.

Oligodendrocytes   **Oligodendrocytes** (ol′i-gō-den′drō-sīt; *oligos* = few) are large cells with a bulbous body and slender cytoplasmic extensions or processes (figure 14.7*d*). The processes of oligodendrocytes ensheathe portions of many different axons, each repeatedly wrapping around part of an axon like electrical tape wrapped around a wire. This protective covering around the axon is called a myelin sheath, which we discuss in a later section.

Glial Cells of the PNS

The two glial cell types in the PNS are satellite cells and neurolemmocytes.

Satellite Cells   **Satellite cells** are flattened cells arranged around neuronal cell bodies in ganglia. (Recall that a ganglion is a collection of neuron cell bodies located outside the CNS.) For example, figure 14.7*e* illustrates how satellite cells surround the cell bodies of sensory neurons located in a specific type of ganglion called a posterior root ganglion. Satellite cells physically separate cell bodies in a ganglion from their surrounding interstitial fluid, and regulate the continuous exchange of nutrients and waste products between neurons and their environment.

Neurolemmocytes   **Neurolemmocytes** (noor-ō-lem′ō-sīt), also called *Schwann cells,* are associated with PNS axons (figure 14.7*f*).

CLINICAL VIEW

Tumors of the Central Nervous System

Neoplasms resulting from unregulated cell growth, commonly known as tumors, sometimes occur in the central nervous system. A tumor that originates within the organ where it is found is called a primary tumor. Because mature neurons do not divide and are incapable of giving rise to tumors, primary CNS tumors originate in supporting tissues within the brain or spinal cord that have retained the capacity to undergo mitosis: the meninges (protective membranes of the CNS) or the glial cells. Glial cell tumors, termed **gliomas,** may be either relatively benign and slow-growing or malignant (capable of metastasizing [spreading] to other areas of the body).

A secondary tumor is a neoplasm that has originated at one site but subsequently spread to some other organ. For example, lung cancer can metastasize to the nervous system and form additional tumors.

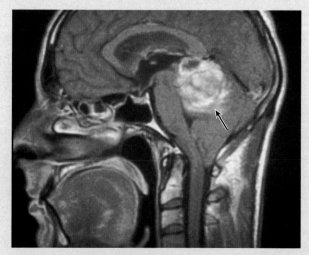

An MRI shows a glioma (arrow).

These cells are responsible for myelinating PNS axons, a process to be discussed in the next section.

Myelination of Axons

Key topics in this section:

- Composition and function of a myelin sheath
- Comparison of saltatory and continuous conduction

The main activity of axons is nerve impulse conduction. A **nerve impulse** is the rapid movement of an electrical charge along a neuron's plasma membrane. A nerve impulse is also known as an *action potential,* because a nerve impulse is caused by an actual voltage (potential) change that moves along the plasma membrane of the axon. The nerve impulse's ability to travel along an axon is affected by a process called myelination.

Myelination

Myelination is the process by which part of an axon is wrapped with a **myelin sheath,** the insulating covering around the axon consisting of concentric layers of myelin. In the CNS, a myelin sheath forms from oligodendrocytes, and in the PNS, it forms from neurolemmocytes, Therefore, myelin mainly consists of the plasma membranes of these glial cells and contains a large proportion of fats and a lesser amount of proteins. The high lipid content of the myelin sheath gives the axon a distinct, glossy-white appearance.

Figure 14.8 illustrates the process of myelinating a PNS axon. The neurolemmocyte starts to encircle a 1 millimeter portion of the axon, much as if you were wrapping a piece of tape around a portion of your pencil. As the neurolemmocyte continues to wrap around the axon, its cytoplasm and nucleus are squeezed to the periphery (the outside edge). The overlapping inner layers of the plasma membrane form the myelin sheath. Sometimes the name **neurilemma** is used to describe this delicate, thin outer membrane of the neurolemmocyte.

In the CNS, an oligodendrocyte can myelinate a 1 millimeter portion of *many* axons, not just one. **Figure 14.9a** shows oligodendrocytes myelinating portions of three different axons. The cytoplasmic extensions of the oligodendrocyte wrap successively around a portion of each axon, and successive plasma membrane layers form the myelin sheath.

In the PNS, a neurolemmocyte can myelinate a 1 millimeter portion of a single axon only (figure 14.9b). Thus, if an axon is longer than 1 millimeter (and most PNS axons are), it takes many neurolemmocytes to myelinate the entire axon. Figure 14.3a shows an axon that has seven neurolemmocytes wrapped around it. The axons in many of the nerves in the body have hundreds or thousands of neurolemmocytes covering their entire length.

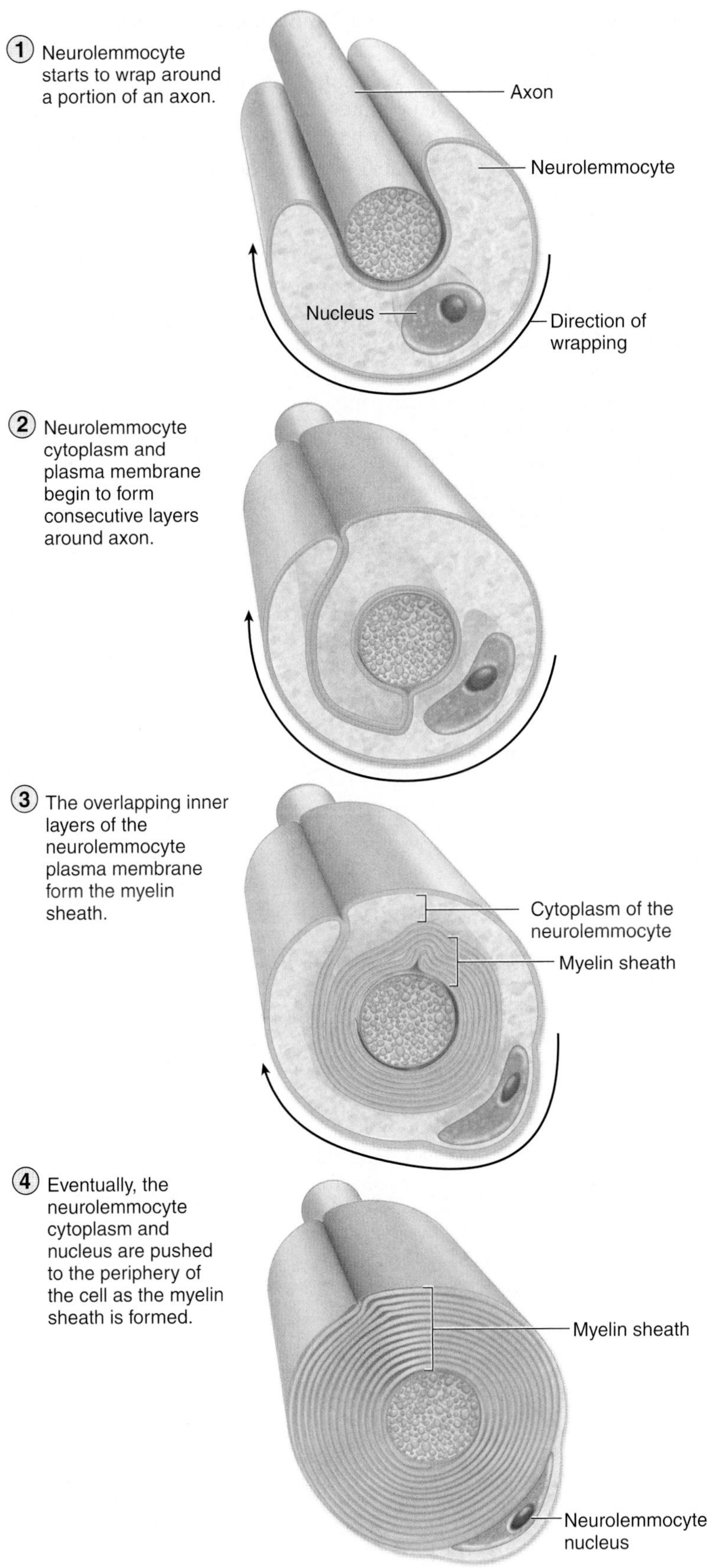

(1) Neurolemmocyte starts to wrap around a portion of an axon.

Axon

Neurolemmocyte

Nucleus

Direction of wrapping

(2) Neurolemmocyte cytoplasm and plasma membrane begin to form consecutive layers around axon.

(3) The overlapping inner layers of the neurolemmocyte plasma membrane form the myelin sheath.

Cytoplasm of the neurolemmocyte

Myelin sheath

(4) Eventually, the neurolemmocyte cytoplasm and nucleus are pushed to the periphery of the cell as the myelin sheath is formed.

Myelin sheath

Neurolemmocyte nucleus

Figure 14.8

Myelination of PNS Axons.   A myelin sheath surrounds most axons. In the PNS, successive adjacent neurolemmocytes form the myelin sheaths along the length of PNS axons.

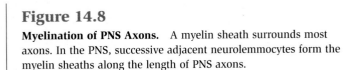

Not all axons are myelinated. **Unmyelinated axons** in the PNS (shown in **figure 14.10**) are associated with a neurolemmocyte, but no myelin sheath covers them. In other words, the axon merely rests in a portion of the neurolemmocyte rather than being wrapped by successive layers of the plasma membrane. In the CNS, unmyelinated axons are not associated with oligodendrocytes.

Nerve Impulse Conduction

The myelin sheath supports, protects, and insulates an axon. Note in figure 14.9 that small spaces interrupt the myelin sheath between adjacent oligodendrocytes or neurolemmocytes. These gaps are called **neurofibril nodes,** or *nodes of Ranvier.* At these nodes, and only at these nodes, can a change in voltage occur across the plasma membrane and result in the movement of a nerve impulse. Thus, in a myelinated axon, the nerve impulse seems to "jump" from neurofibril node to neurofibril node, a process called **saltatory conduction.** In an unmyelinated axon, the nerve impulse must travel the entire length of the axon membrane, a process called **continuous conduction.**

A myelinated axon produces a faster nerve impulse because only the exposed membrane regions are affected as the impulse jumps toward the end of the axon. In an unmyelinated axon, a nerve impulse takes longer to reach the end of the axon because every part of the membrane must be affected by the voltage change. Thus, a myelinated axon also requires less energy in the form of ATP than does an unmyelinated axon. ATP must be used by the cell to reestablish the resting condition that existed prior to the passage of the nerve impulse. Using saltatory conduction, large-diameter, myelinated axons conduct nerve impulses rapidly to the skeletal muscles in the limbs. Using continuous conduction, unmyelinated axons conduct nerve impulses from pain and some cold stimuli.

WHAT DO YOU THINK?

3 If myelinated axons produce faster nerve impulses than unmyelinated axons, why aren't all axons in the body myelinated?

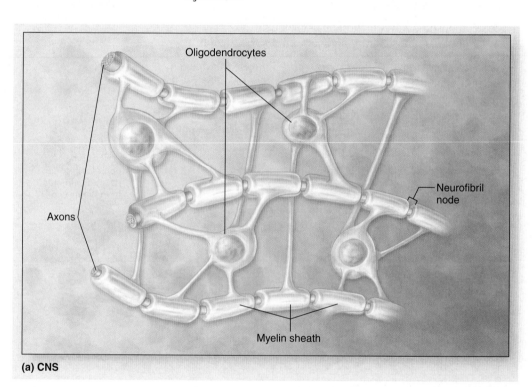

(a) CNS

Figure 14.9

Myelin Sheaths in the CNS and PNS. *(a)* In the CNS, several extensions from an oligodendrocyte wrap around small parts of multiple axons. *(b)* In the PNS, the neurolemmocyte ensheathes only one small part of a single axon.

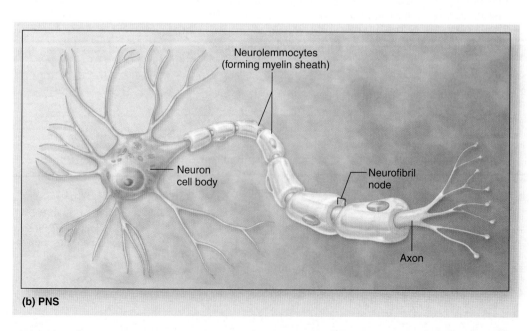

(b) PNS

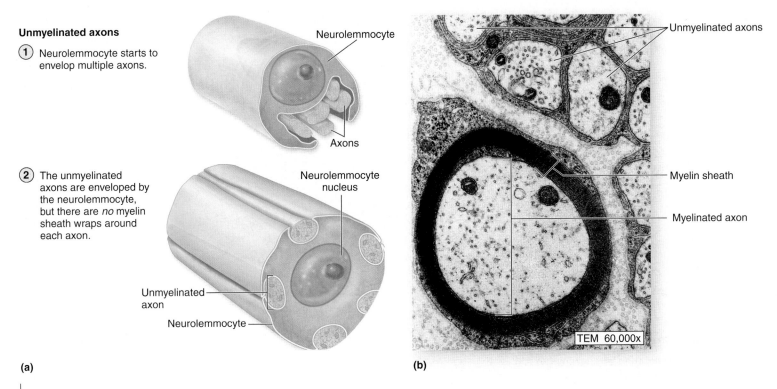

Unmyelinated axons

(1) Neurolemmocyte starts to envelop multiple axons.

Neurolemmocyte

Axons

(2) The unmyelinated axons are enveloped by the neurolemmocyte, but there are *no* myelin sheath wraps around each axon.

Neurolemmocyte nucleus

Unmyelinated axon

Neurolemmocyte

(a)

Unmyelinated axons

Myelin sheath

Myelinated axon

TEM 60,000x

(b)

Figure 14.10

Comparison of Unmyelinated and Myelinated Axons. *(a)* Unmyelinated axons are surrounded by a neurolemmocyte but are *not* wrapped in a myelin sheath. *(b)* An electron micrograph shows a myelinated axon and some unmyelinated axons.

WHAT DID YOU LEARN?

(7) What are some differences in the way axons are myelinated in the PNS versus the CNS?

(8) What are neurofibril nodes, and where are they found?

Axon Regeneration

Key topics in this section:

- Conditions under which axons can regenerate
- Events that occur after injury to a PNS axon

PNS axons are vulnerable to cuts, crushing injuries, and other trauma. However, a damaged axon can regenerate if at least some neurilemma remains. PNS axon regeneration depends upon three factors: (1) the amount of damage; (2) neurolemmocyte secretion of nerve growth factors to stimulate outgrowth of severed axons; and (3) the distance between the site of the damaged axon and the effector organ (as the distance to the effector increases, the possibility of repair decreases).

Neurolemmocytes help repair a damaged axon through a regeneration process called **Wallerian** (waw-ler′ē-an) **degeneration,** illustrated in **figure 14.11** and described here:

1. The axon is severed by some type of trauma.
2. The end of the proximal portion of the severed end seals off by membrane fusion and swells. The swelling is a result of cytoplasm flowing from the neuron cell body

through the axon. The severed distal portion of the axon and its myelin sheath degenerate; macrophages remove the debris by phagocytosis. The neurolemmocytes in the distal region survive.

3. Neurolemmocytes form a regeneration tube in conjunction with the remaining *endoneurium* (an areolar connective tissue wrapping around axons [see next section]) of the severed axon.
4. The axon regenerates, and remyelination occurs. The regeneration tube guides the axon sprout as it begins to grow rapidly through the regeneration tube at a rate of about 5 millimeters per day under the influence of nerve growth factors released by the neurolemmocytes.
5. Innervation is restored as the axon reestablishes contact with its original effector.

Potential regeneration of damaged neurons within the CNS is very limited due to several factors. First, oligodendrocytes do not release a nerve growth factor, and in fact they actively inhibit axon growth by producing and secreting several growth-inhibitory molecules. Second, the large number of axons crowded within the CNS tends to complicate regrowth activities. Finally, both astrocytes and connective tissue coverings may form some scar tissue that obstructs axon regrowth.

WHAT DID YOU LEARN?

(9) What three factors determine PNS axon regeneration?

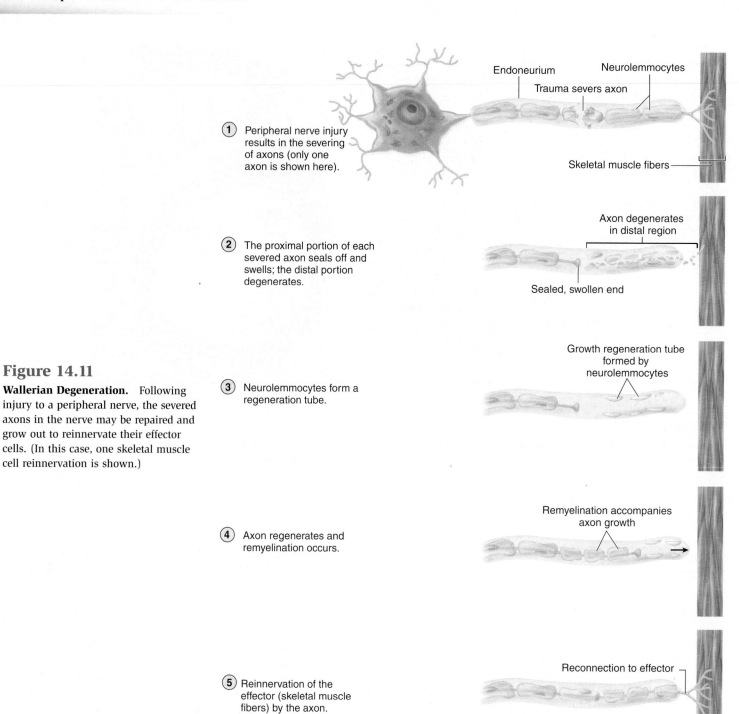

Figure 14.11

Wallerian Degeneration. Following injury to a peripheral nerve, the severed axons in the nerve may be repaired and grow out to reinnervate their effector cells. (In this case, one skeletal muscle cell reinnervation is shown.)

① Peripheral nerve injury results in the severing of axons (only one axon is shown here).

Endoneurium   Neurolemmocytes

Trauma severs axon

Skeletal muscle fibers

② The proximal portion of each severed axon seals off and swells; the distal portion degenerates.

Axon degenerates in distal region

Sealed, swollen end

③ Neurolemmocytes form a regeneration tube.

Growth regeneration tube formed by neurolemmocytes

④ Axon regenerates and remyelination occurs.

Remyelination accompanies axon growth

⑤ Reinnervation of the effector (skeletal muscle fibers) by the axon.

Reconnection to effector

Nerves

Key topic in this section:

- Structure of a nerve

A **nerve** is a cablelike bundle of parallel axons. While a single axon typically must be viewed using a microscope, a nerve tends to be a macroscopic structure. **Figure 14.12** shows a typical nerve. Like a muscle, a nerve has three successive connective tissue wrappings:

- An individual axon in a myelinated neuron is surrounded by neurolemmocytes and then wrapped in the **endoneurium** (en-dō-noo′rē-ŭm; *endon* = within), a delicate layer of areolar connective tissue that separates and electrically isolates each

axon. Also within this connective tissue layer are capillaries that supply each axon.

- Groups of axons are wrapped into separate bundles called **fascicles** (fas′i-kl) by a cellular dense irregular connective tissue layer called the **perineurium** (per-i-noo′rē-ŭm; *peri* = around). This layer supports blood vessels supplying the capillaries within the endoneurium.

- All of the fascicles are bundled together by a superficial connective tissue covering termed the **epineurium** (ep-i-noo′rē-ŭm; *epi* = upon). This thick layer of dense irregular connective tissue encloses the entire nerve, providing both support and protection to the fascicles within the layer.

CLINICAL VIEW

Nerve Regeneration and Spinal Cord Injuries

Spinal cord injuries frequently leave individuals unable to walk or paralyzed from the neck down. At one time, people with a spinal injury at the neck level were doomed to die, chiefly because of inadequate stimulation of the diaphragm and subsequent respiratory failure. Today, aggressive and early treatment of spinal cord injuries helps save lives that would have been lost just 5 years ago. Early use of steroids immediately after the injury appears to preserve some muscular function that might otherwise be lost. Early use of antibiotics has substantially reduced the number of deaths caused by pulmonary and urinary tract infections that accompany spinal cord injuries. There is even hope for repair and reconnection of damaged nerves. Recent research with rats has achieved reconnection and partial restoration of function of severed spinal cords. Researchers in several laboratories have devised a method of creating a "bridge" of nervous tissue that spans the injured area and have used transplanted olfactory nerve tissue as a guide for regrowing severed spinal cord axons capable of reaching the correct targets in the rat brain. This procedure has yet to be tried in humans. In addition, other research indicates that neural stem cells may be able to regenerate CNS axons.

Actor Christopher Reeve, known to many as Superman, was injured in a horse-riding accident and was paralyzed inferior to the level of the second cervical vertebra. Reeve became a tireless proponent and fund-raiser for spinal cord injury research. Up until his death in 2004, his own strides in rehabilitation were remarkable and helped advance spinal cord injury treatments for others. Hopefully, research and clinical medicine will lead to further help for the numerous victims of such traumatic injury.

Actor Christopher Reeve was a pioneer in challenging previous conceptions about neuron regeneration.

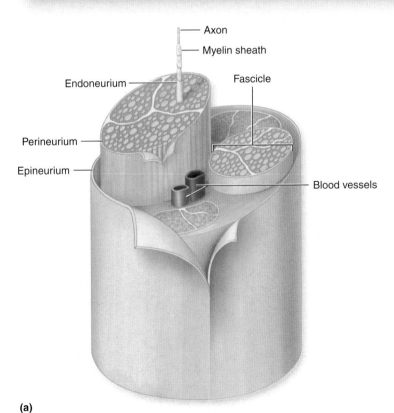

(a)

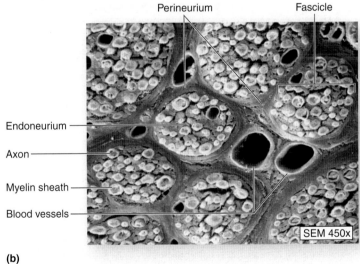

(b)

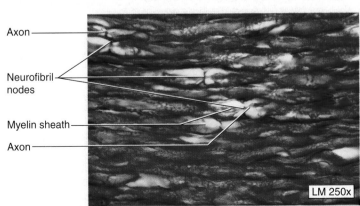

(c)

Figure 14.12

Nerve Structure. *(a)* A nerve is formed from many parallel axons wrapped by successive connective tissue layers. *(b)* SEM shows a cross section of a nerve. *(c)* A photomicrograph shows a longitudinal section of a nerve.

Nerves are a component of the peripheral nervous system. Sensory neurons convey sensory information to the central nervous system, motor neurons convey motor impulses from the central nervous system to the muscles and glands. Mixed nerves convey both types of information.

Synapses

Key topics in this section:

- Components of a synapse
- Conduction of nerve impulses in electrical and chemical synapses

Axons terminate as they contact other neurons, muscle cells, or gland cells at specialized junctions called **synapses** (sin'aps; *syn* = together, *hapto* = to clasp) where the nerve impulse is transmitted to the other cell. **Figure 14.13a** shows an axon transmitting a nerve impulse to another neuron at a synapse. As the axon approaches the cell onto which it will terminate, it generally branches repeatedly into

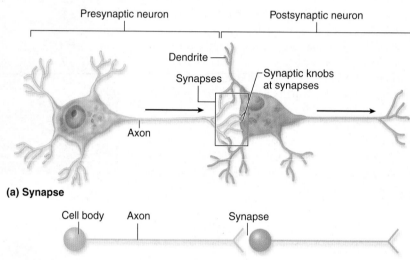

(a) Synapse

(b) Simplified representation of a synapse

Figure 14.13

Synapses. Synapses are intercellular junctions where two excitable cells come in contact to exchange information. *(a)* A synapse occurs where the plasma membrane of a presynaptic neuron synaptic knob comes in close proximity to the plasma membrane of a postsynaptic neuron. Arrows indicate the direction of nerve impulse flow. *(b)* In this simplified representation of a synapse, the spheres represent the cell bodies of the presynaptic and postsynaptic cells, the line represents the axon, and the angled arrow represents the synaptic knobs. *(c)* An axodendritic synapse occurs between an axon and a dendrite; an axosomatic synapse occurs between an axon and a cell body; and an axoaxonic synapse is between an axon and another axon.

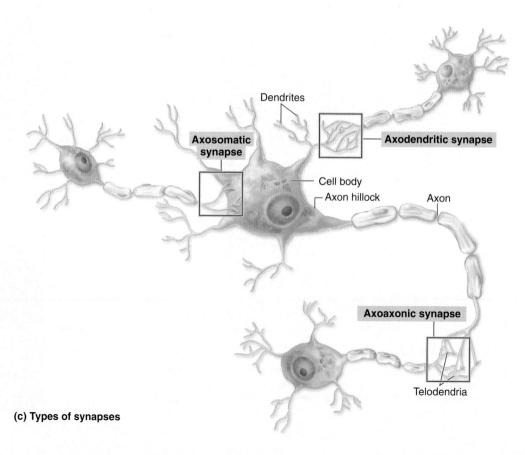

(c) Types of synapses

several telodendria, and each telodendrion loses its myelin covering. Additionally, the synaptic endings usually form swellings called synaptic knobs at the ends of the axon branches. A typical synapse in the CNS consists of the close association of a **presynaptic** (prē-si-nap′tik; *pre* = before) **neuron** and a **postsynaptic** (pōst-si-nap′tik; *post* = after) **neuron** at a region where their plasma membranes are separated by a very narrow space called the **synaptic cleft.** Presynaptic neurons transmit nerve impulses through their axons toward a synapse; postsynaptic neurons conduct nerve impulses through their dendrites and cell bodies away from the synapse.

Figure 14.13*b* shows a simplified diagram of a synapse. Here, the cell body of each neuron is represented by a sphere, and the axon is shown as a straight line. The synaptic knob is represented by an angled arrow attached to the axon, and the space between that angled arrow and the cell body of the next neuron is the synapse.

Axons may establish synaptic contacts with any portion of the surface of another neuron, except those regions covered by a myelin sheath. Three common types of synapses are axodendritic, axosomatic, and axoaxonic (figure 14.13*c*):

- The **axodendritic** (ak′sō-den-drit′ik) **synapse** is the most common type. It occurs between the synaptic knobs of a presynaptic neuron and the dendrites of the postsynaptic neuron. These specific connections occur either on the expanded tips of narrow dendritic spines or on the shaft of the dendrite.
- The **axosomatic** (ak′sō-sō-mat′ik) **synapse** occurs between synaptic knobs and the cell body of the postsynaptic neuron.
- The **axoaxonic** (ak′sō-ak-son′ik) **synapse** is the least common synapse and far less understood. It occurs between

the synaptic knob of a presynaptic neuron and the synaptic knob of a postsynaptic neuron. The action of this synapse appears to influence the activity of the synaptic knob.

Synaptic Communication

Most neurons exhibit both presynaptic and postsynaptic sides and functions. Synapses may be of two types: electrical or chemical.

Electrical Synapses

In an **electrical synapse,** the plasma membranes of the presynaptic and postsynaptic cells are bound tightly together. Electrical synapses are fast and secure, and they permit two-way signaling. At this synapse, **gap junctions** formed by connexons between both plasma membranes (review chapter 4) facilitate the flow of ions, such as sodium ions (Na^+), between the cells **(figure 14.14*a*)**. This causes a local current flow between neighboring cells. Remember that a voltage change caused by movement of charged ions results in a nerve impulse. Thus, these cells act as if they shared a common plasma membrane, and the nerve impulse passes between them with no delay. Electrical synapses are not very common in the brains of mammals. In humans, for example, these synapses occur primarily between smooth muscle cells (such as the smooth muscle in the intestines), where quick, uniform innervation is essential. Electrical synapses are also located in cardiac muscle at the intercalated discs (chapter 4).

Chemical Synapses

The most numerous type of synapse is the **chemical synapse.** This type of synapse facilitates most of the interactions between neurons and all communications between neurons and effectors. At these junctions,

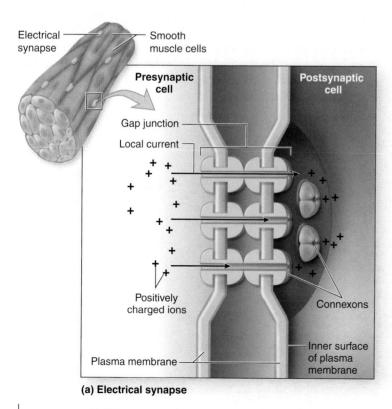

(a) Electrical synapse

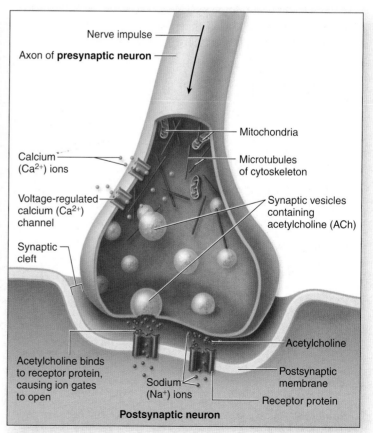

(b) Chemical synapse

Figure 14.14

Electrical and Chemical Synapses. *(a)* In an electrical synapse, ions pass through gap junctions between neurons from the presynaptic to the postsynaptic cell. *(b)* In a chemical synapse, a neurotransmitter is released from the presynaptic neuron to receptors on the membrane of the postsynaptic neuron.

the presynaptic membrane releases a signaling molecule called a **neurotransmitter.** There are many different neurotransmitters, but acetylcholine (ACh) is the most common neurotransmitter and is our example in figure 14.14*b*. Some types of neurons use other neurotransmitters. The neurotransmitter molecules are released only from the presynaptic cell. They then bind to receptor proteins found only in the plasma membrane of the postsynaptic cell, and this causes a brief voltage change across the membrane of the postsynaptic cell. Thus, a unidirectional flow of information and communication takes place; it originates in the presynaptic cell and is received by the postsynaptic cell. A very precise sequence of events is required for the conduction of a nerve impulse from the presynaptic neuron to the postsynaptic neuron:

1. A nerve impulse travels through the axon and reaches its synaptic knob.
2. The arrival of the nerve impulse at the synaptic knob causes an increase in calcium ion (Ca^{2+}) movement into the synaptic knob through voltage-regulated calcium ion channels in the membrane.
3. Entering calcium ions cause synaptic vesicles to move to and bind to the inside surface of the membrane; neurotransmitter molecules within the synaptic vesicles are released into the synaptic cleft by exocytosis.
4. Neurotransmitter molecules diffuse across the synaptic cleft to the plasma membrane of the postsynaptic cell.
5. Neurotransmitter molecules attach to specific protein receptors in the plasma membrane of the postsynaptic cell, causing ion gates to open. Note: The time it takes for neurotransmitter release, diffusion across the synaptic cleft, and binding to the receptor is called the **synaptic delay.**
6. An influx of sodium ions (Na^+) moves into the postsynaptic cell through the open gate, affecting the charge across the membrane.
7. Change in the postsynaptic cell voltage causes a nerve impulse to begin in the postsynaptic cell.
8. The enzyme acetylcholinesterase (AChE) resides in the synaptic cleft and rapidly breaks down molecules of ACh

that are released into the synaptic cleft. Thus, AChE is needed so that ACh will not continuously stimulate the postsynaptic cell.

Once a nerve impulse is initiated, two factors influence the rate of conduction of the impulse: the axon's diameter and the presence (or absence) of a myelin sheath. The larger the diameter of the axon, the more rapidly the impulse is conducted because of less resistance to current flow as charged ions move into the axon. Also, as previously mentioned, an axon with a myelin sheath conducts impulses many times faster than an unmyelinated axon because of the differences between saltatory and continuous conduction.

WHAT DID YOU LEARN?

12 What are the two types of synaptic communication?

13 What factors influence the impulse conduction rate?

Neural Integration and Neuronal Pools

Key topic in this section:

■ The four different neuronal circuits and how each one functions

The nervous system is able to coordinate and integrate nervous activity in part because billions of interneurons within the CNS are grouped in complex patterns called **neuronal pools** (or *neuronal circuits* or *pathways*). Neuronal pools are defined based upon function, not anatomy, into four types of circuits: converging, diverging, reverberating, and parallel-after-discharge (**figure 14.15**). A pool may be localized, with its neurons confined to one specific location, or its neurons may be distributed in several different regions of the CNS. However, all neuronal pools are restricted in their number of input sources and output destinations.

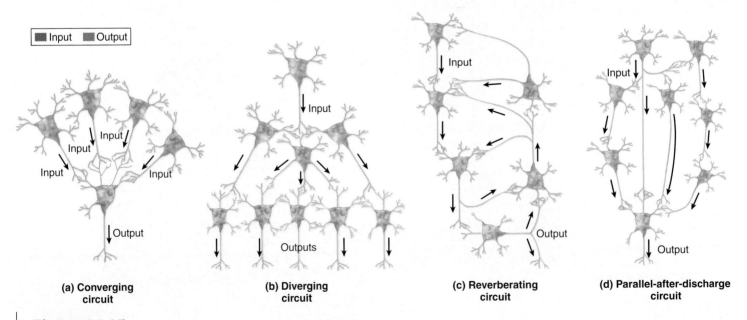

(a) Converging circuit

(b) Diverging circuit

(c) Reverberating circuit

(d) Parallel-after-discharge circuit

Figure 14.15

Neuronal Pools. Neuronal pools are groups of neurons arranged in specific patterns (circuits) through which impulses are conducted and distributed. Four types of neuronal pools are recognized: *(a)* converging circuit, *(b)* diverging circuit, *(c)* reverberating circuit, and *(d)* parallel-after-discharge circuit.

CLINICAL VIEW: In Depth

Nervous System Disorders

Four serious diseases that attack portions of the nervous system are amyotrophic lateral sclerosis, multiple sclerosis, Parkinson disease, and Guillain-Barré syndrome.

Amyotrophic lateral sclerosis (**ALS**; often called *Lou Gehrig disease*) is a well-known motor neuron disease that progresses quickly and is eventually fatal. It affects neurons in both the brain and the spinal cord, leading to progressive degeneration of the somatic motor system. ALS patients generally have weakened and atrophied muscles, especially in the hands and forearms. Additionally, they may experience speech impairment, breathing difficulties, and chewing and swallowing problems that result in choking or excessive drooling. However, the disease does not affect sensory abilities, such as hearing, sight, or smell. No effective treatment or cure exists, and the disease is invariably fatal.

ALS affects both males and females, but it occurs in males more often. About 90% of ALS cases occur in families with no previous history of the disease. In contrast, about 10% of cases are inherited and called *familial* (meaning that more members of the same family are affected than can be accounted for by chance). The inherited form of ALS has been localized to a gene on chromosome 21.

Multiple sclerosis (**MS**) is progressive demyelination of neurons in the central nervous system accompanied by the destruction of oligodendrocytes. As a result, the conduction of nerve impulses is disrupted, leading to impaired sensory perception and motor coordination. Repeated inflammatory events at myelinated sites cause scarring (sclerosis), and in time some function is permanently lost. The disease usually strikes young adults between the ages of 18 and 40. It is five times more prevalent in whites than in blacks.

Although MS is very disabling, it progresses slowly, and most patients lead productive lives, especially during recurring periods of remission. Symptoms are diverse because almost any myelinated site in the brain or spinal cord may be affected. Among the typical symptoms are vision problems, muscle weakness and spasms, urinary infections and bladder incontinence, and drastic mood changes.

The most widely held view of the cause of multiple sclerosis is that the body's immune system attacks its own central nervous system, making MS an autoimmune disorder. Treatment depends to some degree on the stage and severity of the disease. Steroids are useful during periods of acute symptoms, whereas interferons (natural proteins produced by the immune system) are used for prolonged therapy. Recent experiments have shown that one form of interferon lowers the activity of immune cells, reducing the number and severity of attacks.

Parkinson disease (*Parkinsonism* or "shaky palsy") is a slowly progressive disorder affecting muscle movement and balance. The condition is characterized by stiff posture, tremors, and reduced spontaneity of facial expressions. It results from loss of cells that produce the neurotransmitter dopamine in a specific region of the brainstem. For further information about Parkinson disease, see Clinical View: In Depth, "Brain Disorders," in chapter 15, page 470.

Guillain-Barré syndrome (**GBS**) is a disorder of the peripheral nervous system characterized by muscle weakness that begins in the distal limbs, but rapidly advances to involve proximal muscles as well (a condition known as ascending paralysis). At the microscopic level, inflammation causes loss of myelin from the peripheral nerves and spinal nerve roots. Most cases of GBS are preceded by an acute, flulike illness, although no specific infectious agent has ever been identified. In rare instances, the condition may follow an immunization. Even though GBS appears to be an immune-mediated condition, the use of steroids provides little if any measurable improvement. In fact, most people recover almost all neurologic function on their own with little medical intervention. Should hospitalization and treatment be required, therapies fall into four categories: (1) supportive, including breathing assistance if indicated; (2) physical therapy to increase muscle flexibility and strength; (3) injections of high-dose immunoglobulins to "turn off" the production of antibodies causing the disease; and (4) plasmaphoresis, a process of filtering the blood to remove the antibodies that are causing the myelin destruction.

(a)

(b)

Individuals with neurodegenerative diseases must overcome physical challenges in order to carry on the activities of daily life. (a) Amyotrophic lateral sclerosis (scientist and writer Stephen Hawking). (b) Multiple sclerosis.

In a **converging circuit**, nerve impulses converge (come together) at a single postsynaptic neuron (figure 14.15a). This neuron receives input from several presynaptic neurons. For example, multiple sensory neurons synapse on the neurons in the salivary nucleus in the brainstem, resulting in the production of saliva. The various inputs may originate from more than one stimulus—in this example, smelling food, seeing the time on the clock indicating dinnertime, hearing food preparation activities, or seeing pictures of food in a magazine. These multiple inputs lead to a single output, the production of saliva.

A **diverging circuit** spreads information from one presynaptic neuron to several postsynaptic neurons, or from one pool to multiple pools (figure 14.15b). For example, the few neurons in the brain that control the movements of skeletal muscles in the legs during walking also stimulate the muscles in the back that maintain posture and balance while walking. In this case, a single or a few inputs lead to multiple outputs.

Reverberating circuits utilize feedback to produce a repeated, cyclical stimulation of the circuit, or a reverberation (figure 14.15c). Once activated, a reverberating circuit may continue to function until either inhibitory stimuli or synaptic fatigue breaks the cycle. (**Synaptic fatigue** occurs when repeated stimuli cause temporary inability of the presynaptic cell to meet demands of synaptic transmission as a result of a lack of neurotransmitter production.) The repetitive nature of a reverberating circuit ensures that we continue breathing while we are asleep.

In a **parallel-after-discharge circuit,** several neurons or neuronal pools process the same information at one time. A single presynaptic neuron stimulates different groups of neurons, each of which passes the nerve impulse along a pathway that ultimately synapses with a common postsynaptic cell (figure 14.15d). This type of circuit is believed to be involved in higher-order thinking, such as the type needed to perform precise mathematical calculations.

WHAT DID YOU LEARN?

14 How is a diverging circuit different from a reverberating circuit?

Development of the Nervous System

Key topic in this section:

■ Early events in nervous system development

Nervous tissue development begins in the embryo during the third week when a portion of the ectoderm that overlies the notochord thickens. This thickened ectoderm is called the **neural plate,** and the cells of the plate collectively are called the **neuroectoderm.** The neuroectoderm undergoes dramatic changes, called **neurulation** (noor-oo-lā′shŭn), to form nervous tissue structures. The process of neurulation is shown in **figure 14.16** and explained here:

1. The neural plate develops a central longitudinal indentation called the **neural groove.** As this is occurring, cells along the lateral margins of the neural plate proliferate, becoming the thickened **neural folds.** The tips of the neural folds form the neural crest and are occupied by **neural crest cells** (or simply, the *neural crest*).

2. The neural folds elevate and approach one another as the neural groove continues to deepen. The neural crest cells are now at the very highest point of the neural groove. When viewed from a superior angle, the neural folds resemble the

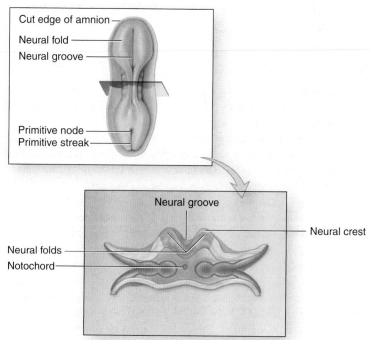

① Neural folds and neural groove form from the neural plate.

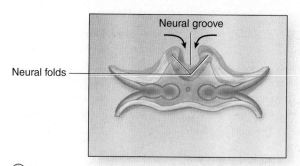

② Neural folds elevate and approach one another.

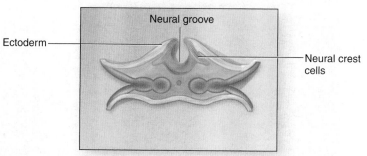

③ Neural crest cells begin to "pinch off" from the neural folds and form other structures.

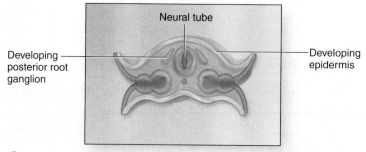

④ Neural folds fuse to form the neural tube.

Figure 14.16

Nervous System Development. The process of neurulation begins in the third week, and the neural tube finishes closing by the end of week 4.

CLINICAL VIEW

Neural Tube Defects

Neural tube defects (NTD) are serious developmental deformities of the brain, spinal cord, and meninges. The two basic categories of NTDs are anencephaly and spina bifida. Both conditions result from localized failure of the developing neural tube to close.

Anencephaly (an'en-sef'ă-lē; *an* = without, enkephalos = brain) is the substantial or complete absence of a brain as well as the bones making up the cranium. Infants with anencephaly rarely live longer than a few hours following birth. Fortunately, neural tube defects of this magnitude are rare, and are easily detected with prenatal ultrasound, thus alerting the parents to the condition.

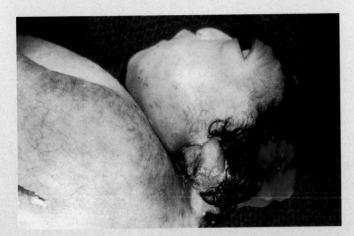

A newborn with anencephaly.

Spina bifida (spī'nă bĭ'fid'ă; *spina* = spine, *bifidus* = cleft in two parts) occurs more frequently than anencephaly. This defect results when the caudal portion of the neural tube, often in the lumbar or sacral region, fails to close. Two forms of spina bifida occur: the more severe spina bifida cystica and the less severe spina bifida occulta. In spina bifida cystica, almost no vertebral arch forms, so the posterior aspect of the spinal cord in this region is left unprotected (figures *a* and *b*). Typically, there is a large cystic structure in the back, filled with cerebrospinal fluid (CSF) and covered by a thin layer of skin or in some cases only by meninges (protective membranes around the spinal cord). Surgery is generally done promptly to close the defect, reduce the risk of infection, and preserve existing spinal cord function. Paralysis of the lower limbs is often part of the spina bifida syndrome. But even with these problems, most children with spina bifida cystica live well into adulthood.

Spina bifida occulta is less serious, but much more common than the cystica variety. This condition is characterized by a partial defect of the vertebral arch, typically involving the vertebral lamina and spinous process (figure *c*). The bony defect is small, and the spinal cord or meninges does not protrude. Often there is a tuft of hair in the region of the bony defect. Most people with this condition are otherwise asymptomatic, and it is generally detected incidentally during an x-ray for an unrelated reason. Estimates of the incidence of spina bifida occulta range as high as 17% of the population in some x-ray studies.

Although the risk of neural tube defects cannot be eliminated, it can be greatly reduced. Researchers have discovered that increased intake of vitamin B_{12} and the B vitamin **folic acid (folate)** by pregnant women is correlated with a decreased incidence of neural tube defects. Both vitamin B_{12} and folic acid are critical to DNA formation and are necessary for cellular division and tissue differentiation. Thus, pregnant women are encouraged to take prenatal vitamins containing high levels of these chemicals, and the food industry has begun fortifying many breads and grains with folate as well.

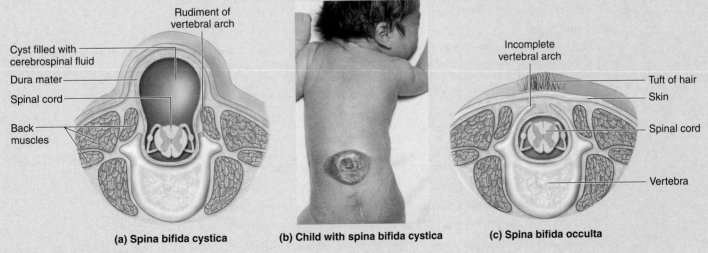

(a) Spina bifida cystica   **(b) Child with spina bifida cystica**   **(c) Spina bifida occulta**

Spina bifida is a neural tube disorder that occurs in two forms: (a, b) spina bifida cystica; (c) spina bifida occulta. (See chapter 7 for a description of vertebral development.)

sides of a hot dog roll, with the neural groove represented by the opening in the roll.

3. The neural crest cells begin to "pinch off" from the neural folds and form other structures.

4. By the end of the third week, the neural folds have met and fused at the midline, and the neural groove starts to form a **neural tube,** which has an internal lumen called the **neural canal.** The neural tube initially fuses at its midline, and later the neural folds slightly superior and inferior to this midline fuse as well. Thus, the neural tube forms as the neural folds "zip" together both superiorly and inferiorly.

For a short time, the neural tube is open at both its ends. These openings, called **neuropores** (noor′ō-pōr), close during the end of the fourth week. The opening closest to the future head is the **cranial neuropore,** while the opening closest to the future buttocks region is the **caudal neuropore.** If these openings do not close, the developing human will have a neural tube defect (see Clinical View). The developing neural tube forms the central nervous system. In particular, the cranial part of the neural tube expands to form the brain (see chapter 15), while the caudal part of the neural tube expands to form the spinal cord (see chapter 16). Also, please refer back to chapter 7 to review vertebral development.

CLINICAL TERMS

demyelination (dē-mī′ĕ-li-nā′shun) Progressive loss or destruction of myelin in the CNS and PNS with preservation of the axons; often leads to loss of sensation and/or motor control.

neuritis (noo-rī′tis) Inflammation of a nerve.

neuropathy (noo-rop′ă-thē) Classical term for a disorder affecting any segment of the nervous system.

neurotoxin (noor-ō-tok′sin) Any poison that acts specifically on nervous tissue.

CHAPTER SUMMARY

Organization of the Nervous System 414	■ The nervous system includes all the nervous tissue in the body.
	Structural Organization: Central and Peripheral Nervous Systems 414
	■ The central nervous system is composed of the brain and the spinal cord.
	■ The peripheral nervous system is composed of the cranial nerves, spinal nerves, and ganglia.
	Functional Organization: Sensory and Motor Nervous Systems 414
	■ The nervous system is functionally subdivided into a sensory nervous system that conveys sensory information to the CNS, and a motor nervous system that conducts motor commands to muscles and glands.
Cytology of Nervous Tissue 416	■ Neurons are excitable cells that transmit nerve impulses, and glial cells completely surround neurons and support them.
	Neurons 416
	■ A generalized neuron has a cell body and processes called dendrites and an axon. They are classified structurally by the number of processes attached to the cell body (unipolar, bipolar, or multipolar) and functionally as sensory neurons, motor neurons, or interneurons.
	Glial Cells 420
	■ Glial cells support neurons in the CNS. Astrocytes help form the blood-brain barrier and regulate tissue fluid composition; ependymal cells line CNS cavities and produce cerebrospinal fluid; microglial cells act as phagocytes in nervous tissue; and oligodendrocytes myelinate CNS axons.
	■ In the PNS, satellite cells support neuron cell bodies in ganglia, and neurolemmocytes myelinate PNS axons.
Myelination of Axons 423	■ A nerve impulse is the rapid movement of a charge along a neuron's plasma membrane.
	Myelination 423
	■ Oligodendrocytes (CNS) and neurolemmocytes (PNS) wrap around axons of neurons, forming a discontinuous myelin sheath along the axon, with small gaps called neurofibril nodes.
	Nerve Impulse Conduction 424
	■ The myelin sheath insulates the axonal membrane, resulting in faster nerve impulse conduction.
	■ Unmyelinated axons are associated with a neurolemmocyte but not ensheathed by it.
Axon Regeneration 425	■ Regeneration of damaged neurons is limited to PNS axons that are able to regrow under certain conditions by a process called Wallerian degeneration.
Nerves 426	■ A nerve is a bundle of many parallel axons organized in three layers: an endoneurium around a single axon, a perineurium around a fascicle, and an epineurium around all of the fascicles.
Synapses 428	■ The specialized junction between two excitable cells where a nerve impulse is transmitted is called a synapse.
	■ Swellings of axons at their end branches are called synaptic knobs.
	■ The space between the presynaptic and postsynaptic cells is the synaptic cleft.
	■ Synapses are classified according to the point of contact between the synaptic knob and the postsynaptic cell as axodendritic, axosomatic, or axoaxonic.

Synapses (continued) 428	**Synaptic Communication** 429
	■ Synapses are termed electrical when a flow of ions passes from the presynaptic cell to the postsynaptic cell through gap junctions; synapses are termed chemical when a nerve impulse causes the release of a chemical neurotransmitter from the presynaptic cell that induces a response, in the postsynaptic cell.
	■ A myelinated axon conducts impulses faster than an unmyelinated axon, and the larger the diameter of the axon, the faster is the rate of conduction.
Neural Integration and Neuronal Pools 430	■ Interneurons are organized into neuronal pools, which are groups of interconnected neurons with specific functions.
	■ In a converging circuit, neurons synapse on the same postsynaptic neuron.
	■ A diverging circuit spreads information to several neurons.
	■ In a reverberating circuit, neurons continue to restimulate presynaptic neurons in the circuit.
	■ A parallel-after-discharge circuit involves parallel pathways that process the same information over different amounts of time and deliver that information to the same output cell.
Development of the Nervous System 432	■ Nervous tissue development begins in the early embryo with the formation of the neural plate. As this plate grows and develops, a neural groove appears as a depression in the plate, prior to the elevation of neural folds along the lateral side of the plate. The fusion of the neural folds gives rise to a neural tube, from which the brain and spinal cord develop.
	■ A neural tube defect can result if part of the neural tube fails to fuse.

CHALLENGE YOURSELF

Matching

Match each numbered item with the most closely related lettered item.

_____ 1. motor nervous system
_____ 2. effector
_____ 3. oligodendrocyte
_____ 4. chromatophilic substance
_____ 5. collaterals
_____ 6. microglial cells
_____ 7. multipolar neurons
_____ 8. interneuron
_____ 9. chemical synapse
_____ 10. dendrite

a. skeletal muscle fiber
b. neuron part that usually receives incoming impulses
c. stain darkly with basic dyes
d. transmits motor information
e. uses a neurotransmitter
f. makes myelin sheaths in CNS
g. neurons with multiple dendrites
h. side branches of axons
i. respond to CNS infection
j. sensory to motor neuron communication

Multiple Choice

Select the best answer from the four choices provided.

_____ 1. The cell body of a mature neuron does not contain
a. a nucleus.
b. ribosomes.
c. a centriole.
d. mitochondria.

_____ 2. Neurons that have only two processes attached to the cell body are called
a. unipolar.
b. bipolar.
c. multipolar.
d. efferent.

_____ 3. Which neurons are located *only* within the CNS?
a. motor neurons
b. unipolar neurons
c. sensory neurons
d. interneurons

_____ 4. A structure or cell that collects sensory information is a
a. motor neuron.
b. receptor.
c. neurolemmocyte.
d. ganglion.

_____ 5. The glial cells that help produce CSF in the CNS are
a. satellite cells.
b. microglial cells.
c. ependymal cells.
d. astrocytes.

_____ 6. Which of the following is *not* a part of the CNS?
a. microglial cell
b. spinal cord
c. neurolemmocyte
d. brain

_____ 7. Which of these cells transmits, transfers, and processes a nerve impulse?
a. neurolemmocyte
b. astrocyte
c. neuron
d. oligodendrocyte

_____ 8. Which type of neuronal pool utilizes nerve impulse feedback to repeatedly stimulate the circuit?
a. converging circuit
b. diverging circuit
c. reverberating circuit
d. parallel-after-discharge circuit

_____ 9. At an electrical synapse, presynaptic and postsynaptic membranes interface through
   a. neurofibril nodes.
   b. gap junctions.
   c. telodendria.
   d. neurotransmitters.

_____ 10. The epineurium is
   a. a thick, dense irregular connective tissue layer enclosing the nerve.
   b. a group of axons.
   c. a delicate layer of areolar connective tissue.
   d. a cellular layer of dense regular connective tissue.

Content Review

1. What are the three structural types of neurons? How do they compare to the three functional types of neurons?

2. What is the function of sensory neurons?

3. Identify the principal types of glial cells, and briefly discuss the function of each type.

4. How does the myelin sheath differ between the CNS and the PNS?

5. Describe the procedure by which a PNS axon may repair itself.

6. Describe the arrangement and structure of the three coverings that surround axons in ANS nerves.

7. Clearly distinguish among the following: a neuron, an axon, and a nerve.

8. What are the differences between electrical and chemical synapses? Which is the more common type of synapse in humans?

9. Discuss the similarities and differences between converging and parallel-after-discharge circuits.

10. What are the basic developmental events that occur during neurulation?

Developing Critical Reasoning

1. Over a period of 6 to 9 months, Marianne began to experience vision problems as well as weakness and loss of fine control of the skeletal muscles in her leg. Blood tests revealed the presence of antibodies (immune system proteins) that attack myelin. Beyond the presence of the antibodies, what was the cause of Marianne's vision and muscular difficulties?

2. Surgeons were able to reattach an amputated limb, sewing both the nerves and the blood vessels back together. After the surgery, which proceeded very well, the limb regained its blood supply almost immediately, but the limb remained motionless and the patient had no feeling in it for several months. Why did it take longer to reestablish innervation than circulation?

ANSWERS TO "WHAT DO YOU THINK?"

1. The term _visceral_ refers to organs, especially thoracic and abdominal organs such as the heart, lungs, and gastrointestinal tract. Therefore, the parts of the sensory and motor nervous systems that innervate these viscera are called the visceral sensory and visceral motor (autonomic) nervous systems.

2. Tumors occur due to uncontrolled mitotic growth of cells. Since glial cells are mitotic and neurons typically are nonmitotic, a "brain tumor" almost always develops from glial cells.

3. A myelinated axon takes up more space than an unmyelinated axon. There simply isn't enough space in the body to hold myelin sheaths for every axon. Thus, the body conserves this space by myelinating only the axons that must transmit nerve impulses very rapidly.

Visit the McKinley/O'Loughlin _Human Anatomy_, 2e website at aris.mhhe.com

15

OUTLINE

NERVOUS SYSTEM

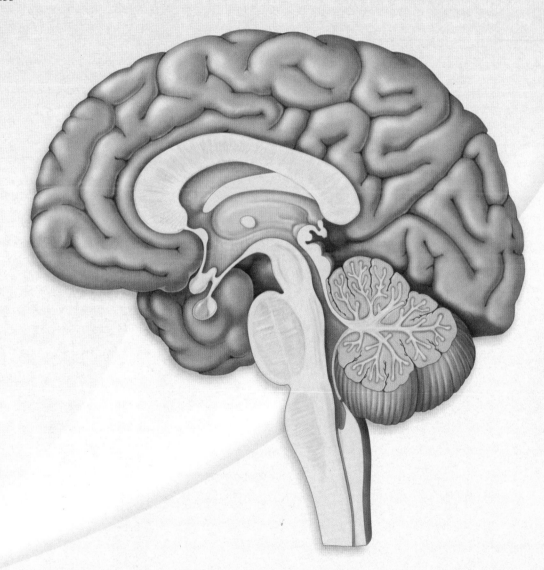

Brain and Cranial Nerves

Approximately 4 to 6 million years ago, when the earliest humans were evolving, brain size was a mere 440 cubic centimeters (cc), not much larger than that of a modern chimpanzee. As humans have evolved, brain size has increased steadily and reached an average volume of 1200–1500 cc and an average weight of 1.35 to 1.4 kilograms. In addition, the texture of the outer surface of the brain (its hemispheres) has changed. Our skull size limits the size of the brain, so the tissue forming the brain's outer surface folded on itself so that more neurons could fit into the space within the skull. Although modern humans display variability in brain size, it isn't the *size* of the brain that determines intelligence, but the number of active synapses among neurons.

The brain is often compared to a computer because they both simultaneously receive and process enormous amounts of information, which they then organize, integrate, file, and store prior to making an appropriate output response. But in some ways this is a weak comparison, because no computer is capable of the multitude of continual adjustments that the brain's neurons perform. The brain can control numerous activities simultaneously, and it can also respond to various stimuli with an amazing degree of versatility.

Brain Development and Tissue Organization

Key topics in this section:

- Embryonic development of the divisions of the brain
- Organization of neural tissue

Figure 15.1 shows the major parts of the adult brain from several views. Our discussion in this chapter focuses on the brain's four major regions: the cerebrum, the diencephalon, the brainstem, and the cerebellum. When viewed superiorly, the cerebrum is divided into two halves, called the left and right cerebral hemispheres. Each hemisphere may be further subdivided into five functional areas called lobes. Four lobes are visible superficially, and one is only

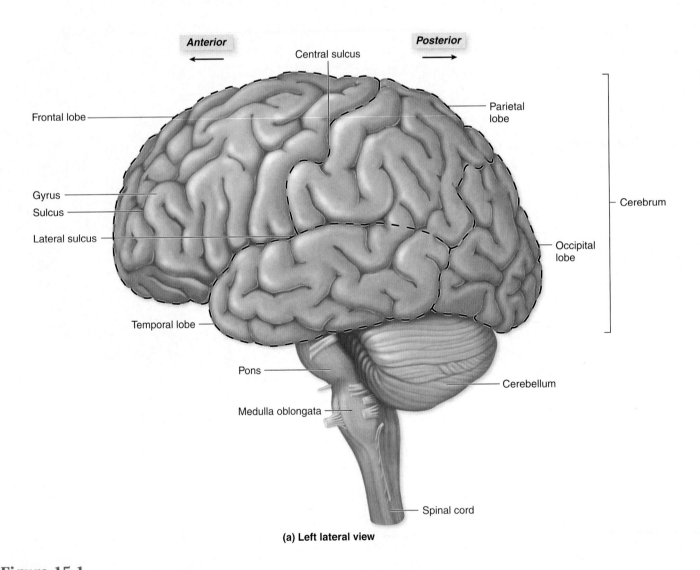

(a) Left lateral view

Figure 15.1

The Human Brain. The brain is a complex organ that has several subdivisions. (*a*) An illustration and a cadaver photo show left lateral views of the brain, revealing the cerebrum, cerebellum, and portions of the brainstem; the diencephalon is not visible.

seen internally (see figure 15.11). The outer surface of an adult brain exhibits folds called **gyri** (jī'rī; sing., *gyrus*; *gyros* = circle) and shallow depressions between those folds called **sulci** (sŭl'sī; sing., *sulcus*; furrow, ditch). The brain is associated with 12 pairs of cranial nerves (see figure 15.24).

Two directional terms are often used to describe brain anatomy. **Anterior** is synonymous with *rostral* (meaning "toward the nose"), and **posterior** is synonymous with *caudal* (meaning "toward the tail").

Embryonic Development of the Brain

In order to understand how the structures of the adult brain are named and connected, it is essential to know how the brain develops. In the human embryo, the brain forms from the cranial (superior) part of the neural tube, which undergoes disproportionate growth rates in different regions. By the late fourth week of development, this growth has formed three **primary brain vesicles**, which eventually give rise to all the different regions of the adult brain.

The names of these vesicles describe their relative positions in the developing head: The forebrain is called the **prosencephalon** (pros-en-sef'a-lon; *proso* = forward, *enkephalos* = brain); the midbrain is called the **mesencephalon** (mez'en-sef'a-lon; *mes* = middle); and the hindbrain is called the **rhombencephalon** (rom-ben-sef'a-lon; *rhombo* = rhomboid) **(figure 15.2a)**.

By the fifth week of development, the three primary vesicles further develop into a total of five secondary brain vesicles (figure 15.2b):

■ The **telencephalon** (tel-en-sef'ă-lon; *tel* = head end) arises from the prosencephalon and eventually forms the cerebrum.

■ The **diencephalon** (dī-en-sef'ă-lon; *dia* = through) derives from the prosencephalon and eventually forms the thalamus, hypothalamus, and epithalamus.

■ The **mesencephalon** (mez-en-sef'ă-lon; *mesos* = middle) is the only primary vesicle that does not form a new secondary vesicle.

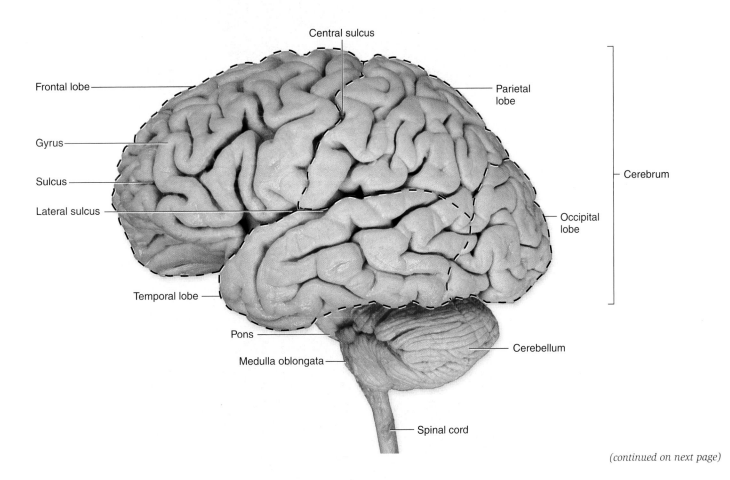

(continued on next page)

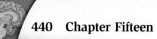

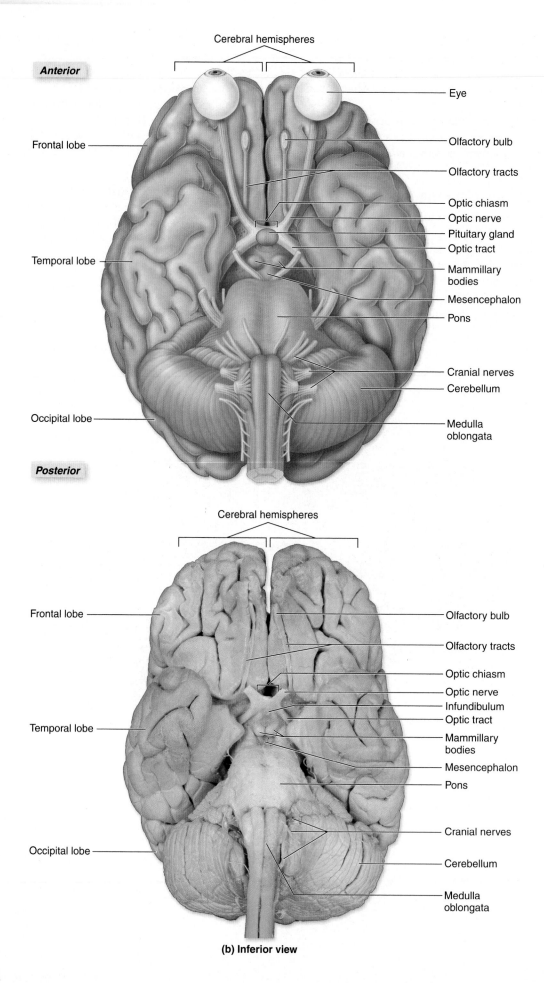

(b) Inferior view

Figure 15.1

The Human Brain (continued). (b) In inferior view, an illustration and a cadaver photo best illustrate the cranial nerves arising from the base of the brain. (c) Internal structures such as the thalamus and hypothalamus are best seen in midsagittal view.

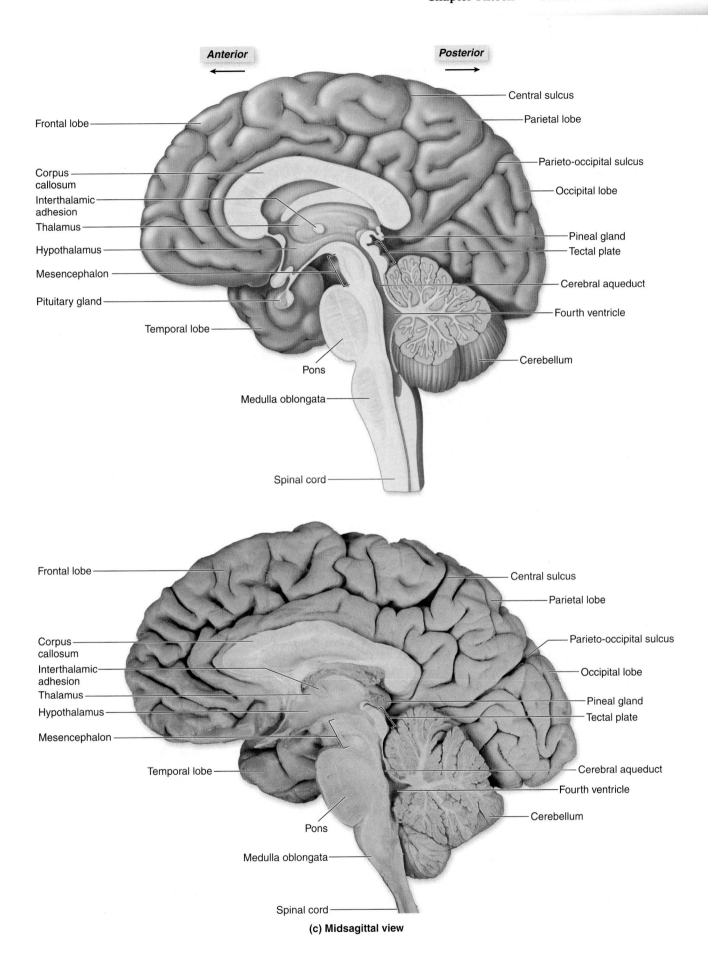

Anterior

Posterior

Frontal lobe

Corpus callosum

Interthalamic adhesion

Thalamus

Hypothalamus

Mesencephalon

Pituitary gland

Temporal lobe

Pons

Medulla oblongata

Spinal cord

Central sulcus

Parietal lobe

Parieto-occipital sulcus

Occipital lobe

Pineal gland

Tectal plate

Cerebral aqueduct

Fourth ventricle

Cerebellum

Frontal lobe

Corpus callosum

Interthalamic adhesion

Thalamus

Hypothalamus

Mesencephalon

Temporal lobe

Pons

Medulla oblongata

Spinal cord

Central sulcus

Parietal lobe

Parieto-occipital sulcus

Occipital lobe

Pineal gland

Tectal plate

Cerebral aqueduct

Fourth ventricle

Cerebellum

(c) Midsagittal view

- The **metencephalon** (met-en-sef′ă-lon; *meta* = after) arises from the rhombencephalon and eventually forms the pons and cerebellum.
- The **myelencephalon** (mī-el-en-sef′ă-lon; *myelos* = medulla) also derives from the rhombencephalon, and it eventually forms the medulla oblongata.

Table 15.1 summarizes the embryonic brain structures and their corresponding structures in the adult brain.

During the embryonic and fetal periods, the telencephalon grows rapidly and envelops the diencephalon. As the future brain develops, its surface becomes folded, especially in the telencephalon, leading to the formation of the adult sulci and gyri (see figure 15.1a). The bends and creases that occur in the developing brain determine the boundaries of the brain's cavities. Together, the bends, creases, and folds in the telencephalon surface are necessary in order to fit the massive amount of brain tissue within the confines of the cranial cavity. Most of the gyri and sulci develop late in the fetal period, so that by the time the fetus is born, its brain closely resembles that of an adult (figure 15.2c–e).

> ## Study Tip!
>
> When reviewing the embryonic development of the brain, note that during the *fifth* week of development, *five* secondary brain vesicles form.

Figure 15.2

Structural Changes in the Developing Brain. *(a)* As early as 4 weeks, the growing brain is bent because of space restrictions in the developing head. *(b)* At 5 weeks, the secondary brain vesicles appear. *(c)* By 13 weeks, the telencephalon grows rapidly and envelops the diencephalon. *(d)* Some major sulci and gyri are present by 26 weeks. *(e)* The features of an adult brain are present at birth.

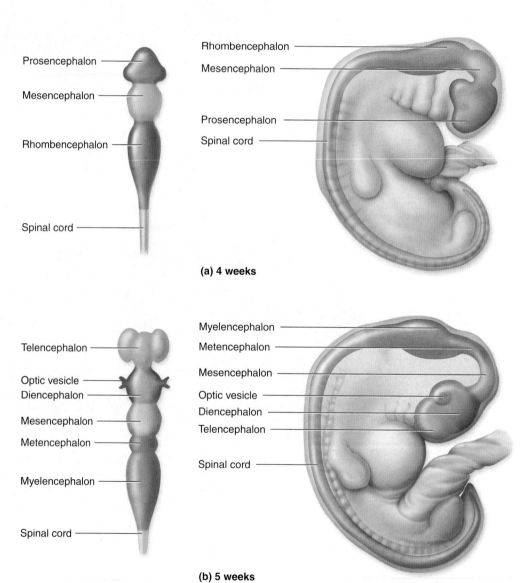

(a) 4 weeks

(b) 5 weeks

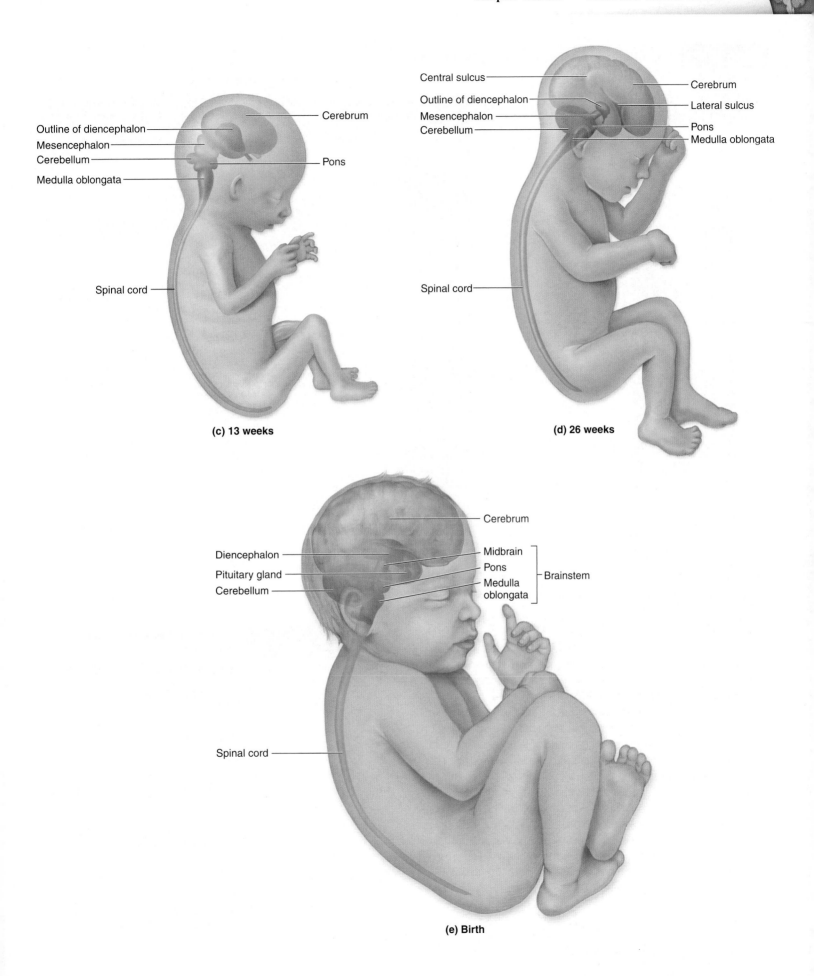

Cerebrum

Outline of diencephalon

Mesencephalon

Cerebellum

Medulla oblongata

Pons

Spinal cord

(c) 13 weeks

Central sulcus

Outline of diencephalon

Mesencephalon

Cerebellum

Cerebrum

Lateral sulcus

Pons

Medulla oblongata

Spinal cord

(d) 26 weeks

Cerebrum

Diencephalon

Pituitary gland

Cerebellum

Midbrain

Pons

Medulla oblongata

Brainstem

Spinal cord

(e) Birth

| Table 15.1 | Major Brain Structures: Embryonic Through Adult |

EMBRYONIC DEVELOPMENT $\longrightarrow$ ADULT STRUCTURE

Neural Tube	Primary Brain Vesicles	Secondary Brain Vesicles (future adult brain regions)[1]	Neural Canal Derivative[2]	Structures Within Brain Region
		Telencephalon	Lateral ventricles	Cerebrum
	Prosencephalon (forebrain)	Diencephalon	Third ventricle	Epithalamus, thalamus, hypothalamus
	Mesencephalon (midbrain)	Mesencephalon	Cerebral aqueduct	Cerebral peduncles, superior colliculi, inferior colliculi
	Rhombencephalon (hindbrain)	Metencephalon	Anterior part of fourth ventricle	Pons, cerebellum
		Myelencephalon	Posterior part of fourth ventricle; central canal	Medulla oblongata

Anterior

Posterior

Neural canal

Neural canal

[1] The embryonic secondary vesicles form the adult brain regions, and so they share the same names.

[2] The neural canal in each specific brain region will form its own named "space."

Organization of Neural Tissue Areas in the Brain

Two distinct tissue areas are recognized within the brain and spinal cord: gray matter and white matter. The **gray matter** houses motor neuron and interneuron cell bodies, dendrites, telodendria, and unmyelinated axons. (Origin of gray color described in chapter 14.) The **white matter** derives its color from the myelin in the myelinated axons. During brain development, an outer, superficial region of gray matter forms from migrating peripheral neurons. As a result, the external layer of gray matter, called the cerebral **cortex** (kor′teks; bark), covers the surface of most of the adult brain (the cerebrum and the cerebellum). The white matter lies deep to the gray matter of the cortex. Finally, within the masses of white matter, the brain also contains discrete internal clusters of gray matter called **cerebral nuclei,** which are oval, spherical, or sometimes irregularly shaped clusters of neuron cell bodies. **Figure 15.3** shows the apportionment of gray matter and white matter in various regions of the brain. **Table 15.2** is a glossary of nervous system structures.

WHAT DID YOU LEARN?

❶ Identify the primary vesicles that form during brain development.

❷ What is the name of a depression between two adjacent surface folds in the telencephalon?

Table 15.2	Glossary of Nervous System Structures
Structure	**Description**
Ganglion	Cluster of neuron cell bodies within the PNS
Center	Group of CNS neuron cell bodies with a common function
Nucleus	Center that displays discrete anatomic boundaries
Nerve	Axon bundle extending through the PNS
Nerve plexus	Network of nerves
Tract	CNS axon bundle in which the axons have a similar function and share a common origin and destination
Funiculus	Group of tracts in a specific area of the spinal cord
Pathway	Centers and tracts that connect the CNS with body organs and systems

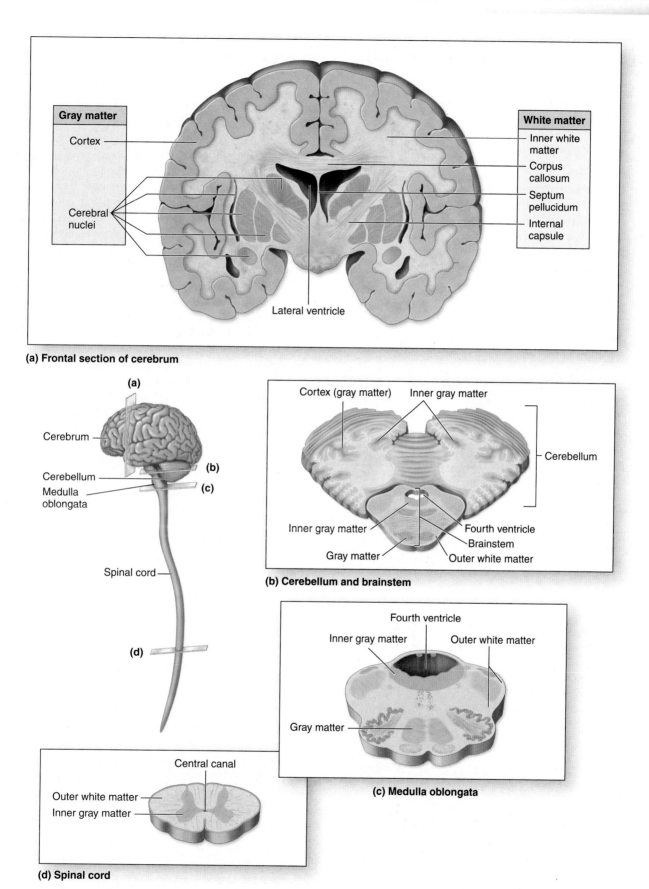

(a) Frontal section of cerebrum

Gray matter
- Cortex
- Cerebral nuclei

White matter
- Inner white matter
- Corpus callosum
- Septum pellucidum
- Internal capsule

Lateral ventricle

Cerebrum

Cerebellum

Medulla oblongata

Spinal cord

Cortex (gray matter)   Inner gray matter

Cerebellum

Inner gray matter   Fourth ventricle

Gray matter   Brainstem

Outer white matter

(b) Cerebellum and brainstem

Fourth ventricle

Inner gray matter   Outer white matter

Gray matter

(c) Medulla oblongata

Central canal

Outer white matter
Inner gray matter

(d) Spinal cord

Figure 15.3

Gray and White Matter in the CNS.   The gray matter represents regions containing neuron cell bodies, dendrites, and telodendria, whereas the white matter derives its color from myelinated axons. The distribution of gray and white matter is compared in *(a)* the cerebrum, *(b)* the cerebellum and brainstem, *(c)* the medulla oblongata, and *(d)* the spinal cord.

Support and Protection of the Brain

Key topics in this section:

- Characteristics of the cranial meninges and the cranial dural septa
- Origin, function, and pattern of cerebrospinal fluid circulation
- Structure of the blood-brain barrier and how it protects the brain

The brain is protected and isolated by multiple structures. The bony cranium provides rigid support, while protective connective tissue membranes called meninges surround, support, stabilize, and partition portions of the brain. Cerebrospinal fluid (CSF) acts as a cushioning fluid. Finally, the brain has a blood-brain barrier to prevent harmful materials from entering the bloodstream.

Cranial Meninges

The **cranial meninges** (mě-nin′jes, mē′nin-jēz; sing., *meninx*, men′ingks, mē′ninks; membrane) are three connective tissue layers that separate the soft tissue of the brain from the bones of the cranium, enclose and protect blood vessels that supply the brain, and contain and circulate cerebrospinal fluid. In addition, some parts of the cranial meninges form some of the veins that drain blood from the brain. From deep (closest to the brain) to superficial (farthest away from the brain), the cranial meninges are the pia mater, the arachnoid, and the dura mater **(figure 15.4)**.

Pia Mater

The **pia mater** (pē′ǎ mah′ter, pī′ǎ mā′ter; *pia* = tender, delicate) is the innermost of the cranial meninges. It is a thin layer of delicate areolar connective tissue that is highly vascularized and tightly adheres to the brain, following every contour of the surface.

Arachnoid

The **arachnoid** (ǎ-rak′noyd), also called the *arachnoid mater* or the *arachnoid membrane*, lies external to the pia mater (figure 15.4). The term arachnoid means "resembling a spider web," and this meninx is so named because it is partially composed of a delicate web of collagen and elastic fibers, termed the **arachnoid trabeculae**. Immediately deep to the arachnoid is the **subarachnoid space**. The arachnoid trabeculae extend through this space from the arachnoid to the underlying pia mater. Between the arachnoid and the overlying dura mater is a potential space, the **subdural space.** The subdural space becomes an actual space if blood or fluid accumulates there, a condition called a subdural hematoma (see Clinical View).

Dura Mater

The **dura mater** (doo′rǎ mā′ter; *dura* = tough, *mater* = mother) is an external tough, dense irregular connective tissue layer composed of two fibrous layers. As its Latin name indicates, it is the strongest of the meninges. Within the cranium, the dura mater is composed of two layers. The **meningeal** (mě-nin′jē-ǎl, men′in-jē′ǎl) **layer** lies deep to the periosteal layer. The **periosteal** (per-ē-os′tē-ǎl; *peri* = around, *osteon* = bone) **layer,** the more superficial layer, forms the periosteum on the internal surface of the cranial bones.

The meningeal layer is usually fused to the periosteal layer, except in specific areas where the two layers separate to form large, blood-filled spaces called **dural venous sinuses.** Dural venous sinuses are typically triangular in cross section, and unlike most other veins, they do not have valves to regulate venous blood flow. The dural venous sinuses are, in essence, large veins that drain blood from the brain and transport this blood to the internal jugular veins that help drain blood circulation to the head.

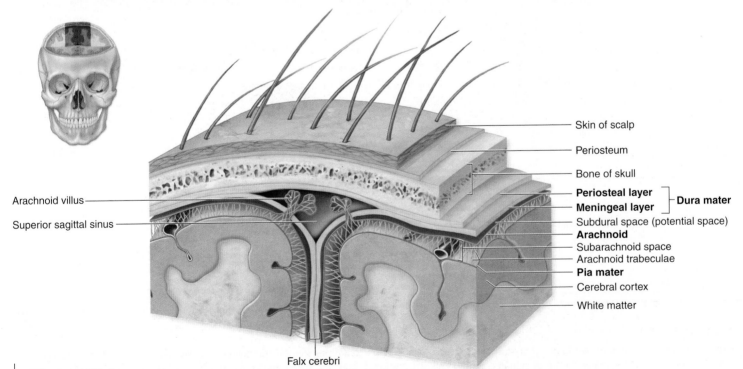

Arachnoid villus
Superior sagittal sinus
Skin of scalp
Periosteum
Bone of skull
Periosteal layer ⎤
Meningeal layer ⎦ Dura mater
Subdural space (potential space)
Arachnoid
Subarachnoid space
Arachnoid trabeculae
Pia mater
Cerebral cortex
White matter
Falx cerebri

Figure 15.4

Cranial Meninges. A coronal section of the head depicts the organization of the three meningeal layers: the dura mater, the arachnoid, and the pia mater. In the midline, folds of the inner meningeal layer of the dura mater form the falx cerebri, which partitions the two cerebral hemispheres. The inner meningeal layer and the outer periosteal layer sometimes separate to form the dural venous sinuses, such as the superior sagittal sinus (shown here), which drain blood away from the brain.

The dura mater and the bones of the skull may be separated by the potential **epidural** (ep-i-doo′răl; *epi* = upon, *durus* = hard) **space,** which contains the arteries and veins that nourish the meninges and bones of the cranium. Under normal (healthy) conditions, the potential space is not a space at all. However, it has the *potential* to become a *real* space and fill with fluid or blood as a result of trauma or disease (see Clinical View, "Epidural and Subdural Hematomas," page 448, for examples). (Note in figure 15.4 that no epidural space is labeled, since this is a potential space, not a real space.)

Cranial Dural Septa

The meningeal layer of the dura mater extends as flat partitions (septa) into the cranial cavity at four locations. Collectively, these double layers of dura mater are called **cranial dural septa.** These membranous partitions separate specific parts of the brain and provide additional stabilization and support to the entire brain. There are four cranial dural septa: the falx cerebri, tentorium cerebelli, falx cerebelli, and diaphragma sellae **(figure 15.5)**.

The **falx cerebri** (falks sĕ-rē′bri; *falx* = sickle, *cerebro* = brain) is the largest of the four dural septa. This large, sickle-shaped vertical fold of dura mater, located in the midsagittal plane, projects into the longitudinal fissure between the left and right cerebral hemispheres. Anteriorly, its inferior portion attaches to the crista galli of the ethmoid bone; posteriorly, its inferior portion attaches to the internal occipital crest. Running within the margins of this dural septa are two dural venous sinuses: the **superior sagittal sinus** and the **inferior sagittal sinus** (see figure 23.11*b*, page 696).

The **tentorium cerebelli** (ten-tō′rē-ŭm ser-e-bel′ī) is a horizontally oriented fold of dura mater that separates the occipital and

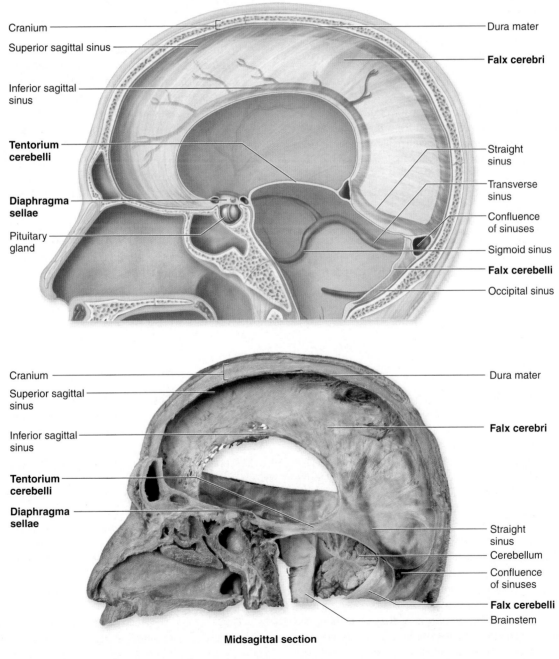

Midsagittal section

Figure 15.5

Cranial Dural Septa. An illustration and a cadaver photo of a midsagittal section of the skull show the orientation of the falx cerebri, falx cerebelli, tentorium cerebelli, and diaphragma sellae.

temporal lobes of the cerebrum from the cerebellum. It is named for the fact that it forms a dural "tent" over the cerebellum. The **transverse sinuses** run within its posterior border. The anterior surface of the tentorium cerebelli has a gap or opening, called the **tentorial notch** (or *tentorial incisure*), to allow for the passage of the brainstem.

Extending into the midsagittal line inferior to the tentorium cerebelli is the **falx cerebelli,** a sickle-shaped vertical partition that divides the left and right cerebellar hemispheres. A tiny **occipital sinus** (another dural venous sinus) runs in its posterior vertical border.

The smallest of the dural septa is the **diaphragma sellae** (dī-ă-frag′mă sel′ē; *sella* = saddle), which forms a "roof" over the sella turcica of the sphenoid bone. A small opening within it allows for the passage of a thin stalk, called the infundibulum, that attaches the pituitary gland to the base of the hypothalamus (described later).

WHAT DO YOU THINK?

1 How does the meningeal layer that provides the most support and physical protection to the brain perform its primary task?

Brain Ventricles

Ventricles (ven′tri-kl; *ventriculus* = little cavity) are cavities or expansions within the brain that are derived from the lumen (opening) of the embryonic neural tube. The ventricles are continuous with one another as well as with the central canal of the spinal cord **(figure 15.6)**.

There are four ventricles in the brain: Two **lateral ventricles** are in the cerebrum, separated by a thin medial partition called the **septum pellucidum** (pe-loo′si-dum; *pellucid* = transparent). Within the diencephalon is a smaller ventricle called the **third ventricle.** Each lateral ventricle communicates with the third ventricle through an opening called the **interventricular foramen**

(formerly called the *foramen of Munro*). A narrow canal called the **cerebral aqueduct** (ak′we-dŭkt; canal) (also called the *mesencephalic aqueduct* and *aqueduct of the midbrain* and formerly called the *aqueduct of Sylvius*), passes through the mesencephalon and connects the third ventricle with the tetrahedron-shaped **fourth ventricle.** The fourth ventricle is located between the pons and the cerebellum. The fourth ventricle narrows at its inferior end before it merges with the slender **central canal** in the spinal cord. All of the ventricles contain cerebrospinal fluid.

Cerebrospinal Fluid

Cerebrospinal (sĕ-rē′brō-spī-năl) **fluid (CSF)** is a clear, colorless liquid that circulates in the ventricles and subarachnoid space. CSF bathes the exposed surfaces of the central nervous system and completely surrounds the brain and spinal cord. CSF performs several important functions:

- **Buoyancy.** The brain floats in the CSF, which thereby supports more than 95% of its weight and prevents it from being crushed under its own weight. Without CSF to support it, the heavy brain would sink through the foramen magnum.
- **Protection.** CSF provides a liquid cushion to protect delicate neural structures from sudden movements. When you try to walk quickly in a swimming pool, your movements are slowed as the water acts as a "movement buffer." CSF likewise helps slow movements of the brain if the skull and/or body move suddenly and forcefully.
- **Environmental stability.** CSF transports nutrients and chemicals to the brain and removes waste products from the brain. Additionally, CSF protects nervous tissue from chemical fluctuations that would disrupt neuron function. The waste products and excess CSF are eventually transported into the venous circulation, where they are filtered from the blood and secreted in urine in the urinary system.

CLINICAL VIEW

Epidural and Subdural Hematomas

A **hemorrhage** is any loss of blood from a vessel. A pooling of blood outside of a vessel is referred to as a **hematoma** (hē-mă-tō′mă, hem′ă-tō′mă; *hemato* = blood, *oma* = tumor). An **epidural hematoma** is a pool of blood that forms in the epidural space of the brain. Recall that the epidural space is normally a potential space. However, as blood oozes into the region, the layers of the potential space become separated, and an actual space filled with blood is produced. Most epidural hematomas are caused by a severe blow to the side of the head, usually at the pterion (tĕ′rē-on), the junction of the temporal, sphenoid, frontal, and parietal bones. Due to the force of impact, the person suffers a fractured skull, resulting in hemorrhage of the middle meningeal artery and the formation of an epidural hematoma. Typically, a person with a fractured skull loses consciousness immediately. Following the period of unconsciousness, the patient wakes up and appears relatively normal, but after a few hours, the adjacent brain tissue becomes distorted and compressed as the hematoma continues to increase in size.

Severe neurologic injury or even death may occur if the bleeding is not stopped and the accumulated blood removed. Treatment involves surgically drilling a hole in the skull, suctioning out the blood, and ligating (tying off) the bleeding vessel.

A **subdural hematoma** is a hemorrhage that occurs in the subdural space between the dura mater and the arachnoid. These hematomas typically result when veins are ruptured due to fast or violent rotational motion of the head. In fact, isolated cases of subdural hematomas have occurred in otherwise healthy individuals who have ridden "monster roller coasters." These people developed a severe headache immediately after the ride, and CT scans showed clear evidence of subdural hematoma. Many neurologists now feel there is a risk (albeit very low) of developing a subdural hematoma from a roller coaster ride involving high-speed turns that jerk and whip the head.

With a subdural hematoma, the blood accumulates relatively slowly, but this injury is still a serious medical emergency and must be treated similarly to an epidural hematoma.

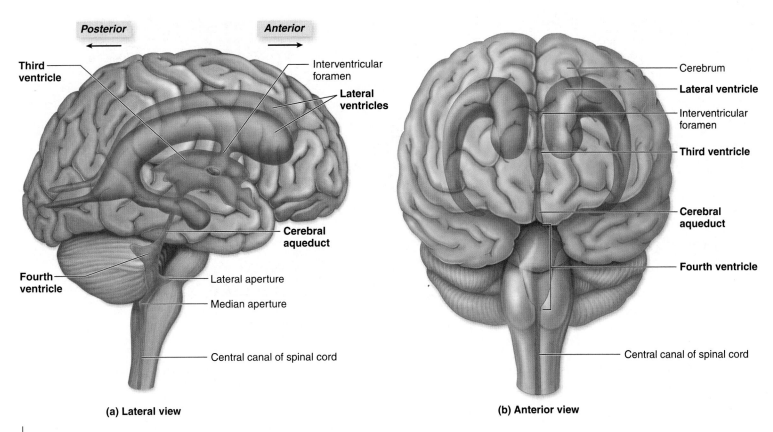

(a) Lateral view

(b) Anterior view

Figure 15.6

Ventricles of the Brain.   The ventricles are formed from the embryonic neural canal. They are sites of production of cerebrospinal fluid (CSF), which transports chemical messengers, nutrients, and waste products. *(a)* Lateral and *(b)* anterior views show the positioning and relationships of the ventricles.

CLINICAL VIEW

Brain Injuries

Brain injuries commonly occur due to head trauma that causes displacement and distortion of nervous tissue. Two common types of brain injury are concussions and contusions.

A **concussion** is the most common brain injury. It is characterized by temporary, abrupt loss of consciousness after a blow to the head or the sudden stop of a moving head. Although a concussion leaves no obvious physical defect or sign of injury in the brain, symptoms including headache, drowsiness, lack of concentration, confusion, and amnesia may occur. Multiple concussions have a cumulative effect, causing the person to lose a small amount of mental ability with each episode. Extreme examples occur in prizefighters in whom repeated head injuries actually can induce Parkinson disease, as exemplified by boxer Muhammad Ali, who experienced numerous concussions and now has this condition.

A **contusion** is visible bruising of the brain due to trauma that causes blood to leak from small vessels. A contusion may result in a torn pia mater, which permits blood to enter the subarachnoid space. Usually, the person immediately loses consciousness (normally for no longer than 5 minutes). Respiration abnormalities and decreased blood pressure sometimes occur as well.

CSF Formation

Cerebrospinal fluid is formed by the **choroid plexus** (ko′royd plek′sŭs; *chorioeides* = membrane, *plexus* = a braid) in each ventricle. The choroid plexus is composed of a layer of **ependymal** (ep-en′di-măl; *ependyma* = an upper garment) **cells** and the capillaries that lie within the pia mater (**figure 15.7**). CSF is produced by secretion of a fluid from the ependymal cells that originates from the blood plasma. CSF is somewhat similar to blood plasma, although certain ion concentrations differ between the two types of fluid.

WHAT DO YOU THINK?

2 What do you think happens if the amount of CSF produced begins to exceed the amount removed or drained at the arachnoid villi?

CSF Circulation

The choroid plexus produces CSF at a rate of about 500 milliliters (ml) per day. The CSF circulates through and eventually leaves the ventricles and enters the subarachnoid space, where the total volume of CSF at any given moment ranges between 100 ml and 160 ml. This means that excess CSF is continuously removed from the subarachnoid space so the fluid will not accumulate and compress and damage the nervous tissue. Fingerlike extensions of the arachnoid project through the dura mater into the dural venous sinuses to form **arachnoid villi** (vil′i; shaggy hair). Collections of arachnoid

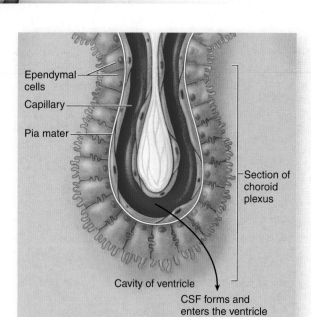

Ependymal cells

Capillary

Pia mater

Section of choroid plexus

Cavity of ventricle

CSF forms and enters the ventricle

(a) Choroid plexus

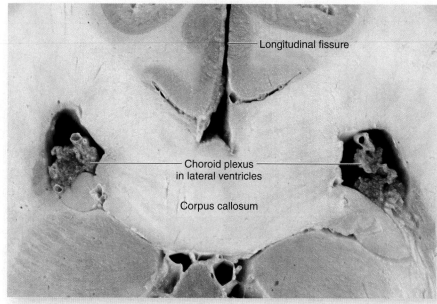

Longitudinal fissure

Choroid plexus in lateral ventricles

Corpus callosum

(b) Coronal section of the brain, close-up

Figure 15.7

Choroid Plexus.   The choroid plexus, which forms the cerebrospinal fluid, is composed of ependymal cells and capillaries within the pia mater. *(a)* The diagram shows a section of the choroid plexus. *(b)* The photograph shows a coronal brain section that includes the choroid plexus in lateral ventricles.

villi form **arachnoid granulations.** Excess CSF moves across the arachnoid membrane at the arachnoid villi to return to the blood within the dural venous sinuses. Within the subarachnoid space, cerebral arteries and veins are supported by the arachnoid trabeculae and surrounded by cerebrospinal fluid.

Figure 15.8 shows the process of CSF production, circulation, and removal, which consists of the following steps:

1. CSF is produced by the choroid plexus in the ventricle.
2. CSF flows from the lateral ventricles and third ventricle through the cerebral aqueduct into the fourth ventricle.
3. Most of the CSF in the fourth ventricle flows into the subarachnoid space by passing through openings in the roof

of the fourth ventricle. These ventricular openings are the paired **lateral apertures** and the single **median aperture.** CSF also fills the central canal of the spinal cord.

4. As it travels through the subarachnoid space, CSF removes waste products and provides buoyancy for the brain and spinal cord.
5. As CSF accumulates within the subarachnoid space, it exerts pressure within the arachnoid villi. This pressure exceeds the pressure of blood in the venous sinuses. Thus, the arachnoid villi extending into the dural venous sinuses provide a conduit for a one-way flow of excess CSF to be returned into the blood within the dural venous sinuses.

CLINICAL VIEW

Hydrocephalus

Hydrocephalus (hī-drō-sef′a-lŭs; *hydro* = water, *kephale* = head) literally means "water on the brain," and refers to the pathologic condition of excessive CSF, which often leads to brain distortion. Most cases of hydrocephalus result from either an obstruction in CSF flow that restricts its reabsorption into the venous bloodstream or some intrinsic problem with the arachnoid villi themselves.

If hydrocephalus develops in a young child, prior to closure of the cranial sutures, the head becomes enlarged, and neurologic damage may result. If hydrocephalus develops after the cranial sutures have closed, the brain may be compressed within the fixed cranium as the ventricles expand, resulting in permanent brain damage.

Severe cases of hydrocephalus are most often treated by inserting a tube called a ventriculoperitoneal (VP) shunt. The shunt drains excess CSF from the ventricles to the abdominopelvic cavity.

Although VP shunts have been used for 30 years, complications such as infection and blockage sometimes occur. A newly developed surgical procedure known as endoscopic third ventriculostomy (ETV) creates a hole in the floor of the third ventricle that directly drains into the subarachnoid space.

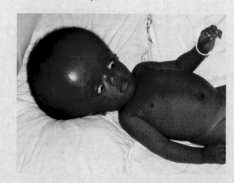

Infant with hydrocephalus.

(1) CSF is produced by the choroid plexus in the ventricles.

(2) CSF flows from the third ventricle through the cerebral aqueduct into the fourth ventricle.

(3) CSF in the fourth ventricle flows into the subarachnoid space by passing through the paired lateral apertures or the single median aperture, and into the central canal of the spinal cord.

(4) As the CSF flows through the subarachnoid space, it removes waste products and provides buoyancy to support the brain.

(5) Excess CSF flows into the arachnoid villi, then drains into the dural venous sinuses. Pressure allows the CSF to be released into the blood without permitting any venous blood to enter the subarachnoid space. The greater pressure on the CSF in the subarachnoid space ensures that CSF moves into the venous sinuses.

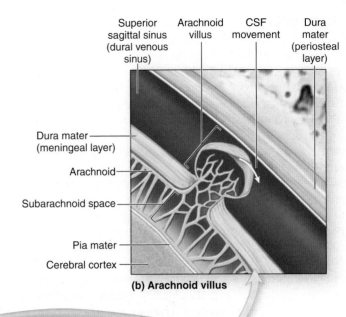

(b) Arachnoid villus

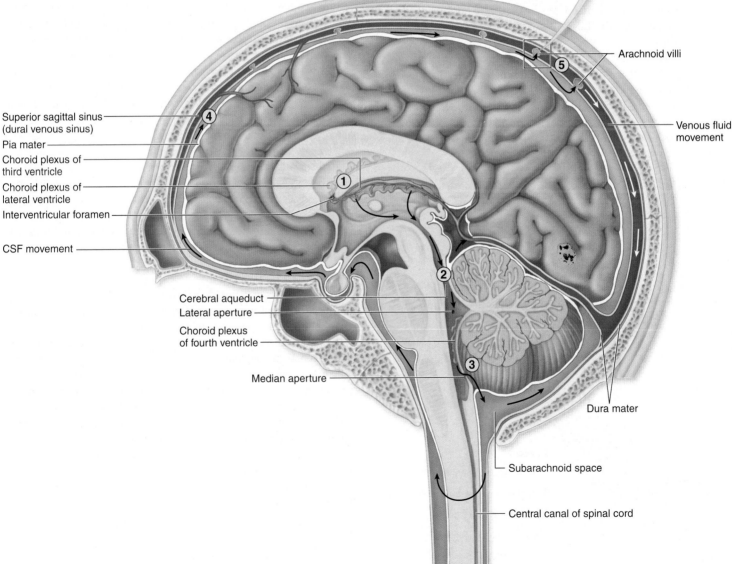

(a) Midsagittal section

Figure 15.8

Production and Circulation of Cerebrospinal Fluid.   (a) A midsagittal section identifies the sites where cerebrospinal fluid (CSF) is formed and the pathway of its circulation toward the arachnoid villi. (b) CSF flows from the arachnoid villi into the dural venous sinuses.

Blood-Brain Barrier

Nervous tissue is protected from the general circulation by the **blood-brain barrier (BBB)**, which strictly regulates what substances can enter the interstitial fluid of the brain (see chapter 14). The blood-brain barrier keeps the neurons in the brain from being exposed to drugs, waste products in the blood, and variations in levels of normal substances (e.g., ions, hormones) that could adversely affect brain function.

Recall that the **perivascular feet** of astrocytes cover, wrap around, and completely envelop capillaries in the brain. Both the capillary endothelial cells and the astrocyte perivascular feet contribute to the blood-brain barrier **(figure 15.9)**. The continuous basement membrane of the endothelial cells is a significant barrier. Tight junctions between adjacent endothelial cells reduce capillary permeability and prevent materials from diffusing across the capillary wall. The astrocytes act as "gatekeepers" that permit materials to pass to the neurons after leaving the capillaries. Even so, the barrier is not absolute. Usually only lipid-soluble (dissolvable in fat) compounds, such as nicotine, alcohol, and some anesthetics, can diffuse across the endothelial plasma membranes and into the interstitial fluid of the CNS to reach the brain neurons.

The blood-brain barrier is markedly reduced or missing in three distinct locations in the CNS: the choroid plexus, the hypothalamus, and the pineal gland. The capillaries of the choroid plexus must be permeable in order to produce CSF. Both the hypothalamus and the pineal gland produce some hormones, which must have ready access to the bloodstream, so their capillaries must be more permeable as well.

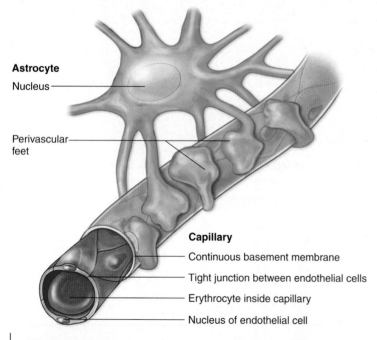

Figure 15.9

Blood-Brain Barrier. The perivascular feet of the astrocytes (when they completely cover the capillary) and the tight endothelial junctions of the capillaries work together to prevent harmful materials in the bloodstream from reaching the brain. (Here we show just a few perivascular feet of astrocytes, so that their structure may be appreciated. Note: The perivascular feet completely surround capillaries in the brain.)

Labels in figure:
- Astrocyte
- Nucleus
- Perivascular feet
- Capillary
- Continuous basement membrane
- Tight junction between endothelial cells
- Erythrocyte inside capillary
- Nucleus of endothelial cell

Cerebrum

Key topics in this section:

- Anatomic structures and functional areas of the cerebrum
- Tracts associated with central white matter of the cerebrum
- Components of the cerebral nuclei and their function

The **cerebrum** is the location of conscious thought processes and the origin of all complex intellectual functions. It is readily identified as the two large hemispheres on the superior aspect of the brain (see figure 15.1*a,b*). Your cerebrum enables you to read and comprehend the words in this textbook, turn its pages, form and remember ideas, and talk about your ideas with your peers. It is the center of your intelligence, reasoning, sensory perception, thought, memory, and judgment, as well as your voluntary motor, visual, and auditory activities.

The cerebrum is formed from the telencephalon. Recall from an earlier section in this chapter that the outer layer of gray matter is called the cerebral cortex and an inner layer is white matter. Deep to the white matter are discrete regions of gray matter called cerebral nuclei. As described earlier, the surface of the cerebrum folds into elevated ridges, called gyri, which allow a greater amount of cortex to fit into the cranial cavity. Adjacent gyri are separated by shallow sulci or deeper grooves called **fissures** (fish'ŭr). The cerebrum also contains a large number of neurons, which are needed for the complex analytical and integrative functions performed by the cerebral hemispheres.

Cerebral Hemispheres

The cerebrum is composed of two halves, called the left and right **cerebral hemispheres** (hem'i-sfēr; *hemi* = half, *sphaira* = ball) **(figure 15.10)**. The paired cerebral hemispheres are separated by a deep **longitudinal fissure** that extends along the midsagittal plane. The cerebral hemispheres are separate from one another, except at a few locations where bundles of axons called **tracts** form white matter regions that allow for communication between them. The largest of these white matter tracts, the **corpus callosum** (kōr'pŭs kal-lō'sŭm; *corpus* = body, *callosum* = hard), connects the hemispheres (see a midsagittal section of the corpus callosum in figure 15.1*c*). The corpus callosum provides the main communications link between these hemispheres.

Three points should be kept in mind with respect to the cerebral hemispheres:

- In most cases, it is difficult to assign a precise function to a specific region of the cerebral cortex. Considerable overlap and indistinct boundaries permit a single region of the cortex to exhibit several different functions. Additionally, some aspects of cortical function, such as memory or consciousness, cannot easily be assigned to any single region.
- With few exceptions, both cerebral hemispheres receive their sensory information from and project motor commands to the opposite side of the body. The right cerebral hemisphere controls the left side of the body, and vice versa.

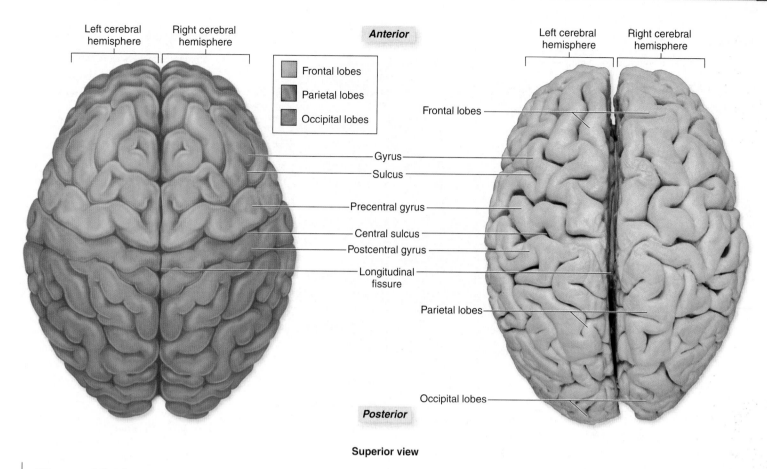

Superior view

Figure 15.10

Cerebral Hemispheres. Superior views comparing an illustration and a cadaver photo show the cerebral hemispheres, where our conscious activities, memories, behaviors, plans, and ideas are initiated and controlled.

- The two hemispheres appear as anatomic mirror images, but they display some functional differences, termed **hemisphere lateralization**. For example, the portions of the brain that are responsible for controlling speech and understanding verbalization are frequently located in the left hemisphere. These differences primarily affect higher-order functions, which are addressed in chapter 17.

WHAT DO YOU THINK?

3 In the past, one treatment for severe epilepsy was to cut the corpus callosum, thus confining epileptic seizures to just one cerebral hemisphere. How would cutting the corpus callosum affect communication between the left and right hemispheres?

Lobes of the Cerebrum

Each cerebral hemisphere is divided into five anatomically and functionally distinct lobes. The first four lobes are superficially visible and are named for the overlying cranial bones: the frontal, parietal, temporal, and occipital lobes **(figure 15.11)**. The fifth lobe, called the insula, is not visible at the surface of the hemispheres. Each lobe exhibits specific cortical regions and association areas.

The **frontal lobe** (lōb) lies deep to the frontal bone and forms the anterior part of the cerebral hemisphere. The frontal lobe ends posteriorly at a deep groove called the **central sulcus** that marks the boundary with the parietal lobe. The inferior border of the frontal lobe is marked by the **lateral sulcus,** a deep groove that

separates the frontal and parietal lobes from the temporal lobe. An important anatomic feature of the frontal lobe is the **precentral gyrus,** which is a mass of nervous tissue immediately anterior to the central sulcus. The frontal lobe is primarily concerned with voluntary motor functions, concentration, verbal communication, decision making, planning, and personality.

The **parietal lobe** lies internal to the parietal bone and forms the superoposterior part of each cerebral hemisphere. It terminates anteriorly at the central sulcus, posteriorly at a relatively indistinct **parieto-occipital sulcus,** and laterally at a lateral sulcus. An important anatomic feature of this lobe is the **postcentral gyrus,** which is a mass of nervous tissue immediately posterior to the central sulcus. The parietal lobe is involved with general sensory functions, such as evaluating the shape and texture of objects being touched.

The **temporal lobe** lies inferior to the lateral sulcus and underlies the temporal bone. This lobe is involved with hearing and smell.

The **occipital lobe** forms the posterior region of each hemisphere and immediately underlies the occipital bone. This lobe is responsible for processing incoming visual information and storing visual memories.

The **insula** (in'soo-lă; inland) is a small lobe deep to the lateral sulcus. It can be observed by laterally reflecting (pulling aside) the temporal lobe. The insula's lack of accessibility has prevented aggressive studies of its function, but it is apparently involved in memory and the interpretation of taste.

Table 15.3 summarizes the lobes of the cerebrum and their subdivisions.

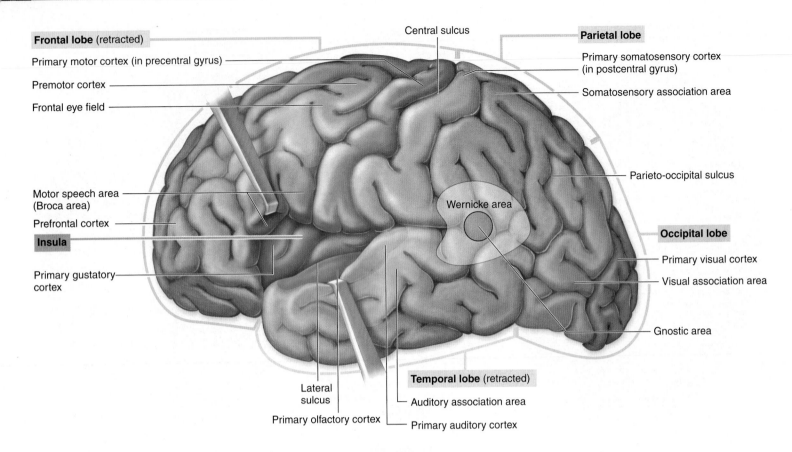

Left lateral view

Figure 15.11

Cerebral Lobes. Each cerebral hemisphere is partitioned into five structural and functional areas called lobes. Within each lobe are specific cortical regions and association areas.

Table 15.3	Cerebral Lobes	
Lobe	**Cortices and Association Areas Within Lobe**	**Primary Functions**
Frontal	Primary motor cortex (located within precentral gyrus) Premotor cortex Motor speech area (Broca area) (usually found only on the left frontal lobe) Frontal eye fields	Higher intellectual functions (concentration, decision making, planning); personality; verbal communication; voluntary motor control of skeletal muscles
Parietal	Primary somatosensory cortex (located within postcentral gyrus) Somatosensory association area Part of Wernicke area Part of gnostic area	Sensory interpretation of textures and shapes; understanding speech and formulating words to express thoughts and emotions
Temporal	Primary auditory cortex Primary olfactory cortex Auditory association area Olfactory association area Part of Wernicke area Part of gnostic area	Interpretation of auditory and olfactory sensations; storage of auditory and olfactory experiences; understanding speech
Occipital	Primary visual cortex Visual association areas	Conscious perception of visual stimuli; integration of eye-focusing movements; correlation of visual images with previous visual experiences
Insula	Primary gustatory cortex	Interpretation of taste; memory

Functional Areas of the Cerebrum

Research has shown that specific structural areas of the cerebral cortex have distinct motor and sensory functions. In contrast, some higher mental functions, such as language and memory, are dispersed over large areas. Three categories of functional areas are recognized: motor areas that control voluntary motor functions; sensory areas that provide conscious awareness of sensation; and association areas that primarily integrate and store information. Although many structural areas have been identified, there is still much that is not known or understood about the brain.

Motor Areas

The cortical areas that control motor functions are housed within the frontal lobes. The **primary motor cortex,** also called the *somatic motor area,* is located within the precentral gyrus of the frontal lobe (figure 15.11). Neurons there control voluntary skeletal muscle activity. The axons of these neurons project contralaterally (to the opposite side) to the brainstem and spinal cord. Thus, the left primary motor cortex controls the right-side voluntary muscles, and vice versa.

The innervation of the primary motor cortex to various body parts can be diagrammed as a **motor homunculus** (hō-mŭngk'ū-lŭs; diminutive man) on the precentral gyrus (**figure 15.12,** *left*). The bizarre, distorted proportions of the homunculus body reflect the amount of cortex dedicated to the motor activity of each body part. For example, the hands are represented by a much larger area of cortex than the trunk, because the hand muscles perform much more detailed, precise movements than the trunk muscles do. From a functional perspective, more motor activity is devoted to the hand in humans than in other animals because our hands are adapted for the precise, fine motor movements needed to manipulate the environment, and many motor units are devoted to muscles that move the hand and fingers.

The **motor speech area,** previously called the *Broca area,* is located in most individuals within the inferolateral portion of the left frontal lobe (see figure 15.11). This region is responsible for controlling the muscular movements necessary for vocalization.

The **frontal eye field** is on the superior surface of the middle frontal gyrus, which is immediately anterior to the premotor cortex in the frontal lobe. These cortical areas control and regulate the eye movements needed for reading and coordinating binocular vision (vision in which both eyes are used together). Some investigators include the frontal eye fields within the premotor area, thus considering the frontal eye fields part of the motor association cortex.

CLINICAL VIEW

Brodmann Areas

In the early 1900s, Korbinian Brodmann studied the comparative anatomy of the mammalian brain cortex. His colleagues encouraged him to correlate physiologic activities with previously determined anatomic locations. He performed his physiologic studies on epileptic patients undergoing surgical procedures and on laboratory rodents. Based on these findings, Brodmann produced a map

that shows the specific areas of the cerebral cortex where certain functions occur. Brodmann developed the numbering system below, which correlates with his map and shows that similar cognitive functions are usually sequential. Recent technological improvements now allow neuroscientists to more precisely pinpoint the location of physiologic activities in the brain cortex, and thus many do not use the Brodmann Area maps. However, for historical perspective and early views of the brain, they do have relevance.

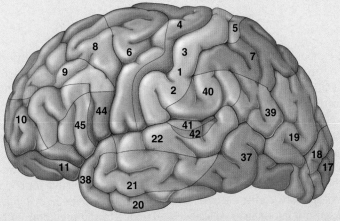

Modern rendition of Korbinian Brodmann's map of the brain, showing selected Brodmann areas.

Area	Function	Area	Function
1, 2, 3	Primary body sensation (somatosensory) in parietal lobe	22	Auditory association area in temporal lobe
4	Primary motor area (precentral gyrus) in frontal lobe	37	Visual association area in temporal lobe
5	Sensory association area in parietal lobe	38	Emotion area in temporal lobe
6	Premotor area in frontal lobe	39	Visual association area in temporal lobe
7	Sensory association area in parietal lobe	40	Sensory association area in parietal lobe
8	Frontal eye field in frontal lobe	41	Primary auditory cortex in temporal lobe
9, 10, 11	Cognitive activities (judgment or reasoning) in frontal lobe	42	Auditory association area in temporal lobe
17	Primary visual cortex in occipital lobe	44	Motor speech area in frontal lobe
18, 19	Visual association area in occipital lobe	45	Motor speech area in frontal lobe
20, 21	Visual association area in temporal lobe		

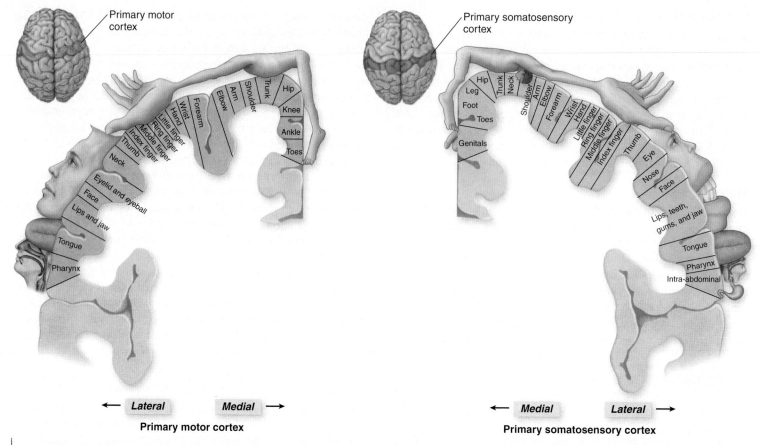

Figure 15.12

Primary Motor and Somatosensory Cortices. Body maps called the motor homunculus and the sensory homunculus illustrate the topography of the primary motor cortex and the primary somatosensory cortex in coronal section. The figure of the body (homunculus) depicts the nerve distributions; the size and location of each body region indicates relative innervation. Each cortex occurs on both sides of the brain but appears on only one side in this illustration.

Sensory Areas

The cortical areas within the parietal, temporal, and occipital lobes typically are involved with conscious awareness of sensation. Each of the major senses has a distinct cortical area.

The **primary somatosensory cortex** is housed within the postcentral gyrus of the parietal lobes. Neurons in this cortex receive general somatic sensory information from touch, pressure, pain, and temperature receptors. We typically are conscious of the sensations received by this cortex. A **sensory homunculus** may be traced on the postcentral gyrus surface, similar to the motor homunculus (figure 15.12, *right*). The surface area of somatosensory cortex devoted to a body region indicates the amount of sensory information collected within that region. Thus, the lips, fingers, and genital region occupy larger portions of the homunculus, while the trunk of the body has proportionately fewer receptors, so its homunculus region is smaller.

Sensory information for sight, sound, taste, and smell arrives at other cortical regions (see figure 15.11). The **primary visual cortex,** located in the occipital lobe, receives and processes incoming visual information. The **primary auditory cortex,** located in the temporal lobe, receives and processes auditory information. The **primary gustatory** (gŭs′tă-tōr-ē; *gustatio* = taste) **cortex** is in the insula and is involved in processing taste information. Finally, the **primary olfactory** (ol-fak′tŏ-rē; *olfactus* = to smell) **cortex,** located in the temporal lobe, provides conscious awareness of smells.

Association Areas

The primary motor and sensory cortical regions are connected to adjacent **association areas** that either process and interpret incoming data or coordinate a motor response (see figure 15.11). Association areas integrate new sensory inputs with memories of past experiences. Following are descriptions of the main association areas.

The **premotor cortex,** also called the *somatic motor association area,* is located in the frontal lobe, immediately anterior to the precentral gyrus. It permits us to process motor information and is primarily responsible for coordinating learned, skilled motor activities, such as reading a book or playing the piano. The effects of damage to the premotor cortex make its function easily apparent. For example, an individual who has sustained trauma to an area that coordinates eye movements still understands written letters and words, but cannot read because his or her eyes cannot follow the lines on a printed page.

The **somatosensory association area** is located in the parietal lobe and lies immediately posterior to the primary somatosensory cortex. It interprets sensory information and is responsible for integrating and interpreting sensations to determine the texture, temperature, pressure, and shape of objects. The somatosensory association area allows us to identify objects while our eyes are closed. For example, we can tell the difference between the coarse feel of a handful of dirt, the smooth, round shape of a marble, and the thin, flat, rounded surface of a coin because those textures have already been stored in the somatosensory association area.

The **auditory association area** is located within the temporal lobe, posteroinferior to the primary auditory cortex. Within this area, the cortical neurons interpret the characteristics of sound and store memories of sounds heard in the past. The next time an annoying song is playing over and over in your head, you will know that this auditory association area is responsible (so try to hear a favorite song before turning off the radio or CD player).

The **visual association area** is located in the occipital lobe and surrounds the primary visual area. It enables us to process visual information by analyzing color, movement, and form, and to use this information to identify the things we see. For example, when we look at a face, the primary visual cortex receives bits of visual information, but the visual association area is responsible for integrating all of this information into a recognizable picture of a face.

A functional brain region acts like a multi-association area between lobes for integrating information from individual association areas. One functional brain region is the **Wernicke area** (see figure 15.11), which is typically located only within the left hemisphere, where it overlaps the parietal and temporal lobes. The Wernicke area is involved in recognizing, understanding, and comprehending spoken or written language. As you may expect, the Wernicke area and the motor speech area must work together in order for fluent communication to occur.

Another functional brain region, called the **gnostic** (nō′stik; *gnōsis* = knowledge) **area** (or *common integrative area*), is composed of regions of the parietal, occipital, and temporal lobes. This region integrates all sensory, visual, and auditory information being processed by the association areas within these lobes. Thus it provides comprehensive understanding of a current activity. For example, suppose you awaken from a daytime nap: The hands on the clock indicate that it is 12:30, you smell food cooking, and you hear your friends talking about being hungry. The gnostic area then interprets this information to mean that it is lunchtime.

Higher-Order Processing Centers

Other association areas are called higher-order processing areas. These centers process incoming information from several different association areas and ultimately direct either extremely complex motor activity or complicated analytical functions in response. Both cerebral hemispheres house higher-order processing centers involving such functions as speech, cognition, understanding spatial relationships, and general interpretation (see chapter 17).

Central White Matter

The **central white matter** lies deep to the gray matter of the cerebral cortex and is composed primarily of myelinated axons. Most of these axons are grouped into bundles called **tracts,** which are classified as association tracts, commissural tracts, or projection tracts (**figure 15.13**).

Association tracts connect different regions of the cerebral cortex within the same hemisphere. Short association tracts are composed of **arcuate** (ar′kū-āt; *arcuatus* = bowed) **fibers;** they connect neighboring gyri within the same lobe. The longer association tracts, which are composed of **longitudinal fasciculi** (fă-sik′-ū-lī; *fascis* = bundle), connect gyri in different lobes of the same hemisphere. An example of an association tract composed of arcuate fibers is the tract that connects the primary motor cortex (of the frontal lobe) with the premotor or motor association area (also of the frontal lobe). An example of a longitudinal fasciculi is the tract that connects the Wernicke area to the motor speech area.

Commissural (kom-i-sūr′ăl; *committo* = combine) **tracts** extend between the cerebral hemispheres through axonal bridges called commissures. The prominent commissural tracts that link the left and right cerebral hemispheres include the large, C-shaped **corpus callosum** and the smaller **anterior** and **posterior commissures**.

Projection tracts link the cerebral cortex to the posterior brain regions and the spinal cord. Examples of projection tracts are the corticospinal tracts that carry motor signals from the cerebrum to the brainstem and spinal cord. The packed group of axons in these tracts passing to and from the cortex between the cerebral nuclei is called the **internal capsule.**

Table 15.4 summarizes the characteristics of the three white matter tracts of the cerebrum.

CLINICAL VIEW

The Case of Phineas Gage

The curious and tragic story of Phineas Gage has bettered our understanding of the brain and its higher-order functions. Mr. Gage lived in the mid-1800s and worked on a railroad construction crew. In September of 1848, while using an iron rod to tamp down blasting powder, the gunpowder exploded, and sent the 13-pound, 3-1/2-foot rod through Phineas's head just below his left eye. The rod ripped through his left frontal lobe, exited the top of his head, and landed 30 yards away. Astonishingly, Gage remained conscious and survived the accident.

Mr. Gage eventually regained his strength, but his personality was completely changed. Whereas before the accident, he was described as capable, well-balanced, and shrewd, after his injury, he became irreverent, fitful, and grossly profane. He was virtually incapable of making decisions, showed no regard for his former co-workers, and was described by his friends as "no longer Gage." Computerized reconstructions of Mr.

[1] Damasio H., et al. 1994. The return of Phineas Gage: Clues about the brain from the skull of a famous patient. *Science* 264(5162):1102–5.

Gage's skull showing the sites of the injury have allowed researchers to determine fairly accurately what parts of his brain were damaged.[1] Based on the case of Phineas Gage, medical science has learned that the frontal lobes are important to the proper functioning of our personalities and that the frontal cortex is intimately linked to the basic elements of decision making.

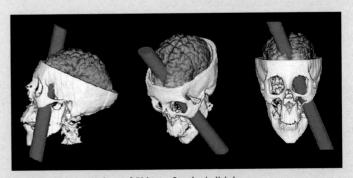

Computer reconstructions of Phineas Gage's skull injury.

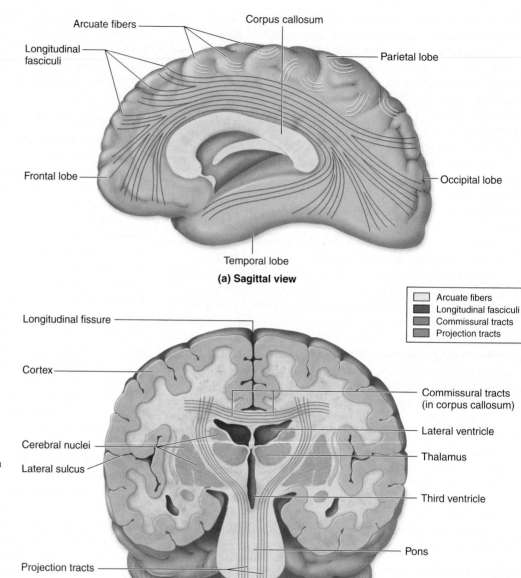

(a) Sagittal view

(b) Coronal section

Legend:
- Arcuate fibers
- Longitudinal fasciculi
- Commissural tracts
- Projection tracts

Figure 15.13

Central White Matter Tracts. White matter tracts are composed of both myelinated and unmyelinated axons. Three major groups of axons are recognized based on their distribution. *(a)* A sagittal view shows arcuate fibers and longitudinal fasciculi association tracts, which extend between gyri within one hemisphere. *(b)* A coronal view shows how commissural tracts extend between cerebral hemispheres, while projection tracts extend between the hemispheres and the brainstem.

Table 15.4	White Matter Tracts in the Cerebrum	
Tracts	**Distribution of Axons**	**Examples**
Association tracts	Connect separate cortical areas within the same hemisphere	
Arcuate fibers	Connect neighboring gyri within a single cerebral lobe	Tracts connecting primary motor cortex (frontal lobe) to motor association area (frontal lobe)
Longitudinal fasciculi	Connect gyri between different cerebral lobes of the same hemisphere	Tracts connecting Wernicke area (parietal/temporal lobes) and motor speech area (frontal lobe)
Commissural tracts	Connect corresponding lobes of the right and left hemispheres	Corpus callosum, anterior commissure, posterior commissure
Projection tracts	Connect cerebral cortex to the diencephalon, brainstem, cerebellum, and spinal cord	Corticospinal tracts (motor axons traveling from cerebral cortex to spinal cord; sensory axons traveling from spinal cord to cerebrum)

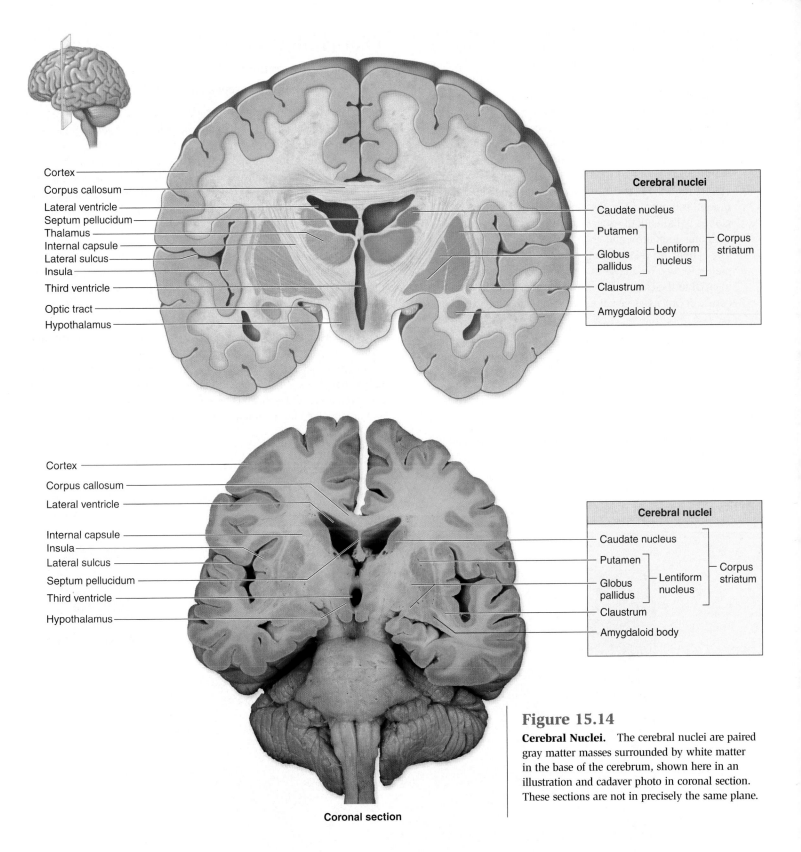

Cortex
Corpus callosum
Lateral ventricle
Septum pellucidum
Thalamus
Internal capsule
Lateral sulcus
Insula
Third ventricle
Optic tract
Hypothalamus

Cerebral nuclei

Caudate nucleus
Putamen — Lentiform nucleus — Corpus striatum
Globus pallidus
Claustrum
Amygdaloid body

Cortex
Corpus callosum
Lateral ventricle

Internal capsule
Insula
Lateral sulcus
Septum pellucidum
Third ventricle
Hypothalamus

Cerebral nuclei

Caudate nucleus
Putamen — Lentiform nucleus — Corpus striatum
Globus pallidus
Claustrum
Amygdaloid body

Coronal section

Figure 15.14

Cerebral Nuclei.   The cerebral nuclei are paired gray matter masses surrounded by white matter in the base of the cerebrum, shown here in an illustration and cadaver photo in coronal section. These sections are not in precisely the same plane.

Cerebral Nuclei

The **cerebral nuclei** (also called the *basal nuclei*) are paired, irregular masses of gray matter buried deep within the central white matter in the basal region of the cerebral hemispheres inferior to the floor of the lateral ventricle (**figure 15.14**; see figure 15.3*a*). (These masses of gray matter are sometimes incorrectly called the basal ganglia. However, the term *ganglion* is best restricted to clusters of neuron cell bodies *outside* the CNS, whereas a *nucleus* is a collection of cell bodies *within* the CNS.)

Cerebral nuclei have the following components:

- The **C**-shaped **caudate** (kaw′dāt; *caud* = tail) **nucleus** has an enlarged head and a slender, arching tail that parallels the swinging curve of the lateral ventricle. When a person begins to walk, the neurons in this nucleus stimulate the appropriate muscles to produce the pattern and rhythm of arm and leg movements associated with walking.

- The **amygdaloid** (ā-mig′dă-loyd; *amygdala* = almond) **body** (often just called the *amygdala*) is an expanded region at the

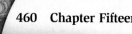

tail of the caudate nucleus. It participates in the expression of emotions, control of behavioral activities, and development of moods (see the section on the limbic system later in this chapter).

■ The **putamen** (pū-tā′men; *puto* = to prune) and the **globus pallidus** (pal′i-dŭs; *globus* = ball, *pallidus* = pale) are two masses of gray matter positioned between the bulging external surface of the insula and the lateral wall of the diencephalon. The putamen and the globus pallidus combine to form a larger body, the **lentiform** (len′ti-fōrm; *lenticula* = lentil, *forma* = shape) **nucleus,** which is usually a compact, almost rounded mass. The putamen functions in controlling muscular movement at the subconscious level, and the globus pallidus both excites and inhibits the activities of the thalamus to control and adjust muscle tone.

■ The **claustrum** (klaws′trŭm; barrier) is a thin sliver of gray matter formed by a layer of neurons located immediately internal to the cortex of the insula and derived from that cortex. It processes visual information at a subconscious level.

The term **corpus striatum** (strī-ā′tŭm; *striatus* = furrowed) describes the striated or striped appearance of the internal capsule as it passes among the caudate nucleus and the lentiform nucleus.

WHAT DID YOU LEARN?

7 What is the function of the corpus callosum?

8 List the five lobes that form each cerebral hemisphere and the function of each lobe.

9 An athlete suffers a head injury that causes loss of movement in his left leg. What specific area of the brain was damaged?

10 What is the function of association areas in the cerebrum?

Diencephalon

Key topic in this section:

■ Divisions of the diencephalon and their functions

The **diencephalon** (dī-en-sef′ă-lon; *dia* = through) is a part of the prosencephalon sandwiched between the inferior regions of the cerebral hemispheres. This region is often referred to as the "in-between brain." The components of the diencephalon include the epithalamus, the thalamus, and the hypothalamus **(figure 15.15)**. The diencephalon provides the relay and switching centers for some sensory and motor pathways and for control of visceral activities.

Epithalamus

The **epithalamus** (ep′i-thal′ă-mŭs) partially forms the posterior roof of the diencephalon and covers the third ventricle. The posterior portion of the epithalamus houses the pineal gland and the habenular nuclei. The **pineal** (pin′ē-ăl; *pineus* = pinecone-like) **gland** (or *pineal body*) is an endocrine gland. It secretes the hormone **melatonin,** which appears to help regulate day-night cycles known as the body's **circadian rhythm.** (Some companies are marketing the sale of melatonin in pill form as a cure for jet lag, although this "cure" has yet to be proven.) The **habenular** (hă-ben′ū-lăr; *habena* = strap) **nuclei** help relay signals from the *limbic system* (described later in this chapter) to the mesencephalon and are involved in visceral and emotional responses to odors.

Thalamus

The **thalamus** (thal′ă-mŭs; bed) refers to paired oval masses of gray matter that lie on each side of the third ventricle **(figure 15.16)**. The thalamus forms the superolateral walls of the third ventricle. When viewed in midsagittal section, the thalamus is located between the

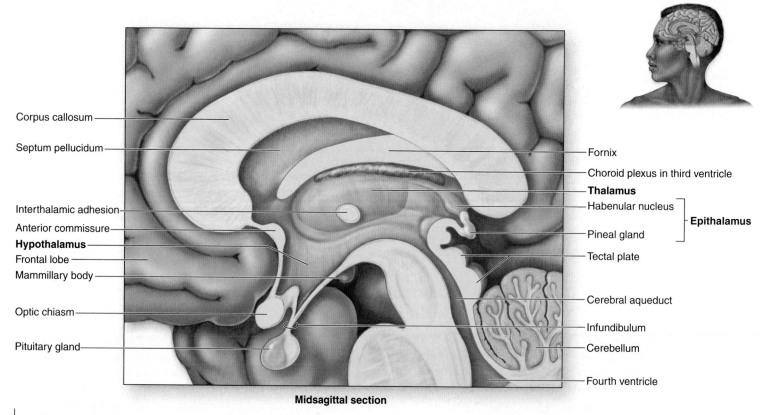

Midsagittal section

Figure 15.15

Diencephalon. The diencephalon encloses the third ventricle and connects the cerebral hemispheres to the brainstem. The right portion of the diencephalon is shown here in midsagittal section.

anterior commissure and the pineal gland. The **interthalamic adhesion** (or *intermediate mass*) is a small, midline mass of gray matter that connects the right and left thalamic bodies.

Each part of the thalamus is a gray matter mass composed of about a dozen major **thalamic nuclei** that are organized into groups; axons from these nuclei project to particular regions of the cerebral cortex (figure 15.16*b*). Sensory impulses from all the conscious senses except olfaction converge on the thalamus and synapse in at least one of its nuclei. The major functions of each group of nuclei are detailed in **table 15.5**.

The thalamus is the principal and final relay point for sensory information that will be processed and projected to the primary somatosensory cortex. Only a relatively small portion of the sensory information that arrives at the thalamus is forwarded to the cerebrum because the thalamus acts as an information filter. For example, the thalamus is responsible for filtering out the sounds and sights in a busy dorm cafeteria when you are trying to study. The thalamus also "clues in" the cerebrum about where this sensory information came from. For example, the thalamus lets the cerebrum know that a nerve impulse it receives came from the eye, indicating that the information is visual.

WHAT DO YOU THINK?

4   If there were no thalamus, how would the cerebrum's interpretation of sensory stimuli be affected?

Hypothalamus

The **hypothalamus** (hī′pō-thal′ă-mŭs; *hypo* = under) is the antero-inferior region of the diencephalon. A thin, stalklike **infundibulum** (in-fŭn-dib′ū-lŭm; funnel) extends inferiorly from the hypothalamus to attach to the pituitary gland **(figure 15.17)**.

Functions of the Hypothalamus

The hypothalamus has numerous functions, which are controlled by specific nuclei as listed in **table 15.6**. Functions of the hypothalamus include:

■ **Master control of the autonomic nervous system.** The hypothalamus is a major autonomic integration center. In essence, it is the "president" of the corporation known as the autonomic nervous system. It projects descending

Table 15.5	Functions Controlled by Thalamic Nuclei
Nuclei Group	**Functions**
Anterior group	Changes motor cortex excitability and modifies mood
Lateral group	Controls sensory flow to parietal lobes and emotional information to cingulate gyrus
Medial group	Sends signals about conscious awareness of emotional states to frontal lobes
Posterior group	Lateral geniculate nuclei: Relay visual information from optic tract to visual cortex and midbrain Medial geniculate nuclei: Relay auditory information from inner ear to auditory cortex Pulvinar nuclei: Integrate and relay sensory information for projection to association areas of cerebral cortex
Ventral group	Ventral anterior nuclei: Relay somatic motor information from cerebral nuclei and cerebellum to primary motor cortex and premotor cortex of frontal lobe Ventral lateral nuclei: Same as ventral anterior nuclei Ventral posterior nuclei: Relay sensory information to primary somatosensory cortex of parietal lobe

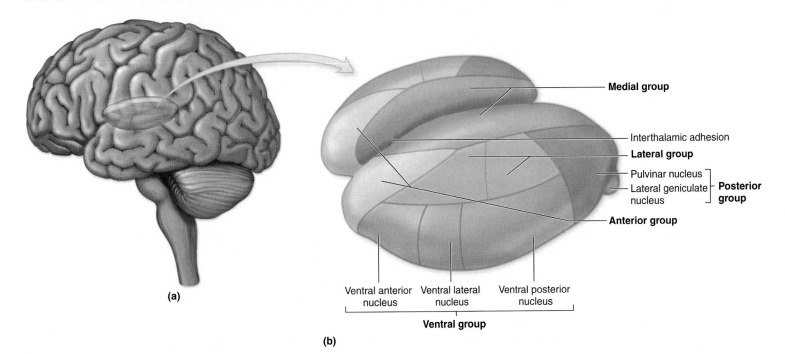

(a)

(b)

Figure 15.16

Thalamus.   (*a*) Lateral view of a transparent brain identifies the approximate location of the thalamus. (*b*) The thalamus is composed of clusters of nuclei organized into groups, as shown in this enlarged view. Not all of the nuclei may be seen from this angle.

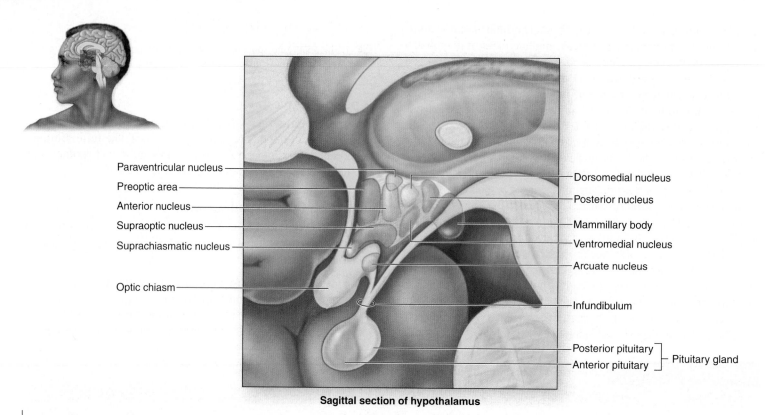

Sagittal section of hypothalamus

Figure 15.17

Hypothalamus. The hypothalamus is the most inferior region of the diencephalon. This sagittal view shows the organization and location of some of the hypothalamic nuclei.

Table 15.6	Functions Controlled by Selected Hypothalamic Nuclei
Nucleus or Hypothalamic Region	**Function**
Anterior nucleus	"Thirst center" (stimulates fluid intake); autonomic control center
Arcuate nucleus	Regulates appetite, release of gonadotropin-releasing hormone, release of growth hormone-releasing hormone, and release of prolactin-inhibiting hormone
Mammillary body	Processes sensations related to smelling; controls swallowing
Paraventricular nucleus	Produces oxytocin and antidiuretic hormone (ADH)
Preoptic area	"Thermostat" (regulates body temperature)
Suprachiasmatic nucleus	Regulates sleep-wake (circadian) rhythm
Supraoptic nucleus	Produces oxytocin and antidiuretic hormone (ADH)
Ventromedial nucleus	"Satiety center" (produces hunger sensations)

axons to autonomic nuclei in the inferior brainstem that influence heart rate, blood pressure, digestive activities, and respiration.

- **Master control of the endocrine system.** The hypothalamus is also "president" of another "corporation"—the endocrine system—overseeing most but not all of that system's functions. The hypothalamus secretes hormones that control secretory activities in the anterior pituitary gland.

In turn, subsequent normal secretions from the pituitary gland control metabolism, growth, stress responses, and reproduction. Additionally, the hypothalamus produces two hormones that are transported through axons in the infundibulum and then stored and released in the posterior pituitary: Antidiuretic hormone reduces water loss at the kidneys, and oxytocin stimulates smooth muscle contractions in the uterus, mammary gland, and prostate gland.

- **Regulation of body temperature.** The body's thermostat is located within the hypothalamus. Neurons in the preoptic area detect altered blood temperatures and signal other hypothalamic nuclei, which control the mechanisms that heat or cool the body (shivering and sweating, respectively).
- **Control of emotional behavior.** The hypothalamus is located at the center of the limbic system, the part of the brain that controls emotional responses, such as pleasure, aggression, fear, rage, contentment, and the sex drive.
- **Control of food intake.** Neurons within the ventromedial nucleus monitor levels of nutrients such as glucose and amino acids in the blood and produce sensations of hunger.
- **Control of water intake.** Specific neurons within the anterior nucleus continuously monitor the blood solute (dissolved substances) concentration. High solute concentration stimulates both the intake of fluid and the production of antidiuretic hormone by neurons in the supraoptic nucleus and paraventricular nucleus (see chapter 20).
- **Regulation of sleep-wake (circadian) rhythms.** The suprachiasmatic nucleus directs the pineal gland when to secrete melatonin. Thus, both work to regulate circadian rhythms.

WHAT DID YOU LEARN?

11 Where is the epithalamus? What is the location and function of the pineal gland in relation to the epithalamus?

12 Describe the structure and the general function of the thalamus.

13 What structure connects the pituitary gland to the hypothalamus?

Brainstem

Key topic in this section:

- Components of the brainstem and their functions

The **brainstem** connects the prosencephalon and cerebellum to the spinal cord. Three regions form the brainstem: the superiorly placed mesencephalon, the pons, and the inferiorly placed medulla oblongata (**figure 15.18**). The brainstem is a bidirectional passageway for all tracts extending between the cerebrum and the spinal cord. It also contains many autonomic centers and reflex centers required for our survival, and it houses nuclei of many of the cranial nerves.

Mesencephalon

The **mesencephalon** (or *midbrain*) is the superior portion of the brainstem. Extending through the mesencephalon is the **cerebral aqueduct** connecting the third and fourth ventricles; it is surrounded by a region called the **periaqueductal gray matter** (**figure 15.19**). The nuclei of two cranial nerves that control some eye movements are housed in the mesencephalon: the oculomotor nerve (CN III) and the trochlear nerve (CN IV). The mesencephalon contains several major regions.

Cerebral peduncles (pĕ'dŭng-kl; *pedunculus* = little foot) are motor tracts located on the anterolateral surfaces of the mesencephalon. Somatic motor axons descend (project inferiorly) from the primary motor cortex, through these peduncles, to the spinal cord. In addition, the mesencephalon is the final destination of the **superior cerebellar peduncles** connecting the cerebellum to the mesencephalon.

The **tegmentum** (teg-men'tŭm; covering structure) is sandwiched between the substantia nigra (described in the next paragraph) and the periaqueductal gray matter. The tegmentum contains the pigmented **red nuclei** and the reticular formation (to be discussed in chapter 17). The reddish color of the nuclei is due to both blood vessel density and iron pigmentation in the neuronal cell bodies. The tegmentum integrates information from the cerebrum and cerebellum and issues involuntary motor commands to the erector spinae muscles of the back to help maintain posture while standing, bending at the waist, or walking.

The **substantia nigra** (sŭb-stan'shē-ă nī'gră; *niger* = black) consists of bilaterally symmetrical nuclei within the mesencephalon. It is best observed in cross section (figure 15.19). Its name derives from its almost black appearance, which is due to melanin pigmentation. The substantia nigra is squeezed between the cerebral peduncles and the tegmentum. The medial lemniscus (see the following section on the medulla oblongata) is a band of axons immediately posterior to the substantia nigra. The substantia nigra houses clusters of neurons that produce the neurotransmitter dopamine, which affects brain processes that control movement, emotional response, and ability to experience pleasure and pain. These neurons are dark-hued due to the melanin they contain. Degeneration of these cells in the substantia nigra is a pathology that underlies Parkinson disease (see Clinical View: In Depth, page 470).

The **tectum** (tek'tŭm; roof) is the posterior region of the mesencephalon dorsal to the cerebral aqueduct. It contains two pairs of sensory nuclei, the superior and inferior colliculi, which are collectively called the **tectal plate** (*quadrigeminal* (kwah'dri-jem'i-năl) *plate* or *corpora quadrigemina*) (see figure 15.18*b*). These nuclei are relay stations in the processing pathway of visual and auditory sensations. The **superior colliculi** (ko-lik'yū-lī; sing., *colliculus*; mound) are the superior nuclei. They are called "visual reflex centers" because they help visually track moving objects and control reflexes such as turning the eyes and head in response to a visual stimulus. For example, the superior colliculi are at work when you think you see a large animal running at you and turn suddenly toward the image. The paired **inferior colliculi** are the "auditory reflex centers," meaning that they control reflexive turning of the head and eyes in the direction of a sound. For example, the inferior colliculi are at work when you hear the loud "BANG!" of a car back-firing and turn suddenly toward the noise.

Pons

The **pons** (ponz; bridge) is a bulging region on the anterior part of the brainstem that forms from part of the metencephalon (**figure 15.20**; see figure 15.18). Housed within the pons are sensory and motor tracts that connect to the brain and spinal cord. In addition, the **middle cerebellar peduncles** are transverse groups of fibers that connect the pons to the cerebellum. The pons also houses two **autonomic respiratory centers:** the **pneumotaxic** (noo-mō-tăk'sik) **center** and the **apneustic** (ap-noo'stik) **center.** These centers regulate the rate and depth of breathing, and both of them influence

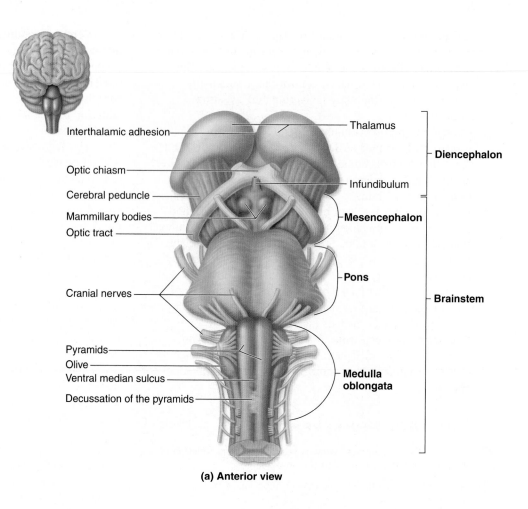

(a) Anterior view

Interthalamic adhesion

Optic chiasm

Cerebral peduncle

Mammillary bodies

Optic tract

Cranial nerves

Pyramids

Olive

Ventral median sulcus

Decussation of the pyramids

Thalamus

Diencephalon

Infundibulum

Mesencephalon

Pons

Brainstem

Medulla
oblongata

Figure 15.18

Brainstem. (a) Anterior and
(b) posterolateral views show the
locations of the mesencephalon, pons, and
medulla oblongata within the brainstem.

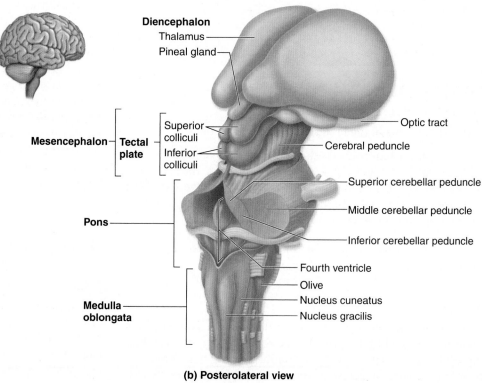

(b) Posterolateral view

Diencephalon
Thalamus
Pineal gland

Mesencephalon **Tectal
plate**

Superior
colliculi

Inferior
colliculi

Pons

**Medulla
oblongata**

Optic tract

Cerebral peduncle

Superior cerebellar peduncle

Middle cerebellar peduncle

Inferior cerebellar peduncle

Fourth ventricle

Olive

Nucleus cuneatus

Nucleus gracilis

Figure 15.19

Mesencephalon. Components of the mesencephalon are shown in cross-sectional view.

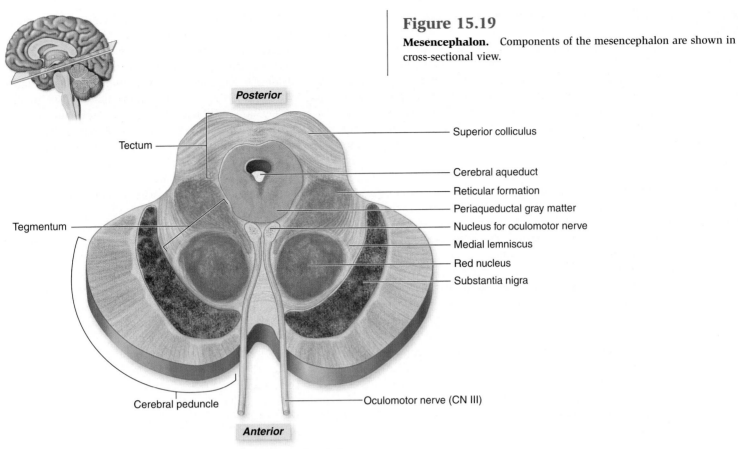

Mesencephalon, cross-sectional view

Labels:
Posterior
Tectum
Superior colliculus
Cerebral aqueduct
Reticular formation
Periaqueductal gray matter
Nucleus for oculomotor nerve
Tegmentum
Medial lemniscus
Red nucleus
Substantia nigra
Cerebral peduncle
Oculomotor nerve (CN III)
Anterior

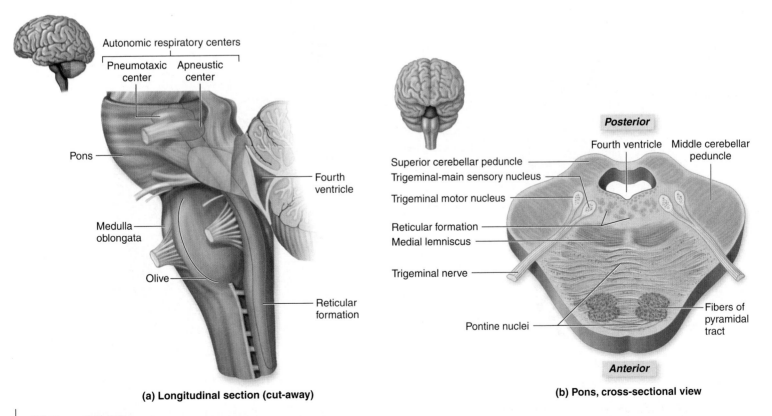

(a) Longitudinal section (cut-away)

Labels:
Autonomic respiratory centers
Pneumotaxic center
Apneustic center
Pons
Medulla oblongata
Olive
Fourth ventricle
Reticular formation

(b) Pons, cross-sectional view

Labels:
Posterior
Fourth ventricle
Middle cerebellar peduncle
Superior cerebellar peduncle
Trigeminal-main sensory nucleus
Trigeminal motor nucleus
Reticular formation
Medial lemniscus
Trigeminal nerve
Pontine nuclei
Fibers of pyramidal tract
Anterior

Figure 15.20

Pons. The pons is a bulge on the ventral side of the hindbrain that contains nerve tracts, nuclei, and part of the reticular formation. *(a)* A partially cut-away longitudinal section identifies two nuclei involved in respiration control, the pneumotaxic and apneustic centers. *(b)* A cross section through the pons shows the pontine nuclei, fiber tracts, and some cranial nerve nuclei.

and modify the activity of the respiratory center in the medulla oblongata.

The pons houses sensory and motor cranial nerve nuclei for the trigeminal (CN V), abducens (CN VI), and facial (CN VII) cranial nerves. Some of the nuclei for the vestibulocochlear cranial nerve (CN VIII) are located there. Additionally, nuclei called the **superior olivary complex** are located in the inferior pons. This nuclear complex receives auditory input and is involved in the pathway for sound localization.

Medulla Oblongata

The **medulla oblongata** (me-dool'ă ob-long-gah'tă; marrow or middle; *oblongus* = rather long), or simply the **medulla,** is formed from the myelencephalon. It is the most inferior part of the brainstem and is continuous with the spinal cord inferiorly. The posterior portion of the medulla resembles the spinal cord with its flattened, round shape and narrow central canal. As the central canal extends anteriorly toward the pons, it enlarges and becomes the fourth ventricle. All communication between the brain and spinal cord involves tracts that ascend or descend through the medulla oblongata **(figure 15.21**; see figures 15.18 and 15.20).

Several external landmarks are visible on the medulla oblongata. The anterior surface exhibits two longitudinal ridges called the **pyramids** (pir'ă-mid), which house the motor projection tracts called the corticospinal (pyramidal) tracts. In the posterior region of the medulla, most of these axons cross to the opposite side of

the brain at a point called the **decussation of the pyramids** (dē-kŭ-să'shŭn; *decussate* = to cross in the form of an **X**). As a result of the crossover, each cerebral hemisphere controls the voluntary movements of the opposite side of the body. Immediately lateral to each pyramid is a distinct bulge, called the **olive,** which contains a large fold of gray matter called the **inferior olivary nucleus.** The inferior olivary nuclei relay ascending sensory impulses, especially proprioceptive information, to the cerebellar cortex. Additionally, paired **inferior cerebellar peduncles** are tracts that connect the medulla oblongata to the cerebellum.

Within the medulla oblongata are additional nuclei that have various functions. The cranial nerve nuclei are associated with the vestibulocochlear (CN VIII), glossopharyngeal (CN IX), vagus (CN X), accessory (CN XI), and hypoglossal (CN XII) cranial nerves. In addition, the medulla oblongata contains the paired **nucleus cuneatus** (kyoo-nē-ā'tŭs; wedge) and the **nucleus gracilis** (gras-i'lis; slender), which relay somatic sensory information to the thalamus. The nucleus cuneatus receives posterior root fibers corresponding to sensory innervation from the arm and hand of the same side. The nucleus gracilis receives posterior root fibers carrying sensory information from the leg and lower limbs of the same side. Bands of myelinated fibers composing a **medial lemniscus** exit these nuclei and decussate in the inferior region of the medulla oblongata. The medial lemniscus projects through the brainstem to the ventral posterior nucleus of the thalamus.

Finally, the medulla oblongata contains several autonomic nuclei, which regulate functions vital for life. Autonomic nuclei

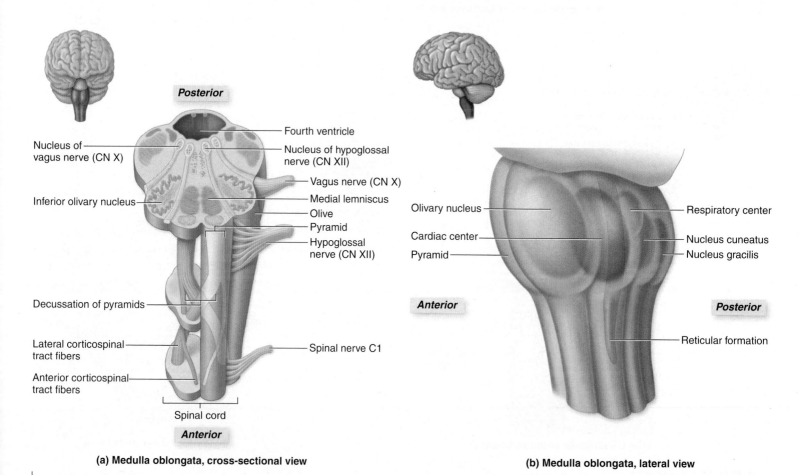

(a) Medulla oblongata, cross-sectional view

(b) Medulla oblongata, lateral view

Figure 15.21

Medulla Oblongata. The medulla oblongata connects the brain to the spinal cord. *(a)* A cross section illustrates important internal structures and decussations of the pyramids. *(b)* The medulla contains several nuclei that are involved in regulating the heart and respiratory rates and in receiving and sending sensory information about upper and lower limb movement.

group together to form centers in the medulla oblongata. Following are the most important autonomic centers in the medulla oblongata and their functions:

- The **cardiac center** regulates both the heart's rate and its strength of contraction.
- The **vasomotor center** controls blood pressure by regulating the contraction and relaxation of smooth muscle in the walls of the smallest arteries (called arterioles) to alter vessel diameter. Blood pressure increases when vessels are constricted and lowers when vessels are dilated.
- The **respiratory center** regulates the respiratory rate. It is influenced by the apneustic and pneumotaxic centers of the pons.
- Other nuclei in the medulla oblongata are involved in coughing, sneezing, salivating, swallowing, gagging, and vomiting.

WHAT DID YOU LEARN?

14 What part of the brain contains paired visual and auditory sensory nuclei?

15 What are the names of the autonomic respiratory centers in the pons?

WHAT DO YOU THINK?

5 Based on your understanding of the medulla oblongata's functions, would you expect severe injury to the medulla oblongata to cause death, or merely be disabling? Why?

Cerebellum

Key topics in this section:

- Structure and function of the cerebellum
- Relationship between the cerebellum and the brainstem

The **cerebellum** (ser-e-bel'ŭm; little brain) is the second largest part of the brain, and it develops from the metencephalon. The cerebellum has a complex, highly convoluted surface covered by a layer of cerebellar cortex. The folds of the cerebellar cortex are called **folia** (fō'lē-ă; *folium* = leaf) **(figure 15.22)**. The cerebellum is composed of left and right **cerebellar hemispheres**. Each hemisphere consists of two lobes, the **anterior lobe** and the **posterior lobe,** which are separated by the **primary fissure**. Along the midline, a narrow band of cortex known as the **vermis** (ver'mis; worm) separates the left and right cerebellar hemispheres (figure 15.22*b*). The vermis receives sensory input reporting torso position and balance. Its output to the vestibular nucleus (see chapter 17) helps maintain balance. Slender **flocculonodular** (flok'ū-lō-nod'ū-lăr; *flocculo* = wool-like tuft) **lobes** lie anterior and inferior to each cerebellar hemisphere (not shown).

The cerebellum is partitioned internally into three regions: an outer gray matter layer of cortex, an internal region of white matter, and the deepest gray matter layer, which is composed of cerebellar nuclei. The white matter of the cerebellum is called the **arbor vitae** (ar'bōr vī'tē; tree of life) because its distribution pattern resembles the branches of a tree.

The cerebellum coordinates and "fine-tunes" skeletal muscle movements and ensures that skeletal muscle contraction follows

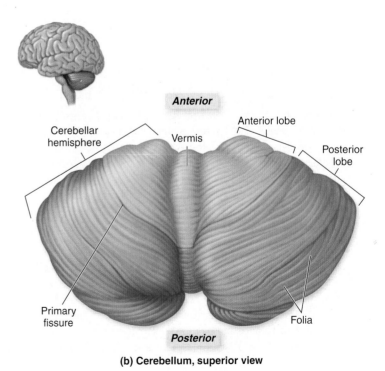

(a) Midsagittal section

Cerebral aqueduct
Pineal gland
Tectal plate
White matter (arbor vitae)
Mammillary body
Mesencephalon
Fourth ventricle
Pons
Medulla oblongata
Folia
Gray matter

Anterior
Anterior lobe
Cerebellar hemisphere
Vermis
Posterior lobe
Primary fissure
Folia
Posterior

(b) Cerebellum, superior view

Figure 15.22

Cerebellum. The cerebellum lies dorsal to the pons and medulla oblongata of the brainstem. *(a)* A midsagittal section shows the relationship of the cerebellum to the brainstem. *(b)* A superior view compares the anterior and posterior lobes of the cerebellum. (Note: The cerebellum has been removed.)

the correct pattern leading to smooth, coordinated movements. The cerebellum stores memories of previously learned movement patterns. This function is performed indirectly, by regulating activity along both the voluntary and involuntary motor pathways at the cerebral cortex, cerebral nuclei, and motor centers in the brainstem. The cerebrum initiates a movement and sends a "rough draft" of the movement to the cerebellum, which then coordinates and fine-tunes it. For example, the controlled, precise movements a pianist makes when playing a concerto are due to fine-tuning by the cerebellum. Without the cerebellum, the pianist's movements would be choppy and sloppy, as in banging an entire hand across the keyboard.

In addition, the cerebellum has several other functions. It adjusts skeletal muscle activity to maintain equilibrium and posture. It also receives proprioceptive (sensory) information from the muscles and joints and uses this information to regulate the body's position. For example, you are able to balance on one foot because the cerebellum takes the proprioceptive information from the body joints and "maps out" a muscle tone plan to keep the body upright. The specifics of how this proprioceptive information travels are covered in detail in chapter 19. Finally, because proprioceptive information from the body's muscles and joints is sent to the cerebellum, the cerebrum knows the position of each body joint and its muscle tone, even if the person is not looking at the joint. For example, if you close your eyes, you are still aware of which body joints are flexed and which are extended because the cerebrum gives you this awareness.

Cerebellar Peduncles

Three thick tracts, called **peduncles**, link the cerebellum with the brainstem (see figure 15.18*b*). The **superior cerebellar peduncles**

connect the cerebellum to the mesencephalon. The **middle cerebellar peduncles** connect the pons to the cerebellum. The **inferior cerebellar peduncles** connect the cerebellum to the medulla oblongata. It is these extensive communications that enable the cerebellum to "fine-tune" skeletal muscle movements and interpret all body proprioceptive movement.

WHAT DID YOU LEARN?

16 What part of the brain contains flocculonodular lobes, folia, and a vermis?

17 What name is given to a thick tract linking the brainstem and cerebellum?

Limbic System

Key topic in this section:

■ Structures and functions of the limbic system

The brain has two important systems that work together for a common function, even though their structures are scattered throughout the brain. These systems are the limbic system and the reticular formation, a loosely organized gray matter core in the brainstem. We will discuss the limbic system here; the reticular formation will be described in chapter 17.

The **limbic** (lim′bik) **system** is composed of multiple cerebral and diencephalic structures that collaboratively process and experience emotions. It is a collective name for the human brain structures that are involved in motivation, emotion, and memory with an emotional association. The limbic system affects memory formation by integrating past memories of physical sensations with emotional states.

The structures of the limbic system form a ring or border around the diencephalon (*limbus* = border). Although neuroanatomists continue to debate the components of the limbic system, the brain structures commonly recognized are shown in **figure 15.23** and listed here:

1. The **cingulate** (sin′gū-lāt; *cingulum* = girdle, to surround) **gyrus** is an internal mass of cerebral cortex located within the longitudinal fissure and superior to the corpus callosum. This mass of tissue may be seen only in sagittal section, and it surrounds the diencephalon. It receives input from the other components of the limbic system.
2. The **parahippocampal gyrus** is a mass of cortical tissue in the temporal lobe. Its function is associated with the hippocampus (described next).
3. The **hippocampus** (hip-ō-kam′pŭs; seahorse) is a nucleus located superior to the parahippocampal gyrus that connects to the diencephalon via a structure called the fornix. As its name implies, this nucleus is shaped like a seahorse. Both the hippocampus and the parahippocampal gyrus are essential in storing memories and forming long-term memory.
4. The **amygdaloid body** connects to the hippocampus. The amygdaloid body is involved in several aspects of emotion, especially fear. It can also help store and code memories based on how a person emotionally perceives them—for example, as related to fear, extreme happiness, or sadness.

CLINICAL VIEW

Effects of Alcohol and Drugs on the Cerebellum

Disorders of the cerebellum are frequently characterized by impaired skeletal muscle function. Typical symptoms include uncoordinated, jerky movements, a condition termed **ataxia** (ă-tak′sē-ă; *a* = without, *taxis* = order), or loss of equilibrium that often presents as uncoordinated walking. A variety of drugs, especially alcohol, can temporarily, and in some cases permanently, impair cerebellar function. For example, drinking too much alcohol leads to the following symptoms of impaired cerebellar function, which are used in the classic sobriety tests performed by police officers:

■ **Disturbance of gait.** A person under the influence of alcohol rarely walks in a straight line, but appears to sway and stagger. In addition, falling and bumping into objects are likely, due to the temporary cerebellar disturbance.

■ **Loss of balance and posture.** When attempting to stand on one foot, a person who is intoxicated usually tips and falls over.

■ **Inability to detect proprioceptive information.** When asked to close the eyes and touch the nose, an intoxicated person frequently misses the mark. This reaction is due to reduced ability to sense proprioceptive information, compounded by uncoordination of skeletal muscles.

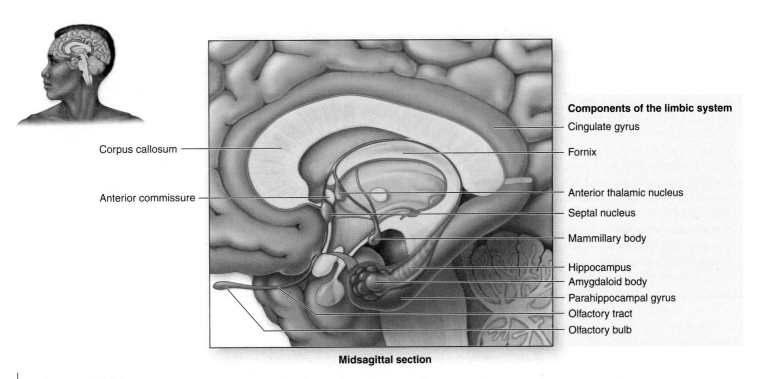

Corpus callosum

Anterior commissure

Components of the limbic system

Cingulate gyrus

Fornix

Anterior thalamic nucleus

Septal nucleus

Mammillary body

Hippocampus

Amygdaloid body

Parahippocampal gyrus

Olfactory tract

Olfactory bulb

Midsagittal section

Figure 15.23

Limbic System.   The components of the limbic system are shown here in midsagittal section with three-dimensional reconstruction. The limbic system affects behavior and emotions.

CLINICAL VIEW

Frontal Lobotomy

Years ago, when no medical treatments were available for the severely mentally ill, the **frontal lobotomy** was touted as a "cure" for people who were violent or profoundly disturbed. Although techniques varied, the basic procedure involved introducing a cutting instrument into the frontal cortex, often through a small hole drilled in the skull in the region of the medial canthus of each eye. The instrument, generally a long, spatula-like blade, was then moved back and forth, severing the frontal cortical connections from the rest of the brain. Surgeons performed thousands of lobotomies from the late 1930s until the early 1950s. In Japan, for instance, the procedure was even performed on children who had simply done poorly in school. In the United States, the procedure was offered to prisoners in exchange for early parole.

In the late 1940s, independent studies showed that the mental conditions of only about one-third of the patients actually improved due to lobotomy, while the remaining two-thirds stayed the same or actually became worse. Also at this time, medications were developed to treat depression and other serious psychiatric problems, obviating the need for such a drastic measure. Thus, the lobotomy passed into medical history in the 1950s, and many states and foreign countries have since passed laws forbidding its use. To add to the irony surrounding the lobotomy,

Dr. Moniz, the father of the procedure, was shot by one of his patients who had undergone the surgery.

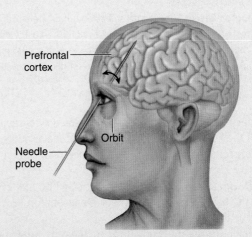

Prefrontal cortex

Orbit

Needle probe

In one type of frontal lobotomy, the cutting instrument is inserted in the medial canthus of each eye and through the thin superior border of the orbit.

CLINICAL VIEW: In Depth

Brain Disorders

A brain disorder may be characterized by a malfunction in sensory gathering or motor expression or by some combination of both activities. Disturbances of the brain include headache, cerebral palsy, encephalitis, epilepsy, Huntington disease, and Parkinson disease.

Headache can occur even though the brain itself is pain-insensitive, due to pressure produced by tumors, hemorrhage, meningitis, or inflamed nerve roots. More typical causes are emotional stress, increased blood pressure, and food allergies, all of which cause blood vessel diameter changes. **Migraine headaches** are severe, recurring headaches that usually affect only one side of the head. Headaches are not a brain disorder, but they may accompany other diseases or brain disorders.

Cerebral palsy (pawl′zē) is actually a group of neuromuscular disorders that usually result from damage to an infant's brain before, during, or immediately after birth. Three forms of cerebral palsy involve impairment of skeletal motor activity to some degree: *athetoid,* characterized by slow, involuntary, writhing hand movements; *ataxic,* marked by lack of muscular coordination; and *spastic,* exhibiting increased muscular tone. Mental retardation and speech difficulties may accompany this disorder.

Encephalitis (en-sef-ă-lī′tis; *enkephalos* = brain, *itis* = inflammation) is an acute inflammatory disease of the brain, most often due to viral infection. Symptoms include drowsiness, fever, headache, neck pain, coma, and paralysis. Death may occur.

Epilepsy (ep′i-lep′sē; *epilepsia* = seizure) is characterized by recurring attacks of motor, sensory, or psychological malfunction, with or without unconsciousness or convulsive movements. During an epileptic seizure, neurons in the brain fire at unpredictable times, even without a stimulus. The term epilepsy does not apply to a specific disease; rather, epilepsy refers to a group of symptoms with many causes. Some epileptic events may be *grand mal* seizures, which affect motor areas of the brain and cause severe spasms and loss of consciousness. Others may be *petit mal* seizures, which affect sensory areas and do not lead to convulsions or prolonged unconsciousness.

Huntington disease is an autosomal dominant hereditary disease that affects the cerebral nuclei. It causes rapid, jerky, involuntary movements that usually start unilaterally in the face, but over months and years progress to the arms and legs. Progressive intellectual deterioration also occurs, including personality changes, memory loss, and irritability. The disease has an onset age of 35–40, and is fatal within 10–20 years.

Parkinson disease is a slow-progressing neurologic condition that affects muscle movement and balance. Parkinson patients exhibit stiff posture, an expressionless face, slow voluntary movements, a resting tremor (especially in the hands), and a shuffling gait. The disease is caused by a deficiency of the neurotransmitter dopamine, which results from decreased dopamine production by degenerating neurons in the substantia nigra. Dopamine deficiency prevents brain cells from performing their usual inhibitory functions within the cerebral nuclei. By the time symptoms develop, the person has lost 80–90% of the cells responsible for producing dopamine.

The causes of Parkinson disease include medication reactions, effects of certain illicit drugs, and genetics. However, most cases occur with no obvious cause and are termed *idiopathic.* Current treatments rely on medications that enhance the amount of dopamine in the remaining cells of the substantia nigra. Although this approach addresses the decreased levels of the neurotransmitter, it does nothing to replace the dead cells. On rare occasions, surgery is done to remove adjacent parts of the brain that are responsible for the jerky movements, but this is a last-ditch effort to alleviate the troubling symptoms.

Boxer Muhammad Ali and actor Michael J. Fox are two famous Parkinson disease patients.

5. The **olfactory bulbs, olfactory tracts,** and **olfactory cortex** are part of the limbic system as well, since particular odors can provoke certain emotions or be associated with certain memories.

6. The **fornix** (fōr′niks; arch) is a thin tract of white matter that connects the hippocampus with other diencephalon limbic system structures.

7. Various nuclei in the diencephalon, such as the **anterior thalamic nuclei,** the **habenular nuclei,** the **septal nuclei,** and the **mammillary** (mam′i-lār-ē; *mammilla* = nipple) **bodies,** interconnect other parts of the limbic system and contribute to its overall function.

WHAT DID YOU LEARN?

18 Describe how the hippocampus and the olfactory structures participate in limbic system function.

Cranial Nerves

Key topics in this section:

- Names and locations of the 12 cranial nerves
- Principal functions of each cranial nerve pair

Cranial nerves are part of the peripheral nervous system and originate on the inferior surface of the brain. There are 12 pairs of cranial nerves. They are numbered according to their positions, beginning with the most anteriorly placed nerve and using Roman numerals, sometimes preceded by the prefix *CN* (**figure 15.24**). The name of each nerve generally has some relation to its function. Thus, the 12 pairs of cranial nerves are the olfactory (CN I), optic (CN II), oculomotor (CN III), trochlear (CN IV), trigeminal (CN V), abducens (CN VI), facial (CN VII), vestibulocochlear (CN VIII), glossopharyngeal (CN IX), vagus (CN X), accessory (CN XI), and hypoglossal (CN XII).

Study Tip!

Developing a code or phrase called a **mnemonic** (nē-mon'ik; *mnemoni-kos* = pertaining to memory) may help you remember the cranial nerves. Mnemonics you devise yourself will be the most relevant to you, but here is a sample mnemonic for the cranial nerves:

Oh	(olfactory)
once	(optic)
one	(oculomotor)
takes	(trochlear)
the	(trigeminal)
anatomy	(abducens)
final	(facial)
very	(vestibulocochlear)
good	(glossopharyngeal)
vacations	(vagus)
are	(accessory)
heavenly!	(hypoglossal)

Each cranial nerve is composed of many axons. Some cranial nerves (e.g., CN XII, hypoglossal nerve) are composed of motor axons only, while other cranial nerves (e.g., CN II, optic nerve) are composed of sensory axons only. Still other cranial nerves (e.g., CN V, trigeminal nerve) are composed of both sensory and motor axons. **Tables 15.7** and **15.8** list whether a cranial nerve has somatic motor, parasympathetic motor, and/or sensory components.

Table 15.7 summarizes the main sensory and motor functions of each cranial nerve. For easier reference, each main function of a nerve is color-coded. Pink represents a sensory function, and blue stands for a somatic motor function; green denotes a parasympathetic motor function (see chapter 18). Table 15.8 lists the individual cranial nerves and discusses their functions, origins, and pathways. The color-coding in table 15.7 carries over to table 15.8, so you can easily determine whether a cranial nerve has sensory and/or motor components.

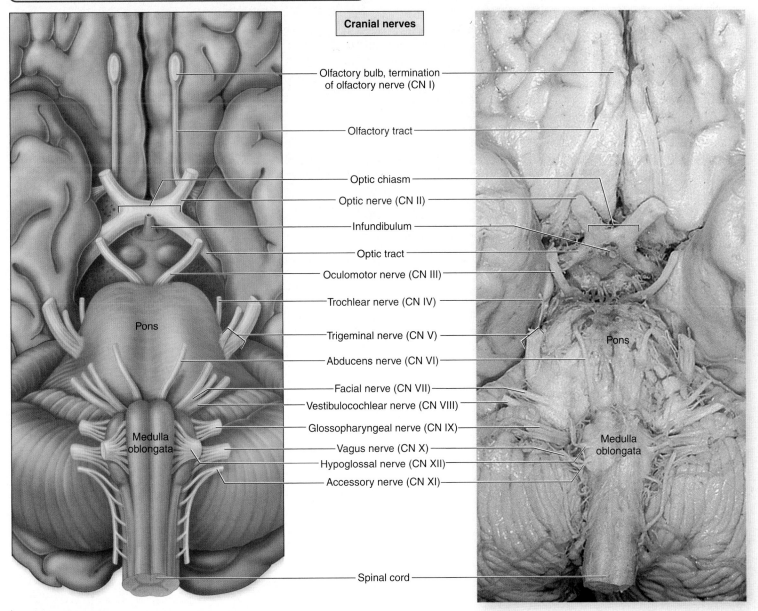

Cranial nerves

- Olfactory bulb, termination of olfactory nerve (CN I)
- Olfactory tract
- Optic chiasm
- Optic nerve (CN II)
- Infundibulum
- Optic tract
- Oculomotor nerve (CN III)
- Trochlear nerve (CN IV)
- Trigeminal nerve (CN V)
- Abducens nerve (CN VI)
- Facial nerve (CN VII)
- Vestibulocochlear nerve (CN VIII)
- Glossopharyngeal nerve (CN IX)
- Vagus nerve (CN X)
- Hypoglossal nerve (CN XII)
- Accessory nerve (CN XI)
- Spinal cord

Pons

Medulla oblongata

Pons

Medulla oblongata

Figure 15.24

Cranial Nerves. A view of the inferior surface of the brain shows the 12 pairs of cranial nerves.

Table 15.7	Primary Functions of Cranial Nerves		
Cranial Nerve	**Sensory Function**	**Somatic Motor Function**	**Parasympathetic Motor (Autonomic) Function**[1]
I (olfactory)	Olfaction (smell)	*None*	*None*
II (optic)	Vision	*None*	*None*
III (oculomotor)	*None*[2]	4 extrinsic eye muscles (medial rectus, superior rectus, inferior rectus, inferior oblique); levator palpebrae superioris muscle (elevates eyelid)	Innervates sphincter pupillae muscle in eye to make pupil constrict; contracts ciliary muscles to make lens of eye more rounded (as needed for near vision)
IV (trochlear)	*None*[2]	Superior oblique eye muscle	*None*
V (trigeminal)	General sensory from anterior scalp, nasal cavity, entire face, most of oral cavity, teeth, anterior two-thirds of tongue; part of auricle of ear	Muscles of mastication, mylohyoid, digastric (anterior belly), tensor tympani, tensor veli palatini	*None*
VI (abducens)	*None*[2]	Lateral rectus eye muscle	*None*
VII (facial)	Taste from anterior two-thirds of tongue	Muscles of facial expression, digastric (posterior belly), stylohyoid, stapedius	Increases secretion from lacrimal gland of eye, submandibular and sublingual salivary glands
VIII (vestibulocochlear)	Hearing (cochlear branch); equilibrium (vestibular branch)	*None*[3]	*None*
IX (glossopharyngeal)	Touch and taste to posterior one-third of tongue, visceral sensory from carotid bodies	One pharyngeal muscle (stylopharyngeus)	Increases secretion from parotid salivary gland
X (vagus)	Visceral sensory information from pharynx, larynx, carotid bodies, heart, lungs, most abdominal organs General sensory information from external acoustic meatus, eardrum, and pharynx	Most pharyngeal muscles; laryngeal muscles	Innervates smooth muscle and glands of heart, lungs, larynx, trachea, most abdominal organs
XI (accessory)	*None*	Trapezius muscle, sternocleidomastoid muscle	*None*
XII (hypoglossal)	*None*	Intrinsic tongue muscles and extrinsic tongue muscles	*None*

[1] The autonomic nervous system contains a parasympathetic division and a sympathetic division. Some cranial nerves carry parasympathetic axons and are listed in this table. Detailed information about these divisions is found in chapter 18.

[2] These nerves do contain some tiny proprioceptive sensory axons in the muscles, but in general, these nerves tend to be described as motor only.

[3] A few motor axons travel with this nerve to the inner ear, but they are not considered a significant component of the nerve.

Table 15.8	**Cranial Nerves**

CN I OLFACTORY NERVE (ol-fak'tŏ-rē; *olfacio* = to smell)

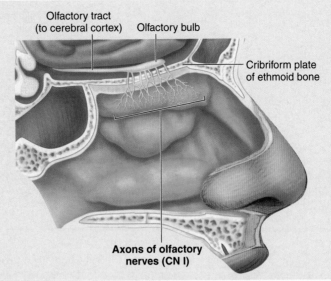

Olfactory tract (to cerebral cortex)

Olfactory bulb

Cribriform plate of ethmoid bone

Axons of olfactory nerves (CN I)

Description	Conducts olfactory (smell) sensations to brain; only type of nervous tissue to regenerate.
Sensory function	Olfaction (smell)
Origin	Receptors (bipolar neurons) in olfactory mucosa of nasal cavity
Pathway	Travels through the cribriform foramina of ethmoid bone and synapses in the olfactory bulbs, which are located in the anterior cranial fossa
Conditions caused by nerve damage	Anosmia (partial or total loss of smell)

CN II OPTIC NERVE (op'tik; *ops* = eye)

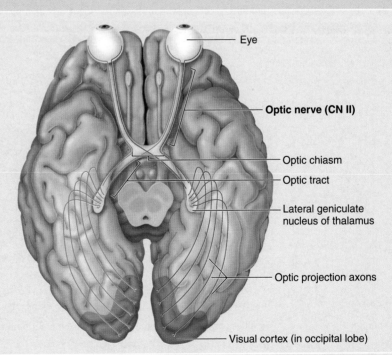

Eye

Optic nerve (CN II)

Optic chiasm

Optic tract

Lateral geniculate nucleus of thalamus

Optic projection axons

Visual cortex (in occipital lobe)

Description	Special sensory nerve of vision that is an outgrowth of the brain; more appropriately called a brain tract
Sensory function	Vision
Origin	Retina of the eye
Pathway	Enters cranium via optic canal of sphenoid bone; left and right optic nerves unite at optic chiasm; optic tract travels to lateral geniculate nucleus of thalamus; finally, information is forwarded to the occipital lobe
Conditions caused by nerve damage	Anopsia (visual defects)

(continued on next page)

Table 15.8 Cranial Nerves (continued)

CN III OCULOMOTOR NERVE (ok'ū-lō-mō'tŏr; *oculus* = eye, *motorius* = moving)

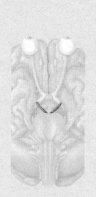

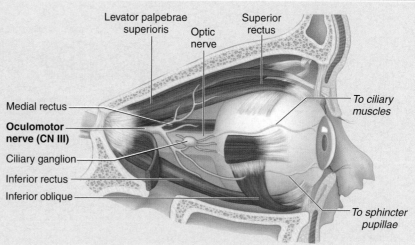

Description	Innervates upper eyelid muscle and four of the six extrinsic eye muscles
Somatic motor function	Supplies four extrinsic eye muscles (superior rectus, inferior rectus, medial rectus, inferior oblique) that move eye Supplies levator palpebrae superioris muscle to elevate eyelid
Parasympathetic motor function	Innervates sphincter pupillae muscle of iris to make pupil constrict Contracts ciliary muscles to make lens of eye more spherical (as needed for near vision)
Origin	Oculomotor and Edinger Westphal nuclei within mesencephalon
Pathway	Leaves cranium via superior orbital fissure and travels to eye and eyelid. (Parasympathetic fibers travel to ciliary ganglion, and postganglionic parasympathetic fibers then travel to iris and ciliary muscle.)
Conditions caused by nerve damage	Ptosis (upper eyelid droop); paralysis of eye muscles, leading to strabismus (eyes not in parallel/deviated improperly), diplopia (double vision), focusing difficulty

CN IV TROCHLEAR NERVE (trok'lē-ar; *trochlea* = a pulley)

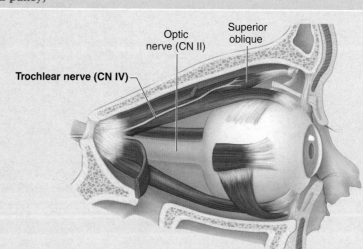

Description	Innervates one extrinsic eye muscle (superior oblique) that loops through a pulley-shaped ligament
Somatic motor function	Supplies one extrinsic eye muscle (superior oblique) to move eye inferiorly and laterally
Origin	Trochlear nucleus within mesencephalon
Pathway	Leaves cranium via superior orbital fissure and travels to superior oblique muscle
Conditions caused by nerve damage	Paralysis of superior oblique, leading to strabismus (eyes not in parallel/deviated improperly), diplopia (double vision)

Table 15.8	Cranial Nerves *(continued)*

CN V TRIGEMINAL NERVE (trī-jem′i-năl; *trigeminous* = threefold)

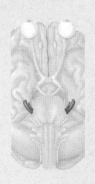

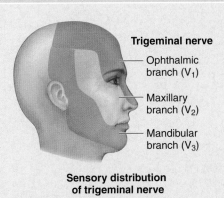

Sensory distribution of trigeminal nerve

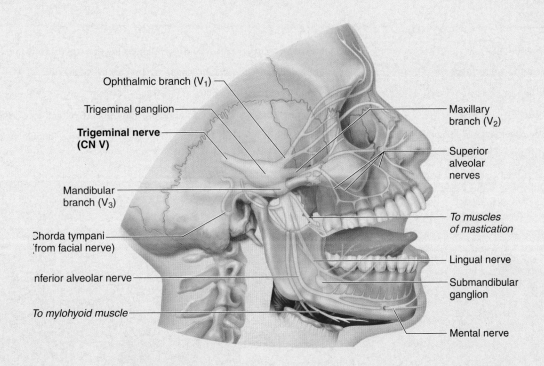

Description	This nerve consists of three divisions: ophthalmic (V_1), maxillary (V_2), and mandibular (V_3); receives sensory impulses from face, oral cavity, nasal cavity, and anterior scalp, and innervates muscles of mastication
Sensory function	Sensory stimuli for this nerve are touch, temperature, and pain. V_1: Conducts sensory impulses from cornea, nose, forehead, anterior scalp V_2: Conducts sensory impulses from nasal mucosa, palate, gums, cheek V_3: Conducts sensory impulses from anterior two-thirds of tongue, skin of chin, lower jaw, lower teeth; one-third from sensory fibers of auricle of ear
Somatic motor function	Innervates muscles of mastication (temporalis, masseter, lateral and medial pterygoids), mylohyoid, anterior belly of digastric, tensor tympani muscle, and tensor veli palatini
Origin	Nucleus in the pons
Pathway	V_1: Sensory fibers enter cranium via superior orbital fissure and travel to trigeminal ganglion before entering pons. V_2: Sensory fibers enter cranium via foramen rotundum and travel to trigeminal ganglion before entering pons. V_3: Motor fibers leave pons and exit cranium via foramen ovale to supply muscles. Sensory fibers travel through foramen ovale to trigeminal ganglion before entering pons.
Conditions caused by nerve damage	Trigeminal neuralgia (tic douloureux) is caused by inflammation of the sensory components of the trigeminal nerve and results in intense, pulsating pain lasting from minutes to several hours

(continued on next page)

Table 15.8	**Cranial Nerves** *(continued)*

CN VI ABDUCENS NERVE (ab-doo'senz; to move away from)

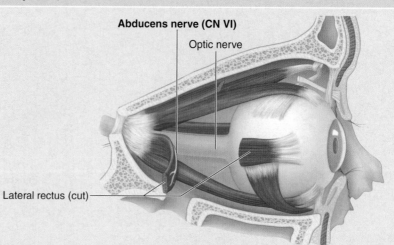

Description	Innervates lateral rectus eye muscle, which abducts the eye ("pulls away laterally")
Somatic motor function	Innervates one extrinsic eye muscle (lateral rectus) for eye abduction
Origin	Pontine (abducens) nucleus in pons
Pathway	Leaves cranium through superior orbital fissure and travels to lateral rectus muscle
Conditions caused by nerve damage	Paralysis of lateral rectus limits lateral movement of eye; diplopia (double vision)

CN VII FACIAL NERVE (fā'shăl; *fascialis* = of the face)

Description	Innervates muscles of facial expression, lacrimal (tear) gland, and most salivary glands; conducts taste sensations from anterior two-thirds of tongue
Sensory function	Taste from anterior two-thirds of tongue
Somatic motor function	The five major motor branches (temporal, zygomatic, buccal, mandibular, and cervical) innervate the muscles of facial expression, the posterior belly of the digastric muscle, and the stylohyoid and stapedius muscles.
Parasympathetic motor function	Increases secretions of the lacrimal gland of the eye as well as the submandibular and sublingual salivary glands
Origin	Nuclei within the pons
Pathway	Sensory fibers travel from the tongue via the chorda tympani branch of the facial nerve through a tiny foramen to enter the skull, and fibers synapse at the geniculate ganglion of the facial nerve. Somatic motor fibers leave the pons and enter the temporal bone through the internal acoustic meatus, project through temporal bone, and emerge through stylomastoid foramen to supply the musculature. Parasympathetic motor fibers leave the pons, enter the internal acoustic meatus, leave with either the greater petrosal nerve or chorda tympani nerve, and travel to an autonomic ganglion before innervating their respective glands.
Conditions caused by nerve damage	Decreased tearing (dry eye) and decreased salivation (dry mouth); loss of taste sensation to anterior two-thirds of tongue and/or facial nerve palsy (sometimes called Bell palsy) characterized by paralyzed facial muscles, eyelid droop, sagging at corner of mouth

Table 15.8   Cranial Nerves (continued)

CN VIII VESTIBULOCOCHLEAR NERVE (ves-tib'ū-lō-kok'lē-ăr; relating to the vestibule and cochlea of the ear)

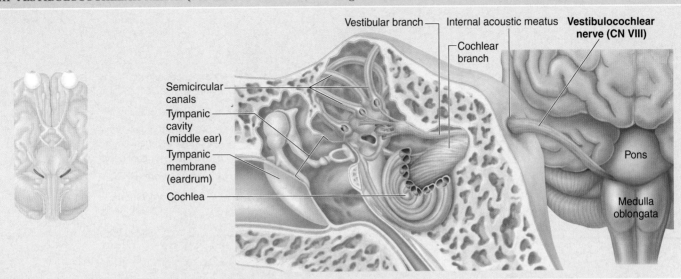

Description	Conducts equilibrium and auditory sensations to brain; formerly called the auditory nerve or acoustic nerve
Sensory function	Vestibular branch conducts impulses for equilibrium while cochlear branch conducts impulses for hearing
Origin	Vestibular branch: Hair cells in the vestibule of the inner ear Cochlear branch: Cochlea of the inner ear
Pathway	Sensory cell bodies of the vestibular branch are located in the vestibular ganglion, while sensory cell bodies of the cochlear branch are located in the spiral ganglion near the cochlea. The vestibular and cochlear branches merge, and together enter cranial cavity through internal acoustic meatus and travel to junction of the pons and the medulla oblongata
Conditions caused by nerve damage	Lesions in vestibular branch produce loss of balance, nausea, vomiting, and dizziness; lesions in cochlear branch result in deafness (loss of hearing)

CN IX GLOSSOPHARYNGEAL NERVE (glos'ō-fǎ-rin'jē-ǎl; *glossa* = tongue)

Description	Receives taste and touch sensations from posterior tongue structures, innervates one pharynx muscle and the parotid salivary gland
Sensory function	General sensation and taste to posterior one-third of tongue; chemoreceptor fibers to carotid bodies (structures on the carotid arteries that detect and monitor O_2 and CO_2 levels in the blood)
Somatic motor function	Innervates stylopharyngeus (pharynx muscle)
Parasympathetic motor function	Increases secretion of parotid salivary gland
Origin	Sensory fibers originate on taste buds and mucosa of posterior one-third of tongue, as well as the carotid bodies. Motor fibers originate in nuclei in the medulla oblongata.
Pathway	Sensory fibers travel from posterior one-third of tongue and carotid bodies along nerve through the inferior or superior ganglion into the jugular foramen, and travel to pons. Somatic motor fibers leave cranium via jugular foramen and travel to stylopharyngeus. Parasympathetic motor fibers travel to otic ganglion and then to parotid gland.
Conditions caused by nerve damage	Reduced salivary secretion (dry mouth); loss of taste sensations to posterior one-third of tongue

Table 15.8	Cranial Nerves *(continued)*

CN X VAGUS NERVE (vā′gŭs; wandering)

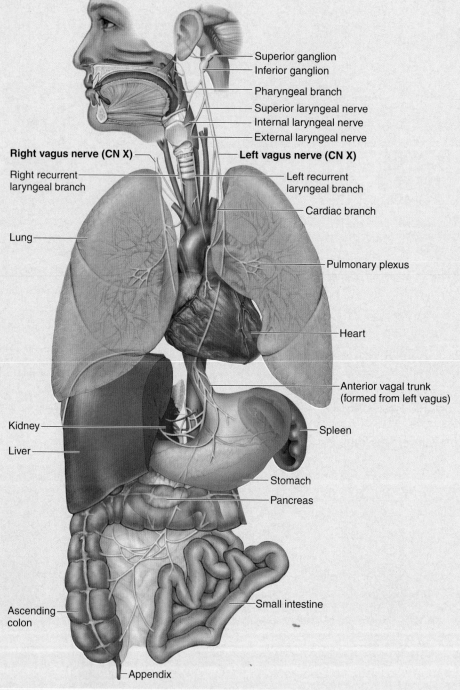

Superior ganglion
Inferior ganglion
Pharyngeal branch
Superior laryngeal nerve
Internal laryngeal nerve
External laryngeal nerve

Right vagus nerve (CN X)
Left vagus nerve (CN X)

Right recurrent laryngeal branch
Left recurrent laryngeal branch

Cardiac branch

Lung

Pulmonary plexus

Heart

Anterior vagal trunk (formed from left vagus)

Kidney
Spleen

Liver

Stomach
Pancreas

Ascending colon

Small intestine

Appendix

Description	Innervates structures in the head and neck and in the thoracic and abdominal cavities
Sensory function	Visceral sensory information from pharynx, larynx, heart, lungs, and most abdominal organs. General sensory information from external acoustic meatus, eardrum, and pharynx
Somatic motor function	Innervates most pharynx muscles and larynx muscles
Parasympathetic motor function	Innervates visceral smooth muscle, cardiac muscle, and glands of heart, lungs, pharynx, larynx, trachea, and most abdominal organs
Origin	Motor nuclei in medulla oblongata
Pathway	Leaves cranium via jugular foramen before traveling and branching extensively in neck, thorax, and abdomen; sensory neuron cell bodies are located in the superior and inferior ganglia associated with the nerve
Conditions caused by nerve damage	Paralysis leads to a variety of larynx problems, including hoarseness, monotone, or complete loss of voice. Other lesions may cause difficulty in swallowing or impaired gastrointestinal system mobility

Table 15.8	Cranial Nerves *(continued)*

CN XI ACCESSORY NERVE (ak-ses′ō-rē; *accedo* = to move toward)

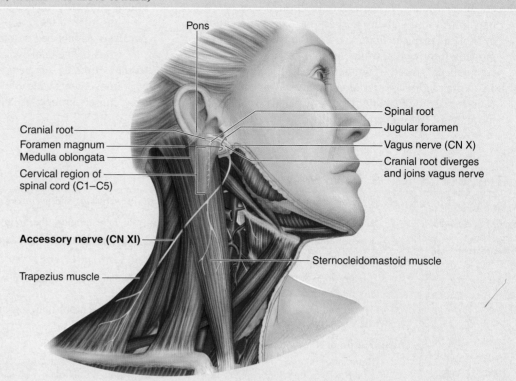

Description	Innervates trapezius, sternocleidomastoid, and some pharynx muscles; formerly called the "spinal accessory nerve"
Somatic motor function	Cranial root: Travels with CN X fibers to pharynx Spinal root: Innervates trapezius and sternocleidomastoid
Origin	Cranial root: Motor nuclei in medulla oblongata Spinal root: Motor nuclei in spinal cord
Pathway	Spinal root travels superiorly to enter skull through foramen magnum; there, cranial and spinal roots merge and leave the skull via jugular foramen. Once outside the skull, cranial root splits to travel with CN X (vagus), and spinal root travels to sternocleidomastoid and trapezius.
Conditions caused by nerve damage	Paralysis of trapezius and sternocleidomastoid, resulting in difficulty in elevating shoulder (trapezius function) or turning head to opposite site (sternocleidomastoid function).

CN XII HYPOGLOSSAL NERVE (hī-pō-glos′ăl; *hypo* = below, *glossus* = tongue)

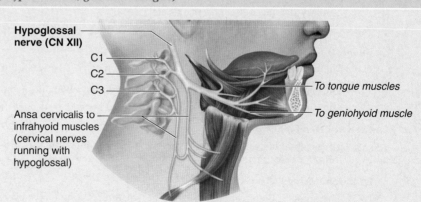

Description	Innervates intrinsic and extrinsic tongue muscles; name means "under the tongue"
Somatic motor function	Innervates intrinsic and extrinsic tongue muscles
Origin	Hypoglossal nucleus in medulla oblongata
Pathway	Leaves cranium via hypoglossal canal; travels inferior to mandible and to inferior surface of tongue
Conditions caused by nerve damage	Swallowing and speech difficulties due to impaired tongue movement; if a single hypoglossal nerve (either left or right) is paralyzed, a protruded (stuck out) tongue deviates to the side of the damaged nerve.

CLINICAL TERMS

aphasia (ă-fā'zē-ă; *a* = without, *phasis* = speech) Loss or impairment in the ability to speak or understand written or spoken language; *nonfluent aphasia* is the inability to properly form words, while *fluent aphasia* is the inability to comprehend speech or written language.

apraxia (ă-prak'sē-ă) Disorder of voluntary movement characterized by inability to perform skilled or purposeful movements, but no paralysis.

encephalopathy (en-sef'ă-lop'ă-thē; *pathos* = suffering) Any disorder of the brain.

paralysis Diminished or total loss of voluntary movement in a muscle due to injury or disease in either the muscle or its nerve.

tremor (trem'er, -ŏr; *tremo* = to tremble) Repetitive oscillatory movements caused by irregular contraction of opposing muscle groups; usually involuntary.

CHAPTER SUMMARY

	▪ The brain is a complex organ that controls all of our activities and performs numerous functions simultaneously.
Brain Development and Tissue Organization 438	▪ The brain has four major regions: the cerebrum, the diencephalon, the brainstem, and the cerebellum.
	▪ Directional terms in brain anatomy are anterior, meaning toward the nose, and posterior, meaning toward the tail.

Embryonic Development of the Brain   439

- ▪ Three primary vesicles (prosencephalon, mesencephalon, and rhombencephalon) form from the neural tube by the late fourth week of development.
- ▪ Five secondary vesicles (telencephalon, diencephalon, mesencephalon, metencephalon, and myelencephalon) form from the primary vesicles by the fifth week of development.

Organization of Neural Tissue Areas in the Brain   444

- ▪ Gray matter is made up of neuron and interneuron cell bodies, dendrites, telodendria, and unmyelinated axons; white matter derives its white color from the myelin in myelinated axons.

Support and Protection of the Brain 446	▪ The brain is protected and isolated by the cranium, the cranial meninges, the cerebrospinal fluid, and a blood-brain barrier.

Cranial Meninges   446

- ▪ The cranial meninges are the pia mater, arachnoid, and dura mater.
- ▪ Dura mater septa project between the major parts of the brain to stabilize its position within the cranium. These septa are the falx cerebri, tentorium cerebelli, falx cerebelli, and diaphragma sellae.

Brain Ventricles   448

- ▪ Fluid-filled spaces in the brain are the paired lateral ventricles, the third ventricle, cerebral aqueduct, and the fourth ventricle.

Cerebrospinal Fluid   448

- ▪ CSF is a clear, colorless fluid that provides buoyancy, protection, and a stable environment for the brain and spinal cord.
- ▪ The choroid plexus produces CSF in each ventricle. It is composed of specialized ependymal cells covered by a capillary-rich layer of connective tissue.
- ▪ CSF enters the subarachnoid space from the ventricles and returns to the venous circulation through the arachnoid villi.

Blood-Brain Barrier   452

- ▪ The blood-brain barrier regulates movement of materials between the blood and the interstitial fluid of the brain.

Cerebrum 452	▪ The cerebrum is the center of our sensory perception, thought, memory, judgment, and voluntary motor actions.
	▪ The cerebral hemispheres contain gyri separated by sulci or deeper grooves called fissures.

Cerebral Hemispheres   452

- ▪ The cerebral hemispheres are separated by a deep longitudinal fissure. Each hemisphere contains five lobes: the frontal, parietal, temporal, and occipital lobes, and the insula.

Functional Areas of the Cerebrum   455

- ▪ The primary motor cortex in the frontal lobe directs voluntary movements. A motor speech area is responsible for the motor functions involved in speaking.
- ▪ The primary somatosensory cortex in the parietal lobe collects somatic sensory information from various general sensory receptors.

Central White Matter   457

- ▪ The central white matter contains three major groups of axons: association tracts, commissural tracts, and projection tracts.

Cerebral Nuclei   459

- ▪ The cerebral nuclei are located within the central white matter in the cerebral hemispheres. They include the caudate nucleus, amygdaloid body, putamen and globus pallidus, and claustrum.

Diencephalon 460	■ The diencephalon has processing and relay centers to integrate the sensory and motor pathways.
	Epithalamus 460
	■ The epithalamus partially forms the posterior roof of the diencephalon; it contains the pineal gland and habenular nuclei.
	Thalamus 460
	■ The thalamus is the principal and final relay point for integrating, assimilating, and amplifying sensory signals sent to the cerebral cortex.
	Hypothalamus 461
	■ The hypothalamus houses control and integrative centers, and oversees endocrine and autonomic nervous system functions.
Brainstem 463	■ The brainstem is composed of three regions: the mesencephalon, pons, and medulla oblongata.
	Mesencephalon 463
	■ The mesencephalon is the superior portion of the brainstem containing the cerebral peduncles, substantia nigra, tegmentum, tectal plate, and nuclei for two cranial nerves.
	Pons 463
	■ The pons is an anterior bulge in the metencephalon containing axon tracts, nuclei for involuntary control of respiration, nuclei for four cranial nerves, and nuclei that process and relay commands arriving through the middle cerebellar peduncles.
	Medulla Oblongata 466
	■ The medulla oblongata connects the brain to the spinal cord. It contains sensory processing centers, autonomic reflex centers, and nuclei for four cranial nerves.
Cerebellum 467	■ The cerebellum helps maintain posture and balance, and establishes patterned muscular contractions.
	Cerebellar Peduncles 468
	■ Three paired cerebellar peduncles (superior, middle, and inferior) are thick fiber tracts that connect the cerebellum with different parts of the brainstem.
Limbic System 468	■ The limbic system and the reticular formation are functional brain systems whose components are widely scattered in the brain, but whose activities combine for a common function.
	■ The limbic system includes a group of cortical structures that surround the corpus callosum and thalamus near the anterior region of the diencephalon. They function in memory and emotional behavior.
Cranial Nerves 470	■ Twelve nerve pairs, called cranial nerves, project from the brain. Each nerve has a specific name and is designated by a Roman numeral.

C H A L L E N G E   Y O U R S E L F

Matching

Match each numbered item with the most closely related lettered item.

_____ 1. mesencephalon

_____ 2. medulla oblongata

_____ 3. falx cerebri

_____ 4. vagus nerve

_____ 5. frontal lobe

_____ 6. temporal lobe

_____ 7. abducens nerve

_____ 8. cerebral nuclei

_____ 9. superior colliculi

_____ 10. thalamus

a. contains the primary auditory cortex

b. innervates most thoracic/abdominal organs

c. has nuclei for CN III and CN IV

d. responsible for involuntary arm swinging

e. contains the motor speech area

f. innervates lateral rectus

g. sensory information relay center

h. autonomic centers for heart rate and respiration

i. dura mater fold between cerebral hemispheres

j. visual reflex centers

Multiple Choice

Select the best answer from the four choices provided.

_____ 1. Which cranial nerve has three divisions (ophthalmic, maxillary, and mandibular)?
   a. accessory (CN XI)
   b. glossopharyngeal (CN IX)
   c. trigeminal (CN V)
   d. hypoglossal (CN XII)

_____ 2. The subdivision of the brain that does not initiate somatic motor movements, but rather coordinates and fine-tunes those movements is the
   a. medulla oblongata.
   b. cerebrum.
   c. cerebellum.
   d. diencephalon.

_____ 3. The visceral reflex center is housed within the
   a. cerebellum.
   b. superior colliculus.
   c. hypothalamus.
   d. pons.

_____ 4. Which of the following is *not* a function of the hypothalamus?
  a. controls endocrine system
  b. regulates sleep-wake cycle
  c. controls autonomic nervous system
  d. initiates voluntary skeletal muscle movement

_____ 5. Which of the following statements is false about the choroid plexus?
  a. It is located within the ventricles of the brain.
  b. It is composed of ependymal cells and capillaries.
  c. It forms the blood-brain barrier.
  d. It produces and circulates cerebrospinal fluid.

_____ 6. The _____ are descending motor tracts on the anterolateral surface of the mesencephalon.
  a. olives
  b. inferior colliculi
  c. cerebral peduncles
  d. tegmenta

_____ 7. Which cerebral lobe is located immediately posterior to the central sulcus and superior to the lateral sulcus?
  a. frontal lobe
  b. temporal lobe
  c. parietal lobe
  d. insula

_____ 8. The primary motor cortex is located in which cerebral structure?
  a. precentral gyrus
  b. postcentral gyrus
  c. motor speech area
  d. prefrontal cortex

_____ 9. The _____ are the isolated, innermost gray matter areas near the base of the cerebrum, inferior to the lateral ventricles.
  a. auditory association areas
  b. cerebral nuclei
  c. substantia nigra
  d. corpus callosum fibers

_____ 10. Which structure contains some autonomic centers involved in regulating respiration?
  a. pons
  b. superior colliculi
  c. cerebellum
  d. thalamus

Content Review

1. List the five secondary brain vesicles, and describe which adult brain structures are formed from each.

2. Describe (a) how and where the cerebrospinal fluid is formed, (b) its subsequent circulation, and (c) how and where it is reabsorbed into the vascular system.

3. Which specific area of the brain may be impaired if you cannot tell the difference between a smooth and a rough surface using your hands only?

4. What activities occur in the visual association area?

5. What are the cerebral nuclei, where are they located, and what is their function?

6. List the functions of the hypothalamus.

7. After a severe blow to the head, a patient suddenly does not produce sensations of hunger and seems unable to tell if he is dehydrated. The attending physician should suspect damage to or a lesion within what general region of the brain?

8. Identify the components of the limbic system.

9. Name the cranial nerve that innervates: (1) the lacrimal glands and most of the salivary glands; (2) the lateral rectus eye muscles; (3) the intrinsic and extrinsic tongue muscles.

10. During surgery to remove a tumor from the occipital lobe of the left cerebrum, a surgeon must cut into the brain to reach the tumor. List in order (starting with the covering skin) all of the layers that must be "cut through" to reach the tumor.

Developing Critical Reasoning

1. Shannon felt strange when she awoke one morning. She could not hold a pen in her right hand when trying to write an entry in her diary, and her muscles were noticeably weaker on the right side of her body. Additionally, her husband noticed that she was slurring her speech, so he took her to the emergency room. What does the ER physician suspect has occurred? Where in the brain might the physician suspect that abnormal activity or perhaps a lesion is located, and why?

2. Parkinson disease is the result of decreased levels of the neurotransmitter dopamine in the brain. However, these patients can't take dopamine in drug form because the drug cannot reach the brain. What anatomic structure prevents the drug from reaching the brain? How could these same anatomic structures be beneficial to an individual under different circumstances?

3. During a robbery at his convenience store, Dustin was shot in the right cerebral hemisphere. He survived, although some specific functions were impaired. Would Dustin have been more likely or less likely to have survived if he had been shot in the brainstem? Why?

ANSWERS TO "WHAT DO YOU THINK?"

1. The dura mater provides the most support and physical protection. This layer is the thickest and most durable of the cranial meninges. It is composed of dense irregular connective tissue, and it anchors the brain "in place" via the cranial dural septa.

2. If CSF production exceeds the amount removed, a pathologic condition called hydrocephalus occurs. This excess volume of CSF often results in brain distortion and subsequent damage to the brain.

3. Cutting the corpus callosum dramatically reduces communication between the right and left cerebral hemispheres, but some communication is maintained via the much smaller anterior and posterior commissures.

4. If there were no thalamus, the cerebrum would still receive sensory stimuli, but the information would not be decoded first. So, the cerebrum would not be able to distinguish taste information from touch or vision information.

5. Severe injury to the medulla oblongata would most likely cause death, because the medulla oblongata is responsible for basic reflex and life functions, including breathing and heartbeat.

Visit the McKinley/O'Loughlin *Human Anatomy,* 2e website at aris.mhhe.com

16

NERVOUS SYSTEM

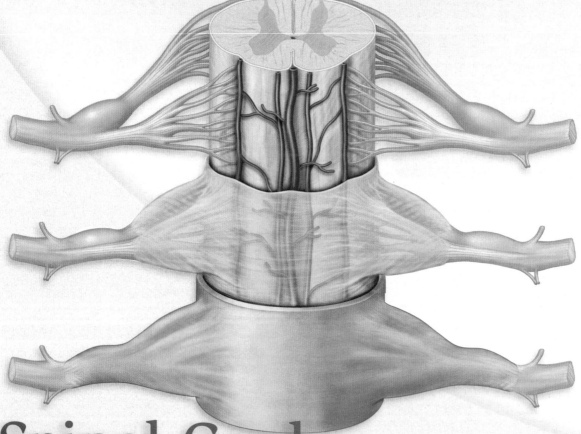

Spinal Cord and Spinal Nerves

The spinal cord provides a vital link between the brain and the rest of the body, and yet it exhibits some functional independence from the brain. The spinal cord and its attached spinal nerves serve two important functions. First, they are a pathway for sensory and motor impulses. Second, the spinal cord and spinal nerves are responsible for reflexes, which are our quickest reactions to a stimulus. In this chapter, we describe the anatomy of the spinal cord and the integrative activities that occur there.

Gross Anatomy of the Spinal Cord

Key topics in this section:

- Structure of the spinal cord
- Basic functions of the spinal cord

A typical adult spinal cord ranges between 42 and 45 centimeters (cm) (16 to 18 inches) in length. It extends inferiorly from the brain through the vertebral canal and ends at the level of the L_1 vertebra. The spinal cord may be subdivided into the following parts (**figure 16.1a**):

- The **cervical part** is the superiormost region of the spinal cord. It is continuous with the medulla oblongata. The cervical part contains motor neurons whose axons contribute to the cervical spinal nerves and receives input from sensory neurons through these spinal nerves (figure 16.1b).
- The **thoracic part** lies inferior to the cervical part. It contains the neurons for the thoracic spinal nerves.
- The **lumbar part** is a shorter segment of the spinal cord that contains the neurons for the lumbar spinal nerves.
- The **sacral part** lies inferior to the lumbar part and contains the neurons for the sacral spinal nerves.
- The **coccygeal** (kok-sij′ē-ăl) **part** (not shown in figure 16.1) is the most inferior "tip" of the spinal cord. (Some texts consider this part a portion of the sacral part of the spinal cord.) One pair of coccygeal spinal nerves arises from this region.

Note that the different *parts* of the spinal cord do not match up exactly with the *vertebrae* of the same name. For example, the lumbar part of the spinal cord is actually closer to the inferior thoracic vertebrae than to the lumbar vertebrae. This discrepancy is due to the fact that the growth of the vertebrae continued longer than the growth of the spinal cord itself. Thus, the spinal cord in an adult is shorter than the vertebral canal that houses it.

The tapering inferior end of the spinal cord is called the **conus medullaris** (kōn′ŭs med-oo-lăr′is; *kōnos* = cone, *medulla* = middle). The conus medullaris marks the official "end" of the spinal cord proper (usually at the level of the first lumbar vertebra). Inferior to this point, groups of axons collectively called the **cauda equina** (kaw′dă ē-kwī′nă) project inferiorly from the spinal cord. These nerve roots are so named because they resemble a horse's tail (*cauda* = tail, *equus* = horse). Within the cauda equina is the **filum terminale** (fī′lŭm ter′mi-năl; *terminus* = end). The filum terminale is a thin strand of pia mater that helps anchor the conus

medullaris to the coccyx. Figure 16.1c shows the conus medullaris and the cauda equina.

Viewed in cross section, the spinal cord is roughly cylindrical, but slightly flattened both posteriorly and anteriorly. Its external surface has two longitudinal depressions: A narrow groove, the **posterior** (or *dorsal*) **median sulcus**, dips internally on the posterior surface, and a slightly wider groove, the **anterior** (or *ventral*) **median fissure**, is observed on its anterior surface.

Cross-sectional views of the spinal cord vary, depending upon the part from which the section was taken (**table 16.1**). These subtle differences make identifying specific spinal cross sections a bit easier. For example, the diameter of the spinal cord changes along its length because the amount of gray matter and white matter and the function of the cord vary in different parts. Therefore, the spinal cord parts that control the upper and lower limbs are larger because more neuron cell bodies are located there, and more space is occupied by axons and dendrites. The **cervical enlargement**, located in the inferior cervical part of the spinal cord, contains the neurons that innervate the upper limbs. The **lumbosacral enlargement** extends through the lumbar and sacral parts of the spinal cord and innervates the lower limbs.

The spinal cord is associated with 31 pairs of spinal nerves that connect the CNS to muscles, receptors, and glands. Spinal nerves are considered *mixed nerves* because they contain both motor and sensory axons. Spinal nerves are identified by the first letter of the spinal cord part to which they attach combined with a number. Thus, each side of the spinal cord contains 8 cervical nerves (called C1–C8), 12 thoracic nerves (T1–T12), 5 lumbar nerves (L1–L5), 5 sacral nerves (S1–S5), and 1 coccygeal nerve (Co1). Spinal nerve names can be distinguished from cranial nerve names (discussed in chapter 15) because cranial nerves are designated by CN followed by a Roman numeral.

Study Tip!

With one exception, the number of spinal nerves matches the number of vertebrae in that region. For example, there are 12 pairs of thoracic spinal nerves and 12 thoracic vertebrae. The sacrum is formed from 5 fused sacral vertebrae, and there are 5 pairs of sacral spinal nerves. The coccygeal vertebrae tend to fuse into one structure, and there is 1 pair of coccygeal nerves. The only exception to this rule is that there are 8 pairs of cervical spinal nerves, but only 7 cervical vertebrae, because the first cervical pair emerges between the atlas (the first cervical vertebra) and the occipital bone.

WHAT DID YOU LEARN?

1 Identify the spinal cord enlargements. What is their function?

2 List the specific names of spinal nerves according to their region and the total number of pairs of spinal nerves.

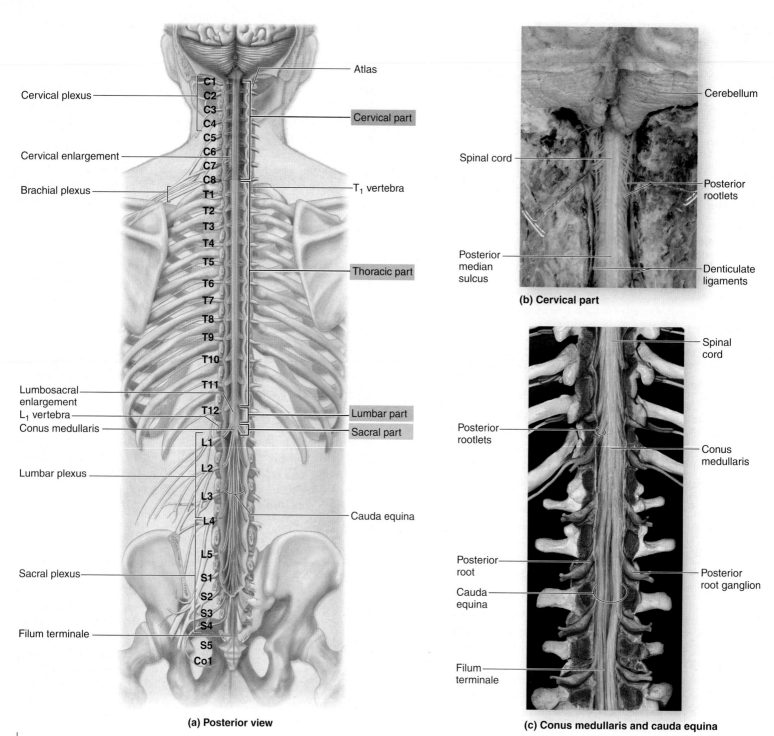

Cervical plexus

Cervical enlargement

Brachial plexus

Lumbosacral enlargement

L₁ vertebra

Conus medullaris

Lumbar plexus

Sacral plexus

Filum terminale

Atlas

Cervical part

T₁ vertebra

Thoracic part

Lumbar part

Sacral part

Cauda equina

(a) Posterior view

Cerebellum

Spinal cord

Posterior rootlets

Posterior median sulcus

Denticulate ligaments

(b) Cervical part

Spinal cord

Posterior rootlets

Conus medullaris

Posterior root

Posterior root ganglion

Cauda equina

Filum terminale

(c) Conus medullaris and cauda equina

Figure 16.1

Gross Anatomy of the Spinal Cord.   The spinal cord extends inferiorly from the medulla oblongata through the vertebral canal. *(a)* The vertebral arches have been removed to reveal the anatomy of the adult spinal cord and its spinal nerves. *(b)* Cadaver photo of the cervical part of the spinal cord. *(c)* Cadaver photo of the conus medullaris and the cauda equina.

Table 16.1	Cross Sections of Representative Parts of the Spinal Cord			
Spinal Cord Part	**General Size (Diameter)**	**General Shape**	**White Matter/Gray Matter Ratio**	**Other Characteristics**
Cervical *Posterior* Posterior median sulcus Anterior median fissure *Anterior*	Largest of all spinal cord parts (13–14 mm in transverse diameter)	Oval; slightly flattened on both anterior and posterior surfaces	Large proportion of white matter compared to gray matter	In superior segments (C1–C5), anterior horns are relatively small, and posterior horns are relatively large In inferior segments (C6–C8), anterior horns are larger and posterior horns are even more enlarged
Thoracic *Posterior* *Anterior*	Smaller than the cervical part (9–11 mm in transverse diameter)	Oval; still slightly flattened anteriorly and posteriorly	Larger proportion of white matter than gray matter	Anterior and posterior horns are enlarged only in first thoracic segment; small lateral horns are visible
Lumbar *Posterior* *Anterior*	Slightly larger than the thoracic part (11–13 mm in transverse diameter)	Less oval, almost circular	Relative amount of white matter is reduced both in proportion to gray matter and in comparison to cervical part	Anterior and posterior horns are very large; small lateral horns present in first two sections of lumbar part only
Sacral *Posterior* *Anterior*	Very small	Almost circular	Proportion of gray matter to white matter is largest in this spinal cord part	Anterior and posterior horns relatively large compared to the size of the cross section

Spinal Cord Meninges

Key topic in this section:

■ Arrangement and functions of the spinal meninges

The spinal cord is protected and encapsulated by **spinal cord meninges,** which are continuous with the cranial meninges described in chapter 15. In addition, spaces between some of the meninges have clinical significance. The structures and spaces (both real and potential) that encircle the spinal cord, listed from outermost to innermost, are as follows: vertebra, epidural space, dura mater, subdural space, arachnoid, subarachnoid space, and pia mater (**figure 16.2**).

The **epidural** (ep-i-doo′răl) **space** lies between the dura mater and the periosteum covering the inner walls of the vertebra, and houses areolar connective tissue, blood vessels, and adipose con-nective tissue. It is in this space that an epidural anesthetic is given. Deep to the epidural space is the most external of the meninges, the **dura mater.** Although the cranial dura mater has an outer periosteal layer and an inner meningeal layer, the spinal dura mater consists of just one meningeal layer. The dura mater provides stability to the spinal cord. In addition, at each intervertebral foramen, the dura mater extends between adjacent vertebrae and fuses with the con-nective tissue layers that surround the spinal nerves.

WHAT DO YOU THINK?

❶ Why doesn't the spinal dura mater have two layers as the cranial dura mater does? What structures that are formed from cranial dura mater must be missing around the spinal cord?

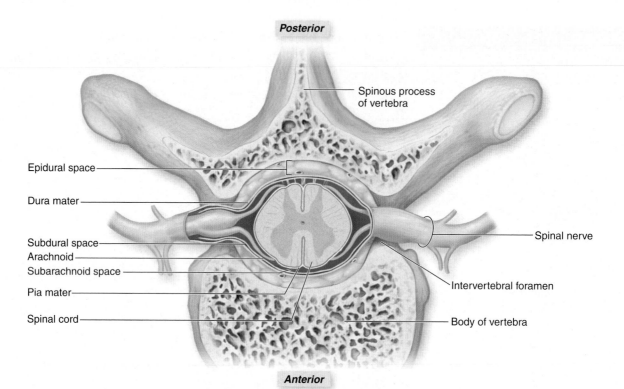

Posterior

Spinous process of vertebra

Epidural space

Dura mater

Subdural space

Arachnoid

Subarachnoid space

Pia mater

Spinal cord

Spinal nerve

Intervertebral foramen

Body of vertebra

Anterior

(a) Cross section of vertebra and spinal cord

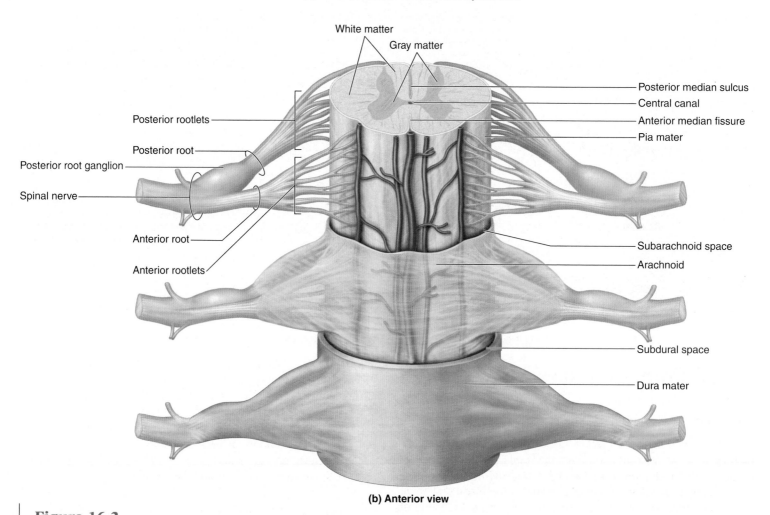

White matter

Gray matter

Posterior rootlets

Posterior root

Posterior root ganglion

Spinal nerve

Anterior root

Anterior rootlets

Posterior median sulcus

Central canal

Anterior median fissure

Pia mater

Subarachnoid space

Arachnoid

Subdural space

Dura mater

(b) Anterior view

Figure 16.2

Spinal Meninges and Structure of the Spinal Cord. *(a)* A cross section of the spinal cord shows the relationship between the meningeal layers and the superficial landmarks of the spinal cord and vertebral column. *(b)* Anterior view shows the spinal cord and meninges.

Lumbar Puncture (Spinal Tap)

It is sometimes necessary to analyze the cerebrospinal fluid (CSF) in order to determine if there is an infection or disorder of the central nervous system. The clinical procedure for obtaining CSF is known as a **lumbar puncture** (commonly referred to as a *spinal tap*). The surface anatomy landmarks and the procedure for performing a lumbar puncture are discussed in a Clinical View in chapter 13. Recall that the adult spinal cord typically ends at the level of the L_1 vertebra, so it is safe to insert the needle between the L_3 and L_4 vertebrae or between the L_4 and L_5 vertebrae. The needle must be inserted through the skin, the subcutaneous layer, back muscles, and ligamentum flavum. Then, the needle must pass through the epidural space, dura mater, arachnoid, and enter the subarachnoid space. Here, approximately 3–9 milliliters of CSF are taken and then analyzed to determine the nature of the nervous system ailment.

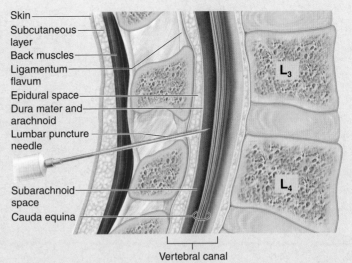

Site of needle insertion for a lumbar puncture.

In most anatomic and histologic preparations, a narrow **subdural space** separates the dura mater from the **arachnoid.** This space is found only in tissue preparations, and in life it is merely a potential space. Deep to the arachnoid is the **subarachnoid space,** which is a real space filled with **cerebrospinal fluid (CSF).** The **pia mater,** deep to the subarachnoid space, is a delicate, innermost meningeal layer composed of elastic and collagen fibers. This meninx directly adheres to the spinal cord and supports some of the blood vessels supplying the spinal cord. **Denticulate** (den-tik′ū-lāt; *dentatus* = toothed) **ligaments** are paired, lateral triangular extensions of the spinal pia mater that attach to the dura mater. These ligaments help suspend and anchor the spinal cord laterally to the dura mater (see figure 16.1*b*).

⚡ WHAT DID YOU LEARN?

❸ What is the function of the denticulate ligaments?

Sectional Anatomy of the Spinal Cord

Key topics in this section:

- Internal anatomy of the spinal cord
- Distribution of gray matter and white matter in the spinal cord
- How gray matter and white matter process information

The spinal cord is partitioned into an inner gray matter region and an outer white matter region **(figure 16.3).** The gray matter is dominated by the dendrites and cell bodies of neurons and glial cells and unmyelinated axons, whereas the white matter is composed primarily of myelinated axons.

Location and Distribution of Gray Matter

The **gray matter** in the spinal cord is centrally located, and its shape resembles a letter H or a butterfly. The gray matter may be subdivided into the following components: anterior horns, lateral horns, posterior horns, and the gray commissure.

Anterior horns are the left and right anterior masses of gray matter. The anterior horns primarily house the *cell bodies of somatic motor neurons*, which innervate skeletal muscle. **Lateral horns** are found in the T1–L2 parts of the spinal cord only. The lateral horns contain the *cell bodies of autonomic motor neurons*, which innervate cardiac muscle, smooth muscle, and glands. **Posterior horns** are the left and right posterior masses of gray matter. The *axons of sensory neurons* and the *cell bodies of interneurons* are located in the posterior horns. (Note that the cell bodies of these sensory neurons are not found in the posterior horns; rather, they are located in the posterior root ganglia, which are mentioned later in this chapter.) The **gray commissure** (kom′i-shūr; *commissura* = a seam) is a horizontal bar of gray matter that surrounds a narrow **central canal**. The gray commissure primarily contains unmyelinated axons and serves as a communication route between the right and left sides of the gray matter.

Study Tip!

Use this analogy to remember the location of somatic motor neurons: The *anterior* horns contain the somatic motor neurons, just as the *anterior* part of a car houses the motor.

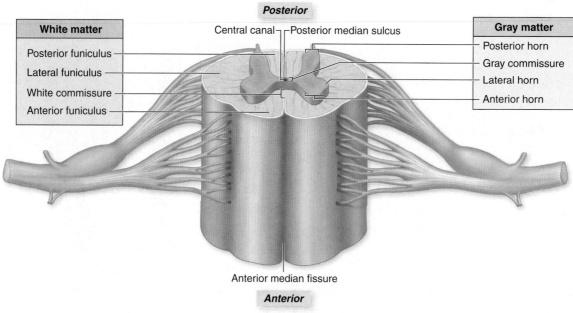

(a) Gray and white matter

Figure 16.3

Gray Matter and White Matter Organization in the Spinal Cord. *(a)* The gray matter is centrally located, and the white matter is externally located. *(b)* Histology of a transverse section of the spinal cord.

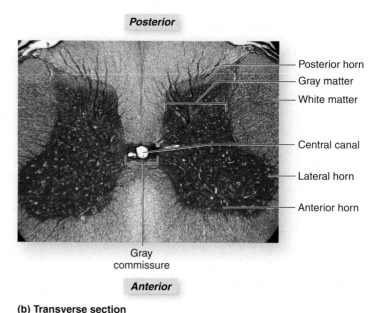

(b) Transverse section

Within these regions of gray matter are various functional groups of neuron cell bodies called **nuclei** (noo′klē-ī) **(figure 16.4)**. **Sensory nuclei** in the posterior horns contain interneuron cell bodies. **Somatic sensory nuclei** receive information from sensory receptors, such as pain or pressure receptors in the skin, while **visceral sensory nuclei** receive information from sensory receptors such as the stretch receptors in the smooth muscle walls of viscera. **Motor nuclei** in the anterior and lateral horns contain motor neuron cell bodies that send nerve impulses to muscles and glands. The **somatic motor nuclei** in the anterior horns innervate skeletal muscle, while the **autonomic motor nuclei** in the lateral horns innervate smooth muscle, cardiac muscle, and glands.

Location and Distribution of White Matter

The white matter of the spinal cord is external to the gray matter. White matter on each side of the cord is also partitioned into three regions, each called a **funiculus** (fū-nik′ū-lŭs; pl., *funiculi*[1], fū-nik′ū-lī; *funis* = cord) (see figure 16.3*a*). A **posterior funiculus** lies between the posterior gray horns on the posterior side of the cord and the posterior median sulcus. The white matter region on each lateral side of the spinal cord is the **lateral funiculus.** The **anterior**

---

[1] Note: Anterior and lateral funiculi were formerly called *columns*. The Federative Committee on Anatomical Terminology (FCAT) now states that the term "column" refers to structures within the gray matter of the spinal cord, while "funiculus" refers to the white matter regions.

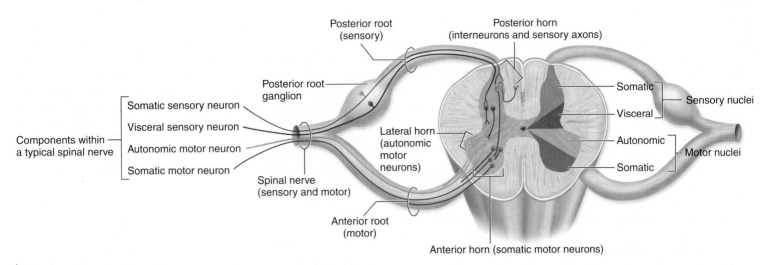

Figure 16.4

Neuron Pathways and Nuclei Locations.   The collections of neuron cell bodies within the CNS form specific nuclei. Neurons are color-coded on the left side of the drawing, while their respective nuclei are color-coded on the right side of the drawing.

funiculus is composed of tracts of white matter that occupy the space on each anterior side of the cord between the anterior gray horns and the anterior median fissure; the anterior funiculi are interconnected by the **white commissure.**

The axons within each white matter funiculus are organized into smaller structural units called **tracts** (*tractus* = a drawing out) or **fasciculi** (fă-sik′ū-lī; *fascis* = bundle), to be described in chapter 17. Individual tracts conduct either sensory impulses (ascending tracts from the spinal cord to the brain) or motor commands (descending tracts from the brain to the spinal cord) only. Each funiculus region (posterior, lateral, and anterior) contains both ascending and descending tracts. Thus, each funiculus contains axons of both motor and sensory nerves.

WHAT DID YOU LEARN?

4 Describe the arrangement of gray matter and white matter in the spinal cord.

5 Compare the components of the anterior horns and the posterior horns of the spinal cord.

6 What are the three groups of funiculi in the white matter of the spinal cord?

Spinal Nerves

Key topics in this section:

- Spinal nerve formation and branches
- Dermatome description and importance
- Structure, distribution, and innervation patterns of the spinal nerve plexuses

The 31 pairs of **spinal nerves** connect the central nervous system to muscles, glands, and receptors. Recall from chapter 14 that a spinal nerve is formed from the union of thousands of motor

and sensory axons, and contains three types of successive connective tissue wrappings: endoneurium, perineurium, and epineurium (see figure 14.12). Motor axons in a spinal nerve originate from the spinal cord (see figures 16.2 and 16.4). Anteriorly, multiple **anterior rootlets** arise from the spinal cord and merge to form a single **anterior root** (or *ventral root*), which contains motor axons only. These motor axons arise from cell bodies in the anterior and lateral horns of the spinal cord. Likewise, the posterior aspect of the spinal cord contains multiple **posterior rootlets** that enter the posterior aspect of the spinal cord. These rootlets were derived from a single **posterior root** (or *dorsal root*), which contains sensory axons only. The cell bodies of these sensory neurons are located in a **posterior root ganglion,** which is attached to the posterior root (see figures 16.2*b* and 16.4).

Each anterior root and its corresponding posterior root unite within the intervertebral foramen to become a spinal nerve. Thus, a spinal nerve contains both motor axons (from the anterior root) and sensory axons (from the posterior root). You can compare a spinal nerve to a cable composed of multiple wires. The "wires" within a spinal nerve are the motor and sensory axons.

Each spinal nerve exits the vertebral canal and travels through an intervertebral foramen superior or inferior to the vertebra of the same number. For example, the second cervical spinal nerve exits the vertebral canal through the intervertebral foramen between the C_1 and the C_2 vertebrae. The eighth cervical spinal nerve is the exception; it leaves the intervertebral foramen between the C_7 and T_1 vertebrae. The spinal nerves inferior to C8 exit below the vertebra of the same number. So, for example, the second thoracic spinal nerve exits the vertebral canal through the intervertebral foramen between the T_2 and T_3 vertebrae.

Because the spinal cord is shorter than the vertebral canal, the roots of the lumbar and sacral spinal nerves have to travel inferiorly to reach their respective intervertebral foramina through which they pass before they can merge and form a spinal nerve. Thus, the anterior and posterior roots of the lumbar and sacral spinal nerves must be much longer than the roots of the other spinal nerves.

Spinal Nerve Distribution

After leaving the intervertebral foramen, a typical spinal nerve almost immediately splits into branches, termed rami (**figure 16.5**). The **posterior** (*dorsal*) **ramus** (rā'mŭs; pl., *rami*, rā'mī; branch) is the smaller of the two main branches. It innervates the deep muscles of the back (e.g., erector spinae and transversospinalis) and the skin of the back. The **anterior** (*ventral*) **ramus** is the larger of the two main branches. The anterior ramus splits into multiple other branches, which innervate the anterior and lateral portions of the trunk, the upper limbs, and the lower limbs. Many of the anterior rami go on to form nerve plexuses, which are described in the next section. Additional rami, called the rami communicantes, are also associated with spinal nerves. These rami contain axons associated with the autonomic nervous system. Each set of rami extends between the spinal nerve and a ball-like structure called the sympathetic trunk ganglion. These ganglia are interconnected and form a beaded necklace-like structure called the sympathetic trunk. The rami communicantes, the sympathetic trunk, and the rest of the autonomic nervous system are described in detail in chapter 18.

WHAT DO YOU THINK?

2 Why is an anterior ramus so much larger than a posterior ramus? (Hint: Think about what structures each innervates.)

Dermatomes

A **dermatome** (der'mă-tōm; *derma* = skin, *tome* = a cutting) is a specific segment of skin supplied by a single spinal nerve. All spi-

nal nerves except C1 innervate a segment of skin, and so each of these nerves is associated with a dermatome. Thus, the skin of the body may be divided into sensory segments that collectively make up a dermatome map (**figure 16.6**). For example, the horizontal segment of skin around the umbilicus (navel) region is supplied by the anterior ramus of the T10 spinal nerve. The dermatome map follows a segmental pattern along the body. Dermatomes are clinically important because they can indicate potential damage to one or more spinal nerves. For example, if a patient experiences **anesthesia** (numbness) along the medial side of the arm and forearm, the C8 spinal nerve may be damaged.

Dermatomes are also involved in **referred visceral pain,** a phenomenon in which pain or discomfort from one organ is mistakenly referred to a dermatome. For example, the appendix is innervated by axons from the T10 regions of the spinal cord, so appendicitis typically causes referred visceral pain to the T10 dermatome in the umbilicus region rather than in the abdominopelvic region of the appendix itself. Thus, pain in a dermatome may arise from an organ nowhere near the dermatome. Referred visceral pain is explored further in chapter 19.

Nerve Plexuses

A **nerve plexus** (plek'sŭs; a braid) is a network of interweaving anterior rami of spinal nerves. The anterior rami of most spinal nerves form nerve plexuses on both the right and left sides of the body. These nerve plexuses then split into multiple "named" nerves that innervate various body structures. The principal

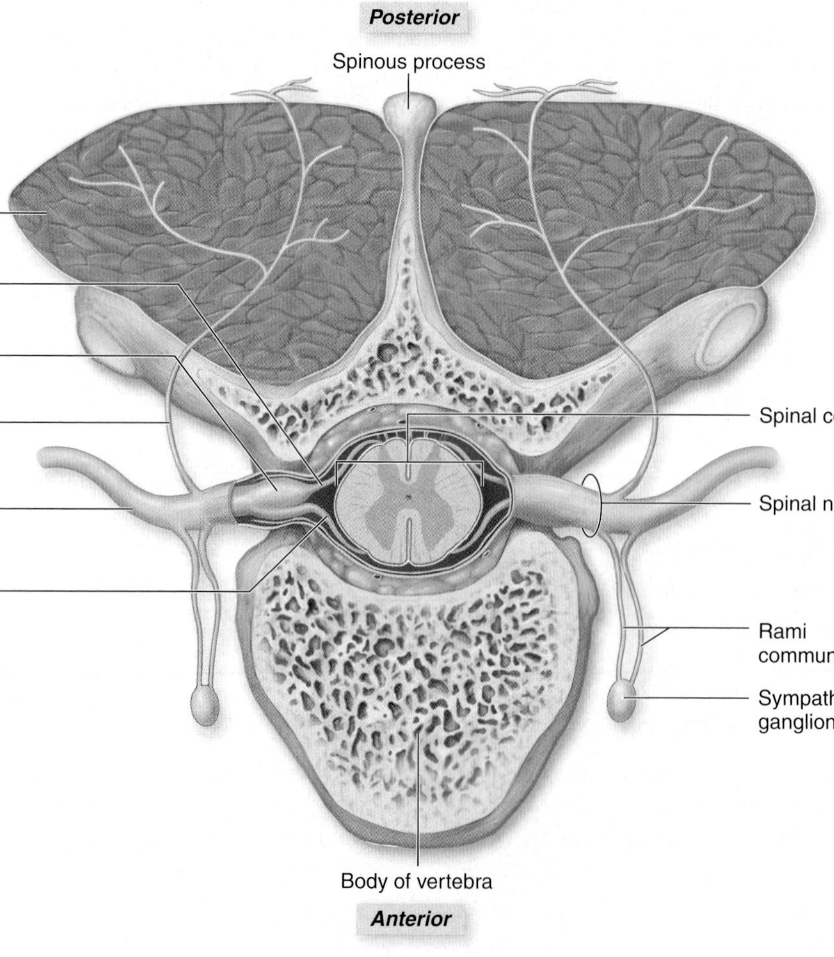

Figure 16.5

Spinal Nerve Branches. The major branches of a spinal nerve are the posterior ramus and the anterior ramus.

Posterior

Spinous process

Deep muscles of back

Posterior root

Posterior root ganglion

Posterior ramus

Anterior ramus

Anterior root

Spinal cord

Spinal nerve

Rami communicantes

Sympathetic trunk ganglion

Body of vertebra

Anterior

CLINICAL VIEW

Shingles

Some adults experience a reactivation of their childhood chickenpox infection, a condition termed **shingles** (shing'glz). During the initial infection, the chickenpox virus (varicella-zoster) sometimes leaves the skin and invades the posterior root ganglia of the spinal cord. There, the virus remains latent until adulthood, when it becomes reactivated and proliferates, traveling through the sensory axons to the dermatome. (The word *shingles* is derived from the latin word *cingulum*, meaning "girdle," reflecting the dermatomal pattern of its spread.) Within the dermatome, the virus continues to proliferate in the skin, giving rise to a rash and blisters, which are often accompanied by intense burning or tingling pain. Sometimes the pain precedes the development of the visible skin changes. Shingles patients are contagious as long as they have the blisters.

About 10% of adults will experience shingles during their lifetime, most after the age of 50. Psychological stress, other infections (such as a cold or the flu), and even a sunburn can trigger the development of shingles. Shingles is usually a self-limiting condition, although the disease can recur, and for some patients, dermatomal pain may last long after the skin changes have resolved.

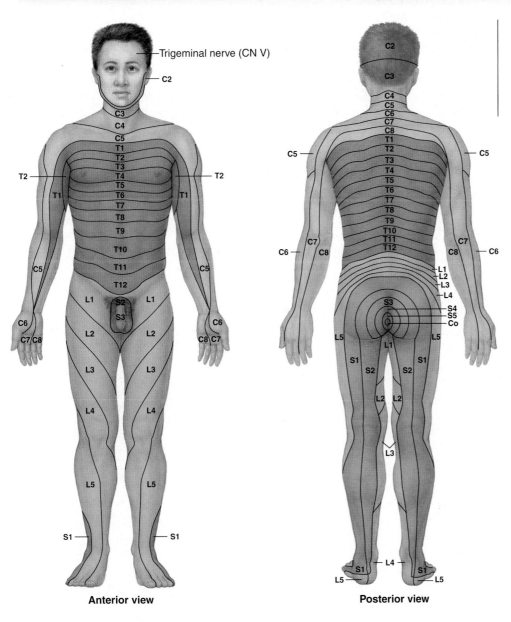

Anterior view   **Posterior view**

Figure 16.6

Dermatome Maps.   A dermatome is an area of skin supplied by a single spinal nerve. These diagrams only approximate the dermatomal distribution.

plexuses are the cervical plexuses, brachial plexuses, lumbar plexuses, and sacral plexuses (see figure 16.1).

⚚ WHAT DO YOU THINK?

3 What is the benefit of having an intricate nerve plexus, rather than a single nerve that innervates a structure?

Nerve plexuses are organized such that axons from each anterior ramus extend to body structures through several different branches. In addition, each terminal branch of the plexus houses axons from several different spinal nerves. Most of the named nerves from a plexus are composed of axons from multiple spinal nerves. Thus, damage to a single segment of the spinal cord or damage to a single spinal nerve generally does not result in complete loss of innervation to a particular muscle or region of skin.

Most of the thoracic spinal nerves, as well as nerves S5–Co1, do not form plexuses. We discuss the thoracic spinal nerves (called intercostal nerves) first, followed by the individual nerve plexuses.

Intercostal Nerves

The anterior rami of spinal nerves T1–T11 are called **intercostal nerves** because they travel in the intercostal space sandwiched between two adjacent ribs (**figure 16.7**). (T12 is called a **subcostal nerve**, because it arises below the ribs, not between two ribs.) The innervation pattern of the T1–T12 nerves is as follows: With the exception of T1, the intercostal nerves do not form plexuses. A portion of the anterior ramus of T1 helps form the brachial plexus, but a branch of it travels within the first intercostal space. The anterior ramus of nerve T2 emerges from its intervertebral foramen and innervates the intercostal muscles of the second intercostal space. Additionally, a branch of T2 conducts sensory impulses from the skin covering the axilla and the medial surface of the arm. Anterior rami of nerves T3–T6 follow the costal grooves of the ribs to innervate the intercostal muscles and receive sensations from the anterior and lateral chest wall. Anterior rami of nerves T7–T12 innervate not only the inferior intercostal spaces, but also the abdominal muscles and their overlying skin.

Cervical Plexuses

The left and right **cervical plexuses** are located deep on each side of the neck, immediately lateral to cervical vertebrae C_1–C_4 (**figure 16.8**). They are formed primarily by the anterior rami of spinal nerves C1–C4. The fifth cervical spinal nerve is not considered part of the cervical plexus, although it contributes some axons to one of the plexus branches. Branches of the cervical plexuses innervate anterior neck muscles as well as the skin of the neck and portions of the head and shoulders. The branches of the cervical plexuses are described in detail in **table 16.2**.

One important branch of the cervical plexus is the **phrenic** (fren′ik; *phren* = diaphragm) **nerve,** which is formed primarily from the C4 nerve and some contributing axons from C3 and C5. The phrenic nerve travels through the thoracic cavity to innervate the diaphragm.

> **Study Tip!**
>
> This mnemonic will help you remember the nerves that innervate the diaphragm: C three, four, and five keep the diaphragm alive.

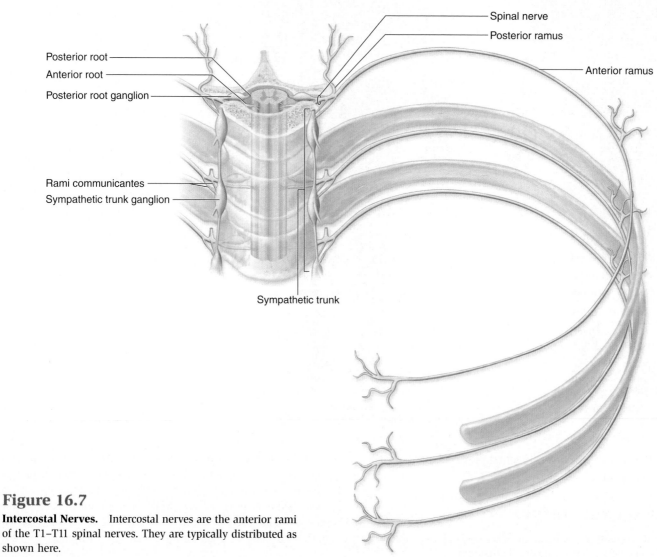

Figure 16.7

Intercostal Nerves. Intercostal nerves are the anterior rami of the T1–T11 spinal nerves. They are typically distributed as shown here.

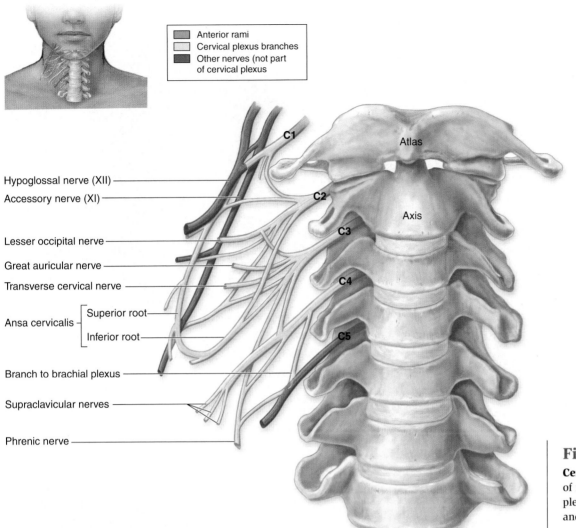

■	Anterior rami
■	Cervical plexus branches
■	Other nerves (not part of cervical plexus)

Hypoglossal nerve (XII)
Accessory nerve (XI)

Lesser occipital nerve

Great auricular nerve

Transverse cervical nerve

Ansa cervicalis — Superior root
Inferior root

Branch to brachial plexus

Supraclavicular nerves

Phrenic nerve

C1
Atlas
C2
Axis
C3
C4
C5

Figure 16.8

Cervical Plexus. Anterior rami of nerves C1–C4 form the cervical plexus, which innervates the skin and many muscles of the neck.

Table 16.2	**Branches of the Cervical Plexuses**	
Nerves	**Anterior Rami**	**Innervation**
MOTOR BRANCHES		
Ansa cervicalis		Geniohyoid; infrahyoid muscles (omohyoid, sternohyoid, sternothyroid, and thyrohyoid)
Superior root	C1, C2	
Inferior root	C3, C4	
Phrenic	C3–C5 (primarily C4)	Diaphragm
Segmental branches	C1–C4	Anterior and middle scalenes
CUTANEOUS BRANCHES		
Greater auricular	C2, C3	Skin on ear; connective tissue capsule covering parotid gland
Lesser occipital	C2	Skin of scalp superior and posterior to ear
Supraclavicular	C3, C4	Skin on superior part of chest and shoulder
Transverse cervical	C2, C3	Skin on anterior part of neck

Note: While CN XII (hypoglossal) travels with the nerves of the cervical plexus, this cranial nerve is not considered part of the plexus.

Brachial Plexuses

The left and right **brachial plexuses** are networks of nerves that supply the upper limb. Each brachial plexus is formed by the anterior rami of spinal nerves C5–T1 (**figure 16.9**). The components of the brachial plexus extend laterally from the neck, pass superior to the first rib, and then continue into the axilla. Each brachial plexus innervates the pectoral girdle and the entire upper limb of one side.

Structurally, each brachial plexus is more complex than a cervical plexus and composed of anterior rami, trunks, divisions, and

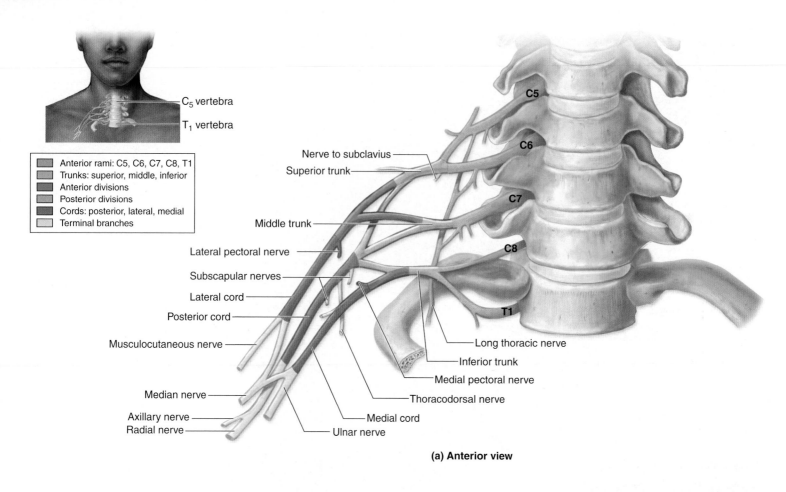

Anterior rami: C5, C6, C7, C8, T1
Trunks: superior, middle, inferior
Anterior divisions
Posterior divisions
Cords: posterior, lateral, medial
Terminal branches

C₅ vertebra
T₁ vertebra

C5
C6
C7
C8
T1

Nerve to subclavius
Superior trunk

Middle trunk

Lateral pectoral nerve
Subscapular nerves
Lateral cord
Posterior cord
Musculocutaneous nerve

Median nerve
Axillary nerve
Radial nerve

Long thoracic nerve
Inferior trunk
Medial pectoral nerve
Thoracodorsal nerve
Medial cord
Ulnar nerve

(a) Anterior view

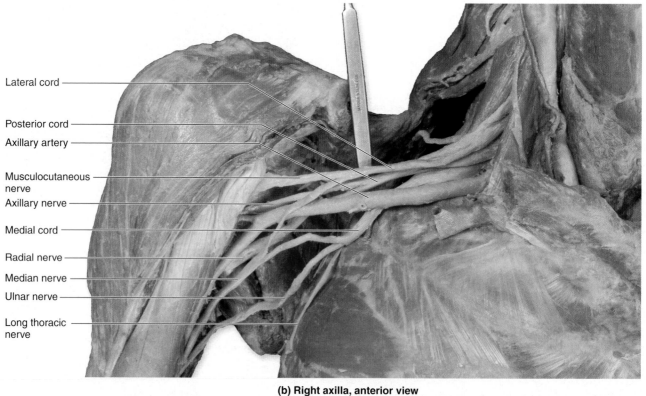

Lateral cord

Posterior cord
Axillary artery

Musculocutaneous nerve
Axillary nerve

Medial cord

Radial nerve
Median nerve
Ulnar nerve

Long thoracic nerve

(b) Right axilla, anterior view

Figure 16.9

Brachial Plexus. Anterior rami of nerves C5–T1 form the brachial plexus, which innervates the upper limb. *(a)* Rami, trunks, divisions, and cords form the subdivisions of this plexus. *(b)* A cadaver dissection identifies major nerves from the right brachial plexus. *(c)* Complete pathways of main brachial plexus branches in the right upper limb.

cords when examined from a medial to lateral perspective. The **anterior rami** (sometimes called *roots*) of the brachial plexus are simply the continuations of the anterior rami of spinal nerves C5–T1. These rami emerge from the intervertebral foramina and travel through the neck. The five roots unite to form the **superior, middle,** and **inferior trunks** in the posterior triangle of the neck. Nerves C5 and C6 unite to form the superior trunk; nerve C7 remains as the middle trunk; and nerves C8 and T1 unite to form the inferior trunk. Portions of each trunk divide inferior to the clavicle into an **anterior division** and a **posterior division,** which primarily contain axons that innervate the anterior and posterior parts of the upper limb, respectively.

> ## Study Tip!
>
> In general, nerves from the *anterior division* of the brachial plexus tend to innervate muscles that *flex* the parts of the upper limb. Nerves from the *posterior division* of the brachial plexus tend to innervate muscles that *extend* the parts of the upper limb. So, if you know a nerve is a branch of the anterior division of the brachial plexus, the nerve likely innervates a group of flexor muscles!

Upon reaching the axilla, the anterior and posterior divisions converge to form three cords, which are named with respect to their position near the axillary artery. The **posterior cord** is posterior to the axillary artery and is formed by the posterior divisions of the superior, middle, and inferior trunks; therefore, it contains portions of C5–T1 nerves. The **medial cord** is medial to the axillary artery and is formed by the anterior division of the inferior trunk; it contains portions of nerves C8–T1. The **lateral cord** is lateral to the axillary artery and is formed from the anterior divisions of the superior and middle trunks; thus, it contains portions of nerves C5–C7.

Finally, five major **terminal branches** emerge from the three cords: the axillary, median, musculocutaneous, radial, and ulnar nerves **(table 16.3).** The **axillary nerve** traverses through the axilla and posterior to the surgical neck of the humerus. The axillary nerve emerges from the posterior cord of the brachial plexus and innervates both the deltoid and teres minor muscles. It receives sensory information from the superolateral part of the arm. The **median nerve** is formed from branches of the medial and lateral cords of the brachial plexus. This nerve travels along the midline of the arm and forearm, and deep to the carpal tunnel in the wrist. It innervates most of the anterior forearm muscles, the thenar muscles, and the lateral two lumbricals. It receives

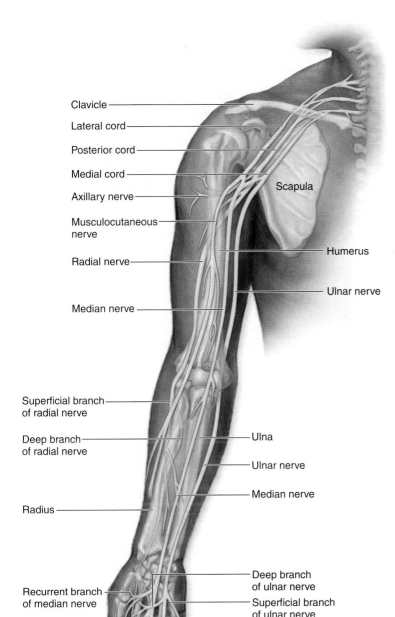

Clavicle

Lateral cord

Posterior cord

Medial cord

Axillary nerve

Musculocutaneous nerve

Radial nerve

Median nerve

Superficial branch of radial nerve

Deep branch of radial nerve

Radius

Recurrent branch of median nerve

Digital branch of median nerve

Scapula

Humerus

Ulnar nerve

Ulna

Ulnar nerve

Median nerve

Deep branch of ulnar nerve

Superficial branch of ulnar nerve

Digital branch of ulnar nerve

(c) Right upper limb, anterior view

Table 16.3		Branches of the Brachial Plexus	
Terminal Branch	**Anterior Rami**	**Motor Innervation**	**Cutaneous Innervation**
Axillary Nerve Formed from posterior cord, posterior division of the brachial plexus Posterior cord Teres minor — Axillary nerve Deltoid	C5, C6	Deltoid (*arm abductor*) Teres minor (*lateral rotator of arm*)	Superolateral arm
Median Nerve Formed from medial and lateral cords, anterior division of the brachial plexus Lateral cord Posterior cord Medial cord **Median nerve** Pronator teres Flexor carpi radialis Palmaris longus Flexor digitorum superficialis Flexor pollicis longus Pronator quadratus Thenar muscles Lateral two lumbricals Flexor digitorum profundus (lateral half)	C5–T1	Most anterior forearm muscles (*pronators, flexors of wrist, digits*) Flexor carpi radialis Flexor digitorum superficialis Pronator teres Pronator quadratus Lateral 1/2 of flexor digitorum profundus Flexor pollicis longus Thenar (thumb) muscles (*move thumb*) Flexor pollicis brevis Abductor pollicis brevis Opponens pollicis Lateral 2 lumbricals (*flex MP joints and extend PIP and DIP joints*)	Palmar aspects and dorsal tips of lateral 3-1/2 digits (*thumb, index finger, middle finger, and 1/2 of ring finger*)

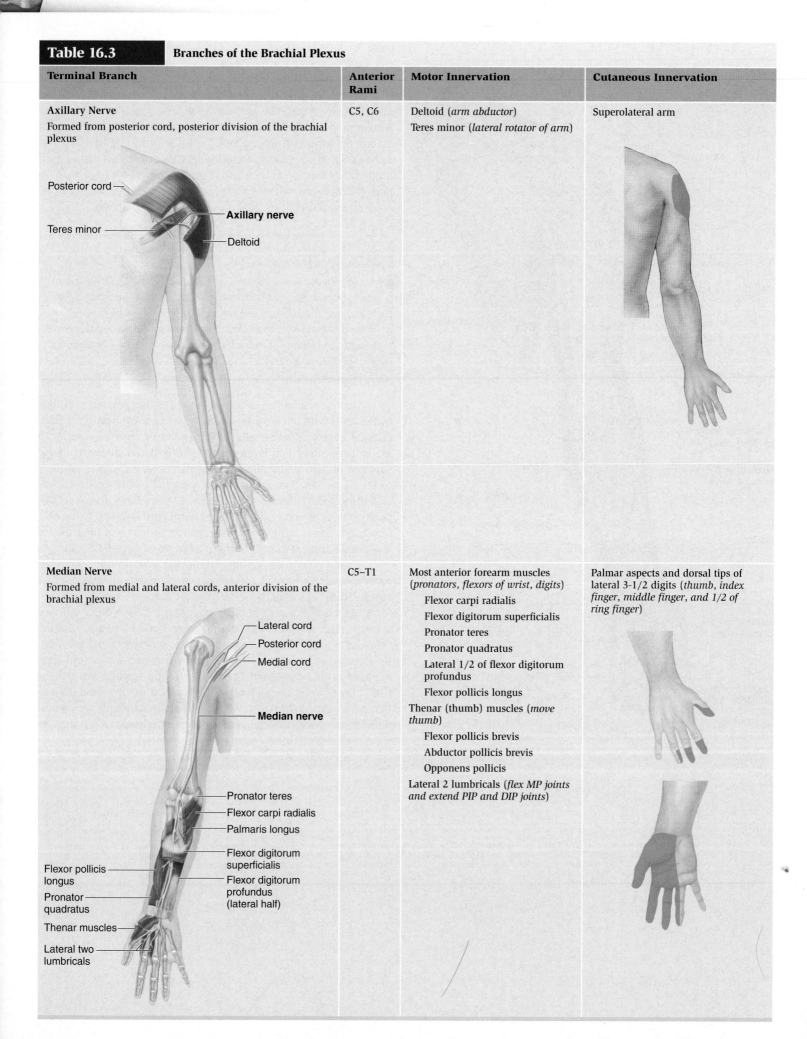

Table 16.3 **Branches of the Brachial Plexus**

<table>
<tr><th>Terminal Branch</th><th>Anterior Rami</th><th>Motor Innervation</th><th>Cutaneous Innervation</th></tr>
<tr><td>Musculocutaneous Nerve
Formed from the lateral cord, anterior division of the brachial plexus
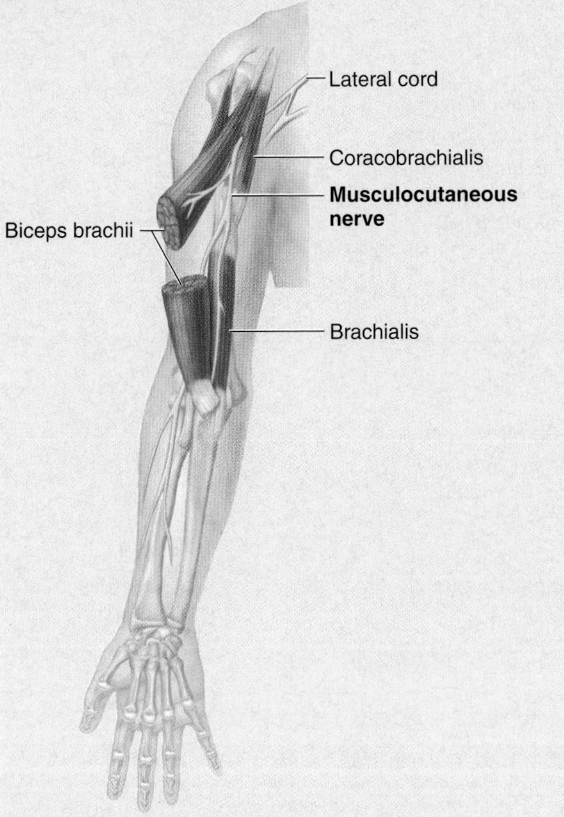
Lateral cord
Coracobrachialis
Musculocutaneous nerve
Biceps brachii
Brachialis</td><td>C5–C7</td><td>Anterior arm muscles (flex humerus, flex elbow joint, supinate forearm)
Coracobrachialis
Biceps brachii
Brachialis</td><td>Lateral region of forearm
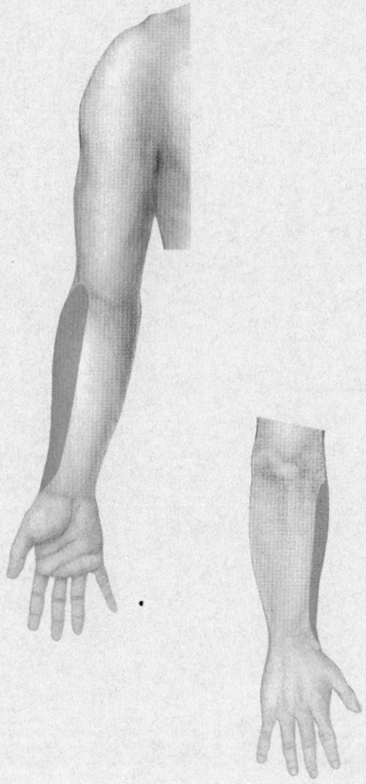</td></tr>
<tr><td>Radial Nerve
Formed from the posterior cord, posterior division of the brachial plexus
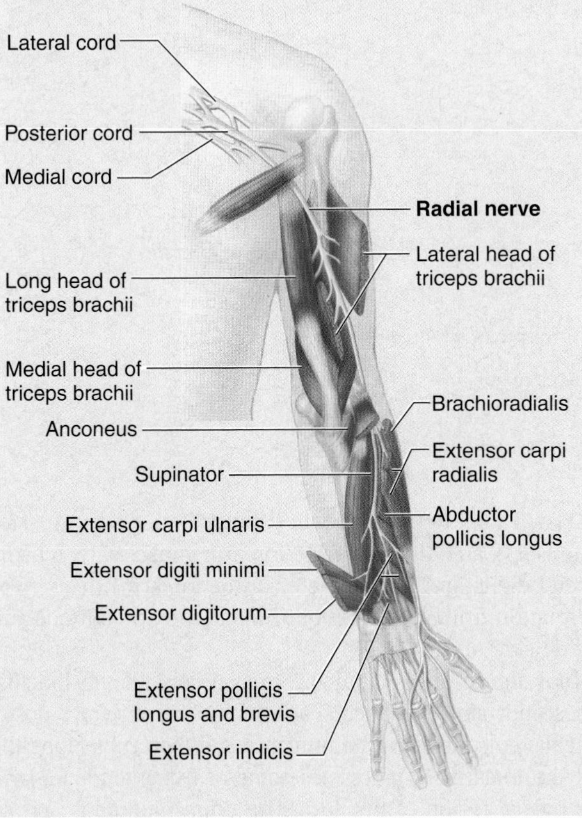
Lateral cord
Posterior cord
Medial cord
Radial nerve
Lateral head of triceps brachii
Long head of triceps brachii
Medial head of triceps brachii
Brachioradialis
Anconeus
Extensor carpi radialis
Supinator
Abductor pollicis longus
Extensor carpi ulnaris
Extensor digiti minimi
Extensor digitorum
Extensor pollicis longus and brevis
Extensor indicis</td><td>C5–T1</td><td>Posterior arm muscles (extend forearm)
Triceps brachii
Anconeus
Posterior forearm muscles (supinate forearm, extend wrist digits, 1 muscle that abducts thumb)
Supinator
Extensor carpi radialis muscles
Extensor digitorum
Extensor carpi ulnaris
Extensor pollicis longus
Extensor pollicis brevis
Abductor pollicis brevis
Extensor digiti minimi
Extensor indicis
Brachioradialis (flexes forearm)</td><td>Posterior region of arm
Posterior region of forearm
Dorsal aspect of lateral 3 digits (except their distal tips)
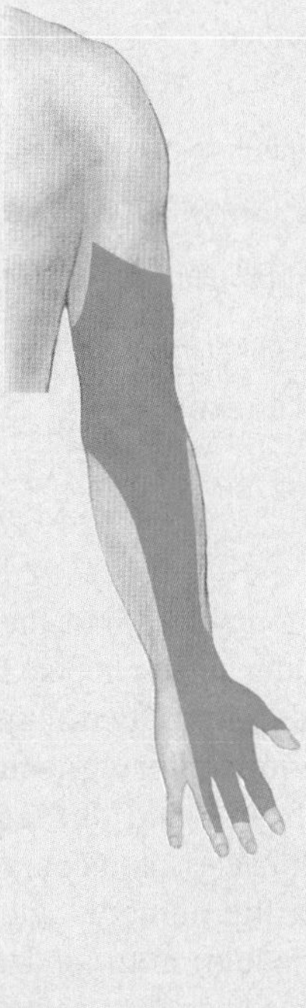</td></tr>
</table>

Table 16.3	Branches of the Brachial Plexus *(continued)*

Terminal Branch	Anterior Rami	Motor Innervation	Cutaneous Innervation
Ulnar Nerve Formed from the medial cord, anterior division of the brachial plexus	C8–T1	Anterior forearm muscles (*flexors of wrist and digits*) 　Medial half of flexor digitorum profundus 　Flexor carpi ulnaris Hypothenar muscles Palmar interossei (*adduct fingers*) Dorsal interossei (*abduct fingers*) Adductor pollicis (*adducts thumb*) Medial 2 lumbricals (*flex MP joints and extend PIP and DIP joints*)	Dorsal and palmar aspects of medial 1-1/2 digits (little finger, medial aspect of ring finger)

Lateral cord
Posterior cord
Medial cord

Ulnar nerve

Flexor carpi ulnaris

Flexor digitorum profundus (medial half)

Adductor pollicis

Hypothenar muscles

Medial two lumbricals

Dorsal and palmar interossei

Smaller Branches of the Brachial Plexus	Anterior Rami	Motor Innervation	Cutaneous Innervation
Dorsal scapular	C5	Rhomboids, levator scapulae	
Long thoracic	C5–C7	Serratus anterior	
Lateral pectoral	C5–C7	Pectoralis major	
Medial pectoral	C8–T1	Pectoralis major Pectoralis minor	
Medial cutaneous nerve of arm	C8–T1		Medial side of arm
Medial cutaneous nerve of forearm	C8–T1		Medial side of forearm
Nerve to subclavius	C5–C6	Subclavius	
Suprascapular	C5–C6	Supraspinatus, infraspinatus	
Subscapular nerves	C5–C6	Subscapularis, teres major	
Thoracodorsal	C6–C8	Latissimus dorsi	

sensory information from the palmar side of the lateral 3-1/2 fingers (thumb, index finger, middle finger, and the lateral one-half of the ring finger) and from the dorsal tips of these same fingers.

The **musculocutaneous** (mŭs′kū-lō-kū-tā′nē-ŭs) **nerve** arises from the lateral cord of the brachial plexus. It innervates the anterior arm muscles (coracobrachialis, biceps brachii, and brachialis), which flex the humerus and flex the forearm. It also receives sensory information from the lateral surface of the forearm. The **radial nerve** arises from the posterior cord of the brachial plexus. It travels along the posterior side of the arm and then along the radial side of

the forearm. The radial nerve innervates the posterior arm muscles (forearm extensors) and the posterior forearm muscles (extensors of the wrist and digits and the supinator of the forearm). It receives sensory information from the posterior arm and forearm surface and the dorsolateral side of the hand.

The **ulnar nerve** arises from the medial cord of the brachial plexus and descends along the medial side of the arm. It wraps posterior to the medial epicondyle of the humerus and then runs along the ulnar side of the forearm. It innervates some of the anterior forearm muscles (the medial region of the flexor digitorum profundus and all

CLINICAL VIEW: In Depth

Brachial Plexus Injuries

Injuries to parts of the brachial plexus are fairly common, especially in individuals aged 18–22. Minor plexus injuries are treated by simply resting the limb. More severe brachial plexus injuries may require nerve grafts or nerve transfers; for very severe injuries, no effective treatment exists. Various nerves of the brachial plexus may be injured.

AXILLARY NERVE INJURY

The axillary nerve can be compressed within the axilla, or it can be damaged if the surgical neck of the humerus is broken (recall that the axillary nerve travels posterior to the surgical neck of the humerus). A patient whose axillary nerve is damaged has great difficulty abducting the arm due to paralysis of the deltoid muscle, as well as anesthesia (lack of sensation) along the superolateral skin of the arm.

RADIAL NERVE INJURY

The radial nerve is especially subject to injury during humeral shaft fractures or in injuries to the lateral elbow. Nerve damage results in paralysis of the extensor muscles of the forearm, wrist, and fingers. A common clinical sign of radial nerve injury is "wrist drop," meaning that the patient is unable to extend his or her wrist. The patient also experiences anesthesia along the posterior arm, the forearm, and the part of the hand normally supplied by this nerve.

POSTERIOR CORD INJURY

The posterior cord of the brachial plexus (which includes the axillary and radial nerves) is commonly injured in the axilla. One cause is improper use of crutches, a condition called *crutch palsy*. Similarly, the posterior cord can be compressed if a person drapes the upper limb over the back of a chair for an extended period of time. Because this can happen if someone passes out in a drunken stupor, this condition is also referred to as *drunkard's paralysis*. Fortunately, full function of these nerves is often regained after a short period of time.

MEDIAN NERVE INJURY

The median nerve may be impinged on or compressed as a result of carpal tunnel syndrome because of the close confines of this narrow passage. Additionally, the nerve may be injured by any deep laceration of the wrist. Median nerve injury often results in paralysis of the thenar group of muscles. The classic sign of median nerve injury is the "ape hand" deformity, which develops over time as the thenar eminence wastes away until the hand eventually resembles that of an ape (apes lack well-developed thumb muscles). The lateral two lumbricals are also paralyzed, and sensation is lost in the part of the hand supplied by the median nerve.

ULNAR NERVE INJURY

The ulnar nerve may be injured by fractures or dislocations of the elbow because of this nerve's close proximity to the medial epicondyle of the humerus. When you "hit your funny bone," you have actually hit your ulnar nerve. Most of the intrinsic hand muscles are paralyzed (including the interossei muscles, the hypothenar muscles, the adductor pollicis, and the medial two lumbricals), making the person unable to adduct or abduct the fingers. In addition, the person experiences sensory loss along the medial side of the hand. A clinician can test for ulnar nerve injury by having a patient hold a piece of paper tightly between the fingers as the doctor tries to pull it away. If the person has weak interosseous muscles, the paper can be easily extracted.

SUPERIOR TRUNK INJURY

The superior trunk of the brachial plexus can be injured by excessive separation of the neck and shoulder, as when a person riding a motorcycle is flipped from the bike and lands on the side of the head. A superior trunk injury affects the C5 and C6 anterior rami, so any brachial plexus branch that has these nerves is also affected to some degree.

INFERIOR TRUNK INJURY

The inferior trunk of the brachial plexus can be injured if the arm is excessively abducted, as when a neonate's arm is pulled too hard during delivery. In children and adults, inferior trunk injuries happen when grasping something above the head in order to break a fall—for example, grabbing a branch to keep from falling out of a tree. An inferior trunk injury involves the C8 and T1 anterior rami, so any brachial plexus branch that is formed from these nerves (such as the ulnar nerve) is also affected to some degree.

of the flexor carpi ulnaris). It also innervates most of the intrinsic hand muscles, including the hypothenar muscles, the palmar and dorsal interossei, and the medial two lumbricals. It receives sensations from the skin of the dorsal and palmar aspects of the medial 1-1/2 fingers (the little finger and the medial half of the ring finger).

The brachial plexus also gives off numerous other nerves that innervate portions of the upper limb and pectoral girdle. These branches are not as large as the terminal branches (see table 16.3).

WHAT DO YOU THINK?

4 Which nerve might you have damaged if you have difficulty abducting your arm and experience anesthesia (lack of sensation) along the superolateral arm?

Lumbar Plexuses

The left and right **lumbar plexuses** are formed from the anterior rami of spinal nerves L1–L4 located lateral to the L_1–L_4 vertebrae and within the psoas major muscle in the posterior abdominal wall (**figure 16.10**). The lumbar plexus is structurally less complex than the brachial plexus.

However, like the brachial plexus, the lumbar plexus is subdivided into an anterior division and a posterior division. The primary nerves of the lumbar plexus are listed in **table 16.4**.

The main nerve of the posterior division of the lumbar plexus is the **femoral nerve.** This nerve supplies the anterior thigh muscles, such as the quadriceps femoris (knee extensor) and the sartorius and iliopsoas (hip flexors). It also receives sensory information from the anterior and inferomedial thigh as well as the medial aspect of the leg. The main nerve of the anterior division is the **obturator nerve,** which travels through the obturator foramen to the medial thigh. There, the nerve innervates the medial thigh muscles (which adduct the thigh) and receives sensory information from the superomedial skin of the thigh. Smaller branches of each lumbar plexus innervate the abdominal wall, the scrotum and the labia, and the inferior portions of the abdominal muscles (table 16.4).

WHAT DO YOU THINK?

5 Which nerve of the lumbar plexus might you have damaged if you have difficulty extending your knee?

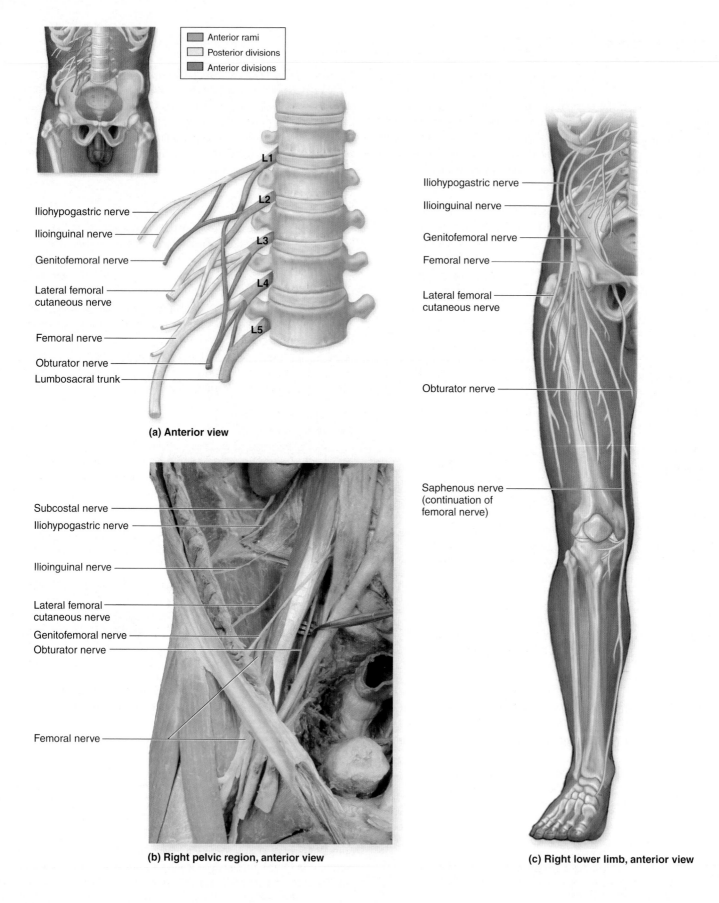

Anterior rami
Posterior divisions
Anterior divisions

L1
L2
L3
L4
L5

Iliohypogastric nerve
Ilioinguinal nerve
Genitofemoral nerve
Lateral femoral cutaneous nerve
Femoral nerve
Obturator nerve
Lumbosacral trunk

(a) Anterior view

Subcostal nerve
Iliohypogastric nerve
Ilioinguinal nerve
Lateral femoral cutaneous nerve
Genitofemoral nerve
Obturator nerve
Femoral nerve

(b) Right pelvic region, anterior view

Iliohypogastric nerve
Ilioinguinal nerve
Genitofemoral nerve
Femoral nerve
Lateral femoral cutaneous nerve
Obturator nerve
Saphenous nerve (continuation of femoral nerve)

(c) Right lower limb, anterior view

Figure 16.10

Lumbar Plexus.   *(a)* Anterior rami of nerves L1–L4 form the lumbar plexus. *(b)* Cadaver photo shows the components of the lumbar plexus. *(c)* Pathways of lumbar plexus nerves.

Table 16.4	Branches of the Lumbar Plexus			

Main Branch		Anterior Rami	Motor Innervation	Cutaneous Innervation

Femoral Nerve
 	
 Iliacus
 Femoral nerve
 Sartorius
 Rectus femoris
 Vastus intermedius
 Vastus lateralis
 	
 L2
 L3
 L4
 Psoas major
 Pectineus
 Vastus medialis
 Sartorius
		L2–L4	Anterior thigh muscles
			   Quadriceps femoris (*knee extensor*)
			   Iliopsoas (*hip flexor*)
			   Sartorius (*hip and knee flexor*)
			 Pectineus[1]
				Anterior thigh
				 Inferomedial thigh
				 Medial side of leg
				 Most medial aspect of foot

Obturator Nerve
 	
 L2
 L3
 L4
 Obturator nerve
 Obturator externus
 Adductor longus
 Adductor brevis
 Adductor longus
 Adductor magnus
 Gracilis
		L2–L4	Medial thigh muscles (*adductors of thigh*)
			   Adductors
			   Gracilis
			   Pectineus[1]
			 Obturator externus (*lateral rotator of thigh*)
				Superomedial thigh

[1] Pectineus may be innervated by the femoral nerve, obturator nerve, or branches from both nerves.

(*continued on next page*)

Table 16.4	Branches of the Lumbar Plexus *(continued)*		
Smaller Branches of the Lumbar Plexus	**Anterior Rami**	**Motor Innervation**	**Cutaneous Innervation**
Iliohypogastric	L1	Partial innervation to abdominal muscles (*flex vertebral column*)	Superior lateral gluteal region Inferior abdominal wall
Ilioinguinal	L1	Partial innervation to abdominal muscles (*flex vertebral column*)	Inferior abdominal wall Scrotum (males) or labia majora (females)
Genitofemoral	L1, L2		Small area in anterior superior thigh Scrotum (males) or labia majora (females)
Lateral femoral cutaneous	L2, L3		Anterolateral thigh

Sacral Plexuses

The left and right **sacral plexuses** are formed from the anterior rami of spinal nerves L4–S4 and are located immediately inferior to the lumbar plexuses **(figure 16.11)**. The lumbar and sacral plexuses are sometimes considered together as the *lumbosacral plexus*. The nerves emerging from a sacral plexus innervate the gluteal region, pelvis, perineum, posterior thigh, and almost all of the leg and foot.

The anterior rami of the sacral plexus organize themselves into an anterior division and a posterior division. The nerves arising from the anterior division tend to innervate muscles that flex (or plantar flex) parts of the lower limb, while the posterior division nerves tend to innervate muscles that extend (or dorsiflex) part of the lower limb. **Table 16.5** lists the main and smaller nerves of the sacral plexus.

The **sciatic** (sī-at′ik) **nerve,** also known as the *ischiadic* (is-kē-at′ik; hip joint) *nerve,* is the largest and longest nerve in the body. This nerve projects from the pelvis through the greater sciatic notch of the os coxae and extends into the posterior region of the thigh. The sciatic nerve is actually composed of two divisions—the *tibial division* and the *common fibular division*—wrapped in a common sheath. Just superior to the popliteal fossa, the two divisions of the sciatic nerve split into two nerves. The **tibial nerve** is formed from the anterior divisions of the sciatic nerve. In the posterior thigh, the tibial division of the sciatic nerve innervates the hamstrings (except for the short head of the biceps femoris) and the hamstring part of the adductor magnus. It travels in the posterior compartment of the leg, where it supplies the plantar flexors of the foot and the toe flexors. In the foot, the tibial nerve splits into the lateral and medial plantar nerves, which innervate the plantar muscles of the foot and conduct sensory impulses from the skin covering the sole of the foot. The **common fibular** (*common peroneal*) **nerve** is formed from the posterior division of the sciatic nerve. As the common fibular division of the sciatic nerve, it supplies the short head of the biceps femoris muscle. Along the lateral knee, as it wraps around the neck of the fibula, this nerve splits into two main branches: the deep fibular nerve and the superficial fibular nerve.

The **deep fibular** (*deep peroneal*) **nerve** travels in the anterior compartment of the leg and terminates between the first and second

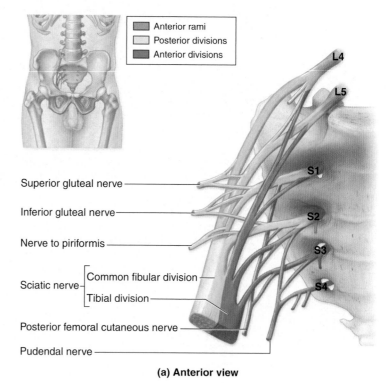

(a) Anterior view

Figure 16.11

Sacral Plexus. Anterior rami of nerves L4, L5, and S1–S4 form the sacral plexus. *(a)* The sacral plexus has six rami and both anterior and posterior divisions. *(b)* A posterior view shows the distribution of nerves of the sacral plexus. *(c, d)* Cadaver photos reveal the major sacral plexus nerves of the right gluteal and popliteal regions.

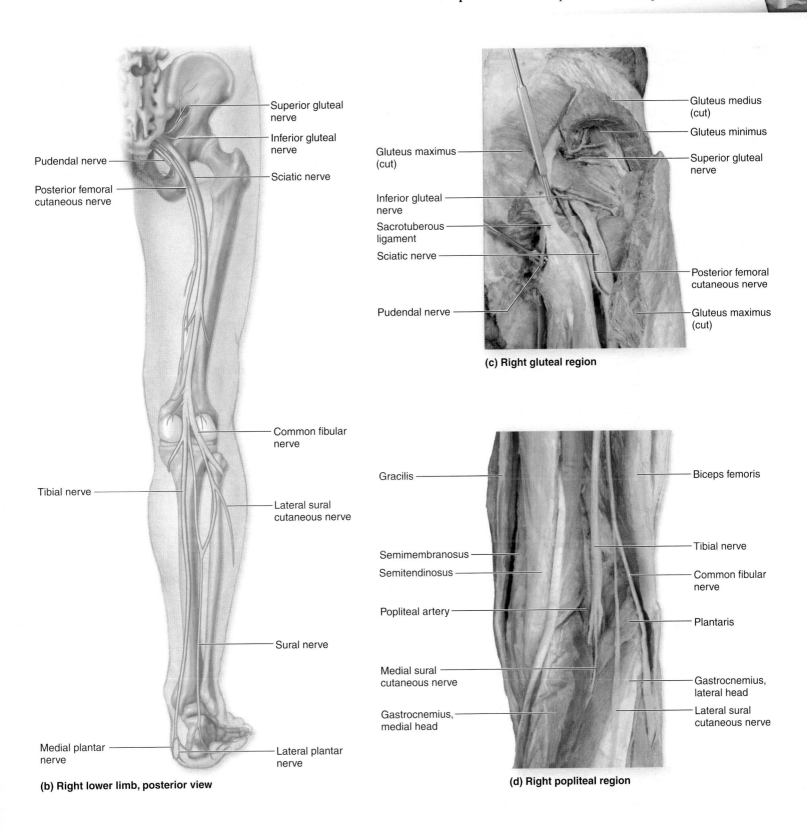

Superior gluteal nerve

Inferior gluteal nerve

Pudendal nerve

Posterior femoral cutaneous nerve

Sciatic nerve

Tibial nerve

Common fibular nerve

Lateral sural cutaneous nerve

Sural nerve

Medial plantar nerve

Lateral plantar nerve

(b) Right lower limb, posterior view

Gluteus maximus (cut)

Inferior gluteal nerve

Sacrotuberous ligament

Sciatic nerve

Pudendal nerve

Gluteus medius (cut)

Gluteus minimus

Superior gluteal nerve

Posterior femoral cutaneous nerve

Gluteus maximus (cut)

(c) Right gluteal region

Gracilis

Semimembranosus

Semitendinosus

Popliteal artery

Medial sural cutaneous nerve

Gastrocnemius, medial head

Biceps femoris

Tibial nerve

Common fibular nerve

Plantaris

Gastrocnemius, lateral head

Lateral sural cutaneous nerve

(d) Right popliteal region

toes. It supplies the anterior leg muscles (which dorsiflex the foot and extend the toes) and the muscles on the dorsum of the foot (which extend the toes). In addition, this nerve receives sensory innervation from the skin between the first and second toes on the dorsum of

the foot. The **superficial fibular** (*superficial peroneal*) **nerve** travels in the lateral compartment of the leg. Just proximal to the ankle, this nerve becomes superficial along the anterior part of the ankle and dorsum of the foot. The superficial fibular nerve innervates the lateral

Table 16.5		Branches of the Sacral Plexus	
Main Branch	**Anterior Rami**	**Motor Innervation**	**Cutaneous Innervation**
Sciatic Nerve (Composed of tibial and common fibular divisions wrapped in a common sheath)	L4–S3	(See tibial and common fibular nerves)	(See tibial and common fibular nerves)

Tibial Nerve
 	
 L4
 L5
 S1
 S2
 S3
 Tibial division of sciatic nerve
 Adductor magnus
 Biceps femoris (long head)
 Semitendinosus
 Semimembranosus
 Tibial nerve
 Gastrocnemius
 Popliteus
 Soleus
 Tibialis posterior
 Flexor digitorum longus
 Flexor hallucis longus
 Medial plantar nerve
 Lateral plantar nerve
	L4–S3	Posterior thigh muscles (*extend thigh and flex leg*)
		    Long head of biceps femoris
		    Semimembranosus
		    Semitendinosus
		    Part of adductor magnus
		 Posterior leg muscles (*plantar flexors of foot, flexors of knee*)
		    Flexor digitorum longus
		    Flexor hallucis longus
		    Gastrocnemius
		    Soleus
		    Popliteus
		    Tibialis posterior (*inverts foot*)
		 Plantar foot muscles (*via medial and lateral plantar nerve branches*)
			Branches to the heel, and via its medial and lateral plantar nerve branches (which supply the sole of the foot)

Common Fibular Nerve (Divides into deep fibular and superficial fibular branches) L4 L5 S1 S2 S3 Common fibular division of sciatic nerve Biceps femoris short head **Common fibular nerve** Fibularis longus Fibularis brevis Tibialis anterior **Superficial fibular nerve** **Deep fibular nerve** Extensor digitorum longus Extensor hallucis longus Fibularis tertius Extensor digitorum brevis Extensor hallucis brevis	L4–S2	Short head of biceps femoris (*knee flexor*); see also deep fibular and superficial fibular nerves	(See deep fibular and superficial fibular nerves)

Table 16.5	Branches of the Sacral Plexus *(continued)*			
Main Branch		**Anterior Rami**	**Motor Innervation**	**Cutaneous Innervation**

Deep Fibular Nerve	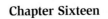 Common fibular nerve — Tibialis anterior — Superficial fibular nerve — **Deep fibular nerve** — Extensor digitorum longus — Extensor hallucis longus — Fibularis tertius — Extensor digitorum brevis — Extensor hallucis brevis	L4–S1	Anterior leg muscles (*dorsiflex foot, extend toes*)
			    Tibialis anterior (*inverts foot*)
			    Extensor hallucis longus
			    Extensor digitorum longus
			    Fibularis tertius
			Dorsum foot muscles (*extend toes*)
			    Extensor hallucis brevis
			    Extensor digitorum brevis
				Dorsal interspace between first and second toes

Superficial Fibular Nerve	Common fibular nerve — Fibularis longus — Fibularis brevis — **Superficial fibular nerve**	L5–S2	Lateral leg muscles (*foot evertors and plantar flexors*)
			    Fibularis longus
			    Fibularis brevis
				Anteroinferior part of leg; most of dorsum of foot

Smaller Branches of the Sacral Plexus	**Anterior Rami**	**Motor Innervation**	**Cutaneous Innervation**
Inferior gluteal nerve	L5–S2	Gluteus maximus (*thigh extensor*)	
Superior gluteal nerve	L4–S1	Gluteus medius, gluteus minimus, and tensor fasciae latae (*abductors of thigh*)	
Posterior femoral cutaneous nerve	S1–S3		Skin on posterior thigh
Pudendal nerve	S2–S4	Muscles of perineum, external anal sphincter, external urethral sphincter	Skin on external genitalia

Sacral Plexus Nerve Injuries

Some branches of the sacral plexus are readily subject to injury. For example, a poorly placed gluteal intramuscular injection can injure the superior or inferior gluteal nerves, and in some cases even the sciatic nerve (see Clinical View in chapter 13). Additionally, a herniated intervertebral disc may impinge on the nerve branches that form the sciatic nerve. Injury to the sciatic nerve produces a condition known as **sciatica** (sī-at'i-kă), which is characterized by extreme pain down the posterior thigh and leg. This pain does not go away unless the injury to the sciatic nerve is remedied. For example, repairing a herniated disc will alleviate the compression on the sciatic nerve.

The common fibular nerve is especially prone to injury due to fracture of the neck of the fibula or compression from a leg cast that is too tight. Compression of the nerve compromises it and its branches (superficial fibular, deep fibular), paralyzing the anterior and lateral leg muscles and leaving the person unable to dorsiflex and evert the foot. One classic sign of fibular nerve injury is "foot drop." As a person lifts the affected foot to take a step, the lack of innervation of the anterior and lateral leg muscles causes the foot to fall into the plantar-flexed position. Because the person can't dorsiflex the foot to walk normally, he or she compensates by flexing the hip to lift the affected area and keep from tripping or stubbing the toes.

compartment muscles of the leg (foot evertors and weak plantar flexors). It also conducts sensory impulses from most of the dorsal surface of the foot and the anteroinferior part of the leg.

WHAT DID YOU LEARN?

7 Where is a posterior root ganglion located, and what does it contain?

8 Identify the nerve plexuses, from superior to inferior.

9 What nerves form the brachial plexus?

10 What are the main nerves of the lumbar and sacral plexuses?

Reflexes

Key topics in this section:

- Properties of a reflex
- Structures and steps involved in a reflex arc
- Reflexes as diagnostic indicators

Reflexes are rapid, automatic, involuntary reactions of muscles or glands to a stimulus. All reflexes have similar properties:

- A *stimulus* is required to initiate a response to sensory input.
- A *rapid response* requires that few neurons be involved and synaptic delay be minimal.
- An *automatic response* occurs the same way every time.
- An *involuntary response* requires no intent or pre-awareness of the reflex activity. Thus, reflexes are usually not suppressed. Awareness of the stimulus occurs after the reflex action has been completed, in time to correct or avoid a potentially dangerous situation.

An example of a reflex occurs when you accidentally touch a hot burner on a stove. Instantly and automatically, you remove your hand from the stimulus (the hot burner), even before you are completely aware that your hand was touching something extremely hot. A reflex is a survival mechanism; it allows us to quickly respond to a stimulus that may be detrimental to our well-being without having to wait for the brain to process the information.

Components of a Reflex Arc

A **reflex arc** is the neural "wiring" of a single reflex. It always begins at a receptor in the PNS, communicates with the CNS, and ends at a peripheral effector, such as a muscle or gland cell. The number of

intermediate steps varies, depending on the complexity of the reflex. Generally, five steps are involved in a simple reflex arc, as illustrated in **figure 16.12** and described here:

1. **Stimulus activates receptor.** Sensory receptors (dendritic endings of a sensory neuron) respond to both external and internal stimuli, such as temperature, pressure, or tactile changes.
2. **Nerve impulse travels through sensory neuron to the CNS.** Sensory neurons conduct impulses from the receptor into the spinal cord.
3. **Information from nerve impulse is processed in the integration center by interneurons.** More complex reflexes may use a number of interneurons within the CNS to integrate and process incoming sensory information and transmit information to a motor neuron. The simplest reflexes do not involve interneurons; rather, the sensory neuron synapses directly on a motor neuron in the anterior gray horn of the spinal cord.
4. **Motor neuron transmits nerve impulse to effector.** The motor neuron transmits a nerve impulse through the anterior root and spinal nerve to the peripheral effector organ.
5. **Effector responds to nerve impulse from motor neuron.** An **effector** (ē-fek'tŏr, -tōr; producer) is a peripheral target organ that responds to the impulse from the motor neuron. This response is intended to counteract or remove the original stimulus.

Reflex arcs may be ipsilateral or contralateral. A reflex arc is termed **ipsilateral** (ip-si-lat'er-ăl; *ipse* = same, *latus* = side) when both the receptor and effector organs of the reflex are on the same side of the spinal cord. For example, an ipsilateral effect occurs when the muscles in your left arm contract to pull your left hand away from a hot object. A reflex arc is **contralateral** (kon-tră-lat'er-ăl; *contra* = opposite) when the sensory impulses from a receptor organ cross over through the spinal cord to activate effector organs in the opposite limb. For example, a contralateral effect occurs when you step on a sharp object with your left foot and then contract the muscles in your right leg to maintain balance as you withdraw your left leg from the damaging object.

Reflexes may also be monosynaptic or polysynaptic (**figure 16.13**). A **monosynaptic** (mon'ō-si-nap'tik; *monos* = single) **reflex** is the simplest of all reflexes. The sensory axons synapse directly on the motor neurons, whose axons project to the effector. Interneurons are not involved in processing this reflex. Very minor synaptic delay is incurred in the single synapse of this reflex arc, resulting in a

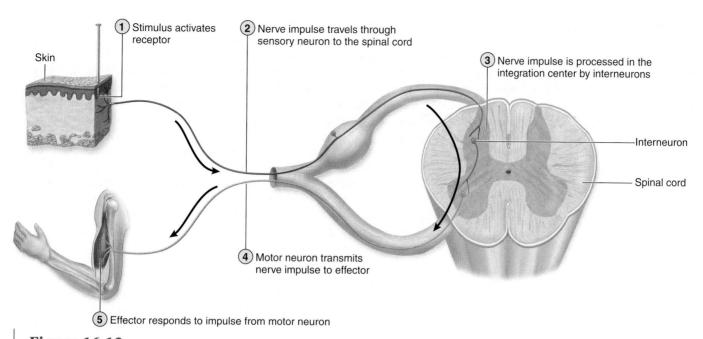

① Stimulus activates receptor

② Nerve impulse travels through sensory neuron to the spinal cord

③ Nerve impulse is processed in the integration center by interneurons

Skin

Interneuron

Spinal cord

④ Motor neuron transmits nerve impulse to effector

⑤ Effector responds to impulse from motor neuron

Figure 16.12

Reflex Arc. A reflex arc is a nerve pathway composed of neurons that control rapid, unconscious, automatic responses to a stimulus.

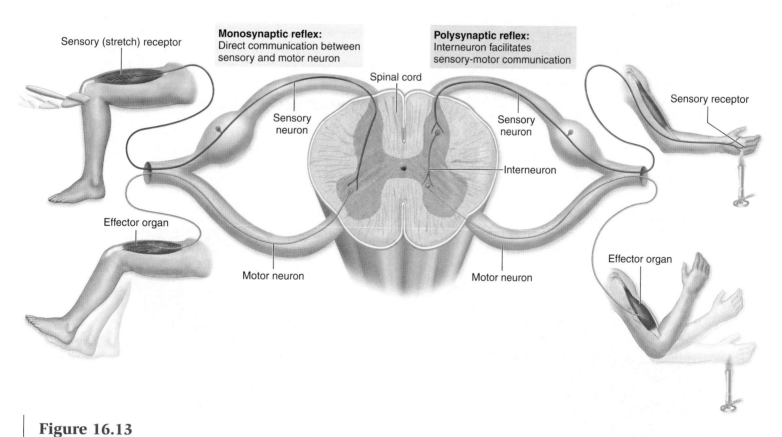

Sensory (stretch) receptor

Monosynaptic reflex: Direct communication between sensory and motor neuron

Polysynaptic reflex: Interneuron facilitates sensory-motor communication

Spinal cord

Sensory neuron

Sensory neuron

Sensory receptor

Interneuron

Effector organ

Motor neuron

Motor neuron

Effector organ

Figure 16.13

Monosynaptic and Polysynaptic Reflexes. (*Left*) The minimal number of neurons and the pathways of a monosynaptic reflex are compared to (*right*) those of a polysynaptic reflex.

very prompt reflex response. An example of a monosynaptic reflex is the patellar (knee-jerk) reflex, that physicians use to assess the functioning of the spinal cord. By tapping the patellar ligament with a reflex hammer, the muscle spindles in the quadriceps muscles are stretched. The brief stimulus causes a reflexive contraction that occurs unopposed and produces a noticeable kick of the leg.

Polysynaptic (pol'ē-si-nap'tik; *polys* = many) **reflexes** have more complex neural pathways that exhibit a number of synapses involving interneurons within the reflex arc. Because this reflex arc has more components, there is a more prolonged delay between stimulus and response. An example of a polysynaptic reflex is the crossed-extensor reflex, which is the reflex that supports postural muscle activity when you withdraw your foot from a painful stimulus, as in the previous example. Contraction of the extensor muscles in the other limb occurs by polysynaptic reflexes to enable you to maintain balance.

Examples of Spinal Reflexes

Some common spinal reflexes are the withdrawal reflex, the stretch reflex, and the Golgi tendon reflex.

A **withdrawal (flexor) reflex** is a polysynaptic reflex arc that is initiated by a painful stimulus, such as touching something very hot. Stimulation of a receptor organ causes the transmission of sensory information to the spinal cord. Interneurons receive the sensory information and stimulate motor neurons to direct flexor muscles to contract in response. Simultaneously, antagonistic (extensor) muscles are inhibited so that the traumatized body part may be quickly withdrawn from the harmful stimulation.

The **stretch reflex** is a monosynaptic reflex that monitors and regulates skeletal muscle length. Stretch in a muscle is monitored by a stretch receptor called a muscle spindle. When a stimulus results in the stretching of a muscle, that muscle reflexively contracts **(figure 16.14)**. The patellar (knee-jerk) reflex is an example

of a stretch reflex. The stimulus (the tap on the patellar ligament) stretches the quadriceps femoris muscle and initiates contraction of the muscle, thereby extending the knee joint.

The **Golgi tendon reflex** is a polysynaptic reflex that prevents skeletal muscles from tensing excessively. Golgi tendon organs are nerve endings located within tendons near a muscle–tendon junction **(figure 16.15)**. As a muscle contracts, force is exerted on its associated tendon, resulting in increased tension in the tendon and activation of the Golgi tendon organ. Nerve impulses in the Golgi tendon organ signal interneurons in the spinal cord, which in turn inhibit the actions of the motor neurons. When the motor neurons that cause the muscle contraction are inhibited, the associated muscle is allowed to relax, thus protecting the muscle and tendon from excessive tension damage.

Reflex Testing in a Clinical Setting

Reflexes can be an important diagnostic tool. Clinicians use them to test specific muscle groups and specific spinal nerves or segments of the spinal cord **(table 16.6)**. Although some variation in reflexes is normal, a consistently abnormal reflex response may indicate damage to the nervous system or muscles.

A reflex response may be normal, hypoactive, or hyperactive. The term **hypoactive reflex** means that a reflex response is diminished or absent. A hypoactive reflex may indicate damage to a segment of the spinal cord, or it may suggest muscle disease or damage to the neuromuscular junction. A **hyperactive reflex** refers to an abnormally strong response. It may indicate damage somewhere in either the brain or spinal cord, especially if it is accompanied by **clonus** (klō'nŭs; tumult), rhythmic oscillations between flexion and extension when the muscle reflex is tested.

WHAT DID YOU LEARN?

11 List the five steps in a reflex arc.

12 What is the major difference between monosynaptic reflexes and polysynaptic reflexes?

Motor
Sensory

Muscle stretch stimulates
sensory nerve impulse
to travel to the CNS

Muscle spindle

Sensory nerve
endings

Motor nerve
endings

Muscle

Stretch

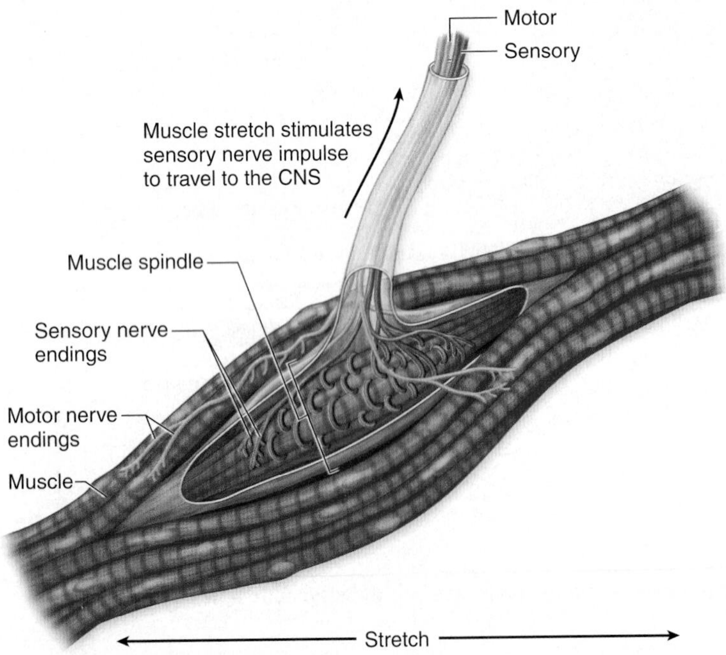

Figure 16.14

Stretch Reflexes. A stretch reflex is a simple monosynaptic reflex involving two neurons. A stretching force detected by a muscle spindle results in the contraction of that muscle.

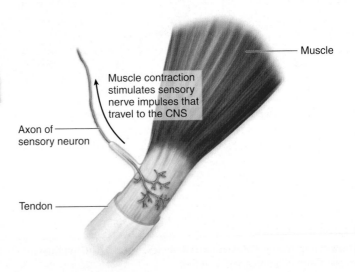

Muscle

Muscle contraction
stimulates sensory
nerve impulses that
travel to the CNS

Axon of
sensory neuron

Tendon

Figure 16.15

Golgi Tendon Reflex. Contraction of a muscle generates tension on its tendons, and may cause a tendon reflex response. This illustration shows a Golgi tendon organ, which detects the contraction force and results in the relaxation of that muscle.

Table 16.6	Some Clinically Important Reflexes	
Reflex	**Spinal Nerve Segments Tested**	**Normal Action of Effector**
Biceps reflex	C5, C6	Flexes elbow when biceps brachii tendon is tapped
Triceps reflex	C6, C7	Extends elbow when triceps brachii tendon is tapped
Abdominal reflexes	T8–T12	Contract abdominal muscles when one side of the abdominal wall is briskly stroked
Cremasteric reflex	L1, L2	Elevates testis (due to contraction of cremaster muscle in scrotum) when medial side of thigh is briskly stroked
Patellar (knee-jerk) reflex	L2–L4	Extends knee when patellar ligament is tapped
Ankle (Achilles) reflex	S1, S2	Plantar flexes ankle when calcaneal tendon is tapped
Plantar reflex	L5, S1	Plantar flexes foot; flexes toes when plantar side of foot is briskly stroked[1]

[1] This is the *normal* reflex response in adults; in adults with spinal cord damage and in normal infants, the **Babinski sign** occurs, which is extension of the great toe and fanning of the other toes.

Development of the Spinal Cord

Key topic in this section:

■ How the spinal cord and spinal nerves develop in the embryo

Recall from previous chapters that the central nervous system forms primarily from the embryonic **neural tube,** while the cranial and spinal nerves form primarily from **neural crest cells** that have split off from the developing neural tube. The cranial (superior) part of the neural tube expands and develops into the brain, while the caudal (inferior) part of the neural tube forms the spinal cord. The following discussion focuses on the caudal part of the neural tube and its nearby neural crest cells.

As the caudal part of the neural tube differentiates and specializes, the spinal cord begins to develop **(figure 16.16)**. However, this developmental process is much less complex than that for the brain. A hollow **neural canal** in the neural tube develops into the central canal of the spinal cord. Note that the neural canal doesn't "shrink" in size; rather, the neural tube around it grows at a rapid rate. Thus, as the neural tube walls grow and expand, the neural canal in the newborn appears as a tiny hole called the central canal.

During the fourth and fifth weeks of embryonic development, the walls of the neural tube start to grow rapidly and unevenly. Part of the neural tube forms the white matter of the spinal cord, while other components form gray matter. By the sixth week of development, a horizontal groove called the **sulcus limitans** (lim′i-tanz;

limes = boundary) forms in the lateral walls of the central canal (figure 16.16b). The sulcus limitans also represents a dividing point in the neural tube as two specific regions become evident on each side: the basal plates and the alar plates.

The **basal plates** lie anterior to the sulcus limitans. The basal plates develop into the anterior and lateral horns, motor structures of the gray matter. They also form the anterior part of the gray commissure. The **alar** (ā′lăr; *ala* = wing) **plates** lie posterior to the sulcus limitans. By about the ninth week of development, the alar plates develop into posterior horns, sensory structures of the gray matter. They also form the posterior part of the gray commissure.

During the embryonic period, the spinal cord extends the length of the vertebral canal. However, during the fetal period, the growth of the vertebral column (and its vertebral canal) outpaces that of the spinal cord. By the sixth fetal month, the spinal cord is at the level of the S_1 vertebra, while a newborn's spinal cord ends at about the L_3 vertebra. By adulthood, the spinal cord length extends only to the level of the L_1 vertebra. This disproportionate growth explains why the lumbar, sacral, and coccygeal regions of the spinal cord and its associated nerve roots do not lie next to their respective vertebrae.

WHAT DID YOU LEARN?

13 From what embryonic structures do most components of the cranial and spinal nerves form?

14 What structures develop from the alar plates?

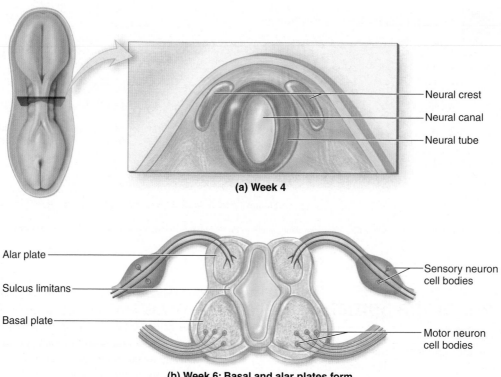

(a) Week 4

Neural crest
Neural canal
Neural tube

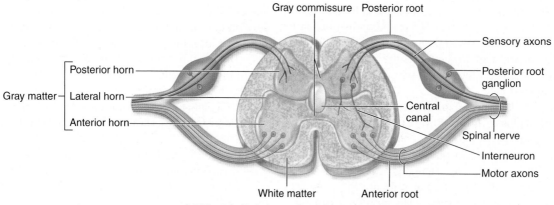

Alar plate

Sulcus limitans

Basal plate

Sensory neuron cell bodies

Motor neuron cell bodies

(b) Week 6: Basal and alar plates form

Gray commissure   Posterior root

Posterior horn

Gray matter — Lateral horn

Anterior horn

Sensory axons

Posterior root ganglion

Central canal

Spinal nerve

Interneuron

Motor axons

White matter   Anterior root

(c) Week 9: Gray horns form from basal and alar plates

Figure 16.16

Spinal Cord Development.   The spinal cord begins development as a tubular extension of the brain. *(a)* A cross section shows the structures of the neural tube of an embryo in week 4 of development. Transverse sections show *(b)* the formation of the basal and alar plates at week 6 and *(c)* the developing spinal cord at week 9.

CLINICAL TERMS

hemiplegia (hem-ē-plē′jē-ă; *hemi* = one-half, *plege* = stroke) Paralysis of the upper and lower limbs on one side of the body only, usually as a result of a stroke (cerebrovascular accident).

myelitis (mī-ĕ-lī′tis; *myelos* = medulla, marrow) Inflammation of the spinal cord.

paraplegia (par-ă-plē′jē-ă) Paralysis that results in loss of motor control in both lower limbs.

quadriplegia (kwah′dri-plē′jē-ă; *quattor* = four) Paralysis of both the upper and the lower limbs.

CHAPTER SUMMARY

	■ The spinal cord and its attached spinal nerves serve as a pathway for sensory and motor impulses and are responsible for reflexes.
Gross Anatomy of the Spinal Cord 485	■ The adult spinal cord extends inferiorly from the brain through the vertebral canal and ends at the level of the L_1 vertebra.
	■ Thirty-one pairs of spinal nerves connect the spinal cord to the body: 8 pairs of cervical nerves, 12 pairs of thoracic nerves, 5 pairs of lumbar nerves, 5 pairs of sacral nerves, and 1 pair of coccygeal nerves.
Spinal Cord Meninges 487	■ An outer epidural space separates the tough dura mater membrane from the inner walls of the vertebral canal.
	■ The subarachnoid space, which is internal to the arachnoid, houses the cerebrospinal fluid.
	■ The pia mater is the innermost meningeal layer. It is bound laterally to the spinal dura mater by paired denticulate ligaments.
Sectional Anatomy of the Spinal Cord 489	■ Gray matter is centrally located and H-shaped; peripheral to the gray matter is the white matter, which is composed primarily of myelinated axons.

Location and Distribution of Gray Matter   489

- ■ Gray matter is composed of three horns: anterior (cell bodies of somatic motor neurons), lateral (cell bodies of autonomic motor neurons), and posterior (sensory axons and interneurons).
- ■ The gray commissure connects the left and right sides of the gray matter and contains the axons of interneurons that extend between opposite sides of the spinal cord.

Location and Distribution of White Matter   490

- ■ The white matter is organized into three pairs of funiculi, each composed of specific tracts. Sensory information moves through ascending tracts to the brain, and motor information is carried by descending tracts from the brain to the spinal cord.

Spinal Nerves 491

- ■ Spinal nerves originate from anterior and posterior rootlets.

Spinal Nerve Distribution   492

- ■ Spinal nerves have two branches: A posterior ramus innervates the skin and deep muscles of the back, and an anterior ramus innervates the anterior and lateral portions of the trunk and the limbs.

Nerve Plexuses   492

- ■ A nerve plexus is a network of interwoven anterior rami.

Intercostal Nerves   494

- ■ The anterior rami of spinal nerves T1–T11 form the intercostal nerves. Nerve T12 is called a subcostal nerve.

Cervical Plexuses   494

- ■ The cervical plexus is formed from the anterior rami of C1–C4. It innervates the neck and portions of the head and shoulders.

Brachial Plexuses   495

- ■ The brachial plexus innervates the upper limb and is formed from the anterior rami of C5–T1.

Lumbar Plexuses   501

- ■ The lumbar plexus innervates the anterior and medial thigh and the skin of the medial leg. It is formed from the anterior rami of L1–L4.

Sacral Plexuses   504

- ■ The sacral plexus innervates most of the lower limb and is formed from the anterior rami of L4–S4.

(continued on next page)

CHAPTER SUMMARY (continued)

Reflexes 508	■ A reflex is a rapid, automatic, involuntary motor response of muscles or glands to a stimulus.
	Components of a Reflex Arc 508
	■ The five steps involved in a reflex arc are (1) activation of a receptor by a stimulus; (2) impulse conduction to the CNS; (3) integration and processing of information by interneurons; (4) stimulation of a motor neuron; and (5) effector organ response.
	■ The simplest reflex arc is a monosynaptic reflex, in which the sensory neuron synapses directly with the motor neuron. A polysynaptic reflex involves a sensory neuron, a motor neuron, and at least one interneuron connecting the sensory and motor neurons.
	Examples of Spinal Reflexes 510
	■ A withdrawal reflex is polysynaptic and activates flexor muscles to immediately remove a body part from a painful stimulus.
	■ A stretch reflex is monosynaptic and regulates skeletal muscle length and tone.
	■ A Golgi tendon reflex is polysynaptic and prevents excessive tension in a muscle by inhibiting the contraction of the muscle, allowing it to relax.
	Reflex Testing in a Clinical Setting 510
	■ Testing differentiates between hypoactive and hyperactive reflexes, and can help diagnose nervous system or muscular disorders.
Development of the Spinal Cord 511	■ The neural tube forms basal plates and alar plates. The basal plates form the anterior horns, lateral horns, and the anterior part of the gray commissure, while the alar plates form the posterior horns and the posterior part of the gray commissure.

CHALLENGE YOURSELF

Matching

Match each numbered item with the most closely related lettered item.

_____ 1. sacral plexus

_____ 2. posterior root

_____ 3. filum terminale

_____ 4. white matter

_____ 5. cervical plexus

_____ 6. basal plate

_____ 7. reflex

_____ 8. lateral horn

_____ 9. dermatome

_____ 10. femoral nerve

a. strand of pia mater that anchors spinal cord to coccyx

b. innervates infrahyoid muscles

c. forms anterior and lateral horns

d. contains axons of sensory neurons

e. contains cell bodies of autonomic motor neurons

f. a segment of skin supplied by a spinal nerve

g. innervates gluteal region and most of lower limb

h. innervates anterior thigh muscles

i. composed of tracts and funiculi

j. rapid, involuntary motor reaction of a muscle

Multiple Choice

Select the best answer from the four choices provided.

_____ 1. The tapered inferior end of the spinal cord is called the
   a. conus medullaris.
   b. filum terminale.
   c. cauda equina.
   d. posterior root.

_____ 2. The anterior root of a spinal nerve contains
   a. axons of both motor and sensory neurons.
   b. axons of sensory neurons only.
   c. interneurons.
   d. axons of motor neurons only.

_____ 3. Identify the meningeal layer immediately deep to the subdural space.
   a. pia mater
   b. arachnoid
   c. epidural space
   d. dura mater

_____ 4. Axons cross from one side of the spinal cord to the other through a region called the
   a. lateral horn.
   b. posterior horn.
   c. gray commissure.
   d. anterior horn.

_____ 5. The radial nerve originates from the _____ plexus.
   a. cervical
   b. lumbar
   c. sacral
   d. brachial

_____ 6. Which structure provides motor innervation to the deep back muscles and receives sensory information from the skin of the back?
    a. anterior ramus
    b. anterior root
    c. posterior ramus
    d. posterior root

_____ 7. Lower limbs are supplied by neurons from the _____ of the spinal cord.
    a. lumbosacral enlargement
    b. thoracic region
    c. cervical enlargement
    d. All of these are correct.

_____ 8. The subarachnoid space contains _____, and the epidural space contains _____.
    a. CSF; fat, connective tissue, and blood vessels
    b. fat, connective tissue, and blood vessels; blood
    c. CSF; pia mater
    d. fat, connective tissue, and blood vessels; CSF

_____ 9. The white matter of the spinal cord is composed primarily of
    a. unmyelinated axons.
    b. neurolemmocytes and satellite cells.
    c. myelinated axons.
    d. cell bodies of neurons.

_____ 10. Which statement is true about intercostal nerves?
    a. They are formed from the posterior rami of spinal nerves.
    b. They form a thoracic plexus of nerves.
    c. They originate from the thoracic part of the spinal cord.
    d. They innervate the deep back muscles of the thoracic region.

Content Review

1. Identify the spinal cord parts, which spinal nerves are associated with them, and their relationship to the corresponding vertebrae.

2. Where is the epidural space? What is housed there?

3. List the three gray matter horns on each side of the spinal cord, and discuss the neuronal composition of each. In addition, list which types of nuclei (motor or sensory) are located in each horn.

4. Where are the cervical plexuses located, and what do they innervate?

5. What are the main terminal branches of the brachial plexus, and what muscles do these terminal branches innervate?

6. What anterior rami form the lumbar plexus, and what are some nerves formed from this plexus?

7. What muscles do the tibial and common fibular nerves innervate?

8. What is a reflex? How does it differ from a muscle movement that you consciously control, as when you consciously contract your biceps brachii muscle?

9. What are the differences between withdrawal, stretch, and Golgi tendon reflexes?

10. Where are the basal plates of the neural tube, and what does each form?

Developing Critical Reasoning

1. Arthur dove off a small cliff into water that was shallower than he expected and hit his head on a submerged object. He is now a quadriplegic, meaning that both his upper and lower limbs are paralyzed. Approximately where is the location of his injury? What is the likelihood that Arthur will recover from this injury? (You may want to review parts of chapter 15 to answer the latter question.)

2. Jessica was knocked off her bicycle and fractured the medial epicondyle of her elbow. The neurologist detected swelling and increased pressure around Jessica's injury, and suspected that a nerve might be damaged as well. What nerve is likely damaged, and what other symptoms would the neurologist expect as a result of this injury?

ANSWERS TO "WHAT DO YOU THINK?"

1. The two layers of the cranial dura mater split to form the dural venous sinuses, which are large veins that drain blood away from the brain. The spinal cord lacks dural venous sinuses, but has smaller veins that travel in the epidural space around the spinal cord.

2. An anterior ramus is larger than a posterior ramus because the posterior rami only innervate deep back muscles and the skin of the back, while the anterior rami innervate almost all other body structures (e.g., the limbs and the anterior and lateral trunk).

3. A nerve plexus houses axons from several different spinal nerves. Thus, damage to a single segment of the spinal cord or damage to a single spinal nerve generally does not result in complete loss of innervation to a particular muscle or region of skin.

4. Anesthesia along the upper lateral arm and difficulty abducting the arm indicate damage to the axillary nerve.

5. Difficulty extending the knee indicates damage to the femoral nerve.

Visit the McKinley/O'Loughlin _Human Anatomy_, 2e website at aris.mhhe.com

17

NERVOUS SYSTEM

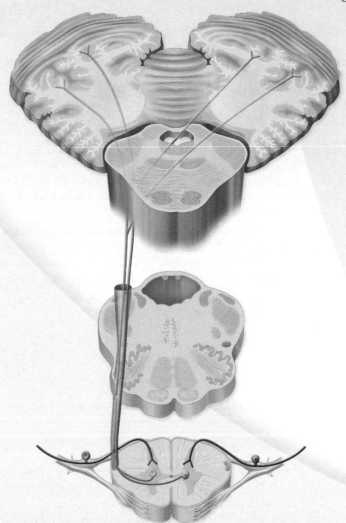

Pathways and Integrative Functions

Ronald Reagan, the fortieth president of the United States, died in June 2004 after a long bout with Alzheimer disease. More than a decade earlier, Mr. Reagan had publicly revealed the onset of his illness by saying, "At the moment, I feel just fine." Alzheimer disease is a progressive dementia that debilitates the functioning of the central nervous system (CNS) and usually affects people in their 60s or over. This neurodegenerative disease causes progressive decline in memory, judgment, and reasoning, as well as disruption of neurologic function within the brain. The cerebral cortex atrophies, and abnormal protein deposits accumulate in the brain. Mr. Reagan's intellectual capacity declined over the ensuing years. As one anonymous individual put it, "His mind just faded away." This chapter focuses on the brain's higher-order activities—such as memory and learning—which depend on the proper functioning of sensory and motor pathways in the nervous system.

General Characteristics of Nervous System Pathways

Key topic in this section:

- Characteristics of sensory and motor pathways in the spinal cord

The CNS communicates with peripheral body structures through **pathways.** These pathways conduct either sensory or motor information; processing and integration occur continuously along them. These pathways travel through the white matter of the brainstem and/or spinal cord as they connect various CNS regions with cranial and spinal nerves.

A pathway consists of a tract and nucleus. **Tracts** are groups or bundles of axons that travel together in the CNS. Each tract may work with multiple nuclei groups in the CNS. A **nucleus** is a collection of neuron cell bodies located within the CNS (see table 15.2).

Nervous system pathways are sensory or motor. **Sensory pathways** are also called *ascending pathways* because the sensory information gathered by sensory receptors ascends through the spinal cord to the brain, while **motor pathways** are also called *descending pathways* because they transmit motor information that descends from the brain through the spinal cord to muscles or glands. Most of the nervous system pathways we discuss in this chapter share several general characteristics:

- Most pathways **decussate** (dē′kŭ-sāt; *decusseo* = to make in the form of an X) (cross over) from one side of the body to the other side at some point in their travels. This crossover process, called decussation, means that the left side of the brain processes information from the right side of the body, and vice versa. For example, when you write with your right hand, the left side of your brain is controlling those right-sided muscles. The term **contralateral** is used to mean the opposite side, whereas the term **ipsilateral** means the same side. Over 90% of all pathways decussate, although the point at which decussation occurs can vary slightly from pathway to pathway.

WHAT DO YOU THINK?

❶ Can you think of a reason why most pathways decussate (cross over) from one side of the body to the other?

- In most pathways, there is a precise correspondence of receptors in body regions, through axons, to specific functional areas in the cerebral cortex. This correspondence is called **somatotopy**

(sō-mă-tot′ō-pē; *soma* = body, *topos* = place). For example, recall the homunculus map in chapter 15 (see figure 15.12), which depicted the surface of the precentral gyrus and showed the parts of the primary motor cortex that control specific body regions. The pathways that connect these parts of the primary motor cortex to a specific body part exhibit somatotopy. Somatotopy is also seen in the sensory homunculus on the primary somatosensory cortex of the postcentral gyrus.

- All pathways are composed of paired tracts. A pathway on the left side of the CNS has a matching tract on the right side of the CNS. Because each tract innervates structures on only one side of the body, both left and right tracts are needed to innervate both the left and right sides of the body.

- Most pathways are composed of a series of two or three neurons that work together. Sensory pathways have primary neurons, secondary neurons, and sometimes tertiary neurons that facilitate the pathway's functioning. In contrast, motor pathways use an upper motor neuron and a lower motor neuron. The cell bodies are located in the nuclei associated with each pathway. We discuss the specific neurons in greater detail later in this chapter.

WHAT DID YOU LEARN?

❶ What is meant by somatotopy?

> ### Study Tip!
>
> Tracts and pathways are named according to their origin and termination. Each has a composite name: The prefix, or first half of the name, indicates its origin, and the suffix, or second half of the name, indicates its destination. For example, sensory pathways usually begin with the prefix *spino-*, indicating that they originate in the spinal cord. So the tract that originates in the spinal cord and terminates in the cerebellum is called the *spinocerebellar tract.* Motor pathways begin with either *cortico-*, indicating an origin in the cerebral cortex, or the name of a brainstem nucleus, such as *rubro-*, indicating an origin within the red nucleus of the mesencephalon. Tracts that terminate in the spinal cord have the suffix *-spinal* as part of their name, Thus, both *corticospinal* and *rubrospinal* denote motor tracts.

Sensory Pathways

Key topics in this section:

- Location and relationship of primary, secondary, and tertiary neurons
- The three major somatosensory pathways

Sensory pathways are ascending pathways that conduct information about limb position and the sensations of touch, temperature, pressure, and pain to the brain. *Somatosensory pathways* process stimuli received from receptors within the skin, muscles, and joints, while *viscerosensory pathways* process stimuli received from the viscera.

The multiple types of body sensations detected by the somatosensory system are grouped into three spinal cord pathways, each with a different brain destination: (1) Discriminative touch permits us to describe textures and shapes of unseen objects and includes pressure, touch, and vibration perception. (2) Temperature and pain allow us to detect those sensations, as well as the sensation of an itch. (3) Proprioception allows us to detect the position of joints, stretch in muscles, and tension in tendons. (Note: Visceral pain pathways will be discussed in chapter 18).

Sensory receptors detect stimuli and then conduct nerve impulses to the central nervous system. Sensory pathway centers within either the spinal cord or the brainstem process and filter the incoming sensory information. These centers determine whether the incoming sensory stimulus should be transmitted to the cerebrum or terminated. Consequently, not all incoming impulses reach the cerebral cortex and our conscious awareness.

Functional Anatomy of Sensory Pathways

Sensory pathways utilize a series of two or three neurons to transmit stimulus information from the body periphery to the brain (**table 17.1**). The first neuron in this chain is the **primary neuron** (or *first-order neuron*). The dendrites of this sensory neuron are part of the receptor that detects a specific stimulus. The cell bodies of primary neurons reside in the posterior root ganglia of spinal nerves or the sensory ganglia of cranial nerves. The axon of the primary neuron projects to a secondary neuron within the CNS. The **secondary neuron** (or *second-order neuron*), the second neuron in this chain, is an interneuron. The cell body of this neuron resides within either the posterior horn of the spinal cord or a brainstem nucleus. The axon of a secondary neuron projects either to the thalamus for conscious sensations or to the cerebellum for unconscious proprioception. The

axon of the secondary neuron arriving in the thalamus synapses with the tertiary neuron, the third neuron in the chain. The **tertiary neuron** (or *third-order neuron*) is an interneuron whose cell body resides within the thalamus. Recall that the thalamus is the central processing and coding center for almost all sensory information; thus, it makes sense (pun intended!) that the last neuron in a sensory pathway chain resides in the thalamus.

The three major types of **somatosensory pathways** are the posterior funiculus–medial lemniscal pathway, the anterolateral pathway, and the spinocerebellar pathway (**figure 17.1**).

Posterior Funiculus–Medial Lemniscal Pathway

The **posterior funiculus–medial lemniscal pathway** (or *posterior column pathway*) projects through the spinal cord, brainstem, and diencephalon before terminating within the cerebral cortex (**figure 17.2**). Its name derives from two components: the tracts within the spinal cord, collectively called the **posterior funiculus** (fū-nik′ū-lŭs; *funis* = cord); and the tracts within the brainstem, collectively called the **medial lemniscus** (lem-nis′kŭs; ribbon). This pathway conducts sensory stimuli concerned with proprioceptive information about limb position and discriminative touch, precise pressure, and vibration sensations.

The posterior funiculus–medial lemniscal pathway uses a chain of three neurons to signal the brain about a specific stimulus. Axons of the primary neurons traveling in spinal nerves reach the CNS through the posterior roots of spinal nerves. Upon entering the spinal cord,

Table 17.1	Sensory Pathway Neurons			
Neuron	**Functional Classification**	**Cell Body Origin**		**Projects To:**
Primary	Sensory neuron	Posterior root ganglia of spinal nerves; sensory ganglia of cranial nerves		Secondary neuron
Secondary	Interneuron	Posterior horn of brainstem nucleus		Thalamus or cerebellum
Tertiary	Interneuron	Thalamus		Cerebral cortex

Figure 17.1

Sensory Pathways in the Spinal Cord. The major sensory (ascending) pathways, shown in various shades of blue, and bilaterally symmetrical tracts. The major motor tracts are indicated in pale red. Note: These colors are used to denote the different sensory pathways only.

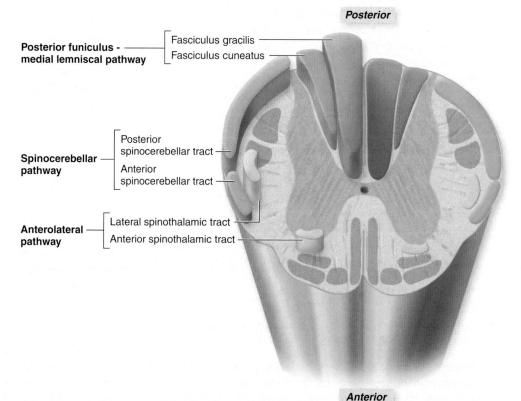

Posterior

Posterior funiculus - medial lemniscal pathway — Fasciculus gracilis / Fasciculus cuneatus

Spinocerebellar pathway — Posterior spinocerebellar tract / Anterior spinocerebellar tract

Anterolateral pathway — Lateral spinothalamic tract / Anterior spinothalamic tract

Anterior

these axons ascend within a specific posterior funiculus, either the **fasciculus cuneatus** (kū′nē-ā-tŭs; *cuneus* = wedge) or the **fasciculus gracilis** (gras′i-lis). The fasciculus cuneatus houses axons from sensory neurons originating in the upper limbs, superior trunk, neck, and posterior region of the head, whereas the fasciculus gracilis carries axons from sensory neurons originating in the lower limbs and inferior trunk. The sensory input into both posterior funiculi is organized somatotopically—that is, there is a correspondence between a receptor's location in a body part and a particular location in the CNS. Thus, the sensory information originating from inferior regions is medially located within the fasciculus, and the sensory information originating at progressively more superior regions is located more laterally.

Sensory axons ascending within the posterior funiculi synapse on secondary neuron cell bodies housed within a posterior funiculus nucleus in the medulla oblongata. These nuclei are either the nucleus cuneatus or the nucleus gracilis, and they correspond to the fasciculus cuneatus and fasciculus gracilis, respectively. These secondary neurons then project axons to relay the incoming sensory information to the thalamus on the opposite side of the brain through the medial lemniscus. Decussation occurs after secondary neuron axons exit their specific nuclei and before they enter the

medial lemniscus. As the sensory information travels toward the thalamus, the same classes of sensory input (touch, pressure, and vibration) that have been collected by cranial nerves CN V (trigeminal), CN VII (facial), CN IX (glossopharyngeal), and CN X (vagus) are integrated and incorporated into the ascending pathways, collectively called the trigeminothalamic tract.

The axons of the secondary neurons synapse on cell bodies of the tertiary neurons within the thalamus. Within the thalamus, the ascending sensory information is sorted according to the region of the body involved (somatotopically). Axons from these tertiary neurons conduct sensory information to a specific location of the primary somatosensory cortex.

Anterolateral Pathway

The **anterolateral pathway** (or *spinothalamic pathway*) is located in the anterior and lateral white funiculi of the spinal cord **(figure 17.3)**. It is composed of the **anterior spinothalamic tract** and the **lateral spinothalamic tract**. Axons projecting from primary neurons enter

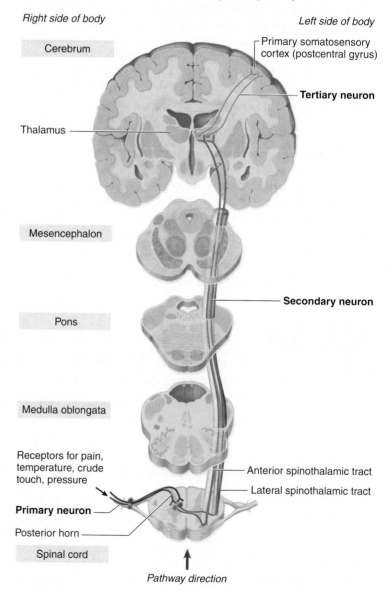

Right side of body   *Left side of body*

Cerebrum

Thalamus

Mesencephalon

Pons

Medulla oblongata

Receptors for pain, temperature, crude touch, pressure

Primary neuron

Posterior horn

Spinal cord

Primary somatosensory cortex (postcentral gyrus)

Tertiary neuron

Secondary neuron

Anterior spinothalamic tract

Lateral spinothalamic tract

Pathway direction

Figure 17.3

Anterolateral Pathway. This pathway conducts crude touch, pressure, pain, and temperature sensations toward the brain. Decussation of axons occurs at the level where the primary neuron axon enters the spinal cord. The primary neuron is purple, the secondary neuron is dark blue, and the tertiary neuron is light green.

Right side of body   *Left side of body*

Cerebrum

Thalamus

Mesencephalon

Medulla oblongata

Receptors for discriminative touch, proprioception, precise pressure, and vibration (from neck, trunk, limbs)

Anterior root

Posterior root

Spinal cord

Primary somatosensory cortex (postcentral gyrus)

Tertiary neuron

Secondary neuron

Medial lemniscus

Nucleus gracilis

Nucleus cuneatus

Medial lemniscus

Decussation prior to entry into the medial lemniscus

Primary neuron

Fasciculus gracilis ⎤ Posterior
Fasciculus cuneatus ⎦ funiculus

Pathway direction

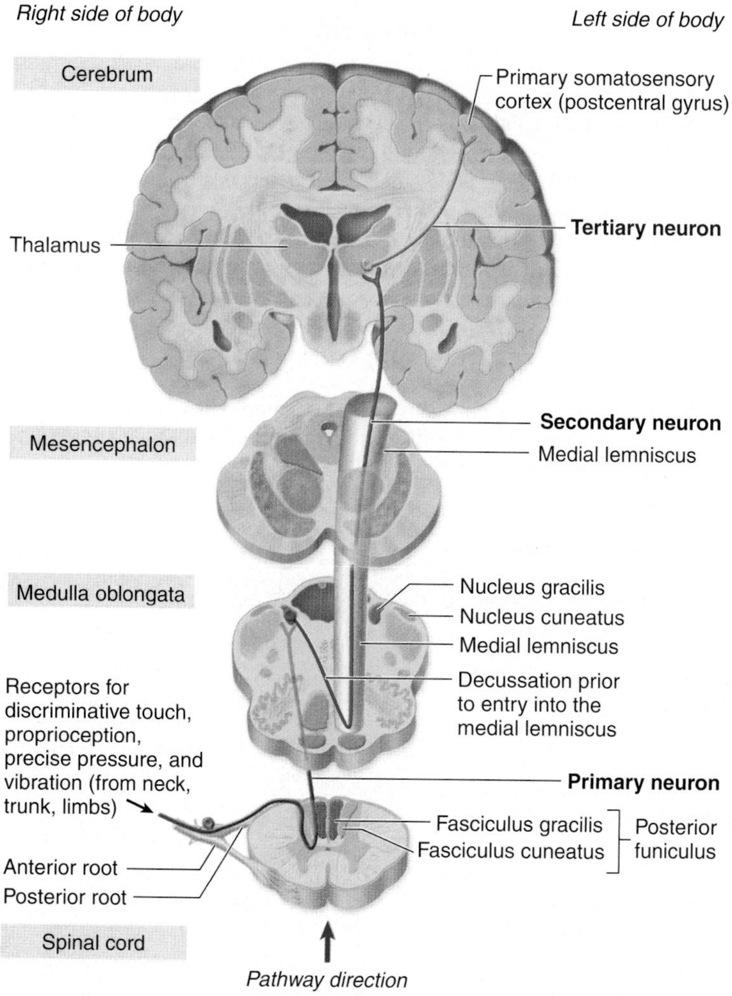

Figure 17.2

Posterior Funiculus–Medial Lemniscal Pathway. This pathway conducts sensory information about limb position, fine touch, precise pressure, and vibration. This pathway is bilaterally symmetrical; to avoid confusion, only sensory input from the right side of the body is shown here. Decussation of axons occurs in the medulla oblongata. The primary neuron is purple, the secondary neuron is dark blue, and the tertiary neuron is light green.

the spinal cord and synapse on secondary neurons within the posterior horns. Axons entering these pathways conduct stimuli related to crude touch and pressure as well as pain and temperature.

Axons of the secondary neurons in the anterolateral pathway cross over to the opposite side of the spinal cord before ascending toward the brain. This decussation occurs through the **anterior white commissure,** located anterior to the gray commissure. The anterior and lateral spinothalamic pathways, like the posterior funiculus–medial lemniscal pathway, are somatotopically organized: Axons transmitting sensory information from more inferior segments of the body are located lateral to those from more superior segments. Secondary neuron axons synapse on tertiary neurons located within the thalamus. Axons from the tertiary neurons then conduct stimulus information to the appropriate region of the primary somatosensory cortex.

Spinocerebellar Pathway

The **spinocerebellar pathway** conducts proprioceptive information to the cerebellum for processing to coordinate body movements. The spinocerebellar pathway is composed of anterior and posterior spinocerebellar tracts; these are the major routes for transmitting postural input to the cerebellum **(figure 17.4)**. Sensory input arriving at the cerebellum through these tracts is critical for regulating posture and balance and for coordinating skilled movements. Note that these spinocerebellar tracts are different from the other sensory pathways in that they do not use tertiary neurons; rather, they only have primary and secondary neurons. Information conducted in spinocerebellar pathways is integrated and acted on at a subconscious level.

Anterior spinocerebellar tracts conduct impulses from the inferior regions of the trunk and the lower limbs. Their axons enter the cerebellum through the superior cerebellar peduncle. **Posterior spinocerebellar tracts** conduct impulses from the lower limbs, the trunk, and the upper limbs. Their axons enter the cerebellum through the inferior cerebellar peduncle.

Specific characteristics and details of these pathways are summarized in **tables 17.2** and **17.3**.

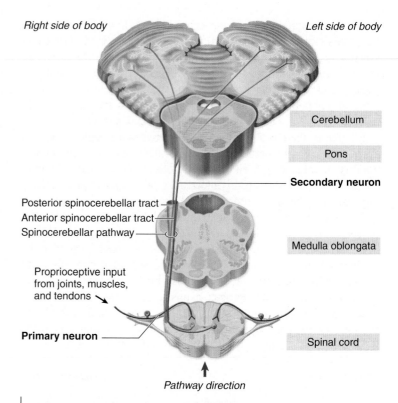

Figure 17.4

Spinocerebellar Pathway. This pathway conducts proprioceptive information to the cerebellum through both the anterior and posterior spinocerebellar tracts. Only some of the axons destined to enter the anterior spinocerebellar pathway decussate at the level where the primary neuron axon enters the spinal cord. Only primary (purple) and secondary (dark blue) neurons are found in this type of pathway.

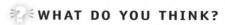

WHAT DO YOU THINK?

2 You have learned that most sensory impulses never reach our conscious awareness. Why? What would be the drawback to being consciously aware of almost all sensory impulses?

Table 17.2		Locations and Functions of Principal Sensory Spinal Cord Pathways		
Pathway	**Funiculus**	**Origin of the Pathway Neuron**	**Termination**	**Function**
POSTERIOR FUNICULUS–MEDIAL LEMNISCAL PATHWAY				
Fasciculus cuneatus	Posterior	Upper limb, superior trunk, neck, posterior head	Nucleus cuneatus of medulla oblongata	Conduct sensory impulses for proprioceptive information about limb position and discriminative touch, precise pressure, and vibration sensation
Fasciculus gracilis	Posterior	Lower limb, inferior trunk	Nucleus gracilis of medulla oblongata	
ANTEROLATERAL PATHWAY				
Anterior spinothalamic	Anterior	Posterior horn interneurons	Thalamus: Tertiary neurons project to primary somatosensory cortex	Conducts sensory impulses for crude touch and pressure
Lateral spinothalamic	Lateral	Posterior horn interneurons	Thalamus: Tertiary neurons project to primary somatosensory cortex	Conducts sensory impulses for pain and temperature
SPINOCEREBELLAR PATHWAY				
Anterior spinocerebellar	Lateral	Posterior horn interneurons	Cerebellum	Conducts proprioceptive impulses from inferior regions of trunk and lower limbs
Posterior spinocerebellar	Lateral	Posterior horn interneurons	Cerebellum	Conducts proprioceptive impulses from lower limbs, regions of trunk, and upper limbs

Table 17.3	Neuron Cell Body Locations and Decussation Sites of the Principal Sensory Spinal Cord Pathways			
Pathway	Location of Neuron Cell Bodies			Structures Involved with Decussation
	Primary Neuron	*Secondary Neuron*	*Tertiary Neuron*	
POSTERIOR FUNICULUS–MEDIAL LEMNISCAL PATHWAY				
Fasciculus cuneatus	Posterior root ganglion	Nucleus cuneatus	Thalamus	Axons of secondary neurons decussate prior to entry into medial lemniscus
Fasciculus gracilis	Posterior root ganglion	Nucleus gracilis	Thalamus	Axons of secondary neurons decussate prior to entry into medial lemniscus
ANTEROLATERAL PATHWAY				
Anterior spinothalamic	Posterior root ganglion	Posterior horn interneurons	Thalamus	Axons of secondary neurons decussate within spinal cord at level of entry
Lateral spinothalamic	Posterior root ganglion	Posterior horn interneurons	Thalamus	Axons of secondary neurons decussate within spinal cord at level of entry
SPINOCEREBELLAR PATHWAY				
Anterior spinocerebellar	Posterior root ganglion	Posterior horn interneurons	None	Some axons decussate in spinal cord and pons, while other axons do not decussate
Posterior spinocerebellar	Posterior root ganglion	Posterior horn interneurons	None	Axons do not decussate

WHAT DID YOU LEARN?

2 What information is conducted by sensory pathways?

3 Compare primary and secondary neurons in the sensory pathways.

4 Which type of sensory pathway conducts proprioceptive information?

Motor Pathways

Key topics in this section:

- Key features and regional anatomy of motor pathways
- Characteristics of direct and indirect motor pathways
- How cerebral nuclei and the cerebellum function in motor activities

Motor pathways are descending pathways in the brain and spinal cord that control the activities of skeletal muscle.

Functional Anatomy of Motor Pathways

Motor pathways are formed from the cerebral nuclei, the cerebellum, descending projection tracts, and motor neurons. **Descending projection tracts** are motor pathways that originate from the cerebral cortex and brainstem (**figure 17.5**). Motor neurons within these tracts either synapse directly on motor neurons in the CNS or on interneurons that, in turn, synapse on motor neurons.

There are at least two motor neurons in the somatic motor pathway: an upper motor neuron and a lower motor neuron (**table 17.4**). These neurons are involved in voluntary movements. The cell body of an **upper motor neuron** is housed within either the cerebral cortex or a nucleus within the brainstem. Axons of the upper motor neuron synapse either directly on lower motor neurons or on interneurons that synapse directly on lower motor neurons. The cell body of a **lower motor neuron** is housed either within the

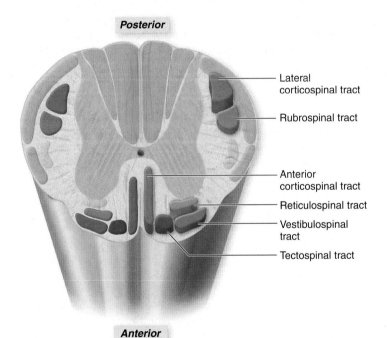

Posterior

Lateral corticospinal tract

Rubrospinal tract

Anterior corticospinal tract

Reticulospinal tract

Vestibulospinal tract

Tectospinal tract

Anterior

Figure 17.5

Descending Projection Tracts. Bilaterally symmetrical motor pathways descend from both the cortex and the brainstem into the spinal cord. Tract names indicate their point of origin and the spinal cord as their destination. Descending motor pathways are shown in shades of red and orange; ascending sensory pathways are shown in pale blue.

Table 17.4	Types of Motor Pathway Neurons		
Neuron	**Cell Body Origin**	**Projects To:**	**Activity**
Upper motor neuron	Cerebral cortex or brainstem nucleus	Lower motor neuron	May be either excitatory or inhibitory
Lower motor neuron	Brainstem nucleus or anterior horn of spinal cord	Skeletal muscle	Excitatory only

anterior horn of the spinal cord or within a brainstem cranial nerve nucleus. Axons of the lower motor neurons exit the CNS and project to the skeletal muscle to be innervated.

The two types of motor neurons perform different activities: The activity of the upper motor neuron either excites or inhibits the activity of the lower motor neuron, but the activity of the lower motor neuron is always excitatory because its axon connects directly to the skeletal muscle fibers. The cell bodies of motor neurons and most interneurons involved in the innervation and control of limb and trunk muscles reside in the spinal cord anterior horn and the gray matter zone between the anterior horn and the posterior horn. The neurons that innervate the head and neck are located in the motor nuclei of cranial nerves and in the reticular formation (introduced in chapter 15 and discussed in this chapter on page 530).

Motor neuron axons form two types of somatic motor pathways: direct pathways and indirect pathways. The direct pathways are responsible for conscious control of skeletal muscle activity, and the indirect pathways are responsible for unconscious control of skeletal muscle activity.

Direct Pathway

The **direct pathway,** also called the **pyramidal** (pi-ram′i-dal) **pathway** or *corticospinal pathway,* originates in the pyramidal cells of the primary motor cortex. The name *pyramidal* is derived from the shape of the upper motor neuron cell bodies, which have a tetrahedral, or pyramid-like, shape. Their axons project either into the brainstem or into the spinal cord to synapse directly on lower motor neurons.

The axons from pyramidal cell upper motor neurons descend through the internal capsule, enter the cerebral peduncles, and ultimately form two descending motor tracts of the direct pathway: corticobulbar tracts and corticospinal tracts.

Corticobulbar Tracts   The **corticobulbar** (kōr′ti-kō-bŭl′bar) **tracts** originate from the facial region of the motor homunculus within the primary motor cortex. Axons of these upper motor neurons extend to the brainstem, where they synapse with lower motor neuron cell bodies that are housed within brainstem cranial nerve nuclei. (Note: The term *bulbar* means resembling a bulb and is used to indicate the rhombencephalon in the brainstem.) Axons of these lower motor neurons help form the cranial nerves. The corticobulbar tracts transmit motor information to control the following movements:

- Eye movements (via CN III, IV, and VI)
- Cranial, facial, pharyngeal, and laryngeal muscles (via CN V, VII, IX, and X)
- Some superficial muscles of the back and neck (via CN XI)
- Intrinsic and extrinsic tongue muscles (via CN XII)

Corticospinal Tracts   The **corticospinal** (kor′ti-kō-spī′năl; *spinalis* = backbone) **tracts** descend from the cerebral cortex through the brainstem and form a pair of thick anterior bulges in the medulla oblongata called the pyramids. Then they continue into the spinal cord to synapse on lower motor neurons in the anterior horn of the

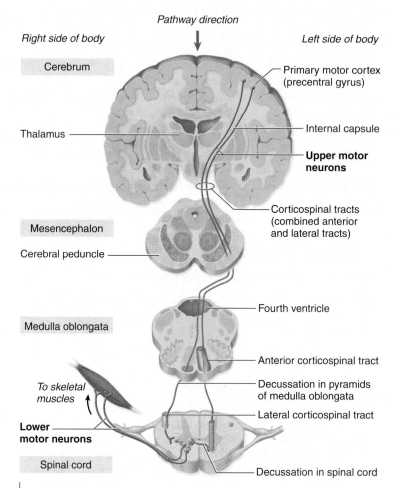

Pathway direction

Right side of body   Left side of body

- Primary motor cortex (precentral gyrus)
- Cerebrum
- Internal capsule
- Thalamus
- **Upper motor neurons**
- Corticospinal tracts (combined anterior and lateral tracts)
- Mesencephalon
- Cerebral peduncle
- Fourth ventricle
- Medulla oblongata
- Anterior corticospinal tract
- To skeletal muscles
- Decussation in pyramids of medulla oblongata
- Lateral corticospinal tract
- **Lower motor neurons**
- Spinal cord
- Decussation in spinal cord

Figure 17.6

Corticospinal Tracts.   Corticospinal tracts originate as collections of motor neurons within the motor cortex of the cerebrum and synapse on motor neurons within the anterior horns of the spinal cord to control voluntary motor activity. The upper motor neuron is dark green, and the lower motor neuron is lavender.

spinal cord **(figure 17.6)**. The corticospinal tracts are composed of two components: lateral and anterior corticospinal tracts.

The **lateral corticospinal tracts** include about 85% of the axons of the upper motor neurons that extend through the medulla oblongata. They decussate within the pyramids of the medulla oblongata and then form the lateral corticospinal tracts in the lateral funiculi of the spinal cord. These tracts contain axons that innervate both lower motor neurons of the anterior horn of the spinal cord and interneurons within the spinal cord. Axons of the lower motor neurons innervate skeletal muscles that control skilled movements in the limbs. Some examples of skilled movements include playing a guitar, dribbling a soccer ball, or typing on your computer keyboard.

The **anterior corticospinal tracts** represent the remaining 15% of the axons of upper motor neurons that extend through the medulla

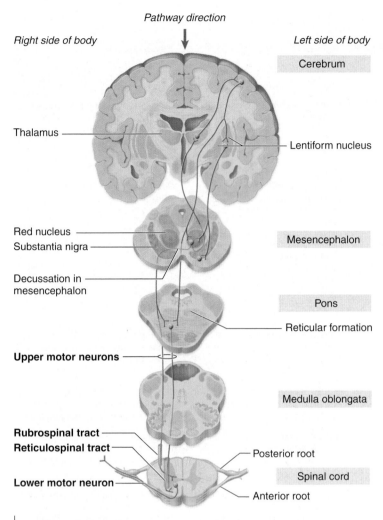

Pathway direction

Right side of body

Left side of body

Cerebrum

Thalamus

Lentiform nucleus

Red nucleus

Substantia nigra

Mesencephalon

Decussation in
mesencephalon

Pons

Reticular formation

Upper motor neurons

Medulla oblongata

Rubrospinal tract
Reticulospinal tract

Posterior root

Spinal cord

Lower motor neuron

Anterior root

Figure 17.7

Indirect Motor Pathways in the Spinal Cord. These motor
pathways originate from neurons housed within the brainstem.
The upper motor neuron is dark green, and the lower motor neuron
is lavender.

oblongata. The axons of these neurons do not decussate at the level
of the medulla oblongata. Instead, they remain on their original side
of the CNS and descend ipsilaterally, meaning "on the same side," to
form the anterior corticospinal tracts in both anterior white funiculi.
At each spinal cord segment, some of these axons decussate through
the median plane in the anterior white commissure. After crossing
to the opposite side, they synapse either with interneurons or lower
motor neurons in the anterior horn of the spinal cord. Axons of the
lower motor neurons innervate axial skeletal muscle.

Indirect Pathway

Several nuclei within the mesencephalon initiate motor commands
for activities that occur at an unconscious level. These nuclei and
their associated tracts constitute the **indirect pathway,** so named
because upper motor neurons originate within brainstem nuclei
(that is, they are not pyramidal cells in the cerebral cortex). The
axons of the indirect pathway take a complex, circuitous route
before finally conducting the motor impulse into the spinal cord.
Motor impulses conducted by axons of the upper motor neurons
in the indirect pathway descend from specific brainstem nuclei into
major tracts of the spinal cord and terminate on either interneurons
or lower motor neurons **(figure 17.7)**.

The indirect pathway modifies or helps control the pattern of
somatic motor activity. This is accomplished by (1) altering motor
neuron sensitivity to incoming impulses in order to control muscles
individually or in groups, and (2) activating feedback loops that
project to the primary motor cortex. This pathway controls some
muscular activity localized within the head, limbs, and trunk of the
body. It is multisynaptic and exhibits a high degree of complexity:
Nerve impulses travel through diverse circuits that involve the primary motor cortex, premotor cortex, cerebral nuclei, thalamus, limbic system, reticular formation, cerebellum, and brainstem nuclei.

Motor signals within the indirect pathway can alter or help
regulate the contraction of skeletal muscles by exciting or inhibiting the lower motor neurons that innervate the muscles. Interaction
among components of these motor pathways occurs both within the
brain and at the level of the motor neurons.

The different tracts of the indirect pathway are grouped
according to their primary functions. The **lateral pathway** regulates
and controls precise, discrete movements and tone in flexor muscles of the limbs—for example, the type of movement required to
gently lay a baby in her crib. (See discussion in chapter 16.) This
pathway consists of the **rubrospinal** (roo′brō-spī′năl; *rubro* = red)
tracts that originate in the red nucleus of the mesencephalon. The
medial pathway regulates muscle tone and gross movements of the
muscles of the head, neck, proximal limb, and trunk. The medial
pathway consists of three groups of tracts: reticulospinal tracts, tectospinal tracts, and vestibulospinal tracts.

- The **reticulospinal** (re-tik-ū-lō-spī′năl) **tracts** originate from
  the reticular formation in the mesencephalon. They help
  control more unskilled automatic movements related to
  posture and maintaining balance.
- The **tectospinal** (tek-tō-spī′năl) **tracts** conduct motor
  commands away from the superior and inferior colliculi in
  the tectum of the mesencephalon to help regulate positional
  changes of the arms, eyes, head, and neck as a consequence
  of visual and auditory stimuli.
- The **vestibulospinal** (ves-tib′ū-lō-spī′năl) **tracts** originate
  within vestibular nuclei of the brainstem. Impulses
  conducted within these tracts regulate muscular activity that
  helps maintain balance during sitting, standing, and walking.

Table 17.5 summarizes the characteristics of the principal
types of motor pathways.

Role of the Cerebral Nuclei

Cerebral nuclei, discussed in chapter 15, are described again here
because they interact with motor pathways in important ways. The
cerebral nuclei receive impulses from the entire cerebral cortex,
including the motor, sensory, and association cortical areas, as well
as input from the limbic system **(figure 17.8)**. Most of the output
from cerebral nuclei goes to the primary motor cortex; cerebral nuclei
do not exert direct control over lower motor neurons. Cerebral nuclei
provide the patterned background movements needed for conscious
motor activities by adjusting the motor commands issued in other
nuclei. For example, when you start walking, you voluntarily initiate
the movement, and the cerebral nuclei then control the continuous
motor commands until you decide to stop walking.

Role of the Cerebellum

The cerebellum plays a key role in movement by regulating the functions of the motor pathways. The cerebellum continuously receives
convergent input from the various sensory pathways and from the

Table 17.5	Principal Motor Spinal Cord Pathways			
Origin of Tract	**Manner of Decussation**	**Destination of Upper Motor Neurons**	**Termination Site**	**Function**
DIRECT PATHWAY				
Corticobulbar tracts	All cranial nerve motor nuclei receive bilateral (both ipsilateral and contralateral) input except CN VI, VII to the lower face, and XII. These receive only contralateral input.	Brainstem only	Cranial nerve nuclei; reticular formation	Voluntary movement of cranial muscles
Lateral corticospinal tracts	All decussate at the pyramids	Lateral funiculus	Gray matter region between posterior and anterior horns; anterior horn; all levels of spinal cord	Voluntary movement of limb muscles
Anterior corticospinal tracts	Decussation occurs in spinal cord at level of lower motor neuron cell body	Anterior funiculus	Gray matter region between posterior and anterior horns; anterior horn; cervical part of spinal cord	Voluntary movement of axial muscles
INDIRECT PATHWAY				
Lateral pathway Rubrospinal tract	Decussate at ventral tegmentum	Lateral funiculus	Lateral region between posterior and anterior horns; anterior horn; cervical part of spinal cord	Regulates and controls precise, discrete movements and tone in flexor muscles of the limbs
Medial pathway Reticulospinal tract	No decussation (ipsilateral)	Anterior funiculus	Medial region between posterior and anterior horns; anterior horn; all parts of spinal cord	Controls more unskilled automatic movements related to posture and maintaining balance
Tectospinal tract	Decussate at dorsal tegmentum	Anterior funiculus	Medial region between posterior and anterior horns; anterior horn; cervical part of spinal cord	Regulates positional changes of the arms, eyes, head, and neck due to visual and auditory stimuli
Vestibulospinal tract	Some decussate (contralateral) and some do not (ipsilateral)	Anterior funiculus	Medial region between posterior and anterior horns; anterior horn; medial tracts to cervical and superior thoracic parts of spinal cord; lateral tracts to all parts of spinal cord	Regulates muscular activity that helps maintain balance during sitting, standing, and walking

Figure 17.8

Cerebral Nuclei and Selected Indirect Motor System Components. A partially cut-away brain diagram shows the general physical location of the cerebral nuclei and some structures of the indirect motor system.

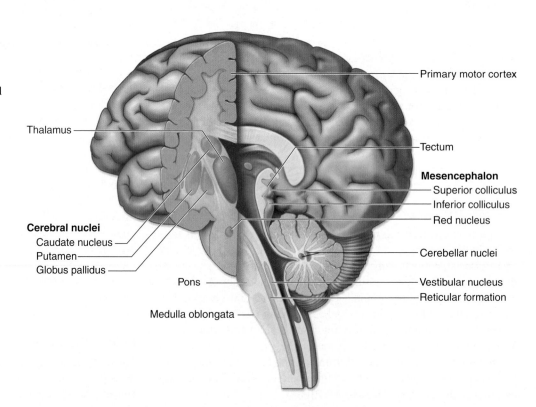

Thalamus

Cerebral nuclei
Caudate nucleus
Putamen
Globus pallidus

Pons

Medulla oblongata

Primary motor cortex

Tectum

Mesencephalon
Superior colliculus
Inferior colliculus
Red nucleus

Cerebellar nuclei

Vestibular nucleus
Reticular formation

CLINICAL VIEW

Cerebrovascular Accident

A **cerebrovascular accident** (**CVA,** or **stroke**) is caused by reduced blood supply to a part of the brain due to a blocked or damaged arterial blood vessel. If a *thrombus* (a blood clot within the blood vessel) forms at a narrowed region of a cerebral artery, it can completely block the lumen of the artery. On occasion, a CVA also results from an *embolus* (a blood clot that formed someplace else) that breaks free, travels through the vascular system, and becomes lodged in a cerebral blood vessel. An especially serious form of stroke results when a weakened blood vessel in the brain ruptures and hemorrhages, quickly leading to unconsciousness and death.

Symptoms of a CVA include loss or blurring of vision, weakness or slight numbness, headache, dizziness, and walking difficulties. Depending on the location of the blockage, the person may experience regional sensory loss, motor loss, or both. For example, a patient who suddenly exhibits speech difficulties and loss of motor control of the right arm may be

experiencing a CVA that affects the left hemisphere precentral gyrus (see chapter 15) in the region of the motor homunculus upper limb and the motor speech area. If the obstruction lasts longer than about 10 minutes, tissue in the brain may die. A massive stroke can leave a person completely paralyzed and without sensation over as much as half the body.

Additionally, elderly people sometimes experience brief episodes of lost sensation or motor ability or "tingling" in the limbs. Such a short-lived episode, called a **transient ischemic attack (TIA)** or "ministroke," results from a temporary plug in a blood vessel that dissolves in a matter of minutes. However, TIAs can indicate substantial risk for a more serious vessel blockage in the future.

The risk for a CVA increases with age, and is also influenced by family history, race, and gender. People can lower their risk of stroke by making lifestyle changes and getting treatment for existing heart conditions, high cholesterol levels, and high blood pressure. People with a history of TIAs should take an agent that inhibits platelet aggregation, such as aspirin.

motor pathways themselves (**figure 17.9**). In this way, the cerebellum unconsciously perceives the state of the body, receives the plan for movement, and then follows the activity to see if it was carried out correctly. When the cerebellum detects a disparity between the intended and actual movement, it may generate an error-correcting signal. This signal is transmitted to both the premotor and primary motor cortices via the thalamus and the brainstem. Descending pathways then transmit these error-correcting signals to the motor neurons. Thus, the cerebellum influences and controls movement by indirectly affecting the excitability of motor neurons. The cerebellum

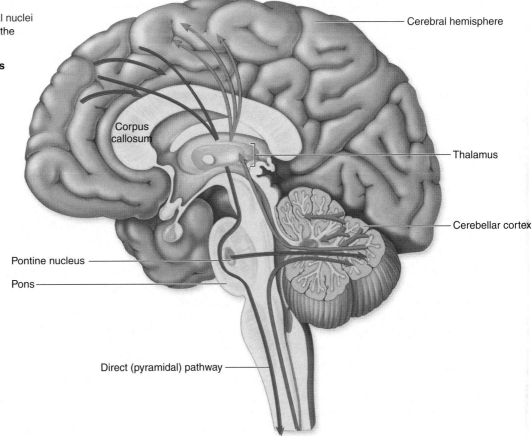

Voluntary movements
The primary motor cortex and the basal nuclei in the forebrain send impulses through the nuclei of the pons to the cerebellum.

Assessment of voluntary movements
Proprioceptors in skeletal muscles and joints report degree of movement to the cerebellum.

Integration and analysis
The cerebellum compares the planned movements (motor signals) against the results of the actual movements (sensory signals).

Corrective feedback
The cerebellum sends impulses through the thalamus to the primary motor cortex and to motor nuclei in the brainstem.

Cerebral hemisphere

Corpus callosum

Thalamus

Cerebellar cortex

Pontine nucleus

Pons

Direct (pyramidal) pathway

Sagittal section

Figure 17.9

Cerebellar Pathways. Input to the cerebellum originates from the motor cortex of the cerebrum and the pons (blue arrows), and the spinocerebellar tracts (purple arrows). Within the cerebellum, the integration and analysis of input information occurs (green arrow). Output from the cerebellum (red arrows) extends through the cerebellar peduncles (not shown).

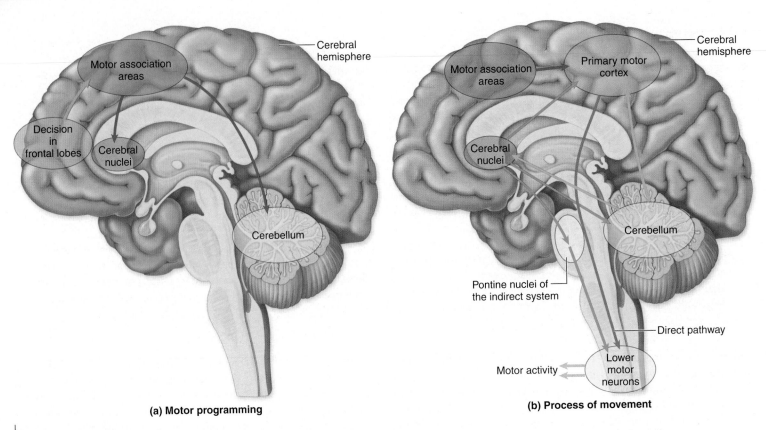

(a) Motor programming

(b) Process of movement

Figure 17.10

Somatic Motor Control. Several regions of the brain participate in the control and modification of motor programming to produce somatic motor activities. *(a)* Motor programs require conscious directions from the frontal lobes to cerebral nuclei and the cerebellum. *(b)* The process of movement is initiated when commands are received by the primary motor cortex from the motor association areas.

is critically important in coordinating movements because it specifies the exact timing of control signals to different muscles. For example, when you randomly run your fingers up and down the strings of a guitar, you are only making noise, not music. It is the cerebellum that directs the precise, exquisite finger movements necessary to produce a recognizable instrumental piece.

Levels of Processing and Motor Control

Simple reflexes that stimulate motor neurons represent the lowest level of motor control. The nuclei controlling these reflexes are located in the spinal cord and the brainstem. Brainstem nuclei also participate in more complex reflexes. Upon receipt of sensory impulses, they initiate motor responses to control motor neurons directly or oversee the regulation of reflex centers elsewhere in the brain. The pattern of feedback, control, and modification between brain regions establishes the motor programming that ultimately produces somatic motor control over the process of movement as illustrated in **figure 17.10**. The most complex unconscious motor patterns are controlled by neurons in the cerebellum, cerebral nuclei, and mesencephalon. Examples of these carefully patterned motor activities include riding a bicycle (cerebellum), swinging the arms while walking (cerebral nuclei), and sudden startled movements due to visual or auditory stimuli (mesencephalon).

Highly variable and complex voluntary motor patterns are controlled by the cerebral cortex and occupy the highest level of processing and motor control. Motor commands may be conducted to specific motor neurons directly, or they may be conveyed indirectly by altering the activity of a reflex control center. Figure 17.10*b* diagrammatically illustrates some steps involved in the interactions between the cerebral nuclei, motor association areas, and cerebel-

lum, and the primary motor cortex, which then issues commands for the programming and execution of a voluntary movement.

WHAT DID YOU LEARN?

5 Identify the CNS components that form the somatic motor pathways.

6 Compare and contrast the upper and lower motor neurons with respect to their cell body origin, what structure(s) they project to, and whether they are excitatory or inhibitory.

7 What is the primary difference between direct and indirect motor pathways?

8 Compare and contrast the influence of cerebral nuclei and the cerebellum on skeletal muscle.

Higher-Order Processing and Integrative Functions

Key topics in this section:

- Locations and functions of the integrative areas of the cerebral cortex
- Cerebral lateralization and functional differences between the hemispheres
- Cerebral centers involved in written and spoken language
- Processes and brain locations related to cognition
- Brain regions and structures involved in memory storage and recall

Higher-order mental functions encompass learning, memory, reasoning, and consciousness. These functions occur within the cortex of the cerebrum and involve multiple brain regions connected by complicated networks and arrays of axons. Both conscious and unconscious processing of information are involved in higher-order mental functions, and they may be continually adjusted or modified.

Development and Maturation of Higher-Order Processing

From infancy on, our motor control and processing capabilities become increasingly complex as we grow and mature. The maturation of the control and processing pathways is reflected in increased structural and functional complexity within the CNS. As previously discussed (chapter 16), the spinal reflex is the most basic level of CNS control. As the CNS continues to develop, many neurons expand their number of connections, providing the increased number of synaptic junctions required for increasingly complex reflex activities and processing. However, even though we are born with a large number of already formed synapses and many more form during childhood and adolescence as our nervous system matures, numerous synapses will degenerate unless we "activate and exercise" our brain to stimulate their use and retention.

During the first year of life, the number of cortical neurons continues to increase. The myelination of most CNS axons continues throughout the first 2 years. The brain grows rapidly in size and complexity so that by the age of 5, brain growth is 95% complete. (The rest of the body doesn't reach its adult size until puberty.) Some CNS axons remain unmyelinated until the teenage years (e.g., some of the axons in the prefrontal cortex). In general, the axons of PNS neurons continue to myelinate past puberty. A person's ability to carry out higher-order mental functions is a direct result of the level of nervous system maturation.

Cerebral Lateralization

Anatomically, the left and right cerebral hemispheres appear identical, but careful examination reveals a number of differences. Humans tend to have shape asymmetry of the frontal and occipital lobes of the brain, called **petalias.** Right-handed individuals tend to have right frontal petalias, meaning that the right frontal lobe projects farther than the left frontal lobe, and left occipital petalias, meaning that the left occipital lobe projects farther than the right occipital lobe. Conversely, left-handed individuals tend to have the reverse pattern (left frontal–right occipital petalias). The hemispheres also differ with respect to some of their functions. Each hemisphere tends to be specialized for certain tasks, a phenomenon called **cerebral lateralization** (lat′er-al-ĭ-ză′shŭn). Higher-order centers in both hemispheres tend to have different but complementary functions.

In most people, the left hemisphere is the **categorical hemisphere.** It usually contains the Wernicke area and the motor speech area. It is specialized for language abilities, and is also important in performing sequential and analytical reasoning tasks, such as those required in science and mathematics. This hemisphere appears to direct or partition information into smaller fragments for analysis. The term "categorical hemisphere" reflects this hemisphere's function in categorization and symbolization.

The other hemisphere (the right in most people) is called the **representational hemisphere**, because it is concerned with visuo-spatial relationships and analyses. It is the seat of imagination and insight, musical and artistic skill, perception of patterns and spatial relationships, and comparison of sights, sounds, smells, and tastes.

Please note that the terms categorical hemisphere and representational hemisphere reflect cognitive localizations. This terminology is not anatomic in nature. In fact, the hemisphere terms are psychology terms. Additionally, it must be recognized that the relative size of the dominant versus the nondominant hemisphere is not appreciated at the gross anatomic level.

Both cerebral hemispheres remain in constant communication through commissures, especially the corpus callosum, which contains hundreds of millions of axons that project between the hemispheres (**figure 17.11**).

Lateralization of the cerebral hemispheres develops early in life (prior to 5–6 years of age). In a young child, the functions of a damaged or removed hemisphere are often taken over by the other hemisphere before lateralization is complete. Some aspects of lateralization differ between the sexes. Women have a thicker posterior part of the corpus callosum due to additional commissural axons in this region. Adult males tend to exhibit more lateralization than females and suffer more functional loss when one hemisphere is damaged.

CLINICAL VIEW

Hemispherectomies and Cerebral Lateralization

Epilepsy is a disorder in which neurons emit nerve impulses too frequently and rapidly, causing seizures that detrimentally affect motor and sensory function. The seizure activity almost always originates on one side of the brain. Most seizures may be controlled by anticonvulsant medications, but if medications are ineffective, surgery may be the next therapy. Surgical removal of the brain part that is the source of the seizures often eliminates seizure episodes. Because the most common source of seizures is the temporal lobe, most patients undergo a temporal lobectomy. This procedure offers hope to epilepsy patients who have been unresponsive to medical therapy. If the seizure source is in a different part of the brain, either an extratemporal lobectomy or a corpus callosotomy may be attempted. In severe cases, a drastic form of therapy is a cerebral **hemispherectomy** (hem′ē-sfēr-ek′tō-mē) in which the side of the brain responsible for the seizure activity is surgically removed. Physicians only pursue a procedure of this magnitude when studies have unequivocally shown which cerebral hemisphere is the source of the seizures. When a hemisphere is removed, additional cerebrospinal fluid fills the space it previously occupied. About 90–95% of epilepsy patients experience long-term seizure control following hemispherectomy. Of that group, 70–85% remain seizure-free, while 10–20% experience at least an 80% reduction in seizure frequency.

Although brain function does not return to complete normalcy following a hemispherectomy, amazingly the remaining hemisphere takes over some of the functions of the missing hemisphere. The younger the individual, the better the chances that the other hemisphere can take over functions previously performed by the missing hemisphere. Hemispherectomy is not without risk. Death from the surgery alone occurs in about 2% of individuals. Long-term complications include displacement of the remaining cerebral hemisphere and problems with CSF flow. But despite these possible adverse developments, most patients' conditions are improved by the surgery.

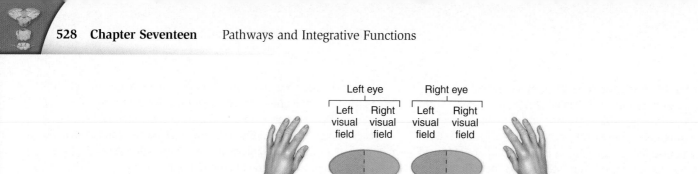

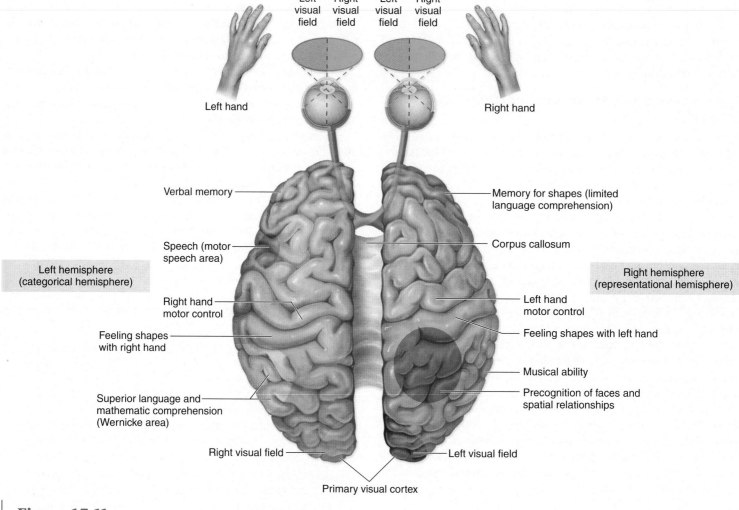

Figure 17.11

Cerebral Lateralization.   The cerebral hemispheres exhibit functional differences as a result of specialization.

Cerebral lateralization is highly correlated with handedness. Right-handed individuals tend to have a slightly different lateralization pattern than those who are left-handed. Neuroanatomic research indicates that the petalia patterns just described differ between left-handed and right-handed individuals. In about 95% of the population, the left hemisphere is the categorical hemisphere, thus correlating with the 90% incidence of right-handed individuals in the population. However, the correlation is not nearly as strict among left-handed people, who may have either hemisphere as their categorical hemisphere. Interestingly, a thicker corpus callosum in left-handers suggests that more signals may be relayed between their hemispheres. Finally, the left hemisphere is the speech-dominant hemisphere; it controls speech in almost all right-handed people as well as in many left-handed ones.

Language

The higher-order processes involved in language include reading, writing, speaking, and understanding words. You may recall from

chapter 15 that two important cortical areas involved in integration are the Wernicke area and the motor speech area (**figure 17.12**). The Wernicke area is involved in interpreting what we read or hear, while the motor speech area receives axons originating from the Wernicke area and then helps regulate the respiratory patterns and precise motor activities required to enable us to speak. Thus, the Wernicke area is central to our ability to recognize written and spoken language. Immediately posterior to the Wernicke area is the angular gyrus, a region that processes the words we read into a form that we can speak (figure 17.12*a*). First, the Wernicke area sends a speech plan to the motor speech area, which initiates a specific patterned motor program that is transmitted to the primary motor cortex. Next, upper motor neurons in the primary motor cortex (pyramidal cells) signal the lower motor neurons, which then innervate the muscles of the cheeks, larynx, lips, and tongue to produce speech.

In most people, the Wernicke area is in the categorical hemisphere (the left). In the representational hemisphere, a cortical

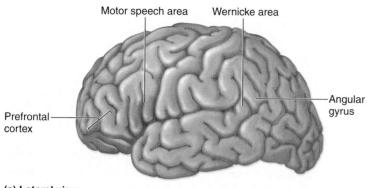

Motor speech area   Wernicke area

Prefrontal cortex

Angular gyrus

(a) Lateral view

Figure 17.12

Functional Areas Within the Cerebral Cortex. *(a)* In most people, the left cerebral hemisphere houses the Wernicke area, the motor speech area, and the prefrontal cortex. *(b)* A PET scan shows the areas of the brain that are most active during speech.

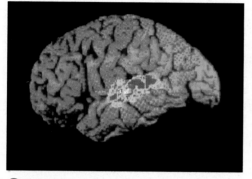

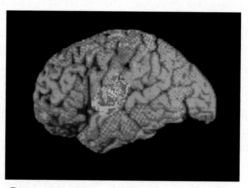

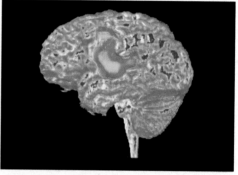

① Auditory information about a sentence travels to the primary auditory cortex. The Wernicke area then interprets the sentence.

(b) PET scans

② Information from the Wernicke area travels to the motor speech area.

③ Information travels from the motor speech area to the primary motor cortex, where motor commands involving muscles used for speech are given.

region opposite the Wernicke area recognizes the emotional content of speech. A lesion in this area of the cerebrum can make a person unable to understand emotional nuances, such as bitterness or happiness, in spoken words. A lesion in the cortical region of the representational hemisphere opposite the motor speech area results in **aprosody**, which causes dull, emotionless speech. (Do not immediately assume that some of your instructors have aprosody, if you frequently experience boring lectures.)

Cognition

Mental processes such as awareness, knowledge, memory, perception, problem solving, decision making, information processing, and thinking are collectively called **cognition** (kog-ni'shŭn; *cognitio* = to become acquainted with). The term cognition is often used to mean the act of knowing, and it may be interpreted in a social or cultural sense to describe knowledge development culminating in thought and action. The association areas of the cerebrum, which form about 70% of the nervous tissue in the brain, are responsible for both cognition and the processing and integration of information between sensory input and motor output areas.

Various studies of individuals suffering from brain lesions (caused by cancer, infection, stroke, and trauma) have helped us determine the functions of the association areas of the cerebrum. For example, the frontal association area (prefrontal cortex) integrates information from the sensory, motor, and association areas to enable the individual to think, plan, and execute appropriate behavior. Thus, an individual with a frontal lobe lesion exhibits personality abnormalities. If a person loses the ability to detect and identify

CLINICAL VIEW

Dyslexia

Dyslexia (dis-lek'sē-ă; *dys* = bad, *lexis* = word) is an inherited learning disability characterized by problems with single-word decoding. Affected individuals not only have trouble reading, but may also have problems writing and spelling accurately. They may be able to recognize letters normally, but their level of reading competence is far below that expected for their level of intelligence. Their writing may be disorganized and uneven, with the letters of words in incorrect order or even completely reversed. Occasionally, the ability to recognize and interpret the meaning of pictures and objects is also impaired. Interestingly, many people seemingly "outgrow" this condition, or at least develop improved reading ability over time. This improvement may reflect neural maturation or retraining of parts of the brain to better decode words and symbols.

Some researchers have postulated that dyslexia is a form of **disconnect syndrome,** in which transfer of information between the cerebral hemispheres through the corpus callosum is impaired. Genetic studies of a Finnish family identified a defective gene on chromosome 15 that appeared related to the transmission of dyslexia from a father to three of his children. Further research is ongoing to determine the environmental and genetic factors involved in dyslexia.

stimuli (termed loss of awareness) on one side of the body or in the limbs on that side, the primary somatosensory area in the hemisphere opposite the affected side of the body has been damaged. An individual who has **agnosia** (ag-nō'zē-ă; *a* = without, *gnosis* = knowledge) displays an inability to either recognize or understand the meaning of various stimuli. For example, a lesion in the temporal lobe may result in the inability to either recognize or understand the meanings of sounds or words. Specific symptoms of agnosia vary, depending on the location of the lesion within the cerebrum.

Memory

Memory is a versatile element of human cognition involving different lengths of time and different storage capacities. Storing and retrieving information requires higher-order mental functions and depends on complex interactions among different brain regions. These brain regions include components of the limbic system, such as the amygdaloid body and hippocampus, the insula lobe, and the frontal cortex. On a broader scale, in addition to memory, information management by the brain entails both learning (acquiring new information) and forgetting (eliminating trivial or nonuseful information).

Neuroscientists and psychologists classify memory in various ways. For example, **sensory memory** occurs when we form important associations based on sensory input from the environment. Sensory memory holds an exact copy of what is heard or seen (auditory and visual). It lasts for fractions of a second (a few seconds at the most) and has unlimited capacity.

Short-term memory (STM) follows sensory memory. It is generally characterized by limited capacity (approximately seven small segments of information) and brief duration (ranging from seconds to hours). Suppose that, in a Friday morning anatomy lecture, your instructor lists the general functions of the cerebral lobes on the board. Unless you study this information over the weekend, you will probably not recall it by Monday's lecture because it was just a small bit of information in STM.

Some psychologists believe that short-term memory represents a group of systems used to temporarily store and manipulate information that together comprise **working memory.** This type of memory is required to perform several different mental activities simultaneously. For example, a newly hired newspaper delivery person must glance at an address list for customer home delivery, drive through the neighborhood, and make judgments as to where to place the delivered paper, such as on the porch, in the mailbox, on the driveway, and at the same time decide the most efficient route to follow.

Once information is placed into **long-term memory (LTM),** it may exist for limitless periods of time. Continuing with the same example, if over the weekend you read and recopy your lecture notes, review the text and figures in the book, and prepare note cards, you may have stored the information about the cerebral lobes as LTM. Not only will you be well prepared for your next exami-

nation, but you may even remember this information for years to come. (However, information in LTM needs to be retrieved occasionally or it can be "lost," and our ability to store and retrieve information declines with age.)

It appears that our brain must organize complex information in short-term memory prior to storing it in long-term memory **(figure 17.13)**. Conversion from STM to LTM is called **encoding,** or *memory consolidation*. Encoding requires the proper functioning of two components of the limbic system: the amygdaloid body and the hippocampus (see chapter 15). When a sensory perception forms in the primary somatosensory cortex, cortical neurons convey impulses along two parallel tracts extending to the amygdaloid body and the hippocampus. Connections from the amygdaloid body to the hypothalamus may link memories to specific emotions, while normal functioning of the hippocampus is required for the formation of STM. Long-term memories are stored in the association areas of the cerebral cortex. For example, voluntary motor activity is housed in the premotor cortex, and memory of sounds is stored in the auditory association area. Because STM and LTM involve different anatomic structures, loss of the ability to form STM does not affect the maintenance or accessibility of LTM.

WHAT DO YOU THINK?

3 What types of study habits best convert short-term memories into long-term memories? Do you practice these habits when you study for your exams?

Consciousness

Consciousness includes an awareness of sensation, voluntary control of motor activities, and activities necessary for higher mental processing. Levels of consciousness exist on a continuum. The highest state of consciousness and cortical activity is **alertness,** in which the individual is responsive, aware of self, and well-oriented to person, place, and time. Normal people alternate between periods of alertness and **sleep,** which is the natural, temporary absence of consciousness from which a person can be aroused by normal stimulation. Cortical activity is depressed during sleep, but functions continue in the vital centers in the brainstem.

Consciousness involves the simultaneous activity of large areas of the cerebral cortex. Projecting vertically through the core of the midbrain, pons, and medulla is a loosely organized core of gray matter called the **reticular formation (figure 17.14)**. The reticular formation extends slightly into the diencephalon and the spinal cord as well. This functional brain system has both motor and sensory components.

The motor component of the reticular formation communicates with the spinal cord and is responsible for regulating muscle tone (especially when the muscles are at rest). The motor component of the reticular formation also assists in autonomic motor functions,

Figure 17.13

Model of Information Processing.
Cognitive psychologists have proposed a model to show the relationships between sensory memory and short-term memory. Long-term memory develops later.

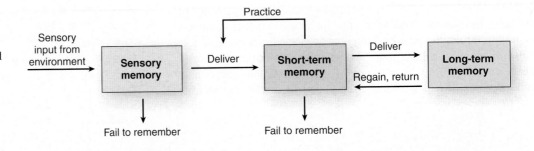

CLINICAL VIEW

Amnesia

Amnesia (am-ne'ze-ă; forgetfulness) refers to complete or partial loss of memory. Most often, amnesia is temporary and affects only a portion of a person's experiences. Causes of amnesia range from psychological trauma to direct brain injury, such as a severe blow to the head or even a CVA.

Because memory processing and storage involve numerous regions of the brain, the type of memory loss that occurs in an episode of amnesia depends on the area of the brain damaged. For example, damage to or loss of sensory association areas in the cerebral cortex can prevent sensory signals from arriving at the primary somatosensory cortex adjacent to the area of injury. The most serious forms of amnesia result from damage to the thalamus and limbic structures, especially the hippocampus. If one or more of these structures is damaged, memory storage and consolidation may be seriously disrupted or completely lacking. The nature of the underlying problem determines whether amnesia is complete or partial, and to what degree recovery, if any, is possible. There are several common types of amnesia.

Anterograde (an'ter-ō-grād; moving ahead) **amnesia** is a form in which a person finds it hard or even impossible to process and/or store ongoing events, although his or her memories from the past are intact and retrievable. Because day-to-day events are forgotten so quickly, the world is always new to the person with anterograde amnesia. Some medications, including general anesthetics, illicit drugs, and even alcohol, temporarily disrupt memory processing, resulting in temporary anterograde amnesia immediately following their use.

In **retrograde** (ret'rō-grād; behind) **amnesia,** the person loses memories of past events. Short-term retrograde amnesia may follow a blow to the head, as might occur in an auto accident or a football game. These patients often experience a memory gap spanning as much as 20 minutes immediately preceding the event.

Posttraumatic amnesia follows a head injury. The duration and extent of the amnesia depend on the severity and location of the brain damage. Posttraumatic amnesia commonly involves features of both retrograde and anterograde amnesia, with the patient exhibiting decreased ability to recall past events as well as to process current events.

Hysterical amnesia covers episodes of memory loss linked to psychological trauma. The condition is usually temporary, with memories of the traumatic event typically returning slowly several days afterward. For some people, however, recall is never complete.

Korsakoff psychosis is a form of impaired memory processing that follows years of alcohol abuse, possibly coupled with thiamine (vitamin B_1) deficiency. Although short-term memory seems intact, the person has serious difficulty recalling simple stories or lists of unrelated words, relating events, identifying common symbols, and even recognizing the faces of friends. This progressive condition is often accompanied by neurologic problems, such as uncoordinated movements and sensory loss in the limbs.

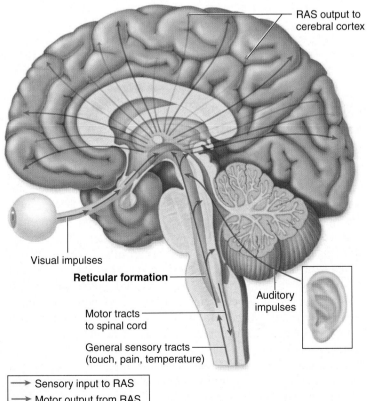

Visual impulses
Reticular formation
Auditory impulses
Motor tracts to spinal cord
General sensory tracts (touch, pain, temperature)
RAS output to cerebral cortex

→ Sensory input to RAS
→ Motor output from RAS
→ RAS output to cerebrum

(a) Reticular formation

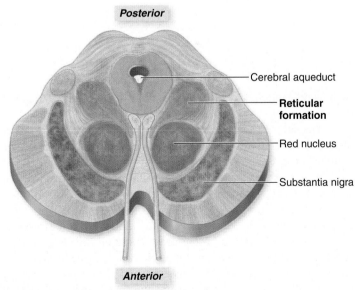

Posterior
Cerebral aqueduct
Reticular formation
Red nucleus
Substantia nigra
Anterior

(b) Cross section of mesencephalon

Figure 17.14

The Reticular Formation. *(a)* The reticular formation is distributed throughout the brainstem. The reticular formation receives and processes various types of stimuli (blue arrow). It participates in cyclic activities such as arousing the cortex to consciousness (purple arrow) and controlling the sleep-wake cycle. Some outputs from the reticular formation influence muscle activity (red arrow). *(b)* A cross section through the mesencephalon shows the position of the reticular formation in the brainstem.

Pathologic States of Unconsciousness

The level of consciousness of a healthy person varies greatly during a 24-hour period, ranging from wide awake and alert to deep sleep. When a person is asleep, he or she is technically unconscious, but not pathologically so. Prior to entering the sleep state, an individual becomes **lethargic** (leth-ar'jik), a normal level of reduced alertness and awareness associated with an inclination to sleep.

Other unconscious conditions are pathologic. A brief loss of consciousness, termed fainting or **syncope** (sin'kŏ-pē), often signals inadequate cerebral blood flow due to low blood pressure, as might follow hemorrhage or sudden emotional stress. **Stupor** (stoo'per; *stupeo* = to be stunned) is a moderately deep level of unconsciousness from which the person can be aroused only by extreme repeated or painful stimuli. A stupor may be associated with metabolic disorders such as low blood sugar, diseases of the liver or kidney, CVA or other brain trauma, or drug use.

A **coma** is a deep and profound state of unconsciousness from which the person cannot be aroused, even by repeated or painful stimuli. A person in a coma is alive, but unable to respond to the environment. A coma may result from severe head injury or CVA, marked metabolic failure (as occurs in advanced liver and kidney disease), very low blood sugar, or drug use. To gauge the depth of a coma, and possibly identify the cause, the physician performs a detailed neurologic exam. The presence or absence of certain reflexes, coupled with a particular type of periodic breathing (Cheyne-Stokes respiration), can help determine the location and nature of the problem.

A **persistent vegetative state (PSV)** is a condition in which the person has lost his or her thinking ability and awareness of the environment, but noncognitive brain functions continue, such as the brainstem's monitoring of heart rate, breathing, and the sleep-wake cycle. Some people in this state exhibit spontaneous movements, such as moving their eyes, grimacing, crying, and even laughing. A persistent vegetative state may follow a coma, and its underlying causes are similar to those producing comas. People in a persistent vegetative state may outwardly appear somewhat normal, but they are unable to speak and do not respond to commands.

We were vividly reminded of the impreciseness of the medical assessment and description of consciousness with the dramatic case of Terri Schiavo. Ms. Schiavo experienced respiratory and cardiac arrest in 1990 at age 26. In 1993 she was diagnosed as being in a persistent vegetative state. In 1998, her husband (guardian) petitioned the courts to remove her feeding tube. A legal, political, and social battle then began, and lasted until March 2005, when the feeding tube was removed and Ms. Schiavo passed away. Many judges, lawyers, politicians, and laypeople weighed in on both sides of this issue; even federal and state legislatures became embroiled in the battle. But although many people had maintained that Ms. Schiavo's condition was capable of improving, autopsy eventually revealed extensive and "irreversible" damage to all regions of her brain, according to the coroner. The meaning of PSV continues to be a controversial issue.

---

such as respiration, blood pressure, and heart rate, by working with the autonomic centers in the medulla and pons.

The sensory component of the reticular formation is responsible for alerting the cerebrum to incoming sensory information. This component, also known as the **reticular activating system (RAS),** contains sensory axons that project to the cerebral cortex. The RAS processes visual, auditory, and touch stimuli and uses this information to keep us in a state of mental alertness. Additionally, the RAS arouses us from sleep. The sound of an alarm clock can awaken us because the RAS receives various sensory stimuli and sends it to the cerebrum, thereby arousing it. Conversely, little or no stimuli (e.g., when you are in bed with the lights out, and no sounds are disturbing you) allow you to sleep, because the RAS is not stimulated to act. The Clinical View examines other levels of consciousness.

WHAT DID YOU LEARN?

9 Distinguish between the activities controlled by the categorical and representational hemispheres.

10 What is the function of the Wernicke area?

11 What is cognition?

12 What is meant by encoding with respect to memory formation?

Aging and the Nervous System

Key topic in this section:

- Effects of aging on the nervous system

Noticeable effects of aging on the brain and nervous system commence at about 30 years of age. Structural changes occur in nervous tissue, in the blood vessels of the brain, and in the brain's gross appearance. However, it must be emphasized that although aging may result in some brain atrophy, recognizable changes in brain size may not necessarily occur. Functional changes in the brain affect its performance and the body's ability to maintain homeostasis. Structural and functional changes overlap somewhat because, for example, a decrease in a blood vessel's carrying capacity (a structural change) often results in diminished metabolic activity in a particular brain region (a functional change).

Structurally, at the tissue level the number of neurons in the brain diminishes. Neurons that die are not replaced; thus, the amount of gray matter decreases. Superficial observation confirms that overall weight is reduced in older individuals. Loss of neurons causes a reduction in the number of synaptic connections, which adversely affects brain function due to diminished communication between neurons. Reduction in the number of sensory neurons

CLINICAL VIEW: In Depth

Alzheimer Disease: The "Long Goodbye"

Alzheimer disease (AD) has become the leading cause of dementia in the developed world. (*Dementia* refers to a general loss of cognitive abilities, including memory, language, and decision-making skills.)

Typically, AD does not become clinically apparent until after the age of 65; its diagnosis is often delayed due to confusion with other forms of cognitive impairment. Symptoms include slow, progressive loss of higher intellectual functions and changes in mood and behavior. AD gradually causes language deterioration, impaired visuospatial skills, indifferent attitude, and poor judgment, while leaving motor function intact. Patients become confused and restless, often asking the same question repeatedly. AD progresses relentlessly over months and years, and thus has come to be known as "the long goodbye." Eventually, it robs its victims of their memory, their former personality, and even the capacity to speak.

We have all heard that mental activity helps us stay sharp. As people age, mental decline often appears to be related to altered or decreased numbers of synapses between neurons. Many different avenues of research suggest that the key to brain vitality is brain activity. It is not necessary to initiate extreme changes to obtain vital brain benefits. Ways to stimulate an active brain include taking a daily walk, playing games, attending plays or lectures, gardening, working crosswords or other puzzles, reading and writing daily, or participating in community groups.

The underlying cause of AD remains a mystery, although both genetics and environment seem to play a role. Postmortem examinations of the brains of AD patients show marked and generalized cerebral atrophy. Microscopic examinations of brain tissue reveal a profound decrease in the number of cerebral cortical neurons, especially those within the temporal and frontal lobes. Surviving neurons contain abnormal aggregations of protein fibers, termed a **neurofibrillary tangle.** In addition, an abnormal protein, **amyloid** (am'i-loyd) **precursor protein (APP),** appears in the brain as well as in the walls of the cerebral arterioles. These histologic changes are most evident in the hippocampus, the region of the brain vital to memory processing. Biochemical alterations also occur, most significantly a decreased level of the neurotransmitter acetylcholine in the cerebrum.

At present, there is no cure for AD, although some medications help alleviate the symptoms and seem to slow the progress of the disease. The most beneficial medications are a class of inhibitors that prevent the enzyme acetylcholinesterase from degrading acetylcholine (ACh), thus allowing higher brain levels of ACh, which we have seen is an important neurotransmitter (see chapter 14). An interesting experimental therapy has proposed trying to trick the immune system into degrading and removing APP. Trial studies were begun in which AD patients were immunized against APP, but these studies were halted because the immune response in some people resulted in generalized brain inflammation, and actually worsened the patients' conditions. However, even though attempts to remove this abnormal protein have not been successful, measuring its amount in the bloodstream may be an effective screening tool for early diagnosis of AD.

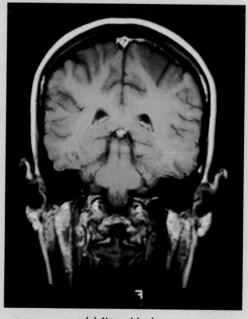

(a) Normal brain

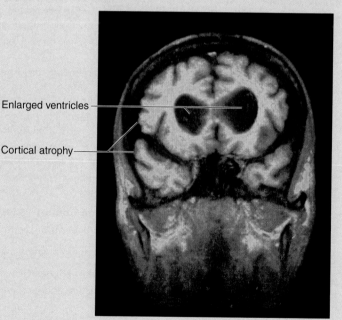

Enlarged ventricles

Cortical atrophy

(b) Alzheimer brain

MRIs show coronal sections of (a) a normal brain and (b) an Alzheimer brain. (Note the large ventricles and wide spaces between gyri in the AD brain.)

leads to diminished ability to detect and discriminate among external stimuli such as pain, light touch, pressure, and postural changes. Because the affected individual cannot appropriately assess the immediate environment, he or she may exhibit reduced control and coordination of movement. Physical injury is also possible, due to the inability to assess and avoid harmful stimuli, such as a hot object or a noxious odor.

In addition, blood flow to specific brain regions decreases because deposited lipids and atherosclerotic plaques often cause narrowing of the internal walls of blood vessels. The resulting diminished nutrient supply affects nervous tissue performance.

Ultimately, any decrease in the number of neurons, the number of interneuronal synapses, or the supply of nutrients and removal of wastes impairs a person's cognitive capacity.

C L I N I C A L   T E R M S

ataxia (ă-tak′sē-ă; *a* = not, *taxis* = order) Inability to perform coordinated body movements.

confusion Condition of reduced awareness in which a person is easily distracted and easily startled by sensory stimuli; alternates between drowsiness and excitability; resembles a state of minor delirium.

delirium (dē-lir′ē-ŭm; *deliro* = to be crazy) Altered state of consciousness consisting of confusion, distractibility, disorientation, disordered thinking and memory, defective perception, hyperactivity, agitation, and autonomic nervous system overactivity.

hemiballismus (hem-ē-bal-iz′mŭs; *ballismos* = jumping about) Jerking, involuntary movement involving proximal limb musculature on only one side of the body.

C H A P T E R   S U M M A R Y

General Characteristics of Nervous System Pathways 517	■ Sensory and motor pathways share the following characteristics: decussation, somatotopy, pairing, and usually two or three neurons within the CNS.
Sensory Pathways 517	■ Pathways of the somatosensory system mediate limb position, touch, temperature, and pain. **Functional Anatomy of Sensory Pathways 518** ■ Primary neurons conduct stimuli to the CNS; secondary neurons are interneurons that synapse either onto a tertiary neuron within the thalamus or in the cerebellum. Decussation occurs by axons of either the primary or secondary neurons. ■ The posterior funiculus–medial lemniscal pathway conducts stimuli of fine touch, precise pressure, and proprioception (posture). ■ The anterolateral pathway (anterior and posterior spinothalamic tracts) carries stimuli related to pain, pressure, temperature, and touch. Decussation of secondary neuron axons occurs in the spinal cord at the level of entry. ■ The spinocerebellar pathway (anterior and posterior spinocerebellar tracts) conducts stimuli to the cerebellum related to tendons, joints, and muscle posture.
Motor Pathways 521	■ The motor pathways of the brain and spinal cord work together to control skeletal muscle. **Functional Anatomy of Motor Pathways 521** ■ Somatic motor pathways involve an upper motor neuron and a lower motor neuron. ■ Somatic motor commands travel through either the direct system (conscious control) or the indirect system (unconscious control). ■ Pyramidal cells are primary motor cortex neurons; the pyramidal pathway provides a rapid, direct mechanism for conscious skeletal muscle control. It consists of two pairs of descending motor tracts: the corticobulbar tracts, the lateral corticospinal tracts, and the anterior corticospinal tracts. ■ The indirect pathway is composed of centers that issue motor commands at an unconscious level: the rubrospinal, reticulospinal, tectospinal, or vestibulospinal tracts. ■ The cerebral nuclei are processing centers for patterned background movement (such as arm swinging while walking). They adjust motor commands issued in other processing centers. ■ The cerebellum helps regulate the functions of the descending somatic motor pathways by influencing and controlling movement. **Levels of Processing and Motor Control 526** ■ The most complex motor patterns are controlled by neurons in the cerebellum, cerebral nuclei, and mesencephalon.
Higher-Order Processing and Integrative Functions 526	■ Higher-order mental functions encompass learning, memory, reasoning, and consciousness. **Development and Maturation of Higher-Order Processing 527** ■ Higher-order functions mature and increase in complexity as development proceeds. **Cerebral Lateralization 527** ■ In most individuals, the left hemisphere is the categorical hemisphere, and the right is the representational hemisphere. **Language 528** ■ The Wernicke area is responsible for recognition of spoken and written language. The motor speech area initiates a specific motor program for the muscles of the cheeks, larynx, lips, and tongue to produce speech. **Cognition 529** ■ Mental processes such as awareness, perception, thinking, knowledge, and memory are collectively called cognition.

Higher-Order Processing and Integrative Functions (continued) 526	**Memory 530** ■ Memory is a higher-order mental function involving the storage and retrieval of information gathered through previous activities. Memory classifications include sensory memory, short-term memory (STM), working memory, and long-term memory (LTM). **Consciousness 530** ■ The highest state of consciousness and cortical activity is alertness. Sleep is a natural, temporary absence of consciousness.
Aging and the Nervous System 532	■ Both structural and functional changes accompany the aging of the brain. Some of these changes are reduced neuron population, including sensory neurons; reduced blood flow through the brain; and reduced cognition.

CHALLENGE YOURSELF

Matching

Match each numbered item with the most closely related lettered item.

_____ 1. aprosody

_____ 2. spinothalamic tract

_____ 3. tertiary neuron

_____ 4. decussation

_____ 5. direct pathway

_____ 6. lower motor neuron

_____ 7. memory

_____ 8. indirect pathway

_____ 9. spinocerebellar

_____ 10. interneuron

a. contains no tertiary neurons

b. exits the CNS

c. axon crossover

d. information storage and retrieval

e. secondary neuron in an ascending pathway

f. originates in the thalamus

g. unconscious control of skeletal muscle

h. absence of emotional speech

i. pyramidal cell

j. detects crude touch, pain, pressure, temperature

Multiple Choice

Select the best answer from the four choices provided.

_____ 1. The fasciculus cuneatus and fasciculus gracilis compose the _____
   a. spinocerebellar tracts.
   b. posterior funiculi.
   c. spinothalamic tracts.
   d. anterior white commissure.

_____ 2. The motor tracts that conduct impulses to regulate the skilled movements of the upper and lower limbs are the
   a. reticulospinal tracts.
   b. corticospinal tracts.
   c. rubrospinal tracts.
   d. tectospinal tracts.

_____ 3. Higher-order mental functions encompass each of the following except
   a. memory.
   b. learning.
   c. reasoning.
   d. coughing.

_____ 4. Which of these are not part of an indirect motor pathway?
   a. rubrospinal tracts
   b. tectospinal tracts
   c. corticobulbar tracts
   d. reticulospinal tracts

_____ 5. Pyramidal cell axons project through corticospinal tracts and synapse at
   a. motor nuclei of cranial nerves.
   b. motor neurons in the anterior horns of the spinal cord.
   c. motor neurons in the posterior horns of the spinal cord.
   d. motor neurons in the lateral horns of the spinal cord.

_____ 6. The right hemisphere tends to be dominant for which functions?
   a. mathematical calculations
   b. motor commands involved with speech
   c. musical and artistic skill
   d. analytic reasoning

_____ 7. Somatotopy is the
   a. relationship between sensory receptors and motor units.
   b. positioning of motor neurons in the cerebellar cortex.
   c. precise correspondence between specific body and CNS areas.
   d. relationship between upper and lower motor neurons.

_____ 8. A loss of consciousness due to fainting is called
   a. lethargy.
   b. syncope.
   c. coma.
   d. sleep.

_____ 9. Which of these is the least likely to affect information transfer from STM (short-term memory) to LTM (long-term memory)?
   a. emotional state
   b. repetition or rehearsal
   c. auditory association cortex
   d. cerebral nuclei

_____ 10. Where are tertiary neurons found?
   a. extending between the posterior horn and the anterior horn
   b. extending between the posterior horn and the brainstem
   c. extending between the thalamus horn and the primary somatosensory cortex
   d. extending between the primary motor cortex and the brainstem

Content Review

1. Discuss the concept of somatotopy as it relates to the motor cortex.

2. Describe the function of primary neurons, secondary neurons, and tertiary neurons in the sensory pathways of the nervous system.

3. Describe the pathway by which the pressure applied to the right hand during a handshake is transmitted and perceived in the left primary somatosensory cortex.

4. A young gymnast has suffered a spinal cord injury and can no longer detect pain sensations in her leg. What spinal tract has been affected?

5. Identify and describe the distribution of the motor pathways that conduct conscious, voluntary motor impulses through the spinal cord.

6. Describe the relationship between the cerebral nuclei and the cerebellum in motor activities.

7. What is meant by the term cerebral lateralization?

8. Distinguish between sensory memory, short-term memory, and long-term memory.

9. Describe the activities of the reticular activating system.

10. What is the consequence of reduction in the number of sensory neurons during aging?

Developing Critical Reasoning

1. Melissa had a horrible headache, restricted movement in her right arm, and slight slurring of her speech. After an MRI was performed, the ER physician suggested that Melissa had suffered a CVA (cerebrovascular accident). What structures in the brain were affected by this incident? What might be the cause of the problem and the expected outcome?

2. Randolph, a college professor, suffered a severe blow to the head while being robbed at an ATM. This trauma caused him to be unable to impart any emotion into his lectures, although he could still speak. Identify Randolph's condition, and locate the area of damage to his brain.

ANSWERS TO "WHAT DO YOU THINK?"

1. No one is quite sure why most pathways decussate. Some researchers have speculated that in more primitive brain systems, most pathways were bilateral (a combination of decussating and undecussating pathways), and as brains evolved, only one of the two pathways remained (typically, the decussating pathway). However, this hypothesis has not been proven and does not explain why the decussating pathways would remain.

2. If our brains were consciously aware of every single sensory stimulus, we would go on "sensory overload." There would be too many sensations to effectively process and interpret, and we would not be able to concentrate on the task at hand. For example, being able to "tune out" many sensory stimuli allows you to study in a crowded student lounge with the music blaring.

3. Study habits that repeat and reinforce the material are best for converting short-term memories into long-term memories. Rewriting your notes, making flashcards, reading the text, and reviewing the material on a regular basis (not just once or twice) all help form long-term memory.

Visit the McKinley/O'Loughlin _Human Anatomy_, 2e website at aris.mhhe.com.

OUTLINE

NERVOUS SYSTEM

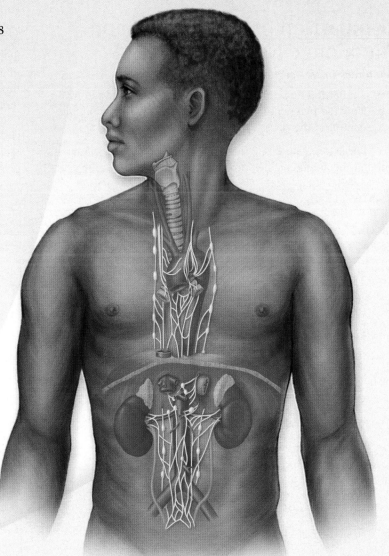

Autonomic Nervous System

On a twisting downhill slope, an Olympic skier is concentrating on controlling his body to negotiate the course faster than anyone else in the world. Compared to the spectators in the viewing areas, his pupils are more dilated, and his heart is beating faster and pumping more blood to his skeletal muscles. At the same time, organ system functions not needed in the race are practically shut down. Digestion, urination, and defecation can wait until the race is over. The skier exhibits a state of heightened readiness, called the "fight-or-flight" response, because the sympathetic division of the autonomic nervous system is dominant.

The **autonomic** (aw-tō-nom′ik; *auto* = self, *nomos* = law) **nervous system (ANS)** is a complex system of nerves that govern involuntary actions. The ANS works constantly with the **somatic nervous system (SNS)** to regulate body organs and maintain normal internal functions. We begin this chapter by comparing the SNS and the ANS.

Comparison of the Somatic and Autonomic Nervous Systems

Key topics in this section:

- Similarities and differences between the SNS and the ANS
- How the two-neuron chain facilitates communication and control in the ANS

Recall from figure 14.2 (page 415) that the somatic nervous system and the autonomic nervous system are part of both the central nervous system and the peripheral nervous system. The SNS operates under our conscious control, as exemplified by voluntary activities such as getting out of a chair, picking up a ball, walking outside, and throwing the ball for the dog to chase. (We have already seen that some SNS activities, such as swinging the arms while walking, occur at the subconscious level.) By contrast, ANS functions are involuntary, and we are usually unaware of them. For example, we are oblivious to the muscular actions of the stomach during digestion or changes in blood vessel diameter to adjust blood pressure.

Both the SNS and the ANS use sensory and motor neurons **(figure 18.1)**. In the SNS, somatic sensory neurons conduct stimulus information from a sensory receptor, such as a tactile receptor in the skin, while somatic motor neurons innervate skeletal muscle fibers. The ANS, by contrast, is activated by visceral sensory neurons. For example, some of these sensory neurons detect pressure by monitoring stretch in blood vessels and organ walls, while others measure carbon dioxide concentration in the blood. Some somatosensory receptors, such as those that detect temperature and light, also activate specific ANS responses (e.g., pupil constriction in response to bright light). In addition, autonomic motor neurons innervate smooth muscle cells, cardiac muscle cells, or glands. These motor neurons can either excite or inhibit cells in the viscera.

The motor neurons of the SNS innervate skeletal muscle fibers, typically causing conscious, voluntary movements. A single lower motor neuron axon extends uninterrupted from the spinal cord to one or more muscle fibers (figure 18.1). The impulses conducted by these

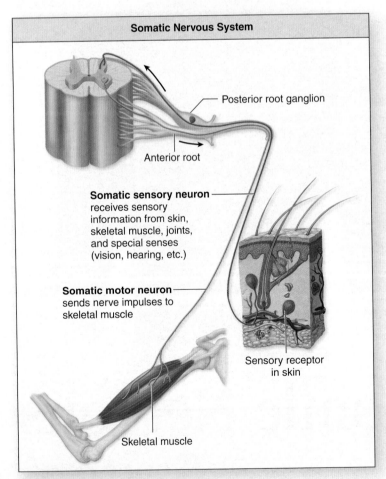

Somatic Nervous System

Posterior root ganglion

Anterior root

Somatic sensory neuron receives sensory information from skin, skeletal muscle, joints, and special senses (vision, hearing, etc.)

Somatic motor neuron sends nerve impulses to skeletal muscle

Sensory receptor in skin

Skeletal muscle

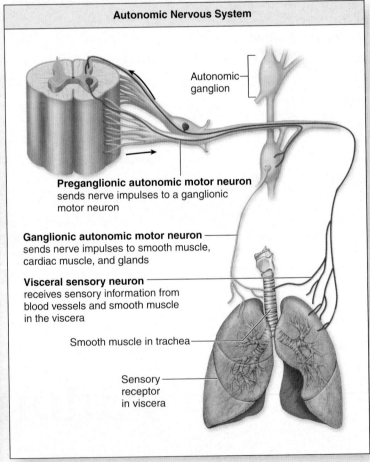

Autonomic Nervous System

Autonomic ganglion

Preganglionic autonomic motor neuron sends nerve impulses to a ganglionic motor neuron

Ganglionic autonomic motor neuron sends nerve impulses to smooth muscle, cardiac muscle, and glands

Visceral sensory neuron receives sensory information from blood vessels and smooth muscle in the viscera

Smooth muscle in trachea

Sensory receptor in viscera

Figure 18.1
Comparison of Somatic and Autonomic Motor Nervous Systems. The somatic nervous system extends a single motor neuron to its effector, while the autonomic nervous system uses two motor neurons, which meet in an autonomic ganglion, to reach its effector. However, both systems use a single sensory neuron to convey impulses to the CNS.

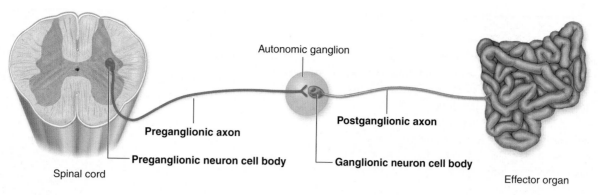

Figure 18.2

Components of the Autonomic Nervous System.   The autonomic nervous system employs a preganglionic neuron, which is housed in the CNS (brain or spinal cord). The preganglionic axon synapses with a ganglionic neuron in an autonomic ganglion. The postganglionic axon (from the ganglionic neuron) travels to the effector.

motor neurons stimulate skeletal muscle fibers, causing them to contract. Contraction continues until neuron impulses cease to stimulate the muscle fiber. By contrast, the ANS uses a pathway that includes a two-neuron chain to innervate muscles and glands (**figure 18.2**). The first of the two ANS motor neurons is the **preganglionic** (prē′gang-lē-on′ik) **neuron.** Its cell body lies within the brainstem or the spinal cord. A **preganglionic axon** extends from this cell body and exits the CNS in either a cranial nerve or a spinal nerve. This axon projects to the cell body of the second neuron, which is housed within an autonomic ganglion in the peripheral nervous system. The second neuron in this pathway is called a **ganglionic neuron,** and a **postganglionic axon** extends from its cell body to effector cells or an effector organ.

WHAT DO YOU THINK?

❶ Why does the autonomic motor nervous system use two neurons (preganglionic and ganglionic) in a chain to an effector? (For the answer, read the next section.)

The two-neuron chain vastly increases communication and control in the ANS. **Neuronal convergence** (kon-ver′jens; *con* = with, *vergere* = to incline) occurs when axons from numerous preganglionic cells synapse (converge) on a single ganglionic cell. In contrast, **neuronal divergence** (di-ver′jens; *di* = apart) occurs when axons from one preganglionic cell synapse on numerous ganglionic cells.

Table 18.1 summarizes the characteristics of the somatic and autonomic nervous systems.

WHAT DID YOU LEARN?

❶ How are motor neurons organizationally different in the ANS versus the SNS?

❷ What organs are innervated by the ANS?

❸ Where is a ganglionic neuron cell body located?

Table 18.1	Comparison of Somatic and Autonomic Motor Nervous Systems	
Feature	**Somatic Nervous System**	**Autonomic Nervous System**
Type of Control	Voluntary control (from cerebral cortex; input from basal nuclei, brainstem, cerebellum, and spinal cord)	Involuntary control (from brainstem, hypothalamus, limbic system, and spinal cord)
Number of Neurons in Pathway	One neuron in pathway; somatic motor neuron axon extends from CNS to effector	Two neurons in pathway; preganglionic neuron in CNS projects preganglionic axon to ganglionic neuron; ganglionic neuron projects postganglionic axon to effector
Ganglia Associated with Motor Neurons	None	Autonomic ganglia: sympathetic trunk ganglia; prevertebral ganglia; terminal or intramural ganglia
Sensory Input	General somatic senses, proprioceptors; special senses	Some somatic and visceral senses
Ganglia Associated with Sensory Input	Posterior root ganglia; sensory ganglia of cranial nerves	Posterior root ganglia; sensory ganglia of cranial nerves
Effector Organs	Skeletal muscle fibers	Cardiac muscle fibers, smooth muscle fibers, glands
Response of Effector	Excitation only	Either excitation or inhibition of effectors
Neurotransmitter Released	Acetylcholine (ACh)	ACh from all preganglionic axons and parasympathetic postganglionic axons, and a few sympathetic postganglionic axons; norepinephrine (NE) from most sympathetic postganglionic axons
Axon Properties	Myelinated, thick; fast conduction	Preganglionic axons are thin, myelinated; postganglionic axons are thinner, unmyelinated, have slow conduction

Overview of the Autonomic Nervous System

Key topic in this section:

- Comparison of the parasympathetic and sympathetic divisions

The ANS is subdivided into the parasympathetic division and the sympathetic division. These two divisions are similar in that they both use a preganglionic neuron and a ganglionic neuron to innervate muscles or glands. Both divisions contain the autonomic ganglia that house the ganglionic neurons. Both divisions are involuntary and are concerned with the body's internal environment in general. However, these two divisions perform dramatically different functions (**figure 18.3**).

The **parasympathetic** (par-ă-sim-pa-thet′ik; *para* = alongside, *sympatheo* = to feel with) **division** is primarily concerned with conserving energy and replenishing nutrient stores. Thus, it is most active when the body is at rest or digesting a meal, and has been nicknamed the "rest-and-digest" division. The parasympathetic division also helps maintain **homeostasis,** a constant internal environment.

The **sympathetic** (sim-pă-thet′ik) **division** is primarily concerned with preparing the body for emergencies. It is often referred to as the "fight-or-flight" division because increased sympathetic activity results in the increased alertness and metabolic activity needed in stressful or frightening situations. During these fight-or-flight events, the sympathetic division exhibits a mass activation response, whereby all components receiving sympathetic innervation get stimulated. (In contrast, the parasympathetic division is discrete and localized, meaning only one or a few structures are innervated at the same time.)

The parasympathetic and sympathetic divisions are similar in that their preganglionic axons are myelinated, while the postganglionic axons are unmyelinated. These two divisions are also distinguished by several anatomic differences. The major difference is that their preganglionic neuron cell bodies are housed in different regions of the CNS. Parasympathetic preganglionic neurons originate in either the brainstem or the lateral gray matter of the S2–S4 spinal cord segments, while sympathetic preganglionic neurons originate in the lateral horns of the T1–L2 spinal cord segments (figure 18.3).

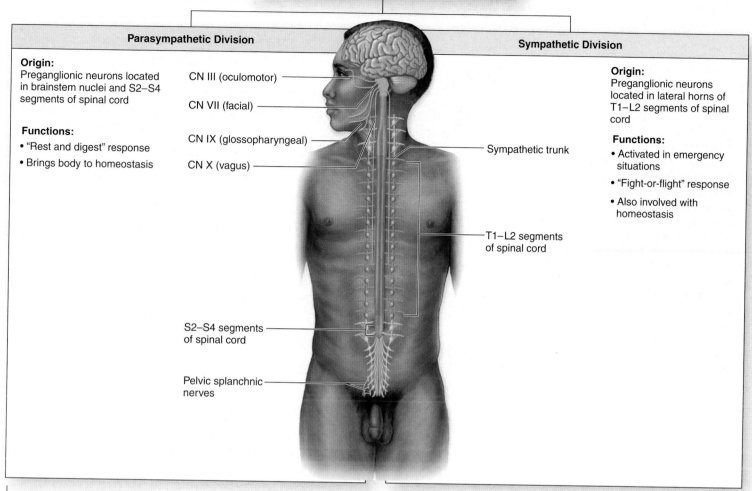

Components of Autonomic Nervous System

Parasympathetic Division

Origin:
Preganglionic neurons located in brainstem nuclei and S2–S4 segments of spinal cord

Functions:
• "Rest and digest" response
• Brings body to homeostasis

CN III (oculomotor)

CN VII (facial)

CN IX (glossopharyngeal)

CN X (vagus)

S2–S4 segments of spinal cord

Pelvic splanchnic nerves

Sympathetic Division

Origin:
Preganglionic neurons located in lateral horns of T1–L2 segments of spinal cord

Functions:
• Activated in emergency situations
• "Fight-or-flight" response
• Also involved with homeostasis

Sympathetic trunk

T1–L2 segments of spinal cord

Figure 18.3

Comparison of Parasympathetic and Sympathetic Divisions.   The parasympathetic and sympathetic divisions of the ANS have the same basic components, but they differ in their origins, locations of the preganglionic cell bodies, axon lengths, and amount of branching.

Figure 18.4 depicts additional anatomic differences: (1) Parasympathetic preganglionic axons are longer, and postganglionic axons are shorter when compared to their counterparts in the sympathetic division. In the sympathetic division, preganglionic axons are shorter and postganglionic axons are longer. (2) Parasympathetic autonomic ganglia are close to or within the wall of the effector organ, while sympathetic autonomic ganglia are relatively close to the vertebral column. (3) The amount of preganglionic axon branching to ganglionic neurons differs between the divisions. Parasympathetic preganglionic

axons tend to have few (less than 4) branches, while sympathetic preganglionic axons tend to have many branches (more than 20).

Table 18.2 summarizes the comparison of the parasympathetic and sympathetic divisions of the autonomic nervous system.

WHAT DID YOU LEARN?

❹ Describe the anatomic differences between the postganglionic axons in the parasympathetic and sympathetic divisions.

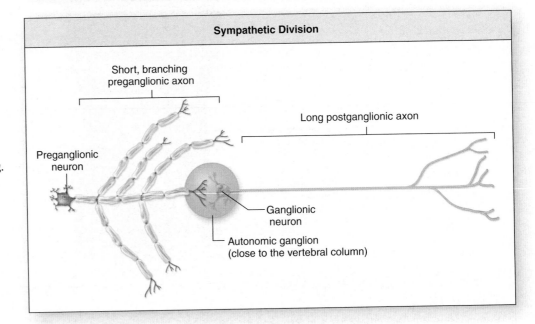

Figure 18.4

Anatomic Differences Between Parasympathetic and Sympathetic Neurons. In both the parasympathetic and sympathetic divisions, preganglionic axons are myelinated and relatively larger in diameter, and postganglionic axons are unmyelinated and relatively smaller in diameter. (*Top*) The parasympathetic division has longer preganglionic axons and shorter postganglionic axons; its preganglionic axons exhibit very little branching. (*Bottom*) The sympathetic division has shorter preganglionic axons and longer postganglionic axons; the preganglionic axons show much branching.

Table 18.2	**Comparison of Parasympathetic and Sympathetic Divisions**	
Feature	**Parasympathetic Division**	**Sympathetic Division**
Function	Conserves energy and replenishes energy stores; maintains homeostasis; "rest-and-digest" division	Prepares body to cope with emergencies and intensive muscle activity; "fight-or-flight" division
Location of Preganglionic Neuron Cell Bodies	Brainstem and lateral gray matter in S2–S4 segments of spinal cord	Lateral horns in T1–L2 segments of spinal cord
Location of Ganglionic Neuron Cell Bodies	Terminal or intramural ganglion	Sympathetic trunk ganglion or prevertebral ganglion
Divergence of Axons	Few (1 axon innervates < 4 ganglionic cell bodies)	Extensive (1 axon innervates > 20 ganglionic cell bodies)
Length of Preganglionic Axon	Long	Short
Length of Postganglionic Axon	Short	Long
Location of Ganglia	Terminal ganglia located close to the target organ; intramural ganglia located within wall of the target organ	Sympathetic trunk (paravertebral) ganglia located on either side of vertebral column; prevertebral (collateral) ganglia located anterior to vertebral column and descending aorta
Rami Communicantes	None	White rami attach to T1–L2 spinal nerves; gray rami attach to *all* spinal nerves

Parasympathetic Division

Key topics in this section:

■ Anatomy of the parasympathetic division.

■ Relationship of the parasympathetic division to the brain, the cranial nerves, and the sacral spinal cord

■ Effects of parasympathetic innervation on effectors

The parasympathetic division of the ANS is structurally simpler than the sympathetic division. The parasympathetic division is also termed the *craniosacral* (krā′nē-ō-sā′krăl) *division* because its preganglionic neurons are housed within nuclei in the brainstem and within the lateral gray matter of the S2–S4 spinal cord segments. The ganglionic neurons in the parasympathetic division are found in either **terminal** (ter′mi-năl; *terminus* = a boundary) **ganglia,** which are located close to the target organ, or **intramural** (in′tră-mū′răl; *intra* = within, *murus* = wall) **ganglia,** which are located within the wall of the target organ.

Cranial Nerves

The cranial nerves associated with the parasympathetic division are the oculomotor (CN III), facial (CN VII), glossopharyngeal (CN IX), and vagus (CN X) (see figure 18.3). The first three of these nerves convey parasympathetic innervation to the head, while the vagus nerve is the source of parasympathetic stimulation for the thoracic and most abdominal organs **(figure 18.5)**.

Review table 15.7 for illustrations of the cranial nerve pathways and the locations of their associated parasympathetic ganglia.

The **oculomotor nerve (CN III)** is formed by axons extending from some cell bodies housed in nuclei in the mesencephalon. The preganglionic axons extend from CN III to the **ciliary** (sil′ē-ar-ē; *ciliaris* = eyelash) **ganglion** within the orbit. Postganglionic axons project from this ganglion to the ciliary muscle and sphincter pupillae muscle of the iris of the eye. Parasympathetic innervation to the ciliary muscle results in lens accommodation, which makes the lens more rounded so that we can see close-up objects. The postganglionic axons that travel to the pupillary constrictor muscle result in pupil constriction when the eye is exposed to bright light.

The **facial nerve (CN VII)** contains parasympathetic preganglionic axons that exit the pons and control the production and secretion of tears, nasal secretions, and saliva. Two branches of parasympathetic preganglionic axons exit the facial nerve and terminate at one of two ganglia. The greater petrosal nerve terminates at the **pterygopalatine** (ter′i-gō-pal′a-tīn) **ganglion** near the junction of the maxilla and palatine bones. Postganglionic axons project to the lacrimal glands and small glands of the nasal cavity, oral cavity, and palate to increase secretion by these glands. The chorda tympani terminates on ganglionic neurons in the **submandibular** (sŭb-man-dib′ū-lăr; *sub* = under) **ganglion** near the angle of the mandible. Postganglionic axons projecting from this ganglion supply the submandibular and sublingual salivary glands in the floor of the mouth, causing an increase in salivary gland secretions. Thus, your mouth waters when you smell an aromatic meal due in part to these parasympathetic axons.

WHAT DO YOU THINK?

❷ The pterygopalatine ganglion is sometimes nicknamed the "hay fever ganglion." Why is this nickname appropriate?

The **glossopharyngeal nerve (CN IX)** innervates the parotid salivary gland. Parasympathetic stimulation exits the brainstem in the glossopharyngeal nerve. From this nerve, the preganglionic parasympathetic axons branch and synapse on ganglionic neurons in the **otic** (ō′tik; *ous* = ear) **ganglion,** which is positioned anterior to the ear near the foramen ovale. Postganglionic axons from the otic ganglion cause an increase in secretion from the parotid salivary glands.

Each **vagus nerve (CN X)** is responsible for supplying parasympathetic innervation to the thoracic organs and most of the abdominal organs, as well as the gonads (ovaries and testes).[1] Almost 80% of all parasympathetic preganglionic axons are transmitted through the vagus nerve. The term *vagus* means "wanderer," which describes the wandering pathway of the vagus nerve as it projects inferiorly through the neck and travels throughout the trunk. Left and right vagus nerves extend multiple branches to the thoracic organs. As these nerves travel inferiorly, their position changes slightly, and they are referred to as the anterior and posterior vagal trunks. In the thoracic cavity, parasympathetic innervation causes increased mucous production and decreased diameter in the airways, as well as decreases in the heart rate and the force of heart contractions. These trunks pass through the diaphragm and associate with the descending abdominal aorta within the abdominal cavity, where they project to their ganglia located immediately adjacent to or within the wall of their target organs. This parasympathetic innervation also causes increased smooth muscle motility and secretory activity in digestive tract organs.

Sacral Spinal Nerves

The remaining preganglionic parasympathetic axons originate from preganglionic neuron cell bodies housed within the lateral gray matter of the S2–S4 spinal cord segments (figure 18.5). These preganglionic parasympathetic axons branch to form the **pelvic splanchnic** (splangk′nik; *visceral*) **nerves,** which contribute to the superior and inferior hypogastric plexus. The preganglionic parasympathetic axons that emanate from each plexus project to the ganglionic neurons within either the terminal or intramural ganglia. The target organs innervated include the distal portion of the large intestine, the rectum, most of the reproductive organs, the urinary bladder, and the distal part of the ureter. This parasympathetic innervation causes increased smooth muscle motility (muscle contraction) and secretory activity in the digestive organs, mentioned above, contraction of smooth muscle in the bladder wall, and erection of the female clitoris and the male penis.

Effects and General Functions of the Parasympathetic Division

The parasympathetic division is most active during times when the body must process nutrients, conserve energy, and attempt to return to homeostasis. The lack of extensive divergence in preganglionic axons prevents the mass activation seen in the sympathetic division. Thus, the effects of the parasympathetic nervous system tend to be discrete and localized. In other words, parasympathetic activity can affect one group of organs without necessarily having to "turn on" all other organs. **Table 18.3** summarizes the effects of parasympathetic innervation.

---

[1] It is unclear what function, if any, these parasympathetic fibers have on the gonads.

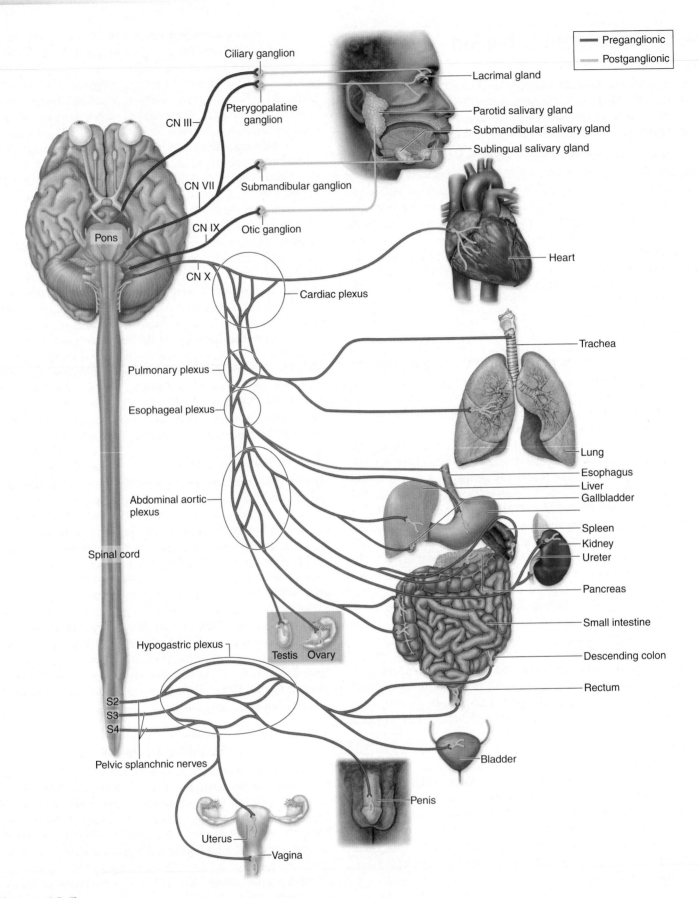

Preganglionic
Postganglionic

Ciliary ganglion
Lacrimal gland
Pterygopalatine ganglion
CN III
Parotid salivary gland
Submandibular salivary gland
Sublingual salivary gland
CN VII
Submandibular ganglion
CN IX
Otic ganglion
Pons
CN X
Cardiac plexus
Heart
Pulmonary plexus
Trachea
Esophageal plexus
Lung
Esophagus
Liver
Gallbladder
Abdominal aortic plexus
Spleen
Kidney
Ureter
Spinal cord
Pancreas
Small intestine
Hypogastric plexus
Testis   Ovary
Descending colon
Rectum
S2
S3
S4
Bladder
Pelvic splanchnic nerves
Penis
Uterus
Vagina

Figure 18.5

Overview of Parasympathetic Pathways. Preganglionic axons from the brain and spinal cord innervate the viscera in the head, neck, and trunk.

Table 18.3	Parasympathetic Division Outflow		
Nerve(s)	CNS Origin of Preganglionic Neuron	Autonomic Ganglion	Effector Organs Innervated
CN III (Oculomotor)	Mesencephalon	Ciliary ganglion	Ciliary muscles to control lens for accommodation; sphincter pupillae muscle of eye to constrict pupil
CN VII (Facial)	Pons	Pterygopalatine ganglion	Lacrimal glands; glands of nasal cavity, palate, oral cavity
		Submandibular ganglion	Submandibular and sublingual salivary glands
CN IX (Glossopharyngeal)	Medulla oblongata	Otic ganglion	Parotid salivary glands
CN X (Vagus)	Medulla oblongata	Multiple terminal and intramural ganglia	Thoracic viscera and most abdominal viscera
Pelvic Splanchnic Nerves	S2–S4 segments of spinal cord	Terminal and intramural ganglia	Some abdominal viscera and most pelvic viscera

WHAT DID YOU LEARN?

5 What are the differences between the terminal and intramural ganglia?

6 Identify the cranial nerves involved in the parasympathetic division of the ANS.

Sympathetic Division

Key topics in this section:

- Anatomy of the sympathetic division
- Relationship of the sympathetic division to the spinal cord and the spinal nerves
- Sympathetic function of the adrenal medulla
- Effects of sympathetic innervation on effectors

The sympathetic division is also called the *thoracolumbar* (thŏr′ă-kō-lŭm′bar) *division* because the preganglionic neuron cell bodies originate and are housed between the first thoracic (T1) and the second lumbar (L2) spinal segments.

Organization and Anatomy of the Sympathetic Division

The sympathetic division is much more complex than the parasympathetic division, both anatomically and functionally (**figure 18.6**; see figure 18.4). The sympathetic preganglionic neuron cell bodies are housed in the **lateral horn** of the T1–L2 segments of the spinal cord. From there, the preganglionic sympathetic axons travel with somatic motor neuron axons to exit the spinal cord and enter first the anterior roots and then the T1–L2 spinal nerves. However, these preganglionic sympathetic axons remain with the spinal nerve for only a short distance before they leave the spinal nerve.

Immediately anterior to the paired spinal nerves are the left and right **sympathetic trunks,** each of which is located immediately lateral to the vertebral column (**figure 18.7**). A sympathetic trunk looks much like a pearl necklace. The "string" of the "necklace" is composed of bundles of axons, while the "pearls" are the **sympathetic trunk ganglia** (*paravertebral* or *chain ganglia*), which house sympa-

thetic ganglionic neuron cell bodies. One sympathetic trunk ganglion is approximately associated with each spinal nerve. However, the cervical portion of each sympathetic trunk is partitioned into only three sympathetic trunk ganglia—the superior, middle, and inferior cervical ganglia—as opposed to the eight cervical spinal nerves. The **superior cervical ganglion** contains postganglionic sympathetic neuron cell bodies whose axons are distributed to structures within the head and neck. These sympathetic postganglionic axons innervate the sweat glands in the head and neck, the smooth muscle in blood vessels of the head and neck, the dilator pupillae muscle of the eye, and the superior tarsal muscle of the eye (which elevates the eyelid). The middle and inferior cervical ganglia house neuron cell bodies that extend postganglionic axons to the thoracic viscera.

Connecting the spinal nerves to each sympathetic trunk are rami communicantes (rā′mī kŏ-mū-ni-kan′tēz; *communico* = to share with someone). **White rami communicantes** (or **white rami**) carry *pre*ganglionic sympathetic axons from the T1–L2 spinal nerves to the sympathetic trunk. Thus, white rami are associated only with the T1–L2 spinal nerves. Since preganglionic axons are myelinated, the white ramus has a whitish appearance (hence, its name). White rami are similar to "entrance ramps" onto a highway. **Gray rami communicantes** (or **gray rami**) carry *post*ganglionic sympathetic axons from the sympathetic trunk to the spinal nerve. Since the postganglionic axons are unmyelinated, the gray rami have a grayish appearance. Gray rami are similar to "exit ramps" off a highway. Gray rami connect to all spinal nerves: the cervical, thoracic, lumbar, sacral, and coccygeal spinal nerves. By these routes, the sympathetic information that started out in the thoracolumbar region can be dispersed to all parts of the body.

Splanchnic nerves are composed of preganglionic sympathetic axons that did not synapse in a sympathetic trunk ganglion. They extend anteriorly from each sympathetic trunk to most of the viscera. (These splanchnic nerves should not be confused with the pelvic splanchnic nerves associated with the parasympathetic division.) Some of the larger splanchnic nerves have specific names:

- The **greater thoracic splanchnic nerve** forms from preganglionic axons extending from the T5–T9 sympathetic trunk ganglia.

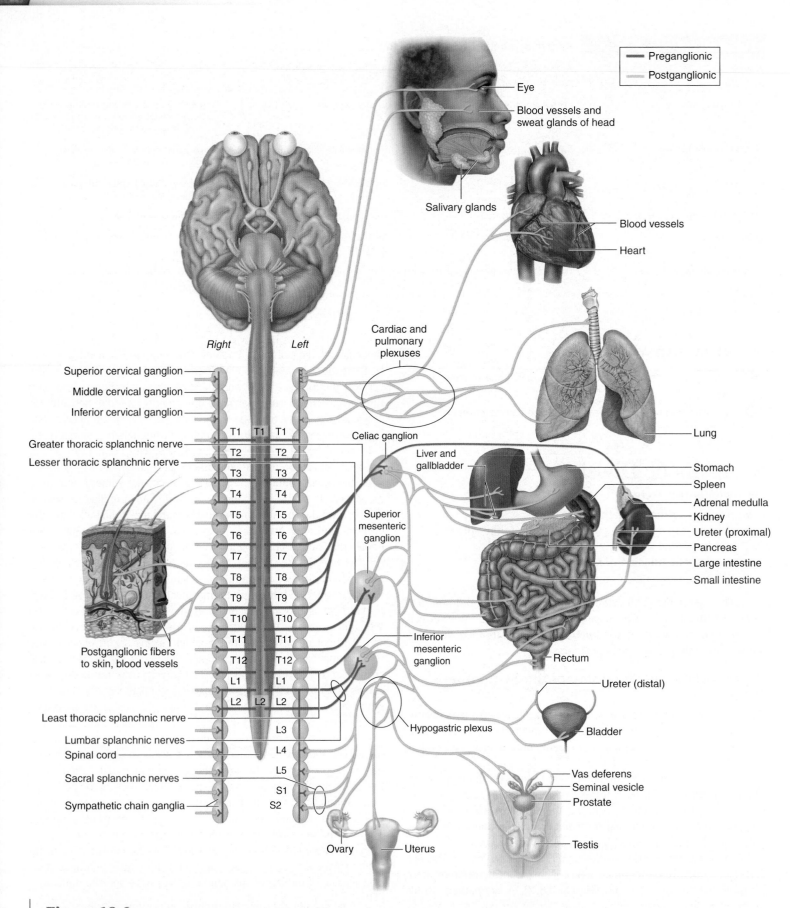

Figure 18.6

Overview of Sympathetic Pathways. The right sympathetic trunk shows the outflow of preganglionic axons and the distribution of postganglionic axons innervating the skin. The left sympathetic trunk illustrates sympathetic postganglionic axon pathways through the gray rami, spinal nerves, and splanchnic nerves. (Note, however, that in reality each sympathetic trunk innervates both the skin and the viscera.)

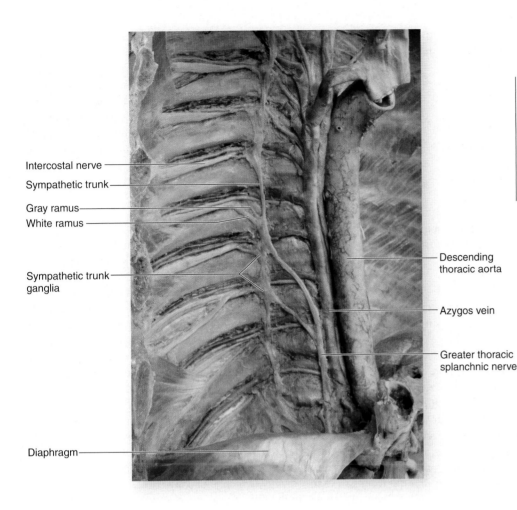

Intercostal nerve

Sympathetic trunk

Gray ramus

White ramus

Sympathetic trunk
ganglia

Diaphragm

Descending
thoracic aorta

Azygos vein

Greater thoracic
splanchnic nerve

Figure 18.7

Sympathetic Trunk. An anterolateral cadaver photo of the right side of the thoracic cavity shows the sympathetic trunk, the gray and white rami communicantes, their attachment to the intercostal nerves, and the greater thoracic splanchnic nerve.

- The **lesser thoracic splanchnic nerve** forms from preganglionic axons extending from the T10–T11 sympathetic trunk ganglia.
- The **least thoracic splanchnic nerve** forms from preganglionic axons extending from the T12 sympathetic trunk ganglia.
- The **lumbar splanchnic nerves** originate from the L1 and L2 sympathetic trunk ganglia.

In addition to these, there also are small sacral splanchnic nerves that originate from the sacral sympathetic ganglia. Splanchnic nerves typically terminate in **prevertebral** (or *collateral*) **ganglia.** These ganglia are called "prevertebral" because they are immediately anterior to the vertebral column on the anterolateral wall of the abdominal aorta. Prevertebral ganglia typically cluster around the origins of the major abdominal arteries and are named for these arteries. For example, the celiac ganglion is located around the origin of the celiac trunk (an artery). Sympathetic postganglionic axons extend away from the ganglionic neuron cell bodies in these ganglia and innervate many of the abdominal organs.

Types of Prevertebral Ganglia

The **prevertebral ganglia** differ from the sympathetic trunk ganglia segments in that (1) they are single structures, rather than paired; (2) they are anterior to the vertebral column (hence the name prevertebral) on the anterolateral surface of the aorta; and (3) they are located only in the abdominopelvic cavity. Prevertebral ganglia include the celiac, superior mesenteric, and inferior mesenteric ganglia.

The **celiac ganglion** is adjacent to the origin of the celiac artery. Its appearance often varies in individuals; thus, it is usually composed of two connected masses, but may also form a single mass. The left and right greater thoracic splanchnic nerves (composed of axons from the T5–T9 segments of the spinal cord) synapse on ganglionic neurons within the celiac ganglion. Postganglionic axons from the celiac ganglion innervate the stomach, spleen, liver, gallbladder, and proximal part of the duodenum and part of the pancreas.

The **superior mesenteric** (mez-en-ter'ik; *mesos* = middle, *enteron* = intestine) **ganglion** is adjacent to the origin of the superior mesenteric artery. The lesser and least thoracic splanchnic nerves project to and terminate in the superior mesenteric ganglion. Thus, this ganglion receives preganglionic sympathetic neurons from the T10–T12 segments of the spinal cord. Postganglionic axons extending from the superior mesenteric ganglion innervate the distal half of the duodenum, part of the pancreas, the remainder of the small intestine, the proximal part of the large intestine, the kidneys, and the proximal parts of the ureters.

The **inferior mesenteric ganglion** is adjacent to the origin of the inferior mesenteric artery. It receives sympathetic preganglionic axons via the lumbar splanchnic nerves, which originate in the

L1–L2 segments of the spinal cord. Its postganglionic axons project to and innervate the distal colon, rectum, urinary bladder, distal parts of the ureters, and most of the reproductive organs.

7 The sympathetic division originates in what area and segments of the spinal cord?

8 Distinguish between the sympathetic trunk ganglia and the prevertebral ganglia.

9 Describe the structural and functional differences between the white and gray rami communicantes. Do these structures contain myelinated or unmyelinated axons? Which carry preganglionic axons, and which carry postganglionic axons?

Sympathetic Pathways

All sympathetic preganglionic neurons originate in the lateral gray horns of the T1–L2 segments of the spinal cord. However, the sympathetic pathways of the axons of these neurons vary, depending upon the location and the type of effector organ being innervated. Recall that preganglionic axons extend from the preganglionic neuron cell bodies via the anterior roots and travel with the T1–L2 spinal nerves. The preganglionic axons immediately leave the spinal nerve and travel through white rami to enter the sympathetic trunk. Once inside the sympathetic trunk, the preganglionic axons may remain at the level of entry, or travel superiorly or inferiorly within the sympathetic trunk.

Axons exit the sympathetic trunk ganglia by one of four pathways (**figure 18.8**). An axon takes the **spinal nerve pathway** if a preganglionic neuron synapses with a ganglionic neuron in a sympathetic trunk ganglion. In this case, the postganglionic axon travels through a gray ramus that is at the same "level" as the ganglionic neuron. For example, if the preganglionic and ganglionic neurons synapse in the L4 sympathetic trunk ganglion, the postganglionic axon travels through the gray ramus at the level of the L4 spinal nerve. After the postganglionic axon travels through the gray ramus, it may enter the spinal nerve and extend to its target organ. The structures in the skin (such as arrector pili muscles and blood vessels) receive their sympathetic innervation via this pathway.

In the **postganglionic sympathetic nerve pathway**, the preganglionic neuron synapses with a ganglionic neuron in a sympathetic trunk ganglion, but the postganglionic axon does *not* leave the trunk via a gray ramus. Instead, the postganglionic axon extends away from the sympathetic trunk ganglion (in the form of a postganglionic sympathetic axon) and goes directly to the effector organ. The esophagus, heart, lungs, and thoracic blood vessels typically receive their sympathetic innervation from this pathway.

The **splanchnic nerve pathway** uses splanchnic nerves, which are preganglionic axons that pass through the sympathetic trunk ganglia without synapsing. These splanchnic nerves extend from the anterior side of the sympathetic trunk ganglia to the prevertebral ganglia. There, the preganglionic axon synapses with a ganglionic neuron. The postganglionic axon then travels to the effector organs. The abdominal and pelvic organs receive their sympathetic innervation via this pathway.

The final pathway is the **adrenal medulla pathway.** In this pathway, the internal region of the adrenal gland, called the **adrenal** (ă-drē'năl) **medulla,** receives preganglionic sympathetic axons. When these preganglionic axons synapse on cells within the adrenal medulla, those cells release hormones that are circulated within the

bloodstream and help prolong the fight-or-flight response. These hormones are **epinephrine** (ep'i-nef'rin; *epi* = upon, *nephros* = kidney) and, to a lesser degree, **norepinephrine** (nŏr-ep-i-nef'rin) (discussed in chapter 20). Both of these hormones potentiate (prolong) the effects of the sympathetic stimulation. For example, if you narrowly miss getting into a car accident, your heart continues to beat quickly, you breathe rapidly, and you feel tense and alert well after the event. In this case, the epinephrine and norepinephrine circulating in your bloodstream are prolonging the effects of the sympathetic stimulation.

Effects and General Functions of the Sympathetic Division

The sympathetic division may innervate a single effector or many effectors. For example, a single effector is involved when smooth muscle controls the diameter of the pupil of the eye, while many effectors respond together, a phenomenon termed **mass activation,** during an emergency or crisis situation. In mass activation, numerous collateral branches of preganglionic sympathetic axons synapse with a large number of ganglionic neurons to stimulate many ganglionic sympathetic neurons and simultaneously activate many effector organs. Mass activation of the sympathetic division causes a heightened sense of alertness due to stimulation of the reticular activation system. **Table 18.4** shows how specific structures are affected by the sympathetic division.

Mass activation often occurs simultaneously with an increase in tonus in skeletal muscle. However, this increased skeletal muscle tension is not due to activation of the ANS, but merely to changes in muscle tone. In addition, the affected individual experiences a feeling of excess energy, which is usually caused by mobilization of energy reserves in the liver. Some obvious systemic changes accompany sympathetic stimulation, including increases in heart rate and blood pressure and parallel increases in depth of respiration and breathing rate. Finally, the pupils dilate due to innervation of the dilator pupillae muscle in the iris of the eye.

3 When a person is very stressed and tense, his or her blood pressure typically rises. What aspect of the sympathetic nervous system causes this rise in blood pressure?

Raynaud Syndrome

Raynaud syndrome, or *Raynaud phenomenon*, is a sudden spasm or constriction of the small arteries of the digits. The immediate decrease in blood flow results in blanching (loss of the red hue) of the skin distal to the area of vascular constriction. The vascular constriction is accompanied by pain, which may even continue for a while after the vessels have dilated and restored the local blood flow. Episodes are typically triggered by exposure to cold, although emotional stress has been known to precipitate a Raynaud attack. Only a few people experience this condition, which is believed to result from an exaggerated local sympathetic response. The severity of this medical condition depends on the frequency and the length of time of each occurrence. Most people affected with Raynaud syndrome must avoid the cold and other triggering circumstances.

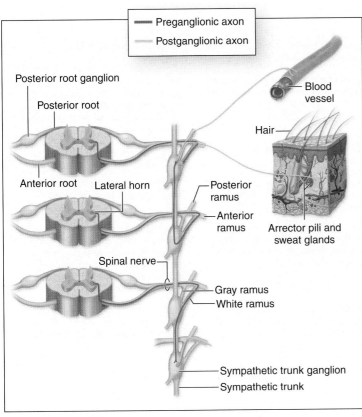

(a) Spinal nerve pathway

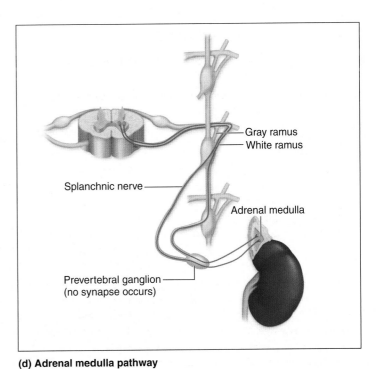

(b) Postganglionic sympathetic nerve pathway

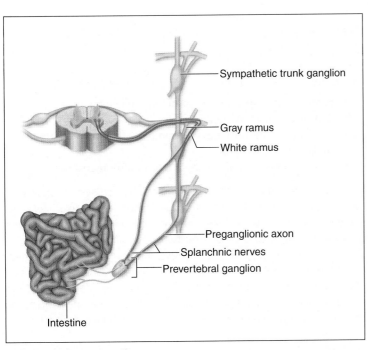

(c) Splanchnic nerve pathway

(d) Adrenal medulla pathway

Figure 18.8

Types of Sympathetic Pathways.   Pathways of *(a)* a spinal nerve, *(b)* a postganglionic sympathetic nerve, *(c)* a splanchnic nerve, and *(d)* the adrenal medulla.

Table 18.4	Sympathetic Division Outflow		
Destination	**Spinal Cord Segment Origin**	**Postganglionic Axon Pathway from Sympathetic Trunk**	**Organs Innervated[1]**
Head and neck	T1–T2 (almost all sympathetic innervation to the head comes from T1)	Via superior cervical ganglion and travel with blood vessels to the head	Blood vessels, sweat glands, and arrector pili muscles of head and neck; dilator pupillae muscle of eye, tarsal glands of eye, superior tarsal muscle of eye
Integumentary structures	T1–L2	Via gray rami to all spinal nerves	Sweat glands and arrector pili muscles, blood vessels in skin
Thoracic organs	T1–T5	Via cervical and thoracic ganglia to autonomic nerve plexuses near organs	Esophagus, heart, lungs, blood vessels within thoracic cavity
Most abdominal organs	T5–T12	Via thoracic splanchnic nerves to prevertebral ganglia (e.g., celiac, superior mesenteric, and inferior mesenteric ganglia)	Abdominal portion of esophagus; stomach, liver, gallbladder, spleen, pancreas, small intestine, most of large intestine, kidneys, ureters, adrenal glands, blood vessels within abdominopelvic cavity
Pelvic organs	T10–L2	Via lumbar and sacral splanchnic nerves to autonomic nerve plexuses that travel to target organ	Distal part of large intestine, anal canal, and rectum; distal part of ureters; urinary bladder, reproductive organs

[1] Sympathetic axons innervate the smooth muscle, cardiac muscle, and glands associated with the organs listed.

WHAT DID YOU LEARN?

10 How can the sympathetic axons stimulate so many effector organs simultaneously?

11 What is the function of splanchnic nerves in the sympathetic division?

12 From what structure are epinephrine and norepinephrine released following sympathetic stimulation?

CLINICAL VIEW

Horner Syndrome

Horner syndrome is a condition caused by damage to the sympathetic innervation to the head. This damage results from impingement, injury, or severing of the cervical sympathetic trunk or the T1 sympathetic trunk ganglion, where postganglionic sympathetic axons traveling to the head originate. The absence of sympathetic innervation on one side of the head leads to certain clinical signs on that side. The patient presents with **ptosis** (tō′sis; a falling), in which the superior eyelid droops because the superior tarsal muscle is paralyzed. Paralysis of the dilator pupillae muscle of the eye results in **miosis** (mī-ō′sis; *meiosis* = lessening), which is a constricted pupil. **Anhydrosis** (an-hǐ-drō′sis; *an* = without, *hidros* = sweat) occurs because the sweat glands no longer receive sympathetic innervation. A fourth symptom is distinct flushing due to lack of sympathetic innervation to blood vessel walls that results in vasodilation.

Other Features of the Autonomic Nervous System

Key topics in this section:

- Structure and location of autonomic plexuses
- Types of neurotransmitters
- Dual innervation by the parasympathetic and sympathetic divisions of the ANS
- How autonomic reflexes help maintain homeostasis

Both divisions of the autonomic nervous system innervate organs through specific axon bundles called autonomic plexuses. Communication between neurons and effectors in the autonomic nervous system is by chemical messengers, called neurotransmitters. These chemical messengers and the receptors on body organs to which they bind are specific in each division of the autonomic nervous system. Most organs are innervated by both divisions of the autonomic nervous system in what is called dual innervation. Autonomic reflexes help us maintain homeostasis. We discuss autonomic plexuses first.

Autonomic Plexuses

Autonomic plexuses are collections of sympathetic postganglionic axons and parasympathetic preganglionic axons, as well as some visceral sensory axons. These sympathetic and parasympathetic axons are close to one another, but they do not interact or synapse with one another. Although these plexuses look like disorganized masses of axons, they provide a complex innervation pattern to their target organs (**figure 18.9**).

In the mediastinum of the thoracic cavity, the **cardiac plexus** consists of postganglionic sympathetic axons that extend from the cervical and thoracic sympathetic trunk ganglia, as well as preganglionic

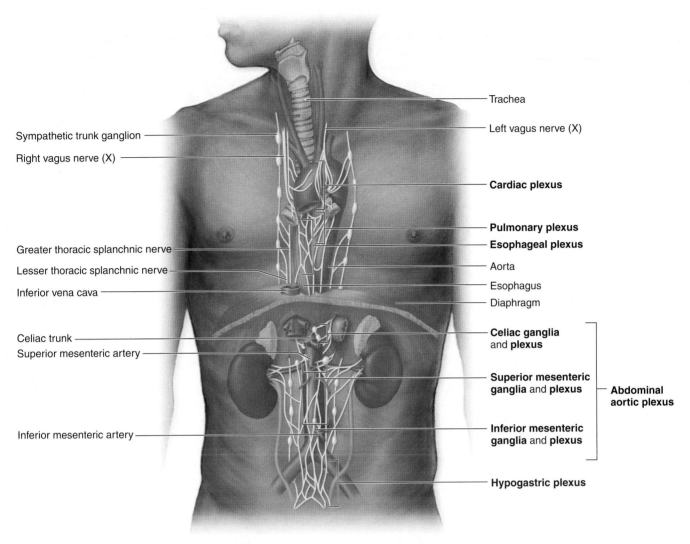

Figure 18.9

Autonomic Plexuses.   Autonomic plexuses are located in both the thoracic and abdominopelvic cavities. This anterior view shows the cardiac, pulmonary, and esophageal plexuses in the thoracic cavity and the abdominal aortic plexus (celiac, superior mesenteric, inferior mesenteric plexuses) in the abdominopelvic cavity.

axons from the vagus nerve. Increased sympathetic activity increases heart rate and blood pressure, while increased parasympathetic activity decreases heart rate.

The **pulmonary plexus** consists of postganglionic sympathetic axons from the thoracic sympathetic trunk ganglia and preganglionic axons from the vagus nerve. The axons project to the bronchi and blood vessels of the lungs. Stimulation of this parasympathetic pathway causes a reduction in the diameter of the bronchi (called bronchoconstriction) and increased secretion from mucous glands of the bronchial tree. Sympathetic innervation causes bronchodilation (increase in the diameter of the bronchi).

The **esophageal plexus** consists of preganglionic axons from the vagus nerve. Smooth muscle activity in the inferior esophageal wall is coordinated by parasympathetic axons that control the swallowing reflex in the inferior region of the esophagus by innervating smooth muscle in the inferior esophageal wall and the cardiac sphincter, a value through which swallowed food and drink must pass.

The **abdominal aortic plexus** consists of the **celiac plexus, superior mesenteric plexus,** and **inferior mesenteric plexus.**

The abdominal aortic plexus is composed of postganglionic axons projecting from the prevertebral ganglia and preganglionic axons from the vagus nerve that enter the abdominopelvic cavity with the esophagus.

The **hypogastric plexus** consists of a complex meshwork of postganglionic sympathetic axons (from the aortic plexus and the lumbar region of the sympathetic trunk) and preganglionic parasympathetic axons from the pelvic splanchnic nerve. Its axons innervate viscera within the pelvic region.

Neurotransmitters and Receptors

Two neurotransmitters are used in the ANS: acetylcholine (ACh) and norepinephrine (NE) **(figure 18.10)**. All preganglionic axons release ACh, which binds specific receptors in the ganglionic plasma membrane and has an excitatory effect on the ganglionic cell. All postganglionic parasympathetic axons and a few postganglionic sympathetic axons release ACh onto the effector. The ACh released from parasympathetic axons has either an excitatory or inhibitory effect on the effector, depending on the receptor on the effector plasma membrane. In contrast, the ACh released from sympathetic axons is excitatory only. Most

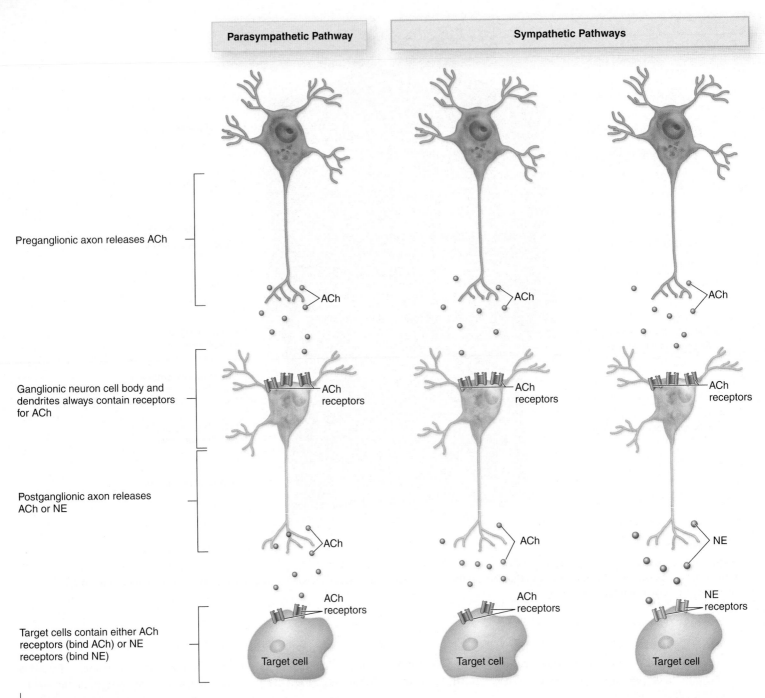

Preganglionic axon releases ACh

Ganglionic neuron cell body and dendrites always contain receptors for ACh

Postganglionic axon releases ACh or NE

Target cells contain either ACh receptors (bind ACh) or NE receptors (bind NE)

Parasympathetic Pathway

Sympathetic Pathways

ACh

ACh

ACh

ACh receptors

ACh receptors

ACh receptors

ACh

ACh

NE

ACh receptors

ACh receptors

NE receptors

Target cell

Target cell

Target cell

Figure 18.10

Neurotransmitters Used in the Autonomic Nervous System. In the parasympathetic pathway, both the preganglionic and postganglionic axons release acetylcholine (ACh). In the sympathetic pathway, all preganglionic axons and a few specific postganglionic axons release ACh. Most postganglionic sympathetic axons release norepinephrine (NE).

postganglionic sympathetic axons release NE onto the effector, which has either an excitatory or an inhibitory effect on the effector, depending on the receptor on the effector plasma membrane.

The axons that release acetylcholine are called *cholinergic*. The axons that release norepinephrine are called *adrenergic*.

Dual Innervation

Many visceral effectors have **dual innervation,** meaning that they are innervated by postganglionic axons from both ANS divisions. The actions of the divisions usually oppose each other, and so they are said to exert antagonistic effects on the same organ. Examples of dual innervation include the following:

- **Control of pupillary diameter.** Sympathetic innervation causes pupil dilation; parasympathetic innervation causes pupil constriction.
- **Control of digestive system activities.** Sympathetic stimulation reduces blood flow to the GI tract; parasympathetic innervation increases activities related to the digestion and processing of ingested food.
- **Control of heart rate.** Sympathetic stimulation increases the heart rate; parasympathetic stimulation decreases the heart rate.

In some ANS effectors, opposing effects are achieved without dual innervation. For example, many blood vessels are innervated by sympathetic axons only. Maintaining sympathetic stimulation holds smooth muscle contraction constant, resulting in blood pressure stability. Increased sympathetic stimulation causes vasoconstriction and results in increased blood pressure, while decreased stimulation causes vasodilation and results in decreased blood pressure. Thus, opposing effects are achieved by increasing or decreasing activity in one division.

Autonomic Reflexes

The autonomic nervous system helps maintain homeostasis through the involuntary activity of **autonomic reflexes,** also termed *visceral reflexes.* Autonomic reflexes consist of smooth muscle contractions, cardiac muscle contractions, or secretion by glands that are mediated by autonomic reflex arcs in response to a specific stimulus. One common autonomic reflex is the micturition reflex, which partly controls the release of urine **(figure 18.11).** Other reflexes include alteration of heart rate, changes in respiratory rate and depth, regulation of digestive system activities, and alteration of pupil diameter. A classic autonomic reflex involves the reduction of blood pressure. When an individual has elevated blood pressure, stretch receptors in the walls of large blood vessels are stimulated. Impulses from these stretch receptors then travel through visceral sensory neurons to the cardiac center in the medulla oblongata. This leads to parasympathetic input to the pacemaker of the heart, resulting in a decrease in heart rate and a concomitant decrease in blood pressure. Autonomic reflexes are comparable to spinal reflexes because they involve a sensory receptor, sensory neurons, interneurons in the CNS, motor neurons, and effector cells.

WHAT DID YOU LEARN?

13 What neurotransmitters are used in the ANS?

14 What is meant by dual innervation?

CLINICAL VIEW

Autonomic Dysreflexia

Autonomic dysreflexia is a potentially dangerous vascular condition that causes blood pressure to rise profoundly, sometimes so high that blood vessels rupture. At greatest risk are the thin-walled cerebral vessels; stroke is a common fatal complication of this condition. Autonomic dysreflexia is caused by hyperactivity of the autonomic nervous system in the weeks and months after a spinal cord injury. The majority of patients are either quadriplegic or have some form of spinal cord lesion superior to the sixth thoracic segment.

Often, the initial reaction to spinal cord trauma or injury is spinal shock, which is characterized by the loss of autonomic reflexes. However, this decrease in reflex activities may suddenly be replaced by autonomic reflex activities that cause certain viscera to respond abnormally to the lack of nerve supply, a phenomenon called **denervation hypersensitivity.** For example, when a person loses the ability to voluntarily evacuate the bladder, the bladder may continue to fill with urine to the point of overdistension. This induces a spinal cord reflex resulting in the involuntary relaxation of the internal urethral sphincter, thus allowing the bladder to empty. Essentially, this is an "override" mechanism designed to prevent rupture of the urinary bladder. Unfortunately, activation of this override mechanism can also stimulate a sympathetic nervous system reflex that causes transient, though marked, blood vessel narrowing due to vasoconstriction. The area of vascular constriction is inferior to the level of the spinal cord injury or lesion. This vasoconstriction produces the profound elevation in blood pressure characteristic of autonomic dysreflexia.

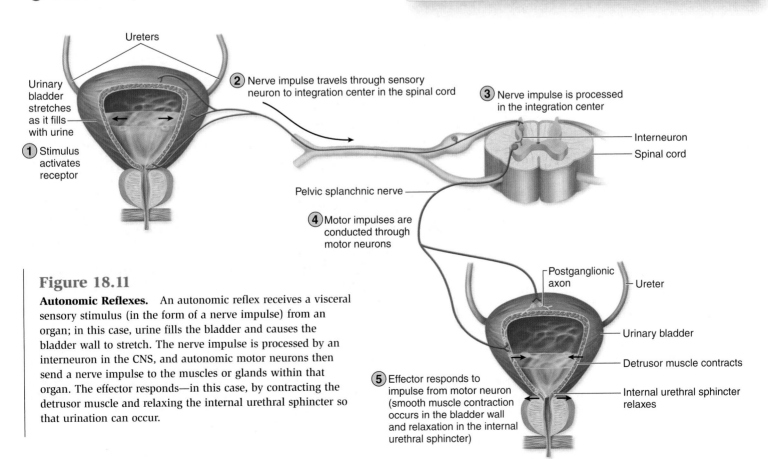

Figure 18.11

Autonomic Reflexes. An autonomic reflex receives a visceral sensory stimulus (in the form of a nerve impulse) from an organ; in this case, urine fills the bladder and causes the bladder wall to stretch. The nerve impulse is processed by an interneuron in the CNS, and autonomic motor neurons then send a nerve impulse to the muscles or glands within that organ. The effector responds—in this case, by contracting the detrusor muscle and relaxing the internal urethral sphincter so that urination can occur.

Ureters

Urinary bladder stretches as it fills with urine

1 Stimulus activates receptor

2 Nerve impulse travels through sensory neuron to integration center in the spinal cord

3 Nerve impulse is processed in the integration center

Interneuron

Spinal cord

Pelvic splanchnic nerve

4 Motor impulses are conducted through motor neurons

Postganglionic axon

Ureter

Urinary bladder

Detrusor muscle contracts

Internal urethral sphincter relaxes

5 Effector responds to impulse from motor neuron (smooth muscle contraction occurs in the bladder wall and relaxation in the internal urethral sphincter)

CNS Control of Autonomic Function

Key topic in this section:

- CNS hierarchy that controls the autonomic nervous system

Several levels of CNS complexity are required to coordinate and regulate ANS function. Thus, despite the name "autonomic," the ANS is a regulated nervous system, not an independent one. Autonomic function is influenced by four CNS regions: the cerebrum, hypothalamus, brainstem, and spinal cord (**figure 18.12**).

ANS activities are affected by conscious activities in the cerebral cortex and subconscious communications between association areas in the cortex with the centers of sympathetic and parasympathetic control in the hypothalamus. Additionally, sensory processing in the thalamus and emotional states controlled in the limbic system directly affect the hypothalamus.

The hypothalamus is the integration and command center for autonomic functions. It contains nuclei that control visceral functions in both divisions of the ANS, and it communicates with other CNS regions, including the cerebral cortex, thalamus, brainstem, cerebellum, and spinal cord. The hypothalamus is the central brain structure involved in emotions and drives that act through the ANS. For example, the sympathetic nervous system fight-or-flight response originates in the sympathetic nucleus in this brain region.

The brainstem nuclei in the mesencephalon, pons, and medulla oblongata mediate visceral reflexes. These reflex centers control accommodation of the lens, blood pressure changes, blood vessel diameter changes, digestive activities, heart rate changes, and pupil size. The centers for cardiac, digestive, and vasomotor functions are housed within the brainstem.

Some autonomic responses, notably the parasympathetic activities associated with defecation and urination, are processed and controlled at the level of the spinal cord without the involvement of the brain. However, the higher centers in the brain may consciously inhibit these reflex activities.

Study Tip!

The analogy of a corporation can help you understand the hierarchy of control of the ANS:

- The **hypothalamus** is the president of the Autonomic Nervous System Corporation. It oversees all activity in this system.
- The **autonomic reflex centers** in the brainstem and spinal cord are the vice presidents of the corporation. They have a lot of control and power in this corporation. Ultimately, though, they must answer to the president (hypothalamus).
- The **preganglionic and ganglionic neurons** are the workers in the corporation. They are ultimately under the control of both the president and vice presidents of the corporation. Also, these workers tend to do most of the real work in the company!

WHAT DID YOU LEARN?

15 What CNS structure is the integration and command center for autonomic function?

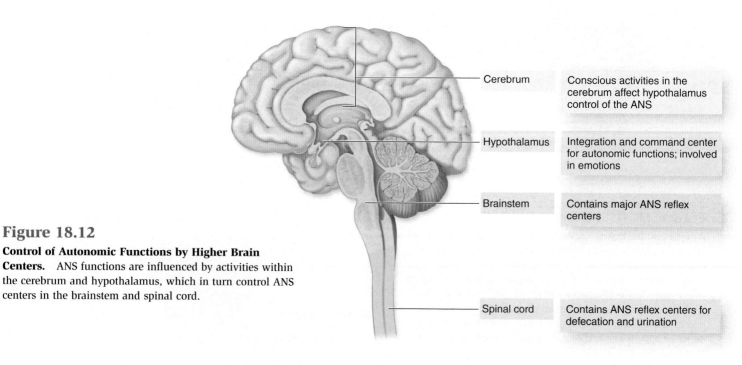

Figure 18.12

Control of Autonomic Functions by Higher Brain Centers. ANS functions are influenced by activities within the cerebrum and hypothalamus, which in turn control ANS centers in the brainstem and spinal cord.

Cerebrum — Conscious activities in the cerebrum affect hypothalamus control of the ANS

Hypothalamus — Integration and command center for autonomic functions; involved in emotions

Brainstem — Contains major ANS reflex centers

Spinal cord — Contains ANS reflex centers for defecation and urination

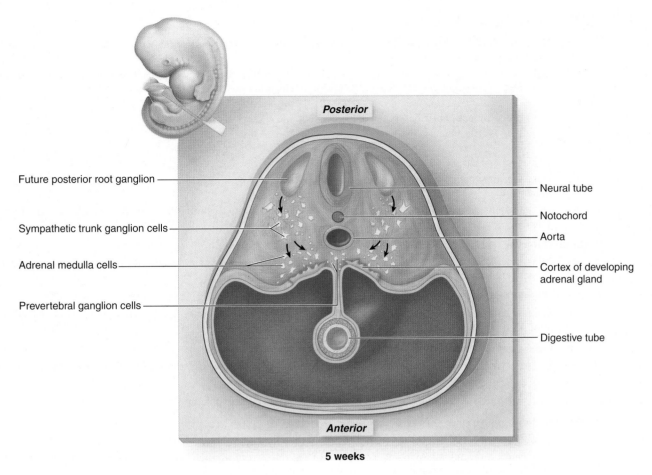

Posterior

Future posterior root ganglion

Sympathetic trunk ganglion cells

Adrenal medulla cells

Prevertebral ganglion cells

Neural tube

Notochord

Aorta

Cortex of developing adrenal gland

Digestive tube

Anterior

5 weeks

Figure 18.13

Neural Crest Cell Derivatives. A transverse section through a 5-week embryo shows structures that develop from migrating neural crest cells, including posterior root ganglia, many of the ANS structures, and the cells of the adrenal medulla.

Development of the Autonomic Nervous System

Key topic in this section:

■ How the autonomic nervous system develops in an embryo

Recall from previous chapters that the embryonic **neural tube** forms the central nervous system structures, while the neural crest cells form most of the peripheral nervous system structures. Because the autonomic nervous system has both CNS and PNS components, it forms from both neural tube and neural crest cells. In general, the neural tube forms the cell bodies of preganglionic neurons (since these structures are housed within the CNS), the hypothalamus (the master control center of the ANS), the autonomic nervous system centers in the brainstem, the white rami, and the autonomic reflex centers within the spinal cord. In general, the neural crest cells form all autonomic ganglia, ganglionic neurons and their postganglionic axons, gray rami communicantes, the sympathetic chain ganglia, and the adrenal medulla.

The neural crest cells begin to migrate during the fourth week of development. Those slated to form ANS structures differentiate soon thereafter. Preganglionic neurons begin to extend axons anteriorly from the neural tube during the fifth week of development (**figure 18.13**). These axons encounter the ganglionic neurons, and the sympathetic trunk begins to form during week 6. By the end of the eighth week, the rami communicantes have formed; the developing heart and lungs begin to receive autonomic innervation in the tenth week of development.

CLINICAL TERMS

Hirschsprung disease (congenital megacolon) Dilation and hypertrophy of the colon due to absence (aganglionosis) or marked reduction (hypoganglionosis) in the number of ganglion cells within the colon.

vagotomy (vā-got′ō-mē; *tome* = incision) Surgical separation or splitting of the vagus nerve, usually performed to reduce gastric acid secretion in ulcer patients when medications have failed.

CHAPTER SUMMARY

Comparison of the Somatic and Autonomic Nervous Systems 538

- The SNS innervates skeletal muscle. The ANS innervates smooth muscle, cardiac muscle, and glands, and controls involuntary motor activities.
- A single motor neuron axon innervates skeletal muscle fibers in the SNS, while the ANS has a two-neuron pathway consisting of preganglionic neurons in the CNS and ganglionic neurons in the PNS.

Overview of the Autonomic Nervous System 540

- The ANS is composed of a parasympathetic division and a sympathetic division.

Parasympathetic Division 543

- The parasympathetic preganglionic neurons are housed either within the brainstem or within the sacral region of the spinal cord.
- The ganglionic neurons in the parasympathetic division are located within either terminal ganglia or intramural ganglia.

Cranial Nerves 543

- Parasympathetic preganglionic axons extend from cell bodies in brainstem nuclei through the oculomotor, facial, glossopharyngeal, and vagus cranial nerves.

Sacral Spinal Nerves 543

- The remaining preganglionic parasympathetic cell bodies are housed within the S2–S4 segments of the spinal cord and form pelvic splanchnic nerves.

Effects and General Functions of the Parasympathetic Division 543

- The parasympathetic division of the ANS alters activities of effector organs to manage and control food processing, energy absorption, and relaxation activities.

Sympathetic Division 545

- The sympathetic division outflow is from the T1–L2 lateral horn segments.

Organization and Anatomy of the Sympathetic Division 545

- Preganglionic neuron cell bodies are housed within the lateral gray horn of the spinal gray matter.
- Myelinated, preganglionic sympathetic axons exit the spinal cord through the anterior root of a spinal nerve and travel through the white rami communicantes to the sympathetic trunk ganglia.

Sympathetic Pathways 548

- In the spinal nerve pathway, the postganglionic axon enters the spinal nerve through the gray ramus and travels to the blood vessels and glands distributed throughout the limbs and body wall of the trunk.
- In the postganglionic sympathetic nerve pathway, the postganglionic axon leaves the sympathetic trunk and extends directly to the target organ.
- In the splanchnic nerve pathway, the preganglionic axon passes through the sympathetic trunk without synapsing and travels to the prevertebral ganglia.
- In the adrenal medulla pathway, the preganglionic axons extend through the autonomic ganglia without synapsing. They synapse on secretory cells in the adrenal medulla that release epinephrine and norepinephrine.

Effects and General Functions of the Sympathetic Division 548

- Sympathetic division pathways prepare the body for fight or flight.

Other Features of the Autonomic Nervous System 550

- Both divisions of the autonomic nervous system innervate organs through specific axon bundles.

Autonomic Plexuses 550

- Autonomic plexuses are meshworks of postganglionic sympathetic axons, preganglionic parasympathetic axons, and visceral sensory neuron axons in the anterior body cavities that merge and intermingle but do not synapse with each other.

Neurotransmitters and Receptors 551

- Two neurotransmitters are used in the ANS: acetylcholine (ACh) and norepinephrine (NE).
- Both the preganglionic and postganglionic axons in the parasympathetic division release acetylcholine; the preganglionic axon and a few postganglionic axons in the sympathetic division release acetylcholine; however, most of the postganglionic axons of the sympathetic division release norepinephrine.

Dual Innervation 552

- Many visceral effectors have dual innervation, meaning they are innervated by axons from both ANS divisions. The actions of the divisions often oppose each other, and thus they exert antagonistic effects on the same organ.

Autonomic Reflexes 553

- Homeostasis in the human body is maintained through the activity of autonomic reflexes. These reflexes result in smooth muscle contractions, cardiac muscle contractions, or secretion by glands.

CNS Control of Autonomic Function 554	▪ Autonomic function is influenced by four CNS regions: cerebrum, hypothalamus, brainstem, and spinal cord.
Development of the Autonomic Nervous System 555	▪ The neural tube gives rise to most of the CNS structures of the ANS. ▪ The neural crest cells give rise to most of the PNS structures of the ANS.

CHALLENGE YOURSELF

Matching

Match each numbered item with the most closely related lettered item.

_____ 1. norepinephrine

_____ 2. autonomic plexus

_____ 3. ganglionic neuron

_____ 4. hypothalamus

_____ 5. sympathetic division

_____ 6. gray ramus

_____ 7. splanchnic nerve

_____ 8. sympathetic trunk ganglia

_____ 9. parasympathetic division

_____ 10. acetylcholine

a. contains sympathetic postganglionic axons only

b. controls entire ANS function

c. hormone secreted by adrenal medulla

d. second ANS neuron

e. neurotransmitter for all preganglionic axons

f. craniosacral division

g. preganglionic axons to prevertebral ganglia

h. network of pre- and postganglionic axons

i. fight-or-flight division

j. lateral to spinal cord

Multiple Choice

Select the best answer from the four choices provided.

_____ 1. A splanchnic nerve in the sympathetic division of the ANS
   a. connects neighboring sympathetic trunk ganglia.
   b. controls parasympathetic functions in the thoracic cavity.
   c. is formed by preganglionic axons that travel to prevertebral ganglia.
   d. travels through parasympathetic pathways in the head.

_____ 2. Some parasympathetic preganglionic neuron cell bodies are housed within the
   a. hypothalamus.
   b. sacral region of the spinal cord.
   c. cerebral cortex.
   d. thoracolumbar region of the spinal cord.

_____ 3. Which of the following is *not* a function of the sympathetic division of the ANS?
   a. increases heart rate and breathing rate
   b. prepares for emergency
   c. increases digestive system motility and activity
   d. dilates pupils

_____ 4. Postganglionic axons from the celiac ganglion innervate which of the following?
   a. stomach
   b. urinary bladder
   c. lung
   d. adrenal medulla

_____ 5. Sympathetic division splanchnic nerves end in the _____ ganglia, which are anterior to the vertebral column and aorta.
   a. intramural
   b. sympathetic trunk
   c. prevertebral
   d. terminal

_____ 6. All parasympathetic division synapses use _____ as a neurotransmitter.
   a. dopamine
   b. norepinephrine
   c. acetylcholine
   d. epinephrine

_____ 7. Which autonomic nerve plexus innervates the pelvic organs?
   a. cardiac plexus
   b. esophageal plexus
   c. hypogastric plexus
   d. inferior mesenteric plexus

_____ 8. Which of the following describes a sympathetic postganglionic axon?
   a. long, unmyelinated axon
   b. short, myelinated axon
   c. short, unmyelinated axon
   d. long, myelinated axon

_____ 9. Neural crest cells form
   a. the hypothalamus.
   b. white rami communicantes.
   c. autonomic ganglia.
   d. autonomic reflex centers.

_____ 10. All of the following cranial nerves carry parasympathetic preganglionic nerve axons except
   a. CN III (oculomotor).
   b. CN V (trigeminal).
   c. CN IX (glossopharyngeal).
   d. CN X (vagus).

Content Review

1. What four CNS regions control the autonomic nervous system?

2. For the following ganglia, identify the location and the division of the ANS each is part of: sympathetic trunk ganglia, prevertebral ganglia, terminal and intramural ganglia.

3. Compare and contrast the postganglionic axons of the parasympathetic and sympathetic divisions. Examine the axon length, myelination (or lack thereof), and the neurotransmitter used.

4. Explain how adrenal medulla stimulation potentiates (prolongs) the effects of the sympathetic division of the autonomic nervous system.

5. Identify and describe the four basic pathways used in the sympathetic division.

6. Are the cell bodies of sympathetic and parasympathetic neurons located in the central nervous system, in the peripheral nervous system, or in both? Explain your answer.

7. Identify the types of axons that compose the gray and white rami communicantes, describe their anatomic arrangement and location, and discuss the reason for the differences in their color.

8. Describe how the general functions of the sympathetic and parasympathetic divisions of the ANS differ.

9. What may occur with the mass activation of the sympathetic division of the ANS?

10. Describe the embryonic components that form ANS structures.

Developing Critical Reasoning

1. Holly takes night classes at the local community college. After her lecture, she walks alone to her car and suddenly hears several dozen screeching birds fly away from the tree she is walking under. Holly immediately feels her heart pounding and notices that her breathing rate has increased. Minutes later, she still feels tense and "on edge." What happened internally to cause Holly's initial response? Why did Holly still feel tense minutes later?

2. Some faculty dislike teaching lecture classes after lunch, complaining that the students do not pay attention at this time. From an anatomic viewpoint, what is happening to these students?

ANSWERS TO "WHAT DO YOU THINK?"

1. Compared to the somatic motor system, the autonomic motor system has a limited number of resources (nerves) to transmit the motor information throughout the body. By using a two-neuron chain, nerve impulses are able to diverge to a larger number of resources. (The study tip comparing the ANS to the airline industry [see page 540] also helps answer this question.)

2. The pterygopalatine ganglion is nicknamed the "hay fever ganglion" because when it is overstimulated, it causes some of the classic allergic reactions, including watery eyes, runny and itchy nose, sneezing, and scratchy throat.

3. Sympathetic innervation causes vasoconstriction of most blood vessels. When blood vessels are constricted, it takes more force and pressure to pump blood through the vessels, so blood pressure rises.

Visit the McKinley/O'Loughlin _Human Anatomy_, 2e website at aris.mhhe.com

NERVOUS SYSTEM

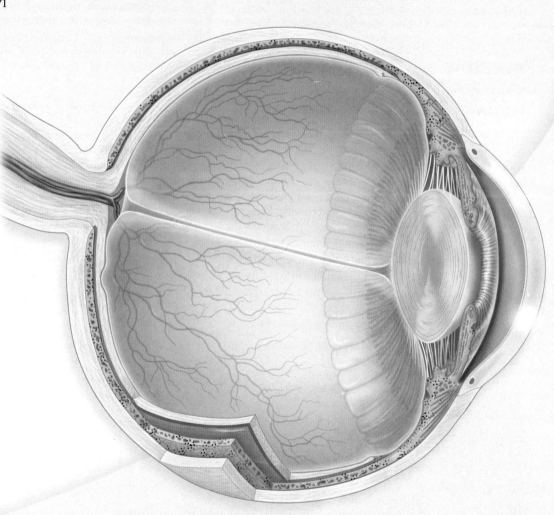

Senses: General and Special

Our bodies are continually exposed to an onslaught of sensory information, both externally from the environment and internally within the body. These changes in the external or internal environment are called **stimuli** (stim′ū-lī; *stimulus* = a goad) Some of these stimuli give us pleasure, as when we watch a funny movie, listen to a CD, touch a baby's face, or eat a tasty meal. Other stimuli alert us to potential dangers, as when we see a stranger lurking in the dark, smell smoke upon walking into a room, or touch a hot pan on the kitchen stove. Still others provide routine, moment-to-moment information about our body's external surroundings and its internal state.

Our conscious awareness of incoming sensory information is called **sensation.** Only a stimulus that reaches the cerebral cortex of the brain results in a sensation of that stimulus. Thus, although our bodies are constantly bombarded by sensory stimuli, we are consciously aware of only a fraction of these stimuli.

Stimuli are detected by receptors. This chapter focuses on the two classes of receptors in our bodies: receptors for the **general senses** (temperature, pain, touch, stretch, and pressure) and receptors for the **special senses** (gustation, olfaction, vision, equilibrium, and hearing).

Receptors

Key topic in this section:

- Properties and classification schemes for receptors

Receptors (*recipio* = to receive) are structures that detect stimuli. They range in complexity from the relatively simple dendritic ending of a sensory neuron to complex structures called sense organs whose nerve endings are associated with epithelium, connective tissue, or muscular tissue. Receptors in the body monitor both external and internal environmental conditions and conduct information about those stimuli to the central nervous system. We are aware of some specific stimuli, other stimuli never reach our consciousness.

The **receptive field** of a receptor is the entire area through which the sensitive ends of the receptor cell are distributed **(figure 19.1)**. There is an inverse relationship between the size of the recep-

tive field and our ability to identify the exact location of a stimulus. If the receptive field is small, precise localization and sensitivity are easily determined. In contrast, a broad receptive field only detects the general region of the stimulus. Although it might seem advantageous for all receptors to have small receptive fields, the number of receptors in the body would have to markedly increase in order to detect environmental stimuli if all receptive fields were very small. The energy costs to maintain activity in such a large number of receptors would be enormous.

All receptors act as **transducers** (trans-doo′ser; *trans* = across, *duco* = to lead), which are structures that transform the energy of one system (for example, heat) into a different form of energy (for example, a nerve impulse). Receptors may be either tonic or phasic **(figure 19.2)**. **Tonic receptors** respond continuously to stimuli at a constant rate; examples are the balance receptors in the ear that keep the head upright. **Phasic receptors** detect a new stimulus or a change in a stimulus that has already been applied, but over time their sensitivity decreases. An example is the tactile receptors of the skin that sense the increased pressure if we are pinched. Phasic receptors can undergo a change called **adaptation** (*adapto* = to adjust), which is a reduction in sensitivity to a continually applied stimulus. For example, upon sitting in a chair, we are immediately aware of the pressure increase wherever our body contacts the chair.

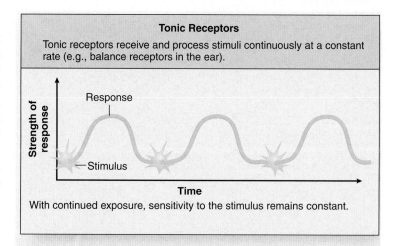

Tonic Receptors

Tonic receptors receive and process stimuli continuously at a constant rate (e.g., balance receptors in the ear).

Strength of response — Response — Stimulus — Time

With continued exposure, sensitivity to the stimulus remains constant.

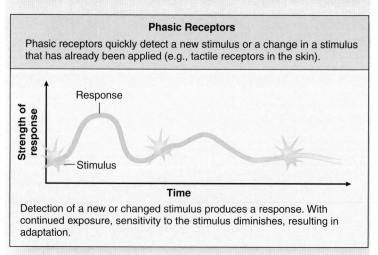

Phasic Receptors

Phasic receptors quickly detect a new stimulus or a change in a stimulus that has already been applied (e.g., tactile receptors in the skin).

Strength of response — Response — Stimulus — Time

Detection of a new or changed stimulus produces a response. With continued exposure, sensitivity to the stimulus diminishes, resulting in adaptation.

Figure 19.1

Receptive Fields.   The specific area monitored by each sensory receptor is called its receptive field. A smaller receptive field (1) offers greater specificity of localization than does a larger receptive field (2).

Figure 19.2

Receptor Responses.   Tonic receptor activity continuously responds to stimuli, whereas phasic receptor activity detects a new or changed stimulus.

But soon, we become unaware of this pressure because adaptation has occurred in the deep pressure receptors of the skin.

Classification of Receptors

General sense receptors are distributed throughout the skin and organs. Special sense receptors are housed within complex organs in the head. Three criteria are used to describe receptors—receptor distribution, stimulus origin, and modality of stimulus **(table 19.1)**.

Receptor Distribution

Receptors can be classified based on their distribution in the body as part of the general senses or the special senses.

The receptors for general senses are subdivided into two categories: (1) **Somatic receptors** are housed within the body wall; they include receptors for external stimuli, including chemicals, temperature, pain, touch, proprioception, and pressure. (2) **Visceral receptors** are located in the walls of the viscera; they respond to chemicals, temperature, and pressure and are sometimes also called *interoceptors* or *visceroceptors*.

The receptors for the special senses are located within sense organs and housed only in the head. The five special senses are gustation (taste), olfaction (smell), vision, equilibrium, and hearing (audition).

In this chapter, we usually use receptor distribution as our criterion when discussing receptors.

Table 19.1	Criteria for Classifying Receptors
Classification	**Description**
RECEPTOR DISTRIBUTION (BODY LOCATION)	
General senses: Structurally simple; distributed throughout the body	
Somatic receptors (found within body wall)	
Chemical	Respond to specific molecules dissolved in fluid
Temperature	Respond to changes in temperature
Pain	Detect damage to tissue
Touch	Detect fine or light touch
Proprioception	Monitor changes in position, and in tension and stretch of skeletal muscle, tendons, and joint capsules
Pressure	Respond to mechanical pressure, vibration, and stretch
Visceral sensory receptors (located within walls of viscera)	
Chemical	Respond to specific molecules dissolved in fluid
Temperature	Respond to heat or cold (cold receptors outnumber heat receptors)
Pressure	Respond to stretch
Special senses: Structurally complex; located only in the head	
Gustation	Perceive tastes
Olfaction	Perceive odors
Vision	Perceive objects by reflected or emitted light
Equilibrium	Maintain coordination and balance
Hearing	Perceive sounds
STIMULUS ORIGIN (STIMULUS LOCATION)	
Exteroceptors	Receptors in the skin or mucous membranes that line cavities that are open to the outside of the body (e.g., nasal cavity, oral cavity, vagina, and anal canal); also include the special senses (gustation, olfaction, vision, equilibrium, and hearing)
Interoceptors (visceroceptors)	Receptors located within the walls of viscera
Proprioceptors	Receptors in skeletal muscles, tendons, and joint capsules that sense position or state of contraction of the muscle
MODALITY OF STIMULUS (STIMULATING AGENT)	
Chemoreceptors	Detect chemicals/specific molecules dissolved in fluid; respond to odors and tastes in fluids (mucus, saliva)
Thermoreceptors	Detect changes in temperature
Photoreceptors	Detect changes in light intensity, color, and movement of light rays
Mechanoreceptors	Detect physical deformation due to touch, pressure, vibration, and stretch (primarily cutaneous tactile receptors)
Baroreceptors	Detect pressure change within body structures
Nociceptors	Detect tissue damage; pain receptors

Stimulus Origin

Based on where the stimulus originates, there are three types of receptors: exteroceptors, interoceptors, and proprioceptors.

Exteroceptors (eks′ter-ō-sep′ter, -tōr; *exterus* = external) detect stimuli from the external environment. For example, the receptors in your skin (generally called *cutaneous receptors*) are exteroceptors because external stimuli typically cause sensations to the skin. Likewise, receptors for your special senses are considered exteroceptors because they usually interpret external stimuli, such as the taste of the food you just ate or the sound of music on the radio. Exteroceptors are also found in the mucous membranes that open to the outside of the body, such as the nasal cavity, oral cavity, vagina, and anal canal.

Interoceptors (in′ter-ō-sep′ter; *inter* = between), also called *visceroceptors,* detect stimuli in internal organs (viscera). These receptors are primarily stretch receptors in the smooth muscle of these organs. Most of the time we are unaware of these receptors, but when the smooth muscle stretches to a certain point (e.g., when eating a large meal stretches our stomach wall), we may become aware of these sensations. Interoceptors also report on pressure, chemical changes in the visceral tissue, and temperature. When we feel pain in our internal organs, it is usually because a tissue has been deprived of oxygen (as in a heart attack), because the smooth muscle has been stretched so much that we are uncomfortable, or because the tissue has suffered trauma, and damaged cells have released chemicals that stimulate specific interoceptors.

Proprioceptors (prō′prē-ō-sep′ter; *proprius* = one's own) are located in muscles, tendons, and joints. They detect body and limb movements, skeletal muscle contraction and stretch, and changes in joint capsule structure. Thus, even if you are not looking at your body joints, you are aware of their positioning and the state of contraction of your skeletal muscles, because proprioceptors send this information to the CNS.

Modality of Stimulus

Receptors also may be classified according to the stimulus they perceive, called the **modality of stimulus**, or the *stimulating agent*. For example, some receptors respond only to temperature changes, while others respond to chemical changes. There are six groups of receptors, based on their modality of stimulus: chemoreceptors, thermoreceptors, photoreceptors, mechanoreceptors, baroreceptors, and nociceptors.

- **Chemoreceptors** (kē′mō-rē-sep′tōr) detect chemicals such as specific molecules dissolved in fluid in our external and internal environments, including ingested food and drink, body fluids, and inhaled air. For example, the receptors in the taste buds on our tongue are chemoreceptors, because they respond to the specific molecules in ingested food.

Likewise, chemoreceptors in some of our blood vessels monitor the concentration of oxygen and carbon dioxide molecules in our blood.

- **Thermoreceptors** (ther′mō-rē-sep′tōr; *therme* = heat) respond to changes in temperature.
- **Photoreceptors** (fō′tō-rē-sep′tōr; *phot* = light) are located in the eye, where they detect changes in light intensity, color, and movement.
- **Mechanoreceptors** (mek′ă-nō-rē-sep′tōr; *mechane* = machine) respond to touch, pressure, vibration, and stretch. Most of the cutaneous receptors are mechanoreceptors, because they respond to pressure and touch on the skin. Mechanoreceptors are also located in the ear for equilibrium and hearing.
- **Baroreceptors** (bar′ō-rē-sep′ter, -tōr; *baros* = weight) detect changes in pressure within body structures. These sensory receptors branch repeatedly within the connective tissues in vessel or organ walls, especially the elastic layers. Any pressure stimulus that causes wall deformation results in a change in the nerve impulse rates from the receptors and a pressure sensation.
- **Nociceptors** (nō-si-sep′ter, -tōr; *noci* = pain) respond to pain caused by either external or internal stimuli. *Somatic nociceptors* detect chemical, heat, or mechanical damage to the body surface or skeletal muscles. For example, exposure to acid, touching a hot pan, or suffering a sprained ankle stimulate somatic nociceptors. *Visceral nociceptors* detect internal body damage within the viscera due to excessive stretching of smooth muscle, oxygen deprivation of the tissue, or chemicals released from damaged tissue.

Study Tip!

Here is one way to remember the differences between somatic nociceptors and visceral nociceptors: When you eat a very spicy meal, your mouth stings and burns because the nociceptors in your mouth are somatic nociceptors. They detect the spicy stimuli and send impulses to the brain where the stimuli are translated into the stinging, burning sensation. When you swallow that spicy meal and it travels through your GI tract, you may not experience the stinging, burning sensation because the nociceptors there are visceral nociceptors that respond only to abnormal muscle stretch, oxygen deprivation, or chemical imbalance in the tissue. Now, when the waste products from that spicy meal are expelled from the body, the anus may burn and sting because the nociceptors around the anus and lower anal canal are somatic nociceptors. Thus, the somatic nociceptors in the mouth and around the anus have similar reactions to the spices!

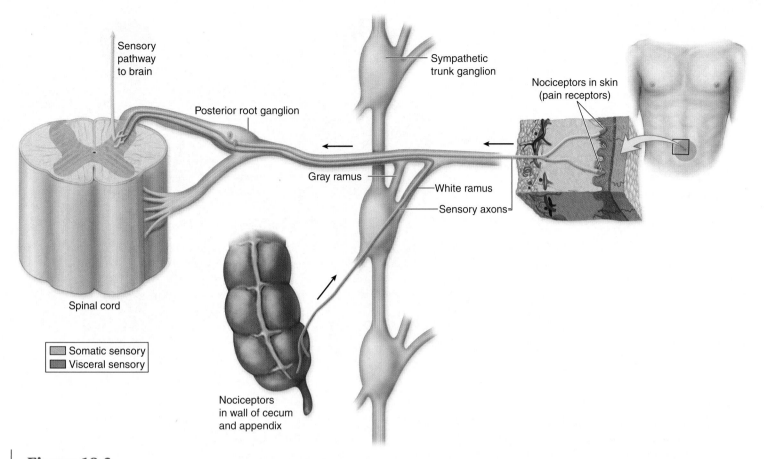

Figure 19.3

Source of Referred Pain.   Misinterpretation of a pain source occurs when sensory impulses from two different organs are conducted to the brain via a common pathway. For example, pain that is perceived as originating from the umbilical region of the skin actually has its source within the appendix.

Clinically Significant Types of Pain   **Phantom** (fan′tŏm) **pain** is a sensation associated with a body part that has been removed. Following the amputation of an appendage, the patient often continues to experience pain that feels like it is coming from the removed part. The stimulation of a sensory neuron pathway from the removed limb anywhere on the remaining intact portion of the pathway propagates nerve impulses and conducts them to the CNS, where they are interpreted as originating in the amputated limb. In other words, the cell bodies of the sensory neurons that provided sensation to the leg remain alive because they were not part of the leg. This so-called *phantom limb syndrome* can be quite debilitating. Some people experience extreme pain, whereas others have an insatiable desire to scratch a nonexistent itch.

Referred pain occurs when impulses from certain viscera are perceived as originating not from the organ, but in dermatomes of the skin (see figure 16.6). Numerous cutaneous and visceral sensory neurons conduct nerve impulses through the same ascending tracts within the spinal cord **(figure 19.3)**. As a result, impulses conducted along ascending pathways may be localized incorrectly. Consequently, the sensory cortex in the brain is unable to differentiate between the actual and false sources of the stimulus.

Clinically, some common sites of referred pain are useful in medical diagnosis **(figure 19.4)**. For example, cardiac problems are often a source of referred pain because the heart receives its sympathetic innervation from the T1–T5 segments of the spinal cord. Pain associated with a myocardial infarction (heart attack)

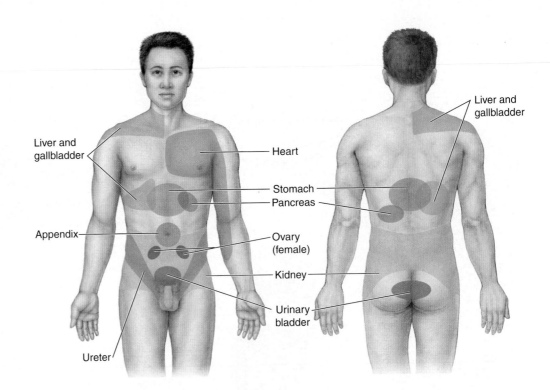

Figure 19.4
Common Sites of Referred Pain.

may be referred to the skin dermatomes innervated by the T1–T5 spinal nerves, which lie along the pectoral region and the medial side of the arm. Thus, some individuals who are experiencing heart problems may feel pain along the medial side of the left arm, where the T1 dermatome lies. By the same token, kidney and ureter pain may be referred along the T10–L2 dermatomes, which typically overlie the inferior abdominal wall in the groin and loin regions.

Visceral pain is usually referred along the sympathetic nerve pathways, but sometimes it can follow parasympathetic pathways as well. For example, referred pain from the urinary bladder can follow the parasympathetic pathways (via the pelvic splanchnic nerves). Since the pelvic splanchnic nerves lie in the S2–S4 segments of the spinal cord, pain may be referred to the S2–S4 dermatomes, which overlie the medial buttocks region.

WHAT DO YOU THINK?

1 Based on stimulus origin, would a cutaneous pain receptor be classified as an exteroceptor, an interoceptor, or a proprioceptor? Based on receptor distribution, would this same cutaneous receptor be classified as a general somatic, general visceral, or special sense? Finally, how would this pain receptor in the skin be classified based on modality of stimulus?

WHAT DID YOU LEARN?

1 What is a sensation? How is a sensation related to a stimulus?

2 What are the three groups of receptors classified by stimulus origin?

3 What stimuli affect mechanoreceptors and thermoreceptors?

Tactile Receptors

Key topic in this section:

■ Types of tactile receptors and their functions

Receptors for general senses are usually simple. Their stimuli include chemicals, temperature, pain, touch, stretch, and pressure. Thus, chemoreceptors, thermoreceptors, nociceptors, mechanoreceptors, and proprioceptors all participate in the general senses. Having discussed these receptors in general in the previous section, now we will examine tactile receptors in depth.

Tactile (tak'til; *tango* = to touch) **receptors** are the most numerous type of receptor. They are mechanoreceptors that react to touch, pressure, and vibration stimuli. They are located in the dermis and the subcutaneous layer **(figure 19.5)**. Tactile receptors range from simple, dendritic ends that have no connective tissue wrapping (called unencapsulated) to complex structures that are wrapped with connective tissue or glial cells (called encapsulated) **(table 19.2)**.

Unencapsulated Tactile Receptors

Unencapsulated receptors have no connective tissue wrapping around them and are relatively simple in structure. The three types of unencapsulated receptors are free nerve endings, root hair plexuses, and tactile discs.

Free nerve endings are terminal branches of dendrites. They are the least complex of the tactile receptors and reside closest to the surface of the skin, usually in the papillary layer of the dermis. Often, some dendritic branches extend into the deepest epidermal strata and terminate between the epithelial cells. These tactile receptors primarily detect pain and temperature stimuli, but some also detect light touch and pressure.

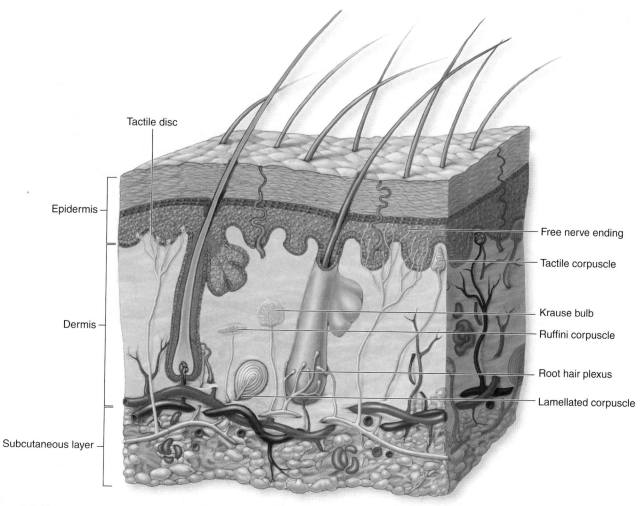

Tactile disc

Epidermis

Dermis

Subcutaneous layer

Free nerve ending

Tactile corpuscle

Krause bulb

Ruffini corpuscle

Root hair plexus

Lamellated corpuscle

Figure 19.5

Tactile Receptors.   Various types of tactile receptors in the skin supply information about touch, pressure, or vibration in our immediate environment.

Root hair plexuses are specialized free nerve endings that form a weblike sheath around hair follicles in the reticular layer of the dermis. Any movement or displacement of the hair changes the arrangement of these branching dendrites, initiating a nerve impulse. These receptors quickly adapt; thus, although we feel the initial contact of a long-sleeved shirt on our arm hairs when we put on the garment, our conscious awareness subsides immediately until we move and the root hair plexuses are restimulated.

Tactile discs, previously called *Merkel discs,* are flattened nerve endings that function as tonic receptors for fine touch. These receptors are important in distinguishing the texture and shape of a stimulating agent. Tactile discs are associated with special **tactile cells** (*Merkel cells*), which are located in the stratum basale of the epidermis. Tactile cells exhibit a small receptive field and communicate directly with the dendrites of a sensory neuron.

Encapsulated Tactile Receptors

Encapsulated receptors are covered either by connective tissue or glial cells. Encapsulated tactile receptors include Krause bulbs, lamellated corpuscles, Ruffini corpuscles, and tactile corpuscles.

Krause bulbs are located near the border of the stratified squamous epithelium in the mucous membranes of the oral cavity, nasal cavity, vagina, and anal canal, where they detect light pressure stimuli and low-frequency vibration.

Lamellated (lam′ē-lāt-ed) **corpuscles,** previously called *Pacinian corpuscles,* are large receptors that detect deep pressure and high-frequency vibration. The center of the receptor houses several dendritic endings of sensory neurons wrapped within numerous concentric layers of flat, fibroblast-like cells. This structure ensures that only deep-pressure stimuli will activate the receptor. Lamellated corpuscles are found deep within the reticular layer of the dermis; in the subcutaneous layer of the palms of the hands, soles of the feet, breasts, and external genitalia; in the synovial membranes of joints; and in the walls of some organs.

Ruffini corpuscles detect both continuous deep pressure and distortion in the skin. These are tonic receptors that do not exhibit adaptation; they are housed within the dermis and subcutaneous layer.

Tactile corpuscles, previously called *Meissner corpuscles,* are physically different from the unencapsulated tactile discs discussed previously. Tactile corpuscles are large, encapsulated oval receptors.

| Table 19.2 | Types of Tactile Receptors |

UNENCAPSULATED RECEPTORS[1]

	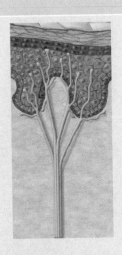	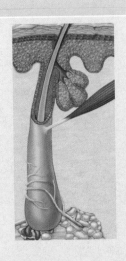	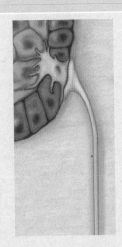
Ending Type	**Free Nerve Ending**	**Root Hair Plexus**	**Tactile Disc**
Location	Widespread in deep epidermis and papillary layer of the dermis	Surrounds hair follicles in the reticular layer of the dermis	Stratum basale of epidermis
Function	Detect pressure, change in temperature, pain, touch	Detect movement of the hair	Detect light touch, textures, and shapes
Modality of Stimulus	Some are nociceptors; others are thermoreceptors or mechanoreceptors	Mechanoreceptors	Mechanoreceptors

ENCAPSULATED RECEPTORS[2]

Ending Type	**Krause Bulb**	**Lamellated Corpuscle**	**Ruffini Corpuscle**	**Tactile Corpuscle**
Location	Mucous membranes of oral cavity, nasal cavity, vagina, and anal canal	Dermis, subcutaneous tissue, synovial membranes, and some viscera	Dermis and subcutaneous layer	Dermal papillae, especially in lips, palms, eyelids, nipples, genitals
Function	Detect light pressure and low-frequency vibration	Detect deep pressure and high-frequency vibration	Detect continuous deep pressure and skin distortion	Detect fine, light touch and texture
Modality of Stimulus	Mechanoreceptors	Mechanoreceptors	Mechanoreceptors	Mechanoreceptors

[1] *Unencapsulated* means there is no connective tissue wrapping.

[2] *Encapsulated* means either wrapped in connective tissue or ensheathed by a glial cell.

They are formed from highly intertwined dendrites enclosed by modified neurolemmocytes, which are then covered with dense irregular connective tissue. Tactile corpuscles are phasic receptors for light touch, shapes, and texture. They are housed within the dermal papillae of the skin, especially in the lips, palms, eyelids, nipples, and genitals.

⚗️ WHAT DID YOU LEARN?

❹ How do unencapsulated receptors differ from encapsulated receptors?

❺ What is the function of Ruffini corpuscles?

Gustation

Key topic in this section:

■ Structure, location, and innervation pathway of gustatory receptors

Our sense of taste, called **gustation** (gŭs-tā′shŭn; *gusto* = to taste), permits us to perceive the characteristics of what we eat and drink. Gustation is also referred to as contact chemoreception because we must come in contact with a substance to experience its flavor. **Gustatory (taste) cells** are taste receptors housed in specialized sensory organs termed **taste buds** on the tongue surface.

On the dorsal surface of the tongue are epithelial and connective tissue elevations called **papillae** (pă-pil′ē; *papula* = a nipple), which are of four types: filiform, fungiform, vallate, and foliate (**figure 19.6**). **Filiform** (fil′i-fōrm; *filum* = thread) **papillae** are short and spiked; they are distributed on the anterior two-thirds of the dorsal tongue surface. These papillae do not house taste buds and, thus, have no role in gustation. **Fungiform** (fŭn′ji-fōrm) **papillae** are blocklike projections primarily located on the tip and sides of the tongue. They contain only a few taste buds each. **Vallate** (val′āt; *vallo* = to surround with) (*circumvallate*) **papillae** are the least numerous yet the largest papillae on the tongue. They are arranged in an inverted V shape on the posterior dorsal surface of the tongue. Each papilla is surrounded by a deep, narrow depression. Most of our taste buds are housed within the walls of these papillae along the side facing the depression. **Foliate** (fō′lē-āt) **papillae** are not well developed on the human tongue. They extend as ridges on the posterior lateral sides and house only a few taste buds during infancy and early childhood.

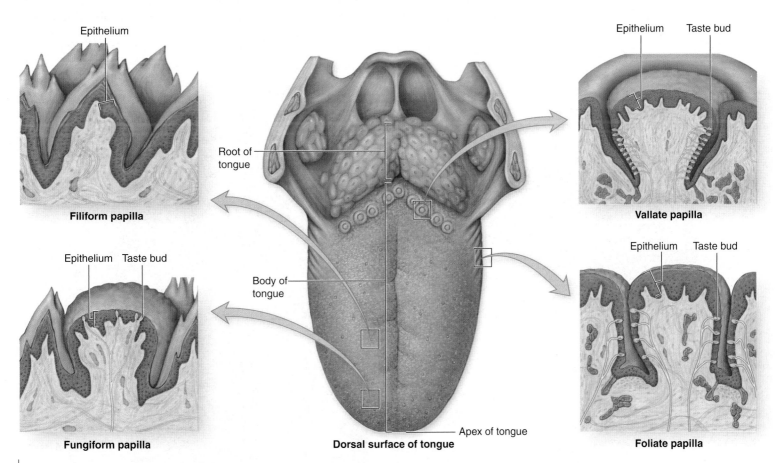

Figure 19.6

Tongue Papillae.   Papillae are small elevations on the tongue surface that may be one of four types: filiform, fungiform, vallate, or foliate. In adults, only fungiform and vallate papillae contain taste buds for gustation; in infants and young children, the foliate papillae have a few taste buds.

Each cylindrical taste bud is composed of numerous taste receptors called **gustatory cells** (or *gustatory receptors*), which are enclosed by **supporting cells (figure 19.7)**. The dendritic ending of each gustatory cell is formed by a slender **gustatory microvillus,** sometimes called a *taste hair*. The gustatory microvillus extends through an opening in the taste bud, called the **taste pore,** to the surface of the tongue. This is the receptive portion of the cell. Within the oral cavity, saliva keeps the environment moist; molecules in food dissolve in the saliva and stimulate the gustatory microvilli. Thus, gustatory receptors are classified as chemoreceptors, because food molecules must be dissolved in saliva before they can be tasted.

The gustatory cells within the taste buds are specialized neuroepithelial cells that have a 7- to 10-day life span. A population of stem cells called **basal cells** constantly replaces the gustatory cells. Beginning at about age 50, our ability to distinguish between different tastes declines due to reduction in both gustatory cell replacement and the number of taste buds. Thus, an elderly person may complain that food is tasteless or very bland.

Gustatory Discrimination

The tongue detects five basic **taste sensations:** salty, sweet, sour, bitter, and umami. Salty tastes are caused by metal ions such as potassium or sodium in food or drink; sweet tastes result from the presence of organic compounds, such as sugar; sour tastes are caused by hydrogen ions from acids, such as those found in lemons, limes, and other fruit; bitter tastes can be caused by alkaloids (such as those in brussel sprouts), toxins, or poisons; and umami is a pleasant taste that has been described as "chicken soup-like" and is caused by amino acids such as glutamate or aspartate.

For the basic taste sensations, the threshold required to stimulate the receptor varies. Our taste receptors are more sensitive to bitter and sour stimuli, since those sensations might indicate something toxic or poisonous in our food. Employing a combination of taste modalities allows us to perceive a wide variety of tastes.

In the past, researchers thought certain tastes were best interpreted along specific regions of the tongue; however, recent research has found that these "taste maps" were incorrect and that taste sensations are spread over broader regions of the tongue than previously thought.

Gustatory Pathways

Once the gustatory cells within taste buds have detected sensory stimuli, nerve impulses conduct this information through CN VII (facial) from the anterior two-thirds of the tongue and CN IX (glossopharyngeal) from the posterior one-third of the tongue **(figure 19.8)**. All gustatory information projects first to the **nucleus solitarius** in the medulla oblongata. This nucleus extends through the length of the medulla oblongata and slightly into the inferior region of the pons. It receives taste sensations from cranial nerves VII and

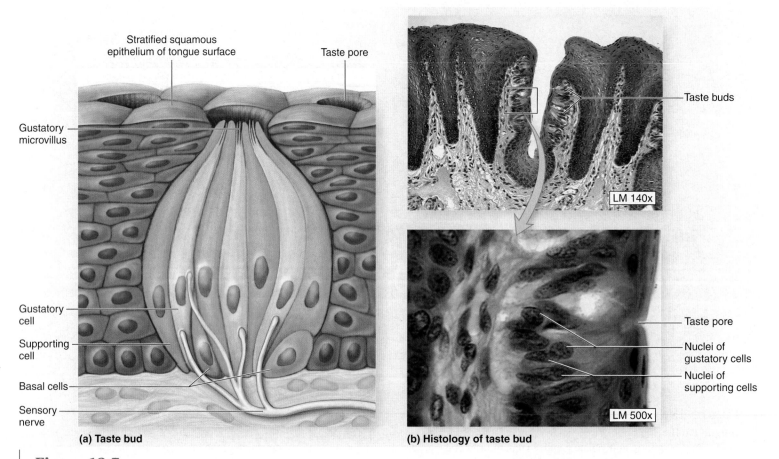

(a) Taste bud

(b) Histology of taste bud

Figure 19.7

Taste Buds. *(a)* The detailed structure of a taste bud and its organization in the wall of a papilla. *(b)* Photomicrographs show the histologic structure of taste buds.

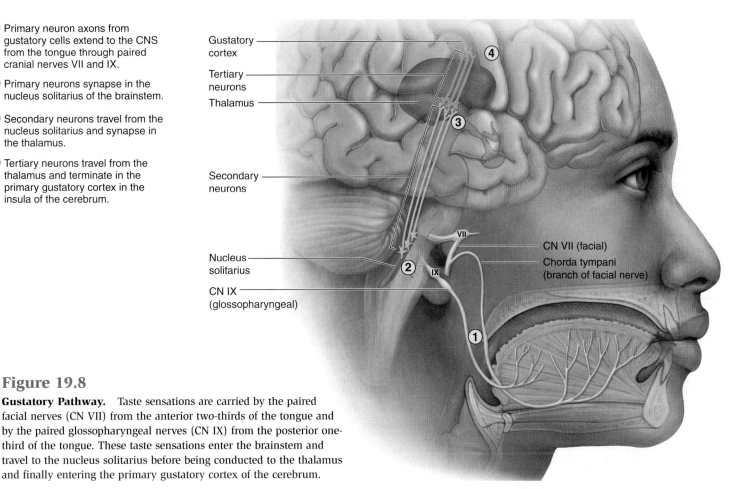

1. Primary neuron axons from gustatory cells extend to the CNS from the tongue through paired cranial nerves VII and IX.

2. Primary neurons synapse in the nucleus solitarius of the brainstem.

3. Secondary neurons travel from the nucleus solitarius and synapse in the thalamus.

4. Tertiary neurons travel from the thalamus and terminate in the primary gustatory cortex in the insula of the cerebrum.

Gustatory cortex

Tertiary neurons

Thalamus

Secondary neurons

Nucleus solitarius

CN IX (glossopharyngeal)

CN VII (facial)

Chorda tympani (branch of facial nerve)

Figure 19.8

Gustatory Pathway.   Taste sensations are carried by the paired facial nerves (CN VII) from the anterior two-thirds of the tongue and by the paired glossopharyngeal nerves (CN IX) from the posterior one-third of the tongue. These taste sensations enter the brainstem and travel to the nucleus solitarius before being conducted to the thalamus and finally entering the primary gustatory cortex of the cerebrum.

IX. Gustatory information then projects to the thalamus for processing. From the thalamus, axons project gustatory information to the primary gustatory cortex in the insula of the cerebrum for analysis (see table 15.3). The conscious perception of taste requires integrating taste sensations with those of temperature, texture, and smell.

WHAT DID YOU LEARN?

6 Which cranial nerves receive taste sensations from the tongue? Which nerves receive general sensory information from the tongue?

WHAT DO YOU THINK?

2 When a person has a stuffy nose from a "cold" or hay fever, he or she typically can't detect tastes as well. Why?

Olfaction

Key topic in this section:

- Structure, location, and innervation pathway of olfactory receptors

Olfaction (ol-fak′shŭn; *olfacio* = to smell) is the sense of smell. Smell is also referred to as *remote chemoreception* because an object can be at a distance and we may still be able to sense its odor. Compared to many other animals, olfaction in humans is not highly

developed because we do not require olfactory information to find food or communicate with others. Within the nasal cavity, paired **olfactory organs** are the organs of smell. They are composed of several components **(figure 19.9)**. An **olfactory epithelium** lines the superior part of the nasal septum (an aggregate area of about 5 square centimeters). This specialized epithelium is composed of three distinct cell types: (1) **olfactory receptor cells** (also called *olfactory neurons*), which detect odors; (2) **supporting cells,** which sandwich the olfactory neurons and sustain and maintain the receptors; and (3) **basal cells,** which function as stem cells to replace olfactory epithelium components.

Internal to the olfactory epithelium is an areolar connective tissue layer called the **lamina propria** (lam′i-nă prō-pri′ă). Included with the collagen fibers and ground substance of this layer are mucin-secreting structures called **olfactory glands** (or *Bowman glands*) and many blood vessels and nerves.

Olfactory Receptor Cells

Olfactory receptor cells are bipolar neurons that have undergone extensive differentiation and modification. At the apical surface of each neuron, the neck and apical head together form a thin, knobby projection that extends into the mucus covering the olfactory epithelium. Projecting from each knob into the overlying mucus are numerous thin, unmyelinated, cilia-like extensions called **olfactory hairs,** which house receptors for airborne molecules. These olfactory hairs are immobile and usually appear as a tangled mass within the mucous layer. Deep breathing causes the inhaled air to mix and swirl, so both fat- and water-soluble odor molecules diffuse into the

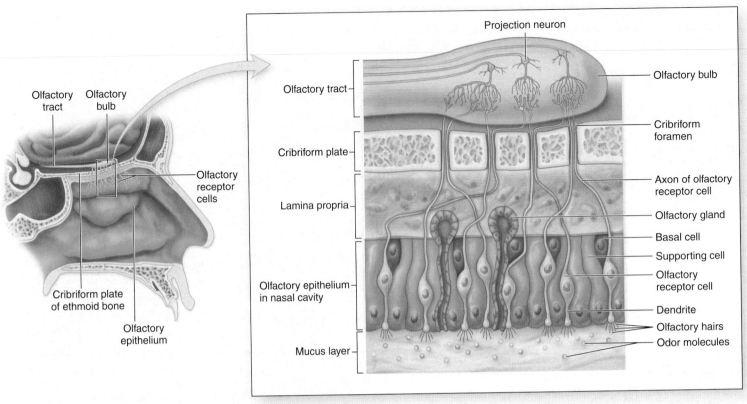

(a) Olfactory receptor cells

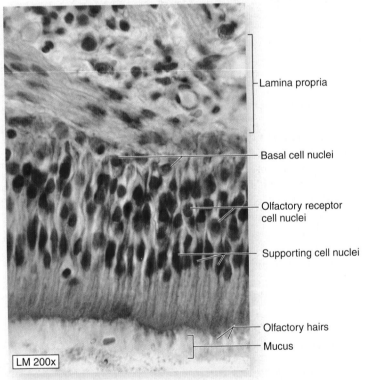

(b) Olfactory epithelium

Figure 19.9

Olfactory Organs. Olfactory organs house olfactory receptor cells that detect chemical stimuli in the air we breathe. (*a*) When olfactory receptor cells within the olfactory epithelium are stimulated, they send nerve impulses in axons that pass through the cribriform plate and synapse on neurons within the olfactory bulb. (*b*) A photomicrograph of the olfactory epithelium.

mucous layer covering the olfactory receptor cells. Receptor proteins on the olfactory hairs detect specific molecules. Airborne molecules dissolved in the mucous lining bind to those receptors. Depending on which receptors are stimulated, different smells will be detected. Once the receptors are stimulated, adaptation occurs rapidly. Thus, an initial strong smell (such as rotting food in a trash can that hasn't been emptied for a week) may seem to dissipate as your olfactory receptors adapt to the foul odor.

Olfactory Discrimination

In contrast to the five basic taste sensations, the olfactory system can recognize as many as eight different primary odors as well as many thousands of other chemical stimuli. Primary odors are those that are detectable by a large number of people, such as camphorous, fishy, malty, minty, musky, and sweaty. Secondary odors are those produced by a combination of chemicals and not detected or recognized by everyone. For example, some flower blossoms exhibit almost 100 odoriferous compounds, and individuals in the general population vary widely in their ability to recognize some or all of these. However, there are no obvious structural differences between olfactory receptor cells.

Olfactory receptor cells are one of the few types of neurons that undergo mitosis to replace aged cells. As with gustatory receptors, the number of olfactory receptor cells declines with age, thus diminishing the sense of olfaction. In addition, the remaining olfactory neurons lose their sensitivity to odors. Thus, an elderly individual has a decreased ability to recognize odor molecules.

Olfactory Pathways

Olfactory nerve (CN I) axons are discrete bundles of olfactory neuron axons that project through foramina in the cribriform plate and enter a pair of **olfactory bulbs** inferior to the frontal lobes of the brain. Neurons within the olfactory bulbs project axon bundles, called **olfactory tracts,** to the primary olfactory cortex in the temporal lobe of the cerebrum (see table 15.3).

The olfactory pathway is so sensitive that only a few stimulating molecules are needed to bind to receptors and initiate olfactory sensation. Olfactory stimuli do not immediately project to the thalamus; instead, they travel to the olfactory cortex in the temporal lobe. Later, olfactory sensations can project from the temporal lobes to the thalamus and then to the frontal lobes for more specific discrimination. However, there are widespread olfactory associations within the hypothalamus and limbic system. Thus, "smells" often initiate behavioral and emotional reactions.

🔅 WHAT DID YOU LEARN?

❼ What are the components and functions of the olfactory organs?

❽ What are the olfactory hairs?

Vision

Key topics in this section:

- Structure and function of the accessory structures of the eye
- Anatomy of the eye
- Vision receptors and their innervation pathways
- Formation of the embryonic eye

Visual stimuli help us form specific detailed visual images of objects in our environment. The sense of vision uses visual receptors (photoreceptors) in the eyes to detect light, color, and movement. Before describing the structures of the eye, we briefly examine the accessory structures that support and protect the eye's exposed surface.

Accessory Structures of the Eye

The accessory structures of the eye provide a superficial covering over its anterior exposed surface (conjunctiva), prevent foreign objects from coming in contact with the eye (eyebrows, eyelashes, and eyelids), and keep the exposed surface moist, clean, and lubricated (lacrimal glands) **(figure 19.10)**.

Conjunctiva

A specialized stratified squamous epithelium termed the **conjunctiva** (kon-jŭnk-tī′vă) forms a continuous lining of the external, anterior surface of the eye (the **ocular conjunctiva**) and the internal surface of the eyelid (the **palpebral** [pal′pē-brăl] **conjunctiva**) (figure 19.10*b*).

The space formed by the junction of the ocular conjunctiva and the palpebral conjunctiva is called the **conjunctival fornix.** Eye movement is controlled by extrinsic muscles of the eye. These muscles were described and illustrated in chapter 11. Please review the text description on "Extrinsic Eye Muscles," figure 11.4, and table 11.2.

The conjunctiva contains numerous goblet cells, which lubricate and moisten the eye. In addition, the conjunctiva houses numerous blood vessels that supply the avascular sclera ("white") of the eye, as well as abundant free nerve endings that detect foreign objects as they contact the eye. There is no conjunctiva on the surface of the cornea (the transparent center of the external eye) because the numerous blood vessels that run through the conjunctiva could interfere with letting light rays through.

The **eyebrows** are slightly curved rows of thick, short hairs at the superior edge of the orbit along the superior orbital ridge. They function primarily to prevent sweat from dripping into the open eyes. **Eyelashes** extend from the margins of the eyelids and prevent large foreign objects from contacting the anterior surface of the eye.

The **eyelids,** also called the *palpebrae* (pal-pē′brĕ), form the movable anterior protective covering over the surface of the eye. Closing the eyelids covers the delicate anterior surface of the eyes and also distributes **lacrimal fluid** (tears) to cleanse and lubricate this surface. Each eyelid is formed by a fibrous core (the **tarsal plate**), tarsal muscles, tarsal glands, the palpebral part of the orbicularis oculi muscle, the palpebral conjunctiva, and a thin covering of skin. **Tarsal glands,** previously called *Meibomian glands,* are sebaceous glands that produce a secretion to prevent tear overflow from the open eye and keep the eyelids from adhering together. The eyelids' free margins are separated by a central **palpebral fissure.**

The eyelids are united at **medial** and **lateral palpebral commissures.** At the medial commissure is a small, reddish body called the **lacrimal caruncle** (kar′ŭng-kl) that houses ciliary glands. **Ciliary glands** are modified sweat glands that form the thick secretory products that contribute to the gritty, particulate material often noticed around the eyelids after awakening.

CLINICAL VIEW

Conjunctivitis

Conjunctivitis (kon-jŭnk-ti-vī′tis) is the most common nontraumatic eye complaint seen by physicians. It presents as inflammation and reddening of the conjunctiva, and is often called "pink eye." An irritant causes dilation of blood vessels and aggregation of inflammatory cells in the connective tissue internal to the conjunctival epithelium. Conjunctivitis frequently results from a viral infection, but bacteria can be a cause as well. Conjunctivitis also can be a response to airborne allergens (pollen or animal dander), chemicals, or physical irritants, such as contact lenses left in too long.

Trachoma is a chronic, contagious form of conjunctivitis caused by *Chlamydia trachomatis.* Infection by this bacterium results in inflammation because of hypertrophy of the conjunctiva. The condition is accompanied by the appearance of newly formed, minute, gray-yellow granules in the conjunctiva. In developing countries, trachoma is a common cause of neonatal blindness; the newborn is infected as it passes through the birth canal. Blindness occurs when the inflammatory process causes scarring and thickening of the conjunctiva.

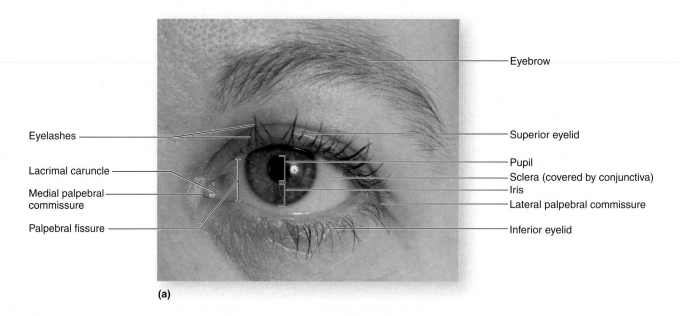

(a)

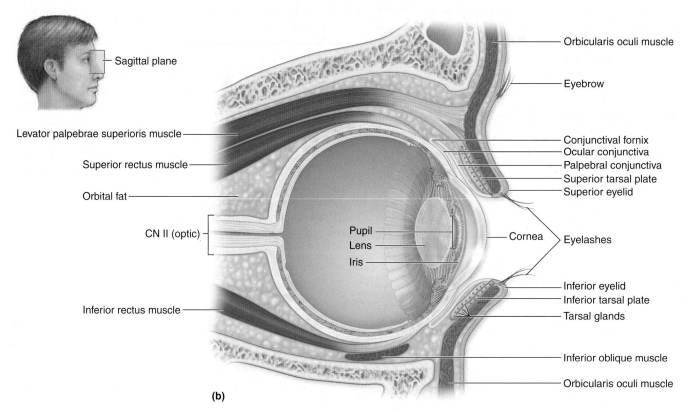

(b)

Figure 19.10

External Anatomy of the Eye and Surrounding Accessory Structures. *(a)* Accessory structures protect the eye. *(b)* A sagittal section shows the eye and its accessory structures.

Lacrimal Apparatus

Each eye is associated with a **lacrimal** (lak′ri-măl; *lacrima* = a tear) **apparatus** that produces, collects, and drains lacrimal fluid, more commonly known as *tears,* from the eye **(figure 19.11)**. Lacrimal fluid lubricates the anterior surface of the eye to reduce friction from eyelid movement; continuously cleanse and moisten the eye surface; and help prevent bacterial infection because they contain an antibiotic-like enzyme called *lysozyme.*

A **lacrimal gland** is located within the superolateral depression of each orbit. It is composed of an orbital or superior part and a palpebral or inferior part. The gland continuously produces tears. The

blinking motion of the eyelids "washes" the lacrimal fluid released from excretory ducts over the eyes. Gradually, the lacrimal fluid is transferred to the **lacrimal caruncle** at the medial surface of the eye. On the superior and inferior sides of the lacrimal caruncle are two small openings called the **lacrimal puncta** (pungk′tă; sing., *punctum;* to prick). (If you examine your own eye, each punctum appears as a "hole" in the lacrimal caruncle.) Each lacrimal punctum has a **lacrimal canaliculus** that drains lacrimal fluid into a rounded **lacrimal sac.** Finally, a **nasolacrimal duct** receives the lacrimal fluid from the lacrimal sac. This duct, which is along the lateral side of the nose, delivers the drained fluid into the nasal cavity, where it mixes with mucus.

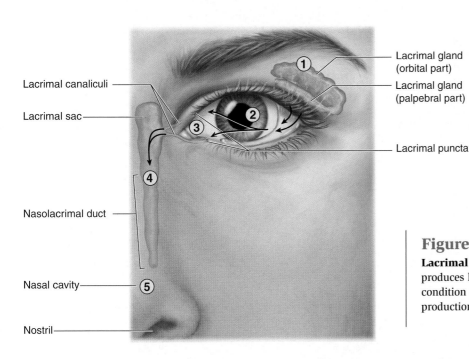

Lacrimal canaliculi

Lacrimal sac

Nasolacrimal duct

Nasal cavity

Nostril

Lacrimal gland (orbital part)

Lacrimal gland (palpebral part)

Lacrimal puncta

1 Lacrimal fluid (tears) is produced in the lacrimal gland.

2 Lacrimal fluid is dispersed across eye surface.

3 Lacrimal fluid enters the lacrimal canaliculi and collects in the lacrimal sac.

4 Lacrimal fluid from the lacrimal sac drains through the nasolacrimal duct.

5 Lacrimal fluid enters the nasal cavity.

Figure 19.11

Lacrimal Apparatus.   The lacrimal apparatus continually produces lacrimal fluid that cleanses and maintains a moist condition on the anterior surface of the eye. Lacrimal fluid production and drainage occur in a series of steps.

WHAT DO YOU THINK?

3 When a person cries, tears spill onto the cheeks, and the person usually gets a temporary runny nose. Based on your knowledge of the lacrimal apparatus, what is happening?

Eye Structure

The eye is an almost spherical organ that measures about 2.5 centimeters in diameter. Nearly 80% of the eye is receded into the orbit of the skull (see chapter 7), a space also occupied by the lacrimal gland, extrinsic eye muscles, numerous blood vessels that supply nutrients, and the cranial nerves that innervate the eye and other structures in the face and orbit. **Orbital fat** (see figure 19.10) cushions the posterior and lateral sides of the eye, providing support, protection, and vasculature. The hollow interior of the eye is called the **vitreous chamber.** It is posterior to the lens and is filled with a gelatinous, viscous fluid called the **vitreous body,** or *vitreous humor.*

Three principal layers form the wall of the eye: the fibrous tunic (outermost layer), the vascular tunic (middle layer), and the retina (innermost layer) **(figure 19.12a).**

Fibrous Tunic

The external layer of the eye wall, called the **fibrous tunic,** is composed of the anterior cornea and the posterior sclera.

The avascular, transparent **cornea** (kōr′nē-a) forms the anterior surface of the fibrous tunic. It exhibits a convex shape, and thus it refracts (bends) light rays coming into the eye. The cornea is composed of an inner simple squamous epithelium, a middle layer of collagen fibers, and an outer stratified squamous epithelium. At its circumferential edge, this epithelium is continuous with the ocular conjunctiva, and it adjoins the sclera. The structural continuity between the cornea and sclera is called the **limbus** (lim′bŭs) or the *corneal scleral junction.*

The cornea contains no blood vessels; thus, both its external and internal epithelial surfaces must obtain nutrients by alternative means. Nutrients and oxygen are supplied to the external corneal epithelium by fluid from the lacrimal glands, whereas the internal epithelium obtains needed gases and nutrients from the fluid in the anterior chamber of the eye.

Most of the fibrous tunic is formed by tough **sclera** (sklĕr′ă; *skleros* = hard), a part of the outer layer that is called the "white" of the eye. It is composed of dense irregular connective tissue that includes both collagen and elastic fibers. The sclera provides for eye shape and protects the eye's delicate internal components.

Vascular Tunic

The middle layer of the eye wall is the **vascular tunic,** also called the *uvea* (oo′vē-ă; *uva* = grape). It is composed of three distinct regions; from posterior to anterior, they are the choroid, the ciliary body, and the iris (figure 19.12). The vascular tunic houses an extensive array of blood vessels, lymph vessels, and the intrinsic muscles of the eye.

The **choroid** (kō′royd; *choroeides* = like a membrane) is the most extensive and posterior region of the vascular tunic. It houses a vast network of capillaries, which supply both nutrients and oxygen to the retina, the inner layer of the eye wall. Cells of the choroid are filled with pigment from the numerous melanocytes in this region. The melanin pigment is needed to absorb extraneous light that enters the eye, thus allowing the retina to clearly interpret the remaining light rays and form a visual image.

The **ciliary** (sil′ē-ar-ē; *cilium* = eyelid) **body** is located anterior to the choroid. The ciliary body is composed of **ciliary muscles** (bands of smooth muscle) and **ciliary processes** (folds of epithelium that cover the ciliary muscles). Extending from the ciliary body to the lens are suspensory ligaments. Relaxation and contraction of the ciliary muscles change the tension on the suspensory ligaments, thereby altering the shape of the lens. In addition, the ciliary body epithelium secretes a fluid called aqueous humor (to be discussed in a later section).

The most anterior region of the vascular tunic is the **iris** (ī′ris; rainbow), which is the colored portion of the eye. In the center of the iris is a black hole called the **pupil** (pū′pil). The peripheral edge of the iris is continuous with the ciliary body. The iris is composed of two layers of pigment-forming cells (anterior and posterior layers), two groups of smooth muscle fibers, and an array of vascular and nervous structures. The iris controls pupil size or diameter—and thus the amount of light entering the eye—using its two smooth muscle layers: the sphincter pupillae and the dilator pupillae

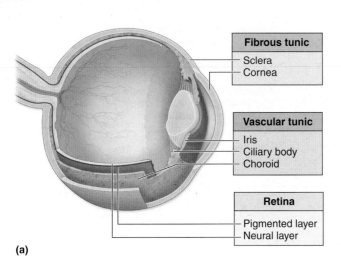

(a)

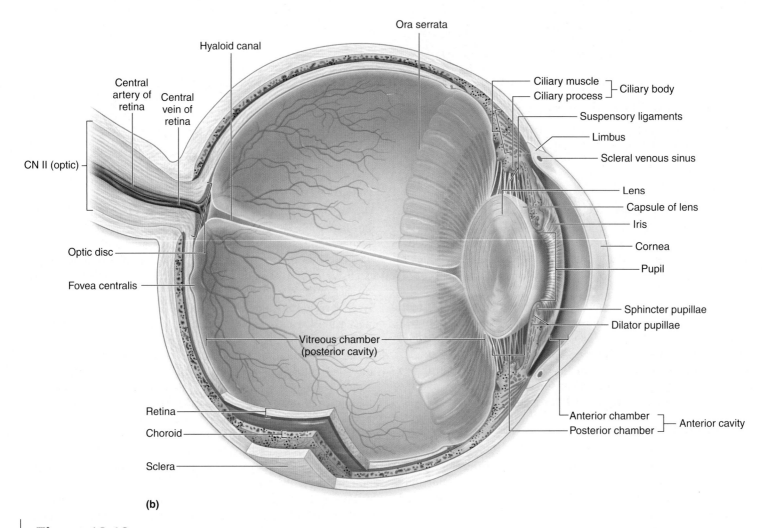

(b)

Figure 19.12

Anatomy of the Internal Eye. Sagittal views depict *(a)* the three tunics of the eye, and *(b)* internal eye structures showing the relationships among the components of the vascular layer, the lens, and the cornea.

muscles **(figure 19.13)**. The **sphincter pupillae** (or *pupillary constrictor*) is arranged in a pattern that resembles concentric circles around the pupil. Under the control of the parasympathetic division of the ANS, it constricts the pupil. The **dilator pupillae** (or *pupillary dilator*) is organized in a radial pattern extending peripherally through the iris. It is controlled

by the sympathetic division of the ANS to dilate the pupil. Only one set of these smooth muscles can contract at any one time. When stimulated by bright lights, parasympathetic innervation causes the sphincter pupillae to contract and thus decrease pupil diameter, whereas low light levels activate sympathetic stimulation to cause pupil dilation.

Pupillary constriction—Contraction of sphincter pupillae (parasympathetic innervation)

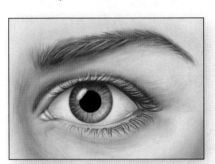

Pupillary dilation—Contraction of dilator pupillae (sympathetic innervation)

Sphincter pupillae contracts

Dilator pupillae relaxes

Dilator pupillae contracts

Sphincter pupillae relaxes

Figure 19.13

Pupil Diameter. Pupillary constriction decreases the diameter of the pupil to reduce the amount of light entering the eye. Pupillary dilation increases pupil diameter to increase light entry into the eye. Pupillary constriction is controlled by the parasympathetic division of the ANS, and pupillary dilation is controlled by the sympathetic division.

Retina

The internal layer of the eye wall, called the **retina** (ret′i-nă; *rete* = a net), or either the *internal tunic* or *neural tunic*, is composed of two layers: an outer pigmented layer and an inner neural layer (**figure 19.14a**). The **pigmented layer** is immediately internal to the choroid and attached to it. This layer provides vitamin A for photoreceptor cells. Light rays that pass through the inner layer are absorbed in this outer layer. The **neural layer** houses all of the photoreceptors and their associated neurons. This layer of the retina is responsible for receiving light rays and converting them into nerve impulses that are transmitted to the brain.

The retina extends posteriorly from the **ora serrata** (ō′ră sĕ-ră′tă; *serratus* = sawtooth) to line the internal posterior surface of the eye wall. The ora serrata is the jagged margin between the photosensitive posterior region of the retina and the nonphotosensitive anterior region of the retina that continues anteriorly to cover the ciliary body and the posterior side of the iris (see figure 19.12b).

Table 19.3 summarizes the characteristics of the three tunics of the eye wall.

Organization of the Neural Layer   Three distinct layers of cells form the neural layer: photoreceptor cells, bipolar cells, and ganglion cells. Incoming light must pass almost through the entire neural layer before reaching the photoreceptors.

The outermost layer of the neural layer is composed of **photoreceptor cells** of two types: **Rods** have a rod-shaped outer part and function in dim light, while **cones** have a cone-shaped outer part and function in high-intensity light and in color vision.

Rods are longer and narrower than cones. Rods are primarily located in the peripheral regions of the neural layer. There are more than 100 million rod cells per eye. Rods are especially important when the light is dim. They detect movement but exhibit poor visual acuity. In addition, rods pick up contrasting dark and light tones, but cannot distinguish color. Thus, when you are trying to see at night, it is primarily the rods that are working. The rods can perceive the object, but your vision may not be particularly sharp, and you may find it difficult to see any color variation.

CLINICAL VIEW

Detached Retina

A **detached retina** occurs when the outer pigmented and inner neural layers of the retina separate. Detachment may result from head trauma (soccer players and high divers are especially susceptible), or it may have no overt cause. Individuals who are nearsighted, due to a more elliptical eyeball, are at increased risk for detachment because their retina is typically thinned or "stretched" more than that of a normal eye. There is also increased risk for retinal detachment in diabetics and older individuals.

Normally, the pigmented and neural layers of the retina are held in close opposition by the pressure of the vitreous body in the posterior cavity against the retina. Retinal layer separation results in fluid accumulation between them. Recall that the pigmented layer is directly internal to and adheres to the choroid, the vascular source of the retina. The layer separation results in deprivation of nutrients for the neurons, photoreceptors, and other cells in the inner neural layer. Degeneration and death of nervous tissue result if the blood supply is not restored.

Symptoms of a detached retina include a large number of "floaters" (small, particle-like objects) in the vision; the appearance of a "curtain" in the affected eye; flashes of light; and decreased, watery, or wavy vision.

The treatment for a detached retina depends on the location of the detachment and its severity. **Pneumatic retinopexy** is a treatment for upper retinal detachment. The physician inserts a needle into the anesthetized eye and injects a gas bubble into the vitreous body. The gas bubble rises and pushes against the upper retinal detachment, forcing the two layers back together again. The gas bubble absorbs and disappears over 1–2 weeks, and then a laser may be used to "tack" the two layers of the retina together. The **scleral buckle** uses a silicone band to press inward on the sclera in order to hold the retina in place. A laser is then used to reattach the retina. The scleral buckle remains as a permanent fixture in the eye.

Individuals who have had a detached retina in one eye are at increased risk for developing a detachment in the other eye as well. Thus, the retinal detachment patient must be extra-vigilant in monitoring his or her vision.

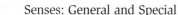

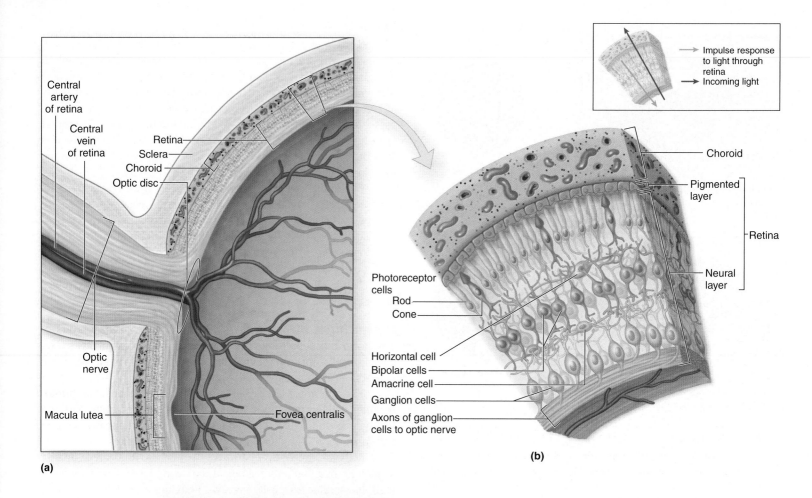

(a)

(b)

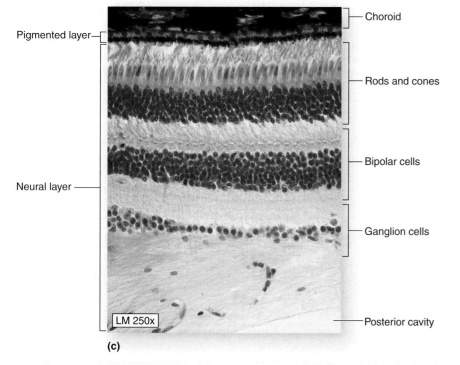

(c)

Figure 19.14

Structure and Organization of the Retina. The retina is composed of two distinct layers: the outer pigmented layer and the inner neural layer. *(a)* The optic nerve is composed of ganglionic cell axons that originate in the neural layer. *(b)* The neural layer is composed of three cellular layers: the outer photoreceptor cells (rods and cones), the middle bipolar cells, and the inner ganglion cells. *(c)* Histologic section of the retina.

In contrast, cones occur at a density of less than 10 million per eye. Cones are activated by high-intensity light and provide precise visual acuity and color recognition. Thus, when you notice the fine details in a colorful picture, the cones of your neural layer are responsible. Cones are concentrated at the posterior part of the neural layer at the visual axis of the eye.

Internal to the photoreceptor layer is a layer of **bipolar cells,** a type of bipolar neuron. Sandwiched between the photoreceptor layer and the bipolar cells is a thin web of **horizontal cells** that form connections between the photoreceptor and bipolar cells. Rods and cones form synapses on the dendrites of the bipolar neurons. There are far fewer bipolar neurons than photoreceptors, and thus

Tunic/Structures	Components	Location	Function
Table 19.3	**Tunics of the Eye Wall**		
FIBROUS TUNIC (*EXTERNAL LAYER*)			
Cornea	Two layers of epithelium with collagen fibers in the middle; avascular	Forms the anterior one-sixth of the external eye layer	Transmits and refracts incoming light
Sclera	Dense regular connective tissue	Posterior covering of the eye (forms the remaining five-sixths of the external eye layer); the "white" of the eye	Supports eye shape and protects it Protects delicate internal structures Extraocular muscle attachment site
VASCULAR TUNIC (*MIDDLE LAYER*)			
Choroid	Areolar connective tissue; highly vascularized	Forms posterior two-thirds of middle wall of eye	Supplies nourishment to retina Pigment absorbs extraneous light
Ciliary body	Ciliary smooth muscles and an inner, secretory epithelium	Between choroid posteriorly and iris anteriorly	Holds suspensory ligaments that attach to the lens and change lens shape for far and near vision Epithelium secretes aqueous humor
Iris	Two layers of smooth muscle (sphincter pupillae and dilator pupillae) with a central opening called the pupil	Anterior region of middle layer	Controls pupil diameter and thus the amount of light entering the eye
RETINA (*INTERNAL LAYER*)			
Pigmented layer	Pigmented epithelial cells and supporting cells	Outermost portion of retina that directly adheres to choroid	Absorbs extraneous light Provides vitamin A for photoreceptor cells
Neural layer	Photoreceptors, bipolar neurons, and ganglion cells	Inner portion of retina	Detects incoming light rays; light rays are converted to nerve impulse and transmitted to brain

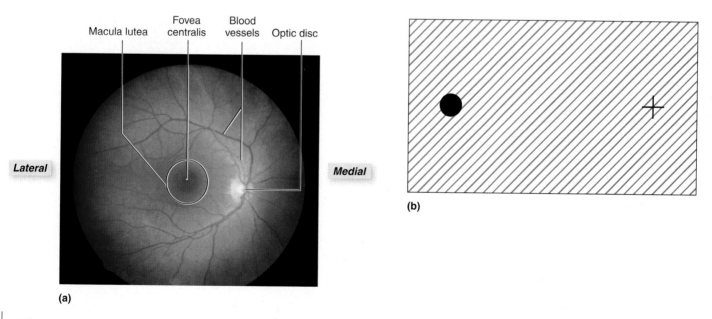

Figure 19.15

Internal View of the Retina Showing the Optic Disc (Blind Spot). An ophthalmoscope is used to view the retina through the pupil. (*a*) Blood vessels travel through the optic nerve to enter the eye at the optic disc. (*b*) Check your blind spot! Close your left eye. Hold this figure in front of your right eye, and stare at the black spot. Move the figure toward your open eye. At a certain point, when the image of the plus sign is over the optic disc, the plus sign seems to disappear.

information must converge as visual signals are directed toward the brain from the stimulated photoreceptors.

Ganglion cells form the innermost layer in the neural layer that is adjacent to the vitreous body in the posterior cavity. Neuronal convergence continues between the bipolar neurons and ganglionic neurons. **Amacrine** (am′ă-krin) **cells** help process and integrate visual information as it passes between bipolar and ganglionic neurons. Axons extend from the ganglionic cells into and through the **optic disc.** The axons of the ganglionic cells converge to form the optic nerve (CN II) as they exit the eye and extend toward the brain. The optic disc lacks photoreceptors, and consequently it is called the **blind spot** because no image forms there (**figure 19.15**).

Macular Degeneration

Macular degeneration, the physical degeneration of the macula lutea of the retina, has become a leading cause of blindness in developed countries. Although the majority of cases are reported in people over 55, the condition may occur in younger people as well. Most non-age-related cases are associated with an aggravating condition such as diabetes, an ocular infection, heredity, or trauma to the eye. People who smoke and those with hypertension are at increased risk for developing macular degeneration.

Macular degeneration occurs in two forms—*dry* or *wet*. About 70% of affected people have the dry form, which is characterized by photoreceptor loss and thinning of the pigmented layer. The remaining 30% of patients have the wet form, which has all of the dry form characteristics plus bleeding, capillary proliferation, and scar tissue formation. The wet form is the most advanced and aggressive type, and it accounts for most of the cases of blindness. The disease is often present in both eyes, although it may progress at different rates in each eye.

The first symptom of macular degeneration is typically the loss of visual acuity. Other common vision problems include: straight lines appearing distorted or wavy; a dark, blurry area of visual loss in the center of the visual field; diminished color perception; "floaters"; dry eyes; and cataracts.

At present, there is no cure for macular degeneration. However, its progression may be slowed. Early detection has become an important element in treatment. To track the progression of the disease, doctors rely heavily on self-monitoring, in which the patient regularly performs a simple visual test using the *Amsler grid* (see figures *c,d*). The grid consists of evenly spaced vertical and horizontal lines with a small dot in the middle. While staring at the dot, the patient looks for wavy lines, blurring, or missing parts of the grid.

If the lines of the grid do not appear straight and regular, or if parts seem to be missing, the patient should be examined for changes in the macula. The most effective treatment for macular degeneration uses laser photocoagulation to destroy the abnormally proliferating vessels in the macula. Although this procedure cannot restore lost vision, it does slow the pace of visual loss. A number of trials involving dietary modifications have been undertaken, including supplementation with vitamins A, C, and E as well as restriction of saturated fats. However, at present no definitive data indicate that these measures reduce the symptoms or decrease the progression of the disease.

(a) Normal vision

(b) The same scene as viewed by a person with macular degeneration

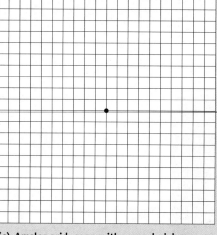

(c) Amsler grid, seen with normal vision

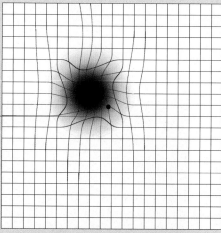

(d) Amsler grid, as viewed by a person with macular degeneration

Comparison showing differences between normal vision (left) and the vision of a person with macular degeneration (right).

Just lateral to the optic disc is a rounded, yellowish region of the neural layer called the **macula** (mak′ū-lă) **lutea** (loo′te-a; saffron-yellow). Within the macula lutea is a pit called the **fovea centralis** (fō′vē-ă, pit; sen′tră′lis, central). This pit is the area of sharpest vision; when you read the words in your text, they are precisely focused here. Although the other regions of the neural layer also receive and interpret light rays, no other region can focus as precisely as can the fovea centralis because the fovea centralis contains the highest proportion of cones and almost no rods.

Lens

The **lens** is a strong, yet deformable, transparent structure bounded by a dense, fibrous, elastic capsule. Internally, it is composed of precisely arranged layers of cells that have lost their organelles and are filled completely by a protein called *crystallin*. The lens focuses incoming light onto the retina, and its shape determines the degree of light refraction.

The **suspensory** (sŭs-pen′sŏ-rē; *suspensio* = to hang up) **ligaments** attach to the lens capsule at its periphery, where they transmit tension that enables the lens to change shape. The tension in the suspensory ligaments comes from contraction of the **ciliary muscles** in the ciliary body. When the ciliary muscles relax, the ciliary body moves posteriorly, away from the lens, and so the tension on the suspensory ligaments increases. This constant tension causes the lens to flatten **(figure 19.16a)**. A flattened lens is necessary for far vision (viewing distant objects). This shape of the lens is the "default" position of the lens.

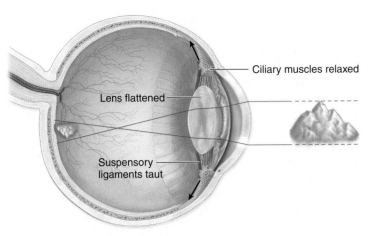

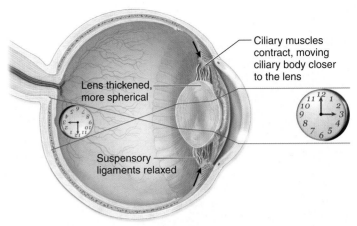

(a) Lens shape for distant vision

(b) Lens shape for near vision (accommodation)

Figure 19.16

Lens Shape in Far Vision and Near Vision. *(a)* To focus a distant object on the retina, the ciliary muscles within the ciliary body relax, which tenses the suspensory ligaments and flattens the lens. *(b)* Accommodation occurs when a nearby object is focused on the retina. The ciliary muscles contract, causing the suspensory ligaments to relax and the lens to thicken (become more spherical).

In contrast, when we wish to see close objects (use our near vision), the lens must become more spherical to properly refract (bend) the light rays and focus an image on the retina. The process of making the lens more spherical in order to view close-up objects is called **accommodation** (ă-kom′ŏ-dā′shŭn; *accommodo* = to adapt) (figure 19.16*b*). Accommodation is controlled by the parasympathetic division of the ANS. Stimulation of the ciliary muscles by parasympathetic axons causes the muscles to contract. When the ciliary muscles contract, the entire ciliary body moves anteriorly and thus moves closer to the lens itself. This process reduces the suspensory ligaments' tension and releases some of their "pull" on the lens, so the lens can become more spherical.

CLINICAL VIEW

Cataracts

Cataracts (kat′ă-rakt) are small opacities within the lens that, over time, may coalesce to completely obscure the lens. Cataracts are a major cause of blindness worldwide. Most cases occur as a result of aging, although other causative factors include diabetes, intraocular infections, excessive ultraviolet light exposure, and glaucoma. The resulting vision problems include difficulty focusing on close objects, reduced visual clarity as a consequence of clouding of the lens, "milky" vision, and reduced intensity of colors.

A cataract needs to be removed only when it interferes with normal daily activities. Newer surgical techniques include **phaco-emulsification** (or **phaco**), a process whereby the opacified center of the lens is fragmented using ultrasonic sound waves, thus making it easy to remove. The destroyed lens is then replaced with an artificial intraocular lens, which becomes a permanent part of the eye. Occasionally, an intraocular lens replacement is not suitable, in which case the person is fitted with a soft contact lens after the cataract has been removed.

Eye without a cataract

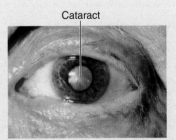

Eye with a cataract

Normal vision

Image seen through cataract

(Left) The appearance of a normal eye and normal vision compared to (right) an eye with a cataract and vision through a cataract.

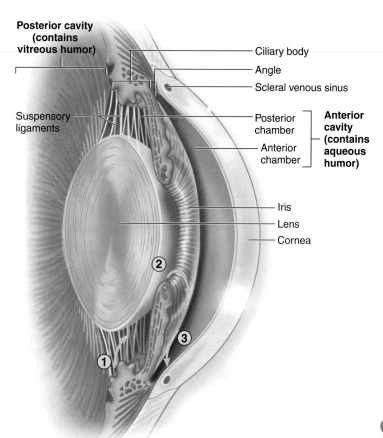

Posterior cavity (contains vitreous humor)

Suspensory ligaments

Ciliary body
Angle
Scleral venous sinus

Posterior chamber
Anterior chamber

Anterior cavity (contains aqueous humor)

Iris
Lens
Cornea

① ② ③

① Epithelial cells covering the ciliary body secrete aqueous humor into the posterior chamber of the anterior cavity.

② Aqueous humor slowly flows through the posterior chamber, around the lens, and through the pupil into the anterior chamber of the anterior cavity.

③ Aqueous humor is drained into the scleral venous sinus and thus transported to the venous bloodstream.

Figure 19.17

Aqueous Humor: Secretion and Reabsorption. Aqueous humor is a watery secretion that flows continuously through the eye, removing waste products and helping to maintain the chemical environment. Its pathway is represented here by numbers 1, 2, and 3.

With increasing age, the lens becomes less resilient and less able to become spherical. Thus, even if the suspensory ligaments relax, the lens may not be able to spring out of the default, flattened position into the more spherical shape needed for near vision. If the lens cannot become spherical, reading close-up words becomes very difficult. This age-related change is called **presbyopia** (prez-bē-ō′pē-ă; *presbys* = old man, *ops* = eye). You may have noticed your parents holding the newspaper at arm's length. They do this because they can only focus on the words if the paper is far enough away that their lenses do not need to be spherical to accurately refract the light rays from the type. As we age, presbyopia may become more severe as the lens loses more of its elasticity, and thus most of us eventually need reading glasses in order to focus on close-up objects.

Cavities of the Eye

The internal space of the eye is subdivided by the lens into two separate cavities: the anterior cavity and the posterior cavity **(figure 19.17)**. The **anterior cavity** is the space anterior to the lens and posterior to the cornea. The iris of the eye subdivides the anterior cavity further into two chambers: The **anterior chamber** is between the iris and cornea, while the **posterior chamber** is between the lens and the iris.

The anterior cavity contains a fluid called **aqueous** (ak′qwē-ŭs; watery) **humor** (hū′mer), which is a filtrate of plasma that resembles CSF and is produced by the epithelium covering the ciliary body. Aqueous humor removes waste products and helps maintain the chemical environment within the anterior and posterior chambers of the eye. The aqueous humor is secreted into the posterior chamber. From the posterior chamber, it flows through the pupil into the anterior chamber. The aqueous humor is continually reabsorbed across the covering epithelium into a vascular space, called the **scleral venous sinus** (previously called the *canal of Schlemm*), located in the limbus between the cornea and the sclera. The scleral venous sinus conducts the reabsorbed aqueous humor to the veins that drain the eye. Thus, as with cerebrospinal fluid, excess aqueous humor is removed from these eye chambers and transported into the venous circulation.

The **posterior cavity** is posterior to the lens and anterior to the retina. It is occupied by the transparent, gelatinous **vitreous** (vit′rē-ŭs; glassy) **humor,** which completely fills the space between the lens and the retina. The vitreous humor helps maintain eye shape,

---

CLINICAL VIEW

Glaucoma

Glaucoma (glaw-kō′ma) is a disease that exists in three forms, all characterized by increased intraocular pressure: angle-closure glaucoma, open-angle glaucoma, and congenital or juvenile glaucoma.

Angle-closure and open-angle glaucoma both involve the angle formed in the anterior chamber of the eye by the union of the choroid and the corneal-scleral junction (see figure 19.17). This angle is the important tract for draining the aqueous humor. If it narrows, fluid and pressure build up within the anterior chamber. About one-third of all cases of glaucoma develop as a direct consequence of the narrowing of this angle, a condition called *primary angle–closure glaucoma*. *Open-angle glaucoma* accounts for about two-thirds of glaucoma cases. In this instance, although the drain angles are adequate, fluid transport out of the anterior chamber is impaired. *Congenital glaucoma* occurs only rarely and is due to hereditary factors or intrauterine infection.

Irrespective of the cause of glaucoma, fluid buildup causes a posterior dislocation of the lens and a substantial increase in pressure in the posterior chamber. Compression of the choroid layer may occur, constricting the blood vessels that nourish the retina. Retinal cell death and increased pressure may distort the axons within the optic nerve. Eventually, the patient may experience such symptoms as reduced field of vision, dim vision, and halos around lights. These symptoms are often unrecognized until it is too late and the damage is irreversible. Furthermore, a sudden closure of the tract that drains aqueous humor can lead to rapid buildup of pressure, with ocular distortion and excruciating pain.

Both medical and surgical treatments are available for arresting the advance of glaucoma, but they cannot restore lost vision. For this reason, early detection and treatment are critical for preserving vision. It's a simple matter for the ophthalmologist or optometrist to measure the anterior chamber pressure, and this screening test should be a part of everyone's periodic medical evaluation.

support the retina, and transmit light from the lens to the retina. Within the vitreous body is a thin **hyaloid** (hī′ă-loyd; *hyalos* = glass) **canal**. This canal is a remnant of embryonic hyaloid blood vessels that once supplied the retina and lens. As the eye developed, the distal parts of the hyaloid vessels regressed, leaving the proximal parts of the vessels to become the retinal vessels. The path left by the distal part of the hyaloid vessels then became this hyaloid canal.

Visual Pathways

Light stimuli are detected by photoreceptors in the retina. Stimulation of photoreceptors by incoming light causes a change in the rods and cones. These photoreceptor cells, in turn, signal the change to the bipolar cells, resulting in the stimulation of the ganglion cells and the generation of a nerve impulse. The visual image that is formed is a result of the processing and summation of information as it is collected and integrated in the retina. Due to this continual processing and integration, there are fewer cells in each layer, from photoreceptor to ganglionic neuron.

Ganglionic axons converge to form the optic nerve. Optic nerves project from each eye through paired optic foramina and converge at the **optic chiasm** immediately anterior to the pituitary gland **(figure 19.18)**. Only the ganglionic axons originating from the medial region of each retina cross to the opposite side of the brain at the optic chiasm. The optic chiasm is a flattened structure anterior to the infundibulum (see chapter 15) that is the location of decussation (or crossing) of optic nerve axons. Ganglionic axons originating from the lateral region of each retina remain on the same side of the brain and do not cross. **Optic tracts** form laterally from the optic chiasm as a composite of ganglionic axons originating from the retinas of each eye.

Upon entry into the brain, some axons within each optic tract project to the **superior colliculi.** Collectively, these projecting axons and the indirect motor pathway they stimulate are called the **tectal system.** This system coordinates the movements of the eyes, head, and neck in responding to visual stimuli.

The remainder of the optic tract axons extend to the thalamus, specifically to the **lateral geniculate** (je-nik′ū-lāt) **nucleus,** where visual information is processed within each thalamic body. Neurons from the thalamus project to the primary visual cortex of the occipital lobe for interpretation of incoming visual stimuli.

Development of the Eye

Eye development begins during the fourth week of development when left and right **optic vesicles** evaginate from the diencephalic portion of the prosencephalon **(figure 19.19)**. Thus, these optic vesicles (the future retina of the eye) are formed from the neural tube. As the optic vesicle nears the surface ectoderm, the vesicle invaginates (indents) so that it forms an **optic cup.** The optic cup contains an inner and outer layer, and is connected to the developing brain by an **optic stalk. Hyaloid vessels** (blood vessels) enter the optic cup through an opening in the optic stalk. The ectoderm overlying the optic cup forms a depression called a **lens pit.** This lens pit further indents and forms a circular structure called the **lens vesicle,** which will develop into the future lens of the eye. As mentioned previously, the hyaloid vessels eventually regress, leaving the hollow cavity known as the hyaloid canal.

During the fifth and sixth weeks of development, the lens vesicle pinches off the surface of the ectoderm and becomes an internal structure. Meanwhile, the inner and outer layers of the optic cup give rise to the inner neural layer and outer pigmented layer of the retina, respectively. Initially, a space exists between these two layers, but that space disappears, and the layers contact each other before birth.

Within the developing optic cup, a gelatinous material called the **primary vitreous body** begins forming; it eventually gives rise to the vitreous humor in the eye. The vitreous humor is not continuously secreted, so the vitreous humor in your eye right now is the same vitreous humor that was produced during this embryonic period.

During the sixth and seventh weeks of development, the mesenchyme around the optic cup forms the choroid and sclera. The iris, ciliary body, and suspensory ligaments form from the choroid beginning in the third month of development. Meanwhile, the cornea develops from the mesenchyme anterior to the lens as well as from part of the overlying ectoderm. These different tissue layers form the multiple tissue layers of the cornea by the eighth week of development.

Finally, the eyelids and the conjunctiva form from both the outer surface ectoderm and an underlying layer of mesenchyme. The lacrimal glands, which are essentially a type of exocrine gland, are formed from ectoderm that burrowed into the deeper connective tissue layers around the superolateral surface of the eye.

WHAT DID YOU LEARN?

9 How is lacrimal fluid spread across the eye and removed from the orbital region?

10 What are the three eye tunics; what is the primary function of each tunic?

11 Identify the three main cell layers in the neural layer of the retina, and discuss the function of each.

12 What is the hyaloid canal, and what does it represent?

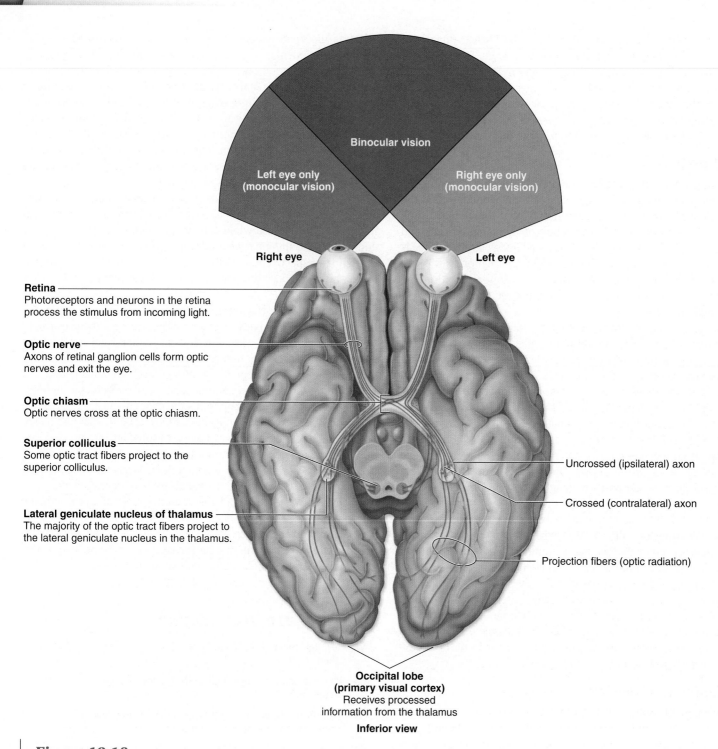

Binocular vision

Left eye only (monocular vision)

Right eye only (monocular vision)

Right eye

Left eye

Retina
Photoreceptors and neurons in the retina process the stimulus from incoming light.

Optic nerve
Axons of retinal ganglion cells form optic nerves and exit the eye.

Optic chiasm
Optic nerves cross at the optic chiasm.

Superior colliculus
Some optic tract fibers project to the superior colliculus.

Lateral geniculate nucleus of thalamus
The majority of the optic tract fibers project to the lateral geniculate nucleus in the thalamus.

Uncrossed (ipsilateral) axon

Crossed (contralateral) axon

Projection fibers (optic radiation)

Occipital lobe (primary visual cortex)
Receives processed information from the thalamus

Inferior view

Figure 19.18

Visual Pathways. Each optic nerve conducts visual stimulus information. At the optic chiasm, some axons from the optic nerve decussate. The optic tract on each side then contains axons from both eyes. Visual stimulus information is processed by the thalamus and then interpreted by visual association areas in the cerebrum.

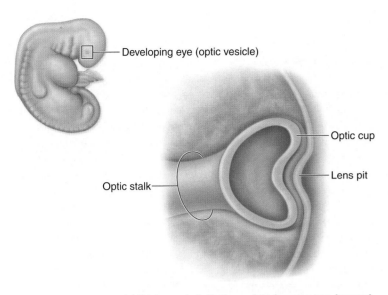

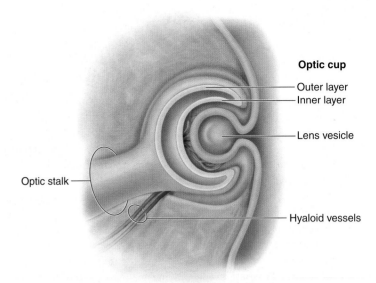

(a) **Early week 4: Optic vesicle forms a two-layered optic cup; overlying ectoderm forms a lens pit.**

(b) **Late week 4: Optic cup deepens and forms inner and outer layers; lens pit forms lens vesicle.**

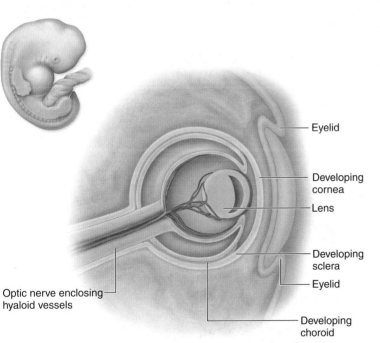

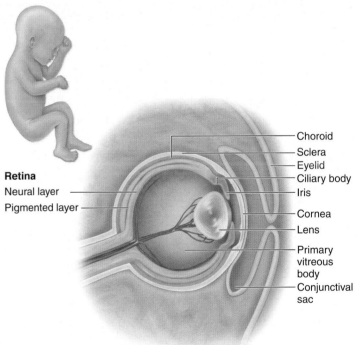

(c) **Week 6: Lens becomes an internal structure; corneas, sclera, and choroid start to form.**

(d) **Week 20: Three tunics of the eye have formed.**

Figure 19.19

Eye Development.   Development of the eye begins in the embryo during week 4 with the formation of optic vesicles lateral to the forebrain. *(a)* Early in week 4 of development, the optic vesicle forms a two-layered optic cup, and the overlying ectoderm forms a lens pit. *(b)* By late week 4, the lens pit becomes a lens vesicle, and hyaloid blood vessels begin to be incorporated into the forming optic nerve. *(c)* At week 6, the cornea, sclera, and choroid begin to develop. *(d)* By week 20, the three tunics of the eye, a lens, and primary vitreous body are present.

CLINICAL VIEW: In Depth

How Vision Can Be Functionally Impaired

Emmetropia (em-ĕ-trō′pē-ă; *emmetros* = according to measure; *ops* = eye) is the condition of normal vision in which parallel rays of light are focused exactly on the retina. Any variation in the curvature of either the cornea or the lens, or in the overall shape of the eye, causes entering light rays to form an abnormal focal point. Conditions that can result include hyperopia, myopia, and astigmatism.

People with **hyperopia** (hī-per-ō′pē-ă) have trouble seeing close-up objects, and so are called "farsighted." In this optical condition, only convergent rays (those that come together from distant points) can be brought to focus on the retina. The cause of hyperopia is a "short" eyeball; parallel light rays from objects close to the eye focus posterior to the retina. By contrast, people with **myopia** (mī-ō′pē-ă; *myo* = to shut) have trouble seeing far-away objects, and so are called "nearsighted."

In myopia, only rays relatively close to the eye focus on the retina. The cause of this condition is a "long" eyeball; parallel light rays from objects at some distance from the eye focus anterior to the retina within the vitreous body. Another variation is **astigmatism** (ă-stig′mă-tizm; *a* = not, *stigma* = a point), which causes unequal focusing and blurred images due to unequal curvatures along different meridians in one or more of the refractive surfaces (cornea, anterior surface, or posterior surface of the lens). Astigmatism can give a person headaches or eyestrain. Vision is distorted or blurred at all distances.

The typical treatment for vision disturbances is eyeglasses. A concave eyeglass lens is used to treat myopia, because the concave lens refracts (bends) the light rays to make the focused image appear directly on the retina, instead of too far in front of it. A convex lens is used to treat both hyperopia and age-related far-sightedness (called presbyopia). The convex lens refracts the light rays so they can be better focused on the retina.

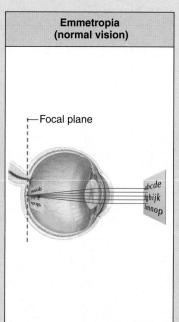

Emmetropia (normal vision)

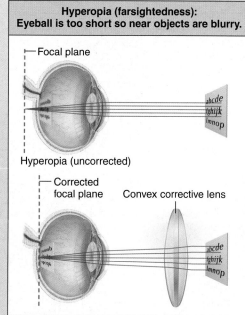

Hyperopia (farsightedness): Eyeball is too short so near objects are blurry.

Hyperopia (uncorrected)

Corrected focal plane · Convex corrective lens

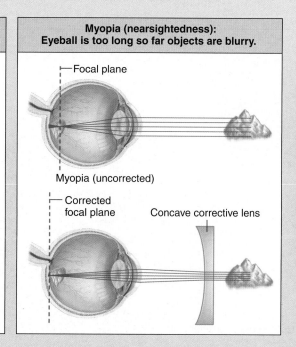

Myopia (nearsightedness): Eyeball is too long so far objects are blurry.

Myopia (uncorrected)

Corrected focal plane · Concave corrective lens

Vision correction using (center) convex and (right) concave lenses.

Equilibrium and Hearing

Key topics in this section:

- Anatomy of the ear
- Detection and processing of equilibrium sensations
- Collection and processing of sound
- Formation of the embryonic ear

The ear is partitioned into three distinct anatomic regions: external, middle, and inner **(figure 19.20)**. The external ear is located mostly on the outside of the body, and the middle and inner areas are housed within the petrous part of the temporal bone. Movements of the inner ear fluid result in the sensations of hearing and equilibrium, or balance.

External Ear

The external ear is a skin-covered, primarily cartilaginous structure called the **auricle,** or *pinna* (pin′ă; wing). The auricle is funnel-shaped, and serves to protect the entry into the ear and to direct sound waves into the bony tube called the **external acoustic meatus,** which extends medially and slightly superiorly from the lateral surface of the head. The external acoustic meatus terminates at the **tympanic** (tim-pan′ik; *tympanon* = drum) **membrane,** or *eardrum,* a delicate, funnel-shaped epithelial sheet that is the partition between the external and middle ear. The tympanic membrane vibrates when sound waves hit it. These vibrations provide the means for transmission of sound wave energy into the middle and inner ear.

The narrow external opening in the external acoustic meatus prevents large objects from entering and damaging the tympanic

In more recent times, surgical techniques have been developed to treat vision problems. **Corneal incision** involves cutting the cornea to change its shape, and thereby change its ability to refract light. One type of corneal incision procedure, called a **radial keratotomy** (ker'ă-tot'ŏ-me; *tome* = incision), or **RK**, treats nearsightedness. The ophthalmologist makes radial-oriented cuts in the cornea, much like the spokes of a wheel. These cuts flatten the cornea, and allow it to refract the light rays so that they focus on the retina.

Currently more common than corneal incision is **laser vision correction,** which uses a laser to change the shape of the cornea. The laser can either flatten the center of the cornea to correct myopia or shape the outer edges of the cornea to treat hyperopia. Two of the more popular types of laser vision correction are photorefractive keratectomy and laser-assisted in situ keratomileusis.

Photorefractive keratectomy (PRK) is called a *photoablation procedure* because the laser removes (ablates) tissue directly from the surface of the cornea. The removal of tissue results in a newly shaped

cornea that can focus better. This procedure is becoming less popular, because its regression rate is high—that is, the epithelial tissue that covers the surface of the cornea can regrow and regenerate, leading to partial return to uncorrected vision.

Laser-assisted in situ keratomileusis (ker'ă-tō-mī-loo'sis) (or **LASIK**) is rapidly becoming the most popular laser vision correction procedure. It can treat nearsightedness, farsightedness, and astigmatism. LASIK removes tissue from the inner, deeper layer of the cornea, which is less likely to regrow than surface tissue, so less vision regression occurs. The accompanying illustration outlines the LASIK procedure.

No vision correction procedure is without risk. Some may not correct vision enough, while others may overcorrect it, making a nearsighted person farsighted. Problems with corneal healing or a poorly placed incision can disrupt vision forever. However, laser vision correction has helped millions of individuals (including one of the authors of this textbook). If you are considering it, you should be aware of the risks and have a thorough eye exam and risk consultation beforehand.

① Cornea is sliced with a sharp knife. Flap of cornea is reflected, and deeper corneal layers are exposed.

② A laser removes microscopic portions of the deeper corneal layers, thereby changing the shape of the cornea.

③ Corneal flap is put back in place, and the edges of the flap start to fuse within 72 hours.

LASIK laser vision correction procedure.

membrane. Near its entrance, fine hairs help guard the opening. Deep within the canal, ceruminous glands produce a waxlike secretion called *cerumen,* which combines with dead, sloughed skin cells to form earwax. This material may help reduce infection within the external acoustic meatus by impeding microorganism growth.

Middle Ear

The middle ear contains an air-filled **tympanic cavity** (figure 19.20). Medially, a bony wall that houses the oval window and round window (discussed later) separates the middle ear from the inner ear. The tympanic cavity maintains an open connection with the atmosphere through the **auditory tube** (also called the *pharyngotympanic tube* or *Eustachian tube*). This passageway opens into the nasopharynx (upper throat) from the middle ear. It has a normally closed,

slitlike opening at its connection to the nasopharynx. Air movement through this tube occurs as a result of chewing, yawning, and swallowing, which allows the pressure to equalize in the middle ear.

WHAT DO YOU THINK?

④ When an airplane descends to a lower altitude, you may feel greater pressure in your ears, followed by a "popping" sensation, before more normal pressure resumes. What do you think happens anatomically?

Auditory Ossicles

The tympanic cavity of the middle ear houses the three smallest bones of the body, called the **auditory ossicles** (os'i-kl)

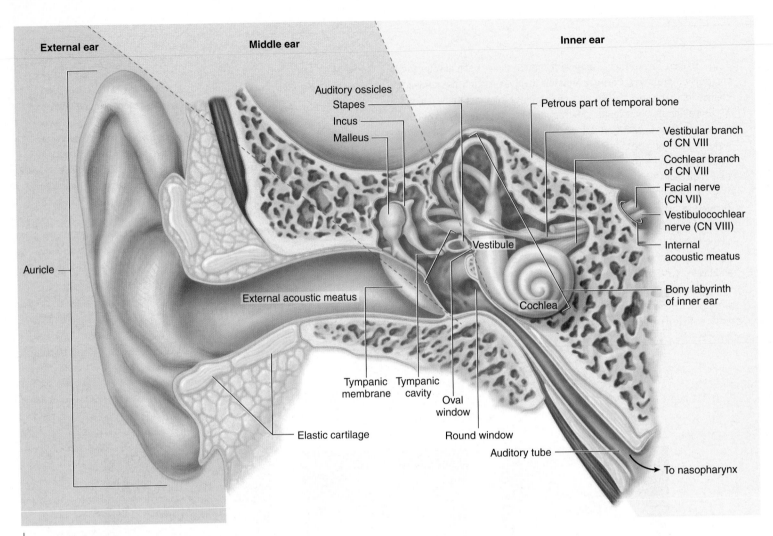

External ear

Middle ear

Inner ear

Auditory ossicles
Stapes
Incus
Malleus

Petrous part of temporal bone

Vestibular branch
of CN VIII

Cochlear branch
of CN VIII

Facial nerve
(CN VII)

Vestibulocochlear
nerve (CN VIII)

Internal
acoustic meatus

Vestibule

Auricle

Bony labyrinth
of inner ear

Cochlea

External acoustic meatus

Tympanic   Tympanic
membrane   cavity
Oval
window

Elastic cartilage

Round window

Auditory tube

To nasopharynx

Figure 19.20

Anatomic Regions of the Ear.   The external, middle, and inner regions are shown in this view of the right ear.

(figure 19.21; see also figure 19.20). These bones are, from lateral to medial, the malleus (hammer), the incus (anvil), and the stapes (stirrup). The **malleus** (mal′ē-ŭs) is attached to the medial surface of the tympanic membrane, and suspended by ligaments bound to the wall of the tympanic cavity. It resembles a large hammer in shape. The **incus** (ing′kŭs) resembles an anvil and is the middle auditory ossicle. The **stapes** (stā′pēz) resembles a stirrup on a saddle. It has a cylindrical, disclike footplate that fits into the oval window, an opening that marks the lateral wall of the inner ear.

The auditory ossicles are responsible for amplifying sound waves and transmitting them into the inner ear via the oval window. When sound waves strike the tympanic membrane, the three middle ear ossicles vibrate along with the tympanic membrane, causing the footplate of the stapes to move in and out of the oval window. The movement of this ossicle initiates pressure waves in the fluid within the closed compartment of the inner ear. Since the tympanic membrane is 20 times greater in diameter than the stapes footplate in the oval window, sounds transmitted across the middle ear are amplified more than 20-fold, and we are able to detect very faint sounds.

Two tiny skeletal muscles, called the **stapedius** and the **tensor tympani,** are located within the middle ear. These muscles restrict ossicle movement when loud noises occur, and thus protect the sensitive receptors in the inner ear.

Inner Ear

The inner ear is located within the petrous part of the temporal bone, where there are spaces or cavities called the **bony labyrinth** (lab′i-rinth; an intricate, mazelike passageway) **(figure 19.22;** see also figure 19.20). Within the bony labyrinth are membrane-lined, fluid-filled tubes and spaces called the **membranous labyrinth.** Receptors for equilibrium and hearing are housed, along with supporting cells, within a sensory epithelium lining part of the membranous labyrinth.

The space between the outer walls of the bony labyrinth and the membranous labyrinth is filled with a fluid called **perilymph** (per′i-limf), which is similar in composition to both extracellular fluid and cerebrospinal fluid (CSF). In the inner ear, the perilymph suspends, supports, and protects the membranous labyrinth from the wall of the bony labyrinth. The membranous labyrinth contains a unique fluid called **endolymph** (en′dō-limf). Endolymph exhibits a low sodium and high potassium concentration similar to that of intracellular fluid.

The bony labyrinth is structurally and functionally partitioned into three distinct regions: the **vestibule** (ves′ti-bool; *vestibulum =*

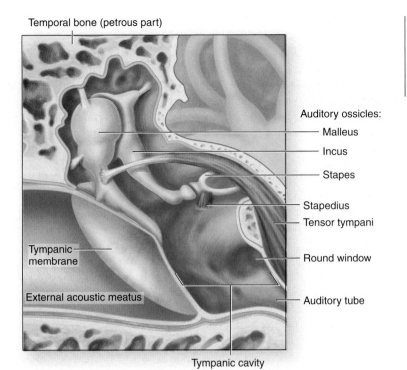

Temporal bone (petrous part)

Auditory ossicles:
- Malleus
- Incus
- Stapes

Stapedius
Tensor tympani

Tympanic membrane

Round window

External acoustic meatus

Auditory tube

Tympanic cavity

Figure 19.21

Auditory Ossicles. The middle ear contains the auditory ossicles and associated structures within the tympanic cavity, as shown in this view of the right middle ear.

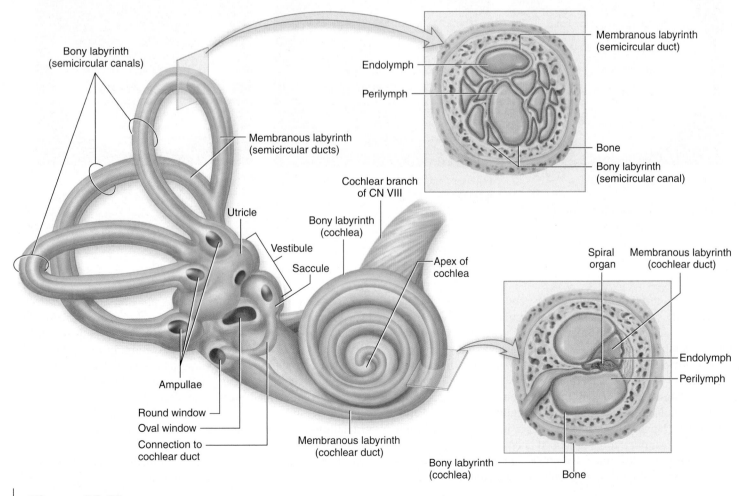

Bony labyrinth (semicircular canals)

Membranous labyrinth (semicircular duct)

Endolymph

Perilymph

Membranous labyrinth (semicircular ducts)

Bone

Bony labyrinth (semicircular canal)

Cochlear branch of CN VIII

Utricle

Vestibule

Saccule

Bony labyrinth (cochlea)

Apex of cochlea

Spiral organ

Membranous labyrinth (cochlear duct)

Endolymph

Perilymph

Ampullae

Round window

Oval window

Connection to cochlear duct

Membranous labyrinth (cochlear duct)

Bony labyrinth (cochlea)

Bone

Figure 19.22

Inner Ear. The inner ear is composed of a bony labyrinth cavity that houses a fluid-filled membranous labyrinth. Within the bony labyrinth are the vestibular organs for equilibrium and balance (saccule, utricle, and semicircular canals) and the cochlea for hearing.

CLINICAL VIEW

Otitis Media

Otitis (ō-tī′tis; *itis* = inflammation) **media** (mē′dē-ă; *medius* = middle) is an infection of the middle ear. It is most often experienced by young children, whose auditory tubes are horizontal, relatively short, and undeveloped. If a young child has a respiratory infection, the causative agent may spread from the throat into the tympanic cavity and the mastoidal air cells. Fluid then accumulates in the middle ear cavity, resulting in pressure, pain, and sometimes impaired hearing.

Classic symptoms of otitis media include fever (sometimes over 104°F), pulling on or holding the affected ear, and general irritability. An **otoscope** (ō′tō-skōp; *skopeo* = to view) is an instrument used to examine the tympanic membrane, which normally appears white and "pearly," but in cases of severe otitis media is red and may even bulge due to fluid pressure in the middle ear.

Some otitis media infections clear up on their own, but pediatricians may prescribe antibiotics to relieve the potential bacterial infection.

After the infection is relieved, material drains from the auditory tube to alleviate middle ear pressure. Otitis media can be a serious health problem because infection can spread to the mastoidal air cells, which may cause meningitis. Additionally, fusion of the auditory ossicles may occur as a consequence of chronic otitis media, resulting in impaired hearing.

Repeated ear infections, or a chronic ear infection that does not respond to antibiotic treatment, usually calls for a surgical procedure called a **myringotomy** (mir-ing-got′ō-mē; *myringa* = membrane), whereby a ventilation tube is inserted into the tympanic membrane. This procedure offers immediate relief from the pressure and allows the infection to heal and the pus and mucus to drain from the middle ear. Eventually, the inserted tube is sloughed, and the tympanic membrane heals.

Once a child is about 5 years old, the auditory tube has become larger, more angled, and better able to drain fluid and prevent infection from reaching the middle ear. Thus, the occurrence rate for ear infections drops dramatically at this time.

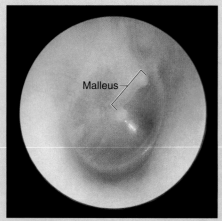

Normal tympanic membrane

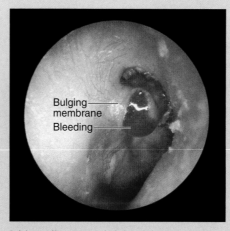

Otitis media present, bulging red tympanic membrane

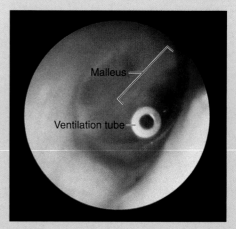

Myringotomy

Middle ear views as seen with an otoscope.

entrance court), the **semicircular canals,** and the **cochlea** (kok′lē-ă = snail shell). The vestibule and semicircular canals compose a general area called the **vestibular complex.** The vestibule of the vestibular complex contains two saclike, membranous labyrinth parts—the **utricle** (oo′tri-kl; *uter* = leather bag) and the **saccule** (sak′ūl; *saccus* = sack)—which are interconnected through a narrow passageway **(figure 19.23a).** Within the semicircular canals of the vestibular complex, the membranous labyrinth is called the **semicircular ducts.** Finally, the cochlea houses a membranous labyrinth called the cochlear duct, to be described later.

The sensory epithelium in the utricle and saccule forms a covering on a small, raised oval area called the **macula** located along the internal wall of both sacs. This epithelium is composed of a mixed layer of hair cells and supporting cells (figure 19.23b). **Hair cells** are the sensory receptors of the inner ear for both equilibrum and hearing. They continuously release neurotransmitter molecules to the sensory neurons that monitor their activity. The apical surface of each hair cell has a covering of numerous (more than 50) long, stiff microvilli, termed **stereocilia** (ster′ē-ō-sil′ē-ă; *stereos* = solid, *cilium* = eyelid). Each hair cell also has one long cilium, termed a **kinocilium** (kī-nō-sil′-ē-ŭm; *kino* = movement) on its apical surface (figure 19.23c). Both the stereocilia and kinocilia may be physically bent or displaced, resulting in changes in the amount and rate of neurotransmitter release from the hair cell.

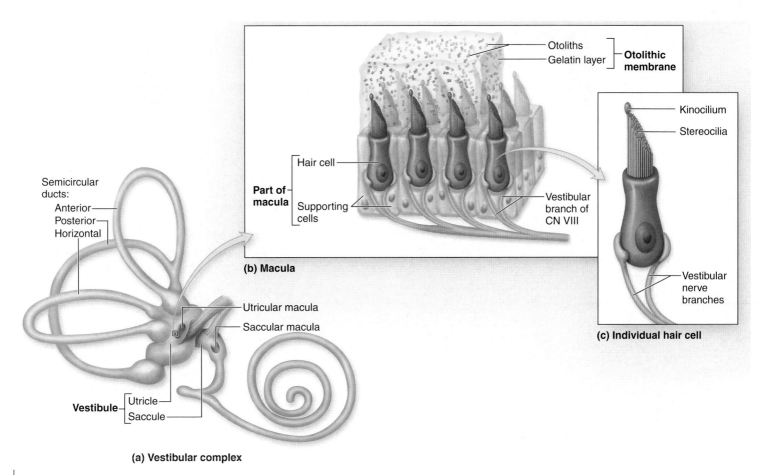

(b) Macula

(c) Individual hair cell

(a) Vestibular complex

Figure 19.23

Macula Structure. *(a)* The maculae are located within the walls of the saccule and utricle in the vestibular complex. Hair cells within the maculae detect both the orientation of the head when the body is stationary and linear acceleration of the head. *(b)* An enlarged view of a macula shows that the apical surface of the hair cells is covered by a gelatinous layer overlaid with otoliths, called the otolithic membrane. *(c)* An individual hair cell has numerous microvilli called stereocilia and a single long kinocilium.

Structures and Mechanism of Equilibrium

Sensory receptors in the utricle and saccule detect the position of the head. Stereocilia and kinocilia projecting from the hair cells embed within a gelatinous mass that completely covers the apical surface of the epithelium. This gelatinous layer is embedded with a bunch of small calcium carbonate crystals called **otoliths** (o′tō-lith; *lithos* = stone), or *statoconia*. Together, the gelatinous layer and the crystals form the **otolithic membrane** (or *statoconic membrane*). The otoliths push on the underlying gelatin, thereby increasing the weight of the otolithic membrane covering the hair cells. The position of the head influences the position of the otolithic membrane **(figure 19.24)**. When the head is held erect, the otolithic membrane applies pressure directly onto the hair cells, and minimal stimulation of the hair cells occurs. However, tilting the head due to either acceleration or deceleration causes the otolithic membrane to shift its position on the macula surface, thus distorting the stereocilia. Bending of the stereocilia results in a change in the amount of neurotransmitter released from the hair cells and a simultaneous change in the stimulation of the sensory neurons. Thus, distortion of the hair cells

in one particular direction causes stimulation, whereas the opposing movement causes inhibition of the sensory neuron. These impulses are sent to the brain to indicate the direction the head has tilted.

The semicircular canals are continuous with the superoposterior region of the utricle. There are three semicircular canals: anterior, posterior, and lateral. The anterior and posterior canals are vertical but at right angles to each other, whereas the lateral canal is slightly off the horizontal plane. Within each semicircular canal is a semicircular duct connected to the utricle. Receptors in the semicircular ducts detect rotational movements of the head. Contained within each semicircular duct is an expanded region, called the **ampulla** (am-pul′lă; pl., *ampullae*, am-pul′lē), located at the end furthest from the utricle connection **(figure 19.25)**. The ampulla contains an elevated region, called the **crista ampullaris** (or *ampullary crest*), that is covered by an epithelium of hair cells and supporting cells. These hair cells embed both their kinocilia and stereocilia into an overlying gelatinous dome called the **cupula** (koo′poo-lă). This accessory structure extends completely across the semicircular duct to the roof over the ampulla.

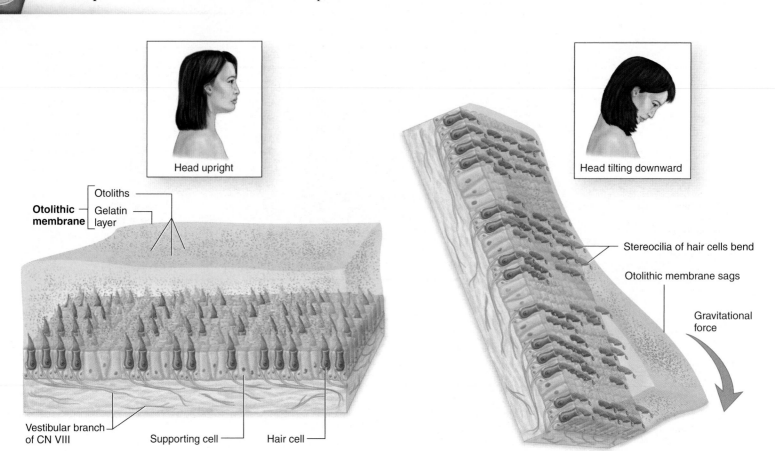

Figure 19.24

Detecting Head Position. When the head is upright, the stereocilia are encased within the otolithic membrane, and their positions indicate to the brain that the head is upright. Tilting the head downward causes the otolithic membrane to move slightly, causing the stereocilia to bend and initiating a nerve impulse that reports the change in head position.

CLINICAL VIEW

Motion Sickness

Motion sickness is a sense of nausea, mild disorientation, and dizziness that some of us have felt while flying in an airplane or riding in an automobile. Although a minor nuisance for most of us, it can be a debilitating problem for some travelers. Motion sickness develops when a person is subjected to acceleration and directional changes, but there is no visual contact with the outside horizon. In this situation, the vestibular complex of the inner ear is sending impulses to the brain that conflict with the visual reference. The eyes tell the brain we are standing still in an airplane or a ship's cabin, but the inner ear is saying something completely different. Fear or anxiety about developing motion sickness can lower the threshold of experiencing symptoms.

Motion sickness may be alleviated or prevented in some individuals. The affected person should seek a place of lesser movement and reestablish the visual reference. Most people who tend to have motion sickness learn not to read while moving. Some people drink a carbonated beverage or eat soda crackers, although the reason this lessens the symptoms isn't clear. Antihistamines are effective in reducing symptoms, and a number of nonprescription preparations are available as well, including Dramamine®, Bonine®, and Marezine®. These medications can be taken orally or administered as a skin patch. People who are apt to develop symptoms of motion sickness are advised to take the medication at least 1 hour prior to departing on a trip.

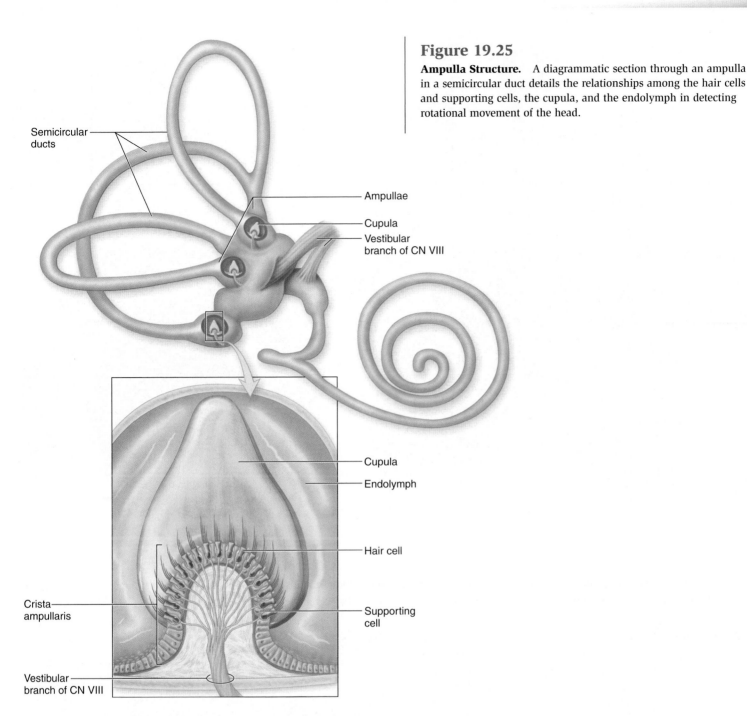

Figure 19.25

Ampulla Structure.   A diagrammatic section through an ampulla in a semicircular duct details the relationships among the hair cells and supporting cells, the cupula, and the endolymph in detecting rotational movement of the head.

The hair cells of the sensory epithelium covering the crista ampullaris detect rotational movements of the head (**figure 19.26**). When the head rotates, it causes endolymph within the semicircular ducts to move and push against the cupula, causing bending of the stereocilia. Stereocilia bending results in altered neurotransmitter release from the hair cells and simultaneous stimulation of the sensory neurons. As previously described, fluid movement in one direction causes distortion of the hair cells, resulting in stimulation, whereas the opposing movement causes distortion resulting in inhibition.

Vestibular Sensation Pathways   CN VIII receives all sensory stimuli from the inner ear. This cranial nerve has two branches: The vestibular branch, also called the vestibular nerve, receives equilibrium stimuli from both the vestibule and the semicircular canals; the cochlear branch of CN VIII, also called the cochlear nerve, receives sound stimuli from the cochlea.

Sensory neurons housed within the vestibular ganglia monitor changes in the activities of the hair cells within the vestibule and the semicircular canals. Vestibular nerve axons project to paired **vestibular nuclei** within the superior region of the medulla oblongata. These nuclei integrate the stimuli related to balance and equilibrium, and project impulses inferiorly through the vestibulospinal tracts to maintain muscle tone and balance. Additionally, the vestibular nuclei send impulses to motor nuclei in the brainstem and spinal cord to control reflexive motor activities associated with eye movements and head and neck functions. Finally, impulses about our changing equilibrium are sent to the cerebellum, the thalamus, and eventually the cerebral cortex for further processing.

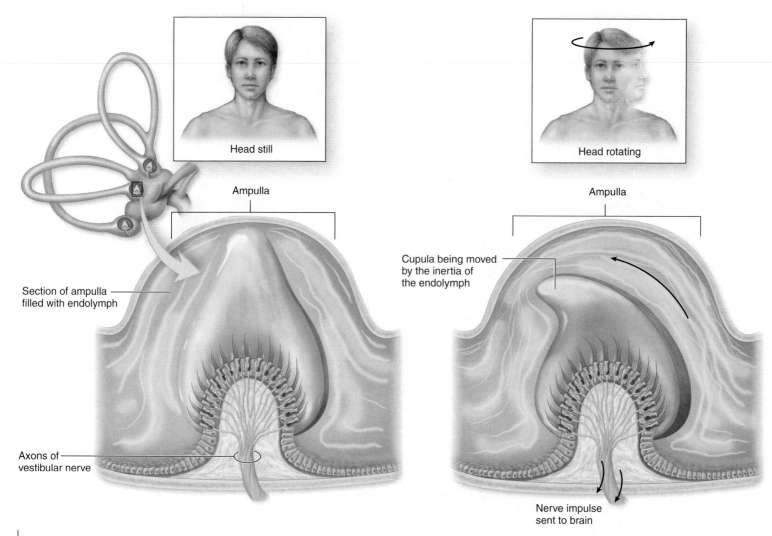

Head still

Head rotating

Ampulla

Ampulla

Section of ampulla filled with endolymph

Cupula being moved by the inertia of the endolymph

Axons of vestibular nerve

Nerve impulse sent to brain

Figure 19.26

Function of the Crista Ampullaris. Rotation of the head causes endolymph within the semicircular duct to push against the cupula covering the hair cells, resulting in bending of their stereocilia and the initiation of a nerve impulse.

Structures for Hearing

Cochlea Hearing organs are housed within the cochlea in both inner ears. The cochlea is a snail-shaped spiral chamber in the bone of the inner ear. It has a spongy bone axis, called the **modiolus** (mō-dī'ō-lŭs; *hub of a wheel*). Protected within the core of the modiolus, the membranous labyrinth houses the **spiral organ** (formerly called the *organ of Corti*), which is responsible for hearing **(figure 19.27)**.

The **cochlear duct,** or *scala media* (skā'lă; *stairway*), is the membranous labyrinth that runs through the cochlea. The roof and floor of the cochlear duct are formed by the vestibular and basilar membranes, respectively. These membranes partition the bony labyrinth of the cochlea into two smaller chambers, both filled with perilymph. The superior chamber is the **scala vestibuli** (*vestibular duct*), and the inferior chamber is the **scala tympani** (*tympanic duct*). The scala vestibuli and scala tympani merge through a small channel called the **helicotrema** (hel'i-kō-trē'mă; *helix* = spiral, *trema* = hole) at the apex of the cochlear spiral apex.

Spiral Organ Within the cochlear duct, the spiral organ is a thick sensory epithelium consisting of both hair cells and supporting cells that rests on the basilar membrane. The hair cells extend stereocilia into an overlying gelatinous structure called the **tectorial** (tek-tōr'ē-ăl; *tectus* = to cover) **membrane.** Any movement of the basilar membrane causes distortion of the stereocilia, with subse-

Cochlear Shape—Why a Spiral?

The spiral shape of the cochlea facilitates its fitting into the temporal bones of the skull, but this shape has functional significance as well. Recent research suggests that our ability to detect low-frequency sound is a direct result of the cochlea's spiral shape. Just as a child on a rotating merry-go-round is forced to the outside of the swirling platform, some investigators suggest that the curve of the cochlea focuses pressure waves in the perilymph (formed as a result of sound waves striking the tympanic membrane) to the outer edge of the spiral where these pressure waves are more readily detected by the vibration-sensitive hair cells in the spiral organ. Low-frequency sounds peak near the narrowing tip of the cochlea, where it appears there is greater sensitivity.

quent stimulation of sensory neurons as the amount of neurotransmitter released by the hair cells changes. The cell bodies of these sensory neurons are housed within the **spiral ganglia** in the modiolus medial to the cochlear duct. The distortion is caused by pressure waves in the fluid of the inner ear as a consequence of sound waves

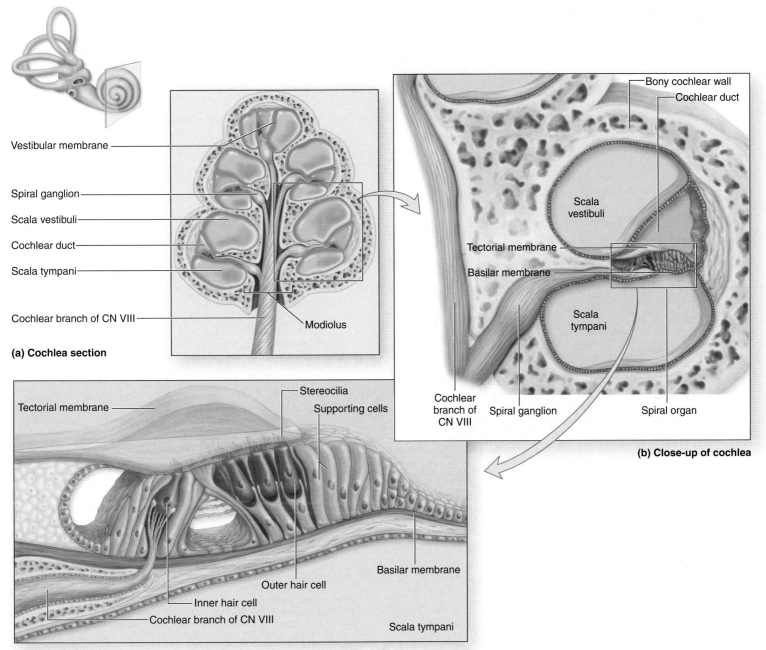

(a) Cochlea section

Vestibular membrane

Spiral ganglion

Scala vestibuli

Cochlear duct

Scala tympani

Cochlear branch of CN VIII

Modiolus

Bony cochlear wall

Cochlear duct

Scala vestibuli

Tectorial membrane

Basilar membrane

Scala tympani

Cochlear branch of CN VIII

Spiral ganglion

Spiral organ

(b) Close-up of cochlea

(c) Spiral organ

Tectorial membrane

Stereocilia

Supporting cells

Inner hair cell

Cochlear branch of CN VIII

Outer hair cell

Basilar membrane

Scala tympani

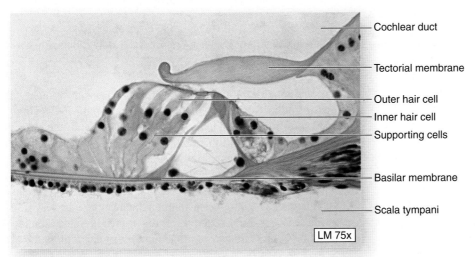

(d) Spiral organ

Cochlear duct

Tectorial membrane

Outer hair cell

Inner hair cell

Supporting cells

Basilar membrane

Scala tympani

LM 75x

Figure 19.27

Structure of the Cochlea and Spiral Organ. The cochlea exhibits a snail-like spiral shape and is composed of three fluid-filled ducts. *(a)* A section through the cochlea details the relationship among the three ducts: the cochlear duct, scala tympani, and scala vestibuli. *(b)* A magnified view of the cochlea. *(c)* Hair cells rest on the basilar membrane of the spiral organ in the cochlear duct. *(d)* Light micrograph of the spiral organ.

Table 19.4	Inner Ear Structures		
Bony Labyrinth Structure	**Membranous Labyrinth Structure Housed in Bony Labyrinth**	**Structure(s) Housing Receptors**	**Function of Receptors**
Vestibule	Utricle, saccule	Maculae	Detect acceleration and deceleration movements of the head
Semicircular canals	Semicircular ducts	Cupulas	Detect rotational movements of the head
Cochlea	Cochlear duct	Spiral organ	Hearing (converting sound waves to a nerve impulse)

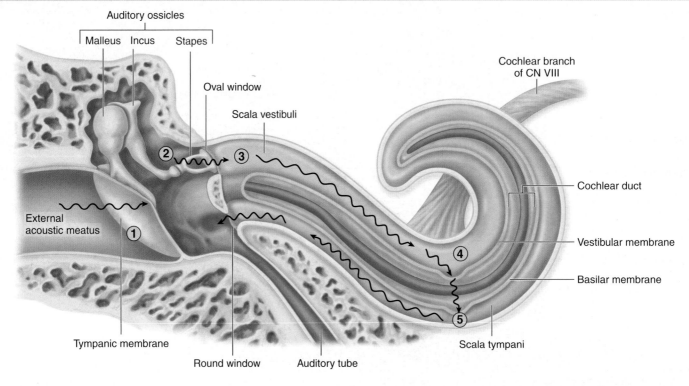

① Sound waves enter external acoustic meatus and make the tympanic membrane vibrate.

② Tympanic membrane vibration causes movement by the auditory ossicles; sound waves are amplified. The stapes moves within the oval window; pressure waves are generated.

③ Pressure waves begin at the oval window and travel through the scala vestibuli.

④ High-frequency and upper medium-frequency pressure waves in the scala vestibuli cause the vestibular membrane to vibrate, resulting in pressure wave formation in the endolymph of the cochlear duct. These pressure waves displace a specific region of the basilar membrane. Hair cells in the spiral organ are distorted, causing a stimulus in the cochlear branch of CN VIII.

⑤ Remaining pressure wave vibrations are transferred to the scala tympani and exit the inner ear via the round window.

Figure 19.28

Sound Wave Pathways Through the Ear. Sound waves enter the external ear, are conducted through the ossicles of the middle ear, and then are detected by a specific region of the spiral organ in the inner ear.

arriving at the tympanic membrane (see the next section). The pressure waves cause the basilar membrane to flutter or "bounce," resulting in distortion of the stereocilia (figure 19.27).

    Table 19.4 summarizes the structures of the inner ear and clarifies which ones are part of the bony labyrinth and which are part of the membranous labyrinth.

Process of Hearing

The sound wave pathway through the ear is shown in **figure 19.28** and described here:

1. Sound waves are collected and funneled by the auricle of the external ear. From there, sound waves enter the external acoustic meatus and make the tympanic membrane vibrate.

2. The vibration of the tympanic membrane causes movement by the auditory ossicles. Sound waves are amplified, allowing even soft sounds to be heard by the ear. The foot of the stapes moves like a piston in the oval window, thereby transmitting the effect of sound waves into pressure waves in the inner ear.

3. Pressure waves originate within the inner ear at the oval window and travel through the perilymph in the scala vestibuli.

4. High-frequency and upper medium-frequency pressure waves in the scala vestibuli cause the vestibular membrane to vibrate, resulting in pressure wave formation in the endolymph of the cochlear duct. Pressure waves in the cochlear duct displace a specific region of the basilar membrane. Hair cells in the spiral organ of this region are distorted, causing a stimulus in the cochlear branch of CN VIII.

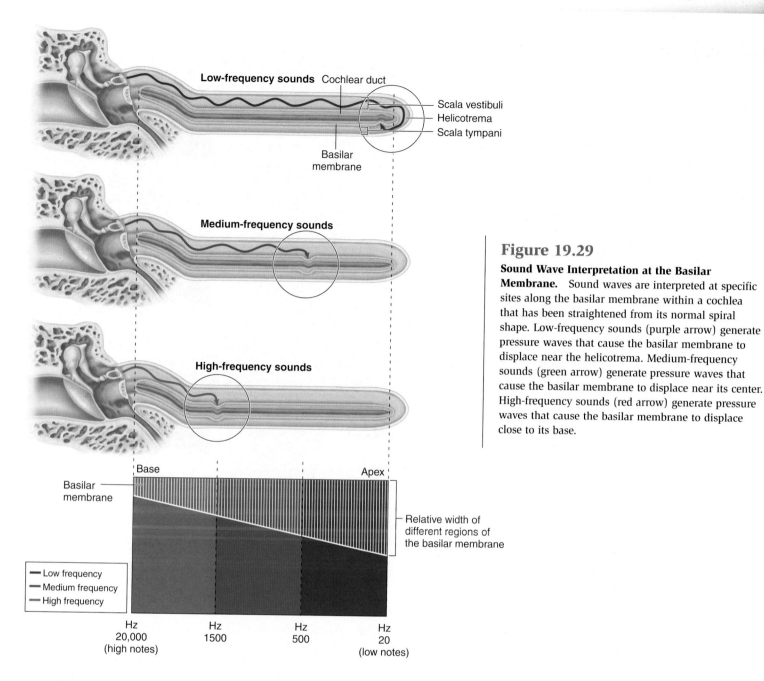

Figure 19.29

Sound Wave Interpretation at the Basilar Membrane. Sound waves are interpreted at specific sites along the basilar membrane within a cochlea that has been straightened from its normal spiral shape. Low-frequency sounds (purple arrow) generate pressure waves that cause the basilar membrane to displace near the helicotrema. Medium-frequency sounds (green arrow) generate pressure waves that cause the basilar membrane to displace near its center. High-frequency sounds (red arrow) generate pressure waves that cause the basilar membrane to displace close to its base.

5. The remaining pressure wave vibrations in the cochlear duct are transmitted to the perilymph of the scala tympani, and they exit the inner ear at the round window. As the pressure waves leave the inner ear, the window bulges slightly.

The sound waves that give rise to pressure waves in the inner ear are characterized by their frequency and intensity. **Frequency** is the number of waves that move past a point during a specific amount of time. Frequency is measured in *hertz* (*Hz*), and is classified as high, medium, or low (**figure 19.29**). **Intensity** refers to a sound's loudness and is measured in units called *decibels* (*dB*). An inaudible sound has no decibels. The region of the spiral organ stimulated by pressure waves in the perilymph varies according to the frequency of the sound: Low-frequency sounds stimulate the spiral organ far away from the oval window, while high-frequency sounds stimulate it near the oval window.

Figure 19.29 shows that high-frequency and upper medium-frequency pressure waves are transmitted across the vestibular membrane to cause the basilar membrane to vibrate, resulting in hair cell stimulation in that region. Low-frequency pressure waves travel around the helicotrema to the scala tympani to cause basilar membrane vibration.

CLINICAL VIEW

Are Rock Concerts Bad for Your Health?

Continuous or repeated exposure to loud noises is known to damage hearing, and recent studies report that the development of a benign tumor, called **acoustic neuroma**, often accompanies such hearing loss. The acoustic neuroma is a tumor of the vestibular branch of CN VIII. Its development was first reported as a consequence of mobile phone use. Further investigations suggest that acoustic neuromas develop in patients exposed to loud noise from other sources, including plant machinery, motors, sporting events, power tools, and music. Unfortunately, the repeated, long-term exposure to very loud music (such as that experienced at rock concerts) provided the greatest possibility for tumor development. An interesting follow-up would be to evaluate the musicians themselves for increased tumor development.

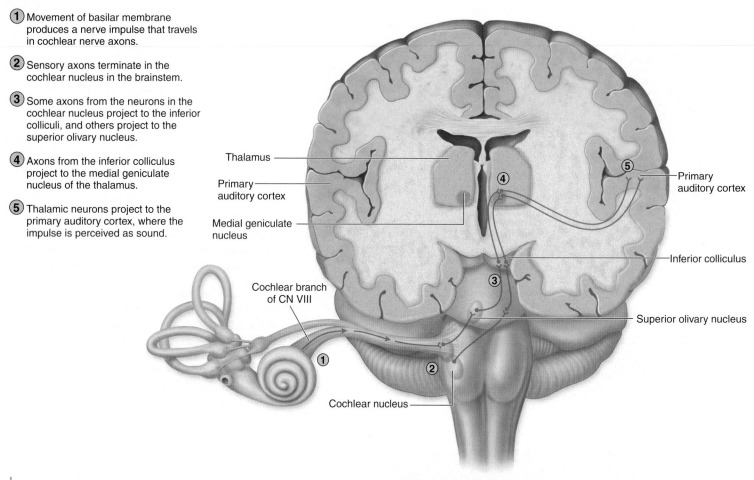

(1) Movement of basilar membrane produces a nerve impulse that travels in cochlear nerve axons.

(2) Sensory axons terminate in the cochlear nucleus in the brainstem.

(3) Some axons from the neurons in the cochlear nucleus project to the inferior colliculi, and others project to the superior olivary nucleus.

(4) Axons from the inferior colliculus project to the medial geniculate nucleus of the thalamus.

(5) Thalamic neurons project to the primary auditory cortex, where the impulse is perceived as sound.

Figure 19.30

Central Nervous System Pathways for Hearing. (1) The cochlear nerve branch of CN VIII conducts impulses generated by the detection of sounds to (2) the cochlear nucleus in the brainstem. (3) Impulses are conducted to both the inferior colliculus and the superior olivary nucleus. (4) The medial geniculate nucleus within the thalamus receives impulses from the inferior colliculus. (5) Following processing, neurons within the thalamus transmit the information to the primary auditory cortex.

Auditory Pathways

In the previous section, we discussed how sound waves travel through the inner ear. But how are those sound waves transferred to the brain, where they are interpreted as sound? This auditory pathway is shown in **figure 19.30** and described here:

1. When the basilar membrane bounces, the stereocilia on spiral organ hair cells bend against the tectorial membrane, producing a nerve impulse. The nerve impulse travels through the cochlear branch that attaches to the hair cells. (Recall that the cell bodies of these sensory neurons are located in the spiral ganglia, medial to the cochlear duct in the modiolus.) The cochlear branch and the vestibular branch merge to form the vestibulocochlear nerve (CN VIII).

2. The sensory axons of the cochlear nerve (the "primary neurons" in this sensory pathway) terminate in the paired **cochlear nuclei** within the brainstem. These sensory neurons synapse with secondary neurons housed within this nucleus.

3. After integration and processing of incoming information within the cochlear nucleus, axons from secondary neurons in this nucleus project to both the inferior colliculi within the mesencephalon and the superior olivary nuclei within the myelencephalon.

4. Thereafter, inferior colliculi neurons extend axons to the medial geniculate nucleus of the thalamus.

5. Neurons reaching the medial geniculate nucleus of the thalamus synapse with tertiary neurons in the thalamus; axons from the tertiary neurons extend to the primary auditory cortex in the temporal lobe. The nerve impulses in the primary auditory cortex are perceived as sounds.

Development of the Ear

Ear development begins during the fourth week **(figure 19.31)**. All embryonic germ layers contribute to ear formation, including the surface ectoderm and the layer that is beginning to specialize as the neuroectoderm. Each part of the ear originates separately.

External and Middle Ear Development

The auricle of the external ear forms from the surface ectoderm. Masses of tissue from the first and second pharyngeal arches come together and fuse, forming the unique shape of the auricle. Each external acoustic meatus forms from an external indentation of the ectoderm called the **first pharyngeal cleft** (or *groove*). The tympanic cavity and the auditory tube of each middle ear are formed from an internal tubelike expansion of the pharynx called the **first pharyngeal pouch.** (Eventually, four pharyngeal clefts and pouches form, but only the first pharyngeal cleft and pouch develop into external and middle ear structures.) Specifically, as the first pharyngeal pouch grows, it develops into a trumpet-shaped structure called the **tubotympanic recess.**

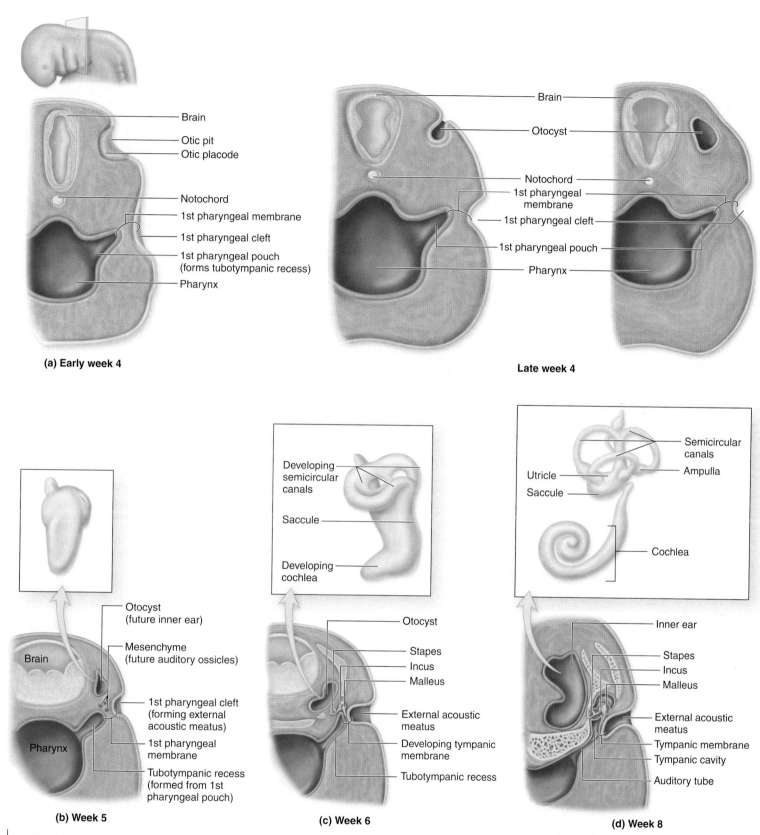

(a) Early week 4

Late week 4

(b) Week 5

(c) Week 6

(d) Week 8

Figure 19.31

Development of the Ear. *(a)* Ear development begins during week 4, when the otic placode invaginates and turns into an otocyst (future inner ear). The first pharyngeal cleft, membrane, and pouch develop and form middle and external ear structures. *(b)* By week 5, a tubotympanic recess forms from the first pharyngeal pouch, and the first pharyngeal cleft begins to form the external acoustic meatus. *(c)* The auditory ossicles form beginning in week 6, and the first pharyngeal membrane forms the tympanic membrane (eardrum). *(d)* By week 8, the otocyst forms a recognizable inner ear, and the tubotympanic recess forms the auditory tube and tympanic cavity of the middle ear.

CLINICAL VIEW

Cochlear Implants

A **cochlear implant** is an electronic device that assists some hearing-impaired people. It is not a hearing aid and does not restore normal hearing; rather, it compensates for damaged or nonfunctioning parts of the inner ear. A cochlear implant can help some deaf individuals learn to speak.

The components of a cochlear implant include: (1) an external microphone to detect sound (typically worn behind one ear), (2) a speech processor to arrange the sounds from the microphone, (3) a transmitter connected to a receiver/stimulator (placed within the cochlea) to convert the processed sound into electrical impulses, and (4) some form of electrode that sends the impulses from the stimulator to the brain for interpretation.

A cochlear implant parallels normal hearing by selecting sounds, processing them into electrical signals, and then sending sound information to the brain for interpretation. However, this type of "hearing" is considerably different from normal hearing. Patients with implants report that voices sound squeaky and high-pitched, similar to those of mouse cartoon characters; however, this limitation does not prevent people with implants from having oral communication capabilities.

As of 2002, 60,000 people worldwide had received cochlear implants, but some controversy accompanies their increasing use. First, any surgical installation procedure is accompanied by a small risk for infection or complication. Second, some implant users never hear well enough to speak, while many others require further assistance with language development. Finally, cochlear implant opponents have expressed concern that the implant threatens the unique, sign-based culture of the deaf person. Others question the use of implants in children, who may not have made the decision for themselves. Cochlear implant supporters, on the other hand, argue that the procedure facilitates oral or manual language development, which is a precursor to cognitive maturation.

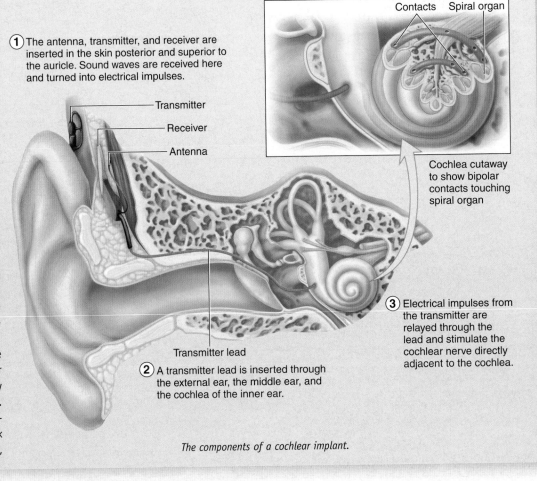

① The antenna, transmitter, and receiver are inserted in the skin posterior and superior to the auricle. Sound waves are received here and turned into electrical impulses.

Transmitter
Receiver
Antenna

Contacts   Spiral organ

Cochlea cutaway to show bipolar contacts touching spiral organ

③ Electrical impulses from the transmitter are relayed through the lead and stimulate the cochlear nerve directly adjacent to the cochlea.

Transmitter lead

② A transmitter lead is inserted through the external ear, the middle ear, and the cochlea of the inner ear.

The components of a cochlear implant.

The narrow part of this "trumpet" forms the auditory tube, while the expanded end forms the tympanic cavity of the middle ear.

As the first pharyngeal pouch and the first pharyngeal cleft grow toward one another, their inner endoderm and outer ectoderm form the **first pharyngeal membrane,** which later thins and forms the tympanic membrane. The auditory ossicles develop from mesenchyme cells that are near, but not inside, the tympanic cavity. During the ninth month of development, the tympanic cavity expands and envelops the ossicles, thereby enclosing these ear bones within the tympanic cavity.

Inner Ear Development

Each inner ear forms from an **otic placode** (plak´ōd; *plax* = flat) that is formed from ectoderm and is at the level of the rhombencephalon (hindbrain). Each otic placode invaginates, forming an **otic pit.** Eventually, each otic pit closes off and becomes an internal structure called an **otocyst** (ō´tō-sist; *kystis* = bladder), or *otic vesicle.*

The otocyst is initially very primitive. During the fifth and sixth weeks of development, the otocyst begins to expand. Its anterior tip elongates and starts to form the future cochlear duct. The utricle and saccule become better differentiated. The hair cells of the spiral organ start to develop by week 7. Also during the seventh week, the three semicircular ducts form along with the ampulla structures within them. The shape of the membranous labyrinth is attained by the end of the eighth week. As the petrous part of the temporal bone (bony labyrinth) develops around the membranous labyrinth, the space between the two labyrinths fills with perilymph.

WHAT DID YOU LEARN?

⑬ What is the function of the auditory ossicles?

⑭ What is the difference between the bony and membranous labyrinths?

⑮ What is the structure of a macula, and what is its function?

⑯ In general, how are the hair cells in the spiral organ stimulated?

CLINICAL TERMS

conduction deafness Deafness due to middle ear conditions that prevent vibration transfer between the tympanic membrane and the oval window.

keratitis (ker-ă-tī′tis; *keras* = horn, *itis* = inflammation) Inflammation of the cornea.

Meniere disease An equilibrium disturbance characterized by vertigo, nausea, vomiting, tinnitus, and progressive hearing loss; it is caused by distortion of the membranous labyrinth of the endolymphatic duct due to excessive accumulation of fluid.

nystagmus (nis-tag′mŭs; a nodding) Involuntary, rhythmic oscillation of the eyeballs; may occur after damage to the brainstem.

otitis externa External acoustic meatus inflammation; often called "swimmer's ear," because water immersion can predispose this condition.

otosclerosis (ō′tō-sklē-rō′sis; *sclerosis* = hardening) Condition in which abnormal, microscopic bone growth in the inner ear traps the stapes in the oval window.

sensorineural deafness (sen′sōr-i-noor′ăl) Deafness caused by problems within the cochlea or along the auditory nerve pathway.

tinnitus (ti-nī′tŭs; *tinnio* = to jingle) Ear noises (ringing, roaring, or hissing).

vertigo (ver′ti-gō; *verto* = to turn) Sensation of rotation, such as spinning or whirling.

CHAPTER SUMMARY

	■ The body detects numerous stimuli continuously. Stimuli may be pleasurable, alert us to danger, or provide information about moment-to-moment changes in our environment.
	■ A stimulus is a change in the internal or external environment that is detected by a receptor and causes a response. Impulses caused by a stimulus arrive in the CNS and are perceived as a sensation.
Receptors 560	■ Receptors detect changes in the external and internal environments, and range in complexity from single-celled to multicellular.
	Classification of Receptors 561
	■ General sense receptors are housed in the skin, musculoskeletal organs, and viscera. Special sense receptors are housed within the head.
	■ Receptors are defined by stimulus origin (exteroceptors, interoceptors, proprioceptors); receptor distribution (general senses and special senses); and modality of stimulus.
	■ Receptors for general senses have a specificity for chemicals (chemoreceptors); temperature (thermoreceptors); light (photoreceptors); touch, stretch, vibration, and pressure (mechanoreceptors and baroreceptors); and pain (nociceptors).
Tactile Receptors 564	■ Tactile receptors react to touch, pressure, and vibration stimuli. Tactile receptors may be either unencapsulated or encapsulated.
	■ Free nerve endings are unencapsulated dendrites that detect pain and temperature stimuli, as well as some light touch and pressure.
	Unencapsulated Tactile Receptors 564
	■ These tactile receptors are relatively simple in structure and have no connective tissue wrapping.
	Encapsulated Tactile Receptors 565
	■ Tactile receptors covered either by connective tissue or glial cells.
Gustation 567	■ Gustation is the chemical sense of taste. Gustatory cells are housed in taste buds that also have supporting cells and basal cells.
	Gustatory Discrimination 568
	■ The five basic taste sensations are salty, sweet, sour, bitter, and umami.
	Gustatory Pathways 568
	■ Cranial nerves CN VII (facial) and CN IX (glossopharyngeal) are involved in gustation.
Olfaction 569	■ Olfaction is the chemical sense of smell. Paired olfactory organs are composed of a lining olfactory epithelium (olfactory receptor cells, supporting cells, and basal cells), a lamina propria, and olfactory glands.
	Olfactory Receptor Cells 569
	■ Olfactory receptor cells are bipolar neurons with olfactory hairs.
	Olfactory Discrimination 570
	■ The olfactory system recognizes different chemical stimuli. The number of olfactory receptors declines as we age.
	Olfactory Pathways 571
	■ Olfactory receptor cell axons enter a pair of olfactory bulbs. Neurons within the olfactory bulbs project axon bundles called olfactory tracts to the olfactory cortex in the cerebrum.

(continued on next page)

CHAPTER SUMMARY (continued)

Vision   571

- The eyes house photoreceptors that detect light, color, and movement.

Accessory Structures of the Eye   571

- Accessory structures that cover and protect the eye include the conjunctiva, eyebrows, eyelashes, eyelids, and lacrimal glands.
- Each lacrimal apparatus includes a lacrimal gland and a series of channels and ducts that distribute and drain thin, watery lacrimal fluid (tears) to cleanse the eye surface and moisten the conjunctiva.

Eye Structure   573

- The eye is an almost spherical, hollow organ that has three principal layers, called tunics.
- The fibrous tunic (external layer), which has an anterior cornea and a posterior sclera, helps focus light on the retina; protects and maintains the shape of the eye; and provides the attachment site for the extrinsic eye muscles (sclera).
- The vascular tunic (middle layer) has three regions: the choroid, which distributes vascular and lymphatic vessels inside the eye; the ciliary body, which assists lens shape changes with the suspensory ligaments and secretes aqueous humor; and the iris, which controls pupil diameter.
- The retina is composed of an outer pigmented layer and an inner neural layer that houses all of the photoreceptors and their associated neurons.
- The three cell layers in the neural layer are (1) photoreceptors (rods and cones) that are stimulated by photons of light; (2) supporting cells and neurons that process and begin to integrate incoming visual stimuli; and (3) an inner layer of ganglion cells that conduct impulses to the brain.
- The retina has a posterior, yellowish region called the macula lutea, which houses a concentration of cones. Vision is most acute at a depression within the center of the macula lutea called the fovea centralis.
- The lens is a hard, deformable, transparent structure bounded by a dense fibrous, elastic capsule.

Visual Pathways   581

- The optic nerves formed by ganglionic axons that project from each eye through the paired optic foramina converge at the optic chiasm.
- The optic tract is the continuation of nerve fibers from the optic chiasm. Each optic tract is a composite of ganglionic axons originating from the retinas of each eye.

Development of the Eye   581

- Eye development begins during the fourth week with the evagination of the forebrain that forms optic vesicles.
- Optic vesicles invaginate and form a double-layered epithelium called the optic cup.
- The lens develops from the lens vesicle.

Equilibrium and Hearing   584

- Receptors in the ear provide for the senses of equilibrium and hearing.

External Ear   584

- The external ear is a skin-covered, cartilage-supported structure called the auricle.
- The external acoustic meatus directs sound waves to the tympanic membrane, a delicate epithelial sheet that partitions the spaces of the outer and middle ear. Its vibrations provide the means for transmission of sound wave energy to the inner ear.

Middle Ear   585

- The tympanic cavity (middle ear) is an air-filled space occupied by three small auditory ossicles: the malleus, the incus, and the stapes, which transmit sound wave energy from the outer to the inner ear.

Inner Ear   586

- The inner ear houses the structures for equilibrium and hearing. Specialized receptors called hair cells are housed in the membranous labyrinth, which lies within a cavernous space in dense bone called the bony labyrinth.
- Hair cells are the sensory receptors for equilibrium and hearing. When their apical surface stereocilia, and sometimes kinocilia, are displaced, a change occurs in neurotransmitter release and the firing rate of the monitoring sensory neuron.
- The saccule and utricle are membranous sacs of the vestibular complex that contain areas of hair cells called maculae, which detect acceleration or deceleration of the head in one direction.
- The semicircular canals of the vestibular complex house the membranous semicircular ducts, each with an ampulla that houses the hair cells for detecting rotational movements of the head.
- Equilibrium stimuli from the vestibule and semicircular canals are conducted through the vestibular branch of the vestibulocochlear nerve (CN VIII).
- Hearing organs are housed within the cochlea.
- Sound is perceived when impinging sound waves cause vibrations of the tympanic membrane, resulting in auditory ossicle vibrations that lead to vibrations of the basilar membrane. Movement of the basilar membrane causes stereocilia distortion, which is perceived as sound.
- Sound waves transmitted in the fluid of the inner ear are detected by hair cells of the spiral organ. Impulses are conducted through the cochlear branch of the vestibulocochlear nerve (CN VIII) to the cochlear nuclei in the brainstem.

Development of the Ear   596

- Ear development begins during the fourth week. All embryonic germ layers contribute to the formation of the ear, including the surface ectoderm and the layer that is beginning to specialize as the neuroectoderm. Each part of the ear (outer, middle, and inner) originates separately.

CHALLENGE YOURSELF

Matching

Match each numbered item with the most closely related lettered item.

_____ 1. proprioceptors
_____ 2. auditory ossicles
_____ 3. endolymph
_____ 4. nociceptor
_____ 5. basilar membrane
_____ 6. choroid
_____ 7. gustation
_____ 8. suspensory ligaments
_____ 9. adaptation
_____ 10. olfaction

a. sensation of taste
b. supports the spiral organ
c. detects pain stimuli
d. sensitivity reduced due to constant stimulus
e. attaches to lens
f. detects tension in tendons, position of joints
g. sensation of smell
h. fluid in the membranous labyrinth
i. malleus, incus, stapes
j. pigmented middle layer of eye

Multiple Choice

Select the best answer from the four choices provided.

_____ 1. Unencapsulated, terminal branches of dendrites are called
  a. lamellated corpuscles.
  b. free nerve endings.
  c. organs of Ruffini.
  d. Krause bulbs.

_____ 2. Baroreceptors are a class of mechanoreceptor that respond to
  a. light touch.
  b. pain stimuli.
  c. increase in muscle tension.
  d. changes in pressure.

_____ 3. Which sensory structure has stereocilia of hair cells embedded in a gelatinous structure called a cupula?
  a. semicircular duct
  b. saccule
  c. cochlear duct
  d. utricle

_____ 4. The photoreceptors that perceive color and sharp vision are
  a. amacrine cells.
  b. rods.
  c. horizontal cells.
  d. cones.

_____ 5. The tarsal glands secrete
  a. a high-salt fluid to prevent endolymph from becoming perilymph.
  b. a low-protein product into the aqueous humor.
  c. a lipid-rich product that prevents the eyelids from sticking together.
  d. a mucus-rich product that maintains the moistness of the olfactory cilia.

_____ 6. The arrangement of tunics in the eye, from the center of the eye to the periphery, is
  a. retina, vascular, fibrous.
  b. vascular, retina, fibrous.
  c. vascular, fibrous, retina.
  d. retina, fibrous, vascular.

_____ 7. Receptors in the walls of blood vessels that respond to discrete changes in gas concentration in the blood are called
  a. gustatory receptors.
  b. chemoreceptors.
  c. thermoreceptors.
  d. nociceptors.

_____ 8. The only sensations to reach the cerebral cortex without first processing through the synapses in the thalamus are
  a. pain.
  b. olfaction.
  c. proprioception.
  d. touch.

_____ 9. The lacrimal glands produce lacrimal fluid for each of the following functions except
  a. cleansing the eye surface.
  b. preventing bacterial infection.
  c. humidifying the eye orbit.
  d. moistening the eye surface.

_____ 10. Which statement is true about the cochlear duct?
  a. It is part of the bony labyrinth.
  b. It is filled with perilymph.
  c. It contains hair cells that convert sound waves into nerve impulses.
  d. It contains a spiral organ that rests on a vestibular membrane.

Content Review

1. What are the classifications of receptors according to modality of stimulus? Give an anatomic example of each.

2. How are visceral nociceptors different from somatic nociceptors, and how do they relate to the phenomenon known as "referred pain"?

3. What are the pathways by which taste sensations reach the brain?

4. Describe the pathway by which olfactory stimuli travel from the nasal cavity to the brain.

5. What structures in the wall of the eye help control the amount of light entering the eye?

6. How is the lens able to focus images from a book that you are reading, and then immediately also focus the image of children playing in your backyard?

7. Discuss the formation, circulation, and reabsorption of aqueous humor in the eye.

8. Where are the tensor tympani and the stapedius located, and what important function do they perform?

9. Briefly describe the structural relationship between the membranous labyrinth and the bony labyrinth.

10. Describe the pathway by which sound waves enter and exit the ear, and how that sound is converted into a nerve impulse.

Developing Critical Reasoning

1. Savannah is an active 3-year-old who began to cough and sniffle. Then she experienced an earache and a marked decrease in hearing acuity. During a physical examination, her pediatrician noted the following signs: elevated temperature, a reddened tympanic membrane with a slight bulge, and some inflammation in the pharynx. What might the pediatrician call this condition, and how would it be treated? How does Savannah's age relate to her illness?

2. After Alejandro quit smoking, he discovered that foods seemed much more flavorful than when he had smoked. How are smoking and taste perceptions linked?

A N S W E R S   T O   " W H A T   D O   Y O U   T H I N K ? "

1. A cutaneous pain receptor would be classified as an exteroceptor (based on stimulus origin), a general somatic receptor (based on receptor distribution), and a nociceptor (based on modality of stimulus).

2. Olfaction plays a major role in detecting tastes. Most taste is due to our perception of the odor, rather than the taste, of the food. If your nose is stuffed up and you can't smell the food, your taste perception is impaired as well.

3. When a person cries, tears flood the lacrimal apparatus. The lacrimal puncta are unable to collect all the tears, which then flow over the lower eyelids. In addition, the nasolacrimal duct pours many more tears than normal into the nasal cavity where they mix with mucus, resulting in the runny nose.

4. Air pressure increases when the airplane descends, which is why you may feel greater pressure in your ears. The "popping" noise results from your auditory tubes opening and releasing some of that pressure in your middle ear.

Visit the McKinley/O'Loughlin *Human Anatomy*, 2e website at aris.mhhe.com

OUTLINE

Anatomy &
Physiology | **REVEALED®**
aprevealed.com

ENDOCRINE SYSTEM

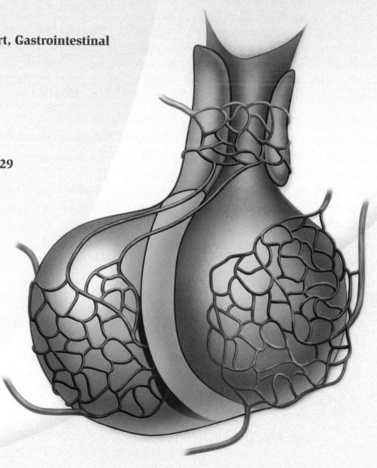

Endocrine System

If you haven't eaten in a while, your blood glucose level becomes low, leaving you groggy and mentally slow. When you eat a meal and it courses through your digestive system, the bloodstream takes up glucose, and your body tissues become newly energized. The **endocrine system** has come to your rescue to help your body achieve homeostasis! Specifically, an endocrine organ called the pancreas has secreted the hormone insulin into the bloodstream to stimulate your body cells to take up glucose for storage or energy production.

This chapter begins with an overview of the endocrine organs (called glands) and the endocrine cells, and then describes how they regulate body functions as well as what happens when they malfunction. We also explore the endocrine functions of other body organs, such as the kidneys, heart, GI tract, and gonads.

Endocrine Glands and Hormones

Key topics in this section:

- Similarities and differences between the endocrine system and the nervous system
- Hormone classification based on chemical structure
- How feedback loops regulate hormone secretion

We have learned in earlier chapters that exocrine glands usually produce secretions that are released into ducts opening onto an epithelial surface. In contrast, **endocrine glands** are ductless organs that secrete their molecular products directly into the bloodstream. All endocrine cells, whether organized into a single organ or scattered in small clusters within other organs, are located within highly vascularized areas to ensure that their products enter the bloodstream immediately.

The endocrine system and the nervous system often work together to bring about homeostasis in the body. Both affect specific target organs. However, their communication methods and their effects differ. For example, the endocrine system uses hormones, while the nervous system uses nerve impulses to transmit information. **Table 20.1** lists several similarities and differences between the endocrine and nervous systems.

Overview of Hormones

The endocrine glands release **hormones** (hōr'mōn; *hormao* = to rouse) into the bloodstream **(figure 20.1)**. Hormones are molecules that have an effect on specific organs. Only cells with specific receptors for the hormone (enabling the hormone to bind to the cell) respond to that hormone. These cells are called **target cells,** and the organs that contain them are called **target organs.** In contrast, organs, tissues, or cells that do not have the specific receptor for a hormone do not bind or attach the hormone and do not respond to its stimulating effects. One example is prolactin, a hormone produced by the anterior pituitary. The target organ for this hormone is the mammary gland; that is, some cells in the mammary gland have receptors that pick up and respond to the prolactin by producing milk to nourish an infant. However, other organs (such as the heart and lungs) do not respond to prolactin because they do not have receptors that bind this hormone. The study of the structural components of the endocrine system, the hormones they produce, and the effects of those hormones on target organs is termed **endocrinology** (en'dō-kri-nol'ō-jē; *endo* = within, *krino* = to separate).

A hormone's structure determines how it interacts with the target cells. Hormones are classified based on their chemical structure into three distinct groups:

- **Peptide** (pep'tīd) **hormones** are formed from chains of amino acids. Most of our body's hormones are peptide hormones. Longer chains are called *protein hormones.* An example is growth hormone.
- **Biogenic amines** (bī'ō-jen-ik ă-mēn', am'in) are small molecules produced by altering the structure of a specific amino acid. An example is thyroid hormone.
- **Steroid** (ster'oyd) **hormones** are a type of lipid derived from cholesterol. An example is testosterone.

Negative and Positive Feedback Loops

Hormone levels are regulated by a self-adjusting mechanism called feedback, meaning that the product of a pathway acts back at an

Table 20.1	**Comparison of the Endocrine System and the Nervous System**	
Features	**Endocrine System**	**Nervous System**
Communication Method	Secretes hormones into bloodstream; hormones are distributed to target cells throughout body	Neurotransmitter release causes nerve impulse between neuron and its target cell
Target of Stimulation	Any cell in the body with a receptor for the hormone	Other neurons, muscle cells, and gland cells
Response Time	Relatively slow reaction time: seconds to minutes to hours	Rapid reaction time: milliseconds or seconds (sometimes minutes)
Effect of Stimulation	Causes metabolic activity changes in target cells	Causes contraction of muscles or secretion from glands
Range of Effect	Typically has widespread, general effects throughout the body	Typically has localized, specific effects in the body
Duration of Response	Long-lasting: minutes to days to weeks; may continue after stimulus is removed	Short-term: milliseconds; terminates with removal of stimulus
Recovery Time	Slow return to prestimulation level of activity	Rapid, immediate return to prestimulation level of activity

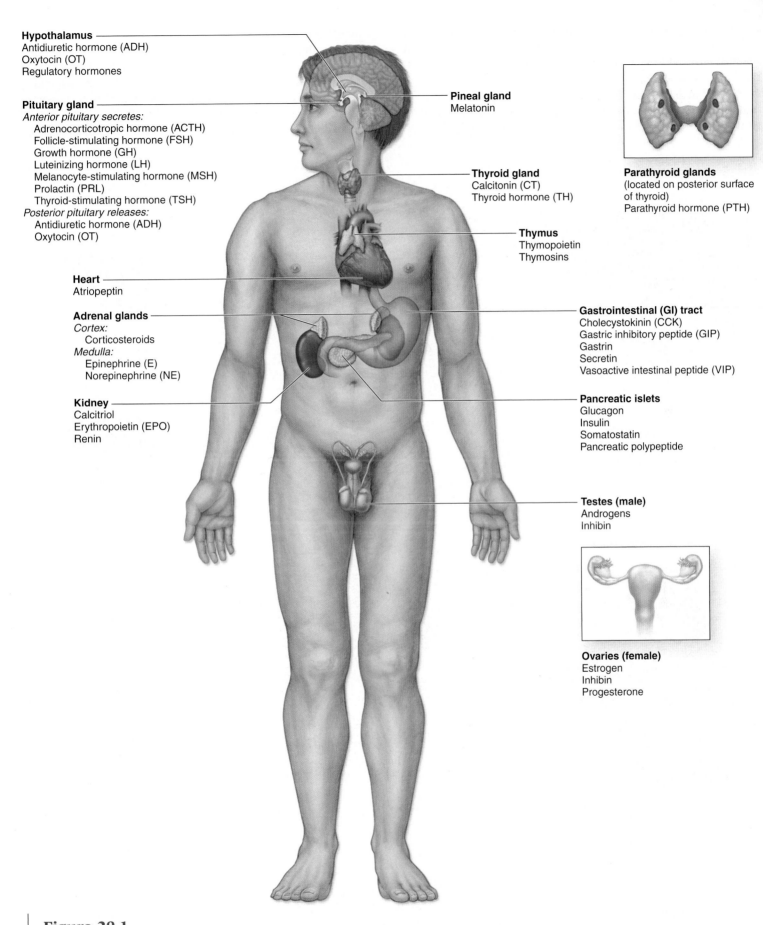

Hypothalamus
Antidiuretic hormone (ADH)
Oxytocin (OT)
Regulatory hormones

Pituitary gland
Anterior pituitary secretes:
    Adrenocorticotropic hormone (ACTH)
    Follicle-stimulating hormone (FSH)
    Growth hormone (GH)
    Luteinizing hormone (LH)
    Melanocyte-stimulating hormone (MSH)
    Prolactin (PRL)
    Thyroid-stimulating hormone (TSH)
Posterior pituitary releases:
    Antidiuretic hormone (ADH)
    Oxytocin (OT)

Heart
Atriopeptin

Adrenal glands
Cortex:
    Corticosteroids
Medulla:
    Epinephrine (E)
    Norepinephrine (NE)

Kidney
Calcitriol
Erythropoietin (EPO)
Renin

Pineal gland
Melatonin

Thyroid gland
Calcitonin (CT)
Thyroid hormone (TH)

Thymus
Thymopoietin
Thymosins

Parathyroid glands
(located on posterior surface
of thyroid)
Parathyroid hormone (PTH)

Gastrointestinal (GI) tract
Cholecystokinin (CCK)
Gastric inhibitory peptide (GIP)
Gastrin
Secretin
Vasoactive intestinal peptide (VIP)

Pancreatic islets
Glucagon
Insulin
Somatostatin
Pancreatic polypeptide

Testes (male)
Androgens
Inhibin

Ovaries (female)
Estrogen
Inhibin
Progesterone

Figure 20.1

Endocrine System.   Endocrine glands and other organs that contain endocrine cells are found throughout the body. They produce and release various types of hormones.

earlier step in the pathway to regulate the pathway's activities. This pattern is circular, so is often called a **feedback loop.** There are two types of feedback loops: negative feedback loops and positive feedback loops **(figure 20.2)**.

In a **negative feedback loop,** a stimulus starts a process, and eventually either the hormone that is secreted or a product of its effects causes the process to slow down or turn off. Most hormonal systems work by negative feedback mechanisms. One example is the regulation of the blood glucose level in the body (figure 20.2a). A normal blood glucose level exists when the body is at homeostasis for energy production. Eating food is a stimulus that eventually causes the blood glucose level to rise. In response, the pancreas secretes a hormone called insulin that helps reduce the blood glucose level. Insulin acts in two ways: It tells target cells

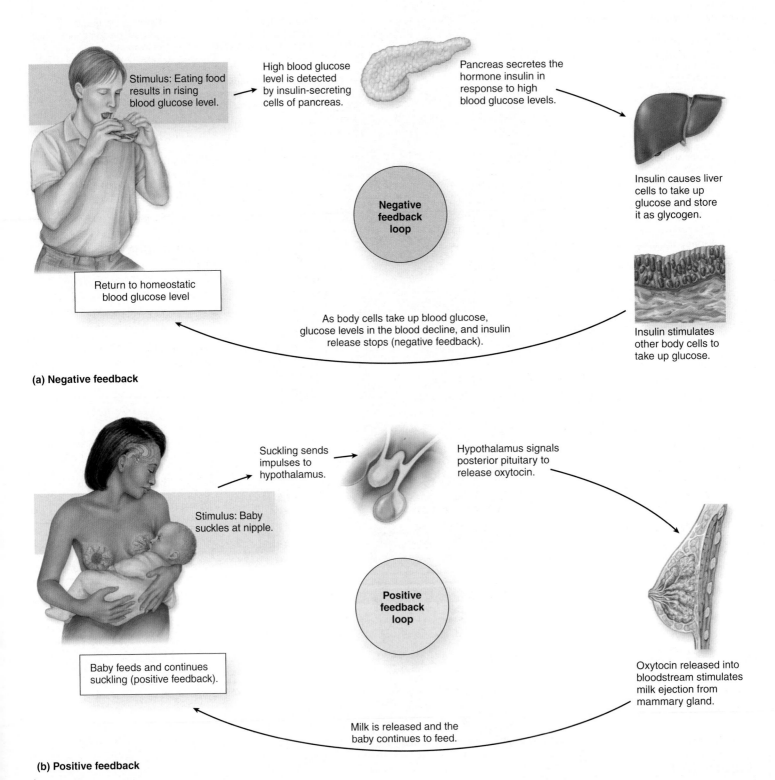

Stimulus: Eating food results in rising blood glucose level.

High blood glucose level is detected by insulin-secreting cells of pancreas.

Pancreas secretes the hormone insulin in response to high blood glucose levels.

Insulin causes liver cells to take up glucose and store it as glycogen.

Negative feedback loop

Return to homeostatic blood glucose level

As body cells take up blood glucose, glucose levels in the blood decline, and insulin release stops (negative feedback).

Insulin stimulates other body cells to take up glucose.

(a) Negative feedback

Suckling sends impulses to hypothalamus.

Hypothalamus signals posterior pituitary to release oxytocin.

Stimulus: Baby suckles at nipple.

Positive feedback loop

Baby feeds and continues suckling (positive feedback).

Oxytocin released into bloodstream stimulates milk ejection from mammary gland.

Milk is released and the baby continues to feed.

(b) Positive feedback

Figure 20.2

Negative and Positive Feedback Loops in the Endocrine System. The initial step in any feedback pathway is the stimulus. (*a*) A negative feedback loop occurs when the end product of a pathway acts to turn off or slow down the pathway. (*b*) A positive feedback loop is involved when the end product of a pathway stimulates further pathway activity.

in the liver to take up glucose in the blood and store it in the form of glycogen, and it stimulates other body target cells to take up the glucose for use in metabolism. As the blood glucose level declines, a normal blood glucose level is reached. Once this homeostatic level is achieved, insulin secretion stops, and glucose uptake by target cells ceases.

A **positive feedback loop** accelerates the original process, either to ensure that the pathway continues to run or to speed up its activities. Only a few positive feedback loops occur in the human body. One example is the process of milk release from the mammary glands (figure 20.2b). The initial stimulus is the baby suckling at the nipple, which sends nerve impulses to the hypothalamus. In turn, the hypothalamus signals the posterior pituitary to release the hormone oxytocin. When oxytocin is released into the bloodstream, it acts on the mammary gland cells and stimulates milk release. As milk is released, the baby continues to suckle, which continues to stimulate the hypothalamus. This process goes on until the baby stops suckling or all the milk is expelled from the breast. Then milk release ceases.

Many of the body's feedback mechanisms are much more complex than the examples just given, usually involving multiple steps or multiple feedback loops. Complex loops are the most common self-adjusting regulatory mechanisms because they permit an exquisite fine-tuning of the process, not just an all-or-none effect.

WHAT DID YOU LEARN?

1 What general name identifies cells that respond to a specific hormone?

2 What is the purpose of a negative feedback loop?

Hypothalamic Control of the Endocrine System

Key topics in this section:

- How the hypothalamus controls the anterior pituitary
- The hypothalamus produces the posterior pituitary hormones
- Stimulation of the adrenal medulla by the hypothalamus

As the master control center of the endocrine system (see chapter 15), the hypothalamus (in the inferior region of the diencephalon) oversees most endocrine activity, and it does so in three ways (**figure 20.3**).

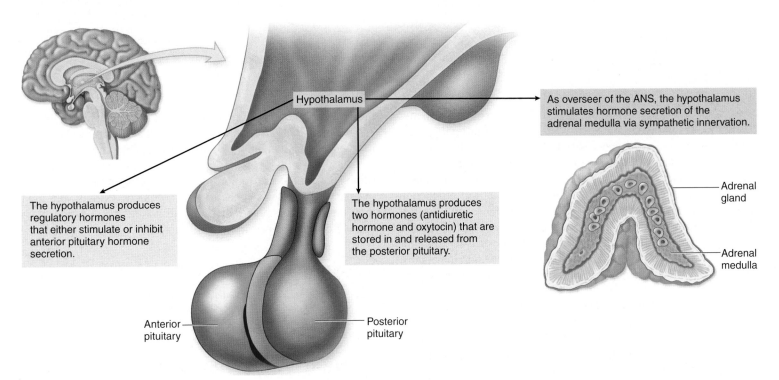

Hypothalamus

As overseer of the ANS, the hypothalamus stimulates hormone secretion of the adrenal medulla via sympathetic innervation.

The hypothalamus produces regulatory hormones that either stimulate or inhibit anterior pituitary hormone secretion.

The hypothalamus produces two hormones (antidiuretic hormone and oxytocin) that are stored in and released from the posterior pituitary.

Adrenal gland

Adrenal medulla

Anterior pituitary

Posterior pituitary

Figure 20.3
How the Hypothalamus Controls Endocrine Function. Hypothalamic influence on the anterior pituitary, posterior pituitary, and adrenal medulla regulates key hormones.

Table 20.2	Regulatory Hormones Secreted by the Hypothalamus			
Hormone	**Source**	**Hormone Target**	**Hormone Effects**	**Related Disorders**
RELEASING HORMONES				
Corticotropin-releasing hormone (CRH)	Neuroendocrine neurons in paraventricular nucleus	Corticotropic cells in pars distalis of anterior pituitary	Increases secretion of adrenocorticotropic hormone (ACTH)	Hyposecretion: Addison disease Hypersecretion: Cushing syndrome, androgenital syndrome
Gonadotropin-releasing hormone (GnRH)	Neuroendocrine neurons in medial preoptic and arcuate nuclei	Gonadotropic cells in pars distalis of anterior pituitary	Increases secretion of follicle-stimulating hormone (FSH) and luteinizing hormone (LH)	Hyposecretion: Kallman syndrome Hypersecretion: Precocious puberty
Growth hormone-releasing hormone (GHRH); also called somatocrinin	Neuroendocrine neurons in arcuate nucleus	Somatotropic cells in pars distalis of anterior pituitary	Increases secretion of growth hormone (GH)	Hyposecretion: Pituitary dwarfism Hypersecretion: Giantism, acromegaly
Prolactin-releasing factor (PRF)	Hypothalamus	Mammotropic cells in pars distalis of anterior pituitary	Increases secretion of prolactin (PRL)	Hyposecretion: Menstrual disorders, delayed puberty, male infertility Hypersecretion: Reduced menstruation, anovulatory infertility, male impotence
Thyrotropin-releasing hormone (TRH)	Neuroendocrine neurons in paraventricular and anterior hypothalamic nuclei	Thyrotropic cells in pars distalis of anterior pituitary	Increases secretion of thyroid-stimulating hormone (TSH)	Hyposecretion: Hypothyroidism Hypersecretion: Hyperthyroidism
INHIBITING HORMONES				
Growth hormone-inhibiting hormone (GHIH); also called somatostatin	Neuroendocrine neurons in periventricular nucleus	Somatotropic cells in pars distalis of anterior pituitary	Decreases secretion of growth hormone (GH)	Hyposecretion: Giantism, acromegaly Hypersecretion: Pituitary dwarfism
Prolactin-inhibiting hormone (PIH)	Neuroendocrine neurons in arcuate nucleus	Mammotropic cells in pars distalis of anterior pituitary	Decreases secretion of prolactin (PRL)	Hyposecretion: Reduced menstruation, anovulatory infertility, male impotence Hypersecretion: Menstrual disorders, delayed puberty, male infertility

First, special cells in the hypothalamus secrete hormones that influence the secretory activity of the anterior pituitary gland. These hormones are called **regulatory hormones** because they are molecules secreted into the blood to regulate secretion of most anterior pituitary hormones. Regulatory hormones fall into one of two groups: **Releasing hormones (RH)** stimulate the production and secretion of specific anterior pituitary hormones, and **inhibiting hormones (IH)** deter the production and secretion of specific anterior pituitary hormones (table 20.2). Because the anterior pituitary secretes hormones that influence many endocrine organs, the hypothalamus has indirect control over these endocrine organs as well.

Second, the hypothalamus produces two hormones that are transported to and stored in the posterior pituitary: antidiuretic hormone (ADH) and oxytocin (OT). The posterior pituitary does not synthesize any hormones. It only releases the two hypothalamic hormones transported to it.

Finally, because the hypothalamus is also the master control center of the autonomic nervous system (ANS), it directly oversees the stimulation and hormone secretion of the adrenal medulla. The adrenal medulla is an endocrine structure that secretes its hormones in response to stimulation by the sympathetic nervous system.

Although the hypothalamus oversees most endocrine activity, some endocrine cells are not under its direct control. For example, the parathyroid glands release their hormones without any input from the hypothalamus; rather, they respond directly to concentrations of chemical levels in the bloodstream.

WHAT DO YOU THINK?

1 If a patient had a damaged hypothalamus, how might this affect the workings of the endocrine system?

WHAT DID YOU LEARN?

3 What types of hormones does the hypothalamus secrete to regulate the functioning of the anterior pituitary?

Pituitary Gland

Key topics in this section:

- Anatomy of the anterior and posterior pituitary
- Hormones of the anterior and posterior pituitary and their functions
- Relationships between the hypothalamus and the pituitary gland

The **pituitary** (pi-too′i-tār-ē) **gland,** or *hypophysis* (hī-pof′i-sis; undergrowth), lies inferior to the hypothalamus **(figure 20.4).** The small, slightly oval gland is housed within the hypophyseal fossa in the sella turcica of the sphenoid bone. The pituitary gland is connected to the hypothalamus by a thin stalk, the **infundibulum** (in-fŭn-dib′ū-lŭm; a funnel). The infundibulum extends from the base of the hypothalamus at the **median eminence,** a conical projection sandwiched between the optic chiasm anteriorly and the mammillary bodies posteriorly. The pituitary gland is covered superiorly by the diaphragma sellae, which is a cranial dural septa that ensheathes the stalk of the infundibulum to restrict pituitary gland movement (see figure 15.5).

The pituitary gland is partitioned both structurally and functionally into an anterior pituitary and a posterior pituitary (sometimes referred to as just *anterior lobes* and *posterior lobes,* respectively). Both regions are derived from different embryonic structures (see later in this chapter).

Anterior Pituitary

Most of the pituitary gland is composed of the **anterior pituitary,** or *adenohypophysis* (ad′ĕ-nō-hī-pof′i-sis; *adenos* = gland), which is the part of the pituitary gland that both produces and secretes hormones. It is partitioned into three distinct areas (figure 20.4). The **pars distalis** is the large anterior portion of the anterior pituitary. A thin **pars intermedia** is a scant region between the pars distalis and the posterior pituitary. The histology of the pituitary gland is shown in **figure 20.5.** The **pars tuberalis** is a thin wrapping around the infundibular stalk (a posterior pituitary structure), which is also called the infundibulum.

Like all endocrine glands, the anterior pituitary has an extensive distribution of many blood vessels to facilitate uptake and transport of hormones.

Control of Anterior Pituitary Secretions

The anterior pituitary is controlled by regulatory hormones secreted by the hypothalamus (see figure 20.3). These regulatory hormones reach the anterior pituitary by traveling through a blood vessel network called the **hypothalamo-hypophyseal** (hī′pō-thal′ă-mō-hī-pō-fiz′ē-ăl) **portal system (figure 20.6).** This portal system is composed of two capillary plexuses interconnected by portal veins. This vein takes blood from one organ to another organ, such as from the hypothalamus to the pituitary, before the blood is returned to the heart.

The hypothalamo-hypophyseal portal system is essentially a "shunt" that takes venous blood carrying regulatory hormones from the hypothalamus directly to the anterior pituitary before the blood returns to the heart. The specific steps of this portal system take place as follows:

1. The median eminence of the hypothalamus has a very porous capillary network called the **primary plexus** (or *primary capillary plexus*). This plexus receives arterial blood from the superior hypophyseal artery and is drained by the **hypophyseal portal veins.**
2. The hypophyseal portal veins extend inferiorly from the median eminence of the hypothalamus through the infundibulum to the anterior pituitary. There, they disperse into a network of capillaries called the **secondary plexus** (or *secondary capillary plexus*). From this plexus, the hypothalamic hormones leave the bloodstream and enter the interstitial spaces around the cells of the anterior pituitary, where they can affect the activities of these cells.
3. Blood from the anterior pituitary (along with the hormones the anterior pituitary releases) is drained by **anterior hypophyseal veins.** These veins carry the blood filled with anterior pituitary hormones to the heart, where it is pumped throughout the body.

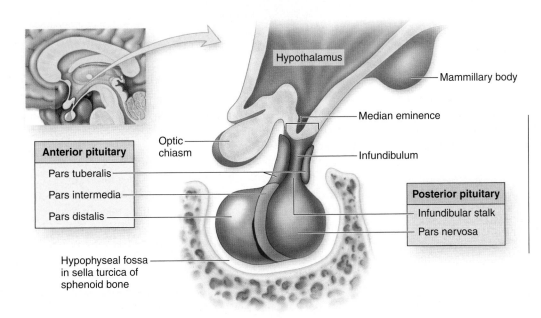

Hypothalamus

Mammillary body

Median eminence

Optic chiasm

Infundibulum

Anterior pituitary

Pars tuberalis

Pars intermedia

Pars distalis

Posterior pituitary

Infundibular stalk

Pars nervosa

Hypophyseal fossa in sella turcica of sphenoid bone

Figure 20.4

Pituitary Gland. The pituitary gland is composed of an anterior and a posterior part, which are attached to the hypothalamus by a stalk called the infundibulum. The pituitary gland occupies the hypophyseal fossa within the sella turcica of the sphenoid bone.

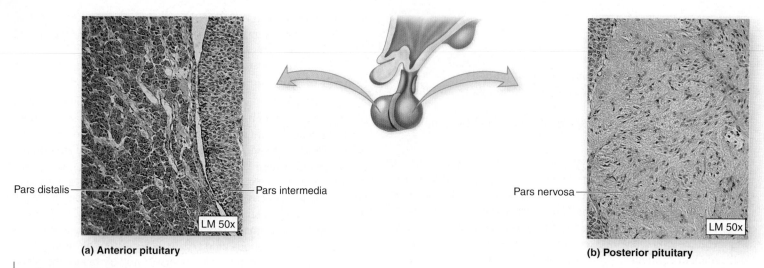

(a) Anterior pituitary

Pars distalis — — Pars intermedia

LM 50x

(b) Posterior pituitary

Pars nervosa —

LM 50x

Figure 20.5

Microscopic Anatomy of the Pituitary Gland.   Comparison of the anterior pituitary (pars distalis and pars intermedia) and the posterior pituitary (pars nervosa).

Figure 20.6

Hypothalamo-Hypophyseal Portal System.   The hypothalamus regulates the function of the anterior and posterior pituitary. The hypothalamo-hypophyseal portal system is the circulatory connection between the hypothalamus (source of regulatory hormones) and the anterior pituitary (location of target cells of the hypothalamic regulatory hormones).

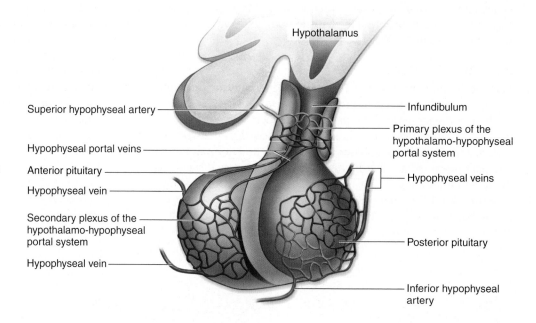

Hypothalamus

Superior hypophyseal artery

Hypophyseal portal veins

Anterior pituitary

Hypophyseal vein

Secondary plexus of the hypothalamo-hypophyseal portal system

Hypophyseal vein

Infundibulum

Primary plexus of the hypothalamo-hypophyseal portal system

Hypophyseal veins

Posterior pituitary

Inferior hypophyseal artery

Thus, the hypothalamo-hypophyseal portal system provides a pathway for hypothalamic hormones to immediately reach the anterior pituitary. In addition, the veins that drain this portal system provide a pathway by which the anterior pituitary hormones may be released into the general bloodstream.

Hormones of the Anterior Pituitary

The anterior pituitary secretes seven major hormones **(figure 20.7)**. Six of the seven hormones are **tropic** (tro′pik) **hormones.** Tropic hormones stimulate other endocrine glands or cells to secrete other hormones. The following specific cells in the anterior pituitary secrete these tropic hormones:

1. **Thyrotropic** (thī-rō-trōp′ik) **cells** in the pars distalis synthesize and secrete **thyroid-stimulating hormone (TSH),** also called *thyrotropin* (thī-rō-trō′pin). TSH regulates the release of thyroid hormone from the thyroid gland. Increased secretion of TSH results from pregnancy, stress, or exposure to low temperatures.

2. **Mammotropic** (mam-ō-trōp′ik) **cells,** also called *lactotropic cells,* in the pars distalis synthesize and secrete **prolactin** (prō-lak′tin; *lac* = milk) **(PRL).** In females, prolactin regulates mammary gland growth and breast milk production. Recent studies suggest that prolactin receptors occur on many body organs, and in males may influence the sensitivity of interstitial cells in the testes to the effects of luteinizing hormone for testosterone secretion.

3. **Corticotropic** (kor′ti-kō-trōp′ik) **cells** are located in the pars distalis. These cells synthesize and secrete **adrenocorticotropic** (ă-drē′nō-kōr′ti-kō-trō′pik) **hormone (ACTH),** also called *corticotropin.* ACTH

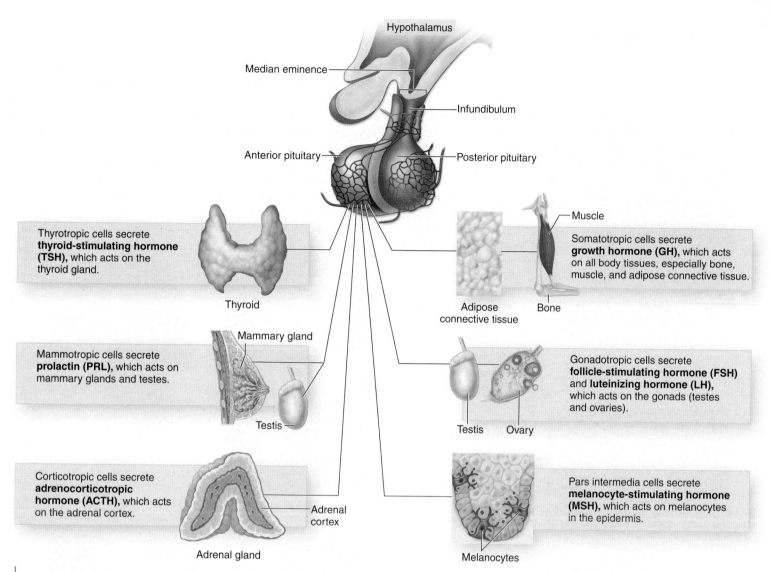

Figure 20.7

Anterior Pituitary Hormones.   The seven anterior pituitary hormones affect different areas of the body.

stimulates the adrenal cortex to produce and secrete its own hormones.

4. **Somatotropic** (sō′mă-tō-trōp′ik) **cells** in the pars distalis synthesize and secrete **growth hormone (GH)**, also called *somatotropin*. GH stimulates cell growth as well as cell division (mitosis) and affects most body cells, including adipose connective tissue, but specifically those of the skeletal and muscular systems. GH exhibits its tropic effects by stimulating the liver to produce **insulin-like growth factor 1 and 2** (IGF-1 and IGF-2), also called **somatomedin** (sō′mă-tō-mē′din), the hormone that stimulates growth at the epiphyseal plates of long bones.

5, 6. **Gonadotropic** (gō′nad-ō-trōp′ik, gon′ă-dō-) **cells** in the pars distalis synthesize and secrete both **follicle-stimulating hormone (FSH)** and **luteinizing** (loo′tē-in-ī-zing) **hormone (LH)**, collectively called **gonadotropins**. These hormones act in both females and males to influence reproductive system activities by regulating

hormone synthesis by the gonads, as well as the production and maturation of gametes in both sexes (see chapter 28).

> ## Study Tip!
> Remember the anterior pituitary tropic hormones with the FLAT PIG mnemonic:
>
> - F is for FSH
> - L is for LH
> - A is for ACTH
> - T is for TSH
>
> - P is for Prolactin
> - I is to be Ignored
> - G is for GH

The remaining hormone of the anterior pituitary is not considered a tropic hormone, because it does not stimulate hormone secretion by other endocrine tissues or glands. Rather, this hormone directly affects other activities in specific cells in the body. **Melanocyte-stimulating**

CLINICAL VIEW

Disorders of Growth Hormone Secretion

Pituitary dwarfism is caused by hyposecretion of GH.

Pituitary dwarfism (dwŏrf'izm) is a condition that exists at birth as a result of inadequate growth hormone production due to a hypothalamic or pituitary problem. Growth retardation is typically not evident until a child reaches 1 year of age, because the influence of growth hormone is minimal during the first 6 to 12 months of life. In addition to short stature, children with pituitary dwarfism often have periodic low blood sugar (hypoglycemia). Injections of human growth hormone over a period of many years can bring about improvement but not normal height.

Too much growth hormone, on the other hand, causes excessive growth and leads to increased levels of blood sugar (hyperglycemia). Oversecretion of growth hormone in childhood causes **pituitary gigantism** (jī'gan-tizm; *gigas* = giant). Beyond extraordinary height (sometimes up to 8 feet), these people have enormous internal organs, a large and protruding tongue, and significant problems with blood glucose management. If untreated, a pituitary giant dies at a comparatively early age, often from complications of diabetes or heart failure.

Excessive growth hormone production in an adult results in **acromegaly** (ak-rō-meg'ă-lē; *megas* = large). The individual does not grow in height, but the bones of the face, hands, and feet enlarge and widen. An increase in mandible size leads to a protruding jaw (prognathism). Internal organs, especially the liver, increase in size, and overproduction and release of glucose lead to the development of diabetes in virtually everyone with acromegaly. Acromegaly may result from loss of "feedback" control of growth hormone at either the hypothalamic or pituitary level, or it may develop because of a GH-secreting tumor of the pituitary. Removal of the pituitary alleviates the effects of acromegaly, but this drastic treatment presents many new problems because of the loss of all pituitary hormones.

Age 9

Age 16

Age 33

Age 52

Acromegaly is caused by secretion of excessive growth hormone (GH) in adulthood. The face and hands are most notably affected, as seen in an individual with acromegaly at ages 9, 16, 33, and 52.

hormone **(MSH)** is secreted by the cells in the pars intermedia of the anterior pituitary. MSH stimulates both the rate of melanin synthesis by melanocytes in the integument and the distribution of melanocytes in the skin. Its secretion has little effect on humans and usually ceases prior to adulthood, except in specific diseases.

Study Tip!

Use the following analogy to understand the complex interrelationships among the hypothalamus, anterior pituitary, and target organs.

The endocrine system is similar to the hierarchy of a corporation:

- The **hypothalamus** is the "president" of the endocrine system. It produces and secretes releasing hormones and inhibiting hormones that act on the anterior pituitary.

- The **anterior pituitary** is the "vice president" of the endocrine system. The vice president is pretty powerful, but usually it must receive orders (hormones) from the president before it can do its job. After receiving regulatory hormones from the hypothalamus, the anterior pituitary secretes hormones that act on the target organs.

- The **target organs** are the "workers" in the endocrine system. They receive orders (hormones) from the vice president. These organs respond to the anterior pituitary hormones (either by secreting their own hormones or by performing some action that helps keep the body at homeostasis).

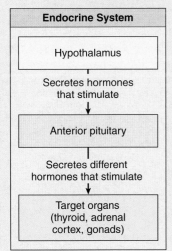

Endocrine System
Hypothalamus
Secretes hormones that stimulate ↓
Anterior pituitary
Secretes different hormones that stimulate ↓
Target organs (thyroid, adrenal cortex, gonads)

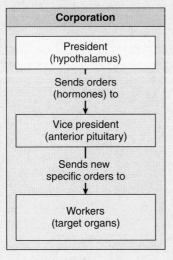

Corporation
President (hypothalamus)
Sends orders (hormones) to ↓
Vice president (anterior pituitary)
Sends new specific orders to ↓
Workers (target organs)

Posterior Pituitary

The **posterior pituitary** (*neurohypophysis*) is the neural part of the pituitary gland because it was derived from nervous tissue at the base of the diencephalon (see figures 20.3 and 20.14). The posterior pituitary is composed of a rounded lobe called the **pars nervosa** and the **infundibular stalk,** also called the **infundibulum.**

The neural connection between the hypothalamus and the posterior pituitary is called the **hypothalamo-hypophyseal tract (figure 20.8)**. The posterior pituitary consists primarily of the endings of unmyelinated axons that extend through the hypothalamo-hypophyseal tract from neuron cell bodies housed in the hypothalamus. These neurons in the hypothalamus are called **neurosecretory** (noor'ō-sē'krĕ-tōr-ē) **cells** because they secrete hormones. The hormones they produce are transported through the unmyelinated axons and housed in their terminals within the posterior pituitary. These hormones may be released from the posterior pituitary when a nerve impulse passes through the neuron to the ending in the posterior pituitary. Instead of releasing a neurotransmitter into a synaptic cleft, the posterior pituitary releases a hormone into the bloodstream.

Two specific hypothalamic nuclei contain the neuron cell bodies whose axons extend into the posterior pituitary. The **supraoptic** (soo-pră-op'tik) **nucleus** is located superior to the optic chiasm, and the **paraventricular** (par-ă-ven-trik'ū-lĕr) **nucleus** is in the anterior-medial region of the hypothalamus adjacent to the third ventricle. The neuron cell bodies in both nuclei produce two closely related peptide hormones, **antidiuretic** (an'tē-dī-ū-ret'ik) **hormone (ADH)** and **oxytocin** (ok-sē-tō'sin; *okytokos* = swift birth). ADH and oxytocin are transported from the hypothalamus to the posterior pituitary via the hypothalamo-hypophyseal tract.

ADH is released from the posterior pituitary in response to various stimuli, including a decrease in blood volume, a decrease in blood pressure, or an increase in the concentration of specific electrolytes (salts) in the blood, all of which indicate that the body is dehydrated. ADH primarily increases water retention from kidney tubules during urine production, resulting in more concentrated urine and conservation of the body's water supply. Another consequence of its release is the vasoconstriction of blood vessels, resulting in increased blood pressure; thus, ADH is also referred to as **vasopressin** (vă-sō-press'in; *vas* = vessel, *pressium* = to press down).

In females, oxytocin stimulates contraction of the uterine wall smooth musculature to facilitate labor and childbirth, and as

Figure 20.8

Hypothalamo-Hypophyseal Tract. The hypothalamus regulates the function of the anterior and posterior pituitary. The hypothalamo-hypophyseal tract is the nervous connection between nuclei in the hypothalamus and axonal endings in the posterior pituitary.

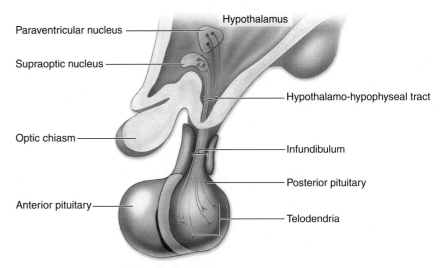

Table 20.3		Pituitary Gland Hormones		
Hormone	**Source**	**Hormone Target**	**Hormone Effects**	**Related Disorders**
HORMONES SECRETED BY THE ANTERIOR PITUITARY				
Adrenocorticotropic hormone (ACTH)	Corticotropic cells of pars distalis	Adrenal cortex	Stimulates production of corticosteroid hormones	Hyposecretion: Addison disease Hypersecretion: Cushing syndrome, androgenital syndrome
Follicle-stimulating hormone (FSH)	Gonadotropic cells of pars distalis	Female: Ovaries Male: Testes	Female: Stimulates growth of ovarian follicles Male: Stimulates sperm production	Hyposecretion: Menstrual problems, impotence, infertility Hypersecretion: Hypergonadism, precocious puberty
Luteinizing hormone (LH)	Gonadotropic cells of pars distalis	Female: Ovaries Male: Testes	Female: Stimulates ovulation, estrogen and progesterone synthesis in corpus luteum of ovary Male: Stimulates androgen synthesis in testes	Hyposecretion: Menstrual problems, impotence, infertility Hypersecretion: Hypergonadism, precocious puberty
Thyroid-stimulating hormone (TSH); also called thyrotropin	Thyrotropic cells of pars distalis	Thyroid gland	Stimulates thyroid hormone synthesis and secretion	Hyposecretion: Hypothyroidism Hypersecretion: Hyperthyroidism
Prolactin (PRL)	Mammotropic cells of pars distalis	Receptors on organs throughout the body Female: Mammary glands Male: Interstitial cells in testes	Female: Stimulates milk production in mammary glands Male: May play a role in the sensitivity of the interstitial cells to LH	Hyposecretion: Menstrual disorders, delayed puberty, male infertility Hypersecretion: Reduced menstruation, female infertility, male impotence
Growth hormone (GH)	Somatotropic cells of pars distalis	Almost every cell in the body	Stimulates increased growth and metabolism in target cells; stimulates synthesis of somatomedin in the liver to stimulate growth at epiphyseal plate	Hyposecretion: Pituitary dwarfism Hypersecretion: Giantism, acromegaly
Melanocyte-stimulating hormone (MSH)	Cells of pars intermedia	Melanocytes	Stimulates synthesis of melanin and dispersion of melanin granules in epidermal cells	Hyposecretion: Obesity Hypersecretion: Skin darkening
HORMONES STORED IN THE POSTERIOR PITUITARY				
Antidiuretic hormone (ADH); also called vasopressin	Supraoptic and paraventricular nuclei of hypothalamus	Kidney, smooth muscle in arteriole walls	Stimulates reabsorption of water from tubular fluid in kidneys; stimulates vasoconstriction in arterioles of body, thereby raising blood pressure	Hyposecretion: Diabetes insipidus Hypersecretion: Water retention
Oxytocin (OT); also called "cuddling hormone"	Supraoptic and paraventricular nuclei of hypothalamus	Female: Uterus, mammary glands Male: Smooth muscle of male reproductive tract	Female: Stimulates smooth muscle contraction in uterine wall; stimulates milk ejection from mammary glands Male: Stimulates contraction of smooth muscle of male reproductive tract	Hyposecretion: Reduced milk release from mammary glands Hypersecretion: Rarely a problem

mentioned earlier, it is also responsible for milk ejection from the mammary gland. In males, oxytocin induces smooth muscle contraction in male reproductive organs, specifically causing the prostate gland to release a component of semen during sexual activity. Recent investigations suggest that oxytocin influences maternal behavior and pair bonding, and thus some researchers call it the "cuddle hormone." Oxytocin receptors have been found throughout the body in a number of organs, including the reproductive organs, pancreas, cardiovascular system, kidney, and brain. One group of studies linked oxytocin to decreases in both heart rate and cardiac output. Interestingly, other studies suggest that oxytocin can modulate stress, social behaviors, and anxiety through the activity of the limbic system.

Table 20.3 summarizes the hormones of the anterior and posterior pituitary gland and the effects of each hormone.

WHAT DID YOU LEARN?

④ Where is the pituitary gland located?

⑤ What are the anterior pituitary hormones and their functions?

⑥ Explain how the hypothalamus controls hormone release from both the anterior and posterior pituitary.

CLINICAL VIEW

Hypophysectomy

The surgical removal of the pituitary gland is called a **hypophysectomy** (hī'pof-i-sek'tō-mē). Although rarely performed today, this surgery was done in years past to treat advanced breast and prostate cancer, two malignancies that depend on hormone stimulation for their growth. Pituitary removal effectively shuts off the hormone source to these tumors, but medications are now available to block the hormone stimulation in these cancers instead.

Currently, a hypophysectomy is performed for tumors in the pituitary gland. Most pituitary tumors cause changes in a person's vision, because the optic chiasm is essentially draped around the anterior pituitary. The preferred route of entry in a hypophysectomy is through the nasal cavity and sphenoidal sinus, directly into the sella turcica. This approach requires very small instruments and allows complete removal of the pituitary with a minimum of trauma.

Thyroid Gland

Key topics in this section:

■ Thyroid gland anatomy and location
■ How thyroid hormones are produced, stored, and secreted

The largest gland entirely devoted to endocrine activities is the **thyroid** (thī'royd; *thyreos* = an oblong shield) **gland (figure 20.9).** The thyroid gland in an adult weighs between 25 and 30 grams and is located immediately inferior to the thyroid cartilage of the larynx and anterior to the trachea. It is covered by a connective tissue capsule.

The thyroid gland exhibits a distinctive "butterfly" shape due to its **left** and **right lobes,** which are connected at the anterior midline by a narrow **isthmus** (is'mŭs; constrict). Both lobes of the thyroid gland are highly vascularized, giving the gland an intense reddish coloration. The gland is supplied by the superior and inferior thyroid arteries. Thyroid veins return venous blood from the thyroid and also transport the thyroid hormones into the general circulation.

Synthesis of Thyroid Hormone by Thyroid Follicles

At the histologic level, the thyroid gland is composed of numerous microscopic, spherical structures called **thyroid follicles.** The wall of each follicle is formed by simple cuboidal epithelial cells, called **follicular cells,** that surround a central lumen. That lumen houses a viscous, protein-rich fluid termed **colloid** (kol'oyd). External to individual follicles are some cells called parafollicular cells, which we discuss later in this chapter.

The follicular cells are instrumental in producing **thyroid hormone,** as follows: Follicular cells synthesize a glycoprotein called **thyroglobulin (TGB)** and secrete it by exocytosis into the colloid. In brief, iodine molecules must be combined with the thyroglobulin in the colloid to produce thyroid hormone precursors, which are TGB molecules that contain immature thyroid hormone within their structure. The precursors are stored in the colloid until the secretion of thyroid hormone is needed. When the thyroid gland is stimulated to secrete thyroid hormone, some of the colloid with thyroid hormone precursors is internalized into the cell by endocytosis. It travels to a lysosome where an enzyme releases the thyroid hormone molecules from the precursor in preparation for its secretion from the follicular cells.

WHAT DO YOU THINK?

❷ Have you ever noticed that the salt you buy in the grocery store is listed as "iodized"? Why is iodine added to our salt?

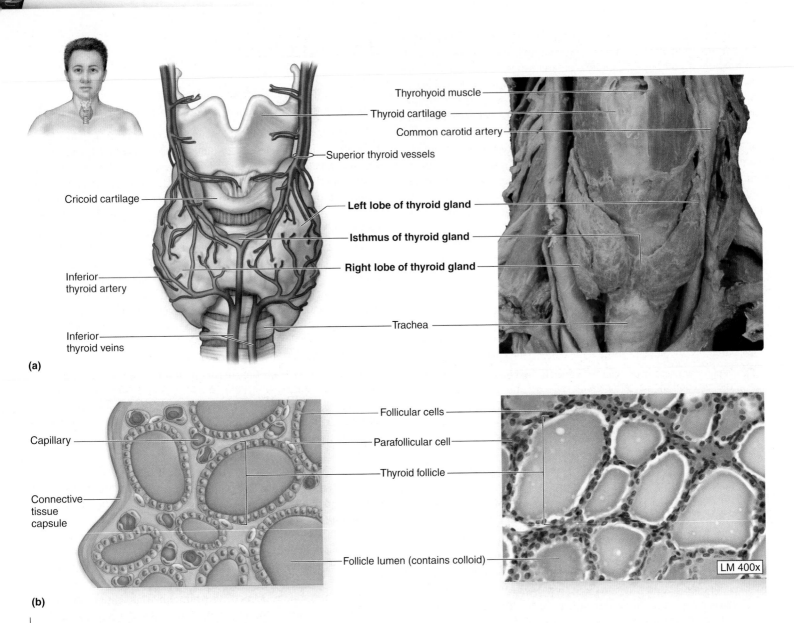

(a)

Thyrohyoid muscle

Thyroid cartilage

Common carotid artery

Superior thyroid vessels

Cricoid cartilage

Left lobe of thyroid gland

Isthmus of thyroid gland

Right lobe of thyroid gland

Inferior thyroid artery

Trachea

Inferior thyroid veins

(b)

Follicular cells

Capillary

Parafollicular cell

Thyroid follicle

Connective tissue capsule

Follicle lumen (contains colloid)

LM 400x

Figure 20.9

Thyroid Gland. *(a)* Location, gross anatomy, and vascular connections of the thyroid gland. *(b)* Microscopic anatomy of thyroid follicles, illustrating the simple cuboidal epithelium of the follicular cells, colloid within the follicle lumen, and the relationship of parafollicular cells to a follicle.

Thyroid Gland–Pituitary Gland Negative Feedback Loop

The regulation of thyroid hormone secretion depends upon a complex thyroid gland–pituitary gland negative feedback process, shown in **figure 20.10** and described here:

1. A stimulus (such as low body temperature or a decreased level of thyroid hormone in the blood) sets the feedback process in motion. This stimulus signals the hypothalamus and causes it to release **thyrotropin-releasing hormone (TRH).**

2. TRH travels through the hypothalamo-hypophyseal portal system to stimulate thyrotropic cells in the anterior pituitary, causing them to synthesize and release **thyroid-stimulating hormone (TSH).**

3. Increased levels of TSH stimulate the follicular cells of the thyroid gland to release **thyroid hormone (TH)** into the bloodstream.

4. TH stimulates its target cells (many of the cells in the body), so cellular metabolism is increased in these cells. Basal body temperature rises as a result of this stimulation.

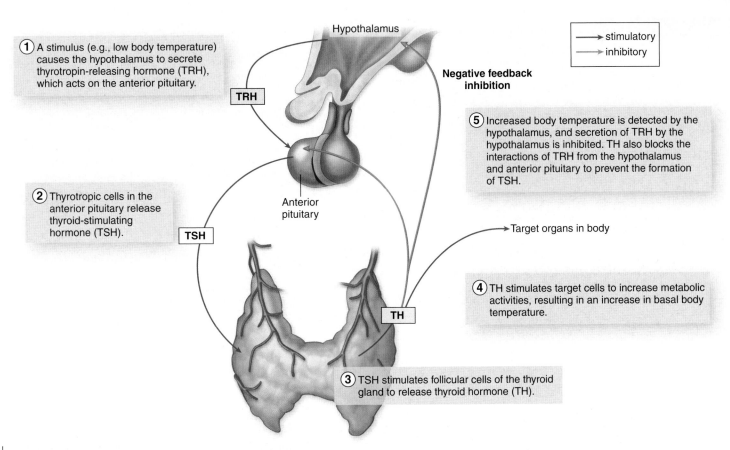

Figure 20.10

Thyroid Gland–Pituitary Gland Negative Feedback Loop. A hypothalamic-releasing hormone stimulates the anterior pituitary gland to secrete thyroid-stimulating hormone (TSH) which causes the thyroid gland to secrete thyroid hormone (TH). The secreted thyroid hormone stimulates target cells and prevents the anterior pituitary from further stimulating the hypothalamus.

5. The increased heat produced as a waste product due to increased metabolism increases body temperature and inhibits the release of TRH by the hypothalamus. Also, TH blocks receptors on the anterior pituitary cells for TRH, thus inhibiting the interactions of the hypothalamus and anterior pituitary. As TH levels rise in the bloodstream, the anterior pituitary stops secreting TSH, and so the thyroid gland is no longer stimulated to release TH.

Parafollicular Cells

External to the thyroid follicles are some larger endocrine cells, called **parafollicular** (par-ă-fo-lik′ū-lăr; *para* = alongside, *folliculus* = a small sac) **cells,** or *C cells,* which occur either individually or in small clusters (see figure 20.9*b*). Parafollicular cells secrete the hormone **calcitonin** (kal-si-tō′nin; *tonos* = stretching) in response to an elevated level of blood calcium. Ultimately, calcitonin acts to reduce the blood calcium level and encourage deposition of calcium into bone. Calcitonin does

this by stimulating osteoblast activity and inhibiting osteoclast activity, resulting in new bone matrix formation, including deposition of calcium salts onto this matrix and a decrease in bone matrix resorption.

Because calcitonin positively affects bone deposition, medical research is looking at developing calcitonin-like medicines to treat bone loss diseases such as osteoporosis. The activity of calcitonin is antagonistic to the effects of parathyroid hormone (PTH), described next.

Table 20.4 lists the disorders that result from abnormal secretion of TH and calcitonin on page 618 and the Clinical View discusses two of these.

WHAT DID YOU LEARN?

7 What is colloid, and where is it located?

8 How does thyroid hormone affect the body?

Table 20.4	Thyroid and Parathyroid Gland Hormones			
Hormone	**Source**	**Hormone Target**	**Hormone Effects**	**Related Disorders**
Thyroid hormone (TH)	Follicular cells of thyroid gland	Most body cells	Increases metabolism, oxygen use, growth, and energy use; supports and increases rate of development	Hypersecretion: Hyperthyroidism, Graves disease Hyposecretion: Hypothyroidism, goiter
Calcitonin	Parafollicular cells of thyroid gland	Bone, kidney	Reduces calcium levels in body fluids; decreases bone resorption and increases calcium deposition in bone	Hypersecretion: Thyroid cancer; some association with lung, breast, and pancreatic cancers; chronic renal failure
Parathyroid hormone (PTH)	Chief cells of parathyroid gland	Bone, small intestine, kidney	Increases calcium levels in blood through bone resorption (so calcium may be delivered to tissues needing calcium ions, such as muscle tissue); increases calcium absorption by small intestine by calcitriol; decreases calcium loss through the kidneys	Hypersecretion: Hyperparathyroidism Hyposecretion: Hypoparathyroidism

CLINICAL VIEW

Disorders of Thyroid Gland Secretion

Thyroid hormone (TH) adjusts and maintains the basal metabolic rate of many cells in our body. In the healthy state, TH release is very tightly controlled, but should the amount vary by even a little, a person may exhibit noticeable symptoms. Disorders of thyroid activity are among the most common metabolic problems clinicians see.

Hyperthyroidism (hī-per-thī'royd-izm) results from excessive production of TH, and is characterized by increased metabolic rate, weight loss, hyperactivity, and heat intolerance. Although there are a number of causes of hyperthyroidism, the more common ones are (1) ingestion of TH supplements (weight control clinics sometimes use TH to increase metabolic activity); (2) excessive stimulation of the thyroid by the pituitary gland; and (3) loss of feedback control by the thyroid itself. This last condition, called **Graves disease**, includes all the symptoms of hyperthyroidism plus a peculiar change in the eyes known as **exophthalmos** (protruding and bulging eyeballs). Hyperthyroidism is treated by removing the thyroid gland, either by surgery or intravenous injections of radioactive iodine. In the latter procedure, the thyroid literally "cooks itself" as it sequesters the radioactive iodine, but other organs are not damaged because they do not store iodine as the thyroid does. (Patients whose thyroid glands have been removed must take hormone supplements on a daily basis.)

Hypothyroidism (hī-pō-thī'royd-izm) results from decreased production of TH. It is characterized by low metabolic rate, lethargy, a feeling of being cold, weight gain (in some patients), and photophobia (an aversion and avoidance of light). Hypothyroidism may be caused by decreased iodine intake, lack of pituitary stimulation of the thyroid, post-therapeutic hypothyroidism (resulting from either surgical removal or radioactive iodine treatments), or destruction of the thyroid by the person's own immune system. Oral replacement of TH is the treatment for this type of hypothyroidism.

A **goiter** refers to enlargement of the thyroid, typically due to an insufficient amount of the dietary iodine needed to produce TH. Although the pituitary releases more TSH in an effort to stimulate the thyroid, the lack of dietary iodine prevents the thyroid from producing the needed TH. The long-term consequence of the excessive TSH stimulation is overgrowth of the thyroid follicles and the thyroid itself. Goiter was a relatively common deformity in the United States until food processors began adding iodine to table salt. It still occurs in parts of the world where iodine is lacking in the diet, and as such is referred to as *endemic goiter*. Unfortunately, goiters do not readily regress once iodine is restored to the diet, and surgical removal is often required.

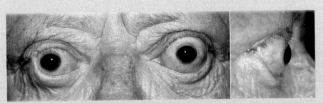

An individual with Graves disease (hyperthyroidism) exhibits exophthalmos, bulging and protruding eyeballs.

Endemic goiter is caused by dietary iodine deficiency.

Parathyroid Glands

Key topics in this section:

- Parathyroid gland anatomy and location
- Hormones secreted by the parathyroid gland and how they function

The small, brownish-red **parathyroid** (par-ă-thī′royd) **glands** are located on the posterior surface of the thyroid gland **(figure 20.11)**. The parathyroid glands are usually four small nodules, but some individuals may have as few as two or as many as six of these glands. The inferior thyroid artery generally supplies all nodules of the parathyroid gland. Rarely, the superior thyroid artery may supply the superior pair of nodules.

There are two different types of cells in the parathyroid gland: chief cells and oxyphil cells. The **chief cells,** or *principal cells,* have a large, spherical nucleus and a relatively clear cytoplasm. They are the source of **parathyroid hormone (PTH). Oxyphil** (ok′sē-fil) **cells** are larger than chief cells, and each oxyphil has a granular, pink cytoplasm. The function of oxyphil cells is not known.

PTH is secreted into the bloodstream in response to decreased blood calcium levels, which may result from normal events such as loss of electrolytes during sweating or urine formation. Because calcium ions are needed for many body functions, including activity at synapses and muscle contraction, inadequate levels of calcium in the blood mean that the body cannot function properly. Thus, parathyroid hormone encourages the release of calcium stores into the blood, ultimately raising blood calcium levels **(figure 20.12)**. PTH influences target organs in several ways.

- PTH stimulates osteoclasts to resorb bone and release calcium ions from bone matrix into the bloodstream.
- PTH stimulates calcitriol synthesis from an inactive form of vitamin D as a result of increased calcium ions in the kidney. Calcitriol is a hormone that promotes calcium absorption of ingested nutrients in the small intestine.
- PTH prevents the loss of calcium ions during the formation of urine in the kidneys. Calcium ions are returned to the bloodstream from filtrate of the blood that is later modified to form urine.

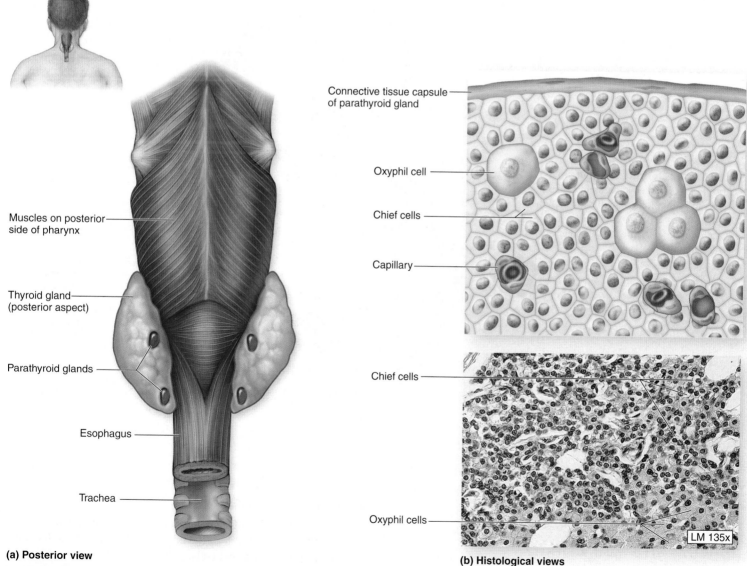

(a) Posterior view

(b) Histological views

Figure 20.11

Parathyroid Glands.   *(a)* The parathyroid glands are four small nodules attached to the capsule of the thyroid gland on its posterior surface.
(b) Each parathyroid gland is enclosed within a connective tissue capsule and composed of chief cells and oxyphil cells.

CLINICAL VIEW

Disorders of Parathyroid Gland Secretion

The parathyroid glands are critical to the control of blood calcium levels. Over- or underactivity of the parathyroids can have serious, if not fatal, consequences. **Hyperparathyroidism** (hī′per-par-ă-thī′royd-izm) refers to overactivity of the parathyroid glands and is the most commonly encountered disorder of the parathyroid glands. Following are the major consequences of hyperparathyroidism: (1) Bones are depleted of calcium (and thus become more subject to fractures); (2) the extra resorption of calcium from the filtrate that becomes urine leads to an increased incidence of kidney stones; (3) high blood calcium levels cause a decrease in bowel motility, which leads to constipation; and (4) high blood calcium levels cause psychological changes, such as depression, decreased mental activity, and eventually coma. Hyperparathyroidism usually results from just one of the four glands increasing in size and beginning to work on its own without control.

Hypoparathyroidism is underactivity of the parathyroid glands. It is a rare condition that results in abnormally low blood calcium levels (hypocalcemia). Most of the symptoms are neuromuscular, and range from mild tingling in the fingers and limbs to marked muscle cramps and contractions (tetany); in severe cases, convulsions may occur, and death may result if left untreated. Hypoparathyroidism is most commonly due to accidental removal or damage to the parathyroid glands during thyroid surgery. Less commonly, hypoparathyroidism occurs because of autoimmune destruction of glands. Currently, no direct replacement for PTH exists, so therapy consists of dietary vitamin D/calcium supplementation.

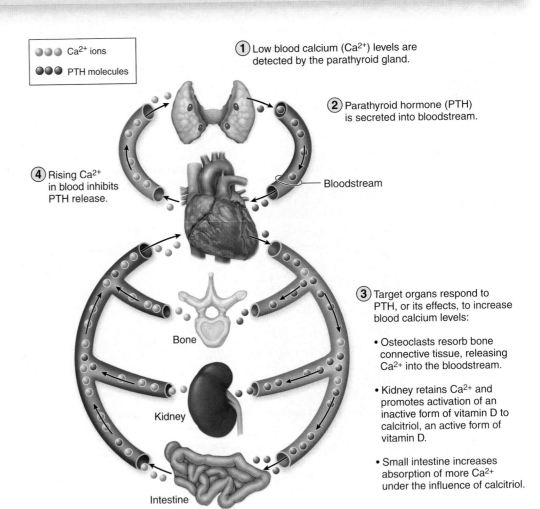

● ● ● Ca²⁺ ions

● ● ● PTH molecules

① Low blood calcium (Ca²⁺) levels are detected by the parathyroid gland.

② Parathyroid hormone (PTH) is secreted into bloodstream.

④ Rising Ca²⁺ in blood inhibits PTH release.

Bloodstream

Bone

Kidney

Intestine

③ Target organs respond to PTH, or its effects, to increase blood calcium levels:

- Osteoclasts resorb bone connective tissue, releasing Ca²⁺ into the bloodstream.

- Kidney retains Ca²⁺ and promotes activation of an inactive form of vitamin D to calcitriol, an active form of vitamin D.

- Small intestine increases absorption of more Ca²⁺ under the influence of calcitriol.

Figure 20.12

Parathyroid Hormone. Parathyroid hormone (PTH) helps maintain calcium levels in body fluids. Low blood calcium levels cause PTH to be released from the parathyroid gland, affecting various target organs.

Table 20.4 lists the hormones secreted by the thyroid and parathyroid glands, and the Clinical View on this page focuses on two disorders that result from abnormal secretion of PTH.

WHAT DID YOU LEARN?

❾ What are the main effects of parathyroid hormone?

Adrenal Glands

Key topics in this section:

- Structure of the adrenal cortex and adrenal medulla
- Hormones produced in the adrenal cortex and medulla and their effects on target cells

The **adrenal** (ă-drē′năl; *ad* = to, *ren* = kidney) **glands** (or *suprarenal glands*) are paired, pyramid-shaped endocrine glands anchored on the superior surface of each kidney (**figure 20.13**). The adrenal

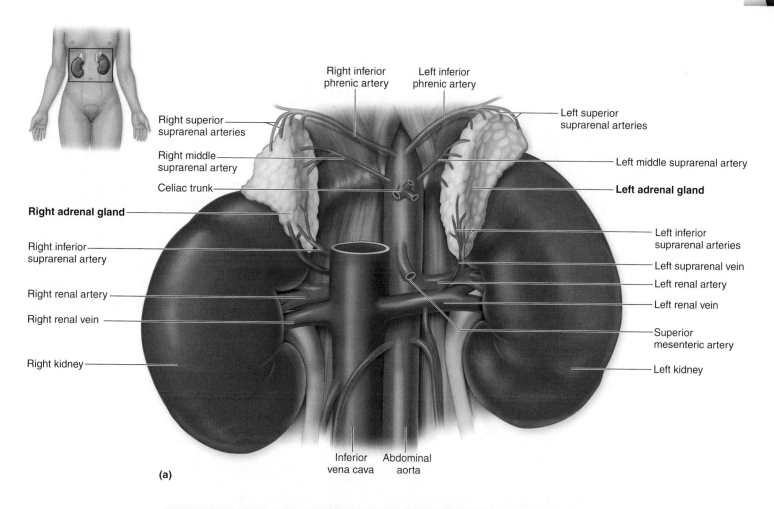

Right inferior phrenic artery

Left inferior phrenic artery

Right superior suprarenal arteries

Left superior suprarenal arteries

Right middle suprarenal artery

Left middle suprarenal artery

Celiac trunk

Left adrenal gland

Right adrenal gland

Right inferior suprarenal artery

Left inferior suprarenal arteries

Left suprarenal vein

Left renal artery

Right renal artery

Left renal vein

Right renal vein

Superior mesenteric artery

Right kidney

Left kidney

Inferior vena cava

Abdominal aorta

(a)

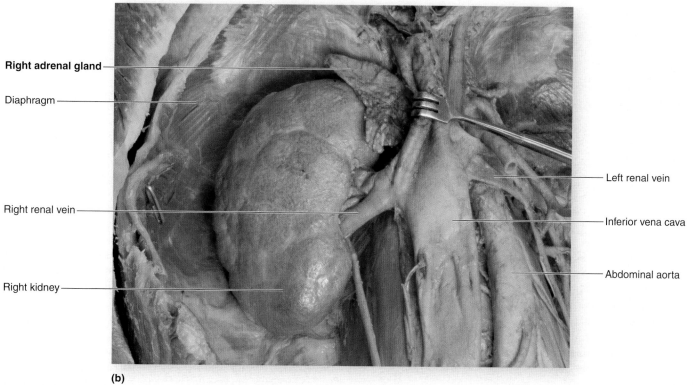

Right adrenal gland

Diaphragm

Left renal vein

Right renal vein

Inferior vena cava

Abdominal aorta

Right kidney

(b)

Figure 20.13

Adrenal Glands. Each adrenal gland is a two-part gland that secretes stress-related hormones. The adrenal cortex produces steroid hormones, and the adrenal medulla produces epinephrine and norepinephrine. (*a*) An anterior view shows the relationships of the kidneys, adrenal glands, and vasculature supplying these glands. (*b*) Cadaver photo. (*continued on next page*)

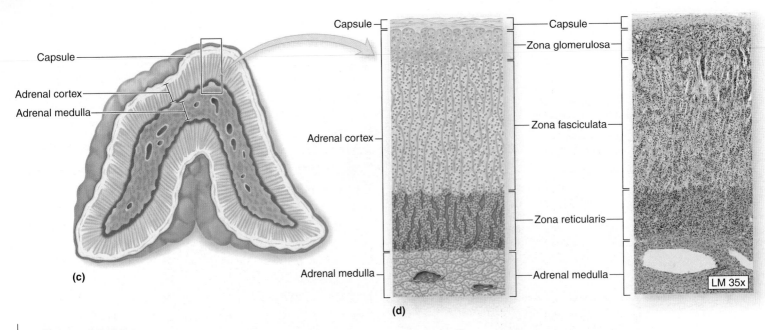

Figure 20.13

Adrenal Glands. (*continued*)   *(c)* A sectioned adrenal gland shows the outer cortex and the inner medulla. *(d)* A diagram and a micrograph illustrate the three zones of the adrenal cortex, as well as the relationship of the cortex to the external capsule and the internal medulla.

Table 20.5	Comparison of Adrenal Cortex and Adrenal Medulla	
Feature	**Adrenal Cortex**	**Adrenal Medulla**
Embryonic Derivative	Intermediate mesoderm	Neural crest cells
Adrenal Gland Location	Outer region of the gland	Inner region of the gland
Mode of Stimulation	Hormonal (stimulated by ACTH from anterior pituitary)	Neural (stimulated by preganglionic axons from sympathetic division of ANS)
Hormones Produced	Corticosteroids: mineralocorticoids, glucocorticoids, gonadocorticoids	Epinephrine, norepinephrine
Effects of Hormones	Mineralocorticoids regulate the balance of electrolytes (e.g., Na$^+$ and K$^+$) in the body Glucocorticoids elevate blood glucose levels during long-term stressful situations (e.g., fasting, injury, anxiety), and stimulate the body to use fats and proteins as energy resources Gonadocorticoids (primarily androgens) are sex hormones	Prolongs fight-or-flight response of the sympathetic division of the ANS

glands are embedded in fat and fascia to minimize their movement. These endocrine glands are supplied by multiple suprarenal arteries that branch from larger abdominal arteries, including the inferior phrenic arteries, the aorta, and the renal arteries. Venous drainage enters the suprarenal veins. The adrenal gland has an outer **adrenal cortex** and an inner central core called the **adrenal medulla**. These two regions secrete different types of hormones (**table 20.5**).

Adrenal Cortex

The adrenal cortex exhibits a distinctive yellow color as a consequence of the stored lipids in its cells. These cells synthesize more than 25 different steroid hormones, collectively called **corticoste-**

roids. Corticosteroid synthesis is stimulated by the ACTH produced by the anterior pituitary. Corticosteroids are vital to our survival; trauma to or removal of the adrenal glands requires corticosteroid supplementation throughout life.

The adrenal cortex is partitioned into three separate regions: the zona glomerulosa, the zona fasciculata, and the zona reticularis (figure 20.13*d*). Different functional categories of steroid hormones are synthesized and secreted in the separate zones.

The **zona glomerulosa** (zō′nă glō-mĕr-ū-lōs′ă; *glomerulus* = ball of yarn) is the thin, outer cortical layer composed of dense, spherical clusters of cells. These cells synthesize **mineralocorticoids** (min′er-al-ō-kōr′ti-koyd), a group of hormones that help regulate the

CLINICAL VIEW

Disorders of Adrenal Cortex Hormone Secretion

Four abnormal patterns of adrenal cortical function are Cushing syndrome, Addison disease, androgenital syndrome, and pheochromocytoma.

Cushing syndrome results from the chronic exposure of the body's tissues to excessive levels of glucocorticoid hormones. This complex of symptoms is seen most frequently in people taking corticosteroids as therapy for autoimmune diseases such as rheumatoid arthritis, although some cases result when the adrenal gland produces too much of its own glucocorticoid hormones. Corticosteroids are powerful immunosuppressant drugs, but they have serious side effects, such as a significant loss of bone mass (osteoporosis), muscle weakness, redistribution of body fat, and salt retention (resulting in overall swelling of the tissues). Cushing syndrome is characterized by body obesity, especially in the face (called "moon face") and back ("buffalo hump"). Other symptoms include hypertension, excess hair growth, kidney stones, and menstrual irregularities.

(a)   (b)

(a) Photo prior to onset of Cushing syndrome. (b) Symptoms resulting from the excessive glucocorticoid secretion in Cushing syndrome include "buffalo hump" and "moon face."

Addison disease is a form of adrenal insufficiency that develops when the adrenal glands fail, resulting in a chronic shortage of glucocorticoids and sometimes mineralocorticoids. Adrenal cortical failure may develop because the pituitary stops supplying ACTH to stimulate the adrenal cortex, or because the adrenal glands themselves are diseased and cannot respond to ACTH. Addison disease is a rare disorder, occurring in about 1 in 100,000 people of all ages and affecting men and women equally. The symptoms include weight loss, general fatigue and weakness, hypotension, and darkening of the skin. Therapy consists of oral corticosteroids taken for the rest of the patient's life. Perhaps the most celebrated person with Addison disease was former President John Fitzgerald Kennedy, who was quietly treated by specialists throughout his presidency. It is now thought that his steroid treatments may have contributed to the president's other medical problems, among them osteoporosis and back pain.

(a)   (b)

(a) John F. Kennedy prior to being treated for Addison disease. (b) In 1960, facial swelling was one of the effects of cortisone treatment for Kennedy's Addison disease.

Adrenogenital syndrome, also known as *androgen insensitivity syndrome* or *congenital adrenal hyperplasia*, first manifests in the embryo and fetus. It is characterized by the inability to synthesize corticosteroids. The pituitary, sensing the deficiency of corticosteroids, releases massive amounts of ACTH in an unsuccessful effort to bring the glucocorticoid content of the blood up to a healthy level. This large amount of ACTH produces *hyperplasia* (increased size) of the adrenal cortex and causes the release of intermediary hormones that have a testosterone-like effect. The result is *virilization* (masculinization) in a newborn. Virilization in girls means the clitoris is enlarged, sometimes to the size of a male penis. The effect may be so profound that the sex of a newborn female is questioned or even mistaken. A virilized male may have an enlarged penis and exhibit signs of premature puberty as early as age 6 or 7. About 75% of affected newborns also have a salt-losing problem, owing to their bodies' inability to manufacture mineralocorticoids and glucocorticoids. Treatment consists of oral corticosteroids, which boost the body's level of corticosteroids and inhibit the pituitary from releasing excessive amounts of ACTH.

A benign tumor called a **pheochromocytoma** (fē'ō-krō'mō-sī-tō'mă; *phaios* = dusky, *chroma* = color) sometimes arises from the chromaffin cells in the adrenal medulla. The tumor is characterized by episodic secretion of large amounts of epinephrine and norepinephrine. The periodic high levels of these hormones cause marked swings in blood pressure, alternating between hypertension and hypotension. Often it's the light-headedness associated with low blood pressure that causes a person to seek medical help. Other complaints include headache, nausea, heart palpitations, and sweating. Beyond tremendous swings in blood pressure, other metabolic problems occur, including abnormally high levels of glucose in the blood (hyperglycemia) and glucose in the urine (glycosuria). If untreated, a pheochromocytoma can lead to fatal brain hemorrhage or heart failure. Surgical removal of the tumor is the only cure.

Table 20.6	Adrenal Gland Hormones			
Hormone	**Source**	**Hormone Target**	**Hormone Effects**	**Related Disorders**
Mineralocorticoids (e.g., aldosterone)	Zona glomerulosa of adrenal cortex	Kidney cells	Regulate electrolyte composition and concentration in body fluids	Hyposecretion: Addison disease Hypersecretion: Hypertension, edema
Glucocorticoids (e.g., cortisol, corticosterone)	Zona fasciculata of adrenal cortex	Liver cells	Stimulate lipid and protein metabolism; regulate blood glucose levels	Hyposecretion: Addison disease Hypersecretion: Cushing syndrome
Gonadocorticoids (e.g., androgens)	Zona reticularis of adrenal cortex	Sex organs	Protein synthesis in sex organ cells	Hyposecretion: Generally no effect: may see effect post-menopause Hypersecretion: Adrenogenital syndrome
Epinephrine and norepinephrine	Chromaffin cells of adrenal medulla	Various cells throughout the body	Work with the sympathetic division of the ANS to stimulate fight-or-flight response	Hyposecretion: Hypertension, hyperglycemia, nervousness, sweating Hypersecretion: Pheochromocytoma (benign tumor in adrenal medulla)

composition and concentration of electrolytes (ions) in body fluids. The principal mineralocorticoid is **aldosterone** (al-dos′ter-ōn), which regulates the ratio of sodium (Na^+) and potassium (K^+) ions in our blood by stimulating Na^+ retention and K^+ secretion. If the ratio of these electrolytes becomes unbalanced, body functions are dramatically affected; severe imbalances can result in death.

The **zona fasciculata** (fă-sik′ū-lă′tă; *fascicle* = bundle of parallel sticks) is the middle layer and the largest region of the adrenal cortex. It is composed of parallel cords of lipid-rich cells that have a bubbly, almost pale appearance. **Glucocorticoids** (gloo-kō-kōr′ti-koyd) are synthesized by these cells. Glucocorticoids stimulate metabolism of lipids and proteins, and help regulate glucose levels in the blood, especially as the body attempts to resist stress and repair injured or damaged tissues. The most common glucocorticoids are **cortisol** (kōr′ti-sol) (*hydrocortisone*) and **corticosterone** (kōr′ti-kos′ter-ōn).

The innermost region of the cortex, called the **zona reticularis** (rĕ-tik′ū-lăr′is; *reticulum* = network), is a narrow band of small, branching cells. These cells are capable of secreting minor amounts of sex hormones called **gonadocorticoids.** The primary gonadocorticoids secreted are **androgens,** which are male sex hormones. In females, the androgens are converted to estrogen. The amount of androgen secreted by the adrenal cortex is small compared to that secreted by the gonads.

Table 20.6 lists the adrenal gland hormones, and the Clinical View on page 623 describes some disorders that result from abnormal secretion of adrenal gland hormones. In addition, the use and abuse of steroids is addressed in the Clinical View in chapter 10, page 304.

Study Tip!

Here is how to remember the adrenal gland hormonal functions:

- The zona glomerulosa regulates the levels of sodium and potassium ions in the blood, so it regulates *salt*.
- The zona fasciculata secretes glucocorticoids, which keep blood glucose levels up, so it regulates *sugar*.
- Finally, the zona reticularis secretes small amounts of androgens, so it regulates *sex* (hormones).

Thus, the adrenal cortex regulates **salt, sugar,** and **sex!**

Adrenal Medulla

The **adrenal medulla** forms the inner core of each adrenal gland (figure 20.13c,d). It has a pronounced red-brown color due to its extensive vascularization. The adrenal medulla primarily consists of clusters of large, spherical cells called **chromaffin** (krō′maf-in; *chroma* = color, *affinis* = affinity) **cells.** These chromaffin cells were formed from neural crest cells, so they are essentially modified ganglionic cells of the sympathetic division of the autonomic nervous system. They are innervated by preganglionic sympathetic axons.

When stimulated by the sympathetic division of the ANS, one population of chromaffin cells secretes the hormone **epinephrine** (ep′i-nef′rin; *epi* = upon, *nephros* = kidney, also called *adrenaline* [ă-dren′ă-lin]). The other population secretes the hormone **norepinephrine** (nōr′ep-i-nef′rin, also called *noradrenaline* [nōr-ă-dren′ă-lin]). These hormones work with the sympathetic division of the autonomic nervous system to prepare the body for an emergency or fight-or-flight situation. Since hormones are released more slowly than nerve impulses and their effects are longer lasting, the secretion of epinephrine and norepinephrine helps prolong the effects of the sympathetic stimulation. Thus, long after the sympathetic axons have ceased to transmit nerve impulses, the sympathetic responses continue to linger because the epinephrine and norepinephrine are still present in the bloodstream and affecting their target cells.

WHAT DID YOU LEARN?

10 What is the primary function of aldosterone?

11 What stimulates the secretion of epinephrine?

WHAT DO YOU THINK?

3 A person who is extremely allergic to bee stings may carry an "Epi-pen," which is an auto-injector of epinephrine. Why would the epinephrine help this individual?

Pancreas

Key topics in this section:

- Pancreas anatomy and location
- Hormones produced by the pancreatic islets and how they function

The **pancreas** (pan′krē-as; *pan* = all, *kreas* = flesh) is an elongated, spongy, nodular organ situated between the duodenum of the small intestine and the spleen (**figure 20.14**) and posterior to the stomach. The pancreas performs both exocrine and endocrine activities, and thus

it is considered a *heterocrine,* or mixed, gland. The pancreas is mostly composed of groups of cells called **pancreatic acini** (sing., *acinus* = grape). Acinar cells produce an alkaline pancreatic juice that is secreted through pancreatic ducts into the duodenum of the small intestine. This pancreatic juice aids digestion. (Pancreas anatomy and its exocrine functions are described further in chapter 26.)

Scattered among the pancreatic acini are small clusters of endocrine cells called **pancreatic islets** (ī′let), also known as *islets of Langerhans* (figure 20.14*b*). Estimates of the number of islets range between 1.5 and 2 million; however, these endocrine cell clusters form only about 1% of the pancreatic volume. A pancreatic islet may be composed of four types of cells: two major types (called alpha cells and beta cells) and two minor types (called delta cells and F cells). Each type produces its own hormone (**table 20.7**).

Alpha cells secrete **glucagon** (gloo′kă-gon) when blood glucose levels drop. Glucagon causes target cells in the liver to break down glycogen into glucose and release glucose into the bloodstream to increase blood glucose levels. Also, this hormone stimulates adipose cells to break down lipid and secrete it into the bloodstream. Thus, if you haven't eaten in a long time and your blood glucose levels are low, the alpha cells secrete glucagon.

Beta cells secrete **insulin** (in′sū-lin; *insula* = island) when blood glucose levels are elevated. Target cells respond to insulin by taking up the glucose, thus lowering blood glucose levels (see figure 20.2*a*). Insulin also promotes glycogen synthesis and lipid storage.

Delta cells are stimulated by high levels of nutrients in the bloodstream. Delta cells synthesize **somatostatin** (sō′mă-tō-stat′in; *stasis* = standing still), also described as **growth hormone-inhibiting hormone (GHIH)** (see table 20.2). Somatostatin slows the release of insulin and glucagon as well as the activity of the digestive organs, thereby also slowing the rate of nutrient entry into the bloodstream. This process gives the other islet hormones the ability to control and coordinate nutrient uptake.

Table 20.7	**Pancreatic Hormones**			
Hormone	**Source**	**Hormone Target**	**Hormone Effects**	**Related Disorders**
Glucagon	Alpha cells of pancreatic islets	Liver, adipose cells	Increases blood glucose levels, glycogen breakdown in liver cells, lipid breakdown in adipose cells	Hyposecretion: Hypoglycemia Hypersecretion: Dehydration
Insulin	Beta cells of pancreatic islets	Liver, body cells	Decreases glucose levels in body fluids, glucose transport into target cells; promotes glycogen and lipid formation and storage	Hyposecretion: Diabetes mellitus Hypersecretion: Hyperinsulism
Somatostatin	Delta cells of pancreatic islets	Alpha and beta cells of pancreatic islets	Slows release of insulin and glucagon to slow rate of nutrient absorption during digestion	Hyposecretion: Giantism, acromegaly Hypersecretion: Suppress insulin and glucagon release
Pancreatic polypeptide	F cells of pancreatic islets	Delta cells of pancreatic islets	Suppresses somatostatin secretion from delta cells	Hyposecretion: Excessive pancreatic enzyme secretion Hypersecretion: Inhibition of gallbladder secretion; suppress pancreas secretion; overstimulates gastric secretion

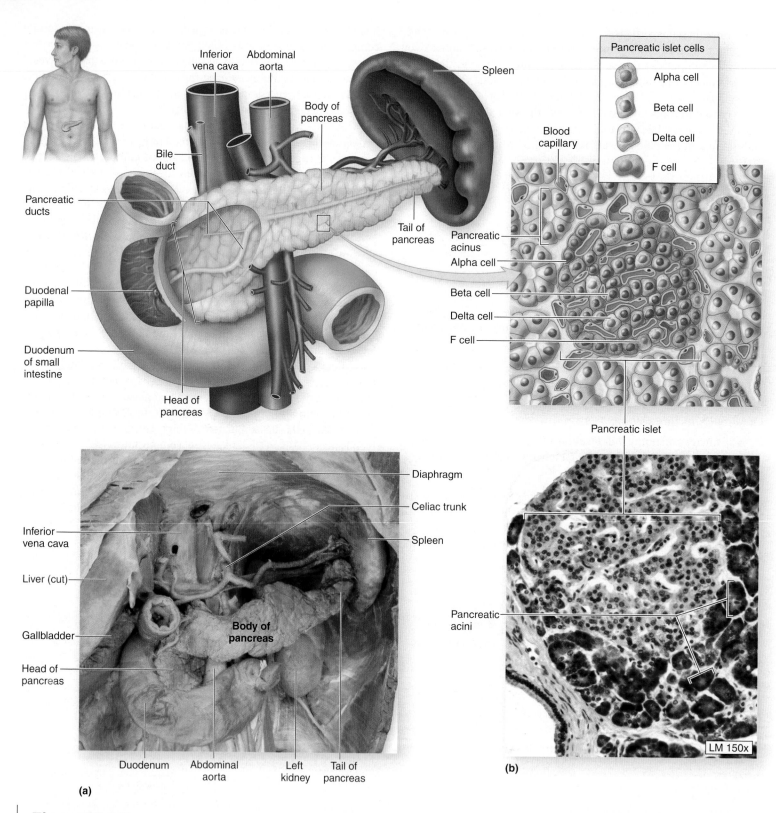

Pancreatic islet cells

- Alpha cell
- Beta cell
- Delta cell
- F cell

Inferior vena cava

Abdominal aorta

Spleen

Body of pancreas

Bile duct

Pancreatic ducts

Duodenal papilla

Duodenum of small intestine

Head of pancreas

Tail of pancreas

Blood capillary

Pancreatic acinus

Alpha cell

Beta cell

Delta cell

F cell

Pancreatic islet

Diaphragm

Celiac trunk

Spleen

Inferior vena cava

Liver (cut)

Gallbladder

Head of pancreas

Body of pancreas

Pancreatic acini

Duodenum

Abdominal aorta

Left kidney

Tail of pancreas

LM 150x

(a)

(b)

Figure 20.14

Pancreas. The pancreas performs both exocrine and endocrine activities. *(a)* A diagram and cadaver photo illustrate the relationship between the pancreas and the small intestine and spleen. *(b)* A diagram and micrograph reveal the histology of a pancreatic islet. Different types of islet cells are distinguished by a specific staining technique called immunohistology.

CLINICAL VIEW: In Depth

Diabetes Mellitus

Diabetes (dī-ă-bē'tez) **mellitus** (me-lī'tŭs; sweetened with honey) is a metabolic condition marked by inadequate uptake of glucose from the blood. The name "diabetes mellitus" is derived from the phrase "sweet urine" because some of the excess glucose is expelled into the urine, resulting in glycosuria (glucose in the urine). Chronically elevated blood glucose levels damage blood vessels, especially the smaller arterioles. Because of its damaging effects on the vascular system, diabetes is the leading cause of retinal blindness, kidney failure, and nontraumatic leg amputations in the United States. Diabetes is also associated with increased incidence of heart disease and stroke. Three categories of diabetes mellitus are type 1 diabetes, type 2 diabetes, and gestational diabetes.

Type 1 diabetes, also referred to as *insulin-dependent diabetes mellitus* (*IDDM*), is characterized by absent or diminished production and release of insulin by the pancreatic islet cells. This type tends to occur in children and younger individuals, and is not directly associated with obesity. Type 1 diabetes develops in a person who harbors a genetic predisposition, although some kind of triggering event is required to start the process. Often, the trigger is a viral infection; then the process continues as an autoimmune condition in which the beta cells of the pancreatic islets are the primary focus of destruction. When the beta cells are destroyed, no insulin is produced, so blood glucose cannot be taken up and utilized by the body tissues. Treatment of type 1 diabetes requires daily injections of insulin. Newer monitoring instruments, such as the insulin pump, now allow rapid and almost continual monitoring of blood glucose. This advancement, coupled with automated delivery of insulin, has greatly improved the treatment and lifestyle of people with IDDM.

Type 2 diabetes, also known as *insulin-independent diabetes mellitus* (*IIDM*), results from either decreased insulin release by the pancreatic beta cells or decreased insulin effectiveness at peripheral tissues. This type of diabetes was previously referred to as *adult-onset diabetes* because it tended to occur in people over the age of 30. However, type 2 diabetes is now rampant in adolescents and young adults. Obesity plays a major role in the development of type 2 diabetes, and more young

people today are considered overweight than ever before. Weight reduction supports the prevention of type 2 diabetes and appears to decrease the symptoms of type 2 diabetes that are already presenting. Most type 2 diabetes patients can be successfully treated with a combination of diet, exercise, and medications that enhance insulin release or increase its sensitivity at the tissue level. In more severe cases, a person with type 2 diabetes must take insulin injections.

Gestational diabetes is seen in some pregnant women, typically in the latter half of the pregnancy. If untreated, gestational diabetes can pose a risk to the fetus as well as increase delivery complications. Most at risk for developing this condition are women who are overweight, African American, Native American, or Hispanic, or those who have a family history of diabetes. While gestational diabetes usually resolves after giving birth, a woman who presents with the condition has a 20–50% chance of developing type 2 diabetes within 10 years.

Until recently, there was no cure for diabetes, but in the past few years pancreas transplants have helped individuals with severe cases of diabetes. Pancreas transplants have several drawbacks, however: They require major surgery, there is a long donor waiting list, and many complications can arise due to the surgery, either from potential rejection of the transplanted organ or the toxic effects of the necessary immunosuppressant antirejection drugs. Recently, a less invasive surgery, called an islet cell transplant, has been developed. In this procedure, the islet cells are removed from a donor pancreas and purified. Then the cells are injected into a vein that enters the liver. Once in the liver, the islet cells embolize (form big clots) and start producing insulin almost immediately. Islet cell transplants are still very new and have many complications. The process of extracting and purifying the islet cells is complicated and can have a high failure rate; bleeding or major blood clots may occur in the vein where the islet cells are transplanted; and the immunosuppressant drugs that must be taken have serious side effects. Furthermore, recent studies have shown that the efficacy of islet cell transplants is only temporary. Most patients need to resume insulin shots within 2 years of transplantation. Thus, islet cell transplant surgery is reserved for patients who have severe forms of diabetes.

F cells are stimulated by protein digestion in the digestive tract. F cells secrete **pancreatic polypeptide (PP)** to suppress and regulate somatostatin secretion from delta cells. Together, these pancreatic hormones provide for orderly uptake and processing of nutrients.

💡 WHAT DID YOU LEARN?

12 What hormones are produced in the pancreatic islets?

Pineal Gland and Thymus

Key topic in this section:

- Anatomy, location, and endocrine functions of the pineal gland and thymus

The **pineal** (pin'ē-al; *pineus* = relating to the pine) **gland,** also called the *pineal body,* is a small, cone-shaped structure attached to the posterior region of the epithalamus (see figure 20.1). It is composed primarily of **pinealocytes,** which secrete **melatonin** (mel-ă-tōn'in; *melas* =

dark hue, *tonas* = contraction), a hormone whose production tends to be cyclic; it increases at night and decreases during the day. Melatonin secretion helps regulate a circadian rhythm (24-hour body clock). It also appears to affect the synthesis of the hypothalamic regulatory hormone responsible for FSH and LH synthesis. The role of melatonin in sexual maturation is not well understood. However, it is known that the removal of the pineal gland in experimental animals results in their premature sexual maturation, while excessive melatonin secretion delays puberty in humans. The pineal gland decreases in size with age.

The **thymus** (thĭ′mŭs; *thymos* = sweetbread) is a bilobed structure located within the mediastinum superior to the heart and immediately posterior to the sternum (see figure 20.1). The size of the thymus varies among individuals; however, it is always relatively large in infants and children. The thymus diminishes in size and activity with age, especially after puberty. The adult thymus is usually only one-third to one-fourth its pre-puberty size and weight. Most of the functional cells of the thymus are lost, and the ensuing space fills with adipose connective tissue.

The thymus functions principally in association with the lymphatic system to regulate and maintain body immunity. It produces **thymopoietin** (thĭ′mō-poy-ē′tin) and **thymosins** (thĭ′mō-sins) (a group of complementary hormones). These hormones act by stimulating and promoting the differentiation, growth, and maturation of a category of lymphocytes called T-lymphocytes (thymus-derived lymphocytes). The thymus will be discussed in greater detail in chapter 24.

 WHAT DID YOU LEARN?

 Where is the pineal gland located?

Endocrine Functions of the Kidneys, Heart, Gastrointestinal Tract, and Gonads

Key topic in this section:

- How hormones secreted by the kidneys, heart, gastrointestinal tract, and gonads help regulate homeostasis

Some of the organs of the urinary, cardiovascular, digestive, and reproductive systems contain their own endocrine cells, which secrete hormones. These hormones help regulate the following: electrolyte levels in the blood, and red blood cell production; blood volume, and blood pressure; digestive system activities; and sexual maturation and activity **(table 20.8)**.

Kidneys

Hormones secreted by endocrine cells in the kidneys help regulate electrolyte concentration in body fluids, the rate of red blood cell production, and an increase in both blood volume and blood pressure. The specific hormones the kidneys secrete are calcitriol and erythropoietin. Renin is an enzyme released by the kidneys.

Calcitriol (kal-si-trī′ol) is a form of vitamin D and is synthesized by specific endocrine cells of the kidneys when they are stimulated by increased blood levels of parathyroid hormone. Calcitriol stimulates the intestinal cells to increase their uptake of both calcium and phosphate, thus increasing calcium levels in body fluids.

The kidneys release **erythropoietin** (ĕ-rith-rō-poy′ĕ-tin) **(EPO)** when the blood's oxygen content is low. EPO causes increases in the rates of erythrocyte (red blood cell) production and maturation. The resulting elevated numbers of erythrocytes increase the blood's ability to transport oxygen.

Renin (rē′nin), an enzyme produced in the kidney, helps form a hormone (angiotensin II) that stimulates specific cells of the adrenal cortex to produce and secrete aldosterone (discussed earlier in this chapter). Renin is released in response to abnormal electrolyte concentration in the fluid being formed into urine in the kidneys. This is a warning sign of possible abnormal electrolyte concentration in the blood.

Heart

Modified cardiac muscle cells in the wall of the right atrium secrete the hormone **atriopeptin** (ā′trē-ō-pep′tin), also called *atrial natriuretic* (nā′trē-ū-ret′ik) *peptide* (*ANP*), in response to excessive stretch in the wall of the heart. This stretching is caused by a marked elevation in either blood volume or blood pressure. Atriopeptin reduces blood volume by causing water loss and sodium excretion from the blood into the urine. With less blood volume, blood pressure drops.

Gastrointestinal Tract

The gastrointestinal (GI) tract, or digestive tract, consists of the organs involved in ingesting and digesting food, which include the stomach, small intestine, large intestine, liver, pancreas, and gallbladder. Some of these organs have their own endocrine cells, which produce hormones that help regulate digestive activities, and stimulate the liver, gallbladder, and pancreas to release secretions needed for efficient digestion. These hormones will be further discussed in chapter 26.

Gonads

The **gonads** (gō′nad; *gone* = seed) are the female and male primary sex organs—the ovaries in females and the testes in males. In addition to producing gametes, the gonads also produce sex hormones. The **ovaries** produce the female sex hormones **estrogen** and **progesterone,** while the **testes** produce the male sex hormones called androgens, many of which are converted into **testosterone.** The gonads also produce **inhibin,** which inhibits follicle-stimulating hormone secretion. The specific functions of these hormones are discussed in chapter 28.

 **WHAT DID YOU LEARN?**

 What hormone is produced by cardiac muscle cells in the heart wall?

Table 20.8	Organs with Endocrine Functions, Their Hormones, and Hormonal Effects	
Gland/Endocrine Cells	**Hormone**	**Hormonal Effects**
Kidney[1]	Calcitriol	Promotes calcium absorption in small intestine
	Erythropoietin	Stimulates erythrocyte production and maturation
Heart	Atriopeptin	Increases sodium and water loss in urine, resulting in decreased blood pressure and volume
GI tract	Various hormones related to digestion	Controls overall secretory activity and motility in GI tract
Ovaries	Estrogen	Stimulates development of female reproductive organs, follicle maturation; regulates menstrual cycle; stimulates growth of mammary glands
	Progesterone	Regulates menstrual cycle; stimulates growth of uterine lining; stimulates growth of mammary glands
	Inhibin	Inhibits secretion of follicle-stimulating hormone (see chapter 28)
Testes	Androgens (primarily testosterone)	Stimulates male reproductive organ development, production of sperm
	Inhibin	Inhibits secretion of follicle-stimulating hormone (see chapter 28)

[1] Renin is released by the kidneys, but it is an enzyme, not a hormone. Renin helps in the production of angiotensin II, a hormone that stimulates the production of aldosterone.

Aging and the Endocrine System

Key topic in this section:

- How endocrine activity changes as people age

Aging reduces the efficiency of endocrine system functions, resulting in decreased secretory activities of endocrine glands and reduced hormone levels in the blood. Many abnormalities or disorders experienced after middle age, such as abdominal weight gain or muscle loss, are directly related to endocrine gland dysfunction or imbalance. For example, the secretion of GH and sex hormones often changes. Reduction in GH levels leads to loss of weight and body mass in the elderly, although continued exercise reduces this effect. In addition, testosterone or estrogen levels decline as males and females age. Often hormone replacement therapy (HRT), attempts to supplement GH and sex hormone levels that have naturally diminished with age.

Development of the Endocrine System

Key topic in this section:

- Events in adrenal, pituitary, and thyroid gland development

Endocrine organs originate at different times during development and are derived collectively from all three germ layers (ectoderm, mesoderm, and endoderm). Here, we describe the development of three representative endocrine organs (the adrenal glands, the pituitary gland, and the thyroid gland), while the development of other endocrine organs is discussed in chapters 24 and 28.

Adrenal Glands

Early in the second month of embryonic development, the two future components of the adrenal gland begin to develop: (1) The adrenal cortex is derived from intermediate mesoderm located in a region called the **urogenital** (ū′rō-jen′i-tal) **ridge.** This urogenital ridge forms the future gonads and some components of the urinary system. (2) The adrenal medulla develops from neural crest cells that migrated from the developing neural tube (see figure 18.13). Thus, the adrenal medulla has a neuroectodermal origin. These neural crest cells migrate to the urogenital ridge region, where they are engulfed by the intermediate mesoderm that forms the developing adrenal cortex. Maturation and development of the adrenal gland continue into early childhood.

Pituitary Gland

The anterior and posterior pituitary have different developmental origins **(figure 20.15)**. The anterior pituitary develops as an outpocketing of the ectoderm from the stomodeum (future opening of the mouth). The posterior pituitary develops as a bud that grows from the developing hypothalamus.

Both parts of the pituitary gland start growing during the third week of development. A growth of tissue called the **hypophyseal pouch** (or *Rathke pouch*) forms in the stomodeum (the future mouth) and grows superiorly toward the future diencephalon. At the same time, a **neurohypophyseal bud** grows from the developing diencephalon toward the stomodeum. This bud later forms most of the adult infundibulum and the posterior pituitary. By late in the second month of development, the hypophyseal pouch loses its connection with the oral cavity as it grows superiorly toward the developing neurohypophyseal bud. Eventually, the hypophyseal pouch and neurohypophyseal bud adhere to one another, and the pituitary gland is formed.

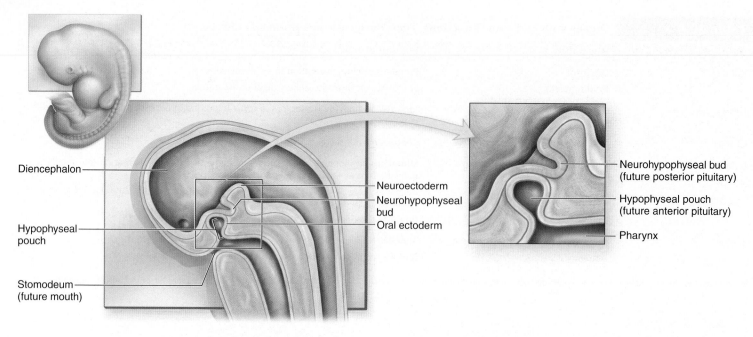

(a) **Week 3: Hypophyseal pouch and neurohypophyseal bud form.**

Diencephalon

Hypophyseal pouch

Stomodeum (future mouth)

Neuroectoderm

Neurohypophyseal bud

Oral ectoderm

Neurohypophyseal bud (future posterior pituitary)

Hypophyseal pouch (future anterior pituitary)

Pharynx

Infundibulum

Hypophyseal pouch

Neurohypophyseal bud

(b) **Late second month: Hypophyseal pouch loses contact with roof of pharynx.**

Anterior pituitary

Pars tuberalis

Pars intermedia

Pars distalis

Posterior pituitary

Median eminence

Pars nervosa

(c) **Fetal period: Anterior and posterior pituitary have formed.**

Figure 20.15

Pituitary Gland Development.   The pituitary gland forms from two separate structures. *(a)* During the third week of development, a hypophyseal pouch (future anterior pituitary) grows from the roof of the pharynx, while a neurohypophyseal bud (future posterior pituitary) forms from the diencephalon. *(b)* By late in the second month, the hypophyseal pouch loses contact with the roof of the pharynx as it merges with the hypophyseal bud. *(c)* During the fetal period, the anterior and posterior pituitary develop fully.

CLINICAL VIEW

Thyroid Gland Developmental Anomalies

Variations occur in how the developing thyroid gland migrates and how the thyroglossal duct regresses. Normally, the thyroid gland should migrate to a position inferior to the larynx, and the thyroglossal duct should regress completely. But sometimes, a portion of the thyroglossal duct persists and forms a variable median-placed **pyramidal lobe** of the thyroid gland.

Another developmental anomaly of the thyroid is an **ectopic thyroid gland,** in which some of the developing thyroid, or the entire

gland, is located along its path of descent. Thus, tissue identifiable as thyroid gland may be found more superiorly in the neck or even along the base of the tongue. This condition is not very common.

Finally, **thyroglossal cysts** may occur if part or all of the thyroglossal duct persists and doesn't form a pyramidal lobe. The remnant of this duct can gradually fill with fluid produced by the cells lining the thyroglossal duct, forming a painless, movable mass in the center of the neck. Thyroglossal duct cysts are typically painless and seen in children at about age 5. Such cysts may require surgical removal.

Thyroid Gland

During the end of the fourth week, the thyroid gland begins to form from endodermal tissue **(figure 20.16)**. The gland first forms as a mass called the **thyroid diverticulum** between the developing anterior two-thirds and the posterior one-third of the tongue. This is called the **foramen cecum** of the tongue. As the thyroid diverticulum develops, it

migrates and descends through the developing neck, initially maintaining a connection with the tongue called the **thyroglossal duct.**

By the seventh week of development, the thyroid gland reaches its final position anterior to the trachea and immediately inferior to the thyroid cartilage of the larynx. The thyroid gland becomes functional by weeks 10–12 of development.

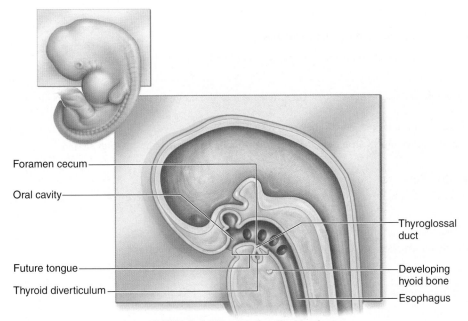

(a) Week 4: Thyroid diverticulum forms.

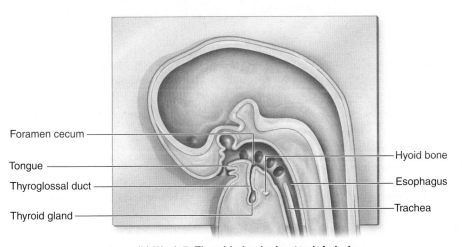

(b) Week 7: Thyroid gland migrates inferiorly.

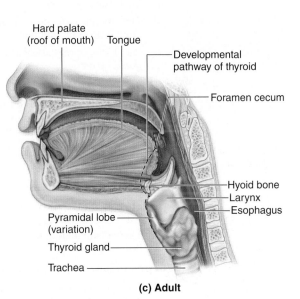

(c) Adult

Figure 20.16

Thyroid Gland Development. The thyroid gland begins to form in the fourth week as an invagination in the floor of the future tongue. The thyroid initially remains attached to the tongue via a thyroglossal duct. *(a)* Inferior extension of the thyroid diverticulum at development week 4. *(b)* By week 7, the thyroid gland migrates inferiorly, and the thyroglossal duct starts to regress. *(c)* In an adult, the thyroid is anterior to the trachea and inferior to the larynx.

CLINICAL TERMS

hypocalcemia (hī′pō-kal-sē′mē-ă) Abnormally low levels of calcium in the circulating blood, resulting in muscle spasms and tetany.

ketoacidosis (kē′tō-as-i-dō′sis) Condition that occurs as a result of diabetes or starvation; an enhanced production of ketones (metabolic acids that accumulate during fatty acid metabolism) in the blood leads to markedly lower blood pH, which can be life-threatening.

myxedema (mik-se-dē′mă; *myxa* = mucus, *oidema* = swelling) Type of hypothyroidism that occurs in adults and is characterized by somnolence, low body temperature, dry skin, and sluggish mental activity.

thyrotoxicosis (thī′rō-tok-si-kō′sis; *thyro* = thyroid, *toxikon* = poison, *osis* = condition) Condition resulting from excessive quantities of thyroid hormone; in its acute form, the patient experiences elevated temperature, rapid heart rate, and abnormal activity in a variety of other body systems.

CHAPTER SUMMARY

■ Homeostasis is maintained by nervous and endocrine system activities.

Endocrine Glands and Hormones 604

■ Endocrine glands are ductless organs that secrete their molecular products directly into the bloodstream.

Overview of Hormones 604

■ Hormones are chemical messengers in the endocrine system. Upon release into the bloodstream, these secretions affect intercellular activities.

■ Target cells have receptors for specific hormones to which they will respond.

■ The three groups of hormones based on chemical structure are peptide hormones, biogenic amines, and steroid hormones.

Negative and Positive Feedback Loops 604

■ Negative feedback inhibits the activity of a pathway, while positive feedback accelerates the activity of the original pathway.

Hypothalamic Control of the Endocrine System 607

■ Regulatory hormones from the hypothalamus control the activity of the anterior pituitary.

■ The hypothalamus produces the two hormones stored in and released from the posterior pituitary.

Pituitary Gland 609

■ The anterior and posterior pituitary parts differ in origin, structure, and function.

Anterior Pituitary 609

■ The anterior pituitary is composed of a large pars distalis, a slender pars intermedia, and the pars tuberalis that wraps around the infundibulum.

■ Regulatory hormones from the hypothalamus travel to the anterior pituitary through the hypothalamo-hypophyseal portal system.

Posterior Pituitary 613

■ The posterior pituitary houses axon endings from some hypothalamic neurons that are part of the hypothalamo-hypophyseal tract. Neurons housed in the supraoptic and paraventricular nuclei of the hypothalamus produce antidiuretic hormone and oxytocin.

Thyroid Gland 615

■ The thyroid gland, located immediately inferior to the thyroid cartilage of the larynx, is composed of two lateral lobes connected medially by a narrow isthmus.

Synthesis of Thyroid Hormone by Thyroid Follicles 615

■ Thyroid follicles are fluid-filled spheres within the thyroid gland; follicle cells synthesize thyroglobulin and store it within the colloid.

■ Thyroid-stimulating hormone from the anterior pituitary stimulates follicle cells to synthesize and release thyroid hormone.

Thyroid Gland–Pituitary Gland Negative Feedback Loop 616

■ A stimulus, such as low body temperature, causes thyroid hormone release. Both the released thyroid hormone and the effects it has on target cells block further release of thyroid hormone.

Parafollicular Cells 617

■ The parafollicular cells produce calcitonin to help lower calcium ion concentrations in body fluids.

Parathyroid Glands 619

■ Four small, nodular parathyroid glands are embedded in the posterior surface of the thyroid gland; they have chief cells that secrete parathyroid hormone in response to decreased levels of calcium ions within body fluids.

■ PTH secretion causes increased osteoclast activity, reduced loss of calcium in the urine, and increased calcium absorption in the intestine.

Adrenal Glands 620

■ Adrenal glands are located along the superior border of each kidney. Each adrenal gland has a peripheral cortex and a medulla core within its capsule.

Adrenal Cortex 622

■ The adrenal cortex synthesizes and secretes steroid hormones collectively called corticosteroids: (1) mineralocorticoids, principally aldosterone; (2) glucocorticoids, primarily cortisol and corticosterone; and (3) relatively minor concentrations of gonadocorticoids.

Adrenal Medulla 624

■ The adrenal medulla is composed of groups of chromaffin cells that secrete either epinephrine or norepinephrine.

Pancreas 625

■ The pancreas is a mixed gland with both exocrine and endocrine functions.

■ Endocrine cells form clusters called pancreatic islets, which are composed of four cell types: Alpha cells secrete glucagon, beta cells produce insulin, delta cells produce somatostatin, and F cells secrete pancreatic polypeptide.

Pineal Gland and Thymus 627	■ The pineal gland has secretory cells called pinealocytes that synthesize melatonin.
	■ The thymus, located in the mediastinum, produces several hormones called thymosins.
Endocrine Functions of the Kidneys, Heart, Gastrointestinal Tract, and Gonads 628	**Kidneys 628** ■ Endocrine cells in the kidneys regulate some homeostatic activities by secreting calcitriol, erythropoietin, and renin. **Heart 628** ■ Modified cardiac muscle cells secrete the hormone atriopeptin, which causes both water and sodium loss from the kidneys. **Gastrointestinal Tract 628** ■ Hormones secreted by endocrine cells in the stomach and small intestine promote activity during digestion. **Gonads 628** ■ Ovaries synthesize and secrete estrogen and progesterone; testes produce androgens, primarily testosterone.
Aging and the Endocrine System 629	■ Usually, the secretory activity of endocrine glands decreases with age, especially the production and activities of GH and estrogen.
Development of the Endocrine System 629	■ Organs of the endocrine system develop individually at different times during the embryonic and fetal stages. **Adrenal Glands 629** ■ The adrenal cortex is derived from intermediate mesoderm, and the adrenal medulla develops from neural crest cells, beginning early in the second month. **Pituitary Gland 629** ■ The anterior pituitary develops as an outpocketing of the ectoderm from the stomodeum, while the posterior pituitary develops as a bud from the developing hypothalamus. **Thyroid Gland 630** ■ The thyroid begins to form from endoderm of the primitive tongue at the end of the fourth week of development.

CHALLENGE YOURSELF

Matching

Match each numbered item with the most closely related lettered item.

_____ 1. oxytocin

_____ 2. prolactin

_____ 3. glucagon

_____ 4. target cells

_____ 5. colloid

_____ 6. thyroid-stimulating hormone

_____ 7. mineralocorticoids

_____ 8. epinephrine

_____ 9. thymosin

_____ 10. hypothalamus

a. contained within the thyroid follicle

b. produces regulatory hormones

c. stimulates milk production

d. stimulates thyroid hormone release

e. involved in maturation of lymphocytes

f. stimulates contraction of uterine wall

g. synthesized by adrenal medulla

h. made by alpha cells of pancreas

i. have receptors for specific hormones

j. electrolyte balance in body fluids

Multiple Choice

Select the best answer from the four choices provided.

_____ 1. Retention of both water and sodium from the kidney occurs as a result of the production and release of
   a. epinephrine.
   b. thyroid hormone.
   c. antidiuretic hormone.
   d. glucocorticoid.

_____ 2. When glucose levels in the blood are elevated,
   a. glucagon is released.
   b. insulin is released.
   c. thyroid hormone is released.
   d. aldosterone is released.

_____ 3. Which of the following is a tropic hormone?
   a. glucagon
   b. epinephrine
   c. melatonin
   d. growth hormone

_____ 4. Follicle-stimulating hormone stimulates the production of
   a. steroids in the adrenal cortex.
   b. gametes (sperm and oocytes) in males and females.
   c. hormones in parafollicular cells of the thyroid gland.
   d. melanin in melanocytes.

_____ 5. Which hormone is *not* synthesized by cells in the pituitary gland?
   a. adrenocorticotropic hormone
   b. growth hormone
   c. luteinizing hormone
   d. thyroid hormone

_____ 6. What is the function of glucocorticoids?
   a. maintain Na^+ levels
   b. develop the circadian rhythm
   c. lower calcium levels in the blood
   d. elevate blood glucose levels during periods of stress

_____ 7. Which hormone is antagonistic to parathyroid hormone?
   a. thyroid hormone
   b. insulin
   c. calcitonin
   d. norepinephrine

_____ 8. Parathyroid hormone is released from the parathyroid gland when blood levels of _____ fall.
   a. sodium
   b. calcium
   c. potassium
   d. iodine

_____ 9. The secretion of epinephrine and norepinephrine from the adrenal medulla is stimulated by
   a. adrenocorticotropic hormone.
   b. increased levels of growth hormone.
   c. sympathetic nerve innervation.
   d. increased levels of glucose in body fluids.

_____ 10. What hormone is released from the kidneys?
   a. follicle-stimulating hormone
   b. erythropoietin
   c. glucagon
   d. growth hormone

Content Review

1. Describe the relationship between hormones and target organs.
2. Identify the three chemical classes of hormones, and give an example of each.
3. Explain how the hypothalamus oversees and controls endocrine system function.
4. Describe the structure of the hypothalamo-hypophyseal portal system, and briefly discuss the advantage of the structural arrangement.
5. What seven hormones are released by the anterior pituitary, what target organs do they affect, and what are the functions of these hormones?
6. What is the function of calcitonin, and under what conditions might the body release it?
7. What are the consequences of insufficient parathyroid hormone (PTH) secretion?
8. Describe the structure of the adrenal cortex. What hormones are secreted in each region?
9. Identify the cell types of the pancreatic islets and the hormones secreted by each type.
10. Explain why the kidneys, heart, GI tract, and gonads are considered part of the endocrine system.

Developing Critical Reasoning

1. The TV news reports that children and teenagers today are heavier than ever and that these populations are at increased risk for type 2 diabetes. Your father asks, "Isn't type 2 diabetes called adult-onset diabetes? If so, then why are children at risk?" How would you answer your father? Compare and contrast type 1 and type 2 diabetes, and discuss how the terms "adult-onset diabetes" and "juvenile diabetes" may no longer be appropriate.

2. Susan is a 35-year-old mother of two who works as an admissions officer at the university. Lately she has been extremely lethargic, sleeps for long periods of time, and feels mentally slow. Additionally, she feels cold most of the time and exhibits muscular weakness. What tests should be performed, and what problem should her physician suspect?

ANSWERS TO "WHAT DO YOU THINK?"

1. Damage to the hypothalamus could affect its secretion of oxytocin and antidiuretic hormone. Also, the hypothalamus might not be able to properly release its regulating hormones, so anterior pituitary function could be adversely affected. Finally, since the hypothalamus oversees the autonomic nervous system, adrenal medulla function (which is controlled by sympathetic innervation) might be affected as well.

2. Iodine is needed for normal thyroid hormone production. Without adequate dietary levels of iodine, an individual is at risk for developing a goiter (an enlarged thyroid gland). In the United States, iodine is added to store-bought salt to ensure that we have adequate dietary levels of iodine.

3. Epinephrine is a treatment for anaphylactic shock, an immediate allergic reaction in which the blood vessels dilate and the airways constrict. Blood pressure drops, and breathing becomes difficult. Epinephrine counteracts these effects by constricting the blood vessels (and thereby raising blood pressure) and opening the airways. People who are extremely allergic may die if not treated promptly.

Visit the McKinley/O'Loughlin *Human Anatomy*, 2e website at aris.mhhe.com

21

Anatomy &
Physiology | **REVEALED**
aprevealed.com

CARDIOVASCULAR SYSTEM

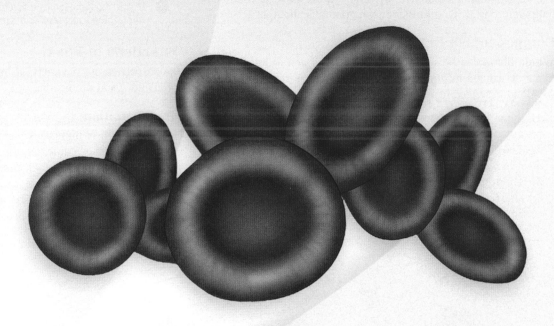

Blood

Within our bodies is a connective tissue so valuable that donating a portion of it to someone else can save that person's life. This tissue regenerates itself continuously and is responsible for transporting the gases, nutrients, and hormones our bodies need for proper functioning. Losing too much of this tissue can kill us, and yet it is something we frequently take for granted.

This valuable connective tissue is blood. Blood is considered a fluid connective tissue because it contains cells, a liquid ground substance (called plasma), and dissolved proteins. In this chapter, we describe the components of blood, its functions, and how the body produces the various types of blood components.

General Composition and Functions of Blood

Key topics in this section:

- Basic components of blood
- How blood functions in transport, regulation, and protection

Blood is a type of fluid connective tissue (see chapter 4). Blood is about four times more viscous than water, meaning that it is thicker and more "goopy." The temperature of blood is about 1°C higher than measured body temperature; thus, if your body temperature is 37°C, your blood temperature is about 38°C.

Components of Blood

Whole blood can be separated into its liquid and cellular components by using a machine called a **centrifuge**, as shown in **figure 21.1** and described here:

1. Blood to be sampled is withdrawn from a vein and collected in a glass tube, called a *centrifuge tube.*
2. The glass tube is placed into the centrifuge, which then spins it in a circular motion for several minutes.
3. The rotational movement separates the blood into liquid and cellular components, thus allowing these elements to be examined separately.

Figure 21.2 shows the three components separated by centrifugation, from bottom to top in the test tube:

- **Erythrocytes** (ĕ-rith′rō-sīt; *erythros* = red, *kytes* = cell), sometimes called *red blood cells*, form the lower layer of the centrifuged blood. They typically average about 44% of a blood sample.
- A **buffy coat** makes up the middle layer. This thin, slightly gray-white layer is composed of cells called **leukocytes** (or *white blood cells*) and cell fragments called **platelets.** The buffy coat forms less than 1% of a blood sample.
- **Plasma** is a straw-colored liquid that lies above the buffy coat in the centrifuge tube; it generally makes up about 55% of blood.

Collectively, the erythrocytes and the components of the buffy coat are called the **formed elements.** It is best not to refer to all of these structures as "cells" because platelets are merely fragments broken off from a larger cell. The formed elements, together with the liquid plasma, compose whole blood (the substance we most commonly refer to simply as "blood").

WHAT DO YOU THINK?

❶ If your body becomes dehydrated, does the plasma percentage in whole blood increase or decrease?

Functions of Blood

Blood carries out a variety of important functions: transportation, regulation, and protection.

Transportation

Blood transports numerous elements and compounds throughout the body. For example, erythrocytes and plasma carry oxygen from the lungs to body cells and then transport the carbon dioxide produced by the cells back to the lungs for expulsion from the body. Blood plasma transports nutrients that have been absorbed from the GI tract. Plasma also transports hormones secreted by

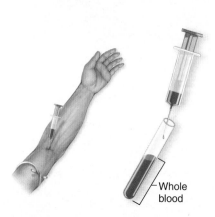

① Withdraw blood into a syringe and place it into a glass centrifuge tube.

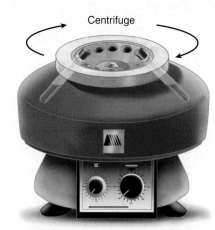

② Place the tube into a centrifuge and spin for about 10 minutes.

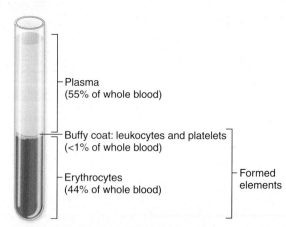

③ Components of blood separate during centrifugation to reveal plasma, buffy coat, and erythrocytes.

Plasma
(55% of whole blood)

Buffy coat: leukocytes and platelets
(<1% of whole blood)

Erythrocytes
(44% of whole blood)

Formed
elements

Figure 21.1

Whole Blood Separation.   A sample of whole blood is used to determine the ratio of plasma to formed elements. The blood sample is drawn from a vein and placed into a glass tube. After centrifugation, the formed elements in the sample remain packed in the bottom of the centrifuge tube.

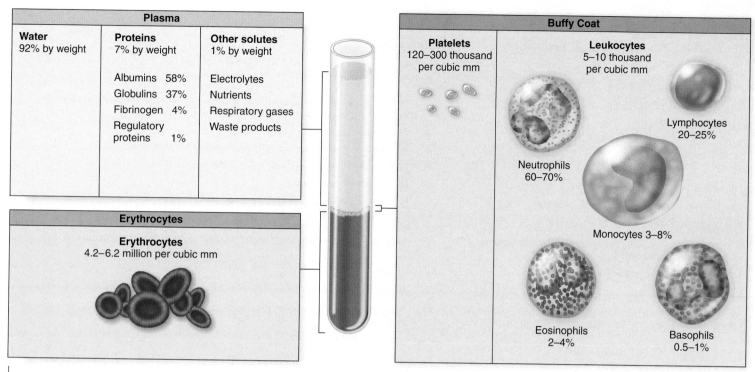

Figure 21.2

Whole Blood Composition.   Whole blood contains plasma (average = about 55%) and formed elements (average = about 45%). (The percentages presented in this figure are average numbers of cells, and the numbers for components of the buffy coat represent average ranges. A cubic millimeter of blood is equivalent to a microliter (μL) of blood.)

the endocrine glands. Finally, plasma carries some waste products from the cells to organs such as the kidneys, where these waste products are removed.

Regulation

Blood regulates body temperature in several ways. Plasma absorbs and distributes heat throughout the body. If the body needs to be cooled, the blood vessels in the dermis dilate and dissipate the excess heat through the integument. Conversely, when the body needs to conserve heat, the dermal blood vessels constrict, and the warm blood is shunted to deeper blood vessels in the body (see chapter 5).

Blood also helps regulate pH levels in the body's tissues. The term **pH** is a measure of how acidic or alkaline a fluid is. A neutral pH (neither acidic nor alkaline fluid, such as water) is measured at exactly 7, while **acidic** fluids (e.g., orange juice) are between 0 and 7, and **alkaline** fluids (e.g., milk) are between 7 and 14. Blood plasma contains compounds and ions that may be distributed to the fluid bathing cells within the tissues (interstitial fluid) to help maintain normal tissue pH. In addition, blood plasma pH is continuously regulated to try to maintain a value of 7.4, which is the pH level required for normal cellular functioning. If the blood pH falls below 7.4 to 7.0, the condition called *acidosis* results, and the central nervous system is depressed; coma and death could occur. If the blood pH rises to 7.8, *alkalosis*, characterized by a hyperexcited nervous system and convulsions, results.

Blood maintains normal fluid levels in the cardiovascular system and prevents fluid loss. A constant exchange of fluid takes place between the blood plasma and the interstitial fluid. If too much fluid is absorbed into the blood, high blood pressure results. If too much fluid escapes the bloodstream and enters the tissues, blood pressure

drops to unhealthily low levels, and the tissues swell with excess fluid. In order to maintain a balance of fluid between the blood and the interstitial fluid, blood contains molecules (such as salts and some proteins) to prevent excess fluid loss from the plasma.

Protection

Leukocytes help guard against infection by mounting an immune response if a pathogen or an **antigen** (an'ti-jen; *anti* = opposite, *gen* = producing) (a substance perceived as foreign to the body) is found. **Antibodies** (an'tē-bod-ē; *body* = main part), which are molecules that can immobilize antigens until a leukocyte can completely kill or remove the antigen, are transported in plasma. In addition, platelets and blood proteins protect the body against blood loss by forming blood clots.

WHAT DID YOU LEARN?

❶ Erythrocytes make up what average percentage of whole blood?

❷ What are the protective functions of the blood?

Blood Plasma

Key topic in this section:

■ Components of plasma

Blood plasma is a complex mixture of water, proteins, and other solutes (**table 21.1**). When the blood cells, platelets, and clotting proteins are removed from plasma, the remaining fluid is termed **serum** (ser'um; whey).

Table 21.1	Composition of Blood Plasma
Plasma Component (Percentage of Plasma)	**Functions**
WATER (~92% OF PLASMA)	Acts as the solvent in which formed elements, solutes, and wastes are suspended
PLASMA PROTEINS (~7% OF PLASMA)	
Albumin (~58% of plasma proteins)	Regulates water movement between the blood and interstitial fluid (and thus the viscosity of blood); transports some fatty acids and hormones
Globulins (~37% of plasma proteins)	Alpha-globulins transport lipids and some metal ions Beta-globulins transport iron ions and lipids in bloodstream Gamma-globulins are antibodies that immobilize pathogens (bacteria, viruses, etc.)
Fibrinogen (~4% of plasma proteins)	Helps with blood clotting
Regulatory proteins (<1% of plasma proteins)	Consists of enzymes, proenzymes, and hormones
OTHER SOLUTES (~1% OF PLASMA)	
Electrolytes (e.g., sodium, potassium, calcium, chloride, iron, bicarbonate, and hydrogen)	Help establish and maintain membrane potentials, maintain pH balance, and regulate osmosis
Nutrients (e.g., amino acids, glucose, cholesterol, vitamins, fatty acids)	Energy source
Respiratory gases	Oxygen and carbon dioxide
Wastes (breakdown products of metabolism) (e.g., lactic acid, creatinine, urea, bilirubin, ammonia)	Waste products are transported to the liver and kidneys where they can be removed from the blood

Water is the most abundant compound in plasma, making up about 92% of plasma's total volume. Water facilitates the transport of materials in the plasma. The next most abundant compounds in plasma are the plasma proteins.

Plasma Proteins

Plasma proteins make up about 7% of the plasma (see figure 21.2). Measured amounts of plasma proteins usually range between 6 and 8 grams of protein in a volume of 100 milliliters of blood (referred to as grams per deciliter [g/dl]). The plasma proteins include albumins, globulins, fibrinogen, and regulatory proteins.

Albumins (al-bū′min; *albumen* = white of egg) are the smallest and most abundant of the plasma proteins, making up approximately 58% of total plasma proteins. They regulate water movement between the blood and interstitial fluid. Secondarily, albumins act as transport proteins that carry ions, hormones, and some lipids in the blood.

Globulins (glob′ū-lin; *globules* = globule) are the second-largest group of plasma proteins, forming about 37% of all plasma proteins. The smaller **alpha-globulins** and the larger **beta-globulins** primarily bind, support, and protect certain water-insoluble or hydrophobic molecules, hormones, and ions. **Gamma-globulins,** also called *immunoglobulins* or antibodies, are soluble proteins produced by some of our defense cells to protect the body against pathogens that may cause disease.

Fibrinogen (fī′brin′ō-jen; *fibra* = fiber) makes up about 4% of all plasma proteins. Fibrinogen is responsible for blood clot formation. Following trauma to the walls of blood vessels, fibrinogen is converted into long, insoluble strands of **fibrin,** which helps form a blood clot.

Regulatory proteins form a very minor class of plasma proteins (less than 1% of total plasma proteins) and include enzymes (proteins that accelerate chemical reactions), proenzymes (inactive precursors of enzymes), and hormones that are being transported to other parts of the body.

Differences Between Plasma and Interstitial Fluid

Plasma is a type of extracellular fluid (ECF), meaning it is a body fluid found outside of (rather than within) cells. Plasma and inter-

stitital fluid (the type of extracellular fluid that bathes the outside of cells) have similar concentrations of most dissolved products, nutrients, and electrolytes. However, the concentration of dissolved oxygen is higher in plasma than in interstitial fluid, because the cells take up and use the oxygen from the interstitial fluid during energy production. This difference in concentration ensures that oxygen will continue to diffuse from blood into the interstitial fluid. Similarly, the concentration of carbon dioxide is lower in blood than in interstitial fluid because cells produce carbon dioxide during energy production, and it diffuses out of the cells into the interstitial fluid. This difference in concentration ensures that carbon dioxide will readily diffuse from the interstitial fluid into the blood, where it will be carried to the lungs and discharged from the body.

WHAT DID YOU LEARN?

3 What are the components of plasma?

4 Identify the four classes of plasma proteins.

Formed Elements in the Blood

Key topics in this section:

- Structural and functional characteristics of erythrocytes
- Life cycle of erythrocytes
- Significance of the ABO and Rh blood groups
- Types of leukocytes and their functions
- Structure of platelets and their role in blood clotting

The formed elements have three components:

- Erythrocytes make up more than 99% of formed elements. Their primary function is to transport respiratory gases in the blood.
- Leukocytes make up less than .01% of formed elements. All leukocytes contribute to mounting an immune response and defending the body against pathogens.

■ Platelets make up less than 1% of formed elements and help with blood clotting.

Table 21.2 summarizes the characteristics of the formed elements.

The percentage of the volume of all formed elements in the blood is called the **hematocrit** (hē′mă-tō-krit, hem′ă-; *hemato* = blood, *krino* = to separate). This medical dictionary definition of the true hematocrit differs from the clinical definition, which equates the hematocrit to the percentage of erythrocytes. The difference between these two numbers is almost negligible, which is why the true hematocrit and the clinical hematocrit are virtually the same. Hematocrit values vary slightly and are dependent on the age and sex of the individual. Adult males tend to have a hematocrit ranging between 42% and 56%, while females' hematocrits range from 38% to 46%. Children's hematocrit ranges also vary in individuals and differ from adult values. In addition, altitude can affect the hematocrit. Let's say a person lives in a cabin high in the Rocky Mountains, where the air is thinner and there is less oxygen. Each time the person breathes at this altitude, she inhales relatively less oxygen than she would inhale at a lower altitude. The person's body compensates by making more erythrocytes; more erythrocytes in the blood can carry more oxygen to the tissues. However, having more erythrocytes means that the blood is relatively more viscous than that of a person living at a lower altitude. Blood that is more viscous is more likely to cause cardiovascular complications, such as major blood clots that lead to heart attacks or strokes.

All of the components of the formed elements can be viewed by preparing a **blood smear,** as shown in **figure 21.3** and described here:

1. A finger is pricked, and a small amount of blood is collected.
2. A blood drop is placed onto a glass slide.
3. A second slide spreads the drop of blood across the first slide, smearing a thin surface of blood along the slide (hence the name "blood smear").
4. The thin layer of blood is stained to provide contrast for viewing after the smear dries. After the stain dries, a glass coverslip is placed over the specimen to protect it.

The prepared slide is then viewed using the light microscope.

Erythrocytes

Although erythrocytes are commonly referred to as red blood cells or RBCs, the term "cell" is a misnomer because mature erythrocytes lack nuclei and organelles. In other words, an erythrocyte is not like other cells in the body, so it is more appropriate to call it a formed element.

CLINICAL VIEW

Blood Doping

In order to enhance their performance in endurance events, some athletes may try to boost their bodies' ability to deliver oxygen to the muscles by increasing the number of erythrocytes in their blood (and thus increasing their hematocrit levels). This can be accomplished in several ways. First, the number of erythrocytes can be increased naturally by living and training at high altitude where the concentration of oxygen in the air is lower. The body compensates for the decreased oxygen concentration in the atmosphere by increasing the rate of erythrocyte production, thus increasing the number of erythrocytes per unit volume of blood.

Another method used by some athletes in recent years is **blood doping.** In this procedure, the athlete essentially donates erythrocytes to himself or herself. Prior to the athletic event, the individual has a unit of blood removed and stored. This causes the body to increase erythrocyte production to make up for the ones just removed. A few days before the competition, the erythrocytes from the donated unit are transfused back into the person's body. The increased number of erythrocytes increases the amount of oxygen transported in the blood and is thought to favorably affect muscle performance, thereby improving athletic performance. However, potential deadly dangers are inherent in blood doping. By increasing the number of erythrocytes per measured volume of blood, blood doping also increases the viscosity of the blood. Thus, the heart must work harder to pump this "thicker," more cellular blood. Eventually, temporary athletic success may be overshadowed by permanent cardiovascular damage that can even lead to death. Therefore, blood doping has now been banned from athletic competitions. In fact, just days before the start of the 2006 Tour de France, this premier cycling event was rocked when some leading contenders were suspended from the race after being charged with blood doping.

Erythrocytes transport oxygen and carbon dioxide to and from the tissues and the lungs, respectively. The fact that erythrocytes lack a nucleus and organelles enables them to carry these respiratory gases more efficiently. A normal, mature erythrocyte is a very small structure, with a diameter of approximately 7.5 micrometers (μm) **(figure 21.4)**. Its unique, biconcave disc structure (at its narrowest point ∼.75 μm and at its widest point ∼2.6 μm) allows respiratory gases to be loaded and unloaded rapidly and efficiently. Erythrocytes line up in single file, termed a **rouleau** (roo-lō′; pl. rouleaux; cylinder), as they pass through small blood vessels. The

Table 21.2	Characteristics of the Formed Elements			
Formed Element	**Size (all measurements are for diameter)**	**Function**	**Life Span**	**Density (average number per mm³ of blood = μl)**
Erythrocytes	7.5 μm	Transport oxygen and carbon dioxide	∼ 120 days	Females: ∼ 4.8 million Males: ∼ 5.4 million
Leukocytes	1.5 to 3 times larger than an erythrocyte; 11.25–22.5 μm	Prepare immune response, defend against antigens	Varies from 12 hours (neutrophil) to years (lymphocyte)	5,000–10,000
Platelets	Less than 1/4 the size of an erythrocyte; ∼ 2 μm	Participate in blood clotting	∼ 8–10 days	120,000–300,000

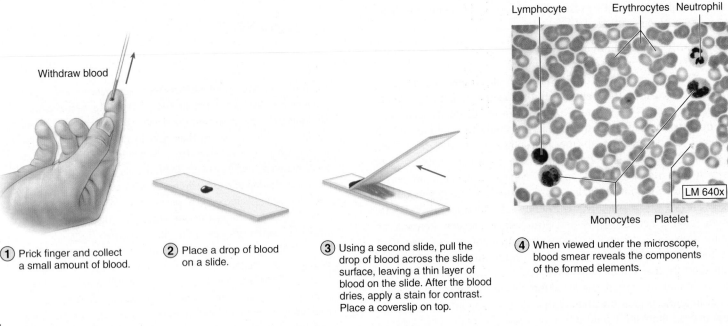

① Prick finger and collect a small amount of blood.

② Place a drop of blood on a slide.

③ Using a second slide, pull the drop of blood across the slide surface, leaving a thin layer of blood on the slide. After the blood dries, apply a stain for contrast. Place a coverslip on top.

④ When viewed under the microscope, blood smear reveals the components of the formed elements.

Figure 21.3

Preparing a Blood Smear.

number of erythrocytes in the bloodstream normally ranges between 4.2 and 6.2 million per cubic millimeter of blood.

Hemoglobin in Erythrocytes

Every erythrocyte is filled with approximately 280 million molecules of a red-pigmented protein called **hemoglobin** (hē-mō-glō′bin; *haima* = blood). Hemoglobin transports oxygen and carbon dioxide, and is responsible for the characteristic bright red color of arterial blood. When blood is maximally loaded with oxygen, it is termed **oxygenated.** Conversely, when some oxygen is lost and carbon dioxide is gained during respiratory gas exchange, blood is called **deoxygenated.** Hemoglobin that contains no oxygen has a deep red color that is perceived as blue because the dark color of the blood within these vessels is observed through the skin and the subcutaneous layer.

Each hemoglobin molecule consists of four protein building blocks, called **globins.** Two of these globins are called **alpha (α) chains,** and the other two, which are slightly different, are called **beta (β) chains (figure 21.5).** All globin chains contain a nonprotein (or **heme**) group that is in the shape of a ring, with an iron ion (Fe^{2+}) in its center. Oxygen binds to these iron ions for transport in the blood. Since each molecule of hemoglobin has four rings, each hemoglobin molecule has four iron ions and is capable of binding four molecules of oxygen. The oxygen binding is fairly weak to ensure rapid attachment and detachment of oxygen with hemoglobin. The result is that oxygen binds to the hemoglobin when the erythrocytes pass through the blood vessels of the lungs, and it leaves the hemoglobin when the erythrocytes pass through the blood vessels of the body tissues. This gas exchange occurs by diffusion as a result of the differences in concentration of oxygen between two areas. For example, oxygen is in higher

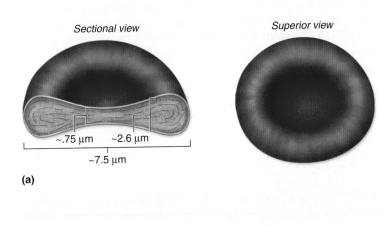

Sectional view

Superior view

~.75 μm   ~2.6 μm

~7.5 μm

(a)

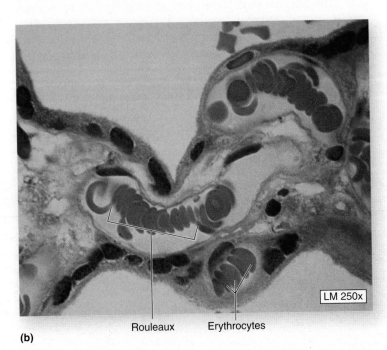

Figure 21.4

Erythrocyte Structure. (*a*) An erythrocyte has the gross structure of a biconcave disc, as shown here in sectional and superior views. (*b*) SEM of erythrocytes shows their three-dimensional structure and a rouleau.

Rouleaux   Erythrocytes

(b)

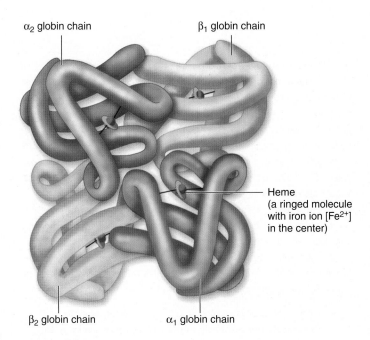

α₂ globin chain — β₁ globin chain

Heme
(a ringed molecule
with iron ion [Fe²⁺]
in the center)

β₂ globin chain — α₁ globin chain

Figure 21.5

Molecular Structure of Hemoglobin. A single molecule of hemoglobin is composed of four protein subunits, called globins, each containing a heme group that holds a single iron ion in its center. Each hemoglobin molecule transports four oxygen molecules that are weakly attracted to the iron ions.

concentration in the lungs compared to the blood, so oxygen diffuses from the lungs into the blood. Conversely, oxygen is in higher concentration in the blood compared to the interstitial fluid around body cells, so oxygen diffuses from the blood to the interstitial fluid. Carbon dioxide and the globin molecule (not the iron ion) have a similar weak attachment relationship for the transport of carbon dioxide molecules.

Erythrocyte Life Cycle

The absence of both a nucleus and cellular organelles comes at a cost to the erythrocyte by reducing its life span. A mature erythrocyte cannot synthesize proteins to repair itself or replace damaged membrane regions. Aging and the wear-and-tear of circulation through blood vessels cause erythrocytes to become more fragile and less flexible. Therefore, the erythrocyte has a finite life span of about 120 days **(figure 21.6)**. Every day, just under 1% of the oldest circulating erythrocytes are removed from circulation. The old erythrocytes are phagocytized in the liver and spleen by cells called macrophages (to be discussed later in this chapter). Some potential erythrocyte components are stored in other organs for recycling, while other components are excreted from the body, as shown in steps 4 and 5 of figure 21.6 and explained here:

- The heme group (minus the iron ion) in hemoglobin is converted first into a green pigment called **biliverdin** (bil-i-ver′din; *bilis* = bile). Biliverdin is eventually converted into a yellow-green pigment called **bilirubin** (bil-i-roo′bin). Bilirubin is a component of a digestive secretion called bile,

Figure 21.6

Recycling the Components of Aged or Damaged Erythrocytes. Erythrocytes have an average life span of about 120 days. Their molecular components are then broken down and recycled or eliminated from the body.

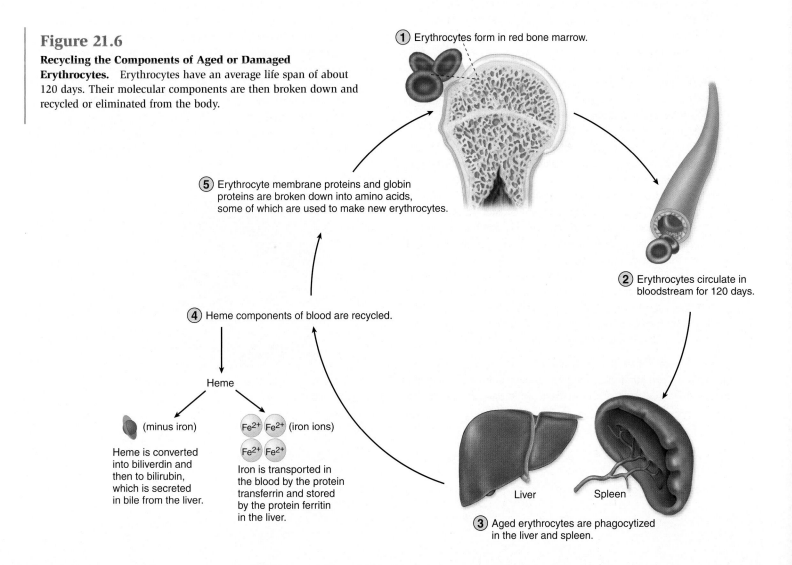

1 Erythrocytes form in red bone marrow.

5 Erythrocyte membrane proteins and globin proteins are broken down into amino acids, some of which are used to make new erythrocytes.

2 Erythrocytes circulate in bloodstream for 120 days.

4 Heme components of blood are recycled.

Heme

(minus iron)

Heme is converted into biliverdin and then to bilirubin, which is secreted in bile from the liver.

Fe²⁺ Fe²⁺ (iron ions)
Fe²⁺ Fe²⁺

Iron is transported in the blood by the protein transferrin and stored by the protein ferritin in the liver.

Liver Spleen

3 Aged erythrocytes are phagocytized in the liver and spleen.

CLINICAL VIEW: In Depth

Erythrocyte Volume Disorders

The number of erythrocytes in a person's blood can vary from the normal range, leading to various clinical disorders. In general, these conditions are classified as either anemia or polycythemia.

Anemia (a-nē'mē-ă; *an* = without) is any condition in which the count of erythrocytes per cubic millimeter of blood is less than the normal range. Anemia occurs due to either inadequate production or decreased survival of erythrocytes. The blood contains fewer erythrocytes than normal, and as a result body tissues are unable to get enough oxygen, so the heart may have to work harder. Symptoms of anemia include lethargy, shortness of breath, pallor of the skin and mucous membranes, fatigue, and heart palpitations. The types of anemia include the following:

- *Aplastic anemia* is characterized by significantly decreased formation of erythrocytes and hemoglobin due to defective red bone marrow. The causes of defective red bone marrow vary, but may include poisons, toxins, or radiation.
- *Congenital hemolytic anemia* occurs when destruction of erythrocytes is more rapid than normal, usually due to a genetic defect. It is caused by the production of abnormal membrane proteins that make the erythrocyte plasma membrane very fragile.
- *Erythroblastic anemia* is characterized by the presence of large numbers of immature, nucleated cells (called *erythroblasts* and *normoblasts*) in the circulating blood. A reduced rate of hemoglobin synthesis causes these immature cells to be present. These cells cannot function normally and thus anemia results.
- *Familial microcytic anemia* is a rare type of inherited anemia associated with a defect in iron uptake and use.
- *Hemorrhagic anemia* results from immediate blood loss due to such factors as chronic ulcers or heavy menstrual flow.
- *Macrocytic anemia* occurs when the average size of circulating erythrocytes is too large. Deficiencies in both vitamin B_{12} and folic acid uptake result in the production of enlarged erythrocytes.
- *Pernicious anemia* is a chronic progressive anemia in adults caused by the body's failure to absorb vitamin B_{12}. A defect in the production of *intrinsic factor* (a glycoprotein secreted by stomach lining cells to enhance B_{12} absorption in the small intestine) leads to pernicious anemia.
- *Sickle-cell disease* is an autosomal recessive anemia that occurs when a person inherits two copies of the sickle-cell gene. Erythrocytes become sickle-shaped, making them unable to flow efficiently through the blood vessels to body tissues and more prone to destruction by rupture (a process called *hemolysis*).

Most anemias are treated by letting the patient's own bone marrow replace the erythrocytes. However, anemia is often a symptom of another disease or problem. For example, while many anemias are due to iron deficiency, the iron deficiency is not due to diet, but rather to chronic blood loss, a process that depletes the body of its iron stores over months or years. The three most common causes of such chronic blood loss are excessive menstrual bleeding, undiagnosed stomach ulcer, and colon cancer. Imagine the magnitude of the mistake a physician could make by simply placing a patient on iron supplements when the underlying cause of the iron deficiency is an undiagnosed cancer of the colon! So, while restoring the patient's erythrocyte count, a physician should also look for any underlying cause of the anemia.

Polycythemia (pol'ē-sī-thē'mē-ă; *poly* = many, *kytos* = cell) is the condition of having too many erythrocytes in the blood (otherwise known as an elevated hematocrit). The affected person has the same total blood volume, but many more erythrocytes than are healthy. The blood becomes thick and viscous, putting a tremendous strain on the heart. Following are some of the different types of polycythemia:

- *Compensatory polycythemia* results from chronic *hypoxia* (inadequate oxygen supply to the body). Smokers develop this condition when long-term exposure to tobacco smoke and chronically high levels of carbon monoxide damage their lungs.
- *Relative polycythemia* is an increase in the number of erythrocytes in the blood per unit volume as a result of a decrease in blood plasma. For example, suppose that a child is severely dehydrated due to a serious case of diarrhea. As the child progressively loses water, his blood becomes more concentrated. This type of polycythemia is a temporary condition, and the ratio of erythrocytes to water in the blood returns to normal when the child becomes rehydrated.
- *Erythrocytosis* is an increase in erythrocytes due to an increase in the level of the hormone erythropoietin (EPO).
- *Polycythemia vera* is a chronic form characterized by an increase in blood volume and the number of erythrocytes. This condition results when erythrocyte growth in the red bone marrow is not regulated. Red cell precursors continue to grow and mature, irrespective of the presence or absence of erythropoietin.

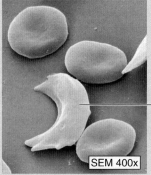

Sickle-shaped erythrocyte

SEM 400x

SEM of blood from a person with sickle-cell disease.

which is produced by liver cells. When bile enters the GI tract, it helps emulsify (break down) fats. The bilirubin in the GI tract is modified into other products that appear in urine from the kidneys and feces from the GI tract.

■ The iron ion component in hemoglobin is removed and transported by a beta-globulin protein, called **transferrin** (trans-fer′in; *trans* = across, *ferrum* = iron), to the liver where the iron ion is passed to another protein, called **ferritin** (fer′i-tin) for storage. Ferritin is stored in the liver and will be transported to the red bone marrow, as needed, for erythrocyte production.

■ Erythrocyte membrane proteins and globin proteins are broken down into free amino acids, some of which the body uses for protein synthesis to make new erythrocytes.

Blood Types

The plasma membrane of an erythrocyte has numerous molecules called **surface antigens** (or *agglutinogens*) that project from the plasma membrane surface. The most commonly identified group of antigens is the **ABO blood group.** This group has two surface antigens, called **A** and **B.** The presence or absence of either the A and/or B surface antigen are the criteria that determine your **ABO blood type,** as shown in **figure 21.7a** and listed here:

■ Blood with erythrocytes having surface antigen A is called **type A** blood.
■ Blood with erythrocytes having surface antigen B is called **type B** blood.

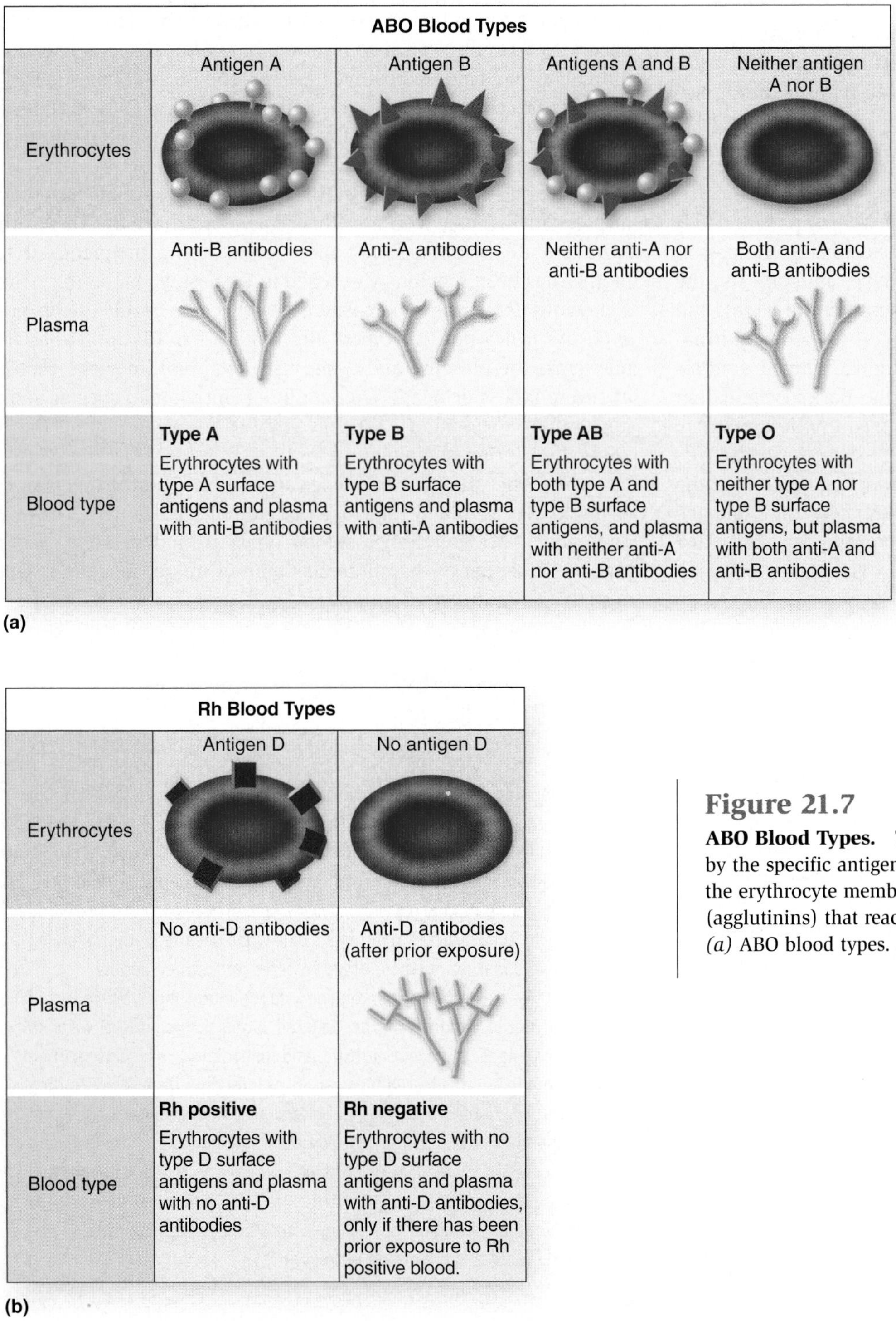

ABO Blood Types				
	Antigen A	Antigen B	Antigens A and B	Neither antigen A nor B
Erythrocytes				
	Anti-B antibodies	Anti-A antibodies	Neither anti-A nor anti-B antibodies	Both anti-A and anti-B antibodies
Plasma				
Blood type	**Type A** Erythrocytes with type A surface antigens and plasma with anti-B antibodies	**Type B** Erythrocytes with type B surface antigens and plasma with anti-A antibodies	**Type AB** Erythrocytes with both type A and type B surface antigens, and plasma with neither anti-A nor anti-B antibodies	**Type O** Erythrocytes with neither type A nor type B surface antigens, but plasma with both anti-A and anti-B antibodies

(a)

Rh Blood Types		
	Antigen D	No antigen D
Erythrocytes		
	No anti-D antibodies	Anti-D antibodies (after prior exposure)
Plasma		
Blood type	**Rh positive** Erythrocytes with type D surface antigens and plasma with no anti-D antibodies	**Rh negative** Erythrocytes with no type D surface antigens and plasma with anti-D antibodies, only if there has been prior exposure to Rh positive blood.

(b)

Figure 21.7

ABO Blood Types.   The blood type of an individual is determined by the specific antigens (agglutinogens) exposed on the surface of the erythrocyte membrane. Likewise, plasma contains antibodies (agglutinins) that react with antigens from outside the body. *(a)* ABO blood types. *(b)* Rh blood types.

- Blood with erythrocytes having surface antigens A and B is called **type AB** blood.
- Blood with erythrocytes having neither surface antigen A nor B is called **type O** blood.

The ABO surface antigens on erythrocytes are accompanied by specific antibodies (or *agglutinins*) that travel in the blood plasma. In general, an antibody is a protein that is produced by a white blood cell (specifically, a *B-lymphocyte*) and designed to recognize and immobilize an antigen it perceives as foreign to the body. An antibody interacts with a specific antigen. The ABO blood group has both **anti-A** and **anti-B antibodies** that react with the surface antigen A and the surface antigen B, respectively. Your blood plasma does not have antibodies that recognize the surface antigens on your erythrocytes. Within the ABO blood group, the following blood types and antibodies are normally associated:

- Type A blood has anti-B antibodies in its blood plasma.
- Type B blood has anti-A antibodies in its blood plasma.
- Type AB blood has *neither* anti-A nor anti-B antibodies in its blood plasma.
- Type O blood has *both* anti-A and anti-B antibodies in its blood plasma.

Blood types become clinically important when a patient needs a blood transfusion (see Clinical View: "Transfusions"). If a person is transfused with blood of an incompatible type, antibodies in the plasma bind to surface antigens of the transfused erythrocytes, and clumps of erythrocytes bind together in a process termed **agglutination** (ă-gloo-ti-nā′shŭn; *ad* = to, *gluten* = glue). Clumped erythrocytes can block blood vessels and prevent the normal circulation of blood. Eventually, some or all of the clumped erythrocytes may rupture, a process called **hemolysis** (hē-mol′i-sis; *lysis* = destruction). The release of erythrocyte contents and fragments into the blood often causes further reactions and ultimately may damage organs. Therefore, compatibility between donor and recipient must be determined prior to blood donations and transfusions using an agglutination test **(figure 21.8)**.

 WHAT DO YOU THINK?

2 Why is an individual with type O blood called a "universal donor"? Likewise, why is an individual with type AB blood called a "universal recipient"?

Another common surface antigen on erythrocyte membranes is part of the Rh blood type. The **Rh blood type** is determined by the presence or absence of the Rh surface antigen, often called either **Rh factor** or **surface antigen D.** When the Rh factor is present, the individual is said to be **Rh positive (Rh⁺).** Conversely, an individual is termed **Rh negative (Rh⁻)** when the surface antigen is lacking from the membranes of his or her erythrocytes (see figure 21.7b).

In contrast to the ABO blood group, where antibodies may be found in the blood even without prior exposure to a foreign antigen, antibodies to the Rh factor appear in the blood only when an Rh negative individual is exposed to Rh positive blood. Often this occurs as a result of an inappropriate blood transfusion. Therefore, individuals who are Rh positive never exhibit Rh antibodies, because they possess the Rh antigen on their erythrocytes. Only individuals who are Rh negative can exhibit Rh antibodies, and that can occur only after exposure to Rh antigens.

The potential presence of Rh antibodies is especially important in pregnant women who are Rh negative and have an Rh positive fetus. An Rh incompatibility may result during pregnancy if the mother has been previously exposed to Rh positive blood (e.g., from a previous fetus with Rh positive blood). As a result of the prior exposure to Rh positive blood, the mother has Rh antibodies that may cross the placenta and destroy the fetal erythrocytes, resulting in severe illness or death. Giving a pregnant woman special immunoglobulins (e.g., RhoGAM) prevents her from developing the Rh antibodies during pregnancy.

The ABO and Rh blood types are usually reported together. For example, types AB and Rh⁺ together are reported as AB⁺. However, remember that ABO and Rh blood types are independent of each other, and neither of them interacts with or influences the presence or activities of the other group.

Study Tip!

— To remember which ABO blood type is associated with which specific antibody, keep in mind that each blood type has an antibody of a different letter:

- Type A blood does *not* have anti-A antibodies (since anti-A antibodies and type A blood start with the same letter); it only has anti-B antibodies.
- Type B blood does *not* have anti-B antibodies. (It can't have anti-B antibodies, because the B antibodies and type B blood start with the same letter); it only has anti-A antibodies.
- Type AB blood has *both* A and B in its name, so it has no anti-A or anti-B antibodies.
- Type O blood has *neither* an A nor a B in its name, so it has both anti-A and anti-B antibodies.

CLINICAL VIEW

Transfusions

Transfusion is the transfer of blood or blood components from a donor to a recipient. Whole blood is almost never transfused today, and is generally not even available. Rather, when you donate a unit of blood, it is almost immediately divided into three components: erythrocytes, clotting proteins, and platelets. When a person needs one of these blood products, the physician administers only what is required, thus allowing a single donation of whole blood to serve up to three people.

To prevent health problems, donor blood must be collected under sterile conditions. The donated blood is first mixed with an anticoagulant to prevent clotting, and immediately refrigerated. Then the donated unit is tested for a variety of infectious diseases, including hepatitis and AIDS, as well as for general liver disease. Finally, the blood is separated into erythrocytes, platelets, and clotting factors. Should leukocytes be needed, they must be collected in a special apparatus that effectively filters the leukocytes from the blood and then returns the blood to the donor. (A donor with healthy red bone marrow can quickly replace the donated leukocytes.)

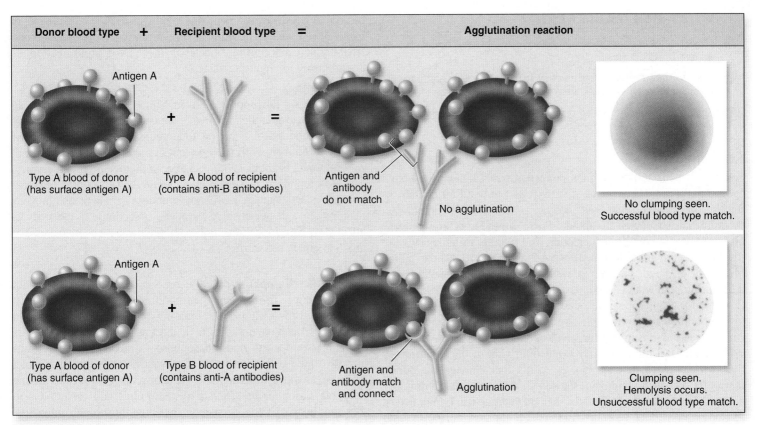

(a) Agglutination test

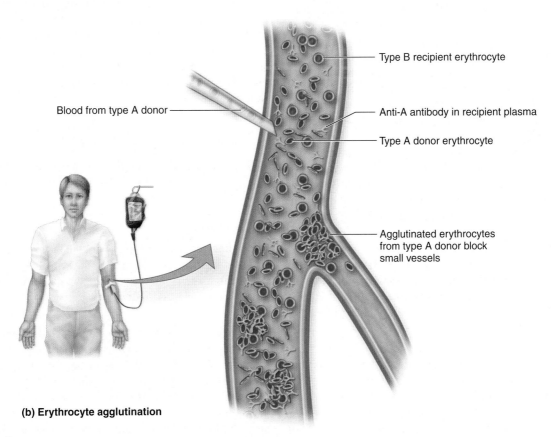

(b) Erythrocyte agglutination

Figure 21.8

Agglutination Reaction. Antibodies in the blood plasma bind to their corresponding surface antigens on the erythrocyte plasma membranes, causing agglutination. *(a)* In a test between plasma and erythrocyte samples, a successful match (no clumping) is compared to an unsuccessful match (clumping). *(b)* If a person receives mismatched blood, erythrocytes agglutinate and block small blood vessels.

5 Why does an erythrocyte lack cellular organelles, and how is this related to its life span?

6 How do transferrin and ferritin participate in recycling erythrocyte components after the cells break down?

7 Should a person with blood type AB donate blood to a person with blood type A? Why or why not?

Leukocytes

Leukocytes help initiate an immune response and defend the body against pathogens. Leukocytes are true "cells" in that they contain a nucleus and cellular organelles. Leukocytes also differ from erythrocytes in that leukocytes are about 1.5 to 3 times larger in diameter and they do not contain hemoglobin. The number of leukocytes in the bloodstream normally ranges between 5000 and 10,000 per cubic millimeter of blood. This range is for adults. Infants normally have a higher number than children or adults.

Abnormal numbers of leukocytes result from various pathologic conditions. For example, a reduced number of leukocytes causes a serious disorder called **leukopenia** (loo-kō-pē′nē-ă; *penia* = poverty). This condition may result from viral or bacterial infection, certain types of leukemia, or toxins that damage the bone marrow. Conversely, **leukocytosis** (loo′kō-sī-tō′sis) results from an elevated leukocyte count (greater than 10,000 per cubic millimeter of blood) and is often indicative of infection, inflammatory reaction, or extreme physiologic stress.

Leukocytes are motile and remarkably flexible. In fact, most leukocytes are found in body tissues (as opposed to the bloodstream). Leukocytes enter the tissue by a process called **diapedesis** (dī′ă-pĕ-dē′sis; *dia* = through, *pedesis* = a leaping), whereby they leave the vessel by squeezing between the endothelial cells of the blood vessel wall. **Chemotaxis** (kē-mō-tak′sis; *taxis* = orderly arrangement) is a process whereby leukocytes are attracted to the site of infection by molecules released by damaged cells, dead cells, or invading pathogens.

The five types of leukocytes are divided into two distinguishable classes—granulocytes and agranulocytes—based upon the presence or absence of visible organelles termed specific granules (**table 21.3**). When a normal blood smear is observed under the microscope, erythrocytes outnumber leukocytes by 500–1000-fold.

Granulocytes

Granulocytes (gran′ū-lō-sīt; *granulum* = small grain) have granules in their cytoplasm that are clearly visible when viewed with a microscope. When a blood smear is stained to provide contrast, three types of granulocytes can be distinguished: neutrophils, eosinophils, and basophils.

Neutrophils   The most numerous leukocyte in the blood is the **neutrophil** (noo′trō-fil; *neuter* = neither), constituting about 60–70% of the total number of leukocytes. The neutrophil is named for its neutral or pale-colored granules within a light lilac cytoplasm. A neutrophil is about 1.5 times larger in diameter than an erythrocyte. Neutrophils exhibit a multilobed nucleus; as many as five lobes are interconnected by thin strands. Because of the various shapes of their nuclei, neutrophils also may be called *polymorphonuclear* (*PMN*) *leukocytes*. Neutrophils usually remain in circulation for about 10 to 12 hours before they exit the blood vessels and enter the tissue spaces, where they phagocytize infectious pathogens, especially bacteria. Specifically, neutrophils target and kill bacteria

by secreting **lysozyme,** an enzyme that helps destroy components of bacterial cell walls. The number of neutrophils in a person's blood rises dramatically during a bacterial infection as more neutrophils are produced to target the bacteria.

Eosinophils   **Eosinophils** (ē-ō-sin′ō-fil; *eos* = dawn) have reddish or pink-orange granules in their cytoplasm. Typically, eosinophils constitute about 2–4% of the total number of leukocytes. Their nucleus is bilobed, with the two lobes connected by a thin strand. An eosinophil is about 1.5 times larger in diameter than an erythrocyte. Eosinophils increase in number when they encounter and react to or phagocytize antigen-antibody complexes or allergens (antigens that initiate a hypersensitive or allergic reaction). If the body is infected by parasitic worms, the eosinophils release chemical mediators that attack the worms.

Basophils   **Basophils** (bā′sō-fil; *basis* = base) are usually about 1.5 times larger in diameter than erythrocytes. They are the least numerous of the granulocytes; typically, basophils constitute about 0.5–1% of the total number of leukocytes. For this reason, it is sometimes difficult to find a basophil on a blood smear. Basophils always exhibit a bilobed nucleus as well as abundant blue-violet granules in the cytoplasm that often obscure the nucleus of the basophil. Basophils are similar to neutrophils and eosinophils in that they may exit the circulation and migrate through interstitial spaces. The primary components of basophil granules are histamine and heparin, which are released during anti-inflammatory or allergic reactions. When histamine is released from these granules, it causes an increase in the diameter of blood vessels (*vasodilation*), resulting in a decrease in blood pressure along with classic allergic symptoms such as swollen nasal membranes, itchy and runny nose, and watery eyes. The release of heparin from basophils inhibits blood clotting (*anticoagulation*).

3 Which type of granulocyte may increase in number if you develop "strep throat" (infection of the throat by *Streptococcus* bacteria)?

Agranulocytes

Agranulocytes (ă-gran′ū-lō-sīt) are leukocytes that have such small granules in their cytoplasm that they are frequently overlooked—hence the name agranulocyte (*a* = without). Agranulocytes include both lymphocytes and monocytes.

Lymphocytes   As their name implies, most **lymphocytes** (lim′fō-sīt) reside in lymphatic organs and structures. Typically, lymphocytes constitute about 20–25% of the total number of leukocytes. Their dark-staining nucleus is usually rounded or slightly indented, and smaller lymphocytes exhibit only a thin rim of blue-gray cytoplasm around the nucleus. When activated, lymphocytes grow larger and have proportionally more cytoplasm. Thus, some of the smaller, nonactivated lymphocytes may have a diameter less than that of an erythrocyte, while activated lymphocytes may be two times the diameter of an erythrocyte.

There are three categories of lymphocytes. **T-lymphocytes** (*T-cells*) manage and direct an immune response; some directly attack foreign cells and virus-infected cells. **B-lymphocytes** (*B-cells*) are stimulated to become plasma cells and produce antibodies.

Table 21.3   Leukocytes

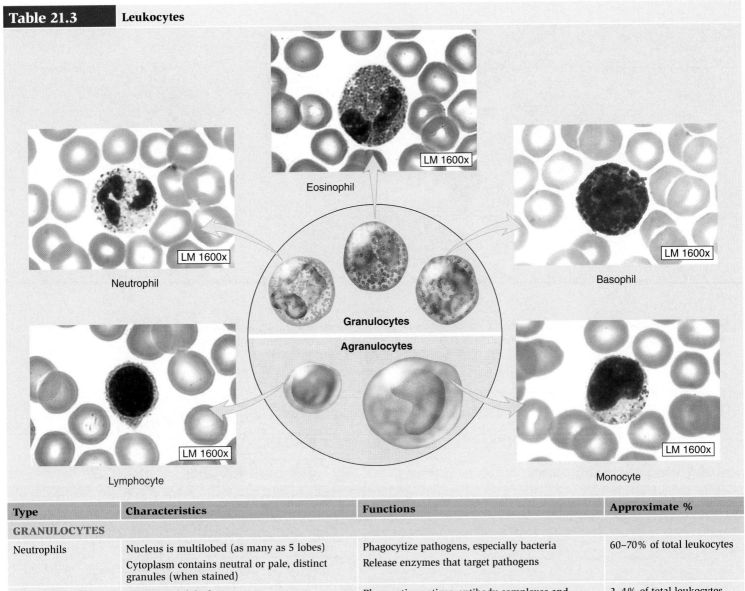

Type	Characteristics	Functions	Approximate %
GRANULOCYTES			
Neutrophils	Nucleus is multilobed (as many as 5 lobes) Cytoplasm contains neutral or pale, distinct granules (when stained)	Phagocytize pathogens, especially bacteria Release enzymes that target pathogens	60–70% of total leukocytes
Eosinophils	Nucleus is bilobed Cytoplasm contains reddish or pink-orange specific granules (when stained)	Phagocytize antigen-antibody complexes and allergens Release chemical mediators to destroy parasitic worms	2–4% of total leukocytes
Basophils	Nucleus is bilobed Cytoplasm contains deep blue-violet specific granules (when stained)	Release histamine (vasodilator) and heparin (anticoagulant) during inflammatory or allergic reactions	0.5–1% of total leukocytes
AGRANULOCYTES			
Lymphocytes	Round or slightly indented nucleus (fills the cell in smaller lymphocytes) Nucleus is usually darkly stained Thin rim of cytoplasm surrounds nucleus (when stained)	Attack pathogens and abnormal/infected cells Coordinate immune cell activity Produce antibodies	20–25% of total leukocytes
Monocytes	Kidney-shaped or C-shaped nucleus Nucleus is generally pale staining Abundant cytoplasm around nucleus (when stained)	Can exit blood vessels and become macrophages Phagocytize pathogens, cellular debris, dead cells	3–8% of total leukocytes

Natural killer cells (*NK cells*) attack abnormal and infected tissue cells. Lymphocytes are examined in detail in chapter 24.

Monocytes   A **monocyte** (mon′ō-sīt; *monos* = single) can be up to three times the diameter of an erythrocyte. Monocytes usually constitute about 3–8% of all leukocytes. The pale-staining nucleus

of a monocyte is kidney-shaped or C-shaped. After approximately 3 days in circulation, monocytes exit blood vessels and take up residence in the tissues, where they change into large phagocytic cells called **macrophages** (mak′rō-fāj; *macros* = large, *phago* = to eat). Macrophages phagocytize bacteria, cell fragments, dead cells, and debris.

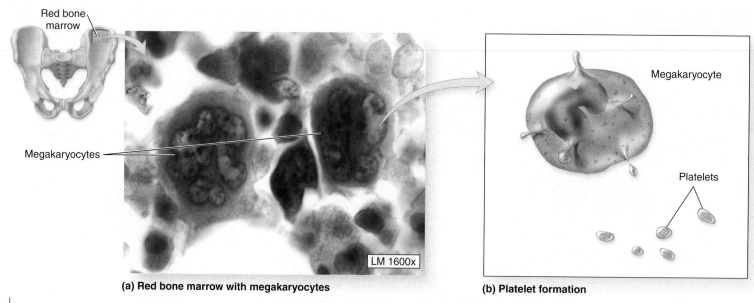

Red bone marrow

Megakaryocytes

LM 1600x

(a) Red bone marrow with megakaryocytes

Megakaryocyte

Platelets

(b) Platelet formation

Figure 21.9

Origin of Platelets. Platelets are derived from megakaryocytes in the red bone marrow. *(a)* Photomicrograph of megakaryocytes in red bone marrow. *(b)* Platelets are formed as membrane-bound cytoplasmic fragments are pinched off from megakaryocytes.

Study Tip!

The mnemonic "Never let monkeys eat bananas" is a simple way to recall the leukocytes in order of their relative abundance:

Never = **N**eutrophil (most abundant)

Let = **L**ymphocyte

Monkeys = **M**onocyte

Eat = **E**osinophil

Bananas = **B**asophil (least abundant)

Platelets

Platelets (plāt′let; *platys* = flat) are irregular, membrane-enclosed cellular fragments that are about 2 micrometers in diameter (less than one-fourth the size of an erythrocyte). In stained preparations, they exhibit a dark central region. Platelets are sometimes called *thrombocytes* (throm′bō-sīt; *thrombos* = clot), although that name is inappropriate because they are cell fragments that never had a nucleus, whereas the suffix *-cyte* implies a complete, nucleated cell.

Platelets are continually produced in the red bone marrow by cells called **megakaryocytes** (meg-ă-kar′ē-ō-sīt; *megas* = big) (**figure 21.9**). Megakaryocytes are easily distinguished both by their large size (about 100 micrometers in diameter) and their dense, multilobed nucleus. As the megakaryocytes shed small volumes of cytoplasm wrapped within plasma membrane, platelets form and immediately enter the general circulation.

Normally, the concentration of platelets in an adult ranges from 120,000 to about 300,000 per cubic millimeter of blood, although the count may rise to as high as 400,000 during times of stress. Severe trauma to a blood vessel causes the blood to coagulate, or clot. A complex process involving components in the plasma produces a web of fibrin that traps erythrocytes and platelets to halt blood flow (**figure 21.10**). If not used to form clots or small platelet

SEM 4100x

Fibrin

Platelets

Erythrocytes

Figure 21.10

Blood Clot. Severe trauma to a blood vessel causes the blood to coagulate, or clot. In a complex process, components in the plasma produce a web of fibrin that traps erythrocytes and platelets and halts blood flow. This SEM shows erythrocytes, fibrin, and platelets within a forming clot.

plugs to stop small vessel leaks, platelets circulate in the blood for 8 to 10 days. Thereafter, they are broken down, and their contents are recycled. An abnormally small number of platelets in circulating blood is termed **thrombocytopenia.**

WHAT DID YOU LEARN?

8 What is meant when a patient is said to have leukopenia?

9 What function do basophils carry out?

10 What are megakaryocytes, and what is their function?

Hemopoiesis: Production of Formed Elements

Key topics in this section:

- Process of hemopoiesis
- Origin and maturation of each type of formed element

Because formed elements have a relatively short life span, new ones are continually produced by the process of **hemopoiesis** (hē′mō-poy-ē′sis; *poiesis* = a making), also called *hematopoiesis*. Hemopoiesis occurs in red bone marrow (see chapter 6). The process starts with hemopoietic stem cells called **hemocytoblasts** (hē′-mō-sī′tō-blast) **(figure 21.11)**. Hemocytoblasts are considered pluripotent cells, meaning that they can differentiate and develop into many different kinds of cells. Hemocytoblasts produce two lines for blood cell development: The **myeloid** (mī′ĕ-loyd; *myelos* = marrow) **line** forms erythrocytes, megakaryocytes, and all leukocytes except lymphocytes; the **lymphoid** (lim′foyd) **line** forms lymphocytes.

A number of hormones and growth factors influence the maturation and division of the blood stem cells. Review figure 21.11 as you read this section to see where each growth factor acts. These so-called **colony-stimulating factors (CSFs)**, or *colony-forming units* (*CFUs*), include the following:

- **Multi-CSF** is a growth factor that increases the formation of erythrocytes, as well as all classes of granulocytes, monocytes, and platelets from myeloid stem cells.
- **GM-CSF** is a growth factor that accelerates the formation of all granulocytes and monocytes from their progenitor cells.
- **G-CSF** is a growth factor that stimulates the formation of granulocytes from myeloblast cells.
- **M-CSF** is a growth factor that stimulates the production of monocytes from monoblasts.
- **Thrombopoietin** is a growth factor that stimulates both the production of megakaryocytes in the bone marrow and the subsequent formation of platelets.
- **Erythropoietin (EPO)** is a hormone produced by the kidneys to increase the rate of production and maturation of erythrocyte progenitor and erythroblast cells.

CLINICAL VIEW

Leukemia

Leukemia (loo-kē′mē-ă) is a malignancy (cancer) in the leukocyte-forming cells. There are several varieties of leukemia, but all are marked by abnormal development and proliferation of leukocytes, both in the bone marrow and the circulating blood. Leukemias are classified based on their duration as either acute or chronic.

Acute leukemia progresses rapidly, and death occurs within a few months after the onset of symptoms (severe anemia, hemorrhages, and recurrent infections). Acute leukemia tends to occur in children and young adults. **Chronic leukemia** progresses more slowly; survival typically exceeds 1 year from the onset of symptoms, which include anemia and a tendency to bleed. Chronic leukemia usually occurs in middle-aged and older individuals.

Leukemia can also be classified based on the type of cell that has become malignant. **Granulocytic leukemia** is characterized by uncontrolled proliferation of immature cells in the myeloid stem cell lines, as well as by the presence of large numbers of immature granulocytes in the circulating blood. **Lymphocytic leukemia** is characterized by increased numbers of malignant lymphocytes and/or lymphocytic precursors (lymphoblasts) in the bone marrow and circulating blood. This type of leukemia often involves lymph nodes and the spleen. **Monocytic leukemia** is a rare form characterized by an increased number of malignant and immature monocytic cells in the bone marrow and circulating blood.

Leukemias represent a malignant transformation of a leukocyte cell line. As abnormal leukocytes increase in number, the erythrocyte and megakaryocytic lines virtually always decrease in number because the proliferating malignant cells literally squeeze them out. This decrease in erythrocyte and platelet production results in the anemia and bleeding that are often the first signs of leukemia. Fortunately, great strides have been made in treating some leukemias over the past 2 decades, especially acute childhood leukemia. Childhood leukemia, once an unquestioned death sentence, stands a good chance of being completely cured today due to improved bone marrow transplant technology in recent years.

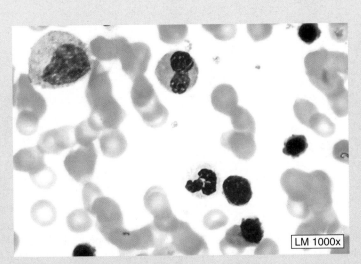

(a) Normal bone marrow sample

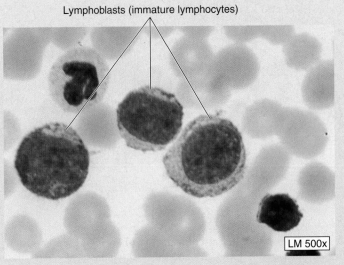

Lymphoblasts (immature lymphocytes)

(b) Bone marrow sample in ALL (acute lymphoblastic leukemia)

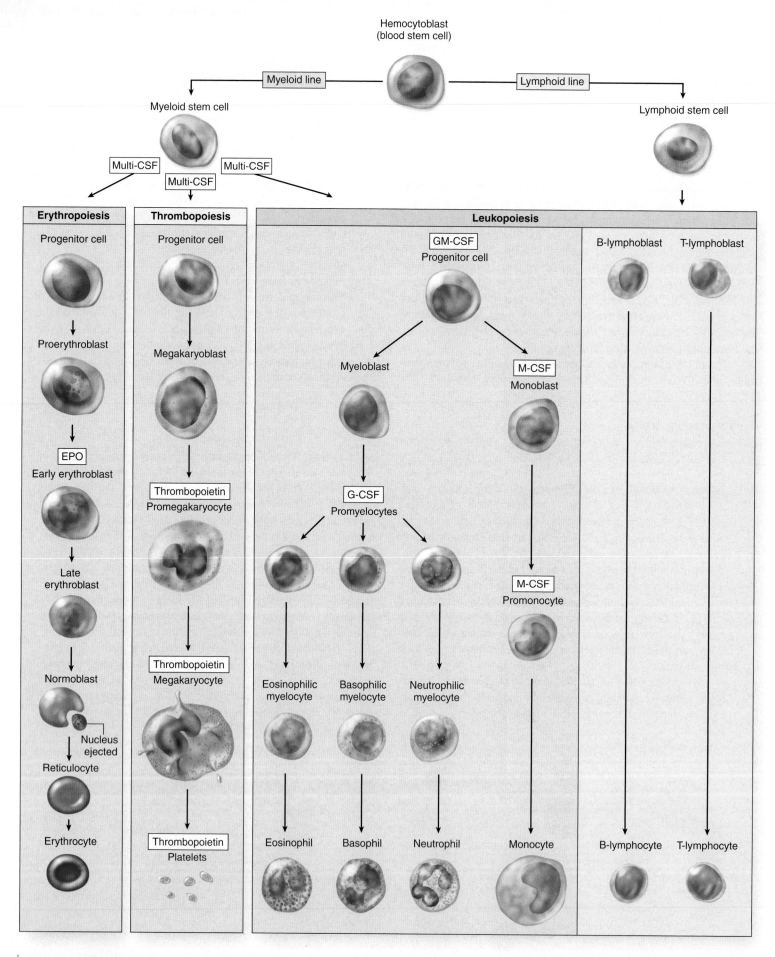

Figure 21.11

Origin, Differentiation, and Maturation of Formed Elements. All formed elements are derived from common hemopoietic stem cells called hemocytoblasts. Both myeloid stem cells and lymphoid stem cells are derived from hemocytoblasts. Myeloid stem cells give rise to erythrocytes, platelets, and to all leukocytes except lymphocytes. Lymphoid stem cells give rise to B- and T-lymphocytes and NK cells (not shown).

Erythropoiesis

Erythrocyte production is called **erythropoiesis** (ĕ-rith′rō-poy-ē′sis). Normally, erythrocytes are produced at the rate of about 3 million per second. The hormone erythropoietin (EPO) controls this rate. Erythropoiesis begins with a **myeloid stem cell** that forms a progenitor cell. The progenitor cell forms a **proerythroblast,** which is a large, nucleated cell. This cell then becomes an **erythroblast,** a slightly smaller cell that is forming hemoglobin in its cytoplasm. The next stage, called a **normoblast,** is a still smaller cell with more hemoglobin in the cytoplasm; its nucleus has been ejected. Eventually, a cell called a **reticulocyte** (re-tik′ū-lō-sīt) is produced. The reticulocyte has lost all organelles except some ribosomes, but it continues to produce hemoglobin. The transformation from myeloid stem cell to reticulocyte takes about 5 days. Some reticulocytes finish maturing while circulating in blood vessels. One to 2 days after entering the circulation, the organelles in the reticulocyte degenerate, and the reticulocyte becomes a mature erythrocyte.

Thrombopoiesis

The production of platelets is called **thrombopoiesis** (throm′bō-poy-ē′sis). From the myeloid stem cell, a committed cell called a **megakaryoblast** is produced. It matures under the influence of thrombopoietin to form a megakaryocyte. Each megakaryocyte then produces thousands of platelets.

Leukopoiesis

The production of leukocytes is called **leukopoiesis** (loo′kō-poy-ē′sis). Leukopoiesis involves three different maturation processes: granulocyte maturation, monocyte maturation, and lymphocyte maturation.

Granulocyte Maturation

All three types of granulocytes (neutrophils, basophils, and eosinophils) are derived from a myeloid stem cell along the myeloid line. This stem cell forms a progenitor cell, which then forms a **myeloblast** (mī′ĕ-lō-blast), which ultimately differentiates into one of the three types of granulocytes.

Monocyte Maturation

Like granulocytes, monocytes are also derived from a myeloid stem cell. In this case, the myeloid stem cell differentiates into a progenitor cell, which then forms a **monoblast** (instead of a myeloblast, as with granulocytes). The monoblast matures into a **promonocyte,** which then forms a **monocyte.**

Lymphocyte Maturation

Lymphocytes are derived from a **lymphoid stem cell** along the lymphoid line. The lymphoid stem cell differentiates into **B-lymphoblasts** and **T-lymphoblasts.** B-lymphoblasts mature into B-lymphocytes, while T-lymphoblasts mature into T-lymphocytes. Some lymphoid stem cells differentiate directly into NK cells.

 WHAT DID YOU LEARN?

11 What are hemocytoblasts?

CLINICAL TERMS

cyanosis (sī-ă-nō′sis; *kyanos* = blue color) Dark bluish or purplish discoloration of the skin and mucous membranes as a result of deficient oxygenation of the blood.

hemoglobinuria (hē′mō-glō-bi-noo′rē-ă; *ouron* = urine) Presence of hemoglobin in the urine.

hemophilia (hē-mō-fil′ē-ă) Inherited disorder characterized by a tendency to hemorrhage uncontrollably because of a defect in the blood-coagulating mechanism.

hemostasis Halt in blood loss.

septicemia (sep-ti-sē′mē-ă; *sepsis* = putrefaction) Systemic disease caused by the spread of microorganisms and their toxins through the circulating blood.

CHAPTER SUMMARY

General Composition and Functions of Blood 636	■ Blood is a fluid connective tissue composed of formed elements, plasma, and dissolved proteins.
	Components of Blood 636
	■ Centrifugation separates whole blood into three components: erythrocytes, a buffy coat composed of leukocytes and platelets, and plasma, a straw-colored fluid.
	Functions of Blood 636
	■ Blood transports nutrients, wastes, and respiratory gases; regulates body temperature, pH, and water levels; and protects the body against the loss of blood volume and the activities of pathogens.
Blood Plasma 637	■ Blood plasma is a mixture of water, proteins, and other solutes.
	Plasma Proteins 638
	■ Plasma proteins are divided into four classes based on their structure and function: Albumins are small proteins that regulate water movement and assist with transport; globulins transport proteins of the immune system; fibrinogen aids in clot formation; and regulatory proteins include enzymes and hormones.
	Differences Between Plasma and Interstitial Fluid 638
	■ Plasma is a component of blood; interstitial fluid bathes body cells.

(continued on next page)

CHAPTER SUMMARY (continued)

Formed Elements in the Blood 638	■ Erythrocytes are the major formed element. Less than 1% of the formed elements consist of leukocytes and cell fragments called platelets. A hematocrit is the percentage of the volume of all formed elements in the blood.

Erythrocytes 639

- ■ Erythrocytes have a biconcave disc structure that facilitates the exchange of respiratory gases into and out of the cells.
- ■ Hemoglobin is a pigmented protein that fills mature erythrocytes; it transports oxygen and carbon dioxide.
- ■ Aged erythrocytes are broken down and their components recycled after about 120 days in the blood.
- ■ Erythrocyte plasma membranes have molecules called surface antigens; blood plasma has antibodies.
- ■ When antibodies bind to surface antigens, agglutination may result in hemolysis of the erythrocytes.
- ■ The Rh blood group is determined by the presence or absence of the Rh surface antigen.

Leukocytes 646

- ■ Leukocytes are white blood cells that defend against invading pathogens, reduce the activities of abnormal cells, and remove damaged cells, debris, and antigen-antibody complexes.
- ■ A disorder in which leukocytes are reduced in number is called leukopenia; a slightly elevated leukocyte count is termed leukocytosis.
- ■ Leukocytes are either granulocytes (neutrophils, eosinophils, and basophils) or agranulocytes (lymphocytes and monocytes) based on the presence (or lack) of specific granules in their cytoplasm.

Platelets 648

- ■ Platelets, or thrombocytes, are the smallest components of the formed elements. These membrane-enclosed packets of cytoplasm are derived from megakaryocytes.
- ■ An abnormal number of platelets causes a condition called thrombocytopenia.

Hemopoiesis: Production of Formed Elements 649	■ Formed elements have a relatively short life span. They are constantly renewed by the process of hemopoiesis. ■ Hemopoietic stem cells are pluripotent cells called hemocytoblasts located in red bone marrow.

Erythropoiesis 651

- ■ Erythropoiesis is the production of erythrocytes. The hormone erythropoietin (EPO) controls the rate of proerythroblast maturation to become an erythrocyte.

Thrombopoiesis 651

- ■ Thrombopoiesis is the production of platelets. A megakaryoblast is produced, and it matures under the influence of thrombopoietin to form a megakaryocyte.

Leukopoiesis 651

- ■ The production of leukocytes is called leukopoiesis. This process begins in the bone marrow with the division of hemocytoblasts.

CHALLENGE YOURSELF

Matching

Match each numbered item with the most closely related lettered item.

_____ 1. monocyte

_____ 2. fibrinogen

_____ 3. neutrophil

_____ 4. erythroblast

_____ 5. hemopoiesis

_____ 6. albumin

_____ 7. antibody

_____ 8. water

_____ 9. erythropoiesis

_____ 10. globulin

a. immature form of erythrocyte with a nucleus

b. the most abundant plasma protein

c. formation of erythrocytes

d. agranulocyte that can develop into macrophage

e. the most abundant compound in plasma

f. the most numerous leukocyte

g. transport protein in plasma

h. blood cell formation and development

i. binds to and immobilizes antigens

j. protein that can be converted into blood clot fibers

Multiple Choice

Select the best answer from the four choices provided.

_____ 1. In the adult, the stem cells for leukocytes reside in the
  a. bloodstream.
  b. red bone marrow.
  c. liver.
  d. muscle.

_____ 2. Which type of leukocyte increases during allergic reactions and parasitic worm infections?
  a. basophil
  b. eosinophil
  c. lymphocyte
  d. neutrophil

_____ 3. This cell forms platelets in the red bone marrow:
  a. lymphocyte
  b. megakaryocyte
  c. eosinophil
  d. reticulocyte

_____ 4. Which of the following is not a function of blood?
   a. prevention of fluid loss
   b. nutrient and waste transport
   c. maintenance of constant pH levels
   d. production of hormones

_____ 5. A person with blood type A has
   a. anti-B antibodies in her blood plasma.
   b. anti-A antibodies in her blood plasma.
   c. both anti-A and anti-B antibodies in her blood plasma.
   d. no antibodies in her blood plasma.

_____ 6. The hematocrit is a measure of
   a. water concentration in the plasma.
   b. the percentage of formed elements in the blood.
   c. the number of platelets in the blood.
   d. antibody concentration in the plasma.

_____ 7. Oxygen attaches to a(n) _____ ion in hemoglobin.
   a. calcium
   b. sodium
   c. iron
   d. potassium

_____ 8. During the recycling of components following the normal destruction of erythrocytes, globin is broken down, and its components are
   a. used to synthesize new proteins.
   b. stored as iron in the liver.
   c. eliminated from the body in the bile.
   d. removed in the urine.

_____ 9. The type of leukocyte that produces antibodies is a (an)
   a. eosinophil.
   b. basophil.
   c. T-lymphocyte.
   d. B-lymphocyte.

_____ 10. Which of the following is *not* a characteristic of a mature erythrocyte?
   a. biconcave disc shape
   b. absence of organelles
   c. life span of about 12 months
   d. filled with hemoglobin

Content Review

1. How does blood help regulate body temperature?

2. What are globulins? What do they do?

3. When blood is centrifuged, a thin, gray-white layer called the buffy coat covers the layer of packed erythrocytes. What are the components of the buffy coat?

4. What is the shape of an erythrocyte, and why is this shape advantageous to its function?

5. How is oxygen carried by erythrocytes?

6. What are the anatomic characteristics of each type of leukocyte? How can you tell these leukocytes apart when viewing a blood smear under the microscope?

7. How do the functions of basophils differ from those of lymphocytes?

8. Briefly describe the origin, structure, and functions of platelets.

9. What is hemopoiesis? What is a hemocytoblast? Briefly describe the two lines of blood cells that develop during hemopoiesis.

10. What are colony-stimulating factors? Where would they be found in the body, and what is their general function?

Developing Critical Reasoning

1. While taking a clinical laboratory class, Marilyn prepared and examined blood smears from several donors. One of the smears had an increased percentage (about 10% of observed leukocytes) of cells containing reddish-orange granules. Discuss the type of cell described and the condition that may have caused this increase in the donor.

2. Abby is a nurse on duty in a hospital emergency room when a critically injured patient is brought in. The physician calls for an immediate blood transfusion, but the patient's blood type is unknown. What blood type should the patient be given and why?

A N S W E R S   T O   " W H A T   D O   Y O U   T H I N K ? "

1. When you become dehydrated, the plasma percentage decreases because the primary component of plasma is water.

2. A person with type O blood is considered a "universal donor" because her erythrocytes have no surface antigens. Without surface antigens, the type O erythrocytes will not be destroyed through agglutination and hemolysis by antibodies in the recipient's plasma. Likewise, a person with type AB blood is considered a "universal recipient" because his blood plasma has no antibodies to the ABO blood types. Thus, the AB recipient may receive any type of blood and not worry about the donor's erythrocytes being destroyed.

3. Neutrophils increase in number when you get "strep throat" because they target bacteria.

Visit the McKinley/O'Loughlin *Human Anatomy*, 2e website at aris.mhhe.com

22

CARDIOVASCULAR SYSTEM

Heart

In chapter 21, we discovered the importance of blood and the myriad of substances it carries. To maintain homeostasis, blood must circulate continuously throughout the body. The continual pumping action of the heart is essential for maintaining blood circulation. If the heart fails to pump adequate volumes of blood, cells are deprived of needed oxygen and nutrients, waste products accumulate, and cell death occurs.

In a healthy, 80-kilogram resting adult, the heart beats about 75 times per minute (about 4500 times per hour or 108,000 times per day). The amount of blood pumped from one ventricle per minute (about 5.25 liters [L] at rest) is called the cardiac output. When the body is more active, and the cells need oxygen and nutrients delivered at a faster pace, the heart can increase its output up to five- or six-fold.

Overview of the Cardiovascular System

Key topics in this section:

- Basic features of the cardiovascular system
- Overview of the pulmonary and systemic circulations
- Position and location of the heart
- Structure and function of the pericardium

As the center of the cardiovascular system, the heart connects to blood vessels that transport blood between the heart and all body tissues. The two basic types of blood vessels are **arteries** (ar′ter-ē), which carry blood *away* from the heart, and **veins** (vān), which carry blood back *to* the heart. The differences between these types of vessels are discussed in chapter 23. Most arteries carry blood high in oxygen (except for the pulmonary arteries, as explained later), while most veins carry blood low in oxygen (except for the pulmonary veins). The arteries and veins entering and leaving the heart are called the **great vessels** because of their relatively large diameter. The heart exhibits several related characteristics and functions:

- The heart's anatomy ensures the **unidirectional flow** of blood through it. Backflow of blood is prevented by valves within the heart.
- The heart acts like two side-by-side pumps that work at the same rate and pump the same volume of blood; one directs blood to the lungs for gas exchange, while the other directs blood to body tissues for nutrient and respiratory gas delivery.
- The heart develops **blood pressure** through alternate cycles of heart wall contraction and relaxation. **Blood pressure** is the force of the blood pushing against the inside walls of the vessels. A minimum blood pressure is essential for pushing blood through the blood vessels.

Pulmonary and Systemic Circulations

The cardiovascular system consists of two circulations: the pulmonary circulation and the systemic circulation **(figure 22.1)**. The **pulmonary** (pŭl′mō-nār-ē; *pulmo* = lung) circulation consists of the chambers on the right side of the heart (right atrium and right ventricle) as well as the pulmonary arteries and veins. This circulation conveys blood to the lungs via **pulmonary arteries** to reduce carbon dioxide and replenish oxygen levels in the blood before returning to

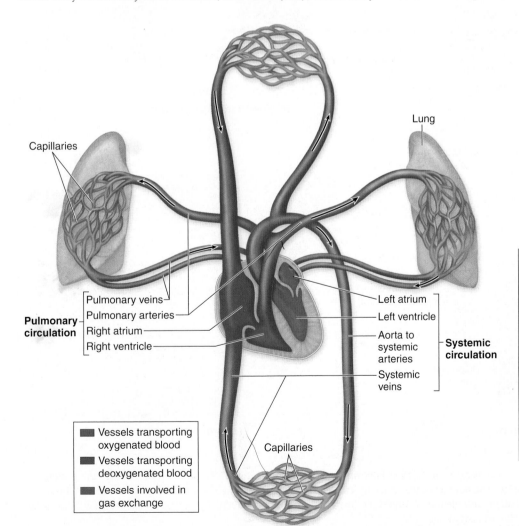

Lung

Capillaries

Pulmonary veins
Pulmonary arteries
Pulmonary circulation
Right atrium
Right ventricle

Left atrium
Left ventricle
Aorta to systemic arteries
Systemic circulation
Systemic veins

Capillaries

- ■ Vessels transporting oxygenated blood
- ■ Vessels transporting deoxygenated blood
- ■ Vessels involved in gas exchange

Figure 22.1

Cardiovascular System.   The cardiovascular system is composed of the pulmonary circulation and the systemic circulation. The pulmonary circulation pumps blood from the right side of the heart through pulmonary vessels, to the lungs, and back to the left side of the heart. The systemic circulation pumps blood from the left side of the heart, through systemic vessels in peripheral tissues, and back to the right side of the heart.

the heart in **pulmonary veins.** Blood returns to the left side of the heart, where it then enters the systemic circulation.

The **systemic** (sis-tem'ik) **circulation** consists of the chambers on the left side of the heart (left atrium and left ventricle), along with all the other named blood vessels. It carries blood to all the peripheral organs and tissues of the body. Blood that is high in oxygen (oxygenated) from the left side of the heart is pumped into the aorta, the largest systemic artery in the body, and then into smaller systemic arteries. Gas is exchanged with tissues from the body's smallest vessels, called capillaries. Systemic veins then carry blood that is low in oxygen (deoxygenated) and high in carbon dioxide and waste products back to the heart. Most veins merge and drain into the **superior** and **inferior venae cavae** (vē'nē ca'vē; sing., *vena cava*), which drain blood into the right atrium. There, the blood enters the pulmonary circulation, and the cycle repeats.

? WHAT DO YOU THINK?

1 We previously mentioned that arteries tend to carry oxygenated blood, but the pulmonary arteries are the exception. Why are the pulmonary arteries carrying deoxygenated blood?

Position of the Heart

The heart is located left of the body midline posterior to the sternum in the mediastinum **(figure 22.2).** The heart is slightly rotated such that its right side or **right border** (primarily formed by the right atrium and ventricle) is located more anteriorly, while its left side or **left border** (primarily formed by the left atrium and ventricle) is located more posteriorly. The posterosuperior surface of the heart, formed primarily by the left atrium, is called the **base**. The pulmonary veins that enter the left atrium border this base. The **superior border** is formed by the great arterial trunks (ascending aorta and

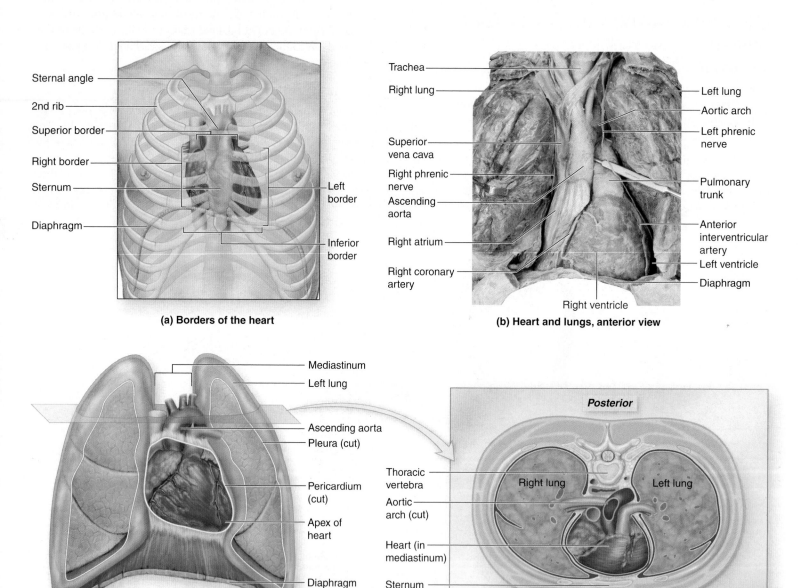

(a) Borders of the heart

(b) Heart and lungs, anterior view

(c) Serous membranes of the heart and lungs

(d) Cross-sectional view

Figure 22.2

Heart Position Within the Thoracic Cavity. The heart is in the mediastinum of the thoracic cavity. (*a*) An anterior view shows the position of the heart posterior to the anterior thoracic cage. The borders of the heart are labeled. (*b*) Cadaver photo of the heart within the mediastinum (anterior view). (*c*) Serous membranes (pericardium and pleura) surround the heart and lungs, respectively. (*d*) A cross-sectional view depicts the heart's relationship to the other organs in the thoracic cavity.

pulmonary trunk) and the superior vena cava. The inferior, conical end is called the **apex** (ā′peks; tip). It projects slightly anteroinferiorly toward the left side of the body. The **inferior border** is formed by the right ventricle.

Characteristics of the Pericardium

The heart is contained within the **pericardium** (per-i-kar′dē-ŭm), a fibrous sac and serous lining **(figure 22.3)**. The pericardium restricts the heart's movements so that it doesn't bounce and move about in the thoracic cavity, and prevents the heart from overfilling with blood.

The pericardium is composed of two parts. The outer portion is a tough, dense connective tissue layer called the **fibrous pericardium.** This layer is attached to both the diaphragm and the base of the great vessels. The inner portion is a thin, double-layered serous membrane called the **serous pericardium.** The serous pericardium may be subdivided into (1) a **parietal layer** of serous pericardium that lines the inner surface of the fibrous pericardium, and (2) a **visceral layer** of serous pericardium (also called the **epicardium**) that covers the outside of the heart. The parietal and visceral layers reflect (fold back) along the great vessels, where these layers become continuous with one another. The thin space between the parietal and visceral layers of the serous pericardium is the **pericardial cavity,** into which serous fluid is secreted in order to lubricate the serous

membranes and facilitate the almost frictionless, continuous movement of the heart when it beats. The pericardial cavity is a potential space with just a thin lining of serous fluid. However, it may become a real space as described in the Clinical View, "Pericarditis."

Study Tip!

To demonstrate the almost frictionless movement of the heart within the pericardial sac, try the following demonstration:

1. Obtain two glass microscope slides. Place them together and then try to slide them back-and-forth past each other. You will find that they stick together (even if they are very clean) and do not move relative to one another very easily!

2. Now, take two similar glass slides and place a very small drop of water onto the surface of one slide. Place the slides together, as before, and then try to slide them back-and-forth past each other.

We predict that you have demonstrated to yourself that glass slides move easily past each other when a water drop is present between them. This ease of movement of two opposing surfaces parallels the sliding movement of the parietal and visceral pericardial surfaces when a thin layer of serous fluid is present between them.

WHAT DID YOU LEARN?

1 What is the basic distinction between arteries and veins?

2 Contrast the pulmonary and systemic circulations.

3 What is the difference between the base of the heart and its apex?

4 Identify the layers of the pericardium. Why is the pericardial cavity described as a potential space?

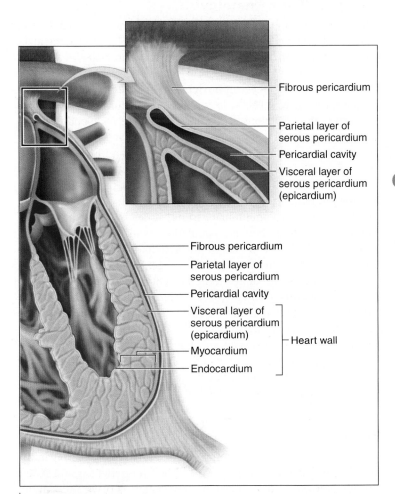

Fibrous pericardium

Parietal layer of serous pericardium

Pericardial cavity

Visceral layer of serous pericardium (epicardium)

Fibrous pericardium

Parietal layer of serous pericardium

Pericardial cavity

Visceral layer of serous pericardium (epicardium)

Myocardium — Heart wall

Endocardium

Figure 22.3

Pericardium.   The pericardium consists of an outer fibrous pericardium and an inner serous pericardium. The serous pericardium consists of a parietal layer, which adheres to the fibrous pericardium, and a visceral layer, which forms the epicardium of the heart. The space between the parietal and visceral layers of the serous pericardium is called the pericardial cavity.

CLINICAL VIEW

Pericarditis

Pericarditis (per′i-kar-dī′tis; *peri* = around, *kardia* = heart, *ites* = inflammation) is an inflammation of the pericardium typically caused by viruses, bacteria, or fungi. Whatever the cause of pericarditis, the pericardium is inflamed. The inflammation causes an increase in capillary permeability. Thus, the capillaries become more "leaky," resulting in fluid accumulation in the pericardial cavity. At this point, the potential space of the pericardial cavity becomes a real space as it fills with fluid and pus. In severe cases, the excess fluid accumulation limits the heart's movement and keeps it from filling with an adequate amount of blood. The heart is unable to pump blood, leading to a medical emergency called **cardiac tamponade** and resulting in heart failure and death.

Pericarditis typically occurs between the ages of 20 and 50. Fever and chest pain are frequent symptoms. Pericarditis pain is located over the center or left side of the chest, and may extend to the neck or left shoulder. Patients often describe the pain as piercing or "knife-like," and say that breathing worsens it. In contrast, pain from a myocardial infarction typically is described as crushing. But although the two conditions are different, the diagnosis of myocardial infarction and pericarditis often may be confused, especially by the patients experiencing the symptoms. A helpful diagnostic finding in pericarditis is **friction rub,** a crackling or scraping sound heard with a stethoscope that is caused by the movement of the inflamed pericardial layers against each other. The inflammation results in the loss of the lubricating action of the serous membranes.

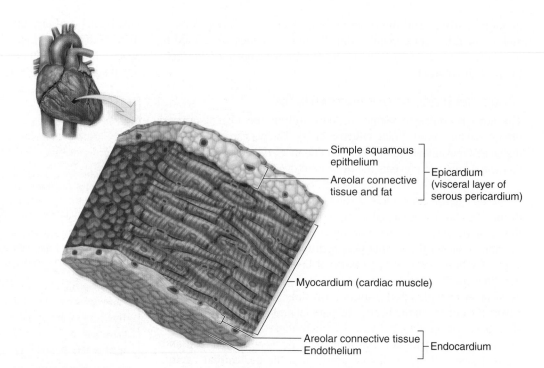

Figure 22.4

Organization of the Heart Wall. The heart wall is composed of an outer epicardium (visceral layer of the serous pericardium), a middle myocardium (cardiac muscle), and an inner endocardium (composed of areolar connective tissue and an endothelium).

Anatomy of the Heart

Key topics in this section:

- External anatomy of the heart and its major vessels
- Internal anatomic characteristics of each heart chamber
- How valves regulate blood flow through the heart

The heart is a relatively small, conical organ approximately the size of a person's clenched fist. In the average normal adult, it weighs about 250 to 350 grams, but certain diseases may cause heart size to increase dramatically.

Heart Wall Structure

The heart wall consists of three distinctive layers: an external epicardium, a middle myocardium, and an internal endocardium (**figure 22.4**).

The **epicardium** (ep-i-kar′dē-ŭm; *epi* = upon, *kardia* = heart) is the outermost heart layer and is also known as the *visceral layer* of the serous pericardium. The epicardium is composed of a serous membrane and areolar connective tissue. As we age, more fat is deposited in the epicardium, and so this layer becomes thicker and more fatty.

The **myocardium** (mī-ō-kar′dē-ŭm; *mys* = muscle) is the middle layer of the heart wall and is composed of cardiac muscle tissue. The myocardium is the thickest of the three heart wall layers. It lies deep to the epicardium and superficial to the endocardium. The myocardial layer is where myocardial infarctions (heart attacks) occur. The arrangement of cardiac muscle in the heart wall permits the compression necessary to pump large volumes of blood out of the heart.

The internal surface of the heart and the external surfaces of the heart valves are covered by **endocardium** (en-dō-kar′dē-ŭm; *endon* = within). The endocardium is composed of a simple squamous epithelium, called an endothelium, and a layer of areolar connective tissue.

External Heart Anatomy

The heart is composed of four hollow chambers: two smaller atria and two larger ventricles (**figure 22.5**). The left and right **atria** (ā′trē-ă; sing., *atrium*; entrance hall) are thin-walled chambers located superiorly. The anterior part of each atrium is a wrinkled, flaplike extension, called an **auricle** (au′ri-kl; *auris* = ear) because it resembles an ear. The atria receive blood returning to the heart through both circulations: The right atrium receives blood from the systemic circulation, and the left atrium receives blood from the pulmonary circulation. Blood that enters an atrium is passed to the ventricle on the same side of the heart. The left and right **ventricles** (ven′tri-kl; *venter* = belly) are the inferior chambers. Two large arteries, the **pulmonary trunk** and the **aorta** (ā-ōr′tă), exit the heart at its superior border. The pulmonary trunk carries blood from the right ventricle into the pulmonary circulation, while the aorta conducts blood from the left ventricle into the systemic circulation. Both ventricles pump the same volume of blood per minute.

The atria are separated from the ventricles externally by a relatively deep **coronary sulcus** (or *atrioventricular sulcus*) that extends around the circumference of the heart. The **anterior interventricular** (in-ter-ven-trik′ū-lăr) **sulcus** and the **posterior interventricular sulcus** are located between the left and right ventricles on the anterior and posterior surfaces of the heart, respectively. These sulci extend inferiorly from the coronary sulcus toward the heart apex. All sulci house blood vessels packed in adipose connective tissue that supply and drain the heart (discussed later in this chapter).

Internal Heart Anatomy: Chambers and Valves

Figure 22.6 depicts the internal anatomy and structural organization of the four heart chambers: the right atrium, right ventricle, left atrium, and left ventricle. Each of these chambers plays a role in the continuous process of blood circulation. Important to their function are valves, epithelium-lined dense connective tissue cusps that permit

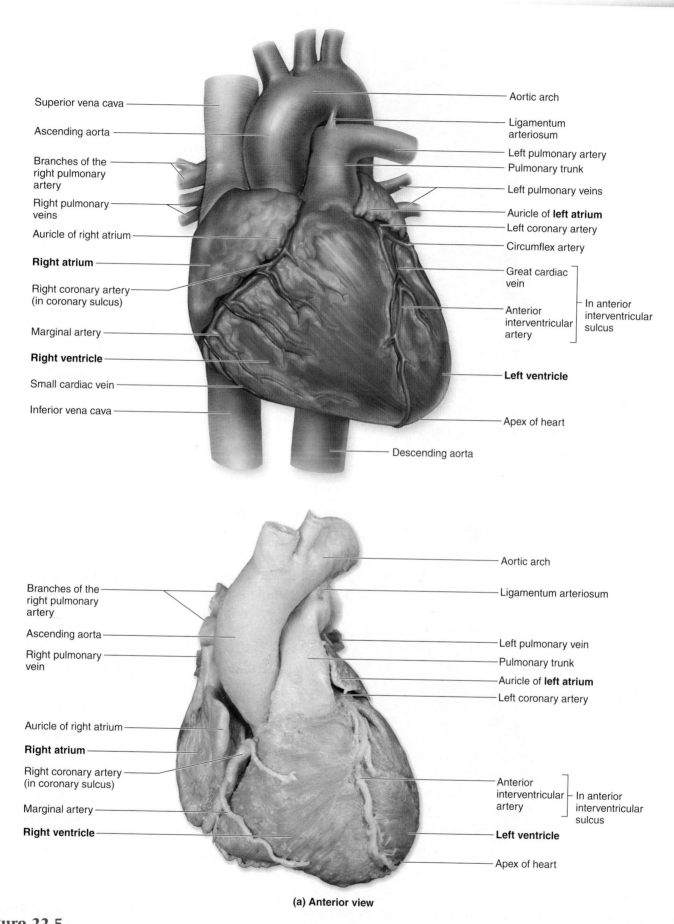

(a) Anterior view

Figure 22.5

External Anatomy and Features of the Heart.   (a) An illustration and a cadaver photo show the heart chambers and associated vessels and the apex of the heart in an anterior view. (*continued on next page*)

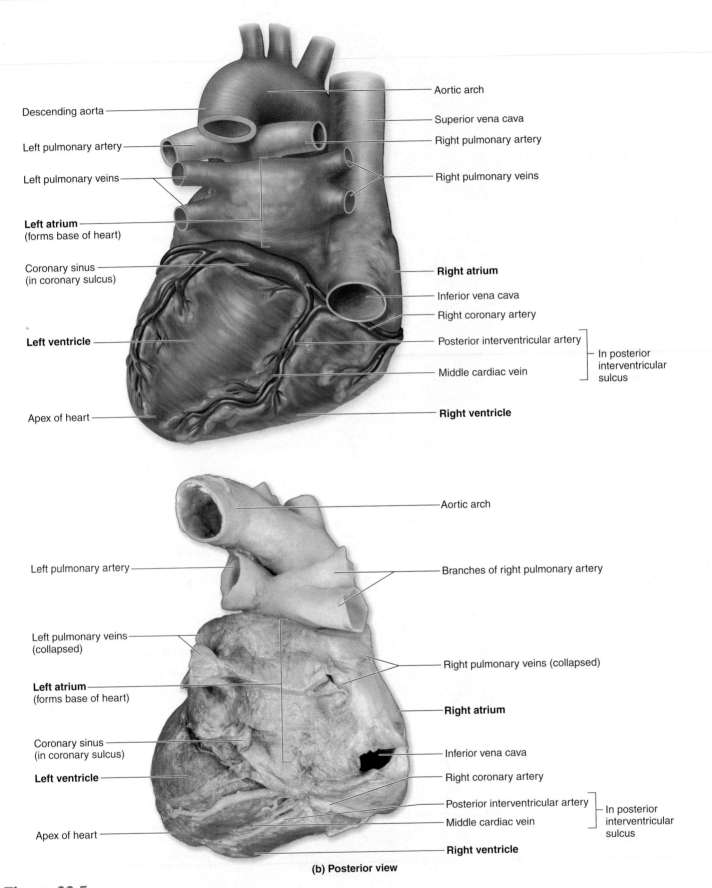

Descending aorta

Left pulmonary artery

Left pulmonary veins

Left atrium
(forms base of heart)

Coronary sinus
(in coronary sulcus)

Left ventricle

Apex of heart

Aortic arch

Superior vena cava

Right pulmonary artery

Right pulmonary veins

Right atrium

Inferior vena cava

Right coronary artery

Posterior interventricular artery

Middle cardiac vein

Right ventricle

In posterior
interventricular
sulcus

Left pulmonary artery

Left pulmonary veins
(collapsed)

Left atrium
(forms base of heart)

Coronary sinus
(in coronary sulcus)

Left ventricle

Apex of heart

Aortic arch

Branches of right pulmonary artery

Right pulmonary veins (collapsed)

Right atrium

Inferior vena cava

Right coronary artery

Posterior interventricular artery

Middle cardiac vein

Right ventricle

In posterior
interventricular
sulcus

(b) Posterior view

Figure 22.5

External Anatomy and Features of the Heart. (*continued*)   *(b)* An illustration and a cadaver photo show the heart chambers and associated vessels and the base of the heart in a posterior view.

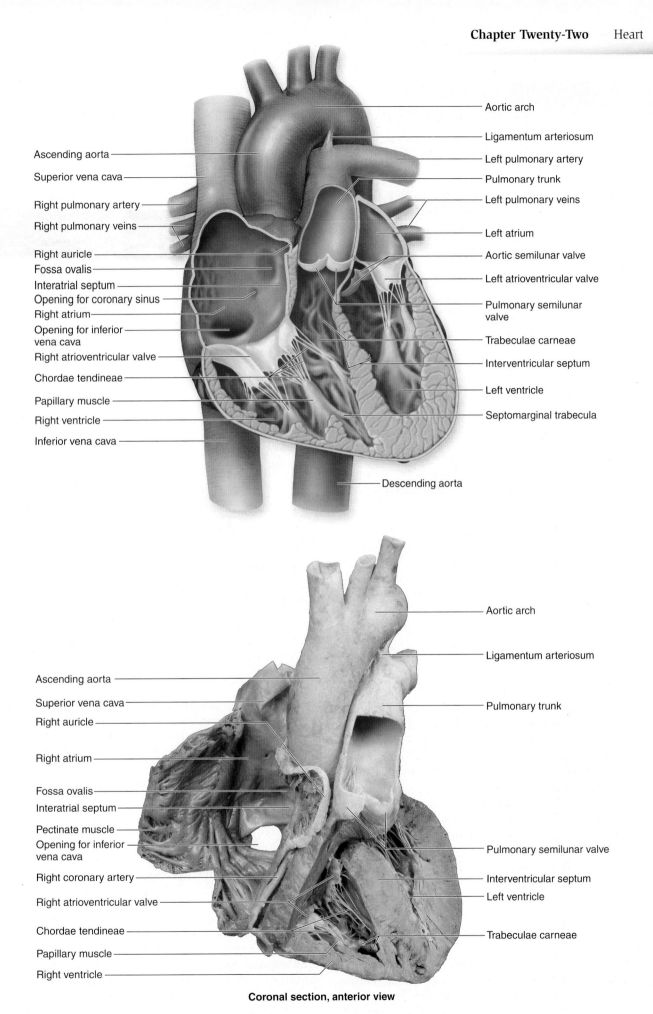

Ascending aorta

Superior vena cava

Right pulmonary artery

Right pulmonary veins

Right auricle
Fossa ovalis
Interatrial septum
Opening for coronary sinus
Right atrium
Opening for inferior
vena cava
Right atrioventricular valve

Chordae tendineae

Papillary muscle

Right ventricle

Inferior vena cava

Aortic arch
Ligamentum arteriosum
Left pulmonary artery
Pulmonary trunk
Left pulmonary veins
Left atrium
Aortic semilunar valve
Left atrioventricular valve
Pulmonary semilunar
valve
Trabeculae carneae
Interventricular septum
Left ventricle
Septomarginal trabecula

Descending aorta

Ascending aorta

Superior vena cava

Right auricle

Right atrium

Fossa ovalis
Interatrial septum

Pectinate muscle
Opening for inferior
vena cava

Right coronary artery

Right atrioventricular valve

Chordae tendineae

Papillary muscle

Right ventricle

Aortic arch

Ligamentum arteriosum

Pulmonary trunk

Pulmonary semilunar valve

Interventricular septum

Left ventricle

Trabeculae carneae

Coronal section, anterior view

Figure 22.6

Internal Anatomy of the Heart.   An illustration and a cadaver photo reveal the internal structure of the heart, including the valves and the musculature of the heart wall.

Table 22.1	Heart Valves		
Heart Valve	**Location**	**Structure**	**Function**
Right atrioventricular valve	Between right atrium and right ventricle	Three triangular-shaped cusps of dense connective tissue covered by endothelium; chordae tendineae attached to free edges	Prevents backflow of blood into right atrium when ventricles contract
Pulmonary semilunar valve	Between right ventricle and pulmonary trunk	Three semilunar cusps of dense connective tissue covered by endothelium; no chordae tendineae	Prevents backflow of blood into right ventricle when ventricles relax
Left atrioventricular valve	Between left atrium and left ventricle	Two triangular-shaped cusps of dense connective tissue covered by endothelium; chordae tendineae attached to free edges	Prevents backflow of blood into left atrium when ventricles contract
Aortic semilunar valve	Between left ventricle and ascending aorta	Three semilunar cusps of dense connective tissue covered by endothelium; no chordae tendineae	Prevents backflow of blood into left ventricle when ventricles relax

the passage of blood in one direction and prevent its backflow (**table 22.1**). Valve cusps are the tapering projection of a cardiac valve, also called *flaps* or *leaflets*.

Fibrous Skeleton

The fibrous skeleton of the heart is located between the atria and the ventricles, and is formed from dense regular connective tissue (**figure 22.7**). The fibrous skeleton performs the following functions:

- Separates the atria and ventricles.
- Anchors heart valves by forming supportive rings at their attachment points.
- Provides electrical insulation between atria and ventricles. This insulation ensures that muscle impulses are not spread randomly throughout the heart, and thus prevents all of the heart chambers from beating at the same time.
- Provides a rigid framework for the attachment of cardiac muscle tissue.

Right Atrium

The **right atrium** receives venous blood from the systemic circulation and the heart muscle itself. Three major vessels empty into the right atrium: (1) The **superior vena cava** drains blood from the head, neck, upper limbs, and superior regions of the trunk; (2) the **inferior vena cava** drains blood from the lower limbs and trunk; and (3) the **coronary sinus** drains blood from the heart wall.

The **interatrial** (in-ter-ā′trē-ăl) **septum** forms a thin wall between the right and left atria. The posterior atrial wall is smooth, but the auricle and anterior wall exhibit obvious muscular ridges, called **pectinate** (pek′ti-nāt; teeth of a comb) **muscles.** The structural differences in the anterior and posterior walls occur because the two walls formed from separate structures during embryonic development. Inspection of the interatrial septum reveals an oval depression called the **fossa ovalis**, also called the *oval fossa*. It occupies the former location of the fetal foramen ovale, which shunted blood from the right atrium to the left atrium during fetal life, as described later in this chapter.

Separating the right atrium from the right ventricle is the **right atrioventricular opening**. This opening is covered by a **right atrioventricular (AV) valve** (also called the *tricuspid valve*, since it has three triangular cusps). Deoxygenated venous blood flows from the right atrium, through the right atrioventricular opening when the valve is open, into the right ventricle. The right AV valve is forced

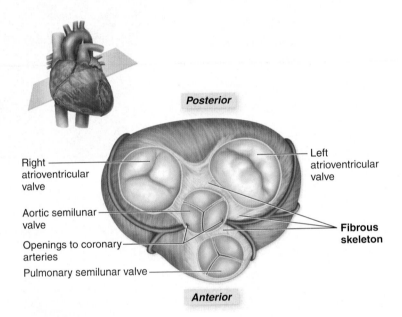

Posterior

Right atrioventricular valve

Left atrioventricular valve

Aortic semilunar valve

Fibrous skeleton

Openings to coronary arteries

Pulmonary semilunar valve

Anterior

Figure 22.7

Fibrous Skeleton of the Heart. Four thick regions of strong dense regular connective tissue encircle the four heart valves and the origins of the pulmonary trunk and the aorta. This fibrous skeleton isolates the atria from the ventricles (so muscle impulses won't randomly pass between them), stabilizes the heart valves, and provides an attachment site for cardiac muscle. This is a superior view of a transverse section.

closed when the right ventricle begins to contract, preventing blood from flowing back into the right atrium.

Right Ventricle

The right ventricle receives deoxygenated venous blood from the right atrium. An **interventricular septum** forms a thick wall between the right and left ventricles. The internal wall surface of each ventricle displays characteristic large, smooth, irregular muscular ridges, called the **trabeculae carneae** (tră-bek′ū-lē; *trabs* = beam; kar′nē-ē; *carne* = flesh) (see figure 22.6).

The right ventricle typically has three cone-shaped, muscular projections called **papillary** (pap′i-lăr-ē; *papilla* = nipple) **muscles,** which anchor numerous thin strands of collagen fibers called

Valve Defects and Their Effects on Circulation

Structural damage to the heart valves can impair blood circulation and lead to serious health problems. Damage may result from developmental abnormalities, infection, hypertension, or other cardiovascular problems.

Valvular insufficiency, also termed *valvular incompetence,* occurs when one or more of the cardiac valves leaks (called "regurgitant flow") because the valve cusps do not close tightly enough. Inflammation or disease may cause the free edges of the valve cusps to become scarred and constricted, allowing blood to regurgitate back through the valve. As the heart works to overcome the effect of the backflow, blood forced through the valve openings may cause heart enlargement. As a result, the heart must work harder to circulate the normal amount of blood.

Valvular stenosis (ste-nō′sis; narrowing) is scarring of the valve cusps so that they become rigid or partially fused and cannot open com-pletely. A stenotic valve is narrowed and presents resistance to the flow of blood, so that output from the affected chamber decreases. Often the affected chamber hypertrophies and dilates—both conditions that may have deleterious consequences. Heart function may become so reduced that the rest of the body cannot receive adequate blood flow. A primary cause of valvular stenosis is rheumatic heart disease.

Rheumatic (roo-mat′ik) **heart disease** may follow a streptococcal infection of the throat. It results when antibodies produced to kill the bacteria cross-react with the body's own connective tissue, thereby initiating an autoimmune disease. All parts of the heart are subject to injury, but the endocardium, the valve cusps, and the left AV valve are typically most affected. Significantly scarred and narrow valves must be surgically repaired or replaced. Patients with a history of rheumatic heart disease must take antibiotics before undergoing dental or medical procedures that are likely to introduce bacteria into the bloodstream.

---

chordae tendineae (kōr′dē ten′di-nē-ē). The chordae tendineae attach to the lower surface of cusps of the right AV valve and prevent the valve from everting and flipping into the atrium when the right ventricle is contracting. A muscle bundle called the **septomarginal trabecula** (see figure 22.6), or *moderator band,* connects the base of the anterior papillary muscle to the interventricular septum. At its superior end, the right ventricle narrows into a smooth-walled, conical region called the **conus arteriosus.** Beyond the conus arteriosus is the **pulmonary semilunar valve,** which marks the end of the right ventricle and the entrance into the pulmonary trunk. The pulmonary trunk divides shortly into right and left pulmonary arteries, which carry deoxygenated blood to the lungs.

Semilunar valves are located within the walls of both ventricles immediately before the connection of the ventricle to the pulmonary trunk and aorta (see figure 22.6). Each valve is composed of three thin, half-moon-shaped, pocketlike semilunar cusps. As blood is pumped into the arterial trunks, it pushes against the cusps, forcing the valves open. When ventricular contraction ceases, blood is prevented from flowing back into the ventricles from the arterial trunk by first entering the pockets of the semilunar valves between the cusps and the chamber wall. This causes the cusps to "fill and expand" and meet at the artery center, effectively blocking blood backflow.

Left Atrium

Once gas exchange occurs in the lungs, the oxygenated blood travels through the pulmonary veins to the left atrium (see figure 22.6). The smooth posterior wall of the left atrium contains openings for approximately four pulmonary veins. Sometimes two of these vessels fuse prior to reaching the left atrium, thus decreasing the number of openings through the atrial wall. Like the right atrium, the left atrium also has pectinate muscles along its anterior wall as well as an auricle.

Separating the left atrium from the left ventricle is the **left atrioventricular opening.** This opening is covered by the **left atrioventricular (AV) valve** (also called the *bicuspid valve,* since it has two triangular cusps). This valve is also sometimes called the *mitral* (mī′trăl) *valve,* because the two triangular cusps resemble a miter (the headpiece worn by a bishop). Oxygenated blood flows from the left atrium, through the left atrioventricular opening when the valve

is open, into the left ventricle. The left AV valve is forced closed when the left ventricle begins to contract, preventing blood backflow into the left atrium.

Left Ventricle

The left ventricular wall is typically three times thicker than the right ventricular wall **(figure 22.8).** The left ventricle requires thicker walls in order to generate enough pressure to force the oxygenated blood that has returned to the heart from the lungs into the aorta and then through the entire systemic circulation. (The right ventricle,

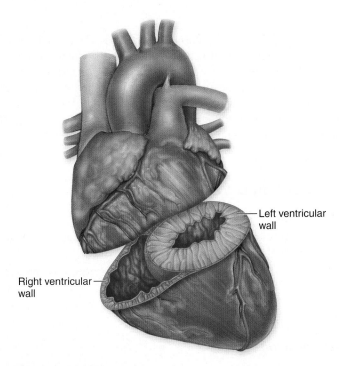

Left ventricular wall

Right ventricular wall

Figure 22.8

Comparison of Right and Left Ventricular Wall Thickness.
The wall of the left ventricle is about three times thicker than that of the right ventricle, because the left ventricle must generate a force sufficient to push blood through the systemic circulation and return it to the heart.

CLINICAL VIEW

Heart Sounds

By using a stethoscope, a physican can discern four normal heart sounds with each contraction: the two familiar sounds referred to as "lubb-dupp" and two minor sounds. The lubb sound signifies the closing of the AV valves, while the dupp sound signifies the closing of the semilunar valves. These heart sounds provide clinically important information about heart activity and the action of heart valves. The minor sounds are caused by contraction of the atria and flow of blood into the ventricles.

The place where sounds from each AV valve and each semilunar valve may best be heard does *not* correspond with the location of the valve (see figure) because some overlap of valve sounds occurs near their anatomic locations.

- The aortic semilunar valve is best heard in the second intercostal space to the right of the sternum.
- The pulmonary semilunar valve is best heard in the second intercostal space to the left of the sternum.
- The right AV valve is best heard by the right side of the inferior end of the sternum.
- The left AV valve is best heard near the apex of the heart (at the level of the left fifth intercostal space, about 9 centimeters from the midline of the sternum).

Often, abnormal heart sounds, generally called a **heart murmur,** are the first indication of heart problems. These sounds may be heard before, during, or after normal heart sounds. A heart murmur is usually the result of turbulence of the blood as it passes through the heart, and may be caused by valvular leakage, decreased valve flex-

ibility or a misshapen valve. Sometimes heart murmurs are of little consequence, but all of them need to be evaluated to rule out a more serious heart problem.

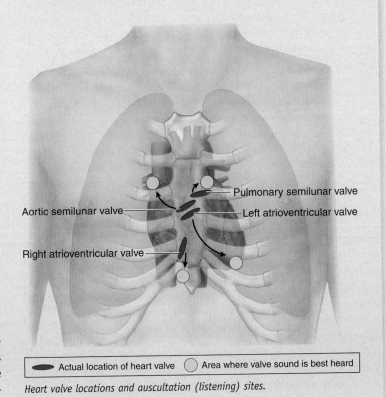

| ⬬ Actual location of heart valve | ◯ Area where valve sound is best heard |

Heart valve locations and auscultation (listening) sites.

---

in contrast, merely has to pump blood to the nearby lungs.) The trabeculae carneae in the left ventricle are more prominent than in the right ventricle. Typically, two large papillary muscles project from the ventricle's inner wall and anchor the chordae tendineae that attach to the cusps of the left AV valve. At the superior end of the ventricular cavity, the **aortic semilunar valve** marks the end of the left ventricle and the entrance into the aorta (see figure 22.7).

⚡ WHAT DID YOU LEARN?

5 Where is the coronary sulcus located?

6 What are two similarities and two differences between the right and left ventricles?

7 What is the composition and function of the chordae tendineae?

Coronary Circulation

Key topic in this section:

- Location, origins, and branches of the coronary blood vessels

Left and **right coronary arteries** travel within the coronary sulcus of the heart to supply the heart wall (**figure 22.9a**). These arteries are the only branches of the ascending aorta. The openings for these arteries are located in the wall of

the ascending aorta immediately superior to the aortic semilunar valve.

The right coronary artery typically branches into the right **marginal artery,** which supplies the right border of the heart, and the **posterior interventricular artery,** which supplies the posterior surface of both the left and right ventricles. The left coronary artery typically branches into the **anterior interventricular artery** (also called the *left anterior descending artery*), which supplies the anterior surface of both ventricles and most of the interventricular septum, and the **circumflex** (ser'kŭm-fleks; around) **artery,** which supplies the left atrium and ventricle. This arterial pattern can vary greatly among individuals. For example, some people may have a posterior interventricular artery that is a branch of the left coronary artery. Knowledge of this variation is essential when treating individuals for coronary artery disease.

The coronary arteries are considered **functional end arteries** (see chapter 23). In other words, while the left and right coronary arteries share some tiny connections, called **anastomoses,** functionally they act like end arteries, which have no anastomoses and are the "end of the line" when it comes to arterial blood flow. The anastomoses allow the coronary arteries to shunt a tiny amount of blood from one artery to another. However, if one of the arteries becomes blocked, as happens with coronary artery disease, these anastomoses are too tiny to shunt sufficient blood from one artery to the other. For example, if a branch of the left coronary artery becomes blocked, the right coronary artery cannot shunt enough blood to the part of

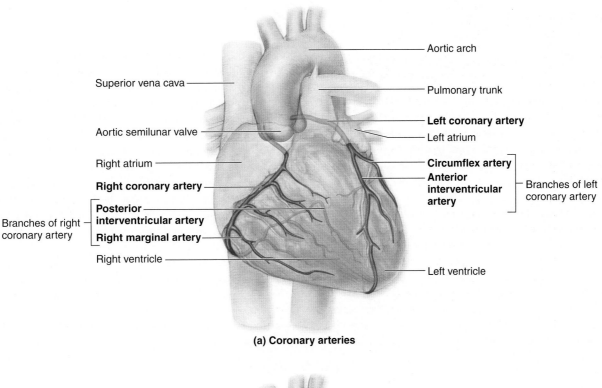

(a) Coronary arteries

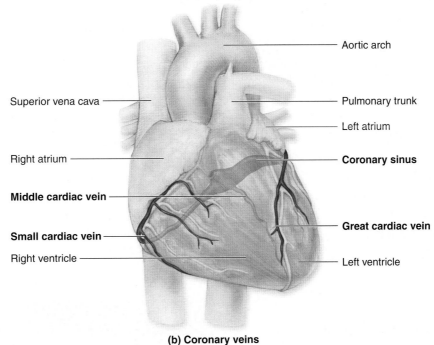

(b) Coronary veins

Figure 22.9

Coronary Circulation.   Anterior view of *(a)* coronary arteries and *(b)* coronary veins that transport blood to and from the cardiac muscle tissue.

the heart supplied by this branch. As a result, the part of the heart wall that was supplied by the branch of the left coronary artery dies due to lack of blood flow to the tissue.

Venous return occurs through one of several cardiac veins (figure 22.9*b*). The **great cardiac vein** runs alongside the anterior interventricular artery; the **middle cardiac vein** runs alongside the posterior interventricular artery; and the **small cardiac vein** travels close to the right marginal artery. These cardiac veins all drain into the **coronary sinus,**

a large vein that lies in the posterior aspect of the coronary sulcus. The coronary sinus drains directly into the right atrium of the heart.

Because the ventricular myocardium is compressed during contraction, most coronary flow occurs during ventricular relaxation. Normally, flow is evenly distributed throughout the thickness of the myocardium. Under certain circumstances, however, coronary flow may be reduced, especially to the regions immediately external to the endocardium. Situations related to inadequate coronary

Angina Pectoris and Myocardial Infarction

The most common cause of death in the United States is **coronary atherosclerosis** (ath'er-ō-skler-ō'sis; *athere* = gruel, *sclerosis* = hardness), or *coronary heart disease* (see Clinical View, "Atherosclerosis," in chapter 23). This condition is characterized by narrowing of the coronary arteries that reduces blood flow to the myocardium and gives rise to chest pain. Coronary atherosclerosis can lead to either angina pectoris or myocardial infarction.

Angina pectoris (an'jĭ-nă, an-ji'na) is a poorly localized pain sensation in the left side of the chest, the left arm and shoulder, or sometimes the jaw and/or the back. Generally it results from strenuous activity, when workload demands on the heart exceed the ability of the narrowed coronary vessels to supply blood. The pain from angina is typically referred along the sympathetic pathways (T1–T5 spinal cord segments), so an individual may experience pain in the chest region or down the left arm, where the T1 dermatome is located. The pain diminishes shortly after the person stops the exertion, and normal blood flow

to the heart is restored. Although many people are successfully treated for years with medications that cause temporary vascular dilation, such as nitroglycerine, the prognosis and long-term therapy for angina depend on the severity of the vascular narrowing.

Myocardial infarction (in-fark'shŭn) **(MI),** commonly called a heart attack, is a potentially fatal condition resulting from sudden and complete occlusion (blockage) of a coronary artery. A region of the myocardium is deprived of oxygen, and some of this tissue may die (necrose). Most people experiencing an MI report a sudden, excruciating and crushing substernal chest pain, although some people have relatively little pain. Other immediate symptoms include weakness, shortness of breath, and marked sweating. Some MI sufferers die immediately from myocardial conduction disturbances that lead to ventricular fibrillation. Others may die within a few hours or days because the heart has been profoundly and suddenly weakened by the loss of a large amount of tissue. (Mature cardiac cells have little or no capacity to regenerate, so when an MI results in cell death, scar tissue forms to fill the gap.) In people who survive an MI, the remaining cardiac muscle eventually may strengthen itself over time as the remaining cardiac muscle fibers hypertrophy.

blood flow include **tachycardia** (tak'i-kar'dē-ă; *tachys* = quick), an increased heart rate that shortens diastole, and **hypotension** (low blood pressure), which reduces the ability of blood to flow through the ventricular myocardium.

WHAT DO YOU THINK?

2 Do the coronary arteries fill with blood when the ventricles contract or when they relax?

WHAT DID YOU LEARN?

8 Why are coronary arteries considered functional end arteries?

How the Heart Beats: Electrical Properties of Cardiac Tissue

Key topics in this section:

- Comparison of cardiac muscle and skeletal muscle
- Conduction of muscle impulses along muscle fibers
- Autorhymicity and the heart's conducting system

The efficient pumping of blood through the heart and blood vessels requires precisely coordinated contractions of the heart chambers. These contractions are made possible by the properties of cardiac muscle tissue and by specialized cells in the heart, known collectively as its conducting system.

Characteristics of Cardiac Muscle Tissue

The myocardium is composed of cardiac muscle, which was described briefly in chapters 4 (table 4.14) and 10 (table 10.6). Recall that cardiac muscle cells are relatively short, branched cells

that usually house one or two central nuclei and numerous mitochondria for ATP supply **(figure 22.10).** Cardiac muscle cells are arranged in spiral bundles and wrapped around and between the heart chambers. Cardiac muscle tissue exhibits some characteristics that are similar to those of skeletal muscle and others that are different **(table 22.2).** For example, the cells in both tissue types are striated, with extensive capillary networks that supply needed nutrients and oxygen. However, cardiac and smooth muscle differ in the following ways:

- The sarcoplasmic reticulum in cardiac muscle is less extensive and not as organized as in skeletal muscle.
- Cardiac muscle has no terminal cisternae, while skeletal muscle does.

Table 22.2	Comparison of Cardiac and Skeletal Muscle	
Cardiac Muscle	**Skeletal Muscle**
Cells are short and branching	Cells are long and cylindrical
1 or 2 nuclei in the center of the cell	Multiple nuclei at the periphery of the cell
Cells joined by intercellular junctions in intercalated discs	Cells do not have specialized intercellular junctions
Functional contractile unit is the sarcomere	Functional contractile unit is the sarcomere
T-tubules overlie Z-discs	T-tubules overlie A band/I band junctions
Composed of thick and thin filaments	Composed of thick and thin filaments
Contains sarcoplasmic reticulum but less than in skeletal muscle	Contains sarcoplasmic reticulum but more than in cardiac muscle
More mitochondria than in skeletal muscle	Fewer mitochondria than in cardiac muscle

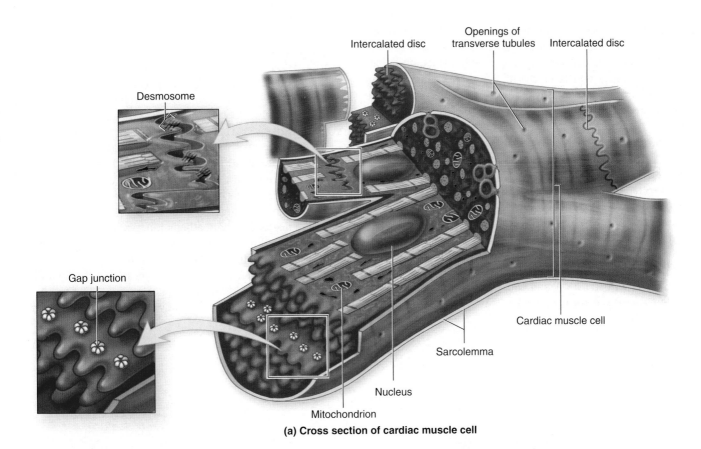

Desmosome

Gap junction

Intercalated disc

Openings of transverse tubules

Intercalated disc

Cardiac muscle cell

Sarcolemma

Nucleus

Mitochondrion

(a) Cross section of cardiac muscle cell

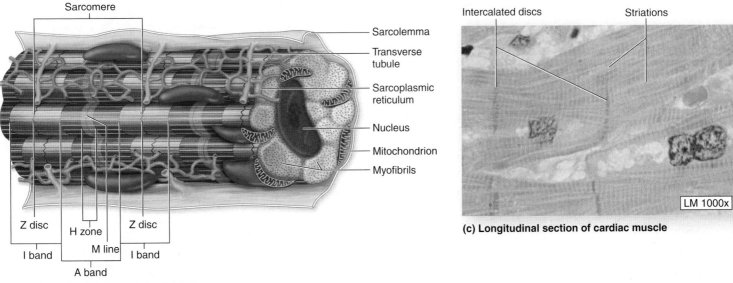

Sarcomere

Sarcolemma

Transverse tubule

Sarcoplasmic reticulum

Nucleus

Mitochondrion

Myofibrils

Z disc

H zone

M line

Z disc

I band

I band

A band

Intercalated discs

Striations

LM 1000x

(c) Longitudinal section of cardiac muscle

(b) Cardiac muscle cell, longitudinal view

Figure 22.10

Organization and Histology of Cardiac Muscle.   Cardiac muscle cells form the myocardium. *(a)* Individual cells are relatively short, branched, and striated. They are connected to adjacent cells by intercalated discs. *(b)* Transverse tubules are invaginations of the sarcolemma that surround myofibrils and overlie the Z disc. The sarcoplasmic reticulum is meager in cardiac muscle compared to its abundance in skeletal muscle. *(c)* Light micrograph of cardiac muscle in longitudinal section.

- Cardiac muscle lacks the extensive association of smooth endoplasmic reticulum (SER) and transverse tubules (T-tubules) present in skeletal muscle.
- In cardiac muscle, T-tubules overlie Z discs instead of the junctions of A bands and I bands as seen in skeletal muscle.
- The T-tubules in cardiac muscle have a less extensive distribution and a reduced association with SER compared to

those in skeletal muscle, allowing for both the delayed onset and prolonged contraction of cardiac muscle tissue.

Contraction of Heart Muscle

Cardiac muscle cells contract as a single unit because muscle impulses (changes in voltage potential across the sarcolemma [see chapter 10]) are distributed immediately and simultaneously throughout all cells of first the atria and then the ventricles. Neighboring cardiac muscle cells in the walls of heart chambers have formed specialized cell–cell contacts called **intercalated discs** (in-ter′kă-lā-ted disk), which electrically and mechanically link the cells together and permit the immediate passage of muscle impulses (figure 22.10).

Within the intercalated discs, gap junctions increase the flow of ions between the cells as the muscle impulse moves along the sarcolemma. The gap junctions of intercalated discs provide a low-resistance pathway across the membranes of adjoining cardiac muscle cells, allowing the unrestricted passage of ions required for the synchronous beating of cardiac muscle cells. Numerous desmosomes (see figure 4.1) within the intercalated discs prevent cardiac muscle cells from pulling apart. Therefore, cardiac muscle cells function as a single, coordinated unit; the precisely timed stimulation and response by cardiac muscle cells of both the atria and the ventricles are dependent upon these structural features.

The Heart's Conducting System

The heart exhibits **autorhythmicity,** meaning that the heart itself (not external nerves) is responsible for initiating the heartbeat. Certain cardiac muscle cells are specialized to initiate and conduct muscle impulses to the contractile muscle cells of the myocardium. Collectively, these specialized cells are called the heart's **conducting system (figure 22.11).**

The heartbeat is initiated by the specialized cardiac muscle cells of the **sinoatrial (SA) node,** which are located in the posterior wall of the right atrium, adjacent to the entrance of the superior vena cava. The cells of the SA node act as the **pacemaker,** the rhythmic center that establishes the pace for cardiac activity. Under the influence of parasympathetic innervation, SA node cells initiate impulses 70–80 times per minute.

The muscle impulse travels from the SA node to the **atrioventricular (AV) node.** The AV node is located in the floor of the right atrium between the right AV valve and the opening for the coronary sinus. The AV node normally slows conduction of the impulse as it

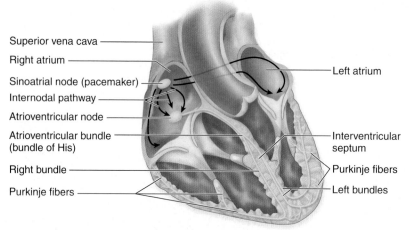

1. Muscle impulse is generated at the sinoatrial node. It spreads throughout the atria and to the atrioventricular node by the internodal pathway.

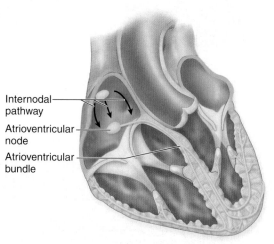

2. Atrioventricular node cells delay the muscle impulse as it passes to the atrioventricular bundle.

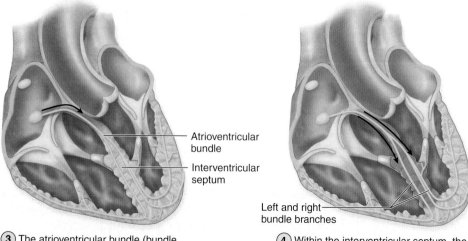

3. The atrioventricular bundle (bundle of His) conducts the muscle impulse into the interventricular septum.

4. Within the interventricular septum, the right and left bundles split from the atrioventricular bundle.

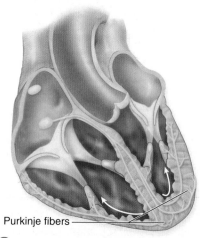

5. The muscle impulse is delivered to Purkinje fibers in each ventricle and distributed throughout the ventricular myocardium.

Figure 22.11

Conducting System of the Heart. Modified cardiac muscle fibers initiate the heartbeat and then spread and conduct the impulse throughout the heart.

CLINICAL VIEW

The Electrocardiogram

Electrical currents within the heart can be detected during a routine physical examination using monitoring electrodes attached to the skin—usually at the wrist, ankles, and six separate locations on the chest. The electrical signals are collected and charted as an **electrocardiogram** (ē-lek-trō-kar′dē-ō-gram; *gramma* = drawing), also called an **ECG** or *EKG*. When readings from the different electrodes are compared, they collectively provide an accurate, comprehensive assessment of the electrical activity of the heart.

An ECG provides a composite tracing of all muscle impulses generated by myocardial cells. It records the wave of change in voltage (potential) across the sarcolemma that (1) originates in the SA node, (2) radiates through both of the atria to the AV node, (3) passes through the AV node and the interventricular septum to the heart apex, and (4) stimulates the Purkinje fibers in the ventricular myocardium.

The progress of impulse transmission through the various parts of the conducting system is mirrored in an electrocardiogram. A typical ECG tracing for one heart cycle has three principal deflections: a P wave above the baseline, a QRS complex that begins (Q) and ends (S) with small downward deflection from the baseline and has a large deflection (R) above the baseline, and a T wave above the baseline (see figure). These waves are indicators of depolarization and repolarization within specific regions of the heart:

1. The **P wave** is generated when the impulse originating in the SA node depolarizes the cells of the atria.

2. The **QRS complex** identifies the beginning of depolarization of the ventricles. Simultaneously, the atria repolarize; however, this repolarization signal is masked by the electrical activity of the ventricles.

3. The **T wave** is a small, rounded peak that denotes ventricular repolarization.

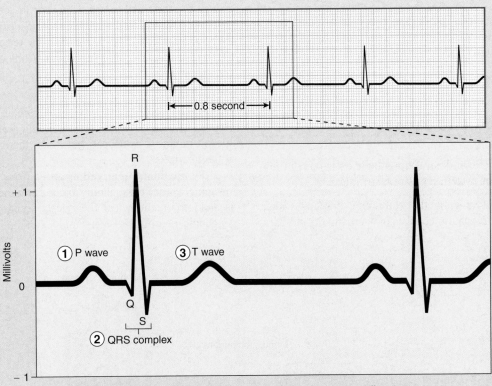

The events of a single cardiac cycle as recorded on an electrocardiogram.

travels from the atria to the ventricles, providing a delay between activation and contraction of the ventricles.

Recall that the fibrous skeleton insulates the atria from the ventricles to prevent random muscle impulses from spreading between the atria and the ventricles. There is an opening in the fibrous skeleton that allows the AV node to communicate with the next part of the conduction system, the **atrioventricular (AV) bundle.** This bundle, also known as the *bundle of His*, receives the muscle impulse from the AV node and extends into the interventricular septum before dividing into **left** and **right bundles.** These bundles conduct the impulse to conduction fibers called **Purkinje** (pŭr-kin′jē) **cells** that begin within the apex of the heart and extend through the walls of the ventricles. Purkinje fibers are larger than other cardiac muscle cells. Muscle impulse conduction along the Purkinje fibers is extremely rapid, consistent with the large size of the cells, and the impulse spreads immediately throughout the ventricular myocardium.

WHAT DID YOU LEARN?

9 What are three of the ways in which cardiac muscle differs from skeletal muscle?

10 Which component of the conducting system is located in the floor of the right atrium, between the right AV valve and the coronary sinus opening?

11 What is meant by the term *autorhythmic* when used to describe the heart?

Cardiac Arrhythmia

Cardiac arrhythmia (ă-rith′mē-ă; *a* = not, *rhythmos* = rhythm), also called *dysrhythmia*, is any abnormality in the rate, regularity, or sequence of the cardiac cycle. Several common arrhythmias have been described:

■ **Atrial flutter** occurs when the atria attempt to beat at a rate of 200 to 400 times per minute, and as a consequence literally bombard the AV node with muscle impulses. Abnormal muscle impulses flow continuously through the atrial condition system, thus stimulating the atrial musculature and AV node over and over. This condition may persist for years, and frequently degenerates into atrial fibrillation.

■ **Atrial fibrillation** (fĭ-bri-lă′shŭn) differs from atrial flutter in that the muscle impulses are significantly more chaotic, leading to an irregular heart rate. The ventricles respond by increasing and decreasing contraction activities, which may lead to serious disturbances in the cardiac rhythm.

■ **Premature ventricular contractions (PVCs)** often result from stress, stimulants such as caffeine, or sleep deprivation. They occur either singly or in rapid bursts due to abnormal impulses

initiated within the AV node or the ventricular conduction system. All of us experience an occasional PVC, and they are not detrimental unless they occur in great numbers. Most PVCs go unnoticed, although occasionally one is perceived as the heart "skipping a beat" and then "jumping" in the chest.

A more serious arrhythmia is **ventricular fibrillation,** a rapid, repetitious movement of the ventricular muscle that replaces normal contraction. This is a life-threatening condition caused by scattered impulses originating at different times and places throughout the entire myocardium. Because the contractions of a heart in fibrillation are uncoordinated, the heart does not pump blood, and blood circulation stops. This cessation of cardiac activity is called **cardiac arrest.** Fibrillation almost certainly results in death unless the normal rhythmic contractions of the heart are promptly restored. To restore normal heart contractions, medical personnel apply a strong electrical shock to the skin of the chest using paddle electrodes. The electrical current passes through the chest wall to completely and immediately depolarize the entire myocardium. This procedure is analogous to pushing the reset button on a computer—and as in rebooting the computer, the hope is that when the heart begins to function again, it will work as intended.

Innervation of the Heart

Key topic in this section:

■ How the sympathetic and parasympathetic divisions of the autonomic nervous system regulate heart rate

The heart is innervated by the autonomic nervous system (**figure 22.12**). This innervation consists of sympathetic and para-

sympathetic components, collectively referred to as the **coronary plexus.** The innervation by autonomic centers in the brainstem doesn't initiate a heartbeat, but it can increase or decrease the rate of the heartbeat.

Sympathetic innervation arises from the T1–T5 segments of the spinal cord. Preganglionic axons enter the sympathetic trunk and ascend into the thoracic and cervical portions, where they synapse on ganglionic neurons. Postganglionic axons project from the superior,

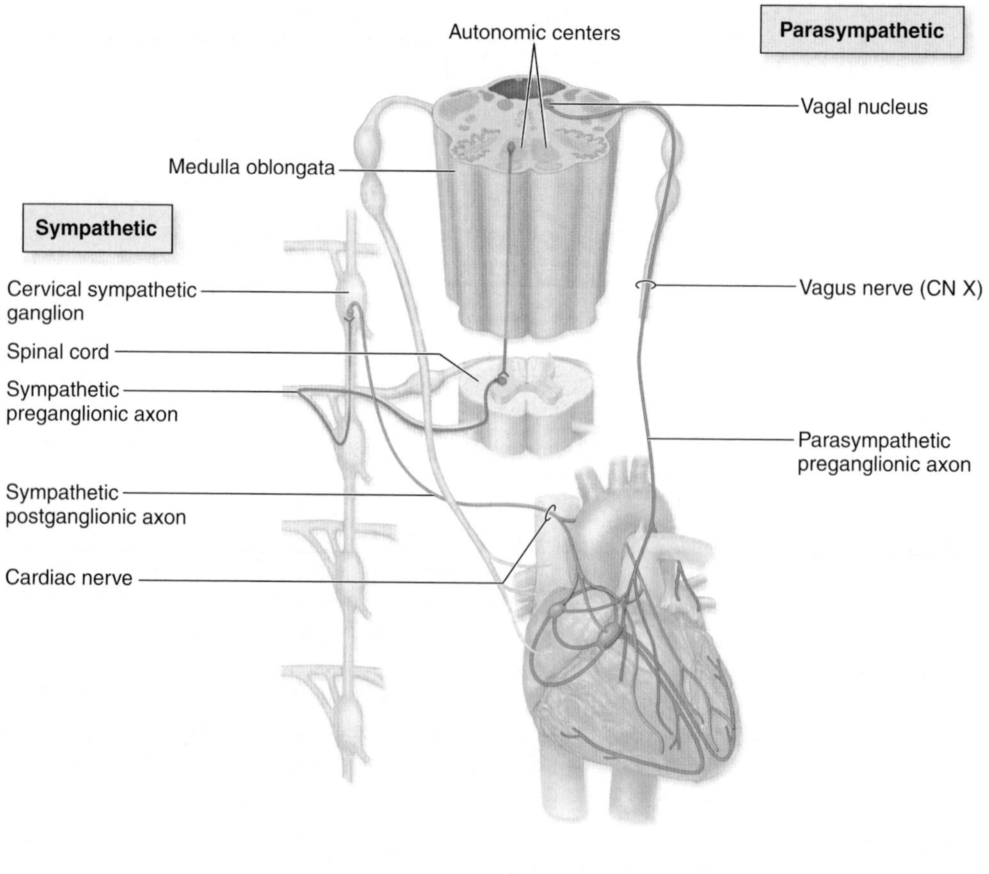

Figure 22.12

Autonomic Innervation of the Heart. The amount of blood pumped from the heart and the heart rate are modified by autonomic centers in the brainstem. (*Left*) Sympathetic stimulation is carried through the cardiac nerves. (*Right*) Parasympathetic stimulation is carried by vagus nerves.

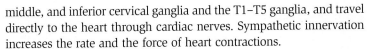

middle, and inferior cervical ganglia and the T1–T5 ganglia, and travel directly to the heart through cardiac nerves. Sympathetic innervation increases the rate and the force of heart contractions.

Parasympathetic innervation comes from the medulla oblongata via the left and right vagus nerves (CN X). As the vagus nerves descend into the thoracic cavity, they give off branches that supply the heart. Parasympathetic innervation decreases the heart rate, but generally tends to have no effect on the force of contractions.

Although there is a particularly rich sympathetic and parasympathetic innervation of the SA and AV nodes, the working myocardial cells are also supplied by both types of autonomic axons.

Study Tip!

Here is one way to help you remember the effects of sympathetic and parasympathetic innervation on the heart:

Sympathetic innervation **S**peeds the heart rate.

Parasympathetic innervation does the o**P**posite (decreases the heart rate).

WHAT DID YOU LEARN?

12 How does the sympathetic division of the autonomic nervous system affect the heart rate?

Tying It All Together: The Cardiac Cycle

Key topics in this section:

- Events in the cardiac cycle
- Pattern of blood flow through the heart

A **cardiac cycle** is the time from the start of one heartbeat to the initiation of the next. During a single cardiac cycle, all chambers within the heart experience alternate periods of contraction and relaxation. The contraction of a heart chamber is called **systole** (sis′tō-lē). During this period, the contraction of the myocardium forces blood either into another chamber (from atrium to ventricle) or into a blood vessel (from a ventricle into the attached large artery). The relaxation phase of a heart chamber is termed **diastole** (dī-as′tō-lē; dilation). During this period between contraction phases, the myocardium of each chamber relaxes, and the chamber fills with blood.

At the beginning of the cardiac cycle, the left and right atria contract simultaneously. When the atria contract (atrial systole), blood is forced into the ventricles through the open AV valves. During this time, blood is still returning to the atria in the superior vena cava, inferior vena cava, and coronary sinus (right atrium) and pulmonary veins (left atrium). After the atria begin to relax (atrial diastole), left and right ventricular contraction (ventricular systole) occurs **(figure 22.13a)**. Thus, only two of the four chambers (either the atria or the ventricles) contract at the same time. When the ventricles contract, the atrioventricular openings close as blood pushes against the cusps of the AV valves, and their edges meet to form a seal. Papillary muscles and the chordae tendineae prevent these valve cusps from everting. The semilunar valves are forced open, and blood enters the pulmonary trunk and the aorta. When the ventricles are relaxing during the cardiac cycle (ventricular diastole), most of the blood flows passively from the relaxing atria into the ventricles through the open AV valves (figure 22.13b). Therefore, for the last half of the cardiac cycle, all four chambers are in diastole together.

WHAT DO YOU THINK?

3 What could happen if the AV valves were to evert into the atria?

Steps in the Cardiac Cycle

The events in a normal cardiac cycle are shown in **figure 22.14** and described here:

1. **Atrial systole** occurs at the beginning of the cardiac cycle. It is a brief contraction of the atrial myocardium initiated by the heart pacemaker. Contraction of the atria finishes filling the ventricles through the open AV valves while the ventricles are in diastole. The semilunar valves remain closed.
2. **Early ventricular systole** is the beginning of the ventricular contraction. The atria remain in diastole. The AV valves are forced closed, and the semilunar valves remain closed.
3. **Late ventricular systole** occurs later in the ventricular contraction. The atria remain in diastole, and the AV valves remain closed. Pressure on blood in the ventricles forces the semilunar valves to open, and blood is ejected into the arterial trunks.
4. **Early ventricular diastole** is the start of ventricular relaxation. The atria remain in diastole. The semilunar valves close to prevent blood backflow into the ventricles. The AV valves remain closed.
5. **Late ventricular diastole** is a continuation of ventricular relaxation and an important time for ventricular filling. The atria remain in diastole. The AV valve opens, and passive filling of the ventricle from the atria begins and continues as most of the ventricular filling occurs. The semilunar valves remain closed.

Summary of Blood Flow During the Cardiac Cycle

As the heart chambers cyclically contract and relax, pressure on the blood in the chambers alternately increases and decreases. **Table 22.3** illustrates the flow of blood through the four heart chambers during the cardiac cycle. As you learn this sequence, keep the following general principles in mind:

- Blood flows from veins into the atria under low pressure.
- Blood only passes from the atria into the ventricles if the AV valves are open. Most of the ventricular filling is passive (about 70%), occurring when both chambers are relaxing (in diastole) and the atrial pressure is greater than the ventricular pressure. Filling of the final 30% of the ventricles occurs when the atria contract (in systole).
- Ventricular contraction (systole) increases pressure on the blood within the ventricles. When ventricular pressure rises significantly, the AV valves close, and the semilunar valves are forced open due to increased pressure in the ventricles, allowing blood to enter the large arterial trunks.

WHAT DID YOU LEARN?

13 Distinguish between systole and diastole.

14 What events occur in the ventricles during late ventricular diastole?

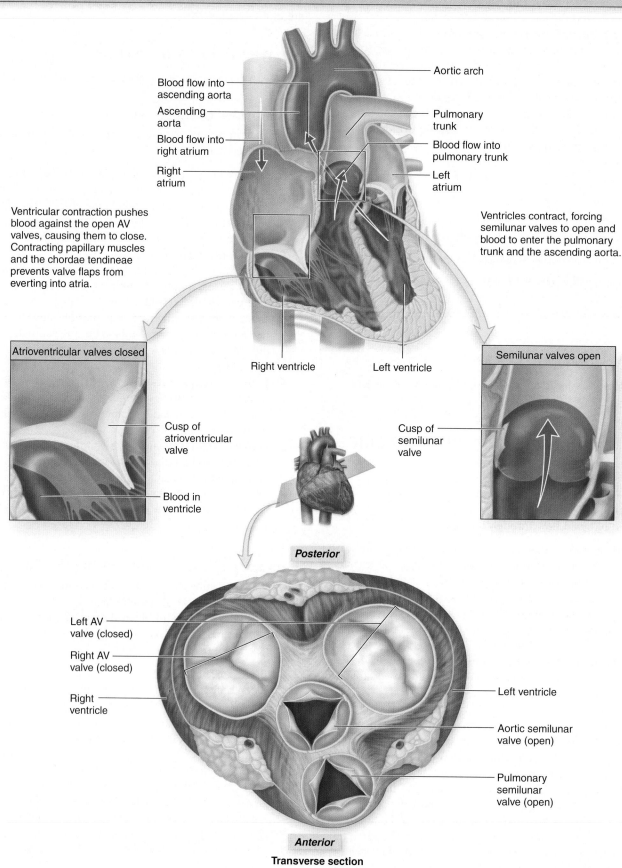

(a) Ventricular Systole (Contraction)

Blood flow into ascending aorta

Ascending aorta

Blood flow into right atrium

Right atrium

Aortic arch

Pulmonary trunk

Blood flow into pulmonary trunk

Left atrium

Ventricular contraction pushes blood against the open AV valves, causing them to close. Contracting papillary muscles and the chordae tendineae prevents valve flaps from everting into atria.

Ventricles contract, forcing semilunar valves to open and blood to enter the pulmonary trunk and the ascending aorta.

Right ventricle   Left ventricle

Atrioventricular valves closed

Cusp of atrioventricular valve

Blood in ventricle

Semilunar valves open

Cusp of semilunar valve

Posterior

Left AV valve (closed)

Right AV valve (closed)

Right ventricle

Left ventricle

Aortic semilunar valve (open)

Pulmonary semilunar valve (open)

Anterior

Transverse section

Figure 22.13

Ventricular Systole and Ventricular Diastole.   *(a)* The semilunar valves open during ventricular systole to allow blood to flow into the large arteries; the AV valves close during ventricular systole to prevent backflow of blood into the atria. *(b)* During ventricular diastole, the AV valves open to allow blood to enter the ventricles from the atria; the semilunar valves remain closed to prevent backflow of blood into the ventricles from the large arteries. Transverse sections in both *(a)* and *(b)* show a superior view.

(b) Ventricular Diastole (Relaxation)

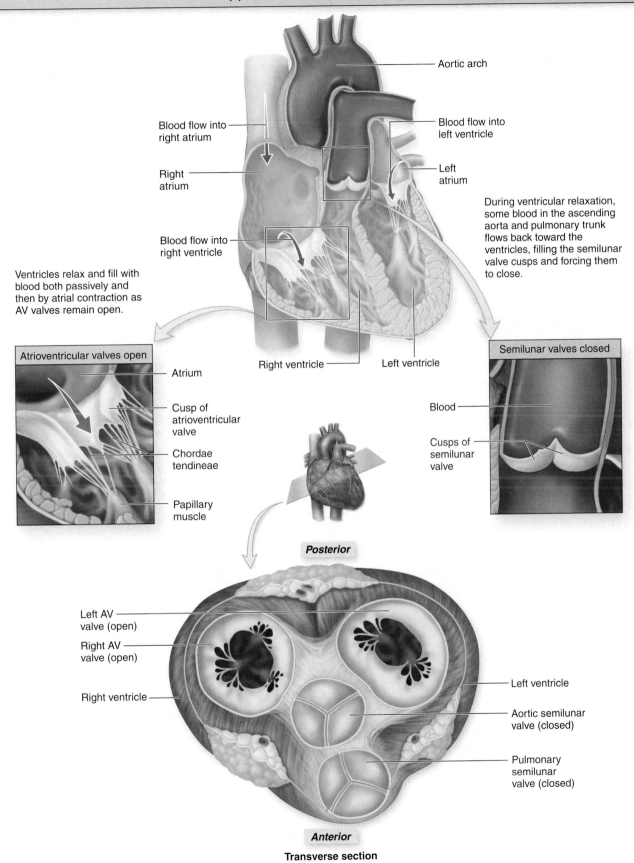

Aortic arch

Blood flow into right atrium

Blood flow into left ventricle

Right atrium

Left atrium

Blood flow into right ventricle

During ventricular relaxation, some blood in the ascending aorta and pulmonary trunk flows back toward the ventricles, filling the semilunar valve cusps and forcing them to close.

Ventricles relax and fill with blood both passively and then by atrial contraction as AV valves remain open.

Right ventricle

Left ventricle

Atrioventricular valves open

Semilunar valves closed

Atrium

Cusp of atrioventricular valve

Chordae tendineae

Papillary muscle

Blood

Cusps of semilunar valve

Posterior

Left AV valve (open)

Right AV valve (open)

Right ventricle

Left ventricle

Aortic semilunar valve (closed)

Pulmonary semilunar valve (closed)

Anterior

Transverse section

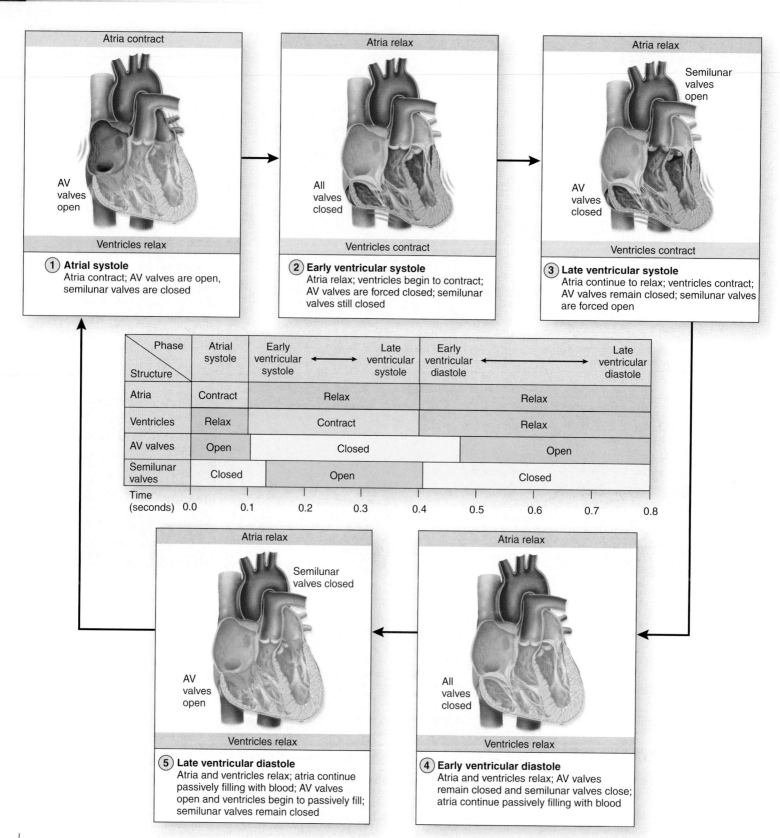

1 **Atrial systole**
Atria contract; AV valves are open, semilunar valves are closed

2 **Early ventricular systole**
Atria relax; ventricles begin to contract; AV valves are forced closed; semilunar valves still closed

3 **Late ventricular systole**
Atria continue to relax; ventricles contract; AV valves remain closed; semilunar valves are forced open

Phase / Structure	Atrial systole	Early ventricular systole ⟷ Late ventricular systole		Early ventricular diastole ⟷ Late ventricular diastole	
Atria	Contract	Relax		Relax	
Ventricles	Relax	Contract		Relax	
AV valves	Open	Closed		Open	
Semilunar valves	Closed	Open		Closed	
Time (seconds)	0.0 0.1	0.2 0.3	0.4	0.5 0.6	0.7 0.8

5 **Late ventricular diastole**
Atria and ventricles relax; atria continue passively filling with blood; AV valves open and ventricles begin to passively fill; semilunar valves remain closed

4 **Early ventricular diastole**
Atria and ventricles relax; AV valves remain closed and semilunar valves close; atria continue passively filling with blood

Figure 22.14

Cardiac Cycle. The cardiac cycle consists of all of the events that occur with a single heartbeat: contraction (systole) and relaxation (diastole) of all four heart chambers.

Table 22.3   **Blood Flow Through the Heart**

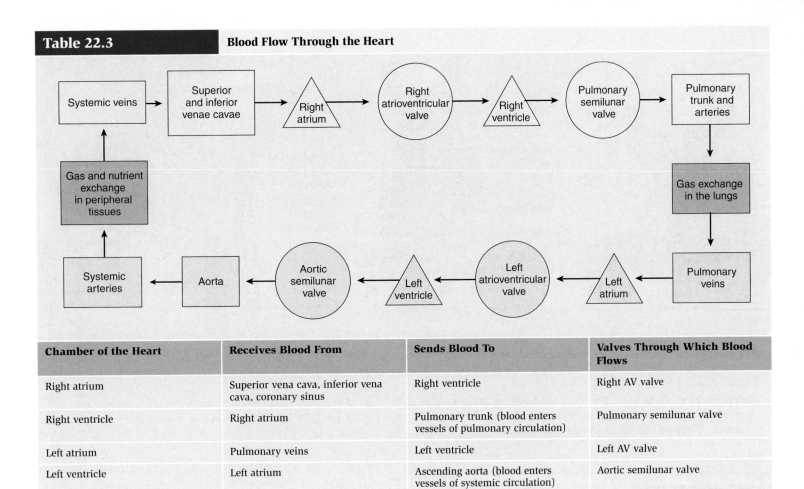

Chamber of the Heart	Receives Blood From	Sends Blood To	Valves Through Which Blood Flows
Right atrium	Superior vena cava, inferior vena cava, coronary sinus	Right ventricle	Right AV valve
Right ventricle	Right atrium	Pulmonary trunk (blood enters vessels of pulmonary circulation)	Pulmonary semilunar valve
Left atrium	Pulmonary veins	Left ventricle	Left AV valve
Left ventricle	Left atrium	Ascending aorta (blood enters vessels of systemic circulation)	Aortic semilunar valve

Aging and the Heart

Key topic in this section:

■ How heart function changes as we age

A healthy heart is capable of quickly and efficiently altering both the heartbeat rate and the volume of blood pumped during either an increase or decrease in activity. Consuming a diet low in saturated fats, abstaining from smoking, and exercising regularly help maintain a strong, vigorous heart. However, the decreased flexibility and elasticity of connective tissue that occur with aging can cause the heart valves to become slightly inflexible. As a result, a heart murmur may develop, and blood flow through the heart may be altered. Decreased conducting system efficiency reduces the heart's ability to pump the extra blood needed during stress and exercise. In addition, the muscular wall of the ventricle increases in thickness when high blood pressure causes the ventricles to work harder to pump blood into the arterial trunks. Consequently, the ventricular myocardium undergoes hypertrophy. Hypertrophy of the heart may have different causes (such as hypertension or narrowing of vessels connected to the heart), but cardiac muscle cells always thicken and lose the normal arrangement of formed cell bundles. Thus, although the heart is enlarged, it works less efficiently.

WHAT DID YOU LEARN?

15 What are some age-related changes that result in altered heart functions?

Development of the Heart

Key topic in this section:

■ Formation of postnatal heart structures from the primitive heart tube

Development of the heart commences in the third week, when the embryo becomes too large to receive its nutrients through diffusion alone. At this time, the embryo needs its own blood supply, heart, and blood vessels for transporting oxygen and nutrients through its growing body. The steps involved in heart development are complex, because the heart must begin working before its development is complete.

By day 19 (middle of week 3), two **heart tubes** (or *endocardial tubes*) form from mesoderm in the embryo. By day 21, these paired tubes fuse, forming a single primitive heart tube **(figure 22.15)**. This tube develops the following named expansions that ultimately give rise to postnatal heart structures (listed from inferior to superior): **sinus venosus, primitive atrium, primitive ventricle,** and **bulbus cordis** (bŭl′bŭs kōr′dis). The sinus venosus and primitive atrium form parts of the left and right atria. The primitive ventricle forms most of the left ventricle. The bulbus cordis may be further subdivided into a trabeculated part of the right ventricle, which forms most of the right ventricle; the **conus cordis,** which forms the outflow tracts for the ventricles; and the **truncus arteriosus,** which forms the ascending aorta and pulmonary trunk. **Table 22.4** lists these primitive heart tube components and the structures they develop into.

By day 22, the primitive heart begins to beat and begins its process of bending and folding. The heart folding is complete by the end of the

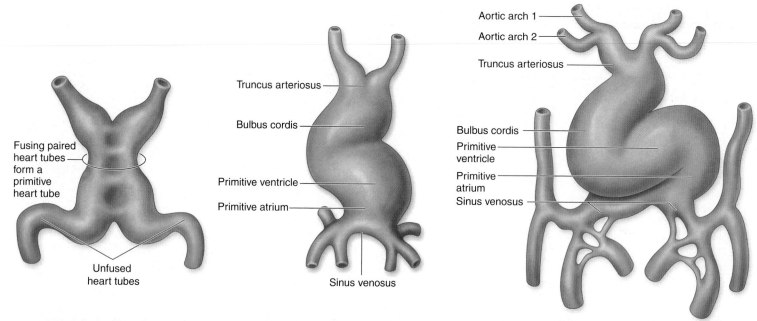

(a) 21 days: Paired heart tubes fuse.   **(b) 22 days: Primitive heart tube begins to fold.**   **(c) 28 days: S-shaped heart tube completes folding.**

Figure 22.15

Development of the Heart.   The heart develops from mesoderm. By day 19, paired heart tubes are present in the cardiogenic region of the embryo. (*a*) These paired tubes fuse by day 21 to form a primitive heart tube. (*b*) The primitive heart tube bends and folds upon itself, beginning on day 22. (*c*) By day 28, the heart tube is S-shaped.

Table 22.4	Primitive Heart Tube Components and Their Postnatal Structures
Heart Tube Component	**Postnatal Derivative**
Sinus venosus	Superior vena cava, coronary sinus, smooth posterior wall of right atrium
Primitive atrium	Anterior muscular portions of left and right atria
Primitive ventricle	Most of left ventricle
Bulbus cordis	
Trabeculated part of right ventricle	Most of right ventricle
Conus cordis	Outflow tracts from ventricles to aorta and pulmonary trunk
Truncus arteriosus	Ascending aorta, pulmonary trunk

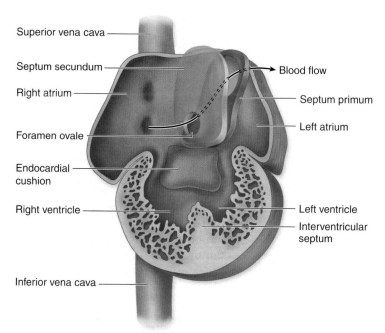

Early week 7 (43 days)

Figure 22.16

Interatrial Septum.   The interatrial septum is formed by two overlapping septa (primum and secundum). The foramen ovale is a passageway that detours blood away from the pulmonary circulation into the systemic circulation prior to birth.

week (figure 22.15*b*). The bulbus cordis is pulled inferiorly, anteriorly, and to the embryo's right, while the primitive ventricle moves left to reposition. The primitive atrium and sinus venosus reposition superiorly and posteriorly. Thus, by day 28 the heart tube is S-shaped (figure 22.15*c*).

The next major steps in heart development occur during weeks 5–8, when the single heart tube becomes partitioned into four chambers (two atria and two ventricles), and the main vessels entering and leaving the heart form. The common atrium is subdivided into a left and right atrium by an interatrial septum, which consists of two parts (**septum primum** and **septum secundum**) that partially overlap. These two parts connect to tissue masses called **endocardial cushions.** An opening in the septum secundum (which is covered by the septum primum) is called the **foramen ovale** (ō-val′ē) (**figure 22.16**). Because the embryonic lungs are not functional, much of the blood is shunted away from the lungs and to the rest of the body. Since it is the right side of the heart that sends blood to the lungs, blood is shunted from the right atrium to the left atrium by traveling through the foramen ovale and pushing the septum primum to the left. Blood cannot flow back from the left atrium to the right, because the septum primum's movement to the right is stopped when it comes against the septum secundum. Thus, the septum primum acts as a unidirectional flutter valve. When the baby is born and the lungs are fully functional, the blood from the left atrium pushes the septum primum and secundum together, creating a closed interatrial septum. The only remnant of the embryonic opening is an oval-shaped depression in the interatrial septum called the **fossa** (fos′ă; trench) **ovalis.**

Left and right ventricles are partitioned by an interventricular septum that grows superiorly from the floor of the ventricles. The AV valves, papillary muscles, and chordae tendineae all form from portions of the ventricular walls as well. The superior part of the interventricular septum develops from the **aorticopulmonary septum,** which is a spiral-shaped mass that also subdivides the truncus arteriosus into the pulmonary trunk and the ascending aorta.

Many congenital heart malformations result from incomplete or faulty development during these early weeks. For example, in an **atrial septal defect** the postnatal heart still has an opening between the left and right atria. Thus, blood from the left atrium (the higher-pressure system) is shunted to the right atrium (the lower-pressure system). This can lead to enlargement of the right side of the heart. **Ventricular septal defects** can occur if the interventricular septum is incompletely formed. A common malformation called **tetralogy of Fallot** occurs when the aorticopulmonary septum divides the truncus arteriosus unevenly. As a result, the patient has a ventricular septal defect, a very narrow pulmonary trunk (pulmonary stenosis), an aorta that overlaps both the left and right ventricles, and an enlarged right ventricle (right ventricular hypertrophy). A good knowledge of heart development is essential in understanding these congenital heart malformations.

CLINICAL TERMS

bradycardia (brad-ē-kar′dē-ă; *bradys* = slow) Slowing of the heartbeat, usually described as less than 50 beats per minute.

cardiomyopathy (kar′dē-ō-mī-op′ă-thē) Another term for disease of the myocardium; causes vary and include thickening of the ventricular septum (hypertrophy), secondary disease of the myocardium, or sometimes a disease of unknown cause.

endocarditis (en′dō-kar-dī′tis) Inflammation of the endocardium. Types include bacterial (caused by the direct invasion of bacteria), chorditis (affecting the chordae tendineae), infectious (caused by

microorganisms), mycotic (due to infection by fungi), and rheumatic (due to endocardial involvement as part of rheumatic heart disease).

ischemia (is-kē′mē-ă; *ischo* = to keep back) Inadequate blood flow to a structure caused by obstruction of the blood supply, usually due to arterial narrowing or disruption of blood flow.

myocarditis (mī′ō-kar-dī′tis) Inflammation of the muscular walls of the heart. This uncommon disorder is caused by viral, bacterial, or parasitic infections, exposure to chemicals, or allergic reactions to certain medications.

CHAPTER   SUMMARY

Overview of the Cardiovascular System 655	▪ Arteries carry blood away from the heart, and veins return blood to the heart. ▪ Heart functions include one-directional blood flow through the cardiovascular system, coordinated side-by-side pumps, and generation of pressure to drive blood through blood vessels. **Pulmonary and Systemic Circulations 655** ▪ The pulmonary circulation conveys blood to and from the lungs, and the systemic circulation carries blood to and from all the organs and tissues. **Position of the Heart 656** ▪ The heart is located posterior to the sternum in the mediastinum. ▪ Its base is the posterosuperior surface formed primarily by the left atrium; the apex is in the conical, inferior end. **Characteristics of the Pericardium 657** ▪ The pericardium that encloses the heart has an outer fibrous portion and an inner serous portion. ▪ The pericardial cavity is a thin space between the layers of the serous pericardium. Pericardial fluid produced by the serous membranes lubricates the surfaces to reduce friction.
Anatomy of the Heart 658	▪ The heart is a small, cone-shaped organ. **Heart Wall Structure 658** ▪ The heart wall has an epicardium (visceral pericardium), myocardium (thick layer of cardiac muscle), and endocardium (thin endothelium and areolar connective tissue). **External Heart Anatomy 658** ▪ The heart has four chambers: Two smaller atria receive blood returning to the heart, and two larger ventricles pump blood away from the heart. ▪ Around the circumference of the heart, a deep coronary sulcus separates the atria and ventricles. Shallow sulci extend from the coronary sulcus on the anterior and posterior surfaces between the ventricles. Coronary vessels lie within the sulci.

(continued on next page)

CHAPTER SUMMARY (continued)

Anatomy of the Heart (continued) 658	**Internal Heart Anatomy: Chambers and Valves** 658
	▪ Dense regular connective tissue forms the fibrous skeleton of the heart that separates the atria and the ventricles. This skeleton (1) separates atria and ventricles, (2) anchors heart valves, (3) electrically insulates atria and ventricles, and (4) provides for attachment of cardiac muscle tissue.
	▪ The right atrium receives blood from the superior vena cava, the inferior vena cava, and the coronary sinus.
	▪ Atria are separated by an interatrial septum, and ventricles are separated by an interventricular septum.
	▪ Four pulmonary veins empty into the left atrium.
	▪ Thick-walled ventricles receive blood from the atria through open AV valves. The free edges of the AV valve cusps are prevented from everting into the atria by the chordae tendineae.
	▪ Trabeculae carneae are large, irregular muscular ridges on the inside of the ventricle wall.
	▪ Papillary muscles are cone-shaped projections on the inner surface of the ventricles that anchor the chordae tendineae.
	▪ Semilunar valves are located within the wall of each ventricle near its connection to a large artery. The right ventricle houses the pulmonary semilunar valve, and the left ventricle houses the aortic semilunar valve.
Coronary Circulation 664	▪ The major coronary artery branches are the anterior interventricular and circumflex arteries from the left coronary artery, and the right marginal and posterior interventricular arteries from the right coronary artery.
	▪ Venous return is through the cardiac veins into the coronary sinus, which drains into the right atrium of the heart.
How the Heart Beats: Electrical Properties of Cardiac Tissue 666	**Characteristics of Cardiac Muscle Tissue** 666
	▪ Cardiac muscle cells are small and branched. They form sheets of cardiac muscle tissue arranged into spiral bundles wrapped around and between the heart chambers.
	▪ Intercalated discs tightly link the muscle cells together and permit the immediate passage of muscle impulses.
	Contraction of Cardiac Muscle 667
	▪ The heart exhibits autorhythmicity. Its stimulus to contract is initiated by cells of the sinoatrial (SA) node in the roof of the right atrium.
	▪ Muscle impulses travel to the atrioventricular (AV) node and then through the AV bundle within the interventricular septum to Purkinje fibers in the heart apex.
	The Heart's Conducting System 668
Innervation of the Heart 670	▪ Sympathetic and parasympathetic innervation of the heart have opposing influences on heart rate. Autonomic innervation increases or decreases the rate of the heartbeat, but does not initiate it.
Tying It All Together: The Cardiac Cycle 671	▪ A cardiac cycle is the period of time from the start of one heartbeat to the beginning of the next.
	▪ All chambers within the heart experience alternate periods of contraction (systole) and relaxation (diastole) during a single cycle.
	Steps in the Cardiac Cycle 671
	Summary of Blood Flow During the Cardiac Cycle 671
	▪ Cyclic contraction and relaxation of heart chambers result in the pumping of blood out of the chambers and the subsequent refilling of the chambers with blood for the next cycle.
Aging and the Heart 675	▪ Decreased flexibility of heart structures as we age reduces the efficiency of cardiac function. Increased blood pressure results in hypertrophy of the myocardium.
Development of the Heart 675	▪ Development of the heart commences in the third week; two heart tubes form and fuse to become a single primitive heart tube.
	▪ The primitive heart tube develops expansions that later form postnatal heart structures.

CHALLENGE YOURSELF

Matching

Match each numbered item with the most closely related lettered item.

_____ 1. right marginal artery

_____ 2. pulmonary veins

_____ 3. left AV valve

_____ 4. diastole cells

_____ 5. SA node

_____ 6. venae cavae

_____ 7. intercalated disc

_____ 8. right AV valve

_____ 9. circumflex artery

_____ 10. systole

a. veins that carry blood to right atrium

b. period of relaxation

c. contains three cusps; also known as tricuspid valve

d. specialized junction between cardiac muscle cells

e. veins that carry blood to left atrium

f. period of contraction

g. branch of right coronary artery

h. origin of heartbeat

i. branch of left coronary artery

j. also known as bicuspid or mitral valve

Multiple Choice

Select the best answer from the four choices provided.

_____ 1. Muscle impulses are spread rapidly between cardiac muscle cells by
  a. sarcomeres.
  b. intercalated discs.
  c. chemical neurotransmitters.
  d. AV valves.

_____ 2. Venous blood from the heart wall enters the right atrium through the
  a. superior vena cava.
  b. coronary sinus.
  c. inferior vena cava.
  d. pulmonary veins.

_____ 3. How is blood prevented from flowing into the right ventricle from the pulmonary trunk?
  a. closing of the right AV valves
  b. opening of the pulmonary semilunar valve
  c. contraction of the right atrium
  d. closing of the pulmonary semilunar valve

_____ 4. Which of the following is the correct circulatory sequence for blood to pass through part of the heart?
  a. R. atrium → right AV valve → R. ventricle → pulmonary semilunar valve
  b. R. atrium → left AV valve → R. ventricle → pulmonary semilunar valve
  c. L. atrium → right AV valve → L. ventricle → aortic semilunar valve
  d. L. atrium → left AV valve → L. ventricle → pulmonary semilunar valve

_____ 5. The pericardial cavity is located between the
  a. fibrous pericardium and the parietal layer of the serous pericardium.
  b. parietal and visceral layers of the serous pericardium.
  c. visceral layer of the serous pericardium and the epicardium.
  d. myocardium and the visceral layer of the serous pericardium.

_____ 6. In the developing heart, the atria form from the primitive atrium and the
  a. sinus venosus.
  b. bulbus cordis.
  c. primitive ventricle.
  d. conus cordis.

_____ 7. The irregular muscular ridges in the ventricular walls are the
  a. papillary muscles.
  b. trabeculae carneae.
  c. chordae tendineae.
  d. moderator bands.

_____ 8. Sympathetic innervation of cardiac muscle originates from
  a. CN X (vagus nerve).
  b. L1–L2 segments of the spinal cord.
  c. the AV node.
  d. T1–T5 segments of the spinal cord.

_____ 9. When the ventricles contract, all of the following occur except
  a. closing of the AV valves.
  b. blood ejecting into the pulmonary trunk and aorta.
  c. closing of the semilunar valves.
  d. opening of the semilunar valves.

_____ 10. The thickest part of the heart wall is the
  a. pericardium.
  b. epicardium.
  c. myocardium.
  d. endocardium.

Content Review

1. What are the differences between the pulmonary and systemic circulations?

2. What chamber walls primarily form the anterior side of the heart? What chamber walls form the posterior side of the heart?

3. Compare the structure, location, and function of the parietal and visceral layers of the serous pericardium.

4. Where is the fibrous skeleton of the heart located? What are its functions?

5. Why are the chordae tendineae required for the proper functioning of the AV valves?

6. Explain why the walls of the atria are thinner than those of the ventricles, and why the walls of the right ventricle are relatively thin when compared to the walls of the left ventricle.

7. Identify and compare the branches of the right and left coronary arteries. In general, what portions of the heart does each branch supply?

8. Compare cardiac and skeletal muscle. In what ways are these muscle types similar? In which ways are they different?

9. Describe the functional differences in the effects of the sympathetic and parasympathetic divisions of the autonomic nervous system on the activity of cardiac muscle.

10. What are the phases of the cardiac cycle? Describe each phase with respect to which heart chambers are in systole or diastole, which valves are open or closed, and blood flow into or out of the chambers.

Developing Critical Reasoning

1. It was the end of the semester, and Huang had begun to prepare for his final examinations. Unfortunately, he still had to work his full-time job. In order to find sufficient time to study, he stayed up late and drank large amounts of coffee to stay alert. One evening during a very late study session, Huang felt a pounding in his chest and thought he was having a heart attack. His roommate took Huang to the emergency room. After an examination and interview by the physician, Huang was told that he probably had a cardiac arrhythmia. What was the most probable cause of the arrhythmia?

2. Josephine is a 55-year-old overweight woman who has a poor diet and does not exercise. One day while walking briskly, she experienced pain in her chest and down her left arm. Her doctor told her she was experiencing angina due to heart problems. Josephine asks you to explain what causes angina, and why she was feeling pain in her arm even though the problem was with her heart. What do you tell her?

ANSWERS TO "WHAT DO YOU THINK?"

1. The pulmonary arteries carry blood low in oxygen to the lungs because the lungs are responsible for replenishing the oxygen supplies in the blood. After deoxygenated blood travels to the lungs, the pulmonary capillaries are involved in gas exchange—that is, carbon dioxide is removed from the blood, and oxygen enters the blood. Then the pulmonary veins carry the newly oxygenated blood back to the heart.

2. When the ventricles contract, they also constrict the coronary arteries. Thus, the coronary arteries can fill with blood only when the ventricles are relaxed.

3. If the AV valves were to evert into the atria, some of the blood from the ventricles would be pushed into the atria, resulting in inefficient pumping of blood. When the heart has to work harder to pump the blood out of the heart, clinical problems result (see Clinical View: "Valve Defects and Their Effects on Circulation" for further information).

Visit the McKinley/O'Loughlin *Human Anatomy*, 2e website at aris.mhhe.com

23

OUTLINE

CARDIOVASCULAR SYSTEM

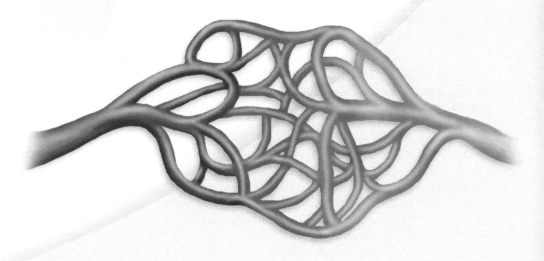

Vessels and Circulation

Blood vessels are analogous to highways—they are an efficient mode of transport for oxygen, carbon dioxide, nutrients, hormones, and waste products to and from body tissues. The heart is the mechanical pump that propels the blood through the vessels. Together, the heart and blood vessels form a closed-loop system, whereby blood is continuously pumped to and from all areas of the body.

Blood vessels are not rigid and immobile; rather, they can pulsate and change shape in accordance with the body's needs. This chapter examines the basic histologic structure common to all blood vessels and traces the flow of blood through all of the different body regions.

Study Tip!

— Here are some tips to help you remember the names and kinds of blood vessels:

1. Blood vessels often share names with either the body region they traverse or the bone next to them. For example, the radial artery travels near the radius, and the axillary artery is in the axillary region.

2. Some blood vessels are named for the structure they supply. For example, the renal arteries supply the kidneys, the gonadal arteries supply the gonads, and the facial arteries supply the face.

3. Arteries and veins that travel together (called companion vessels) sometimes share the same name. For example, the femoral artery is accompanied by the femoral vein.

4. Writing out your own simplified flowchart of blood vessels for each body region will help you better understand and remember the pattern of blood flow in that region.

Anatomy of Blood Vessels

Key topics in this section:

- Structure of arteries, capillaries, and veins
- How the different types of vessels interconnect to transport blood

The **systemic circulation** consists of the blood vessels that extend to all body regions. The **pulmonary circulation** consists of the vessels that take the blood to and from the lungs for the purpose of gas exchange. Both circulations work continuously and in tandem with each other.

The three classes of blood vessels are arteries, capillaries, and veins. In the systemic circulation, as the ventricles of the heart contract, arteries convey blood away from the heart to the body. Arteries branch into smaller and smaller vessels until they feed into the capillaries, where gas and nutrient exchange occurs. From the capillaries, veins return blood to the heart.

Arteries become progressively smaller as they branch and extend farther from the heart, while veins become progressively larger as they merge and come closer to the heart. The site where

two or more arteries (or two or more veins) converge to supply the same body region is called an **anastomosis** (ă-nas'tō-mō'sis; pl., *anastomoses*). Arterial anastomoses provide alternate blood supply routes to body tissues or organs. (For an example, see figure 23.12, which shows anastomoses among the superior and inferior epigastric arteries.) Some arteries do not form anastomoses; these so-called **end arteries** provide only one pathway through which blood can reach an organ. Examples of end arteries include the renal artery of the kidney and the splenic artery of the spleen. Other arteries (such as the coronary arteries in the heart wall) are called **functional end arteries,** meaning that their anastomoses are so tiny that the arteries may almost be considered end arteries. Veins tend to form many more anastomoses than do arteries.

Often, an artery travels with a corresponding vein. These vessels are called **companion vessels** because they supply the same body region and tend to lie next to one another.

Blood Vessel Tunics

Both artery and vein walls have three layers, called **tunics** (too'nik; *tunica* = coat). The tunics surround the **lumen** (loo'men), or inside space, of the vessel through which blood flows. These tunics are the tunica intima, tunica media, and tunica externa **(figure 23.1)**.

The innermost layer of a blood vessel wall is the **tunica intima** (too'ni-kă in-ti'mă; *intimus* = inmost), or *tunica interna*. It is composed of an **endothelium** (a simple squamous epithelium lining the blood vessel lumen) and a subendothelial layer made up of a thin layer of areolar connective tissue.

The **tunica media** (mē'dē-ă; *medius* = middle) is the middle layer of the vessel wall. It is composed of circularly arranged layers of smooth muscle cells. Sympathetic innervation causes the smooth muscle to contract, resulting in **vasoconstriction** (vā'sō-kon-strik'shŭn), or narrowing of the blood vessel lumen. When the fibers relax, **vasodilation** (vā'sō-dī-lā'shŭn), or widening of the blood vessel lumen, results.

The **tunica externa** (eks-ter'nă; *externe* = outside), or *tunica adventitia*, is the outermost layer of the blood vessel wall. It is composed of areolar connective tissue that contains elastic and collagen fibers. The tunica externa helps anchor the vessel to other structures. Very large blood vessels require their own blood supply to the tunica externa in the form of a network of small arteries called the **vasa vasorum** (vā-să vā-sŏr'ŭm; vessels of vessels). The vasa vasorum extend through the tunica externa.

In arteries, the thickest layer is the tunica media, while veins have a thicker tunica externa. The lumen in an artery is narrower than in its companion vein of the same size **(figure 23.2)**. Further, arteries tend to have more elastic and collagen fibers in all their tunics, which means that artery walls remain open (patent), can spring back to shape, and can withstand changes in blood pressure. In contrast, vein walls tend to collapse if there is no blood in them. **Table 23.1** compares the characteristics of arteries and veins.

Finally, capillaries contain only the tunica intima, but this layer consists of a basement membrane and endothelium only. Intercellular clefts are the thin spaces between adjacent cells in the

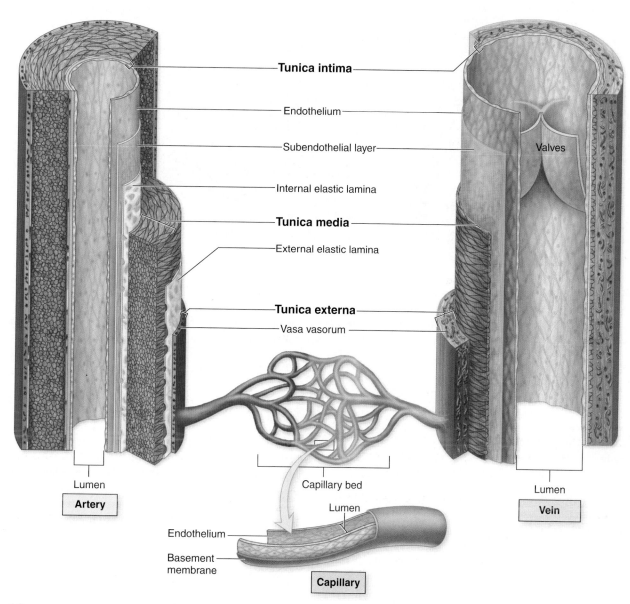

Figure 23.1

Walls of an Artery, a Capillary, and a Vein.   Both arteries and veins have a tunica intima, tunica media, and tunica externa. However, an artery has a thicker tunica media and a relatively smaller lumen, while a vein's thickest layer is the tunica externa, and it has a larger lumen. Some veins also have values. Capillaries typically have only a tunica intima, but they do not have a subendothelial layer—just the endothelium and a basement membrane.

capillary wall. Having only the tunica intima, without connective tissue and muscle layers, allows for rapid gas and nutrient exchange between the blood and the tissues.

Arteries

Arteries (ar′ter-ē) transport blood away from the heart. The arteries in the systemic circulation carry oxygenated blood to the body tissues. In contrast, the pulmonary arteries (part of the pulmonary circulation) carry deoxygenated blood to the lungs.

The three basic types of arteries are elastic arteries, muscular arteries, and arterioles (**figure 23.3**, *right*). In general, as an artery's diameter decreases, there is a corresponding decrease in the amount of elastic fibers and a relative increase in the amount of smooth muscle.

Elastic Arteries

Elastic arteries are the largest arteries, with diameters ranging from 2.5 to 1 centimeter. They are also called *conducting arteries* because they conduct blood away from the heart to the smaller

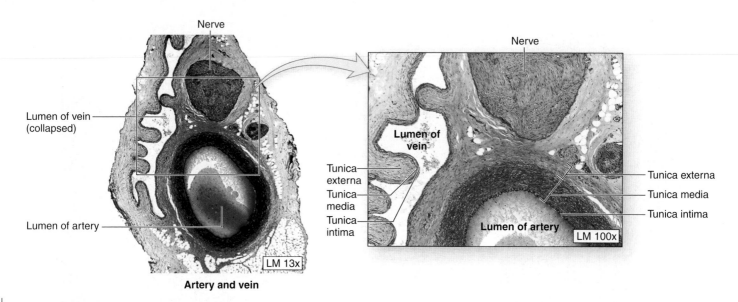

Artery and vein

Figure 23.2

Microscopic Comparison of Arteries and Veins. Arteries generally maintain their shape in tissues as a result of the thicker tunica media layer and more elastic fibers in their walls. The walls of neighboring veins often collapse when they are not filled with blood.

Table 23.1	Comparison of Arteries and Veins	
Characteristic	**Artery**	**Vein**
Lumen Diameter	Narrower than vein lumen	Wider than artery lumen; often appears collapsed when cut in cross section
General Wall Thickness	Thicker than companion vein	Thinner than companion artery
Cross-Sectional Shape	Retains its circular cross-sectional shape	Cross section tends to flatten out (collapse) if no blood is in the vein
Thickest Tunic	Tunica media	Tunica externa
Elastic and Collagen Fibers in Tunics	More than in vein	Less than in artery
Valves	None	Present in most veins
Blood Pressure	Higher than in veins (larger arteries typically > 90 mm Hg)	Lower than in arteries (approximately 2 mm Hg)
Blood Flow	Transports blood away from heart	Transports blood to heart
Blood Oxygen Levels	Systemic arteries carry blood high in O_2	Systemic veins carry blood low in O_2
	Pulmonary arteries carry blood low in O_2	Pulmonary veins carry blood high in O_2

muscular arteries. As their name suggests, these arteries have a large proportion of elastic fibers throughout all three tunics, especially in the tunica media (**figure 23.4a**). The abundant elastic fibers allow the elastic artery to stretch when a heart ventricle ejects blood into it. In this manner, the elastic arteries are able to withstand the strong pulsations of the ejected blood as well as reduce the force of the pulsations somewhat, so that the pressure of the arterial blood equalizes slightly as it reaches the smaller arteries and eventually the capillaries. Examples of elastic arteries include the aorta and the pulmonary, brachiocephalic, common carotid, subclavian, and common iliac arteries. Elastic arteries branch into muscular arteries.

Muscular Arteries

Muscular arteries typically have diameters ranging from 1 centimeter to 3 millimeters. These medium-sized arteries are also called *distributing arteries* because they distribute blood to body organs and tissues. Unlike elastic arteries, the elastic fibers in muscular arteries are confined to two circumscribed rings: The **internal elastic lamina** (lam′i-nă) separates the tunica intima from the tunica media, and the **external elastic lamina** separates the tunica media from the tunica externa (figure 23.4b).

Muscular arteries have a proportionately thicker tunica media, with multiple layers of smooth muscle fibers. The greater amount of

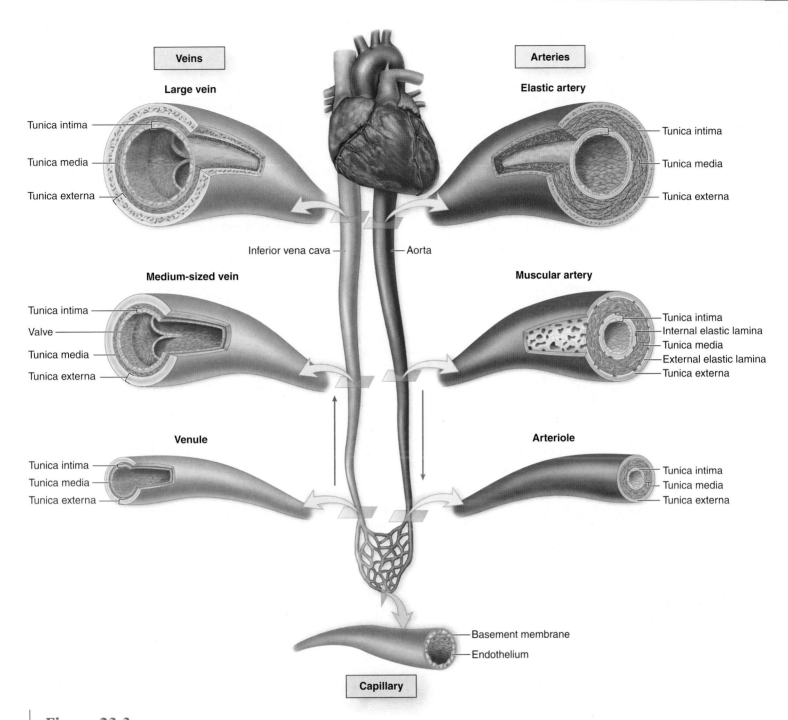

Figure 23.3

Comparison of Companion Vessels.   Arteries and veins supplying the same region are called companion vessels. The thickness of the tunics differs, depending on the size of the vessels.

muscle and lesser amount of elastic fibers result in less distensibility but better ability to vasoconstrict and vasodilate. Most of the named blood vessels (such as the brachial, anterior tibial, coronary, and inferior mesenteric arteries) are examples of muscular arteries. Muscular arteries branch into arterioles.

Arterioles

Arterioles (ar-tēr'ē-ōl) are the smallest arteries, with diameters ranging from 3 millimeters to 10 micrometers. In general, arterioles have less than six layers of smooth muscle in their tunica media (figure 23.4c). Larger arterioles have all three tunics, whereas the smallest arterioles may have an endothelium surrounded by a single layer of smooth muscle fibers. Sympathetic innervation

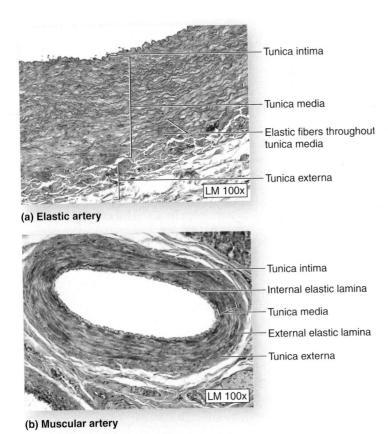

— Tunica intima

— Tunica media

— Elastic fibers throughout tunica media

— Tunica externa

LM 100x

(a) Elastic artery

— Tunica intima
— Internal elastic lamina
— Tunica media
— External elastic lamina
— Tunica externa

LM 100x

(b) Muscular artery

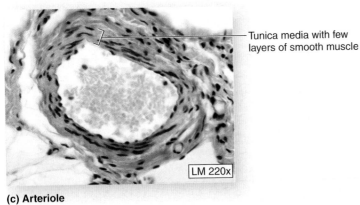

— Tunica media with few layers of smooth muscle

LM 220x

(c) Arteriole

Figure 23.4

Types of Arteries.   *(a)* Elastic arteries have vast arrays of elastic fibers in their tunica media. *(b)* Muscular arteries have a tunica media composed of numerous layers of smooth muscle flanked by elastic laminae. *(c)* Arterioles typically have a tunica media composed of six or fewer layers of smooth muscle cells.

causes contraction in the smooth muscle of the arteriole wall and results in vasoconstriction of the arteriole, which raises blood pressure. Arteriole vasoconstriction decreases blood flow into the capillaries, whereas arteriole vasodilation increases blood flow into the capillaries.

> ## Study Tip!
>
> Is that a muscular artery or an arteriole you are examining under the microscope? One easy way to figure this out is to count the smooth muscle layers in the tunica media:
>
> - Arterioles have six or fewer layers of smooth muscle in their tunica media.
> - Muscular arteries usually have many more layers of smooth muscle in their tunica media.

Capillaries

Capillaries (kap'-i-lar-ē; *capillaris* = relating to hair), the smallest blood vessels, connect arterioles to venules (see figure 23.3). The average capillary diameter is 8–10 micrometers, just slightly larger than the diameter of a single erythrocyte. The narrow vessel diameter means erythrocytes must travel in single file (called a rouleau; see chapter 21) through each capillary. Most capillaries consist solely of a tunica intima composed of a very thin, single layer of endothelium and a basement membrane; there is no subendothelial layer. The thin wall and the narrow vessel diameter are optimal for diffusion of gases and nutrients between blood in the capillaries and body tissues. Oxygen and nutrients from the blood can pass through the endothelial lining to the tissues, while carbon dioxide and waste products may be removed from the tissues and enter the cardiovascular system for removal. Some capillaries have additional functions; for example, the capillaries in the small intestine mucosa are also responsible for receiving digested nutrients. Thus, capillaries are called the **functional units** of the cardiovascular system.

The extensive capillaries in a capillary bed create an increase in the total surface area of these blood vessels (especially when compared to a single arteriole or venule). The increase in total surface area results in slower blood flow through the capillaries, thus allowing sufficient time for nutrient, gas, and waste exchange between the tissues and the blood. The capillary bed architecture, along with this slower blood flow, results in a change from a pulsatile flow to a steady flow of blood.

Capillaries do not function independently; rather, a group of capillaries (10–100) functions together and forms a **capillary bed (figure 23.5)**. A capillary bed is fed by a **metarteriole** (met'ar-tēr'ē-ōl; *meta* = after), which is a vessel branch of an arteriole. The proximal part of the metarteriole is encircled by scattered smooth muscle fibers, while the distal part of the metarteriole (called the **thoroughfare channel**) has no smooth muscle fibers. The thoroughfare channel connects to a **postcapillary venule** (ven'ool, vē'nool), which drains the capillary bed. Vessels called **true capillaries** branch from the metarteriole and make up the bulk of the capillary bed. At the origin of each true capillary, a smooth muscle ring called the **precapillary sphincter** controls blood flow into the true capillaries. Sphincter relaxation permits blood to flow into the true capillaries, whereas sphincter contraction causes blood to flow directly from the metarteriole into

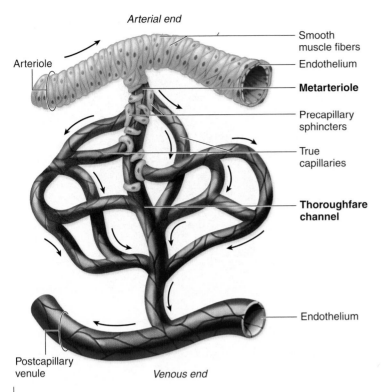

Arterial end

Arteriole

Smooth muscle fibers

Endothelium

Metarteriole

Precapillary sphincters

True capillaries

Thoroughfare channel

Endothelium

Postcapillary venule

Venous end

Figure 23.5

Capillary Bed Structure.   A capillary bed originates from a metarteriole. The distal part of the metarteriole is called the thoroughfare channel, and it merges with the postcapillary venule. True capillaries branch from the metarteriole, and blood flow into these true capillaries is regulated by the precapillary sphincters.

the postcapillary venule via the thoroughfare channel. The precapillary sphincters open when the tissue needs nutrients, and they close when the tissue's needs have been met. These precapillary sphincters go through cycles of contracting and relaxing at a rate of about 5–10 cycles per minute. This cyclical process is called **vasomotion.**

The three basic kinds of capillaries are continuous capillaries, fenestrated capillaries, and sinusoids (**figure 23.6**). In **continuous**

capillaries, the most common type, endothelial cells form a complete, continuous lining and are connected by tight junctions (see chapter 4). Materials can pass through the endothelial cells or the intercellular clefts via either simple diffusion or pinocytosis (a type of endocytosis whereby droplets of fluid are packaged in pinocytotic vesicles, see chapter 2). Continuous capillaries are found in muscle, skin, the thymus, the lungs, and the CNS.

Fenestrated (fen'es-trā'ted; *fenestra* = window) **capillaries** have **fenestrations** (or pores) within each endothelial cell. These fenestrations measure 10–100 nanometers in diameter. The basement membrane remains continuous. Fenestrated capillaries are seen where a great deal of fluid transport between the blood and interstitial tissue occurs, such as in the small intestine (intestinal villi), the ciliary process of the eye, most of the endocrine glands, and the kidney.

Sinusoids (si'nŭ-soyd; *sinus* = cavity, *eidos* = appearance), or *discontinuous capillaries,* have larger gaps than fenestrated capillaries, and their basement membrane is either discontinuous or absent. Sinusoids tend to be wider, larger vessels with openings that allow for transport of larger materials, such as proteins or cells (e.g., blood cells). Sinusoids are found in bone marrow, the anterior pituitary, the parathyroid glands, the adrenal glands, the spleen, and the liver.

Veins

Veins (vān) drain capillaries and return the blood to the heart. Compared with a corresponding artery, vein walls are relatively thin, and the vein lumen is larger (see figure 23.3, *left*). Systemic veins carry deoxygenated blood to the right atrium of the heart, while pulmonary veins carry oxygenated blood to the left atrium of the heart. Because blood pressure gradually decreases as blood travels through smaller arteries and into capillaries, the pressure has been substantially reduced by the time blood reaches the veins. At rest, the body's veins hold about 60% of the body's blood. Thus, veins function as **blood reservoirs.**

Venules

Venules are the smallest veins, measuring from 8 to 100 micrometers in diameter. Venules are companion vessels with arterioles since both supply the same areas and generally are of similar size. The smallest ones, called **postcapillary venules,** drain capillaries (see figure 23.5). Postcapillary venules resemble continuous capillaries in structure,

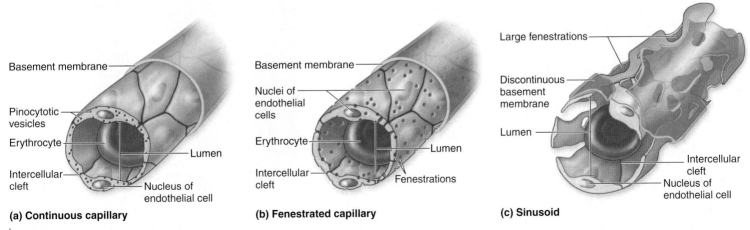

Basement membrane

Pinocytotic vesicles

Erythrocyte

Intercellular cleft

Lumen

Nucleus of endothelial cell

(a) Continuous capillary

Basement membrane

Nuclei of endothelial cells

Erythrocyte

Intercellular cleft

Lumen

Fenestrations

(b) Fenestrated capillary

Large fenestrations

Discontinuous basement membrane

Lumen

Intercellular cleft

Nucleus of endothelial cell

(c) Sinusoid

Figure 23.6

Types of Capillaries.   (*a*) Continuous capillaries have tight junctions between the endothelial cells that permit minimal fluid leakage. (*b*) Fenestrated capillaries have fenestrations (pores) in the endothelial cells to permit small molecules to move out of the vessel. (*c*) Sinusoids have big gaps between the endothelial cells and a discontinuous basement membrane that promotes transport of larger molecules.

except that they have a slightly wider lumen. The mechanism of **dia-pedesis** (dī'ă-pĕ-dē'sis), by which leukocytes migrate from blood vessels into interstitial fluid, occurs primarily in the postcapillary venules. Smaller venules merge to form larger venules. The largest venules have all three tunics. Venules merge to form veins.

Veins

A venule becomes a vein when its diameter is greater than 100 micrometers. Smaller and medium-sized veins typically travel with muscular arteries, while the largest veins travel with (correspond to) elastic arteries (see figure 23.3). Blood pressure in veins is too low to overcome the forces of gravity. To prevent blood from pooling in the limbs and assist blood moving back to the heart, most veins contain numerous **valves** formed primarily of tunica intima and strengthened by elastic and collagen fibers. Thus, as blood flows superiorly in the limbs, these one-way valves close to prevent backflow (**figure 23.7**).

In addition to valves, many deep veins pass between skeletal muscle groups. As the skeletal muscles contract, veins are squeezed to help pump the blood toward the heart. This process is called the **skeletal muscle pump**. When skeletal muscles are more active—for example, when a person is walking—blood is pumped more quickly and efficiently toward the heart. Conversely, inactivity (as when

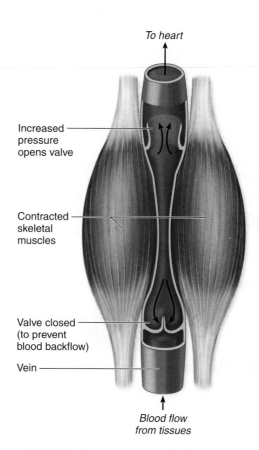

Figure 23.7

Valves in Veins. Valves are one-way flaps that prevent pooling and backflow of venous blood to ensure that blood flows toward the heart. Particularly in the lower limbs, the contraction of skeletal muscles squeezes the veins passing between muscles and forces blood toward the heart.

CLINICAL VIEW

Varicose Veins

Varicose (var'i-kōs; *varix* = dilated vein) **veins** are dilated, tortuous (having many curves or twists) veins. The valves in these veins have become non-functional, causing blood to pool in one area and the vein to swell and bulge. Varicose veins are most common in the superficial veins of the lower limbs. They may be a result of genetic predisposition, aging, or some form of stress on the venous system that inhibits venous return (such as standing for long periods of time, obesity, or pregnancy). Varicose veins may become inflamed and painful, especially if fluid leaks from them into the tissues.

Symptoms of varicose veins may be alleviated by elevating the affected body part or wearing compression stockings (to promote blood movement in the lower limbs). In a procedure called sclero-therapy, an irritant is injected into small varicose veins to make them scar and seal off. Typically, a patient needs multiple sclerotherapy sessions before optimal results are seen. For larger varicose veins, an outpa-tient surgical procedure called stripping or vein removal (phlebectomy) is necessary. These veins can be removed without affecting the circulation, since the blood may be shunted to other veins that are not varicose. Even after treatment, it is possible for varicose veins to recur.

Varicose veins in the anorectal region are called **hemorrhoids** (hem'ŏ-royd). Hemorrhoids occur due to increased intra-abdominal pressure, as when a person strains to have a bowel movement or is in labor during childbirth. Hemorrhoids may need to be surgically excised if they become too painful or bleed excessively.

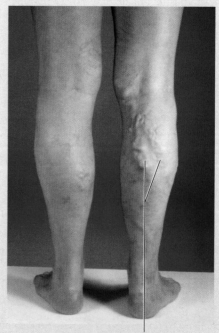

Varicose veins

CLINICAL VIEW

Deep Vein Thrombosis

Deep vein thrombosis (throm-bō′sis; a clotting) **(DVT)** refers to a **thrombus**, which is a blood clot in a vein. The most common site for the thrombus is a vein in the calf (sural) region; the femoral region is another common site. The blood clot partially or completely blocks the flow of blood in the vein. DVT typically occurs in individuals with heart disease or those who are inactive or immobile for a long period of time, such as bedridden patients or those who have been immobilized in a cast. Even healthy individuals who have been on a long airline trip may develop DVT. In fact, DVT is sometimes called "economy class syndrome" in reference to the reduced amount of leg room in economy class seating on airlines. DVT may also be a complication in pregnancy, where fluid accumulation in the legs and impingement of the fetus on the inferior vena cava may prevent efficient blood flow back to the heart. For inactive individuals, the leg muscles (e.g., gastrocnemius and soleus) do not contract as often and can't help propel blood through the deep veins, thus allowing the blood to pool and potentially to clot.

Initial signs of DVT include fever, tenderness and redness in the affected area, severe pain and swelling in the areas drained by the affected vein, and rapid heartbeat. A person experiencing these symptoms should seek immediate medical attention. The most serious complication of DVT is a **pulmonary embolus** (em′bō-lŭs; a plug), in which a blood clot breaks free within the vein and travels through vessels to the lung, eventually blocking a branch of the pulmonary artery and potentially causing respiratory failure and death. If a DVT is diagnosed, the patient is given anticoagulation medication (such as low-molecular-weight heparin) to help prevent further clotting and break up the existing clot.

To reduce the risk for DVT, a person should maintain a healthy weight, stay active, and treat medical conditions that may increase the risk for DVT. On a long airline flight or car trip, stretching the legs and moving the feet frequently assist venous circulation in the legs. Bedridden individuals may wear full-length compression stockings to assist circulation in the lower limbs.

---

a person has been sitting for a long time or is bedridden) leads to blood pooling in the leg veins.

WHAT DID YOU LEARN?

1 What is the structure and function of an anastomosis?

2 Describe the capillary types. List a location for each in the body.

3 Explain how valves and muscular pumps help veins propel blood back to the heart.

Blood Pressure

Key topics in this section:

- Blood pressure measurement
- Comparison of diastolic and systolic blood pressure

- How blood pressure changes as blood travels through arteries, capillaries, and veins

The rhythmic pumping of blood through the heart produces a rhythmic pulsation of blood through the arteries. When you check your **pulse** (pŭls; *pulsus* = a stroke), you are feeling these pulsations of blood. Arterial walls contain elastic connective tissue, which allows them to expand and recoil in response to the pulsating blood. Due to this pulsating blood, larger arteries tend to exhibit high pressure, which gradually decreases as blood travels into smaller and smaller arteries.

Blood pressure is the force per unit area that blood places on the inside wall of a blood vessel and is measured in millimeters of mercury (mm Hg). Arterial blood pressure is measured with a **sphygmomanometer** (sfig′mō-mă-nom′ĕ-ter; *sphygmos* = pulse, *manos* = thin, *metron* = measure) **(figure 23.8a)**. A cuff is wrapped around the arm such that when inflated it completely

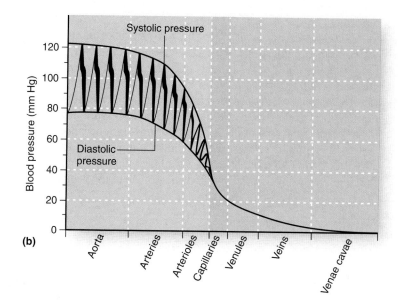

(a)   **(b)**

Figure 23.8

Blood Pressure. *(a)* A sphygmomanometer is used to measure blood pressure in the brachial artery. *(b)* A tracing compares the blood pressure changes as blood flows through the different vessels in the cardiovascular system.

CLINICAL VIEW

Hypertension: The "Silent Killer"

Hypertension is chronically elevated blood pressure, defined as a systolic pressure greater than 140 mm Hg and/or a diastolic pressure greater than 90 mm Hg. About 90–95% of all hypertension cases are **essential hypertension,** in which the cause is idiopathic (unknown). **Secondary hypertension** accounts for the other 5–10% of cases, meaning that the high blood pressure is caused by another condition, usually renal disease or an adrenal gland tumor.

Hypertension has many serious effects on the body. It causes changes in the blood vessel walls, making them prone to further injury. Increased damage makes the blood vessels more likely to develop atherosclerosis (see Clinical View: In Depth on page 706). In addition, hypertension causes undue stress on arterioles, resulting in thickening of the arteriole walls and reduction in luminal diameter, a condition called **arteriolosclerosis.** Furthermore, hypertension causes thickening of the renal arteries, leading to renal failure, and it can greatly damage the cerebral arteries, making them prone to rupture, which results in a fatal brain hemorrhage or stroke. Finally, hypertension is a major cause of heart failure owing to the extra workload placed on the heart.

Since hypertension is initially asymptomatic (meaning no noticeable symptoms are present), it has been dubbed the "silent killer." Everyone is encouraged to have their blood pressure checked early in life, and regularly, to make sure they are not suffering from hypertension. Mild hypertension may be controlled by losing weight, eating a healthy diet, exercising regularly, and not smoking. Stress is also associated with hypertension; thus, reducing stress, or learning to manage it, is important in treatment. In many instances, however, medication may be needed to control hypertension. Diuretics increase urine output, thereby reducing salt and water retention and consequently lowering blood volume. Beta-blockers slow the heart rate and lower heart output, while ACE (angiotensin-converting enzyme) inhibitors block angiotensin II (a protein that constricts arterioles), thereby increasing vasodilation.

---

compresses the brachial artery, temporarily stopping the flow of blood. Then air is gradually released from the cuff while a health-care practitioner places a stethoscope just distal to the compressed artery and listens for the sound of blood flow. Once the first pulsation is heard through the stethoscope, the sphygmomanometer is read. The reading on the dial at this time is the **systolic blood pressure,** the pressure in the vessel during ventricular systole (ventricular contraction). The health-care practitioner continues to listen to artery pulsations, and when the pulsations stop (because blood is flowing evenly through the blood vessel), the sphygmomanometer is read again for a measurement of **diastolic blood pressure,** the pressure during ventricular diastole (ventricular relaxation).

Systolic pressure is greater than diastolic pressure due to the greater force from ventricular contraction. Blood pressure is expressed as a ratio, in which the numerator (upper number) is the systolic pressure and the denominator (lower number) is the diastolic pressure. The average adult has a blood pressure of 120/80 mm Hg.

Blood pressure is produced in the ventricles of the heart, so pressure is highest in the arteries closest to the heart, such as the aorta. As the arteries branch into smaller vessels and travel greater distances from the ventricles, blood pressure decreases (figure 23.8b). By the time the blood reaches the capillaries, fluctuations between systolic and diastolic blood pressure disappear. Blood pressure at this point is about 40 mm Hg, and once blood leaves the capillaries, pressure is below 20 mm Hg. Blood pressure drops from 20 mm Hg in the venules to almost 0 mm Hg by the time blood travels through the venae cavae to the right atrium of the heart.

Several factors can influence arterial blood pressure. Increased blood volume and increased cardiac output (the amount of blood pumped out of a ventricle in one minute) both can increase blood pressure. Vasoconstriction raises blood pressure, while vasodilation reduces blood pressure. Some medicines and drugs can either increase or decrease pressure as well. People who are overweight or less healthy tend to have increased blood pressure.

WHAT DO YOU THINK?

❶ Nicotine stimulates the heart and raises cardiac output. It also causes arteriole vasoconstriction. Given this information, would you expect the blood pressure of a smoker to be high or relatively low?

WHAT DID YOU LEARN?

❹ What is blood pressure? What are the similarities and differences between systolic pressure and diastolic pressure?

Systemic Circulation

Key topic in this section:

- Major blood vessels involved in blood flow to and from all the body tissues

The systemic circulation consists of blood vessels that extend to all body regions. In this section, we discuss arterial and venous systemic blood flow according to body region. As you read these descriptions, it may help you to refer to **figure 23.9**, which shows the locations of the major arteries and veins.

General Arterial Flow Out of the Heart

Oxygenated blood is pumped out of the left ventricle of the heart and enters the **ascending aorta.** The **left** and **right coronary arteries** emerge immediately from the wall of the ascending aorta and supply the heart. The ascending aorta curves toward the left side of the body and becomes the **aortic arch** (*arch of the aorta*). Three main arterial branches emerge from the aortic arch: (1) the **brachiocephalic** (brā′kē-ō-se-fal′ik) **trunk,** which bifurcates into the **right common carotid** (ka-rot′id) **artery** supplying arterial blood to right side of the head and neck, and the **right subclavian** (sŭb-klā′vē-an; *sub* = beneath; clavicle) **artery** supplying the right upper limb and some thoracic structures; (2) the **left common carotid artery** supplying the left side of the head and neck; and (3) the **left subclavian artery** supplying the left upper limb and some thoracic structures.

The aortic arch curves and projects inferiorly as the **descending thoracic aorta** that extends several branches to supply the thoracic wall. When this artery extends inferiorly through the aortic opening (hiatus) in the diaphragm, it is renamed the **descending abdominal aorta.** In the abdomen, arterial branches originate from the aorta wall to supply the abdominal wall and organs.

At the level of the fourth lumbar vertebra, the descending abdominal aorta bifurcates into **left** and **right common iliac** (il′ē-ak; *ileum* = groin) **arteries.** Each of these arteries further divides into an **internal iliac artery** (to supply pelvic and perineal structures) and an **external iliac artery** (to supply the lower limb).

General Venous Return to the Heart

Once oxygenated blood is distributed throughout the body, it returns to the heart through veins that often share the same names as their corresponding arteries (figure 23.9). The veins that drain the head, neck, and upper limbs merge to form the **left** and **right brachiocephalic veins,** which in turn merge to form the **superior vena cava.** The superior vena cava drains directly into the right atrium. The veins inferior to the diaphragm merge to collectively form the **inferior vena cava.** The inferior vena cava lies to the right side of the descending abdominal aorta, and is responsible for carrying venous blood toward the heart from the lower limbs, pelvis and perineum, and abdominal structures. The inferior vena cava extends through the caval opening in the diaphragm and also drains blood directly into the right atrium.

Blood Flow Through the Head and Neck

The left and right common carotid arteries supply most of the blood to the head and neck. They travel parallel immediately lateral to either side of the trachea **(figure 23.10a).** At the superior border of the thyroid cartilage, each artery divides into an **external carotid artery** that supplies structures external to the skull, and an **internal carotid artery** that supplies internal skull structures. Prior to this bifurcation, the common carotid artery contains a structure called the carotid sinus, which is a receptor that detects changes in blood pressure.

Blood Flow Through the Neck and Superficial Head Structures

The external carotid artery supplies blood to several branches: (1) The **superior thyroid artery** supplies the thyroid gland, larynx, and some anterior neck muscles; (2) the **ascending pharyngeal** (fă-rin′jē-ăl; *pharynx* = throat) **artery** supplies the pharynx; (3) the **lingual** (ling′gwăl) **artery** supplies the tongue; (4) the **facial artery** supplies most of the facial region; (5) the **occipital artery** supplies the posterior portion of the scalp; and (6) the **posterior auricular artery** supplies the ear and the scalp around the ear. Thereafter, the external carotid artery divides into the **maxillary artery,** which supplies the teeth, gums, nasal cavity, and meninges, and the **superficial temporal artery,** which supplies the side of the head and the parotid gland.

Venous return is through smaller veins that merge to form the **facial, superficial temporal,** and **maxillary** (mak′si-lār-ē) **veins** (figure 23.10b). Some of these veins merge and drain into either the **internal jugular vein** or the **external jugular vein** that drains into the subclavian vein and then into the brachiocephalic vein. Figure 23.10c shows the major arteries and veins of the head and neck.

Blood Flow Through the Cranium

The internal carotid artery branches only after it enters the skull through the carotid canal. Once inside the skull, it forms multiple branches, including the **anterior** and **middle cerebral arteries,** which supply the brain, and the **ophthalmic** (op′thal′mik; *ophthalmos* = eye) **arteries,** which supply the eyes **(figure 23.11)**.

The **vertebral arteries** emerge from the subclavian arteries and travel through the transverse foramina of the cervical vertebrae before entering the skull through the foramen magnum, where they merge to form the **basilar** (bas′i-lār; *basis* = base) **artery.** The basilar artery travels immediately anterior to the pons and extends many branches prior to subdividing into the **posterior cerebral arteries,** which supply the posterior portion of the cerebrum.

The **cerebral arterial circle** (*circle of Willis*) is an important anastomosis of arteries around the sella turcica. The circle is formed from posterior cerebral arteries, **posterior communicating arteries** (branches of the posterior cerebral arteries), internal carotid arteries, **anterior cerebral arteries,** and **anterior communicating arteries** (which connect the two anterior cerebral arteries). This arterial circle equalizes blood pressure in the brain and can provide collateral channels should one vessel become blocked.

WHAT DO YOU THINK?

❷ If both common carotid arteries were blocked, would any blood be able to reach the brain? Why or why not?

Some cranial venous blood is drained by the **vertebral veins** that extend through the transverse foramina of the cervical vertebrae and drain into the brachiocephalic veins. However, most of the venous blood of the cranium drains through several large veins

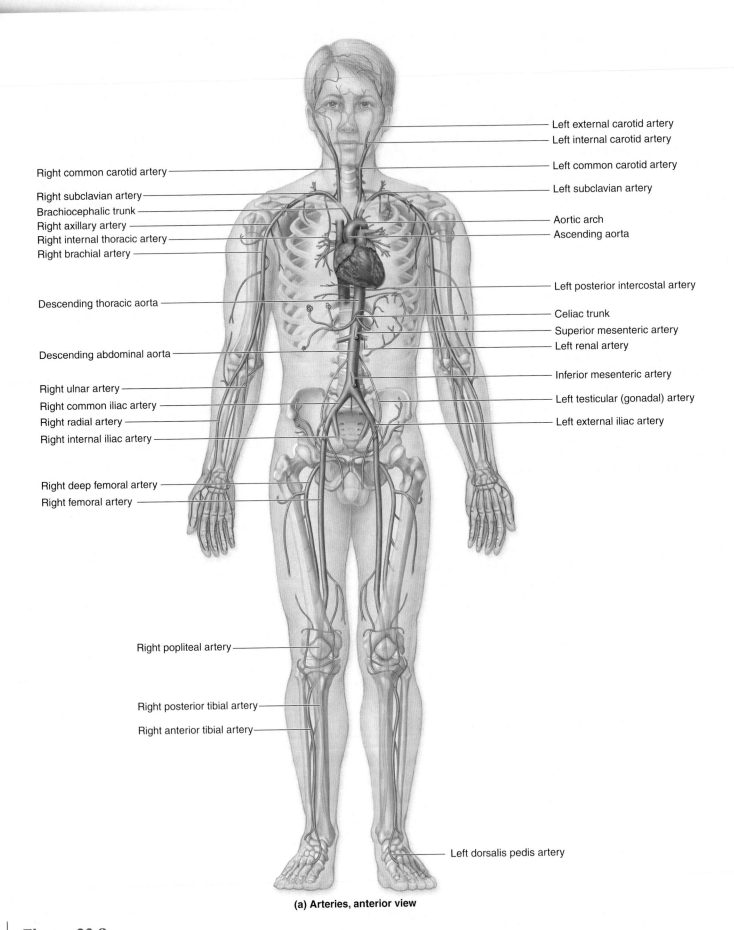

Left external carotid artery

Left internal carotid artery

Right common carotid artery

Left common carotid artery

Right subclavian artery

Left subclavian artery

Brachiocephalic trunk

Right axillary artery

Aortic arch

Right internal thoracic artery

Ascending aorta

Right brachial artery

Left posterior intercostal artery

Descending thoracic aorta

Celiac trunk

Superior mesenteric artery

Left renal artery

Descending abdominal aorta

Inferior mesenteric artery

Right ulnar artery

Left testicular (gonadal) artery

Right common iliac artery

Right radial artery

Left external iliac artery

Right internal iliac artery

Right deep femoral artery

Right femoral artery

Right popliteal artery

Right posterior tibial artery

Right anterior tibial artery

Left dorsalis pedis artery

(a) Arteries, anterior view

Figure 23.9

General Vascular Distribution.   Arteries in the systemic circulation carry blood from the heart to systemic capillary beds; systemic veins return this blood to the heart. *(a)* Anterior view of the systemic arteries. *(b)* Anterior view of the systemic veins.

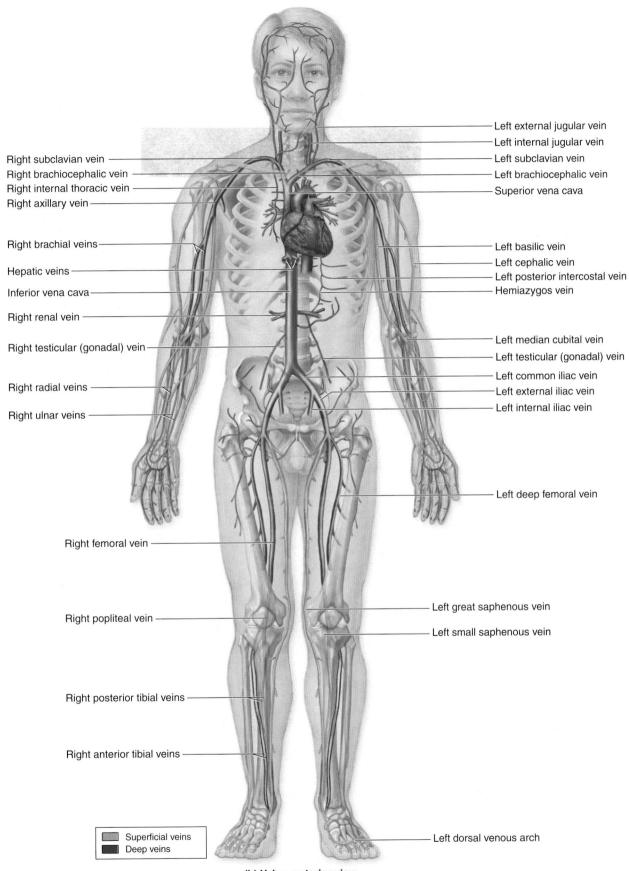

Right subclavian vein

Right brachiocephalic vein

Right internal thoracic vein

Right axillary vein

Right brachial veins

Hepatic veins

Inferior vena cava

Right renal vein

Right testicular (gonadal) vein

Right radial veins

Right ulnar veins

Right femoral vein

Right popliteal vein

Right posterior tibial veins

Right anterior tibial veins

Left external jugular vein

Left internal jugular vein

Left subclavian vein

Left brachiocephalic vein

Superior vena cava

Left basilic vein

Left cephalic vein

Left posterior intercostal vein

Hemiazygos vein

Left median cubital vein

Left testicular (gonadal) vein

Left common iliac vein

Left external iliac vein

Left internal iliac vein

Left deep femoral vein

Left great saphenous vein

Left small saphenous vein

Left dorsal venous arch

Superficial veins

Deep veins

(b) Veins, anterior view

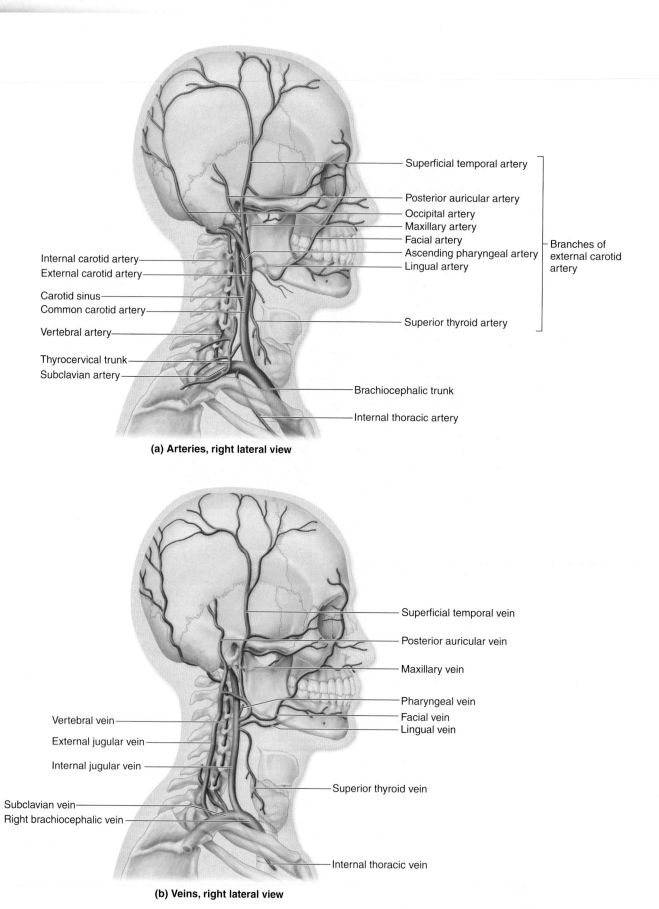

(a) Arteries, right lateral view

Superficial temporal artery
Posterior auricular artery
Occipital artery
Maxillary artery
Facial artery
Ascending pharyngeal artery
Lingual artery
Superior thyroid artery

Branches of external carotid artery

Internal carotid artery
External carotid artery
Carotid sinus
Common carotid artery
Vertebral artery
Thyrocervical trunk
Subclavian artery

Brachiocephalic trunk
Internal thoracic artery

(b) Veins, right lateral view

Superficial temporal vein
Posterior auricular vein
Maxillary vein
Pharyngeal vein
Facial vein
Lingual vein
Superior thyroid vein

Vertebral vein
External jugular vein
Internal jugular vein
Subclavian vein
Right brachiocephalic vein

Internal thoracic vein

Figure 23.10

Blood Flow to the External Head and Neck.   Right lateral views show *(a)* major arteries and *(b)* major veins of the head and neck. *(c)* A cadaver photo shows the major vessels of the head and neck.

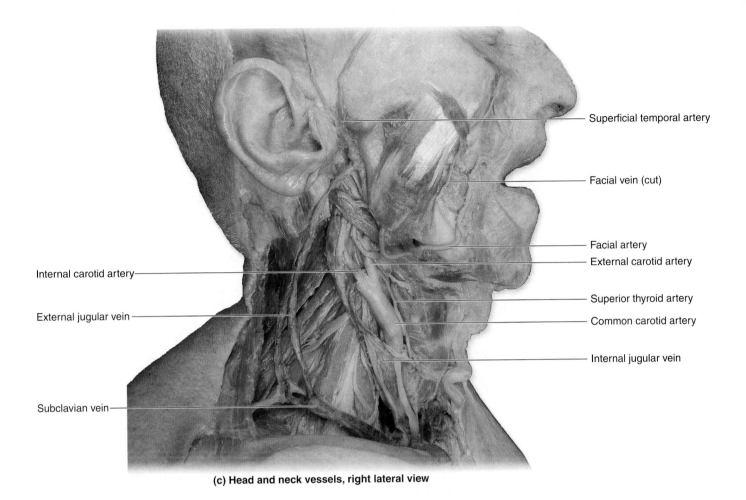

(c) Head and neck vessels, right lateral view

Labels on figure:
- Superficial temporal artery
- Facial vein (cut)
- Facial artery
- External carotid artery
- Superior thyroid artery
- Common carotid artery
- Internal jugular vein
- Internal carotid artery
- External jugular vein
- Subclavian vein

collectively known as the **dural** (doo-răl) **venous sinuses** (figure 23.11*b*). Recall that these large veins are formed between the two layers of dura mater and also receive excess CSF. There are no valves in the dural venous sinus system, so potentially the blood can flow in more than just one direction.

The **dural venous sinus system** has several components (see also figure 15.5). The **superior sagittal sinus** is located superior to the longitudinal fissure of the brain; it drains into one of the transverse sinuses (usually the right one). The **inferior sagittal sinus** occupies the inferior free edge of the falx cerebri. The **straight sinus** is formed by the merging of the inferior sagittal sinus and the great cerebral vein; this sinus drains into **left** and **right transverse sinuses** that run horizontally along the internal margin of the occipital bone. Finally, the S-shaped **left** and **right sigmoid** (sig′moyd; *sigma* = letter S) **sinuses** are a continuation of the transverse sinuses; they drain into the internal jugular veins. The internal jugular veins and subclavian veins merge to form the brachiocephalic veins that drain into the superior vena cava. Additional components of the dural venous sinus system include the **occipital** and **marginal sinuses,** the **superior** and **inferior petrosal sinuses,** and the **cavernous sinuses.**

Blood Flow Through the Thoracic and Abdominal Walls

The arterial flow to the thoracic and abdominal walls is supplied by several pairs of arteries **(figure 23.12)**. A left and right **internal thoracic artery** emerge from each subclavian artery to supply the mammary gland and anterior thoracic wall. Each internal thoracic artery has the following branches: the first six **anterior intercostal arteries** that supply the anterior intercostal spaces, and a **musculophrenic** (mŭs′kū-lō-fren′ik; *phren* = diaphragm) **artery** that divides into anterior intercostal arteries 7–9. The internal thoracic artery then becomes the **superior epigastric** (ep-i-gas′trik; *epi* = upon, *gastric* = stomach) **artery,** which carries blood to the superior abdominal wall. The **inferior epigastric artery,** a branch of the external iliac artery, supplies the inferior abdominal wall. Together, the superior and inferior epigastric arteries form extensive anastomoses.

The left and right **costocervical** (kos′tō-ser′vi-kal) **trunks** and **thyrocervical trunks** emerge from each subclavian artery. The costocervical trunk has a branch called the **supreme intercostal artery,** which branches into the first and second **posterior intercostal arteries.** Posterior intercostal arteries 3–11 are branches of the descending

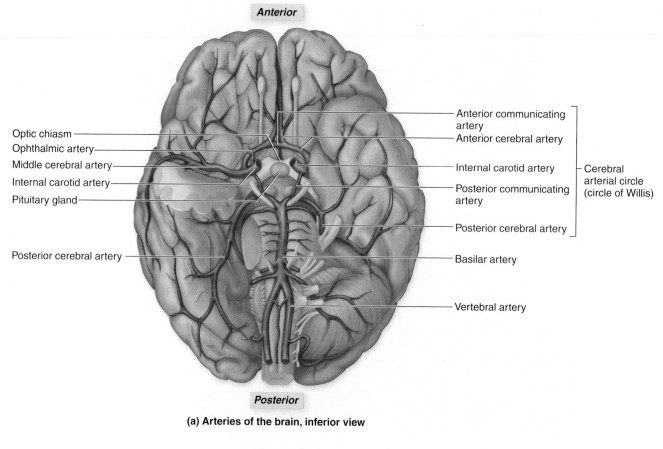

Anterior

Optic chiasm

Ophthalmic artery

Middle cerebral artery

Internal carotid artery

Pituitary gland

Posterior cerebral artery

Anterior communicating artery

Anterior cerebral artery

Internal carotid artery

Posterior communicating artery

Posterior cerebral artery

Basilar artery

Vertebral artery

Cerebral arterial circle (circle of Willis)

Posterior

(a) Arteries of the brain, inferior view

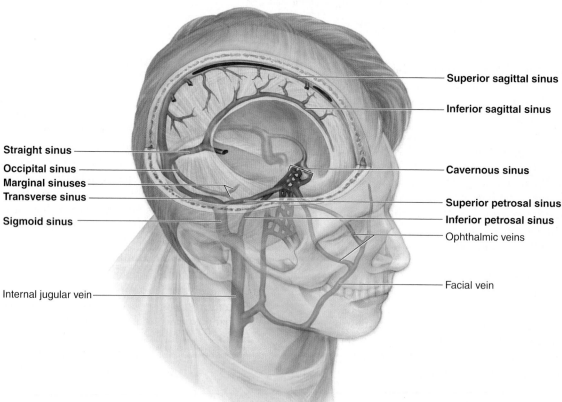

Straight sinus

Occipital sinus

Marginal sinuses

Transverse sinus

Sigmoid sinus

Internal jugular vein

Superior sagittal sinus

Inferior sagittal sinus

Cavernous sinus

Superior petrosal sinus

Inferior petrosal sinus

Ophthalmic veins

Facial vein

(b) Cranial and facial veins, right superior anterolateral view

Figure 23.11

Blood Flow Through the Cranium.   The internal carotid and vertebral arteries supply blood to the cranium, and blood is drained from the cranium by the internal jugular veins. *(a)* The arterial circulation is revealed in an inferior view of the brain with a portion of the right temporal lobe and right cerebellar hemisphere removed. *(b)* Venous drainage of the cranium is shown from a superior anterolateral view. The dural venous sinuses are labeled in bold.

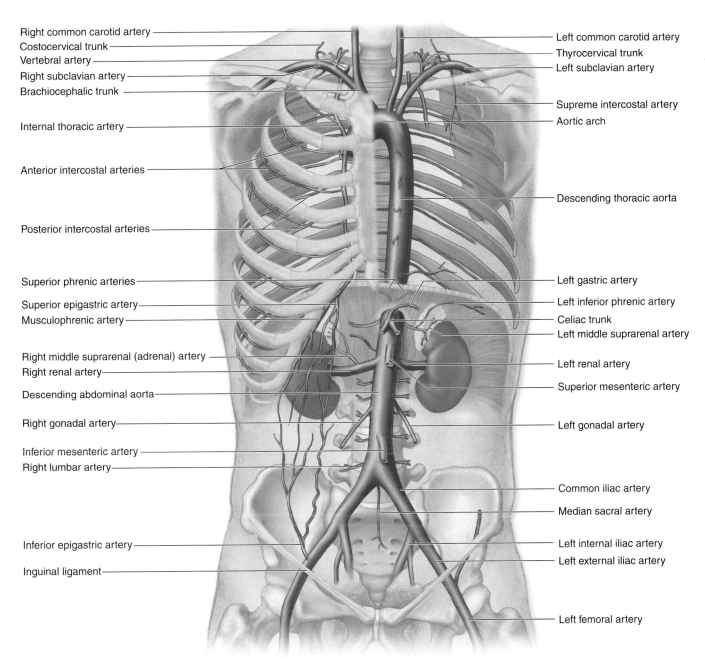

Right common carotid artery
Costocervical trunk
Vertebral artery
Right subclavian artery
Brachiocephalic trunk

Internal thoracic artery

Anterior intercostal arteries

Posterior intercostal arteries

Superior phrenic arteries
Superior epigastric artery
Musculophrenic artery

Right middle suprarenal (adrenal) artery
Right renal artery

Descending abdominal aorta

Right gonadal artery

Inferior mesenteric artery
Right lumbar artery

Inferior epigastric artery

Inguinal ligament

Left common carotid artery
Thyrocervical trunk
Left subclavian artery

Supreme intercostal artery
Aortic arch

Descending thoracic aorta

Left gastric artery
Left inferior phrenic artery
Celiac trunk
Left middle suprarenal artery

Left renal artery
Superior mesenteric artery

Left gonadal artery

Common iliac artery
Median sacral artery

Left internal iliac artery
Left external iliac artery

Left femoral artery

Figure 23.12

Arterial Circulation to the Thoracic and Abdominal Body Walls.   The torso is supplied by arteries that have extensive anastomoses. In this illustration, the more anteriorly placed arteries as well as the "vessel arcs" formed by the anterior and posterior intercostal arteries are shown on the right side of the body. The left side of the body illustrates the deeper, posteriorly placed arteries only.

thoracic aorta. The posterior and anterior intercostal arteries anastomose, and each pair forms a horizontal vessel arc that spans a segment of the thoracic wall.

Finally, five pairs of **lumbar arteries** branch from the descending abdominal aorta to supply the posterolateral abdominal wall.

In addition to the paired vessels just described, an unpaired **median sacral artery** arises at the bifurcation of the aorta in the pelvic region to supply the sacrum and coccyx.

Venous drainage of the thoracic and abdominal walls is a bit more complex than the arterial pathways **(figure 23.13)**. **Anterior intercostal veins**, a **superior epigastric vein**, and a **musculophrenic vein** all merge into the **internal thoracic vein**. Each internal thoracic vein drains into its respective brachiocephalic vein. The

inferior epigastric vein merges with the **external iliac vein** that eventually drains into the inferior vena cava. The first and second **posterior intercostal veins** then merge with the **supreme intercostal vein** that drains into the brachiocephalic vein. Remember, the brachiocephalic veins merge to form the superior vena cava.

The **lumbar veins** and **posterior intercostal veins** drain into the azygos system of veins along the posterior thoracic wall. The **hemiazygos** (hem′ē-az′ī-gos) and **accessory hemiazygos veins** on the left side of the vertebrae drain the left-side veins. The **azygos vein** drains the right-side veins and also receives blood from the hemiazygos veins. The azygos vein also receives blood from the esophageal veins, bronchial veins, and pericardial veins. The azygos vein merges with the superior vena cava.

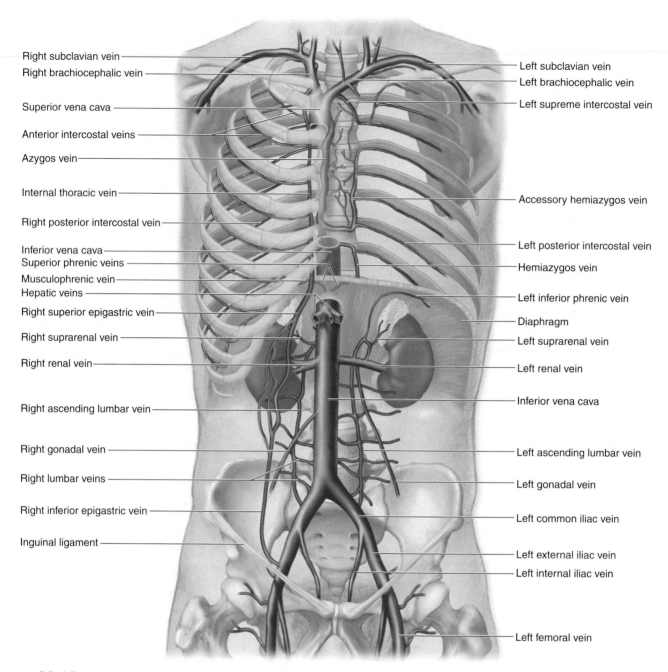

Right subclavian vein
Right brachiocephalic vein

Superior vena cava

Anterior intercostal veins

Azygos vein

Internal thoracic vein

Right posterior intercostal vein

Inferior vena cava
Superior phrenic veins
Musculophrenic vein
Hepatic veins
Right superior epigastric vein

Right suprarenal vein

Right renal vein

Right ascending lumbar vein

Right gonadal vein

Right lumbar veins

Right inferior epigastric vein

Inguinal ligament

Left subclavian vein
Left brachiocephalic vein
Left supreme intercostal vein

Accessory hemiazygos vein

Left posterior intercostal vein
Hemiazygos vein

Left inferior phrenic vein
Diaphragm
Left suprarenal vein
Left renal vein

Inferior vena cava

Left ascending lumbar vein

Left gonadal vein

Left common iliac vein

Left external iliac vein
Left internal iliac vein

Left femoral vein

Figure 23.13

Venous Circulation to the Thoracic and Abdominal Body Walls. The right side of this body primarily illustrates the more anteriorly placed veins as well as the "vessel arcs" formed by the anterior and posterior intercostal veins. The left side of the body illustrates the deeper, posteriorly placed veins only.

Figure 23.14 shows the arteries and veins of the posterior thoracic and abdominal walls.

Blood Flow Through the Thoracic Organs

The main thoracic organs include the heart, the lungs, the esophagus, and the diaphragm. The vessels of the heart were described in chapter 22; the vessels of the other thoracic organs are discussed here (see figures 23.12 and 23.13).

Lungs

The bronchi, bronchioles, and connective tissue of the lungs are supplied by three or four small **bronchial arteries** that emerge as

tiny branches from the anterior wall of the descending thoracic aorta. Left and right **bronchial veins** drain into the azygos system of veins. The rest of the lung receives its oxygen via diffusion directly from the tiny air sacs (alveoli) of the lungs.

Esophagus

Several small **esophageal** (ē-sof′ă-jē′ăl, ē′sŏ-faj′ē-ăl) **arteries** emerge from the anterior wall of the descending thoracic aorta and supply the esophagus. Additionally, the **left gastric artery** forms several **esophageal branches** that supply the abdominal portion of the esophagus. **Esophageal veins** drain the esophageal wall, and may take either of two routes: into the azygos vein or into the left gastric vein. The latter merges with the hepatic portal vein.

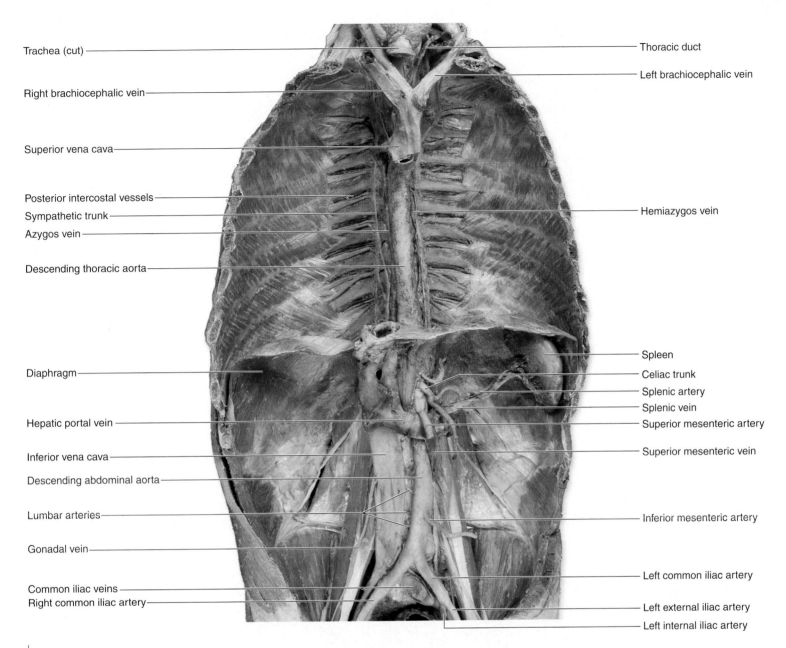

Trachea (cut)

Right brachiocephalic vein

Superior vena cava

Posterior intercostal vessels
Sympathetic trunk
Azygos vein

Descending thoracic aorta

Diaphragm

Hepatic portal vein

Inferior vena cava
Descending abdominal aorta

Lumbar arteries

Gonadal vein

Common iliac veins
Right common iliac artery

Thoracic duct
Left brachiocephalic vein

Hemiazygos vein

Spleen
Celiac trunk
Splenic artery
Splenic vein
Superior mesenteric artery

Superior mesenteric vein

Inferior mesenteric artery

Left common iliac artery

Left external iliac artery
Left internal iliac artery

Figure 23.14

Vessels of the Posterior Body Wall.   A cadaver photo illustrates the venous and arterial supply to the posterior thoracic and abdominal walls.

Diaphragm

Arterial blood is supplied to the diaphragm by paired vessels. **Superior phrenic** (fren′ik; *phren* = diaphragm) **arteries** emerge from the descending thoracic aorta; both **musculophrenic arteries** and pericardiacophrenic arteries arise from the internal thoracic artery; and **inferior phrenic arteries** emerge from the descending abdominal aorta to supply the diaphragm. Superior phrenic and inferior phrenic veins merge with the inferior vena cava, while the musculophrenic and pericardiacophrenic veins drain into the internal thoracic veins that merge with the brachiocephalic veins.

🔦 WHAT DID YOU LEARN?

5 What are the three main branches of the aortic arch, and what main body regions does each supply?

6 Which arteries supply the brain?

7 Describe venous drainage into the azygos system.

Blood Flow Through the Gastrointestinal Tract

Arterial Supply to the Abdomen

Three unpaired arteries emerge from the anterior wall of the descending abdominal aorta to supply the gastrointestinal (GI) tract. From superior to inferior, these arteries are the celiac trunk, superior mesenteric artery, and inferior mesenteric artery.

The **celiac** (sē′lē-ak; *koilia* = belly) **trunk** is located immediately inferior to the aortic opening (hiatus) of the diaphragm **(figure 23.15a)**. It supplies the stomach, part of the duodenum, the liver, part of the pancreas, and the spleen. Three branches emerge from this arterial trunk: the left gastric, splenic, and common hepatic arteries. (1) The **left gastric artery** supplies the lesser curvature of the stomach, and extends some esophageal branches. The left gastric artery anastomoses with the right gastric artery. (2) The **splenic** (splen′ik) **artery** supplies the spleen part of the stomach (via branches called the **left gastroepiploic** [gas′trō-ep′i-plō′ik]

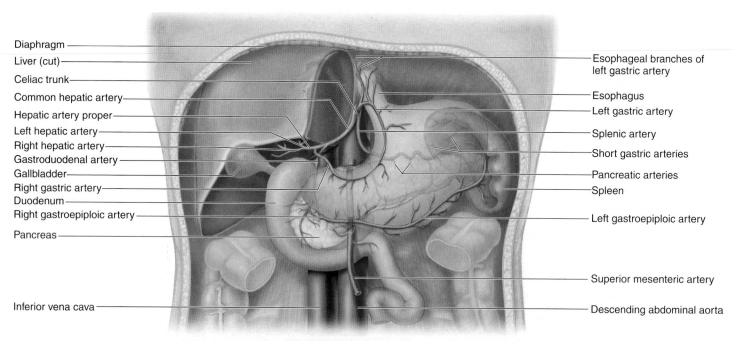

Diaphragm
Liver (cut)
Celiac trunk
Common hepatic artery
Hepatic artery proper
Left hepatic artery
Right hepatic artery
Gastroduodenal artery
Gallbladder
Right gastric artery
Duodenum
Right gastroepiploic artery
Pancreas

Inferior vena cava

Esophageal branches of
left gastric artery
Esophagus
Left gastric artery
Splenic artery
Short gastric arteries
Pancreatic arteries
Spleen
Left gastroepiploic artery

Superior mesenteric artery

Descending abdominal aorta

(a) Celiac trunk branches

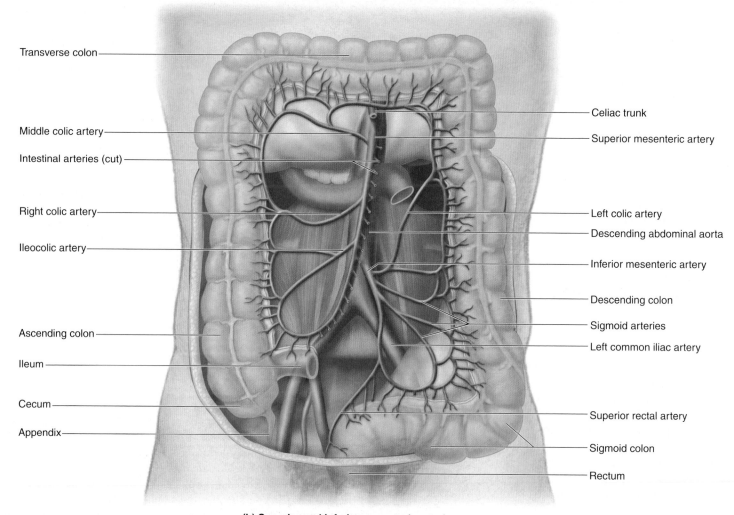

Transverse colon

Middle colic artery

Intestinal arteries (cut)

Right colic artery

Ileocolic artery

Ascending colon

Ileum

Cecum

Appendix

Celiac trunk

Superior mesenteric artery

Left colic artery
Descending abdominal aorta
Inferior mesenteric artery
Descending colon
Sigmoid arteries
Left common iliac artery

Superior rectal artery
Sigmoid colon
Rectum

(b) Superior and inferior mesenteric arteries

Figure 23.15

Arterial Supply to the Gastrointestinal Tract and Abdominal Organs. The celiac trunk, superior mesenteric artery, and inferior mesenteric artery supply most of the abdominal organs. (a) Branches of the celiac trunk supply part of the esophagus, stomach, spleen, pancreas, liver, and gallbladder. (b) Branches of the superior mesenteric and inferior mesenteric arteries primarily supply the intestines.

artery and short gastric arteries), and the pancreas (via branches called pancreatic arteries). (3) The last branch of the celiac trunk is the **common hepatic** (he-pat′ik; *hepat* = liver) **artery,** which extends to the right side of the body where it divides into a hepatic artery proper and a gastroduodenal artery. The **hepatic artery proper** supplies the liver (via **left** and **right hepatic arteries**), the gallbladder (via the **cystic artery**), and part of the stomach (via the **right gastric artery**). The **gastroduodenal** (gas′trō-doo′ō-dē′năl) **artery** supplies the greater curvature of the stomach (via the **right gastroepiploic artery**), the duodenum, and the pancreas (via the superior pancreaticoduodenal arteries).

The **superior mesenteric** (mez-en-ter′ik; *mesos* = middle, *enteron* = intestine) **artery** is located immediately inferior to the celiac trunk (figure 23.15*b*). The superior mesenteric artery supplies blood to most of the small intestine, the pancreas, and the proximal part of the large intestine. Its branches include: (1) the inferior pancreaticoduodenal arteries that anastomose with the superior pancreaticoduodenal arteries to supply the duodenum and pancreas; (2) 18–20 **intestinal arteries** that supply the jejunum and ileum; (3) the **ileocolic** (il′ē-ō-kol′ik) **artery** that supplies the ileum, cecum, and appendix; (4) the **right colic artery** that typically supplies the ascending colon; and (5) the **middle colic artery** that supplies most of the transverse colon.

The **inferior mesenteric artery** is the most inferior of the three unpaired arteries that arise from the descending abdominal aorta. It emerges approximately 5 centimeters superior to bifurcation of the aorta at about the level of vertebra L_3. The branches of the inferior mesenteric artery include: (1) the **left colic artery** that supplies the distal part of the transverse colon and part of the descending colon; (2) the **sigmoid arteries** that supply the inferior descending colon and the sigmoid colon; and (3) the **superior rectal** (rek′tăl; *rectus* = straight) **artery** that is a continuation of the inferior mesenteric artery and supplies the rectum.

Venous Return from the Abdomen

In contrast to most arteries and veins that run parallel to one another, the celiac artery and the common hepatic artery do not have corresponding veins of the same name. Rather, the veins of the gastrointestinal tract all merge into some part of the hepatic portal system.

Hepatic Portal System

The **hepatic portal** (pōr′tăl; *porta* = gate) **system** is a venous network that drains the GI tract and shunts the blood to the liver for absorption and processing of transported materials **(figure 23.16)**. Following nutrient absorption, the blood exits the liver through **hepatic veins** that merge with the inferior vena cava.

The hepatic portal system is needed because the GI tract absorbs digested nutrients, and these nutrients must be absorbed

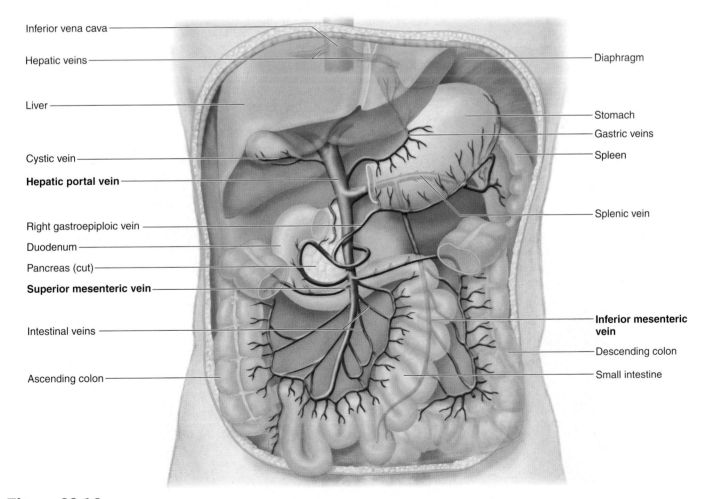

Figure 23.16

Hepatic Portal System.   The hepatic portal system is a network of veins that transports venous blood from the GI tract to the liver for nutrient processing.

and processed in the liver. The liver also detoxifies harmful agents that have been absorbed by the gastrointestinal blood vessels. The most efficient route for handling and processing the absorbed nutrients and any absorbed harmful substances is a system of vessels that drains the GI tract directly into the liver, rather than distributing these materials throughout the entire cardiovascular system. The hepatic portal system also receives products of erythrocyte destruction from the spleen, so that the liver can "recycle" some of these components.

The **hepatic portal vein** is the large vein that receives deoxygenated (oxygen-poor) but nutrient-rich blood from the gastrointestinal organs. Three main venous branches merge to form this vein: (1) The **inferior mesenteric vein,** a vertically positioned vein draining the distal part of the large intestine, receives blood from the superior rectal vein, sigmoid veins, and left colic vein. The inferior mesenteric vein typically (but not always) drains into the splenic vein. (2) The **splenic vein,** a horizontally positioned vein draining the spleen, receives blood from pancreatic veins, short gastric veins, and the right gastroepiploic vein. (3) The **superior mesenteric vein,** another vertically positioned vein on the right side of the body, drains the small intestine and part of the large intestine. It receives blood from the intestinal veins, pancreaticoduodenal veins, ileocolic vein, and right and middle colic veins. Some small veins, such as the left and right gastric veins, drain directly into the hepatic portal vein.

Figure 23.17 is a cadaver photo showing the arteries and veins of the posterior abdominal wall region. Note in this example that the inferior mesenteric vein drains into the superior mesenteric vein, not the splenic vein. This figure illustrates that the hepatic portal system can show great variation in some individuals.

Study Tip!

Although the pattern of the veins of the hepatic portal system can vary, together they typically resemble the side view of a chair. The front leg of the "chair" represents the inferior mesenteric vein, while the back leg represents the superior mesenteric vein. The seat of the chair is the splenic vein, while the back represents the hepatic portal vein.

Hepatic portal vein — Splenic vein

Superior mesenteric vein — Inferior mesenteric vein

The configuration of the veins of the hepatic portal system resembles the side view of a chair.

The venous blood in the hepatic portal vein flows through the sinusoids of the liver for absorption, processing, and storage of nutrients. In these sinusoids, the venous blood mixes with some arterial blood entering the liver in the hepatic arteries. Thus, liver

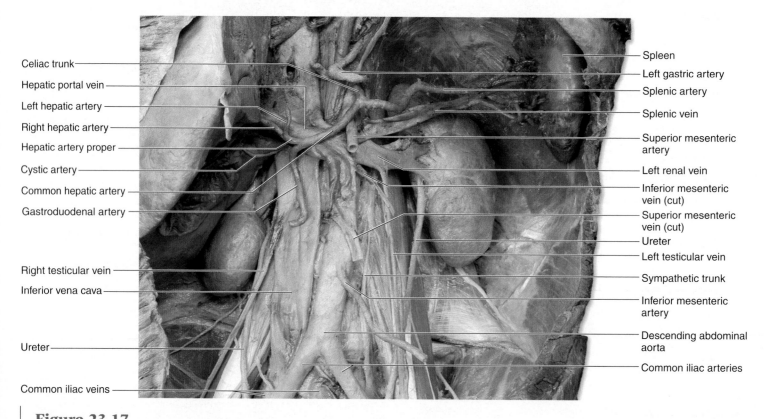

Celiac trunk
Hepatic portal vein
Left hepatic artery
Right hepatic artery
Hepatic artery proper
Cystic artery
Common hepatic artery
Gastroduodenal artery

Right testicular vein
Inferior vena cava

Ureter

Common iliac veins

Spleen
Left gastric artery
Splenic artery
Splenic vein
Superior mesenteric artery
Left renal vein
Inferior mesenteric vein (cut)
Superior mesenteric vein (cut)
Ureter
Left testicular vein
Sympathetic trunk
Inferior mesenteric artery
Descending abdominal aorta
Common iliac arteries

Figure 23.17

Major Vessels of the Posterior Abdominal Wall. In this cadaver photo, note that the inferior mesenteric vein varies from the "average" hepatic portal system pattern by draining into the superior mesenteric vein.

cells also receive oxygenated blood. Once nutrient absorption has occurred, blood leaves the liver through **hepatic veins** that merge with the inferior vena cava.

Blood Flow Through the Posterior Abdominal Organs, Pelvis, and Perineum

In addition to the arteries already mentioned, three other paired arterial branches emerge from the sides of the descending abdominal aorta: (1) The **middle suprarenal artery** supplies each adrenal gland; (2) the **renal** (rē'năl) **artery** supplies each kidney; and (3) the **gonadal** (gō-nad'ăl) **artery** supplies each gonad (testes in males, ovaries in females) (see figure 23.12). (Note: The right middle suprarenal artery may branch from the right renal artery in some individuals.) Subsequently, these organs are drained by veins having the same name as the arteries. **Left suprarenal** and **gonadal veins** typically merge with and drain into the **left renal vein.** The left renal vein, **right renal vein, right suprarenal vein,** and **right gonadal vein** merge directly into the inferior vena cava.

The primary arterial supply to the pelvis and perineum is from the **internal iliac artery**, one of the two main branches of the common iliac artery **(figure 23.18)**. Some branches of the internal iliac artery include: (1) the **superior** and **inferior gluteal** (gloo'tē-ăl) **arteries** that supply the gluteal region; (2) the **obturator** (ob'too-rā-tŏr) **artery** that supplies medial muscles of the thigh; (3) the **internal pudendal** (pū-den'dăl; *pudeo* = to feel ashamed) **artery** that supplies the anal canal and perineum; (4) the **middle rectal artery** that supplies the lower portion of the rectum; and (5) the **uterine** (ū'ter-in) **artery** and **vaginal artery** (in females) that supply the uterus and vagina. Some minor branches of the internal iliac artery include the iliolumbar arteries that supply the posterior abdominal wall muscles, the lateral sacral arteries that supply the region around the sacrum, and the vesical arteries that supply the bladder and the prostate (in males).

The pelvis and perineum are drained by veins with the same name as the supplying arteries. The veins merge with the **internal iliac vein** that merges with the **common iliac vein,** which subsequently drains into the inferior vena cava.

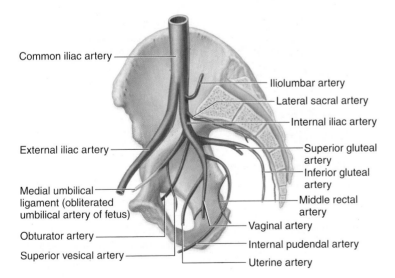

Figure 23.18

Arterial Supply to the Pelvis.   Branches of the right internal iliac artery distribute blood to the pelvic organs. Shown is a female pelvis; a male pelvis would have no uterine artery, and instead of a vaginal artery, would have an inferior vesical artery.

Labels in figure:
- Common iliac artery
- Iliolumbar artery
- Lateral sacral artery
- Internal iliac artery
- External iliac artery
- Superior gluteal artery
- Inferior gluteal artery
- Medial umbilical ligament (obliterated umbilical artery of fetus)
- Middle rectal artery
- Obturator artery
- Vaginal artery
- Internal pudendal artery
- Superior vesical artery
- Uterine artery

> **WHAT DID YOU LEARN?**
>
> ⑧ What three main branches arise from the celiac trunk?
>
> ⑨ Describe the hepatic portal system. Which veins drain into it? What is the main function of the hepatic portal system?
>
> ⑩ What branches of the internal iliac artery are seen in females only?

Blood Flow Through the Upper Limb

Blood flow through the upper limb closely mirrors its flow through the lower limb in many respects. Both the upper and lower limbs are supplied by a main arterial vessel: the subclavian artery for the upper limb and the femoral artery for the lower limb. This artery bifurcates at the elbow or knee. Arterial and venous arches are seen in both the hand and foot. Finally, both the upper limb and the lower limb have superficial and deep networks of veins.

Arterial Flow

A **subclavian artery** supplies blood to each upper limb **(figure 23.19a)**. The left subclavian artery emerges directly from the aortic arch, while the right subclavian artery is a division of the brachiocephalic trunk. The subclavian artery extends multiple branches: (1) the vertebral artery, (2) the thyrocervical trunk (to supply the thyroid gland and some neck and shoulder muscles), (3) the costocervical trunk, and (4) the internal thoracic artery.

After the subclavian artery passes over the lateral border of the first rib, it is renamed the **axillary** (ak'sil-ār-ē) **artery.** The axillary artery extends many branches to the shoulder and thoracic region, including the supreme thoracic artery (not to be confused with the supreme intercostal artery), the thoracoacromial artery, the lateral thoracic artery, the humeral circumflex arteries, and the subscapular artery. When the axillary artery passes the inferior border of the teres major muscle, it is renamed the **brachial** (brā'kē-ăl) **artery.** The brachial artery travels alongside the humerus. One of its branches is the **deep brachial artery** (also known as the *profunda brachii artery* or *deep artery of the arm*), which supplies blood to most brachial (arm) muscles. In the cubital fossa, the brachial artery divides into an **ulnar** (ŭl'năr) **artery** and a **radial** (rā'dē-ăl) **artery.** Both arteries supply the forearm and wrist before they anastomose and form two arterial arches in the palm: the **superficial palmar** (pawl'măr) **arch** (formed primarily from the ulnar artery) and the **deep palmar arch** (formed primarily from the radial artery). **Digital arteries** emerge from the arches to supply the fingers.

> **WHAT DO YOU THINK?**
>
> ❸ If the left ulnar artery were cut, would any blood be able to reach the left hand and fingers? Why or why not?

Venous drainage of the upper limb is through two groups of veins: superficial and deep.

Superficial Venous Drainage of the Upper Limb

On the dorsum of the hand, a **dorsal venous network** (or *arch*) of veins drains into both the **basilic** (ba-sil'ik) **vein** and the **cephalic** (se-fal'ik) **vein** (figure 23.19b). The basilic vein runs adjacent to the medial surface of the upper limb and eventually helps form the axillary vein. The cephalic vein runs alongside the lateral aspect of the upper limb, enters the deltopectoral triangle, and drains into the axillary vein. These superficial veins have perforating branches that allow them to connect to the deeper veins.

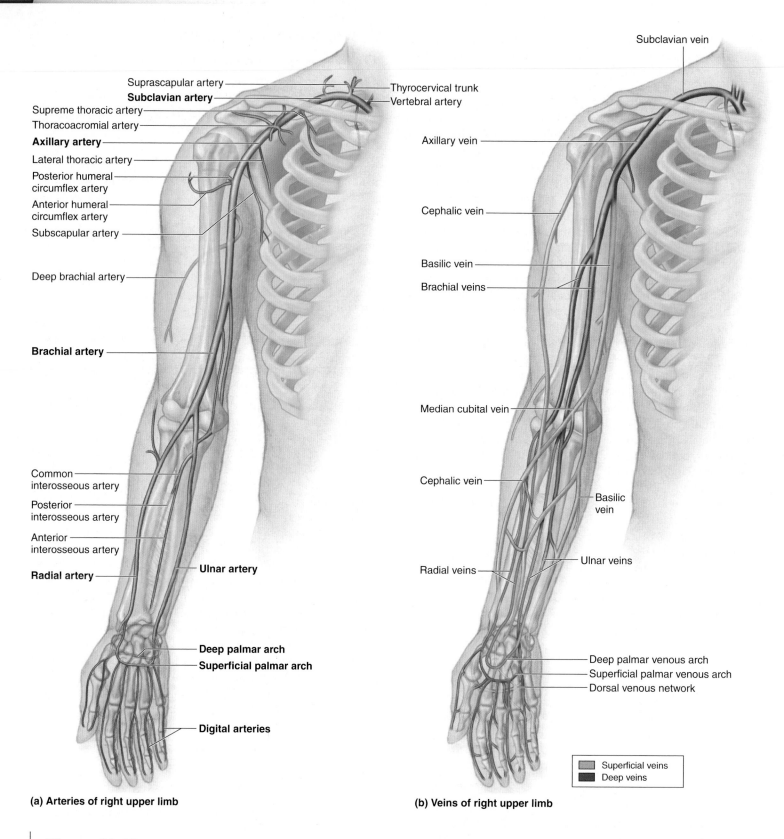

(a) Arteries of right upper limb

(b) Veins of right upper limb

Superscapular artery
Subclavian artery
Supreme thoracic artery
Thoracoacromial artery
Axillary artery
Lateral thoracic artery
Posterior humeral circumflex artery
Anterior humeral circumflex artery
Subscapular artery
Deep brachial artery
Brachial artery
Common interosseous artery
Posterior interosseous artery
Anterior interosseous artery
Radial artery
Ulnar artery
Deep palmar arch
Superficial palmar arch
Digital arteries

Thyrocervical trunk
Vertebral artery

Subclavian vein
Axillary vein
Cephalic vein
Basilic vein
Brachial veins
Median cubital vein
Cephalic vein
Basilic vein
Radial veins
Ulnar veins
Deep palmar venous arch
Superficial palmar venous arch
Dorsal venous network

Superficial veins
Deep veins

Figure 23.19

Vascular Supply to the Upper Limb.   The subclavian artery carries oxygenated blood to the upper limb; veins merge to return deoxygenated blood to the heart. *(a)* Arteries that supply the upper limb. *(b)* Veins that return blood from the upper limb. *(c)* Cadaver photo of the major vessels of the right arm. *(d)* Cadaver photo of the major vessels of the right forearm.

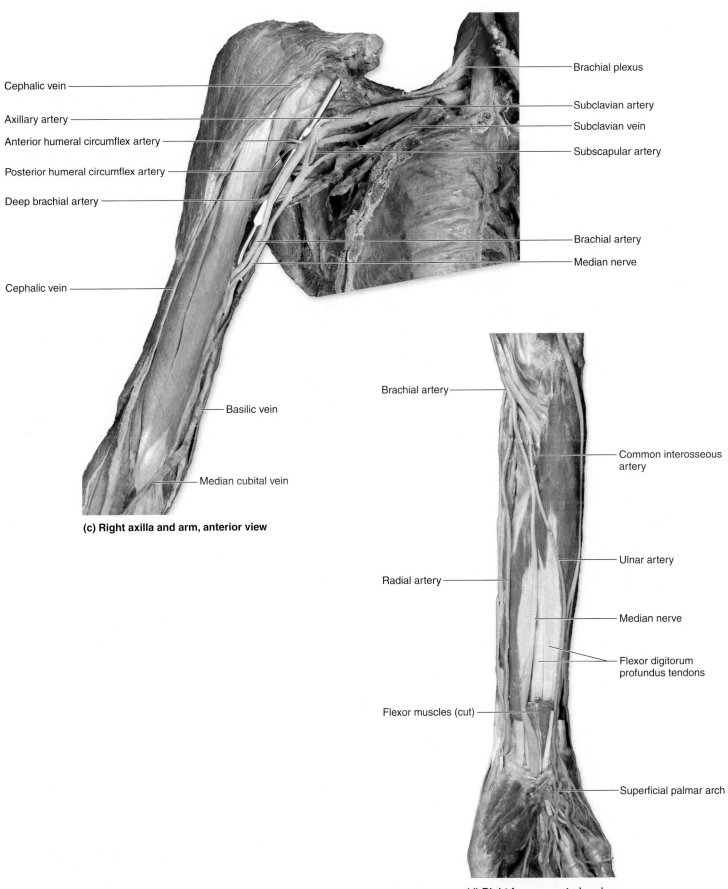

Cephalic vein

Axillary artery

Anterior humeral circumflex artery

Posterior humeral circumflex artery

Deep brachial artery

Cephalic vein

Brachial plexus

Subclavian artery

Subclavian vein

Subscapular artery

Brachial artery

Median nerve

Basilic vein

Median cubital vein

(c) Right axilla and arm, anterior view

Brachial artery

Radial artery

Flexor muscles (cut)

Common interosseous artery

Ulnar artery

Median nerve

Flexor digitorum profundus tendons

Superficial palmar arch

(d) Right forearm, anterior view

CLINICAL VIEW: In Depth

Atherosclerosis

Atherosclerosis (ath′er-ō-skler-ō′sis, *athere* = gruel, *sclerosis* = hardening) is a progressive disease of the elastic and muscular arteries. It is characterized by the presence of an **atheroma** (or *atheromatous plaque*), which leads to thickening of the tunica intima and narrowing of the arterial lumen. Atherosclerosis is linked to over 50% of all deaths in the United States, and it is a leading cause of morbidity and mortality in other countries of the western world as well.

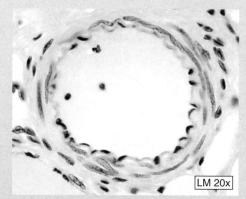

(a) Normal artery

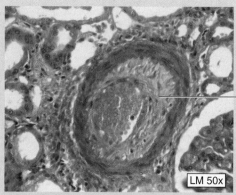

Atheroma

(b) Atherosclerotic artery

ETIOLOGY OF ATHEROSCLEROSIS

Although the cause of atherosclerosis is not completely understood, the **response-to-injury hypothesis** is the most widely accepted. This proposal states that injury (especially repeated injury) to the endothelium of an arterial wall results in an inflammatory reaction that eventually leads to the development of an atheroma. The injury could be caused by infection, trauma to the vessel, or hypertension. The injured endothelium becomes more permeable, which encourages leukocytes and platelets to adhere to the lesion and initiate an inflammatory response. Low-density lipoproteins (LDLs and VLDLs) enter the tunica intima, combine with oxygen, and remain stuck to the vessel wall. This oxidation of lipoproteins attracts monocytes, which adhere to the endothelium and migrate into the wall. As the monocytes migrate into the wall, they digest the lipids and develop into structures called **foam cells**. Eventually, smooth muscle cells from the tunica media migrate into the atheroma and proliferate, causing further enlargement of the atheroma. Atherosclerotic plaques cause narrowing of the lumen of the blood vessel, thereby restricting blood flow to the regions the artery supplies. In addition, the plaque may rupture, causing a blood clot to form and completely block the artery.

Atherosclerosis is a progressive disease. The plaques begin to develop in early adulthood and grow and enlarge as we age. People are unaware of the plaques until they become large enough to restrict blood flow in an artery and cause vascular complications.

RISK FACTORS FOR ATHEROSCLEROSIS

Some individuals are genetically prone to atherosclerosis. **Hypercholesterolemia** (an increased amount of cholesterol in the blood), which also tends to run in families, has been positively associated with the rate of development and severity of atherosclerosis. In addition, males tend to be affected more than females, and symptomatic atherosclerosis increases with age. Finally, smoking and hypertension both cause vascular injury, which increases the risk.

TREATMENT AND PREVENTION OPTIONS

If an artery is occluded (blocked) in one or just a few areas, a treatment called **angioplasty** (an′jē-ō-plas-tē; *angeion* = vessel, *plastos* = formed) is used. A physician inserts a balloon-tip catheter into an artery, and positions it at the site where the lumen is narrowed. Then the balloon is inflated, forcibly expanding the narrowed area. To ensure that the area remains open, a stent may be placed at the site. A stent is a piece of wire-mesh that springs open to keep the vessel lumen open. The stent becomes a permanent part of the vessel. For occluded coronary arteries, a much more invasive treatment known as **coronary bypass surgery** may be needed. A vein (e.g., great saphenous vein) or artery (e.g., internal thoracic artery) is detached from its original location and grafted from the aorta to the coronary artery system, thus bypassing the area(s) of atherosclerotic narrowing.

Ideally, the best treatment for atherosclerosis is to try to prevent it by:

- Maintaining a healthy diet and watching your cholesterol level. High cholesterol levels can be treated with drugs called statins.
- Not smoking.
- Monitoring your blood pressure regularly. If you have hypertension, seek treatment as soon as possible.

Balloon catheter   Atheroma   Artery

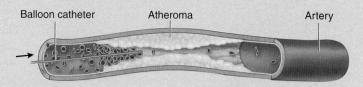

(1) Balloon catheter is used to carry an uninflated balloon to the area in artery that is obstructed.

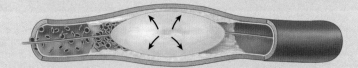

(2) Balloon inflates, compressing the atheroma.

(3) Balloon is deflated following lumen widening, and then catheter is withdrawn. A stent may be placed in the artery as well.

Angioplasty is used to expand the narrowed region of an artery.

In the cubital region, an obliquely positioned **median cubital vein** connects the cephalic and basilic veins. The median cubital vein is a common site for **venipuncture,** in which a vein is punctured to draw blood or inject a solute. All of these superficial veins are highly variable among individuals and have multiple superficial tributaries draining into them.

Deep Venous Drainage of the Upper Limb

The digital veins and **superficial** and **deep palmar venous arches** drain into pairs of **radial veins** and **ulnar veins** that run parallel to arteries of the same name. (Paired veins that run alongside an artery are collectively known as *venae commitantes.*) At the level of the cubital fossa, the radial and ulnar veins merge to form a pair of **brachial veins** that travel with the brachial artery. Brachial veins and the basilic vein merge to form the **axillary vein.** Superior to the lateral border of the first rib, the axillary vein is renamed the **subclavian vein.** When the subclavian vein and jugular veins of the neck merge, they form the brachiocephalic vein. As we have seen, the left and right brachiocephalic veins form the superior vena cava.

The combined arteries and veins of the upper limbs are shown in cadaver photos in figure 23.19c,d.

Blood Flow Through the Lower Limb

The arterial and venous blood flow of the lower limb is very similar to that of the upper limb. As we discuss lower limb blood flow, compare it with that of the upper limb.

Arterial Flow

The main arterial supply for the lower limb is the external iliac artery, which is a branch of the common iliac artery **(figure 23.20a).** The external iliac artery travels inferior to the inguinal ligament, where it is renamed the **femoral** (fem'ŏ-răl) **artery.** The **deep femoral artery** (*profunda femoris artery* or *deep artery of the thigh*) emerges from the femoral artery to supply the hip joint (via medial and lateral femoral circumflex arteries) and many of the thigh muscles, before traversing posteromedially along the thigh. The femoral artery passes through an opening in the adductor magnus muscle and enters the posteriorly placed popliteal fossa, where the vessel is renamed the **popliteal** (pop-lit'ē-ăl, pop-li-tē'ăl) **artery.** The popliteal artery supplies the knee joint and muscles in this region.

The popliteal artery divides into an **anterior tibial** (tib'ē-ăl) **artery** that supplies the anterior compartment of the leg, and a **posterior tibial artery** that supplies the posterior compartment of the leg. The posterior tibial artery extends a branch called the **fibular** (fib'ū-lăr) **artery,** which supplies the lateral compartment leg muscles.

The posterior tibial artery continues to the plantar side of the foot, where it branches into **medial** and **lateral plantar arteries.** The anterior tibial artery crosses over the anterior surface of the ankle, where it is renamed the **dorsalis pedis artery.** The dorsalis pedis artery and a branch of the lateral plantar artery unite to form the **plantar** (plan'tăr) **arch** of the foot. **Digital arteries** extend from the plantar arch and supply the toes.

WHAT DO YOU THINK?

4 If the right femoral artery were blocked, would any blood be able to reach the right leg? Why or why not?

Venous drainage of the lower limb is through two groups of veins: superficial and deep.

Superficial Venous Drainage of the Lower Limb

On the dorsum of the foot, a **dorsal venous arch** drains into the **great saphenous** (să-fē'nŭs) **vein** and the **small saphenous vein** (figure 23.20b). The great saphenous vein originates in the medial ankle and extends adjacent to the medial surface of the entire lower limb before it drains into the femoral vein. The small saphenous vein extends adjacent to the lateral ankle and then travels along the posterior calf, before draining into the popliteal vein. These superficial veins have perforating branches that connect to the deeper veins. If the valves in these veins (or the perforating branches) become incompetent, varicose veins develop (see the Clinical View on page 688).

Deep Venous Drainage of the Lower Limb

The digital veins and deep veins of the foot drain into pairs of **medial** and **lateral plantar veins.** These veins drain into a pair of **posterior tibial veins.** A pair of **fibular veins** travel alongside the fibular artery and drain into the posterior tibial veins. On the dorsum of the foot and ankle, deep veins drain into a pair of **anterior tibial veins,** which traverse alongside the anterior tibial artery. The anterior and posterior tibial veins merge to form a **popliteal vein** that accompanies the popliteal artery in the popliteal fossa. This vein curves to the anterior portion of the thigh and is renamed the **femoral vein.** Once this vein passes superior to the inguinal ligament, it is renamed once again as the **external iliac vein.** The external and internal iliac veins merge in the pelvis, forming the **common iliac vein.** Left and right common iliac veins then merge to form the inferior vena cava.

Figure 23.21 shows arteries and veins of the lower limb in the femoral region.

WHAT DID YOU LEARN?

11 What are the superficial veins that help drain the upper limb?

12 How does arterial blood travel through the lower limb? List the branching that occurs, beginning with the external iliac artery.

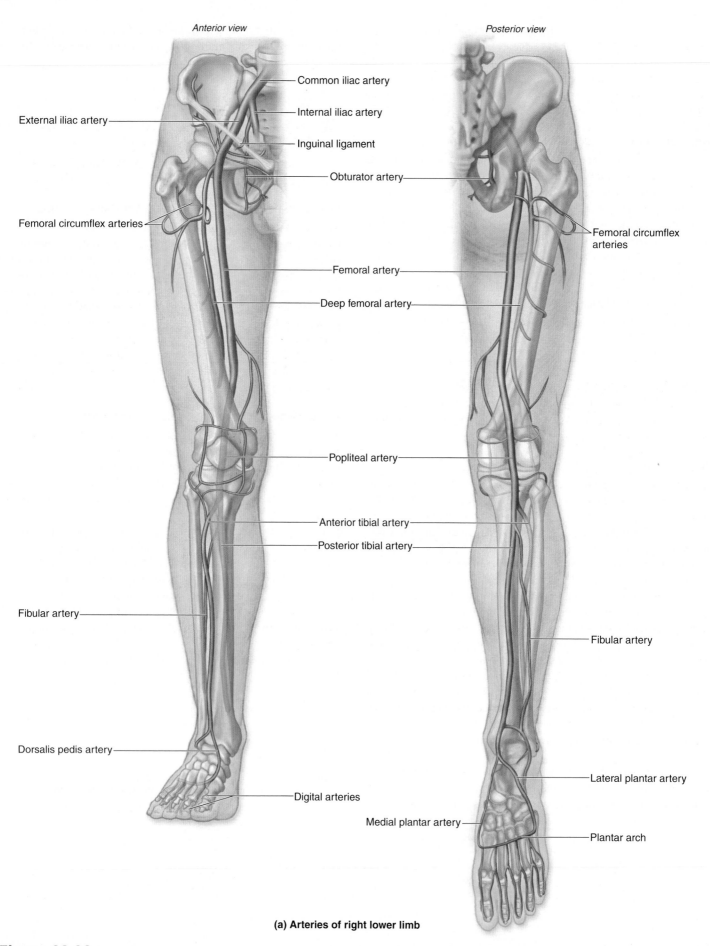

Anterior view

Posterior view

Common iliac artery

Internal iliac artery

External iliac artery

Inguinal ligament

Obturator artery

Femoral circumflex arteries

Femoral circumflex arteries

Femoral artery

Deep femoral artery

Popliteal artery

Anterior tibial artery

Posterior tibial artery

Fibular artery

Fibular artery

Dorsalis pedis artery

Lateral plantar artery

Digital arteries

Medial plantar artery

Plantar arch

(a) Arteries of right lower limb

Figure 23.20

Vascular Supply to the Lower Limb. The external iliac artery carries oxygenated blood to the lower limb; veins merge to return deoxygenated blood to the heart. *(a)* Anterior and posterior views of arteries distributed throughout the right lower limb. *(b)* Anterior and posterior views of veins that return blood from the right lower limb.

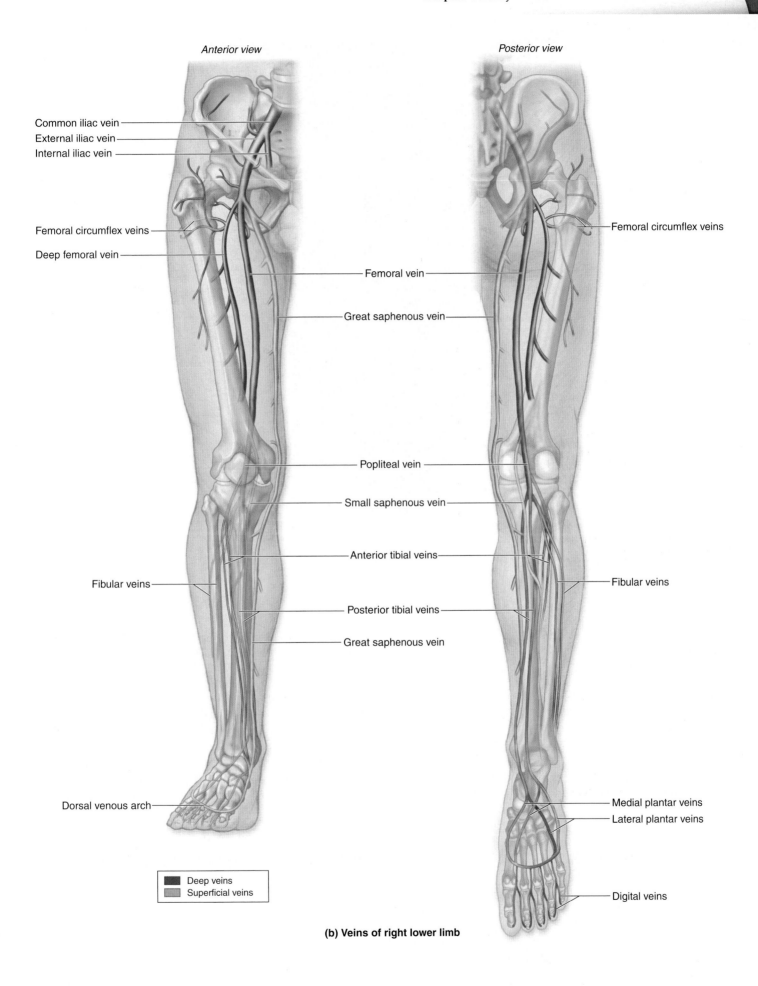

Anterior view

Posterior view

Common iliac vein

External iliac vein

Internal iliac vein

Femoral circumflex veins

Deep femoral vein

Femoral circumflex veins

Femoral vein

Great saphenous vein

Popliteal vein

Small saphenous vein

Anterior tibial veins

Fibular veins

Fibular veins

Posterior tibial veins

Great saphenous vein

Dorsal venous arch

Medial plantar veins

Lateral plantar veins

Deep veins
Superficial veins

Digital veins

(b) Veins of right lower limb

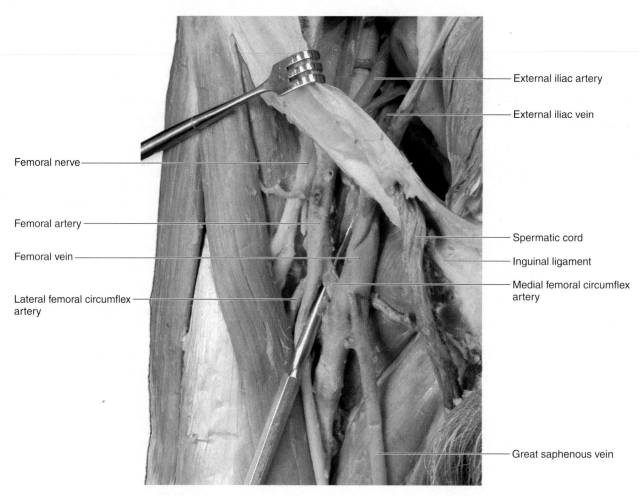

Figure 23.21

Vascular Supply to the Lower Limb. A cadaver photo shows major vessels of the anterior thigh.

Pulmonary Circulation

Key topic in this section:

■ Pulmonary circulation vessels and their pathways

The pulmonary circulation is responsible for carrying deoxygenated blood from the right side of the heart to the lungs, and then returning the newly oxygenated blood to the left side of the heart **(figure 23.22)**. Blood low in oxygen is pumped out of the right ventricle into the **pulmonary trunk.** This vessel travels superiorly and slightly to the left before it bifurcates into a **left pulmonary artery** and a **right pulmonary artery** that go to the lungs. The pulmonary arteries divide into smaller arteries that continue to subdivide to form arterioles. These arterioles finally branch into pulmonary capillaries, where gas exchange occurs. Carbon dioxide is removed from the blood and enters the tiny air sacs (alveoli) of the lungs, while oxygen moves in the opposite direction, from the air sacs into the blood. The

capillaries merge to form venules and then the **pulmonary veins.** Typically, two left and two right pulmonary veins carry the newly oxygenated blood to the left atrium of the heart.

Compared to the systemic circulation, the vessels that make up the pulmonary circulation are relatively short. Blood doesn't need to be pumped as far in the pulmonary circulation, since the lungs are close to the heart. In addition, the pulmonary arteries have less elastic connective tissue and wider lumens than systemic arteries. As a result, blood pressure is lower in the pulmonary arteries than in the systemic arteries, and pressure is correspondingly lower on the right side of the heart than on the left side.

WHAT DID YOU LEARN?

13 Compare the bronchial arteries and veins (discussed earlier in this chapter) with the pulmonary arteries and veins.

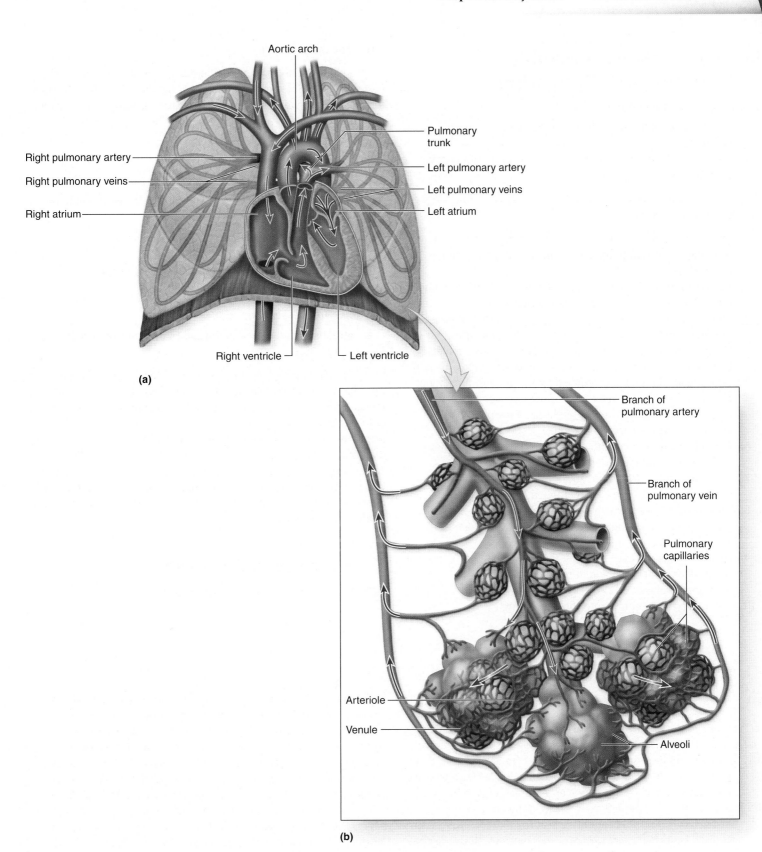

Aortic arch

Pulmonary
trunk

Right pulmonary artery

Right pulmonary veins

Right atrium

Left pulmonary artery

Left pulmonary veins

Left atrium

Right ventricle

Left ventricle

(a)

Branch of
pulmonary artery

Branch of
pulmonary vein

Pulmonary
capillaries

Arteriole

Venule

Alveoli

(b)

Figure 23.22

Pulmonary Circulation.   The pulmonary circulation conducts blood from the heart to and from the gas exchange surfaces of the lungs. Blood circulation through the heart is indicated by colored arrows (blue = deoxygenated blood; red = oxygenated blood). *(a)* Deoxygenated blood is pumped from the right ventricle to the lungs through the pulmonary arteries. Oxygenated blood returns to the heart from the lungs within the pulmonary veins. *(b)* At the microscopic level, pulmonary capillaries are associated with the alveoli of the lungs.

Review of Heart, Systemic, and Pulmonary Circulation

Key topics in this section:

- Oxygenation of blood in the systemic and pulmonary circulations
- Heart chambers involved in the systemic and pulmonary circulations

A simplified flowchart of blood circulation is outlined in **figure 23.23** and described here:

1. The systemic circulation begins when oxygenated blood flows from the left atrium to the left ventricle and then is pumped into the aorta.
2. Blood in the aorta enters elastic arteries and then flows into muscular arteries before entering arterioles.
3. Blood enters the systemic capillaries from arterioles; gases, nutrients, and wastes are exchanged in the capillaries.
4. Deoxygenated blood that is low in nutrients exits capillary beds. It drains into venules that merge to form veins.
5. Deoxygenated blood is conducted by the venous circulation to either the superior vena cava or the inferior vena cava for entry into the right atrium of the heart. Now the blood has entered the pulmonary circulation.
6. Blood flows from the right atrium into the right ventricle, and then it is pumped into the pulmonary trunk.
7. The pulmonary trunk bifurcates into left and right pulmonary arteries that carry deoxygenated blood to the lungs.
8. This blood passes through a series of smaller and smaller arteries before entering pulmonary capillaries, where gas exchange occurs.
9. Oxygenated blood exits the lung through a series of progressively larger veins that merge to form the pulmonary veins.
10. Pulmonary veins empty into the left atrium.
11. The cycle repeats.

Keep in mind that this outline is an oversimplification of complex events. For example, remember that both ventricles contract together. Thus, while some blood is traveling through the systemic vessels, blood is also traveling through the pulmonary vessels.

WHAT DID YOU LEARN?

14 Oxygenated blood leaves what chamber of the heart and travels through which major vessel?

Aging and the Cardiovascular System

Key topic in this section:

- Structure, function, and durability of blood vessels during aging

As adults get older, the heart and blood vessels become less resilient. Many of the elastic arteries are less able to withstand the forces from the pulsating blood. Systolic blood pressure may increase with age, exacerbating this problem. As a result, older individuals are more apt to develop an **aneurysm** (an'ū-rizm; *aneurysma* = a dilation), whereby part of the arterial wall thins and balloons out. This wall is more prone to rupture, which can cause massive bleeding and may lead to death. In addition, as we grow older, the incidence and severity of atherosclerosis increases, at least for people living in the developed world.

WHAT DID YOU LEARN?

15 How are aging and blood pressure related?

CLINICAL VIEW

Abdominal Aortic Aneurysm

An aneurysm is a localized, abnormal dilation of a blood vessel. Although an aneurysm can form in any type of vessel, aneurysms are particularly common in arteries, especially the aorta, because of the higher blood pressure on the arterial side of the circulation. After being initiated by a weakness in the wall of the vessel, an aneurysm tends to increase in size over a period of weeks or months until it ruptures.

Abdominal aortic aneurysm is a relatively common medical problem, and is most often a consequence of atherosclerosis. Most abdominal aortic aneurysms develop between the level of the renal arteries and the point near where the aorta bifurcates into the common iliac arteries. Since no pain fibers are associated with the aorta, an aneurysm can increase in size and reach the point of rupture without the patient ever being aware of it. A ruptured aorta is a surgical emergency that few people survive. An abdominal aortic aneurysm may be detected during a physical exam as a pulsating abdominal mass. X-ray and ultrasound studies can confirm the diagnosis and determine the size and extent of the aneurysm. For a number of years, aortic aneurysm repair has involved removing the dilated segment of aorta and replacing it with an artificial vascular prosthesis. This risky surgical procedure requires making a large abdominal incision to gain access to the dilated segment of aorta. More recently, stents have been developed that can be inserted through an incision in the femoral artery, positioned in the area of the aneurysm using x-ray guidance, and then expanded to reinforce the weakened and dilated area of the aortic wall. This procedure is less invasive and traumatic than major abdominal surgery. Unfortunately, the stent does not always lead to a complete cure, and complications are still possible.

Systemic circulation (black arrows)

① Oxygenated blood flows from the left atrium to the left ventricle and then is pumped into the aorta.

② Blood passes from the aorta into elastic arteries and then into muscular arteries before entering arterioles.

③ Blood in arterioles enters systemic capillaries for exchange of gases and nutrients.

④ Deoxygenated blood exits capillary beds into venules and then into veins.

⑤ Deoxygenated blood is conducted to either the superior or inferior vena cava and then enters the right atrium of the heart.

Pulmonary circulation (yellow arrows)

⑥ Blood flows from the right atrium to the right ventricle and is then pumped into the pulmonary trunk.

⑦ The pulmonary trunk conducts deoxygenated blood into pulmonary arteries to the lungs.

⑧ The blood passes through smaller and smaller arteries before entering pulmonary capillaries for gas exchange.

⑨ Oxygenated blood exits the lung via a series of progressively larger veins that merge to form the pulmonary veins.

⑩ Pulmonary veins drain into the left atrium.

⑪ The cycle repeats.

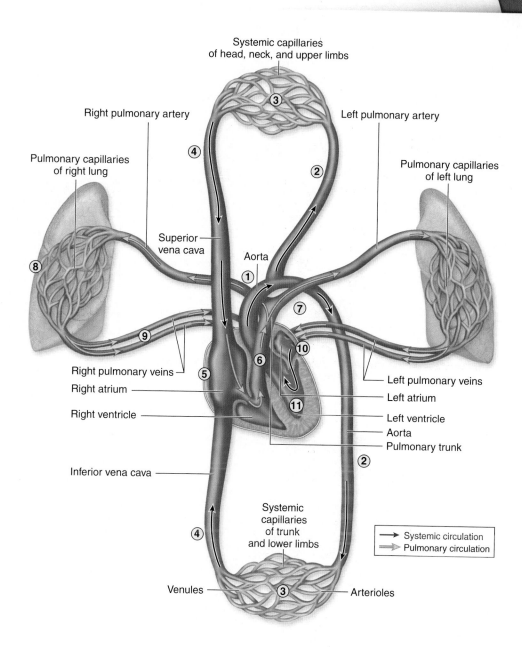

Figure 23.23

Cardiovascular System Circulatory Routes.   Blood travels through two routes: the systemic circulation and the pulmonary circulation. In the systemic circulation (black arrows), blood is pumped into the arteries, through the systemic capillary beds, and then back to the heart in systemic veins. In the pulmonary circulation (yellow arrows), blood is pumped through the pulmonary arteries to pulmonary capillary beds in the lungs and then back to the heart in pulmonary veins.

Blood Vessel Development

Key topics in this section:

- Developmental fates of the embryonic vessels
- Comparison of the fetal and postnatal circulatory patterns

The heart and its blood vessels begin to develop in the embryo during the third week. The blood vessels form by a process called **vasculogenesis** (vas′kū-lō-jen′ĕ-sis; *vasculum* = small vessel, *gene-sis* = production), whereby some of the mesoderm forms cells called

angioblasts, and these angioblasts connect to form the first primitive blood vessels. These vessels then grow and invade developing tissues throughout the embryo.

Artery Development

The embryo initially has both a **left** and **right dorsal aorta** (**figure 23.24a**). These two vessels remain separate until the level of the fourth thoracic vertebra, where they fuse to form a **common dorsal aorta** that supplies blood to the inferior part of the body. Eventually, the superior part of the right dorsal aorta degenerates and disappears, leaving the left

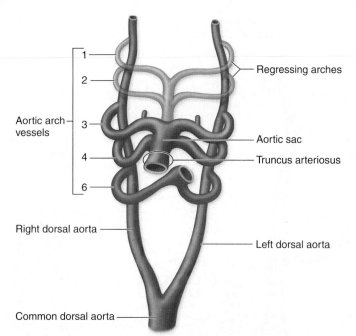

(a) Late week 4 to early week 5: Paired aortic arch vessels connect to left and right dorsal aortae

Aortic Arch Vessel	Postnatal Structure Formed by Vessels
1	Small part of maxillary arteries
2	Small part of stapedial arteries
3	Left and right common carotid arteries
4	Right vessel: proximal part of right subclavian artery Left vessel: aortic arch (connects to the left dorsal aorta)
6	Right vessel: right pulmonary artery Left vessel: left pulmonary artery and ductus arteriosus

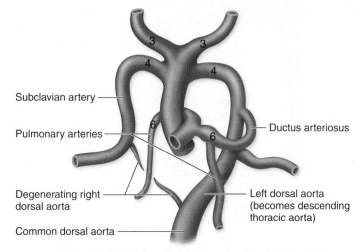

(b) Week 7: Right dorsal aorta degenerates; left dorsal aorta becomes descending thoracic aorta

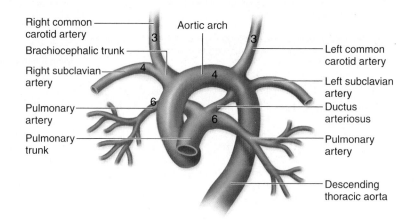

(c) Week 8: Aortic arch and branches formed

Figure 23.24

Thoracic Artery Development.　Aortic arch vessels form most of the major thoracic and head and neck arteries. *(a)* By late week 4/early week 5, paired aortic arch vessels arise from the truncus arteriosus and attach to paired left and right dorsal aortae. *(b)* During week 7, the right dorsal artery starts to degenerate, and the left dorsal aorta becomes the descending thoracic aorta. *(c)* By week 8, the aortic arch vessels have undergone remodeling to form the aortic arch, major branches of the arch, and the pulmonary arteries.

dorsal aorta and common dorsal aorta (figure 23.24b,c). The left dorsal aorta and common dorsal aorta form the descending thoracic aorta.

Beginning the fourth week, the truncus arteriosus of the heart connects to the left and right dorsal aortae by a series of paired **aortic arch vessels** (figure 23.24a). These vessels are numbered 1–6. Aortic arch vessels 1 and 2 primarily regress, vand only a small portion of them remain to form small segments of arteries in the head. Aortic arch vessel 5 never forms in humans. The remaining aortic arch vessels—3, 4, and 6—develop into adult arteries (figure 23.24b,c). In addition, the most superior part of the truncus arteriosus (called the **aortic sac**) forms the brachiocephalic trunk. The dorsal aorta (now part of the descending aorta) develops vascular "sprouts" that form

many of the blood vessels in the body. These blood vessel sprouts grow and migrate to the areas that need vascularization.

Vein Development

The venous system of the embryo develops from three venous systems: the **vitelline** (vī-tel′in; *vitellus* = yolk) **system,** the **umbilical** (ŭm-bil′i-kăl) **system,** and the **cardinal** (kar′di-năl; *cardinalis* = principal) **system.** All three systems initially are bilateral and connect to the sinus venosus of the heart. However, these three systems are eventually remodeled so that venous blood return is shifted to the right side of the heart. Each system is responsible for a specific body area: The vitelline system drains the gastrointestinal region;

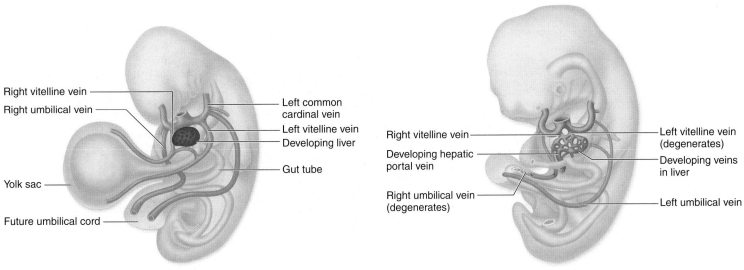

Right vitelline vein

Right umbilical vein

Yolk sac

Future umbilical cord

Left common cardinal vein

Left vitelline vein

Developing liver

Gut tube

(a) Week 4: Bilateral vitelline and umbilical arteries in place

Right vitelline vein

Developing hepatic portal vein

Right umbilical vein (degenerates)

Left vitelline vein (degenerates)

Developing veins in liver

Left umbilical vein

(b) Week 5: Asymmetric remodeling of the veins occurs

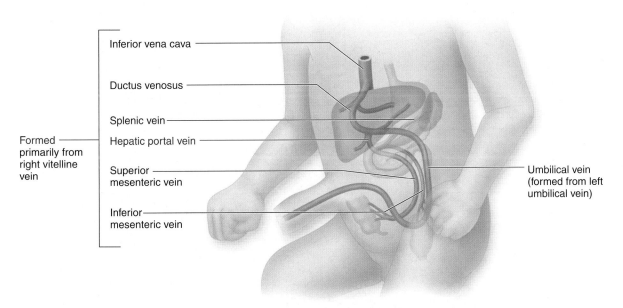

Inferior vena cava

Ductus venosus

Splenic vein

Hepatic portal vein

Formed primarily from right vitelline vein

Superior mesenteric vein

Inferior mesenteric vein

Umbilical vein (formed from left umbilical vein)

(c) Week 12: Hepatic portal system (right vitelline vein) and a single umbilical vein (left umbilical vein)

Figure 23.25

Development of Vitelline and Umbilical Veins.   The vitelline veins carry blood from the yolk sac, while the umbilical veins carry oxygenated blood to the embryo. *(a)* At week 4, the bilateral vessels are present. *(b)* By week 5, the vitelline vessels form the blood vessels in the liver, and the right vitelline vein forms the hepatic portal vein. *(c)* By week 12, the right vitelline vein forms most of the hepatic portal system, while the left vitelline vein regresses. Conversely, the right umbilical vein regresses, and the left umbilical vein persists as the single umbilical vein.

the umbilical system carries oxygenated blood from the placenta; and the cardinal system forms most of the veins of the head, neck, and body wall. (Limb veins are formed from separate venous plexuses that interconnect with the cardinal system.)

The vitelline system of veins consists of **left** and **right vitelline veins**, which are apparent through the fourth week **(figure 23.25)**. Beginning in the fifth week and continuing through the twelfth week, the left vitelline vein primarily degenerates, and the right vitelline vein forms the hepatic portal system, the sinusoids of the liver, and the portion of the inferior vena cava between the liver and the heart. Also

formed from the right vitelline vein is the **ductus venosus** (dŭk′tŭs vē-nō′sŭs), which connects the umbilical vein to the inferior vena cava and heart, and shunts blood away from the fetal liver.

The umbilical system of veins originally begins with a **left** and **right umbilical vein.** However, by the second month of development, the right umbilical vein disappears, and the left umbilical vein connects directly to the ductus venosus. Thus, within the umbilical cord are one umbilical vein and a pair of umbilical arteries.

The cardinal system of veins consists of a series of paired veins: the **anterior cardinal veins, posterior cardinal veins, supracardinal**

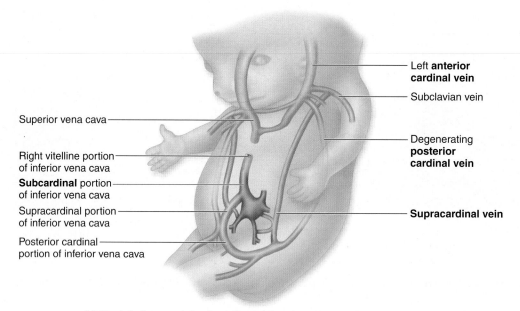

Superior vena cava

Right vitelline portion
of inferior vena cava

Subcardinal portion
of inferior vena cava

Supracardinal portion
of inferior vena cava

Posterior cardinal
portion of inferior vena cava

Left **anterior
cardinal vein**

Subclavian vein

Degenerating
**posterior
cardinal vein**

Supracardinal vein

(a) Week 8: Supra- and subcardinal veins undergo asymmetric remodeling

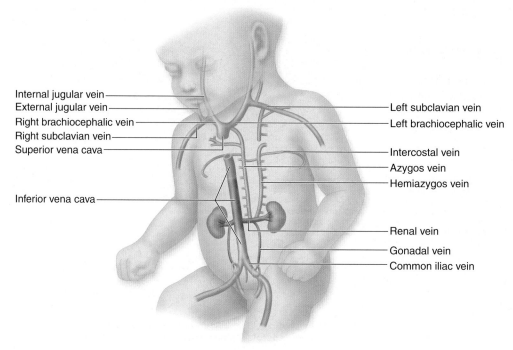

Internal jugular vein
External jugular vein
Right brachiocephalic vein
Right subclavian vein
Superior vena cava

Inferior vena cava

Left subclavian vein
Left brachiocephalic vein

Intercostal vein
Azygos vein
Hemiazygos vein

Renal vein
Gonadal vein
Common iliac vein

(b) Birth: Mature vessel pattern formed

Figure 23.26

Cardinal Vein Development.   The primary venous drainage for the embryo is initially formed by the cardinal veins. *(a)* By week 8, some of these cardinal veins undergo asymmetric remodeling and begin to form some of the named veins in the body. *(b)* At birth, the cardinal veins have been remodeled to form the inferior vena cava and the posterior thoracic and abdominal wall vessels.

veins, and **subcardinal veins (figure 23.26)**. The anterior cardinal veins develop into the veins of the head and neck and the veins superior to the heart. By the eighth week of development, the posterior cardinal veins degenerate and are largely replaced by the supracardinal and subcardinal veins. The supracardinal and subcardinal veins undergo asymmetrical remodeling, whereby venous blood flow is shifted to the right side of the body. The subcardinal veins form veins that drain the posterior abdominal wall, while the supracardinal veins form the hemiazygos and azygos system of veins. The inferior vena

cava is formed from parts of the right vitelline, right subcardinal, right supracardinal, and right posterior cardinal veins. The mature vessel pattern is formed well before birth and shown in figure 23.26*b*.

Comparison of Fetal and Postnatal Circulation

The cardiovascular system of the fetus is structurally and functionally different from that of the newborn. Whereas the fetus receives oxygen and nutrients directly from the mother through the placenta, its postnatal cardiovascular system is independent. In addition, since the fetal lungs are not functional, the blood pressure in the pulmonary arteries and right side of the heart is greater than the pressure in the left side of the heart. Finally, several fetal vessels help shunt blood directly to the organs in need and away from the organs that are not yet functional. As a result, the fetal cardiovascular system has some structures that are modified or that cease to function once the human is born. **Figure 23.27** compares the fetal and postnatal circulation patterns.

Fetal circulation occurs as follows:

1. Oxygenated blood from the placenta enters the body of the fetus through the **umbilical vein**.
2. The blood from the umbilical vein is shunted away from the liver and directly toward the inferior vena cava through the **ductus venosus**.
3. Oxygenated blood in the ductus venosus mixes with deoxygenated blood in the inferior vena cava.
4. Blood from the superior and inferior venae cavae empties into the right atrium.
5. Since pressure is greater on the right side of the heart (compared to the left side), most of the blood is shunted from the right atrium to the left atrium via the **foramen ovale**. This blood flows into the left ventricle and then is pumped out through the aorta.
6. A small amount of blood enters the right ventricle and pulmonary trunk, but much of this blood is shunted from the pulmonary trunk to the aorta through a vessel detour called the **ductus arteriosus** (ar-tēr′ē-ō′sŭs).
7. Blood travels to the rest of the body, and the deoxygenated blood returns to the placenta through a pair of **umbilical arteries.**
8. Nutrient and gas exchange occurs at the placenta, and the cycle repeats.

At birth, the fetal circulation begins to change into the postnatal pattern. When the baby takes its first breath, pulmonary resistance drops, and the pulmonary arteries dilate. As a result, pressure on the right side of the heart decreases so that the pressure is greater on the left side of the heart, which handles the systemic circulation.

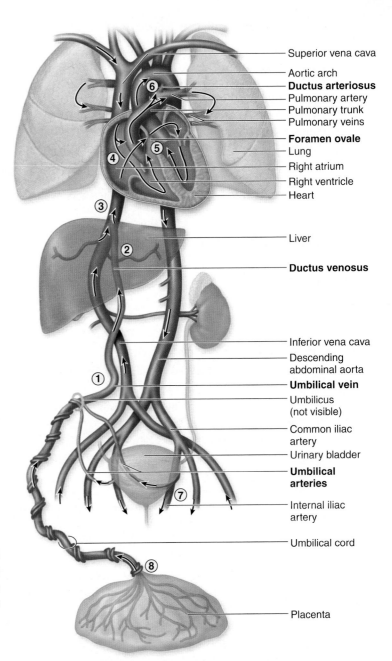

Fetal Cardiovascular Structure	Postnatal Structure
Ductus arteriosus	Ligamentum arteriosum
Ductus venosus	Ligamentum venosum
Foramen ovale	Fossa ovalis
Umbilical arteries	Medial umbilical ligaments
Umbilical vein	Round ligament of liver (ligamentum teres)

Figure 23.27

Fetal Circulation.   Structural changes in both the heart and blood vessels accommodate the different needs of the fetus and the newborn. The pathway of blood flow is indicated by black arrows. The numbers correspond to the steps listed in the text. The chart at the bottom of the drawing summarizes the fate of each of the fetal cardiovascular structures.

CLINICAL VIEW

Patent Ductus Arteriosus

In some infants (especially premature infants), the ductus arteriosus fails to constrict and close after birth. This open (patent) ductus arteriosus occurs in approximately 8 per 10,000 births. Since the systemic circulation is under higher pressure than the pulmonary circulation, a patent ductus arteriosus serves as a conduit through which blood from the aorta can enter the pulmonary system. If left untreated, this shunting will, over a period of several years, result in high blood pressure in the pulmonary circuit. This pulmonary hypertension then leads to failure of the right ventricle. Because circulating chemicals called prostaglandins help keep the ductus arteriosus open during fetal life, the first form of treatment for a patent ductus arteriosus is prostaglandin-inhibiting medication. In the uncommon instance that medication does not work, the ductus arteriosus is surgically repaired.

The postnatal changes occur as follows:

- The umbilical vein and umbilical arteries constrict and become nonfunctional. They turn into the **round ligament of the liver** (or *ligamentum* [lig′ă-men′tŭm; band] *teres*) and the **medial umbilical ligaments,** respectively.
- The ductus venosus ceases to be functional and constricts, becoming the **ligamentum venosum** (vē-nō′sŭm).
- Since pressure is now greater on the left side of the heart, the two flaps of the interatrial septum close off the foramen ovale. The only remnant of the foramen ovale is a thin, oval depression in the wall of the septum called the **fossa ovalis**.

- Within 10–15 hours after birth, the ductus arteriosus closes and becomes a fibrous structure called the **ligamentum arteriosum.**

WHAT DID YOU LEARN?

16 The hepatic portal system is formed primarily from what embryonic vein system?

17 Each medial umbilical ligament is a remnant of what embryonic vessel?

CLINICAL TERMS

hypotension Low blood pressure.

vasculitis (vas-kū-lī′tis) Inflammation of any type of blood vessel. If only arteries are inflamed, the condition is called *arteritis*; if only veins are inflamed, it is called *phlebitis*.

CHAPTER SUMMARY

	▪ Blood vessels form a closed supply system to transport oxygen and nutrients to body tissues, and remove waste products from these tissues.
Anatomy of Blood Vessels 682	▪ Arteries conduct blood away from the heart; capillaries exchange gases, nutrients, and wastes with body tissues; and veins conduct blood to the heart.
	Blood Vessel Tunics 682
	▪ The tunica intima (innermost layer) is composed of an endothelium, a basement membrane, and a layer of areolar connective tissue.
	▪ The tunica media (middle layer) is composed of smooth muscle. This is the largest tunic in an artery.
	▪ The tunica externa (outermost layer) is composed of areolar connective tissue and adipose connective tissue. This is the largest tunic in a vein.
	▪ Capillaries have a tunica intima, composed of an endothelial layer and a basement membrane only.
	Arteries 683
	▪ Elastic arteries have the largest diameter and the greatest proportion of elastic fibers in their walls.
	▪ Muscular arteries are medium-sized arteries with more smooth muscle and fewer elastic fibers to ensure vasodilation and vasoconstriction.
	▪ Arterioles are the smallest arteries.
	Capillaries 686
	▪ Capillaries, the smallest blood vessels, connect arterioles with venules. Gas and nutrient exchange occurs in the capillaries.
	▪ The three types of capillaries are continuous capillaries, fenestrated capillaries, and sinusoids.
	Veins 687
	▪ Venules are small veins that merge into larger veins. Blood pressure is low in the veins, which act as reservoirs and hold about 60% of the body's blood at rest.
	▪ One-way valves prevent blood backflow in veins.
Blood Pressure 689	▪ Blood pressure is the force exerted by the blood on the vessel wall. Systolic blood pressure is a measure of pressure during ventricular contraction, and diastolic pressure is a measure of pressure during ventricular relaxation.
Systemic Circulation 690	▪ The systemic circulation conducts oxygenated blood to and deoxygenated blood from peripheral capillary beds.
	General Arterial Flow Out of the Heart 691
	▪ The ascending aorta gives off the left and right coronary arteries to supply the heart.
	▪ The aortic arch has three branches: the brachiocephalic trunk, the left common carotid artery, and the left subclavian artery.
	▪ The descending thoracic aorta extends several branches to supply the thoracic wall.
	▪ The descending abdominal aorta bifurcates into common iliac arteries; these vessels divide into internal and external iliac arteries.

General Venous Return to the Heart 691

 ■ Deoxygenated blood returns to the heart via the superior and inferior venae cavae.

Blood Flow Through the Head and Neck 691

 ■ Common carotid arteries branch into the internal and external carotid arteries, which supply most of the blood to the head and neck.

 ■ The cerebral arterial circle is an arterial anastomosis that supplies the brain.

 ■ Vertebral veins and the dural venous sinuses drain the cranium.

Blood Flow Through the Thoracic and Abdominal Walls 695

 ■ The thoracic and abdominal walls are supplied by paired arteries.

 ■ Hemiazygos and accessory hemiazygos veins drain the left side of the thorax, and the azygos vein drains the right side of the thorax.

Blood Flow Through the Thoracic Organs 698

 ■ Bronchial arteries and bronchial veins supply the connective tissue, bronchi, and bronchioles of the lung.

 ■ Esophageal arteries and veins supply the esophagus.

 ■ Superior phrenic arteries, the musculophrenic arteries, and the inferior phrenic arteries and veins supply the diaphragm.

Blood Flow Through the Gastrointestinal Tract 699

 ■ Three unpaired arteries supply the gastrointestinal tract organs: the celiac trunk, the superior mesenteric artery, and the inferior mesenteric artery.

 ■ The hepatic portal vein is a large vein that receives oxygen-poor but nutrient-rich blood from the gastrointestinal organs and takes it to the liver. Blood exits the liver via hepatic veins.

Blood Flow Through the Posterior Abdominal Organs, Pelvis, and Perineum 703

 ■ Paired branches of the descending abdominal aorta supply the posterior abdominal organs and the pelvis and perineum. Venous drainage is by veins of the same name as the arteries.

Blood Flow Through the Upper Limb 703

 ■ The subclavian artery continues as the axillary artery and then becomes the brachial artery. The brachial artery divides into an ulnar artery and a radial artery.

 ■ Anastomoses of ulnar and radial arteries form the superficial palmar arch and the deep palmar arch; digital arteries emerge from the arches to supply the fingers.

 ■ The superficial group of veins contains the basilic, median cubital, and cephalic veins that drain into the axillary vein. The deep group of veins contains veins that bear the same names as the arteries.

Blood Flow Through the Lower Limb 707

 ■ The external iliac artery extends inferior to the inguinal ligament and is renamed the femoral artery. It enters the popliteal fossa, and then becomes the popliteal artery before dividing into anterior and posterior tibial arteries. The posterior tibial artery gives off a fibular artery. The posterior tibial artery branches into medial and lateral plantar arteries.

 ■ The superficial group of veins includes the great saphenous vein and the small saphenous vein. The deep group of veins consists of veins that bear the same names as the corresponding arteries.

Pulmonary Circulation 710

 ■ The pulmonary circulation carries deoxygenated blood to the lungs and returns oxygenated blood to the heart.

Review of Heart, Systemic, and Pulmonary Circulation 712

 ■ Oxygenated blood is pumped from the left ventricle through the systemic circulation and back to the right side of the heart. This deoxygenated blood is pumped from the right ventricle through the pulmonary circulation and returns as oxygenated blood to the left side of the heart.

Aging and the Cardiovascular System 712

 ■ As adults get older, the heart and blood vessels become less resilient, systolic blood pressure may rise, and the incidence and severity of atherosclerosis increase.

Blood Vessel Development 713

 ■ The blood vessels form by a process called vasculogenesis beginning in the third week of development.

Artery Development 713

 ■ The right dorsal aorta regresses, and the left dorsal aorta (plus the common aorta) form the descending aorta.

 ■ Aortic arch vessels 1, 2, 3, 4, and 6 develop into parts of adult arteries in the head, neck, and thorax.

Vein Development 714

 ■ Three bilateral venous systems connect the sinus venosus of the heart: the vitelline system, the umbilical system, and the cardinal system. These are the origin of the venous system.

Comparison of Fetal and Postnatal Circulation 716

 ■ The fetal cardiovascular system contains some structures that are modified or cease to function once the baby is born.

CHALLENGE YOURSELF

Matching

Match each numbered item with the most closely related lettered item.

_____ 1. hepatic portal vein

_____ 2. capillary

_____ 3. median cubital vein

_____ 4. common iliac artery

_____ 5. dural venous sinus

_____ 6. azygos vein

_____ 7. hemiazygos vein

_____ 8. popliteal artery

_____ 9. brachiocephalic trunk

_____ 10. pulmonary vein

a. common site for venipuncture

b. drains venous blood from the brain

c. drains right posterior intercostal veins

d. sends oxygenated blood to right upper limb

e. continuation of femoral artery

f. drains directly into left atrium

g. composed of endothelium and basement membrane only

h. bifurcation of descending abdominal aorta

i. superior mesenteric vein drains into it

j. left posterior intercostal veins drain into it

Multiple Choice

Select the best answer from the four choices provided.

_____ 1. Which of the following is not a type of capillary?
a. continuous
b. sinusoid
c. elastic
d. fenestrated

_____ 2. Some venous blood from the upper limb drains through the
a. cephalic vein.
b. great saphenous vein.
c. external jugular vein.
d. inferior vena cava.

_____ 3. All of the following are direct branches of the celiac trunk except the
a. splenic artery.
b. right gastric artery.
c. left gastric artery.
d. common hepatic artery.

_____ 4. Which type of vessel has a large number of smooth muscle cell layers in its tunica media as well as elastic tissue confined to an internal elastic lamina and external elastic lamina?
a. elastic artery
b. muscular artery
c. arteriole
d. venule

_____ 5. Which statement is true about veins?
a. Veins always carry deoxygenated blood.
b. Veins drain into smaller vessels called venules.
c. The largest tunic in a vein is the tunica externa.
d. The lumen of a vein tends to be smaller than that of a comparably sized artery.

_____ 6. Circle the correct pathway that blood follows through the upper limb arteries:
a. subclavian → axillary → ulnar → radial → brachial
b. subclavian → axillary → brachial → cephalic → basilic
c. subclavian → ulnar → brachial → radial
d. subclavian → axillary → brachial → radial and ulnar

_____ 7. Which of the following veins typically does not drain directly into the inferior vena cava?
a. renal
b. hepatic portal
c. common iliac
d. right gonadal

_____ 8. After birth, the umbilical vein becomes the
a. medial umbilical ligament.
b. ligamentum venosum.
c. ligamentum arteriosum.
d. round ligament of the liver.

_____ 9. The left fourth aortic arch vessel in an embryo becomes the
a. left common carotid artery.
b. left subclavian artery.
c. aortic arch.
d. left pulmonary artery.

_____ 10. Vasa vasorum are found in the tunica _____ of a large blood vessel.
a. intima
b. media
c. externa
d. All of these are correct.

Content Review

1. List and describe the three tunics in most blood vessels.

2. Compare and contrast arteries and veins with respect to function, tunic size, and lumen size.

3. Describe the three types of arteries, and give an example of each.

4. What is the main function of capillaries? What are the three kinds of capillaries?

5. Is blood pressure higher in arteries or veins? What are the consequences of hypertension?

6. Identify the three main branches of the aortic arch that receive oxygenated blood, and identify the areas of the body they supply.

7. How is blood flow through the upper and lower limbs similar?

8. Compare the systemic and pulmonary circulations. Discuss the function of arteries and veins in each system.

9. How does aging affect blood vessel anatomy and function?

10. What postnatal changes occur in the heart and blood vessels? Why do these occur?

Developing Critical Reasoning

1. Two 50-year-old men are trying to determine their risk for developing atherosclerosis. John jogs three times a week, maintains a healthy weight, and eats a diet low in saturated fats. Thomas rarely exercises, is overweight, and only occasionally eats healthy meals. Based on your knowledge of the cardiovascular system and atherosclerosis, which man do you think is more at risk for developing the disease? What other factors could put a person at risk for atherosclerosis?

2. Arteries tend to have a lot of vascular anastomoses around body joints (such as the elbow and knee). Propose a reason why this would be beneficial.

3. The internal thoracic artery is frequently used as a coronary bypass vessel (a replacement artery for a blocked coronary artery). What makes this vessel a good choice for this surgery? Will blood flow to the thoracic wall be compromised as a result? Why or why not?

A N S W E R S   T O   " W H A T   D O   Y O U   T H I N K ? "

1. A smoker would have elevated blood pressure, since nicotine increases cardiac output and causes vasoconstriction.

2. Blood could still reach the brain through the vertebral arteries. However, it is unlikely that these arteries could provide sufficient blood to the entire brain and head.

3. If the left ulnar artery were cut, the left hand and fingers could still receive blood via the left radial artery.

4. If the right femoral artery were blocked, blood flow to the right leg would be cut off; in other words, the popliteal artery and the branches to the leg would not receive any blood.

Visit the McKinley/O'Loughlin *Human Anatomy*, 2e website at aris.mhhe.com

24

LYMPHATIC SYSTEM

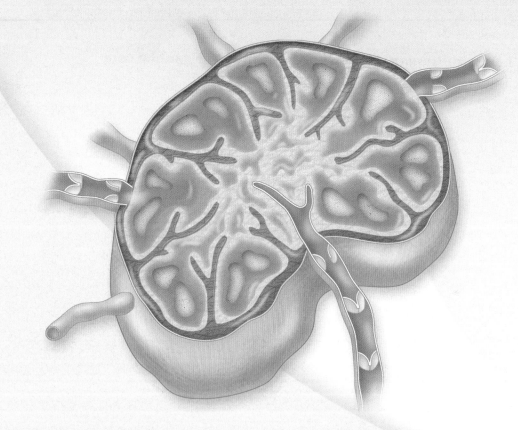

Lymphatic System

Whave seen in chapters 21–23 how the cardiovascular system transports blood throughout the body, where it exchanges gases and nutrients with the tissues. Another body system, called the lymphatic system, assists the cardiovascular system by transporting excess interstitial fluid through lymph vessels. Once this fluid enters the vessels, the fluid is renamed lymph. Along the way, lymph is filtered and checked for foreign or pathologic material, such as bacteria and cancer cells. Lymphatic structures contain certain cells that initiate an immune response to abnormal materials and perform other functions essential to homeostasis and survival. Without the primary immune response by the lymphatic system, the body would be unable to fight infection and keep itself healthy.

In this chapter we examine the lymph vessels, lymphatic structures, and lymphatic organs of the body, and learn how each of these components plays an important role in keeping us healthy.

Functions of the Lymphatic System

Key topic in this section:

■ How the lymphatic system aids homeostasis and guards the health of body cells and tissues

The **lymphatic** (lim-fat'ik) **system** involves several organs as well as a system of lymphatic cells and lymph vessels located throughout the body **(figure 24.1)**. Together, these structures transport fluids and help the body fight infection. However, not all of these components of the lymphatic system are involved in each function.

At the arterial end of a capillary bed, blood pressure forces fluid from the blood into the interstitial spaces around cells. This fluid is called **interstitial fluid** (not to be confused with *extracellular fluid*, a term that encompasses both interstitial fluid and plasma [see chapter 2]). Most of this fluid is reabsorbed at the venous end of the capillaries, but an excess of about 3 liters of fluid per day remains in the interstitial spaces. A network of **lymph vessels** (figure 24.1) reabsorbs this excess fluid and returns it to the venous circulation. If this excess fluid were not removed, body tissues would swell, a condition called **edema** (e-dě'mă; *oidema* = a swelling). Further, this excess fluid would accumulate outside the bloodstream, causing blood levels to drop precipitously. Thus, the lymphatic system prevents interstitial fluid levels from rising out of control and helps maintain blood volume levels.

Lymph vessels also transport dietary lipids. Although most nutrients are absorbed directly into the bloodstream, some larger materials, such as lipids and lipid-soluble vitamins, are unable to enter the bloodstream directly from the gastrointestinal (GI) tract. These materials are transported through tiny lymph vessels called lacteals, which drain into larger lymph vessels and eventually into the bloodstream.

Lymphatic organs house lymphocytes. While some lymphocytes circulate in the bloodstream, most are located in the lymphatic structures and organs. Some lymphatic organs assist in these cells' maturation, while others serve as a site for lymphocyte replication (mitosis).

Finally, the lymphatic system generates an immune response and increases the lymphocyte population when necessary. Lymphatic structures contain T-lymphocytes, B-lymphocytes, and macrophages (monocytes that have migrated from the bloodstream into other tissues). These cells are constantly monitoring the blood and the interstitial fluid for **antigens** (an'ti-gen; *anti*(body) + *gen* = producing), which are any substances perceived as abnormal to the body, such as bacteria, viruses, and even cancer cells. If antigens are discovered, lymphatic cells initiate a systematic defense against the antigens, called an **immune** (i-mūn') **response**. Some of the cells produce

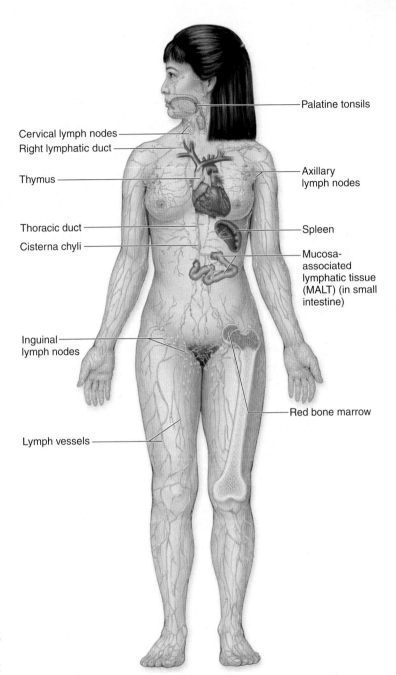

Figure 24.1

Lymphatic System.   The lymphatic system consists of lymph vessels, lymphatic cells, and lymphatic organs that work together to pick up and transport interstitial fluid back to the blood and to mount an immune response when needed.

soluble proteins called **antibodies** that bind to and immobilize the foreign or abnormal agent, thus damaging it or identifying it to other elements of the immune system. Other cells attack and destroy the antigen directly. Still other cells become memory cells, which remember the past antigen encounters and initiate an even faster and more powerful response should the same antigen appear again.

WHAT DID YOU LEARN?

❶ What is the "immune response," and how does the lymphatic system initiate it?

Lymph and Lymph Vessels

Key topics in this section:

- Components of lymph
- Path of lymph from interstitial tissues to the bloodstream

Excess interstitial fluid and solutes are returned to the bloodstream through a lymph vessel network. When the combination of interstitial fluid, solutes, and sometimes foreign material enters the lymph vessels, the liquid mixture is called **lymph** (limf; *lympha* = clear spring water). The lymph vessel network is composed of increasingly larger vessels, as follows (from smallest to largest in diameter): lymphatic capillaries, lymphatic vessels, lymphatic trunks, and lymphatic ducts. Thus, the term "lymph vessel" is a general term to describe all of these specific lymphatic capillaries, lymphatic vessels, trunks, and ducts.

Lymphatic Capillaries

The lymph vessel network begins with microscopic vessels called **lymphatic capillaries.** Lymphatic capillaries are closed-ended tubes that are interspersed among most blood capillary networks **(figure 24.2)**, except those within the red bone marrow and central nervous system. In addition, avascular tissues (such as epithelia) lack lymphatic capillaries. A lymphatic capillary is similar to a blood capillary in that its wall is an endothelium. However, lymphatic capillaries tend to be larger in diameter, lack a basement membrane, and have overlapping endothelial cells. **Anchoring filaments** help hold these endothelial cells to the nearby structures. These overlapping endothelial cells act as one-way flaps; when interstitial fluid pressure rises, the margins of the endothelial cells push into the lymphatic capillary lumen and allow interstitial fluid to enter. When the pressure increases

in the lymphatic capillary, the cell wall margin pushes back into place next to the adjacent endothelial cell. The fluid that is now "trapped" in the lymph capillary cannot be released back into the interstitial spaces. This process is analogous to the movement of the entryway door to your house or apartment. Imagine that the door is unlocked and the knob is turned. Putting pressure on the outside of the door (like the pressure of interstitial fluid on the outside of the lymphatic capillary wall) causes it to open to the inside so you can enter. Once inside, pressure applied to the inside surface of the door (or fluid pressure against the inside lymphatic capillary surface) causes it to close.

The small intestine (part of the GI tract) contains special types of lymphatic capillaries called **lacteals** (lak'tē-ăl; *lactis* = milk). Lacteals pick up not only interstitial fluid, but also dietary lipids and lipid-soluble vitamins (vitamins that must be dissolved in lipids before they can be absorbed). The lymph from the GI tract has a milky color due to the lipid, and for this reason the GI tract lymph is also called **chyle** (kīl; *chylos* = juice).

Lymphatic Vessels

Lymphatic capillaries merge to form larger structures called **lymphatic vessels.** Lymphatic vessels resemble small veins, in that both contain three tunics (intima, media, and externa) and both have **valves** within the lumen. Since the lymph vessel network is a low-pressure system, valves prevent lymph from pooling in the vessel and help prevent lymph backflow **(figure 24.3)**. These valves are especially important in areas where lymph flow is against the direction of gravity. Contraction of nearby skeletal muscles also helps move lymph through the vessels.

Some lymphatic vessels connect directly to lymphatic organs called lymph nodes. **Afferent lymphatic vessels** bring lymph to a lymph node where it is filtered for foreign or pathogenic material.

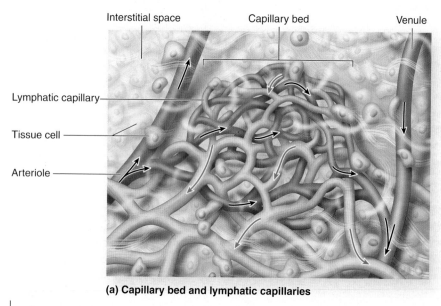

Interstitial space · Capillary bed · Venule · Lymphatic capillary · Tissue cell · Arteriole

(a) Capillary bed and lymphatic capillaries

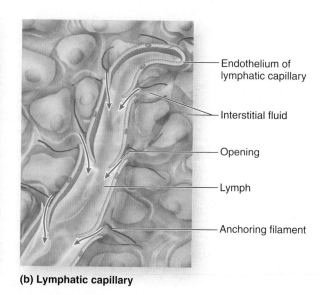

Endothelium of lymphatic capillary · Interstitial fluid · Opening · Lymph · Anchoring filament

(b) Lymphatic capillary

Figure 24.2

Lymphatic Capillaries. *(a)* Lymphatic capillaries arise as blind-ended vessels in connective tissue spaces among most blood capillary networks. Here, the black arrows show blood flow and the green arrows show lymph flow. *(b)* A lymphatic capillary takes up interstitial fluid through one-way flaps in its endothelial lining.

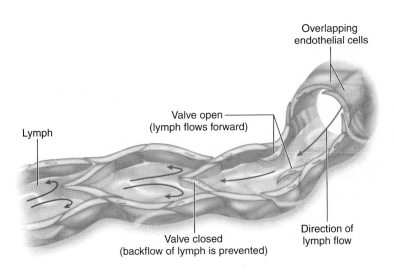

Overlapping
endothelial cells

Valve open
(lymph flows forward)

Lymph

Direction of
lymph flow

Valve closed
(backflow of lymph is prevented)

(a) Lymphatic vessel, longitudinal section

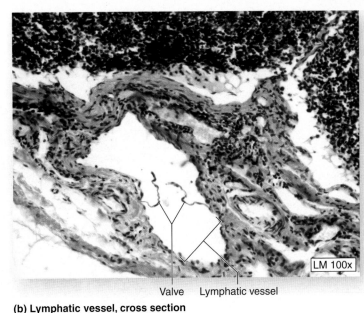

LM 100x

Valve   Lymphatic vessel

(b) Lymphatic vessel, cross section

Figure 24.3

Lymphatic Vessels and Valves. *(a)* Lymphatic vessels contain valves to prevent backflow of lymph. *(b)* Histologic cross section of a lymphatic vessel.

Once filtered, the lymph exits the lymph node via **efferent lymphatic vessels**. Lymph nodes are often found in clusters, so after one lymph node receives and filters lymph, the lymph is passed to another lymph node in the cluster, then to another lymph node, and so on. Thus, lymph is repeatedly examined for the presence of foreign or pathogenic materials.

Lymphatic Trunks

Left and right **lymphatic trunks** form from merging lymphatic vessels **(figure 24.4)**. Each lymphatic trunk drains lymph from a major body region, as follows:

- **Jugular trunks** drain lymph from the head and neck.
- **Subclavian trunks** drain lymph from the upper limbs, breasts, and superficial thoracic wall.
- **Bronchomediastinal trunks** drain deep thoracic structures.
- **Intestinal trunks** drain most abdominal structures.
- **Lumbar trunks** drain the lower limbs, abdominopelvic wall, and pelvic organs.

Lymphatic Ducts

Lymphatic trunks drain into the largest vessels, called **lymphatic ducts.** The two lymphatic ducts empty lymph back into the venous circulation. The **right lymphatic duct** is located near the right clavicle. The right lymphatic duct returns the lymph into the junction of the right subclavian vein and the right internal jugular vein. It receives lymph from the lymphatic trunks that drain the right

side of the head and neck, right upper limb, and right side of the thorax. The **thoracic duct** is the largest lymphatic vessel, with a length of about 37.5–45 centimeters (15–18 inches). At the base of the thoracic duct and anterior to the L_2 vertebra is a rounded, saclike structure called the **cisterna chyli** (sis-ter′nă kī′lī; cistern). The cisterna chyli gets its name from the milky lymph called chyle it receives from the small intestine. Left and right intestinal and lumbar trunks drain into the cisterna chyli. The thoracic duct travels superiorly from the cisterna chyli and lies directly anterior to the vertebral bodies. It passes through the aortic opening of the diaphragm, and then it ascends to the left of the vertebral body midline. It drains lymph into the junction of the left subclavian vein and left internal jugular vein. The thoracic duct receives lymph from most regions of the body, including the left side of the head and neck, left upper limb, left thorax, and all body regions inferior to the diaphragm (including the right lower limb and right side of the abdomen).

WHAT DID YOU LEARN?

2 What is lymph?

3 Describe the structure of lymphatic capillaries. Into what structures do they drain?

4 Which major body regions drain lymph to the right lymphatic duct?

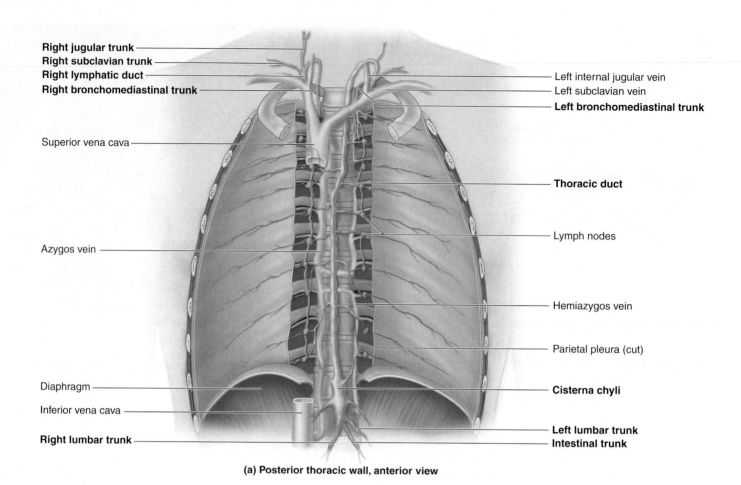

Right jugular trunk
Right subclavian trunk
Right lymphatic duct
Right bronchomediastinal trunk

Left internal jugular vein
Left subclavian vein
Left bronchomediastinal trunk

Superior vena cava

Thoracic duct

Lymph nodes

Azygos vein

Hemiazygos vein

Parietal pleura (cut)

Diaphragm

Cisterna chyli

Inferior vena cava

Left lumbar trunk
Intestinal trunk

Right lumbar trunk

(a) Posterior thoracic wall, anterior view

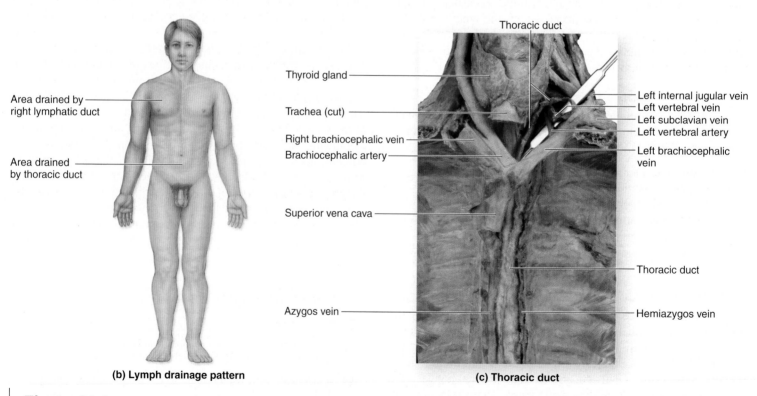

Thoracic duct

Area drained by
right lymphatic duct

Thyroid gland

Left internal jugular vein
Left vertebral vein
Left subclavian vein
Left vertebral artery

Trachea (cut)

Area drained
by thoracic duct

Right brachiocephalic vein
Brachiocephalic artery

Left brachiocephalic
vein

Superior vena cava

Thoracic duct

Azygos vein

Hemiazygos vein

(b) Lymph drainage pattern

(c) Thoracic duct

Figure 24.4

Lymphatic Trunks and Ducts. Lymph drains from lymphatic trunks into lymphatic ducts that each empty into the junctions of the internal jugular and subclavian veins. *(a)* An anterior view of the posterior thoracic wall illustrates the major lymphatic trunks and ducts. *(b)* Pattern of lymph drainage into the right lymphatic duct and the thoracic duct. *(c)* A cadaver photo of the thoracic duct.

Lymphedema

Lymphedema (limf′e-dē′mă) refers to an accumulation of interstitial fluid that occurs due to interference with lymphatic drainage in a part of the body. As the interstitial fluid accumulates, the affected area swells and becomes painful. If lymphedema is left untreated, the protein-rich interstitial fluid may interfere with wound healing and can even contribute to an infection by acting as a growth medium for bacteria.

Most cases of lymphedema are *obstructive,* meaning they are caused by blockage of lymph vessels. There are several causes of obstructive lymphedema:

- Any surgery that requires removal of a group of lymph nodes (e.g., breast cancer surgery when the axillary lymph nodes are removed) puts an individual at increased risk for lymphedema.
- The spread of malignant tumors within the lymph nodes and/or lymph vessels can obstruct lymphatic drainage.
- Radiation therapy may cause scar formation that interferes with lymphatic drainage.
- Trauma or infection of the lymph vessels obstructs lymphatic drainage.

In addition, millions of individuals in Southeast Asia and Africa have developed lymphedema as a result of infection by threadlike parasitic filarial worms. **Lymphatic filariasis** (fil-ă-rī′ă-sis; *filum* = thread) is a type of lymphedema whereby filarial worms lodge in the lymphatic system, live and reproduce there for years, and eventually obstruct lymphatic drainage. Some filarial worms gain entrance to the body through cracks in the skin of the foot, which is why many cases of lymphedema in the foot are seen. However, mosquitoes are the most common vector for transmitting filariasis. Once the mature worms have entered the body, they become permanent "residents." An affected body part can swell to many times its normal size. In these extreme cases, the condition also is known as **elephantiasis** (el-ĕ-fan-tī′ă-sis; *elephas* = elephant). Patients are treated with medications to kill the filarial worms, although the damage to the lymphatic system may be irreversible.

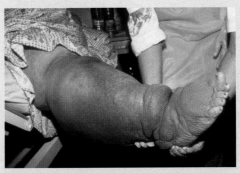

Elephantiasis (lymphatic filariasis) of the lower limb.

Lymphedema has no cure, but it can be controlled. Patients may wear compression stockings or other compression garments to reduce swelling and assist interstitial fluid return to the circulation. Certain exercise regimens may improve lymphatic drainage as well. Ideally, an individual with any symptoms of lymphedema, such as swelling and pain in a body region or skin feeling tight, should seek medical assistance quickly in order for treatment to be most effective.

Lymphatic Cells

Key topics in this section:

- Types of lymphatic cells
- Function of lymphocytes in the body's immune response
- Lymphocyte formation

Lymphatic cells (also called *lymphoid cells*) are located in both the lymphatic system and the cardiovascular system. The lymphatic cells work together to elicit an immune response. Among the types of lymphatic cells are macrophages, some epithelial cells, dendritic cells, and lymphocytes.

Macrophages (mak′rō-faj; *macros* = large, *phago* = to eat) are monocytes that have migrated into the lymphatic system from the bloodstream; they are responsible for phagocytosis of foreign substances. Special epithelial cells called **nurse cells** are found in the thymus, where they secrete thymic hormones. **Dendritic** (den-drit′ik) **cells** are found in the lymphatic nodules of a lymph node; they internalize antigens from the lymph and present them to other lymphatic cells. (Recall from chapter 5 that dendritic cells within the skin epidermis perform the same function.) **Lymphocytes** are the most abundant cells in the lymphatic system. There are three types of lymphocytes, and each has a specific job in the overall immune response.

Types and Functions of Lymphocytes

The three types of lymphocytes are T-lymphocytes (also called *T-cells*), B-lymphocytes (also called *B-cells*), and NK cells. All three types migrate through the lymphatic system and search for antigens.

> ### Study Tip!
> Lymphocytes are identified according to the tissue or organ where they mature:
>
> **T**-lymphocytes mature in the **T**hymus.
>
> **B**-lymphocytes mature in the **B**one marrow.

T-lymphocytes

T-lymphocytes make up about 70–85% of body lymphocytes. The lymphocyte plasma membrane contains a coreceptor that can recognize a particular antigen. (Coreceptors are named with the letters "CD" followed by a number.) There are several types of T-lymphocytes, each with a particular kind of coreceptor. The two main groups are helper T-lymphocytes and cytotoxic T-lymphocytes **(figure 24.5)**.

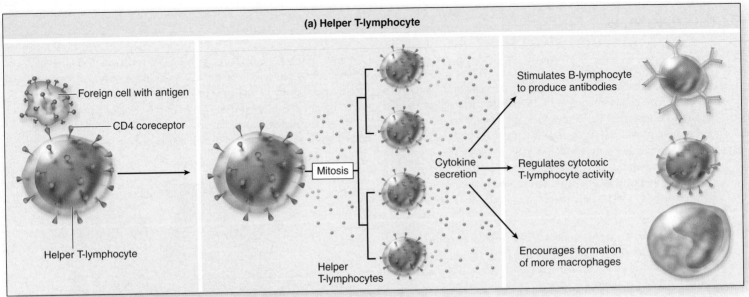

(a) Helper T-lymphocyte

Foreign cell with antigen

CD4 coreceptor

Helper T-lymphocyte

Mitosis

Helper T-lymphocytes

Cytokine secretion

Stimulates B-lymphocyte to produce antibodies

Regulates cytotoxic T-lymphocyte activity

Encourages formation of more macrophages

1 Helper T-lymphocyte recognizes antigen.

2 Helper T-lymphocyte secretes cytokines and begins to undergo mitosis to form more helper T-lymphocytes.

3 Cytokines secreted by helper T-lymphocytes initiate and control the immune response.

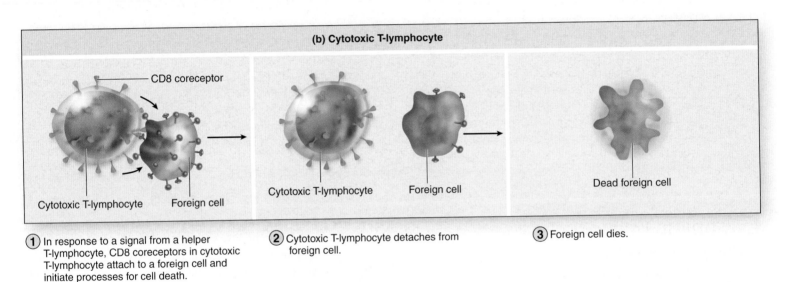

(b) Cytotoxic T-lymphocyte

CD8 coreceptor

Cytotoxic T-lymphocyte   Foreign cell

Cytotoxic T-lymphocyte   Foreign cell

Dead foreign cell

1 In response to a signal from a helper T-lymphocyte, CD8 coreceptors in cytotoxic T-lymphocyte attach to a foreign cell and initiate processes for cell death.

2 Cytotoxic T-lymphocyte detaches from foreign cell.

3 Foreign cell dies.

Figure 24.5

Types of T-lymphocytes and Their Role in the Immune Response. *(a)* Helper T-lymphocytes recognize antigens and then secrete cytokines to initiate both the maturation of immune defense cells and the immune response. *(b)* Cytotoxic T-lymphocytes recognize foreign antigens and directly attack and kill foreign cells, thereby reducing threats by pathogens.

Helper T-lymphocytes are needed to begin an effective defense against antigens. They primarily contain the CD4 coreceptor. For this reason, helper T-lymphocytes are also called *CD4+ cells*, or *T4 cells*. There are many kinds of helper T-lymphocytes in the body, and each is activated by and responds to one type of antigen only. For example, one type of helper T-lymphocyte may respond to the chickenpox virus, but this same helper T-lymphocyte will not be activated if it comes across *Streptococcus* bacteria. In essence, helper T-lymphocytes initiate and oversee the immune response; in other words, they are the "conductors" in the immune response "symphony." Helper T-lymphocytes regulate the immune response using two methods. The first method is to present an antigen to other lymphatic cells. The second method is to secrete **cytokines** (sī′tō-

kin; *kinesis* = movement), which are chemical signals that bind to receptors on other lymphatic cells and activate them.

Cytotoxic T-lymphocytes, also called *CD8+ cells* or *T8 cells*, primarily contain the CD8 coreceptor. These lymphocytes come in direct contact with infected or foreign cells and kill them. Each type of cytotoxic T-lymphocyte responds to one type of antigen only. Cytotoxic T-lymphocytes can kill in either of two ways: by secreting substances into abnormal cells that cause unregulated entry of material into the cell, which may cause cell swelling and bursting, or by triggering cell death directly. A cytotoxic T-lymphocyte acts only after it is activated by a helper T-lymphocyte.

In addition to the two main groups, other subsets of T-lymphocytes include memory T-lymphocytes and suppressor T-lymphocytes.

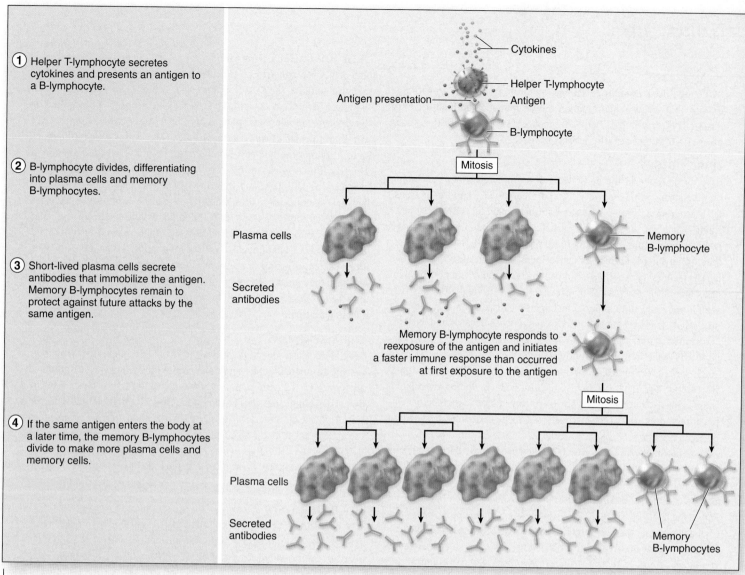

① Helper T-lymphocyte secretes cytokines and presents an antigen to a B-lymphocyte.

② B-lymphocyte divides, differentiating into plasma cells and memory B-lymphocytes.

③ Short-lived plasma cells secrete antibodies that immobilize the antigen. Memory B-lymphocytes remain to protect against future attacks by the same antigen.

④ If the same antigen enters the body at a later time, the memory B-lymphocytes divide to make more plasma cells and memory cells.

Cytokines
Helper T-lymphocyte
Antigen presentation
Antigen
B-lymphocyte
Mitosis
Plasma cells
Memory B-lymphocyte
Secreted antibodies

Memory B-lymphocyte responds to reexposure of the antigen and initiates a faster immune response than occurred at first exposure to the antigen

Mitosis
Plasma cells
Secreted antibodies
Memory B-lymphocytes

Figure 24.6

B-lymphocytes and Their Role in the Immune Response.   B-lymphocytes are activated by helper T-lymphocytes when presented with an antigen. Their response to primary and secondary exposure to an antigen is shown here in a series of steps.

Memory T-lymphocytes arise from some cytotoxic T-lymphocytes that previously destroyed a foreign cell. They patrol the body, and if they encounter the same antigen again, they mount an even faster immune response than occurred at the first exposure to the antigen. **Suppressor** (soo-pres′ŏr) **T-lymphocytes** often contain the CD8 coreceptor and appear to "turn off" the immune response once it has been activated to help regulate its performance. Some other T-lymphocyte types not mentioned here also populate the body.

B-lymphocytes

B-lymphocytes make up about 15–30% of the lymphocytes in the body. B-lymphocytes contain antigen receptors that respond to one particular antigen and stimulate the production of **immunoglobulins (Ig)** (im′ū-nō-glob′ū-lin), or **antibodies,** that respond to that particular antigen. There are five main classes of immunoglobulins based on the order of amino acids in their composition. These classes, from most common in the plasma to least common, are IgG, IgA, IgM,

IgD, and IgE. These immunoglobulins are released by the specific B-lymphocytes to immobilize or neutralize specific antigens. Typically, a B-lymphocyte cannot be activated until a helper T-lymphocyte presents it with an antigen. Once it is activated, the B-lymphocyte undergoes cell division and differentiates into one of two types of B-lymphocytes: plasma cells or memory B-lymphocytes (**figure 24.6**).

Most of the activated lymphocytes become **plasma** (plaz′ma) **cells,** mature cells that produce and secrete large amounts of antibodies. The antibodies then immobilize, neutralize, and destroy the particular antigen. Plasma cells may be either short-lived or long-lived. The short-lived plasma cells have a life span of less than a week, while long-lived plasma cells can live for months or years.

A few of the activated B-lymphocytes do not differentiate into plasma cells and instead become **memory B-lymphocytes**. These cells "remember" the initial antigen attack and stand guard to mount a faster, more efficient immune response should the same antigen strike again. If the antigen does strike again, the body responds so

CLINICAL VIEW: In Depth

HIV and AIDS

AIDS (acquired immunodeficiency syndrome) is a life-threatening disease that results from infection by the **human immunodeficiency virus (HIV)**. HIV targets helper T-lymphocytes; the loss of these cells gives rise to the devastating effects of AIDS.

EPIDEMIOLOGY

HIV can be found in the body fluids of an infected person, including blood, semen, vaginal secretions, and breast milk. The virus is transmitted during activities that allow intimate contact with these body fluids, such as unprotected sexual or anal intercourse, sharing hypodermic needles with other intravenous drug users, or breast-feeding an infant. Current evidence indicates that HIV is *not* spread by casual kissing, sharing eating utensils, using a public toilet, or other nonintimate types of physical contact. Although HIV was first seen in the early 1980s among the homosexual male population and IV drug users, it is now a major disease among heterosexual populations. The United Nations program on AIDS (UNAIDS) estimates that 90% of all HIV infections are currently transmitted heterosexually. Prior to 1985, before HIV and AIDS were well known, HIV could be transmitted through the donated blood supply. Individuals who received blood transfusions sometimes received HIV-infected blood, thereby becoming infected as well. This discovery led to more stringent screening of blood donors.

Since the early 1980s, over 60 million people have become infected with HIV, and more than 12 million have died. The incidence of AIDS is increasing throughout the world, but the disease is particularly rampant on the continents of Africa and Asia. Sub-Saharan Africa has been hit especially hard: 20% of all individuals are infected in South Africa, while Zimbabwe and Kenya have infection rates of 33%. The AIDS epidemic in Africa has led to massive numbers of deaths, and children are frequently orphaned as both parents succumb to the disease. Asian countries are also seeing a surge of new HIV and AIDS cases in recent years. Health officials are concerned that these numbers will quickly multiply unless preventive measures are taken soon.

PREVENTION

The key to limiting the spread of HIV infection is to refrain from behaviors that allow the virus to be transmitted. Unprotected intercourse (especially anal intercourse) and oral sex can spread HIV, so individuals should either practice abstinence or protect themselves by using condoms. (Other contraceptives, such as birth control pills, do *not* protect an individual from HIV infection.) Both partners in a monogamous relationship should be tested for the HIV virus via a simple blood test before engaging in sexual intercourse. Intravenous drug users should not share needles. As a precaution, health-care workers should wear gloves and be careful around patients' body fluids. HIV-infected pregnant women need special prenatal care to keep from transmitting the virus to their fetuses, and HIV-infected mothers are discouraged from breast-feeding, because the virus is present in breast milk.

HOW HIV CAUSES DAMAGE

The HIV virus consists of two identical copies of a single strand of genetic material (RNA) surrounded by an outer protein coat. A small part of this protein coat binds to the CD4 coreceptor on a helper T-lymphocyte. (Some macrophages also have a CD4 coreceptor, so HIV can bind to them as well.) After HIV attaches to the helper T-lymphocyte, it enters the helper T-lymphocyte, the protein coat is shed, and the HIV RNA is released into the helper T-lymphocyte. A DNA copy is made of the HIV RNA, and then the HIV DNA is incorporated into the helper T-lymphocyte's DNA. Thus, the helper T-lymphocyte becomes an "HIV factory" as it divides and produces new HIV viruses that will travel through the body and destroy other helper T-lymphocytes. Since helper T-lymphocytes oversee the body's immune response, their decrease results in a loss of normal immune function. Thus, the infected individual is prone to certain types of cancer and opportunistic infections, diseases that would normally be eradicated by a healthy immune system.

EARLY SYMPTOMS

Several weeks to several months after initial HIV infection, many individuals experience flulike symptoms, while others have no symptoms at all. Typically, the early symptoms disappear after a few weeks. Healthy helper T-lymphocytes divide to replace the cells that were lost in the initial phase of infection. However, in the long run, HIV continues to replicate at a faster rate than the immune system can replace the dying infected cells. Over a period of months to years, the population of helper T-lymphocytes drops to a dangerous level, setting the stage for AIDS.

HIV BLOOD TESTS

HIV blood tests detect the presence of HIV antibodies in the blood. It can take as long as 6 months for antibody levels in the blood to rise to a point where they can be detected by the blood test. Thus, individuals who have been exposed to HIV, but are tested within the first 6 months, may receive a false-negative result simply because the antibodies have not yet reached the detectable level. Even though the antibody test is negative at this early stage, the person can still infect others.

WHEN DOES HIV BECOME AIDS?

HIV is diagnosed as AIDS when a person's helper T-lymphocyte count drops to below 200 cells per cubic milliliter, when an opportunistic infection or related illness develops, or when a particular type of malignancy develops. Common opportunistic infections include pneumocystic pneumonia and histoplasmosis. Malignancies that tend to occur in people whose immune systems are weakened include Kaposi's sarcoma and non-Hodgkin lymphoma. Opportunistic infections and malignancies account for up to 80% of all AIDS-related deaths. In addition, many AIDS patients have some form of CNS complications, including meningitis, encephalitis, neurologic deficits, and neuropathies.

TREATMENT OPTIONS

HIV infection is a lifelong illness; there is no cure. Current pharmaceutic treatments alleviate symptoms or help prevent the spread of HIV infection in the body, but they cannot eradicate HIV from an infected individual. In addition, most of these drugs have numerous unpleasant side effects.

The first HIV drug treatment was AZT (Zidovudine), which helps prevent the HIV RNA from being transcribed into viral DNA. AZT can help contain the infection, but the HIV virus often develops resistance to it. Other newer HIV drugs target other cellular activities of HIV, helping prevent HIV from replicating in the helper T-lymphocytes. Combinations of these different drugs (called "drug cocktails") are often given to HIV patients to retard the development of drug resistance and to ensure more effective elimination of the viral copies from the blood (called viral load). Patients taking a triple combination of drugs typically experience a dramatic reduction in viral load and even a slight rise in their helper

T-lymphocyte count. However, they must take these drugs for life, or else the HIV and AIDS will progress.

Unfortunately, HIV drugs are expensive and not widely available in developing countries, where the need for them is greatest. One hopeful sign is that pharmaceutical companies are negotiating with the governments of developing countries to make cheaper forms of these drugs available. In addition, pharmaceutical companies are starting to work together to create better and easier-to-use medications. For example, in July 2006, the FDA approved an HIV medication that combines three HIV drugs in the "cocktails" in a single pill. This pill is marketed under the brand name *Atripla* and was produced through the collaboration of several drug companies. The single daily dose medication will make treating HIV in foreign countries much easier, since these drugs will be easier to distribute than multiple pills and patients will be more likely to comply with the simpler dosing schedule.

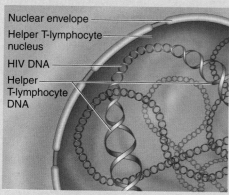

① HIV targets and attaches to CD4 coreceptor on helper T-lymphocyte.

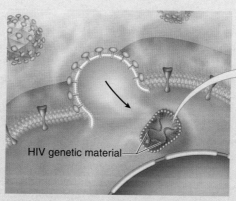

② HIV releases its genetic material into helper T-lymphocyte.

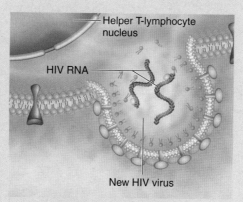

③ HIV DNA is made from HIV RNA.

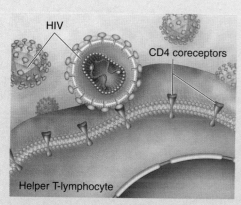

④ HIV DNA incorporates itself into the helper T-lymphocyte DNA.

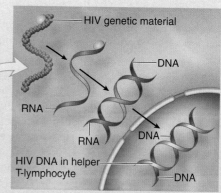

⑤ The helper T-lymphocyte becomes an "HIV-factory," producing HIV viruses that will be released from the helper T-lymphocyte and travel throughout the body.

HIV (human immunodeficiency virus) targets helper T-lymphocytes in a multistep process.

Table 24.1	Types of Lymphocytes	
Cell Type	**Function**	**Type of Antigen Response**
T-LYMPHOCYTE		
Helper T-lymphocyte	Initiates and oversees the immune response	Responds to a single antigen
Cytotoxic T-lymphocyte	Directly kills foreign cells; must be activated by a helper T-lymphocyte first	Responds to a single antigen
Memory T-lymphocyte	A type of cytotoxic T-lymphocyte that has already killed; patrols the body looking for the same antigen again	Responds to a single antigen
Suppressor T-lymphocyte	Helps "turn off" the immune response once it has been activated	Responds to a single antigen
B-LYMPHOCYTE		
Plasma cell	Produces and secretes antibodies	Responds to a single antigen
Memory B-lymphocyte	Remembers an initial antigen attack and mounts a faster, more efficient response should the same antigen type attack again	Responds to a single antigen
NK (NATURAL KILLER) CELL		
NK (natural killer) cell	Kills a wide variety of infected and cancerous cells	Responds to multiple antigens

quickly that no symptoms may occur. Memory B-lymphocytes have a much longer life span than plasma cells; some live for months or even years.

Some vaccines (e.g., polio vaccine, flu vaccine) introduce modified or dead forms of an antigen so that memory cells may be formed and the body can fight and eliminate the illness before any symptoms ever develop. Depending upon the life span of the particular memory B-lymphocytes, the vaccine may provide lifelong immunity, or periodic **booster shots** may be needed to ensure continued protection against the antigen.

✎? WHAT DO YOU THINK?

❶ Tetanus (commonly known as *lockjaw*) is a severe illness that causes painful muscle spasms and convulsions, and can lead to death. If adults are advised to get a tetanus booster shot about once every 10 years, what is the probable life span of the tetanus-detecting memory B-lymphocytes?

NK Cells

NK (natural killer) cells, also called *large granular lymphocytes,* make up the remaining small percentage of body lymphocytes. NK cells tend to have CD16 receptors. Unlike T- and B-lymphocytes, which respond to one particular antigen only, NK cells can kill a wide variety of infected cells and some cancerous cells.

Table 24.1 reviews the main types of lymphocytes and their functions.

Lymphopoiesis

Lymphopoiesis (lim-fō-poy-ē′sis) is the process of lymphocyte development and maturation. When a lymphocyte fully matures, it becomes **immunocompetent,** meaning that the lymphocyte is fully able to participate in the immune response. Immature lymphocytes cannot participate in the immune response. All lymphocyte types originate in the red bone marrow, but their maturation sites differ (**figure 24.7**).

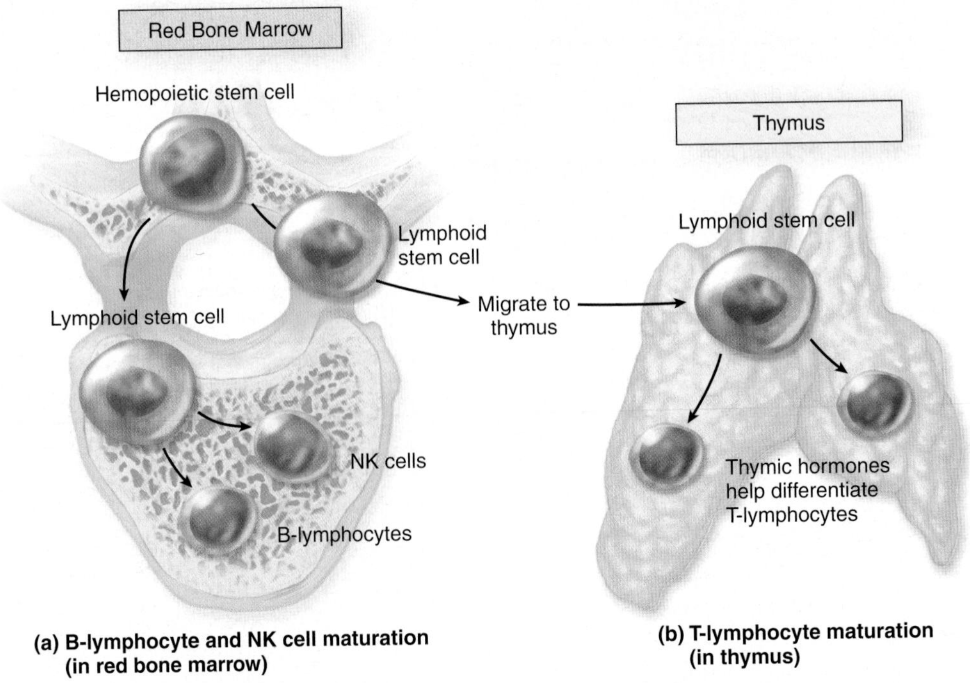

Figure 24.7

Lymphopoiesis. *(a)* B-lymphocytes and NK cells mature in the red bone marrow. *(b)* T-lymphocytes mature and differentiate in the thymus under the influence of thymic hormones.

(a) B-lymphocyte and NK cell maturation (in red bone marrow)

(b) T-lymphocyte maturation (in thymus)

In the red bone marrow, a hemopoietic stem cell gives rise to several types of immature blood precursor cells, including **lymphoid** (lim'foyd) **stem cells** (a slightly more differentiated type of stem cell). Some lymphoid stem cells remain in the red bone marrow and mature into B-lymphocytes and NK cells. Once the B-lymphocytes and NK cells mature, they migrate from the bone marrow, travel through the bloodstream, and enter lymphatic structures and lymphatic organs.

Other lymphoid stem cells leave the red bone marrow and migrate to the thymus for subsequent maturation. Under the influence of thymic hormones, these stem cells mature and differentiate into specific types of T-lymphocytes. Once the maturation and differentiation process is complete, the T-lymphocytes migrate to the other lymphatic structures in the body. The T-lymphocyte maturation process primarily takes place from childhood until puberty. Thereafter, the thymus regresses and becomes almost nonfunctional in the adult.

Most lymphocytes have long life spans, and some can live for many years. Once lymphocytes leave their maturation sites, they can proliferate through cell division. However, note that each type of lymphocyte can only replicate its own kind—that is, a B-lymphocyte cannot produce a T-lymphocyte, only other B-lymphocytes. In addition, mature helper T-lymphocytes can only divide into other mature helper T-lymphocytes, not other types of T-lymphocytes.

WHAT DID YOU LEARN?

5 List the main types of lymphatic cells.

6 List and describe the functions of the different types of T-lymphocytes and B-lymphocytes.

7 How and where are lymphocytes formed?

Lymphatic Structures

Key topics in this section:

- Structure and functions of lymphatic nodules
- Organs of the lymphatic system and their functions

Besides the lymphatic vessels, the lymphatic system consists of lymphatic nodules and various lymphatic organs (see figure 24.1).

Lymphatic Nodules

Lymphatic nodules (nod'ūl; *nodulus* = knot), or *lymphatic follicles*, are oval clusters of lymphatic cells with some extracellular connective tissue matrix that are *not* surrounded by a connective tissue capsule. The pale center of a lymphatic nodule is called the **germinal** (jer'mi-năl; *germen* = bud) **center;** it contains proliferating B-lymphocytes and some macrophages. T-lymphocytes are located outside the germinal center. Lymphatic nodules filter and attack antigens. Individually, a lymphatic nodule is small. However, in some areas of the body, many lymphatic nodules group together to form larger structures, such as mucosa-associated lymphatic tissue (MALT) or tonsils.

MALT (Mucosa-Associated Lymphatic Tissue)

Large collections of lymphatic nodules are located in the lamina propria of the mucosa of the gastrointestinal, respiratory, genital, and urinary tracts. Together, these collections of lymphatic nodules are called **MALT (mucosa-associated lymphatic tissue).** As food, air, and urine enter their respective tracts, the lymphatic cells in the MALT detect antigens and initiate an immune response. MALT is very prominent in the mucosa of the small intestine, primarily in the ileum. There, collections of lymphatic nodules called **Peyer patches** can become quite large and bulge into the gut lumen (**figure 24.8a**).

Tonsils

Tonsils (ton'sil; *tonsilla* = a stake) are large clusters of lymphatic cells and extracellular connective tissue matrix that are not completely surrounded by a connective tissue capsule (figure 24.8b,c). Tonsils consist of multiple germinal centers and have invaginated outer edges called crypts. Crypts help trap material and facilitate its identification by lymphocytes. Several groups of tonsils are found in the pharynx (throat): **Pharyngeal tonsils,** or *adenoids* (ad'ĕ-noydz; *aden* = gland), are in the posterior wall of the nasopharynx; **palatine tonsils** are in the posterolateral region of the oral cavity; and **lingual tonsils** are along the posterior one-third of the tongue.

WHAT DO YOU THINK?

2 If your tonsils are removed, how does your body develop an immune response against antigens in the throat? Are any other sources of lymphatic cells or structures located there?

CLINICAL VIEW

Tonsillitis and Tonsillectomy

Because the tonsils are designed to protect the pharynx from infection, they frequently become inflamed and infected, a condition called **acute tonsillitis** (ton'si-lī'tis). The palatine tonsils are most commonly affected. The tonsils redden and enlarge—in severe cases, to the point that they partially obstruct the pharynx and may cause respiratory distress.

Tonsils may be infected by viruses (such as adenoviruses) or bacteria (most commonly *Streptococcus*). Streptococcal tonsillitis often results in very red tonsils that have whitish specks (called whitish exudate). The symptoms of tonsillitis include fever, chills, sore throat, and difficulty swallowing. Bacterial tonsillitis (e.g., "strep throat") is successfully treated with antibiotics such as penicillin or amoxicillin. If tonsillitis is caused by a virus, measures to relieve the inflammation (such as pain medication and/or gargling) are advised, since standard antibiotics are not effective against viruses.

Persistent or recurrent infections can lead to permanent enlargement of the tonsils and a condition known as **chronic tonsillitis**. If medical treatment does not help the chronic tonsillitis, surgical removal of the tonsils **(tonsillectomy)** may be indicated. Typically, medical guidelines suggest performing a tonsillectomy only if the person has had 6–7 tonsillar infections in 1 year, or 2–3 infections per year for several years running. Research indicates that tonsillectomy does not significantly affect the body's response to new infections.

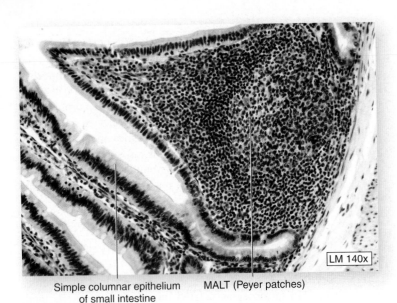

Simple columnar epithelium   MALT (Peyer patches)
of small intestine

(a) MALT in small intestine

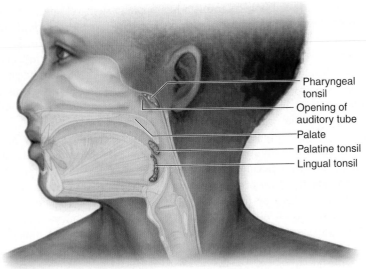

Pharyngeal tonsil
Opening of auditory tube
Palate
Palatine tonsil
Lingual tonsil

(b) Tonsils

Figure 24.8

Lymphatic Nodules. *(a)* MALT (mucosa-associated lymphatic tissue) in the ileum of the small intestine is called Peyer patches. *(b)* Tonsils reside in the wall of the pharynx and are composed of *(c)* lymphatic nodules.

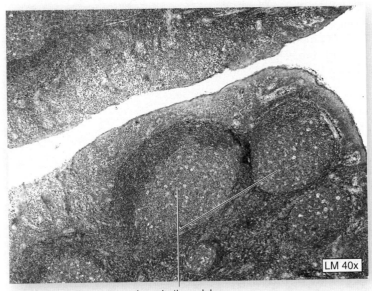

Lymphatic nodules

(c) Histology of tonsil

Lymphatic Organs

A lymphatic organ consists of lymphatic cells within an extracellular connective tissue matrix, and is completely surrounded by a connective tissue capsule. The main lymphatic organs are the thymus, lymph nodes, and spleen.

Thymus

The **thymus** (thī′mŭs;) is a bilobed organ located in the anterior mediastinum. In infants and young children, the thymus is quite large and extends into the superior mediastinum as well **(figure 24.9a)**. The thymus continues to grow until puberty, when it reaches a maximum weight of 30–50 grams. After reaching this size, cells of the thymus regress, and much of the functional thymus is eventually replaced by adipose connective tissue. In adults, the thymus atrophies and becomes almost nonfunctional.

In its prime, the thymus consists of two fused **thymic lobes,** each surrounded by a connective tissue capsule. Fibrous extensions of the capsule, called **trabeculae** (tră-bek′ū-lē) or *septa,* subdivide the thymic lobes into **lobules;** each lobule has an outer **cortex** and

an inner **medulla** (figure 24.9*b*). These distinct zones support different stages of T-lymphocyte development. The cortex contains immature T-lymphocytes, nurse cells, and some macrophages. The medulla contains epithelial cells and T-lymphocytes that have completed maturation. In addition, the medulla contains **thymic corpuscles** (or *Hassall corpuscles*), which are circular aggregations of aged, degenerated nurse cells (figure 24.9*c*).

The thymus functions as a site for T-lymphocyte maturation and differentiation. Immature lymphocytes migrate to the thymus during embryonic development. These immature cells then reside in the cortex of each lobule. The nurse cells in the cortex secrete several **thymic hormones** that stimulate T-lymphocyte maturation and differentiation. T-lymphocytes within the thymus do not participate in the immune response and are protected from antigens in the body by a well-formed **blood-thymus barrier** around the blood vessels in the cortex. When the T-lymphocytes differentiate (e.g., mature into helper T-lymphocytes or cytotoxic T-lymphocytes), they migrate to the medulla of each lobule. No blood-thymus barrier is present in the medulla, so the mature T-lymphocytes enter the bloodstream and migrate to other lymphatic system structures. The T-lymphocyte

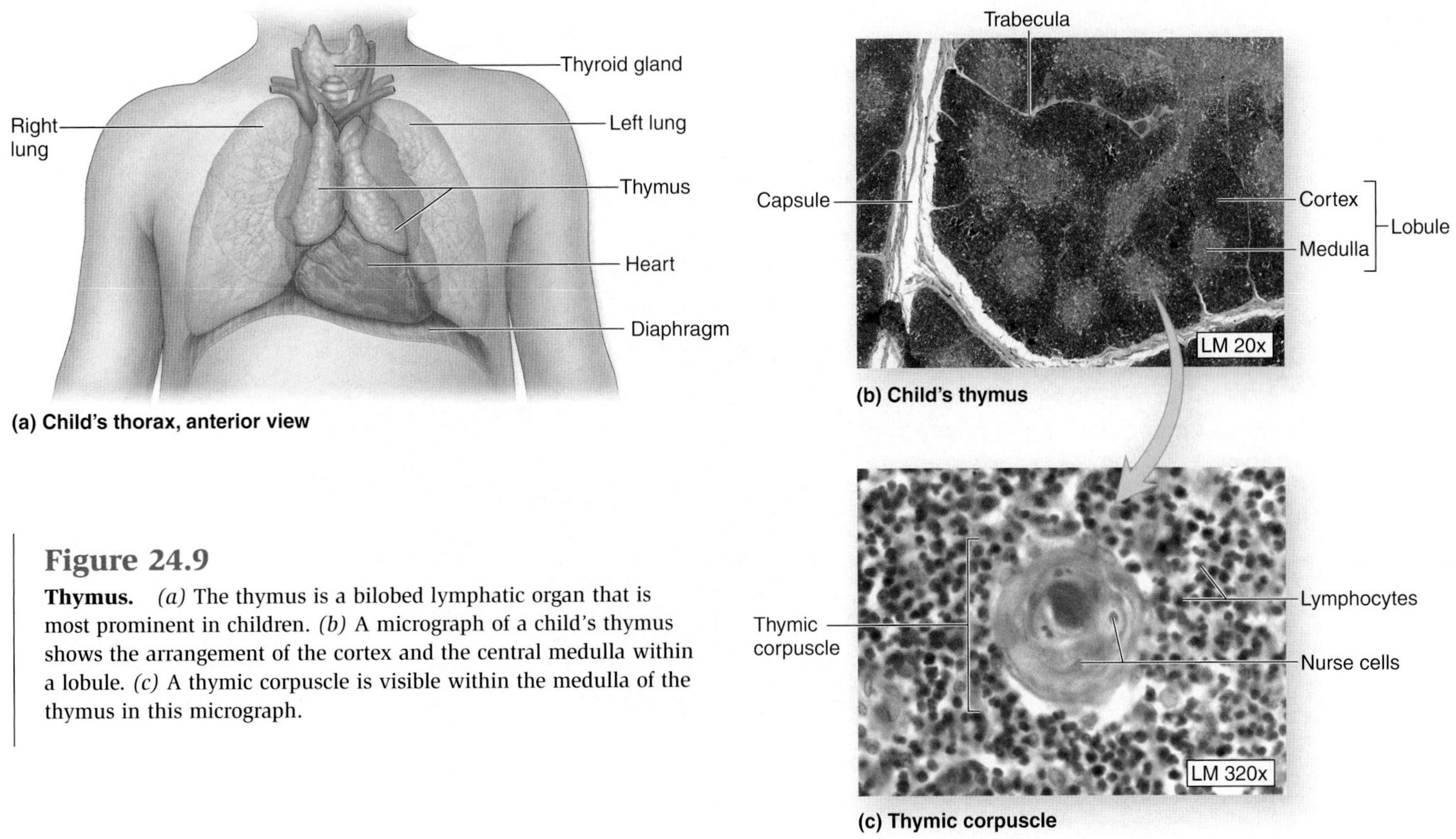

(a) Child's thorax, anterior view

Right lung
Thyroid gland
Left lung
Thymus
Heart
Diaphragm

(b) Child's thymus

Trabecula
Capsule
Cortex
Medulla
Lobule
LM 20x

(c) Thymic corpuscle

Thymic corpuscle
Lymphocytes
Nurse cells
LM 320x

Figure 24.9

Thymus. *(a)* The thymus is a bilobed lymphatic organ that is most prominent in children. *(b)* A micrograph of a child's thymus shows the arrangement of the cortex and the central medulla within a lobule. *(c)* A thymic corpuscle is visible within the medulla of the thymus in this micrograph.

maturation and differentiation process occurs primarily when we are young. Once adulthood is reached, differentiated T-lymphocytes can be produced by cell division only, not by maturation of new cells in the thymus.

Lymph Nodes

Lymph nodes are small, round or oval structures located along the pathways of lymph vessels (see figure 24.1). They range in length from 1 to 25 millimeters, and typically are found in clusters that receive lymph from selected body regions. For example, the cluster of lymph nodes in the armpit, called the **axillary lymph nodes,** receives lymph from the breast, axilla, and upper limb. Lymph nodes clustered in the groin, called **inguinal lymph nodes,** receive lymph from the lower limb and pelvis. **Cervical lymph nodes** receive lymph from the head and neck. In addition to these clusters, lymph nodes are found individually throughout the body.

Each lymph node is surrounded by a tough connective tissue **capsule** (cap′sool) **(figure 24.10a)**. The capsule projects internal extensions called **trabeculae** into the node, subdividing the node into compartments. The trabeculae also provide a pathway through which blood vessels and nerves may enter the lymph node. Lymphatic cells surround the trabeculae, and tiny open channels called lymphatic sinuses provide a pathway through which lymph flows.

The lymph node regions deep to the capsule are subdivided into an outer **cortex** and an inner **medulla.** The cortex consists of lymphatic nodules and lymphatic sinuses called **cortical sinuses**. Remember that lymphatic nodules contain an outer region of T-lymphocytes surrounding an inner germinal center that houses proliferating B-lymphocytes and some macrophages. In addition, dendritic cells within these lymph nodes collect antigens from the

lymph and present them to the T-lymphocytes. The lymph node medulla has strands of lymphatic cells (primarily B-lymphocytes and macrophages) supported by connective tissue fibers called **medullary cords.** The medulla also contains lymphatic sinuses called **medullary sinuses.**

The primary function of a lymph node is to filter antigens from lymph and initiate an immune response when necessary. Afferent lymphatic vessels carry lymph to the lymph node, where it slowly percolates through the cortical sinuses and then the medullary sinuses. Macrophages line the lymphatic sinuses and remove foreign debris from the lymph. Lymph then exits the lymph node by way of one or two efferent lymphatic vessels. Efferent lymphatic vessels originate at the indented portion of the lymph node called the **hilum** (hī′lŭm) or *hilus*.

If antigens are presented to lymphocytes, an immune response is generated. The lymphocytes undergo cell division (especially in the germinal centers), and these new lymphocytes eventually travel to the bloodstream, where they can help fight infection. When a person is sick (e.g., influenza or strep throat), some of the lymph nodes are often swollen and tender to the touch. This is a sign that lymphocytes are proliferating and beginning to fight the infection.

Cancerous cells from other areas of the body can travel easily through the lymphatic system (a process called **metastasis**), and become entrapped in lymph nodes. These cancerous cells can proliferate and also contribute to enlarged lymph nodes. A lymph node enlarged by cancer tends to be firm and nontender, as opposed to a lymph node that is swollen and tender due to an infection. If an individual is diagnosed with a cancer, the lymph nodes that drain the affected organ or body region are examined to determine if the cancer has spread. For example, if cancer is detected in a breast, the axillary lymph nodes are examined.

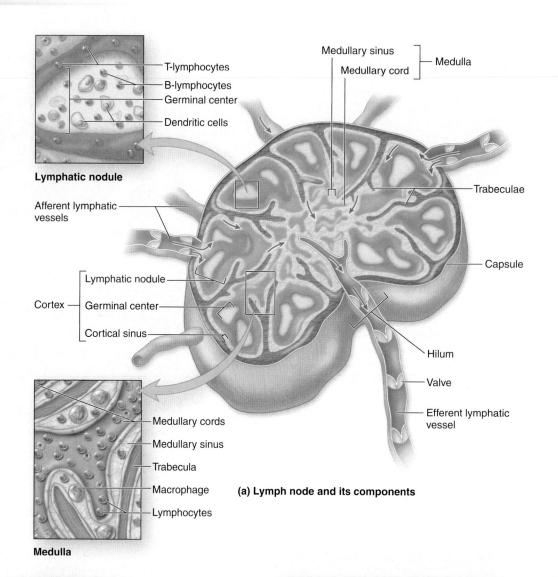

T-lymphocytes
B-lymphocytes
Germinal center
Dendritic cells

Lymphatic nodule

Afferent lymphatic
vessels

Cortex
Lymphatic nodule
Germinal center
Cortical sinus

Medullary sinus
Medullary cord ⎤ Medulla

Trabeculae

Capsule

Hilum

Valve

Efferent lymphatic
vessel

(a) Lymph node and its components

Medullary cords
Medullary sinus
Trabecula
Macrophage
Lymphocytes

Medulla

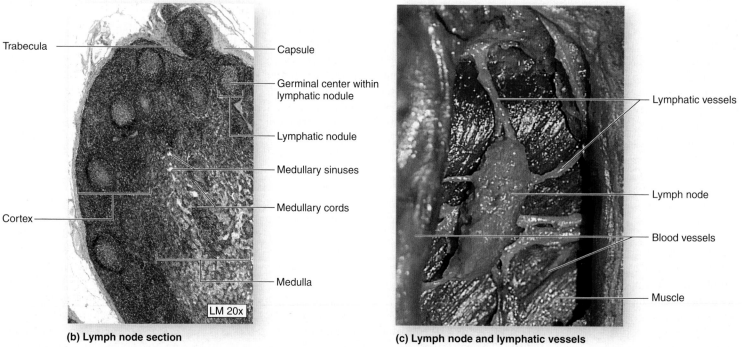

Trabecula
Capsule

Germinal center within
lymphatic nodule

Lymphatic nodule

Medullary sinuses

Cortex

Medullary cords

Medulla

LM 20x

(b) Lymph node section

Lymphatic vessels

Lymph node

Blood vessels

Muscle

(c) Lymph node and lymphatic vessels

Figure 24.10

Lymph Nodes.   Lymph nodes are small, encapsulated structures that filter the lymph in lymphatic vessels. *(a)* Green arrows indicate the direction of lymph flow into and out of the lymph node. *(b)* A micrograph of a lymph node shows the cortex and medulla. *(c)* A cadaver photo of a lymph node and lymphatic vessels.

CLINICAL VIEW

Lymphoma

A **lymphoma** (lim-fō′mă; *oma* = tumor) is a malignant neoplasm that develops from lymphatic structures. Usually (but not always) a lymphoma presents as a nontender, enlarged lymph node, often in the neck or axillary region. Some patients have no further symptoms, while others may experience night sweats, fever, and unexplained weight loss in addition to the nodal enlargement. Lymphomas are grouped into two categories: Hodgkin lymphoma and non-Hodgkin lymphoma.

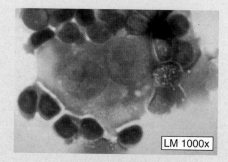

Reed-Sternberg cell, a characteristic of Hodgkin lymphoma.

LM 1000x

Hodgkin lymphoma (or *Hodgkin disease*) is characterized by the presence of the **Reed-Sternberg cell,** a large cell whose two nuclei resemble owl eyes, surrounded by lymphocytes within the affected lymph node. Hodgkin lymphoma affects young adults (ages 16–35) and people over 60. It arises in a lymph node and then spreads to other nearby lymph nodes. If caught early, Hodgkin lymphoma can be treated and cured by excision of the tumor, radiation, and/or chemotherapy.

Non-Hodgkin lymphomas are much more common than Hodgkin lymphomas. These lymphomas typically develop in lymphatic structures, usually from abnormal B-lymphocytes, and less commonly from T-lymphocytes. Some kinds of non-Hodgkin lymphoma are aggressive and often fatal, whereas others are slow-growing and indolent. Treatment depends on the type of non-Hodgkin lymphoma, the extent of its spread at the time of discovery, and the rate of progression of the malignancy.

The AIDS epidemic is associated with a significant rise in aggressive B-lymphocyte non-Hodgkin lymphomas, prompting the Centers for Disease Control to revise the definition of AIDS to include HIV-infected patients who have this type of lymphoma. AIDS-related non-Hodgkin lymphomas are aggressive and difficult to treat, putting the patient's health in further jeopardy.

Spleen

The largest lymphatic organ in the body is the **spleen,** which is located in the left upper quadrant of the abdomen, inferior to the diaphragm and adjacent to ribs 9–11. This deep red organ lies lateral to the left kidney and posterolateral to the stomach. The spleen can vary considerably in size and weight, but typically is about 12 centimeters long and 7 centimeters wide. The spleen's posterolateral aspect (called the diaphragmatic surface) is convex and rounded, while the concave anteromedial border (the *visceral surface*) contains the **hilum** (or *hilus*), where blood vessels and nerves enter and leave the spleen (**figure 24.11a,d**). A **splenic** (splen′ik) **artery** delivers blood to the spleen, while blood returns to the circulation by way of a **splenic vein.**

The spleen is surrounded by a dense irregular connective tissue **capsule.** The capsule sends extensions called **trabeculae** into the organ. Within these trabeculae extend branches of the splenic artery and vein called **trabecular vessels.** The spleen lacks a cortex and medulla. Rather, the cells around the trabeculae are subdivided into white pulp and red pulp. Red pulp surrounds each cluster of white pulp (figure 24.11b,c).

The **white pulp** is associated with the arterial supply of the spleen and consists of circular clusters of lymphatic cells (T-lymphocytes, B-lymphocytes, and macrophages). In the center of each cluster is a **central artery.** As blood enters the spleen and flows through the central arteries, the white pulp lymphatic cells monitor the blood for foreign materials, bacteria, and other antigens. If antigens are found, the T- and B-lymphocytes elicit an immune response. Thus, while lymph nodes monitor *lymph* for antigens, the spleen monitors *blood* for antigens.

The **red pulp** is associated with the venous supply of the spleen, since blood that enters the spleen in the central arteries then travels through blood vessels in the red pulp. Red pulp consists of splenic cords and splenic sinusoids. The **splenic cords** (*cords of*

Bilroth) contain erythrocytes, platelets, macrophages, and some plasma cells. One of the functions of the spleen is to serve as a blood reservoir, whereby the formed elements are stored in these splenic cords. In situations where more erythrocytes (and thus, greater oxygen delivery) are needed, such as during exercise, these erythrocytes reenter the bloodstream. Since the spleen contains a large amount of blood, severe trauma to the spleen results in massive hemorrhage.

Among the splenic cords, the **splenic sinusoids** act like enlarged capillaries that carry blood. These vessels have a discontinuous basal lamina, so blood cells can easily enter and exit across the vessel wall. Macrophages lining the sinusoid lumen phagocytize (1) bacteria and foreign debris from the blood, and (2) old and defective erythrocytes and platelets (a process called **hemolysis** (hē-mol′i-sis; *lysis* = destruction). Old erythrocytes can rupture or become trapped in the sinusoids, making them a target for these macrophages. Sinusoids merge to form small veins, and eventually the filtered blood leaves the spleen via the splenic vein.

In summary, the spleen performs the following functions:

- Initiates an immune response when antigens are found in the blood (a white pulp function)
- Serves as a reservoir for erythrocytes and platelets (a red pulp function)
- Phagocytizes old, defective erythrocytes and platelets (a red pulp function)
- Phagocytizes bacteria and other foreign materials

Table 24.2 summarizes the lymphatic structures and organs and their functions.

WHAT DO YOU THINK?

❸ If your spleen were removed (splenectomy), would you be able to fight off illness and infections effectively? Why or why not?

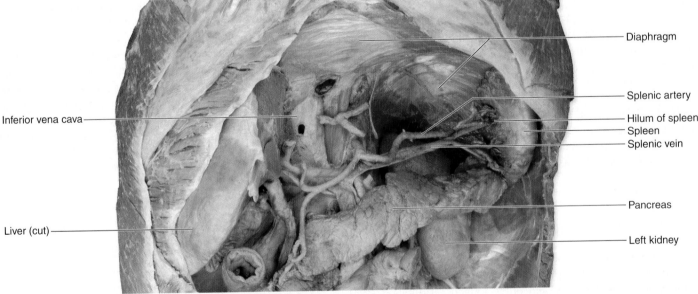

(a) Anterior view of spleen

Diaphragmatic surface

Diaphragm

Visceral surface

Splenic artery

Splenic vein

Hilum

(b) Red and white pulp of spleen

Central artery

White pulp

Red pulp

Splenic sinusoids

Trabecula

Splenic cords

Capsule

(c) Histology of spleen

Trabeculae

Red pulp

Central artery

White pulp

Capsule

LM 40x

(d) Abdominal cavity, anterior view

Inferior vena cava

Liver (cut)

Diaphragm

Splenic artery

Hilum of spleen

Spleen

Splenic vein

Pancreas

Left kidney

Figure 24.11

Spleen. *(a)* An anterior view illustrates the hilum as well as the splenic artery and vein. *(b)* A diagram depicts the microscopic arrangement of blood vessels, the red pulp, and the white pulp. *(c)* A micrograph of the spleen shows areas of white pulp and red pulp. *(d)* A cadaver photo shows the spleen and its relationship to the diaphragm, pancreas, and kidney. In this photo, the pancreas has been moved inferiorly to show the splenic vessels more clearly.

Table 24.2	Lymphatic Structures and Organs		
Component	**Structure or Organ[1]**	**Functions**	**Location**
Lymphatic nodules	Structure	Filter and attack antigens	Throughout body
MALT (mucosa-associated lymphatic tissue)	Structure	Filter and attack antigens in food, air, or urine	Within walls of GI, respiratory, genital, and urinary tracts
Tonsils	Structure	Protect against inhaled and ingested materials	Within pharynx
Thymus	Organ	Site of T-lymphocyte maturation and differentiation; stores maturing lymphocytes	Superior mediastinum (in adults); anterior and superior mediastinum (in children)
Lymph nodes	Organ	Filter lymph; mount immune response	Throughout body; frequently in clusters in the axillary, inguinal, and cervical regions
Spleen	Organ	Filters blood and recycles aged erythrocytes and platelets; serves as a blood reservoir; houses lymphocytes; mounts immune response to foreign antigens in the blood	In left upper quadrant of abdomen, near 9th–11th ribs and inferior to diaphragm

[1] A lymphatic structure is unencapsulated or has an incomplete connective tissue capsule, while a lymphatic organ has a complete connective tissue capsule encircling it.

WHAT DID YOU LEARN?

8 What is MALT, and where is it found?

9 Describe the main function of the thymus, and explain how the blood-thymus barrier supports this function.

10 Describe the basic structure and function of a lymph node.

11 How do the white pulp and red pulp of the spleen differ with respect to both cell population and function?

Aging and the Lymphatic System

Key topic in this section:

■ How aging affects the lymphatic system

Some lymphatic system functions are not affected by aging. For example, the body effectively continues to transport lymph back to the bloodstream and absorb dietary lipids from the small intestine. However, other functions change as we age. First, when an individual reaches adulthood, the thymus no longer matures and differentiates T-lymphocytes. New T-lymphocytes can be produced only by division (mitosis) of pre-existing lymphocytes. Second, the lymphatic system's ability to provide immunity and fight disease decreases as we get older. Helper T-lymphocytes do not respond to antigens as well, and do not always reproduce rapidly. Reduced numbers of helper T-lymphocytes lead to fewer B-lymphocytes and other kinds of T-lymphocytes. Therefore, the body's ability to acquire immunity and resist infection decreases, making elderly people more susceptible to illnesses and more likely to become sicker than younger adults. Older individuals are advised to get a pneumococcal vaccine or yearly influenza vaccinations because of their increased risk of developing *Streptococcus pneumoniae* infections or the flu, respectively. The faltering immune system may also be less able to target and eliminate malignant cells, suggesting one reason why the elderly tend to be more prone to cancer.

WHAT DID YOU LEARN?

12 Why are elderly individuals more prone to illnesses?

Development of the Lymphatic System

Key topic in this section:

■ Lymphatic system formation in the developing embryo and fetus

The origin of the lymph vessels is poorly understood. Some anatomists believe they originate from the endothelium of veins, while others support the theory that they originate from local mesoderm. Despite this conflict, we know that the first lymphatic structures (called **primary lymph sacs**) appear during the sixth week of development. A total of six primary lymph sacs form: Two **jugular lymph sacs** develop near each junction of the subclavian and future internal jugular veins; two **posterior lymph sacs** develop near each junction of the external and internal iliac veins; a **retroperitoneal** (re′trō-per′i-tō-nē′ăl) **lymph sac** forms in the digestive system mesentery; and a **cisterna chyli** forms dorsal to the aorta **(figure 24.12a)**.

Paired lymph vessels connect the lymph sacs by the ninth week (figure 24.12b). Eventually, portions of the paired lymph vessels are obliterated, and a single thoracic duct forms that travels from the cisterna chyli, along the bodies of the vertebrae, and empties into the union of the left subclavian and left internal jugular veins. The right lymphatic duct is formed from some other lymph vessels. Additional smaller lymph vessels form during and after the embryonic period.

During the fetal period, connective tissue subdivides the lymph sacs (except the cisterna chyli) into rounded structures that later become the lymph nodes (figure 24.12c). Lymphocytes within the developing lymph sacs eventually form the cortex and medulla of each lymph node.

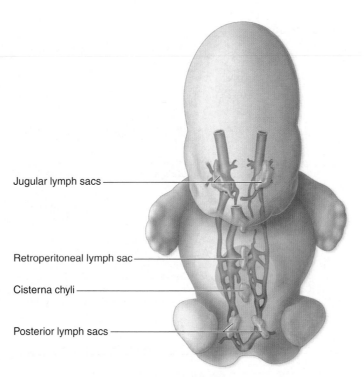

Jugular lymph sacs

Retroperitoneal lymph sac

Cisterna chyli

Posterior lymph sacs

(a) Week 6: Primary lymph sacs form

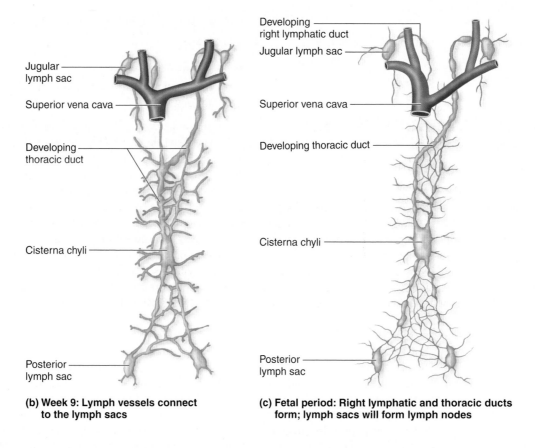

Developing
right lymphatic duct

Jugular lymph sac

Superior vena cava

Developing thoracic duct

Cisterna chyli

Posterior
lymph sac

Jugular
lymph sac

Superior vena cava

Developing
thoracic duct

Cisterna chyli

Posterior
lymph sac

**(b) Week 9: Lymph vessels connect
to the lymph sacs**

**(c) Fetal period: Right lymphatic and thoracic ducts
form; lymph sacs will form lymph nodes**

Figure 24.12

Development of the Lymphatic System. *(a)* Lymph sacs appear by week 6 of development. *(b)* Paired lymph vessels develop by week 9.
(c) During the fetal period, some of the lymph vessels enlarge and form the right lymphatic and thoracic ducts. Lymph sacs eventually develop into
lymph nodes.

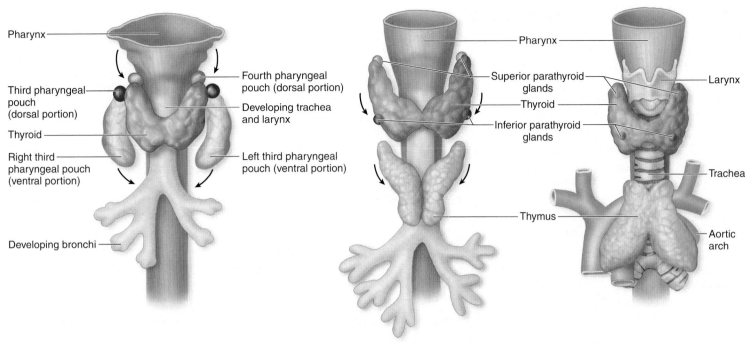

(a) Week 5: Ventral portions of left and right third pharyngeal pouches migrate inferiorly

(b) Week 7: Left and right third pharyngeal pouches fuse to form the bilobed thymus

(c) Fetal period: Thymus is positioned in mediastinum

Figure 24.13

Development of the Thymus.   The thymus originates from the ventral portions of the third pharyngeal pouches. *(a)* These pouches branch off the pharynx and migrate inferiorly beginning in the fifth week. *(b)* The left and right pouches fuse in the future mediastinum during the seventh week, forming the bilobed thymus. *(c)* By the fetal period, the thymus gland is large and is positioned within the mediastinum.

The thymus forms from the *ventral* portions of the left and right third pharyngeal pouches **(figure 24.13)**. These pouches are the internal, endodermal portions of the third pharyngeal arches. During weeks 4–7 of development, the left and right third pouches migrate inferiorly from the pharynx to their final position posterior to the sternum. There, the two ventral portions fuse to form a single thymus gland. Lymphocytes begin infiltrating the gland shortly after its formation. By the fetal period, the thymus gland is large and located within the mediastinum, superior to the heart and anterior to the aortic arch and trachea.

The spleen is formed from mesoderm that condenses in the future greater omentum during the fifth week of development. Initially, the spleen functions solely as a blood cell-producing organ.

During the second trimester, T-lymphocytes and B-lymphocytes enter the organ, helping form the characteristic red pulp and white pulp.

The palatine tonsils form from endoderm from the left and right second pharyngeal pouches and the second pharyngeal arch mesoderm. During months 3–5 of development, lymphatic cells enter the tonsils. Other tonsils (e.g., pharyngeal tonsils, lingual tonsils) and lymphatic structures such as MALT form from aggregations of lymphatic cells and connective tissue.

WHAT DID YOU LEARN?

13 When does the spleen begin to develop?

CLINICAL TERMS

autoimmune disease Disease in which the body's immune system mistakenly attacks its own healthy tissues. Examples include systemic lupus erythematosus (SLE), multiple sclerosis (MS), rheumatoid arthritis, type 1 diabetes mellitus, and scleroderma.

lymphadenectomy (lim-fad-ě-nek′tō-mē; *aden* = gland) Removal or excision of lymph nodes.

lymphangitis (*angeion* = vessel) Inflammation of the lymph vessels.

splenomegaly (splē-nō-meg′ă-lē; *mega* = large) Enlarged spleen, often seen in association with infection (e.g., mononucleosis).

C H A P T E R   S U M M A R Y

Functions of the Lymphatic System 723	■ The lymphatic system carries interstitial fluid back to the bloodstream, transports dietary lipids, houses and develops lymphocytes, and generates an immune response.
Lymph and Lymph Vessels 724	■ Lymph is interstitial fluid containing solutes and sometimes foreign material that is transported through lymph vessels to the blood.
	■ There are many types of lymph vessels. From smallest to largest, they are lymphatic capillaries, lymphatic vessels, lymphatic trunks, and lymphatic ducts.

Lymphatic Capillaries   724

■ Lymphatic capillaries, the smallest lymph vessels, are endothelium-lined vessels with overlapping internal edges of endothelial cells that regulate lymph entry.

■ Lacteals are lymphatic capillaries in the small intestine; they pick up and transport the lymph (called chyle) from the intestine.

Lymphatic Vessels   724

■ Lymphatic vessels form from merging lymphatic capillaries. They have valves to prevent lymph backflow.

■ Afferent lymphatic vessels conduct lymph to lymph nodes, and efferent lymphatic vessels conduct lymph away from lymph nodes.

Lymphatic Trunks   725

■ Lymphatic trunks form from merging lymphatic vessels; each trunk drains a major body region into a lymphatic duct.

Lymphatic Ducts   725

■ The right lymphatic duct drains the right side of the head and neck, the right upper limb, and the right side of the thorax. It drains into the junction of the right subclavian vein and the right internal jugular vein.

■ The thoracic duct drains lymph from the left side of the head and neck, the left upper limb, the left thorax, and all body regions inferior to the diaphragm. It drains into the junction of the left subclavian vein and left internal jugular vein.

Lymphatic Cells 727	■ Lymphatic cells include macrophages that phagocytize foreign substances, epithelial cells that secrete thymic hormones, dendritic cells that filter antigens from lymph, and lymphocytes that perform specific functions in the immune response.

Types and Functions of Lymphocytes   727

■ Helper T-lymphocytes respond to one type of antigen only, and secrete cytokines, which are chemical signals that activate other lymphatic cells.

■ Cytotoxic T-lymphocytes kill infected or foreign cells following direct contact with them.

■ Memory T-lymphocytes arise from cytotoxic T-lymphocytes that previously killed a foreign cell, and cause a faster immune response than the first time.

■ Suppressor T-lymphocytes often "turn off" the immune response once it has been activated.

■ Activated B-lymphocytes respond to one particular antigen; they proliferate and differentiate into either plasma cells or memory B-lymphocytes.

■ Plasma cells produce and secrete large numbers of antibodies.

■ Memory B-lymphocytes mount an even faster and more powerful immune response upon reexposure to an antigen.

■ NK cells respond to multiple antigens; they destroy infected cells and some cancerous cells.

Lymphopoiesis   732

■ Some hemopoietic stem cells remain in the red bone marrow and mature into B-lymphocytes and NK cells. Other stem cells exit the marrow and migrate to the thymus for subsequent maturation into T-lymphocytes.

Lymphatic Structures 733	■ Lymphatic structures include lymphatic nodules and various lymphatic organs.

Lymphatic Nodules   733

■ Lymphatic nodules are oval clusters of lymphatic cells and extracellular connective tissue matrix that are not contained within a connective tissue capsule.

■ MALT (mucosa-associated lymphatic tissue) is composed of lymphatic nodules housed in the walls of the GI, respiratory, genital, and urinary tracts.

■ Tonsils are large clusters of partially encapsulated lymphatic cells and extracellular connective matrix.

Lymphatic Organs   734

■ The lymphatic organs are composed of lymphatic structures completely surrounded by a connective tissue capsule.

■ The thymus is where T-lymphocytes mature and differentiate under stimulation by thymic hormones.

■ Lymph nodes are small structures that filter lymph.

■ The spleen is partitioned into white pulp (consists of clusters of lymphatic cells that generate an immune response when exposed to antigens in the blood) and red pulp (consists of splenic cords that store blood and sinusoids containing macrophages that phagocytize foreign debris, old erythrocytes, and platelets).

Aging and the Lymphatic System 739	■ The lymphatic system's ability to provide immunity and fight disease decreases as we get older.
Development of the Lymphatic System 739	■ The primary lymph sacs eventually give rise to lymph nodes. ■ The thymus forms from the ventral portions of the left and right third pharyngeal pouches. ■ The spleen forms from mesodermal condensations during week 5 of development. ■ The palatine tonsils are derived from the second pharyngeal pouches.

CHALLENGE YOURSELF

Matching

Match each numbered item with the most closely related lettered item.

_____ 1. lymph node
_____ 2. antibody
_____ 3. helper T-lymphocyte
_____ 4. spleen
_____ 5. red bone marrow
_____ 6. macrophage
_____ 7. lymphatic capillary
_____ 8. thymus
_____ 9. lymphatic vessel
_____ 10. thoracic duct

a. receives lymph from some lymphatic trunks

b. former monocyte that phagocytizes foreign debris

c. B-lymphocytes mature and differentiate here

d. smallest type of lymph vessel

e. immobilizes an antigen

f. T-lymphocytes mature and differentiate here

g. filters lymph

h. drains directly into and out of lymph nodes

i. removes old and defective erythrocytes

j. cell type that regulates the immune response

Multiple Choice

Select the best answer from the four choices provided.

_____ 1. Lymph from which of the following body regions drains into the thoracic duct?
a. right side of the thorax
b. right upper limb
c. right lower limb
d. right side of the head

_____ 2. Which type of lymphatic cell is responsible for producing antibodies?
a. macrophage
b. helper T-lymphocyte
c. plasma cell
d. NK cell

_____ 3. Which statement is false about lymphatic nodules?
a. The center has proliferating B-lymphocytes and some macrophages.
b. T-lymphocytes are located along the periphery.
c. Lymphatic nodules are completely surrounded by a connective tissue capsule.
d. Lymphatic nodules in the ileum of the small intestine are called Peyer patches.

_____ 4. What is the function of the blood-thymus barrier?
a. It protects maturing T-lymphocytes from antigens in the blood.
b. It filters the blood and starts an immune response when necessary.
c. It subdivides the thymus into a cortex and a medulla.
d. It forms thymic corpuscles.

_____ 5. Which type of lymph vessel consists solely of an endothelium and has one-way flaps that allow interstitial fluid to enter?
a. lymphatic vessel
b. lymphatic capillary
c. lymphatic duct
d. lymphatic trunk

_____ 6. Which statement is true about lymph nodes?
a. Cancerous lymph nodes are swollen and tender to the touch.
b. The medulla of a lymph node contains lymphatic nodules.
c. Lymph enters the lymph node through afferent lymphatic vessels.
d. Lymphatic sinuses are located in the cortex of a lymph node only.

_____ 7. In an early _Streptococcus_ infection of the throat, all of the following structures may swell except the
a. pharyngeal tonsil.
b. spleen.
c. cervical lymph node.
d. palatine tonsil.

_____ 8. Which of the following is a function of the white pulp of the spleen?
   a. phagocytizes erythrocytes
   b. serves as a blood cell reservoir
   c. elicits an immune response if antigens are detected in the blood
   d. serves as a site for hemopoiesis during fetal life

_____ 9. The primary lymph sacs form the
   a. thoracic duct.
   b. right lymphatic duct.
   c. spleen.
   d. lymph nodes.

_____ 10. What change occurs to the adult lymphatic system as we get older?
   a. The body produces and transports less lymph.
   b. Greater numbers of B-lymphocytes are produced.
   c. Helper T-lymphocytes do not respond as well to antigens.
   d. The lymph nodes enlarge.

Content Review

1. List the functions of the lymphatic system

2. Describe what lymph is, and draw a flowchart that illustrates what structures the lymph travels through to return to the bloodstream.

3. Which body regions have their lymph drained to the thoracic duct?

4. Compare and contrast the types of lymphatic cells with respect to their appearance, function, and place of maturation.

5. Describe how the lymphatic cells elicit an immune response.

6. Describe the basic composition of a lymphatic nodule, and give examples in the body where lymphatic nodules may be found.

7. Describe how the thymus's anatomy and function change from infancy on.

8. Describe the basic anatomy of a lymph node, how lymph enters and leaves the node, and the functions of this organ.

9. Compare and contrast the red and white pulp of the spleen with respect to the anatomy and functions of each.

10. Describe how lymph vessels and lymph nodes develop.

Developing Critical Reasoning

1. Arianna was diagnosed with mononucleosis, an infectious disease that targets B-lymphocytes. When she went to the doctor, he palpated her left side, just below the rib cage. The doctor told Arianna she was checking to see if a certain organ was enlarged, a complication that can occur with mononucleosis. What lymphatic organ was the doctor checking, and why would it become enlarged? Include some explanation of the anatomy and histology of this organ in your answer.

2. Why is HIV infection so devastating to the body? In your answer, explain what cells are infected and why the body cannot produce more mature, noninfected cells. Also explain how AIDS affects the way the body fights infection, and give some examples of ailments that are common among AIDS patients.

3. Jordan has an enlarged lymph node along the side of his neck, and he is worried that the structure may be a lymphoma. What are some criteria to help distinguish between infected lymph nodes and malignant lymph nodes? If the lymph node is cancerous, how would a physician determine if the cancer has spread to other parts of the body?

ANSWERS TO "WHAT DO YOU THINK?"

1. If you need a booster shot about once every 10 years, that indicates a maximum 10-year life span of tetanus-detecting memory B-lymphocytes. Many of these cells may die off before the 10 years is up, which is why a physician gives you a tetanus booster earlier if you have been exposed to tetanus—for example, by piercing your foot on a rusty nail that could carry tetanus bacteria.

2. If the tonsils are removed, other lymphatic tissue and lymphatic organs in the head and neck, such as lymph nodes, can mount an immune response to antigens in the throat. Also, the lymphocytes circulating in the bloodstream can detect antigens in the throat.

3. Most individuals who have their spleens removed can fight off illness and infections effectively, because the other lymphatic tissues and organs take over the immune functions previously handled by the spleen. However, the risk for severe bacterial infection is greater, since there is no spleen to filter bacteria from the blood. For this reason, individuals who have undergone splenectomies may need to be vaccinated against certain bacteria and undergo long-term (years or even lifelong) antibiotic therapy.

Visit the McKinley/O'Loughlin _Human Anatomy_, 2e website at aris.mhhe.com

OUTLINE

RESPIRATORY SYSTEM

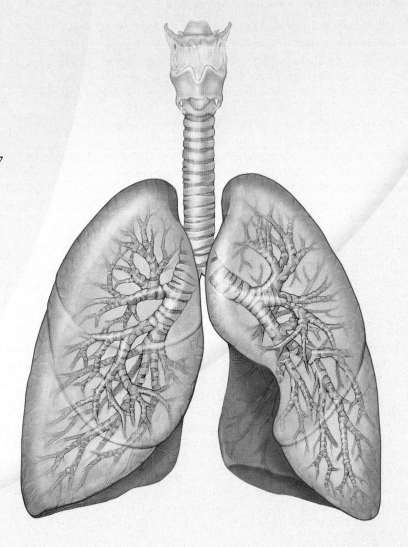

Respiratory System

The **respiratory** (res′pi-ră-tōr′ē; *respiro* = to breathe) **system** provides the means for gas exchange required by living cells. Oxygen must be supplied without interruption, and carbon dioxide, a waste product generated by the cells, must be continuously expelled. The respiratory and cardiovascular systems are inseparable partners. While the respiratory system exchanges gases between the atmosphere and the blood, the cardiovascular system transports those gases between the lungs and the body cells. This chapter examines the cells, tissues, and organs involved in the complex and vital process of respiration.

General Organization and Functions of the Respiratory System

Key topics in this section:

- Components of the conducting and respiratory portions of the respiratory system
- Comparison of external and internal respiration
- Other functions of the respiratory system

Anatomically, the respiratory system consists of an upper respiratory tract and a lower respiratory tract (**figure 25.1**). Functionally, it can be divided into a conducting portion, which transports air, and a respiratory portion, where gas exchange with the blood occurs. The **conducting portion** includes the nose, nasal cavity, and pharynx of the upper respiratory tract and the larynx, trachea, and progressively smaller airways (from the primary bronchi to the terminal bronchioles) of the lower respiratory tract. The **respiratory portion** is composed of small airways called respiratory bronchioles and alveolar ducts as well as air sacs called alveoli in the lower respiratory tract.

Respiratory System Functions

The primary function most of us associate with the respiratory system is breathing, also termed pulmonary ventilation. Breathing consists of two cyclic phases: **inhalation** (in-hă-lā′shŭn), also called *inspiration,* and **exhalation,** also called *expiration* (eks-pi-rā′shŭn). Inhalation draws gases into the lungs, and exhalation forces gases out of the lungs.

Gas Exchange

The continuous movement of gases into and out of the lungs is necessary for the process of gas exchange. There are two types of gas exchange: external respiration and internal respiration. **External respiration** involves the exchange of gases between the atmosphere and the blood. Oxygen in the atmosphere is inhaled into the lungs. It diffuses from the lungs into the blood within the cardiovascular system at the same time carbon dioxide diffuses from the blood into the lungs in order to be exhaled. **Internal respiration** involves the exchange of gases between the blood and the cells of the body. Blood transports oxygen from the lungs to the body cells and transports carbon dioxide produced by the body cells to the lungs.

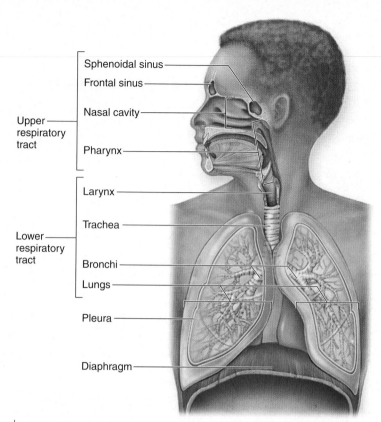

Figure 25.1

Gross Anatomy of the Respiratory System. The major components of the respiratory system are organized into the upper and lower respiratory tracts.

In addition to gas exchange, the respiratory system also functions in gas conditioning, sound production, olfaction, and defense.

Gas Conditioning

As inhaled gases pass through conducting airways, the gases are "conditioned" prior to reaching the gas exchange surfaces of the lungs. Specifically, the gases are warmed to body temperature, humidified (moistened), and cleansed of particulate matter through contact with the respiratory epithelium and its sticky mucous covering. Conditioning is facilitated by the twisted pathways through the nasal cavity and paranasal sinuses, which cause the inhaled air to become very turbulent during inhalation. This swirling of inhaled gases means the air remains in the nasal cavity and paranasal sinuses for a relatively longer time, providing greater opportunity for conditioning.

Sound Production

As air is forced out of the lungs and moves through the larynx, sound may be produced, such as speech or singing. Other anatomic structures aid sound production, including the nasal cavity, paranasal sinuses, teeth, lips, and tongue.

Olfaction

The superior region of the nasal cavity is covered with olfactory epithelium, which contains receptors for the sense of smell (see chapter 19). These receptors are stimulated when airborne molecules are inhaled and dissolved in the mucus covering this olfactory epithelium.

Defense

Finally, both the structure of the respiratory system and some of the cells within the respiratory epithelium protect the body against infection by airborne molecules. The entrance to the respiratory system (the nose) is inferiorly directed, is lined with coarse hairs, and has twisted passageways to prevent large particles, microorganisms, and insects from entering. Additionally, numerous goblet cells are dispersed throughout the pseudostratified ciliated columnar epithelium lining much of the upper respiratory tract. Mucous glands housed within the lamina propria deep to the epithelium contribute to the layer of mucus covering the epithelium and keep it from drying out. Mucous glands also secrete lysozyme, an enzyme that helps defend against inhaled bacteria. The layer of sticky mucus traps inhaled dust, dirt particles, microorganisms, and pollen. If we are exposed to airborne allergens, large quantities of small particulate material, irritating gases, or pathogens, the rate of mucin production increases.

WHAT DID YOU LEARN?

❶ Explain the functions carried out by the respiratory system in addition to gas exchange.

❷ How does mucus help with respiratory system functions?

CLINICAL VIEW

Cystic Fibrosis

Cystic fibrosis (sis'tik fī-brō'sis) is the most common serious genetic disease in Caucasians, occurring with a frequency of 1 in 3200 births. The condition is inherited as an autosomal recessive trait, and is rare among people of Asian and African descent. The name cystic fibrosis refers to the characteristic scarring and cyst formation within the pancreas, first recognized in the 1930s. Cystic fibrosis affects the organs that secrete mucin, tears, sweat, digestive juices, and saliva. A defective gene produces an abnormal plasma membrane protein involved in chloride ion transport, so individuals with cystic fibrosis cannot secrete chloride. This lack of chloride secretion causes sodium and water to move from the mucus back into the secretory cell itself, thus dehydrating the mucus covering the epithelial surface. The mucus becomes thick and sticky, obstructing the airways of the lungs and the ducts of the pancreas and salivary glands. In the lungs, the mucus becomes so thick it results in airway obstruction. Pulmonary infections, secondary to airway obstruction, are common and can be life-threatening. In the case of the pancreas, the obstructed ducts lead to a backup of digestive enzymes that eventually destroy the pancreas itself.

Interestingly, the normal chloride transport protein works in the opposite direction in the sweat glands of the skin. Chloride and sodium are not reabsorbed from the sweat, and so they become concentrated on the skin in individuals with cystic fibrosis. Mothers of babies with cystic fibrosis often find that the baby tastes "salty" when kissed. Thus, clinically elevated chloride levels in sweat are one method of diagnosing the disease.

The primary treatment for cystic fibrosis involves agents that break up the thick mucus in the lungs. In addition, antibiotics for pulmonary infections are required chronically, because prevention and early treatment of infection are vital to reducing long-term complications. Absorption problems caused by pancreatic damage are treated with orally administered digestive enzymes, vitamins, and caloric supplements. Since the gene responsible for cystic fibrosis has been identified, scientists have been investigating ways to insert copies of the healthy gene into the epithelial cells of the respiratory tracts of cystic fibrosis patients. In the most promising method found thus far, the healthy gene is transmitted via a modified adenovirus.

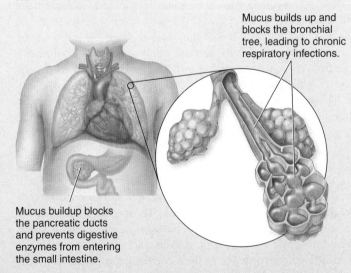

Mucus builds up and blocks the bronchial tree, leading to chronic respiratory infections.

Mucus buildup blocks the pancreatic ducts and prevents digestive enzymes from entering the small intestine.

Cystic fibrosis results in thickened mucus that obstructs both the respiratory passageways and ducts of glands such as the pancreatic ducts.

Upper Respiratory Tract

Key topics in this section:

- Organization and functions of upper respiratory tract organs
- Comparison of the regions of the pharynx

The **upper respiratory tract** is composed of the nose and nasal cavity, paranasal sinuses, pharynx, and associated structures. These structures are all part of the conducting portion of the respiratory system. When an individual has an upper respiratory tract infection, some or all of these structures are involved.

Nose and Nasal Cavity

The **nose** is the main conducting airway for inhaled air. The nose is supported superiorly by paired **nasal bones** that form the bridge of the nose. Anteroinferiorly from the bridge is the fleshy, cartilaginous **dorsum nasi.** The dorsum nasi is supported by one pair of **lateral cartilages** and two pairs of **alar cartilages.** Paired **nostrils,** or *nares* (nā′res; sing., nā′ris), open on the inferior surface of the nose.

The internal surface of the nose leads to the **nasal cavity (figure 25.2).** The nasal cavity is continuous posteriorly with the nasopharynx via paired openings called **choanae** (kō′an-ē; sing., *choana*), or *internal nares.* The frontal bone, nasal bones, cribriform plate of the ethmoid, and sphenoid bone form the roof of the nasal cavity. The palatine process of the maxillae and the horizontal plate of the palatine bones form the hard palate, which is the nasal cavity floor. The anterior region of the nasal cavity, near the nostrils, is called the **vestibule.**

The nasal cavity is lined with pseudostratified ciliated columnar epithelium. Within this epithelium are numerous goblet cells that produce mucin, and immediately deep to this epithelium is an extensive vascular network. Near the vestibule are coarse hairs called **vibrissae** (vī-bris′ē; sing., *vibrissa; vibro* = to quiver) that help trap larger particles before they pass through the nasal cavity. The most superior part of the nasal cavity contains the **olfactory epithelium,** which is composed of both a pseudostratified ciliated columnar epithelium and olfactory receptor cells.

The **nasal septum** divides the nasal cavity into left and right portions. It is formed anteriorly by **septal nasal cartilage.** A thin, bony sheet formed by the perpendicular plate of the ethmoid bone (superiorly) and the vomer bone (inferiorly) forms the posterior part of the nasal septum.

Along the lateral walls of the nasal cavity are three paired, bony projections: the **superior, middle,** and **inferior nasal conchae** (kon′kē; sing., *concha*; a shell). These conchae subdivide the nasal cavity into separate air passages, each called a **nasal meatus** (mē-ā′tŭs). The superior, middle, and inferior meatuses are located immediately inferior to their corresponding nasal conchae.

As inhaled air passes over constricted, narrow grooves in each meatus, the inhaled air becomes turbulent. Increased turbulence ensures that the air remains in the nasal cavity for a longer time, so that the air becomes warmed and humidified. Because the conchae help produce this turbulence, they are sometimes called the "turbinate" bones.

Besides functioning in filtration, conditioning, and olfaction, the nasal cavity is a resonating chamber that contributes to sound production, discussed later in this chapter.

WHAT DO YOU THINK?

1. What does it mean if someone has a "deviated septum"? What kinds of problems can arise with a deviated septum?

Paranasal Sinuses

Four bones of the skull contain paired air spaces called the **paranasal** (par-ă-nā′săl; *para* = alongside) **sinuses,** which together decrease skull bone weight. These spaces are named for the bones in which they are housed; thus, from a superior to inferior direction, they are the **frontal, ethmoidal sphenoidal,** and **maxillary sinuses (figure 25.3;** see also chapter 7). All sinuses communicate with the nasal cavity by ducts and are lined with the same pseudostratified ciliated columnar epithelium as the nasal cavity.

Study Tip!

The nasal cavity and the paranasal sinuses are the primary structures that warm and humidify the air we inhale. To illustrate this, breathe through your mouth instead of your nose on a cold day. Your throat and trachea may feel "raw," and you may cough because the air entering the lungs from your mouth is not being properly conditioned. Repeat this experiment by breathing through your nose. Once the air is warmed and humidified in the nasal cavity, it is much easier to breathe.

Pharynx

The common space used by both the respiratory and digestive systems is the **pharynx** (far′ingks), commonly called the throat (figure 25.2). The pharynx is funnel-shaped, meaning that it is slightly wider superiorly and narrower inferiorly. The pharynx originates posterior to the nasal and oral cavities and extends inferiorly to the level of the bifurcation of the larynx and esophagus. For most of its length, the pharynx is the common pathway for both inhaled and exhaled air (the respiratory system) and ingested food (the digestive system).

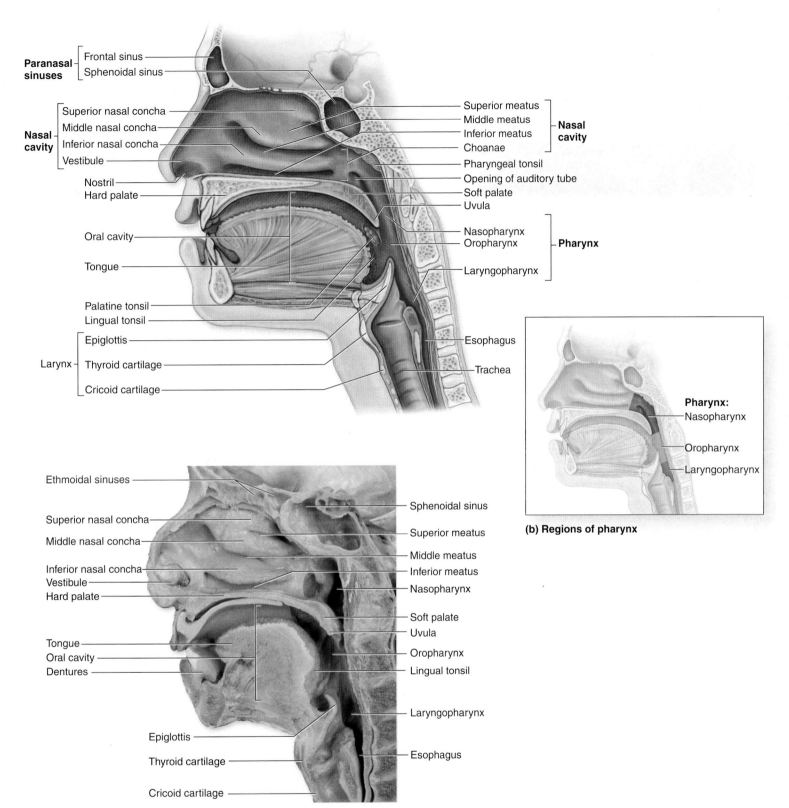

Paranasal sinuses
- Frontal sinus
- Sphenoidal sinus

Nasal cavity
- Superior nasal concha
- Middle nasal concha
- Inferior nasal concha
- Vestibule

- Nostril
- Hard palate
- Oral cavity
- Tongue

- Palatine tonsil
- Lingual tonsil

Larynx
- Epiglottis
- Thyroid cartilage
- Cricoid cartilage

Nasal cavity
- Superior meatus
- Middle meatus
- Inferior meatus
- Choanae

- Pharyngeal tonsil
- Opening of auditory tube
- Soft palate
- Uvula

Pharynx
- Nasopharynx
- Oropharynx
- Laryngopharynx

- Esophagus
- Trachea

(b) Regions of pharynx

Pharynx:
- Nasopharynx
- Oropharynx
- Laryngopharynx

- Ethmoidal sinuses
- Superior nasal concha
- Middle nasal concha
- Inferior nasal concha
- Vestibule
- Hard palate
- Tongue
- Oral cavity
- Dentures
- Epiglottis
- Thyroid cartilage
- Cricoid cartilage

- Sphenoidal sinus
- Superior meatus
- Middle meatus
- Inferior meatus
- Nasopharynx
- Soft palate
- Uvula
- Oropharynx
- Lingual tonsil
- Laryngopharynx
- Esophagus

(a) Sagittal section

Figure 25.2

Anatomy of the Upper Respiratory Tract. The upper respiratory tract includes the nose, nasal cavity, paranasal sinuses, and pharynx. *(a)* A diagrammatic sagittal section and cadaver photo of the head show the upper respiratory tract structures and their relationship to the larynx, trachea, and esophagus. *(b)* The three specific regions of the pharynx (nasopharynx, oropharynx, and laryngopharynx) are highlighted in a diagrammatic sagittal section.

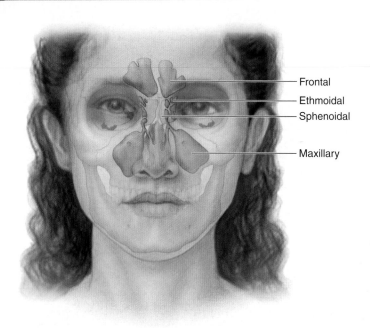

- Frontal
- Ethmoidal
- Sphenoidal
- Maxillary

Figure 25.3

Paranasal Sinuses. The paranasal sinuses are air-filled cavities named for the bones in which they are found: frontal, ethmoidal, sphenoidal, and maxillary.

The pharynx is lined by a mucosa and contains skeletal muscles that are primarily used for swallowing. Its flexible lateral walls are distensible in order to force swallowed food into the esophagus. The pharynx is partitioned into three adjoining regions: the nasopharynx, oropharynx, and laryngopharynx (**table 25.1**). Figure 25.2*b* color-codes these regions of the pharynx for your reference.

Nasopharynx

The **nasopharynx** (nā'zō-far'ingks) is the superiormost region of the pharynx. The nasopharynx is located directly posterior to the nasal cavity and superior to the soft palate, which separates it from the posterior part of the oral cavity. It is lined with a pseudostratified ciliated columnar epithelium.

Normally, only air passes through the nasopharynx. Material from the oral cavity and oropharynx is typically blocked from entering the nasopharynx by the soft palate, which elevates when we swallow. However, sometimes an accident occurs, and food or drink enters the nasopharynx and the nasal cavity. For example, if a person is swallowing and laughing at the same time, the soft palate cannot form as good a seal for the nasopharynx. The force from the laughing and the lack of a good seal may propel some of the material into the nasal cavity. In other instances, severe vomiting can propel material so forcibly that the vomitus is expelled through both the oral cavity and the nasal cavity.

In the lateral walls of the nasopharynx, paired **auditory tubes** (*eustachian tubes* or *pharyngotympanic tubes*) connect the nasopharynx to the middle ear. Recall that the auditory tubes equalize air pressure between the middle ear and the atmosphere by allowing excess air pressure to be released into the nasopharynx. The posterior nasopharynx wall also houses a single **pharyngeal tonsil** (commonly called the *adenoids* [ad'ĕ-noydz; *aden* = gland, *eidos* = resemblance]).

Oropharynx

The middle pharyngeal region, the **oropharynx** (ōr'ō-far'ingks), is immediately posterior to the oral cavity. The oropharynx is bounded by the edge of the soft palate superiorly and by the hyoid bone inferiorly. It is a common respiratory and digestive pathway through which both air and swallowed food and drink pass. Nonkeratinized stratified squamous epithelium lines the oropharynx because this epithelium is strong enough to withstand the abrasion of swallowed food.

The **fauces** (faw'sēz; throat) is the opening that represents the threshold for entry into the oropharynx from the oral cavity. Two pairs of muscular arches, the anterior palatoglossal arches and the posterior palatopharyngeal arches, form the entrance into the oropharynx from the oral cavity.

Lymphatic organs in the oropharynx provide the "first line of defense" against ingested or inhaled foreign materials. The **palatine tonsils** are on the lateral wall between the arches, and the **lingual tonsils** are at the base of the tongue.

Laryngopharynx

The inferior, narrowed region of the pharynx is the **laryngopharynx** (lă-ring'gō-far'ingks). It extends inferiorly from the hyoid bone and is continuous with the larynx and esophagus. The laryngopharynx terminates at the superior border of the esophagus, which is equivalent to the inferior border of the cricoid cartilage in the larynx. In

Table 25.1	Regions of the Pharynx		
	Function	**Epithelial Lining**	**Characteristics**
Nasopharynx	Conducts air	Pseudostratified ciliated columnar epithelium	Posterior to nasal cavity Pharyngeal tonsil on posterior wall Auditory tubes open into nasopharynx to equalize air pressure in the middle ear
Oropharynx	Conducts air; serves as passageway for food and drink	Nonkeratinized stratified squamous epithelium	Posterior to oral cavity Paired palatine tonsils on lateral walls Lingual tonsils on base of tongue (and thus in anterior region of oropharynx) Extends between soft palate and level of hyoid bone
Laryngopharynx	Conducts air; serves as passageway for food and drink	Nonkeratinized stratified squamous epithelium	Extends from level of hyoid bone to beginning of esophagus (posterior to level of cricoid cartilage in larynx)

fact, the larynx forms the anterior wall of this part of the pharynx. The laryngopharynx is lined with a nonkeratinized stratified squamous epithelium since it permits passage of both food and air.

WHAT DID YOU LEARN?

3 What changes occur in inhaled gases as they travel through the respiratory system?

4 What is the function of the nasal conchae?

5 How is swallowed food prevented from entering the nasopharynx?

Lower Respiratory Tract

Key topics in this section:

- Organization and functions of lower respiratory tract organs and regions
- Characteristics of the respiratory membrane

The **lower respiratory tract** is made up of conducting airways (larynx, trachea, bronchi, bronchioles, and their associated structures) as well as the respiratory portion of the respiratory system (respiratory bronchioles, alveolar ducts, and alveoli) **(table 25.2;** see figure 25.1). A lower respiratory tract infection affects some or all of these structures.

Larynx

The **larynx** (lar′ingks), also called the *voice box*, is a short, somewhat cylindrical airway **(figure 25.4)**. It is continuous superiorly with the laryngopharynx, and inferiorly with the trachea; it is anterior to the esophagus. The superior aspect of the larynx is lined with a nonkeratinized stratified squamous epithelium. Inferior to the vocal cords, the larynx lining becomes a pseudostratified ciliated columnar epithelium. The larynx conducts air into the lower respiratory tract, and produces sounds.

The larynx is supported by a framework of nine pieces of cartilage (three individual pieces and three cartilage pairs) that are held in place by ligaments and muscles. The largest cartilage is the **thyroid cartilage,** which forms only the anterior and lateral walls of the larynx. It has no posterior component and it is formed from hyaline cartilage. A dense connective tissue band called the **thyrohyoid membrane** attaches the superior border of the thyroid cartilage to the hyoid bone. The V-shaped anterior projection of the thyroid cartilage is called the **laryngeal** (lă-rin′jē-ăl) **prominence** (commonly referred to as the "Adam's apple" in males). The overall growth of the thyroid cartilage is stimulated by testosterone; thus, the Adam's apple is usually prominent and larger in males following puberty.

The ring-shaped **cricoid** (krī′koyd; *kridos* = a ring) **cartilage** forms the inferior base of the larynx and connects to the trachea inferiorly. The cricoid cartilage is composed of hyaline cartilage. It

Table 25.2	Structures of the Lower Respiratory Tract			
Structure[1]	**Anatomic Description**	**Wall Support**	**Epithelial Lining**	**Function**
Larynx	Connects to pharynx and trachea; composed of cartilage, skeletal muscle, and laryngeal ligaments; also called the voice box	Nine pieces of cartilage; supported by ligaments and skeletal muscle	Nonkeratinized stratified squamous epithelium superior to vocal folds; pseudostratified ciliated columnar epithelium inferior to vocal folds	Conducting: air; produces sound
Trachea	Flexible, but semirigid tubular organ connecting larynx to primary bronchi; incomplete, C-shaped cartilages keep trachea patent (open)	C-shaped cartilage rings	Pseudostratified ciliated columnar epithelium	Conducting: air
Bronchi	Largest airways of the bronchial tree; consist of primary, secondary, tertiary, and smaller bronchi	Incomplete rings and irregular plates of cartilage; some smooth muscle	Larger bronchi lined by pseudostratified ciliated columnar epithelium; smaller bronchi lined by simple columnar epithelium	Conducting: air
Bronchioles	Smaller conducting airways of bronchial tree; larger bronchioles branch into smaller bronchioles; terminal bronchioles are the last part of the conducting portion	No cartilage; proportionately greater amounts of smooth muscle in walls of terminal bronchioles	Epithelium ranges from simple columnar (for the largest bronchioles) to simple cuboidal (for smaller bronchioles)	Conducting: air; smooth muscle in walls allows for bronchoconstriction and bronchodilation
Respiratory bronchioles	Smallest conducting airways; begin the respiratory portion	No cartilage; smooth muscle is scarce in their walls	Simple cuboidal epithelium	Respiratory: gas exchange
Alveolar ducts	Tiny airways that branch off respiratory bronchioles; multiple alveoli found along walls of alveolar duct	No cartilage, no smooth muscle	Simple squamous epithelium	Respiratory: gas exchange
Alveoli	Tiny microscopic air sacs	No cartilage, no smooth muscle	Simple squamous epithelium	Respiratory: gas exchange

[1] Structures are listed in the order that air passes through them during inhalation.

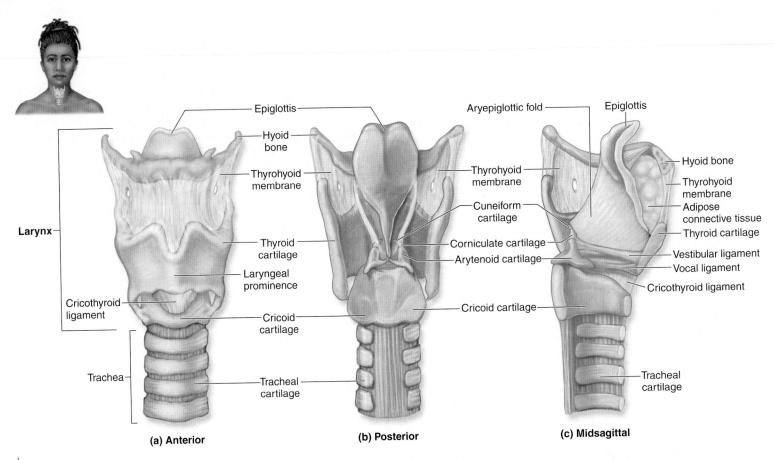

(a) Anterior **(b) Posterior** **(c) Midsagittal**

Figure 25.4

Larynx. The larynx functions primarily to prevent food and fluid from entering the lower respiratory tract. Its secondary function is sound production. Laryngeal anatomy and its relationship to the hyoid bone and trachea are compared in *(a)* anterior, *(b)* posterior, and *(c)* midsagittal views.

has a narrow anterior region, but its posterior region is wide to support the posterior larynx. A dense regular connective tissue band called the **cricothyroid ligament** attaches the cricoid cartilage to the inferior edge of the thyroid cartilage.

The large, spoon- or leaf-shaped **epiglottis** (ep-i-glot′is; *epi* = on, *glottis* = mouth of windpipe) is formed primarily of elastic cartilage. The epiglottis projects superiorly into the pharynx from its attachment to the thyroid cartilage. When a person swallows, the larynx moves anteriorly and superiorly, causing the epiglottis to close over the laryngeal opening and prevent materials from entering the larynx. After swallowing, the larynx returns to its normal position, and the epiglottis elevates and returns to its original position.

The paired **arytenoid** (ar-i-tē′noyd) **cartilages** have a pyramidal shape, and they rest on the superoposterior border of the cricoid cartilage. The paired **corniculate** (kōr-nik′ū-lāt; *corniculatus* = horned) **cartilages** attach to the superior surface of the arytenoid cartilages. The paired **cuneiform** (kū′nē-i-fōrm; *cuneus* = wedge) **cartilages** do not directly attach to any other cartilages. Instead, they are supported within a mucosa-covered connective tissue sheet called the **aryepiglottic fold.** The aryepiglottic fold extends between

the lateral sides of each arytenoid cartilage and the epiglottis to support some of the laryngeal soft tissue structures.

Two groups of laryngeal muscles are located within the larynx. The **intrinsic muscles** attach to the arytenoid and corniculate cartilages. They cause the arytenoid cartilages to pivot, and regulate tension on the vocal folds. The **extrinsic muscles** are the infrahyoid muscles (see chapter 11) that attach the hyoid bone to the thyroid cartilage. They normally stabilize the larynx and help move it during swallowing.

Sound Production

Certain structures of the larynx function specifically in sound production. Two pairs of ligaments extend from the posterior surface of the thyroid cartilage to the arytenoid cartilages. The inferior ligaments, called **vocal ligaments,** are covered by a mucous membrane. These ligaments together with their mucosa are called the **vocal folds (figure 25.5).** Vocal folds are "true vocal cords" because they produce sound when air passes between them. The superior ligaments are called **vestibular ligaments** (see figure 25.4*c*). Together with the mucosa covering them, they are called the **vestibular folds**

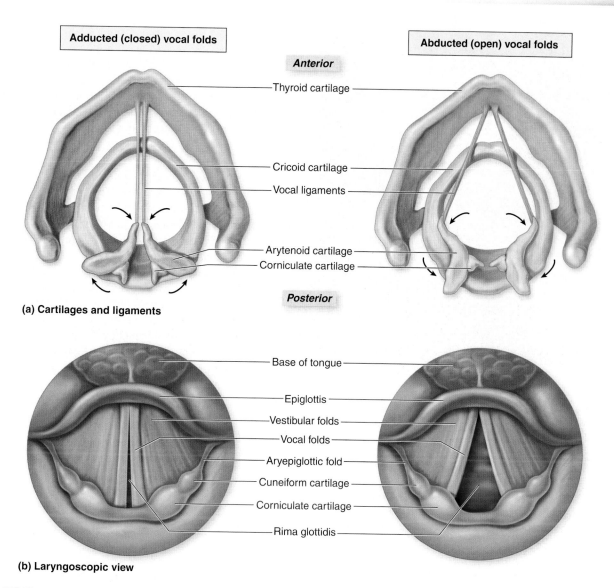

| Adducted (closed) vocal folds | Abducted (open) vocal folds |

Anterior

- Thyroid cartilage
- Cricoid cartilage
- Vocal ligaments
- Arytenoid cartilage
- Corniculate cartilage

Posterior

(a) Cartilages and ligaments

- Base of tongue
- Epiglottis
- Vestibular folds
- Vocal folds
- Aryepiglottic fold
- Cuneiform cartilage
- Corniculate cartilage
- Rima glottidis

(b) Laryngoscopic view

Figure 25.5

Vocal Folds.   The vocal folds (true vocal cords) are epithelium-covered elastic ligaments extending between the thyroid and arytenoid cartilages. These folds surround the rima glottidis and are involved in sound production. Adducted (closed) and abducted (open) vocal folds are shown in *(a)* a superior view of the cartilages and ligaments only and *(b)* a diagrammatic laryngoscopic view of the coverings around these cartilages and ligaments.

(**figure 25.6**). These folds are "false vocal cords" because they have no function in sound production, but protect the vocal folds. The vestibular folds attach to the corniculate cartilages.

When intrinsic muscles of the larynx make the arytenoid cartilages pivot, they can abduct or adduct the vocal folds. The opening between the vocal folds is called the **rima glottidis** (rī′mă; slit; glo-tī′-dis). This opening widens if the vocal folds are abducted and becomes narrower if the vocal folds are adducted (see figure 25.5). The term **glottis** (glot′is) refers to the rima glottidis plus the vocal folds.

When air is forced through the rima glottidis, the vocal folds begin to vibrate, and this vibration produces sound. The nonkeratinized stratified squamous epithelium lining the vocal folds with-

stands this abrasive contact between the two vocal folds and their vibrational activity during sound production. The length, tension, and position of the vocal folds determine the characteristics of the sound, as follows:

- The range of a voice (be it soprano or bass) is determined by the length of the vocal folds. Longer vocal folds produce lower sounds than shorter vocal folds. As we grow, our vocal folds increase in length, which is why our voices become deeper as we mature into adults. Also, both the growth of the thyroid cartilage and the longer and thicker vocal folds in mature males help explain why men typically have deeper voices than females.

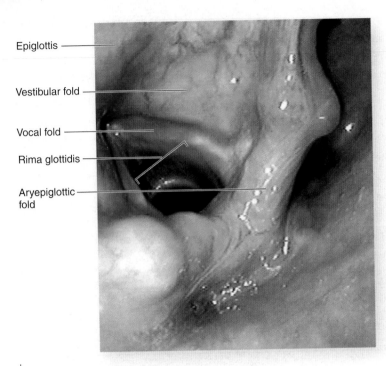

Epiglottis

Vestibular fold

Vocal fold

Rima glottidis

Aryepiglottic fold

Figure 25.6

Laryngoscopic View of the Larynx. A superior laryngoscopic view shows the vestibular folds, the vocal folds, and the rima glottidis opening into the trachea.

- Pitch refers to the frequency of sound waves, and is determined by the amount of tension or tautness on the vocal folds as regulated by the intrinsic laryngeal muscles. Increasing the tension on the vocal folds causes the vocal folds to vibrate more when air passes by them and produces a higher sound. Conversely, the less taut the vocal folds, the less they vibrate and the lower the pitch of the sound.
- Loudness depends on the force of the air passing across the vocal folds. A lot of air forced through the rima glottidis produces a loud sound; a little air forced through the rima glottidis produces a soft sound. When you whisper, only the most posterior portion of the rima glottidis is open, and the vocal folds do not vibrate. Since the vocal folds are not vibrating, the whispered sounds are all of the same pitch.

Study Tip!

The vocal folds are comparable to the strings of a harp. The short strings produce the high notes, while the long strings produce the low notes. Thus, shorter vocal folds produce higher notes than longer vocal folds.

Keep in mind that recognizable speech also requires the participation of the pharynx, nasal and oral cavities, paranasal sinuses, lips, teeth, and tongue. If you have a stuffy nose, the quality of your voice changes to a more nasal tone. Try this experiment: Hold your nose and then speak. You will notice that your voice sounds quite different when air doesn't pass through the nasal cavity. A sinus infection can also cause the sound of the voice to change as fluid accumulation leads to decreased space in the paranasal sinuses for

sound resonance. In addition, young children tend to have high, nasal-like voices because their sinuses are not yet well-developed, so they lack large "chambers" where sounds can resonate. A child also has shorter, smaller vocal folds, which produce a higher voice. When a male goes through puberty, his laryngeal cartilages and vocal folds grow rapidly, producing the "cracking" voice that eventually leads to a deeper voice at maturity.

Trachea

The **trachea** (trā′kē-ă; rough) is a flexible, slightly rigid tubular organ often referred to as the "windpipe." The trachea extends through the mediastinum and lies immediately anterior to the esophagus, inferior to the larynx, and superior to the primary bronchi of the lungs. The trachea averages approximately 2.5 centimeters in diameter and 12 to 14 centimeters in length.

The anterior and lateral walls of the trachea are supported by 15 to 20 C-shaped **tracheal cartilages (figure 25.7a)**. These cartilage "rings" reinforce and provide some rigidity to the tracheal wall to ensure that the trachea remains open (patent) at all times. The cartilage rings are connected by elastic connective tissue sheets called **anular** (an′ū-lăr; *anulus* = ring) **ligaments**. The open ends of each C-shaped piece are bound together by the **trachealis muscle** and an elastic, ligamentous membrane (figure 25.7b). The trachealis distends during swallowing and bulges into the lumen of the trachea to allow for expansion of the esophagus to accommodate larger materials being swallowed. During coughing, it contracts to reduce trachea diameter, thus facilitating the more rapid expulsion of air

CLINICAL VIEW

Laryngitis

Laryngitis (lar-in-jī′tis) is inflammation of the larynx that may extend to the surrounding structures. Viral or bacterial infection is the number one cause of laryngitis. Less frequently, laryngitis follows overuse of the voice, such as yelling for several hours at a football game. Symptoms include hoarse voice, sore throat, and sometimes fever. In severe cases, the inflammation and swelling can extend to the epiglottis. In children, whose airways are proportionately smaller, a swollen epiglottis may lead to sudden airway obstruction and become a medical emergency.

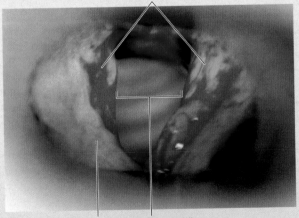

Inflamed vocal folds

Vestibular fold   Rima glottidis

A laryngoscopic view shows the inflamed, reddened vocal folds characteristic of laryngitis.

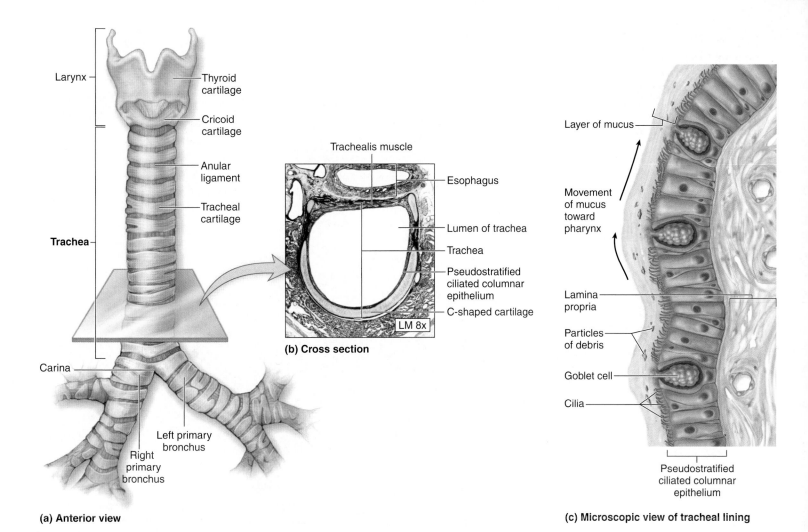

Figure 25.7

Trachea. *(a)* The trachea connects to the larynx superiorly and the primary bronchi inferiorly. *(b)* A cross-sectional photomicrograph shows the relationship of the trachea (anteriorly) and the esophagus (posteriorly). The wall of the trachea is supported by C-shaped rings of cartilage. *(c)* The trachea is lined with a pseudostratified ciliated columnar epithelium that propels mucus and debris away from the lungs and toward the pharynx.

and helping to loosen material (foreign objects or food materials) from the air passageway.

The mucosa lining the trachea is a pseudostratified ciliated columnar epithelium with numerous mucin-secreting goblet cells and an underlying lamina propria that houses mucin-secreting glands (figure 25.7c). The movement of cilia propels mucus laden with dust and dirt particles toward the larynx and the pharynx, where it is swallowed. A submucosal layer deep to the mucosa contains many submucous glands.

At the level of the sternal angle, the trachea bifurcates into two smaller tubes, called the right and left **primary bronchi** (brong′kī; *bronchos* = windpipe) (or *main bronchi*). Each primary bronchus projects inferiorly and laterally toward a lung. The most inferior tracheal cartilage separates the primary bronchi at their origin and forms an internal ridge called the **carina** (kă-rī′nă; keel).

WHAT DO YOU THINK?

2 In chronic smokers, the lining of the trachea and bronchi changes from a pseudostratified ciliated columnar epithelium to a stratified squamous epithelium. Why do you think this change occurs? What are some consequences of this epithelium in the trachea?

Bronchial Tree

The **bronchial tree** is a highly branched system of air-conducting passages that originate from the left and right primary bronchi and progressively branch into narrower tubes as they diverge throughout the lungs before ending in terminal bronchioles **(figure 25.8)**. Incomplete rings of hyaline cartilage support the walls of the primary bronchi to ensure that they remain open. The right primary bronchus is shorter, wider, and more vertically oriented than the left primary bronchus; thus, foreign particles are more likely to lodge in the right primary bronchus.

The primary bronchi enter the hilum of each lung together with the pulmonary vessels, lymphatic vessels, and nerves. Each primary bronchus then branches into **secondary bronchi** (or *lobar bronchi*). The left lung has two secondary bronchi since it has two lobes; the right lung has three lobes and three secondary bronchi. Secondary bronchi are smaller in diameter than primary bronchi. They further divide into **tertiary bronchi**. The right lung is supplied by 10 tertiary bronchi, and the left lung is supplied by 8 to 10 tertiary bronchi. (The difference in the number depends upon whether some of the left lung tertiary bronchi are combined or separate structures.) Each tertiary bronchus is called a *segmental bronchus* because it

Figure 25.8

Bronchial Tree.   (a) The bronchial tree is composed of conducting passageways that originate at the primary bronchi and end at the terminal bronchioles (not visible in this view). (b) The major components of the bronchial tree are color-coded.

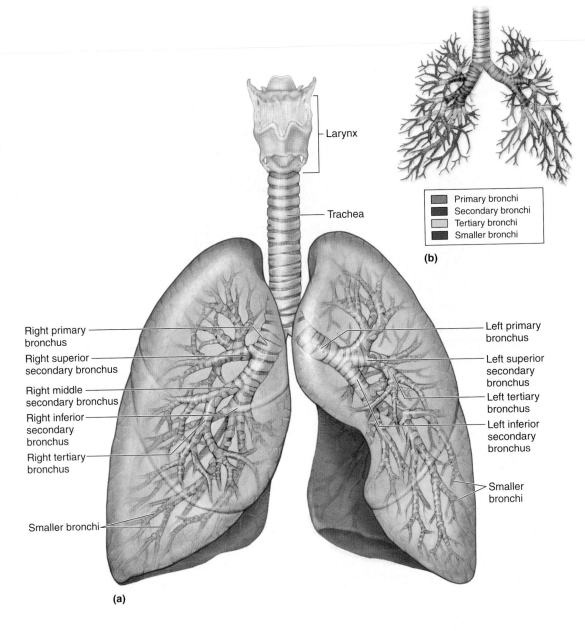

— Larynx

— Trachea

Primary bronchi
Secondary bronchi
Tertiary bronchi
Smaller bronchi

(b)

Right primary bronchus

Right superior secondary bronchus

Right middle secondary bronchus

Right inferior secondary bronchus

Right tertiary bronchus

Smaller bronchi

Left primary bronchus

Left superior secondary bronchus

Left tertiary bronchus

Left inferior secondary bronchus

Smaller bronchi

(a)

supplies a part of the lung called a bronchopulmonary segment, discussed later in this chapter.

There are approximately 9 to 12 different levels of bronchial branch division (not all shown in figure 25.8). Thus, the primary, secondary, and tertiary bronchi are the first, second, and third generations of bronchi, respectively. All types of bronchi exhibit some common characteristics:

- Incomplete rings of cartilage in the walls become less numerous and smaller, eventually consisting only of scattered pieces of cartilage as the bronchi continue to divide and decrease in diameter.
- The largest branches of bronchi are lined by a pseudostratified ciliated columnar epithelium, whereas smaller branches of bronchi are lined by a ciliated columnar epithelium.

- A complete ring of smooth muscle is found between the mucosa of the airways and the cartilaginous support in the wall.

The bronchi branch into smaller and smaller tubules that eventually reach a diameter of less than 1 millimeter. These smaller tubules, called **bronchioles** (brong′kē-ōl), are no longer lined with pseudostratified ciliated columnar epithelium, but with simple columnar or simple cuboidal epithelium. Unlike the smallest bronchi, which have irregular plates of cartilage in their walls, the walls of bronchioles contain no cartilage, since their small diameter alone prevents collapse. Instead, they have a thicker layer of smooth muscle than do bronchi, a characteristic that helps regulate airway constriction or dilation. Contraction of the smooth muscle narrows bronchioles (called **bronchoconstriction**), whereas relaxation of

Bronchitis

Bronchitis (brong-kī′tis) is inflammation of the bronchi caused by viruses or bacteria, or by inhaling vaporized chemicals, particulate matter, or cigarette smoke from the air. Clinically, bronchitis is divided into two categories, acute and chronic.

Acute bronchitis develops rapidly either during or after an infection, such as a cold. Symptoms include cough, wheezing, pain upon inhalation, and fever. Most cases of acute bronchitis resolve completely within 10 to 14 days.

Chronic bronchitis results from long-term exposure to irritants such as chemical vapors, polluted air, or cigarette smoke. Medically, chronic bronchitis is defined as the production of large amounts of mucus, associated with a cough lasting 3 continuous months. If exposure to the irritant persists, permanent changes to the bronchi occur, including (1) thickened bronchial walls with subsequent narrowing of their lumens, (2) overgrowth (hyperplasia) of the mucin-secreting cells of the bronchi, and (3) accumulation of lymphocytes within the bronchial walls. These long-term changes in the bronchi increase the likelihood of bacterial infections, and chronic bronchitis greatly increases the chance of developing pneumonia as well.

the smooth muscle dilates bronchioles (called **bronchodilation**). Constriction or dilation of the bronchioles regulates the amount of air traveling through the bronchial tree.

The **terminal bronchioles** are the final segment of the conducting pathway. They conduct air into the respiratory portion of the respiratory system.

WHAT DO YOU THINK?

3 Can you think of any reasons why you would want your bronchioles to constrict? Why wouldn't you want your bronchioles fully dilated all the time?

Respiratory Bronchioles, Alveolar Ducts, and Alveoli

As previously stated, the respiratory portion of the respiratory system consists of respiratory bronchioles, alveolar ducts, and pulmonary alveoli **(figure 25.9)**. Within this respiratory portion, the epithelium is much thinner than in the conducting portion, thus facilitating gas diffusion between pulmonary capillaries and the respiratory structures.

Terminal bronchioles branch to form the **respiratory bronchioles.** Subsequent partitioning of the respiratory bronchioles forms smaller respiratory bronchioles. Eventually, the smallest respiratory bronchioles subdivide into thin airways called **alveolar ducts,** which are lined with a simple squamous epithelium. The distal end of an alveolar duct terminates as a dilated **alveolar sac.**

Both of these airways—respiratory bronchioles and alveolar ducts—contain small saccular outpocketings called **alveoli** (al-vē′ō-lī; sing., al-vē′ō-lŭs; *alveus* = hollow sac). An alveolus is about 0.25 to 0.5 millimeter in diameter. Its thin wall is specialized to promote diffusion of gases between the alveolus and the blood in the pulmonary capillaries **(figure 25.10)**. The lungs contain approximately 300–400 million alveoli. Alveoli abut one another, so their sides become slightly flattened. Thus, an alveolus in cross section actually looks more hexagonal or polygonal in shape than circular (see figure 25.9*a*). The small openings in the walls between adjacent alveoli, called **alveolar pores,** provide for collateral ventilation of alveoli.

The spongy nature of the lung is due to the packing of millions of alveoli together.

Two cell types form the alveolar wall. The predominant cell is an **alveolar type I cell,** also called a *squamous alveolar cell*. This simple squamous epithelial cell promotes rapid gas diffusion across the alveolar wall. The **alveolar type II cell,** called a *septal cell*, is part of a smaller population of cells within the alveolar wall. Typically, it displays an almost cuboidal shape. Alveolar type II cells secrete **pulmonary surfactant** (ser-fak′tănt), a fluid that coats the inner alveolar surface to reduce surface tension and prevent the collapse of the alveoli. **Alveolar macrophages,** or *dust cells*, may be either fixed or free. Fixed alveolar macrophages remain within the connective tissue of the alveolar walls, while free alveolar macrophages are migratory cells that continually crawl within the alveoli, engulfing any microorganisms or particulate material that has reached the alveoli. The alveolar macrophages are able to leave the lungs either by entering alveolar lymphatics or by being coughed up in sputum (matter from the respiratory tract, such as mucus mixed with saliva) and then expectorated from the mouth.

The **respiratory membrane** is the thin wall between the alveolar lumen and the blood (see figure 25.10). It consists of the plasma membranes of an alveolar type I cell, and an endothelial cell of a capillary, and their fused basement membranes. Oxygen diffuses from the lumen of the alveolus across the respiratory membrane into the pulmonary capillary, thereby allowing the erythrocytes in the blood to become oxygenated again. Conversely, carbon dioxide diffuses from the blood in the capillary through the respiratory membrane to enter the alveolus. Once in the alveoli, carbon dioxide is exhaled from the respiratory system into the external environment.

WHAT DID YOU LEARN?

6 What function is served by the vocal folds?

7 What are some anatomic differences between bronchi and bronchioles?

8 How do terminal and respiratory bronchioles differ in structure and function?

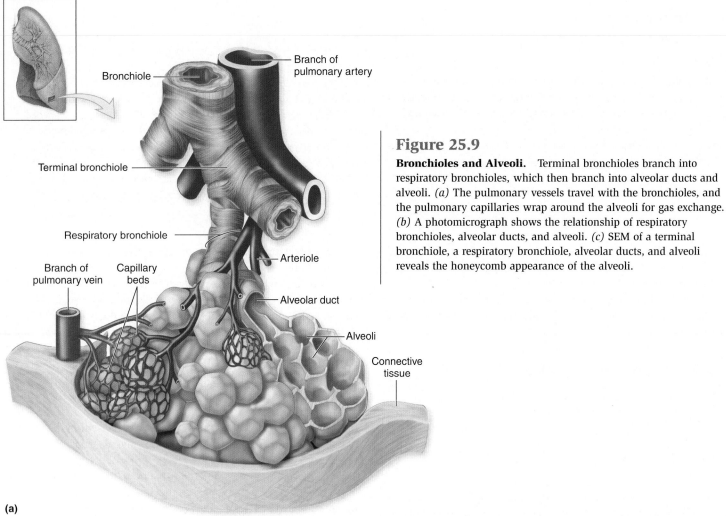

(a)

Figure 25.9

Bronchioles and Alveoli. Terminal bronchioles branch into respiratory bronchioles, which then branch into alveolar ducts and alveoli. (a) The pulmonary vessels travel with the bronchioles, and the pulmonary capillaries wrap around the alveoli for gas exchange. (b) A photomicrograph shows the relationship of respiratory bronchioles, alveolar ducts, and alveoli. (c) SEM of a terminal bronchiole, a respiratory bronchiole, alveolar ducts, and alveoli reveals the honeycomb appearance of the alveoli.

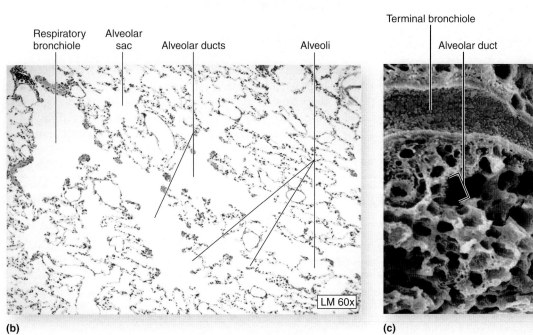

(b)

(c)

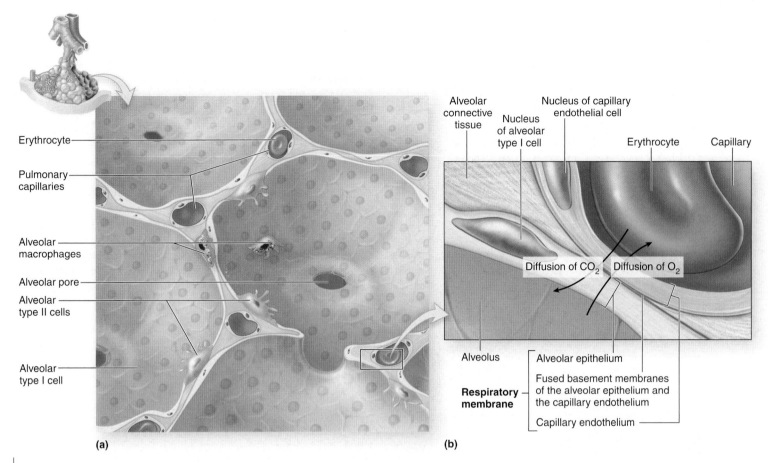

Figure 25.10

Alveoli and the Respiratory Membrane.   Gas exchange between the alveoli and the pulmonary capillaries occurs across a thin respiratory membrane. *(a)* A diagram shows the structural arrangement of several adjacent alveoli. *(b)* The respiratory membrane consists of an alveolar type I cell, an endothelial cell of a capillary, and their fused basement membranes. Oxygen diffuses from alveoli into the blood within the capillary, and carbon dioxide diffuses in the opposite direction. (Note: The pulmonary surfactant covering layer is not shown here.)

Lungs

Key topics in this section:

- Structure and function of the pleura
- Anatomy of the lungs

The lungs house the bronchial tree and the respiratory portion of the respiratory system. The lungs are located on the lateral sides of the thoracic cavity and separated from each other by the mediastinum.

WHAT DO YOU THINK?

4 As you will soon learn, the left lung is physically smaller than the right lung. Why is the left lung smaller?

Pleura and Pleural Cavities

The outer lung surfaces and the adjacent internal thoracic wall are lined by a serous membrane called **pleura** (ploor'ă), which is formed from simple squamous epithelium called a mesothelium. The outer surface of each lung is tightly covered by the **visceral pleura,** while the internal thoracic walls, the lateral surfaces of the mediastinum, and the superior surface of the diaphragm are lined by the **parietal pleura (figure 25.11)**. (These pleural layers may also be viewed in the transverse section of the thoracic cavity shown in figure 22.2*d*.) The visceral and parietal pleural layers are continuous at the hilum of each lung. Between these serous membrane layers is a **pleural cavity.** When the lungs are fully inflated, the pleural cavity is a potential space because the visceral and parietal pleurae are almost in contact with each other. The pleural membranes produce an oily, serous fluid that acts as a lubricant, ensuring that opposing pleural membrane surfaces slide by each other with minimal friction during breathing.

Gross Anatomy of the Lungs

The paired, spongy lungs are the primary organs of respiration. Each lung has a conical shape. Its wide, concave **base** rests inferiorly upon the muscular diaphragm, and its relatively blunt superior

CLINICAL VIEW

Pneumothorax

Pneumothorax (noo-mō-thōr'aks; *pneuma* = air) is a condition that occurs when free air gets into the pleural cavity, the space between the parietal and visceral pleura. A pneumothorax may develop in one of two ways. Air may be introduced externally from a penetrating injury to the chest, such as a knife wound or gunshot, or it may originate internally as when a broken rib lacerates the surface of the lung.

The presence of free air in the pleural space sometimes causes the affected lung or a portion of it to deflate, a condition termed **atelectasis** (at-ē-lek'tă-sis; *ateles* = incomplete, *ektasis* = extension). The collapsed portion of the lung remains down until the air has been removed from the pleural space. If a pneumothorax is small, the air exits naturally within a few days. However, a large pneumothorax is a medical emergency requiring insertion of a tube into the pleural space to suck out the free air. After the air has been removed, an airtight bandage is placed over the entry site to prevent air from reentering the pleural space.

A particularly dangerous condition is **tension pneumothorax,** in which a hole in the chest or lung allows air to enter and acts as a one-way valve. As the patient struggles to breathe, air is pulled in through the wound but cannot escape. Air pressure within the pleural space becomes greater, causing atelectasis of the lung and eventually displacing the heart and mediastinal structures. Both lungs then become compressed, and respiratory distress and death occur unless the tension pneumothorax is promptly treated.

In addition to air, fluid can also accumulate in the pleural space. For example, blood may collect (**hemothorax**) due to a lacerated artery, a blood vessel that leaks as a result of surgery, heart failure, or certain tumors. An accumulation of serous fluid within the pleural cavity is called **hydrothorax,** and an accumulation of pus, as occurs with pneumonia, is called **empyema.**

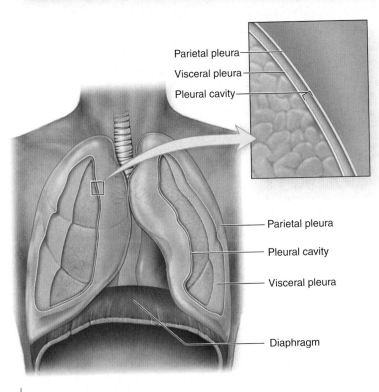

Parietal pleura
Visceral pleura
Pleural cavity

Parietal pleura

Pleural cavity

Visceral pleura

Diaphragm

Figure 25.11

Pleural Membranes.   The serous membranes associated with the lungs are called the pleura. The parietal pleura lines the inner surface of the thoracic cavity, and the visceral pleura covers the outer surface of the lungs. The thin space between these layers is called the pleural cavity.

region, called the **apex** (or *cupola*), projects superiorly to a point that is slightly superior and posterior to the clavicle (**figure 25.12**). Both lungs are bordered by the thoracic wall anteriorly, laterally, and posteriorly, and supported by the rib cage. Toward the midline, the lungs are separated from each other by the mediastinum.

The relatively broad, rounded surface in contact with the thoracic wall is called the **costal surface** of the lung. The **mediastinal surface** of the lung is directed medially, facing the mediastinum and slightly concave in shape. This surface houses the vertical, indented

hilum through which the bronchi, pulmonary vessels, lymph vessels, and nerves pass. Collectively, all structures passing through the hilum are termed the **root** of the lung.

The right and left lungs exhibit some obvious structural differences. Since the heart projects into the left side of the thoracic cavity, the left lung is slightly smaller than the right lung. The left lung has a medial surface indentation, called the **cardiac impression,** that is formed by the heart. The left lung also has an anterior indented region called the **cardiac notch.** The descending thoracic aorta forms a groovelike impression on the medial surface of the left lung.

The right lung is subdivided into the **superior** (*upper*), **middle,** and **inferior** (*lower*) **lobes** by two fissures. The **horizontal fissure** separates the superior from the middle lobe, while the **oblique fissure** separates the middle from the inferior lobe. The left lung has only two lobes, superior and inferior, which are subdivided by an oblique fissure. The **lingula** of the left lung is located on the superior lobe. The lingula is homologous to the middle lobe of the right lung.

The left and right lungs may be partitioned into **bronchopulmonary segments**—10 in the right lung, and typically 8 to 10 in the left lung (**figure 25.13**). (The discrepancy in bronchopulmonary segment number for the left lung comes from the merging or lumping of some left bronchopulmonary segments into combined ones by some anatomists.) Each bronchopulmonary segment is supplied by its own tertiary bronchus and a branch of the pulmonary artery and vein. In addition, each segment is surrounded by connective tissue, thereby encapsulating one segment from another and ensuring that each bronchopulmonary segment is an autonomous unit. Thus, if a portion of a lung is diseased, a surgeon can remove the entire bronchopulmonary segment that is affected, while the remaining healthy bronchopulmonary segments continue to function as before.

Blood Supply To and From the Lungs

Both the pulmonary circulation and the bronchial circulation supply the lungs. Recall from chapter 23 that the **pulmonary circulation** conducts blood to and from the gas exchange surfaces of the lungs to replenish its depleted oxygen levels and get rid of excess carbon dioxide (see figures 23.22 and 23.23). Deoxygenated blood is

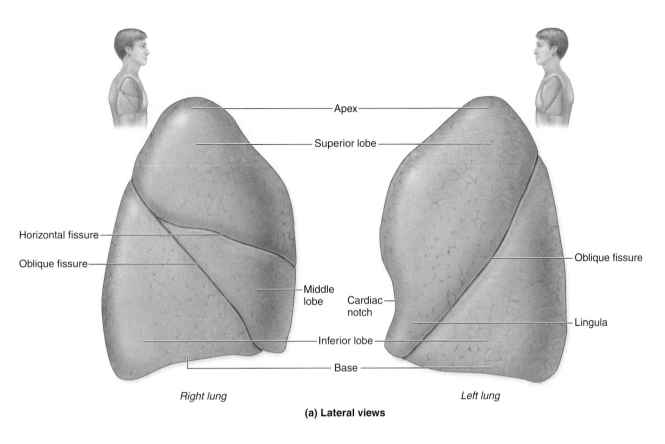

Right lung   *Left lung*

(a) Lateral views

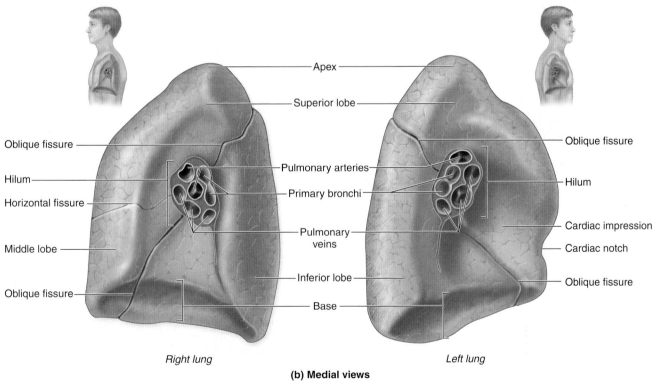

Right lung   *Left lung*

(b) Medial views

Figure 25.12

Gross Anatomy of the Lungs. The lungs are composed of lobes separated by distinct depressions called fissures. *(a)* Lateral views show the three lobes of the right lung and the two lobes of the left lung. *(b)* Medial views show the hilum of each lung, where the pulmonary vessels and bronchi enter and leave the lung.

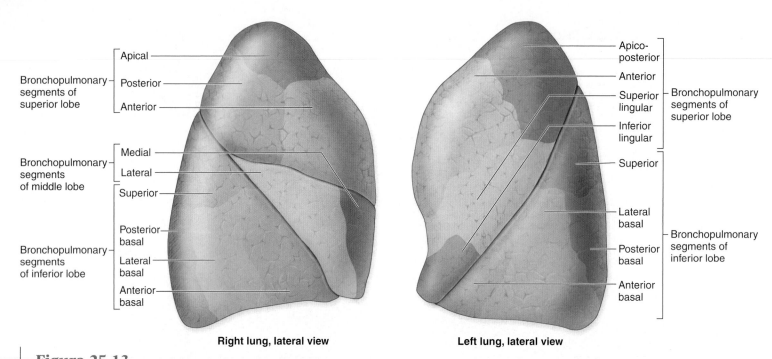

Right lung, lateral view   **Left lung, lateral view**

Figure 25.13

Bronchopulmonary Segments of the Lungs.   The portion of each lung supplied by each tertiary bronchus (represented by different colors) is a bronchopulmonary segment. (The medial basal bronchopulmonary segment cannot be seen from this view.)

pumped from the right ventricle through the pulmonary trunk into pulmonary arteries, which enter the lung. Thereafter, continuous branching of these vessels leads to pulmonary capillaries that encircle all alveoli. The deoxygenated blood that enters these capillaries becomes oxygenated before it returns to the left atrium through a series of pulmonary venules and veins.

The **bronchial circulation** is a component of the systemic circulation. The bronchial circulation consists of tiny bronchial arteries and veins that supply the bronchi and bronchioles of the lung. This part of the circulation system is much smaller than the pulmonary system, because most tiny respiratory structures (alveoli and alveolar ducts) exchange respiratory gases directly with the inhaled air. Approximately three or four tiny **bronchial arteries** branch from the

anterior wall of the descending thoracic aorta and divide to form capillary beds to supply structures in the bronchial tree. Increasingly larger **bronchial veins** collect venous blood and drain into the azygos and hemiazygous systems of veins.

Lymphatic Drainage

Lymph nodes and vessels are located within the connective tissue of the lung as well as around the bronchi and pleura **(figure 25.14)**. The lymph nodes collect carbon, dust particles, and pollutants that were not filtered out by the pseudostratified ciliated columnar epithelium. The lymph from the lung is conducted first to **pulmonary lymph nodes** within the lungs. Lymphatic ves-

Figure 25.14

Lung Lymphatic Drainage. Lymph vessels conduct lymph to the pulmonary, bronchopulmonary, and tracheobronchial lymph nodes. Lymph is then drained by the bronchomediastinal trunks into the right lymphatic duct or the thoracic duct.

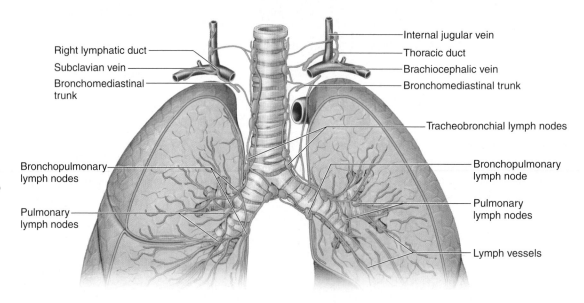

CLINICAL VIEW

Pneumonia

Pneumonia (noo-mō′nē-ă) is an infection of the alveoli of the lung. Common causative agents include viruses and bacteria, and sometimes fungi. The infection may involve an entire lung or just one lobe. Pneumonia results in tissue swelling and accumulated leukocytes in the affected area, thus greatly diminishing the capacity for gas exchange.

Pneumonia is a contagious disease that is usually spread by respiratory droplets. Symptoms include cough, fever, and rapid breathing. In addition, the bronchi produce and expel sputum (mucus and other matter), which may be rust- or green-tinged.

Diagnosis of pneumonia depends on symptoms and characteristic changes seen on a chest x-ray. A sputum culture is often helpful in identifying the specific organism. Treatment may include antibiotics, respiratory support, and medications to relieve symptoms. Patients with severe cases of pneumonia or those with coexisting lung diseases, such as chronic bronchitis or emphysema, may require supplemental oxygen.

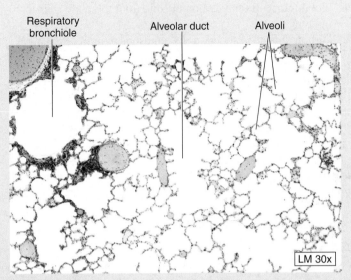

Respiratory bronchiole | Alveolar duct | Alveoli

LM 30x

Normal lung tissues.

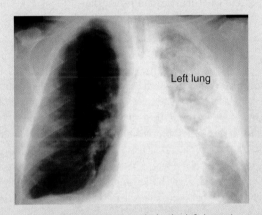

Left lung

Chest x-ray of a patient with pneumonia in the left lung. A normal lung appears as a black space on an x-ray because its spongy structure is not dense. In contrast, a pneumonia lung appears white or opaque on an x-ray due to accumulation of fluid and cells.

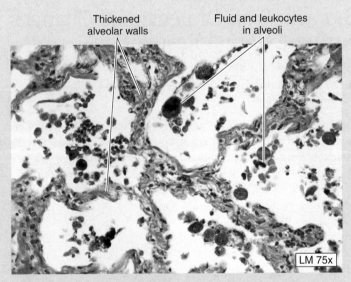

Thickened alveolar walls | Fluid and leukocytes in alveoli

LM 75x

Tissues within a lung affected by pneumonia.

sels exit these lymph nodes and conduct lymph to **bronchopulmonary lymph nodes** located at the hilum of the lung. These vessels drain first into **tracheobronchial lymph nodes** and then into the **left** and **right bronchomediastinal trunks** (discussed in chapter 24). The right bronchomediastinal trunk drains into the right lymphatic duct, while the left bronchomediastinal trunk drains into the thoracic duct.

WHAT DO YOU THINK?

5 The lymph nodes of the lung become black and darkened over time in both smokers and nonsmokers. Why do these lymph nodes turn black?

WHAT DID YOU LEARN?

9 What is the hilum of the lung, and how does it function?

Pulmonary Ventilation

Key topic in this section:

■ The process of ventilation

Breathing, also known as **pulmonary ventilation,** is the movement of air into and out of the respiratory system. At rest, a normal adult breathes about 16 times per minute, and approximately 500 milliliters of air are exchanged with the atmosphere per breath. The airflow exchange is caused by the muscular actions associated with inhalation and exhalation, as well as by differences in atmospheric air pressure and lung (intrapulmonary) air pressure. Gases are exchanged in the following cycle:

■ Oxygen in the atmospheric air is drawn into the lungs by inhalation.
■ Oxygen is transported to the body cells from the lungs by blood circulating through the cardiovascular system.

- Cells use the oxygen and generate carbon dioxide as a waste product.
- Blood transports the carbon dioxide from the body cells to the lungs.
- Carbon dioxide is added to the atmosphere during exhalation.

The movement of gases into and out of the respiratory system follows **Boyle's law,** which states, "The pressure of a gas decreases if the volume of the container increases, and vice versa." Thus, when the volume of the thoracic cavity increases even slightly during inhalation, the intrapulmonary pressure decreases slightly, and air flows into the lungs through the conducting airways. Therefore, air flows from a region of higher pressure (the atmosphere) into a region of lower pressure within the lungs (the intrapulmonary region). Similarly, when the volume of the thoracic cavity decreases during exhalation, the intrapulmonary pressure increases and forces air out of the lungs into the atmosphere.

WHAT DID YOU LEARN?

10 What is pulmonary ventilation?

Thoracic Wall Dimensional Changes During External Respiration

Key topic in this section:

- How the thoracic cavity changes in size and shape during respiration

As you inhale, the dimensions of your thoracic cavity generally increase, forming a larger space for the expanding lungs. During exhalation, your thoracic cavity dimensions return to their original size. Thus, the thoracic cavity becomes larger during inhalation and smaller during exhalation, as diagrammed in **figure 25.15**. Vertical dimensional changes occur with movements of the diaphragm, which forms the rounded "floor" of the thoracic cavity. The diaphragm contracts, causing its depression—that is, its dome-shaped central portion flattens and moves inferiorly to press against the abdominal viscera, resulting in inhalation. When you exhale, the diaphragm relaxes and returns to its original position.

Lateral dimensional changes occur with rib movements. Elevation of the ribs increases the lateral dimensions of the thoracic cavity, while depression of the ribs decreases the lateral dimensions of the thoracic cavity.

Figure 25.16 shows that several muscles of external respiration move the ribs:

- The **scalene muscles** help increase thoracic cavity dimensions by elevating the first and second ribs during forced inhalation.
- The **external intercostal muscles** extend from a superior rib inferomedially to the adjacent inferior rib. The ribs elevate upon contraction of the external intercostals, thereby increasing the transverse dimensions of the thoracic cavity during inhalation.
- The **internal intercostal muscles** lie at right angles to the external intercostals and deep to them. Contraction of the internal intercostals depresses the ribs, but this only occurs during forced exhalation. Normal exhalation requires no active muscular effort.
- A small **transversus thoracis** (see also figure 11.13) extends across the inner surface of the thoracic cage and attaches to ribs 2–6. It helps depress the ribs.

Two posterior thoracic muscles also assist with external respiration. These muscles are located deep to the trapezius and latissimus dorsi, but superficial to the erector spinae muscles (see also figure 11.11). The **serratus posterior superior** elevates ribs 2–5 during inhalation, and the **serratus posterior inferior** depresses ribs 8–12 during exhalation.

In addition, some accessory muscles assist with external respiration activities. The pectoralis minor and sternocleidomastoid help with forced inhalation, while the abdominal muscles (external and internal obliques, transversus abdominis, and rectus abdominis) assist in active exhalation. (Researchers are still debating the effects of some of the external respiration muscles.)

Finally, a slight anterior-posterior dimensional change occurs in the thoracic cavity due to movement of the inferior portion of the sternum. When you inhale, the inferior portion of the sternum moves anteriorly, slightly increasing the anterior-posterior dimensions of the thorax. When you exhale, the inferior portion of the sternum moves posteriorly and returns to its original position.

WHAT DID YOU LEARN?

11 What types of dimensional changes occur to the thorax when you inhale, and what muscles are responsible?

Study Tip!

— To visualize rib movement during external respiration, think of the thoracic cavity as a bucket and the ribs as the bucket handles. When the bucket handles are lifted up, they move relatively farther away from the edges of the bucket. Thus, the measurement from the bucket handles (ribs) to the bucket (thoracic cavity) increases, just as the lateral dimensions of the thoracic cavity increase. When the bucket handles are depressed, they move next to the edges of the bucket, and so the distance from the bucket handle (ribs) to the bucket (thoracic cavity dimension) decreases.

Inhalation: Ribs (bucket handles) elevated, lateral dimension increased

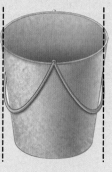

Exhalation: Ribs (bucket handles) depressed, lateral dimension decreased

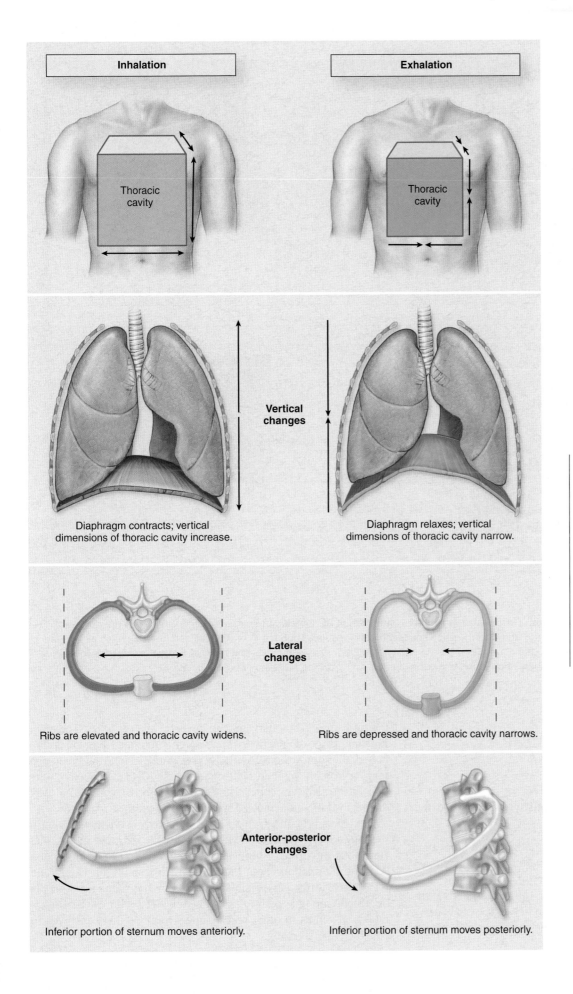

Inhalation	Exhalation

Thoracic cavity

Diaphragm contracts; vertical dimensions of thoracic cavity increase.

Diaphragm relaxes; vertical dimensions of thoracic cavity narrow.

Vertical changes

Ribs are elevated and thoracic cavity widens.

Ribs are depressed and thoracic cavity narrows.

Lateral changes

Anterior-posterior changes

Inferior portion of sternum moves anteriorly.

Inferior portion of sternum moves posteriorly.

Figure 25.15

Thoracic Cavity Dimensional Changes Associated with Breathing.   The boxlike thoracic cavity changes size upon inhalation and exhalation. During inhalation, the box increases in vertical, lateral, and anterior-posterior dimensions due to movement of the sternum, ribs, and diaphragm, respectively. Upon exhalation, these dimensions decrease, and the thoracic cavity becomes smaller.

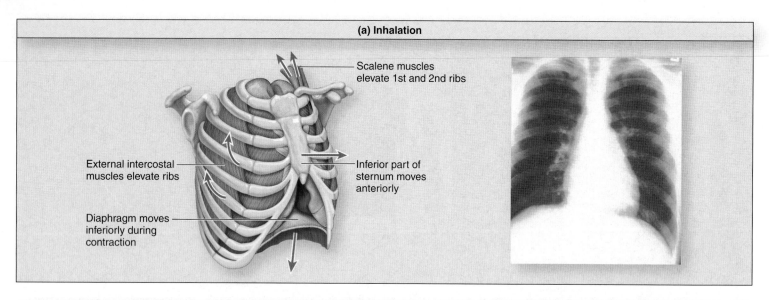

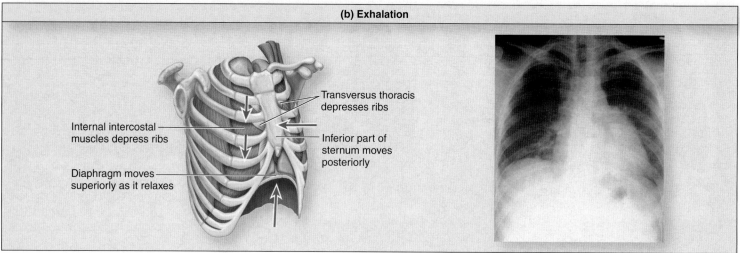

Figure 25.16

Muscles Involved in External Respiration. *(a)* Inhalation requires contraction of the external intercostal muscles (to elevate the ribs) and the diaphragm (which moves inferiorly during contraction). Forced inhalation also requires contraction of the scalene muscles, which elevate the first and second ribs. *(b)* During exhalation, these muscles relax. Additionally, the transversus thoracis and internal intercostal muscles contract to depress the ribs during forced exhalation. Companion x-rays show the thoracic cavity during inhalation and exhalation.

Innervation of the Respiratory System

Key topic in this section:

■ Components of the autonomic nervous system that regulate ventilation

The larynx, trachea, bronchial tree, and lungs are innervated by the autonomic nervous system. The autonomic nerves that innervate the heart also send branches to these respiratory structures (see figure 22.13 for a review of these nerves). The vagus nerve is the primary innervator of the larynx. Damage to one of the vagus nerve branches going to the larynx can cause a person to have a monotone or a permanently hoarse voice.

Sympathetic innervation to the lungs originates from the T1–T5 (or occasionally T2–T5) segments of the spinal cord. These preganglionic fibers enter the sympathetic trunk and synapse with ganglionic neurons. The postganglionic sympathetic fibers (called

the cardiac nerves) innervate both the heart and the lungs. The main function of the sympathetic innervation is to open up or dilate the bronchioles (bronchodilation). Parasympathetic innervation to the lungs is from the left and right vagus nerves (CN X). The main function of the parasympathetic innervation is to decrease the airway diameter of the bronchioles (bronchoconstriction).

Collectively, the sympathetic and parasympathetic fibers form the **pulmonary plexus,** a weblike network of nerve fibers that surrounds the primary bronchi and enters the lungs at the hilum. Sensory information about the "stretch" in smooth muscle around the bronchial tree is typically conducted by the vagus nerve to the brainstem and then relayed to centers involved with external respiration as well as to other reflex centers, such as those involved in coughing and sneezing.

WHAT DO YOU THINK?

6 When an asthma inhaler provides relief for bronchoconstriction, is it mimicking sympathetic or parasympathetic stimulation?

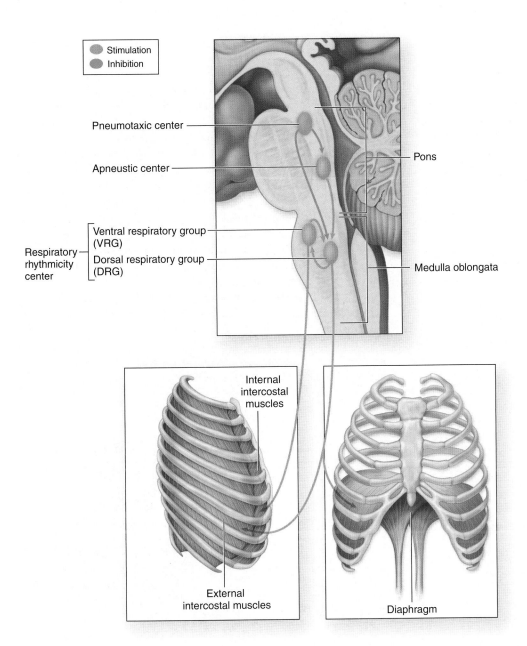

Legend:
- Stimulation
- Inhibition

Pneumotaxic center

Apneustic center

Respiratory rhythmicity center
- Ventral respiratory group (VRG)
- Dorsal respiratory group (DRG)

Pons

Medulla oblongata

Internal intercostal muscles

External intercostal muscles

Diaphragm

Figure 25.17

Respiratory Control Centers in the Brainstem. The dorsal respiratory group (DRG) and ventral respiratory group (VRG) within the medulla oblongata regulate normal ventilation rate. The pons houses the pneumotaxic and apneustic centers, which influence the DRG and VRG. The pneumotaxic center is inhibitory to both respiration and the apneustic center. The apneustic center stimulates the DRG.

Ventilation Control by Respiratory Centers of the Brain

The involuntary, rhythmic activities that deliver and remove respiratory gases are regulated in the brainstem. Regulatory respiratory centers are located within the reticular formation through both the medulla oblongata and the pons. The regulatory centers are composed of specific nuclei, called the respiratory rhythmicity center, the apneustic center, and the pneumotaxic center **(figure 25.17)**.

The **respiratory rhythmicity center** in the medulla oblongata establishes the rate and depth of breathing. Two distinct autonomic nuclei form this center. The **dorsal respiratory group (DRG)** is the inspiratory center that controls inhalation. It controls the motor neurons that stimulate the muscles of inspiration. The **ventral respiratory group (VRG)** is the expiratory center for forced exhalation. It functions only during forced exhalation. During normal quiet breathing, the VRG is inactive, and exhalation is a passive event that does not require nervous stimulation. When the VRG is activated, its neurons stimulate accessory respiratory muscles to cause maximal, rapid exhalation—for example, when you exercise and are breathing deeply and forcibly. The DRG is activated during both normal inhalation and forced inhalation. The DRG sends impulses through both the phrenic and intercostal nerves to stimulate the diaphragm and external intercostal muscles.

The **apneustic** (ap-noo'stik) **center** and the **pneumotaxic** (noo-mō-tak'sik) **center** are nuclei housed within the pons. Both areas influence the breathing rate by modifying the activity of the respiratory rhythmicity center. The apneustic center stimulates inspiration through the DRG; the pneumotaxic center inhibits both the activity of the DRG and that of the apneustic center. By inhibiting the DRG, the VRG is able to function and initiate the process of forced exhalation. For example, during vigorous exercise when your respiratory rate must be increased, respiratory gases must be exchanged more frequently than when at rest. Thus, the DRG, once stimulated, must be inhibited fairly quickly, with the simultaneous activation of the VRG, so that forced exhalation can occur and the next inhalation can begin.

WHAT DID YOU LEARN?

12 What is the main function of sympathetic innervation to the lungs?

13 Compare the activities of the DRG and the VRG in the brain's respiratory centers.

CLINICAL VIEW

Asthma

Asthma (az'mă) is a chronic condition characterized by episodes of bronchoconstriction and wheezing, coughing, shortness of breath, and excess pulmonary mucus. Its incidence is increasing among young people, particularly those living in urban areas where airborne industrial pollutants and tobacco smoke are abundant. In most cases, the affected person develops a sensitivity to an airborne agent such as pollen, smoke, mold spores, dust mites, or particulate matter. Upon reexposure to this triggering substance, a localized immune reaction occurs in the bronchi and bronchioles, resulting in bronchoconstriction, swollen submucosa, and increased production of mucus. Episodes typically last an hour or two. Continual exposure to the triggering agent increases the severity and frequency of asthma attacks. Eventually, the walls of the bronchi and bronchioles may become permanently thickened, leading to chronic and unremitting airway narrowing and shortness of breath. If airway narrowing is extreme during a severe asthma attack, death could occur.

Today, the primary treatment for asthma consists of administering inhaled steroids (cortisone-related compounds) to reduce the inflammatory reaction, combined with bronchodilators to alleviate the bronchoconstriction. Avoidance of the triggering agent is also very important. For some patients, allergy shots have proven helpful. In cases of severe asthma, oral doses of steroids may control the allergic hyper-response and reduce the inflammation.

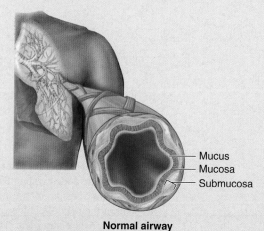

— Mucus
— Mucosa
— Submucosa

Normal airway

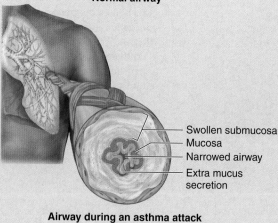

— Swollen submucosa
— Mucosa
— Narrowed airway
— Extra mucus secretion

Airway during an asthma attack

Individuals suffering from asthma may need to use inhaled medications to dilate their constricted bronchioles.

Aging and the Respiratory System

Key topic in this section:

- Age-related respiratory system changes

The respiratory system becomes less efficient with age due to several structural changes. First, aging results in a decrease in elastic connective tissue in the lungs and the thoracic cavity wall. This loss of elasticity reduces the amount of gas that can be exchanged with each breath and results in a decrease in the ventilation rate. In addition, a condition such as emphysema may cause a loss of alveoli or a decrease in their size or functionality. The resulting reduced capacity for gas exchange can cause an older person to become "short of breath" upon exertion.

Finally, as we get older, carbon, dust, and pollution material gradually accumulate in our lymph nodes and lungs. If a person also smokes regularly, the lungs become even darker and blacker throughout because of the deposition of carbon particles in the cells. Two distinct diseases, emphysema and chronic bronchitis, together encompass **chronic obstructive pulmonary disease (COPD),** which is often related to tobacco use. The condition is characterized by lung structural abnormalities resulting from inflammation. The resulting airflow obstruction makes it hard for the patient to exhale.

WHAT DID YOU LEARN?

14 What are some ways that aging can affect the respiratory system?

Smoking, Emphysema, and Lung Cancer

Smoking is one of the most important modifiable factors contributing to disease and premature death in the United States. It significantly increases the risk and severity of atherosclerosis, and is directly related to the development of cancers of the lung, esophagus, stomach, and urinary bladder. Current studies also indicate an association between secondhand smoke exposure and an increased risk of bronchitis, ear infections, and asthma in children. Secondhand smoke is a mixture of the gases and particulate materials released by the burning of tobacco in cigarettes, cigars, and pipes, as well as exhaled by smokers. Unfortunately, second-hand smoke is inhaled by everyone within the environment exposed to it. Potential health-care risks include cancer, asthma, and infections in the respiratory system. The most common smoking-related diseases are emphysema and several types of lung cancer.

Emphysema (em-fi-zē′mă; *en* = in, *physema* = a blowing) is an irreversible loss of pulmonary gas exchange areas due to inflammation of the terminal bronchioles and alveoli, in conjunction with the widespread destruction of pulmonary elastic connective tissue. These combined events lead to an increase in the diameter or dilation of individual alveoli, resulting in a decrease in the total number of alveoli, and the subsequent loss of gas exchange surface area. A person with advanced emphysema has a larger than normal chest circumference because air is trapped within the abnormally expanded and nonfunctional alveoli. The patient is unable to exhale effectively, so that stagnant, oxygen-poor

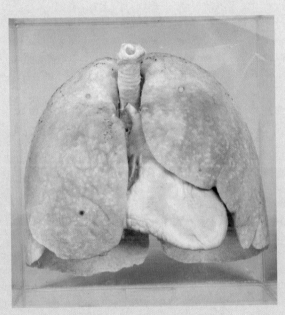

Nonsmoker's lungs.

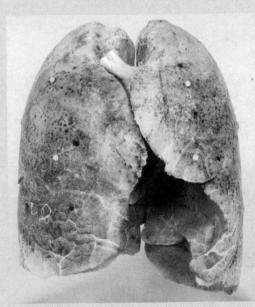

Smoker's lungs.

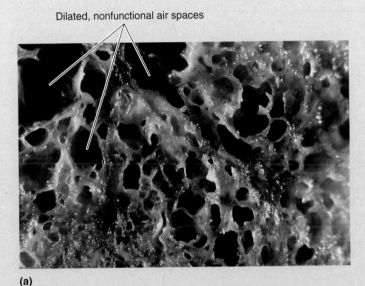

Dilated, nonfunctional air spaces

(a)

Dilated, nonfunctional alveoli

LM 15x

(b)

Emphysema causes dilation of the alveoli and loss of elastic tissue, resulting in poorly functioning alveoli. (a) A gross section of an emphysemic lung shows the dilated alveoli. (b) Microscopically, the alveoli are abnormally large and nonfunctional.

(continued on next page)

CLINICAL VIEW: In Depth

air builds up within the abnormally large (but numerically diminished) alveoli. Most cases of emphysema result from damage caused by smoking. Once the tissue in the lung has been destroyed, it cannot regenerate, and thus there is no cure for emphysema. The best therapy for an emphysema patient is to stop smoking and try to get optimal use from the remaining lung tissue by using a bronchodilator, seeking prompt treatment for pulmonary infections, and taking oxygen supplementation if necessary.

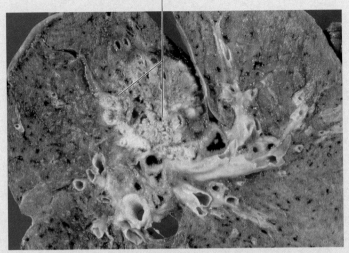

Squamous cell carcinoma

Gross section of a lung with squamous cell carcinoma (speckled white and black regions).

An individual with advanced emphysema must rely on a portable oxygen tank, such as this backpack tank.

Adenocarcinoma is less common than the squamous cell type. An adenocarcinoma of the lung arises from the mucin-producing glands in the respiratory epithelium. It begins when DNA injury causes one of these cells to become malignant and begin to divide uncontrollably. Histologically, an adenocarcinoma displays some of the microscopic features of the gland from which it arose, thereby making it distinguishable from the other forms of lung cancer.

Small-cell carcinoma is a less common type of lung cancer that originates in the primary bronchi and eventually invades the mediastinum. This type of cancer is especially known for its early metastasis to other organs. Small-cell carcinoma arises from the small neuroendocrine cells in the larger bronchi; their secretions help regulate muscle tone in the bronchi and vessels. As a consequence of their endocrine heritage, some of these tumors secrete hormones. For example, a small-cell cancer of the lung occasionally releases ACTH, producing symptoms of Cushing syndrome.

Lung cancer is a highly aggressive and frequently fatal malignancy that originates in the epithelium of the respiratory system. It claims over 150,000 lives annually in the United States. Smoking causes about 85% of all lung cancers. Metastasis, the spread of cancerous cells to other tissues, occurs early in the course of the disease, making a surgical cure unlikely for most patients. Pulmonary symptoms include chronic cough, coughing up blood, excess pulmonary mucus, and increased likelihood of pulmonary infections. Some people are diagnosed based on symptoms that develop after the cancer has already metastasized to a distant site. For example, lung cancer commonly spreads to the brain, so in some cases lung cancer is not discovered until the patient seeks treatment for a seizure disorder related to cancer in the brain.

Lung cancers are classified by their histologic appearance into three basic patterns: squamous cell carcinoma, adenocarcinoma, and small-cell carcinoma.

Squamous cell carcinoma (kar-si-nō′mă; *karkinos* = cancer, *oma* = tumor) is the most common form of lung cancer. At the microscopic level, the pseudostratified ciliated columnar epithelium lining the lungs changes to a sturdier stratified squamous epithelium in order to withstand the chronic inflammation and injury caused by tobacco smoke. If the chronic injury continues, these transformed epithelial cells may accumulate enough genetic damage to become overtly malignant. The malignant cells divide uncontrollably, invade the surrounding tissue, and then spread to distant sites.

Small-cell carcinoma

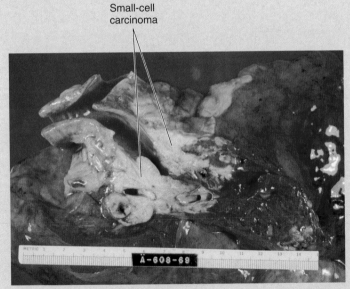

Gross section of a lung with small-cell carcinoma (white regions) around a bronchus.

Development of the Respiratory System

Key topic in this section:

■ How the respiratory system forms in the embryo and fetus

Early in the fourth week of development, a ventral outgrowth extends from the developing pharynx. This endodermal outgrowth is called the **respiratory diverticulum** (dī-ver-tik′ū-lŭm; byroad), or *lung bud*, and it initially maintains communication with the pharynx (**figure 25.18a**). By late in the fourth week, a septum forms between the pharynx and the respiratory diverticulum, partitioning them into two separate tubes. The respiratory diverticulum, formed from endoderm, undergoes intricate branching to form the respiratory tree. Surrounding the respiratory diverticulum is mesoderm, which later differentiates into the vasculature, muscle, and cartilage of each lung.

The respiratory diverticulum grows inferiorly and forms the future trachea. At the end of the fourth week, the respiratory diverticulum branches into a **left** and **right primary bronchial bud.** Each bud forms the rudiments of the left and right primary bronchi, respectively. Growing branches of the pulmonary arteries and veins travel with these developing bronchial buds. By the fifth week of development, the primary bronchial buds branch into **secondary bronchial buds** (figure 25.18b). The secondary bronchial buds form the secondary bronchi of each lung. Thus, the left primary bronchial

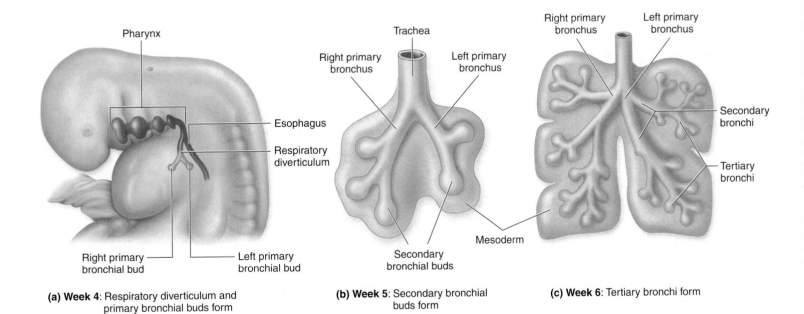

(a) **Week 4**: Respiratory diverticulum and primary bronchial buds form

(b) **Week 5**: Secondary bronchial buds form

(c) **Week 6**: Tertiary bronchi form

Figure 25.18

Development of the Respiratory System. The respiratory system forms as an outgrowth (called the respiratory diverticulum) from the developing pharynx beginning at week 4. *(a)* Primary bronchial buds appear later during week 4. *(b)* Secondary bronchial buds branch from the primary bronchi during week 5 and grow into the surrounding mesoderm. *(c)* By week 6, the tertiary bronchi of the left and right lungs have formed.

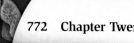

bud branches into two secondary bronchial buds (since the left lung has two secondary bronchi), and the right primary bronchial bud branches into three secondary bronchial buds (for the three secondary bronchi of the right lung).

In the following weeks, the secondary bronchial buds undergo further branching. Week 6 is marked by the development of tertiary bronchi and the rudiments of the bronchopulmonary segments (figure 25.18*c*). From week 6 to week 16, the respiratory tree forms smaller branches until finally the terminal bronchioles form. Thus, the conducting portion of the respiratory system has developed by week 16.

Weeks 16–28 mark the branching and development of respiratory bronchioles from the terminal bronchioles. From week 28 of development until birth, **primitive alveoli** (also called *terminal sacs*) develop more profusely. These primitive alveoli have a thick epithelial lining that must thin into simple squamous epithelium in order for the alveoli to become functional. It isn't until after week 28 that this epithelium becomes sufficiently thinned for respiration. In addition, by about week 28 the alveolar type II cells start to secrete pulmonary surfactant (described earlier in this chapter),

which helps keep the alveoli patent (open) and facilitates inflation. Without sufficient surfactant, the alveoli collapse upon exhalation, and it becomes difficult to reinflate them. Prematurely born infants sometimes experience respiratory distress due to inadequate production of surfactant.

Prior to birth, the respiratory system is nonfunctional because gas exchange occurs between fetal blood and maternal blood at the placenta. The lungs and pulmonary vessels of the fetus are collapsed, and so most of the blood is shunted away from the lungs to the fetus's systemic circulation. At birth, the first contraction of the external intercostal muscles and diaphragm fills the lungs with air. (The pressure changes within the thoracic cavity drive the commencement of pulmonary circulation.) Blood is sent to the lungs, where gas exchange occurs, and the newborn relies on its own lungs (instead of the mother's placenta) for respiratory gas exchange.

Even after we are born, our lungs continue to produce additional primitive alveoli. Some research has indicated that alveoli continue to develop until we are about 8 years old, by which time each lung has approximately 300 to 400 million alveoli. **Table 25.3** summarizes the events in respiratory system development.

Table 25.3	Summary of Respiratory System Development
Week of Development/Age	**Respiratory System Structure Formed**
Early week 4	Respiratory diverticulum forms
Late week 4	Primary bronchial buds form
Week 5	Secondary bronchial buds form
Week 6	Tertiary bronchial buds form
Week 6–Week 16	Successive branching of tertiary bronchial buds forms smaller bronchi and bronchioles; eventually, terminal bronchioles form; conducting portion of respiratory system is complete
Week 16–Week 28	Terminal bronchioles branch into respiratory bronchioles
Week 28–birth	Primitive alveoli form; pulmonary surfactant begins to be produced
Birth–8 years	Alveoli continue to develop; eventually, adult number of alveoli (300–400 million per lung) is attained

CLINICAL TERMS

epistaxis (ep′i-stak′sis; *epi* = on, *stazo* = to fall in drops) Bleeding from the nose; may be caused by allergies, hypertension, infection, or nasal trauma. Also called nosebleed or nasal hemorrhage.

Heimlich maneuver A potentially life-saving technique in which external pressure applied to the abdominal region dislodges and expels a foreign object from the larynx or trachea.

hyaline (hī′ă-lin; *hyalos* = glass) **membrane disease** Disease seen especially in premature neonates with reduced amounts of lung surfactant.

pulmonary embolism (em′bō-lizm) Obstruction or occlusion of a pulmonary vessel by an embolus (foreign material or blood clot).

CHAPTER SUMMARY

General Organization and Functions of the Respiratory System 746	■ The respiratory system has a conducting portion to convey gas to and from the lungs and a respiratory portion for gas exchange with the blood.

Respiratory System Functions 746

- ■ Respiratory system functions include gas exchange, gas conditioning, sound production, olfaction, and defense.

Upper Respiratory Tract 748	■ The conducting airways of the upper respiratory tract are the nose, nasal cavity, paranasal sinuses, pharynx, and their associated structures.

Nose and Nasal Cavity 748

- ■ The nasal cavity is the primary site for conditioning inhaled air. It houses three pairs of nasal conchae that cause turbulence in inhaled air, which passes posteriorly into the nasopharynx through the choanae.

Paranasal Sinuses 748

- ■ Paranasal sinuses are paired air spaces in the frontal, ethmoidal, and sphenoidal bones, and the maxillae. They decrease skull bone weight, help condition inhaled air, and contribute to sound resonance.

Pharynx 748

- ■ The pharynx is composed of (1) the nasopharynx, with paired auditory openings on the lateral wall and a pharyngeal tonsil on the posterior wall; (2) the oropharynx, with paired palatine tonsils on the lateral walls and lingual tonsils at the base of the tongue; and (3) the laryngopharynx, which is continuous with the larynx and esophagus.

Lower Respiratory Tract 751	■ The conducting airways of the lower respiratory tract include the larynx, trachea, bronchi, bronchioles to the terminal bronchioles, and their associated structures. Its respiratory portions include respiratory bronchioles, alveolar ducts, and alveoli.

Larynx 751

- ■ The larynx conducts air into the trachea and lower respiratory tract, and produces sound.
- ■ The larynx is composed of cartilage and has paired vocal folds that produce sound when air passes between them.

Trachea 754

- ■ The trachea is lined by pseudostratified ciliated columnar epithelium and has C-shaped tracheal cartilage rings that support the tracheal wall and prevent its collapse.

Bronchial Tree 755

- ■ The bronchial tree conducts respiratory gases from the primary bronchi to the terminal bronchioles.
- ■ Bronchial tree passageways have cartilage and/or smooth muscle bands to support the walls. The passageway sequence is (1) primary bronchi, (2) secondary bronchi, (3) tertiary bronchi, (4) bronchioles, and (5) terminal bronchioles.

Respiratory Bronchioles, Alveolar Ducts, and Alveoli 757

- ■ Respiratory bronchioles branch from terminal bronchioles and have alveoli outpocketings in their walls.
- ■ An alveolus is a small, thin sac with two types of cells in its wall.
- ■ Alveolar type I cells promote rapid gas diffusion; alveolar type II cells secrete pulmonary surfactant.
- ■ Alveolar macrophages remove inhaled particulate materials from alveolar surfaces.
- ■ The respiratory membrane consists of alveolar type I cells, an endothelial cell of a capillary, and their fused basement membranes.

Lungs 759	■ The lungs are lateral to the mediastinum in the thoracic cavity.

Pleura and Pleural Cavities 759

- ■ The visceral pleura covers the lung outer surface, and the parietal pleura lines the internal thoracic walls; a pleural cavity is sandwiched between the pleural layers. The pleural membranes produce serous fluid.

Gross Anatomy of the Lungs 759

- ■ Lung surfaces include the base (upon the diaphragm), the apex (superior surface), the costal surface (against the thoracic wall), and the mediastinal surface (facing the mediastinum).
- ■ The hilum is a medial opening through which bronchi, pulmonary vessels, lymph vessels, and nerves enter the lungs.

Blood Supply To and From the Lungs 760

- ■ The pulmonary circulation transports blood to and from the gas exchange surfaces of the lungs, and the bronchial circulation supplies the bronchi and bronchioles.

Lymphatic Drainage 762

- ■ The connective tissue in the lung houses lymph nodes and lymph vessels.

(continued on next page)

CHAPTER SUMMARY (continued)

Pulmonary Ventilation 763	■ Breathing, called pulmonary ventilation, is the movement of air into and out of the respiratory tract. Change in air pressure between the atmosphere and the alveoli drives ventilation.
Thoracic Wall Dimensional Changes During External Respiration 764	■ Inhalation causes the thoracic cavity space to increase vertically, laterally, and in an anterior-posterior direction; exhalation causes it to return to its original size. ■ During external respiration, the primary muscles that move the ribs are: for inhalation, (1) scalene, (2) external intercostal, and (3) serratus posterior superior; for forced exhalation, (4) internal intercostal, (5) transversus thoracis, and (6) serratus posterior inferior.
Innervation of the Respiratory System 766	■ Sympathetic stimulation causes bronchodilation; parasympathetic stimulation causes bronchoconstriction. **Ventilation Control by Respiratory Centers of the Brain** 767 ■ The respiratory center in the medulla oblongata has a dorsal respiratory group (DRG) for inspiration and a ventral respiratory group (VRG) for forced expiration. ■ The apneustic and pneumotaxic centers within the pons influence the respiration rate by modifying the activity of the DRG.
Aging and the Respiratory System 768	■ The respiratory system becomes less efficient with age due to loss of elasticity and loss or decreased size and functionality of alveoli.
Development of the Respiratory System 771	■ Early in the fourth week of development, a respiratory diverticulum forms and leads to primary bronchial buds by late in that same week. ■ By the fifth week of development, the primary bronchial buds branch into secondary bronchial buds. These bronchial buds undergo further branching until terminal bronchioles are formed by week 16. Respiratory bronchioles form from the terminal bronchioles during weeks 16–28. ■ Alveoli continue to form from week 28 of development until about 8 years of age, when the adult number of 300–400 million alveoli is attained.

CHALLENGE YOURSELF

Matching

Match each numbered item with the most closely related lettered item.

_____ 1. nasopharynx

_____ 2. bronchiole

_____ 3. nasal meatus

_____ 4. left lung

_____ 5. cricoid cartilage

_____ 6. primary bronchus

_____ 7. alveolar type II cell

_____ 8. arytenoid cartilage

_____ 9. alveolar macrophage

_____ 10. epiglottis

a. solid ring of hyaline cartilage

b. branches directly from the trachea

c. has a cardiac notch and cardiac impression

d. phagocytic cell in alveoli

e. contains pharyngeal tonsil

f. covers laryngeal opening during swallowing

g. causes air turbulence in nasal cavity

h. produces pulmonary surfactant

i. lacks cartilage but has significant amounts of smooth muscle in wall

j. vocal folds attach to it

Multiple Choice

Select the best answer from the four choices provided.

_____ 1. The visceral pleura covers the
   a. outer surface of the lung.
   b. gas exchange surface of the alveoli.
   c. inner wall of the thoracic cavity.
   d. lining of the bronchi and bronchioles only.

_____ 2. An area common to both the respiratory and digestive systems through which food, drink, and air pass is the
   a. nasopharynx.
   b. trachea.
   c. oropharynx.
   d. glottis.

_____ 3. Which statement is *false* about the trachea?
   a. It is lined with a nonkeratinized stratified squamous epithelium.
   b. It is continuous superiorly with the larynx.
   c. It bifurcates into left and right primary bronchi at the level of the sternal angle.
   d. It contains C-shaped cartilage rings.

_____ 4. Which structure is the last, smallest portion of the conducting portion of the respiratory system?
   a. nasopharynx
   b. terminal bronchiole
   c. respiratory bronchiole
   d. alveolus

_____ 5. Which is *not* a function of the paranasal sinuses?
    a. warm inhaled air
    b. responsible for sound resonance
    c. gas exchange
    d. humidify inhaled air

_____ 6. The _____ cartilage of the larynx forms the laryngeal prominence.
    a. arytenoid
    b. cuneiform
    c. thyroid
    d. cricoid

_____ 7. The C-shaped cartilages in the trachea
    a. serve as a point of attachment for some muscles of expiration.
    b. support muscular attachments to the thyroid cartilage and epiglottis.
    c. prevent the trachea from collapsing.
    d. attach the trachea to the esophagus posteriorly.

_____ 8. Which of the following is *not* a muscle of inspiration?
    a. diaphragm
    b. external intercostals
    c. rectus abdominis
    d. scalene

_____ 9. The epithelium lining the alveolus is composed of a
    a. simple squamous epithelium.
    b. pseudostratified ciliated columnar epithelum.
    c. simple cuboidal epithelium.
    d. transitional epithelium.

_____ 10. The apneustic center is involved in
    a. inhibition of the pneumotaxic area.
    b. stimulation of DRG.
    c. stimulation of the pneumotaxic area.
    d. inhibition of VRG.

Content Review

1. What is the function of the mucous lining of the epithelium in the respiratory tract?

2. What type of epithelium is found in the oropharynx, and why is it well suited to this location?

3. What must happen to the vocal folds in order to produce a higher-pitched sound? A lower-pitched sound? What produces a louder sound?

4. What are the components of the bronchial tree, from largest to smallest?

5. Why is cartilage unnecessary in the walls of the bronchioles?

6. Why are alveolar type II cells important in maintaining the inflation of the lungs?

7. How do the left and right lungs differ anatomically?

8. How do the dimensions of the thoracic cavity change when we inhale and exhale? What muscles assist with these dimensional changes?

9. Name the autonomic nervous system respiratory centers in the pons and the medulla oblongata, and describe their functions.

10. Contrast the functions and interactions of the DRG and the VRG in the medulla oblongata.

Developing Critical Reasoning

1. Charlene has had a bad cold for the last few days. While preparing a presentation for her speech class, she records her talk so that she can critique it later. When she listens to the recording, her young daughter exclaims, "Mommy that doesn't even sound like you. What happened to your voice?" How is Charlene's cold related to the changes in her voice?

2. Your best friend George is an athletic 20-year-old who smokes regularly. George tells you, "Smoking doesn't affect me—I can still run and do the sports I like. All that talk about smoking being dangerous doesn't apply to me." Do you agree with George? What would you tell him about the dangers of smoking and some of the conditions he may expect to have later in life?

ANSWERS TO "WHAT DO YOU THINK?"

1. A "deviated septum" is off-center, so one side of the nasal cavity is larger than the other. This alters the normal flow of air through the nose, and if the narrower side becomes blocked, nasal congestion or sinus problems may result.

2. The epithelium changes because a stratified squamous epithelium is more sturdy and protective against smoke than a pseudostratified ciliated columnar epithelium. Unfortunately, since stratified squamous epithelium lacks cilia and goblet cells, less mucus is produced, and no cilia are present to propel particles away from the bronchi toward the pharynx. Thus, the main way to eliminate these particles is by coughing, leading to the chronic "smoker's cough."

3. The constriction of the bronchioles allows for a more forceful expulsion of air from the lungs, which may help dislodge accumulated mucus or inhaled foreign particulate materials.

4. The left lung is smaller because the heart projects into the left side of the thoracic cavity.

5. Lymph nodes darken and turn black as they accumulate the dust, particles, and pollution we inhale over a lifetime.

6. An asthma inhaler mimics sympathetic stimulation because it causes bronchodilation.

Visit the McKinley/O'Loughlin *Human Anatomy*, 2e website at aris.mhhe.com

26

Digestive System

Each time we eat a meal and drink fluids, our bodies take in the nutrients necessary for survival. However, these nutrients must be digested and processed—broken down both mechanically and chemically—into components small enough for our cells to use. Once the nutrients are broken down sufficiently, they are absorbed from the digestive system into the bloodstream and transported to the body tissues.

Not everything we eat can be used by the body. After the nutrients from foods are absorbed, some materials remain that cannot be digested or absorbed, such as cellulose and fiber. These materials must be expelled from the body via a process called defecation. All of these functions are the responsibility of the digestive system.

General Structure and Functions of the Digestive System

Key topics in this section:

- GI tract organs and accessory digestive organs
- Basic functions of the digestive system
- Mechanical and chemical digestion
- The processes of peristalsis and segmentation

The **digestive** (di-jes′tiv, dī-; *digestus* = to force apart, divide, dissolve) **system** includes the organs that ingest the food, transport the ingested material, digest the material into smaller usable components, absorb the necessary digested nutrients into the bloodstream, and expel the waste products from the body. When you eat, you put food into your mouth and it mixes with saliva as you chew. The chewed food mixed with saliva is called a **bolus** (bō′lŭs; lump), and it is the bolus that is swallowed. The stomach processes the bolus and turns it into a pastelike substance called **chyme.** Hereafter in the chapter, we will simplify our discussion by referring to the ingested contents as "material."

The digestive system is composed of two separate categories of organs: digestive organs and accessory digestive organs (**figure 26.1**). The **digestive organs** collectively make up the **gastrointestinal (GI) tract,** also called the *digestive tract* or *alimentary* (al-i-men′ter-ē; *alimentum* = nourishment) *canal.* The GI tract organs are the oral cavity, pharynx, esophagus, stomach, small intestine, and large intestine. These organs form a continuous tube from the mouth to the anus. The contraction of muscle in the GI tract wall propels materials through the tract.

Accessory digestive organs are not part of the long GI tube, but often develop as outgrowths from and are connected to the GI tract. The accessory digestive organs assist the GI tract in the digestion of material. Accessory digestive organs include the teeth, tongue, salivary glands, liver, gallbladder, and pancreas.

Digestive System Functions

The digestive system performs six main functions: ingestion, digestion, propulsion, secretion, absorption, and elimination of wastes.

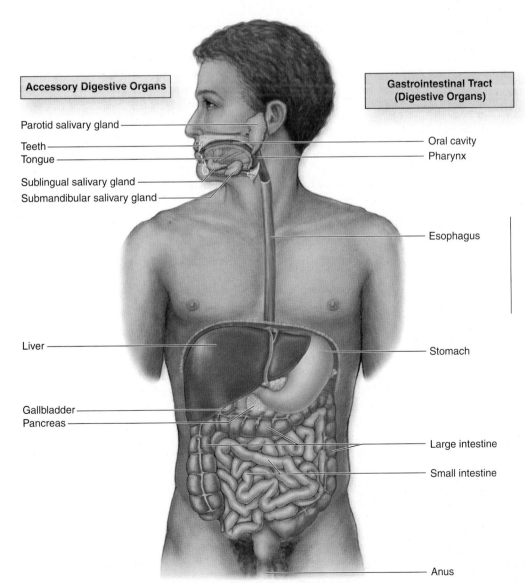

Accessory Digestive Organs
Parotid salivary gland
Teeth
Tongue
Sublingual salivary gland
Submandibular salivary gland
Liver
Gallbladder
Pancreas

Gastrointestinal Tract (Digestive Organs)
Oral cavity
Pharynx
Esophagus
Stomach
Large intestine
Small intestine
Anus

Figure 26.1

Digestive System. The digestive system is composed of the gastrointestinal (GI) tract and accessory digestive organs that assist the GI tract in the process of digestion.

Ingestion (in-jes´chŭn; *ingero* = to carry in) is the introduction of solid and liquid materials into the oral cavity. It is the first step in the process of digesting and absorbing nutrients.

Digestion is the breakdown of large food items into smaller structures and molecules. There are two aspects to digestion: **Mechanical digestion** physically breaks down ingested materials into smaller pieces, and **chemical digestion** breaks down ingested material into smaller molecules by using enzymes. (Review enzymes in chapter 2.) The first part of mechanical digestion is **mastication** (mas´ti-kā´shŭn; *mastico* = to chew), the chewing of ingested material by the teeth in the oral cavity.

After the materials are swallowed, they move through the GI tract by a process termed **propulsion** (prō-pŭl´shŭn; *propello* = to move forth). Two types of movement are involved in propulsion: peristalsis and segmentation. **Peristalsis** (per-i-stal´sis; *stalsis* = constriction) is the process of muscular contraction that forms ripples along part of the GI tract and forces material to move further along the tract (**figure 26.2a**). Peristalsis is like pushing toothpaste through a toothpaste tube: As you push on the flat end of the tube, a portion of toothpaste moves toward the opening. As you push the middle of the tube, the segment of toothpaste moves even closer to the opening. Churning and mixing movements in the small intestine, called **segmentation** (seg´men-tā´shŭn), help disperse the material being digested and combine it with digestive organ secretions (figure 26.2b). The material within the lumen of the small intestine is moved back-and-forth to mix it with the secretory products that are being introduced into that region of the GI tract. Thus, contraction and relaxation of the digestive wall musculature creates a "blender" effect, mixing ingested materials with digestive enzymes and mechanically breaking down larger digested particles into smaller ones.

Secretion (se-krē´shŭn; *secerno* = to separate) is the process of producing and releasing mucin or fluids such as acid, bile, and digestive enzymes. When these products are secreted into the lumen of the GI tract, they facilitate chemical digestion and the passage of material through the GI tract. Some of these products (e.g., acid, bile, digestive enzymes) help digest food. Mucin secretions serve a protective function. Mucin mixes with water to form mucus, and the mucus coats the GI wall to protect and lubricate it against acidic secretions and abrasions by passing materials.

Absorption (ab-sōrp´shŭn; *absorptio* = to swallow) involves either passive movement or active transport of electrolytes, digestion products, vitamins, and water across the GI tract epithelium and into GI tract blood and lymph vessels.

The final function of the digestive system is the **elimination** of wastes. Our bodies utilize most, but not all, of the components of what we eat. All undigestible materials as well as the waste products secreted by the accessory organs into the GI tract are compacted into **feces** (fē´sēz; *faex* = dregs), or *fecal material*, and then eliminated from the GI tract by the process of **defecation** (def-ĕ-kā´shŭn; *defaeco* = to remove the dregs).

WHAT DID YOU LEARN?

❶ What structures compose the GI tract?

❷ How do secretion and absorption differ?

Oral Cavity

Key topic in this section:

■ Structure and function of the tongue, salivary glands, and teeth

The **oral cavity**, or *mouth*, is the entrance to the GI tract (**figure 26.3**). The mouth is the initial site of mechanical digestion (via mastication) and chemical digestion (via an enzyme in saliva). The epithelial lining of the oral cavity is a nonkeratinized stratified squamous epithelium that protects against the abrasive activities associated with digestion. This lining is moistened continually by the secretion of saliva.

The oral cavity is bounded anteriorly by the teeth and lips and posteriorly by the oropharynx. The superior boundary of the oral cavity is formed by the hard and soft palates. The floor, or inferior surface, of the oral cavity is formed by the mylohyoid muscle covered with a mucous membrane. The tongue attaches to the floor of the oral cavity.

The oral cavity has two distinct regions: The **vestibule** is the space between the cheeks or lips and the gums. The **oral cavity proper** lies central to the alveolar processes of the mandible and maxillae.

Cheeks, Lips, and Palate

The lateral walls of the oral cavity are formed by the cheeks, which are covered externally by the integument and contain the buccinator muscles. The buccinator muscles compress the cheeks against the teeth to hold solid materials in place during chewing. The cheeks terminate at the fleshy **lips** (or *labia*), which form the anterior wall of the oral cavity. The lips are formed primarily by the orbicularis oris muscle and covered with keratinized stratified squamous epithelium. Lips have a reddish hue because of their abundant supply of superficial blood vessels and the reduced amount of keratin within their outer epithelial layer. The **gingivae** (jin´ji-vē), or *gums*, are composed of dense irregular connective tissue, with an overlying nonkeratinized stratified squamous epithelium that covers the alveolar processes of the upper and lower jaws and surrounds the necks of the teeth. The internal surfaces of the superior and inferior lips each are attached to the gingivae by a thin mucosa fold in the midline, called the **labial** (lā´bē-ăl) **frenulum** (fren´ū-lŭm; *frenum* = bridle).

The **palate** (pal´ăt) forms the roof of the oral cavity and acts as a barrier to separate it from the nasal cavity. The anterior two-thirds

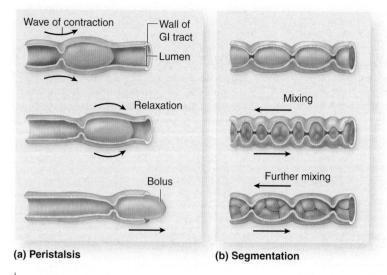

(a) **Peristalsis**   (b) **Segmentation**

Labels in figure: Wave of contraction; Wall of GI tract; Lumen; Relaxation; Bolus; Mixing; Further mixing

Figure 26.2

Peristalsis and Segmentation. A swallowed bolus is propelled through the GI tract by the coordinated contraction and relaxation of the musculature in the GI tract wall. (*a*) Peristalsis is a wave of contraction that moves material ahead of the wave through the GI tract toward the anus. (*b*) Segmentation is a back-and-forth movement in the small intestine whereby ingested material mixes with secretory products to increase the efficiency of digestion and absorption.

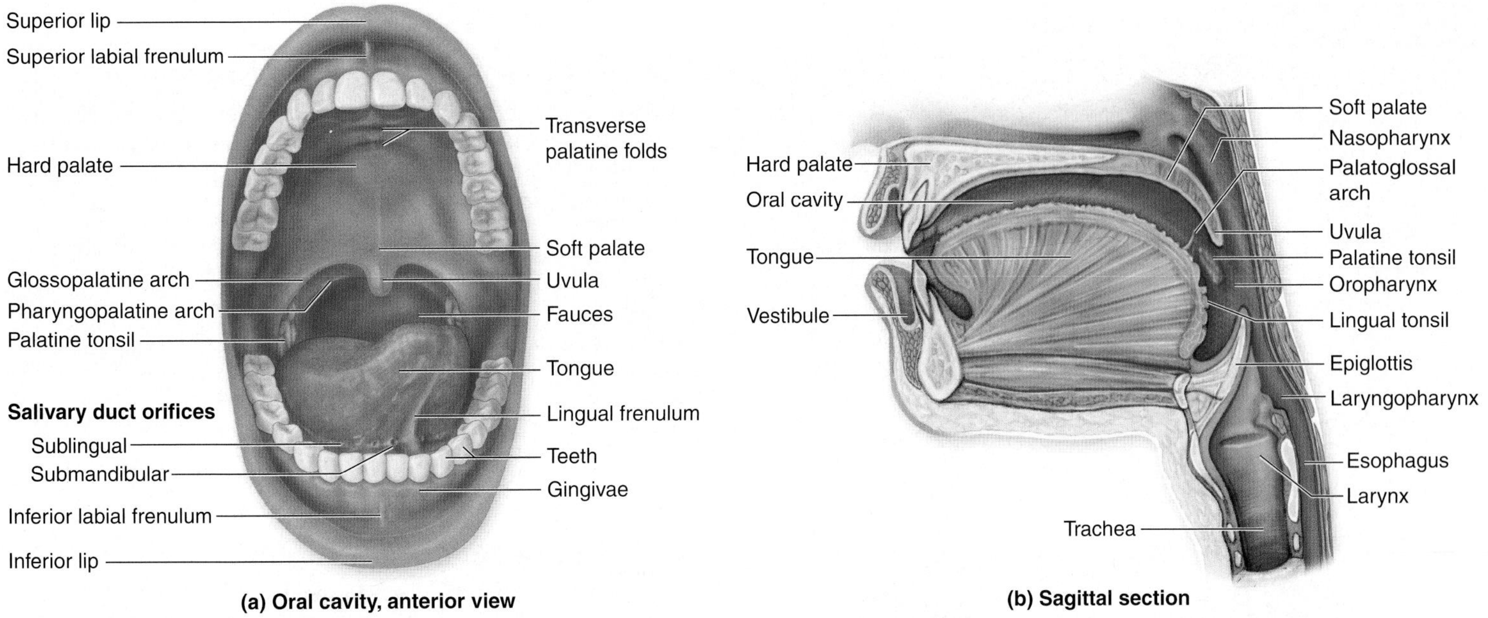

(a) Oral cavity, anterior view

(b) Sagittal section

Figure 26.3

Oral Cavity.   *(a)* Ingested food and drink enter the GI tract through the oral cavity, shown here in anterior view. *(b)* A diagrammatic sagittal section shows the structures of the oral cavity and the pharynx.

of the palate is hard and bony (called the hard palate), while the posterior one-third is soft and muscular (called the soft palate). The **hard palate** is formed by the palatine processes of the maxillae and the horizontal plates of the palatine bones. It is covered with dense connective tissue and nonkeratinized stratified squamous epithelium and exhibits prominent **transverse palatine folds,** or *friction ridges,* that assist the tongue in manipulating ingested materials prior to swallowing. The arching **soft palate** is primarily composed of skeletal muscle and covered with nonkeratinized stratified squamous epithelium. Extending inferiorly from the posterior part of the soft palate is a conical median projection called the **uvula** (ū′vū-lă; *uva* = grape). When you swallow, the soft palate and the uvula elevate to close off the posterior entrance into the nasopharynx and prevent ingested materials from entering the nasal region.

The **fauces** represent the opening between the oral cavity and the oropharynx. The fauces are bounded by paired muscular folds: the **glossopalatine arch** (anterior fold) and the **pharyngopalatine arch** (posterior fold). The **palatine tonsils** are housed between the arches (see chapter 24). These tonsils serve as an "early line of defense" as they monitor ingested food and drink for antigens, and initiate an immune response when necessary.

Tongue

The **tongue** (tŭng) is an accessory digestive organ that is formed primarily from skeletal muscle and covered with stratified squamous epithelium. As described in chapter 19, numerous small projections, termed **papillae** (pă-pil′ē; sing., *papilla; papula* = pimple), cover the superior (dorsal) surface of the tongue. The tongue's stratified squamous epithelium is keratinized over the filiform papillae but nonkeratinized over the rest of the tongue. Chapter 25 described the tongue as a participant in sound production. In addition, the tongue manipulates and mixes ingested materials during chewing and helps compress the materials against the palate to turn them into a bolus. The tongue also performs important functions in swallowing. The inferior surface of the tongue attaches to the floor of the oral cavity by a thin, vertical mucous membrane, the **lingual frenulum** (figure 26.3*a*). In addition,

the posteroinferior surface of the tongue contains **lingual tonsils.** Two categories of skeletal muscles move the tongue (see chapter 11).

Salivary Glands

The salivary glands collectively produce and secrete **saliva** (să-lī′vă), a fluid that assists in the initial activities of digestion. The volume of saliva secreted daily ranges between 1.0 and 1.5 liters. Most saliva is produced during mealtime, but smaller amounts are produced continuously to ensure that the oral cavity mucous membrane remains moist. Water makes up 99.5% of the volume of saliva and is its primary ingredient. Saliva also contains a mixture of other components **(table 26.1)**.

Saliva has various functions. It moistens ingested food and helps it become a slick, semisolid bolus that is more easily swallowed. Saliva also moistens, cleanses, and lubricates the oral cavity structures. The first step in chemical digestion occurs when amylase in saliva begins to break down carbohydrates. Saliva contains antibodies and an antibacterial substance called lysozyme that help inhibit bacterial growth in the oral cavity. Finally, saliva is the watery medium into which food molecules are dissolved so that taste receptors on the tongue can be stimulated.

Three pairs of multicellular salivary glands are located external to the oral cavity: the parotid, submandibular, and sublingual glands **(figure 26.4*a*)**.

The **parotid** (pă-rot′id; *para* = beside, *ot* = ear) **salivary glands** are the largest salivary glands. Each parotid gland is located anterior and slightly inferior to the ear, partially overlying the masseter muscle. The parotid salivary glands produce about 25–30% of the saliva, which is conducted through the **parotid duct** to the oral cavity. The parotid duct extends from the gland, parallel to the zygomatic arch, before penetrating the buccinator muscle and opening into the vestibule of the oral cavity near the second upper molar.

As their name suggests, the **submandibular** (sŭb-man-dib′ū-lăr) **salivary glands** are inferior to the body of the mandible. They produce most of the saliva (about 60–70%). A **submandibular duct** transports saliva from each gland through a papilla in the floor of the mouth on the lateral sides of the lingual frenulum.

Table 26.1	Saliva Characteristics			
Production Rate	**pH Range**	**Composition of Saliva**	**Solute Components**	**Neural Control of Saliva Secretion**
1–1.5 L/day	Slightly acidic (pH 6.4 to 6.8)	99.5% water; 0.5% solutes	Ions (e.g., Na^+, K^+, chloride, bicarbonate) Immunoglobulin A (helps decrease bacterial infections) Lysozyme (antibacterial enzyme) Mucin Salivary amylase (enzyme that breaks down carbohydrates)	Parasympathetic axons in CN IX stimulate parotid salivary gland secretions Parasympathetic axons in CN VII stimulate submandibular and sublingual salivary gland secretions Sympathetic stimulation from cervical ganglia stimulates mucin secretion

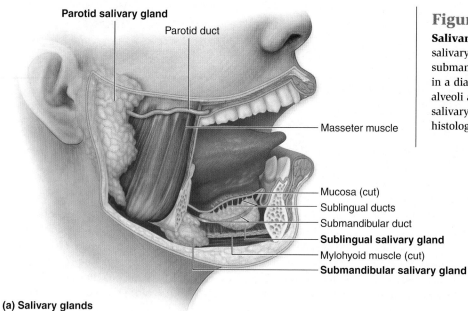

(a) Salivary glands

Figure 26.4

Salivary Glands. Saliva is produced by three pairs of salivary glands. (a) The relative locations of the parotid, submandibular, and sublingual salivary glands are shown in a diagrammatic sagittal section. (b) Serous and mucous alveoli are shown in a diagrammatic representation of salivary gland histology. (c) A photomicrograph reveals the histologic detail of the submandibular salivary gland.

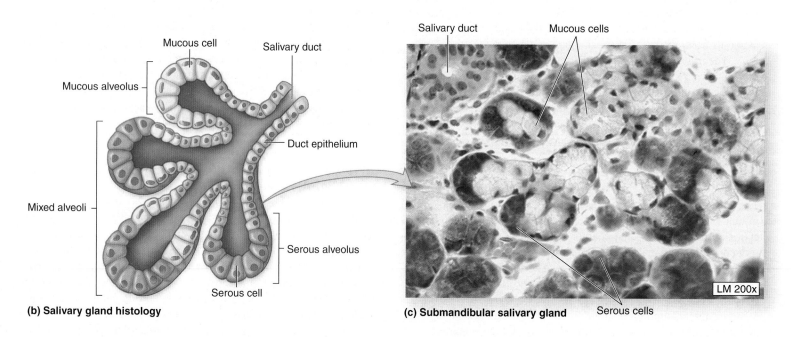

(b) Salivary gland histology

(c) Submandibular salivary gland

Table 26.2	Salivary Gland Characteristics		
Salivary Gland	**Structure and Location**	**Types of Secretion**	**Percentage of Saliva Produced**
Parotid	Largest of the salivary glands; located anterior and slightly inferior to ears; parotid duct conducts secretions into the vestibule near upper 2nd molar	Only serous secretions	25–30%
Submandibular	Located inferior to mandibular body; submandibular duct opens lateral to lingual frenulum	Both mucus and serous secretions	60–70%
Sublingual	Smallest of the salivary glands; located inferior to tongue; tiny sublingual ducts open into floor of oral cavity	Both mucus and serous secretions	3–5%

The **sublingual** (sŭb-ling′gwăl) **salivary glands** are inferior to the tongue and internal to the oral cavity mucosa. Each sublingual salivary gland extends multiple tiny **sublingual ducts** that open onto the inferior surface of the oral cavity, posterior to the submandibular duct papilla. These tiny glands contribute only about 3–5% of the total saliva.

Two types of secretory cells are housed in the salivary glands: mucous cells and serous cells (figure 26.4b,c). **Mucous cells** secrete mucin, which forms mucus upon hydration, while **serous cells** secrete a watery fluid containing ions, lysozyme, and salivary amylase. The proportion of mucous cells to serous cells varies among the three types of salivary glands. The submandibular and sublingual glands produce both serous and mucus secretions, whereas the parotid glands produce only serous secretions. **Table 26.2** summarizes the structure of the salivary glands and their secretions.

The salivary glands are primarily innervated by the parasympathetic division of the autonomic nervous system. In particular, the facial nerve (CN VII) innervates the submandibular and sublingual glands, while the glossopharyngeal nerve (CN IX) innervates the parotid gland. Parasympathetic innervation stimulates salivary gland secretion, which is why your mouth "waters" when you see a delicious dinner in front of you. Your salivary glands are preparing the body for the start of the digestion process. In contrast, sympathetic stimulation inhibits normal secretion from these glands, which is why you may experience a dry mouth after a fight-or-flight response.

WHAT DO YOU THINK?

1 Research suggests that a "dry mouth" (inadequate production of saliva) is correlated with an increase in dental problems, such as cavities. What are the possible reasons for this correlation?

Teeth

The **teeth** are collectively known as the **dentition** (den-tish′ŭn; *dentition* = teething). Teeth are responsible for mastication, the first part of the mechanical digestion process. A tooth has an exposed **crown,** a constricted **neck,** and one or more **roots** that anchor it to the jaw **(figure 26.5)**. The roots of the teeth fit tightly into **dental alveoli,** which are sockets within the alveolar processes of both the maxillae and the mandible. Collectively, the roots, the dental alveoli, and the **periodontal ligaments** that bind the roots to the alveolar processes form a gomphosis joint (described in chapter 9).

Each root of a tooth is ensheathed within hardened material called **cementum** (se-men′tŭm; rough quarry stone). A tough, durable layer of **enamel** (ē-nam′ĕl) forms the crown of the tooth. Enamel, the hardest substance in the body, is composed primarily of

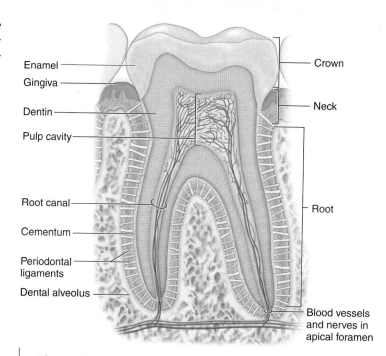

Enamel — Crown

Gingiva

Dentin — Neck

Pulp cavity

Root canal

Cementum — Root

Periodontal ligaments

Dental alveolus

Blood vessels and nerves in apical foramen

Figure 26.5

Anatomy of a Molar. Ingested material is chewed by the teeth in the oral cavity.

calcium phosphate crystals. **Dentin** (den′tin) forms the primary mass of a tooth. Dentin is comparable to bone but harder, and is deep to the cenentum and the enamel. The center of the tooth is a **pulp cavity** that contains a connective tissue called **pulp.** A **root canal** opens into the connective tissue through an opening called the **apical foramen** and is continuous with the pulp cavity. Blood vessels and nerves pass through the apical foramen and are housed in the pulp.

The **mesial** (mē′zē-ăl; *mesos* = middle) **surface** of the tooth is the surface closest to the midline of the mouth, while the **distal surface** of the tooth is farthest from the mouth midline. Other tooth surfaces include: the **buccal surface,** adjacent to the internal surface of the cheek; the **labial surface,** adjacent to the internal surface of the lip; the **lingual surface,** facing the tongue; and the **occlusal** (ŏ-kloo′zăl) **surface,** where the teeth from the opposing superior and inferior arches meet.

Two sets of teeth develop and erupt during a normal lifetime. In an infant, 20 **deciduous** (dē-sid′ū-ŭs; *deciduus* = falling off) **teeth,** also called *milk teeth,* erupt between 6 months and 30 months after birth. These teeth are eventually lost and replaced by 32 **permanent teeth.** As **figure 26.6** shows, the more anteriorly placed permanent teeth tend to

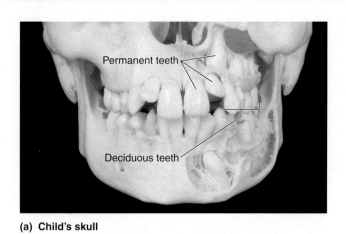

(a) Child's skull

Central incisor (7–9 mos.)
Lateral incisor (9–11 mos.)
Canine (18–20 mos.)
1st molar (14–16 mos.)
2nd molar (24–30 mos.)

Upper teeth

2nd molar (20–22 mos.)

Lower teeth

1st molar (12–14 mos.)

Canine (16–18 mos.)
Lateral incisor (7–9 mos.)
Central incisor (6–8 mos.)

(b) Deciduous teeth

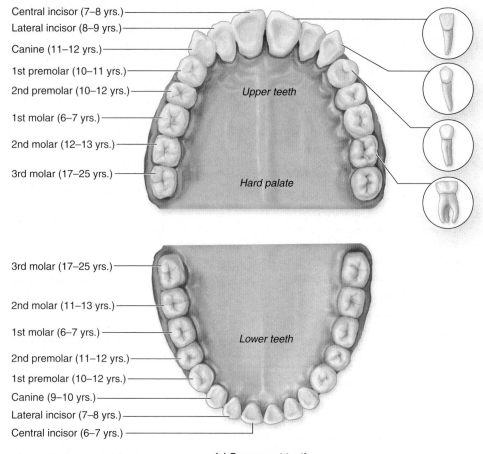

Central incisor (7–8 yrs.)
Lateral incisor (8–9 yrs.)
Canine (11–12 yrs.)
1st premolar (10–11 yrs.)
2nd premolar (10–12 yrs.)
1st molar (6–7 yrs.)
2nd molar (12–13 yrs.)
3rd molar (17–25 yrs.)

Upper teeth

Hard palate

3rd molar (17–25 yrs.)
2nd molar (11–13 yrs.)
1st molar (6–7 yrs.)
2nd premolar (11–12 yrs.)
1st premolar (10–12 yrs.)
Canine (9–10 yrs.)
Lateral incisor (7–8 yrs.)
Central incisor (6–7 yrs.)

Lower teeth

(c) Permanent teeth

Figure 26.6

Teeth. *(a)* Both deciduous teeth and unerupted permanent teeth are visible in this cut-away section of a child's skull. *(b,c)* Comparison of the dentition of deciduous and permanent teeth, including the approximate age at eruption for each tooth.

Table 26.3	Oral Cavity Structures and Their Functions	
Structure	Description	Function
Gingivae	Composed of dense irregular connective tissue and nonkeratinized stratified squamous epithelium	Surround necks of teeth and cover alveolar processes
Hard palate	Anterior roof of mouth; bony shelf covered by dense connective tissue and nonkeratinized stratified squamous epithelium	Forms anterior two-thirds of roof of mouth; separates oral cavity from nasal cavity
Lips	Form part of anterior walls of oral cavity; covered with keratinized stratified squamous epithelium	Close oral cavity during chewing
Salivary glands	Three pairs of large multicellular glands: parotid glands, sublingual glands, and submandibular glands	Produce saliva
Soft palate	Posterior roof of mouth formed from skeletal muscle and covered with nonkeratinized stratified squamous epithelium; the uvula hangs from it	Forms posterior one-third of roof of mouth; helps close off entryway to nasopharynx when swallowing
Teeth	Hard structures projecting from the alveolar processes of the maxillae and mandible: incisors, canines, premolars, and molars	Mastication (chewing food)
Tongue	Composed primarily of skeletal muscle and covered by stratified squamous epithelium; surface covered by papillae	Manipulates ingested material during chewing; pushes material against palate to turn it into a bolus; detects tastes (via taste buds)
Tonsils	Aggregates of partially encapsulated lymphatic tissue	Detect antigens in swallowed food and drink and initiate immune response if necessary
Uvula	Conical, median, muscular projection extending from the soft palate	Assists soft palate in closing off entryway to nasopharynx when swallowing
Vestibule	Space between cheek and gums	Space between lips/cheeks and gums where ingested materials are mixed with saliva and mechanically digested

appear first, followed by the posteriorly placed teeth. (The major exception to this rule are the first molars, which appear at about age 6 and are sometimes referred to as the "6-year molars.") The last teeth to erupt are the third molars, often called *wisdom teeth*, in the late teens or early 20s. Often the jaw lacks space to accommodate these final molars, and they may either emerge only partially or grow at an angle and become impacted (wedged against another structure). Impacted teeth cannot erupt properly because of the angle of their growth.

WHAT DO YOU THINK?

2 If the deciduous teeth are eventually replaced by permanent teeth, why do humans have deciduous teeth in the first place?

The most anteriorly placed permanent teeth are called **incisors** (in-sī′zŏr; *incido* = to cut into). They are shaped like a chisel and have a single root. They are designed for slicing or cutting into food. Immediately posterolateral to the incisors are the **canines** (kā′nīn; *canis* = dog), which have a pointed tip for puncturing and tearing food. **Premolars** are located posterolateral to the canines and anterior to the molars. They have flat crowns with prominent ridges called **cusps** that are used to crush and grind ingested materials. Premolars may have one or two roots. The **molars** (mō′lăr; *molaris* = millstone) are the thickest and most posteriorly placed teeth. They have large, broad, flat crowns with distinctive cusps, and three or more roots. Molars are also adapted for grinding and crushing ingested materials. If the mouth is divided into quadrants, each quadrant contains the following number of permanent teeth: 2 incisors, 1 canine, 2 premolars, and 3 molars.

Table 26.3 summarizes the structures of the oral cavity and their functions.

WHAT DID YOU LEARN?

3 What are the main components of saliva, and what functions do they serve?

4 What are the types of permanent teeth, and what is each tooth's main function in mastication?

Pharynx

Key topic in this section:

■ Structure of the pharynx and action of the pharyngeal constrictors

The common space used by both the respiratory and digestive systems is the **pharynx** (see figure 26.1 and chapter 25). The non-keratinized stratified squamous epithelial lining of the oropharynx and the laryngopharynx provides protection against the abrasive activities associated with swallowing ingested materials.

Three skeletal muscle pairs, called the superior, middle, and inferior **pharyngeal constrictors,** form the wall of the pharynx. Sequential contraction of the pharyngeal constrictors decreases the diameter of the pharynx, beginning at its superior end and moving toward its inferior end, thus pushing swallowed material toward the esophagus. As the pharyngeal constrictors constrict, the epiglottis of the larynx closes over the laryngeal opening to prevent ingested materials from entering the larynx and trachea.

The vagus nerves (CN X) innervate most pharyngeal muscles. The principal arteries supplying the pharynx are branches of the external carotid arteries. The pharynx is drained by the internal jugular veins.

💡 **WHAT DID YOU LEARN?**

5 How do pharyngeal constrictors move swallowed material to the esophagus?

General Arrangement of Abdominal GI Organs

Key topics in this section:

- Peritoneum location and function
- Derivation of specific mesenteries
- The four tunics in the GI tract wall
- Blood vessels, lymphatic structures, and nerves that supply the GI tract

The abdominal organs of the GI tract are supported by serous membranes, and the walls of these organs have specific layers, called tunics.

Peritoneum, Peritoneal Cavity, and Mesentery

The abdominopelvic cavity is lined with moist serous membranes **(figure 26.7)**. The portion of the serous membrane that lines the

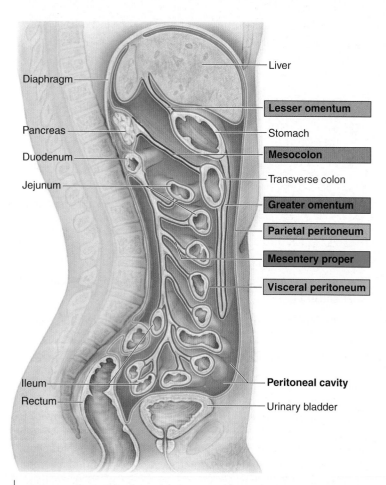

Liver

Diaphragm

Lesser omentum

Pancreas — Stomach

Mesocolon

Duodenum

Jejunum — Transverse colon

Greater omentum

Parietal peritoneum

Mesentery proper

Visceral peritoneum

Ileum

Rectum **Peritoneal cavity**

Urinary bladder

Figure 26.7

Peritoneum and Mesenteries. Many abdominal organs are held in place by double-layered serous membrane folds called mesenteries. The peritoneum is the serous membrane lining the internal abdominal wall (parietal layer) and covering the outer surface of the abdominal organs (visceral layer). This sagittal view through the abdominopelvic cavity shows the relationship between the peritoneal membranes and the abdominal organs they ensheathe.

inside surface of the body wall is called the **parietal peritoneum** (per′i-tō-nē′ŭm; *periteino* = to stretch over). The portion of the serous membrane that reflects to cover the surface of internal organs is called the **visceral peritoneum.** Between these two layers is the **peritoneal cavity,** a potential space where the peritoneal layers that face each other secrete a lubricating serous fluid. The thin layer of fluid in the peritoneal cavity lubricates both the body wall and the internal organ surfaces, allowing the abdominal organs to move freely, and reducing any friction resulting from this movement.

> ## Study Tip!
>
> The serous peritoneum is very similar to the serous pleura and the serous pericardium. These membranes have an outer parietal layer that lines the body wall and a visceral layer that covers the organ. The space between the parietal and visceral layers is where serous fluid is secreted, and this fluid acts as a lubricant to prevent friction as the organ moves. So, if you remember the basics about the pleura and the pericardium, you will know the basics about the peritoneum too.

Within the abdomen, organs that are completely surrounded by visceral peritoneum are called **intraperitoneal** (in′tră-per′i-tō-nē′ăl) **organs.** They include the stomach, part of the duodenum of the small intestine, the jejunum and the ileum of the small intestine, the cecum, the appendix, and the transverse and sigmoid colon of the large intestine. **Retroperitoneal** (re-trō-per′i-tō-nē′ăl) **organs** typically lie directly against the posterior abdominal wall, so only their anterolateral portions are covered with peritoneum. Retroperitoneal organs include most of the duodenum, the pancreas, the ascending and descending colon of the large intestine, and the rectum.

The **mesenteries** (mes′en-ter-ē; *mesos* = middle, *enteron* = intestine) are folds of peritoneum that support and stabilize the intraperitoneal GI tract organs. Blood vessels, lymph vessels, and nerves are sandwiched between the two folds and supply the digestive organs. There are several different types of mesenteries. The **greater omentum** (ō-men′tŭm) extends inferiorly like an apron from the greater curvature of the stomach and covers most of the abdominal organs **(figure 26.8a)**. It often accumulates large amounts of adipose connective tissue. The **lesser omentum** connects the lesser curvature of the stomach and the proximal end of the duodenum to the liver. The lesser omentum may be subdivided into a hepatogastric ligament, which runs from the liver to the stomach, and a hepatoduodenal ligament, which runs from the liver to the duodenum.

The **mesentery proper** is a fan-shaped fold of peritoneum that suspends most of the small intestine from the internal surface of the posterior abdominal wall (figure 26.8b). The peritoneal fold that attaches parts of the large intestine to the internal surface of the posterior abdominal wall is called the **mesocolon** (mez′ō-kō′lon). Essentially, a mesocolon is a mesentery for parts of the large intestine. There are several distinct sections of the mesocolon, each named for the portion of the colon it suspends. For example, transverse mesocolon is associated with the transverse colon, while sigmoid mesocolon is associated with the sigmoid colon.

A **peritoneal ligament** is a peritoneal fold that attaches one organ to another organ, or attaches an organ to the anterior or lateral abdominal wall. Some examples of peritoneal ligaments include: the **coronary ligament,** a peritoneal fold attaching the superior surface of the liver to the diaphragm at the margins of the bare area of the liver; the **falciform** (fal′si-fōrm; *falx* = sickle)

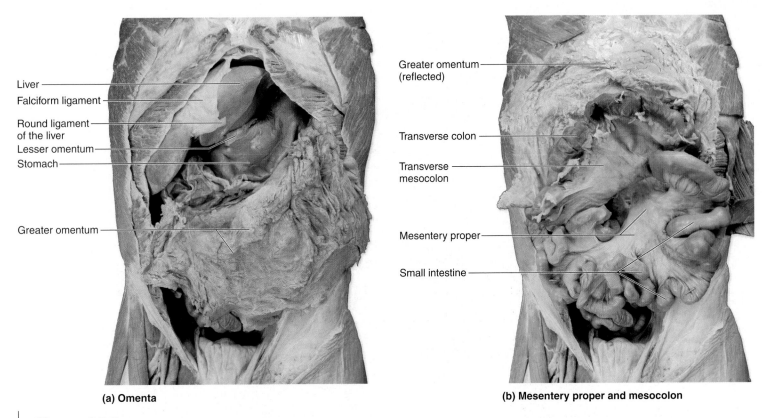

(a) Omenta

(b) Mesentery proper and mesocolon

Figure 26.8

Omenta and Mesentery. Cadaver photos show anterior views of *(a)* the greater and lesser omenta, and *(b)* the mesentery proper and some of the mesocolon.

ligament, a peritoneal fold that attaches the liver to the anterior internal abdominal wall; and the lienorenal ligament, a fold of peritoneum between the spleen and the kidney.

General Histology of GI Organs (Esophagus to Large Intestine)

The GI tract from the esophagus through the large intestine is a tube composed of four concentric layers, called tunics. From deep (the lining of the lumen) to superficial (the external covering), these tunics are the mucosa, the submucosa, the muscularis, and the adventitia or serosa **(figure 26.9)**. The general pattern of the tunics is described next. There are variations in the general pattern, which will be described in detail when we discuss the specific organs in which they appear.

Mucosa

The **mucosa** (mū-kō′să), or *mucous membrane,* has three components: (1) a superficial epithelium lining the lumen of the GI tract; (2) an underlying areolar connective tissue, called the **lamina propria;** and (3) a relatively thin layer of smooth muscle, termed the **muscularis mucosae.**

For most of the abdominal GI tract organs, the lining epithelium is a simple columnar epithelium. The exception to this rule is the esophagus, which is lined with nonkeratinized stratified squamous epithelium.

Submucosa

The **submucosa** is composed of either areolar or dense irregular connective tissue and has far fewer cells than the lamina propria.

Submucosa components include: accumulations of lymphatic tissue in some submucosal regions; mucin-secreting glands that project ducts across the mucosa and open into the lumen of the tract in the esophagus and duodenum; many large blood vessels and lymph vessels; and nerves that extend fine branches into both the mucosa and the muscularis. These nerve fibers and their associated ganglia are collectively referred to as the **submucosal nerve plexus** (*or Meissner plexus*).

Muscularis

The **muscularis** (mŭs-kū′lā′ris) typically contains two layers of smooth muscle. Exceptions to this pattern include the esophagus (which contains a mixture of skeletal and smooth muscle) and the stomach (which contains three layers of smooth muscle). The fibers of the inner layer of smooth muscle are oriented circumferentially around the GI tract, and are called the **inner circular layer.** The fibers of the outer layer are oriented lengthwise along the GI tract, and are called the **outer longitudinal layer.**

If you think of the GI tract as a "tube," then contractions of the circular layer constrict the diameter of the tube lumen, while contractions of the longitudinal layer shorten the tube. At specific locations along the GI tract, the inner circular muscle layer is greatly thickened to form a **sphincter.** A sphincter closes off the lumen opening at some point along the GI tract, and in so doing it can help control the movement of materials through the GI tract. The nerve fibers and the associated ganglia located between the two layers of smooth muscle control its contractions and are collectively referred to as the **myenteric** (mī-en-ter′ik; *mys* = muscle, *enteron* = intestine) **nerve plexus** (or *Auerbach plexus*).

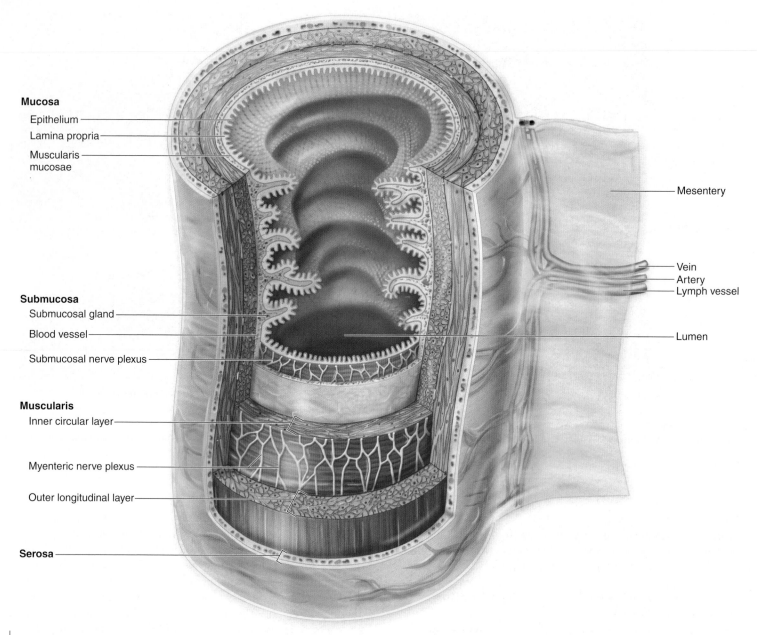

Mucosa
Epithelium
Lamina propria
Muscularis mucosae

Mesentery

Submucosa
Submucosal gland
Blood vessel
Submucosal nerve plexus

Vein
Artery
Lymph vessel

Lumen

Muscularis
Inner circular layer

Myenteric nerve plexus

Outer longitudinal layer

Serosa

Figure 26.9

Tunics of the Abdominal GI Tract. The wall of the abdominal GI tract has four tunics. From the lining of its lumen to the external covering, the tunics are the mucosa, submucosa, muscularis, and adventitia or serosa.

Adventitia or Serosa

The outermost tunic may be either an adventitia or a serosa. An **adventitia** (ad-ven-tish'ă) is composed of areolar connective tissue with dispersed collagen and elastic fibers. A **serosa** (se-rō'să) has the same components as the adventitia, but is covered by a visceral peritoneum. Intraperitoneal organs have a serosa, because they are completely surrounded by visceral peritoneum. Retroperitoneal organs primarily have an adventitia, since these organs are only partially covered by visceral peritoneum. For example, the ascending colon (which is retroperitoneal) has an adventitia, while the stomach (which is intraperitoneal) has a serosa.

Study Tip!

You have just learned that the "default" pattern of the tunics is as follows:

1. Mucosa (typically lined with simple columnar epithelium)
2. Submucosa
3. Muscularis (typically formed from two layers of smooth muscle)
4. Adventitia or serosa

As you learn the basic GI organs, determine how these organs follow or deviate from this default pattern. For example:

- The esophagus has these tunics, but it deviates from the pattern in two ways: Its mucosa has a stratified squamous epithelium, and its muscularis has a skeletal muscle in its superior region, skeletal and smooth muscle in the middle region, and smooth muscle in its inferior region.
- The stomach has these tunics, but it deviates from the pattern in that its muscularis has three layers of smooth muscle, not two.
- The small intestine follows the basic "default" pattern of tunics.
- The large intestine has these tunics, but in its muscularis, the outer longitudinal layer of muscle forms three distinct bands called teniae coli (described later in this chapter).

Knowing the basic tunic pattern, and then figuring out how an organ may deviate from this pattern, will help you better distinguish the histology of the esophagus, stomach, small intestine, and large intestine.

Blood Vessels, Lymphatic Structures, and Nerve Supply

The GI tract has a rich blood and nerve supply. In addition, extensive lymphatic structures along the length of the GI tract act as sentinels to monitor for antigens that may have been ingested. The blood vessels, lymphatic structures, and nerves enter the GI tract from either the surrounding structures (e.g., nearby organs, abdominal wall) or via the supporting mesentery.

Blood Vessels

Branches of the **celiac trunk, superior mesenteric artery,** and **inferior mesenteric artery** supply the abdominal GI tract (see chapter 23). These artery branches split into smaller branches that extend throughout the walls of the GI organs. Branches travel within the tunics, and the mucosa contains capillaries that have fenestrated endothelial cells to promote absorption. The veins arising in the mucosa form anastomoses in the submucosa before exiting the wall of the GI tract adjacent to their companion arteries. Eventually, the veins merge to form the hepatic portal system of veins, to be discussed later in this chapter.

Lymph Vessels and Tissues

Lymphatic capillaries arise as blind tubes in the mucosa of the GI tract. In the small intestine, each villus usually contains a single, blind-ended, central lymphatic capillary called a **lacteal.** Recall from chapter 24 that lacteals are responsible for absorbing dietary lipids and lipid-soluble vitamins (vitamins that can be absorbed only if they are dissolved in lipids first). Outside the organ walls, lymphatic capillaries merge to form lymphatic vessels. These vessels enter and exit the many lymph nodes scattered near the organs and within the mesentery. Eventually, this lymph will be transported to the cisterna chyli, which drains into the thoracic duct.

The lymphatic structures within the GI tract lie primarily in the lamina propria of the mucosa. Lymphatic structures called MALT (mucosa-associated lymphatic tissue) are found in the small intestine and appendix (see chapter 24). In the small intestine, these aggregate nodules are called **Peyer patches.** They appear to the naked eye as oval bodies about the size of a pea, but are often much larger structures. Less commonly, lymphatic structures are found external to the simple epithelium throughout the stomach and intestines where they are known as diffuse lymphatic tissue. Also, solitary lymphatic nodules may occur in the esophagus, the pylorus of the stomach, and along the entire length of the small and large intestines.

Nerves

The nerves associated with the GI tract consist of both autonomic motor and sensory axons. There are three main groups of autonomic plexuses:

- The **celiac plexus** contains sympathetic axons (from the T5–T9 segments of the spinal cord) and parasympathetic axons (from the vagus nerve). This plexus supplies structures that receive their blood supply from branches of the celiac trunk.
- The **superior mesenteric plexus** contains sympathetic axons (from the T8–T12 segments of the spinal cord) and parasympathetic axons (from the vagus nerve). This plexus transmits autonomic innervation to structures that receive their blood supply from branches of the superior mesenteric artery.
- The **inferior mesenteric plexus** contains sympathetic axons (from the L1–L2 segments of the spinal cord) and parasympathetic axons (from the pelvic splanchnic nerves). This plexus supplies structures that receive blood from branches of the inferior mesenteric artery.

In general, parasympathetic innervation promotes digestive system activity by stimulating GI gland secretions and peristalsis, and by relaxing GI sphincters. These changes in activity induce vasodilation and an increase in GI blood flow. Sympathetic innervation inhibits digestive system activity, and it tends to do the opposite of parasympathetic innervation. So sympathetic innervation inhibits some GI gland secretions, inhibits peristalsis, closes the GI sphincters, and vasoconstricts the blood vessels to the GI tract.

WHAT DID YOU LEARN?

6 Compare the terms intraperitoneal and retroperitoneal, and give examples of each type of organ.

7 What are the four main tunics of the abdominal GI tract, and what is the "default" pattern in each tunic?

Esophagus

Key topic in this section:

- Structure and function of the esophagus

The **esophagus** (ē-sof′ă-gŭs; gullet) is a tubular passageway for swallowed materials being conducted from the pharynx to the stomach (see figure 26.1). The inferior region of the esophagus

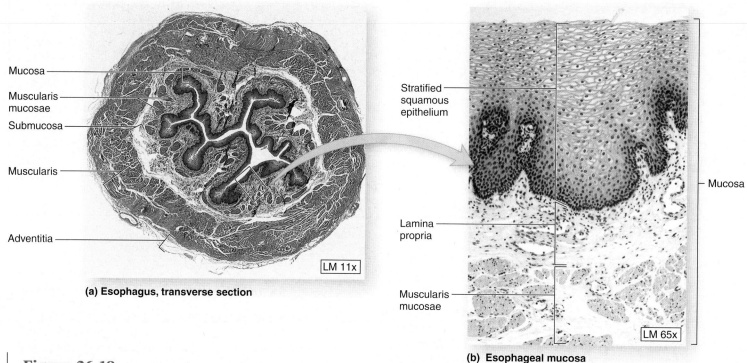

Figure 26.10

Histology of the Esophagus. The esophagus extends inferiorly from the pharynx and conducts swallowed materials to the stomach. *(a)* A photomicrograph of a transverse section through the esophagus identifies the tunics in its wall. *(b)* A photomicrograph shows the esophageal mucosa.

connects to the stomach by passing through an opening in the diaphragm called the **esophageal hiatus** (hī-ā′tŭs; to yawn).

Gross Anatomy

The esophageal wall is thick and composed of concentric tunics that are continuous superiorly with those of the pharynx and inferiorly with those of the stomach. In the living adult human, the esophagus is about 25 centimeters long, and most of its length is within the thorax, directly anterior to the vertebral bodies. Only the last 1.5 centimeters of the esophagus are located in the abdomen. The empty esophagus is flattened; thus, there is no continuously open lumen. Only a passing bolus of food slightly expands the esophagus.

Histology

The esophageal mucosa is different from that of the abdominal GI tract organs in that it is composed of thick, nonkeratinized stratified squamous epithelium **(figure 26.10)**. The stratified squamous epithelium is better suited to withstand the abrasions of the bolus as it moves through the esophagus. Since the esophagus does not absorb any nutrients, this thicker, protective epithelium supports its function.

The submucosa is thick and composed of abundant elastic fibers that permit distension during swallowing. It houses numerous mucous glands that provide a thick, lubricating mucus for the epithelium.

The muscularis is composed of an inner circular layer and an outer longitudinal layer. The muscularis of the esophagus is unique in that it contains a blend of both skeletal and smooth muscle fibers.

The two layers of muscle in the superior one-third of the muscularis layer are skeletal, rather than smooth, to ensure that the swallowed material moves rapidly out of the pharynx and into the esophagus before the next respiratory cycle begins. (Remember that skeletal muscle contracts more rapidly than does smooth muscle.) Striated and smooth muscle fibers intermingle in the middle one-third of the esophagus, and only smooth muscle is found within the wall of the inferior one-third. This transition marks the beginning of a continuous smooth muscle muscularis that extends throughout the stomach and the small and large intestines to the anus. The outermost layer of the esophagus is an adventitia.

At the superior end of the esophagus, the **superior esophageal sphincter** (or *pharyngoesophageal sphincter*) is a thickened ring of circular skeletal muscle marking the area where the esophagus and the pharynx meet. This sphincter is closed during inhalation of air, so that air doesn't enter the esophagus and enters the larynx and trachea instead. The orifice between the esophagus and the stomach is bounded by a thin band of circular smooth muscle, the **inferior esophageal sphincter** (*esophagealgastric* or *cardiac sphincter*). This sphincter isn't strong enough alone to prevent materials from refluxing back into the esophagus; instead, the esophageal opening of the diaphragm assumes the duty of the "stronger sphincter" to prevent materials from regurgitating from the stomach into the esophagus.

WHAT DID YOU LEARN?

8 How do the esophageal tunics differ from the "default" tunic pattern?

CLINICAL VIEW

Reflux Esophagitis and Gastroesophageal Reflux Disease

The inferior esophageal sphincter and the diaphragm usually prevent acidic stomach contents from regurgitating into the esophagus. However, sometimes acidic chyme refluxes into the esophagus, causing the burning pain and irritation of **reflux esophagitis.** Because the pain is felt posterior to the sternum and can be so intense that it is mistaken for a heart attack, this condition is commonly known as "heartburn." Unlike the stomach epithelium, the esophageal epithelium is poorly protected against acidic contents and easily becomes inflamed and irritated. Other symptoms include abdominal pain, difficulty in swallowing, increased belching, and sometimes bleeding.

Reflux esophagitis can occur in anyone, but is seen most frequently in overweight individuals, smokers, those who have eaten a very large meal (especially just before bedtime), and/or people with **hiatal hernias** (hī-ā'tăl her'nē-ă; rupture), in which a portion of the stomach protrudes through the diaphragm into the thoracic cavity. Eating spicy foods or ingesting too much caffeine may exacerbate the symptoms in people affected by reflux esophagitis. Preventive treatment includes lifestyle changes such as limiting meal size, not lying down until at least 2 hours after eating, quitting smoking, and losing weight. Sleeping with the head of the bed elevated 4–6 inches, so that the body lies at an angle rather than flat, appears to alleviate symptoms.

Chronic reflux esophagitis can lead to **gastroesophageal reflux disease (GERD).** In this condition, frequent gastric reflux erodes the esophageal tissue, so over a period of time, scar tissue builds up in the esophagus, leading to narrowing of the esophageal lumen. An **endoscope** (an optical tubular instrument used to visualize the lumina of internal organs) may be inserted into the larynx and esophagus to visualize the damage (see photo). In advanced cases of GERD, the esophageal epithelium may change from stratified squamous to columnar secretory epithelium, a condition known as **Barrett esophagus.** The specific reasons why Barrett esophagus develops are not known. Barrett esophagus is associated with an increased risk of cancerous growths in the esophagus, as these new secretory cells can develop into cancer. The continual reflux and injury may also lead to esophageal ulcers.

GERD can be treated with a series of medications. Proton pump inhibitors (e.g., Prilosec®, Nexium®) limit acid secretion in the stomach by acting on the proton (hydrogen ion) "pumps" that help produce acid. Histamine (H₂) blockers (e.g., Pepcid®, Axid®, Zantac®) also help limit acid secretion in the stomach. Antacids (e.g., Tums®, Rolaids®) help neutralize stomach acid.

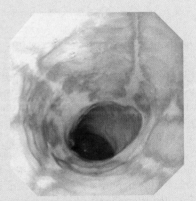

An endoscopic view of the esophagus shows the signs of reflux esophagitis.

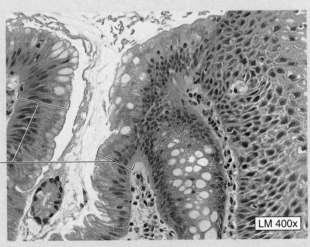

Columnar secretory epithelium

LM 400x

The condition known as Barrett esophagus.

The Swallowing Process

Key topic in this section:

■ The three phases of swallowing

Swallowing, also called *deglutition* (dē-gloo-tish'ŭn; *degluto* = to swallow), is the process of moving ingested materials from the oral cavity to the stomach. There are three phases of swallowing: the voluntary phase, the pharyngeal phase, and the esophageal phase **(figure 26.11)**.

The **voluntary phase** occurs after ingestion. Food and saliva mix in the oral cavity. Chewing forms a bolus that is mixed and manipulated by the tongue and then pushed superiorly against the hard palate. Transverse palatine folds in the hard palate help direct the bolus posteriorly toward the oropharynx.

The appearance of the bolus at the entryway to the oropharynx initiates the **pharyngeal phase,** when tactile sensory receptors trigger the swallowing reflex, which is controlled by the swallowing center in the medulla oblongata. The bolus passes quickly and involuntarily through the pharynx to the esophagus. During this phase, (1) the soft palate and uvula elevate to block the passageway between the nasopharynx and oropharynx; (2) the bolus enters the oropharynx; and (3) the larynx and laryngeal opening elevate toward the epiglottis, ultimately covering and sealing the glottis to prevent swallowed materials from entering the trachea. Pharyngeal constrictors are skeletal muscle groups that contract involuntarily and sequentially to ensure that the swallowed bolus moves quickly (1 second elapses in this phase) through the pharynx and into the esophagus.

The **esophageal phase** is involuntary. It is the time (about 5 to 8 seconds) during which the bolus passes through the esophagus

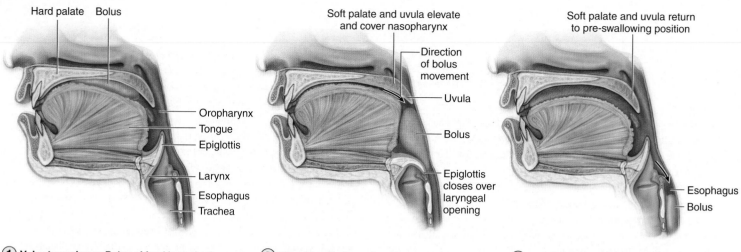

① **Voluntary phase:** Bolus of food is pushed by tongue against hard palate and then moves toward oropharynx.

② **Pharyngeal phase** (involuntary)**:** As bolus moves into oropharynx, the soft palate and uvula close off nasopharynx, and the larynx elevates so the epiglottis closes over laryngeal opening.

③ **Esophageal phase** (involuntary)**:** Peristaltic contractions of the esophageal muscle push bolus toward stomach; soft palate, uvula, and larynx return to their pre-swallowing positions.

Figure 26.11

Phases of Swallowing.   Swallowing, or deglutition, occurs as a result of coordinated muscular activities that force the bolus (1) into the pharynx from the oral cavity, (2) through the pharynx, and (3) into the esophagus on the way to the stomach.

and into the stomach. This phase begins when the superior esophageal sphincter relaxes to allow ingested materials into the esophagus. The presence of the bolus within the lumen of the esophagus stimulates peristaltic waves of muscular contraction that assist in propelling the bolus toward the stomach. The soft palate, uvula, and larynx return to the pre-swallowing positions.

WHAT DID YOU LEARN?

❾ Briefly describe the three phases of swallowing.

Stomach

Key topics in this section:

- Gross anatomy of the stomach
- Histology of the stomach wall
- Secretions of the stomach

The **stomach** (stŭm′ŭk; *stomachus* = belly) is a muscular, J-shaped sac that occupies the left upper quadrant of the abdomen, immediately inferior to the diaphragm **(figure 26.12)**. It continues the mechanical and chemical digestion of the bolus. After the bolus has been completely processed in the stomach, the product is called **chyme** (kīm; *chymos* = juice). Chyme has the consistency of a pastelike soup. The stomach facilitates mechanical digestion by the contractions of its thick muscularis layer, which churns and mixes the bolus and the gastric secretions. The stomach facilitates chemical digestion through its gastric secretions of acid and enzymes.

Gross Anatomy

The stomach is composed of four regions:

- The **cardia** (kar′dē-ă; heart) is a small, narrow, superior entryway into the stomach lumen from the esophagus. The

internal opening where the cardia meets the esophagus is called the **cardiac orifice.**
- The **fundus** (fŭn′dŭs; bottom) is the dome-shaped region lateral and superior to the esophageal connection with the stomach. Its superior surface contacts the diaphragm.
- The **body** is the largest region of the stomach; it is inferior to the cardiac orifice and the fundus.
- The **pylorus** (pī-lōr′ŭs; gatekeeper) is a narrow, medially directed, funnel-shaped pouch that forms the terminal region of the stomach. Its opening with the duodenum of the small intestine is called the **pyloric orifice.** Surrounding this pyloric orifice is a thick ring of circular smooth muscle called the **pyloric sphincter.** The pyloric sphincter regulates the entrance of chyme into the small intestine by closing upon sympathetic innervation and opening upon parasympathetic innervation.

The inferior convex border of the stomach is the **greater curvature,** while the superior concave border is the **lesser curvature.** The greater omentum attaches to the greater curvature edge of the stomach, and the lesser omentum extends between the lesser curvature and the liver.

Internally, the stomach lining is composed of numerous **gastric folds,** or *rugae* (roo′gē; *ruga* = wrinkle). These gastric folds, which are only observed when the stomach is empty, allow the stomach to expand greatly when it fills and then return to its normal J-shape when it empties.

Histology

The stomach is lined by a simple columnar epithelium, although little absorption occurs in the stomach. This epithelium does not contain goblet cells; instead, surface mucous cells (described later) secrete mucin onto the epithelial lining. Another feature that distinguishes the stomach mucosa from the default pattern is that the stomach lining is indented by numerous depressions called **gastric**

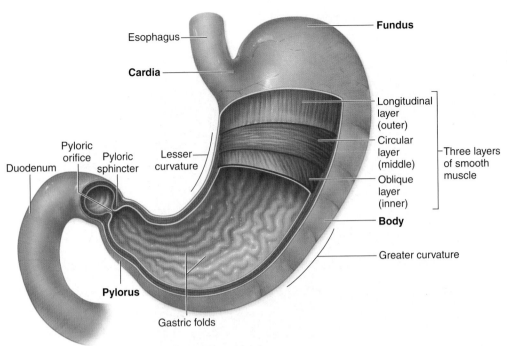

(a) Stomach regions, anterior view

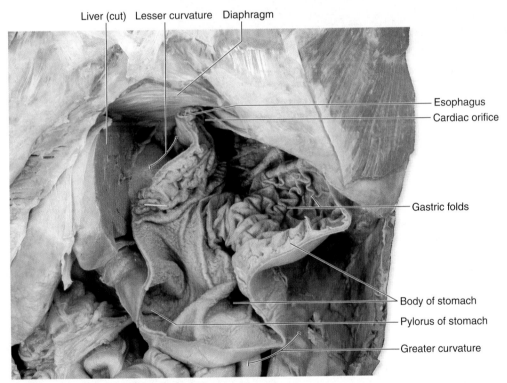

(b) Gross anatomy of stomach (cut open)

Figure 26.12

Gross Anatomy of the Stomach. The stomach is a muscular sac where mechanical and chemical digestion of the bolus occurs. *(a)* The major regions of the stomach are the cardia, the fundus, the body, and the pylorus. Three layers of smooth muscle make up the muscularis. *(b)* A cadaver photo shows an anterior section of the stomach, illustrating the gastric folds and the cardiac orifice.

pits. At the base of each pit are the openings of several branched tubular glands, called **gastric glands,** which extend through the length of the mucosa to its base **(figure 26.13)**. The mucosa of the stomach is only 1.5 millimeters at its thickest point (about the thickness of a nickel). The muscularis mucosae lies between the base of the gastric glands and the submucosa, and it helps expel gastric gland secretory products when it contracts.

The muscularis of the stomach varies from the default pattern in that it is composed of three smooth muscle layers instead of two: an inner oblique layer, a middle circular layer, and an outer longi-

tudinal layer. The oblique layer is best-developed at the cardia and the body of the stomach. The presence of a third layer of smooth muscle assists the continued churning and blending of the bolus. The stomach is intraperitoneal, so its outermost tunic is a serosa.

WHAT DO YOU THINK?

3 The stomach secretes gastric juices, which are highly acidic and can break down and chemically digest food. What prevents the gastric juices from eating away at the stomach itself?

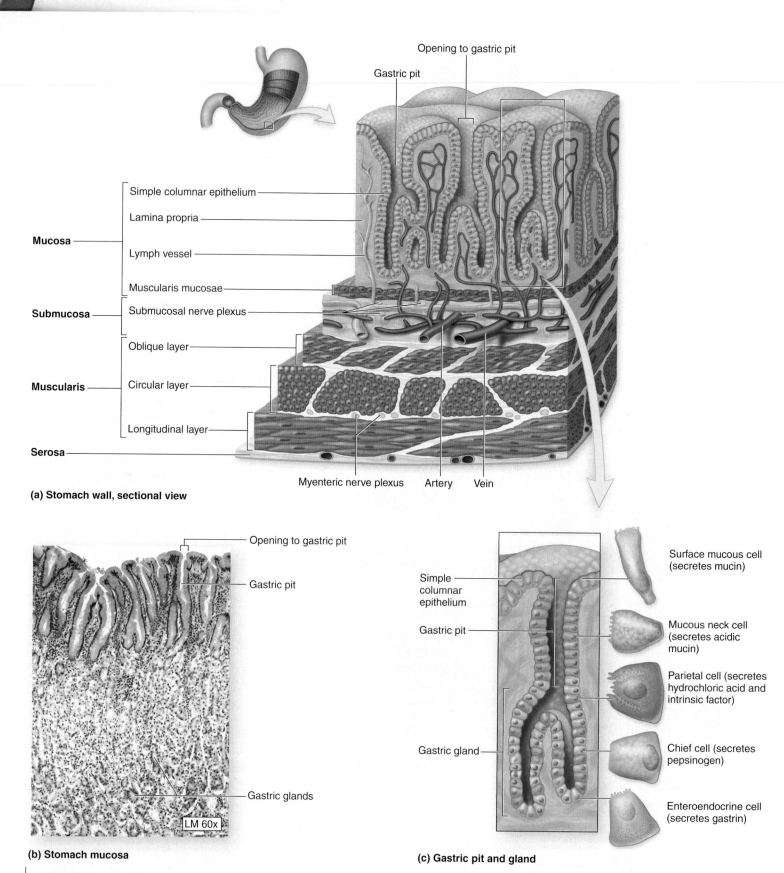

Opening to gastric pit

Gastric pit

Mucosa
- Simple columnar epithelium
- Lamina propria
- Lymph vessel
- Muscularis mucosae

Submucosa
- Submucosal nerve plexus

Muscularis
- Oblique layer
- Circular layer
- Longitudinal layer

Serosa

Myenteric nerve plexus   Artery   Vein

(a) Stomach wall, sectional view

Opening to gastric pit

Gastric pit

Gastric glands

LM 60x

(b) Stomach mucosa

Simple columnar epithelium

Gastric pit

Gastric gland

Surface mucous cell (secretes mucin)

Mucous neck cell (secretes acidic mucin)

Parietal cell (secretes hydrochloric acid and intrinsic factor)

Chief cell (secretes pepsinogen)

Enteroendocrine cell (secretes gastrin)

(c) Gastric pit and gland

Figure 26.13

Histology of the Stomach Wall.   *(a)* The stomach lumen contains invaginations within the mucosa called gastric pits that lead into gastric glands. *(b)* A photomicrograph shows gastric glands and the cells lining the gastric pit. *(c)* A diagrammatic section of a gastric pit and gland shows their structure and the distribution of different types of secretory cells.

Gastric Secretions

Five types of secretory cells form the gastric epithelium (figure 26.13b,c).

Surface mucous cells line the stomach lumen and extend into the gastric pits. They continuously secrete mucin onto the gastric luminal surface to prevent ulceration of the lining upon exposure to the high acidity of the gastric fluid and protect the epithelium from gastric enzymes.

Mucous neck cells are located immediately deep to the base of the gastric pit and are interspersed among the parietal cells. These

---

CLINICAL VIEW

Peptic Ulcers

Normally there is a balance in the stomach between the acidic gastric juices and the protective regenerative nature of the mucosa lining. When this balance is thrown off, the stage is set for the development of a **peptic ulcer,** which is a chronic, solitary erosion of a portion of the lining of either the stomach or the duodenum. Annually, over 4 million people in the United States are diagnosed with an ulcer. Gastric ulcers are peptic ulcers that occur in the stomach, whereas duodenal ulcers are peptic ulcers that occur in the superior part of the duodenum, which is the first segment of the small intestine. Duodenal ulcers are common because the first part of the duodenum receives the chyme from the stomach but has yet to receive the alkaline pancreatic juice that can neutralize chyme's acidic content. Symptoms of an ulcer include a gnawing, burning pain in the epigastric region, which may be worse after eating, as well as nausea, vomiting, and extreme belching. Bleeding may also occur, and the partially digested blood results in dark and tarry stools. If left untreated, an ulcer may erode the entire organ wall and cause **perforation,** which is a medical emergency.

Irritation of the gastric mucosa (**gastritis**) has been linked to many cases of peptic ulcer. Nonsteroidal anti-inflammatory drugs (NSAIDS), such as ibuprofen and aspirin, are a common cause of gastritis, and these drugs also impair healing of the gastric lining. However, the major player in peptic ulcer formation is a bacterium called *Helicobacter pylori,* which is present in over 70% of gastric ulcer cases and well over 90% of duodenal ulcer cases. *H. pylori* resides in the stomach and produces enzymes that break down the components in the gastric mucus, weakening its protective effects. As leukocytes enter the stomach to destroy the bacteria, they also destroy the mucous neck cells. This further irritates the stomach lining and creates an ideal environment for continued *H. pylori* colonization. Thus, the bacteria initiate a cascade of events that lead to erosion of the gastric lining and eventual perforation of the stomach wall if not treated.

Treatment for a gastric ulcer involves eliminating the bacteria as well as reducing gastric acidity to promote healing. Categories of medications that help include an antibiotic taken for 2 weeks to eradicate *H. pylori,* an antacid (Tums®, Rolaids®) to help neutralize stomach acid, a proton-pump inhibitor (e.g., Prilosec®, Nexium®), and/or a histamine (H₂) blocker (e.g., Pepcid®, Axid®, Zantac®) to help limit acid secretion in the stomach.

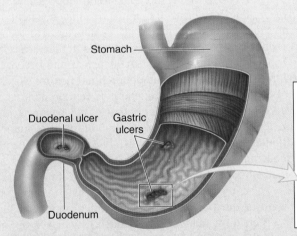

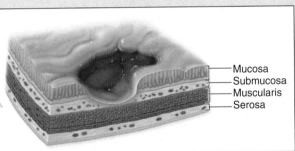

(a) Common locations of gastric and duodenal ulcers

(a) A sectioned view of the stomach and duodenum illustrates some common locations for gastric and duodenal ulcers. (b) A gross anatomic specimen shows a perforated gastric ulcer.

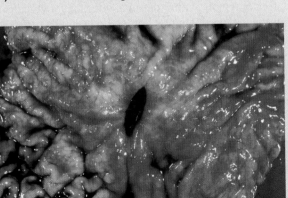

(b) Perforated gastric ulcer

cells produce an acidic mucin that differs structurally and functionally from the mucin secreted by the surface mucous cells. The acidic mucin helps maintain the acidic conditions resulting from the secretion of hydrochloric acid by parietal cells.

Parietal cells (also called *oxyntic cells*) are located primarily in the proximal and middle parts of the gastric gland. Their distinctive features are small intracellular channels called canaliculi, which are lined by microvilli. Hydrochloric acid secreted across the parietal cells' vast internalized surface helps denature proteins to facilitate chemical digestion. Parietal cells also produce **intrinsic factor,** a molecule that binds vitamin B_{12} in the stomach lumen and assists in B_{12} absorption in the ileum of the small intestine.

Chief cells (also called *zymogenic cells* or *peptic cells*) are housed primarily in the distal part of the gastric gland. These cells synthesize and secrete enzymes, primarily inactive pepsinogen, into the lumen of the stomach. The acid content of the stomach then converts inactive pepsinogen into the active enzyme **pepsin,** which chemically digests denatured proteins in the stomach into smaller fragments.

Enteroendocrine (en′ter-ō-en′dō-krin; *enteron* = gut, intestine) **cells** are endocrine cells widely distributed in the gastric glands of the stomach. These cells secrete **gastrin,** a hormone that enters the blood and stimulates the secretory activities of the chief and parietal cells and the contractile activity of gastric muscle. Enteroendocrine cells also produce other hormones, such as **somatostatin,** that modulate the function of nearby enteroendocrine and exocrine cells.

Saliva, mucin, and gastric secretions contribute substantially to the volume of ingested material. In fact, if a person eats 1 liter of food, the amount of chyme that enters the small intestine is 3–4 liters, due to the volume of the secretions.

WHAT DID YOU LEARN?

10 What are the four regions of the stomach, and where is each located?

11 What are the five types of secretory cells in the stomach, and what does each secrete?

Small Intestine

Key topics in this section:

- Gross anatomy of the small intestine
- Comparison of the three regions of the small intestine
- Microscopic structure of the small intestine

The small intestine finishes the chemical digestion process and is responsible for absorbing most of the nutrients. Ingested nutrients spend at least 12 hours in the small intestine as chemical digestion and absorption are completed.

Gross Anatomy and Regions

The **small intestine,** also called the *small bowel,* is a coiled, thin-walled tube about 6 meters (20 feet) in length in the unembalmed cadaver. (It is much shorter in a living individual due to muscle tone. We do not know the exact length in a living human because it would be physically difficult to measure the intestine, and would cause severe discomfort to a living individual.) It extends from the pylorus of the stomach to the cecum of the large intestine, and thus occupies a significant portion of the abdominal cavity. The small intestine receives its blood supply primarily from branches of the superior

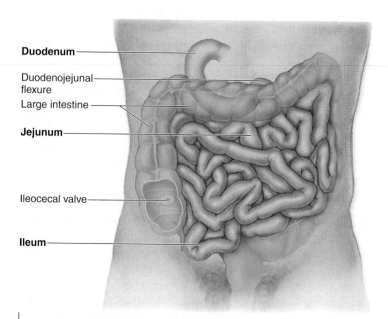

Figure 26.14

Gross Anatomy of the Small Intestine.   The three regions of the small intestine—duodenum, jejunum, and ileum—are continuous and "framed" within the abdominal cavity by the large intestine.

mesenteric artery, and it is innervated by the superior mesenteric plexus. The small intestine consists of three specific segments: the duodenum, the jejunum, and the ileum **(figure 26.14)**.

The **duodenum** (doo-ō-dē′nŭm, doo-od′ĕ-nŭm; breadth of twelve fingers) forms the initial or first segment of the small intestine. It is approximately 25 centimeters (10 inches) long and originates at the pyloric sphincter. The duodenum is arched into a C-shape around the head of the pancreas and becomes continuous with the jejunum at the **duodenojejunal flexure** (flek′sher; *fleksura* = bend). Most of the duodenum is retroperitoneal, although the very proximal part is intraperitoneal and connects to the liver by the lesser omentum. Within the wall of the duodenum is the **major duodenal papilla,** through which bile and pancreatic juice enter the duodenum. A minor duodenal papilla also receives an additional small amount of pancreatic juice via an accessory pancreatic duct.

The **jejunum** (jĕ-joo′nŭm; *jejunus* = empty) is the middle region of the small intestine. Extending approximately 2.5 meters (7.5 feet in an unembalmed cadaver), it makes up approximately two-fifths of the small intestine's total length. The jejunum is the primary region within the small intestine for chemical digestion and nutrient absorption. It is intraperitoneal and suspended in the abdomen by the mesentery proper.

The **ileum** (il′ē-ŭm; *eiles* = twisted) is the final or last region of the small intestine. At about 3.6 meters (10.8 feet) in length in an unembalmed cadaver, the ileum forms approximately three-fifths of the small intestine. Its distal end terminates at the **ileocecal** (il′ē-ō-sē′kăl) **valve,** a sphincter that controls the entry of materials into the large intestine. The ileum is intraperitoneal and suspended in the abdomen by the mesentery proper.

Internally, the mucosal and submucosal tunics of the small intestine are thrown into **circular folds** (also called *plicae* [pli′kē; fold] *circulares*) **(figure 26.15a,b)**. Circular folds, which can be seen with the naked eye, help increase the surface area through which nutrients can be absorbed. In addition, the circular folds act like "speed bumps" to slow down the movement of chyme and

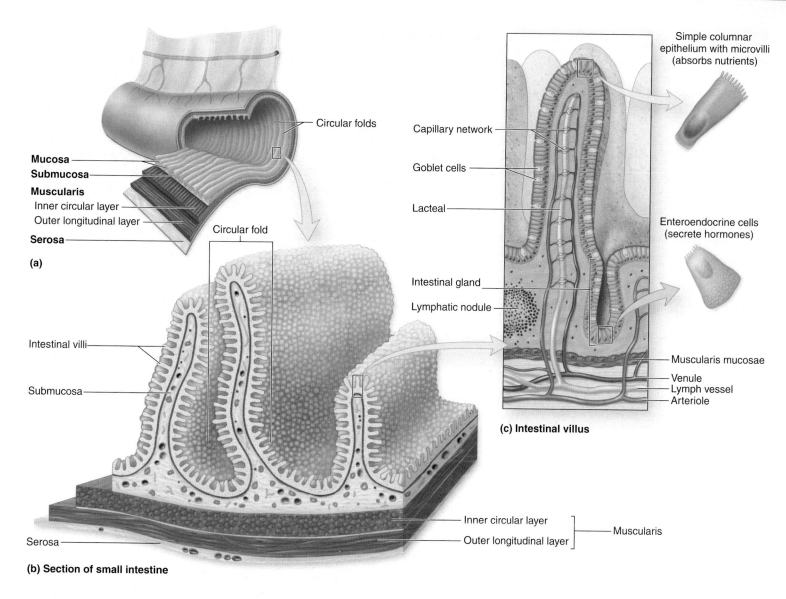

(a)

- Mucosa
- Submucosa
- Muscularis
  - Inner circular layer
  - Outer longitudinal layer
- Serosa

Circular folds

Circular fold

Intestinal villi

Submucosa

Serosa

Inner circular layer
Outer longitudinal layer — Muscularis

(b) Section of small intestine

Simple columnar epithelium with microvilli (absorbs nutrients)

Capillary network

Goblet cells

Lacteal

Intestinal gland

Lymphatic nodule

Enteroendocrine cells (secrete hormones)

Muscularis mucosae
Venule
Lymph vessel
Arteriole

(c) Intestinal villus

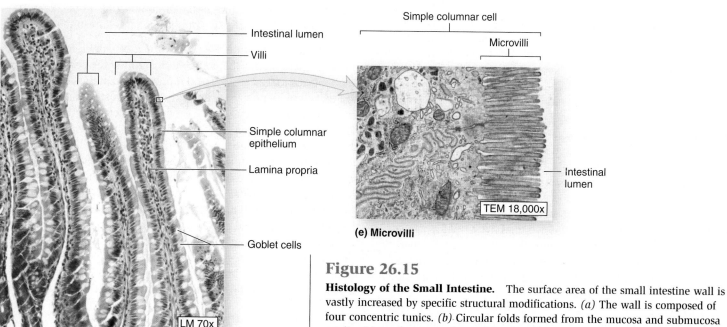

Intestinal lumen

Villi

Simple columnar epithelium

Lamina propria

Goblet cells

(d) Intestinal villi

LM 70x

Simple columnar cell

Microvilli

Intestinal lumen

TEM 18,000x

(e) Microvilli

Figure 26.15

Histology of the Small Intestine. The surface area of the small intestine wall is vastly increased by specific structural modifications. (*a*) The wall is composed of four concentric tunics. (*b*) Circular folds formed from the mucosa and submucosa are lined by a dense covering of fingerlike projections called intestinal villi, which are formed from mucosa only. (*c*) The structure of a single villus. (*d*) A photomicrograph shows the internal structure of villi projecting into the intestinal lumen. (*e*) The plasma membrane along the apical surface of the epithelium contains microvilli.

Table 26.4	Small Intestine Structures Involved in Digestion and Absorption	
Structure	**Anatomy**	**Function**
Circular folds	Circular folds of mucosa and submucosa	Slow down the passage of materials undergoing digestion; increase surface area for both absorption and chemical digestion
Villi	Fingerlike projections of mucosa	Increase surface area for both absorption and chemical digestion
Microvilli	Folded, fingerlike projections of plasma membrane on apical surface of columnar epithelial cells	Increase surface area for both absorption and chemical digestion
Intestinal glands	Invaginations into mucosa between villi	Increase surface area for both absorption and chemical digestion; enteroendocrine cells lining intestinal glands secrete digestive hormones
Submucosal glands	Coiled tubular glands within submucosa with ducts opening into intestinal lumen	Secrete alkaline mucin to protect and lubricate lining of small intestine

ensure that it remains within the small intestine for maximal nutrient absorption. Circular folds are more numerous in the duodenum and jejunum, and least numerous in the ileum.

WHAT DO YOU THINK?

4 Why are the circular folds much more numerous in the duodenum and least numerous in the ileum? How does the abundance of circular folds relate to the main functions of the duodenum and ileum?

Histology

When circular folds are viewed at the microscopic level, smaller, fingerlike projections of mucosa only, called **villi,** can be seen along their surface. These villi further increase the surface area for absorption and secretion. Increasing the absorptive surface area even further are **microvilli** (mī-krō-vil′ī; *mikros* = small) along the free surface of the simple columnar cells (figure 26.15*e*). Individual microvilli are not clearly visible in light micrographs of the small intestine; instead, they collectively appear as a **brush border** that resembles a brightly staining surface on the apical end of the simple columnar cells (figure 26.15*d*). Each villus contains an arteriole and a venule, with a rich capillary network between them. The capillaries absorb most nutrients. In the center of the villus is a lacteal, which is responsible for absorbing lipids and lipid-soluble vitamins, which are too large to be absorbed by the capillaries.

Between some of the intestinal villi are invaginations of mucosa called **intestinal glands** (also known as *intestinal crypts* or *crypts of Lieberkuhn*). These glands extend to the base of the mucosa and slightly resemble the gastric glands of the stomach. Lining them are simple columnar epithelial cells (with goblet cells) and enteroendocrine cells. These enteroendocrine cells release hormones such as **secretin, cholecystokinin,** and **gastric inhibitory peptide.** Some of these hormones temporarily slow down digestive activities as material from the stomach begins to enter the small intestine, thereby prolonging the time for stomach emptying into the small intestine. The goblet cells produce mucin to lubricate and protect the intestinal lining as materials being digested pass through.

Distinctive histologic features characterize the three small intestine regions. The proximal duodenum contains **submucosal glands** (or *Brunner glands*), which produce a viscous, alkaline mucus that protects the duodenum from the acidic chyme. Circular folds are best-developed in the jejunum and nearly absent in the ileum. Additionally, villi are larger and more numerous in the jeju-

num. The number of goblet cells in the mucosa increases from the duodenum to the ileum, due to the increased need for lubrication as digested materials are absorbed and undigested materials are left behind. Peyer patches are abundant primarily in the ileum.

Table 26.4 summarizes all of the small intestine structures that aid in digestion and absorption.

WHAT DID YOU LEARN?

12 Compare and contrast the gross anatomic and histologic characteristics that distinguish the duodenum, jejunum, and ileum.

Large Intestine

Key topics in this section

- Gross anatomy of the large intestine
- Comparison of the large intestine regions
- Microscopic structure of the large intestine
- Movement of material through the large intestine

The **large intestine,** also called the *large bowel,* forms a three-sided perimeter in the abdominal cavity around the centrally located small intestine **(figure 26.16)**. From its origin at the ileocecal junction to its termination at the anus, the large intestine has an approximate length of 1.5 meters (5 feet) and a diameter of 6.5 centimeters (2.5 inches). It is called the "large" intestine because its diameter is greater than that of the small intestine. Recall that the small intestine absorbs much, but not all, of the digested material. On average, about 1 liter of remaining material passes from the small intestine to the large intestine daily.

The large intestine absorbs most of the water and ions from the remaining digested material. In so doing, the watery material that first enters the large intestine soon solidifies and becomes **feces.** The large intestine stores the feces until the body is ready to defecate (expel the feces). The large intestine also absorbs a very small percentage of nutrients still remaining in the digested material.

Gross Anatomy and Regions

The initial or first region of the large intestine is a blind sac called the **cecum** (sē′kŭm; *caecus* = blind), which is located in the right lower abdominal quadrant. This pouch extends inferiorly from the **ileocecal valve,** which represents the attachment of the distal end

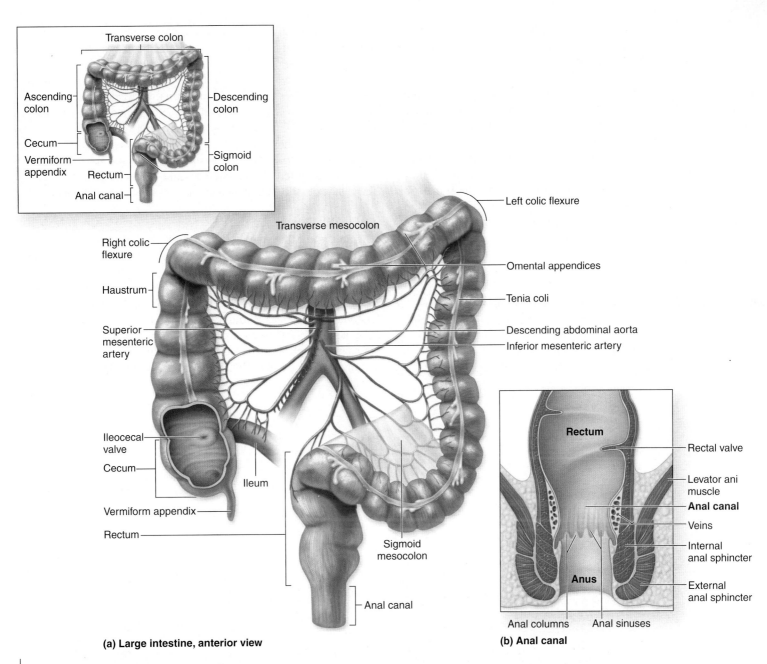

(a) Large intestine, anterior view

(b) Anal canal

Figure 26.16

Gross Anatomy of the Large Intestine. *(a)* Anterior view of the large intestine, which forms the distal end of the GI tract. *(b)* Details of the anal canal.

of the small intestine to the proximal region of the large intestine. Projecting inferiorly from the posteromedial region of the cecum is the **vermiform** (ver′mi-fōrm; *vermis* = worm) **appendix** (ă-pen′diks; appendage), a thin, hollow, fingerlike sac lined by lymphocyte-filled lymphatic nodules. Both the cecum and the vermiform appendix are intraperitoneal organs.

At the level of the ileocecal valve, the **colon** begins and forms an inverted U-shaped arch. The colon is partitioned into four segments: the ascending colon, transverse colon, descending colon, and sigmoid colon.

The **ascending colon** originates at the ileocecal valve and extends superiorly from the superior edge of the cecum along the right lateral border of the abdominal cavity. The ascending colon is retroperitoneal, since its posterior wall directly adheres to the pos-

terior abdominal wall, and only its anterior surface is covered with peritoneum. As it approaches the inferior surface of the liver, the ascending colon makes a 90-degree turn toward the left side of the abdominal cavity. This bend in the colon is called the **right colic** (kol′ik) **flexure,** or the *hepatic flexure* (figure 26.16a).

The **transverse colon** originates at the right colic flexure and curves slightly anteriorly as it projects horizontally across the anterior region of the abdominal cavity. A type of mesentery called the **transverse mesocolon** connects the transverse portion of the large intestine to the posterior abdominal wall. Hence, the transverse colon is intraperitoneal. As the transverse colon approaches the spleen in the left upper quadrant of the abdomen, it makes a 90-degree turn inferiorly. The resulting bend in the colon is called the **left colic flexure,** or the *splenic* (splen′ik) *flexure.*

The **descending colon** is retroperitoneal and found along the left side of the abdominal cavity. It originates at the left colic flexure and descends vertically until it terminates at the sigmoid colon. The **sigmoid** (sig'moyd; resembling letter S) **colon** originates at the **sigmoid flexure,** where the descending colon curves and turns inferomedially into the pelvic cavity. The sigmoid colon is intraperitoneal and has a mesentery called the **sigmoid mesocolon.** The sigmoid colon terminates at the rectum.

The **rectum** (rek'tŭm; *rectus* = straight) is a retroperitoneal structure that connects to the sigmoid colon. The rectum is a muscular tube that readily expands to store accumulated fecal material prior to defecation. Three thick, transverse folds of the rectum, called **rectal valves,** ensure that fecal material is retained during the passing of gas. The rectum then terminates at the anal canal.

> ## Study Tip!
>
> Many of the segments of the colon are named for the direction in which materials travel. So material ascends (travels superiorly) in the *ascending colon,* material travels across in the *transverse colon,* and material descends (travels inferiorly) in the *descending colon.*

The terminal few centimeters of the large intestine are called the **anal** (ā'năl) **canal** (figure 26.16b). The anal canal passes through an opening in the levator ani muscles of the pelvic floor and terminates at the anus. The internal lining of the anal canal contains relatively thin longitudinal ridges, called **anal columns,** between which are small depressions termed **anal sinuses.** As fecal material passes through the anal canal during defecation, pressure exerted on the anal sinuses causes their cells to release excess mucin. As a result, this extra mucus lubricates the anal canal during defecation. At the base of the anal canal are the **internal** and **external anal sphincters,** which close off the opening to the anal canal and relax (open) during defecation.

Histology

The mucosa of the large intestine is lined with simple columnar epithelium and goblet cells **(figure 26.17)**. Unlike the small intestine, the large intestine mucosa lacks intestinal villi; however, it contains numerous intestinal glands that extend to the muscularis mucosae. The glands' goblet cells secrete mucin to lubricate the undigested material as it passes through, and the simple columnar epithelial cells continue to absorb nutrients that were not absorbed during passage through the small intestine. Many lymphatic nodules and lymphatic cells occupy the lamina propria of the large intestine.

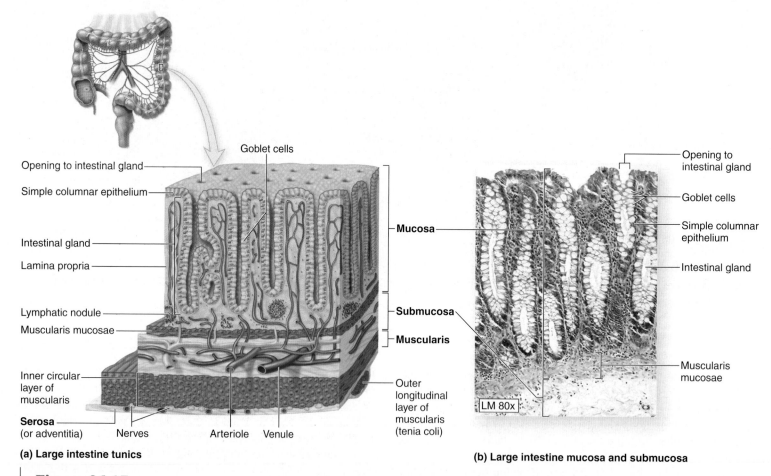

(a) Large intestine tunics

(b) Large intestine mucosa and submucosa

Figure 26.17

Histology of the Large Intestine. *(a)* The luminal wall of the large intestine is composed of several layers. *(b)* A photomicrograph shows the histology of the wall of the large intestine.

CLINICAL VIEW: In Depth

Colorectal Cancer

Colorectal cancer is the second most common type of cancer in the United States, with over 140,000 cases occurring annually and 60,000 of those resulting in death. In recent years, high-profile individuals such as baseball players Eric Davis and Darryl Strawberry and Supreme Court Justice Ruth Bader Ginsburg have been treated for colorectal cancer. These well-publicized cases have helped increase awareness of this deadly disease.

The term **colorectal cancer** refers to a malignant growth anywhere along the colon or rectum. The majority of colorectal cancers appear in the rectum, sigmoid colon, and distal descending colon, which are the segments of the large intestine that have the longest contact with fecal matter before it is expelled from the body. Most colorectal cancers arise from **polyps** (pol'ip; *polys* = many), which are outgrowths from the colon mucosa. Note, however, that colon polyps are very common, and most of them never become cancerous. Low-fiber diets have also been implicated in increasing the risk of colon cancer, because decreased dietary fiber leads to decreased stool (section of fecal matter) bulk and longer time for stools to remain in the large intestine, thus theoretically exposing the large intestine mucosa to toxins in the stools for longer periods of time. Other risk factors include a family history of colorectal cancer, personal history of ulcerative colitis, and older age (since most patients are over the age of 40).

Initially, most patients are asymptomatic. Later, they may notice rectal bleeding (often evidenced as blood in the stool or on the toilet paper) and a persistent change in bowel habits (typically constipation). For some, the bleeding may not be noticeable, but can be detected in a stool sample. Eventually, the person may experience abdominal pain, fatigue, unexplained weight loss, and anemia.

A cancerous growth in the colon must be removed surgically, and sometimes radiation and/or chemotherapy are used as well. Colorectal cancers that are limited to the mucosa have a 5-year survival rate, but the prognosis is poor for cancers that have spread into deeper colon wall tunics or metastasized to the lymph nodes.

The key to an increased survival rate for colorectal cancer is early detection. If caught early, colorectal cancer is very treatable. People should see their doctor if they experience rectal bleeding or any persistent change in bowel habits. By age 50 (or earlier, if you have symptoms or a family history of colorectal cancer), individuals should take advantage of the following screening methods:

1. Take a yearly **fecal occult** (ŏ-kŭlt', ok'ŭlt) **blood test,** which checks for the presence of blood in the stools.
2. Every 5 years, have a **sigmoidoscopy** (sig'moy-dos'kŏ-pē). In this procedure, which is done in the doctor's office, an endoscope is inserted into the anus, rectum, and sigmoid colon to check for polyps or cancer.
3. Every 10 years, undergo a **colonoscopy** (kō-lon-os'kŏ-pē; *skopeo* = to view). A colonoscopy is more extensive than a sigmoidoscopy. The endoscope is inserted through the anus and into the large intestine at least up to the right colic flexure of the colon, and sometimes as far proximally as the ileocecal valve.

In addition, people are advised to eat a high-fiber diet to reduce the risk of developing colorectal cancer.

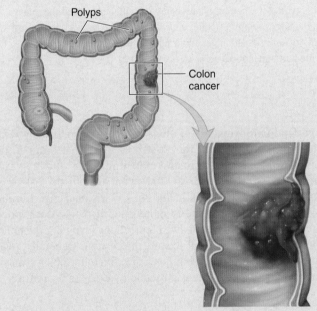

Polyps in the large intestine sometimes lead to colorectal cancer.

The muscularis of the colon and cecum has two layers of smooth muscle, but the outer longitudinal layer does not completely surround the colon and cecum. Instead, these longitudinal smooth muscle fibers form three thin, distinct, longitudinal bundles called **teniae** (tē'nē-ē; ribbons, band) **coli** (kō'lī). The teniae coli act like elastic in a waistband—they help bunch up the large intestine into many sacs, collectively called **haustra** (haw'stră; sing., *haustrum; haustus* = to drink up) (see figure 26.16*a*). Hanging off the external surface of the haustra are lobules of fat called **omental appendices,** or *epiploic* (ep'i-plō'ik; membrane-covered) *appendages.*

Control of Large Intestine Activity

Large intestine movements are regulated by local reflexes in the autonomic nervous system. With the ingestion of more food, peristaltic movements in the ileum increase, as does the frequency of the opening of the ileocecal valve. This so-called **gastroileal reflex** results in the accumulation of more chyme in the cecum and ascending colon. Segmentation movements are infrequent in the colon. Instead, three types of movements are typically associated with the passage of digested material through the large intestine, the absorption of fluid and ions, and the packaging of waste materials for defecation: peristaltic movements, haustral churning, and mass movement.

Appendicitis

Inflammation of the appendix is called **appendicitis** (ă-pen-di-sī'tis). Most cases of appendicitis occur because fecal matter obstructs the appendix, although sometimes an appendix becomes inflamed without any obstruction. As the tissue in its wall becomes inflamed, the appendix swells, the blood supply is compromised, and bacteria may proliferate in the wall. Untreated, the appendix may burst and spew its contents into the peritoneum, causing a massive infection called **peritonitis,** and possibly leading to death.

During the early stages of acute appendicitis, the smooth muscle wall contracts and goes into spasms. Because this smooth muscle is innervated by the autonomic nervous system, pain is referred to the T10 dermatome around the umbilicus.

As the inflammation worsens and the parietal peritoneum becomes inflamed as well, the pain becomes sharp and localized to the right lower quadrant of the abdomen. Individuals with appendicitis typically experience nausea or vomiting, abdominal tenderness in the inferior right quadrant, a low fever, and an elevated leukocyte count. An inflamed appendix is surgically removed in a procedure called an appendectomy.

Peristaltic movements of the large intestine are usually weak and sluggish, but otherwise they resemble those that occur in the wall of the small intestine. **Haustral churning** occurs after a relaxed haustrum fills with digested or fecal material until its distension stimulates reflex contractions in the muscularis, causing churning and movement of the material to more distal haustra. **Mass movements** are powerful, peristaltic-like contractions involving the teniae coli that propel fecal material toward the rectum. Generally, mass movements occur two or three times a day, often during or immediately after a meal. This is the **gastrocolic** (gas'trō-kol'ik) **reflex.**

13 Identify the retroperitoneal and intraperitoneal large intestine structures.

14 What specific movements and reflexes propel material through the large intestine?

Accessory Digestive Organs

Key topics in this section:

- Liver anatomy and blood supply
- Bile secretion
- Gross anatomy and microanatomy of the pancreas
- Pancreatic acinar cell function
- How secretory products travel through the biliary apparatus

The accessory digestive organs produce secretions that facilitate the chemical digestive activities of GI tract organs. Important accessory digestive organs are the liver, the gallbladder, and the pancreas.

Liver

The **liver** (liv'er) lies in the right upper quadrant of the abdomen, immediately inferior to the diaphragm. Weighing 1–2 kilograms, it constitutes approximately 2% of an adult's body weight. The liver is covered by a connective tissue capsule and a layer of visceral peritoneum, except for a small region on its diaphragmatic surface called the bare area.

Gross Anatomy

The liver is composed of four incompletely separated lobes and supported by two ligaments **(figure 26.18)**. The major lobes are the **right lobe** and the **left lobe.** The right lobe is separated from the smaller left lobe by the falciform ligament, a peritoneal fold that secures the liver to the anterior abdominal wall. In the inferior free edge of the falciform ligament lies the **round ligament of the liver** (or *ligamentum teres*), which represents the remnant of the fetal umbilical vein. Subdivisions of the right lobe include the **caudate**

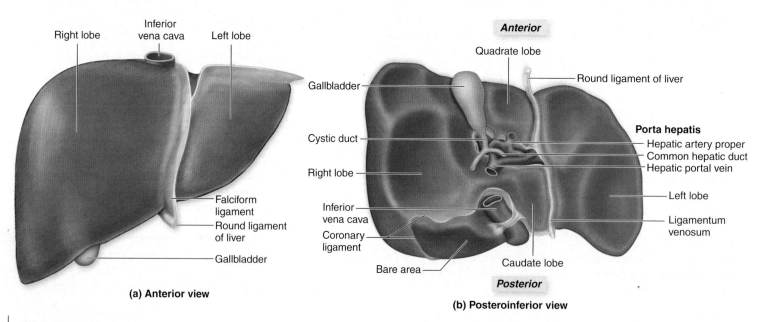

(a) Anterior view

(b) Posteroinferior view

Figure 26.18

Gross Anatomy of the Liver.   The liver is in the upper right quadrant of the abdomen. *(a)* Anterior and *(b)* posteroinferior views show the four lobes of the liver, as well as the gallbladder and the porta hepatis.

(kaw'dāt; *cauda* = tail) **lobe** and the **quadrate** (kwah'drāt; *quadratus* = square) **lobe.** The caudate lobe is adjacent to the inferior vena cava, and the quadrate lobe is adjacent to the gallbladder (figure 26.18*b*).

Along the inferior surface of the liver are several structures that collectively resemble the letter H. The gallbladder and the round ligament of the liver form the vertical superior parts of the H; the **inferior vena cava** and the **ligamentum venosum** form the vertical inferior parts. (Recall from chapter 23 that the ligamentum venosum was a remnant of the ductus venosus in the embryo. This vessel shunted blood from the umbilical vein to the inferior vena cava.) Finally, the **porta** (pōr'tă; gate) **hepatis** (hep'ă-tis) represents the horizontal crossbar of the H and is where blood and lymph vessels, bile ducts, and nerves enter and leave the liver. In particular, the hepatic portal vein and branches of the hepatic artery proper enter at the porta hepatis.

Histology

A connective tissue capsule branches through the liver and forms septa that partition the liver into thousands of small, polyhedral

hepatic lobules, which are the classic structural and functional units of the liver **(figure 26.19).** Within hepatic lobules are liver cells called **hepatocytes** (hep-a'tō-sīt). At the periphery of each lobule are several **portal triads,** composed of branches of the hepatic portal vein, the hepatic artery, and the bile duct.

A dual blood supply serves the liver. The **hepatic portal vein** carries blood from the capillary beds of the GI tract, spleen, and pancreas. It brings approximately 75% of the blood volume to the liver. This blood is rich in nutrients and other absorbed substances but relatively poor in oxygen. The **hepatic artery proper,** a branch of the celiac trunk, splits into **left** and **right hepatic arteries.** These arteries carry well-oxygenated blood to the liver. Blood from branches of the hepatic arteries and hepatic portal vein mixes as it passes to and through the hepatic lobules. At the center of each lobule is a **central vein** that drains the blood from the lobule. Central veins collect venous blood and merge throughout the liver to form numerous **hepatic veins** that eventually empty into the inferior vena cava.

In cross section, a hepatic lobule looks like a side view of a bicycle wheel. The hub of the wheel is represented by the central vein. At the circumference of the wheel where the tire would be are

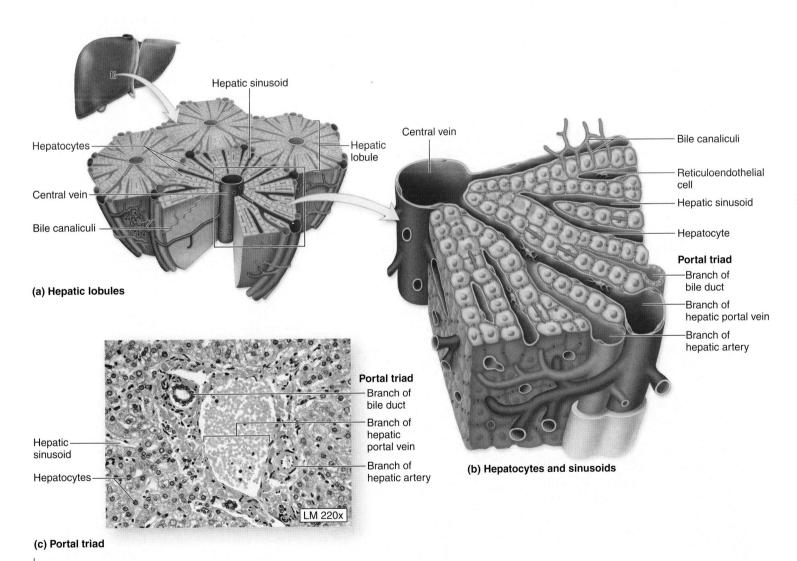

(a) Hepatic lobules

(c) Portal triad

(b) Hepatocytes and sinusoids

Figure 26.19

Histology of the Liver.   *(a)* The functional units of the liver are called hepatic lobules. *(b)* A central vein projects through the center of a hepatic lobule, and several portal triads define its periphery. *(c)* A photomicrograph depicts the portal triad and hepatocytes.

several portal triads that are usually equidistant apart. The numerous spokes of the wheel are the **hepatic sinusoids** (si′nū-soyd; *sinus* = cavity), which are bordered by cords of hepatocytes. Hepatic sinusoids are thin-walled, porous or "leaky" capillaries where venous and arterial blood are mixed and then flow slowly through the hepatic lobule toward the central vein. The sinusoids are lined with stellate cells called **reticuloendothelial cells** (or *Kupffer cells*), which are phagocytic cells that have an immune function.

Hepatocytes absorb nutrients from the sinusoids, and they also produce **bile,** a greenish fluid that breaks down fats to assist in their chemical digestion. Between each cord of hepatocytes is a

small **bile canaliculus.** Bile canaliculi conduct bile from the hepatocytes to the bile duct in the portal triad.

Liver Function

The liver has numerous functions. Besides producing bile, hepatocytes detoxify drugs, metabolites, and poisons. Hepatocytes also store excess nutrients and vitamins and release them when they are needed. Finally, hepatocytes synthesize blood plasma proteins such as albumins, globulins, and proteins required for blood clotting. Reticuloendothelial cells in the liver sinusoids phagocytize debris in the blood as well as help break down and recycle

CLINICAL VIEW

Cirrhosis of the Liver

Chronic injury to the liver inevitably leads to **liver cirrhosis** (sir-rō′sis; *kirrhos* = yellow). Liver cirrhosis results when hepatocytes have been destroyed and are replaced by fibrous scar tissue. This scar tissue often surrounds isolated nodules of regenerating hepatocytes. The fibrous scar tissue also compresses the blood vessels and bile ducts in the liver, leading to **hepatic portal hypertension** (high blood pressure in the hepatic portal venous system) and bile flow impediments.

Liver cirrhosis is caused by chronic injury to the hepatocytes, as may result from chronic alcoholism, liver disease, or certain drugs or toxins. Chronic hepatitis (hep-ă-tī′tis) is a long-term inflammation of the liver that leads to necrosis of liver tissue. Most frequently, viral infections from either **hepatitis B** or **hepatitis C** produce chronic hepatitis. Other disorders that result in liver cirrhosis include some inherited diseases, chronic biliary obstruction, and biliary cirrhosis.

Early stages of liver cirrhosis may be asymptomatic. However, once liver function begins to falter, the patient complains of fatigue, weight loss, and nausea, and may have pain in the right hypochondriac region. During a physical, the doctor may palpate an abnormally small and hard liver. To confirm the diagnosis, a liver biopsy is done by obtaining a small portion of liver tissue through a needle passed into the liver and then examining the cells.

The fibrosis and scarring of liver cirrhosis are irreversible. However, further scarring can be slowed or prevented by treating the cause of the cirrhosis (hepatitis, alcoholism, etc.). Advanced liver cirrhosis may have a variety of complications:

- **Jaundice** (yellowing of the skin and sclerae of the eyes) occurs when the liver can't eliminate enough bilirubin. (Bilirubin is a yellowish product formed when aged erythrocytes are broken down. See chapter 21 for further information.) It is accompanied by darkening of the urine.
- **Edema** (accumulation of fluid in body tissues) and **ascites** (ā-sī′tēz) (fluid accumulation in the abdomen) develop because of decreased albumin production.
- Intense itching occurs when bile products are deposited in the skin.
- Toxins in the blood and brain accumulate because the liver cannot effectively process them.
- Hepatic portal hypertension can lead to dilated veins of the inferior esophagus (esophageal varices).

End-stage liver cirrhosis can be treated only with a liver transplant. Otherwise, death results either from progressive liver failure or from the complications. In the United States, over 25,000 people die each year of liver cirrhosis.

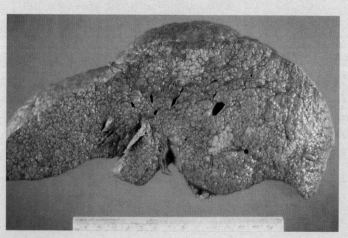

(a) Nodular cirrhosis of the liver

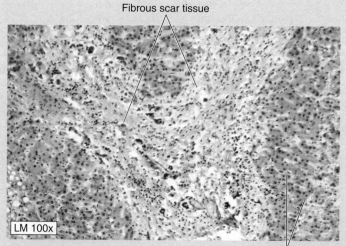

(b) Histology of liver cirrhosis

(a) This gross specimen depicts a type of nodular cirrhosis of the liver. (b) A photomicrograph shows how fibrous scar tissue infiltrates and replaces hepatocytes.

components of aged erythrocytes and damaged or worn-out formed elements.

Gallbladder

Attached to the inferior surface of the liver, a saclike organ called the **gallbladder** concentrates bile produced by the liver and stores this concentrate until it is needed for digestion. The **cystic** (sis'tik; *cysto* = bladder) **duct** connects the gallbladder to the common bile duct. The gallbladder can hold approximately 40 to 60 milliliters of concentrated bile. At the neck of the gallbladder, a sphincter valve controls the flow of bile into and out of the gallbladder. The gallbladder has three tunics: an inner mucosa, a middle muscularis, and an external serosa. Folds in the mucosa permit distention of the wall as the gallbladder fills with bile.

WHAT DO YOU THINK?

5 If your gallbladder were surgically removed, how would this affect your digestion of fatty meals? What diet alterations might you have to make after this surgery?

Pancreas

The **pancreas** is referred to as a mixed gland because it exhibits both endocrine and exocrine functions. The endocrine functions are performed by the cells of the pancreatic islets (see chapter 20). Exocrine activity results in the secretion of digestive enzymes and bicarbonate, collectively called **pancreatic juice,** into the duodenum.

The pancreas is a retroperitoneal organ that extends horizontally from the medial edge of the duodenum toward the left side of the abdominal cavity, where it touches the spleen. It exhibits a wide **head** adjacent to the curvature of the duodenum, a central, elongated **body** projecting toward the left lateral abdominal wall, and a **tail** that tapers as it approaches the spleen **(figure 26.20).**

The pancreas contains modified simple cuboidal epithelial cells called **acinar cells.** These cells, which are organized into large clusters termed **acini** (sing., *acinus*), or *lobules,* secrete the mucin and digestive enzymes of the pancreatic juice. The simple cuboidal epithelial cells lining the pancreatic ducts secrete bicarbonate (alkaline fluid) to help neutralize the acidic chyme arriving in the duodenum from the stomach. Most of the pancreatic juice travels through ducts that merge to form the **main pancreatic duct,** which

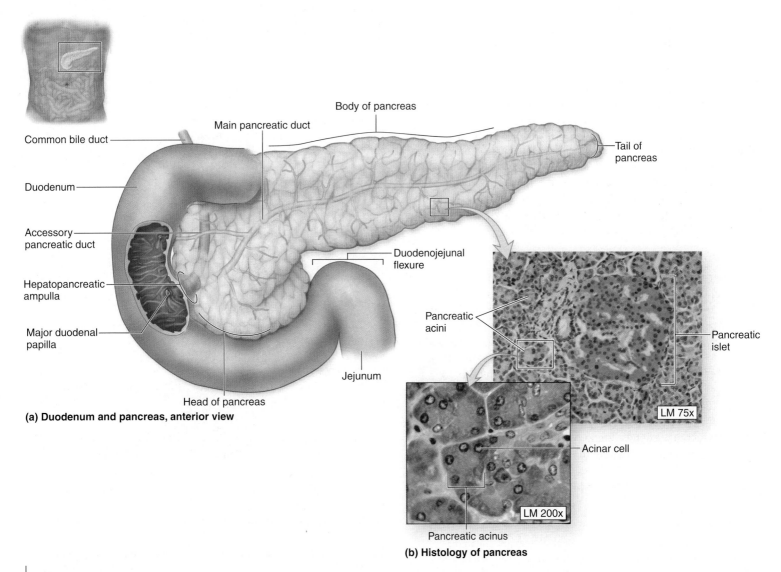

(a) Duodenum and pancreas, anterior view

(b) Histology of pancreas

Figure 26.20

Anatomy and Histology of the Pancreas. *(a)* The components of the pancreas are shown in relationship to the pancreatic duct and the duodenum. *(b)* Photomicrographs depict the histology of acinar cells within the pancreas.

CLINICAL VIEW

Gallstones (Cholelithiasis)

High concentration of certain materials in the bile can lead to the eventual formation of **gallstones.** Gallstones occur twice as frequently in women as in men, and are more prevalent in developed countries. Obesity, increasing age, female sex hormones, Caucasian ethnicity, and lack of physical activity are all risk factors for developing gallstones.

The term **cholelithiasis** (kŏ′lē-li-thī′ă-sis; *chole* = bile, *lithos* = stone, *iasis* = condition) refers to the presence of gallstones in either the gallbladder or the biliary apparatus. Gallstones are typically formed from condensations of either cholesterol or calcium and bile salts. These stones can vary from the tiniest grains to structures almost the size of golf balls. The majority of gallstones are asymptomatic until a gallstone becomes lodged in the neck of the cystic duct, causing the gallbladder to become inflamed **(cholecystitis)** and dilated. The most common symptom is severe pain (called biliary colic) perceived in the right hypochondriac region or sometimes in the area of the right shoulder. Nausea and vomiting may occur, along with indigestion and bloating. Symptoms are typically worse after eating a fatty meal. Treatment consists of surgical removal of the gallbladder, called **cholecystectomy** (kŏ′lē-sis-tek′tō-mĕ; *kystis* = bladder, *ektome* = excision).

Following surgery, the liver continues to produce bile, even in the absence of the gallbladder, but there is no means of concentrating the bile, so further gallstones are unlikely.

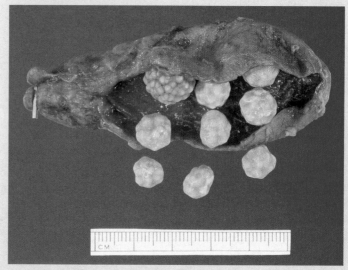

Photo of gallstones in a gallbladder.

drains into the major duodenal papilla in the duodenum. A smaller accessory pancreatic duct drains a small amount of pancreatic juice into a minor duodenal papilla in the duodenum.

Both hormonal and neural mechanisms control pancreatic juice secretions. Enteroendocrine cells in intestinal glands release cholecystokinin to promote the secretion of pancreatic juices from acinar cells, and secretin to stimulate the release of alkaline fluid from pancreatic duct cells. Pancreatic juice secretion is also stimulated by parasympathetic (vagus nerve) activity, while sympathetic activity inhibits pancreatic juice secretion.

Biliary Apparatus

The **biliary** (bil′ē-ār-ē; *bilis* = bile) **apparatus** is a network of thin ducts that carry bile from the liver and gallbladder to the duodenum **(figure 26.21).** The left and right lobes of the liver drain bile into the **left** and **right hepatic ducts,** respectively. The left and right hepatic ducts merge to form a single **common hepatic duct.** The **cystic duct** attaches to the common hepatic duct and carries bile to and from the gallbladder. Bile travels from the common hepatic duct through the cystic duct to be stored in the gallbladder; stored bile travels back through the cystic duct for conduction to the small intestine. The union of the cystic duct and the common hepatic duct forms the **common bile duct** that extends inferiorly to the duodenum.

The **hepatopancreatic ampulla** is a posteriorly placed swelling on the duodenal wall where the common bile duct and main pancreatic duct merge and pierce the duodenal wall. Bile and pancreatic

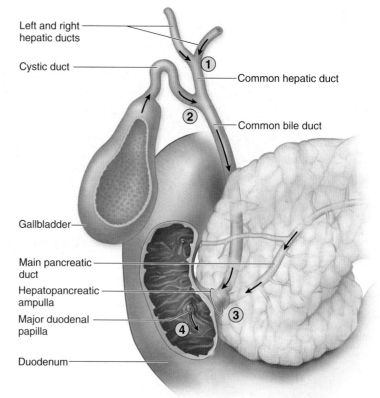

Figure 26.21

Biliary Apparatus.   Bile flows through the biliary apparatus (green), and pancreatic juice flows through the main pancreatic duct until the two vessels merge at the hepatopancreatic ampulla. Numbers indicate the pathway sequence for bile and pancreatic juice.

(1) Left and right hepatic ducts merge to form a common hepatic duct.

(2) Common hepatic and cystic ducts merge to form a common bile duct.

(3) Pancreatic duct merges with common bile duct at the hepatopancreatic ampulla.

(4) Bile and pancreatic juices enter duodenum at the major duodenal papilla.

juice mix in the hepatopancreatic ampulla prior to emptying into the duodenum via the **major duodenal papilla.**

WHAT DID YOU LEARN?

⑮ Identify the structural components of a portal triad.

⑯ What is the function of the gallbladder?

⑰ What is the function of pancreatic acini?

Aging and the Digestive System

Key topic in this section:

■ How the digestive system changes as we age

Age-related changes in digestion system function usually progress slowly and gradually. Overall, reduced secretions accompany normal aging. Mucin secretion decreases, reducing the thickness and amount of mucus in the layer that protects the GI tract and making internal damage to digestive organs more likely. Decreased

CLINICAL VIEW

Intestinal Disorders

Celiac disease (also known as *celiac sprue* or *gluten-sensitive enteropathy*) is an autoimmune disorder of the small intestine. This disease runs in families and may be triggered after surgery, pregnancy, or a viral infection. Affected individuals cannot tolerate a protein called **gluten,** which is found in all forms of wheat, rye, and barley. When an individual with celiac disease ingests food containing gluten, the body's immune cells attack the small intestine mucosa and damage the villi, severely impairing nutrient absorption. Symptoms of celiac disease vary, but may include gas and bloating, diarrhea, fatigue, and weight loss (due to malnutrition). Because these symptoms are similar to those of other bowel disorders, diagnosis may be difficult. The only treatment for celiac disease is to follow a gluten-free diet, but this is challenging because many commercially prepared foods use gluten as a preservative or stabilizer. Thus, the patient with celiac disease must carefully read all ingredient lists on prepared foods.

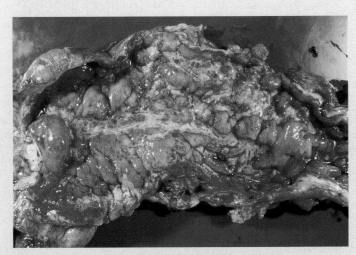

A section of small intestine showing the signs of Crohn disease.

The term **inflammatory bowel disease (IBD)** applies to two autoimmune disorders, Crohn disease and ulcerative colitis. In both of these disorders, selective regions of the intestine become inflamed.

Crohn disease is a condition of young adults characterized by intermittent and relapsing episodes of intense abdominal cramping and diarrhea. Although any region of the gastrointestinal tract, from esophagus to anus, may be involved, the distal ileum is the most frequently and severely affected site. Inflammation involves the entire thickness of the intestinal wall, extending from the mucosa to the serosa. For reasons that are not clear, lengthy regions of the intestine having no trace of injury or inflammation may be followed abruptly by several inches of markedly

diseased intestine. In some patients scarring of the tissue can lead to bowel obstruction. Other patients develop selective malabsorption of certain vitamins.

The age distribution and symptoms of **ulcerative colitis** are similar to those of Crohn disease, but ulcerative colitis involves only the large intestine. The rectum and descending colon are the first to show signs of inflammation and are generally the most severely affected. Also, in ulcerative colitis the inflammation is confined to the mucosa, instead of the full thickness of the intestinal wall. Finally, unlike Crohn disease, ulcerative colitis is associated with a profoundly increased risk of colon cancer. Historically, patients who have had ulcerative colitis for more than 10 years develop colon cancer at a frequency 20 or 30 times that seen in the rest of the population.

A section of large intestine showing the signs of ulcerative colitis.

Treatment of either Crohn disease or ulcerative colitis is complex. Anti-inflammatory drugs, as well as stress reduction and possibly nutritional supplementation, help control symptoms. Surgery may be necessary in both forms of inflammatory bowel disease.

Crohn disease, ulcerative colitis, and celiac disease are distinctly different from a much more common disorder called **irritable bowel syndrome.** Irritable bowel syndrome is characterized by abnormal function of the colon with symptoms of crampy abdominal pain, bloating, constipation, and/or diarrhea. Irritable bowel syndrome occurs in approximately one in every five people in the United States, and is more common in women than men. Irritable bowel syndrome may be diagnosed if a medical evaluation has ruled out Crohn disease and ulcerative colitis. Although neither a cause nor a cure for irritable bowel syndrome is known, most people can control their symptoms by reducing stress, changing their diet, and using certain medicines.

secretion of enzymes and acid results in diminished effectiveness of chemical digestion, which in turn can diminish nutrient absorption. The reduction of glandular secretion is compounded by decreased replacement of epithelial cells.

Another change that occurs with age is a reduction in the thickness of the smooth muscle layers in the muscularis. Consequently, both muscular tone and GI tract motility may be lowered significantly. Additionally, smooth muscle ensheathing the GI tract may adversely affect the function of sphincters or valves that help move materials through the GI tract. As a result, significant changes in nutrient absorption, as well as gastroesophageal reflux disease (GERD), may occur.

Improper dental care or sometimes simply aging can result in either the loss of teeth or periodontal disease. Both conditions adversely affect normal eating habits and nutrition. As food intake dwindles, changes occur in GI tract motility, secretion, and absorption. Marked dietary alterations may also increase the chance of developing cancer in either the stomach or the intestines.

Finally, a direct result of aging is a reduction in olfactory and gustatory sensations. When a person loses the ability to smell or taste food, his or her dietary intake usually suffers.

WHAT DID YOU LEARN?

18 How might the decrease in mucin secretion due to aging affect GI tract organs?

Development of the Digestive System

Key topic in this section:

- Major events in digestive system development

At the end of the third week of development, the disc-shaped embryo undergoes lateral folding, which transforms the endoderm and some lateral plate mesoderm into a cylindrical primitive **gut tube.** The gut tube eventually connects to the primitive mouth and the developing anus, while still maintaining a connection to the yolk sac via a thin stalk called the **vitelline** (vī-tel′in, -ēn) **duct.** A double-layered membrane called **dorsal mesentery** wraps around the gut tube and helps anchor it to the posterior body wall. A portion of the gut tube called the abdominal foregut has **ventral mesentery** that attaches it to the anterior body wall.

The gut tube may be divided into three components: a superior or cephalic **foregut,** a centrally located **midgut,** and an inferior or caudal **hindgut (figure 26.22a). Table 26.5** lists the structures derived from the foregut, midgut, and hindgut. Portions of the gut tube expand to form some of the abdominal organs. Some accessory digestive system organs develop as buds, or outgrowths, from the gut tube. Many of these structures rotate or shift before assuming their final positions in the body.

Stomach, Duodenum, and Omenta Development

By the fourth week of development, a portion of the foregut expands and forms a spindle-shaped dilation that will become the stomach. Between the fifth and seventh weeks, the posterior (dorsal) part of this dilation grows faster than the anterior (ventral) part, so the anterior wall becomes the concave lesser curvature, and the posterior wall becomes convex greater curvature (figure 26.22b). The dorsal mesentery that attaches to the stomach becomes thinner and stretches out from the greater curvature. This mesentery will form the greater omentum, while the ventral mesentery that attaches to the lesser curvature will form the lesser omentum.

During the seventh week, the developing stomach rotates 90 degrees clockwise about a longitudinal axis, as viewed from the superior aspect of the organ (figure 26.22c). Thus, the posterior side of the stomach (the greater curvature) rotates so that it faces the left side of the body, while the anterior side (the lesser curvature) rotates and faces the right side. One week later, the stomach and duodenum rotate about an anterior-posterior axis. This rotation pulls the pyloric region of the stomach and the duodenum superiorly while moving the fundus and the cardia of the stomach slightly inferiorly (figure 26.22d).

Liver, Gallbladder, and Pancreas Development

The liver parenchyma, gallbladder, pancreas, and biliary apparatus develop from buds or outgrowths from the endoderm of the duodenum. (The liver stroma or connective tissue framework is formed from nearby mesoderm.) The **liver bud** appears first, during the third week, develops within the ventral mesentery, and grows superiorly toward the developing diaphragm. By the fourth week, the liver continues to grow, and its attachment to the duodenum becomes a thinned tube that forms the common bile duct. An outgrowth forms off the bile duct that becomes the gallbladder. The connection to the gallbladder and bile duct thins and becomes the cystic duct. As the liver enlarges, the ventral mesentery thins and becomes the falciform ligament.

Table 26.5	Anatomic Derivatives of the Primitive Gut Tube		
Organs Formed	**Foregut**	**Midgut**	**Hindgut**
Digestive organs	Pharynx, esophagus, stomach, proximal half of duodenum	Distal half of duodenum; jejunum, ileum, cecum, appendix, ascending colon, proximal (right) two-thirds of transverse colon	Distal (left) one-third of transverse colon; descending colon, sigmoid colon, rectum, superior portion of anal canal
Accessory digestive organs	Most of liver; gallbladder, biliary apparatus, pancreas		
Other organs	Lungs		Urinary bladder epithelium, urethra epithelium

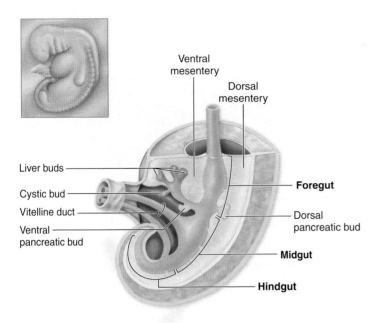

(a) Week 4: Liver, gallbladder, and pancreatic buds develop

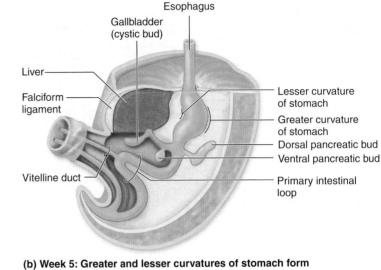

(b) Week 5: Greater and lesser curvatures of stomach form

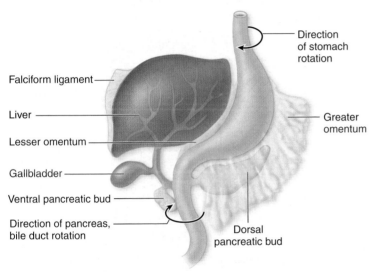

(c) Weeks 6–7: Rotation of stomach, pancreatic buds

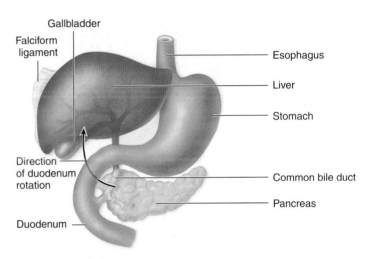

(d) Week 8: Postnatal position of organs attained

Figure 26.22

Development of the Digestive System Foregut.   *(a)* The foregut of the digestive system begins to develop from the primitive gut during week 4 of development. Stages of foregut development continue at *(b)* week 5, *(c)* weeks 6–7, and *(d)* week 8.

The pancreas starts to form at week 4 from two separate outgrowths of the bile duct, called the **ventral pancreatic bud** and the **dorsal pancreatic bud.** By week 6, the ventral pancreatic bud and the biliary apparatus rotate behind the duodenum. The ventral and dorsal pancreatic buds fuse to form the pancreas. Due to this rotation, much of the biliary apparatus now lies posterior to the duodenum and drains into the major duodenal papilla.

Intestine Development

The small and large intestines are formed from the midgut and hindgut, as shown in **figure 26.23**. During the fifth week, the midgut rapidly elongates and forms a **primary intestinal loop.** The cranial portion of the loop forms the jejunum and most of the ileum, while the caudal portion of the loop forms the very distal part of the ileum and much of the large intestine. The loop apex connects to the yolk sac via the vitelline duct.

Due to rapid growth of the liver and the resulting limited space in the developing abdominal cavity, the primary intestinal loop herniates into the umbilicus during the sixth week. As this loop herniates, it undergoes a 90-degree counterclockwise rotation, as viewed from the anterior surface of the body. The cranial loop rotates to the right of the body, while the caudal loop rotates to the left of the body. For the next several weeks, the intestinal loop grows and expands in the umbilicus since there is no available space inside the abdominal cavity.

By weeks 10 and 11, the midgut retracts into the abdomen as the abdominal cavity becomes spacious enough to house all the

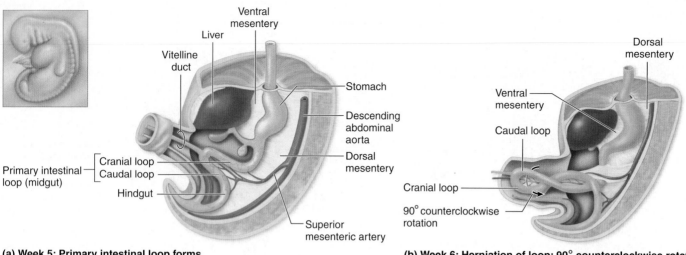

(a) Week 5: Primary intestinal loop forms

(b) Week 6: Herniation of loop; 90° counterclockwise rotation

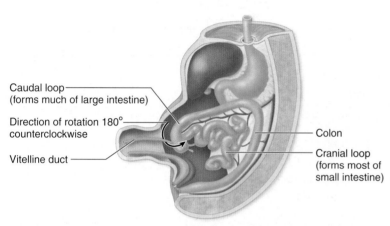

(c) Weeks 10–11: Retraction of intestines back into abdominal cavity; 180° counterclockwise rotation

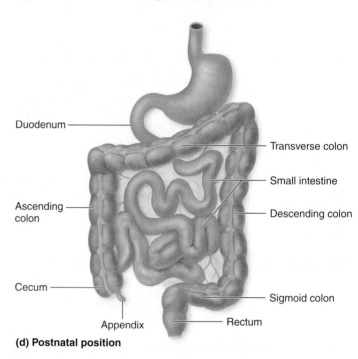

(d) Postnatal position

Figure 26.23

Development of the Digestive System Midgut. The midgut of the digestive system forms almost all of the small intestine and the proximal region of the large intestine. Development of these structures is shown at *(a)* week 5, *(b)* week 6, and *(c)* weeks 10–11. *(d)* The postnatal position of the digestive organs.

intestines. As the midgut retracts, it undergoes another 180-degree counterclockwise rotation, as viewed from the front of the body. Now the cranial loop is on the left side of the body, and the caudal loop is on the right. Thus, the net intestinal rotation is 270 degrees counterclockwise. The cranial loop (forming the jejunum and most of the ileum) retracts and positions itself in the left side of the abdominal cavity. The caudal loop (forming the distal ileum and the proximal part of the large intestine) is pulled to the right side of

the abdominal cavity. The vitelline duct begins to regress and disappears before the baby is born.

WHAT DID YOU LEARN?

19 What structures develop from endoderm outgrowths of the duodenum?

CLINICAL TERMS

bariatric surgery (gastric bypass) Surgically narrowing or blocking off a large portion of the stomach; typically performed in severely obese people who have had trouble losing weight by more traditional methods.

caries (kār′ez; dry rot) Cavities in the tooth enamel caused by plaque and bacterial buildup on the tooth.

diarrhea (dī-ă-rē′ă; *dia* = through, *rhoia* = a flow) Frequent, watery stools.

dysentery (dis-en-tēr-ē; *dys* = bad, *entera* = bowels) Painful, bloody diarrhea due to an infectious agent.

gastroenteritis (gas′trō-en-ter-ī′tis) Inflammation and irritation of the GI tract, often associated with vomiting and diarrhea.

hemorrhoids (hem′ŏ-roydz; *rhoia* = flow) Dilated, tortuous veins around the rectum and/or anus.

intussusception (in-tŭs-sŭ-sep′shŭn; *intus* = within, *suscipio* = to take up) Type of intestinal obstruction in which one part of the intestine constricts and gets pulled into the immediately distal segment of intestine; most commonly seen near the ileocecal junction. Symptoms include bloody stools and severe abdominal pain.

mumps (a lump) Viral infection of the parotid glands, resulting in painful swelling of the glands.

pancreatitis (pan′krē-ă-tī′tis) Inflammation of the pancreas; most often caused by alcoholism or biliary apparatus disease (including gallstones).

volvulus Twisting or torsion of an intestinal segment around itself. If left untreated, obstruction of digestive materials results, and the segment necroses due to poor blood supply. Symptoms include intense abdominal pain and vomiting.

CHAPTER SUMMARY

	■ The digestive system breaks down ingested food into usable nutrients.
General Structure and Functions of the Digestive System 777	■ The digestive organs are the oral cavity, pharynx, esophagus, stomach, small intestine, and large intestine. The accessory digestive organs are the teeth, tongue, salivary glands, liver, gallbladder, and pancreas.

Digestive System Functions 777

- ■ Mechanical digestion is the physical breakdown of ingested materials, and chemical digestion is the enzymatic breakdown of molecules.
- ■ Propulsion moves materials through the GI tract. Peristalsis moves materials from the mouth toward the anus; segmentation mixes ingested materials.
- ■ Secretion is the production and release of fluids; absorption is the movement or transport of materials across the wall of the GI tract.

Oral Cavity 778	■ The oral cavity is the entryway into the GI tract.

Cheeks, Lips, and Palate 778

- ■ The cheeks form the lateral walls of the oral cavity, and the lips frame the anterior opening formed by the orbicularis oris muscle.
- ■ The palate is the oral cavity roof: The hard palate is the anterior, bony portion, and the soft palate is the posterior, muscular portion.

Tongue 779

- ■ The tongue moves and mixes ingested materials; its dorsal surface has papillae.

Salivary Glands 779

- ■ Saliva is a fluid secreted by three pairs of multicellular salivary glands.

Teeth 781

- ■ Adults have 32 permanent teeth of four types: incisors, canines, premolars, and molars.

Pharynx 783	■ Pharyngeal constrictors contract sequentially during swallowing.

General Arrangement of Abdominal GI Organs 784	**Peritoneum, Peritoneal Cavity, and Mesentery 784**

- ■ Parietal peritoneum lines the internal body wall; visceral peritoneum covers abdominal organs.
- ■ Intraperitoneal organs are completely surrounded by visceral peritoneum; retroperitoneal organs are only partially ensheathed.
- ■ Mesenteries are double layers of peritoneum.

General Histology of GI Organs (Esophagus to Large Intestine) 785

- ■ The mucosa is composed of epithelium, an underlying lamina propria, and the muscularis mucosae.
- ■ The submucosa is a dense irregular connective tissue layer containing blood vessels, lymph vessels and nerves.
- ■ The muscularis usually has two smooth muscle layers: an inner circular layer and an outer longitudinal layer.
- ■ The adventitia is the outermost covering; when covered by visceral peritoneum, it is called a serosa.

Blood Vessels, Lymphatic Structures, and Nerve Supply 787

- ■ GI tract organs have extensive blood vessels, lymphatic structures, and nerves.

Esophagus 787	**Gross Anatomy 788**

- ■ The esophagus conducts swallowed materials from the pharynx to the stomach.

(continued on next page)

C H A P T E R S U M M A R Y *(c o n t i n u e d)*

Esophagus (continued) 787	**Histology 788**
	▪ The esophagus is lined by nonkeratinized stratified squamous epithelium.
	▪ The muscularis has all skeletal fibers superiorly, changes to intermingled skeletal and smooth fibers in the middle, and contains all smooth fibers inferiorly.
The Swallowing Process 789	▪ Swallowing has three phases: The voluntary phase moves a bolus into the pharynx; the pharyngeal phase moves the bolus through the pharynx into the esophagus; and the esophageal phase conducts the bolus to the stomach.
Stomach 790	**Gross Anatomy 790**
	▪ Mechanical digestion and chemical digestion occur in the stomach.
	▪ The four stomach regions are the cardia, fundus, body, and pylorus.
	Histology 790
	▪ The stomach mucosa is composed of a simple columnar epithelium containing gastric pits and gastric glands.
	▪ The muscularis has three smooth muscle layers: inner oblique, middle circular, and outer longitudinal.
	Gastric Secretions 793
	▪ Five types of secretory cells form the gastric glands: surface mucous cells (secrete mucin), mucous neck cells (secrete acidic mucin), parietal cells (secrete hydrochloric acid and intrinsic factor), chief cells (secrete pepsinogen), and enteroendocrine cells (secrete gastrin and other hormones).
Small Intestine 794	**Gross Anatomy and Regions 794**
	▪ The small intestine finishes chemical digestion and absorbs most of the nutrients.
	▪ The small intestine is divided into the duodenum, jejunum, and ileum.
	Histology 796
	▪ The small intestine has circular folds with fingerlike villi projecting from them.
Large Intestine 796	**Gross Anatomy and Regions 796**
	▪ The large intestine absorbs fluids and ions, and compacts undigestible wastes.
	▪ The large intestine is composed of the cecum, ascending colon, transverse colon, descending colon, sigmoid colon, rectum, and anal canal.
	Histology 798
	▪ The large intestine mucosa lacks villi, but contains intestinal glands.
	▪ The outer longitudinal layer of the muscularis forms three discrete bands called teniae coli. The teniae coli cause the wall of the colon to bunch into sacs called haustra.
	Control of Large Intestine Activity 799
	▪ Local autonomic reflexes regulate large intestine movements.
Accessory Digestive Organs 800	**Liver 800**
	▪ The liver receives venous blood from the hepatic portal vein and oxygenated blood from the hepatic artery proper.
	▪ Liver functions are bile production, drug detoxification, storage of nutrients and glycogen, and synthesis of plasma proteins.
	Gallbladder 803
	▪ The gallbladder stores and concentrates bile produced by the liver.
	Pancreas 803
	▪ Pancreatic acini produce pancreatic juice that neutralizes acidic chyme.
	Biliary Apparatus 804
	▪ The hepatic ducts drain bile into a single common hepatic duct, and the cystic duct merges with the common hepatic duct to form the common bile duct.
	▪ The common bile duct and the pancreatic duct merge to empty their contents into the duodenum via the major duodenal papilla.
Aging and the Digestive System 805	▪ Age-related changes in the digestive system lead to reduced secretions and diminished absorption of nutrients.

Development of the Digestive System 806	■ The gut tube has a superior foregut, a central midgut, and an inferior hindgut.

Stomach, Duodenum, and Omenta Development   806

■ Stomach formation begins by week 4 of development. Differential growth rates cause the greater and lesser curvatures of the stomach to form.

Liver, Gallbladder, and Pancreas Development   806

■ The liver parenchyma, gallbladder, and pancreas develop from endoderm outgrowths of the duodenum.

Intestine Development   807

■ The small and large intestines form from the midgut and hindgut beginning in week 5.

CHALLENGE YOURSELF

Matching

Match each numbered item with the most closely related lettered item.

_____ 1. circular folds

_____ 2. falciform ligament

_____ 3. pyloric sphincter

_____ 4. haustra

_____ 5. segmentation

_____ 6. simple columnar

_____ 7. muscularis

_____ 8. stratified squamous

_____ 9. Peyer patches

_____ 10. submucosa

a. typically contains two layers of smooth muscle

b. epithelium lining small intestine

c. contains dense irregular connective tissue and blood vessels

d. lymphatic nodules in wall of ileum

e. attaches liver to anterior abdominal wall

f. restricts chyme entry into small intestine

g. epithelium lining the esophagus

h. sacs of large intestine wall

i. increase surface area of small intestine

j. process to mix digested materials in small intestine

Multiple Choice

Select the best answer from the four choices provided.

_____ 1. Which organ is located in the right upper quadrant of the abdomen?
   a. liver
   b. spleen
   c. descending colon
   d. appendix

_____ 2. The _____ cells of the stomach secrete hydrochloric acid (HCl).
   a. chief
   b. parietal
   c. mucous
   d. enteroendocrine

_____ 3. Material leaving the ascending colon next enters the
   a. cecum.
   b. descending colon.
   c. sigmoid colon.
   d. transverse colon.

_____ 4. Which of these organs is retroperitoneal?
   a. stomach
   b. transverse colon
   c. descending colon
   d. ileum

_____ 5. Sympathetic innervation of the GI tract is responsible for
   a. closing the pyloric sphincter.
   b. stimulating peristalsis.
   c. stimulating secretion of the pancreatic acinar cells.
   d. vasodilating the major digestive system blood vessels.

_____ 6. The _____ is derived from the cranial part of the primary intestinal loop.
   a. jejunum
   b. cecum
   c. ascending colon
   d. appendix

_____ 7. The main pancreatic duct merges with the _____, and their contents empty into the duodenum through the major duodenal papilla.
   a. left hepatic duct
   b. common bile duct
   c. cystic duct
   d. common hepatic duct

_____ 8. Which statement is false about pancreatic juice?
   a. It is secreted through the main pancreatic duct into the duodenum.
   b. It is responsible for emulsifying (breaking down) fats.
   c. It is produced by the acinar cells of the pancreas.
   d. The juice has an alkaline pH.

_____ 9. The "living" part of a tooth is the
   a. dentin.
   b. cementum.
   c. pulp.
   d. enamel.

_____ 10. Most of the chemical digestion of our food occurs within the
   a. large intestine.
   b. pancreas.
   c. small intestine.
   d. esophagus.

Content Review

1. What initial stages of digestion occur in the oral cavity?

2. The GI tract from the esophagus to the anal canal is composed of four tunics. Describe the general histology of the tunics and the specific features of the stomach tunics.

3. Compare and contrast gastric juice and pancreatic juice with respect to their composition and function.

4. Compare the anatomy and functions of the circular folds, villi, and microvilli in the small intestine.

5. What are the teniae coli and haustra, and how are they associated?

6. What is the function of each structure in the portal triad of the hepatic lobule, and how do they work together to contribute to the overall functioning of the liver?

7. What is the function of the gallbladder, and what role does it play in digestion?

8. List in order the organs of the GI tract through which ingested material travels, and describe the type(s) of digestion (mechanical/chemical) that takes place in each organ. At what point is the material called a bolus, chyme, and feces?

9. Why are there so many mucin-producing glands along the length of the GI tract?

10. Describe how the small and large intestine form.

Developing Critical Reasoning

1. Alexandra experienced vomiting and diarrhea, and was diagnosed with gastroenteritis (stomach flu). What specific digestive system organs were affected by the illness, and how did the illness interfere with each organ's function?

2. Most cases of colorectal cancer occur in the most distal part of the large intestine (the rectum, sigmoid colon, and descending colon). Why do fewer instances of colon cancer tend to occur in the proximal part of the large intestine? Include the anatomy and function of the colon components in your explanation.

A N S W E R S   T O   " W H A T   D O   Y O U   T H I N K ? "

1. Saliva cleanses the mouth in a variety of ways. As saliva washes over the tongue and teeth, it helps remove foreign materials and buildup. Lysozyme in saliva is an antibacterial enzyme. A person who has a dry mouth is not able to cleanse the mouth well and is more likely to develop dental problems as a result of built-up bacterial and foreign material.

2. A young child's mouth is too small to support adult-sized teeth. Deciduous teeth are much smaller and fit a child's mouth better. As the mouth increases in size by about age 6, it is able to support the first adult-sized teeth.

3. The stomach is lined with specialized surface mucous cells that secrete mucus onto the lining. This mucus prevents the acid from eating away the stomach wall. In addition, the stomach epithelial lining is constantly regenerating to replace any cells that are damaged.

4. The duodenum has many circular folds so that the digested materials can be slowed down and adequately mixed with pancreatic juice and bile. The jejunum also has many circular folds because the slowing down of the materials aids in absorbing the maximum amount of nutrients. By the time chyme reaches the distal ileum, most nutrient absorption has occurred, and so the circular folds are not as essential.

5. When the gallbladder is removed, bile is still produced by the liver, but it is secreted in a "slow drip." A fatty meal requires much more bile for digestion than the amount supplied by a "slow drip," causing the patient to experience gas, pain, bloating, and diarrhea. Thus, surgical patients are initially told to limit their fat intake. The body gradually adjusts so that more bile is secreted, and thus a person may eventually be able to introduce more fat back into his or her diet.

Visit the McKinley/O'Loughlin _Human Anatomy,_ 2e website at aris.mhhe.com

OUTLINE

URINARY SYSTEM

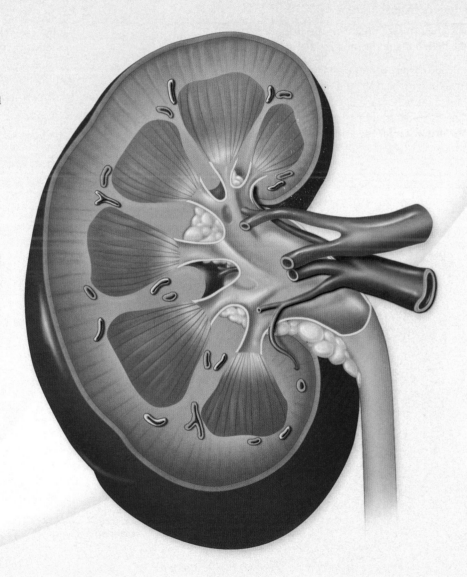

Urinary System

In the course of carrying out their specific functions, the cells of all body systems produce waste products, and these waste products end up in the bloodstream. In this case, the bloodstream is analogous to a river that supplies drinking water to a nearby town. The river water may become polluted with sediment, animal waste, and motorboat fuel—but the town has a water treatment plant that removes these waste products and makes the water safe to drink. The urinary system is the body's "water treatment plant." Without it, waste products could accumulate in the blood and kill us. This chapter focuses on the organs of the urinary system and how they work together to remove waste products from the blood and help maintain homeostasis.

General Structure and Functions of the Urinary System

Key topics in this section:

- The primary organs of the urinary system
- Functions performed by the urinary system

The organs of the **urinary** (ūr′i-nār-ē) **system** are the kidneys, ureters, urinary bladder, and urethra **(figure 27.1)**. The kidneys filter waste products from the bloodstream and convert the filtrate into **urine** (ūr′in). The ureters, urinary bladder, and urethra are collectively known as the **urinary tract** because they transport the urine out of the body.

Besides removing waste products from the bloodstream, the urinary system performs the following other functions:

- **Storage of urine.** Urine is produced continuously, but it would be quite inconvenient if we were constantly excreting urine. The urinary bladder is an expandable, muscular sac that can store as much as 1 liter of urine.
- **Excretion of urine.** The urethra transports urine from the urinary bladder and expels it outside the body. Later in this chapter, we discuss the expulsion of urine from the bladder, which is called micturition or urination.
- **Regulation of blood volume.** The kidneys control the volume of interstitial fluid and blood under the direction of certain hormones. Also, changes in blood volume affect blood pressure, so the kidneys indirectly affect blood pressure.
- **Regulation of erythrocyte production.** As the kidneys filter the blood, they are also indirectly measuring the oxygen level in the blood. If blood oxygen levels are reduced, cells in the kidney secrete a hormone called **erythropoietin** (ĕ-rith-rō-poy′-ĕ-tin) (**EPO**; described in chapter 20). Erythropoietin acts on stem cells in the bone marrow to increase erythrocyte production. Having more erythrocytes allows the blood to transport more oxygen.

Table 27.1 summarizes the organs of the urinary system and their functions.

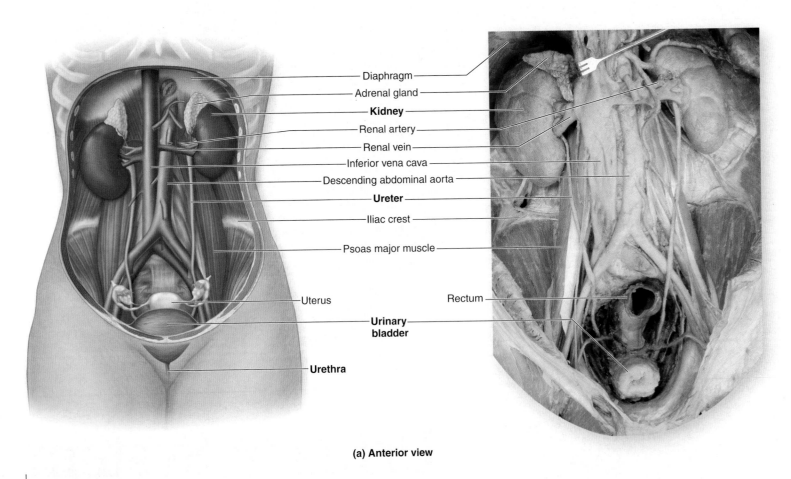

Diaphragm
Adrenal gland
Kidney
Renal artery
Renal vein
Inferior vena cava
Descending abdominal aorta
Ureter
Iliac crest
Psoas major muscle
Uterus
Rectum
Urinary bladder
Urethra

(a) Anterior view

Figure 27.1

Urinary System. The urinary system is composed of two kidneys, two ureters, a single urinary bladder, and a single urethra, shown here in *(a)* anterior and *(b)* posterior views. Dotted lines in *(b)* show the position of the ureters.

Table 27.1	Urinary System Organs and Their Functions		
Organ	**Location**	**Description/Characteristics**	**Function**
Kidneys	Posterior abdominal wall; right kidney is inferior to left kidney	Paired, bean-shaped organs; composed of outer cortex and inner medulla	Filters blood and processes filtrate into tubular fluid, then urine
Ureters	Extend from kidneys to trigone of bladder, along posterior abdominopelvic wall	Paired thin, fibromuscular tubes composed of inner mucosa, middle muscularis of smooth muscle, and an outer adventitia	Transport urine from kidney to urinary bladder via peristalsis
Urinary bladder	Pelvic cavity, posterior to pubic symphysis (when full, it extends into inferior part of abdominal cavity)	Muscular distensible sac composed of inner mucosa, a submucosa, a muscularis, and an outer adventitia or serosa The neck of the bladder is the inferior constricted region where bladder and urethra meet; contains internal urethral sphincter	Reservoir for urine until urination (micturition) occurs
Urethra	Inferior to neck of urinary bladder; extends through muscles of pelvic floor and opens into perineum	Single muscular tube; 3–5 cm long in females; 18–20 cm long in males	Transports urine from urinary bladder to outside of body

WHAT DID YOU LEARN?

1 What organs make up the urinary tract, and what is the main function of the urinary tract?

2 Describe the mechanisms by which the kidneys regulate blood volume and erythrocyte production.

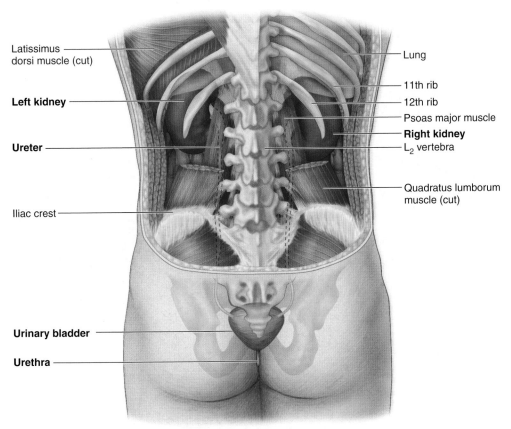

Latissimus dorsi muscle (cut)
Left kidney
Ureter
Iliac crest
Urinary bladder
Urethra

Lung
11th rib
12th rib
Psoas major muscle
Right kidney
L₂ vertebra
Quadratus lumborum muscle (cut)

(b) Posterior view

Kidneys

Key topics in this section:

- Anatomy of the kidneys
- Fundamentals of filtration, tubular reabsorption, and tubular secretion
- Components of a nephron and their roles in urine formation

The kidneys are two symmetrical, bean-shaped, reddish-brown organs located along the posterior abdominal wall, lateral to the vertebral column (figure 27.1). Each kidney weighs approximately 100 grams and measures about 12 centimeters (cm) in length, 6.5 cm in width, and 2.5 cm in thickness.

WHAT DO YOU THINK?

1 A person must have at least one functioning kidney in order to survive. Why do you think a lack of kidneys is deadly?

Gross and Sectional Anatomy of the Kidney

The kidneys are retroperitoneal, since only their anterior surface is covered with peritoneum and the posterior aspect lies directly against the posterior abdominal wall. The **superior pole** (also called *superior extremity*) of the left kidney is at about the level of the T_{12} vertebra, and its **inferior pole** (also called *inferior extrem-*

ity) is at about the level of the L_3 vertebra. The right kidney is positioned about 2 cm inferior to the left kidney to accommodate the large size of the liver. An adrenal gland rests on the superior pole of each kidney.

The kidney has a concave medial border called the **hilum** (hī′lŭm; *hilum* = a small bit), where vessels, nerves, and the ureter connect to the kidney. The hilum is continuous with an internal space within each kidney called the **renal** (rē′năl; *ren* = kidney) **sinus.** The renal sinus houses renal arteries, renal veins, lymphatic vessels, nerves, the renal pelvis, renal calyces, and a variable amount of adipose connective tissue. The kidney's lateral border is convex.

Each kidney is surrounded and supported by several tissue layers. From innermost (closest to the kidney) to outermost, these layers are the fibrous capsule, perinephric fat, renal fascia, and paranephric fat (**figure 27.2**):

- The **fibrous capsule** (kap′sool; *capsa* = box) or *renal capsule,* is composed of dense irregular connective tissue that covers the outer surface of the kidney. The fibrous capsule maintains the kidney's shape, protects it from trauma, and helps prevent infectious pathogens from entering the kidney.
- The **perinephric fat,** or *adipose capsule,* is external to the fibrous capsule and formed by a layer of adipose

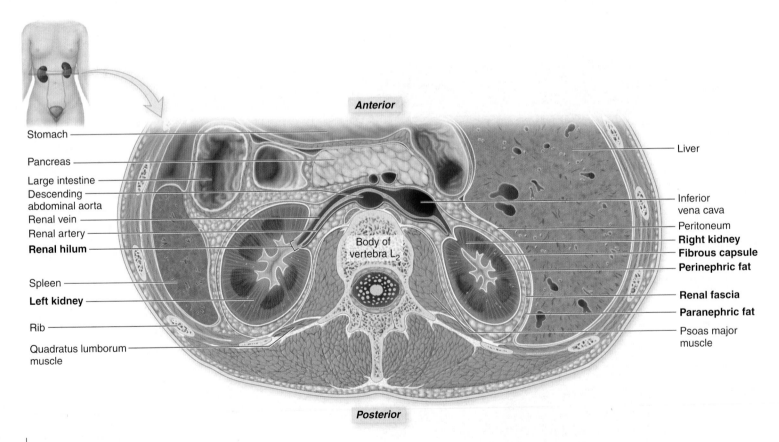

Figure 27.2

Position and Stabilization of the Kidneys. The kidneys lie along the posterior abdominal wall. A cross-sectional view shows that the kidneys are surrounded by four concentric tissue layers: fibrous capsule, perinephric fat, renal fascia, and paranephric fat.

connective tissue that varies in thickness. This layer, which is also called the *perirenal fat,* completely surrounds the kidney and offers cushioning and insulation.

■ The **renal fascia** (fash'ē-ǎ; a band) is external to the perinephric fat and is composed of dense irregular connective tissue. It anchors the kidney to the posterior abdominal wall and peritoneum.

■ The **paranephric fat** is the outermost layer surrounding the kidney. It is composed of adipose connective tissue and lies between the renal fascia and the peritoneum.

When a kidney is sectioned along a coronal plane, an outer **renal cortex** and an inner **renal medulla** can be seen **(figure 27.3).** The medulla tends to be a darker shade than the cortex. Extensions of the cortex, called **renal columns,** project into the medulla and subdivide the medulla into **renal pyramids** (or *medullary pyramids*). An adult kidney typically contains 8 to 15 renal pyramids. The wide base of a renal pyramid lies at the external edge of the medulla where the cortex and medulla meet, called the **corticomedullary junction.** The apex (tip) of the renal pyramid (called the **renal papilla**) projects toward the renal sinus.

Each renal papilla projects into a funnel-shaped space called the **minor calyx** (kā'liks; pl., *calyces,* or *calices,* kal'i-sēz;

cup of a flower). There are between 8 and 15 minor calyces—in other words, one minor calyx for each renal papilla. Several minor calyces merge to form larger spaces called **major calyces**—each kidney typically contains two or three major calyces. Urine from the renal pyramids is collected by the minor calyces and then drained into the major calyces. The major calyces merge to form a large, funnel-shaped **renal pelvis,** which collects urine and transports it into the ureter.

A human kidney is divided into 8 to 15 **renal lobes.** A renal lobe consists of a medullary pyramid and some cortical substance from the renal columns adjacent to it on either side, as well as the cortex external to the pyramid base.

WHAT DID YOU LEARN?

3 Where are the renal cortex and the renal medulla located?

4 Why is the right kidney more inferiorly placed than the left kidney?

Blood Supply to the Kidney

Since the kidneys' primary function is to filter the blood, at least 20% to 25% of the resting cardiac output (see chapter 22)

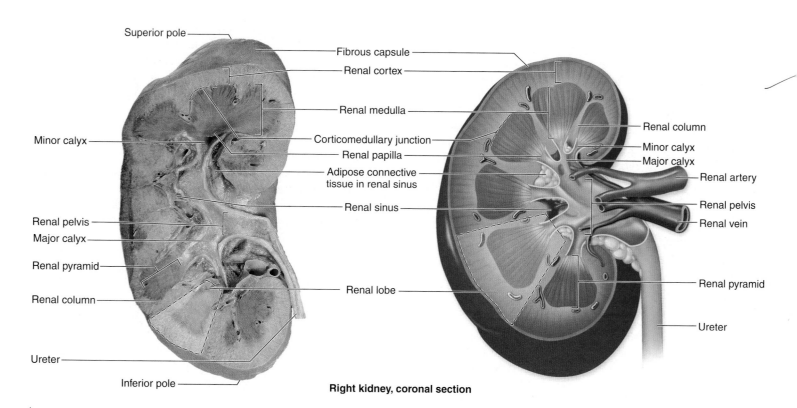

Right kidney, coronal section

Figure 27.3

Gross Anatomy of the Kidney. The bean-shaped kidney is covered by a fibrous capsule and contains an external cortex and an internal medulla.

CLINICAL VIEW

Intravenous Pyelogram

Sometimes a physician needs to visualize the kidneys, ureters, and urinary bladder, especially when the flow of urine from one of the kidneys into the bladder becomes blocked. A prime example is the obstruction of a ureter by a kidney stone, a painful condition that requires prompt treatment. To learn the location of the blockage and to better understand the condition, a physician may ask for an x-ray study known as an intravenous **pyelogram** (pī′el-ō-gram; *gram* = recording).

An intravenous pyelogram is produced by injecting a small amount of radiopaque dye into a vein. The dye is formulated to be water-soluble and quickly cleared by the kidneys. Following injection of the dye, a series of abdominal x-rays are taken over a period of about an hour and a half. As the dye passes through the kidneys and is cleared into the urine, the sequential x-rays provide a "time-lapse" view of urinary system function. The dye allows the kidneys, ureters, and urinary bladder to be shown on x-ray. A problem with filling or movement of urine through one of the kidneys or ureters appears as a narrowing or blockage on the x-ray image.

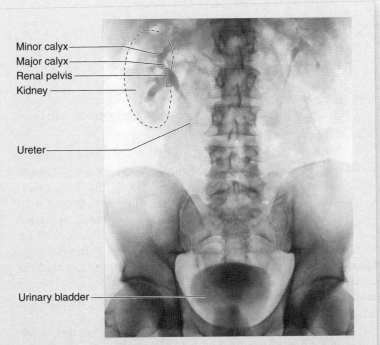

A pyelogram enables physicians to visualize the urinary system organs and identify the location of any blockages.

normally flows through the kidneys via the pathway shown in **figure 27.4**. Blood is carried to a kidney in a **renal artery,** which generally arises and extends from the lateral region of the descending abdominal aorta at the level of the first or second lumbar vertebra. Up to five **segmental** (seg-men′tăl) **arteries** branch from the renal artery within the renal sinus. While still in the renal sinus, the segmental arteries further branch to form the **interlobar** (in-ter-lō′bar; *inter* = between, *lobos* = lobe) **arteries.** Interlobar arteries travel through the renal columns toward the corticomedullary junction, where they branch to form **arcuate** (ar′kū-āt; *arcuatus* = bowed) **arteries.** These arcuate arteries project parallel to the base of the medullary pyramid at the corticomedullary junction. The arcuate arteries give off branches called **interlobular** (in-ter-lob′ū-lăr) **arteries** that project peripherally into the cortex.

As the interlobular arteries enter the cortex, they extend small branches called **afferent** (af′er-ent; *ad* = toward, *ferre* = to lead) **arterioles** (afferent glomerular arteriole). An afferent arteriole then enters a structure called a renal corpuscle and forms a capillary network called the **glomerulus** (glō-mār′ū-lŭs; *glomus* = ball of yarn, *ulus* = small). Some blood plasma is filtered across

the glomerulus into the capsular space within the renal corpuscle. Once some of the blood plasma has been filtered, the remaining blood leaves the glomerulus and enters an **efferent** (ef′er-ent; *efferens* = to bring out) **arteriole** (efferent glomerular arteriole). Note that the efferent arteriole is still carrying oxygenated blood because gas and nutrient exchange with cells of the kidney has not yet occurred.

The efferent arterioles branch into one of two types of capillary networks: peritubular capillaries or vasa recta (figure 27.4). These capillary networks are responsible for the actual exchange of gases, nutrients, and waste materials within the kidney. **Peritubular capillaries** are associated with the convoluted tubules and primarily reside in the cortex of the kidney. **Vasa recta** (vā′să rek′tă; *vasculum* = small vessel, *rectus* = straight) are associated with the nephron loop and primarily reside in the medulla of the kidney.

The peritubular capillaries and vasa recta then drain into a network of veins. The smallest of these veins are the **interlobular veins,** which travel alongside the interlobular arteries. Interlobular veins merge to form **arcuate veins** that project parallel to the base of each medullary pyramid near the corticomedullary junction.

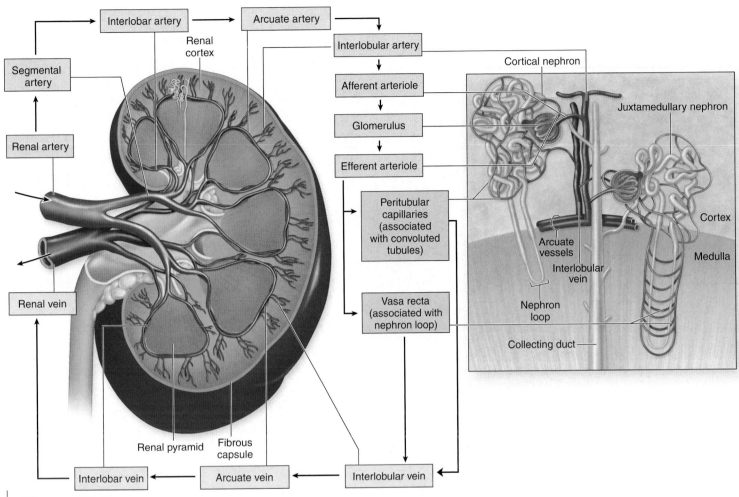

Figure 27.4

Blood Supply to the Kidneys. (*Left*) A coronal view depicts kidney circulation. (*Right*) An expanded view shows circulation to some nephrons. Pink boxes indicate vessels with arterial blood; lavender boxes indicate vessels where reabsorbed materials reenter the blood; blue boxes indicate vessels returning blood to the general circulation.

Arcuate veins merge to form **interlobar veins** that travel through the renal columns from the corticomedullary junction toward the renal sinus. Interlobar veins merge in the renal sinus to form the **renal vein.** (Although many veins are associated with arteries of the same name, there are no segmental veins; rather, the interlobar veins directly form the renal vein.) The renal veins leave the kidney at its hilum and traverse horizontally to drain into the inferior vena cava.

Blood filtration occurs at the glomerulus, and the blood remains highly oxygenated until it reaches the peritubular capillaries and vasa recta, where exchange of gases occurs.

Study Tip!

The names of the blood vessels in the kidney can give you a clue as to their location or appearance:

- Interlobar vessels are located *between* ('inter") the lobes of the kidney.
- Arcuate vessels form vessel "arcs" at the corticomedullary junction.
- Interlobular vessels are located between the smaller lobules of the kidney cortex.
- Afferent arterioles carry blood *to* the glomerulus (remember, "afferent" means "toward").
- Efferent arterioles take blood *away from* the glomerulus (remember, "efferent" means to take away, or "exit").
- Peritubular capillaries are *around* ("peri") the tubules (proximal and distal convoluted tubules).
- Vasa recta means "straight vessels," and these vessels run parallel to the long, straight tubules of the nephron loop.

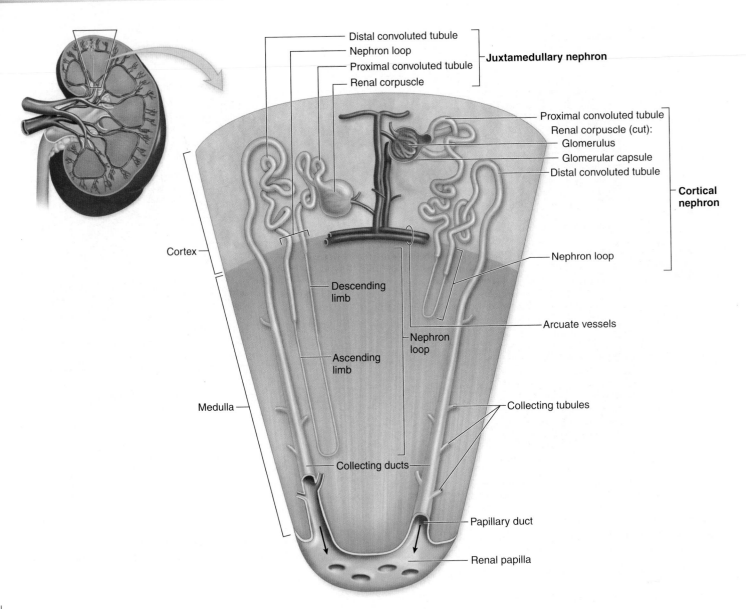

Figure 27.5

Nephron Structure.   The bulk of cortical nephrons are located in the cortex, and their nephron loops barely penetrate the medulla. The juxtamedullary nephrons lie close to the corticomedullary junction, and their relatively long nephron loops extend deep into the medulla.

Nephrons

The functional filtration unit in the kidney is the **nephron** (nef′ron), which consists of the following components: a renal corpuscle (composed of a glomerulus and a glomerular capsule), a proximal convoluted tubule, a nephron loop, and a distal convoluted tubule **(figure 27.5)**. The proximal convoluted tubule, nephron loop, and distal convoluted tubule are collectively known as the **renal tubule.**

Together, both kidneys house approximately 2.5 million nephrons. These microscopic structures measure less than 5 centimeters in total length. There are two types of nephrons: cortical nephrons and juxtamedullary nephrons (see figure 27.4, *right*). Approximately 85% of the nephrons are termed **cortical nephrons** because the bulk of the nephron resides solely in the cortex, and their relatively short nephron loops just barely penetrate the medulla. The remaining 15% of nephrons are called **juxtamedullary** (jŭks′tă-med′ŭ-lăr-ē; *juxta* = close to) **nephrons,** since their renal corpuscles lie close to the corticomedullary junction, and their long nephron loops extend deep into the medulla.

Urine Formation

The nephrons form urine through three interrelated processes: filtration, reabsorption, and secretion.

1. **Filtration** is the process by which water and some dissolved solutes in the blood plasma passively move out of the glomerulus into the capsular space of the renal corpuscle due to pressure differences across the filtration membrane. This water and its dissolved solutes are called **filtrate.**

2. **Tubular reabsorption** occurs when substances in the filtrate move by diffusion or active transport across the wall of the convoluted tubules and the nephron loop to return to the blood. Once filtrate begins to be modified, it is called **tubular fluid.** Usually, all needed solutes and most water that formed the filtrate are reabsorbed by the blood. As reabsorption occurs, some excess solutes, water, and waste products remain within the tubular fluid.

3. **Tubular secretion** is the active transport of solutes out of the blood into the tubular fluid.

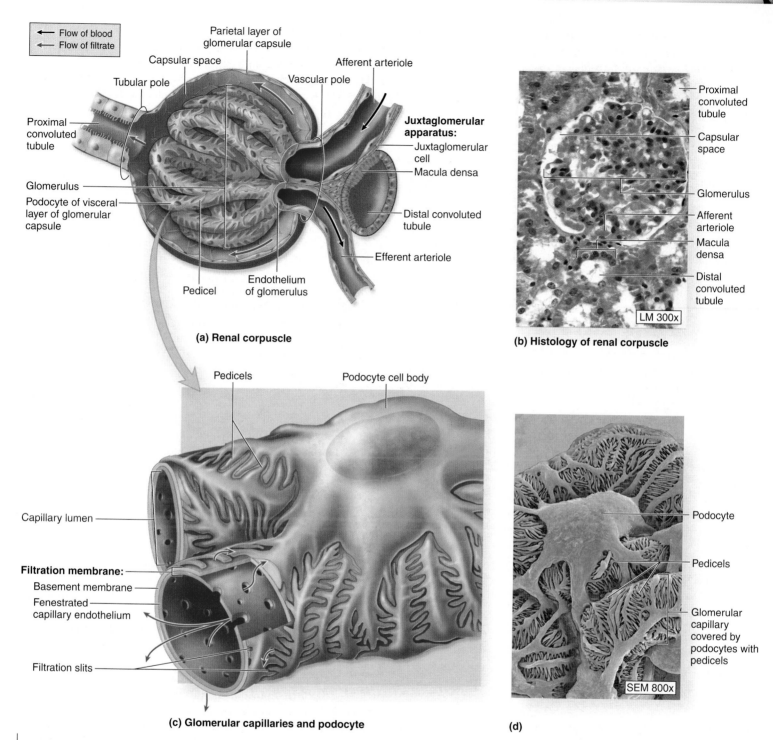

Flow of blood
Flow of filtrate

Parietal layer of glomerular capsule

Capsular space

Tubular pole

Vascular pole

Afferent arteriole

Proximal convoluted tubule

Juxtaglomerular apparatus:
Juxtaglomerular cell
Macula densa

Glomerulus
Podocyte of visceral layer of glomerular capsule

Distal convoluted tubule

Efferent arteriole

Pedicel

Endothelium of glomerulus

(a) Renal corpuscle

Proximal convoluted tubule
Capsular space
Glomerulus
Afferent arteriole
Macula densa
Distal convoluted tubule

LM 300x

(b) Histology of renal corpuscle

Pedicels

Podocyte cell body

Capillary lumen

Filtration membrane:
Basement membrane
Fenestrated capillary endothelium

Filtration slits

Podocyte
Pedicels
Glomerular capillary covered by podocytes with pedicels

SEM 800x

(c) Glomerular capillaries and podocyte

(d)

Figure 27.6

Renal Corpuscle. *(a)* The renal corpuscle is composed of a capillary meshwork called the glomerulus housed within the bulbous glomerular capsule. The afferent arteriole supplies blood to the glomerular capillaries; blood leaves the glomerulus in the efferent arteriole. *(b)* Photomicrograph of a renal corpuscle. *(c)* Filtrate is produced in the renal corpuscle when blood plasma is forced across the glomerular capillary wall under pressure. *(d)* SEM shows the podocytes covering the glomerular capillaries.

Once the tubular fluid exits the collecting duct, it is called urine. Urine consists of solutes and water that have been filtered as filtrate and secreted into the tubular fluid.

Next we examine how the specific components of the nephron carry out these processes.

Renal Corpuscle
A **renal corpuscle** (kōr′pus-l; *corpus* = body, *cle* = tiny) is an enlarged, bulbous region of a nephron **(figure 27.6a)**. It is composed of two structures: the thick tangle of capil-

laries called the **glomerulus,** and an epithelial capsule surrounding the glomerulus called the **glomerular capsule** (*Bowman capsule*). The glomerular capsule has two layers: a **visceral layer** that directly overlies the glomerulus and a **parietal layer.** The visceral layer is composed of specialized cells called podocytes (to be discussed later), while the parietal layer is formed from a simple squamous epithelium. Between these two layers is a **capsular space.** As blood flows through the glomerulus, the solutes and water within the blood are filtered from the glomerulus into the capsular space.

The renal corpuscle has two opposing poles: a **vascular pole,** where the glomerular afferent and efferent arterioles are found, and a **tubular pole,** where the proximal convoluted tubule attaches to the renal corpuscle.

The visceral layer of the glomerular capsule is composed of specialized cells called **podocytes** (pod′ō-sīt; *podos* = foot), which have long processes called **pedicels** (ped′ĭ-sel; *pedicellus* = little foot) or feet, that wrap around the glomerular capillaries to support the capillary wall but not completely ensheathe it. The pedicels are separated by thin spaces called **filtration slits,** which allow materials from the blood plasma to pass into the capsular space (figure 27.6c). Finally, the glomerular capillaries are fenestrated capillaries, which means they have fenestrations or pores that allow materials to be filtered. The filtration slits of the podocytes and the fenestrated glomerulus endothelium collectively make up the **filtration membrane** of the glomerulus.

Several factors cause materials to be filtered from the glomerulus. First, the afferent arteriole is wider in diameter than the efferent arteriole, so the blood in the afferent arteriole enters the glomerulus under high pressure, which helps filter the plasma and solutes out of the glomerulus. Second, the glomerulus is designed to be "leaky" to allow some plasma and solutes to be filtered. Third, the filtration slits between pedicles permit the ready passage of filtered material into the capsular space.

The filtration membrane of the glomerulus, which includes the thick basement membrane of the glomerular endothelium, can only crudely filter the blood. The thick basement membrane also is designed to work with the fenestrations of the capillary in selectively filtering materials from the blood plasma. Thus, too much water and some vital nutrients, ions, plasma proteins, and vitamins escape into the filtrate as well. Therefore, the renal tubule (proximal convoluted tubule, nephron loop, and distal convoluted tubule) becomes responsible for refining and modifying this filtrate. The filtrate first enters the proximal convoluted tubule, where it begins to be modified and becomes tubular fluid.

WHAT DID YOU LEARN?

5 Compare and contrast the juxtamedullary nephron and cortical nephron.

6 What is the difference between filtration and reabsorption?

Proximal Convoluted Tubule The **proximal convoluted tubule (PCT)** originates at the tubular pole of the renal corpuscle (**figure 27.7**; see figure 27.6a). It is lined by a simple cuboidal epithelium with tall microvilli that markedly increase its reabsorption capacity. When viewed under a microscope, the lumen of the proximal convoluted tubule looks fuzzy due to the brush border formed by the tall microvilli. In addition, the cytoplasm of these cells stains brightly, due to an abundance of mitochondria in the cytoplasm.

The cells of the proximal convoluted tubule actively reabsorb almost all nutrients (e.g., glucose and amino acids), ions (especially Na^+), vitamins, and any plasma proteins from the tubular fluid. Approximately 60–65% of the water in the tubular fluid is reabsorbed by osmosis. The reabsorbed solutes and water enter the blood in the peritubular capillaries and are returned to the general circulation in the vascular system.

Nephron Loop The **nephron loop** (or *loop of Henle*) originates at a sharp bend in the proximal convoluted tubule and projects internally toward and into the medulla (see figure 27.5). Each loop has two limbs: a **descending limb** extending from the cortex toward and into the medulla, and an **ascending limb** that returns back to the renal cortex. The descending limb is first lined with a simple cuboidal epithelium (this segment is called the thick descending limb) and then a simple squamous epithelium (this segment is called the thin descending limb). The ascending limb is first lined by a simple squamous epithelium (this segment is called the thin ascending limb) and then by a simple cubodial epithelium (this segment is called the thick ascending limb). The ascending limb of the nephron loop returns to the renal cortex and terminates at the distal convoluted tubule.

The primary function of the nephron loop is to facilitate reabsorption of water and solutes from the tubular fluid. From the ascending limb, sodium ions (Na^+) and chloride ions (Cl^-) are the primary solutes reabsorbed, although other solutes may be reabsorbed as well. When these materials are reabsorbed from the nephron loop, they are returned to the blood via the vasa recta capillary network.

Distal Convoluted Tubule The **distal convoluted tubule (DCT)** originates in the renal cortex at the end of the thick ascending limb of the nephron loop and contacts the afferent arteriole wall at the vascular pole (figure 27.7). Like the proximal convoluted tubule, the distal convoluted tubule is also lined with simple cuboidal epithelium. However, the distal convoluted tubule cells are smaller and contain only sparse, short microvilli (figure 27.7c). Thus, the distal convoluted tubule lumen does not appear fuzzy. In addition, the distal convoluted tubule cytoplasm stains more lightly than that of the proximal convoluted tubule, because the distal convoluted tubule has fewer mitochondria.

The primary function of the distal convoluted tubule is to secrete ions such as potassium (K^+) and hydrogen (H^+) into the tubular fluid. However, reabsorption of water also occurs here, primarily under the influence of aldosterone (see chapter 20). **Aldosterone** is a hormone secreted by the adrenal cortex in response to low blood volume or low solute concentration in the tubular fluid within the kidney. This hormone causes the distal convoluted tubule cells to increase sodium and water reabsorption from tubular fluid. Once the tubular fluid leaves the distal convoluted tubule, it enters the collecting tubules.

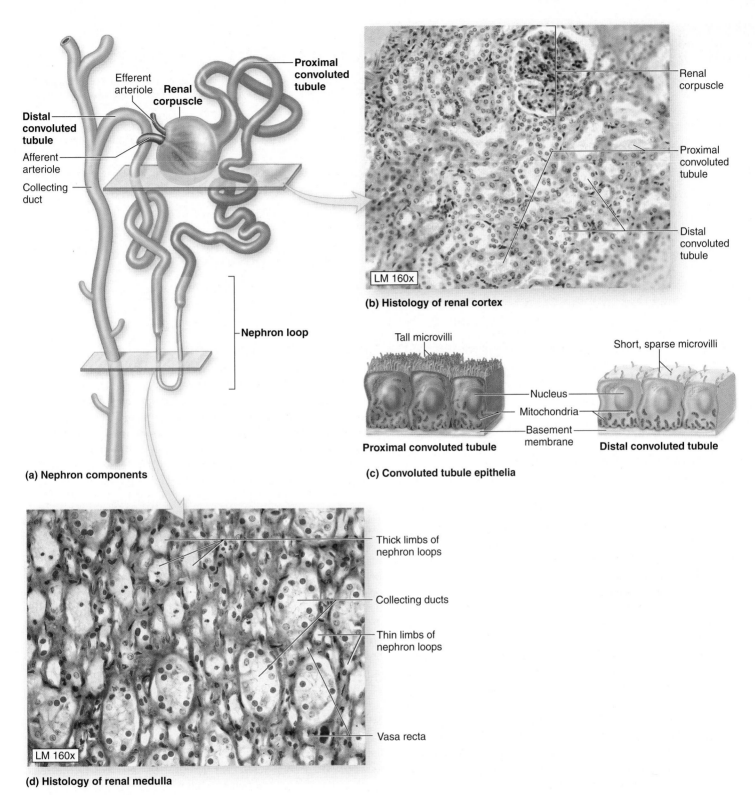

Efferent arteriole

Renal corpuscle

Proximal convoluted tubule

Distal convoluted tubule

Afferent arteriole

Collecting duct

Nephron loop

(a) Nephron components

Renal corpuscle

Proximal convoluted tubule

Distal convoluted tubule

LM 160x

(b) Histology of renal cortex

Tall microvilli

Short, sparse microvilli

Nucleus

Mitochondria

Basement membrane

Proximal convoluted tubule

Distal convoluted tubule

(c) Convoluted tubule epithelia

Thick limbs of nephron loops

Collecting ducts

Thin limbs of nephron loops

Vasa recta

LM 160x

(d) Histology of renal medulla

Figure 27.7

The Convoluted Tubules and Nephron Loop.   The convoluted tubules and nephron loop carry tubular fluid that is being modified to form urine. *(a)* In this drawing, each of the components of the nephron, including the convoluted tubules and nephron loop, is distinguished by a different color. *(b)* A photomicrograph of a section through the cortex compares transverse sections of the proximal and distal convoluted tubules. *(c)* Comparisons of the simple cuboidal epithelium lining the proximal and distal convoluted tubules show differences in the sizes and numbers of microvilli. *(d)* A photomicrograph of a transverse section through the medulla compares nephron loops and collecting ducts.

Table 27.2	Parts of a Nephron and Their Functions	
Structure	**Description**	**Function**
Renal corpuscle	Capillary ball or tuft covered by podocytes and surrounded by an epithelial capsule; capsular space is between the two layers of the capsule	Produces a filtrate of blood that must be modified as it passes through the convoluted tubules and nephron loop
Proximal convoluted tubule	Tubule lined with simple cuboidal epithelium; has a prominent brush border (microvilli); cytoplasm tends to stain more brightly than DCT cells	Reabsorbs ions (especially Na^+), nutrients (glucose and amino acids), plasma proteins, vitamins, and water; secretes some H^+ ions
Nephron loop	Tubule that forms a loop; has thick and thin ascending and descending portions; the most distal part of the loop often extends into the medulla Thick limbs are lined with simple cuboidal epithelium Thin limbs are lined with simple squamous epithelium	Reabsorbs water in tubular fluid; also reabsorbs Na^+ and Cl^-; secretes some H^+ ions
Distal convoluted tubule	Tubule lined with simple cuboidal epithelium with only a sparse brush border; cytoplasm of cells tends to be paler than that of PCT cells	Secretes H^+ and K^+ into tubular fluid; reabsorbs Na^+ and water from tubular fluid

Table 27.2 summarizes the parts of a nephron and their functions.

 WHAT DO YOU THINK?

 Is ADH secreted when the body is dehydrated or well hydrated?

How Tubular Fluid Becomes Urine

When the tubular fluid leaves the distal convoluted tubules, it must travel through a series of small **collecting tubules** that empty into **collecting ducts.** Collecting tubules and collecting ducts project through the renal medulla toward the renal papilla (see figures 27.5 and 27.7d). Both collecting tubules and collecting ducts are lined by a simple epithelium. The epithelial cells are cuboidal in the tubules, but very tall columnar cells in the ducts near the renal papilla.

The collecting ducts are the last structures that have the capacity to modify the tubular fluid further, and can do so under the influence of **antidiuretic hormone (ADH)** (see chapter 20). Antidiuretic hormone secretion by the posterior pituitary results in increased water absorption from the tubular fluid in the collecting ducts, thereby reducing water loss from the kidneys. ADH is secreted in response to either a rise in the concentration of ions in the blood or a fall in blood volume, as when the body is dehydrated and needs to conserve water. If an individual is well hydrated, the collecting ducts merely transport the tubular fluid and do not modify it. However, if an individual is dehydrated, water conservation must occur, and more-concentrated urine is produced. ADH may act on the collecting duct epithelium, making it more able to absorb water from the tubular fluid.

Once the tubular fluid leaves the collecting duct, it may be called urine. The remaining structures in the kidney simply transport the urine and cannot modify this fluid further. Several collecting ducts merge to empty into a papillary duct that opens at the edge of the renal papilla into minor calyx. Urine leaves the renal papilla and enters the minor calyces, which then transport the urine to major calyces, which in turn transport the urine to the renal pelvis. The renal pelvis conducts urine into the ureter.

Juxtaglomerular Apparatus

Associated with the nephron are some structures collectively referred to as the **juxtaglomerular** (jŭks'tă-glō-mer'ū-lăr; *juxta* = near) **apparatus** (see figure 27.6a,b). Components of the juxtaglomerular apparatus are juxtaglomerular cells and a macula densa. **Juxtaglomerular cells** are modified smooth muscle cells of the afferent arteriole located near the entrance to the renal corpuscle. The **macula** (mak'ū-lă; spot) **densa** (den'să; dense) is a group of modified epithelial cells in a distal convoluted tubule that touch the juxtaglomerular cells. These cells, which are located only on the tubule side next to the afferent arteriole, are narrower and taller than other distal convoluted tubule epithelial cells.

The structures of the juxtaglomerular apparatus work together to regulate blood pressure. The macula densa cells continuously monitor ion concentration in tubular fluid. Thus, if either blood volume or solute concentration is reduced, tubular fluid reflects this reduction, and the macula densa cells detect this change and stimulate the juxtaglomerular cells to release renin. Recall from chapter 20 that renin activates the renin-angiotensin pathway, resulting in aldosterone production, which causes increases in blood ion concentrations and blood volume. This arrangement is important in maintaining blood volume and blood pressure homeostasis in the body.

Innervation of the Kidney

Each kidney is innervated by a mass of sensory and autonomic nervous system fibers collectively called the **renal plexus.** The renal plexus accompanies each renal artery and enters the renal sinus of the kidney through the hilum. The renal plexus contains sympathetic innervation from the T10–T12 segments of the spinal cord and parasympathetic innervation from CN X (vagus nerve). The sympathetic innervation is responsible for renal blood vessel vasoconstriction that results in decreased glomerular blood flow and filtrate formation. Increased sympathetic stimulation may also stimulate juxtaglomerular cells to release renin. Pain from the kidneys is typically referred via the sympathetic pathway to the T10–T12 dermatomes. The parasympathetic innervation to the kidney has no known effect.

Renal Failure, Dialysis, and Kidney Transplants

Renal failure refers to greatly diminished or absent renal function caused by the destruction of about 90% of the tissue in the kidney. Renal failure often results from a chronic disease that affects the glomerulus or the small blood vessels of the kidney. Autoimmune conditions, high blood pressure, and diabetes are the three chronic problems accounting for the majority of progressive renal failure. Once the kidneys have been destroyed, there is no chance they will regenerate or begin functioning again. Thus, the two main treatments are dialysis or a kidney transplant.

The term **dialysis** (dī-al′i-sis; *dialyo* = to separate) comes from a Greek word meaning "to separate agents or particles on the basis of their size." Two forms of dialysis are commonly used today: peritoneal dialysis and hemodialysis. In **peritoneal dialysis,** a catheter is permanently placed in the peritoneal cavity, with a bag of dialysis fluid attached to the external end. As the fluid enters the peritoneal cavity and sloshes around, the harmful waste products in the blood are transferred, or *dialyzed,* into the fluid. After several hours, the patient connects a collection bag to the external end of the catheter, positions his or her body so the fluid will flow out of the peritoneal cavity, and opens the catheter valve; the dialysis fluid is then removed.

In **hemodialysis,** the patient's blood is cycled through a machine that filters the waste products across a specially designed membrane. To facilitate this type of dialysis, a vascular connection (an arteriovenous fistula) is made between a conveniently located superficial artery and vein. The patient must remain stationary for the time it takes to cycle the blood through the dialysis unit while the metabolic waste products are removed. Hemodialysis is needed several times a week.

A kidney transplant from a genetically similar person may successfully restore renal function. The replacement kidney is attached to an artery and vein in the inferior abdominopelvic region, where it is relatively easy to establish a vascular connection. The new kidney rests on the superior surface of or immediately lateral to the urinary bladder. Because a pelvic artery and vein connect the donor kidney to the patient's

blood supply, having the kidney near the bladder means only a short segment of ureter is needed for the bladder connection. There is no need to remove the damaged kidney. The transplanted kidney is a foreign tissue, and so the immune system will attack and try to reject this foreign tissue. Therefore, immunosuppressant drugs are administered to suppress the activity of the transplant recipient's immune system. It may be necessary to take these drugs for the remainder of the patient's life.

Donor kidneys may come through an organ procurement program or from a living person. Since most people have two kidneys, and normal renal function actually requires only one, some people have donated a kidney (while still alive) to a relative or even to a total stranger. If successfully matched to the genetic requirements of the recipient, a transplanted kidney assumes all the jobs of the patient's functionless kidneys.

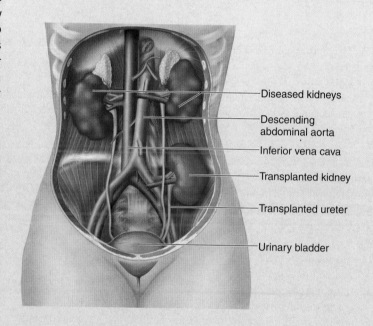

Location of a transplanted kidney in the abdominopelvic cavity.

Labels (top to bottom): Diseased kidneys; Descending abdominal aorta; Inferior vena cava; Transplanted kidney; Transplanted ureter; Urinary bladder

---

WHAT DID YOU LEARN?

7 What are the components of the nephron, and how do they modify the filtrate?

8 Where is ADH produced, and how specifically does it affect water concentration in the urine?

9 What is the structure and function of the juxtaglomerular apparatus?

Urinary Tract

Key topics in this section:

- Anatomy and location of the ureters, urinary bladder, and urethras
- Blood vessels and nerves that supply the organs of the urinary tract

The urinary tract consists of the ureters, urinary bladder, and urethra.

Ureters

The **ureters** (ū-rē′ter, ū′rē-ter) are long, fibromuscular tubes that conduct urine from the kidneys to the urinary bladder. Each tube averages 25 centimeters in length and is retroperitoneal. The ureters originate at the renal pelvis as it exits the hilum of the kidney, and then extend inferiorly to enter the posterolateral wall of the base of the urinary bladder. The wall of the ureter is composed of three concentric tunics. From innermost to outermost, these tunics are the mucosa, muscularis, and adventitia (**figure 27.8**). (Note that the ureter does not have a submucosa.)

The **mucosa** is formed from transitional epithelium, which is both distensible (stretchy) and impermeable to the passage of urine. External to the transitional epithelium of the mucosa is the

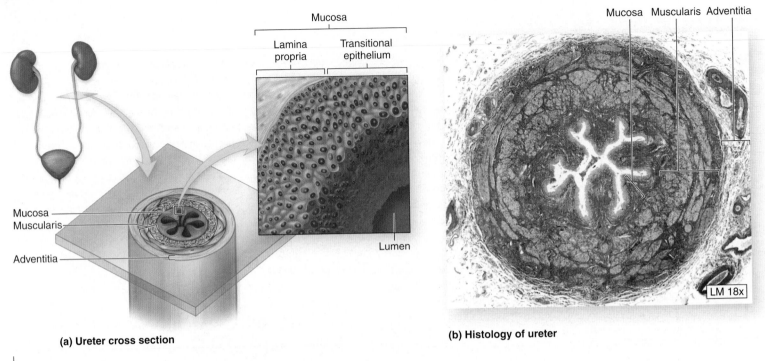

(a) Ureter cross section

(b) Histology of ureter

Figure 27.8

Ureters.   The ureters conduct urine from the kidneys to the urinary bladder for storage prior to voiding from the body. *(a)* Features of a ureter in cross-sectional view. *(b)* A photomicrograph of a ureter in cross section shows its mucosal folds and thick muscularis.

lamina propria, composed of a fairly thick layer of dense irregular connective tissue. The continuous production of urine ensures that the ureters are rarely completely empty, but as peristaltic waves propel urine through the ureter, the ureter may be temporarily empty at specific places along its length. At these locations, the mucosa folds to fill the lumen. Thereafter, when the ureter is distended, the mucosa can be stretched; this folding of the mucosa allows for considerable increase in the luminal diameter when needed.

The middle **muscularis** consists of two smooth muscle layers: an inner longitudinal layer and an outer circular layer. (Note that the arrangement of muscle layers is opposite that in the GI tract, where the circular layer is the inner layer.) The presence of urine within the renal pelvis causes these muscle layers to produce peristaltic waves that propel the urine through the ureters into the urinary bladder.

WHAT DO YOU THINK?

3 Why do the ureters use peristalsis to actively pump urine to the urinary bladder? Why don't they rely on gravity to move the urine to the inferiorly located bladder?

The external layer of the ureter wall is the **adventitia,** which is formed from areolar connective tissue. Some extensions of this areolar connective tissue layer also anchor the ureter to the posterior abdominal wall.

The ureters project through the posteroinferior bladder wall obliquely, and some smooth muscle fibers of the inner longitudinal layer of the muscularis insert into the lamina propria of the bladder. Because of the oblique course of the ureters through the bladder wall, the ureteral walls are compressed as the bladder distends, decreasing the likelihood of urine refluxing into the ureters from the bladder.

Multiple blood vessels supply blood to the ureters. In general, a segment of ureter receives its blood supply from a branch of the nearest artery. Thus, the superior aspects of the ureter are supplied by the renal arteries, while the more inferior aspects may receive their arterial supply from the aorta, common iliac, and/or internal iliac arteries. Venous drainage is through the companion veins.

The ureters are innervated by the autonomic nervous system. Parasympathetic fibers come from CN X (which supplies the superior region of the ureter) and from the pelvic splanchnic nerves (which supply the inferior region of the ureter). There are no known effects of this innervation. Sympathetic fibers come from the T11–L2 segments of the spinal cord. Pain from the ureter (e.g., due to a kidney stone lodged in the ureter) is referred to the T11–L2 dermatomes. These dermatomes are along a "loin-to-groin" region, so "loin-to-groin" pain typically means ureter and/or kidney discomfort.

WHAT DID YOU LEARN?

10 Describe the structure and function of the middle tunic of the ureter.

11 What nerves innervate the ureters?

Urinary Bladder

The **urinary bladder** is an expandable, muscular container that serves as a reservoir for urine **(figure 27.9)**. The bladder is positioned immediately posterior to the pubic symphysis. In females, the urinary bladder is anteroinferior to the uterus and directly anterior to the vagina; in males, the bladder is anterior to the rectum and superior to the prostate gland. The urinary bladder is a retroperitoneal organ, since only its superior surface is covered with peritoneum.

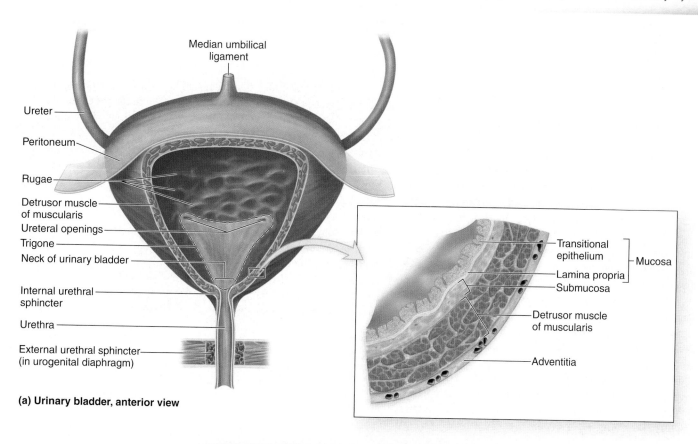

(a) Urinary bladder, anterior view

Median umbilical ligament

Ureter

Peritoneum

Rugae

Detrusor muscle of muscularis

Ureteral openings

Trigone

Neck of urinary bladder

Internal urethral sphincter

Urethra

External urethral sphincter (in urogenital diaphragm)

Transitional epithelium ⎤ Mucosa

Lamina propria ⎦

Submucosa

Detrusor muscle of muscularis

Adventitia

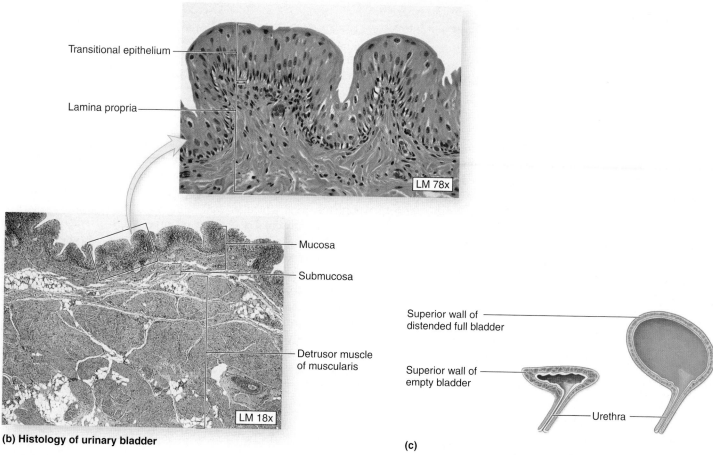

Transitional epithelium

Lamina propria

LM 78x

Mucosa

Submucosa

Detrusor muscle of muscularis

LM 18x

(b) Histology of urinary bladder

Superior wall of distended full bladder

Superior wall of empty bladder

Urethra

(c)

Figure 27.9

Urinary Bladder. *(a)* The urinary bladder is an expandable, muscular sac. This view depicts a female bladder. *(b)* Photomicrographs of the urinary bladder wall show its tunics. *(c)* Sagittal views show that the urinary bladder expands superiorly and becomes more oval in shape as it fills with urine.

CLINICAL VIEW

Renal Calculi

A **renal calculus** (kal'kū-lus, pl., *calculi*; pebbles), or *kidney stone*, is formed from crystalline minerals that build up in the kidney. Normally, urine contains chemicals that prevent these minerals from solidifying, but in some individuals these minerals condense to form hard, rocky structures. Renal calculi can be formed from a variety of different minerals. Over 75% of them contain calcium, in combination with either oxalate or phosphate. The second most common type are struvite (stroo'vīt) stones formed from ammonium and phosphate.

Causes and risk factors for kidney stone formation include inadequate fluid intake and dehydration, reduced urinary flow and volume, and certain abnormal chemical or mineral levels in the urine. **Hypercalcuria,** a high level of calcium in the urine, leads to calcium stones. Frequent urinary tract infections may predispose a person to struvite stones. Certain medical conditions as well as diet can also be correlated with renal calculi formation. In general, males tend to develop stones more often than females.

If the renal calculi are very small, they are asymptomatic, and the person excretes them without ever realizing it. However, a larger stone can become obstructed in the kidney, renal pelvis, or ureter. The term **urolithiasis** (ū-rō-li-thī'ă-sis; *lithos* = stone) refers to the presence of renal calculi anywhere along the urinary tract. Symptoms include severe, cramping pain along the "loin-to-groin" region and possibly nausea and vomiting. The epithelium of the ureter becomes inflamed as it tries to push the stone along its path, resulting in blood in the urine, called **hematuria** (hē-mă-too'rē-a).

Most stones smaller than about 4 millimeters in diameter eventually pass through the urinary tract on their own once the patient drinks plenty of water (2–3 quarts per day) to assist movement. But if the stone is too large (more than 8 millimeters in diameter) and doesn't pass on its own, medical intervention is required. The most common treatment is **lithotripsy** (lith'ō-trip-sē; *tresis* = boring), whereby ultrasound or shock waves are directed toward the stones to pulverize them into smaller particles that can be expelled in the urine. Alternatively, using **ureteroscopy,** a scope is inserted from the urethra into the urinary bladder and ureter to break up and remove the stone. If these treatments aren't viable, traditional surgery may be required.

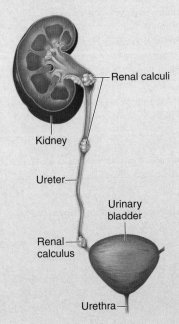

Renal calculi may become lodged at various sites along the urinary tract.

When empty, the urinary bladder exhibits an upside-down pyramidal shape. Filling with urine distends it superiorly until it assumes an oval shape (figure 27.9c). A fibrous, cordlike **median umbilical ligament** extends toward the umbilicus from its origin on the anterosuperior border of the urinary bladder. It is a remnant of the embryologic structure called the urachus, which extends from the superior part of the bladder to the umbilicus. Ureters enter the posterolateral wall of the urinary bladder through the oblique **ureteral openings.** The constricted **neck** of the bladder is located inferiorly and connected to the urethra.

A posteroinferior triangular area of the urinary bladder wall, called the **trigone** (trī'gōn; *trigonum* = triangle), is formed by imaginary lines connecting the two ureteral openings and the urethral opening. The trigone does not move and remains in place as the urinary bladder fills and evacuates. It functions as a funnel to direct the stored urine into the urethra as the bladder wall contracts. The trigone is embryologically different from the rest of the urinary bladder. While the urinary bladder forms from a structure called the cloaca, the trigone forms from the distal parts of the ureters, which become incorporated into the posterior wall of the urinary bladder.

The four tunics that form the wall of the bladder are the mucosa, submucosa, muscularis, and adventitia. The **mucosa** lines the bladder lumen; it is formed by a transitional epithelium that accommodates the shape changes occurring with distension, and by a highly vascularized lamina propria that supports the mucosa.

Additionally, mucosal folds, called **rugae,** allow for even greater distension. Within the trigone region, the mucosa is smooth, thick, and lacking rugae. The **submucosa** lies immediately external to the mucosa and is formed by dense irregular connective tissue that supports the urinary bladder wall.

The **muscularis** consists of three layers of smooth muscle, collectively called the **detrusor** (dē-troo'ser, -sōr; *detrudo* = to drive away) **muscle.** These smooth muscle bundles exhibit such complex orientations that it is difficult to delineate individual layers in random histologic sections. At the neck of the urinary bladder, an involuntary **internal urethral sphincter** is formed by the smooth muscle that encircles the urethral opening.

The **adventitia** is the outer layer of areolar connective tissue of the urinary bladder. A peritoneal membrane covers only the superior surface of the urinary bladder, and in this superior region, the peritoneum plus the connective tissue forms a serosa.

Arterial blood vessels extend to the urinary bladder and penetrate its wall from branches of the internal iliac artery. Venous blood drains into the internal iliac veins.

Micturition

The expulsion of urine from the bladder is called **micturition** (mik-choo-rish'ŭn; *micturio* = to desire to make water) or **urination** (ūr'i-nā'shŭn). Micturition is initiated by a complex sequence of events called the **micturition reflex.** The bladder is supplied by both parasympathetic and sympathetic nerve fibers of the autonomic nervous system. The parasympathetic fibers come from the micturition reflex center located in spinal cord segments S2–S4. The pelvic splanchnic nerves relax the internal urethral sphincter so that urine can pass through and stimulate contraction of the detrusor muscle. Thus, the parasympathetic fibers stimulate micturition. The sympathetic fibers are from the T11–L2 segments of the spinal cord. These fibers cause contraction of the internal urethral sphincter and inhibit contraction of the detrusor muscle. Thus, sympathetic fibers inhibit micturition.

The micturition reflex occurs in a series of steps:

1. When the bladder fills with urine and becomes distended, stretch receptors in the bladder wall are activated, and they signal the micturition reflex center.
2. Impulses within the parasympathetic division of the autonomic nervous system travel to both the internal urethral sphincter and the detrusor muscle.
3. The smooth muscle in the internal urethral sphincter relaxes, and the smooth muscle in the detrusor muscle contracts.
4. The person's conscious decision to urinate causes relaxation of the external urethral sphincter.
5. In addition to the squeezing action of the detrusor muscle on the volume of the urinary bladder, the expulsion of urine is facilitated by contraction of muscles in the abdominal wall and expiratory muscles.
6. Upon emptying of the urinary bladder, the detrusor muscle relaxes, and the neurons of the micturition reflex center are inactivated.

CLINICAL VIEW

Urinary Tract Infections

A **urinary tract infection (UTI)** occurs when bacteria (most commonly *E. coli*) or fungi enter and multiply within the urinary tract. Women are more prone to UTIs because they have a short urethra that is close to the anus, allowing bacteria from the GI tract to more readily enter the female urethra. Sexual intercourse also increases the risk of UTIs in both sexes. In addition, UTIs are associated with the medical use of a urinary catheter, a tube inserted through the urethra into the urinary bladder to help void urine. People who have had more than one UTI are at greater risk for developing others.

A UTI often develops first in the urethra, an inflammation called **urethritis** (ū-rē-thrī′tis). If the infection spreads to the urinary bladder, **cystitis** (sis-tī′tis) results. Occasionally, bacteria from an untreated UTI can spread up the ureters to the kidneys, a condition termed **pyelonephritis** (pī′ĕ-lō-ne-frī′tis).

Symptoms of a UTI include difficult and painful urination, called **dysuria** (dis-ū′rē-ă; *dys* = bad, difficult); the need to urinate frequently; and a feeling of uncomfortable pressure in the pubic region. Some people experience little pain or discomfort, but if the infection spreads to the kidneys, sharp back and flank pain, fever, and occasionally nausea and vomiting occur. A UTI can be diagnosed through **urinalysis** (ū-ri-nal′i-sis), a test of the urine that can reveal the presence of inflammatory cells, blood, and bacteria or fungi. Appropriate antibiotic therapy cures most UTIs.

Urethra

The **urethra** (ū-rē′thră) is a fibromuscular tube that originates at the neck of the urinary bladder and conducts urine to the exterior of the body (**figure 27.10**). The luminal lining of the urethra is a protective mucous membrane that houses clusters of mucin-producing cells called urethral glands. Bundles of smooth muscle fibers surround the mucosa and help propel urine to the outside of the body.

Two urethral sphincters restrict the release of urine until the pressure within the urinary bladder is high enough and voluntary activities needed to release the urine are activated. The **internal urethral sphincter** is the involuntary, superior sphincter surrounding the neck of the bladder, where the urethra originates. This sphincter is a circular thickening of the detrusor muscle and is controlled by the autonomic nervous system. The **external urethral sphincter** is inferior to the internal urethral sphincter and is formed by skeletal muscle fibers of the urogenital diaphragm. This sphincter is a voluntary sphincter controlled by the somatic nervous system. This is the muscle children learn to control when they become "toilet-trained."

The male and female urethras differ slightly in length and morphology.

Female Urethra

The female urethra has a single function: to transport urine to the exterior of the body. The lumen of the female urethra is primarily lined with a stratified squamous epithelium. The urethra is 3 to 5 centimeters long, and opens to the outside of the body at the external urethral orifice located in the female perineum.

Male Urethra

The male urethra has two functions—urinary and reproductive—because it serves as a passageway for both urine and semen. It is approximately 18 to 20 centimeters long and is partitioned into three segments: the prostatic urethra, the membranous urethra, and the spongy urethra.

The **prostatic** (pros-tat′ik) **urethra** is approximately 3 to 4 centimeters long and is the most dilatable portion of the urethra. It extends through the prostate gland, immediately inferior to the male bladder, where multiple small prostatic ducts enter it. The urethra in this region is lined by a transitional epithelium, with many blood vessels in the underlying dense irregular connective tissue. Two smooth muscle bundles surround the mucosa: an internal longitudinal bundle and an external circular bundle. The external circular muscular bundles are a continuation of the thickened circular region of smooth muscle forming the internal urethral sphincter at the bladder outlet.

The **membranous** (mem′bră-nus) **urethra** is the shortest and least dilatable portion of the male urethra. It extends from the inferior surface of the prostate gland through the urogenital diaphragm. As a result, it is surrounded by striated muscle fibers that form the external urethral sphincter of the urinary bladder. The epithelium in this region is often either stratified columnar or pseudostratified columnar.

The **spongy** (spŭn′jē) **urethra** is the longest part (15 centimeters) of the male urethra. It is encased within a cylinder of erectile tissue in the penis called the corpus spongiosum, and extends to the **external urethral orifice.** The proximal part of the spongy urethra is lined by a pseudostratified columnar epithelium, while the distal part has a lining of stratified squamous epithelium.

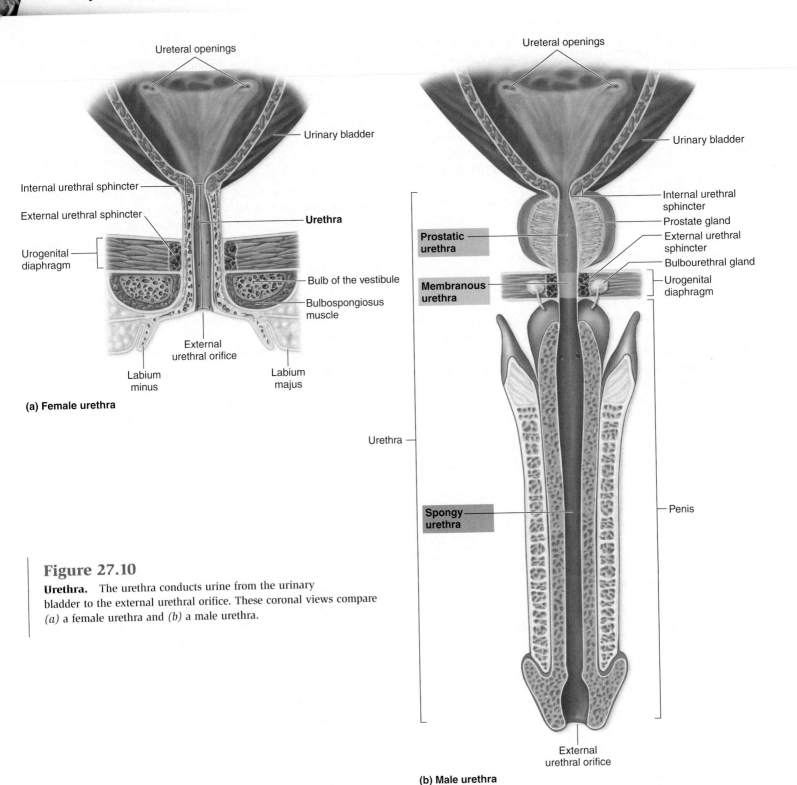

Ureteral openings

Urinary bladder

Internal urethral sphincter

External urethral sphincter

Urethra

Urogenital diaphragm

Bulb of the vestibule

Bulbospongiosus muscle

External urethral orifice

Labium minus

Labium majus

(a) Female urethra

Ureteral openings

Urinary bladder

Internal urethral sphincter

Prostate gland

Prostatic urethra

External urethral sphincter

Bulbourethral gland

Membranous urethra

Urogenital diaphragm

Urethra

Penis

Spongy urethra

External urethral orifice

(b) Male urethra

Figure 27.10

Urethra.   The urethra conducts urine from the urinary bladder to the external urethral orifice. These coronal views compare *(a)* a female urethra and *(b)* a male urethra.

WHAT DID YOU LEARN?

12 Where is the trigone located? How does it assist urinary bladder function?

13 What is the initiating stimulus that activates the micturition reflex?

14 Which portion of the male urethra is the shortest and least dilatable?

Aging and the Urinary System

Key topic in this section:

■ Changes in the urinary system due to aging

Changes in the size and functioning of the kidneys normally begin in our third decade of life and continue thereafter. With age comes a reduced blood flow to the kidneys and a decrease in the number of functional nephrons. The reduction in blood flow contributes to a decreased glomerular filtration rate; consequently, both reabsorption and secretion are reduced. The loss of nephrons results in a diminished ability to filter and cleanse the blood as well as a decrease in the number of target cells within the kidneys that are capable of responding to stimulation by aldosterone or antidiuretic hormone. Thus, the ability to control blood volume and blood pressure is reduced.

Structural changes in the urinary bladder also affect the storage and voiding of urine. The bladder decreases in size, and smooth muscle tone in the bladder wall gradually diminishes, which may lead to more frequent urination. Sometimes, there is a delay in noticing the urge to urinate. Additionally, control of the urethral sphincters—and eventually even control of micturition—may be lost.

The inability to control the expulsion of urine is called **incontinence.** It most commonly occurs among the elderly, and most frequently in women. The extent of the incontinence ranges from occasional urine leakage to complete inability to retain urine. There are two categories of incontinence: (1) Stress incontinence often occurs during vigorous exercise or strenuous coughing, and (2) urge incontinence results from immediate bladder contraction after feeling a strong need to urinate. Causes of incontinence include severe weight gain, complications following pelvic surgery, coincidence with uncontrolled diabetes, severe constipation, pelvic prolapse (in females) as a result of pregnancy, and enlarged prostate (in males).

WHAT DID YOU LEARN?

15 What are the possible consequences of aging on urine production?

Development of the Urinary System

Key topic in this section:

■ Embryonic and fetal development of the urinary system

Most of the urinary system and the reproductive system (discussed in chapter 28) form from the **intermediate mesoderm** of the embryo. Intermediate mesoderm is located between the somites (paraxial mesoderm) and the lateral plate mesoderm. Thus, the urinary system and the reproductive system are linked in both their development and their postnatal functions.

Kidney and Ureter Development

Some of the intermediate mesoderm forms bilateral longitudinal ridges, each called a **urogenital ridge.** The urogenital ridges are located on either side of the vertebral column and give rise to three sets of excretory organs, which become increasingly more advanced. The first excretory organs, each called a **pronephros** (prō-nef′ros; before kidney; pl., *pronephroi*), appear in the cervical region of the embryo and seem to be vestigial—that is, they have no known function in humans. They appear in the early fourth week and quickly degenerate by the end of that same week.

The second set of organs, each called a **mesonephros** (pl., *mesonephroi*), appears just before the pronephroi degenerate. These structures are formed from tissue comprising the urogenital ridge in the thoracic and lumbar regions. Each mesonephros is composed of multiple saclike segments. A **mesonephric** (mez-ō-nef′rik) **duct** drains urine from each mesonephros to the developing urinary

bladder. The mesonephros (or "intermediate" kidney) persists until about week 10 of development.

Early in the fifth week, the final set of excretory organs, each called a **metanephros** (or *permanent kidney*), forms. Each metanephros (pl., *metanephroi*) forms a functional adult kidney and takes over urine production by week 10 **(figure 27.11)**. The components of the metanephros form from two specific structures: the ureteric bud and the metanephric mesoderm **(table 27.3)**. The **ureteric** (ū-rē-ter′ik) **bud** forms as a separate outgrowth from the caudal end of the mesonephric duct and gives rise to the ureter, renal pelvis, calyces, and collecting ducts of the kidney. The **metanephric** (met-ă-nef′rik) **mesoderm** develops from intermediate mesoderm in the sacral region of the embryo, and forms the nephron components of the kidney: the glomerular capsule, the proximal convoluted tubule, the nephron loop, and the distal convoluted tubule. The metanephric mesoderm will not grow and develop unless the ureteric bud grows into and merges with it.

As the ureteric bud grows and merges with the metanephric mesoderm, the ureteric bud undergoes a branching pattern (figure 27.11b). The first set of branches forms the major calyces in week 6, and the second set of branches forms the minor calyces in week 7. This branching pattern continues until the tiny collecting ducts are formed in the kidney, so that by week 32, well over 1 million collecting tubules have formed. As the calyces and collecting ducts form, they signal the metanephric mesoderm to grow and develop into the nephron components. By week 10 of development, the metanephros is able to both produce and expel urine. Fetal urine supplements amniotic fluid production.

During weeks 6–9, the developing kidneys migrate from their position in the pelvic cavity to a more superior position in the lumbar portion of the abdominal cavity (figure 27.11c). The mechanisms for this ascent are not known, but may be related to differential growth of the embryo. As the kidneys ascend, they obtain temporary blood vessels from the nearby vasculature. When the kidneys migrate further, the older temporary blood vessels regress and degenerate, and new blood vessels for the kidney form. Eventually, by week 9, the kidneys complete migration to the lumbar region and acquire their permanent renal arteries branching from the descending abdominal aorta.

Urinary Bladder and Urethra Development

The urinary bladder and urethra develop from the distal part of the hindgut called the **cloaca** (klō-ā′kă; sewer). The cloaca forms not only the epithelium of the urinary bladder and urethra, but also the rectum and anal canal. The cloaca is separated from the outside of the body by a thin **cloacal membrane.** Extending superiorly from

Table 27.3	**Embryonic Derivations of Ureter and Kidney Structures**
Ureteric Bud Origin	**Metanephric Mesoderm Origin**
Ureter	Renal corpuscles
Renal pelvis	Proximal convoluted tubules
Major calyces	Nephron loops
Minor calyces	Distal convoluted tubules
Collecting tubules and ducts	

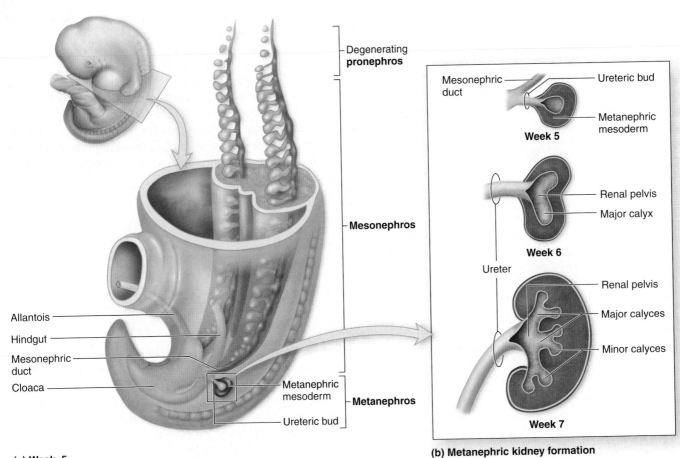

Allantois
Hindgut
Mesonephric duct
Cloaca

Degenerating **pronephros**

Mesonephros

Metanephric mesoderm
Ureteric bud

Metanephros

(a) Week 5

Mesonephric duct
Ureteric bud
Metanephric mesoderm
Week 5

Renal pelvis
Major calyx
Week 6

Ureter

Renal pelvis
Major calyces
Minor calyces
Week 7

(b) Metanephric kidney formation

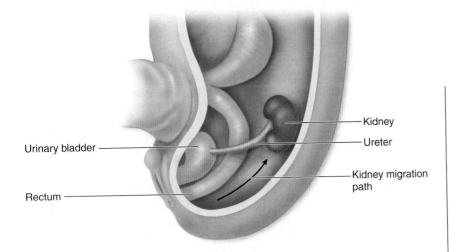

Urinary bladder

Rectum

Kidney
Ureter
Kidney migration path

(c) Weeks 6–9: Kidney migrates from pelvis to lumbar region

Figure 27.11

Kidney Development. The kidneys form from the urogenital ridges of the embryo. (*a*) By week 5, the pronephroi have regressed, and the mesonephroi begin producing urine. The metanephric kidneys, which eventually form the permanent kidneys, also begin to form. (*b*) The development of the metanephric kidney is shown between weeks 5 and 7. (*c*) Between weeks 6 and 9, the kidney migrates superiorly from its initial location in the pelvic cavity to the lumbar region on the posterior abdominal wall.

Kidney Variations and Anomalies

Because the kidneys develop from multiple embryologic structures and must migrate from the pelvic cavity to the lumbar region of the abdominal cavity, variations in both kidney development and placement are common.

Renal agenesis is the failure of a kidney to develop. Kidney development ceases if the ureteric bud and the metanephric mesoderm do not grow toward each other and meet. Failure of one kidney to develop, called *unilateral renal agenesis,* occurs in about 1 per 1000 births, while *bilateral renal agenesis* occurs in about 1 per 3000 births. Unilateral renal agenesis is often asymptomatic, whereas bilateral renal agenesis is invariably fatal.

A **pelvic kidney** can be present if the developing kidney fails to migrate from the pelvic cavity to the abdominal cavity. The pelvic kidney receives its blood supply from branches of the common iliac artery, as opposed to a renal artery that branches from the aorta. A pelvic kidney usually has normal function and causes no problems for the individual.

A **horseshoe kidney** develops when the inferior parts of the left and right kidneys fuse as they try to ascend from the pelvic cavity into the abdominal cavity. The single, large kidney looks like a horseshoe. The kidney's superior migration is usually halted as it gets stuck around the origin of the inferior mesenteric artery. Horseshoe kidneys are fairly common, occurring in about 1 per 600 births. Like a pelvic kidney, a horseshoe kidney typically is asymptomatic and functions normally.

Supernumerary (soo-per-noo′mer-ār-ē) **kidneys** are extra kidneys that develop. They appear to be caused by ureteric bud duplication. However, supernumerary kidneys are very rare. It is more common to see a **duplicated** or **bifid ureter,** from a duplicated ureteric bud, traveling to a single kidney. Typically, these have no clinical significance.

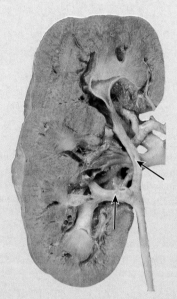

Bifid ureter (arrows).

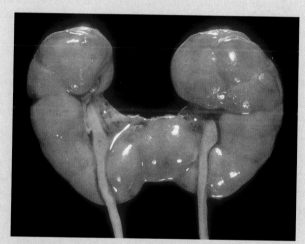

Horseshoe kidney is a fairly common variation.

Finally, because the kidneys must migrate from the pelvic cavity into the abdominal cavity, it is fairly common for **multiple renal vessels** to supply the kidney. Normally, as the kidney migrates, it acquires temporary vessels from the nearby blood supply, and its older, inferior vessels regress and degenerate. If these temporary vessels fail to degenerate, the kidney is left with multiple vessels.

the cloaca to the umbilicus is a thin, tubular hindgut extension called the **allantois** (ă-lan′tō-is). (By week 6, the allantois becomes a fibrous cord called the **urachus,** and in adults the remnant of this cord is called the median umbilical ligament.)

Between weeks 4 and 7, a mass of mesodermal tissue called the **urorectal** (u′rō-rek′tăl) **septum** grows through the cloaca and toward the cloacal membrane **(figure 27.12)**. This septum subdivides the cloaca into an anterior **urogenital** (ū′rō-jen′i-tăl) **sinus** and a posterior **anorectal** (ā′nō-rek′tăl) **canal.** The urogenital sinus (formed from endoderm) and some surrounding mesoderm develop into the urinary bladder and urethra. The urorectal septum grows toward and attaches to the cloacal membrane, subdividing it into an anterior **urogenital membrane** and a posterior **anal membrane.** Both of these membranes rupture by week 8, allowing the urethra and the anal canal to each communicate with the outside of the body.

WHAT DID YOU LEARN?

16   What structures are formed by the cloaca?

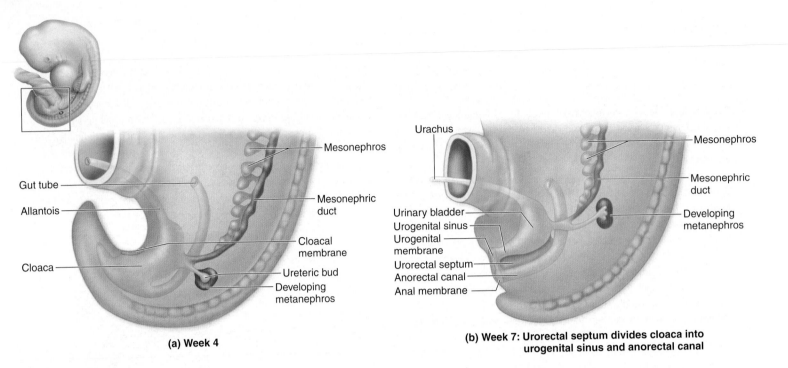

(a) Week 4

(b) Week 7: Urorectal septum divides cloaca into urogenital sinus and anorectal canal

Figure 27.12

Bladder Development. *(a)* A common cloaca unites the distal part of the gut tube with the future urinary bladder. The cloaca is separated from the outside of the body by a thin cloacal membrane. *(b)* A tissue mass called the urorectal septum grows inferiorly and partitions the cloaca into an anteriorly located urogenital sinus (which forms the urinary bladder and urethra) and an anorectal canal. The cloacal membrane is also subdivided into an anteriorly located urogenital membrane and a posteriorly located anal membrane.

CLINICAL TERMS

diuretic (dī-ū-ret′ik; *dia* = throughout) Any chemical or substance that increases urine output. Examples are alcohol, caffeinated drinks, and diuretic drugs prescribed for high blood pressure.

nocturia (nok-too′rē-ă; *nox* = night, *ouron* = urine) Excessive urinating at night.

polyuria (pol-ē-ū′rē-ă; *polys* = much) Excessive secretion of urine resulting in frequent, copious urination.

CHAPTER SUMMARY

General Structure and Functions of the Urinary System 814	▪ The urinary system is composed of the kidneys and the urinary tract (the ureters, the urinary bladder, and the urethra).
	▪ The urinary system filters the blood, transports and stores urine, excretes urine, helps maintain blood volume, and regulates erythrocyte production.
Kidneys 816	**Gross and Sectional Anatomy of the Kidney 816**
	▪ Usually the right kidney is slightly inferior to the left kidney because of the space occupied by the liver.
	▪ The medial concave border has an opening called the renal hilum through which blood vessels, lymph vessels, and nerves pass into a space within the kidney called the renal sinus.
	▪ The kidneys are surrounded by a strong fibrous capsule, perinephric fat, renal fascia, and paranephric fat.
	▪ A longitudinal section of the kidney shows the renal cortex (peripheral) and the renal medulla (central). The renal medulla is subdivided into renal pyramids, each with an apex called a renal papilla that projects into the funnel-shaped minor calyx.
	▪ Urine deposited within the minor calyx flows into the major calyx and then into the renal pelvis.

Kidneys (continued)　816

Blood Supply to the Kidney　817

- Blood is transported to the kidneys in renal arteries, which divide into segmental arteries that form interlobar arteries, which give rise to arcuate arteries. Arcuate arteries branch into interlobular arteries, which supply the glomerulus via the afferent arteriole.
- Blood leaving the glomerulus follows this path: efferent arteriole, peritubular capillaries or vasa recta, interlobular veins, arcuate veins, interlobar veins, and renal vein.

Nephrons　820

- The nephron is the structural and functional unit of the kidney. Nephrons are composed of a renal corpuscle and a renal tubule.
- The renal corpuscle is the site of blood filtration where water and some dissolved solutes move from the blood plasma into the capsular space of the renal corpuscle.
- The proximal convoluted tubule and nephron loop reabsorb almost all plasma proteins, nutrients, and ions from the tubular fluid.
- The distal convoluted tubules originate from the nephron loop.

How Tubular Fluid Becomes Urine　824

- Distal convoluted tubules empty into small collecting tubules that empty into collecting ducts, which merge to empty into the papillary duct that opens at the edge of the renal papilla and releases urine into the minor calyx.

Juxtaglomerular Apparatus　824

- The juxtaglomerular apparatus has juxtaglomerular cells (smooth muscle cells of the afferent arteriole) and a macula densa (distal convoluted tubule epithelial cells adjacent to the afferent arteriole). It releases renin to regulate blood pressure.

Innervation of the Kidney　824

- The renal plexus innervates each kidney with sympathetic and parasympathetic fibers.

Urinary Tract　825

- The urinary tract consists of paired ureters, a urinary bladder, and a single urethra.

Ureters　825

- The ureters are retroperitoneal, fibromuscular tubes that conduct urine from the kidneys to the urinary bladder.
- The ureteral mucosa has a transitional epithelium; the muscularis consists of two layers of smooth muscle that contract to squeeze urine to the urinary bladder.
- The ureters project through the bladder wall obliquely, causing the ureteral walls to compress as the bladder distends.

Urinary Bladder　826

- The urinary bladder stores urine; it has a trigone formed by imaginary lines connecting the posterior ureteral openings and anterior urethral opening. It does not change shape as the bladder fills and empties.
- The mucosa is formed by a transitional epithelium; mucosal folds called rugae support bladder distension.
- The submucosa is formed by dense irregular connective tissue, and the muscularis is composed of three layers of smooth muscle bundles (the detrusor muscle).
- The adventitia is the external layer of the urinary bladder.
- Parasympathetic stimulation promotes micturition, and sympathetic stimulation inhibits it.

Urethra　829

- The urethra conducts urine from the bladder to the exterior of the body.
- An internal urethral sphincter (involuntary) and an external urethral sphincter (voluntary) control the voiding of urine.
- The female urethra is relatively short, and has a single function: to transport urine.
- The male urethra serves as a passageway for both urine and semen. It is about 18 to 20 centimeters long and is partitioned into the prostatic urethra, the membranous urethra, and the spongy urethra.

Aging and the Urinary System　830

- Structural changes in the kidneys and urinary bladder affect the production, storage, and voiding of urine.
- Incontinence is the inability to control urine expulsion.

Development of the Urinary System　831

- Most of the early urinary system and the reproductive system form from the intermediate mesoderm.

Kidney and Ureter Development　831

- The intermediate mesoderm condenses along the posterior body wall and forms a urogenital ridge on each side of the vertebral column.
- Three successive sets of excretory organs form: the pronephros, the mesonephros, and the metanephros (or permanent kidney).

Urinary Bladder and Urethra Development　831

- The urinary bladder and urethra develop from the distal part of the hindgut called the cloaca.

CHALLENGE YOURSELF

Matching

Match each numbered item with the most closely related lettered item.

_____ 1. distal convoluted tubule
_____ 2. urethra
_____ 3. ureter
_____ 4. peritubular capillaries
_____ 5. glomerulus
_____ 6. efferent arteriole
_____ 7. urinary bladder
_____ 8. renal pyramid
_____ 9. cortex
_____ 10. renal pelvis

a. location of renal corpuscle
b. expels urine outside the body
c. major calyces empty into this funnel-shaped region
d. most secretion occurs in this nephron segment
e. stores urine until it is voided
f. structural units that constitute the medulla
g. conducts blood out of the glomerulus
h. vessels involved in reabsorption
i. site of plasma filtration
j. conducts urine from kidney to bladder

Multiple Choice

Select the best answer from the four choices provided.

_____ 1. Which organ is responsible for filtering the blood?
a. ureter
b. urinary bladder
c. kidney
d. urethra

_____ 2. Which statement is _true_ about the urinary bladder?
a. The bladder neck is surrounded by the external urethral sphincter.
b. The detrusor muscle contains only two layers of smooth muscle.
c. The bladder is lined with transitional epithelium.
d. The bladder receives urine from the kidneys via the two urethras.

_____ 3. Tubular fluid from the proximal convoluted tubule next travels to the
a. capsular space.
b. collecting duct.
c. distal convoluted tubule.
d. nephron loop.

_____ 4. The apex of a renal pyramid is called the renal
a. calyx.
b. papilla.
c. column.
d. capsule.

_____ 5. The arteries located at the corticomedullary junction of the kidney are the
a. arcuate arteries.
b. segmental arteries.
c. interlobar arteries.
d. renal arteries.

_____ 6. Which statement is _false_ about the kidneys?
a. The right kidney is positioned more inferiorly than the left kidney.
b. The cortex is subdivided into renal pyramids.
c. The renal artery, renal vein, and ureter connect to the kidney at its hilum.
d. The kidney is covered by a fibrous capsule.

_____ 7. Urine in a major calyx of the kidney next travels to the
a. ureter.
b. minor calyx.
c. urinary bladder.
d. renal pelvis.

_____ 8. Which structure is not controlled by the autonomic nervous system?
a. muscularis of the ureter
b. external urethral sphincter
c. internal urethral sphincter
d. detrusor muscle of the urinary bladder

_____ 9. Reabsorption is the movement of fluid and solutes from the
a. filtrate into the glomerular capillaries.
b. tubular fluid into the capsular space.
c. tubular fluid into the peritubular capillaries.
d. blood vessels into the collecting ducts.

_____ 10. The micturition reflex controls
a. urine formation.
b. voiding of the filled bladder.
c. reabsorption of glucose from filtrate.
d. filling of the urinary bladder.

Content Review

1. What are the basic functions of the urinary system?

2. Describe the connective tissue coverings that surround the kidney, from internal to external. Why are these coverings especially important to kidney structure and function?

3. Map the flow of blood into and out of the kidney. List which structures carry oxygenated blood and which carry deoxygenated blood. In addition, list the structures responsible for gas exchange and reabsorption of materials from the filtrate.

4. Describe the anatomic structure of the glomerulus and the visceral layer of the glomerular capsule.

5. Why are microvilli prominent on the apical surface of the proximal convoluted tubule epithelium but not in the distal convoluted tubule?

6. What do the cells of the juxtaglomerular apparatus secrete? What function does this product perform?

7. What prevents urine stored in the urinary bladder from being forced back through the ureters to the kidney?

8. Describe the innervation of the ureters and urinary bladder.

9. Trace the course of fluid movement, beginning with the production of filtrate in the renal corpuscle and ending with the expulsion of urine from the urethra.

10. What is the cause of a urinary tract infection? Why are these infections more common in women?

Developing Critical Reasoning

1. While drinking many beers one night, Jason noticed that he had to urinate more frequently. The following morning, Jason's mouth felt dry, and he had a headache. A friend told Jason that his symptoms were the result of dehydration. Based on your knowledge of the urinary system, how and why did Jason become dehydrated? What hormone normally regulates the amount of water in the urine, and how did the alcohol interfere with this hormone's function?

2. Males who suffer from either benign prostatic hypertrophy (noncancerous prostate gland enlargement) or prostate cancer often have problems with urination. Based on your knowledge of the male urethra, hypothesize why these urination problems occur.

A N S W E R S   T O   " W H A T   D O   Y O U   T H I N K ? "

1. Without functioning kidneys, the blood would not be able to be filtered, so waste products would accumulate. This accumulation of toxic material in the blood leads to death unless the materials are filtered out.

2. ADH is secreted when the body is dehydrated, so the body can conserve what remaining water it has.

3. When we are lying down, gravity is unable to passively transport urine to the urinary bladder. Thus, peristalsis is also needed so that urine can be actively pumped from the ureters to the urinary bladder no matter what position the body is in.

Visit the McKinley/O'Loughlin *Human Anatomy*, 2e website at aris.mhhe.com

28

REPRODUCTIVE SYSTEM

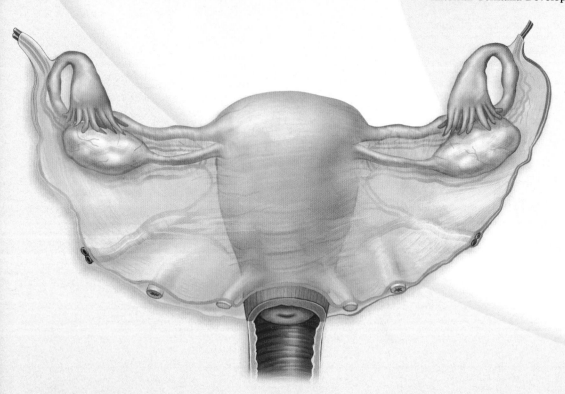

Reproductive System

The female and male reproductive systems provide the means for the sexual maturation of each individual and produce the special cells necessary to propagate the next generation. In this chapter, we first discuss the general similarities between the two reproductive systems and then focus on the specific structures and functions of each system.

Comparison of the Female and Male Reproductive Systems

Key topics in this section:

- Similarities between the female and male reproductive systems
- The events of puberty in females and males
- Components of the perineum in females and males

Besides their obvious differences, the female and male reproductive systems share several general characteristics. For example, some mature reproductive system structures are derived from common developmental structures (primordia) and serve a common function in adults. Such structures are called **homologues** (hōm′ō-log; *homo* = same or alike, *logos* = relation) **(table 28.1)**. The structures listed in this table are described in detail later in this chapter.

Both reproductive systems have primary sex organs called **gonads** (gō′nad; *gone* = seed)—ovaries in females and testes in males. The gonads produce sex cells called **gametes** (gam′ēt; husband or wife), which unite to form a new individual. Female gametes are called oocytes, whereas male gametes are called sperm. In addition, the gonads produce large amounts of **sex hormones** (estrogen and progesterone in the female and androgens in the male), which affect maturation, development, and changes in the activity of the reproductive system organs.

Both reproductive systems have accessory reproductive organs, including ducts to carry gametes away from the gonads toward the site of fertilization (in females) or simply to the outside of the body (in males). Fertilization occurs when female and male gametes fuse. The sexual union between a female and a male is known as **copulation** (kop-ū-lā′shŭn; *copulatio* = a joining), **coitus** (kō′i-tŭs; to come together), or **sexual intercourse.** If fertilization occurs, then the support, protection, and nourishment of the developing human occurs within the female reproductive tract.

Both the female and male reproductive systems are primarily nonfunctional and "dormant" until a time in adolescence known as puberty. At **puberty** (pū′ber-tē; *puber* = grown up), external sex characteristics become more prominent, such as breast enlargement in females and pubic hair in both sexes, and

Table 28.1	Reproductive System Homologues	
Female Organ	**Male Organ Homologue**	**Common Function**
Ovaries	Testes	Produce gametes and sex hormones
Clitoris	Glans of penis	Contain autonomic nervous system axons that stimulate feelings of arousal and sexual climax
Labia majora	Scrotum	Protect and cover some reproductive structures
Vestibular glands	Bulbourethral glands	Secrete mucin for lubrication

the reproductive organs become fully functional. Also, the gametes begin to mature, and the gonads start to secrete their sex hormones. Puberty is initiated when the hypothalamus begins secreting **GnRH (gonadotropin-releasing hormone)** (see chapter 20). GnRH acts on specific cells in the anterior pituitary and stimulates them to release **FSH (follicle-stimulating hormone)** and **LH (luteinizing hormone).** (Prior to puberty, FSH and LH are virtually nonexistent in boys and girls.) As levels of FSH and LH increase, the gonads produce significant levels of sex hormones and start the processes of gamete maturation and sexual maturation.

> ## Study Tip!
>
> A simplified flowchart of the endocrine pathway in puberty is as follows:
>
> **GnRH** (from hypothalamus) → **FSH** and **LH** (from anterior pituitary) → Sex hormone release and gamete maturation (in the gonads).

Both reproductive systems produce gametes. However, the female reproductive tract typically produces and releases a single gamete (oocyte) monthly, while the male reproductive tract produces large numbers (100,000,000 per day) of gametes (sperm) daily. These male gametes are stored within the male reproductive tract for a short time, and if they are not expelled from the body within that period, they are resorbed.

Perineum

In both females and males, the **perineum** (per′i-nē′ŭm) is a diamond-shaped area between the thighs that is circumscribed anteriorly by the pubic symphysis, laterally by the ischial tuberosities, and

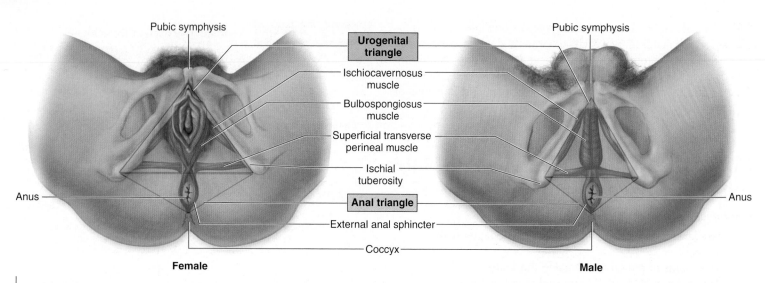

Figure 28.1

Perineum. In both females and males, the perineum is the diamond-shaped area between the thighs extending from the pubis anteriorly to the coccyx posteriorly, and bordered laterally by the ischial tuberosities. An imaginary horizontal line extending from the ischial tuberosities subdivides the perineum into a urogenital triangle anteriorly and an anal triangle posteriorly.

posteriorly by the coccyx (**figure 28.1**). Two distinct triangle bases are formed by an imaginary horizontal line extending between the ischial tuberosities of the ossa coxae. Both triangles house specific structures in the floor of the trunk:

- The anterior triangle, called the **urogenital triangle,** contains the clitoris and the urethral and vaginal orifices in females and the base of the penis and the scrotum in males. Within the urogenital triangle are the muscles that surround the external genitalia, called the ischiocavernosus, bulbospongiosus, and superficial transverse perineal muscles.
- The posterior triangle, called the **anal triangle,** is the location of the anus in both sexes. Surrounding the anus is the external anal sphincter.

Review table 11.12 and figure 11.15 as well when learning these structures.

WHAT DID YOU LEARN?

1 What is puberty?

2 Compare the structures in the female and male urogenital triangles.

Anatomy of the Female Reproductive System

Key topics in this section:

- Gross and microscopic anatomy of the ovaries
- Follicle development, the ovarian cycle, and the process of ovulation
- Anatomy of the uterine tubes and their function
- Regions of the uterus and the uterine cycle
- Anatomy of the vagina and the external genitalia
- Gross and microscopic anatomy of the mammary glands

A sagittal section through the female pelvis illustrates the internal reproductive structures and their relationships to the urinary bladder and rectum (**figure 28.2**). As the peritoneum folds around the various pelvic organs, it produces two major dead-end recesses, or pouches. The anterior **vesicouterine** (ves′i-kō-ū′ter-in; *vesica* = bladder, *utero* = uterus) **pouch** forms the space between the uterus and the urinary bladder, and the posterior **rectouterine** (rek-tō-ū′ter-in) **pouch** forms the space between the uterus anteriorly and the rectum posteriorly.

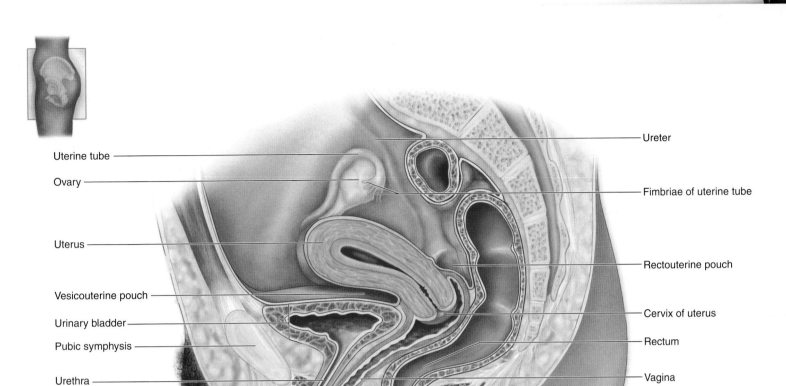

Figure 28.2

Sagittal Section of the Female Pelvic Region.   A sagittal section of the female pelvis illustrates the position of the uterus with respect to the rectum and urinary bladder.

The primary sex organs of the female are the ovaries. The accessory sex organs include the uterine tubes, uterus, vagina, clitoris, and mammary glands.

Ovaries

The **ovaries** are paired, oval organs located within the pelvic cavity lateral to the uterus **(figure 28.3)**. In an adult, the ovaries are slightly larger than an almond—about 2 to 3 cm (centimeters) long, 2 cm wide, and 1 to 1.5 cm thick. Their size usually varies during each menstrual cycle as well as during pregnancy.

The ovaries are anchored within the pelvic cavity by specific cords and sheets of connective tissue. A double fold of peritoneum, called the **mesovarium** (mez′ō-vā′rē-ŭm; *mesos* = middle, *ovarium* = ovary), attaches to each ovary at its **hilum,** which is the

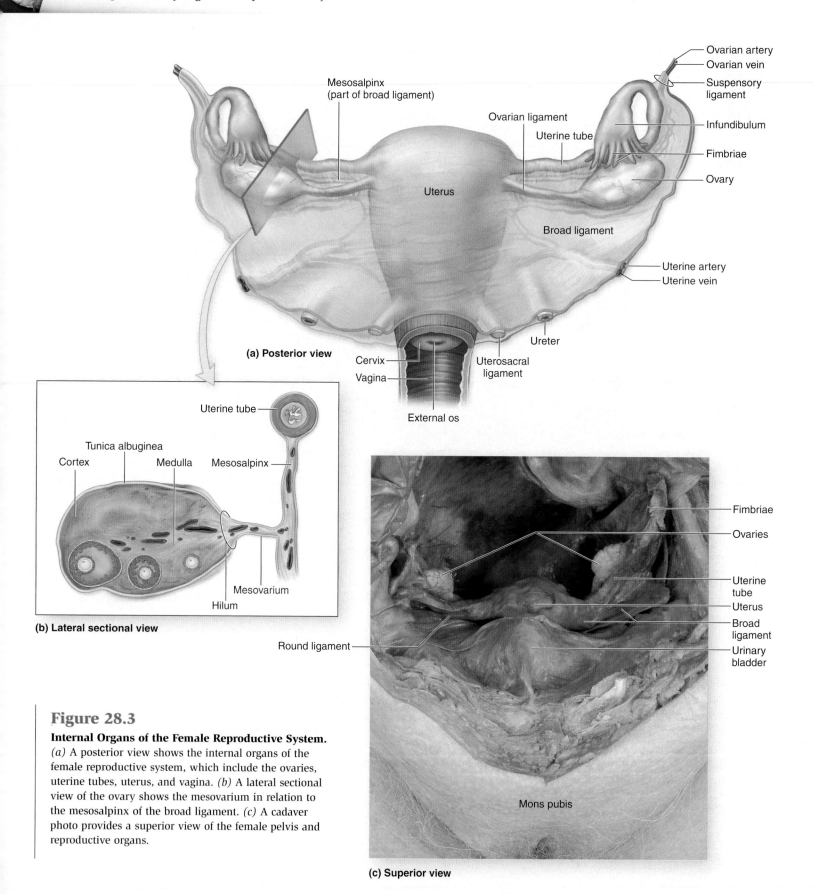

(a) Posterior view

(b) Lateral sectional view

(c) Superior view

Figure 28.3

Internal Organs of the Female Reproductive System.
(a) A posterior view shows the internal organs of the
female reproductive system, which include the ovaries,
uterine tubes, uterus, and vagina. *(b)* A lateral sectional
view of the ovary shows the mesovarium in relation to
the mesosalpinx of the broad ligament. *(c)* A cadaver
photo provides a superior view of the female pelvis and
reproductive organs.

anterior surface of the ovary where its blood vessels and nerves enter. The mesovarium secures each ovary to a **broad ligament,** which is a drape of peritoneum that hangs over the uterus. Each ovary is anchored to the lateral aspect of the uterus by an **ovarian ligament,** which is the superior portion of the round ligament of the uterus. Finally, a **suspensory ligament** attaches to the lateral edge of each ovary and projects superolaterally to the pelvic wall. The ovarian blood vessels and nerves are housed within each suspensory ligament, and they join the ovary at its hilum. Smooth muscle fibers within both the mesovarium and the suspensory ligament contract at the time of ovulation to bring the ovaries into close proximity with the uterine tube openings.

Each ovary is supplied by an **ovarian artery** and an **ovarian vein.** The ovarian arteries are direct branches off the aorta, immediately inferior to the renal vessels. The ovarian veins exit the ovary and drain into either the inferior vena cava or one of the renal veins. Traveling with the ovarian artery and vein are autonomic nerves. Sympathetic axons come from the T10 segments of the spinal cord, whereas parasympathetic axons come from CN X (vagus nerve).

When an ovary is sectioned and viewed microscopically, many features are visible **(figure 28.4)**. Surrounding the ovary is a thin, simple cuboidal epithelial layer called the **germinal epithelium,** so named because early anatomists erroneously thought it was the origin of the female germ (sex) cells. Deep to the germinal epithelium is a connective tissue capsule called the **tunica albuginea** (al-bū-jin′ē-ă; *albugo* = white spot), which is homologous to the tunica albuginea of the testis. Deep to the tunica albuginea, the ovary can be partitioned into an outer **cortex** and an inner **medulla.** The cortex contains ovarian follicles (described next), while the medulla is composed of areolar connective tissue and contains branches of the ovarian blood vessels, lymph vessels, and nerves.

Ovarian Follicles

Within the cortex are thousands of **ovarian follicles.** Ovarian follicles consist of an **oocyte** surrounded by **follicle cells** (or *granulosa cells*), which support the oocyte. There are several different types of ovarian follicles, each representing a different stage of development (figure 28.4):

1. A **primordial follicle** is the most primitive type of ovarian follicle. Each primordial follicle consists of a **primary oocyte** surrounded by a single layer of *flattened* follicle cells. A primary oocyte is an oocyte that is arrested in the first meiotic prophase. About 1.5 million of these types of follicles are present in the ovaries at birth.

2. A **primary follicle** forms from a maturing primordial follicle. Each primary follicle consists of a primary oocyte surrounded by one or more layers of *cuboidal* follicular cells. Each primary follicle secretes estrogen as it continues to mature. The estrogen stimulates changes in the uterine lining.

3. A **secondary follicle** forms from a primary follicle. Each secondary follicle contains a primary oocyte, many layers of follicular cells, and a fluid-filled space called an **antrum**. Within the antrum is a serous fluid that increases in volume as ovulation nears. Surrounding the primary oocyte are two protective structures, the zona pellucida and the corona radiata. The **zona** (zō′nă; zone) **pellucida** (pe-lū′sĭd; *pellucidus* = allowing the passage of light) is a translucent structure that contains glycoproteins. External to the zona pellucida is the **corona** (kō-rō′nă; crown) **radiata** (rā-dē-ā′tă; radiating), which is the innermost layer of follicle cells.

4. A **vesicular follicle** (also called a *mature follicle* or *Graafian follicle*) forms from a secondary follicle. A vesicular follicle contains a **secondary oocyte** (surrounded by a zona pellucida and the corona radiata), numerous layers of follicular cells, and a large, fluid-filled, crescent-shaped antrum. A secondary oocyte has completed meiosis I and is arrested in the second meiotic metaphase. Vesicular follicles become quite large and can be distinguished by their overall size as well as by the size of the antrum.

5. When a vesicular follicle ruptures and expels its oocyte (in a process called ovulation), the remnants of the follicle remaining in the ovary turn into a yellowish structure called the **corpus luteum** (loo-tē′ŭm; *luteus* = saffron-yellow). The corpus luteum does not contain an oocyte. However, the corpus luteum secretes the sex hormones **progesterone** (prō-jes′ter-ōn; *pro* = before; gestation) and **estrogen** (es′trō-jen; *oistrus* = estrus, *gen* = producing). These hormones stimulate the continuing buildup of the uterine lining and prepare the uterus for possible implantation of a fertilized oocyte.

6. When a corpus luteum regresses (breaks down), it turns into a white, connective tissue scar called the **corpus albicans** (al′bi-kanz; white). Most corpus albicans structures are completely resorbed, and only a few may remain within an ovary.

Table 28.2 summarizes the different structures that develop during a female's monthly cycle.

Table 28.2	Ovarian Follicles and Structures That Develop in the Ovary		
Ovarian Structure	**Type of Oocyte**	**Anatomic Characteristics**	**Time of First Appearance**
Primordial follicle	Primary oocyte	Single layer of flattened follicular cells surround an oocyte	Fetal period
Primary follicle	Primary oocyte	Single or multiple layers of cuboidal follicular cells surround an oocyte	Puberty
Secondary follicle	Primary oocyte	Multiple layers of follicular cells surround the oocyte and a small, fluid-filled antrum	Puberty
Vesicular follicle	Secondary oocyte	Many layers of follicular cells surround the oocyte and a very large antrum	Puberty
Corpus luteum	No oocyte	Yellowish, collapsed folds of follicular cells	Puberty
Corpus albicans	No oocyte	Whitish connective tissue scar, remnant of a degenerated corpus luteum	Puberty

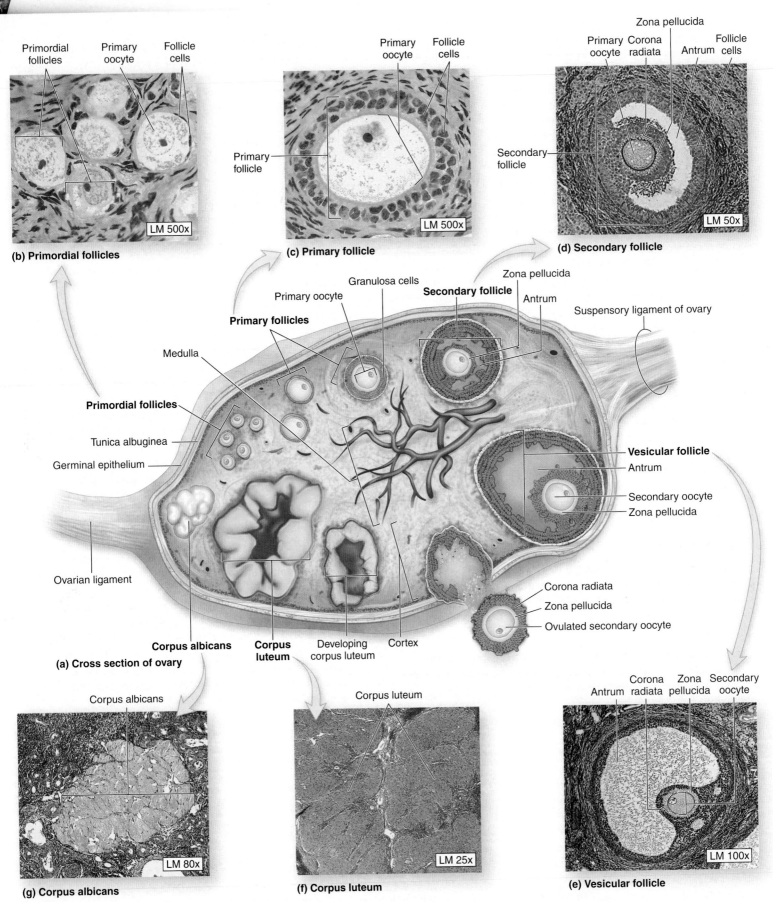

Primordial follicles   Primary oocyte   Follicle cells

(b) Primordial follicles   LM 500x

Primary oocyte   Follicle cells

Primary follicle

(c) Primary follicle   LM 500x

Zona pellucida   Primary oocyte   Corona radiata   Antrum   Follicle cells

Secondary follicle

(d) Secondary follicle   LM 50x

Granulosa cells   Primary oocyte   Zona pellucida   Secondary follicle   Antrum   Suspensory ligament of ovary

Primary oocyte   **Primary follicles**   **Secondary follicle**

Medulla

Primordial follicles

Tunica albuginea

Germinal epithelium

Vesicular follicle   Antrum   Secondary oocyte   Zona pellucida

Ovarian ligament

Corona radiata   Zona pellucida   Ovulated secondary oocyte

Corpus albicans   **Corpus luteum**   Developing corpus luteum   Cortex

(a) Cross section of ovary

Corpus albicans

(g) Corpus albicans   LM 80x

Corpus luteum

(f) Corpus luteum   LM 25x

Corona   Zona   Secondary
Antrum   radiata   pellucida   oocyte

(e) Vesicular follicle   LM 100x

Figure 28.4

Stages of Follicle Development Within an Ovary.   The ovary produces and releases both female gametes (secondary oocytes) and sex hormones. *(a)* A coronal view of the ovary contents depicts the different stages of follicle maturation, ovulation, and corpus luteum development and degeneration. Note that all of the follicles and structures shown in this image would appear at different times during the ovarian cycle—they do not occur simultaneously. Further, the follicles do not migrate through the ovary; rather, all follicles are shown together merely for comparative purposes. Histologic sections identify *(b)* primordial follicles, *(c)* a primary follicle, *(d)* a secondary follicle, and *(e)* part of a vesicular follicle. After ovulation, the remnant of the follicle forms *(f)* the corpus luteum, which then degenerates into *(g)* the corpus albicans.

Oogenesis and the Ovarian Cycle

Oogenesis is the maturation of a primary oocyte to a secondary oocyte and is illustrated in **figure 28.5**.

Before Birth The process of oogenesis begins in a female fetus before birth. At this time, the ovary contains primordial germ cells called **oogonia** (ō-ō-gō′nē-ă; sing., *oogonium; oon* = egg), which are diploid cells, meaning they have 23 pairs of chromosomes. During the fetal period, the oogonia start the process of meiosis, but they are stopped at prophase I. At this point, the cells are called primary oocytes. At birth, the ovaries of a female child are estimated to contain approximately 1.5 million primordial follicles within its cortex. The primary oocytes in the primordial follicles remain arrested in prophase I until after puberty.

Study Tip!

To distinguish a primary oocyte from a secondary oocyte, remember that:

- A **primary** oocyte is arrested in prophase **I** (the term "primary" also means "one").
- A **secondary** oocyte is arrested in metaphase **II** (the term "secondary" means "two").

Additionally, remember that the *only* ovarian follicle containing a secondary oocyte is a vesicular follicle—all other ovarian follicles have primary oocytes only.

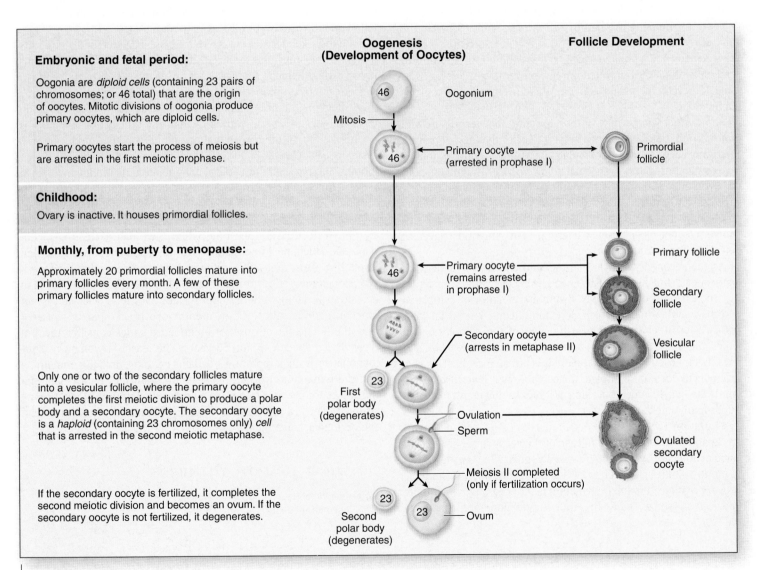

Oogenesis (Development of Oocytes)

Follicle Development

Embryonic and fetal period:

Oogonia are *diploid cells* (containing 23 pairs of chromosomes; or 46 total) that are the origin of oocytes. Mitotic divisions of oogonia produce primary oocytes, which are diploid cells.

Primary oocytes start the process of meiosis but are arrested in the first meiotic prophase.

46 — Oogonium

Mitosis

46 ← Primary oocyte (arrested in prophase I) → Primordial follicle

Childhood:

Ovary is inactive. It houses primordial follicles.

Monthly, from puberty to menopause:

Approximately 20 primordial follicles mature into primary follicles every month. A few of these primary follicles mature into secondary follicles.

46 ← Primary oocyte (remains arrested in prophase I) → Primary follicle / Secondary follicle

Only one or two of the secondary follicles mature into a vesicular follicle, where the primary oocyte completes the first meiotic division to produce a polar body and a secondary oocyte. The secondary oocyte is a *haploid* (containing 23 chromosomes only) *cell* that is arrested in the second meiotic metaphase.

Secondary oocyte (arrests in metaphase II) → Vesicular follicle

23 — First polar body (degenerates)

Ovulation → Ovulated secondary oocyte
Sperm

If the secondary oocyte is fertilized, it completes the second meiotic division and becomes an ovum. If the secondary oocyte is not fertilized, it degenerates.

Meiosis II completed (only if fertilization occurs)

23 — Second polar body (degenerates)

23 — Ovum

Figure 28.5

Oogenesis. Oogenesis begins in a female fetus, when primary oocytes develop in primordial follicles. The ovary and these follicles remain inactive during childhood. At puberty, a select number of primordial follicles each month undergoes maturation and produces a female gamete.

Childhood During childhood, a female's ovaries are inactive, and no follicles develop. In fact, the main event that occurs during childhood is **atresia** (ă-trē′zē-ă; *a* = not, *tresis* = a hole), in which some primordial follicles regress. By the time a female child reaches puberty, only about 400,000 primordial follicles remain in the ovaries.

From Puberty to Menopause When a female child reaches puberty, the hypothalamus releases GnRH (gonadotropin-releasing hormone), which stimulates the anterior pituitary to release FSH (follicle-stimulating hormone) and LH (luteinizing hormone). The levels of FSH and LH vary in a cyclical pattern and produce a monthly sequence of events in follicle development called the **ovarian cycle.** The three phases of the ovarian cycle are the follicular phase, ovulation, and the luteal phase **(figure 28.6)**.

The **follicular phase** occurs during days 1–13 of an approximate 28-day ovarian cycle. At the beginning of the follicular phase, FSH and LH stimulate about 20 primordial follicles to mature into primary follicles. It is unclear why some of the primordial follicles in the ovary are stimulated to mature into primary follicles, while the remainder remain unaffected by the FSH and LH secretion. As the follicles develop, their follicular cells release the hormone **inhibin,** which helps inhibit FSH production, thus preventing excessive ovarian follicle development and allowing the current primary follicles to mature.

Shortly thereafter, a few of these primary follicles mature and become secondary follicles. The primary follicles that do not mature undergo atresia. Late in the follicular phase, one secondary follicle in an ovary matures into a vesicular follicle. Under the influence of LH, the volume of fluid increases within the antrum, and the oocyte is forced toward one side of the follicle, where it is surrounded by a cluster of follicle cells termed the **cumulus** (kū′mū-lŭs; heap) **oophorus** (ō-of′ōr-ŭs; *phorus* = bearing). The innermost layer of these cells is the corona radiata.

As the secondary follicle matures into a vesicular follicle, its primary oocyte finishes meiosis I, and two cells form (see figure 28.5). One of these cells receives a minimal amount of cytoplasm and forms a **polar body,** which is a nonfunctional cell that later regresses. The other cell receives the bulk of the cytoplasm and becomes the secondary oocyte, which continues to develop and reaches metaphase II of meiosis before it is arrested again. This secondary oocyte does not complete meiosis unless it is fertilized by a sperm. If the oocyte is never fertilized, it breaks down and regresses about 24 hours later.

Ovulation (ov′ū′lā′shŭn) occurs on day 14 of a 28-day ovarian cycle and is defined as the release of the secondary oocyte from a vesicular follicle (figure 28.6). Typically, only one ovary ovulates each month—that is, the left ovary ovulates one month, and the right ovary ovulates the next. Ovulation is induced only when there is a peak in LH secretion. As the time of ovulation approaches, the follicle cells in the vesicular follicle increase their rate of fluid secretion, forming a larger antrum and causing further swelling within the follicle. The edge of the follicle that continues to expand at the ovarian surface becomes quite thin and eventually ruptures, expelling the secondary oocyte.

The **luteal phase** occurs during days 15–28 of the ovarian cycle, when the remaining follicle cells in the ruptured vesicular follicle turn into a corpus luteum. The corpus luteum secretes progesterone and estrogen that stabilize and build up the uterine lining, and prepare for possible implantation of a fertilized oocyte.

The corpus luteum has a life span of about 10–13 days if the secondary oocyte is not fertilized. After this time, the corpus luteum regresses and becomes a corpus albicans. As the corpus luteum regresses, its levels of secreted progesterone and estrogen drop, causing the uterine lining to be shed as **menstruation** (men-stroo-ā′shŭn), also called **menses** or a **period.** This marks the end of the luteal phase. A female's first menstrual cycle, called **menarche** (me-nar′kē; *men* = month, *arche* = beginning), is the culmination of female puberty and typically occurs around age 11–12.

If the secondary oocyte is fertilized and if it successfully **implants** in the uterine lining, this fertilized structure (now a preembryo) begins its own development (as discussed in chapter 3). The pre-embryo starts secreting **human chorionic gonadotropin (hCG),** a hormone that enters the mother's bloodstream and acts on the corpus luteum. Essentially, hCG lets the corpus luteum know that implantation has occurred and that the corpus luteum should continue producing progesterone, which will build and stabilize the uterine lining. After 3 months, the placenta of the developing fetus starts producing its own progesterone and estrogen, so by the end of the third month, the corpus luteum has usually regressed and formed a corpus albicans.

After Menopause The time when a woman is nearing menopause is called **perimenopause.** During perimenopause, estrogen levels begin to drop, and a woman may experience irregular periods, skip some periods, or have very light periods. When a woman has stopped having monthly menstrual cycles for 1 year and is not pregnant, she is said to be in **menopause** (men′ō-pawz; *pauses* = cessation). The age at onset of menopause varies considerably, but typically is between 45 and 55 years. The reason a woman reaches menopause is that there are virtually no more follicles remaining and the remaining ovarian follicles stop maturing. As a result, significant amounts of estrogen and progesterone are no longer being secreted. Thus, a woman's endometrial lining does not grow, and she no longer has a menstrual period.

WHAT DO YOU THINK?

❶ If a woman has one ovary surgically removed, can she still become pregnant? Why or why not?

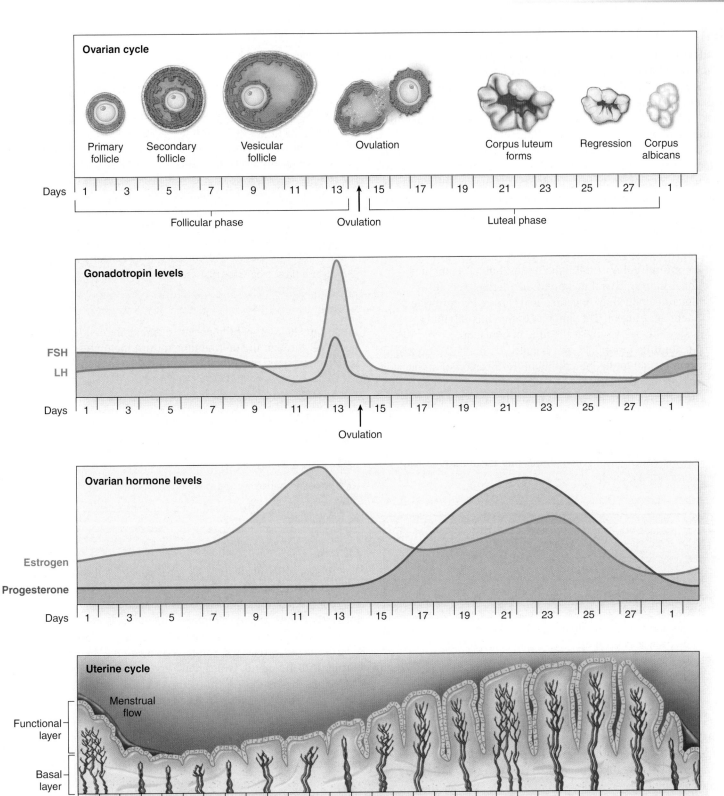

Figure 28.6

Hormonal Changes in the Female Reproductive System.   Cyclic changes in gonadotropins affect ovarian hormone production. FSH causes development of estrogen-producing ovarian follicles during the follicular phase of the ovarian cycle. Estrogen stimulates the proliferative phase in the uterine cycle. Estrogen levels spike as ovulation approaches. High levels of LH promote ovulation at the midpoint of the ovarian cycle. The corpus luteum becomes functional after ovulation, and it produces both progesterone and estrogen to promote uterine lining development. If fertilization does not occur, the corpus luteum degenerates, and menstrual flow begins at the start of the next uterine cycle.

Uterine Tubes

The **uterine tubes,** also called the *fallopian* (fa-lō′pē-an) *tubes* or *oviducts* (ō′vi-dŭkt; *duco* = to lead), extend laterally from both sides of the uterus toward the ovaries (**figure 28.7**). In the lateral part of these tubes, the secondary oocyte is fertilized, and the pre-embryo begins to develop as it travels toward the uterus. Usually it takes the pre-embryo about 3 to 4 days to reach the lumen of the uterus.

The uterine tubes are small in diameter, and reach their maximum length of between 10 and 12 centimeters after puberty. These tubes are covered and suspended by the **mesosalpinx** (mez′ō-sal′pinks; *salpinx* = trumpet), a specific superior part of the broad ligament of the uterus (see figure 28.3*a*). Each uterine tube is composed of contiguous segments that are distinguishable in both gross examination and histologic sections:

- The **infundibulum** (in-fŭn-dib′ū-lŭm; funnel) is the free, funnel-shaped, lateral margin of the uterine tube. Its numerous individual fingerlike folds are called **fimbriae** (fim′brē-ē; fringes). The fimbriae of the infundibulum enclose the ovary only at the time of ovulation.
- The **ampulla** (am-pul′lă; two-handled bottle) is the expanded region medial to the infundibulum. Fertilization of an oocyte typically occurs there.
- The **isthmus** (is′mus) extends medially from the ampulla toward the lateral wall of the uterus. It forms about one-third of the length of the uterine tube.
- The **uterine part** (*intramural part* or *interstitial segment*) extends medially from the isthmus and is continuous with the wall of the uterus.

Study Tip!

One way to remember the segments of the uterine tubes is as follows:

- The infundibulum (the only segment with an "F" in it) has the fimbriae.
- The **a**mpulla is the **a**rm of the uterine tube.
- The **is**thmus **is** the longest.
- The uter**in**e part is **in** the uterus.

The wall of the uterine tube consists of a mucosa, a muscularis, and a serosa. The **mucosa** is formed from a ciliated columnar epithelium and a layer of areolar connective tissue. The mucosa is thrown into linear folds, which reduce the size of the uterine tube lumen. After ovulation, the cilia on the apical surface of the epithelial cells of both the infundibulum and the ampulla begin to beat in the direction of the uterus. This causes a slight current in the fluid within the uterine tube lumen, drawing the ovulated oocyte into the uterine tube and moving it toward the uterus.

The **muscularis** is composed of an inner circular layer and an outer longitudinal layer of smooth muscle cells. The muscular layer increases in relative thickness as the uterine tube approaches the lateral wall of the uterus. Some peristaltic contractions in the

CLINICAL VIEW

Tubal Pregnancy

In a **tubal pregnancy** (or *ectopic pregnancy*), the fertilized oocyte implants in the uterine tube, rather than traveling to the uterus for implantation. The main danger in a tubal pregnancy is that the uterine tube is unable to expand as the embryo grows. Thus, the embryo can remain viable no later than week 8, at which time it has become too large for the confines of the uterine tube. The woman experiences severe cramping, and the uterine tube may rupture if the embryo is not surgically removed. If the uterine tube ruptures, a massive hemorrhage into the abdominopelvic cavity can endanger the life of the mother. Unfortunately, there currently is no way to treat a tubal pregnancy that can spare the developing embryo.

muscularis help propel the oocyte, or pre-embryo if fertilization has occurred, through the uterine tube toward the uterus. The **serosa** is the external serous membrane covering the uterine tube.

WHAT DID YOU LEARN?

- **③** What are the types of ovarian follicles?
- **④** What is ovulation?
- **⑤** What is menarche, and when does it occur?
- **⑥** What type of epithelium lines the uterine tubes, and what is its function?

Uterus

The **uterus** (ū′ter-ŭs; womb) is a pear-shaped, thick-walled muscular organ within the pelvic cavity. It has a lumen (internal space) that connects to the uterine tubes superolaterally and to the vagina inferiorly (figure 28.7*a*). Normally, the uterus is angled anterosuperiorly across the superior surface of the urinary bladder, a position referred to as **anteverted** (an-te-vert′ed; *ante* = before, *versio* = a turning). If the uterus is positioned posterosuperiorly (so that it is projecting toward the rectum), this position is called **retroverted** (re′trō-ver-ted). In older women, the uterus may shift from anteverted to retroverted.

The uterus serves many functions. Following fertilization, the pre-embryo makes contact with the uterine lining and implants in the inner uterine wall. The uterus then supports, protects, and nourishes the developing embryo/fetus by forming a vascular connection that later develops into the placenta. The uterus ejects the fetus at birth after maternal oxytocin levels increase to initiate the uterine contractions of labor. If an oocyte is not fertilized, the muscular wall of the uterus contracts and sheds its inner lining as menstruation.

The uterus is partitioned into the following regions:

- The **fundus** (fŭn′dŭs) is the broad, curved superior region extending between the lateral attachments of the uterine tubes.

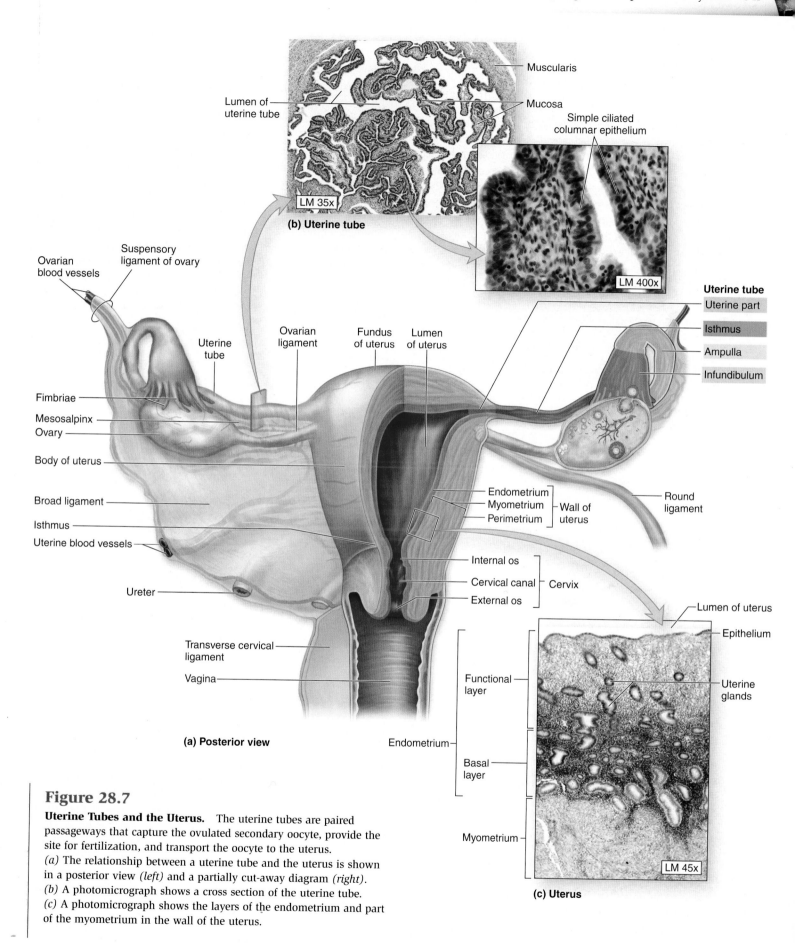

(b) Uterine tube

Muscularis

Lumen of uterine tube

Mucosa

Simple ciliated columnar epithelium

LM 35x

LM 400x

Ovarian blood vessels

Suspensory ligament of ovary

Uterine tube

Ovarian ligament

Fundus of uterus

Lumen of uterus

Uterine tube

Uterine part

Isthmus

Ampulla

Infundibulum

Fimbriae

Mesosalpinx

Ovary

Body of uterus

Broad ligament

Isthmus

Uterine blood vessels

Ureter

Transverse cervical ligament

Vagina

Endometrium

Myometrium

Perimetrium

Wall of uterus

Round ligament

Internal os

Cervical canal

External os

Cervix

Lumen of uterus

Epithelium

Uterine glands

Functional layer

Endometrium

Basal layer

Myometrium

LM 45x

(a) Posterior view

(c) Uterus

Figure 28.7

Uterine Tubes and the Uterus.   The uterine tubes are paired passageways that capture the ovulated secondary oocyte, provide the site for fertilization, and transport the oocyte to the uterus.
(a) The relationship between a uterine tube and the uterus is shown in a posterior view *(left)* and a partially cut-away diagram *(right)*.
(b) A photomicrograph shows a cross section of the uterine tube.
(c) A photomicrograph shows the layers of the endometrium and part of the myometrium in the wall of the uterus.

CLINICAL VIEW

Cervical Cancer

Cervical cancer is one of the most common malignancies of the female reproductive system. It is estimated that over 12,000 new cases of invasive cervical cancer and four times that many noninvasive cervical cancer cases are diagnosed each year. Approximately 4000 women die from cervical cancer annually. Risk factors for cervical cancer include increased age, HIV infection, and low socioeconomic status. However, the most important risk factor is previous human papillomavirus (HPV) infection. Some classes of HPV are considered "high risk" because they are frequently associated with genital and anal cancers in men and women and are sexually transmitted.

The **Papanicolaou (Pap) smear** has become a very effective method of detecting cervical cancer in its early and curable stage. The test is done in the doctor's office as follows:

1. The health-care professional inserts a metal or plastic instrument called a **speculum** (spek'ū-lŭm; mirror) into the vagina to keep the vagina open in order to examine the cervix.

2. Epithelial cells are scraped from the edge of the cervix and placed (smeared) on a microscope slide.

3. The slide is sent to a lab, where a cytologist stains the cells and examines them under the microscope, noting any abnormal cellular development (termed dysplasia).

If dysplastic cells are detected, the health-care professional will likely request a follow-up Pap smear and possibly even a biopsy. Sometimes, dysplastic cells are a result of irritation, infection, or some undetermined cause, and are not cancerous. But if advanced dysplasia is seen, most physicians immediately recommend a biopsy of the cervix, and may even insist on HPV testing. If cervical cancer is present, surgery is indicated. To treat a cancer that is localized, a portion of the cervix may be removed, a procedure known as a **cone biopsy**. In the case of invasive cancer, removal of the entire uterus, called a **hysterectomy** (his-ter-ek'tō-mē; *hystera* = womb), is indicated. Researchers recently have developed a vaccine (™Gardasil) for the most common types of HPV that cause cervical cancer. However, in order for the vaccine to be effective, it must be given to girls who are not yet sexually active. Several states are debating whether girls aged 12 should be required to receive the vaccine.

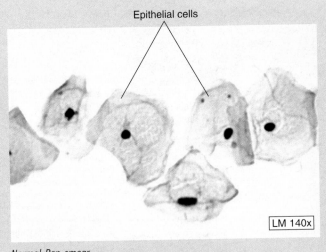

Epithelial cells

LM 140x

Normal Pap smear.

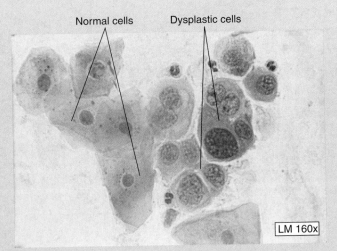

Normal cells   Dysplastic cells

LM 160x

Abnormal Pap smear.

- The major part of the uterus is its middle region, called the **body**, which is composed of a thick wall of smooth muscle.
- A narrow, constricted inferior region of the body that is superior to the cervix is called the **isthmus.**
- The **cervix** is the narrow inferior portion of the uterus that projects into the vagina.

Within the cervix is a narrow channel called the **cervical canal**, which connects to the vagina inferiorly. The superior opening of this canal is the **internal os** (*os* = mouth). The inferior opening of the cervix into the lumen of the vagina is the **external os.** The external os is covered with nonkeratinized stratified squamous epithelium. The cervix contains mucin-secreting glands that help form a thick mucus plug at the external os. This mucus plug is suspected to be a physical barrier that prevents pathogens from invading the uterus from the vagina. The mucus plug thins considerably around the time of ovulation, so sperm may more easily enter the uterus.

Support of the Uterus

Several structures support the uterus. The muscles of the pelvic floor (**pelvic diaphragm** and **urogenital diaphragm**) (see figure 11.15) hold the uterus and vagina in place and help resist intra-abdominal pressure exerted inferiorly on the pelvis. The **round ligaments** (figure 28.7) of the uterus extend from the lateral sides of the uterus, through the inguinal canal and attach to the labia majora. These ligaments help keep the uterus in an anteverted position. The **transverse cervical ligaments** (or *cardinal ligaments*) run from the sides of the cervix and superior vagina laterally to the pelvic wall. They help restrict inferior movements of the uterus. The **uterosacral ligaments** (or *sacrocervical ligaments*; not shown in figure 28.7) connect the inferior portion of the uterus posteriorly to the sacrum.

Many of these ligaments travel between the folds of the broad ligament. Weakness in either the pelvic floor muscles or these ligaments can lead to **prolapse** (prō-laps'; *prolapsus* = a failing) of the uterus, in which the uterus starts to protrude through the vagina.

Despite its name, the broad ligament is not a strong support for the uterus, but rather a peritoneal drape over the uterus.

Blood Supply

Each internal iliac artery extends a branch called a **uterine artery** through the broad ligament to the lateral wall of the uterus (see figure 28.7). Numerous smaller branches from the uterine artery then penetrate the muscular wall of the uterus and further diverge into arcuate arteries. Thereafter, each arcuate artery gives rise to smaller vessels, called radial arteries, which extend into the innermost layer (endometrium) of the uterus. Here they branch into spiral arteries, which swirl throughout the endometrium, extending between and throughout the uterine glands (described below) toward the mucosal surface.

Wall of the Uterus

The uterine wall is composed of three concentric tunics: the perimetrium, myometrium, and endometrium (figure 28.7). The outer tunic of most of the uterus is a serosa called the **perimetrium** (per-i-mē′trē-ŭm; *metra* = uterus). The perimetrium is continuous with the broad ligament. The **myometrium** (mī′ō-mē′trē-ŭm; *mys* = muscle) is the thick, middle tunic of the uterine wall formed from three intertwining layers of smooth muscle. In the nonpregnant uterus, the muscle cells are less than 0.25 millimeters in length. During the course of a pregnancy, smooth muscle cells increase both in size (**hypertrophy**; hī-per′trō-fē) and in number (**hyperplasia**; hi-per-plā′zhē-ă). Some cells may exceed 5 millimeters in length by the end of gestation. The innermost tunic of the uterus, called the **endometrium** (en′dō-mē′trē-ŭm), is an intricate mucosa composed of a simple columnar epithelium and an underlying lamina propria. The lamina propria is filled with compound tubular glands (also called **uterine glands**), which enlarge during the uterine cycle.

Two distinct layers form the endometrium. The deeper layer is the **basal layer,** also called the *stratum basalis* (bā-sā′lis). The basal layer is immediately adjacent to the myometrium, and is a permanent layer that undergoes few changes during each uterine cycle. The more superficial of the two endometrial layers is the **functional layer,** or *stratum functionalis* (fŭnk-shŭn-ăl′is). Beginning at puberty, the functional layer grows from the basal layer under the influence of estrogen and progesterone secreted from the ovarian follicles. If fertilization and implantation do not occur, this lining is shed as menses.

Uterine (Menstrual) Cycle and Menstruation

The cyclical changes in the endometrial lining occur under the influence of estrogen and progesterone secreted by the ovary. The **uterine cycle** (or *menstrual cycle*) consists of three distinct phases of endometrium development: the menstrual phase, proliferative phase, and secretory phase (see figure 28.6, *bottom*).

The **menstrual** (men′stroo-ăl; *menstrualis* = monthly) **phase** occurs approximately during days 1–5 of the cycle. This phase is marked by sloughing of the functional layer and lasts through the period of menstrual bleeding. The **proliferative** (prō-lif′er-ă-tiv; *proles* = offspring, *fero* = to bear) **phase** follows, spanning approximately days 6–14. The initial development of the functional layer of the endometrium overlaps the time of follicle growth and estrogen secretion by the ovary. The last phase is the **secretory** (se-krēt′ĕ-rē, sē′krĕ-tōr-ē) **phase,** which occurs at approximately days 15–28. During the secretory phase, increased progesterone secretion from the corpus luteum results in increased vascularization and development of uterine glands. If the oocyte is not fertilized, the corpus luteum degenerates, and the progesterone level drops dramatically. Without progesterone, the functional layer lining sloughs off, and the next uterine cycle begins with the menstrual phase.

Table 28.3 compares the uterine cycle with the ovarian cycle discussed previously, and **table 28.4** summarizes the hormones that influence the ovarian and uterine cycles. The day ranges listed on figure 28.6 and table 28.3 assume that the woman has a 28-day cycle, meaning she ovulates at day 14 and has a menstrual period about every 28 days. If a woman has a longer cycle (say, she menstruates about every 35 days), her menstrual phase and/or her proliferative phase is longer than average. Typically, a woman ovulates 14 days before menstruation, so the secretory phase day ranges do not vary as much.

WHAT DO YOU THINK?

2 What factors could influence the length and timing of a woman's monthly uterine cycle?

Vagina

The **vagina** (vă-jī′nă) is a thick-walled, fibromuscular tube that forms the inferiormost region of the female reproductive tract and measures about 10 centimeters in length in an adult female (see figure 28.2). The vagina connects the uterus with the outside of

CLINICAL VIEW

Endometriosis

Endometriosis (en′dō-mē-trē-ō′sis) occurs when part of the endometrium is displaced onto the external surface of organs within the abdominopelvic cavity. Scientists think that during the regular uterine (menstrual) cycle of some women, a small amount of endometrium may be expelled from the uterine tubes and become implanted on the surface of the ovaries, uterine tubes, urinary bladder, and intestines. If this displaced endometrium remains viable, it responds to hormone stimulation during each menstrual growth phase. Unfortunately, at the end of the monthly cycle, this displaced endometrium cannot slough and be expelled. Thus, the ensuing hemorrhage and breakdown of the displaced endometrium cause considerable pain and eventually scarring that often leads to deformities of the uterine tubes. Treatments include the use of hormones designed to retard the growth and cycling of the displaced endometriotic tissue, as well as surgical removal of the ectopic endometrium.

Table 28.3	Comparison of Ovarian and Uterine Cycle Phases	
Day[1]	Ovarian Cycle Phase	Uterine Cycle Phase
1–5	Follicular phase	Menstrual phase
6–13		Proliferative phase
14	Ovulation	
15–28	Luteal phase	Secretory phase

[1] This table assumes a 28-day cycle between menstrual periods. If a woman has a longer or shorter cycle, the day ranges for the follicular, menstrual, and proliferative phases will vary.

Table 28.4   Influence of Hormones on the Ovarian and Uterine Cycles

Hormone	Primary Source of Hormone	Effects
Gonadotropin-releasing hormone (GnRH)	Hypothalamus	Stimulates anterior pituitary to produce and secrete FSH and LH
Follicle-stimulating hormone (FSH)	Anterior pituitary	Stimulates development and maturation of ovarian follicles
Luteinizing hormone (LH)	Anterior pituitary	Stimulates ovulation (when there is a peak in LH)
Estrogen	Ovarian follicles (before ovulation), corpus luteum (after ovulation), or placenta (during pregnancy)	Initiates and maintains growth of the functional layer of the endometrium
Progesterone	Corpus luteum or placenta (during pregnancy)	Primary hormone responsible for functional layer growth after ovulation; causes increase in blood vessel distribution, uterine gland size, and nutrient production
Inhibin	Ovarian follicles	Inhibits FSH secretion, so as to prevent excessive follicular development

the body anteroventrally, and thus functions as the birth canal. The vagina is also the copulatory organ of the female, as it receives the penis during intercourse, and it serves as the passageway for menstruation.

The vaginal wall is heavily invested with both blood vessels and lymphatic vessels. Arterial supply comes from the vaginal arter-ies, and venous drainage is via vaginal veins. The lumen of the vagina is flattened anteroposteriorly. The vagina's relatively thin, distensible wall consists of three tunics: an inner mucosa, a middle muscularis, and an outer adventitia.

The **mucosa** consists of a nonkeratinized stratified squamous epithelium and a highly vascularized lamina propria (**figure 28.8**).

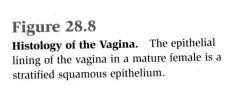

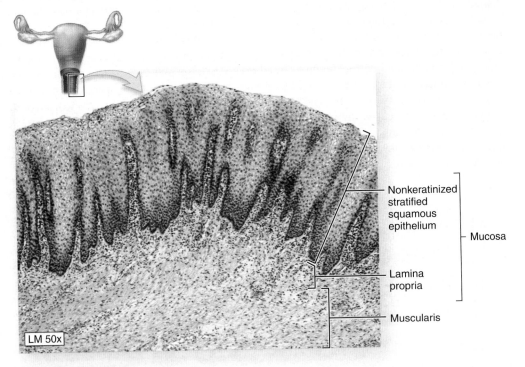

Figure 28.8

Histology of the Vagina. The epithelial lining of the vagina in a mature female is a stratified squamous epithelium.

Nonkeratinized stratified squamous epithelium

Mucosa

Lamina propria

Muscularis

LM 50x

Superifical cells of the vaginal epithelium produce an acidic secretion that helps prevent bacterial and other pathogenic invasion. During each menstrual phase, large numbers of lymphocytes and granulocytes invade the lamina propria to help prevent infection during menstruation. The inferior region of the vaginal mucosa contains numerous transverse folds, or rugae. Near the external opening of the vagina, called the **vaginal orifice,** these mucosal folds project into the lumen to form a vascularized, membranous barrier called the **hymen** (hī′men; membrane). The hymen typically is perforated during the first instance of sexual intercourse, but also may be perforated by tampon use, medical exams, or very strenuous physical activity.

The **muscularis** of the vagina has both outer and inner layers of smooth muscle. The outer layer is composed of bundles of longitudinal smooth muscle cells that are continuous with corresponding muscle cells in the myometrium. The smooth muscle cells of the inner circular layer are interwoven with the outer longitudinal muscle fibers at the point where the two muscle layers meet. Near the vaginal orifice are some skeletal muscle fibers of the muscularis layer that cause partial narrowing of the vaginal orifice. The **adventitia** contains some inner elastic fibers and an outer layer of areolar connective tissue.

External Genitalia

The external sex organs of the female are termed the **external genitalia** or **vulva** (vŭl′vă; a covering) **(figure 28.9)**. The **mons** (monz; mountain) **pubis** is an expanse of skin and subcutaneous connective tissue immediately anterior to the pubic symphysis. The mons pubis is covered with pubic hair in postpubescent females. The **labia majora** (lā′bē-ă mă-jŏr′ă; sing., *labium majus*; *labium* = lip, *majus* = larger) are paired, thickened folds of skin and connective tissue. The labia majora are homologous to the scrotum of the male. In adulthood, their outer surface is covered with coarse pubic hair; they contain numerous sweat and sebaceous glands. The **labia minora** (mī-nŏr′ă; sing., *labium minus*; *minus* = smaller) are paired folds immediately internal to the labia majora. They are devoid of hair and contain a highly vascular layer of areolar connective tissue. Sebaceous glands are located in these folds, as are numerous melanocytes, resulting in enhanced pigmentation of the folds.

The space between the labia minora is called the **vestibule.** Within the vestibule are the **urethral opening** and the vaginal orifice. On either side of the vaginal orifice is an erectile body called the **bulb of the vestibule** (see figure 27.10), which becomes erect and increases in sensitivity during sexual intercourse. A pair of **greater vestibular glands** (previously called *glands of Bartholin*) are housed within the posterolateral walls of the vestibule. These are tubuloacinar glands that secrete mucin, which forms mucus to

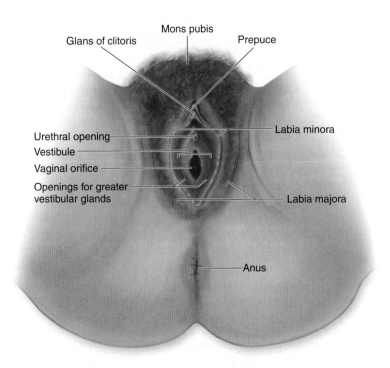

Figure 28.9

Female External Genitalia. Inferior view of the external genitalia, illustrating the urethral opening and the vaginal orifice, which are within the vestibule and bounded by the labia minora.

act as a lubricant for the vagina. Secretion increases during sexual intercourse, when additional lubrication is needed. These secretory structures are homologous to the male bulbourethral glands.

The **clitoris** (klit′ō-ris) is a small erectile body, usually less than 2 centimeters in length, located at the anterior regions of the labia minora. It is homologous to the penis of the male. Two small erectile bodies called **corpora cavernosa** form the **body** of the clitoris. Extending from each of these bodies posteriorly are elongated masses, each called the **crus** (kroos) of the clitoris, which attach to the pubic arch. Capping the body of the clitoris is the **glans** (glanz; acorn). The many specialized sensory nerve receptors housed in the clitoris provide pleasure to the female during sexual intercourse. The **prepuce** (prē′poos; foreskin) is an external fold of the labia minora that forms a hoodlike covering over the clitoris.

Mammary Glands

Each **mammary gland,** or *breast,* is located within the anterior thoracic wall and is composed of a compound tubuloalveolar exocrine

CLINICAL VIEW: In Depth

Contraception Methods

The term **contraception** (kon-tră-sep'shun) refers to birth control, or the prevention of pregnancy. A wide range of birth control methods are available, and they have varying degrees of effectiveness.

Abstinence (ab'sti-nens; *abstineo* = to hold back) means refraining from sexual intercourse. Abstinence is the only 100% proven way *not* to become pregnant.

The **rhythm method** requires avoiding sexual intercourse during the time when a woman is ovulating. Because sperm can live for several days in the female reproductive tract, it is best to avoid intercourse both a few days prior and a few days after ovulating. The rhythm method requires that a woman know when she is ovulating, which may be difficult to determine consistently. As a result, this method has a high failure rate.

Barrier methods of birth control use a physical barrier to prevent sperm from reaching the uterine tubes. Barrier methods include the following:

■ **Condoms**, when used properly, collect the sperm and prevent them from entering the female reproductive tract. They are also the only birth control method that helps protect against sexually transmitted viruses and diseases, such as human papillomavirus, herpes, and HIV. Condoms for males fit snugly on the erect penis,

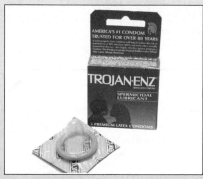

(a) Condoms

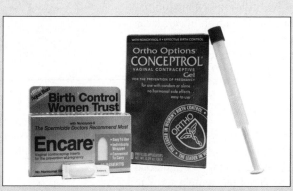

(b) Spermicidal foams

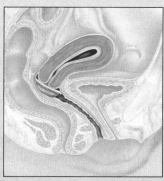

(c) Diaphragm

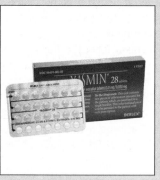

(d) Oral contraceptive

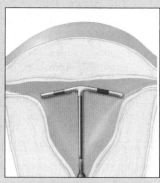

(e) Intrauterine device (IUD)

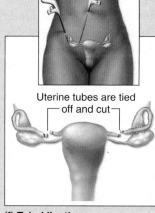

Uterine tubes are tied off and cut

(f) Tubal ligation

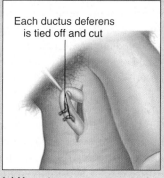

Each ductus deferens is tied off and cut

(g) Vasectomy

Contraception includes barrier, chemical, and surgical methods.

gland (**figure 28.10**). The gland's complex secretory product (called breast milk) contains proteins, fats, and a sugar to provide nutrition to infants.

The **nipple** (nip'l; beak) is a cylindrical projection on the center of the breast. It contains multiple tiny openings of the excretory ducts that transport breast milk. The **areola** (ă-rē'ō-lă; small area) is the pigmented, rosy or brownish ring of skin around the nipple. Its surface often appears uneven and grainy due to the numerous sebaceous glands, called **areolar glands,** immediately internal to the surface. The color of the areola may vary, depending upon whether or not a woman has given birth. In a **nulliparous** (nŭl-ip'ă-rŭs; *nullus* = none, *pario* = to bear) woman (a woman who has never given birth), the areola is rosy or light brown in color. In a **parous** (par'ŭs) woman (a woman who has given birth), the areola may change to a darker rose or brown color.

Internally, the breasts are supported by fibrous connective bands called **suspensory ligaments.** These thin bands extend from the skin and attach to the deep fascia overlying the pectoralis major muscle. Thus, the breast and the pectoralis major muscle are structurally linked.

The mammary glands are subdivided into **lobes,** which are further subdivided into smaller compartments called **lobules.** Lobules contain secretory units termed **alveoli** that produce milk in the lactating female. Alveoli become more numerous and larger while **vaginal condoms** are placed in the vagina prior to sexual intercourse.

- **Spermicidal foams and gels** are chemical barrier methods that kill sperm before they travel to the uterine tubes. They are inserted into the vagina and/or placed on the penis prior to sexual intercourse. Foams and gels are not the most effective method of birth control; rather, they should be used in conjunction with a physical barrier method.
- **Diaphragms** and **cervical caps** are circular, rubbery structures that are inserted into the vagina and placed over the cervix prior to sexual intercourse. Spermicidal gel is used around the edges to help prevent sperm from entering the cervix. Some women find it difficult to correctly place the diaphragm or cervical cap, and incorrect placement can result in pregnancy.

Lactation (nursing a baby) can prevent ovulation and menstruation for several to many months after childbirth *if* a woman nurses her child *constantly* (i.e., much more than five times a day!). The frequent lactation sends signals to the hypothalamus to prevent FSH and LH from being secreted, thus preventing ovulation. Many U.S. women do not nurse a child constantly, so lactation is not a reliable birth control method for them. If a woman is lactating, she should always use another form of birth control as well, because she will not know when her ovulation cycle begins again.

Intrauterine devices (IUDs) are T-shaped, flexible plastic structures inserted into the uterus by a health-care provider. Once in place, the IUD prevents fertilization from occurring, although researchers aren't sure specifically how. The IUD may contain copper or a synthetic progestin, both of which help prevent implantation. An IUD made of copper may stay in place for up to 10 years, while one made of progestin must be removed and replaced each year. Rarely, the IUD may be expelled from the uterus or displaced, in which case a pregnancy can occur.

Chemical methods of birth control are very effective if used properly. They include the following:

- **Oral contraceptives**, commonly called *birth control pills,* typically come in 28-day packets. The first 21 days of pills contain low levels of estrogen and/or progestins, and the last 7 days are sugar pills. (*Note:* progesterone is one type of progestin.) The low levels of estrogen and progestins prevent the LH "spike" needed for ovulation. Thus, oral contraceptives prevent ovulation. During the 7 days of sugar pills, the circulating levels of estrogen and progestins drop, and menstruation occurs. Typically, menstrual flow is much lighter when a woman takes oral contraceptives

because the circulating levels of hormones were low to begin with, so the uterine lining does not build up much. Oral contraceptives require a woman to take a pill a day, at about the same time each day. If she misses one or more days of pills, ovulation may occur.

- **Estrogen/progestin** (prō-jes'tin; *pro* = before; gestation) **patches** are alternatives to the daily oral contraceptive. A patch placed on the body delivers a regular amount of estrogen/progestin through the skin (transdermally). The patch is replaced each week.
- **Implanted/injected progestins** help prevent pregnancy by preventing ovulation and thickening the mucus around the cervix (thus creating a slight physical barrier to the sperm). Depo-Provera is an injectable contraceptive given once every 3 months, while Implanon is an implantable contraceptive that lasts for up to 3 years. The drawback is that ovulation may not occur for many months after stopping the injections.
- **Morning-after pills** (e.g., Plan B) can be taken within 72 hours after having unprotected intercourse. These pills work by inhibiting ovulation, altering the menstrual cycle to delay ovulation, or irritating the uterine lining to prevent implantation.
- **Mifepristone** (Mifiprex in the United States; RU486 in Europe) was approved in 2000 by the U.S. Food and Drug Administration for use during the first 7 weeks of pregnancy. Mifepristone blocks progesterone receptors, so progesterone cannot attach to these receptors and thereby maintain a pregnancy. When taken with a prostaglandin drug, mifepristone induces a miscarriage. Mifepristone's existence is very politically charged, with both sides of the abortion debate arguing for or against its use.

The **surgical methods** of contraception are **tubal ligation** for females and **vasectomy** (va-sek'tō-mē) for males. In a tubal ligation, both uterine tubes are cut, and the ends are tied off or cauterized shut. Thus, tubal ligation prevents both sperm from reaching the oocyte and the oocyte from reaching the uterus.

A vasectomy is an outpatient procedure whereby each ductus deferens is cut, a short segment is removed, and then the ends are tied off. Sperm cannot leave the testis and thus are broken down and resorbed.

Both surgeries are very effective birth control methods, but they are usually permanent and irreversible, so they are not options for people who wish to have more children.

---

during pregnancy. Tiny ducts drain milk from the alveoli and lobules. The tiny ducts of the lobules merge and form 10 to 20 larger channels called **lactiferous** (lak-tif'er-ŭs; *lact* = milk, *fero* = to bear) **ducts.** A lactiferous duct drains breast milk from a single lobe. As each lactiferous duct approaches the nipple, its lumen expands to form a **lactiferous sinus,** a space where milk is stored prior to release from the nipple.

Breast milk is released by a process called **lactation** (lak-tā'shŭn; *lactatio* = to suckle), which occurs in response to a complex sequence of internal and external stimuli. Normally, a woman starts to produce breast milk when she has recently given birth. When a woman is pregnant, the levels of estrogen, progesterone,

and prolactin rise dramatically. Recall from chapter 20 that the hormone **prolactin** is produced in the anterior pituitary and is responsible for milk production. Thus, when the amount of prolactin increases, the mammary gland grows and forms more expanded and numerous alveoli.

While prolactin stimulates production of breast milk, the hormone **oxytocin** is responsible for milk ejection. Recall from chapter 20 that oxytocin is secreted by the hypothalamus and stored in the posterior lobe of the pituitary, and is also responsible for uterine contractions during labor. In response to a stimulus, such as a baby crying or sucking the nipple, oxytocin is released, and milk is ejected from the nipple. As milk drains from the alveoli, increased levels of prolactin

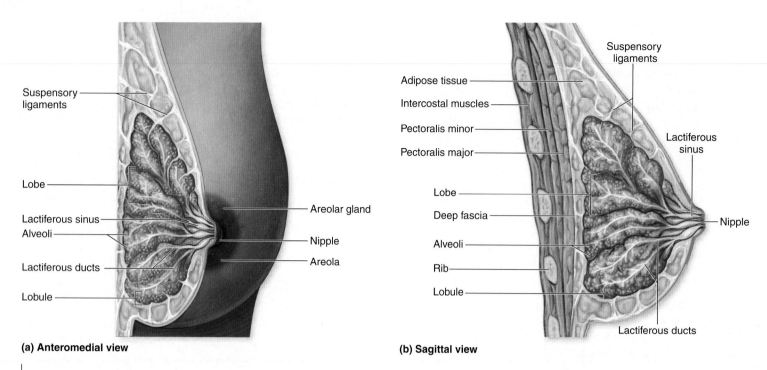

(a) Anteromedial view

(b) Sagittal view

Figure 28.10

Mammary Glands. The mammary glands are composed of glandular tissue and a variable amount of fat. *(a)* An anterior view is partially cut away to reveal internal structures. *(b)* A diagrammatic sagittal section of a mammary gland shows the distribution of alveoli within lobules and the extension of ducts to the nipple.

CLINICAL VIEW

Breast Cancer

Breast cancer affects approximately 1 in every 8 women in the United States, and it also occurs in males, although infrequently. The incidence of breast cancer is rare before age 20. Then it rises steadily to peak at about the age of menopause. Some well-documented risk factors for breast cancer include: maternal relatives with breast cancer, longer reproductive span (early menarche coupled with delayed menopause), obesity, nulliparity (never having been pregnant), late age at first pregnancy, and the presence of specific breast cancer genes (BRCA1 and BRCA2). Except for the genetic influence, all of the risk factors are related to increased exposure to estrogen over a long period of time.

Breast cancers arise from the duct epithelium, not the actual milk-producing cells. Monthly self-examination has proved to be one of the most important means of early detection of breast malignancies. Mammography, which is an x-ray of the breast that can detect small areas of increased tissue density, can identify many small malignancies that are not yet palpable in a self-examination. Recommendations vary, but most physicians agree that women over the age of 40 should have a mammogram done every 1 to 2 years. Women with a family history of breast cancer should consider regular mammography before the age of 40.

Because the lymph drainage from the breast goes predominately to the axilla, the axillary lymph nodes on the side with the cancer must be examined to see if the malignancy has spread. If it has, treatment depends upon the stage of the malignancy, but usually includes surgery and/or chemotherapy. Patients often take drugs that block the effect of the estrogen receptor (e.g., Tamoxifen® and Raloxifene®) for years after the surgery.

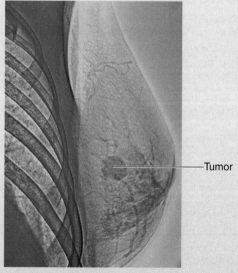

Mammogram showing a cancerous tumor.

are released so the breast can produce more milk. Once a baby stops nursing, the levels of oxytocin drop, and milk ejection ceases.

During the first few weeks of breast-feeding, the mother may experience uterine contractions called **afterpains,** which are caused by the increased levels of oxytocin in her bloodstream. These uterine contractions help shrink the uterus to its prepregnancy size. Afterpains typically become less noticeable and cease a few weeks after birth, by which time the uterus has shrunk considerably.

WHAT DID YOU LEARN?

7 Identify and describe the ligaments that support the uterus and hold it in place.

8 What name is given to the innermost tunic of the uterus? What are the two distinct layers that form this tunic?

9 What mammary gland structures produce and drain milk from each lobe?

Anatomy of the Male Reproductive System

Key topics in this section:

■ Gross and microscopic anatomy of the testes
■ Spermatogenesis and spermiogenesis

■ The male reproductive duct system and the function of each component
■ Anatomy and function of the male accessory glands
■ Components of the penis

In the male, the primary sex organs are the **testes** (tes′tēz; sing., *testis*). The accessory sex organs include a complex set of ducts and tubules leading from the testes to the penis, a group of male accessory glands, and the penis, which is the organ of copulation (**figure 28.11**).

Scrotum

The ideal temperature for producing and storing sperm is about 3° Celsius lower than internal body temperature. The **scrotum** (skrō′tŭm), which is a skin-covered sac between the thighs, provides the cooler environment needed for normal sperm development and maturation (**figure 28.12**). The scrotum is homologous to the labia majora in the female.

Externally, the scrotum contains a distinct, ridgelike seam at its midline, called the **raphe** (rā′fē; *rhaphe* = seam). The raphe persists in an anterior direction along the inferior surface of the penis and in a posterior direction to the anus. The wall of the scrotum is composed of an external layer of skin, a thin layer of superficial fascia immediately internal to the skin, and a layer of smooth muscle, the **dartos** (dar′tos; skinned) **muscle,** immediately internal to the fascia.

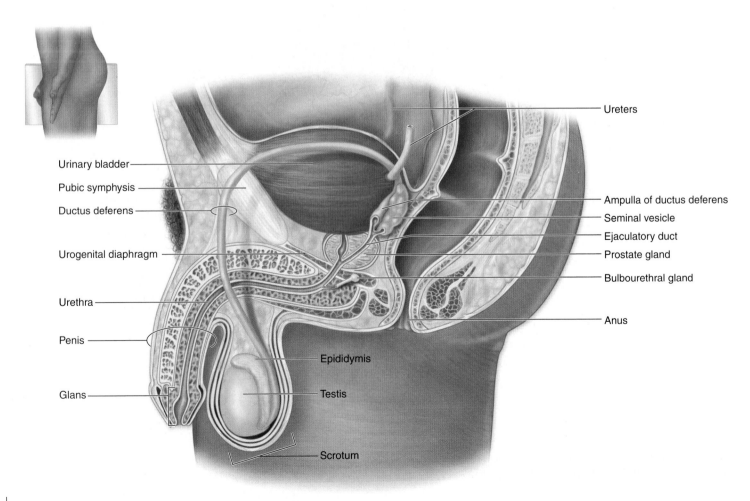

Urinary bladder	Ureters
Pubic symphysis	
Ductus deferens	Ampulla of ductus deferens
	Seminal vesicle
	Ejaculatory duct
Urogenital diaphragm	Prostate gland
	Bulbourethral gland
Urethra	
	Anus
Penis	
	Epididymis
Glans	
	Testis
	Scrotum

Figure 28.11

Male Pelvic Region. A diagrammatic sagittal section shows the locations and relationships of the male pelvic structures.

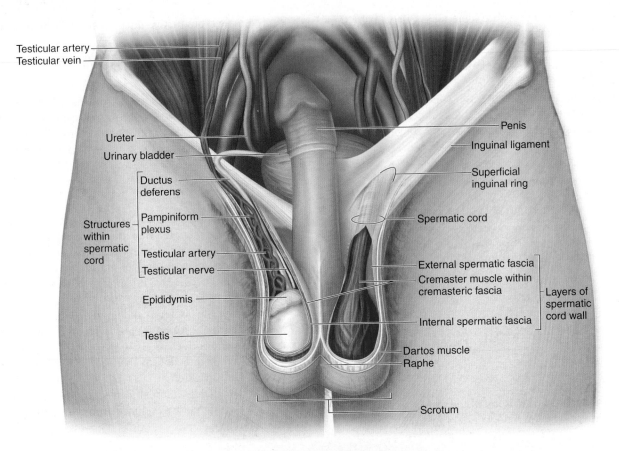

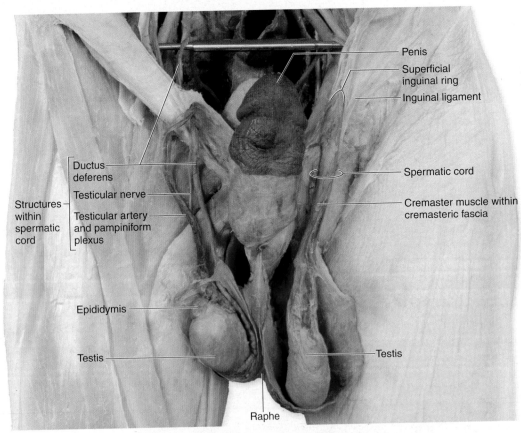

Figure 28.12

Scrotum and Testes. A diagram and a cadaver photo show the scrotum, a skin-covered sac that houses the testes, in anterior view. A multilayered spermatic cord houses the blood vessels, nerves, and a sperm-carrying duct (ductus deferens) for each testis.

When the testes are exposed to elevated temperatures, the dartos muscle relaxes, which unwrinkles the skin of the scrotum and allows the testes to move further away from the body. This inferior movement away from the body cools the testes. At the same time, another muscle (the cremaster muscle) also relaxes to allow the testes to move inferiorly away from the body. The opposite occurs if the testes are exposed to cold. In this case, the dartos and cremaster muscles contract, pulling the testes and scrotum closer to the body in order to conserve heat.

Spermatic Cord

The blood vessels and nerves to the testis travel from within the abdomen to the scrotum in a multilayered structure called the **spermatic cord** (figure 28.12; see **figure 28.13a**). The spermatic cord originates in the **inguinal canal,** a tubelike passageway through the inferior abdominal wall. The spermatic cord wall consists of three layers:

- An **internal spermatic fascia** is formed from fascia deep to the abdominal muscles.
- The **cremaster** (krē-mas′ter; a suspender) **muscle** and **cremasteric fascia** are formed from muscle fiber extensions of the internal oblique muscle and its aponeurosis, respectively.
- An **external spermatic fascia** is formed from the aponeurosis of the external oblique muscle.

Within the spermatic cord is a singular **testicular artery** that is a direct branch from the abdominal aorta. The testicular artery is surrounded by a plexus of veins called the **pampiniform** (pam-pin′i-form; *pampinus* = tendril, *forma* = form) **plexus.** This venous plexus is a means to provide thermoregulation by pre-cooling arterial blood prior to reaching the testes. Autonomic nerves travel with these vessels and innervate the testis.

Testes

In the adult human male, each testis is a relatively small, oval organ housed within the scrotum (figure 28.12). Each weighs approximately 10–12 grams, and displays average dimensions of 4 centimeters (cm) in length, 2 cm in width, and 2.5 cm in antero-posterior diameter. The testes produce sperm and androgens (male sex hormones).

Each testis is covered both anteriorly and laterally by a serous membrane, the **tunica vaginalis** (văj-in-ăl′ĭs; ensheathing).

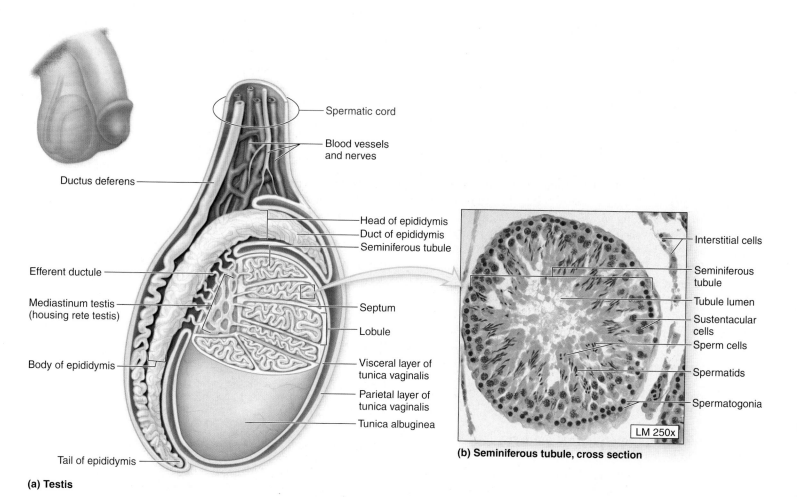

(a) Testis

(b) **Seminiferous tubule, cross section**

Figure 28.13

Testes and Seminiferous Tubules.   (*a*) The gross anatomy of a testis is shown diagrammatically in a cut-away, partial sagittal section. (*b*) A photomicrograph reveals a cross section of a seminiferous tubule in the testis.

This membrane is derived from the peritoneum of the abdominal cavity. The tunica vaginalis has an outer **parietal layer** and an inner **visceral layer** that are separated by serous fluid. A thick, whitish, fibrous capsule called the **tunica albuginea** covers the testis and lies immediately deep to the visceral layer of the tunica vaginalis (figure 28.13a). At the posterior margin of the testis, the tunica albuginea thickens and projects into the interior of the organ as the **mediastinum testis.** Blood vessels, a system of ducts, lymph vessels, and some nerve fibers enter or leave each testis through the mediastinum.

The tunica albuginea projects internally into the testis and forms delicate connective tissue **septa,** which subdivide the internal space into about 250 separate **lobules.** Each lobule contains up to four extremely convoluted, thin and elongated **seminiferous** (sem'i-nif'er-ŭs; *semen* = seed, *fero* = to carry) **tubules.** The seminiferous tubules contain two types of cells: (1) a group of nondividing support cells called the **sustentacular** (sŭs-ten-tak'ū-lăr; *sustento* = to hold upright) (*Sertoli* or *nurse*) **cells,** and (2) a population of dividing germ cells that continuously produce sperm cells beginning at puberty.

The sustentacular cells assist with sperm development. These cells provide a protective environment for the developing sperm cells, and their cytoplasm helps nourish the developing sperm (figure 28.13b). In addition, the sustentacular cells will release the hormone **inhibin** when sperm count is high. Inhibin inhibits FSH secretion, and thus regulates sperm production. (Conversely, when sperm count declines, inhibin secretion decreases.)

The sustentacular cells are secured together by tight junctions, which form a **blood-testis barrier** that is similar to the blood-brain barrier. The blood-testis barrier helps protect developing sperm cells from materials in the bloodstream. It also protects the sperm from the body's leukocytes, which may perceive the sperm as foreign since they have different chromosome numbers and arrangements from the male's other body cells.

The spaces surrounding the seminiferous tubules are called **interstitial spaces.** Within these spaces reside the **interstitial cells** (or *Leydig cells*). Luteinizing hormone stimulates the interstitial cells to produce hormones called **androgens** (an'drŏ-jen; *andros* = male human). There are several types of androgens, the most common one being testosterone. Although the adrenal cortex secretes a small amount of androgens, the vast majority of androgen release is via the interstitial cells in the testis, beginning at puberty. These hormones cause males to develop the classic characteristics of axillary and pubic hair, deeper voice, and sperm production.

WHAT DO YOU THINK?

3 If a male's testes were removed, would he still be able to produce androgens?

Development of Sperm: Spermatogenesis and Spermiogenesis

Spermatogenesis (sper'mă-tō-jen'ĕ-sis; *genesis* = origin) is the process of sperm development that occurs within the seminiferous tubule of the testis. Spermatogenesis does not begin until puberty, when significant levels of FSH and LH stimulate the testis to begin gamete development.

The process of spermatogenesis is shown in **figure 28.14a.** All sperm form from primordial germ (stem) cells called **spermatogonia** (sper'mă-tō-gō'nē-ă; sing., *spermatogonium*; *sperma* = seed, *gone* = generation). Spermatogonia are diploid cells (meaning they have 23 pairs of chromosomes for a total of 46). These cells lie near the base of the seminiferous tubule, surrounded by the cytoplasm of a sustentacular cell. To produce sperm, spermatogonia first divide by mitosis. One of the cells produced is a new spermatogonium (a new germ cell), to ensure that the numbers of spermatogonia never become depleted, and the other cell is a "committed cell" called **primary spermatocyte.** Primary spermatocytes are diploid and an exact copy of spermatogonia. It is the primary spermatocytes that undergo meiosis.

When a primary spermatocyte undergoes meiosis I, the two cells produced are called **secondary spermatocytes.** Secondary spermatocytes are haploid, meaning they have 23 chromosomes only. These cells are still surrounded by the sustentacular cell cytoplasm, but they are relatively closer to the lumen of the seminiferous tubule (as opposed to the base of the seminiferous tubule).

Secondary spermatocytes complete meiosis (go through meiosis II) and form **spermatids** (sper'mă-tid). A spermatid is a haploid cell and is surrounded by the sustentacular cell cytoplasm, very near to the lumen of the seminiferous tubule. The spermatids still have a circular appearance, rather than the sleek shape of mature sperm.

In the final stage of spermatogenesis, a process called **spermiogenesis,** the newly formed spermatids differentiate to become anatomically mature **spermatozoa** (sing., *spermatozoon*; sper'mă-to-zo'on) or **sperm** (figure 28.14b). During spermiogenesis, the spermatid sheds its excess cytoplasm, and the nucleus elongates. A structure called the **acrosome** (ak'rō-sōm; *akros* = tip, *soma* = body) **cap** forms over the nucleus. This acrosome cap contains digestive enzymes that help penetrate the secondary oocyte for fertilization. As the spermatid elongates, a **tail** (*flagellum*) forms from the organized microtubules. The tail attaches to a **midpiece** (*neck*) region containing mitochondria and a centriole. These mitochondria provide the energy to move the tail.

Although the sperm look mature, they do not yet have all of the characteristics needed to successfully travel through the female reproductive tract and fertilize an oocyte. The sperm must leave the seminiferous tubule through a network of ducts (described next) and reside in the epididymis for a period of time in order to become fully motile.

Table 28.5 summarizes the stages of spermatogenesis.

Study Tip!

Use these hints to help identify the various testis cells under the microscope:

1. Spermatogonia are closest to the seminiferous tubule base. As spermatogenesis occurs and the cells mature, more mature cells are found closer to the lumen of the seminiferous tubule. Note how close the spermatozoa are to the lumen.

2. Sometimes it is difficult to distinguish an entire sustentacular cell because its cytoplasm is pale and surrounds the developing sperm. You can identify sustentacular cells by their nucleus, which is oval or flattened and usually has a prominent nucleolus.

3. The interstitial cells are not located within the seminiferous tubule but external to it, usually in clumps of three or more cells.

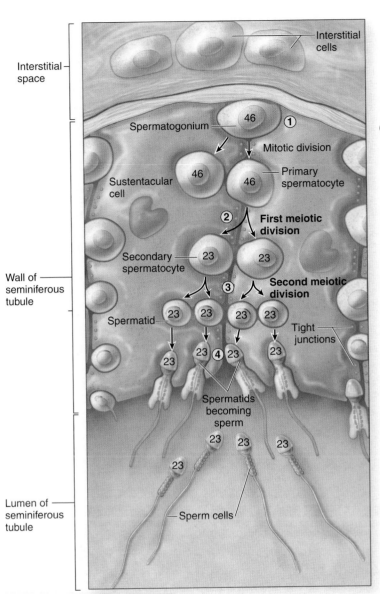

(1) Germ cells that are the origin of sperm cells are *diploid cells* (containing 46 chromosomes, or 23 pairs) called spermatogonia. Mitotic divisions of these cells produce a new germ cell and a committed cell. The committed cell is a primary spermatocyte.

(2) The first meiotic division begins in the *diploid* primary spermatocytes. The *haploid cells* (containing 23 chromosomes only) produced by the first meiotic division are called secondary spermatocytes.

(3) The second meiotic division originates with the secondary spermatocytes and produces spermatids.

(4) The process of spermiogenesis begins with spermatids and results in morphological changes needed to form sperm that will be motile.

(a) Spermatogenesis

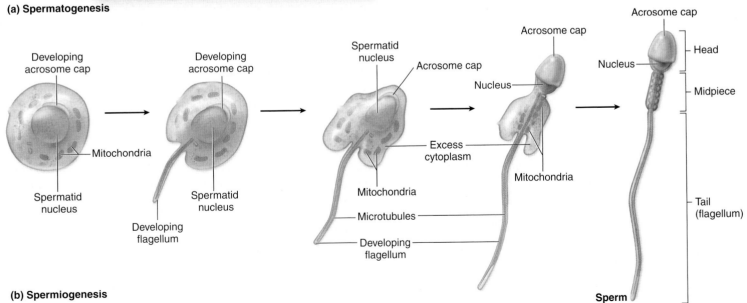

(b) Spermiogenesis

Figure 28.14

Spermatogenesis and Spermiogenesis.   *(a)* The processes of spermatogenesis and spermiogenesis take place within the wall of the seminiferous tubule. *(b)* Structural changes occur during spermiogenesis as a sperm forms from a spermatid.

Table 28.5	Stages of Spermatogenesis		
Cell Type	**Number of Chromosomes**	**Haploid or Diploid**	**Action**
Spermatogonium	23 pairs (46)	Diploid	Divides by mitosis to produce a new spermatogonium and a primary spermatocyte
Primary spermatocyte	23 pairs (46)	Diploid	Completes meiosis I to produce secondary spermatocytes
Secondary spermatocyte	23	Haploid	Completes meiosis II to produce spermatids
Spermatid	23	Haploid	Undergoes spermiogenesis, where most of its cytoplasm is shed, and a midpiece, tail, and head form
Spermatozoon (sperm)	23	Haploid	Leaves seminiferous tubule and matures in epididymis

WHAT DID YOU LEARN?

10 What is the scrotum? It is homologous to what structure in the female?

11 What is the function of the interstitial cells, and where are they located within the testis?

12 Describe the process of spermatogenesis, and mention when it first occurs.

Ducts in the Male Reproductive System

The left and right testes each have their own set of ducts. These ducts store and transport sperm as they mature and pass out of the male body **(figure 28.15)**.

Ducts Within the Testis

The **rete** (rē'tē; net) **testis** is a meshwork of interconnected channels in the mediastinum testis that receive sperm from the seminiferous tubules. The rete testis is lined by simple cuboidal epithelium with short microvilli covering its luminal surface. The channels of the rete testis merge to form the efferent ductules (see figure 28.13).

Approximately 12–15 **efferent ductules** (duk'tool) connect the rete testis to the epididymis. They are lined with both ciliated columnar epithelia that gently propel the sperm toward the epididymis and nonciliated columnar epithelia that absorb excess fluid secreted by the seminiferous tubules. The efferent ductules drain into the epididymis.

Epididymis

The **epididymis** (ep-i-did'i-mis; pl., *epididymides*; *epi* = upon, *didymis* = twin) is a comma-shaped structure composed of an internal duct and an external covering of connective tissue. Its **head** lies on the superior surface of the testis, while the **body** and **tail** are on the posterior surface of the testis (see figure 28.13*a*). Internally, the epididymis contains a long, convoluted **duct of the epididymis,** which is approximately 4 to 5 meters in length and lined with pseudostratified columnar epithelium that contains stereocilia (long microvilli) (figure 28.15*c*).

The epididymis stores sperm until they are fully mature and capable of being motile. Just as a newborn has the anatomic char-

acteristics of an adult, but cannot move as an adult, the sperm that first enter the epididymis look like mature sperm but can't move like mature sperm. If they are expelled too soon, they lack the ability to be motile, which is necessary to travel through the female reproductive tract and fertilize a secondary oocyte. If sperm are not ejected from the male reproductive system in a timely manner, the old sperm degenerate and are resorbed by cells lining the duct of the epididymis.

Ductus Deferens

When sperm leave the epididymis, they enter the **ductus deferens** (děf'er-ens; carry away), also called the *vas deferens*. The ductus deferens is a thick-walled tube that travels within the spermatic cord, through the inguinal canal, and then within the pelvic cavity before it nears the prostate gland (figure 28.15*a*). The wall of the ductus deferens is composed of an inner **mucosa** (lined with pseudostratified ciliated columnar epithelium), a middle **muscularis,** and an outer **adventitia** (figure 28.15*b*). The muscularis contains three layers of smooth muscle: an inner longitudinal, middle circular, and outer longitudinal layer. Contraction of the muscularis is necessary to move sperm cells through the ductus deferens, since sperm do not exhibit motility until they are ejaculated from the penis.

When the ductus deferens travels through the inguinal canal and enters the pelvic cavity, it separates from the other spermatic cord components and extends posteriorly along the superolateral surface of the bladder. It then travels inferiorly and terminates close to the region where the bladder and prostate gland meet. As the ductus deferens approaches the superoposterior edge of the prostate gland, it enlarges and forms the **ampulla** of the ductus deferens (figure 28.15*a*). The ampulla of the ductus deferens unites with the proximal region of the seminal vesicle to form the terminal portion of the reproductive duct system, called the ejaculatory duct.

Ejaculatory Duct

Each **ejaculatory duct** is between 1 and 2 centimeters long. The epithelium of the ejaculatory duct is a pseudostratified ciliated columnar epithelium. The ejaculatory duct conducts sperm (from the ductus deferens) and a component of seminal fluid (from the seminal vesicle) toward the urethra. Each ejaculatory duct opens into the prostatic urethra.

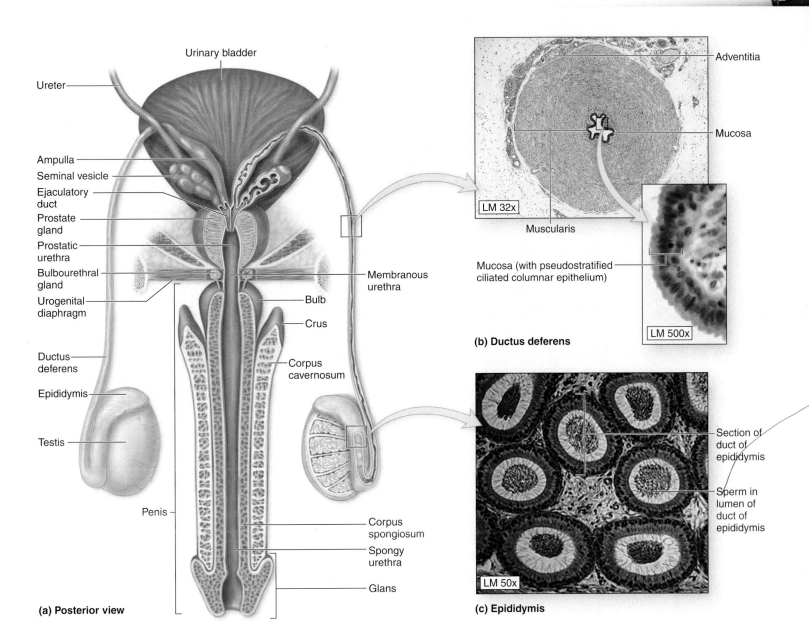

Figure 28.15

Duct System in the Male Reproductive Tract.   (a) A posterior view depicts the structural components of the male reproductive ducts. (b) Micrographs show a cross section through the ductus deferens. (c) A micrograph shows a cross section through the epididymis.

Urethra

The **urethra** transports semen from both ejaculatory ducts to the outside of the body. Recall from chapter 27 that the urethra is subdivided into a **prostatic** (pros-tat′ik) **urethra** that extends through the prostate gland (see figure 28.15), a **membranous urethra** that travels through the urogenital diaphragm, and a **spongy urethra** that extends through the penis.

Accessory Glands

Recall from earlier in this chapter that the vagina has a highly acidic environment to prevent bacterial growth. Sperm cannot survive in this type of environment, so an alkaline secretion called **seminal** (sem′i-nal) **fluid** is needed to neutralize the acidity of the vagina. In addition, as the sperm travel through the female reproductive tract (a process that can take hours to several days), they are nourished

by nutrients within the seminal fluid. The components of seminal fluid are produced by accessory glands: the seminal vesicles, the prostate gland, and the bulbourethral glands.

Seminal Vesicles

The paired **seminal vesicles** are located on the posterior surface of the urinary bladder lateral to the ampulla of the ductus deferens (figure 28.15a). Each seminal vesicle is an elongated, hollow organ approximately 5–8 centimeters long. The wall of each vesicle contains mucosal folds of pseudostratified columnar epithelium (**figure 28.16a**). It is the medial (proximal) portion of the seminal vesicle that merges with a ductus deferens to form the ejaculatory duct.

The seminal vesicles secrete a viscous, whitish-yellow, alkaline fluid containing both fructose and prostaglandins. The fructose is a sugar that nourishes the sperm as they travel through the female

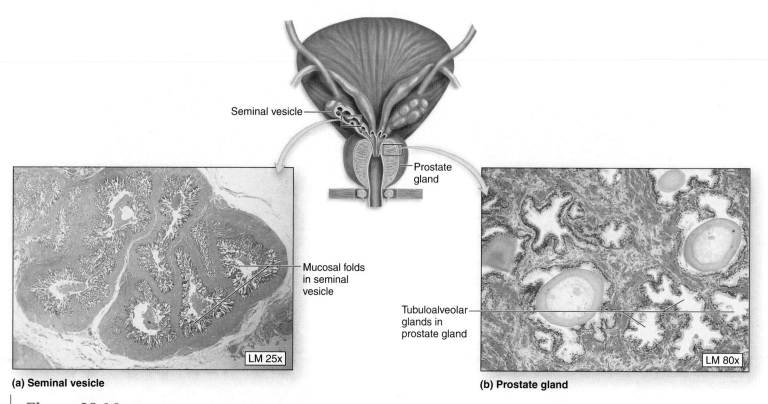

Seminal vesicle

Prostate gland

Mucosal folds in seminal vesicle

LM 25x

(a) Seminal vesicle

Tubuloalveolar glands in prostate gland

LM 80x

(b) Prostate gland

Figure 28.16

Seminal Vesicles and Prostate Gland.   A drawing depicts the relative locations of the seminal vesicles and the prostate gland. Micrographs show sections through *(a)* a seminal vesicle and *(b)* the prostate gland.

reproductive tract. Prostaglandins are hormonelike substances that promote the widening and slight dilation of the external os of the cervix, which facilitates sperm entry into the uterus.

Prostate Gland

The **prostate** (pros′tāt; one who stands before) **gland** is a compact, encapsulated organ that weighs about 20 grams and is shaped like a walnut, measuring approximately 2 cm by 3 cm by 4 cm. It is located immediately inferior to the bladder. The prostate gland includes submucosal glands that produce mucin and more than 30 tubuloalveolar glands that open directly through numerous ducts into the prostatic urethra (figure 28.16*b*). Together, these glands contribute a component to the seminal fluid.

The prostate gland secretes a slightly milky fluid that is weakly acidic and rich in citric acid, seminalplasmin, and prostate-specific antigen (PSA). The citric acid is a nutrient for sperm health, the seminalplasmin is an antibiotic that combats urinary tract infections in the male, and the PSA acts as an enzyme to help liquify semen following ejaculation. (Note that the slightly acidic secretion of the prostate does not cause the seminal fluid to be acidic, and thus the seminal fluid still functions to neutralize the acidity of the vagina.)

Bulbourethral Glands

Paired, pea-shaped **bulbourethral** (bŭl′bō-ū-rē′thrăl) **glands** (or *Cowper glands*) are located within the urogenital diaphragm on each side of the membranous urethra (see figures 28.11 and 28.15). Each gland has a

short duct that projects into the bulb (base) of the penis and enters the spongy urethra. Bulbourethral glands are tubuloalveolar glands that have a simple columnar and pseudostratified columnar epithelium. Their secretory product is a clear, viscous mucin that forms mucus. As a component of the seminal fluid, this mucus protects the urethra and serves as a lubricant during sexual intercourse.

Semen

Seminal fluid from the accessory glands combines with sperm from the testes to make up **semen** (sē′men; seed). When released during intercourse, semen is called the **ejaculate** (ē-jak′ū-lāt), and it normally measures about 3 to 5 milliliters in volume and contains approximately 200 to 500 million spermatozoa. In a sexually active male, the average transit time of human spermatozoa—from their release into the lumen of the seminiferous tubules, passage through the duct system, and appearance in the ejaculate—is about 2 weeks. Since semen is composed primarily of seminal fluid, a male who is very active sexually may have a reduced sperm count because there are fewer sperm to be released from the epididymis; however, the total semen volume remains close to normal for that individual.

WHAT DO YOU THINK?

4   If a male has a vasectomy, is he still able to produce sperm? If so, what happens to those sperm? How is the composition of semen changed in an individual who has had a vasectomy?

CLINICAL VIEW

Benign Prostatic Hyperplasia and Prostate Cancer

Benign prostatic hyperplasia (BPH) is a noncancerous enlargement of the prostate gland. BPH is a common disorder in older men; in fact, its incidence is greater than 90% for men over 80 years of age. Hormonal changes in aging males are the cause of the enlargement.

In BPH, large, discrete nodules form within the prostate and compress the prostatic urethra. Thus, the patient has difficulty starting and stopping a stream of urine, and often complains of **nocturia** (excessive urinating at night), **polyuria** (more-frequent urination), and **dysuria** (painful urination). Some drug regimens help inhibit hormones that cause prostate enlargement, but when medications are no longer effective, surgical removal of the prostatic enlargement is indicated. The most commonly performed surgical procedure is called a **TURP (transurethral resection of the prostate),** in which an instrument called a **resectoscope** (rē-sek′tō-skōp) is inserted into the urethra to cut away the problematic enlargement.

Prostate cancer is one of the most common malignancies among men over 50, and the risk of developing it increases with age. Prostate cancer forms hard, solid nodules, most often in the posterior part of the prostate gland. Early stages of the cancer are generally asymptomatic,

but as it progresses, urinary symptoms may develop. Untreated prostate cancer can metastasize to other body organs.

Early diagnosis and treatment of prostate cancer are vital for cure and long-term survival. A very effective screening tool is a **digital rectal exam,** whereby a physician inserts a finger into the rectum and palpates adjacent structures (including the prostate gland). In addition, most physicals for men over the age of 50 now include a test for **prostate-specific antigen (PSA)** in the blood. The PSA level in a healthy man is typically less than 4 ng/mL. An elevated PSA level can indicate either benign prostatic hyperplasia or prostate cancer. A needle biopsy of the prostate tissue can confirm the diagnosis of cancer.

Several treatment options are available, depending on the stage of the cancer. For earlier stages of the disease, radiation therapy may be beneficial—either traditional external-beam radiation or **interstitial radiotherapy,** in which radioactive palladium or iodine "seeds" are permanently implanted in the prostate. For patients with a more aggressive cancer, the entire prostate and some surrounding structures are surgically removed, a procedure called a **radical prostatectomy.** No matter what the form of treatment, the physician continues to measure PSA levels in the patient's blood to make sure all the cancerous structures have been removed and to check for recurrence.

Penis

The **penis** (pē′nis; tail) and the scrotum form the external genitalia in males **(figure 28.17a)**. Internally, the attached portion of the penis is the **root,** which is dilated internal to the body surface, forming both the **bulb** and the **crus of the penis.** The bulb attaches the penis to the bulbospongiosus muscle in the urogenital triangle, and the crus attaches the penis to the pubic arch. The **body,** or *shaft,* of the penis is the elongated, movable portion. The tip of the penis is called the **glans,** and it contains the **external urethral orifice.** The skin of the penis is thin and elastic. At the distal end of the penis, the skin is attached to the raised edge of the glans and forms a circular fold called the **prepuce** (*foreskin*) (see Clinical View, p. 866).

Within the shaft of the penis are three cylindrical erectile bodies (figure 28.17b). The paired **corpora cavernosa** (kav′er-nō-să; sing., *corpus cavernosum; caverna* = grotto) are located dorsolaterally. Ventral to them in the midline is the single **corpus spongiosum** (spŭn′jē-ō-sŭm), which contains the spongy urethra. Each corpus cavernosum terminates in the shaft of the penis, while the corpus spongiosum continues within the glans. The erectile bodies are ensheathed by the tunica albuginea, which also provides an attachment to the skin over the shaft of the penis.

The erectile bodies are composed of a complex network of **venous spaces** surrounding a central artery. During sexual excitement, blood enters the erectile bodies via the central artery and fills in the venous spaces. As these venous spaces become engorged with blood, the erectile bodies become rigid, a process called **erection** (ē-rek′shŭn; *erecto* = to set up). The rigid erectile bodies compress the veins that drain blood away from the venous spaces. Thus, the spaces fill with blood, but the blood cannot leave the erectile bodies until the sexual excitement ceases. Parasympathetic innervation (via

the pelvic splanchnic nerves) is responsible for increased blood flow and thus the erection of the penis.

Ejaculation (ē-jak-ū-lā′shŭn; *eiaculatus* = to shoot out) is the process by which semen is expelled from the penis with the help of rhythmic contractions of the smooth muscle in the wall of the urethra. Sympathetic innervation (from the lumbar splanchnic nerves) is responsible for ejaculation.

Although in most body systems sympathetic and parasympathetic innervation tend to perform opposite functions, the male reproductive system is an exception. Here, parasympathetic innervation is necessary to achieve an erection, while sympathetic innervation promotes ejaculation. Relaxation of autonomic activity after sexual excitement reduces blood flow to the erectile bodies and shunts most of the blood to other veins, thereby returning the penis to its flaccid condition.

Study Tip!

One way to remember the autonomic innervation for the penis is this phrase: "Point and Shoot!" The **p** in **p**oint (erection) also stands for **p**arasympathetic innervation, while the **s** in **s**hoot (ejaculation) stands for **s**ympathetic innervation.

WHAT DID YOU LEARN?

13 What two structures unite to form the ejaculatory duct?

14 What is the composition of semen, and what structures contribute to semen?

15 Specifically, how do both parasympathetic and sympathetic innervation work on penile function during sexual arousal?

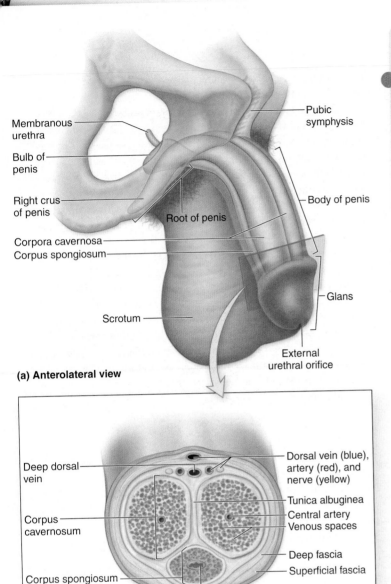

Membranous urethra

Bulb of penis

Right crus of penis

Root of penis

Corpora cavernosa
Corpus spongiosum

Scrotum

Pubic symphysis

Body of penis

Glans

External urethral orifice

(a) Anterolateral view

Deep dorsal vein

Corpus cavernosum

Corpus spongiosum

Skin   Spongy urethra

Dorsal vein (blue), artery (red), and nerve (yellow)

Tunica albuginea
Central artery
Venous spaces

Deep fascia
Superficial fascia

(b) Cross section

Figure 28.17

Anatomy of the Penis. *(a)* An anterolateral view of the circumcised penis. *(b)* A diagrammatic, transverse section shows the arrangement of the erectile bodies.

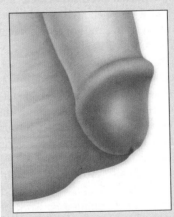
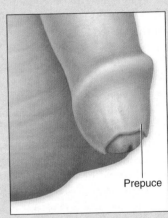

CLINICAL VIEW

Sexually Transmitted Diseases

Sexually transmitted diseases (STDs), also known as *venereal diseases,* are a group of infectious diseases that are usually transmitted via sexual contact. The incidence of STDs has been rising in recent years because individuals are having sexual intercourse at younger ages, and may have multiple sexual partners in their lifetime. Many times, the symptoms of STDs are not immediately noticeable, so infected individuals may spread the disease to someone else without realizing it. Mothers also may transmit STDs to their newborns, either directly across the placenta or at the time of delivery. Condoms have been shown to help prevent the spread of STDs, but they are not 100% effective. Individuals with multiple sex partners should consider being tested routinely for some of the more common STDs, because they may unknowingly be spreading one or more of these conditions.

STDs are a leading cause of **pelvic inflammatory disease** in women, in which the pelvic organs (uterus, uterine tubes, and ovaries) become infected. Should bacteria from an STD infect the uterus and uterine tubes, scarring is likely to follow, leading to blockage of the tubes and infertility. We've already discussed two types of STDs, human papillomavirus (see Clinical View, "Cervical Cancer," earlier in this chapter) and AIDS (in chapter 24). We now explore other common STDs.

Chlamydia (kla-mid'ē-ă) is the most frequently reported bacterial STD in the United States. The responsible agent is *Chlamydia trachomatis.* Most infected people are asymptomatic, while the rest develop symptoms within 1 to 3 weeks after exposure. These symptoms include abnormal vaginal discharge, painful urination (in both males and females), and low back pain. Chlamydia is treated with antibiotics.

Genital herpes (her'pēz; *herpo* = to creep) is caused by herpes simplex virus type 1 (HSV-1) or type 2 (HSV-2). Infected individuals undergo cyclic outbreaks of blister formation in the genital and anal regions; the blisters are filled with fluid containing millions of infectious viruses. The blisters then break and turn into tender sores that remain for 2–4 weeks. Typically, future cycles of blistering are less severe and shorter in duration than the initial episode. There is no cure for herpes, but antiviral medications can lessen the severity and length of an outbreak.

Gonorrhea (gon-ō-rē'ă) is caused by the bacterium *Neisseria gonorrhoeae,* and is spread either by sexual contact or from mother to newborn at the time of delivery. Symptoms include painful urination and/or a yellowish discharge from the penis or vagina. Gonorrhea is treated with antibiotics, although in recent years many gonorrhea strains have become resistant to some antibiotics. If untreated, women may develop pelvic inflammatory disease, and men may develop epididymitis, a painful condition of the epididymis that can lead to infertility. If a newborn acquires the disease, then blindness, joint problems, and/or a life-threatening blood infection may result.

Syphilis (sif'i-lis) is caused by the corkscrew-shaped bacterium *Treponema pallidum.* The bacterium is spread sexually via contact with a syphilitic sore (called a **chancre**), or a newborn may acquire it *in utero.* Babies can acquire congenital syphilis from their mothers and are often stillborn, but if they live, they have a high incidence of skeletal malformities and neurologic problems. Syphilis can be treated with antibiotics. A person can become reinfected with the disease if reexposed to the syphilitic sores.

Aging and the Reproductive Systems

Key topic in this section:

■ Age-related changes in the female and male reproductive systems

Our reproductive systems are basically nonfunctional for several years following birth. When we reach puberty, hormonal changes in the hypothalamus and anterior pituitary stimulate the gonads to begin producing sex hormones. Thereafter, changes occur in many body structures, the reproductive organs mature, and the gonads begin to produce gametes. The time of onset of puberty varies among individuals, but it clearly occurs at a younger average age in females and males today than it did 40 or 50 years ago.

After reaching sexual maturity, the female and male reproductive systems exhibit marked differences in their response to aging. Gametes typically stop maturing in females by their 40s or 50s, and menopause occurs. A reduction in hormone production that accompanies menopause causes some atrophy of the reproductive organs and the breasts. The vaginal wall thickness decreases, as do glandular secretions for maintaining a moist, lubricated lining. The uterus shrinks and atrophies, becoming much smaller than it was before puberty.

The lack of significant amounts of estrogen and progesterone in a menopausal woman also affects other organs and body systems. Women may experience "hot flashes," in which their bodies perceive periodic elevations in body temperature, and they may develop thinning scalp hair and/or an increase in facial hair. Menopausal women are at greater risk for osteoporosis (thinning, brittle bones) and heart disease due to the drop in estrogen and progesterone levels. Formerly, **hormone replacement therapy (HRT),** in the form of estrogen and progesterone supplements, was given to menopausal women to help diminish these symptoms and risks. However, recent studies suggest that the risks associated with HRT (such as breast cancer, lack of protection against heart disease, etc.) may outweigh the benefits in older women. As a result, many physicians have stopped prescribing HRT for menopausal symptoms.

In contrast, males do not experience the relatively abrupt change in reproductive system function that females do. A slight decrease in the size of the testes parallels a reduction in the size of the seminiferous tubules and the number of interstitial cells. As a consequence of the reduced number of interstitial cells, decreased testosterone levels in males in their 50s signal a change called the **male climacteric** (klī-mak′ter-ik, klī-mak-ter′ik). However, males generally do not stop producing gametes as females do following menopause.

Most males experience prostate enlargement (either benign or cancerous) as they age. This prostate enlargement can interfere with sexual and urinary functions. Also associated with aging are **erectile dysfunction** and **impotence,** which refer to the inability to achieve or maintain an erection. Besides aging, other risk factors for this condition include heart disease, diabetes, smoking, and prior prostate surgery. Recently, many drugs have entered the market (e.g., Viagra®) that treat erectile dysfunction by prolonging vasodilation of the penile arteries and thus inhibiting relaxation of the erectile bodies.

WHAT DID YOU LEARN?

16 What are some female body changes that accompany menopause?

Development of the Reproductive Systems

Key topics in this section:

- Development of the female and male reproductive systems
- Common embryonic structures and the hormones that influence their development

The female and male reproductive structures originate from the same basic primordia, which differentiate into female or male structures, depending upon the signals the primordia receive. To better explain this process, we must first distinguish between the genetic and phenotypic sex of an individual.

Genetic Versus Phenotypic Sex

Genetic sex (or *genotypic sex*) refers to the sex of an individual based on her or his sex chromosomes. An individual with two X chromosomes is a genetic female, while an individual with one X

and one Y chromosome is a genetic male. Genetic sex is determined at fertilization.

In contrast, **phenotypic** (fē′nō-tip′ik, fen′ō-) **sex** refers to the appearance of an individual's internal and external genitalia. A person with ovaries and female external genitalia (labia) is a phenotypic female, whereas a person with testes and male external genitalia (penis, scrotum) is a phenotypic male. Phenotypic sex starts to become apparent no earlier than the seventh week of development.

How does the primordial tissue know whether to develop into female reproductive organs or male reproductive organs? In males, a **testis-determining factor (TDF) gene** is located on a specific part of the Y chromosome called the sex-determining region. If the Y chromosome is present, the TDF gene produces proteins to stimulate the production of other hormones (e.g., testosterone and other androgens) that initiate male phenotypic development. If a Y chromosome is absent (e.g., the individual is a genetic female), or if the Y chromosome is either lacking or has an abnormal TDF gene, a female phenotypic sex results. The female phenotypic sex may be thought of as the organism's default pattern. This pattern is not changed unless TDF is present.

Formation of Indifferent Gonads and Genital Ducts

Early in the fifth week of embryonic development, paired **genital ridges** (or *gonadal ridges*) form from **intermediate mesoderm.** The genital ridges will form the gonads. These longitudinal ridges are medial to the developing kidneys at about the level of the tenth thoracic vertebra (**figure 28.18**, *top*). Between weeks 5 and 6, primordial **germ cells** migrate from the yolk sac to the genital ridges. These germ cells will form the future gametes (either sperm or oocytes). Shortly thereafter, two sets of duct systems are formed:

- The **mesonephric** (mez-ō-nef′rik) **ducts** (or *Wolffian ducts*) form most of the male duct system. Recall that the mesonephric ducts also connect the mesonephros (intermediate kidney) to the developing urinary bladder.
- The **paramesonephric ducts** (*Müllerian ducts*) form most of the female duct system, including the uterine tubes, uterus, and superior part of the vagina. These ducts appear lateral to the mesonephric ducts.

CLINICAL VIEW

True Hermaphroditism and Pseudohermaphroditism

The term **hermaphrodite** (her-maf′rō-dīt) is derived from the Greek name Hermaphroditus, the mythological son of the Greek god Hermes and the goddess Aphrodite. In general, a hermaphrodite is an individual with both male and female sex characteristics. **True hermaphroditism** refers to an individual with both ovarian and testicular structures and ambiguous (or female) external genitalia. The person may be a genetic male (XY) or a genetic female (XX). True hermaphroditism is very rare, and typically the ovarian and testicular structures are not functional.

Pseudohermaphroditism (soo′dō-her-maf′rō-dī-tizm; *pseudes* = false) refers to an individual whose genetic sex and phenotypic sex do not match. A **male pseudohermaphrodite** is a genetic male (XY) whose external genitalia resemble those of a female (female phenotypic sex). These individuals usually have testes, but the structures that form the scrotum do not fuse completely, so the structure looks more like labia

majora. Male pseudohermaphroditism usually results from a reduction in male hormones (e.g., testosterone) during development; thus, the testis-determining factor (TDF) gene on the Y chromosome is present, but its proteins are insufficient in the absence of testosterone to masculinize the external genitalia.

A **female pseudohermaphrodite** is a genetic female (XX) with external genitalia that resemble those of a male (male phenotypic sex). Although the ovaries and internal genitalia (e.g., uterine tubes and uterus) are female, the external genitalia (clitoris and labia) resemble male external sex organs. The clitoris enlarges to look like a small penis, and/or the two labia may become partially fused to resemble a scrotum. Female pseudohermaphroditism may result if the female fetus is exposed to excessive androgens (e.g., if the pregnant mother was given certain medications to help prevent miscarriage). More commonly, female pseudohermaphroditism is caused by **congenital adrenal hyperplasia**, in which the fetus's adrenal glands produce excessive amounts of androgens.

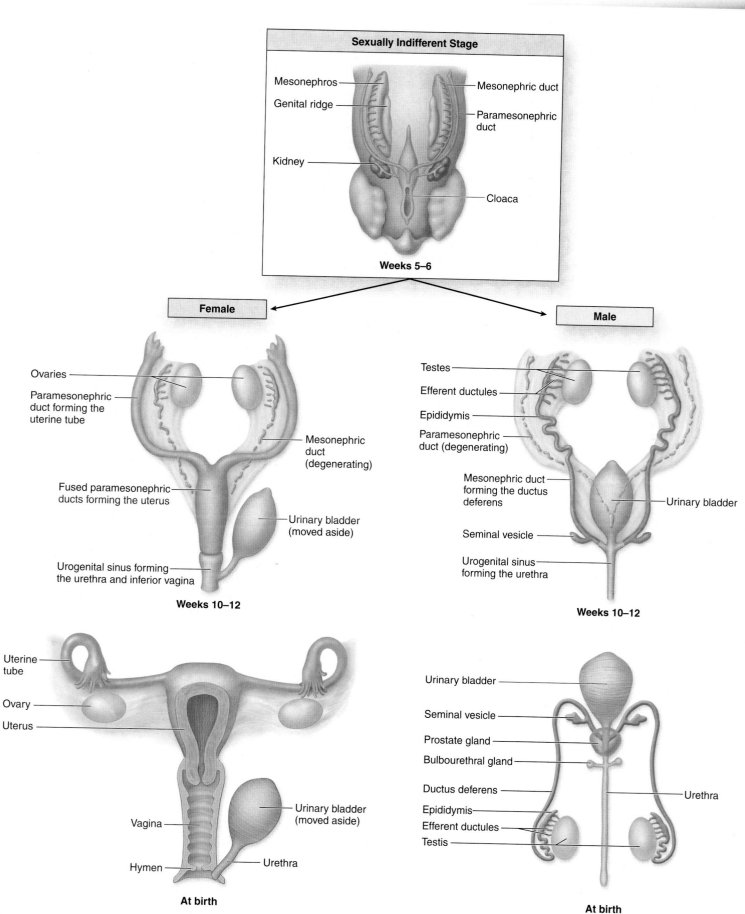

Figure 28.18

Embryonic Development of the Female and Male Reproductive Tracts. Through the first 6 weeks of development, the embryo is termed "sexually indifferent." Thereafter, genetic expression determines sex differentiation.

All human embryos develop both duct systems, but only one of the duct systems remains in the fetus. If the embryo is female, the paramesonephric ducts develop, and the mesonephric ducts degenerate. If the embryo is male, the mesonephric ducts grow and differentiate into male reproductive structures, while the paramesonephric ducts degenerate.

Internal Genitalia Development

The development of the female internal reproductive structures is traced in figure 28.18, *left*. Because no TDF proteins are produced in the female, the mesonephric ducts degenerate. Between weeks 8 and 20, the paramesonephric ducts develop and differentiate. The caudal (inferior) ends of the paramesonephric ducts fuse, forming the uterus and the superior part of the vagina. The cranial (superior) parts of the paramesonephric ducts remain separate and form two uterine tubes. The remaining inferior part of the vagina is formed from the urogenital sinus (which also forms the urinary bladder and urethra).

By about week 7 of development, the TDF gene on the Y chromosome begins influencing the indifferent gonad to become a testis, which then forms sustentacular cells and interstitial cells. Once the sustentacular cells form, they begin secreting **anti-Müllerian hormone (AMH)** (also known as *Müllerian inhibiting substance*), which inhibits the development of the paramesonephric ducts (see figure 28.18, *right*). These paramesonephric ducts degenerate, and between weeks 8 and 12, the mesonephric ducts form the male duct system—efferent ductules, epididymes, vasa deferens, seminal vesicles, and ejaculatory ducts.

The prostate and bulbourethral glands do not form from the mesonephric ducts. Instead, they begin to form as endodermal "buds" or outgrowths of the developing urethra between weeks 10 and 13. As the prostate gland and bulbourethral glands develop, they incorporate mesoderm into their structures as well.

Finally, note that the indifferent gonad originates near the level of thoracic vertebra T_{10}. Throughout prenatal development, the developing testis descends from the abdominal region toward the developing scrotum. A thin band of connective tissue called the **gubernaculum** (goo'ber-nek'ū-lŭm; helm) attaches to the testis and pulls it from the abdomen, through the developing inguinal canal, to its placement in the scrotum. As the embryo grows (but the gubernaculum remains the same length), the testis is passively pulled into the scrotum. This process is slow, beginning in the third month and not completed until the ninth month. It is common for premature male babies to have undescended testes because they were born before the testes had fully descended into the scrotum. Their testes usually descend shortly after birth.

External Genitalia Development

As with the internal genitalia, female and male external genitalia develop from the same primordial structures **(figure 28.19)**. By the sixth week, the following external structures are seen:

- The **urogenital folds** (or *urethral folds*) are paired, elevated structures on either side of the urogenital membrane, a thin partition separating the urogenital sinus from the outside of the body (see chapter 27).
- The **genital tubercle** is a rounded structure anterior to the urogenital folds.
- The **labioscrotal swellings** (or *genital swellings*) are paired elevated structures lateral to the urethral folds.

The external genitalia appear very similar between females and males until about week 12 of development, and they do not become clearly differentiated until about week 20. In the absence of testosterone, female external genitalia develop. The genital tubercle becomes the clitoris. The urogenital folds do not fuse, but become the labia minora. Finally, the labioscrotal folds also remain unfused and become the labia majora. In the male, production and circulation of testosterone cause the primitive external structures to differentiate. The genital tubercle enlarges and elongates, forming the glans of the penis and part of the dorsal side of the penis. The urogenital folds grow and fuse around the developing urethra and form the ventral side of the penis. Finally, the labioscrotal swellings fuse at the midline, forming the scrotum.

Study Tip!

You may be aware that the sex of an unborn child can typically be determined with an ultrasound sometime between weeks 18 and 22. You now know why the physician waits until this time—it is when the external genitalia first become clearly distinguishable.

WHAT DID YOU LEARN?

17 What is the difference between genetic and phenotypic sex?

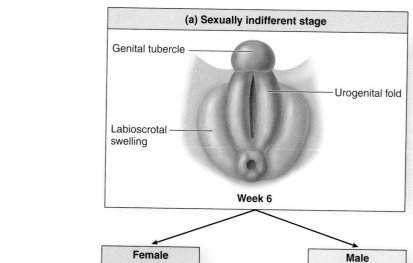

(a) Sexually indifferent stage

Genital tubercle

Urogenital fold

Labioscrotal swelling

Week 6

Female

Male

Figure 28.19

Development of External Genitalia. *(a)* At 6 weeks of development, the external genitalia are undifferentiated. *(b)* By 12 weeks, the urogenital folds begin to fuse in the male and remain open in the female. *(c)* By 20 weeks, external genitalia are well differentiated.

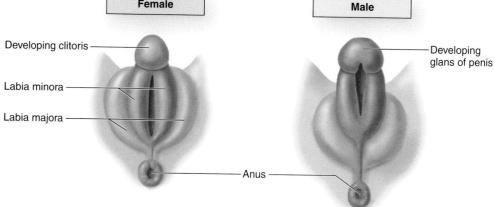

Developing clitoris

Labia minora

Labia majora

Developing glans of penis

Anus

(b) Week 12: Urogenital folds begin to fuse in the male

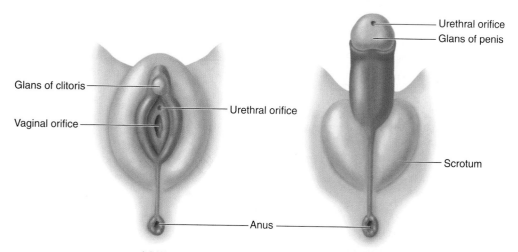

Glans of clitoris

Vaginal orifice

Urethral orifice

Urethral orifice

Glans of penis

Scrotum

Anus

(c) Week 20: External genitalia well differentiated

CLINICAL TERMS

castration (kas-trā′shŭn; *castro* = to deprive of generative power) Removal of the testes or ovaries.

cryptorchidism (krip-tōr′ki-dizm; *krypto* = hidden, *orchis* = testis) A testis that has not descended completely into the scrotum.

dysmenorrhea (dis-men-ōr-rē′ă; *dys* = bad, *men* = month, *rhoia* = flow) Difficult and painful menstruation.

salpingitis Inflammation of the uterine tubes.

CHAPTER SUMMARY

Comparison of the Female and Male Reproductive Systems 839	■ Both reproductive systems have gonads that produce gametes and sex hormones, and a duct system to transport the gametes.
	Perineum 839
	■ In both females and males, the perineum is a diamond-shaped area between the thighs that contains the urogenital and anal triangles.
Anatomy of the Female Reproductive System 840	■ Female internal reproductive organs include paired ovaries and uterine tubes, a uterus, and a vagina.
	Ovaries 841
	■ The cortex of the ovary houses ovarian follicles that consist of an oocyte surrounded by follicle cells.
	■ Changing levels of FSH and LH cause a primordial follicle to mature into a primary follicle. A secondary follicle matures from a primary follicle, and a vesicular follicle matures from a secondary follicle.
	■ A peak in LH causes the secondary oocyte to be released from the vesicular follicle at ovulation; remaining follicular cells become the hormone-producing corpus luteum.
	■ The ovarian cycle consists of the follicular phase, ovulation, and the luteal phase.
	Uterine Tubes 848
	■ The uterine tubes are the site of fertilization. They have an infundibulum, ampulla, isthmus, and uterine part.
	■ The uterine tube wall is composed of an inner mucosa (ciliated columnar epithelium), middle muscularis (two smooth muscle layers), and an external serosa.
	Uterus 848
	■ The uterus is a thick-walled muscular organ that functions as the site of pre-embryo implantation, supports and nourishes the embryo/fetus, ejects the fetus at birth, and is the site of menstruation.
	■ The uterine wall consists of an inner mucosa, the endometrium; a thick-walled middle muscular layer, the myometrium; and an outer serosa, the perimetrium.
	■ The endometrium has a functional layer that is sloughed off as menses and a deeper basal layer that regenerates a new functional layer during the next uterine cycle.
	■ Three distinct phases of endometrium development occur during the uterine cycle: menstrual phase, proliferative phase, and secretory phase.
	Vagina 851
	■ The vagina is a fibromuscular tube that serves as the birth canal for the fetus, a receptacle for the penis during intercourse, and the passageway for menstrual discharge.
	External Genitalia 853
	■ The external female sex organs, collectively called the vulva, include the mons pubis, labia majora, labia minora, and the clitoris.
	Mammary Glands 853
	■ The paired mammary glands produce breast milk.
	■ Prolactin is responsible for milk production; oxytocin is responsible for milk ejection.
Anatomy of the Male Reproductive System 857	■ The primary male reproductive system organs are the testes; accessory sex organs include ducts, male accessory glands, and the penis.
	Scrotum 857
	■ The scrotum houses the testes outside the male body, where the lower temperature is needed to form functional sperm cells.
	Spermatic Cord 859
	■ The spermatic cord transmits blood vessels and nerves from the abdominal cavity to the testis.
	Testes 859
	■ The testis contains up to four seminiferous tubules. Between the seminiferous tubules are interstitial cells, which produce androgens.
	■ Seminiferous tubules contain sustentacular cells and developing sperm cells.
	■ Sustentacular cells nourish developing sperm cells.
	■ Spermatogenesis is the meiotic process that forms haploid spermatids.
	■ Spermiogenesis is the process by which spermatids differentiate into sperm.

Ducts in the Male Reproductive System 862

- ▪ Ducts store and transport sperm. They include the rete testis, the efferent ductules, the epididymis, the ductus deferens, and the ejaculatory duct.
- ▪ The male urethra carries urine or semen at any one time.

Accessory Glands 863

- ▪ Accessory glands produce seminal fluid, a nutrient-rich, alkaline fluid that supports sperm cells. Accessory glands include seminal vesicles, the prostate gland, and the bulbourethral glands.

Semen 864

- ▪ Semen is a mixture of seminal fluid and sperm.

Penis 865

- ▪ The penis is the copulatory organ.
- ▪ The body of the penis contains three parallel erectile bodies and the urethra.

Aging and the Reproductive Systems 867

- ▪ Females undergo a change in reproductive structure and fertility called menopause. Males undergo a male climacteric in which the production of testosterone is measurably reduced.

Development of the Reproductive Systems 868

- ▪ Both male and female reproductive structures originate from the same basic primordia. Gene expression determines how they differentiate.

Genetic Versus Phenotypic Sex 868

- ▪ Genetic sex is based on chromosome type; phenotypic sex refers to the appearance of the internal and external genitalia.

Formation of Indifferent Gonads and Genital Ducts 868

- ▪ Early in development, genital ridges form from intermediate mesoderm. Primordial germ cells migrate from the yolk sac to the genital ridges and form the future gametes.

Internal Genitalia Development 870

- ▪ In the absence of testis-determining factor (TDF), the female reproductive pattern develops. The male reproductive pattern develops as a result of TDF.

External Genitalia Development 870

- ▪ The external genitalia appear very similar until about week 12; external genitalia become fully differentiated about week 20.

CHALLENGE YOURSELF

Matching

Match each numbered item with the most closely related lettered item.

_____ 1. vagina

_____ 2. uterus

_____ 3. clitoris

_____ 4. testes

_____ 5. ovary

_____ 6. prostate gland

_____ 7. scrotum

_____ 8. uterine tube

_____ 9. penis

_____ 10. semen

a. houses the testes

b. produces follicles and sex hormones

c. secretion is milky; contains citric acid

d. contains three erectile bodies

e. normal site for implantation of a pre-embryo

f. fertilization normally occurs here

g. composed of both sperm and seminal fluid

h. birth canal

i. produces spermatozoa

j. contains two erectile bodies

Multiple Choice

Select the best answer from the four choices provided.

_____ 1. In the male, what cells produce androgens?
   a. spermatogonia
   b. interstitial cells
   c. sustentacular cells
   d. All of these are correct.

_____ 2. All of the following organs produce a component of seminal fluid except the
   a. bulbourethral glands.
   b. testes.
   c. seminal vesicles.
   d. prostate gland.

_____ 3. Spermatogonia divide by mitosis to form a new spermatogonium and
   a. a sperm.
   b. spermatids.
   c. a primary spermatocyte.
   d. zygotes.

_____ 4. Sperm are stored in the _____, where they remain until they are fully mature and capable of motility.
   a. epididymis   c. ductus deferens
   b. seminiferous tubule d. rete testis

_____ 5. The female homologue to the penis is the
   a. labia majora.
   b. labia minora.
   c. clitoris.
   d. vagina.

_____ 6. Ovulation occurs due to a dramatic "peak" in which hormone?
   a. progesterone
   b. LH
   c. FSH
   d. prolactin

_____ 7. Which statement is true about the uterus?
   a. The basal layer of the endometrium is shed each month during menses.
   b. The myometrium is composed of several layers of skeletal muscle.
   c. The cervix projects into the vagina.
   d. The round ligament is peritoneum that drapes over the uterus.

_____ 8. Which structure contains a primary oocyte, several layers of follicle cells, and an antrum?
   a. primordial follicle
   b. primary follicle
   c. secondary follicle
   d. vesicular follicle

_____ 9. The most anteriorly placed structure in the female perineum is the
   a. vaginal orifice.
   b. cervix.
   c. labia minora.
   d. mons pubis.

_____ 10. The paramesonephric ducts in the embryo form which of the following?
   a. uterine tubes and uterus
   b. ovary
   c. ductus deferens
   d. seminal vesicle

Content Review

1. What are some similarities between the male and female reproductive systems? What are the anatomic homologues between these systems?

2. What hormones are associated with the female reproductive system, and what is the function of each hormone?

3. Identify the regions of the uterine tube.

4. List the uterine wall layers, and describe the basic anatomy of each layer.

5. Compare and contrast the ovarian cycle phases and the uterine cycle phases. When do they occur? What specific events are associated with each phase?

6. Describe these parts of the mammary gland: nipple, areola, lobe, lobule, and alveoli.

7. What is the function of sustentacular cells in the production of spermatozoa?

8. Describe the process of spermatogenesis, including which cells are diploid and which are haploid.

9. Compare the secretions of the seminal vesicles, the prostate gland, and the bulbourethral glands.

10. What changes occur in the penis to allow a male to attain an erection?

Developing Critical Reasoning

1. Caitlyn had unprotected sex with her fiancé approximately 2 weeks after her last period, and is worried that she might have become pregnant. She asks her physician if there are times during her monthly menstrual cycle when she might be more likely to become pregnant. She also asks how birth control pills prevent a woman from becoming pregnant. What will the physician tell Caitlyn?

2. If parents wish to know the sex of their unborn baby, they usually have to wait until weeks 18–22 of development before a sonogram determining the sex can be performed. Based on your knowledge of reproductive system development, explain why the sex of the unborn baby can't be determined easily before this time.

A N S W E R S   T O   " W H A T   D O   Y O U   T H I N K ? "

1. A woman can become pregnant as long as she has one remaining functioning ovary. However, she may not ovulate every month, since typically each ovary "takes turns" ovulating each month.

2. Stress, age, medications, and body weight all can affect a woman's monthly uterine (menstrual) cycle. Stress and excessively lean body mass can lead to amenorrhea (absence of periods).

3. If a male's testes were removed, the adrenal glands could still produce a small amount of androgens. However, since the testes produce the overwhelming majority of androgens, the small amount produced by the adrenal glands would have little effect on the male.

4. If a male has a vasectomy, sperm still form in the seminiferous tubule and then mature in the epididymis. However, since the sperm are not ejaculated, they die, and their components are broken down and resorbed in the epididymis. An individual who has had a vasectomy ejaculates seminal fluid only, not semen (which contains sperm).

Visit the McKinley/O'Loughlin _Human Anatomy_, 2e website at aris.mhhe.com

Appendix

The Content Review questions from each Challenge Yourself question set are answered on the *Human Anatomy* Online Learning Center at www.mhhe.com/mckinley.

CHAPTER ONE

Challenge Yourself

Matching

1. i	6. d
2. j	7. c
3. g	8. a
4. h	9. f
5. b	10. e

Multiple Choice

1. d	6. a
2. b	7. c
3. b	8. d
4. b	9. a
5. c	10. b

Developing Critical Reasoning

1. The level of anatomic study would be microscopic—either cytologic or histologic—or pathologic. Cytologic studies would examine the blood itself, while histologic studies would examine muscle tissue removed during biopsy. Cytologic studies and histologic studies look in blood and muscle samples for pathogens as well as any increase in white blood cells, results of inflammatory responses, and the presence of cellular debris. Pathologic studies look for anatomic changes in organs as a result of infection.

2. Antebrachial = forearm; patellar = kneecap; sacral = posterior region between the hips; brachial region = the arm.

3. The best techniques would be computed tomography and MRI, but they are expensive. Sonography may provide some information; it would probably not be very precise, but it is inexpensive. Radiography would only provide some information if used in conjunction with radiopaque dyes.

CHAPTER TWO

Challenge Yourself

Matching

1. e	6. j
2. g	7. c
3. h	8. i
4. b	9. a
5. d	10. f

Multiple Choice

1. b	6. d
2. c	7. b
3. a	8. b
4. b	9. c
5. c	10. a

Developing Critical Reasoning

1. The cells have lost their water by the process of osmosis. Water from inside the cell has moved by diffusion across the plasma membrane to the outside of the cell. As a result of the loss of intracellular water, the cells have shrunk.

2. Enclosure of some organelles within a membrane ensures that the contents of the organelle are isolated from the rest of the cytoplasm. Thus, each membrane-bound organelle is able to maintain its specific organization and conduct its own specific function(s) without interference from other cellular activities.

CHAPTER THREE

Challenge Yourself

Matching

1. e	6. i
2. h	7. c
3. b	8. j
4. f	9. g
5. a	10. d

Multiple Choice

1. b	6. b
2. d	7. a
3. b	8. b
4. a	9. c
5. c	10. c

Developing Critical Reasoning

1. The procedure used to check for trisomy in the fetus is called amniocentesis, and it is usually performed during the fourth month of pregnancy. Analysis of the chromosomes in cells from the amniotic fluid may reveal the presence of trisomy. The most common cause of this abnormality is nondisjunction of chromosomes, meaning that sister chromatids for chromosome 21 did not separate during anaphase II of meiosis, thus producing a gamete with 24, not 23, chromosomes.

2. Thalidomide was used to alleviate morning sickness in newly pregnant women, so it affected organogenesis during the embryonic period, especially limb bud formation (see table 3.2).

3. The young woman should be concerned because her pregnancy was likely in the embryonic period when she engaged in binge drinking. Exposure to alcohol during the embryonic period can result in the malformation of some or all organ systems or the death of the embryo.

CHAPTER FOUR

Challenge Yourself

Matching

1.	j	6.	g
2.	h	7.	i
3.	e	8.	b
4.	c	9.	d
5.	f	10.	a

Multiple Choice

1.	d	6.	a
2.	b	7.	d
3.	b	8.	b
4.	c	9.	c
5.	b	10.	b

Developing Critical Reasoning

1. The type of tissue specimen being observed is an areolar (loose) connective tissue. The "open spaces" are apparently due to abundant ground substance. The different types of protein fibers include both collagen (long, straight, and unbranched) and elastic (small and branched) fibers. The different connective tissue cell types probably include fibroblasts, mesenchymal cells, and macrophages.

2. The pain in the knee may be due to degeneration of the cartilage on the surfaces of the bones that form the knee joint. By taking chemical supplements that may help in the regrowth or support of cartilage, the man with the knee joint problems may obtain relief.

CHAPTER FIVE

Challenge Yourself

Matching

1.	g	6.	b
2.	e	7.	c
3.	i	8.	d
4.	f	9.	h
5.	j	10.	a

Multiple Choice

1.	c	6.	b
2.	d	7.	b
3.	a	8.	b
4.	b	9.	c
5.	c	10.	d

Developing Critical Reasoning

1. When the body is cold and needs to conserve heat, the blood vessels in the dermis vasoconstrict—that is, the vessel diameter narrows to allow for less blood flow. In an effort to conserve heat, blood is shunted to deeper body tissues, and relatively less blood flow occurs in the dermal blood vessels. The skin appears pale. When the body becomes too warm, it needs to dissipate heat. Blood vessels in the dermis respond by undergoing vasodilation—that is, the vessel diameter becomes larger to allow for more blood flow. As relatively more blood flows through these vessels, the warmth from the blood dissipates through the skin, and the body cools off. Due to relatively more blood in these dermal blood vessels, your face becomes flushed.

2. Teri's acne is probably caused by plugged sebaceous ducts. At puberty, increased hormone levels stimulate sebaceous gland secretion, making the pores more prone to blockage and leading to various types of acne lesions.

3. John has been chronically overexposed to UV rays from sunlight. This exposure may have damaged the DNA in his epidermal cells and accelerated the process of aging. Because excessive UV exposure is the predominant factor in the development of nearly all skin cancers, the growths removed from John's face were probably cancerous.

CHAPTER SIX

Challenge Yourself

Matching

1.	b	6.	d
2.	j	7.	e
3.	i	8.	g
4.	a	9.	h
5.	c	10.	f

Multiple Choice

1.	b	6.	a
2.	c	7.	a
3.	b	8.	b
4.	b	9.	b
5.	d	10.	b

Developing Critical Reasoning

1. The bone in Marty's leg was undergoing the final phase of fracture repair. The hard callus formed earlier in the repair process persists for several months as an obvious bump composed of excess bony material. Eventually, osteoclasts remove most of the excess bony material from both the external and internal surfaces of the bone.

2. Elise most likely has a low level of calcium in her blood as a result of lack of vitamin D. Her depressed calcium level will naturally slow the bone building process because mineralization of new bone lags behind the formation of osteoid by osteoblasts. Vitamin D is required to stimulate the absorption and transport of calcium from the gastrointestinal tract into the blood. The source of vitamin D is either diet or synthesis in the skin upon exposure to UV radiation. Elise's nutrition was poor, and she tended to remain indoors, away from normal amounts of UV light, making lack of vitamin D likely in her case.

3. The physician was concerned because the fracture apparently occurred across the epiphyseal plate, the site for growth of the long bones of the limbs. If the epiphyseal plate is altered or disrupted, bones do not grow normally. Using screws to align the fractured bone ensures that the growth plate has the correct structure and that normal growth will continue.

CHAPTER SEVEN

Challenge Yourself

Matching

1. b
2. e
3. h
4. j
5. i
6. a
7. d
8. g
9. f
10. c

Multiple Choice

1. c
2. a
3. c
4. c
5. a
6. b
7. b
8. c
9. b
10. a

Developing Critical Reasoning

1. The cause of the lower back pain is probably a herniated disc in the lumbar region. The construction worker twists and flexes the vertebral column and lifts heavy objects, putting him at risk for herniation. The overweight teen has probably overextended the lumbar curve due to altered posture, resulting in a herniation.

2. At birth, the cranium is not completely formed. The flat bones of the skull are connected by flexible areas of dense regular connective tissue called fontanelles. This flexibility allows the cranial bones to overlap as the skull is temporarily deformed to facilitate passage through the birth canal. Fontanelles also allow the skull to return to normal shape in a few days. The cranial bones continue to grow, and the fontanelles have usually been replaced by bone by the age of 15 months.

3. The forensic anthropologist could examine various features of the skull to determine the sex of the individual. A larger, more robust skull with a prominent external occipital protuberance and a squared-off mental protuberance would indicate a male skull. A female skull may be smaller, more gracile, have a more triangular mental protuberance, a more rounded external occipital protuberance, and less pronounced superciliary arches.

CHAPTER EIGHT

Challenge Yourself

Matching

1. b
2. g
3. h
4. j
5. c
6. i
7. f
8. a
9. e
10. d

Multiple Choice

1. b
2. c
3. c
4. b
5. a
6. c
7. c
8. b
9. c
10. b

Developing Critical Reasoning

1. She should not be prescribed thalidomide. If a pregnant female takes thalidomide during weeks 4–8 of embryonic development (the time when the limbs are developing), limb development can be severely disrupted.

2. There are several differences between the pelvis of a female and that of a male. The female ilium flares more laterally, while the male ilium projects more superiorly (thus giving males narrower hips). The female pelvis is wider, the acetabulum projects more laterally (instead of anteriorly), and the greater sciatic notch is much wider. The sacrum tends to be shorter and wider in females. The female pelvis has a wider, larger pelvic inlet and pelvic outlet. The body of the pubis in females is much longer and almost rectangular in shape, compared to the shorter, triangular-shaped appearance of the male pubic body. The female subpubic angle is wider and convex, while the male subpubic angle is much narrower and typically does not extend past 90 degrees. The relative age can be determined from the os coxae. Between the ages of 13 and 15 years, the ilium, ischium, and pubis fuse to form the os coxae of the pelvic girdle. Further, the morphology of the pubic symphysis can be used to estimate the age at death.

3. The young male has a condition called "flat feet," meaning that the medial longitudinal arch is flattened and the entire sole may touch the ground. As a result, long walks or running result in very painful feet because the structures on the sole of the foot are compressed. Since the Army requires standing for a long time and perhaps long marches, individuals with pes planus are ineligible for service.

CHAPTER NINE

Challenge Yourself

Matching

1. i
2. c
3. h
4. j
5. g
6. b
7. f
8. a
9. e
10. d

Multiple Choice

1. c
2. c
3. b
4. c
5. c
6. c
7. a
8. b
9. d
10. d

Developing Critical Reasoning

1. Erin experienced a dislocation of the glenohumeral joint. Immediately following the initial blow, the head of the humerus pushed into the inferior part of the capsule. Next, the head of the humerus tore the inferior part of the capsule and dislocated the humerus, so the humerus was then inferior to the glenoid cavity. Once the humeral head became dislocated from the glenoid fossa, the pulling action of the chest muscles displaced it superiorly and medially, causing the humeral head to lie just inferior to the coracoid process.

2. The knee joint is at risk in "clipping" in a football game. The parts of the joint injured may include the tibial collateral ligament, the medial meniscus, and the anterior cruciate ligament (ACL). When all three are affected, the condition is known as the "unhappy triad" of injuries. First, the leg is forcibly abducted and laterally rotated. If the blow is severe enough, the tibial collateral ligament tears, followed by tearing of the medial meniscus, since these two structures are connected. The force that tears the tibial collateral ligament and medial meniscus is thus transferred to the ACL. Since the ACL is relatively weak, it tears as well.

CHAPTER TEN

Challenge Yourself

Matching

1.	d	6.	a
2.	b	7.	i
3.	g	8.	c
4.	j	9.	h
5.	e	10.	f

Multiple Choice

1.	d	6.	c
2.	d	7.	c
3.	d	8.	a
4.	b	9.	b
5.	c	10.	d

Developing Critical Reasoning

1. Paul experienced atrophy in the muscles of his left forearm due to the lack of stimulation those muscles received during the 6 weeks his arm was in the cast. Atrophy is a reduction in muscle size, tone, and power. If a muscle experiences a marked decrease in use, the muscle becomes flaccid, and its fibers decrease in size and become weaker.

2. Myofilaments are protein filaments consisting of the contracting proteins actin and myosin and some regulatory proteins. The actin proteins and regulatory proteins form the thin filaments, while the myosin proteins make up the thick filaments. The thin filaments insert into and extend between thick filament proteins within a band. The proximity of thick and thin filament proteins permits the interaction of myofilaments within myofibrils when the muscle fiber is stimulated. Stimulation of the muscle fiber results in the shortening of the entire cell. During a contraction, crossbridges form, binding thick filaments to thin, and the myosin heads pivot toward the M line in the center of the sarcomere. This action pulls the thin filaments toward the center of the sarcomere, causing the sarcomere to shorten. Upon the shortening of all sarcomeres, the skeletal muscle fiber itself is shortened.

3. Savannah's 59-year-old grandfather has less tolerance for exercise than Savannah due to many factors. Older individuals experience muscle atrophy, caused by a decrease in cell proliferation, cell volume, and cell number. The decreased muscle strength is compounded by the tendency toward rapid fatigue. The atrophy of muscle secondarily results in a lesser amount of tension generated by that muscle. Consequently, the muscle fatigues more readily while performing a specific amount of work. As a result, the individual has a reduced ability to eliminate the heat generated during muscular contraction. In addition, each skeletal muscle fiber decreases in diameter. The consequences of decreased fiber diameter include: a cutback in oxygen storage capacity because of less myoglobin, reduced glycogen reserves, and a decreased ability to produce ATP. Overall, muscle strength and endurance are impaired, and the individual tends to become fatigued quickly. Depreciated cardiovascular performance also occurs as a result of aging, and thus blood flow to active muscles does not increase with exercise as rapidly as it does in younger people. Finally, after exercising, the muscles in an older individual have a reduced capacity to recover. As we get older, the number of satellite cells in skeletal muscle steadily decreases, resulting in diminished repair capabilities as well.

CHAPTER ELEVEN

Challenge Yourself

Matching

1.	d	6.	i
2.	g	7.	j
3.	a	8.	b
4.	c	9.	f
5.	h	10.	e

Multiple Choice

1.	c	6.	b
2.	a	7.	c
3.	c	8.	c
4.	b	9.	b
5.	a	10.	c

Developing Critical Reasoning

1. When Albon strained to lift the heavy furniture, rising pressure in his abdominopelvic cavity provided the force to push a segment of the intestine into the inguinal canal, causing the inguinal hernia. The inguinal region is one of the weakest areas of the abdominal wall. Males are more likely to develop inguinal hernias than females, because their inguinal canals, and superficial inguinal rings, are larger to allow room for the spermatic cord. Albon's hernia was most likely a direct inguinal hernia, in which herniation occurs through the superficial inguinal ring, but not down the entire length of the inguinal canal. It typically is seen in middle-aged males with poorly developed abdominal muscles and protruding abdomens.

2. Pat might have injured her external urethral sphincter, which assists in urination in females and in both urination and ejaculation in males. This sphincter is under voluntary control, and injury to this muscle could result in the loss of voluntary control over urination.

CHAPTER TWELVE

Challenge Yourself

Matching

1.	b	6.	d
2.	a	7.	f
3.	i	8.	c
4.	g	9.	e
5.	j	10.	h

Multiple Choice

1.	b	6.	b
2.	c	7.	d
3.	b	8.	c
4.	d	9.	c
5.	b	10.	b

Developing Critical Reasoning

1. Most of the muscles in the anterior compartment of the thigh cause extension of the knee, but three muscles cause flexion of the hip. Paralysis of these muscles would therefore compromise both knee extension and hip flexion.

2. Two movements are associated with the elbow joint: extension and flexion. If working out in an exercise facility, Karen could do the following exercises. (1) Arm curls: While standing, pick up a dumbbell. With the arm fully extended at the side, lift the weight toward the shoulder. This would improve flexion. Then slowly let the weight return to the side. (2) Barbell lift: While standing, lift a barbell and curl it up to the chest. Then push the barbell over the head. This would improve flexion and extension. All exercises are performed for 10–12 repetitions, followed by a short rest and a repeat of the exercises three times.

3. The prime mover in elbow flexion is the brachialis. Its movement is not affected by pronation or supination of the forearm. However, the biceps brachii, which is also a forearm flexor, is ineffective in forearm flexion if the forearm is pronated. Thus, pronation of the forearm prevents the biceps brachii from helping with forearm flexion.

CHAPTER THIRTEEN

Challenge Yourself

Matching

1.	g	6.	d
2.	c	7.	a
3.	f	8.	e
4.	i	9.	b
5.	j	10.	h

Multiple Choice

1.	c	6.	c
2.	b	7.	c
3.	a	8.	b
4.	c	9.	c
5.	a	10.	b

Developing Critical Reasoning

1. The health-care worker may have accidentally pierced Marcie's sciatic nerve. Usually, injections are administered in the upper lateral quadrant of the buttock, because the sciatic nerve and gluteal nerves and vessels are located primarily in the medial and inferior lateral part of the buttock.

2. The doctor was able to tell which rib Javier had fractured by locating the sternal angle at the level of the second rib, which facilitates the identification and counting of individual ribs.

3. The doctor was palpating Louisa's cervical lymph nodes, which can become swollen as a result of an infection such as the flu.

CHAPTER FOURTEEN

Challenge Yourself

Matching

1.	d	6.	i
2.	a	7.	g
3.	f	8.	j
4.	c	9.	e
5.	h	10.	b

Multiple Choice

1.	c	6.	c
2.	b	7.	c
3.	d	8.	c
4.	b	9.	b
5.	c	10.	a

Developing Critical Reasoning

1. Marianne may be showing symptoms of multiple sclerosis, in which progressive demyelination of neurons in the central nervous system (accompanied by the destruction of oligodendrocytes) interferes with nerve impulse conduction. The result is impaired sensory perception and motor coordination.

2. The lining of the blood vessels is an epithelium, which continually replaces itself, but nervous tissue does not replace itself readily. Growth of new nervous tissue in the amputated limb must occur by wallerian degeneration, a relatively slow process.

CHAPTER FIFTEEN

Challenge Yourself

Matching

1.	c	6.	a
2.	h	7.	f
3.	i	8.	d
4.	b	9.	j
5.	e	10.	g

Multiple Choice

1.	c	6.	c
2.	c	7.	c
3.	b	8.	a
4.	d	9.	b
5.	c	10.	a

Developing Critical Reasoning

1. Shannon apparently survived a cerebrovascular accident (stroke) in the left frontal lobe of her cerebrum. Damage occurred to the specific regions in the primary motor cortex that control movement of the right upper limb and also to areas of the speech center.

2. Surrounding the brain is the blood-brain barrier. It protects the nervous tissue of the brain from drugs, waste products, and other solutes in the blood by restricting or preventing the movement of these materials out of the blood capillaries into the nervous tissue of the brain. The drug dopamine could alleviate the symptoms of Parkinson disease, but it is not allowed past the blood-brain barrier. The barrier that keeps the drug from reaching the brain is formed by the perivascular feet of astrocytes and the endothelial cells of the capillaries. On the other hand, these same barrier structures prevent materials that could adversely affect the brain from reaching its nervous tissue.

3. Dustin would have been less likely to survive if he had been shot in the brainstem because numerous brainstem structures are involved in involuntary control of heart rate, blood pressure, and respiration. Without the control afforded by these vital structures, survival is not possible.

CHAPTER SIXTEEN

Challenge Yourself

Matching

1.	g	6.	c
2.	d	7.	j
3.	a	8.	e
4.	i	9.	f
5.	b	10.	h

Multiple Choice

1.	a	6.	c
2.	d	7.	a
3.	b	8.	a
4.	c	9.	c
5.	d	10.	c

Developing Critical Reasoning

1. Because Arthur is paralyzed in both the upper and lower limbs, his injury must have occurred in the cervical region of the spinal cord, superior to the cervical enlargement. The likelihood of recovery is not good, since severed axons within the CNS generally do not regenerate (review wallerian degeneration in chapter 15).

2. Jessica may have damaged her ulnar nerve. Ulnar nerve injuries often accompany elbow fractures because the nerve is close to the medial epicondyle of the humerus. Damage to the ulnar nerve may paralyze most of the intrinsic hand muscles, so Jessica may be unable to adduct or abduct her fingers.

CHAPTER SEVENTEEN

Challenge Yourself

Matching

1.	h	6.	b
2.	j	7.	d
3.	f	8.	g
4.	c	9.	a
5.	i	10.	e

Multiple Choice

1.	b	6.	c
2.	b	7.	c
3.	d	8.	b
4.	c	9.	c
5.	b	10.	c

Developing Critical Reasoning

1. Melissa has suffered a cerebrovascular accident, or stroke, due to reduced supply of blood to a specific part of the brain. The diminished supply of nutrients quickly results in damage or death to neurons in the area, with a coincidental loss of functions normally controlled by that brain region. The areas affected in Melissa's brain would be in the left frontal lobe: the motor speech area and the region of the precentral gyrus that controls motor activity of the right upper limb. The extent of Melissa's recovery depends on the location and duration of the cut-off blood supply.

2. Randolph has apparently incurred damage to the cortical region of the representational hemisphere opposite the motor speech area in the categorical hemisphere. Damage to this part of the representational hemisphere results in aprosody, which causes dull, emotionless speech.

CHAPTER EIGHTEEN

Challenge Yourself

Matching

1. c	6. a
2. h	7. g
3. d	8. j
4. b	9. f
5. i	10. e

Multiple Choice

1. c	6. c
2. b	7. c
3. c	8. a
4. a	9. c
5. c	10. b

Developing Critical Reasoning

1. Holly has had a fight-or-flight experience resulting from the sudden, unexpected screeching of the birds. An immediate increase in sympathetic division activity caused noticeable increases in heart rate and respiratory rate, as well as heightened alertness, pupil dilation, and increased metabolic activities. She continues to experience some of these effects even though the noise is gone and she no longer feels frightened, because the initial response to activation of the fight-or-flight process caused the adrenal medulla to release epinephrine and norepinephrine into her bloodstream. These hormones prolong the effects of the sympathetic division stimulation.

2. Students typically have a hard time paying attention in class immediately after a meal because the parasympathetic division of the autonomic nervous system is dominant at this time. This is the "rest-and-digest" division. Energy stores are being replenished, and digestive activities are predominant. A person tends to be less alert during parasympathetic-dominant times.

CHAPTER NINETEEN

Challenge Yourself

Matching

1. f	6. j
2. i	7. a
3. h	8. e
4. c	9. d
5. b	10. g

Multiple Choice

1. b	6. a
2. d	7. b
3. a	8. b
4. d	9. c
5. c	10. c

Developing Critical Reasoning

1. The pediatrician would call this condition otitis media, a middle ear infection often experienced by young children, whose auditory tubes are horizontal, relatively short, and undeveloped. As a child ages, the auditory tube becomes larger and more angled, which helps drain the middle ear; thus, infections decrease.

2. Gustation and olfaction are linked, so the smell of a food contributes more to taste than the food itself. When olfaction is impaired (as occurs with chronic smoking), gustation is also impaired since a person can't detect the odors from the food. When a person quits smoking, the sense of smell gradually improves, and so taste perceptions also improve.

CHAPTER TWENTY

Challenge Yourself

Matching

1. f	6. d
2. c	7. j
3. h	8. g
4. i	9. e
5. a	10. b

Multiple Choice

1. d	6. d
2. b	7. c
3. d	8. b
4. b	9. c
5. d	10. b

Developing Critical Reasoning

1. Type 2 diabetes is also known as insulin-independent diabetes mellitus (IIDM). It results from either decreased insulin release by the pancreatic beta cells or decreased insulin effectiveness in target cells. This type of diabetes was previously called adult-onset diabetes because it tended to occur in people over the age of 30. However, type 2 diabetes is now rampant in adolescents and young adults, since obesity plays a major role in the development of IIDM and many young people are overweight. Type 1 diabetes, also referred to as insulin-dependent diabetes mellitus (IDDM), is characterized by absent or diminished production and release of insulin by the pancreatic islet cells. This type tends to occur in children and younger individuals, and is not directly associated with obesity. Type 1 diabetes develops in a person who harbors a genetic predisposition, although some kind of triggering event is required to start the process. Often, the trigger is a viral infection, and then the process continues as an autoimmune condition in which the beta cells of the pancreatic islets are the primary focus of destruction.

2. Susan's physician should suspect that she has hypothyroidism, a condition that results from decreased production of thyroid hormone and is characterized by low metabolic rate, lethargy, and a feeling of being cold. Her physician should run tests to check for levels of thyroid hormone in the blood and to ensure that iodine is being taken up by the cells of the thyroid gland.

CHAPTER TWENTY-ONE

Challenge Yourself

Matching

1. d   6. b
2. j   7. i
3. f   8. e
4. a   9. c
5. h   10. g

Multiple Choice

1. b   6. b
2. b   7. c
3. b   8. a
4. d   9. d
5. a   10. c

Developing Critical Reasoning

1. Marilyn observed an increased number of eosinophils in the blood smear. Usually these cells are present in concentrations of 2–4% of the blood. The higher percentage may indicate either an increase in the number of antigen-antibody complexes in the body or the presence of a parasitic worm.

2. With no prior knowledge of the blood type of the recipient, the physician should call for a transfusion with type O negative blood. This blood type does not exhibit surface antigens of either the ABO system or the Rh system. Thus, the chance of a transfusion reaction would be minimal.

CHAPTER TWENTY-TWO

Challenge Yourself

Matching

1. g   6. a
2. e   7. d
3. j   8. c
4. b   9. i
5. h   10. f

Multiple Choice

1. b   6. a
2. b   7. b
3. d   8. d
4. a   9. c
5. b   10. c

Developing Critical Reasoning

1. Huang probably suffered from premature ventricular contractions, which cause the heart to skip beats and/or make it feel as if it is jumping in the chest. The premature ventricular contractions most likely resulted from the stress of studying and working, sleep deprivation, and caffeine, the stimulant in the coffee Huang was drinking.

2. Josephine should be told that angina is a poorly localized pain sensation in the left side of the chest, left arm and shoulder, or sometimes the jaw and/or the back. Strenuous activity generally causes angina when workload demands on the heart exceed the ability of the narrowed coronary vessels to supply blood. Her pain diminished shortly after the brisk walking stopped because normal blood flow to the heart had been restored.

CHAPTER TWENTY-THREE

Challenge Yourself

Matching

1. i   6. c
2. g   7. j
3. a   8. e
4. h   9. d
5. b   10. f

Multiple Choice

1. c   6. d
2. a   7. b
3. b   8. d
4. b   9. c
5. c   10. c

Developing Critical Reasoning

1. Thomas is at greater risk for developing atherosclerosis. Males tend to be affected more than females, and although we are comparing the risks for two males, the risk for Thomas is greater because he is overweight and does not follow a healthy diet. Without a healthy diet, the possibility of developing hypertension and vascular injury increases, thus increasing the risk for atherosclerosis. Thomas's diet may also result in increased cholesterol in the blood, which is associated with the rate of development and severity of atherosclerosis.

2. An anastomosis occurs where two or more arteries (or veins) merge and supply the same body region. Vascular anastomoses occur in joints to ensure that the tissues and organs beyond the joint have an uninterrupted supply of blood. When the joint is flexed, some vessels may be pinched. If only one supplying vessel went through a joint (elbow or knee), the part of the limb beyond the joint could potentially lose its blood supply if the joint were flexed and maintained in that state. The anastomosis ensures that there will always be another pathway through which the blood can move beyond the joint, no matter the position of the joint.

3. The internal thoracic artery is an ideal coronary bypass vessel for many reasons. Unlike the great saphenous vein (another vessel sometimes used for bypass surgery), the internal thoracic artery is an artery (not a vein), so it is more structurally similar to the coronary arteries. The internal thoracic artery is immediately adjacent to the heart, so a surgeon does not have to create a completely new incision to remove this artery from its original location. Once this artery has been removed and attached to the heart, blood flow to the thoracic wall is not compromised, because there are extensive anastomoses in this region. The thoracic wall may be supplied by the superior epigastric artery (via its anastomoses with the inferior epigastric artery) as well as by the anterior intercostal arteries (which have anastomoses with the posterior intercostal arteries that extend from the descending thoracic aorta).

CHAPTER TWENTY-FOUR

Challenge Yourself

Matching

1. g	6. b
2. e	7. d
3. j	8. f
4. i	9. h
5. c	10. a

Multiple Choice

1. c	6. c
2. c	7. b
3. c	8. c
4. a	9. d
5. b	10. c

Developing Critical Reasoning

1. Antigens associated with mononucleosis are found within the blood. Arianna's doctor was palpating her spleen to see if it was enlarged. When antigens are present in the blood, lymphatic cells in the white pulp of the spleen detect those antigens, and T-lymphocytes and B-lymphocytes initiate an immune response. One result is that the number of lymphocytes increases dramatically to fight the infection. This increase in lymphocyte number causes enlargement of the spleen.

2. HIV, the human immunodeficiency virus, binds to and then infects helper T-lymphocytes. This causes the helper T-lymphocytes to become HIV factories and prevents them from initiating and overseeing immune responses in the body. Other lymphatic cells target and destroy the infected helper T-lymphocytes, decreasing their numbers. Over a period of time, the helper T-lymphocyte population decreases, and the infected individual is unable to mount immune responses to normal infections. Opportunistic infections (such as pneumocystic pneumonia and histoplasmosis) and cancers (such as Kaposi sarcoma and non-Hodgkin lymphoma) often occur and may result in death.

3. Both cancerous and infected lymph nodes can become enlarged. However, while an infected lymph node is tender or painful to the touch, a cancerous lymph node tends to be hard and painless. A biopsy (in which cells are taken from the lymph node) could confirm the diagnosis. If the lymph node is cancerous, a physician could check to see if other lymph nodes are also hard, nontender, and enlarged. Biopsies of suspicious-looking nodes can help determine if the cancer has spread.

CHAPTER TWENTY-FIVE

Challenge Yourself

Matching

1. e	6. b
2. i	7. h
3. g	8. j
4. c	9. d
5. a	10. f

Multiple Choice

1. a	6. c
2. c	7. c
3. a	8. c
4. b	9. a
5. c	10. b

Developing Critical Reasoning

1. Charlene's cold has had several effects that cause changes in her voice. Inflammation in the nasal cavity alters the volume of space available for sound resonance. Likewise, the accumulation of mucus and fluid in the paranasal sinuses alters sound resonance for voice production. Finally, inflammation in the larynx affects the vocal folds, making the voice sound hoarse and gravely.

2. Smoking is always dangerous. It significantly increases the risk and severity of atherosclerosis and is directly related to the development of cancers of the lung, esophagus, and stomach. The most common smoking-related diseases are emphysema and cancer. Emphysema is an irreversible loss of pulmonary gas exchange area due to inflammation of the terminal bronchioles and alveoli, in conjunction with widespread destruction of pulmonary elastic connective tissue. Proper exhalation fails, and once these lung structures have been destroyed, no regeneration is possible.

Smoking accounts for about 85% of all lung cancers. Metastasis, the spread of cancerous cells to other tissues, occurs early in the course of the disease, making surgical cure unlikely for most patients. Pulmonary symptoms include chronic cough, coughing up blood, excess pulmonary mucous production, and increased likelihood of pulmonary infections. Some people are diagnosed only after the lung cancer has metastasized to distant sites.

CHAPTER TWENTY-SIX

Challenge Yourself

Matching

1. i	6. b
2. e	7. a
3. f	8. g
4. h	9. d
5. j	10. c

Multiple Choice

1. a	6. a
2. b	7. b
3. d	8. b
4. c	9. c
5. a	10. c

Developing Critical Reasoning

1. Due to Alexandra's vomiting, the epithelial lining of the esophagus, pharynx, and oral cavity was exposed to acidic materials that could "burn" it. Additionally, the volume of fluid loss may have contributed to dehydration. The diarrhea caused rapid expulsion of materials from the intestines through the anal canal, so that appropriate nutrient and water absorption did not occur.

2. Colorectal cancer is a malignant growth anywhere along the large intestine or rectum. However, most cases occur in the distal descending colon, the sigmoid colon, and the rectum, regions that are in contact with fecal material longer than the regions of the proximal part of the large intestine. Small outgrowths from the mucosal lining, called polyps, are often the focal point for cancer origin. Often, nutrition may increase a patient's risk for cancer. A reduced-fiber diet results in decreased stool bulk, a greater concentration of toxins in the stools, and longer transit time for fecal materials. An individual's genetic history may also increase the risk for colorectal cancer development.

CHAPTER TWENTY-SEVEN

Challenge Yourself

Matching

1. d   6. g
2. b   7. e
3. j   8. f
4. h   9. a
5. i   10. c

Multiple Choice

1. c   6. b
2. c   7. d
3. d   8. b
4. b   9. c
5. a   10. b

Developing Critical Reasoning

1. Jason experienced more frequent urination because his body was "clearing" the excess fluid from the numerous beers he had consumed. Normally, as our body loses fluid through urination or sweating, we produce ADH to prevent excess fluid loss that leads to dehydration. However, alcohol inhibits the secretion of ADH from the posterior pituitary, and thus excess fluid is lost during urination. This causes the "dry mouth" that is indicative of dehydration.

2. The prostatic urethra extends through the prostate gland, immediately inferior to the male bladder. A patient with benign prostatic hypertrophy has an enlarged prostate that causes constriction of the prostatic urethra, and thus urine cannot be voided normally.

CHAPTER TWENTY-EIGHT

Challenge Yourself

Matching

1. h   6. c
2. e   7. a
3. j   8. f
4. i   9. d
5. b   10. g

Multiple Choice

1. b   6. b
2. b   7. c
3. c   8. c
4. a   9. d
5. c   10. a

Developing Critical Reasoning

1. If Caitlyn has regular 28-day ovarian and uterine cycles, then at 2 weeks since her last period she would have been close to the time of ovulation. Sperm can live for several days in the female reproductive tract, and thus viable sperm may have been present either before, during, or immediately after ovulation, increasing her risk of becoming pregnant significantly. The physician would tell Caitlyn that birth control pills typically come in 28-day packets. The first 21 days of pills contain low levels of estrogen and/or progestins, which prevent the LH "spike" needed for ovulation. Thus, birth control pills prevent ovulation.

2. The external genitalia appear very similar in males and females until about week 12 of development, and do not become clearly differentiated until about week 20. An ultrasound is typically delayed until sometime between weeks 18 and 22 because the external genitalia are not clearly distinguishable until then.

Glossary

Pronunciation Key

Pronouncing a word correctly is as important as knowing its spelling and its contextual meaning. The mastery of all three allows a student to take ownership of the word. The system employed in this text is basic and consists of the following conventions.

1. Vowels marked with a line above the letter are pronounced as follows:
   - ā day, base
   - ē be, feet
   - ī pie, ivy
   - ō so, pole
   - ū unit, cute

2. Vowels marked with the breve (˘) are pronounced as follows:
   - ă above, about
   - ĕ genesis, bet
   - ĭ pit, sip
   - ŏ collide
   - ŭ cut, bud

3. Vowels not marked are pronounced as follows:
   - a mat
   - o not, ought
   - e term

4. Other phonetic symbols used include:
   - ah father
   - aw fall
   - oo food
   - ow cow
   - oy void

5. For consonants, the following key was employed:
   - b   bad
   - ch child
   - d   dog
   - dh this
   - f   fit
   - g   got
   - h   hit
   - j   jive
   - k   keep
   - ks tax
   - kw quit
   - l   learn
   - m   mice
   - n   no
   - ng ring
   - p   put
   - r   right
   - s   so
   - sh shoe
   - t   tight
   - th thin
   - v   very
   - w   wet
   - y   yes
   - z   zero
   - zh measure

6. The principal stressed syllables are followed by a prime (′); single-syllable words do not have a stress mark. Nonstressed syllables are separated by a hyphen.

7. Acceptable alternate pronunciations are given as needed.

A

abduction (ab-dŭk′shŭn) Movement of a body part away from the median plane of the body.

achondroplasia (ā-kon-drŏ-plā′zē-ă) Disease characterized by abnormal conversion of cartilage to bone.

acrosome (ak′rŏ-sōm) Membranous cap on the anterior two-thirds of the sperm nucleus that contains digestive enzymes for penetrating an oocyte.

adaptation (ad-ap-tā′shŭn) Advantageous change of an organ or tissue to meet new conditions.

adduction (ăd-dŭk′shŭn) Medial movement of a body part toward the midline.

adipocyte (ad′i-pō-sīt) Fat storage cell.

adrenergic (ad-rĕ-ner′jik) Relating to nerve cells that use norepinephrine as their neurotransmitter.

afferent (af′er-ent) Inflowing or going toward a center.

agglutination (ă-gloo-ti-nā′shŭn) Process by which erythrocytes adhere and form into clumps that can block blood vessels and disrupt normal blood circulation.

agglutinin (ă-gloo′ti-nin) Antibody that causes clumping with stimulating cells that contain the reactive antigen.

agglutinogen (ă-gloo-tin′ō-jen) Antigenic substance that can stimulate agglutinin formation.

agonist (ag′on-ist) Muscle that contracts to produce a particular movement; also called prime mover.

albumin (al-bū′min) Plasma protein important in regulating fluid balance.

alimentary (al-i-men′ter-ē) Relating to food or nutrition.

alopecia (al-ō-pē′shē-ă) Thinning or loss of hair.

alveolus (al-vē′ō-lŭs; pl., alveoli, -ō-lī) Small hollow or cavity. Air sac in the lungs; also, a milk-secreting portion of a mammary gland.

amblyopia (am-blē-ō′pē-ă) Suppression of central vision or visual acuity in one eye due to the two eyes pointing in different directions so that the two scenes cannot be fused into a single image.

amnion (am′nē-on) Extraembryonic membrane that envelops the embryo.

amphiarthrosis (am′fi-ar-thrŏ′sis) A slightly movable joint.

ampulla (am-pul′lă; pl., ampullae, -lē) Saccular dilation of a canal or duct, such as the ductus deferens in the male reproductive system.

anastomosis (ă-nas′tŏ-mō′sis; pl., anastomoses, -sēz) Union of two structures, such as blood vessels, to supply the same region.

androgen (an′drō-jen) Generic term for a hormone that stimulates the activity of accessory male sex organs or the development of male sex characteristics.

anemia (ă-nē′mē-ă) Any condition in which the number of erythrocytes is below normal.

angioplasty (an′jē-ō-plas′tē) The reopening of a blood vessel through a variety of means.

antagonist (an-tag′ŏ-nist) Muscle that opposes or resists the action of another.

antebrachium (an-te-brā′kē-ŭm) Forearm.

antigen (an′ti-jen) Substance that causes a state of sensitivity and/or responsiveness and reacts with antibodies and/or immune cells of the affected subject.

aorta (ā-ōr′tă) Main trunk of the systemic arterial system, beginning at the left ventricle and ending when it divides to form the common iliac arteries.

aperture (ap′er-choor) Open gap or hole.

apex (ā′peks) Extremity of a conical or pyramidal structure; e.g., the inferior, conical end of the heart.

apical (ap′i-kăl) Related to the tip or extremity of a conical or pyramidal structure.

aponeurosis (ap′ō-noo-rō′sis; pl., aponeuroses, -sēz) Fibrous sheet or flat, expanded tendon.

apoptosis (ap′op-tō′sis) Programmed cell death.

appendicular (ap′en-dik′-ū-lăr) Relating to an appendage or limb; e.g., the appendicular skeleton.

appositional (ap-ō-zish′ŭn-ăl) Being placed or fitted together; e.g., appositional growth of bone.

arcuate (ar′kū-āt) Having a shape that is arched or bowed.

artery (ar′ter-ē) Blood vessel conveying blood away from the heart.

arthrology (ar-throl′ō-jē) Branch of anatomy concerned with joints.

articular (ar-tik′ū-lăr) Relating to a joint.

articulation (ar-tik-ū-lā′shŭn) Joint or connection between bones.

astrocyte (as′trō-sīt) Largest and most abundant glial cell of the nervous system.

atlas (at′las) The first cervical vertebra.

atrium (ā′trē-ŭm; pl., atria, ā′trē-ă) Chamber or cavity to which are connected other chambers or passageways; e.g., the heart has a left atrium and a right atrium.

atrophy (at′rŏ-fē) Wasting of tissues, organs, or the entire body.

auricle (aw′ri-kl) The external ear; also called the pinna.

auscultation (aws-kŭl-tā′shŭn) Listening to the sounds made by various body structures to help diagnose medical conditions.

autolysis (aw-tol′i-sis) Digestion of cells by enzymes present within the cell itself.

autophagy (aw-tof′ă-jē) Segregation and disposal of damaged organelles within a cell.

axial (ak′sē-ăl) Relating to or situated in the central part of the body—the head, neck, and trunk; e.g., the axial skeleton.

axis (ak′sis) The second cervical vertebra.

axolemma (ak′sō-lem′ă) Plasma membrane of an axon.

axon (ak′son) Process of a nerve cell that conducts the impulse away from the cell body.

axoplasm (ak′sō-plazm) Cytoplasm within the axon.

B

baroreceptor (bar′ō-rē-sep′ter, -tōr) Any sensor of pressure changes.

benign (bē-nīn′) Term that denotes the mild character of an illness or the nonmalignant character of a neoplasm.

bile (bīl) Fluid secreted by the liver and discharged from the gallbladder into the duodenum.

biomechanics (bī-ō-me-kan′iks) Analysis of biological systems in mechanical terms.

bolus (bō′lŭs) A single quantity of something, such as a mass of food swallowed.

bursa (ber′să; pl., bursae, ber′sē) Closed, fluid-filled sac lined with a synovial membrane; usually found in areas subject to friction.

C

calcification (kal′si-fi-kā′shŭn) Process in which structures in the body become hardened as a result of deposited calcium salts; normally occurs only in the formation of bone and teeth.

callus (kal′ŭs) Composite mass of cells and extracellular matrix that forms at a fracture site to establish continuity between the bone ends.

calyx (kā′liks; pl., calyces or calices, kal′i-sēz) Cup-shaped structure.

canaliculus (kan-ă-lik′ū-lŭs; pl., canaliculi, -lī) Small canal or channel.

capacitation (kă-pas'i-tā'shŭn) A period of conditioning whereby the sperm cell coat is modified while in the female reproductive tract prior to being able to fertilize the secondary oocyte.

carotene (kar'ō-tēn) Class of yellow-red pigments widely distributed in plants and animals.

carpal (kar'păl) Relating to the wrist.

cartilaginous (kar-ti'laj'i-nŭs) Relating to or consisting of cartilage.

cataract (kat'ă-rakt) Complete or partial opacity of the lens.

cauda equina (kaw'dă ē-kwī'nă) Groups of axons within the vertebral canal inferior to the tapered inferior end of the spinal cord proper.

cecum (sē'kŭm) Blind pouch forming the first part of the large intestine.

centriole (sen'trē-ōl) Organelle that participates in the separation of chromosome pairs during cell division.

centromere (sen'trō-mēr) The nonstaining constriction of a chromosome that is the point of attachment of the spindle fiber.

cerebellum (ser-e-bel'ŭm) The second largest part of the brain; develops posteriorly to the pons in the metencephalon.

cerumen (sĕ-roo'men) Soft, waxy secretion of the ceruminous gland; found in the external auditory meatus.

chemotaxis (kē-mo-tak'sis) Movement in response to chemicals.

choana (kō'an-ă) Opening into the nasopharynx on either side of the nasal cavity.

cholecalciferol (kō'lē-kal-sif'er-ol) Vitamin D of animal origin found in human skin and in the fur and feathers of animals and birds exposed to sunlight.

cholinergic (kol-in-er'jik) Relating to nerve cells that use acetylcholine as their neurotransmitter.

chondroblast (kon'drō-blast) Actively mitotic form of a matrix-forming cell found in growing cartilage.

chondrocyte (kon'drō-sīt) Mature, nondividing cartilage cell.

chordamesoderm (kōr-dă-mes'ō-derm) Mesodermal cells that form the notochord.

chorion (kō're-on) Multilayered, outermost fetal membrane through which nourishment passes and attachment to the uterus occurs.

chromatid (krō'mă-tid) One of the two strands of a chromosome joined by a centromere.

chromatin (krō'ma-tin) Genetic material of the nucleus.

chromosome (krō'mō-sōm) The most organized level of genetic material; a single long molecule of DNA and associated proteins; becomes visible only when cell is dividing.

chyle (kil) Lymph drained from gastrointestinal tract.

cilium (sil'ē-ŭm; pl., cilia, -ă) Motile extension of a cell surface,

containing cytoplasm and microtubules.

circumduction (ser-kŭm-dŭk'shŭn) Movement of a body part in a circular direction.

cisterna (sis-ter'nă; pl., cisternae, -ter'nē) Ultramicroscopic spaces between the membranes of the flattened sacs of various organelles.

cleavage (klēv'ij) Series of mitotic cell divisions occurring in the zygote immediately following its fertilization.

clonus (klō'nŭs) Movement marked by rapid, successive contractions and relaxations of a muscle.

coccyx (kok'siks) The end of the vertebral column, formed from four fused vertebrae; the tailbone.

cognition (kog-ni'shŭn) Mental activities associated with thinking, learning, and memory.

coitus (kō'i-tŭs) Sexual union between a male and a female.

commissure (kom'i-shūr) Bundle of axons passing from one side to the other in the brain or spinal cord.

conjunctiva (kon-jŭnk-tī'vă) Mucous membrane covering the anterior surface of the eyeball and the posterior surface of the eyelids.

connexon (kon-neks'-on) Complex protein assembly that traverses the lipid bilayer of the plasma membrane and may join with those of another cell to form a gap junction.

contralateral (kon-tră-lat'er-ăl) Relating to the opposite side.

conus medullaris (kō'nŭs med-oo-lăr'is) Inferior end of the spinal cord.

cornea (kōr'nē-ă) Transparent structure making up the anterior surface of the eye.

coronal (kōr'o-năl) A vertical plane that divides the body into anterior and posterior parts; also called frontal plane.

coronary (kōr'o-nār-ē) Denoting the blood vessels or other structures and activities related to the heart.

corpus callosum (kōr'pŭs kal-lō-sŭm) White matter tract between the cerebral hemispheres.

costal (kos'tăl) Relating to a rib.

craniosynostosis (krā'nē-ō-sin'os-tō'sis) Premature ossification of the skull and obliteration of one or more sutures.

cubital (kū'bi-tal) Relating to the elbow.

cutaneous (kū-tā'nē-ŭs) Relating to the skin.

cytokine (sī'tō-kīn) Protein that regulates the intensity and duration of an immune response.

cytokinesis (sī'tō-ki-nē'sis) Division of the cytoplasm during cell division.

cytology (sī-tol'ō-jē) Study of cells.

cytoplasm (sī'tō-plazm) All the materials contained between the plasma membrane and the nucleus; includes cytosol, organelles, and inclusions.

cytosol (sī'tō-sol) The viscous, syruplike fluid of the cytoplasm.

D

deciduous (dē-sid'ū-ŭs) Not permanent; e.g., deciduous teeth.

decussation (dē-kŭ-sā'shŭn) Any crossing over or intersection of parts.

defecation (def-ĕ-kā'shŭn) Discharge of feces from the rectum.

deglutition (dē-gloo-tish'ŭn) Swallowing.

dendrite (den'drīt) Process of a neuron that conducts impulses toward the cell body.

dentition (den-tish'ŭn) Natural teeth in the dental arch, considered collectively.

depression (dē-presh'ŭn) Downward or inward displacement of a body part.

dermatology (der-ma-tol'ō-jē) Branch of medicine concerned with all aspects of the skin.

dermatome (der'mă-tōm) Area of skin supplied by a single spinal nerve. Also, during embryonic development, the cells that form the connective tissue of the skin.

dermis (der'mis) Layer of skin internal to the epidermis; contains blood and lymph vessels, nerves and nerve endings, glands, and usually hair follicles.

desmosome (dez'mō-sōm) One type of adhesion between two epithelial cells; a type of intercellular junction.

detrusor (dē-troo'ser, -sōr) Muscle that acts to expel urine from the bladder.

diapedesis (dī'ă-pĕ-dē'sis) Passage of blood or its formed elements through the intact blood vessel wall.

diaphysis (dī-af'i-sis; pl., diaphyses, -sēz) Elongated, rodlike part of a long bone between its two ends; the shaft of a long bone.

diarthrosis (dī-ar-thrō'sis; pl., diarthroses, -sēz) A freely movable (synovial) joint.

diastole (dī-as'tō-lē) The relaxation phase of a heart chamber.

diffusion (di-fū'zhŭn) Random movement of molecules or particles down their concentration gradient.

diploë (dip'lō-ē) Central layer of spongy bone between the two layers of compact bone (outer and inner plates) of the flat cranial bones.

diploid (dip'loyd) State of a cell containing pairs of homologous chromosomes. In humans, the diploid number of chromosomes is 46, 23 pairs.

dislocation (dis-lō-kā'shŭn) Complete displacement of an organ (e.g., a bone) from its normal position.

diuretic (dī-ū-ret'ik) Agent that increases the excretion of urine.

dorsiflexion (dōr-si-flek'shŭn) Upward movement of the foot or toes, or of the hand or fingers.

duodenum (doo-ō-dē'nŭm, doo-od'ĕ-nŭm) First section of small intestine.

dura mater (doo'ră mā'ter) Tough, fibrous membrane forming the outer

covering of the central nervous system.

dyskinesia (dis-ki-nē'zē-ă) Difficulty performing voluntary movements.

E

ectoderm (ek'tō-derm) Outermost of the three primary germ layers of the embryo.

ectopic (ek-top'ik) Out of place; e.g., in an ectopic pregnancy, the pre-embryo implants in the uterine tube rather than in the uterus.

effector (ē-fek'tŏr, -tōr) Peripheral tissue or organ that receives nerve impulses and then reacts.

efferent (ef'er-ent) Outgoing or moving away from a center.

ejaculation (ē-jak-ū-lā'shŭn) Expulsion of semen from the penis.

elevation (el-ĕ-vā'shŭn) Superior movement of a body part.

embryo (em'brē-ō) Organism in the early stages of development; in humans, the embryonic stage extends from the third to the eighth week of development.

embryology (em-brē-ol'ō-jē) Study of the origin and development of the organism, from fertilization of the secondary oocyte until birth.

enamel (ē-nam'ĕl) Hard substance covering the exposed portion of the tooth.

endocardium (en-dō-kar'dē-ŭm) Innermost covering of the heart wall.

endocrine (en'dō-krin) Hormonal secretions that are transported by the circulation.

endocrinology (en'dō-kri-nol'ō-jē) Medical specialty concerned with hormonal secretions and their physiologic and pathologic consequences.

endocytosis (en'dō-sī-tō'sis) Movement of substances from the extracellular environment into the cell.

endoderm (en'dō-derm) Innermost of the three primary germ layers of the embryo.

endolymph (en'dō-limf) Fluid within the membranous labyrinth of the inner ear.

endometrium (en'dō-mē'trē-ŭm) Mucous membrane forming the inner layer of the uterine wall.

endomysium (en'dō-miz'ē-ŭm, -mis'ē-ŭm) Areolar connective tissue layer surrounding a muscle fiber.

endoneurium (en-dō-noo'rē-ŭm) Areolar connective tissue of a peripheral nerve that surrounds the axons.

endosteum (en-dos'tē-ŭm) Layer of cells lining the inner surface of bone in the medullary cavity.

ependymal (ep-en'di-mal) Relating to the cellular lining of the brain ventricles and central canal of the spinal cord.

epicardium (ep-i-kar'dē-ŭm) The visceral layer of serous pericardium.

epidermis (ep-i-derm'is) Superficial epithelial portion of the skin.

epimysium (ep-i-mis′ē-ŭm) Fibrous connective tissue envelope surrounding a skeletal muscle.

epineurium (ep-i-noo′rē-ŭm) Outermost supporting connective tissue layer of peripheral nerves.

epiphyseal line (ep-i-fiz′ē-ăl) Portion of the epiphyseal plate that remains when long bone growth ceases.

epiphysis (e-pif′i-sis; pl., epiphyses, -sēz) Expanded, knobby region at the end of a long bone.

eponychium (ep-ō-nik′ē-ŭm) Corneal layer of epidermis found proximal and lateral to the nail plate.

erection (ē-rek′shŭn) Erectile tissues in the penis fill with blood and cause the penis to enlarge and become firm.

erythrocyte (ĕ-rith′rō-sīt) Mature red blood cell.

erythropoiesis (ĕ-rith′rō-poy-ē′sis) Formation of erythrocytes.

erythropoietin (ĕ-rith-rō-poy′ĕ-tin) Protein that stimulates erythropoiesis.

esophagus (ē-sof′ă-gŭs) Portion of the gastrointestinal tract between the pharynx and the stomach.

eversion (ē-ver′zhŭn) Turning the sole of the foot outward or laterally.

exocrine (ek′sō-krin) Glandular secretions delivered to an apical or luminal surface through a duct.

exocytosis (ek′sō-sī-to′sis) Process whereby secreting granules or droplets are released from a cell.

expiration (eks-pi-rā′shŭn) Forcing air from the lungs.

extension (eks-ten′shŭn) Movement that increases the articulating angle.

exteroceptor (eks′ter-ō-sep′ter, -tōr) Peripheral end organ of the afferent nerves in the skin or mucous membrane that responds to external stimulation.

F

facet (fas′et, fă-set′) Small, smooth area on a bone.

falciform (fal′si-fōrm) Having a sickle shape.

falx cerebri (falks se-rē′brē) Portion of the dura mater septa that projects into the longitudinal fissure between the right and left hemispheres of the brain.

fascia (fash′ē-ă; pl., fasciae, -ē-ē) Sheath of fibrous tissue that envelops the body beneath the skin, encloses muscles, and separates their various layers or groups.

fascicle (fas′i-kl) Band or bundle of muscle or nerve fibers.

fauces (faw′sēz) Space between the oral cavity and the pharynx.

feces (fē′sēz) Material discharged from the GI tract during defecation.

fertilization (fer′til-i-zā′shŭn) Process of sperm penetration of the secondary oocyte.

fetal (fē′tăl) Relating to the fetus; in humans, the fetal period extends from the eighth week after conception until birth.

fibroblast (fī′brō-blast) Large, flat, connective tissue cells with tapered ends that produce the fibers and

ground substance components of the extracellular matrix.

fibrosis (fī-brō′sis) Formation of fibrous connective tissue as a repair or reactive process.

filtrate (fil′trāt) Whatever materials get through a filter.

fimbria (fim′brē-ă; pl., fimbriae, -brē-ē) Any fringelike structure; e.g., the fimbriae of the infundibulum enclose the ovary at the time of ovulation.

fissure (fish′ur) Deep furrow, cleft, or slit.

flagellum (flă-jel′-ŭm; pl., flagella, -ă) Whiplike locomotory organelle that arises from within the cell and extends outside it.

flexion (flek′shŭn) Movement that decreases the angle between the articulating bones.

flexure (flek′sher) Bend in an organ or structure.

folliculitis (fŏ-lik-ū-lī′tis) Inflammation of a hair follicle.

fontanelle (fon′tă-nel′) One of several membranous intervals at the margins of the cranial bones in an infant.

fornix (fōr′niks; pl., fornices, -ni-sēz) Arch-shaped structure.

fossa (fos′ă; pl., fossae, -fos′ē) Depression, often more or less longitudinal in shape, below the level of the surface of a part.

furuncle (fū′rŭng-kl) Localized pyrogenic infection; also called a boil.

G

gamete (gam′ēt) A cell with the haploid number of chromosomes.

gametogenesis (gam′ĕ-tō-jen′ĕ-sis) Formation and development of gametes (sex cells).

ganglion (gang′glē-on; pl., ganglia, -glē-ă) Group of nerve cell bodies in the peripheral nervous system.

gastrulation (gas-troo-lā′shŭn) Formation of the three primary germ layers.

genotype (jen′ō-tīp) The genetic constitution of an individual that, along with environmental influences, contributes to the phenotype.

glabella (glă-bel′ă) The smooth part of the frontal bone that is sandwiched between the superciliary arches.

glycocalyx (glī-kō-kā′liks) Filamentous coating on the apical surface of certain epithelial cells.

gomphosis (gom-fō′sis; pl., gomphoses, -sēz) Fibrous joint in which a peglike process fits into a hole.

gonad (gō′nad) Organ that produces sex cells and sex hormones; the ovaries in a female and the testes in a male.

gustation (gŭs-tā′shŭn) Sense of taste.

gyrus (jī′rŭs; pl., gyri, -rī) Prominent rounded surface elevations that help form the cerebral hemispheres.

H

hapalonychia (hap′ă-lō-nik′ē-ă) Thinning of nails that results in bending and breaking of their free edge.

haploid (hap′loyd) The number of chromosomes in a sperm cell or secondary oocyte. In humans, the haploid number is 23.

hemangioma (he-man′jē-ō′mă) Congenital anomaly characterized by a proliferation of blood vessels forming a neoplasmlike mass.

hematocrit (hē′mă-tō-krit, hem′ă-) Percentage of whole blood attributed to erythrocytes.

hemoglobin (hē-mō-glō′bin) A red pigmented protein that transports oxygen and carbon dioxide; responsible for characteristic bright red color of arterial blood.

hemolysis (hē-mol′i-sis) Alteration or destruction of erythrocytes.

hemopoiesis (hē′mō-poy-ē′sis) Formation and development of blood cells.

hemorrhage (hem′ŏ-rij) Abnormal escape of blood from a blood vessel.

hernia (her′nē-ă) Protrusion of a part through the structures normally surrounding or containing it.

hiatus (hī-ā′tŭs) Opening.

hillock (hil′lok) Any small elevation or prominence.

hilum (hī′lŭm) The part of an organ where structures such as blood vessels, nerves, and lymph vessels enter or leave.

histology (his-tol′ō-jē) Study of tissues formed by cells and cell products.

homeostasis (hō′mē-ō-stā′sis, -os′tă-sis) State of equilibrium in the body with respect to various functions and the chemical composition of fluids and tissues.

homologous (hō-mol′ō-gŭs) Alike in certain critical attributes.

hormone (hōr′mōn) Chemical formed in one part of the body and carried by the blood to another where it can affect cellular activity.

hyaline (hī′a-lin, -lēn) Clear, homogeneous substance; a type of cartilage.

hydroxyapatite (hī-drok′sē-ap-ă-tīt) Natural mineral structure that the crystal lattice of bone and teeth closely resembles.

hyperextension (hī′per-eks-ten′shŭn) Extension of a body part beyond 180 degrees.

hypertonia (hī-per-tō′nē-ă) Extreme tension of muscles or arteries.

hypogastric (hī-pō-gas′trik) Relating to the hypogastrium, or pubic region.

hypotonia (hī-pō-tō′nē-ă) Condition of reduced tension in a body part; e.g., loss of muscle tone.

I

idiopathic (id′ē-ō-path′ik) Describing a disease of unknown cause.

infundibulum (in-fŭn-dib′ū-lŭm) Funnel-shaped structure or passage.

ingestion (in-jes′chŭn) Introduction of food and drink into the GI tract.

inhalation (in-hă-lā′shŭn) Drawing in a breath.

innervation (in-er-vā′shŭn) Supply of axons functionally connected with a part.

insertion (in-ser′shŭn) The usually distal and more movable attachment of a muscle.

integument (in-teg′ū-ment) Cover enveloping the body; includes the epidermis, dermis, and all derivatives of the epidermis; also called skin or the cutaneous membrane.

intercalated disc (in-ter′kă-lā-ted disk) Complex junction that interconnects cardiac muscle cells in the heart wall.

interneuron (in′ter-noo′ron) Type of neuron that resides within the CNS and coordinates activity between sensory and motor neurons.

interoceptor (in′ter-ō-sep′ter) Small sensory receptors within the walls of the viscera.

interstitial (in-ter-stish′ăl) Relating to spaces within a tissue or organ, but not a body cavity; e.g., interstitial fluid occupies the extracellular matrix of tissues.

intervertebral (in-ter-ver′te-brăl) Between vertebrae.

intramembranous ossification (in′tră-mem′bră-nŭs) Bone formation that takes place within a membrane; the formation of flat bones.

intramural (in′tră-mū′răl) Within the substance of the wall of a cavity or hollow organ.

intraperitoneal (in′tră-per′i-tō-nē′ăl) Location of a structure or organ that is completely covered with visceral peritoneum.

invaginate (in-vaj′i-nāt) To infold or insert a structure within itself.

inversion (in-ver′zhŭn) Turning inward of the foot.

isometric (ī-sō-met′rik) Condition in which the ends of a contracting muscle are held fixed and the muscle generates tension but does not shorten.

isotonic (ī-sō-ton′ik) Condition in which a contracting muscle generates tension and then shortens.

K

keratinization (ker′ă-tin-i-zā′shŭn) Development of a horny layer due to cells filled with the protein keratin.

keratinocyte (ke-rat′i-nō-sīt) The most abundant cell type in the epidermis and found throughout the epidermis; cells that produce keratin.

kyphosis (kī-fō′sis) Anteriorly concave curvature of the vertebral column.

L

lacrimal (lak′ri-măl) Relating to tears.

lactation (lak-tā′shŭn) Production of milk.

lacteal (lak′tē-al) Tiny vessel that transports lymph with lipids and lipid-soluble vitamins from the small intestine.

lacuna (lă-koo'nă; pl., lacunae, -koo'nē) Small space, cavity, or depression.

lamella (lă-mel'ă; pl., lamellae, -mel'ē) Layer of bone connective tissue.

lanugo (lă-noo'gō) Fine, soft, unpigmented fetal hair.

larynx (lar'ingks) Organ of voice production that lies between the pharynx and the trachea.

lateralization (lat'er-al-ī-zā'shŭn) Process whereby certain asymmetries of structure and function occur.

lemniscus (lem-nis'kŭs; pl., lemnisci, -nis'ī) Bundle of axons ascending from sensory relay nuclei to the thalamus.

lesion (lē'zhŭn) Pathologic change in a tissue.

lethargy (leth'ar-jē) Mild impairment of consciousness characterized by reduced awareness and alertness.

leukocyte (loo'kō-sīt) Any one of several types of white blood cells.

leukopoiesis (loo'kō-poy-ē'sis) Formation and development of leukocytes.

ligament (lig'ă-ment) Band or sheet of dense regular fibrous tissue that connects bones, cartilage, or other structures.

lobe (lōb) Subdivision of an organ, bounded by some structural demarcation.

lumen (loo'men) The space inside a structure, such as where blood is transported within a blood vessel.

lunula (loo'noo-lă) Pale, arched area at the proximal end of the nail plate.

lymph (limf) Usually transparent fluid found in lymph vessels; derived from interstitial fluid.

lymphoblast (lim'fō-blast) Immature cell that becomes a lymphocyte.

lymphopoiesis (lim-fō-poy-ē'sis) Formation of lymphatic cells.

lysosome (lī'sō-sōm) Organelle containing digestive enzymes.

M

malignant (mă-lig'nant) In reference to a neoplasm, having the property of invasiveness and spread.

mastication (mas-ti-kā'shŭn) Chewing food in preparation for swallowing.

matrix (mā'triks; pl., matrices, mā'tri-sēz) Surrounding substance within which cells or structures are contained or embedded.

mechanoreceptor (mek'ă-nō-rē-sep'tŏr) Receptor that responds to touch, vibration, pressure, or distortion.

mediastinum (me'dē-as-tī'nŭm) Median space of the thoracic cavity.

meiosis (mī-ō'sis) Process of sex cell division that results in four gametes, each with the haploid number of chromosomes.

melanin (mel'ă-nin) Any of the dark brown to black pigments that occur in the skin, hair, and retina.

melanocyte (mel'ă-nō-sīt) Pigment-producing cell in the basal layer of the epidermis.

menarche (me-nar'kē) Time of the first menstrual period.

meninx (mē'ninks; pl., meninges, -jēz) Any one of the membranes covering the brain and spinal cord.

meniscus (mĕ-nis'kŭs; pl., menisci, mĕ-nis'sī) Crescent-shaped fibrocartilage found in certain joints.

menopause (men'ō-pawz) Permanent cessation of menses.

menses (men'sēz) Periodic physiologic hemorrhage from the uterine mucous membrane; commonly referred to as the menstrual period.

menstruation (men-stroo-ā'shŭn) Cyclic endometrial shedding and discharge of bloody fluid.

mesenchyme (mez'en-kīm) An embryonic connective tissue.

mesoderm (mez'ō-derm) The middle of the three primary germ layers of the embryo.

mesothelium (mez-ō-thē'lē-ŭm) The simple squamous epithelium that lines serous cavities.

metaphysis (mĕ-taf'i-sis; pl., metaphyses, -sēz) Section of a long bone between the epiphysis and the diaphysis.

metaplasia (met-ă-plā'zē-ă) Abnormal change in a tissue.

metastasis (mĕ-tas'tă-sis) Spread of a disease from one part of the body to another.

metopic (me-tō'pik, me-top'ik) Relating to the forehead.

microfilament (mī-krō-fil'ă-ment) Smallest structural element of the cytoskeleton.

microglial cell (mī-krog'lē-ă) Category of small glial cells in the central nervous system.

microscopy (mī-kros'kŏ-pē) Investigation of very small objects by means of a microscope.

microtubule (mī-krō-too'bŭl) Hollow, cylindrical cytoplasmic element.

micturition (mik-choo-rish'ŭn) Urination.

mitochondrion (mī-tō-kon'drē-on; pl., mitochondria, -ă) Organelle associated with the production of ATP.

mitosis (mī-tō'sis) Process of somatic cell division.

mons (monz) Anatomic prominence or slight elevation above the general level of the surface; e.g., mons pubis.

mucin (mū'sin) An often protective secretion containing carbohydrate-rich glycoproteins.

mucosa (mū-kō'să) Mucous membrane that lines various body structures.

muscularis (mŭs-kū-lā'ris) Muscular layer in the wall of a hollow organ or tubular structure.

myelin (mī'ĕ-lin) Lipoproteinaceous material of the myelin sheath.

myenteron (mī-en'ter-on) Muscle coat of the intestine; also called the tunica muscularis.

mylohyoid (mī'lō-hī-oyd) Pertaining to the molar teeth or the posterior portion of the mandible and hyoid.

myoblast (mī'ō-blast) Undifferentiated muscle cell with the potential of becoming a muscle fiber.

myocardium (mī-ō-kar'dē-ŭm) Middle layer of the heart wall, consisting of cardiac muscle.

myofibril (mī-ō-fī'bril) One of the longitudinal fibrils found in skeletal and cardiac muscle fibers.

myofilament (mī-ō-fil'ă-ment) A protein filament that makes up the myofibrils in skeletal muscle.

myoglobin (mī-ō-glō'bin) Oxygen-carrying and -storing molecule in muscle.

myometrium (mī-ō-mē'trē-ŭm) Middle tunic (muscular wall) of the uterus.

myotome (mī'ō-tōm) Part of the somite that develops into skeletal muscle.

N

naris (nā'ris; pl., nares, -res) Anterior opening to the nasal cavity.

necrosis (nĕ-krō'sis) Pathologic death of cells or a portion of a tissue or organ.

neoplasia (nē-ō-plā'zē-ă) Process that results in the formation of a neoplasm or abnormal growth.

neoplasm (nē'ō-plazm) An abnormal tissue that grows by cell proliferation more rapidly than normal.

nephron (nef'ron) Convoluted tubular functional unit of the kidney.

neurilemma (noor-i-lem'a) The delicate outer membrane sheath around an axon.

neurofibril (noor-o-fi'bril) Filamentous structure in a neuron.

neurofilament (noor-o-fil'a-ment) Group of intermediate-sized filaments in a neuron.

neuromuscular (noor-o-mus'ku-lar) Relationship between a nerve and a muscle; e.g., a neuromuscular junction.

neuron (noor'on) Functional unit of the nervous system.

neuropore (noor'ō-por) Opening in the embryo leading from the central canal of the neural tube to the exterior.

neurulation (noor-oo-lā'shun) Formation of the neural plate and its closure to form the neural tube.

nociceptor (nō-si-sep'ter, -tŏr) Peripheral sensory receptor for the detection of painful stimuli.

nucleolus (noo'klē'ō-lŭs; pl., nucleoli, -lī) Spherical, dark body within the nucleus where subunits of ribosomes are made.

O

oblique (ob-lēk') Slanted, at an angle.

occult (ŏ-kŭlt', ok'ŭlt) Hidden or concealed.

olfaction (ol-fak'shun) Sense of smell.

oligodendrocyte (ol'i-gō-den'drō-sīt) Category of large glial cells in the central nervous system.

omentum (ō-men'tŭm) Fold of peritoneum from the stomach to another abdominal organ.

oogenesis (ō-ō-jen'ĕ-sis) Formation and development of oocytes.

oogonium (ō-ō-gō'nĕ-ŭm; pl., oogonia, -ă) Primitive germ cell of the oocyte.

opposition (op'pō-si'shŭn) Movement of the thumb across the palm to touch the palmar side of the fingertips.

organelle (or'gă-nel) Complex, organized structures in the cytoplasm of a cell with unique characteristic shapes; called "little organs."

organogenesis (ŏr'gă-nō-jen'ĕ-sis) Formation of organs during development.

orifice (or'i-fis) Aperture or opening.

origin (ōr'i-jin) The less movable of the two points of attachment of a muscle.

os (os; pl., ossa, os'ă) An opening into a hollow organ or canal.

osmosis (os-mō'sis) Process by which a solvent tends to move through a semipermeable membrane from a hypotonic solution to a hypertonic one.

osseous (os'ē-ŭs) Bony.

ossicle (os'i-kl) Small bone.

osteoblast (os'tē-ō-blast) Bone-forming cell.

osteoclast (os'tē-ō-klast) Large cell type that functions in the absorption and removal of bone connective tissue.

osteocyte (os'tē-ō-sīt) Bone cell in a lacuna.

osteogenesis (os'tē-ō-jen'ĕ-sis) Production of new bone.

osteolysis (os-tē-ol'i-sis) Softening, absorption, and destruction of bone connective tissue.

osteon (os'tē-on) Functional unit of compact bone tissue; also called a Haversian system.

osteopenia (os'tē-ō-pē'nē-ă) Decreased calcification or density of bone.

osteoporosis (os'tē-ō-pō-rō'sis) Medical condition characterized by decreased bone mass and increased susceptibility to fracture.

osteoprogenitor (os'tē-ō-prō-jen'i-ter) Precursor to bone cells.

ovulation (ov'ū-lā'shŭn) Release of a secondary oocyte from the ovarian follicle.

ovum (ō'vŭm; pl., ova, -vă) Female sex cell that has been fertilized and has completed meiosis II.

P

palate (pal'ăt) Partition between the oral and nasal cavities; the roof of the mouth.

palpation (pal-pā'shŭn) Using the sense of touch to identify or examine internal body structures.

papilla (pă-pil′ă; pl., papillae, -pil′ē) Small, nipplelike process.

parenchyma (pă-reng′ki-mă) Specific functional cells of a gland or organ.

parietal (pă-rī′ĕ-tăl) Relating to the wall of any cavity.

patellar (pa-tel′ăr) Relating to the kneecap.

pathologic (path-ō-loj′-ik) Pertaining to disease.

pedal (ped′ăl) Relating to the foot.

percussion (per-kŭsh′ŭn) Using sounds produced by tapping the surface of the body to help diagnose medical conditions.

pericardium (per-i-kar′dē-ŭm) Fibroserous membrane covering the heart.

perichondrium (per′-i-kon′drē-ŭm) Layer of dense irregular connective tissue around the surface of cartilage.

perilymph (per′i-limf) Fluid within the osseous labyrinth, surrounding and protecting the membranous labyrinth.

perimetrium (per-i-mē′trē-ŭm) Serous coat of the uterus.

perimysium (per-i-mis′ē-ŭm, -miz′ē-ŭm) Fibrous sheath enveloping each of the fascicles of skeletal muscle fibers.

perineurium (per-i-noo′rē-ŭm) Fibrous sheath of peripheral nerves that surrounds the nerve fascicles.

periosteum (per-ē-os′tē-ŭm) Thick, fibrous membrane covering the entire external surface of a bone, except for the articular cartilage.

peritoneum (per′i-tō-nē′ŭm) Serous sac that lines the abdominopelvic cavity and covers most of the viscera within.

peroxisome (per-ok′si-sōm) Membrane-bound organelle containing oxidative enzymes.

phagocytosis (făg′ō-sī-tō′sis) Process by which cells ingest and digest solid substances.

phalanx (fā′langks; pl., phalanges, -jēz) Long bone of a digit.

phenotype (fē′nō-tīp) Observable characteristics of an individual.

philtrum (fil′trŭm) Groove in the midline of the upper lip.

pinna (pin′ă) Another name for the auricle or outer ear.

pinocytosis (pin′ō-sī-tō′sis) Cellular process of actively engulfing liquid.

placenta (plă-sen′tă) Organ of exchange between the embryo or fetus and the mother.

plasmalemma (plaz-mă-lem′ă) The plasma (cell) membrane.

platelet (plāt′let) Irregularly shaped cell fragment that participates in blood clotting.

pleura (ploor′ă; pl., pleurae, ploor′ē) Serous membranes enveloping the lungs and lining the walls of the pleural cavity.

plexus (plek′sŭs) Network of nerves, blood vessels, or lymph vessels.

pollex (pol′eks) Thumb.

polycythemia (pol′ē-sī-thē′mē-ă) More than the normal number of erythrocytes.

polyspermy (pol′ē-sper-mē) Entrance of more than one sperm into the oocyte.

postsynaptic (pōst-si-nap′tik) On the distal side of a synaptic cleft.

prepuce (prē′poos) Free fold of skin that covers the glans penis.

presynaptic (prē-si-nap′tik) On the proximal side of a synaptic cleft.

pronation (prō-nā′shŭn) Rotational movement of the forearm causing the hand to be directed posteriorly or inferiorly.

proprioceptor (prō′prē-ō-sep′ter) Sensory end organ that senses position or state of contraction.

protraction (prō-trak′shun) Movement of a body part anteriorly in a horizontal plane.

protuberance (prō-too′ber-ans) Swelling or knoblike outgrowth.

pulmonary (pŭl′mō-nār-ē) Relating to the lung.

pulse (pŭls) Rhythmic dilation of an artery resulting from increased blood volume during heart contraction.

pupil (pū′pl) Circular hole in the center of the iris of the eye.

R

radiopaque (rā-dē-ō-pāk′) Relatively impenetrable to x-rays or other forms of radiation.

ramus (rā′mŭs; pl., rami, rā′mī) One of the primary divisions of a nerve or blood vessel. Also, part of an irregularly shaped bone that forms an angle.

raphe (rā′fē) Line of union between two contiguous, bilateral symmetrical structures; e.g., the raphe of the scrotum.

reticulocyte (re-tik′ū-lō-sīt) Young erythrocyte.

retraction (rē-trak′shŭn) Posteriorly directed movement of a body part.

retroperitoneal (re′trō-per′i-tō nē′ăl) External or posterior to the peritoneum.

ribosome (rī′bō-sōm) Ribonucleoprotein structure that is the site of protein synthesis; organelle in the cytoplasm.

rotation (rō-tā′shŭn) Movement of a part around its axis.

rouleau (roo-lō′; pl., rouleaux) Aggregation of erythrocytes in single file.

ruga (roo′gă; pl., rugae, roo′gē) Fold, ridge, or crease.

S

sacroiliac (sā-krō-il′ē-ak) Pertaining to the sacrum and the ilium.

sacrum (sā′krŭm) Next-to-last bone of the vertebral column; formed of five fused vertebrae.

sagittal (saj′i-tăl) Being in an anteroposterior direction.

sarcolemma (sar′kō-lem′ă) Plasma membrane of a muscle fiber.

sarcomere (sar′kō-mēr) Functional unit of skeletal muscle.

sarcoplasm (sar′kō-plazm) Cytoplasm of a muscle fiber.

sclera (skler′ă) Portion of the fibrous layer forming the outer covering of the eyeball, except for the cornea.

sclerotome (skler′ō-tōm) The portion of a somite that gives rise to vertebrae and ribs.

scoliosis (skō-lē-ō′sis) Abnormal lateral and rotational curvature of the vertebral column.

sebum (sē′bŭm) Secretion of a sebaceous gland.

semen (sē′men) Secretion composed of sperm and seminal fluid.

semipermeable (sem′ē-per′mē-ă-bl) Freely permeable to some molecules, but relatively impermeable to other molecules.

sensation (sen-sā′shun) Translation into consciousness of the effects of a stimulus.

serosa (se-rō′să) Outermost coat of a visceral structure that lies in a closed body cavity.

serous (sēr′ŭs) Producing a substance that has a watery consistency.

serum (sēr′ŭm) Clear, watery fluid that remains after clotting proteins are removed from plasma.

sinus (sī′nŭs) Cavity or hollow space; also, a channel for the passage of blood or lymph.

sinusoid (sī′nŭ-soyd) Resembling a sinus; a microscopic space or passage for blood in certain organs, such as the liver or the spleen.

skeleton (skel′ĕ-tŏn) Bony framework of the body; collectively, all the bones of the body.

somatomedin (sō′mă-tō-mē′din) Peptide that stimulates certain anabolic processes in bone and cartilage.

somatotopy (sō-mă-tot′ō-pē) Point-by-point correspondence between a body area and the CNS.

somite (sō′mīt) One of the paired segments consisting of cell masses formed in the early embryonic mesoderm on the sides of the neural tube.

sonography (sō-nog′ră-fē) Radiographic technique using ultrasound waves.

spermatid (sper′mă-tid) Late stage of the development of the sperm.

spermatogenesis (sper′mă-tō-jen′ĕ-sis) Process by which stem cells become sperm.

spermatogonium (sper′mă-tō-gō′nē-ŭm; pl., spermatogonia, -ă) Diploid parent or stem cell that produces sperm.

spermiogenesis (sper′mē-ō-jen′ĕ-sis) Stage of spermatogenesis during which immature spermatids become sperm cells, or spermatozoa.

sphincter (sfingk′ter) Muscle that encircles a duct, tube, or orifice such that its contraction constricts the lumen or orifice.

sphygmomanometer (sfig′mō-mă-nom′ĕ-ter) Instrument for measuring arterial blood pressure.

strabismus (stra-biz′mŭs) Lack of parallelism of the visual axes of the eye.

stria (strī′ă; pl., striae, strī′ē) Stripes, bands, streaks, or lines distinguished by a difference in color, texture, or elevation from surrounding tissue.

stupor (stoo′per) State of impaired consciousness from which the individual can be aroused only by continual stimulation.

subcutaneous (sŭb-kū-tā′nē-ŭs) Beneath the skin.

subluxation (sŭb-lŭk-sā′shŭn) Incomplete dislocation of a body part from its normal position.

sulcus (sool′kŭs; pl., sulci, sŭl′sī) Groove on the surface of the brain.

supernumerary (soo-per-noo′mer-ār-ē) Exceeding the normal number.

supination (soo′pi-nā′shŭn) Rotation of the forearm such that the hand faces anteriorly or superiorly.

suppressor (soo-pres′ŏr) A compound or immune cell that "turns off" a pathway or course of events.

suture (soo′choor) Synarthrosis in which bones are united by a dense regular connective tissue membrane.

symphysis (sim′fi-sis; pl., symphyses, -sēz) Cartilaginous joint in which the two bones are separated by a pad of fibrocartilage.

synapse (sin′aps) Functional contact of a nerve cell with another nerve cell, effector, or receptor.

synarthrosis (sin′ar-thrō′sis; pl., synarthroses, -sēz) Immovable joint.

synchondrosis (sin′kon-drō′sis; pl., synchondroses, -sēz) Cartilaginous joint in which the bones are joined by hyaline cartilage.

syncytiotrophoblast (sin-sish′ē-ō-trō′fō-blast) Outer layer of the trophoblast that produces the hormone human chorionic gonadotropin (hCG).

syndesmosis (sin′dez-mō′sis; pl., syndesmoses, -sēz) Fibrous joint in which the opposing surfaces of articulating bones are united by ligaments.

synergist (sin′er-jist) Structure, muscle, agent, or process that aids the action of another.

synostosis (sin-os-tō′sis; pl., synostoses, -sēz) Osseous union between two bones that were initially separate.

systemic (sis-tem′ik) Relating to the entire organism as opposed to any of its individual parts.

systole (sis′tō-lē) Contraction of the heart.

T

tactile (tak′til) Relating to touch.

tendon (ten′dŏn) Cord of dense regular connective tissue that connects muscle to bone.

teratology (ter-ă-tol′ō-jē) Branch of science concerned with congenital malformations.

testis (tes′tis; pl., testes, -tēz) Male organ for producing gametes and androgens.

thenar (thē′nar) A structure related to the base of the thumb or the thumb's underlying connective components.

thermoreceptor (ther′mō-rē-sep′ter, -tōr) Receptor that is sensitive to heat.

thoracic (thō-ras′ik) Relating to the thorax, the area between the neck and the abdomen.

thrombopoiesis (throm′bō-poy-ē′sis) Formation of blood platelets.

tomography (tō-mog′rǎ-fē) A radiographic image of a plane constructed by means of reciprocal linear or curved motion of the x-ray tube and film cassette; used in producing a CT scan.

trabecula (trǎ-bek′ū-lǎ; pl., trabeculae, -lē) Meshwork; a small piece of the spongy substance of bone, usually interconnected with other similar pieces.

transducer (trans-doo′ser) Device or organ designed to convert energy from one form to another.

transudate (tran′soo-dāt) Fluid that has passed through a membrane.

trophoblast (trof′ō-blast, trō′fō-blast) Cell layer covering the blastocyst that will allow the embryo to receive nourishment from the mother.

tubercle (too′ber-kl) Nodule or slight elevation from the surface of a bone that allows attachment of a tendon or ligament.

tuberosity (too′ber-os′i-tē) Large tubercle or rounded elevation, as on the surface of a bone.

tunic (too′nik) One of the covering layers of a body part, especially a blood vessel or other tubular structure.

U

ultrastructure (ŭl-trǎ-strŭk′choor) Cell structure viewable via the electron microscope.

umbilical (ŭm-bil′i-kǎl) Relating to the umbilicus (navel).

ureter (ū-rē′ter, ū′rē-ter) Tubes that connect the kidney to the urinary bladder.

urethra (ū-rē′thrǎ) Canal leading from the bladder to externally discharge urine.

urinalysis (ū-ri-nal′i-sis) Analysis of the urine to help assess the state of a person's health.

urine (ūr′in) Fluid and dissolved substances excreted by the kidney.

V

vasculogenesis (vas′kū-lō-jen′ĕ-sis) Formation of the vascular system.

vein (vān) Blood vessel carrying blood toward the heart.

vellus (vel′ŭs) Fine, nonpigmented hair covering most of the fetal body.

ventricle (ven′tri-kl) Cavity within an organ such as the heart or the brain.

vertebra (ver′tĕ-brǎ; pl., vertebrae, -brē) Segment of the spinal column.

vertebral (ver′tĕ-brǎl) Relating to a vertebra or the vertebral column.

vesicle (ves′i-kl) Closed cellular structure in the cytoplasm surrounded by a single membrane.

vibrissa (vī-bris′ǎ; pl., vibrissae, vī-bris′ē) Hair in the vestibule of the nose.

viscera (vis′er-ǎ) Organs of the digestive, respiratory, urinary, and endocrine systems.

vulva (vŭl′vǎ) External genitalia of the female.

Z

zygote (zī′gōt) Diploid cell resulting from the union of a sperm cell and a secondary oocyte.

Credits

Photographs

Chapter 1

Figure 1.1a: © Bettmann/Corbis; **1.1b:** © Tim Shaffer/Reuters/Corbis; **1.2a:** © Kathy Talaro/Visuals Unlimited; **1.2b:** © Meiji/Visuals Unlimited; **1.5:** © The McGraw-Hill Companies, Inc./Photo by Jw Ramsey; **Page 18 left:** © Ralph T. Hutchings/Visuals Unlimited; **Page 18 top right:** © ATL/Photo Researchers, Inc.; **Page 18 bottom right:** © SIU BioMed/Custom Medical Stock Photo; **Page 19 left:** © Athenais/Phototake; **Page 19 top right:** © Alfred Pasieka/SPL/Photo Researchers, Inc.; **Page 19 bottom right:** © Hank Morgan/Photo Researchers, Inc.

Chapter 2

Figure 2.2a: © The McGraw-Hill Companies, Inc./Photo by Dr. Alvin Telser; **2.2b:** © VVG/SPL/Photo Researchers, Inc.; **2.2c:** © Eye of Science/Photo Researchers, Inc.; **Table 2.1 Epidermal cells:** © Ed Reschke/Peter Arnold; **Table 2.1 Epithelial cells:** © The McGraw-Hill Companies, Inc./Photo by Dr. Alvin Telser; **Table 2.1 Fat cells:** © The McGraw-Hill Companies, Inc./Photo by Dr. Alvin Telser; **Table 2.1 Muscle cells:** © The McGraw-Hill Companies, Inc./Photo by Dr. Alvin Telser; **Table 2.1 Collagen:** © Ed Reschke/Peter Arnold; **Table 2.1 Lymphocytes:** © The McGraw-Hill Companies, Inc./Photo by Dr. Alvin Telser; **Table 2.1 Nerve cells:** © Carolina Biological Supply Company/Phototake; **Table 2.1 Sperm cells:** © Jason Burns/Phototake; **2.8, 2.9:** © Dennis Kunkel/Phototake; **2.10:** © David M. Phillips/Visuals Unlimited; **2.11:** © D. Friend and Donald Fawcett/Visuals Unlimited; **2.12:** © Donald Fawcett/Visuals Unlimited; **2.13b:** © Gopal Murti/Visuals Unlimited; **2.15:** © Dr. Don W. Fawcett/Visuals Unlimited; **2.16a:** © Eye of Science/Photo Researchers, Inc.; **2.16b:** © Dr. Richard Kessel and Dr. C. Shih/Visuals Unlimited; **2.17:** © Donald Fawcett/Visuals Unlimited; **2.20a:** © Michael Abbey/Photo Researchers, Inc.; **2.20b-d:** © Carolina Biological Supply Company/Phototake; **2.20e:** © Michael Abbey/Photo Researchers, Inc.

Chapter 3

Page 59b: © Hattie Young/Photo Researchers, Inc.; **3.3c:** © David M. Phillips/Visuals Unlimited.

Chapter 4

Table 4.3a-c: © The McGraw-Hill Companies, Inc./Photo by Dr. Alvin Telser; **Table 4.3d:** © Ed Reschke; **Table 4.4a-d, Table 4.5a & b :** © The McGraw-Hill Companies, Inc./Photo by Dr. Alvin Telser; **4.3a:** © VVG/SPL/Photo Researchers, Inc.; **Table 4.6a:** © The McGraw-Hill Companies, Inc./Photo by Dr. Alvin Telser; **Table 4.6b:** © Ed Reschke; **Page 100:** © National Marfan Foundation; **Table 4.9a & b, Table 4.9c:** © The McGraw-Hill Companies, Inc./Photo by Dr. Alvin Telser; **Table 4.10a:** © Ed Reschke; **Table 4.10b:** © The McGraw-Hill Companies, Inc./Dennis Strete, photographer; **Table 4.10c:** © The McGraw-Hill Companies, Inc./Photo by Dr. Alvin Telser; **Table 4.11a:** © Ed Reschke/Peter Arnold; **Table 4.11b:** © Ed Reschke; **Table 4.11c:** © The McGraw-Hill Companies, Inc./Photo by Dr. Alvin Telser; **Table 4.12:** © The McGraw-Hill Companies, Inc./Dennis Strete, photographer; **Table 4.13, Table 4.14a-c:** © The McGraw-Hill Companies, Inc./Photo by Dr. Alvin Telser; **Table 4.15:** © Carolina Biological Supply Company/Phototake; **Page 113 left:** © Dr P. Marazzi/Photo Researchers, Inc.; **Page 113 right:** © Stevie Grand/Photo Researchers, Inc.

Chapter 5

Figure 5.2a: © Ed Reschke/Peter Arnold; **5.3a:** © Carolina Biological Supply Company/Phototake; **5.3b:** © Ed Reschke/Peter Arnold; **5.4b:** © John Burbidge/Photo Researchers, Inc.; **5.9b:** © The McGraw-Hill Companies, Inc./Photo by Dr. Alvin Telser; **5.9c:** © SPL/Photo Researchers, Inc.; **Page 131a:** © Dr. P. Marazzi/Photo Researchers, Inc.; **Page 131b:** © Logical Images, Inc.; **Page 131c:** © John Radcliffe Hospital/Photo Researchers, Inc.; **Page 131d:** © Dr. P. Marazzi/Photo Researchers, Inc.; **5.10b:** © Collection CNRI/Phototake; **5.10c:** © G.W. Willis, MD/Visuals Unlimited; **5.10d:** © The McGraw-Hill Companies, Inc./Photo by Dr. Alvin Telser; **Page 137a:** © Sheila Terry/Photo Researchers, Inc.; **Page 137b:** © Dr. P. Marazzi/Photo Researchers, Inc.; **Page 137c:** © John Radcliffe Hospital/Photo Researchers, Inc.; **Table 5.4 top:** © Dr. Ken Greer/Visuals Unlimited; **Table 5.4 center:** © Dr. P. Marazzi/Photo Researchers, Inc.; **Table 5.4 bottom:** © James Stevenson/Photo Researchers, Inc.

Chapter 6

Figure 6.1b-d, 6.2: © The McGraw-Hill Companies, Inc./Photo by Dr. Alvin Telser; **6.4a:** © The McGraw-Hill Companies, Inc./Photo by Christine Eckel; **6.4b:** © Ralph T. Hutchings/Visuals Unlimited; **6.6c:** © The McGraw-Hill Companies, Inc./Photo by Dr. Alvin Telser; **Page 153:** © SPL/Photo Researchers, Inc.; **6.9a:** © Dr. Richard Kessel & Dr. Randy Kardon/Tissues and Organs/Visuals Unlimited; **6.9b:** © Carolina Biological Supply Company/Phototake; **6.9c:** © Leonard Lessin/SPL/Photo Researchers, Inc.; **Page 159:** © David Hunt/Smithsonian Institution; **6.12a:** © The McGraw-Hill Companies, Inc./Photo by Dr. Alvin Telser; **6.12b:** © Image Shop/Phototake; **Page 163:** Courtesy of Dr. George P. Bogumill, Georgetown University Medical Center, Department of Orthopaedic Surgery; **Page 163a & b:** © Scott Camazine/Photo Researchers, Inc.; **Page 167a:** © SPL/Photo Researchers, Inc.; **Page 167b:** © P. Motta/Photo Researchers, Inc.

Chapter 7

Figure 7.4, 7.5a & b, 7.6, 7.7, 7.8: © The McGraw-Hill Companies, Inc./Photo by Christine Eckel; **Page 181:** Courtesy of Dr. John A. Jane, Sr., David D. Weaver Professor of Neurosurgery, Department of Neurological Surgery, University of Virginia Health System, Charlottesville, Virginia; **7.9, 7.10, 7.11, 7.12a & b, 7.13a & b, 7.14a & b, 7.16b, 7.21a & b, 7.22:** © The McGraw-Hill Companies, Inc./Photo by Christine Eckel; **Page 196 left:** © Dr. M. A. Ansary/SPL/Photo Researchers, Inc.; **Page 196 right:** © NMSB/Custom Medical Stock Photo; **7.23b:** © The McGraw-Hill Companies, Inc./Photo and Dissection by Christine Eckel; **Table 7.4 (all):** © Ralph T. Hutchings/Visuals Unlimited; **7.27a & b, Table 7.5 (all), 7.30a-c, 7.31a & b, 7.32, 7.33c:** © The McGraw-Hill Companies, Inc./Photo by Christine Eckel.

Chapter 8

Figure 8.2c: © The McGraw-Hill Companies, Inc./Photo by Christine Eckel; **8.2d:** © B. Bates/Custom Medical Stock Photo; **8.3a-c, 8.4a-d, 8.5a, 8.5d-f, 8.6a & b:** © The McGraw-Hill Companies, Inc./Photo by Christine Eckel; **8.8:** © Mediscan/Visuals Unlimited; **8.9a & b:** © The McGraw-Hill Companies, Inc./Photo by Christine Eckel; **Table 8.1 top:** © David Hunt/Smithsonian Institution; **Table 8.1 bottom:** © L. Bassett/Visuals Unlimited; **8.11a-d, 8.13a-e:** © The McGraw-Hill Companies, Inc./Photo by Christine Eckel; **8.13f:** © Ralph T. Hutchings/Visuals Unlimited; **8.13g, 8.14a & b:** © The McGraw-Hill Companies, Inc./Photo by Christine Eckel; **Page 244 top left:** © John Watney; **Page 244 bottom left:** © SPL/Photo Researchers, Inc.; **Page 244 top right:** © Mediscan/Visuals Unlimited; **Page 244 center right:** © Bart's Medical Library/Phototake; **Page bottom right:** © ISM/Phototake; **Page 246:** © Science Photo Library/Photo Researchers, Inc.

Chapter 9

Figure 9.7a: © The McGraw-Hill Companies, Inc./Photo by Eric Wise; **9.7b & c:** © The McGraw-Hill Companies, Inc./Photo by Jw Ramsey; **9.7d:** © The McGraw-Hill Companies, Inc./Photo by Eric Wise; **9.7e, 9.8a-c:** © The McGraw-Hill Companies, Inc./Photo by Jw Ramsey; **9.8d:** © The McGraw-Hill Companies, Inc./Photo by Eric Wise; **9.9a & b, 9.10 (all):** © The McGraw-Hill Companies, Inc./Photo by Jw Ramsey; **9.11a:** © The McGraw-Hill Companies, Inc./Photo by Eric Wise; **9.11b & c:** © The McGraw-Hill Companies, Inc./Photo by Jw Ramsey; **9.11d & e:** © The McGraw-Hill Companies, Inc./Photo by Eric Wise; **9.15a:** © The McGraw-Hill Companies, Inc./Photo and Dissection by Christine Eckel; **Page 271a:** © & Courtesy of Dr. Mike Langran/www.ski-injury.com; **Page 271b:** © Science Photo Library, Ltd./Phototake; **9.18d, 9.19c & d, 9.20b & c:** © The McGraw-Hill Companies, Inc./Photo and Dissection by Christine Eckel; **Page 281a:** © John Watney; **Page 281b:** © CNRI/Photo Researchers, Inc.

Chapter 10

Figure 10.6c: © James Dennis/Phototake; **10.7a-c:** © Dr. H. E. Huxley; **10.8a:** © Jean Claude Revy-ISM/Phototake; **10.12:** © Gladden Willis/Visuals Unlimited; **Table 10.6 top:** © Eric Grave/Phototake; **Table 10.6 center:** © John Cunningham/Visuals Unlimited; **Table 10.6 bottom:** © Carolina Biological Supply Company/Phototake; **Page 314:** Courtesy of Muscular Dystrophy Association (www.mda.org).

Chapter 11

Figure 11.2a & b: © The McGraw-Hill Companies, Inc./Photo and Dissection by Christine Eckel; **Page 326:** © Dr. P. Marazzi/SPL/Photo Researchers, Inc.; **11.3a-f:** © The McGraw-Hill Companies, Inc./Photo by Jw Ramsey; **11.8b, 11.10, 11.11:** © The McGraw-Hill Companies, Inc./Photo and Dissection by Christine Eckel; **Page 341:** © AP/Wide World Photos; **11.13b, 11.14b:** © The McGraw-Hill Companies, Inc./Photo and Dissection by Christine Eckel.

Chapter 12

Figure 12.4a & b: © The McGraw-Hill Companies, Inc./Photo and Dissection by Christine Eckel; **12.5c left:** © Elsa/Getty Images; **12.5c center:** © AP/Wide World Photos; **12.5c right:** © Mike Fiala/Getty Images; **12.7a, 12.8a, 12.11a:** © The

McGraw-Hill Companies, Inc./Photo and Dissection by Christine Eckel; **12.12, Page 369:** © The McGraw-Hill Companies, Inc./Photo by Jw Ramsey; **12.13a, 12.14a, 12.15c, 12.17a, 12.18a, 12.20a, 12.21a, 12.22b:** © The McGraw-Hill Companies, Inc./Photo and Dissection by Christine Eckel.

Chapter 13

Opener, 13.1a & b, 13.2, 13.3a & b: © The McGraw-Hill Companies, Inc./Photo by Jw Ramsey; **Page 400 all, Table 13.2b & c:** © The McGraw-Hill Companies, Inc./Photo by Chris Hammond; **13.4a & b, 13.5, 13.6, 13.7, 13.8a & b, 13.9a & b, 13.10a & b, 13.11a-c, 13.12a & b:** © The McGraw-Hill Companies, Inc./Photo by Jw Ramsey.

Chapter 14

Figure 14.3b: © Ed Reschke; **Page 422:** © Simon Fraser/Photo Researchers, Inc.; **14.10b:** © Dr. D. W. Fawcett/Visuals Unlimited; **14.12b:** © Dr. Richard Kessel and Dr. Randy Kardon/Tissues and Organs/Visuals Unlimited; **14.12c:** © Ed Reschke; **Page 427:** © AP/Wide World Photos; **Page 431a:** © AP Photo/The Plain Dealer, Scott Shaw; **Page 431b:** © Don Emmert/AFP/Getty Images; **Page 433 top:** © OJ Staats/Custom Medical Stock Photo; **Page 433b:** © NMSB/Custom Medical Stock Photo.

Chapter 15

Figure 15.1a-c, 15.5b, 15.7b: © The McGraw-Hill Companies, Inc./Photo and Dissection by Christine Eckel; **Page 450:** © Dr. M.A. Ansary/Custom Medical Stock Photo; **15.10:** © The McGraw-Hill Companies, Inc./Photo and Dissection by Christine Eckel; **Page 457:** Damasio H., Grabowski T., Frank R., Galaburda A.M., Damasio A.R. "The Return of Phineas Gage: Clues about the Brain from the Skull of a Famous Patient. *Science* 264 (May 20, 1994), page 1102-5. © 1994 American Association for the Advancement of Science; **15.14:** © The McGraw-Hill Companies, Inc./Photo and Dissection by Christine Eckel; **Page 470:** © AP/Wide World Photos; **15.24:** © The McGraw-Hill Companies, Inc./Photo by Rebecca Gray.

Chapter 16

Figure 16.1b: © The McGraw-Hill Companies, Inc./Photo and Dissection by Christine Eckel; **16.1c:** From Anatomy & Physiology Revealed, © The McGraw-Hill Companies, Inc./The University of Toledo, photography and dissection; **16.3b:** © Dr. David Phillips/Visuals Unlimited; **16.9b, 16.10b, 16.11c & d:** © The McGraw-Hill Companies, Inc./Photo and Dissection by Christine Eckel.

Chapter 17

Figure 17.12b(1 & 2): © Wellcome Dept. of Cognitive Neurology/Photo Researchers, Inc.; **17.12b(3):** © RDF/Visuals Unlimited; **Page 533 left:**

© Stevie Grand/Photo Researchers, Inc.; **Page 533 right:** © J Cavallini/Custom Medical Stock Photo.

Chapter 18

Figure 18.7: © The McGraw-Hill Companies, Inc./Photo and Dissection by Christine Eckel.

Chapter 19

Figure 19.7b(1 & 2): © The McGraw-Hill Companies, Inc./Photo by Dr. Alvin Telser; **19.9b:** © John Cunningham/Visuals Unlimited; **19.10a:** © The McGraw-Hill Companies, Inc./Photo by Jw Ramsey; **19.14c:** © The McGraw-Hill Companies, Inc./Photo by Dr. Alvin Telser; **19.15a:** © Al Blum/Visuals Unlimited; **Page 578a & b:** © PhotoDisc/Getty Images; **Page 580 top left:** © James P. Gilman, CRA/Phototake; **Page 580 top right:** © Dr. P. Marazzi/Photo Researchers, Inc.; **Page 580 bottom:** © PhotoDisc/Getty Images; **Page 588 (all):** © ISM/Phototake; **19.27d:** © Science VU/Visuals Unlimited.

Chapter 20

Figure 20.5a & b: © Educational Images/Custom Medical Stock Photo; **Page 613 top left:** © Frank Trapper/Corbis; **Page 613 bottom left:** © Eric Robert/Corbis; **Page 613 right:** Clinical Pathological Conference on Acromegaly, Diabetes, Hypemetabolism, Protein Use & Heart Failure, American Journal of Medicine 20: 133 (1956). © 2002, with permission from Elsevier; **20.9a:** © The McGraw-Hill Companies, Inc./Photo and Dissection by Christine Eckel; **20.9b:** © The McGraw-Hill Companies, Inc./Photo by Dr. Alvin Telser; **Page 618 left:** © Chris Barry/Phototake; **Page 618 right:** © Scott Camazine/Photo Researchers, Inc.; **20.11b:** © Victor Eroschenko; **20.13b:** © The McGraw-Hill Companies, Inc./Photo and Dissection by Christine Eckel; **20.13d:** © The McGraw-Hill Companies, Inc./Photo by Dr. Alvin Telser; **Page 623 left:** Kathy Carbone; **Page 623 right a:** © AP/Wide World Photos; **Page 623 right b:** © Corbis; **20.14a:** © The McGraw-Hill Companies, Inc./Photo and Dissection by Christine Eckel; **20.14b:** © The McGraw-Hill Companies, Inc./Photo by Dr. Alvin Telser.

Chapter 21

Figure 21.3(4): © The McGraw-Hill Companies, Inc./Photo by Dr. Alvin Telser; **21.4b:** © Ed Reschke; **Page 642:** © Eye of Science/Photo Researchers, Inc.; **21.8a:** © Jean Claude Revy-ISM/Phototake; **Table 21.3 (all):** © The McGraw-Hill Companies, Inc./Photo by Dr. Alvin Telser; **21.9b:** © The McGraw-Hill Companies, Inc./Photo by Dr. Alvin Telser; **21.10:** © Dr. John W. Weisel. *Nature*, Vol. 431, 4 October 2001. © 2001 Nature Publishing Group; **Page 649a:** © The McGraw-Hill Companies, Inc./Photo by Dr. Alvin Telser; **Page 649b:** © Ed Reschke.

Chapter 22

Figure 22.2b: © The McGraw-Hill Companies, Inc./Photo and Dissection by Christine Eckel; **22.5a & b:** © The McGraw-Hill Companies, Inc./Photo and Dissection by Christine Eckel; **22.6:** © The McGraw-Hill Companies, Inc./Photo and Dissection by Christine Eckel; **22.10c:** © Dennis Drenner/Visuals Unlimited.

Chapter 23

Figure 23.2 & inset: © The McGraw-Hill Companies, Inc./Photo by Dr. Alvin Telser; **23.4a-c:** © The McGraw-Hill Companies, Inc./Photo by Dr. Alvin Telser; **Page 688:** © Bart's Medical Library/Phototake; **23.8a:** © RF/Corbis; **23.10c:** © The McGraw-Hill Companies, Inc./Photo and Dissection by Christine Eckel; **23.14:** © The McGraw-Hill Companies, Inc./Photo and Dissection by Christine Eckel; **23.17:** © The McGraw-Hill Companies, Inc./Photo and Dissection by Christine Eckel; **23.19c & d:** © The McGraw-Hill Companies, Inc./Photo and Dissection by Christine Eckel; **Page 706a:** © The McGraw-Hill Companies, Inc./Photo by Dr. Alvin Telser; **Page 706b:** © CNRI/Photo Researchers, Inc. **23.21:** © The McGraw-Hill Companies, Inc./Photo and Dissection by Christine Eckel.

Chapter 24

Figure 24.3b: © The McGraw-Hill Companies, Inc./Photo by Dr. Alvin Telser; **24.4c:** © The McGraw-Hill Companies, Inc./Photo and Dissection by Christine Eckel; **Page 727:** © Andy Crumo, TDR, WHO/Photo Researchers, Inc.; **24.8a & c, 24.9b & c, 24.10b:** © The McGraw-Hill Companies, Inc./Photo by Dr. Alvin Telser; **24.10c:** Dr. Kent Van De Graaff; **Page 737:** © NYU Franklin Research Fund/Phototake; **24.11c:** © The McGraw-Hill Companies, Inc./Photo by Dr. Alvin Telser; **24.11d:** © The McGraw-Hill Companies, Inc./Photo and Dissection by Christine Eckel.

Chapter 25

Figure 25.2a: © The McGraw-Hill Companies, Inc./Photo and Dissection by Christine Eckel; **25.6:** © ISM/Phototake; **Page 754 right:** © ISM/Phototake; **25.7b:** © Science VU/Visuals Unlimited; **25.9b:** © The McGraw-Hill Companies, Inc./Photo by Dr. Alvin Telser; **25.9c:** © Dr. David Phillips/Visuals Unlimited; **Page 763 left:** © Collection CNRI/Phototake; **Page 763 top right:** © Carolina Biological/Visuals Unlimited; **Page 763 bottom right:** © The McGraw-Hill Companies, Inc./Photo by Dr. Alvin Telser; **25.16a & b:** © SIU/Visuals Unlimited; **Page 768:** © Coneyl Jay/Photo Researchers, Inc.; **Page 769 top & bottom:** © Ralph T. Hutchings/Visuals Unlimited; **Page 769a:** © CNRI/Photo Researchers, Inc.; **Page 769b:** © The McGraw-Hill Companies, Inc./Photo by Dr. Alvin Telser; **Page 770 left:** © CHAD

Therapeutics, Inc.; **Page 770 top right:** © Dr. E. Walker/Photo Researchers, Inc.; **Page 770 bottom right:** © Javier Domingo/Phototake.

Chapter 26

Figure 26.4c: © The McGraw-Hill Companies, Inc./Photo by Dr. Alvin Telser; **26.6a:** © The McGraw-Hill Companies, Inc./Photo by Christine Eckel; **26.8a & b:** © The McGraw-Hill Companies, Inc./Photo and Dissection by Christine Eckel; **26.10a:** © Alfred Pasieka/Peter Arnold; **26.10b:** © The McGraw-Hill Companies, Inc./Photo by Dr. Alvin Telser; **Page 789 left:** © David M. Martin, M.D./Photo Researchers, Inc.; **Page 789 right:** © Dr. Edward Lee, Howard University; **26.12b:** © The McGraw-Hill Companies, Inc./Photo and Dissection by Christine Eckel; **26.13b:** © The McGraw-Hill Companies, Inc./Photo by Dr. Alvin Telser; **Page 793b:** © Javier Domingo/Phototake; **26.15d:** © The McGraw-Hill Companies, Inc./Photo by Dr. Alvin Telser; **26.15e:** © Dr. Lee Peachey; **26.17b:** © The McGraw-Hill Companies, Inc./Photo by Dr. Alvin Telser; **26.19c:** © The McGraw-Hill Companies, Inc./Photo by Dr. Alvin Telser; **Page 802a:** © Dr. Joseph William/Phototake; **Page 802b:** © Ida Wyman/Phototake; **26.20b:** © Carolina Biological Supply Company/Phototake; **26.20b (inset):** © Carolina Biological/Visuals Unlimited; **Page 804:** © Gladden Willis/Visuals Unlimited; **Page 805 left & right:** © Javier Domingo/Phototake.

Chapter 27

Figure 27.1a: © The McGraw-Hill Companies, Inc./Photo and Dissection by Christine Eckel; **27.3:** © Ralph T. Hutchings/Visuals Unlimited; **Page 818:** © SPL/Photo Researchers, Inc.; **27.6b:** © The McGraw-Hill Companies, Inc./Photo by Dr. Alvin Telser; **27.6d:** © Dr. Dennis Kunkel/Dennis Kunkel Microscopy, Inc.; **27.7b & d, 27.8b & 27.9b:** © The McGraw-Hill Companies, Inc./Photo by Dr. Alvin Telser; **27.9b (inset):** © Dr. Frederick Skvara/Visuals Unlimited; **Page 833 left:** © J. Siebert/Custom Medical Stock Photo; **Page 833 right:** © The McGraw-Hill Companies, Inc./Photo and Dissection by Christine Eckel.

Chapter 28

Figure 28.3c: © The McGraw-Hill Companies, Inc./Photo and Dissection by Christine Eckel; **28.4b & c:** © The McGraw-Hill Companies, Inc./Photo by Dr. Alvin Telser; **28.4d & e:** © Ed Reschke/Peter Arnold; **28.4f & g:** © The McGraw-Hill Companies, Inc./Photo by Dr. Alvin Telser; **28.7b & inset & 28.7c:** © The McGraw-Hill Companies, Inc./Photo by Dr. Alvin Telser; **Page 850 left:** © Carolina Biological/Visuals Unlimited; **Page 850 right:** © Parviz M. Pour/Photo Researchers, Inc.; **28.8:** © The McGraw-Hill Companies, Inc./Photo by Dr. Alvin Telser; **Page 854a, b & d:** © The McGraw-Hill Companies, Inc./Photo

Line Art

Electronic Publishing Services Inc. Illustration Team

Art Director: Kim E. Moss

Christine Armstrong: Figures 1.4, 2.3, 2.18, 2.19, 2.20, 3.1, 3.2, 3.3, 3.4, 3.5, 3.6, 3.7, 3.8, 3.9, 3.11, 3.12, ua3.2, ua3.3, ua3.4, ua3.5, ua3.6, ua3.7, ua3.8, ua3.9, ua3.10, ua3.11, ua3.12, 3.13, 4.2, ua4.2, ua4.12, ua4.27, 5.11, 6.2, 6.10, 7.17, 8.16, 11.2, 11.8, 14.1, 14.4, 14.9, 14.11, 14.12, 14.13, 14.14, 14.15, 14.16, 15.1, 15.2, 15.3, 15.4, 15.5, 15.6, 15.7, 15.8, 15.9, 15.10, 15.11, 15.13, 15.14, 15.15, 15.16, 15.17, 15.18, 15.19, 15.20, 15.21, 15.22, 15.23, 15.24, ua15.1, ua15.3, ua15.5, ua15.6, 16.2, 16.3, 16.4, 16.5, 16.8, 16.9, 16.10, 16.11, ua16.1, 17.2, 17.3, 17.4, 17.6, 17.7, 17.9, 17.10, 17.11, 17.14, 18.2, 18.3, 18.4, 18.5, 18.6, 18.8, 18.9, 18.10, 18.11, 18.12, 18.13, 19.18, 19.19, 19.30, 20.1, 24.1, 25.1, 25.2, 25.4, 25.5, 25.7, 25.8, 25.9, 25.10, 25.11, 25.12, 25.13, 25.14, 25.15, 25.17, 25.18, ua25.1, ua25.6, 27.3, 27.4, 27.5, 27.6, 27.7, 27.8, 27.9, 27.10, ua27.2, 28.1, 28.9, 28.16

Rachel Bedno Robinson: Figures 1.4, 1.9, 1.10, ua3.14, 4.1, 4.4, 4.5, 4.6, 4.7, ua4.9, ua4.15, ua4.18, 6.1, 6.3, 6.11, 6.13, 6.16, 7.1, 7.24, 7.29, 7.30, ua7.7, 8.1, 8.3, 8.6, 8.9, 8.10, 8.14, 8.15, 9.1, 9.2, 9.3, 9.20, 9.21, ua9.1, 10.13, 10.15, 12.3, 12.21, ua13.6, 15.2, ua15.6, 18.5, 18.6, 18.8, 19.6,

19.7, 19.8, 20.1, 20.2, 20.7, 20.12, 20.14, 21.6, 22.2, 22.6, 22.13, 22.14, ua22.1, 23.9, 23.11, 23.22, 24.1, 24.11, 25.1, 25.2, 25.3, 25.8, 25.11, 25.15, ua25.6, 26.3, 26.7, 28.10

Raychel Ciemma: Figures 4.6, 6.3, 7.10, 7.11, 7.19, 7.26, 7.27, 7.28, 7.29, 7.30, 7.32, 7.33, 8.2, 8.3, 8.6, 8.12, 8.14, 8.15, 9.1, 9.2, 9.3, 9.5, 9.6, 9.14, 9.17, 9.20, 9.21, ua9.1, ua9.3, 11.4, 12.9, 12.11, 12.13, 12.14, ua12.1, ua15.6, 19.9, 19.10, 19.12, 19.14, 19.16, 19.17, ua19.4, 20.11, 25.15, 26.1, 26.2, 26.5, 26.7, 26.9, 26.12, 26.13, 26.14, 26.15, 26.16, 26.17, 26.18, 26.19, 26.20, 26.21, ua26.3, ua26.4, 28.17, 28.18, 28.19, ua28.2, ua28.4

Sherra Cook: Figures ua7.14, ua7.15, 17.1, 17.2, 17.3, 17.4, 17.5, 17.6, 17.7, 17.8, 17.9, 17.10, 17.11, 17.12

Barbara Cousins: Figures ua9.4, 20.3, 20.7, 20.10, 20.14, 20.15

Jennifer Gentry: Figures 1.4, 7.3, 7.15, 7.16, 7.20, 7.31, 10.5, 10.6, 10.7, 10.14, ua10.5, 11.2, 11.5, 11.8, 11.9, 15.12, ua15.3, ua15.6, 20.9, 20.11, 20.12, 20.13, 24.1, 24.4, 24.9, 24.12, 24.13, ua25.1, 26.3, 26.4, 26.6, 26.11, 27.1, 27.2, 28.10, 28.12, 28.14

Jonathan Higgins: Figures 21.1, 21.2, 21.3, 21.4, 21.5, 21.6, 21.7, 21.8, 21.9, 21.11, 24.5, 24.7, 26.9, 26.16, 26.17

Kellie Marsh Holoski: Figures 1.3, 1.4, 1.6, 1.7, 1.8, 1.11, 2.1, 2.16, ua2.18, 3.1, 3.12, 4.3, ua4.1, ua4.2, ua4.3, ua4.4, ua4.5, ua4.6, ua4.7, ua4.8, ua4.9, ua4.10, ua4.11, ua4.13, ua4.14, ua4.15, ua4.16, ua4.17, ua4.18, ua4.19, ua4.20, ua4.21, ua4.22, ua4.23, ua4.24, ua4.25, ua4.26, ua4.27, 5.1, 5.6, 5.9, 5.10, ua5.2, ua5.7, 6.1, 6.2, 6.3, 6.15, 6.17, 7.1, 7.2, 7.5, 7.6, 7.7, 7.8, 7.9, 7.10, 7.12, 7.13, 7.14, 7.15, 7.17, 7.21, 7.22, 7.23, 7.24, 7.25, 8.1, 8.4, 8.5, 8.7, 8.9, 8.11, 8.13, 9.1, 9.2, 9.3, 9.5, 9.6, 9.12, 9.13, 9.15, 9.16, 9.18, 9.19, ua9.1, ua9.3, 11.6, 11.7, 11.10, 11.11, 11.12, 11.13, 11.14, 11.15, 12.3, 12.5, 12.7, 12.8, ua13.1,

ua13.5, 14.1, 14.3, 14.6, 14.7, 14.8, ua15.6, 16.6, 16.7, 16.14, 16.15, 18.5, 18.6, 18.8, 19.1, 19.4, 19.5, 19.11, 19.13, 19.14, 19.20, 19.21, 19.22, 19.23, 19.24, 19.25, 19.26, 19.27, 19.28, 19.29, 19.30, 19.31, 19.32, ua19.1, ua19.5, ua19.7, 20.1, 20.2, 20.7, 21.6, 22.1, 22.2, 22.3, 22.4, 22.6, 22.7, 22.8, 22.9, 22.10, 22.11, 22.12, 22.13, 22.14, 22.15, 22.16, ua22.1, 23.1, 23.3, 23.5, 23.6, 23.7, 23.9, 23.10, 23.12, 23.13, 23.15, 23.16, 23.18, 23.19, 23.20, 23.22, 23.23, 23.24, 23.25, 23.26, 23.27, ua23.2, ua23.4, 24.1, 25.1, 25.11, 26.1

Peter Jurek: Figures 12.19, 15.13

Richard S. LaRocco: Figures 12.15, 12.17, 12.18, 12.20, 12.21, 12.22, 25.16

Joyce Lavery Hall: Figures ua4.8, ua4.11, 5.12, 5.13, 7.2, 7.6, 7.7, 7.9, 7.26, 7.34, 7.35, 7.36, 8.5, 8.10, 8.13, 8.16, 9.4, 9.17, 9.22, 10.17, 11.4, 11.15, 12.22, 12.23, 14.9, ua14.4, ua15.06, 16.1, 16.16, 19.12, 20.15, 20.16, 26.22, 26.23, 27.10, 27.11, 27.12, 28.2, 28.11, 28.13, 28.15, 28.16, ua28.2

Robert Margulies: Figures 16.1, 16.7, 16.8, 16.9, 16.10, 16.11, ua16.2, ua16.3, ua16.4, ua16.5

Matthew McAdams: Figures 1.5, 2.20, 7.23, 10.2, ua10.15, ua10.17, 10.8, 14.7, 14.8, 14.10, 16.4, 16.6, 16.13, ua18.1, 19.2, 19.3, 19.4, 19.11, 19.29, ua19.1, ua19.2, 21.1, 21.2, 21.3, 21.4, 21.5, 21.6, 21.7, 21.8, 21.11, 22.3, 25.15, ua25.5

Kim E. Moss: Cover Art, Figures 1.4, 2.1, 2.3, 2.4, 2.5, 2.6, 2.7, 2.9, 2.10, 2.13, 2.14, ua2.9, ua2.13, ua2.14, ua2.15, 3.5, 3.10, 3.12, ua4.16, ua4.22, ua4.24, 5.2, 5.4, 5.6, 5.7, 5.8, 6.4, 6.5, 6.8, 6.11, 6.14, 6.15, 7.1, 7.3, 7.4, 7.8, 7.10, 7.14, 7.15, 7.25, 7.31, 8.11, 10.1, 10.3, 10.8, 10.14, ua10.1, 11.1, 11.2, 11.14, 12.1, 12.2, 12.4, 12.7, 12.8, 12.15, 12.17, 12.18, 12.20, 12.21, 12.22, 20.1, 20.2, 20.3, 20.4, 20.6, 20.7, 20.9,

20.10, 20.12, 21.11, ua21.2, 24.2, 24.3, 24.5, 24.6, 24.7, 24.10, 24.11, ua24.2, 26.20, 26.21

Evelyn Pence: Figures 5.2, 5.4, 5.10, ua5.1, 14.8, 14.10, 14.11, 19.12, 19.14, 19.16, 19.17, 19.18, 26.12, 26.22, 26.23, ua26.3, ua26.4

Curtis W. Perone: Figures 2.10, 25.16

Marie Rossettie: Figure ua19.6

Brook Wainwright: Figures 10.9, 10.16, ua10.7, 12.6, 12.10, 12.16, 12.19, 12.23, ua15.5, ua16.2, 23.15, 23.16

Kristen Workman: Figure 1.4

Megan E. Rojas: Figures 7.3, 10.2, 10.3, 10.4, 10.11, 10.13, 10.15, 10.16, ua10.8, ua10.10, ua10.11, ua10.13, ua10.17, ua10.21, 12.6, 12.10, 12.15, 12.16, 12.19, 28.3, 28.4, 28.5, 28.6 28.7

Travis Vermilye: Figures ua4.10, ua4.12, ua7.5

Craig Zuckerman: Figures 12.3, 12.5, 12.7, 12.8

Contributing Illustrators:
The following illustrators worked with the illustration team to enhance the artwork that appears throughout this book. Their talents and contributions were invaluable to the production of this art program.

Leigh Campbell
Jonathan Higgins
Michael Hortens
Joyce Lavery Hall
Matthew McAdams
Evelyn Pence
Curtis W. Perone
Wendy Beth Jackelow
Krista Townsend
Brook Wainwright
Kristen Workmen
Alison Wright
Gene Wright
Aysha Venjara

Index

Page numbers followed by an "f" indicate figures; "t" indicates tabular material.

Roots, Combining Forms, Prefixes, and Suffixes

Many terms used in the biological sciences are compound words; that is, words made up of one or more word roots and appropriate prefixes and/or suffixes. Less than 400 roots, prefixes, and suffixes make up more than 90% of the medical vocabulary. These combining forms are most often derived from the ancient Latin or Greek. Prefixes are placed before the root term and suffixes are added after. The following list includes the most common forms used in anatomy and medicine and an example for each. This list, and the word origin information found throughout the text, is intended to facilitate learning an often unnecessarily complex-sounding vocabulary. Exclusively a learning tool, the entries are by intention brief. If you learn them, you will find your progress in your anatomy course swift, steady, and strong (the three "s'es" of *success*).

a-	without, lack of	asymptomatic (absence of symptoms)
ab-	away from	abstinence (to hold back from)
acou-	hearing	acoustics (science of sound)
-ac, -al	pertaining to	cardiac (the heart), myocardial (heart muscle)
ad-	to, toward, near to	adduction (move toward midline)
aden-, adeno-	gland	adenoma (tumor of a gland)
af-	toward	afferent (moving toward)
albi-	white	albinuria (passing of pale or white urine)
-algia	painful condition	myalgia (muscle pain)
an-	without, lack of	anesthesia (absence of pain)
andro-	male	androgens (male hormones)
angi-, angio-	vessel	angiopathy (disease of blood vessels)
ante-	before	antepartum (before birth)
anti-	against	anticoagulant (prevents blood clotting)
apo-	separated from, off	apodia (congenital absence of feet)
arthr-, arthro-	joint	arthritis (inflammation of a joint)
-ary	associated with	urinary (associated with urine)
-asis, -asia	condition or state of	homeostasis (state of metabolic balance)
audio-	hearing	auditory (belonging to the hearing sense)
auri-	ear	auricle (ear-shaped structure)
auto-	self	autolysis (self breakdown)
baro-	weight, pressure	baroreceptor (receptor for pressure changes)
bi-	twice, double	bicuspid (two cusps)
-blast	germ, bud	chondroblast (cartilage-producing cell)
brachi-	arm	brachial (of the arm)
brady-	slow	bradycardia (slow heart rate)
bucc-	cheek	buccal cavity (inside cheek region)
callo-	thick	callosity (thickening of keratinized layer of epidermis)
carcin-	cancer	carcinogenic (causing cancer)
cardio-	heart	cardiogram (register of heart activity)
caud-	tail	caudal (by the tail)
cephal-	head	cephalic (by the head)
cerebro-	brain	cerebrospinal (of the brain and spinal cord)
chondro-	cartilage, gristle	chondrocyte (cartilage cell)
-cide	kill	spermicide (agent that kills sperm)
circum-	around	circumduction (movement forming a circle)
-clast	break	osteoclast (cell that breaks down bone)
co-, com-	with, together	cooperate, gray commissure (connects right/left horns)
contra-	against, opposite	contralateral (opposite side)
cost-	rib	intercostals (between the ribs)
crani-	skull	cranial cavity (where the brain is)
cune-	wedge	cuneiform (wedge shaped)
cuti-	skin	subcutaneous (under the skin)
cyan-	blue color	cyanosis (bluish discoloration of skin)
cysti-, cysto-	sac, bladder	cystoscope (instrument for examining inside of bladder)
-cyte, ctyo-	cell	erythrocyte (red blood cell), cytology (study of cells)
demi-	half	costal demifacet (half-moon facet on vertebra for rib articulation)
derm-	skin	dermatology (study of skin)
di-, diplo-	two	diploid (two sets of chromosomes)

duct-, -duct	lead, draw	ovarian duct, adduct (to lead away from)
dur-	hard	dura mater (tough menix of CNS)
dys-	painful, difficult, bad	dysuria (painful urination)
e-, ec-, ef-, ex-	out, from	efferent (carries away from), excretion (eliminate from)
ecto-	outside, outer	ectocardia (displacement of heart)
-ectomy	to cut out	appendectomy (removal of appendix)
ede-, -edem	swelling	myoedema (muscle swelling)
-el, -elle	small	organelle (tiny structure that performs specific cellular functions)
endo-	within	endocardium (lining within heart chambers)
entero-	intestine	enteritis (inflammation of intestines)
epi-	upon, on	epicardium (membrane covering heart)
ex-, exo-	outside	exhale (breathe out); exocrine (gland that secretes to outside)
extra-	outside	extracellular (outside the cell)
-ferent	carry	afferent (carries toward)
-form	resembling, shape of	fusiform (spindle-shaped)
gastr-, gastro-	stomach	gastric ulcer (stomach ulcer)
-genesis, -genic	produce, origin	gluconeogenesis (glucose from another molecule), carcinogenic (causes cancer)
gloss-, glosso-	tongue	hypoglossal (under the tongue)
glyco-	sugar, sweet	glycolysis (breakdown of glucose)
gyn-	female, woman	gynecology (treatment of female reproductive organs)
hapto-	single	haploid (single set of chromosomes)
hem-, hemato-	blood	hematology (study of blood)
hepato-	liver	hepatitis (inflammation of the liver)
hetero-	different	heterosexual (involving different sexes)
hist-, histo-	tissue	histology (study of tissues)
holo-	whole, entire	hologynic (manifests only in females), holocrania (absence of all bones of skull vault)
homo-, homeo-	same	homeostasis (constancy of body parameters)
hydro-	water	hydroadipsia (absence of thirst for water)
hyper-	over, above	hypertrophy (overgrowth of cells or part)
hypo-	below, under	hypoglycemia (low blood sugar)
idio-	self, distinct	idiopathic (disease of unknown cause)
infra-	below	infraspinatus (below the spine of scapula)
inter-	between	interosseous (between two bones)
intra-	within	intracellular (within the cell)
-issimus	greatest	latissimus (widest)
iso-	equal, same	isotonic (same concentration)
-itis	inflammation	neuritis (inflammation of nerve)
juxta-	near	juxtaglomerular (near the glomerulus)
labi-	lip	labia major (thickened folds of skin and connective tissue in female external genitalia)
lacto-	milk	lactose (milk sugar)
leuko-	white	leukocyte (white blood cell)
lip-	fat	lipid (an operational term denoting solubility characteristics; "fat soluble")
-logy	study	urology (study of urinary system)
-lysis	breaking up, dissolve	hemolysis (breaking up erythrocytes)
macro-	large	macrophage (certain large leukocytes)
mamm-, mast-	breast	mammary glands, mastectomy (breast removal)
medi-	middle	medial (toward the midline)
melano-	black	melanocyte (dark pigment-producing cell)
-mers, -meres	parts	polymers (larger molecules made of monomers)
meta-	after, beyond	metastasis (beyond the original position)
micro-	small	microorganism (very small organism)
mono	one, single	monomer (a single part); monosaccharide (a simple or single sugar)
morph-	form, shape	morphology (study of shape)
myo-	muscle	myometrium (muscular wall of uterus)
necro-	dead	necrotic (being of dead tissue)
neo-	new	neonatal (newborn)
nephro-	kidney	nephrology (study of kidneys)
neuro-	nerve	neurilemma (nerve cell membrane)